GREEN BIOCATALYSIS

GREEN BIOCATALYSIS

Edited by
Ramesh N. Patel

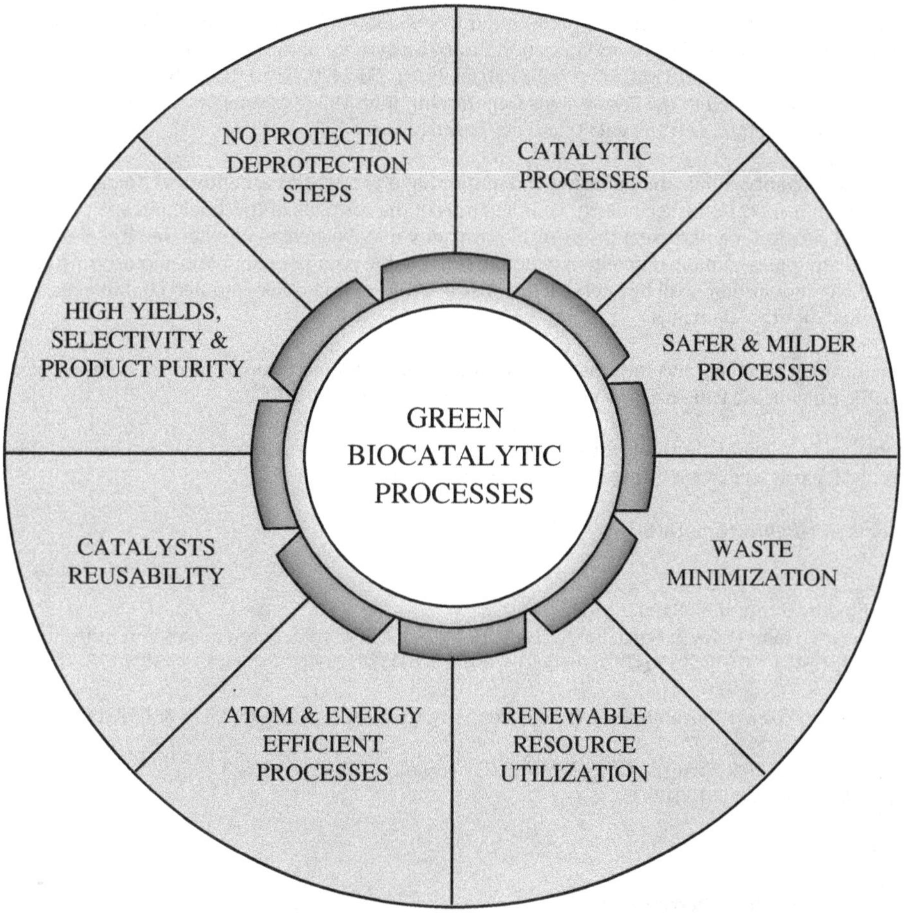

WILEY

Published by John Wiley & Sons, Inc., Hoboken, New Jersey
Published simultaneously in Canada

Library of Congress Cataloging-in-Publication Data:

Names: Patel, Ramesh N., 1942–
Title: Green biocatalysis / edited by Ramesh N. Patel.
Description: Hoboken, New Jersey : John Wiley & Sons, Inc., 2016. | Includes bibliographical references and index.
Identifiers: LCCN 2015047535 (print) | LCCN 2015047720 (ebook) | ISBN 9781118822296 (hardback) | ISBN 9781118822357 (pdf) | ISBN 9781118822364 (epub)
Subjects: LCSH: Enzymes–Biotechnology. | Biocatalysis. | Green chemistry. | BISAC: TECHNOLOGY & ENGINEERING / Chemical & Biochemical.
Classification: LCC TP248.65.E59 G735 2016 (print) | LCC TP248.65.E59 (ebook) | DDC 660–dc23
LC record available at http://lccn.loc.gov/2015047535

Cover image courtesy of Wiley

Set in 10/12pt Palatino by SPi Global, Pondicherry, India

Printed in the United States of America

10 9 8 7 6 5 4 3 2 1

Contents

Preface

Microorganisms, and biocatalysts derived from them, have enormous potential for carrying out the conversion of a variety of synthetic chemicals to useful products in a highly chemoselective, enantioselective, and regioselective manner. Green biocatalysis provides a technology that uses milder and safer processes. Bioprocesses are carried out under ambient temperature and atmospheric pressure in aqueous conditions, which do not cause any racemization, epimerization, or rearrangement of compounds and thus avoid any protection and deprotection steps during synthesis and shorten the synthetic processes. It prevents, eliminates, or minimizes waste rather than offering waste remediation. Green biocatalysis also uses renewable resources and less hazardous materials than chemical processes. Advances in directed evolution under process conditions have provided novel, efficient, and stable biocatalysts for the development of efficient and economical processes under desired operational conditions. One can achieve sustainable development goals using green biocatalytic processes.

Demand for chiral compounds continues to increase, mainly for use in pharmaceuticals but also in other industries such as flavor, fragrance, cosmetics, and agricultural chemicals. Chiral active pharmaceutical ingredients (APIs) were previously usually formulated as racemates; the preference now is for single enantiomers. The switch from a racemic compound to a single enantiomer of API is required to extend life-cycle management and also to improve the bioavailability and efficacy of drugs.

Since the mid-2000s, progress in the high-level expression of enzymes, in *Escherichia coli*, *Pichia pastoris*, and other microbial systems, and improvement in fermentation technology has led to an increase in cell yields in a much shorter time. Advances in protein purification technology, determination of the structure of proteins along with molecular cloning, and the random and directed evolution of biocatalysts have opened up unlimited access to a variety of efficient enzymes and microbial cultures as tools in organic synthesis. The development of efficient immobilization techniques for biocatalysts and the reusability of biocatalysts have provided highly economical and energy-efficient processes for the synthesis of key intermediates and drug products in the pharmaceutical, agrochemical, flavor, fragrance, cosmetics, and pesticide industries.

Various chapters in this book are contributed by internationally renowned scientists and professors from industry and from well known universities, with many years of experience in the design and development of green processes and the industrial application of various processes in the production of fine chemicals and pharmaceuticals. The book contains 29 chapters with over 3000 references and more than 700 equations, drawings, and tables. It examines the use of a variety of classes of enzymes in the development of green processes to generate chiral compounds useful in chemoenzymatic synthesis of drug and agricultural products, cosmetics, and flavor compounds. All the chapters are well presented and cover the following key aspects of biocatalysis:

- A discussion of biocatalysis and green chemistry covers the principles of green chemistry. The concepts of sustainable development and green chemistry are introduced and defined. Green chemistry metrics, such as E factors and the atom economy, together with environmental impact and sustainability metrics, are discussed. The many attractive features of biocatalysis in the context of green chemistry are also discussed.
- There is discussion of the enzymatic synthesis of chiral primary, secondary, and tertiary amines using ω-transaminases, amine oxidases, and the berberine bridge enzyme is presented. New enzyme variants have been created with altered substrate specificity, improved activity and stability with transaminases and oxidases to demonstrate one-pot cascade reactions for efficient multistep transformations.
- Enzymatic applications in decarboxylation and racemization reactions of unnatural compounds using artificial enzyme variants derived from aryl-malonate decarboxylase are presented.
- Green enzymatic processes for the synthesis of key intermediates for the development of drugs are discussed using enzymatic reduction, reductive amination, epoxidation, hydroxylation, and transamination reactions to develop anticancer, antihypertensive, antiviral, anticholesterol, anti-anxiety, antidiabetic, and anti-inflammatory drugs.
- The dynamic kinetic resolution (DKR) of alcohols, amines, and amino acids, and its applications in industrial intermediates are presented. The procedures include the DKR of secondary alcohols, secondary amines, and primary amines by enzyme-metal combinations to access a range of products of excellent enantiomeric excesses with high yields approaching 100%.
- Enzymatic sulfoxidation by flavin-based biocatalysts to obtain chiral sulfoxides, and the application of epoxide hydrolases, chloroperoxidases, and enzymes such as cytochrome P450s, styrene monooxygenases, and alkane monooxygenases are described in detail. Stereoselective desymmetrization of *meso*-epoxides and the enantioconvergent hydrolysis of racemic epoxides is presented, providing enantiopure diol products at 100% yield. The scope of biocatalytic approaches as well as protein engineering for enzymes with better characteristics (stability, activity under process conditions) for bioepoxidation is described.
- The enantioselective acylation of alcohols, and amine reactions by lipases and esterases in organic synthesis with examples of classical and dynamic kinetic resolutions of racemates, are shown, giving attention to the desymmetrization of *meso*-compounds.
- Recent advances in enzyme-catalyzed aldol addition reactions for biocatalytic carbon–carbon bond formation by means of aldolases are described. They offer an exceptionally stereoselective and green tool for asymmetric framework construction and preparation of innovative molecules with engineered enzymes.
- The enzymatic, asymmetric reduction of carbonyl compounds is described, emphasizing directed evolution and protein engineering, to develop novel biocatalysts for reduction and reductive amination processes, with isolated enzymes and whole-cell biocatalysis.
- The application of nitrile-converting enzymes, nitrilases, and hydroxynitrile lyases in the synthesis of chiral compounds and cyanohydrins is covered.

- The discovery and engineering of enzymes for peptide synthesis and activation and synthesis of chiral compounds by ene-reductases (ERs) and flavin-dependent enzymes that catalyze the asymmetric reduction of electronically activated carbon–carbon double bonds are described. These enzymes have attracted increased attention from the industry as the enzymatic reduction usually proceeds with enhanced chemoselectivity and elevated stereoselectivity.
- The application of aromatic hydrocarbon dioxygenases in the preparation of chiral diols, and the biocatalytic introduction of chiral hydroxy groups using oxygenases and hydratases, are very well covered.
- Biocatalysis for drug discovery and development with an industrial perspective, and biocatytic cascade reactions with the integration of biocatalysts with one or more additional reaction steps, and multistep biocatalytic reaction sequences and multienzyme-catalyzed conversions, are presented.
- Recent developments in D-aminopeptidases, racemases, and oxidases and their application in the synthesis of chiral molecules, microbial transformations of pentacyclic triterpenes, and yeast-mediated enantioselective biocatalysis are described.
- Biotechnology for the production of chemicals, pharmaceutical ingredients, flavors and fragrances, food and feed, cosmetics and personal care, polymers, surfactants and lubricants, commodity chemicals, and energy industries is also described.

It is my pleasure to express my sincere appreciation to all the authors for their excellent contribution to *Green Biocatalysis*. My interest in biocatalysis and green processes was developed and stimulated by Drs. David Gibson, Derek Hoare, Nicholas Ornston, Allen Laskin, Ching Hou, Laszlo Szarka, Christopher Cimarusti, John Scott, Richard Mueller, Mauricio Futran, and Robert Waltermire, and by many of my colleagues and collaborators at the University of Texas, Yale University, Exxon Research and Engineering, and Bristol-Myers Squibb. I acknowledge their continuous support and encouragement during my 45-year career. I would like express my sincere appreciation to Antia Lekhwani, Purvi Patel, and Vishnu Priya.R at Wiley for their continuous support and help during the preparation of this book. I would like to give my sincere thanks to my wife, Lekha, and daughter, Sapana, for their support and encouragement while I worked on the book.

About the Editor

Ramesh N. Patel has over 45 years of experience in pharmaceutical and chemical industries. His experience includes building and leading multidisciplinary scientific teams in biochemistry, microbiology, molecular biology, and biochemical engineering. His research interest focused on the development of green biocatalytic processes for the synthesis of chiral compounds and fine chemicals for the development of pharmaceuticals and value-added products.

Dr. Patel received his Ph.D. (1971) in microbiology from the University of Texas at Austin, and he completed NIH and ACS postdoctoral research fellowships (1971–1974) from Yale University, New Haven. He joined the corporate research laboratories of Exxon Research and Engineering Company in 1975, working toward the development of value-added chemical products from cheap hydrocarbon raw materials using various bioprocesses with microbes and isolated enzymes (1975–1987).

Dr. Patel transitioned to the pharmaceutical industry with an interest in biocatalysis and biotechnology and their implications in the development of drugs. During his tenure at Bristol-Myers Squibb (1987–2008), he successfully started and led groups in biocatalysis and biotechnology bringing together multiple scientific disciplines in various project teams. These teams have developed and scaled up over 70 efficient and economical processes for the preparation of core chiral intermediates, drug metabolites, therapeutic proteins, and active pharmaceutical ingredients (APIs). APIs and recombinant proteins were supplied during various drug development stages (toxicology, phase I, phase II, phase III clinical studies), and process technology was transferred to the manufacturing division.

Dr. Patel's achievements include over 170 original publications, 78 process patents, and over 140 invited/external presentations. He is the recipient of a 2004 Biotechnology Lifetime Achievement Award from the American Oil Chemists Society, the 2008 Biocat Industrial Research Award from the International Congress on Biocatalysis (Germany), and the 2012 Distinction of Academician Award from the World Academy of Biocatalysis and Agricultural Biotechnology. He has received three BMS presidential awards for the development of commercial biocatalytic processes for the synthesis of key intermediates for captopril, paclitaxel, and saxagliptin. He has edited two books: *Stereoselective Biocatalysis* (2002, Marcel Dekker), and *Biocatalysis in the Pharmaceutical and Biotechnology Industries* (2007, CRC Press). He is also an editor of the journal *Biocatalysis and Agricultural Biotechnology* published by Elsevier.

Contributors

Andrés R. Alcántara
Organic and Pharmaceutical Chemistry Department, Faculty of Pharmacy, Complutense University of Madrid, Madrid, Spain

Ee Lui Ang
Metabolic Engineering Research Laboratory, Science and Engineering Institutes, Agency for Science, Technology and Research, Singapore, Singapore

Alain Archelas
Institut des Sciences Moléculaires de Marseille, Université d'Aix-Marseille, Marseille, France

Muhammad I. Arif
Biochemical Laboratory, Groningen Biomolecular Sciences and Biotechnology Institute, University of Groningen, Groningen, the Netherlands

Yasuhisa Asano
Asano Active Enzyme Molecule Project, ERATO, JST
and
Department of Biotechnology and Biotechnology Research Center, Toyama Prefectural University, Imizu, Japan

Robert Azerad
Laboratoire de Chimie et Biochimie Pharmacologiques et Toxicologiques, Université Paris Descartes, Paris
and
BDM Department, Bertin Pharma, Montigny-le-Bretonneux, France

Patricia B. Brondani
Groningen Biomolecular Sciences and Biotechnology Institute, University of Groningen, Groningen, the Netherlands

Eduardo Busto
Department of Chemistry, Organic and Bioorganic Chemistry, University of Graz, Graz, Austria

Yoon Kyung Choi
Department of Chemistry, Pohang University of Science and Technology, Pohang, South Korea

Pere Clapés
Department of Biological Chemistry and Molecular Modeling, Instituto de Química Avanzada de Cataluña, IQAC-CSIC, Barcelona, Spain

René Csuk
Organische Chemie, Martin-Luther Universität Halle-Wittenberg, Halle, Germany

Gonzalo de Gonzalo
Departamento de Química Orgánica, Universidad de Sevilla, Sevilla, Spain

Sarah-Marie Dold
Institute of Process Engineering in Life Sciences, Section II: Technical Biology, Karlsruhe Institute of Technology (KIT), Karlsruhe, Germany

Marco W. Fraaije
Groningen Biomolecular Sciences and Biotechnology Institute, University of Groningen, Groningen, the Netherlands

Jin-Song Gong
School of Pharmaceutical Science, Jiangnan University, Wuxi, P.R. China

Vicente Gotor
Departamento de Química Orgánica e Inorgánica, Instituto Universitario de Biotecnología de Asturias, Universidad de Oviedo, Oviedo, Spain

Vicente Gotor-Fernández
Departamento de Química Orgánica e Inorgánica, Instituto Universitario de Biotecnología de Asturias, Universidad de Oviedo, Oviedo, Spain

Harald Gröger
Faculty of Chemistry, Bielefeld University, Bielefeld, Germany

María J. Hernáiz
Organic and Pharmaceutical Chemistry Department, Faculty of Pharmacy, Complutense University of Madrid, Madrid, Spain

Makoto Hibi
Laboratory of Industrial Microbiology, Graduate School of Agriculture, Kyoto University, Kyoto, Japan

Pilar Hoyos
Organic and Pharmaceutical Chemistry Department, Faculty of Pharmacy, Complutense University of Madrid, Madrid, Spain

Werner Hummel
Faculty of Chemistry, Bielefeld University, Bielefeld, Germany

Gilles Iacazio
Institut des Sciences Moléculaires de Marseille, Université d'Aix-Marseille, Marseille, France

Dick B. Janssen
Biochemical Laboratory, Groningen Biomolecular Sciences and Biotechnology Institute, University of Groningen, Groningen, the Netherlands

Mahn-Joo Kim
Department of Chemistry, Pohang University of Science and Technology, Pohang, South Korea

Shigenobu Kishino
Division of Applied Life Science, Graduate School of Agriculture, Kyoto University, Kyoto, Japan

Tanja Knaus
Faculty of Life Sciences, Manchester Institute of Biotechnology, University of Manchester, Manchester, UK

Michael Kotik
Institute of Microbiology, Academy of Sciences of the Czech Republic, Prague, Czech Republic

Wolfgang Kroutil
Department of Chemistry, Organic and Bioorganic Chemistry, University of Graz, Graz, Austria

Rajesh Kumar
Biocatalysis Center of Emphasis, Chemical R&D, PharmaTherapeutics Pharmaceutical Science, Pfizer Worldwide R&D, Groton, CT, USA

Jusuk Lee
Department of Chemistry, Pohang University of Science and Technology, Pohang, South Korea

Youyun Liang
Metabolic Engineering Research Laboratory, Science and Engineering Institutes, Agency for Science, Technology and Research, Singapore, Singapore

Hui Lin
Key Laboratory of Environmental and Applied Microbiology & Environmental Microbiology Key Laboratory of Sichuan Province, Chengdu Institute of Biology, Chinese Academy of Sciences, Chengdu, China

Yan Liu
Key Laboratory of Environmental and Applied Microbiology & Environmental Microbiology Key Laboratory of Sichuan Province, Chengdu Institute of Biology, Chinese Academy of Sciences, Chengdu, China
and
University of the Chinese Academy of Sciences, Beijing, China

Van Martin
Biocatalysis Center of Emphasis, Chemical R&D, PharmaTherapeutics Pharmaceutical Science, Pfizer Worldwide R&D, Groton, CT, USA

Carlos Martinez
Biocatalysis Center of Emphasis, Chemical R&D, PharmaTherapeutics Pharmaceutical Science, Pfizer Worldwide R&D, Groton, CT, USA

Ludmila Martínková
Laboratory of Biotransformation, Institute of Microbiology, Academy of Sciences of the Czech Republic, Prague, Czech Republic

Tomoko Matsuda
Department of Bioengineering, Tokyo Institute of Technology, Yokohama, Japan

Hans-Peter Meyer
Institute of Life Technologies, University of Applied Sciences and Arts Western Switzerland, Sion, Switzerland

Kenji Miyamoto
Department of Biosciences and Informatics, Keio University, Yokohama, Japan

Kenji Mori
Department of Applied Biological Chemistry, Faculty of Agriculture, The University of Tokyo, Tokyo, Japan

Kaoru Nakamura
Science Shop, Kobe University, Kobe, Japan

Jun Ogawa
Division of Applied Life Science, Graduate School of Agriculture, Kyoto University, Kyoto, Japan

Seiji Okazaki
Asano Active Enzyme Molecule Project, ERATO, JST
and
Department of Biotechnology and Biotechnology Research Center, Toyama Prefectural University, Imizu, Japan

Vittorio Pace
Department of Pharmaceutical Chemistry, Faculty of Life Sciences, University of Vienna, Vienna, Austria

Rebecca E. Parales
Department of Microbiology and Molecular Genetics, College of Biological Sciences, University of California, Davis, CA, USA

Jaiwook Park
Department of Chemistry, Pohang University of Science and Technology, Pohang, South Korea

Ramesh N. Patel
SLRP Associates, Consultation in Biotechnology, Bridgewater, NJ, USA

Nina Richter
Department of Chemistry, Organic and Bioorganic Chemistry, University of Graz, Graz, Austria

Jens Rudat
Institute of Process Engineering in Life Sciences, Section II: Technical Biology, Karlsruhe Institute of Technology (KIT), Karlsruhe, Germany

Helmut Schwab
acib GmbH and
Institute of Molecular Biotechnology, Graz, Austria

Nigel S. Scrutton
Faculty of Life Sciences, Manchester Institute of Biotechnology, University of Manchester, Manchester, UK

Roger A. Sheldon
Department of Biotechnology, Delft University of Technology, Delft, the Netherlands
and
Molecular Sciences Institute, School of Chemistry, University of the Witwatersrand, Johannesburg, South Africa

Jin-Song Shi
School of Pharmaceutical Science, Jiangnan University, Wuxi, P.R. China

Robert C. Simon
Department of Chemistry, Organic and Bioorganic Chemistry, University of Graz, Graz, Austria

Kerstin Steiner
acib GmbH, Graz, Austria

Christoph Syldatk
Institute of Process Engineering in Life Sciences, Section II: Technical Biology, Karlsruhe Institute of Technology (KIT), Karlsruhe, Germany

Watumesa A. Tan
Department of Microbiology and Molecular Genetics, College of Biological Sciences, University of California, Davis, CA, USA
and
Faculty of Biotechnology, Atma Jaya Catholic University of Indonesia, Jenderal Sudirman, Jakarta, Indonesia

Helen S. Toogood
Faculty of Life Sciences, Manchester Institute of Biotechnology, University of Manchester, Manchester, UK

Ana Toplak
Biochemical Laboratory, Groningen Biomolecular Sciences and Biotechnology Institute, University of Groningen, Groningen, the Netherlands

Severin Wedde
Faculty of Chemistry, Bielefeld University, Bielefeld, Germany

Romana Wiedner
acib GmbH, Graz, Austria

John Wong
Biocatalysis Center of Emphasis, Chemical R&D, PharmaTherapeutics Pharmaceutical Science, Pfizer Worldwide R&D, Groton, CT, USA

John M. Woodley
Department of Chemical and Biochemical Engineering, Technical University of Denmark (DTU), Kongens Lyngby, Denmark

Bian Wu
Biochemical Laboratory, Groningen Biomolecular Sciences and Biotechnology Institute, University of Groningen, Groningen, the Netherlands

Zhong-Liu Wu
Key Laboratory of Environmental and Applied Microbiology & Environmental Microbiology Key Laboratory of Sichuan Province, Chengdu Institute of Biology, Chinese Academy of Sciences, Chengdu, China

Meng-Yu Xu
Key Laboratory of Environmental and Applied Microbiology Environmental Microbiology Key Laboratory of Sichuan Province, Chengdu Institute of Biology, Chinese Academy of Sciences, Chengdu, China
and
University of the Chinese Academy of Sciences, Beijing, China

Zheng-Hong Xu
School of Pharmaceutical Science, Jiangnan University, Wuxi, P.R. China

Rio Yamanaka
Faculty of Pharmaceutical Science, Himeji Dokkyo University, Himeji, Japan

Kazuyuki Yasukawa
Asano Active Enzyme Molecule Project, ERATO, JST
and
Department of Biotechnology and Biotechnology Research Center, Toyama Prefectural University, Imizu, Japan

Mingzi M. Zhang
Metabolic Engineering Research Laboratory, Science and Engineering Institutes, Agency for Science, Technology and Research, Singapore, Singapore

Huimin Zhao
Metabolic Engineering Research Laboratory, Science and Engineering Institutes, Agency for Science, Technology and Research, Singapore, Singapore
and
Department of Chemical and Biomolecular Engineering, University of Illinois at Urbana-Champaign, Urbana, IL, USA

CHAPTER 1

Biocatalysis and Green Chemistry

Roger A. Sheldon

Department of Biotechnology, Delft University of Technology, Delft, the Netherlands

Molecular Sciences Institute, School of Chemistry, University of the Witwatersrand, Johannesburg, South Africa

1.1 INTRODUCTION TO SUSTAINABLE DEVELOPMENT AND GREEN CHEMISTRY

The publication in 1987 of the report *Our Common Future* by the World Commission on Environment and Development, otherwise known as the Brundtland Report [1], marked the advent of the concept of sustainable development. The report recognized the necessity for industrial and societal development to provide a growing global population with a satisfactory quality of life, but that such development must also be sustainable over time. Sustainable development was defined as *development that meets the needs of the present generation without compromising the needs of future generations to meet their own needs.* In order to be sustainable, it must fulfill two conditions: (i) natural resources should be used at rates that do not unacceptably deplete supplies over the long term, and (ii) residues should be generated at rates no higher than can be assimilated readily by the natural environment [2]. It is abundantly clear, for example, that a society based on nonrenewable fossil resources—oil, coal, and natural gas—is not sustainable in the long term. Sustainability consists of three components: societal, ecological, and economic, otherwise referred to as the three P's—people, planet, and profit.

At the same time, in the mid-1980s, there was a growing concern regarding the copious amounts of waste being generated by the chemical industry. Clearly, a paradigm shift was needed from traditional concepts of reaction efficiency and selectivity, which focus largely on chemical yield, to one that assigns value to maximization of raw materials utilization, elimination of waste, and avoiding the use of toxic and/or hazardous substances [3]. By the same token, there was a pressing need for alternative, cleaner chemistry in order to minimize these waste streams. It led to the emergence of the concepts of waste minimization, zero waste plants, and green chemistry [4]. The latter can be succinctly defined as [5]:

> Green chemistry efficiently utilizes (preferably renewable) raw materials, eliminates waste and avoids the use of toxic and/or hazardous reagents and solvents in the manufacture and application of chemical products.

Originally it was referred to as "clean chemistry" [6]. The now widely accepted term "green chemistry" was introduced in the mid-1990s by Anastas and colleagues [7] of the US Environmental Protection Agency (EPA). The guiding principle is *benign by design* [8] as embodied in the 12 principles of green chemistry of Anastas and Warner:

Green Biocatalysis, First Edition. Edited by Ramesh N. Patel.

The 12 principles of green chemistry are as follows:

1. Waste prevention instead of remediation
2. Atom efficiency
3. Less hazardous materials
4. Safer products by design
5. Innocuous solvents and auxiliaries
6. Energy efficient by design
7. Preferably renewable raw materials
8. Shorter synthesis (avoid derivatization)
9. Catalytic rather than stoichiometric reagents
10. Design products for degradation
11. Analytical methodologies for pollution prevention
12. Inherently safer processes

Green chemistry eliminates waste at source; that is, it is primary pollution prevention rather than end-of-pipe waste remediation, as is inherent in the first principle of green chemistry: prevention is better than cure. Since the mid-1990s, the concept of green chemistry has been widely embraced in both industrial and academic circles [9]. One could say that sustainable development is our ultimate common goal and green chemistry is a means to achieving it.

1.2 GREEN CHEMISTRY METRICS

In order to know whether one process or product is greener than another one, we need meaningful metrics to measure greenness. The most widely accepted metrics of the environmental impact of chemical processes are, probably not coincidentally, the two most simple ones: the E factor [3–6, 10, 11], defined as the mass ratio of waste to desired product, and the atom economy (AE), defined as the molecular weight of the desired product divided by the sum of the molecular weights of all substances produced in the stoichiometric equation, expressed as a percentage [12, 13]. Knowledge of the stoichiometric equation enables one to predict, without performing any experiments, the theoretical amount of waste that will be formed. In Figure 1.1, for example, the AE of the classical chlorohydrin route to ethylene oxide is compared with that of catalytic oxidation with dioxygen. It is interesting to note that the former process produces, on a weight basis, more calcium chloride than ethylene oxide.

The AE is a theoretical number that is based on the assumption that a chemical yield of 100% of the theoretical yield is obtained and that reactants are used in exactly stoichiometric amounts. Furthermore, it disregards substances, such as solvents and acids or bases used in work-up, which do not appear in the stoichiometric equation. The E factor, in contrast, is the actual amount of waste produced in the process, defined as everything but the desired product. It takes the chemical yield into account and includes all reagents, solvent losses, all process aids, and, in principle, even the energy consumed.

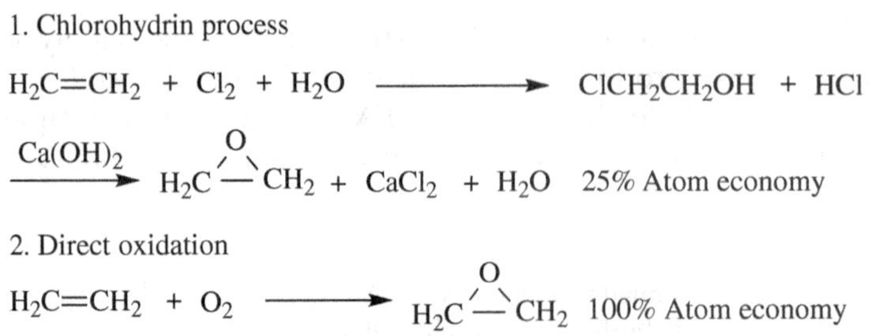

FIGURE 1.1
Atom efficiencies of two processes for ethylene oxide.

Originally [3] water was excluded from the calculation of the E factor as it was thought that its inclusion would lead to exceptionally high E factors in many cases and make meaningful comparisons of processes difficult. There is a definite trend, however, especially in the pharmaceutical industry, toward the inclusion of water in the E factor. The ideal E factor is zero, that is, zero waste. A higher E factor means more waste and, consequently, greater negative environmental impact. Alternatively, one can view the E factor as kilograms of raw materials minus kilograms of desired product, divided by kilograms of product out. It is easily calculated from knowledge of the number of tons of raw materials purchased and the number of tons of product sold. This method of calculation automatically excludes water used in the process, but not the water formed. Typical E factors for various segments of the chemical and allied industries, which we originally estimated in 1992, indicated that the fine chemical and pharmaceutical industries created a substantial waste burden [3]. A modified version of the original presentation, in which the oil-refining segment has been replaced by basic petrochemical hydrocarbon feedstocks, is shown in Table 1.1. This table also contains ranges of (average) annual product tonnages, which allow the annual tonnages of waste to be estimated. Such calculations could have been done in the original presentation, but we refrained from doing so because the relatively low figures for the annual waste tonnages for the pharmaceutical industry could be construed as a reason for inaction on the part of this industry segment, whereas E factors clearly show the need for action.

The substantial increase in E factors on moving downstream from bulk chemicals to fine chemicals and, particularly, pharmaceuticals is a reflection of the increasing molecular complexity of the products and associated multistep syntheses, which can be expected to generate more waste. Consequently, waste generation can be reduced by developing processes that are more step economic as advocated by Wender [14]. In bulk chemicals and basic hydrocarbon feedstock manufacture, in contrast, target molecules are simpler and require a smaller number of steps for their synthesis. This is not the whole story, however. The high E factors in pharma and fine chemicals are also a direct consequence of the widespread use of stoichiometric inorganic and organic reagents in these industry segments. In bulk chemicals manufacture, in contrast, because of the enormous production volumes, the use of stoichiometric reagents is economically prohibitive. We also note that E factors for the production of therapeutic proteins (biopharmaceuticals) on a commercial scale are even higher [15]. The E factor has been widely adopted by the chemical industry—in particular by the pharmaceutical industry [16], as a useful metric for assessing the environmental impact of manufacturing processes [17, 18] and has been shown to be predictive of reductions in manufacturing costs [19].

The number of green metrics subsequently proliferated [20–23]. They can be divided into two types: (i) metrics that are a refinement of the AE concept and (ii) metrics that are variations of the E factor (see Figure 1.2). Examples of the former are reaction mass efficiency (RME) and carbon efficiency (CE) introduced by Constable and coworkers [24] at GlaxoSmithKline (GSK). The RME is defined as the mass of

TABLE 1.1 E Factors in the Chemical and Allied Industries

Industry Segment	Annual Product Tonnage	E Factor (kg waste/ kg product)	Total Annual Waste Tonnage
Basic petrochemicals (ethylene, propylene, butadiene, ethylbenzene)	10 000 000–100 000 000	~0.1	10 000 000
Bulk chemicals (propylene oxide, caprolactam)	10 000–1 000 000	<1 to 5	5 000 000
Fine chemicals (flavors and fragrances, cosmetic ingredients)	100–10 000	5 to >50	500 000
Pharmaceuticals	10–1 000	25 to >100	100 000

E factor

$$E = \frac{\text{Total mass of waste}}{\text{Mass of final product}}$$

Atom efficiency (AE)

$$AE\ (\%) = \frac{\text{m.w of product} \times 100}{\Sigma\ \text{m.w. of reactants}}$$

Mass intensity (MI)

$$MI = \frac{\text{Total mass used in a process}}{\text{Mass of product}}$$

Reaction mass efficiency (RME)

$$RME\ (\%) = \frac{\text{Mass of product C} \times 100}{\text{Mass of A} + \text{Mass of B}}$$

FIGURE 1.2
Green chemistry metrics.

product obtained divided by the total mass of reactants in the stoichiometric equation, expressed as a percentage. It is a refinement of the AE that takes the chemical yield of the product and the actual quantities of reactants used into account. A disadvantage compared to AE is the requirement for experimental data to calculate the RME, which, therefore, cannot be used for rapid analysis of different processes prior to experimental work being performed. The CE is similar to RME but takes only carbon into account, that is, it is the mass of carbon in the product obtained divided by the total mass of carbon present in the reactants.

An example of the second type is mass intensity (MI) [25], defined as the total mass of materials used in a process divided by the mass of product obtained, that is, MI = E factor + 1. An analogous metric, the effective mass yield (EMY), is defined as the mass of the desired product divided by the total mass of nonbenign reactants used in its preparation [26]. The EMY does not include so-called environmentally benign compounds, such as NaCl and acetic acid, but defining nonbenign is difficult and arbitrary.

The AE and E factor are complementary: the former can be used for a quick assessment, before conducting any experiments, while the latter is a measure of the total waste that is actually formed in practice. None of the alternative metrics offer any particular advantage over the AE and E factors for assessing how wasteful a process is. The ideal E factor is zero, which is a better reflection of the ultimate goal of zero waste manufacturing plants than the ideal MI of 1. Moreover, the E factor concept is mathematically simpler since *step E factor contributions are additive while step PMI contributions are not*, because the PMI does not discount the step product from the step mass balance [27].

1.3 ENVIRONMENTAL IMPACT AND SUSTAINABILITY METRICS

Sustainability encompasses the conservation of the Earth's natural resources and minimization of the effect of industrial activities on the health of its inhabitants and the natural environment in addition to economic viability. Green chemistry embodies essentially the same two elements: (i) efficient utilization of raw materials and elimination of waste and (ii) health, safety, and environmental aspects of chemicals and their manufacturing processes but without the economic component. However, the metrics discussed in the preceding section take only the mass of waste generated into account, whereas the environmental impact of this waste is also determined by its nature. Hence, we introduced [6] the term "environmental quotient" (EQ) to take the nature of the waste into account. EQ is the product of the E factor and an unfriendliness multiplier, Q. The latter is dependent on various factors such as toxicity, ease of recycling, etc., and can also be influenced by both the production volume and the location of the facilities. For example, the generation of 100–1000 tons per annum of sodium chloride is unlikely to present a waste problem, but 10 000 tons per annum, in contrast, may already present a disposal problem, thus warranting an increase in Q. Ironically, when very large quantities of sodium chloride are generated, the Q value could decrease again as recycling by electrolysis becomes viable, for example, in propylene oxide manufacture via the chlorohydrin route (see earlier). Hence, the Q value of a particular waste will be determined by, *inter alia*, its ease of disposal or recycling. Generally speaking, organic waste is more easily remediated than inorganic waste, which can be important when considering the green metrics of biocatalytic processes.

Since the mid-2000's, several groups have addressed the problem of quantifying *Q*. For example, Eissen and Metzger [28] developed the Environmental Assessment Tool for Organic Syntheses (EATOS) software in which metrics related to health hazards and persistence and bioaccumulation and ecotoxicity were used to determine the environmental index of the input (substrates, solvents, etc.) and the output (product and waste). Similarly, Saling and coworkers at BASF [29–31] introduced eco-efficiency analyses, which took both economic and environmental aspects into account, including energy, raw materials, emissions, toxicity, hazards, and land use.

The basis for such an analysis is life-cycle assessment (LCA) [32, 33], which is used to assess the environmental impact and sustainability of products and processes within defined domains, for example, cradle to gate, cradle to grave, and gate to gate, on the basis of quantifiable environmental impact indicators, such as energy usage, global warming, ozone depletion, acidification, eutrophication, smog formation, and ecotoxicity, in addition to waste generated. Jessop and coworkers [34], for example, used a combination of eight LCA environmental impact indicators—acidification, ozone depletion, smog formation, global warming, human toxicity by ingestion and inhalation, persistence, bioaccumulation, and abiotic resource depletion—in a gate-to-gate assessment of the greenness of alternative routes to a particular product. The outcome of an LCA resembles an EQ in that it constitutes an integration of the amount of waste with quantifiable environmental indicators based on the nature of the waste.

1.4 SOLVENTS

Typically, solvents constitute more than half of the materials used in a chemical process to produce a drug substance [35]. Consequently, a major source of waste in chemicals manufacture, and an important contributor to high E factors in pharma, is solvent losses, which generally end up in the atmosphere or in groundwater. Moreover, there are health and/or safety issues associated with many traditional organic solvents, such as chlorinated hydrocarbons.

The FDA has issued guidelines [36] for solvent use in the pharmaceutical industry and divided them into four classes:

Class 1 solvents should be avoided in the manufacture of drug substances because of their unacceptable toxicity or deleterious environmental effects. They include benzene and various chlorinated hydrocarbons.

Class 2 solvents should be used only sparingly in pharmaceutical processes because of inherent toxicity. They include acetonitrile, dimethyl formamide, methanol, and dichloromethane.

Class 3 solvents are regarded as less toxic and of lower risk to human health and are, hence, preferred. They include many lower alcohols, esters, ethers, and ketones.

Class 4 solvents, for which no adequate data are available, include diisopropyl ether, methyl tetrahydrofuran, and isooctane.

Consequently, industrial attention is focused both on minimizing overall solvent use and in replacing many traditional organic solvents, such as chlorinated and aromatic hydrocarbons, by more environmental-friendly alternatives such as lower alcohols, esters, and some ethers such as methyl tert-butyl ether (MTBE). Several pharmaceutical companies have produced solvent selection guides to help their chemists in selecting greener, more sustainable solvents [37]. Pfizer, for example, classified solvents into three categories: preferred, usable, and undesirable with an advice regarding substitution of undesirable solvents [16, 38]. Sanofi scientists [39] divided solvents into four categories based on safety, health, and environmental hazards and other industrial issues: (i) recommended, (ii) substitution advisable, (iii) substitution requested, and (iv) banned. Similarly, GSK has a similar guide, with two safety criteria, one health criterion, three environmental criteria including life cycle scoring, and additional red flags, for example, for solvents governed by regulations [40, 41]. Solvents derived from renewable feedstocks, such as ethanol, ethyl lactate, and methyl tetrahydrofuran [42], are becoming popular reaction media as they are seen as "natural" and sustainable.

In the original inventory of E factors of various processes, we assumed [3], if data were not available, that solvents would be recycled by distillation and that this would involve a 10% loss. However, this was probably overoptimistic, certainly for the pharma industry where the widespread use of different solvents for the various steps in multistep syntheses makes recycling difficult owing to cross contamination. The best solvent is no solvent, but if a solvent is needed, it should be safe to use and there should be provisions for its efficient removal from the product and reuse.

1.5 THE ROLE OF CATALYSIS

The waste generated in the manufacture of fine chemicals and pharmaceuticals is largely due to the use of stoichiometric inorganic and organic reagents that are partially incorporated or not incorporated into the product. Typical examples include oxidations with inorganic oxidants such as chromium (VI) salts, permanganates, manganese dioxide, and stoichiometric reductions with metals (Na, Mg, Zn, Fe) and metal hydrides ($LiAlH_4$, $NaBH_4$). Similarly, stoichiometric amounts of mineral acids (H_2SO_4, HF, and H_3PO_4) and Lewis acids ($AlCl_3$, $ZnCl_2$, BF_3) are major sources of waste. The solution is evident: the substitution of antiquated stoichiometric methodologies with cleaner catalytic alternatives [43–45]. This is true elegance and efficiency in organic synthesis [46]. For example, catalytic hydrogenation, oxidation, and carbonylation are highly atom-efficient processes. Similarly, the use of recyclable solid (heterogeneous) acids and bases as catalysts results in substantial reductions in waste in industrial organic synthesis [47, 48]. Indeed, several pharma companies have developed reagent guides for particular reaction types with the aim of improving the greenness and sustainability of their processes [41].

The ultimate in step and AE is the development of catalytic cascade processes whereby several catalytic steps are integrated in one-pot procedures without the need for isolation of intermediates [49]. Such "telescoping" of multistep syntheses into catalytic cascades has several advantages—fewer unit operations, less solvent and reactor volume, shorter cycle times, higher volumetric and space-time yields, and less waste (lower E factor)—that afford substantial economic and environmental benefits. Furthermore, coupling of reactions can be used to drive equilibria toward product, thus avoiding the need for excess reagents.

1.6 BIOCATALYSIS AND GREEN CHEMISTRY

Biocatalysis has many attractive features in the context of green chemistry and sustainable development:

1. The catalyst (an enzyme) is derived from renewable resources and is biocompatible (sometimes even edible), biodegradable, and essentially nonhazardous, that is, it fulfills the criteria of sustainability remarkably well.

2. Biocatalysis avoids the use of, and contamination of products by, scarce precious metals such as palladium, platinum, and rhodium. The long-term commercial viability of many "endangered" elements, such as various noble metals, is questionable. Moreover, the costs of removing traces of noble metals, to an acceptable level, from end products can be substantial.
3. Reactions are performed in an environmentally compatible solvent (water) under mild conditions (physiological pH and ambient temperature and pressure).
4. Reactions of multifunctional molecules proceed with high activities and chemo-, regio-, and stereoselectivities and generally without the need for functional group activation, protection, and deprotection steps required in traditional organic syntheses. This affords processes that are more step economic and more efficient in energy and raw material consumption, generate less waste, and are, therefore, both environmentally and economically more attractive than conventional routes.
5. As a direct result of the higher selectivities and milder reaction conditions, biocatalytic processes often afford products in higher purity than traditional chemical or chemo-catalytic processes.
6. Enzymatic processes (but not fermentations) can be conducted in standard multipurpose batch reactors and, hence, do not require any extra investment, for example, for high-pressure equipment.
7. Biocatalytic reactions are conducted under roughly the same conditions of temperature and pressure, and, hence, it is relatively easy to integrate multiple reactions into eco-efficient catalytic cascade processes [50].

In short, biocatalysis fits very well with the principles of green chemistry and sustainability. As Barry Commoner, the doyen of industrial ecology, observed [51]: "in nature there is no such thing as waste, everything is recycled." As shown in Table 1.2, biocatalysis conforms with 10 of the 12 principles of green chemistry and is not really relevant for the other two (principles 4 and 10), which are concerned with the design of safer, biodegradable products. Consequently, since the mid-1990's, biocatalysis has emerged as an important technology for meeting the growing demand for green and sustainable chemical manufacture [52, 53], particularly in the pharmaceutical industry [54, 55].

Thanks to advances in biotechnology and protein engineering techniques such as *in vitro* evolution [56], it is now possible to produce most enzymes for commercially

TABLE 1.2 Biocatalysis and the Principles of Green Chemistry

Green Chemistry Principles	Biocatalysis
1. Waste prevention	Enables more sustainable routes with significantly reduced waste
2. Atom economy	Enables more atom and step economic routes
3. Less hazardous syntheses	Generally low toxicity
4. Design for safer products	Not relevant
5. Safer solvents and auxiliaries	Usually performed in water or Class 3 solvents
6. Energy efficient	Mild conditions are conducive with energy efficiency
7. Renewable feedstocks	Enzymes are renewable
8. Reduce derivatization	Biocatalysis obviates the need for protection/deprotection
9. Catalysis	Enzymes are catalysts
10. Design for degradation	Not really relevant but enzymes themselves are biodegradable
11. Real-time analysis for pollution prevention	Can be applicable in biocatalytic processes
12. Inherently safer processes	Performed under mild and safe conditions

acceptable prices and to manipulate them such that they exhibit the desired properties with regard to, *inter alia*, substrate specificity, activity, selectivity, stability, and pH optimum [57, 58]. This has made it eminently feasible to optimize the enzyme to fit a predefined optimum process that is genuinely benign by design. Furthermore, the development of effective immobilization techniques has paved the way for optimizing the storage and operational stability and the recovery and recycling of enzymes [59]. In addition, the coimmobilization of two or more enzymes can afford multifunctional solid biocatalysts capable of catalyzing biocatalytic cascade processes [60].

Biocatalytic processes are performed with isolated enzymes or as whole-cell biotransformations. Isolated enzymes have the advantage of not being contaminated with other enzymes present in the cell. The use of whole cells, on the other hand, is less expensive as it avoids the separation and purification of the enzyme. In the case of dead cells, E factors of the two methods are essentially the same: the waste cell debris is separated before or after the biotransformation, respectively. In contrast, substantial amounts of waste biomass can be generated when using growing microbial cells in the fermentation processes. We note, however, that this waste is generally easy to dispose of, for example, as animal feed or can, in principle, be used as a source of energy for the process. Many fermentation processes also involve the formation of copious amounts of inorganic salts that may even be the major contributor to waste. E factors have generally not been calculated for fermentations, but published data [61] regarding mass balances can be used to calculate E factors. The E factor for the bulk fermentation product—citric acid, for example—is 1.4, which compares well with the E factor range of <1–5 typical of bulk petrochemicals. Interestingly, ca. 75% of the waste is accounted for by an inorganic salt, calcium sulfate. If water is included in the calculation, the E factor becomes 17. In contrast, small-volume fermentation processes for low-volume, high-added-value biopharmaceuticals can have extremely high E factors, even when compared with those observed in the production of small-molecule drugs. The fermentative production of recombinant human insulin [15], for example, involves an E factor of ca. 6600 and inclusion of water affords an astronomical E factor of 50 000! In contrast, biocatalysis with isolated enzymes tends to involve significantly higher substrate concentrations and combines a higher productivity with a lower water usage compared to fermentations.

1.7 EXAMPLES OF GREEN BIOCATALYTIC PROCESSES

1.7.1 A Chemoenzymatic Process for Pregabalin

Pfizer scientists have described [62] a second-generation chemoenzymatic process (Figure 1.3) for the manufacture of pregabalin, the active ingredient of the CNS drug Lyrica. It represented a dramatic improvement in process efficiency compared to earlier routes. The stereocenter was set early in the synthesis in accordance with the golden rule of chirotechnology [63], and the wrong enantiomer could be easily racemized and reused. The key enzymatic step was conducted with an inexpensive, readily available laundry detergent lipase at a staggering substrate concentration of 765 g/l. Organic solvent usage was dramatically reduced in a largely aqueous process. Compared to the first-generation manufacturing process, the new process afforded a higher yield and a fivefold reduction in the E factor from 86 to 17.

1.7.2 A Three-Enzyme Process for Atorvastatin Intermediate

Codexis scientists developed and commercialized a green-by-design, three-enzyme process for the synthesis of a key intermediate (Figure 1.4) in the manufacture of atorvastatin, the active ingredient of the cholesterol-lowering drug Lipitor [64, 65]. In the first step, ethyl-4-chloroacetoacetate undergoes highly enantioselective reduction

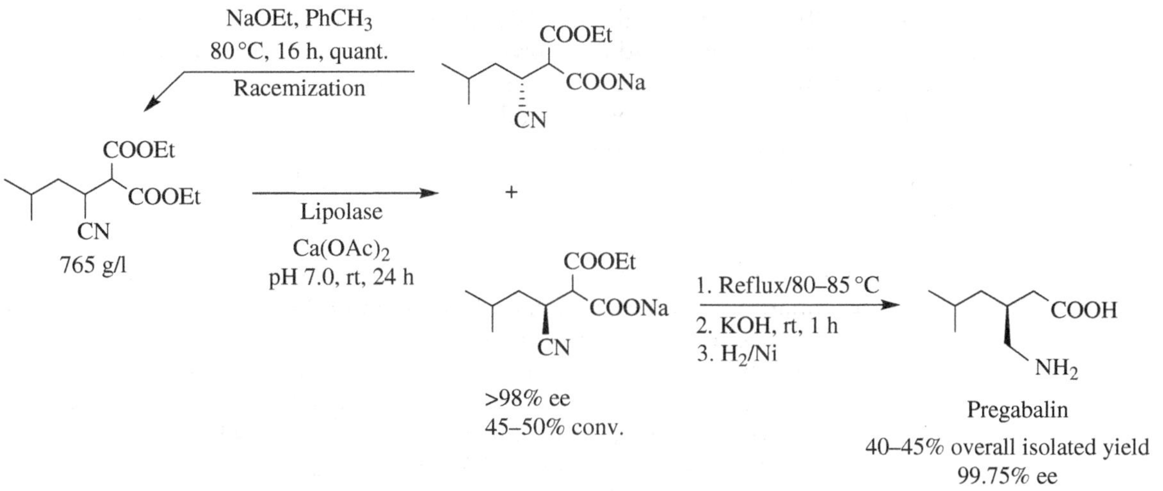

FIGURE 1.3

Chemoenzymatic process for pregabalin.

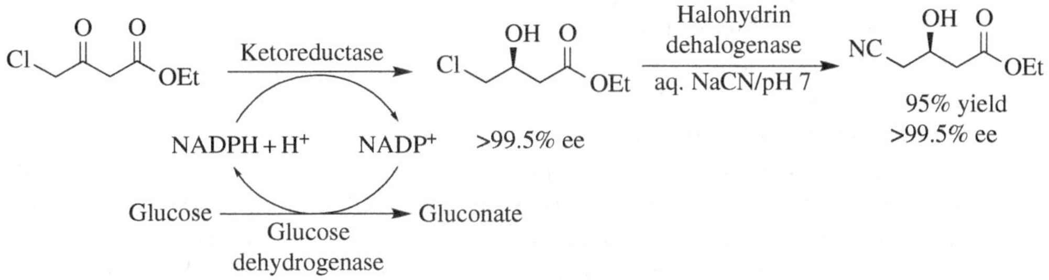

FIGURE 1.4

A two-step three-enzyme process for atorvastatin intermediate.

catalyzed by a ketoreductase (KRED). Cofactor regeneration was achieved with glucose as the hydrogen donor and an NADP-dependent glucose dehydrogenase (GDH) as the catalyst. The (*S*) ethyl-4-chloro-3-hydroxybutyrate product was obtained in 96% isolated yield and >99.5% ee. In the second step, a halohydrin dehalogenase (HHDH) was employed to catalyze a nucleophilic substitution of chloride by cyanide using HCN at neutral pH and ambient temperature.

All previous manufacturing routes to the hydroxynitrile product employed, as the final step, a standard S_N2 substitution of halide with cyanide ion in alkaline solution at elevated temperatures. This resulted in extensive by product formation owing to the base sensitivity of both substrate and product. Since the product is high-boiling oil, troublesome and expensive high-vacuum fractional distillation is required to recover product of acceptable quality, resulting in further yield losses and more waste. Hence, the key to designing an economically and environmentally attractive process was to conduct the cyanation reaction at ambient temperature and neutral pH using the enzyme, HHDH as the catalyst. Overall this afforded an elegant two-step, three-enzyme process for the hydroxynitrile product.

Unfortunately, the wild-type KRED and GDH exhibited prohibitively low activities, and large enzyme loadings were required to obtain an economically viable reaction rate. This resulted in troublesome emulsion formation and associated yield losses in downstream processing. Fortunately, the enzyme loadings could be drastically reduced by employing *in vitro* evolution via DNA shuffling [66] to improve the activity and stability of KRED and GDH. The GDH activity was improved by a factor of 13 and the KRED activity by a factor of 7 while maintaining the nearly perfect enantioselectivity (>99.5%) of the wild-type KRED. With the improved enzymes, the reaction was complete in 8 h with a substrate loading of 160 g/l and phase separation required <1 min, providing the chlorohydrin in >95% isolated yield and >99.9% ee.

Similarly, the activity of the wild-type HHDH in the nonnatural cyanation reaction was extremely low, and the enzyme exhibited severe product inhibition and poor stability under operating conditions. However, after many iterative rounds of DNA shuffling, the inhibition was largely overcome and the HHDH activity was increased more than 2500-fold compared to the wild-type enzyme.

The greenness of process was assessed according to the 12 principles of green chemistry:

Principle 1—waste prevention: The highly selective biocatalytic reactions afforded a substantial reduction in waste, and by avoiding by product formation, the need for yield-sacrificing fractional distillation was circumvented. The butyl acetate and ethyl acetate solvents, used in the extraction of the product from the aqueous layer in the first and second steps, respectively, were recycled with an efficiency of 85%. The E factor for the overall process is 5.8 if process water is excluded (2.3 for the reduction and 3.5 for the cyanation). If process water is included, the E factor for the whole process is 18 (6.6 for the reduction and 11.4 for the cyanation). The main contributors to the E factor (Table 1.3) are solvent losses (51%), sodium gluconate (25%), and the innocuous inorganic salts, NaCl and Na_2SO_4 (combined ca. 22%). The three enzymes and the NADP cofactor account for <1% of the waste. Furthermore, the main waste streams are aqueous and directly biodegradable.

Principle 2—AE: The use of glucose as the reductant for cofactor regeneration is cost-effective, but the AE is poor (45%). However, glucose is an inexpensive renewable raw material and the gluconate coproduct is fully biodegradable.

Principle 3—less hazardous chemical syntheses: The reduction reaction uses essentially nontoxic starting materials and avoids the use of potentially hazardous hydrogen and heavy metal catalysts obviating concern for their removal from waste streams and/or contamination of the product. While cyanide must be used in the second step, as in all practical routes to the product, it is used more efficiently (higher yield) and under less harsh conditions compared to previous processes.

Principle 4—design safer chemicals: This is not applicable as the hydroxynitrile product is the target molecule.

Principle 5—safer solvents and auxiliaries: Safe and environmentally acceptable ethyl acetate and butyl acetate are used, together with water, as cosolvent in the biocatalytic reduction reaction and extraction of the hydroxynitrile product. No auxiliaries are needed.

Principles 6 and 9—design for energy efficiency and catalysis: The process constitutes very efficient biocatalysis with turnover numbers of $>10^5$ for KRED and GDH and $>5\times10^4$ for HHDH. In contrast with previous processes, which employ elevated temperatures for the cyanation step and high-pressure hydrogenation

TABLE 1.3 E Factor of the Process for Atorvastatin Intermediate

Waste	Quantity (kg/kg product)	% of E (Excl. Water)	% of E (Incl. Water)
Substrate losses (8%)	0.09	<2	<1
Triethanolamine	0.04	<1	<1
NaCl and Na_2SO_4	1.29	22	ca. 7
Na-gluconate	1.43	ca. 25	ca. 9
BuOAc (85% recycle)	0.46	ca. 8	ca. 3
EtOAc (85% recycle)	2.50	ca. 43	ca. 14
Enzymes	0.023	<1	<1
NADP	0.005	0.1	<0.1
Water	12.250	—	67
Total waste (E factor)	5.8 kg (18 with H_2O)		

for the reduction step, both steps in the biocatalytic process are run at or close to ambient temperature and pressure and pH7, and the very high energy demands of high-vacuum distillation are dispensed with altogether, resulting in substantial energy savings.

Principles 7 and 10—the use of renewable feedstocks and design for degradation: The enzyme catalysts and the glucose cosubstrate are derived from renewable raw materials and are completely biodegradable. The by-products of the reaction are gluconate, NADP (the cofactor), residual glucose, enzyme, and minerals, and the waste water is directly suitable for biotreatment.

Principle 8—reduce derivatization: The process avoids derivatization steps, that is, it is step economic and involves fewer unit operations than earlier processes, most notably by obviating the troublesome product distillation or bisulfite-mediated separation of dehydrated byproducts.

Principles 11 and 12—real-time analysis for pollution prevention and inherently safer chemistry: The reactions are run in pH-stat mode at neutral pH by computer-controlled addition of base. Gluconic acid generated in the first reaction is neutralized with an aq. NaOH, and HCl generated in the second step is neutralized with feed-on-demand aq. NaCN, regenerating HCN ($pK_a \sim 9$) *in situ*. This minimizes the overall concentration of HCN, affording an inherently safer process. The pH and the cumulative volume of added base are recorded in real time.

In short, the Codexis process is an excellent example of a *benign by design* biocatalytic process for the synthesis of an important pharmaceutical intermediate whereby successful commercialization is enabled by employing modern protein engineering to optimize enzyme performance.

1.7.3 Enzymatic Synthesis of Sitagliptin

Another relevant example is provided by the enzymatic synthesis of the antidiabetic, sitagliptin, which was codeveloped by Merck and Codexis workers [67] to replace a rhodium-catalyzed, high-pressure, asymmetric hydrogenation of an enamine. It involves an overall enantioselective reductive amination of a ketone using an (*R*)-transaminase-catalyzed reaction with isopropylamine (Figure 1.5). The starting point was an (*R*)-selective transaminase, which showed no activity with the ketone substrate. *In silico* studies were employed to identify what was needed to be able to fit the

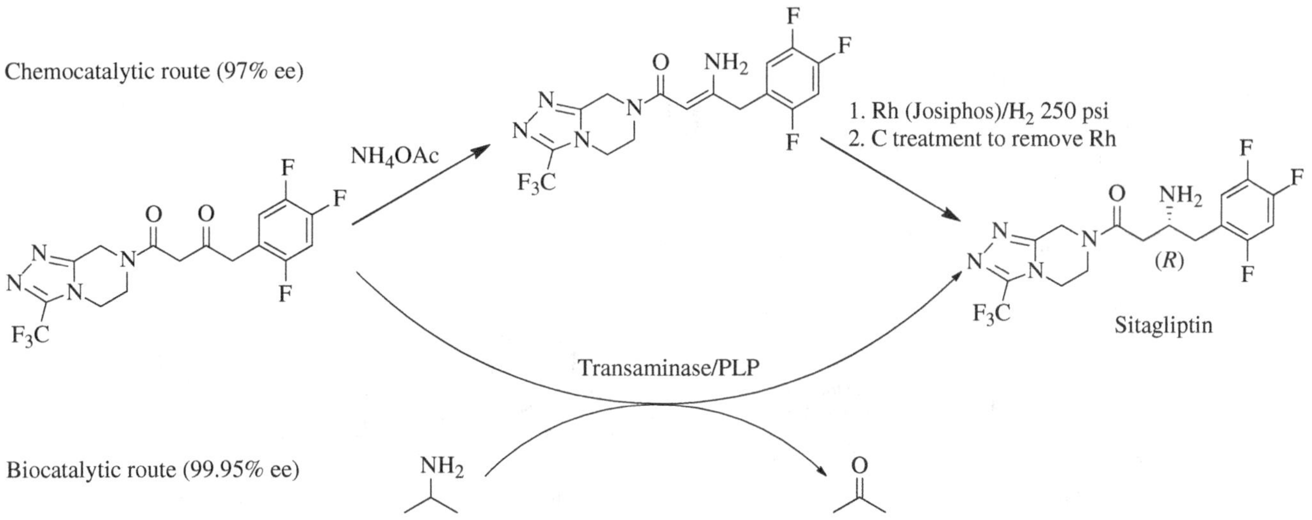

FIGURE 1.5
Two processes for sitagliptin.

ketone into the binding pocket of the enzyme. The amino residues surrounding the binding pocket were then engineered to provide the extra space leading to an enzyme with low activity, which was further improved up to a commercially viable level using *in vitro* evolution. Under optimized conditions, 6 g/l of the best variant in 50% aq. DMSO converted 200 g/l of the ketone substrate to sitagliptin of >99.95% ee with 92% yield. Compared with the rhodium-catalyzed asymmetric hydrogenation, the biocatalytic process displayed a 10–13% increase in overall yield and a 53% increase in productivity (kg/l/day). This resulted in a 19% reduction in total waste with the elimination of all heavy metals combined with a reduction in total manufacturing costs. Furthermore, the enzymatic reaction is run in multipurpose vessels, circumventing the need for specialized high-pressure hydrogenation equipment.

1.7.4 Biocatalytic Synthesis of the Fragrance Chemical (–) Ambrox (Ambrafuran)

The terpenoid molecule, (–)-8,12-epoxy-13,14,15,16-tetranorlabdane, so-called (–) ambrafuran or ambrox (a trade name of Firmenich SA), is one of the most valuable constituents of tincture of ambergris, a substance excreted by the sperm whale (*Physeter catodon* L.). It is much sought after by the perfumery industry and is produced commercially in a hemisynthesis from the diterpenic alcohol, sclareol [68]. The latter is readily extracted in sufficient quantities from clary sage (*Salvia sclarea* L.). The chemical synthesis (Figure 1.6) consists of seven steps involving long reaction times and hazardous reagents such as peracetic acid, lithium aluminum hydride, and butyl lithium; a stoichiometric oxidation with sodium periodate; and the generation of

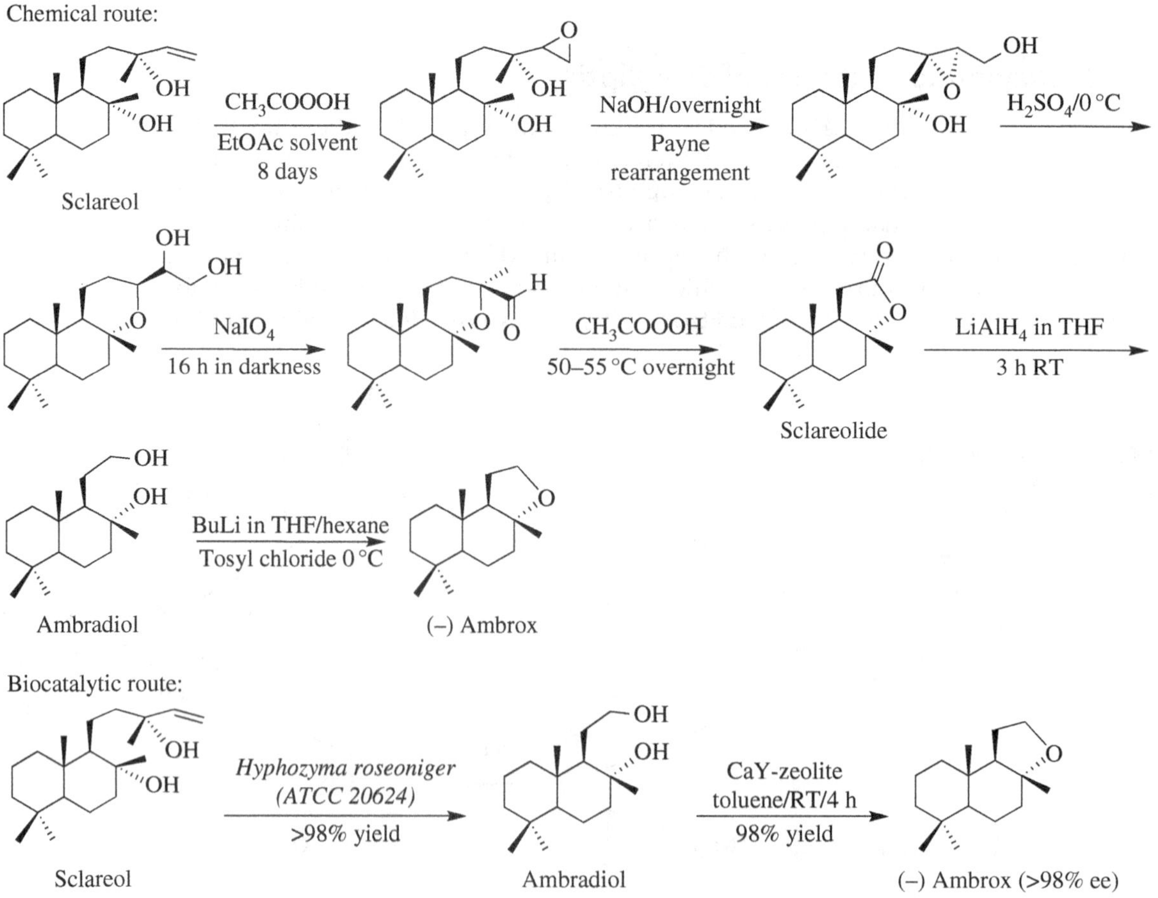

FIGURE 1.6
Two processes for (–) ambrafuran.

copious amounts of waste in addition to the 76% yield of the desired product. In stark contrast, a green, two-step process has been reported [69], which involves the conversion of sclareol to ambradiol as shown in Figure 1.6, catalyzed by whole cells of *Hyphozyma roseoniger*, followed by cyclization to (−) ambrafuran over a Ca-Y zeolite at ambient temperature, both steps proceeding in 98% yield.

1.8 CONCLUSIONS AND FUTURE PROSPECTS

Meaningful metrics for measuring greenness and sustainability are a *conditio sine qua non* for monitoring progress in the trend toward green manufacturing in the chemical and pharmaceutical industries. The widely accepted and complementary concepts of AE and E factors, together with an LCA, provide a sound basis for assessing the greenness and sustainability of different processes and products. Biocatalysis offers numerous benefits in this context. Reactions are conducted under mild conditions employing a catalyst that is biocompatible and biodegradable and derived from renewable resources, thus avoiding the scarcity and product contamination issues associated with the use of noble metal catalysts. Furthermore, processes are step economic and highly selective, resulting in higher product quality and reduced waste generation. In short, biocatalytic processes are green and sustainable, that is, they are more environmentally attractive and more cost-effective compared to classical chemical processes.

REFERENCES

[1] Brundtland, C. G., *Our Common Future*, The World Commission on Environmental Development, Oxford University Press, Oxford, 1987.
[2] Graedel, T. E., in Clark, J., Macquarrie, D. J., eds., *Handbook of Green Chemistry and Technology*, John Wiley & Sons, Inc., New York, 2002, pp. 56–61.
[3] (a) Sheldon, R. A., *Chem. Ind. (Lond.)*, 1992, 903–906; (b) Sheldon, R. A., *Chem. Ind. (Lond.)*, 1997, 12–15.
[4] Sheldon, R. A., *Chem. Soc. Rev.*, 2012, **41**, 1437–1451.
[5] Sheldon, R. A., *C. R. Acad. Sci. Paris IIc: Chim.*, 2000, **3**, 541–551.
[6] (a) Sheldon, R. A., *Chemtech*, March 1994, 38–47; (b) Sheldon, R. A., *J. Chem. Technol. Biotechnol.*, 1997, **68**, 381–388.
[7] Anastas, P. T., Warner, J. C., eds., *Green Chemistry: Theory and Practice*, Oxford University Press, Oxford, 1998.
[8] Anastas, P. T., Farris, C. A., eds., *Benign by Design: Alternative Synthetic Design for Pollution Prevention*, ACS Symposium Series No. 577, American Chemical Society, Washington, DC, 1994.
[9] Lipshutz, B. S., Ghorai, S., *Green Chem.*, 2014, **16**, 3660–3679.
[10] Sheldon, R. A., *Pure Appl. Chem.*, 2000, **72**, 1233–1246.
[11] Sheldon, R. A., *Green Chem.*, 2007, **9**, 1273–1283.
[12] Trost, B. M., *Science*, 1991, **254**, 1471–1477.
[13] Trost, B. M., *Angew. Chem. Int. Ed.*, 1995, **34**, 259–281.
[14] (a) Wender, P. A., Croatt, M. P., Witulski, B., *Tetrahedron*, 2006, **62**, 7505–7511; (b) Wender, P. A., Handy, S. T., Wright, D. L., *Chem. Ind. (Lond.)*, 1997, 765–769.
[15] Ho, S. V., in Dunn, P. J., Wells, A. S., Williams, M. T., eds., *Green Chemistry in the Pharmaceutical Industry*, Wiley-VCH, Weinheim, 2010, pp. 311–331.
[16] Alfonsi, K., Collberg, J., Dunn, P. J., Fevig, T., Jennings, S., Johnson, T. A., Kleine, H. P., Knight, C., Nagy, M. A., Perry, D. A., Stefaniak, M., *Green Chem.*, 2008, **10**, 31–36.
[17] Ritter, S. K., *Chem. Eng. News*, August 18, 2008, **86**, 59.
[18] Thayer, A. N., *Chem. Eng. News*, August 6, 2007, **85**, 11.
[19] Leahy, D. K., Tucker, J. L., Mergelsberg, I., Dunn, P. J., Kopach, M. E., Purohit, V. C., *Org. Process Res. Dev.*, 2013, **17**, 1099–1109.
[20] CalvoFlores, F. G., *ChemSusChem*, 2009, **2**, 905–919.
[21] Auge, J., *Green Chem.*, 2008, **10**, 225–231.
[22] Andraos, J., *Org. Process Res. Dev.*, 2005, **9**, 149–163.

[23] Lapkin, A., Constable, D. J. C., eds., *Green Chemistry Metrics: Measuring and Monitoring Sustainable Processes*, John Wiley & Sons, Ltd, Chichester, 2008.

[24] Constable, D. J. C., Curzons, A. D., Cunningham, V. L., *Green Chem.*, 2002, **4**, 521–527.

[25] (a) Curzons, A. D., Constable, D. J. C., Mortimer, N., Cunningham, V. L., *Green Chem.*, 2001, **3**, 1–6; (b) Jimenez-Gonzalez, C., Ponder, S., Broxterman, Q. B., Manley, J. B., *Org. Process Res. Dev.*, 2011, **15**, 912–917.

[26] Hudlicky, T., Frey, D. A., Koroniak, L., Claeboe, C. D. Brammer, L. E., *Green Chem.*, 1999, **1**, 57–59.

[27] Roschangar, F., Sheldon, R. A. C. H., Senanayake, C. H., *Green Chem.*, 2014, **16**, DOI:10.1039/c4gc01563k.

[28] Eissen, M., Metzger, J. O., *Chem. Eur. J.*, 2002, **8**, 3580–3585.

[29] Landsiedel, R., Saling, P., *Int. J. Life Cycle Assess*, 2002, **5**, 261–268.

[30] Saling, P., *Appl. Microbiol. Biotechnol.*, 2005, **68**, 1–8.

[31] (a) Shonnard, D. R., Kircherer, A., Saling, P., *Environ. Sci. Technol.*, 2003, **37**, 5340–5348; (b) Wall-Markowski, C. A., Kicherer, A., Saling, P., *Environ. Prog.*, 2004, **23**(4), 329–333.

[32] Jimenez-Gonzalez, C., Curzons, A. D., Constable, D. J. C., Cunningham, V. L., *Int. J. Life Cycle Assess*, 2004, **9**, 115–121.

[33] Moretz-Sohn Monteiro, J. G., de Queiroz Fernandes Araujo, O., de Medeiros, J. L., *Clean Technol. Environ. Policy*, 2009, **11**, 209–214 and 459–472.

[34] Mercer, S. M., Andraos, J., Jessop, P. G., *J. Chem. Educ.*, 2012, **89**, 215–220.

[35] Jimenez-Gonzalez, C., Ponder, S., Broxterman, Q. B., Manley, J. B., *Org. Process Res. Dev.*, 2011, **15**, 912–917.

[36] Manual of Policies and Procedures. http://www.fda.gov/downloads/AboutFDA/CentersOffices/CDER/ManualofPoliciesProcedures/ucm124407.pdf (accessed January 5, 2016).

[37] Prat, D., Hayler, J., Wells, A., *Green Chem.*, 2014, **16**, 4546–4551.

[38] Dunn, P. J., *Chem. Soc. Rev.*, 2012, **41**, 1452–1461.

[39] Prat, D., Pardigon, O., Flemming, H. W., Letestu, S., Ducandas, V., Isnard, P., Guntrum, E., Senac, T., Ruisseau, S., Cruciani, P., Hosek, P., *Org. Process Res. Dev.*, 2013, **17**, 1517–1525.

[40] (a) Henderson, R. K., Jimenez-Gonzalez, C., Constable, D. J. C., Alston, S. R., Inglis, G. G. A., Fisher, G., Sherwood, J., Binks, S. P., Curzons, A. D., *Green Chem.*, 2011, **13**, 854–862; (b) Jimenez-Gonzalez, C., Curzons, A. D., Constable, D. J. C., Cunningham, V. L., *Clean Technol. Environ. Policy*, 2005, **7**, 42–50; (c) Curzons, A. D., Constable, D. J. C., Cunningham, V. L., *Clean Prod. Process.*, 1999, **1**, 82–90.

[41] (a) MacMillan, D. S., Murray, J., Sneddon, H. F., Jamieson, C., Watson, A. J. B., *Green Chem.*, 2013, **15**, 596–600; (b) McGonagle, F. I., MacMillan, D. S., Murray, J., Sneddon, H. F., Jamieson, C., Watson, A. J. B., *Green Chem.*, 2013, **15**, 1159–1165.

[42] Pace, V., Hoyos, P., Castoldi, L., Dominguez de Maria, P., Alcantara, A. R., *ChemSusChem*, 2012, **5**, 1369–1379.

[43] Li, C. J., Trost, B. M., *Proc. Natl. Acad. Sci. U. S. A.*, 2008, **105**, 13197–13202.

[44] Sheldon, R. A., Arends, I. W. C. E., Handfed, U., *Green Chemistry and Catalysis*, Wiley-VCH, Weinheim, 2007.

[45] Sheldon, R. A., *Chem. Commun.*, 2008, 3352–3365.

[46] Noyori, R., *Chem. Commun.*, 2005, 1807–1811.

[47] Sheldon, R. A., van Bekkum, H., eds., *Fine Chemicals Through Heterogeneous Catalysis*, Wiley-VCH, Weinheim, 2001, Chapters 3–7.

[48] (a) Thomas, J. M., Hernandez-Garrido, J. C., Bell, R. G., *Top. Catal.*, 2009, **52**, 1630–1639; (b) Kaneda, K., Mizugaki, T., *Energy Environ. Sci.*, 2009, **2**, 655–673.

[49] Bruggink, A., Schoevaart, R., Kieboom, T., *Org. Process. Res. Dev.*, 2003, **7**, 622–640.

[50] Sheldon, R. A., in *Multi-Step Enzyme Catalysis: Biotransformations and Chemoenzymatic Synthesis*, Garcia-Junceda, E., ed., Wiley-VCH, Weinheim, 2008, pp. 109–135.

[51] Commoner, B., *The Closing Circle*, Bantam Books, New York, 1971.

[52] Tao, J., Kazlaukas, R. J., *Biocatalysis for Green Chemistry and Chemical Process Development*, John Wiley & Sons, Inc., Hoboken, NJ, 2011.

[53] Wohlgemuth, R., *Curr. Opin. Biotechnol.*, 2010, **21**, 713–724.

[54] (a) Hoyos, P., Pace, V., Hernaiz, M. J., Alcantara, A. R., *Curr. Green Chem.*, 2014, **1**(2), 155–181; (b) Munoz Solano, D., Hoyos, P., Hernaiz, M. J., Alcantara, A. R., Sanchez-Montero, J. M., *Bioresour. Technol.*, 2012, **115**, 196–207.

[55] (a) Tao, J., Xu, J.-H., *Curr. Opin. Chem. Biol.*, 2009, **13**, 43–50; (b) Zheng, G.-W., Xu, J.-H., *Curr. Opin. Biotechnol.*, 2011, **22**, 784–792; (c) Huisman, G. W., Collier, S. J., *Curr. Opin. Chem. Biol.*, 2013, **17**, 284–292.

[56] (a) Tracewell, C. A., Arnold, F. A., *Curr. Opin. Chem. Biol.*, 2009, **13**, 3–9; (b) Turner, N. J., *Nat. Chem. Biol.*, 2009, **5**, 567–573; (c) Reetz, M. T., *J. Org. Chem.*, 2009, **74**, 5767–5778; (d) Luetz, S., Giver, L., Lalonde, J., *Biotechnol. Bioeng.*, 2008, **101**, 647–653; (e) Woodley, J. M., *Curr. Opin. Chem. Biol.*, 2013, **17**, 310–316.

[57] Bornscheuer, U. T., Huisman, G. W., Kazlauskas, R. J., Lutz, S., Moore, J. C., Robins, K., *Nature*, 2012, **485**, 185–194.

[58] Illanes, A., Cauerhff, A., Wilson, L., Castro, G. R., *Bioresour. Technol.*, 2012, **115**, 48–57.

[59] (a) Sheldon, R. A., van Pelt, S., *Chem. Soc. Rev.*, 2013, **42**, 6223–6235; (b) Sheldon, R. A., *Adv. Synth. Catal.*, 2007, **349**, 1289–1307.

[60] (a) Chmura, A., Rustler, S., Paravidino, M., van Rantwijk, F., Stolz, A., Sheldon, R. A., *Tetrahedron: Asymmetry*, 2013, **24**, 1225–1232; (b) van Pelt, S., van Rantwijk, F., Sheldon, R. A., *Adv. Synth. Catal.*, 2009, **351**, 397–404.

[61] Petrides, B., in Harrison, R. G., Todd, P. W., Rudge, S. R., Petrides, D., eds., Bioprocess design in *Bioseparations Science and Engineering*, Oxford University Press, New York, 2003, pp. 319–364.

[62] Martinez, C. A., Hu, S., Dumond, Y., Tao, J., Kelleher, P., Tully, L., *Org. Process Res. Dev.*, 2008, **12**, 392–398.

[63] Sheldon, R. A., *Chirotechnology: the Industrial Synthesis of Optically Active Compounds*, Marcel Dekker, New York, 1993.

[64] Fox, R. J., Davis, S. C., Mundorff, E. C., Newman, L. M., Gavrilovic, V., Ma, S. K., Chung, L. M., Ching, C., Tam, S., Muley, S., Grate, J., Gruber, J., Whitman, J. C., Sheldon, R. A., Huisman, G. W., *Nat. Biotechnol*, 2007, **25**, 338–344.

[65] Ma, S. K., Gruber, J., Davis, S. C., Newman, L., Gray, D., Wang, A., Grate, J., Huisman G. W., Sheldon, R. A., *Green Chem.*, 2010, **12**, 81–86.

[66] Stemmer, W. P., *Nature*, 1994, **370**, 389–391.

[67] Savile, C. K., Janey, J. M., Mundorff, E. C., Moore, J. C., Tam, S., Jarvis, W. R., Colbeck, J. C., Krebber, A., Fleitz, F. J., Brands, J., Devine, P. N., Huisman, G. W., Hughes, G. J., *Science*, 2010, **329**, 305–309.

[68] (a) Moulines, J., Bats, J.-P., Lamidey, A.-M., Da Silva, N., *Helv. Chim. Acta*, 2004, **87**, 2695–2705; (b) Moulines, J., Lamidey, A.-M., Desvergnes-Breuil, V., *Synth. Commun.*, 2001, **31**(5), 749–758.

[69] Steenkamp, L. H., Taka, M., US Patent Appl, 2010, US 2010/0248316 A1 to CSIR.

CHAPTER 2

Enzymatic Synthesis of Chiral Amines using ω-Transaminases, Amine Oxidases, and the Berberine Bridge Enzyme

Eduardo Busto, Robert C. Simon, Nina Richter, and Wolfgang Kroutil

Department of Chemistry, Organic and Bioorganic Chemistry, University of Graz, Graz, Austria

2.1 INTRODUCTION

Chiral amines can be found in a broad range of naturally occurring compounds such as alkaloids, hormones, amino acids (AAs), or proteins displaying a broad range of biological activities [1, 2]. Various synthetic analogs play a central role as key intermediates in the preparation of valuable chiral compounds such as pharmaceuticals, agrochemicals, organocatalysts, etc. [3]. As a consequence, there is a demand for the chemical industry to prepare efficiently a broad range of structurally diverse amines in an optically pure form.

Consequently, the development of novel stereoselective methods has increased since the mid-2000's. [4] including conventional chemical routes such as asymmetric hydrogenation using transition metals [5, 6] or organocatalyzed reactions [7]. The global market is expected to grow substantially in the near future; therefore, improved methodologies are required to meet the strict regulations in terms of process quality and safety for the pharmaceutical industry. Additionally, process optimization in terms of cost, waste production, time, and energy consumption is necessary to provide the target compounds in an economical and environmental-friendly fashion [8].

Biocatalysis has become a well-accepted synthetic tool for the production of chiral compounds, showing impressive levels of chemo-, regio-, and stereoselectivities [9]. Reactions are usually performed under mild and eco-friendly reaction conditions not only in aqueous buffer but also in neoteric or organic solvents meeting the principles of green chemistry [10, 11]. Protein engineering and high-throughput screening methods allow the development of customized protein variants specifically designed for a target substrate or given reaction conditions (high pH and/or temperature, and/or stability toward organic solvents, and/or high substrate concentration) suitable for industrial applications [12–14].

Green Biocatalysis, First Edition. Edited by Ramesh N. Patel.

While first examples regarding the biocatalytic preparation of chiral amines mainly involved kinetic resolution processes using hydrolases [15], in the last years, the identification of novel biocatalysts including amine oxidases (AOs), amino acid oxidases (AAOs), lyases, or ω-transaminases (ω-TAs) has provided new catalytic tools for the production of chiral amines with high stereoselectivity [16–18].

For this chapter, outstanding examples were selected where ω-TAs, amine oxidases, and the berberine bridge enzyme (BBE) were used as catalysts for the preparation of chiral amines. The account is divided into three different sections according to the type of enzyme—namely, ω-TAs, AOs or AAOs, and the BBE. The chapter focuses on the scope and practical applications, but it also discusses the present limitations of each approach.

2.2 SYNTHESIS OF CHIRAL AMINES USING ω-TRANSAMINASES

2.2.1 ω-Transaminases: Definition and General Facts

Transaminases or aminotransferases (EC 2.6.1.X) represent a broad group of enzymes, which catalyze the transfer of an amino group from an amine donor to a carbonyl compound (an acceptor). Reactions are in general reversible. Transaminases are usually classified depending on the substrate specificity as α- or ω-TAs: α-transaminases (α-TAs) are highly specific enzymes for the amination of exclusively α-keto acids [19, 20]. These enzymes have been successfully used for the preparation of natural and nonnatural AAs [4]. On the other hand, ω-TAs display a broader substrate scope being able to accept ketone or aldehyde as amine acceptor without requiring the presence of an adjacent carboxylic acid moiety, making them very interesting from a synthetic point of view. Both groups of enzymes require a cofactor, namely, pyridoxal-5′-phosphate (PLP) as the electron shuttle between the amine donor and the amine acceptor (Scheme 2.1). First, the amino group is transferred from the amine donor to PLP, affording the reductively aminated cofactor pyridoxamine-5′-phosphate (PMP) and the formally oxidized/deaminated coproduct. The amine donor is usually used in over stoichiometric amounts. Next, the amino group is formally transferred from PMP to the carbonyl compound, thereby regenerating PLP and closing the catalytic cycle (Scheme 2.1).

2.2.2 Stereoselective Transformations Involving ω-TAs

The synthesis of chiral amines using ω-TA can be performed by three different pathways (Scheme 2.2) [19]: (i) kinetic resolution of amines by oxidative deamination, (ii) stereoselective synthesis via amination of prochiral ketones, and (iii) deracemization combining (i) kinetic resolution and (ii) stereoselective synthesis [19].

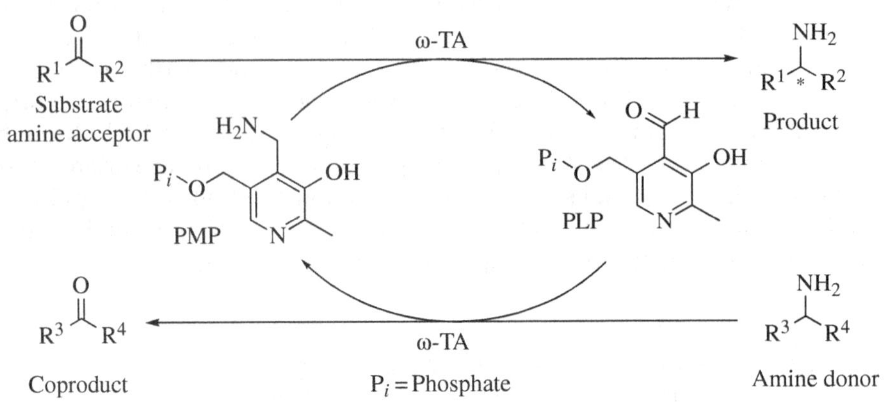

SCHEME 2.1
Catalytic cycle for the ω-TA-catalyzed amination. PLP, pyridoxal-5′-phosphate; PMP, pyridoxamine-5′-phosphate.

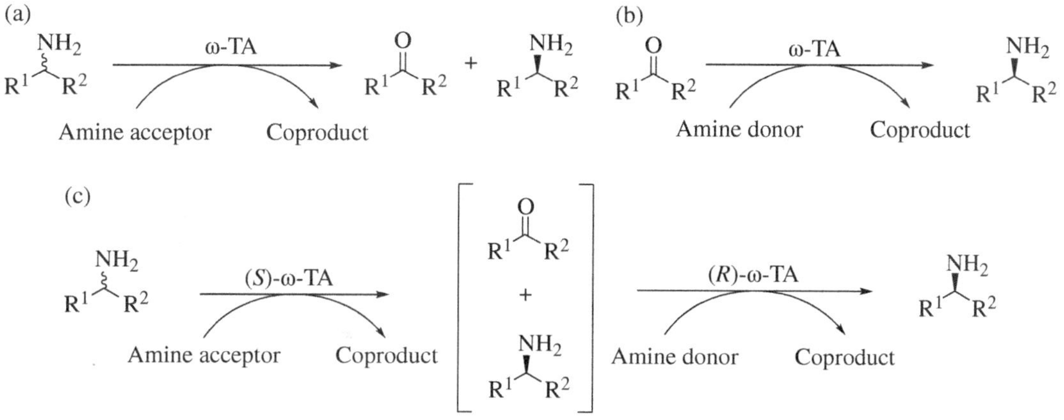

SCHEME 2.2
Stereoselective transformations involving ω-TAs. (a) Kinetic resolution, (b) stereoselective synthesis via asymmetric amination, and (c) deracemization.

The kinetic resolution of amines was exploited first with ω-TAs since the reaction is thermodynamically favored using pyruvate as acceptor [21]. Unfortunately, the inherent maximum 50% yield limits the practical applications of this approach and, therefore, will not be considered in this contribution. Thus, the other methodologies (stereoselective synthesis and deracemization) are preferred since a theoretical yield of 100% is possible. The asymmetric amination allows the preparation of enantiomerically pure amines at the expense of an amine donor, usually an amine or an AA (Scheme 2.2b). On the other hand, the combination of two stereocomplementary ω-TAs enables to establish a deracemization process (Scheme 2.2c). Deracemization strategies are recommended when amines are readily available or the corresponding carbonyl compounds lack stability.

2.2.3 Asymmetric Amination of Ketones

As already mentioned, the asymmetric reductive amination of ketones provides an excellent approach for the production of enantiomerically pure amines with 100% theoretical yield. However, the amination is usually thermodynamically uphill for most amine donors, meaning that the equilibrium is usually displaced to the side of the reagents. As a consequence, different strategies have been established to overcome this limitation. For instance, the use of high excesses of amine donor combined with removal of the carbonyl coproduct by evaporation is a good option when employing 2-propylamine or 2-butylamine as amine donors (Scheme 2.3) [22–24]. However, only a few ω-TAs are able to accept linear amines as amine donors, and most of them prefer alanine. In this case, the thermodynamic barrier can be overcome by recycling/removing the coproduct pyruvate with a coupled multienzyme network; in most cases a lactate dehydrogenase (LDH) and an alanine dehydrogenase (AlaDH) are employed. By using an LDH, pyruvate is reduced to lactate at the expense of NADH, which is recycled using, for example, a glucose dehydrogenase (GDH) at the expense of glucose, usually providing the corresponding amines in excellent conversions [25]. In an alternative approach, an AlaDH catalyzes the recycling of pyruvate to alanine using ammonia and NADH [26]. The oxidized cofactor is recycled, for example, by formate dehydrogenase (FDH) or alternatively glucose and GDH. Alternatively, 1-phenylethylamine shows favorable thermodynamics and is widely accepted by ω-TAs although the stereoselective enzymes consume exclusively one enantiomer; thus 1-phenylethylamine has to be provided in optically pure form [27, 28]. In case the racemic donor is used, kinetic resolution of the donor takes place.

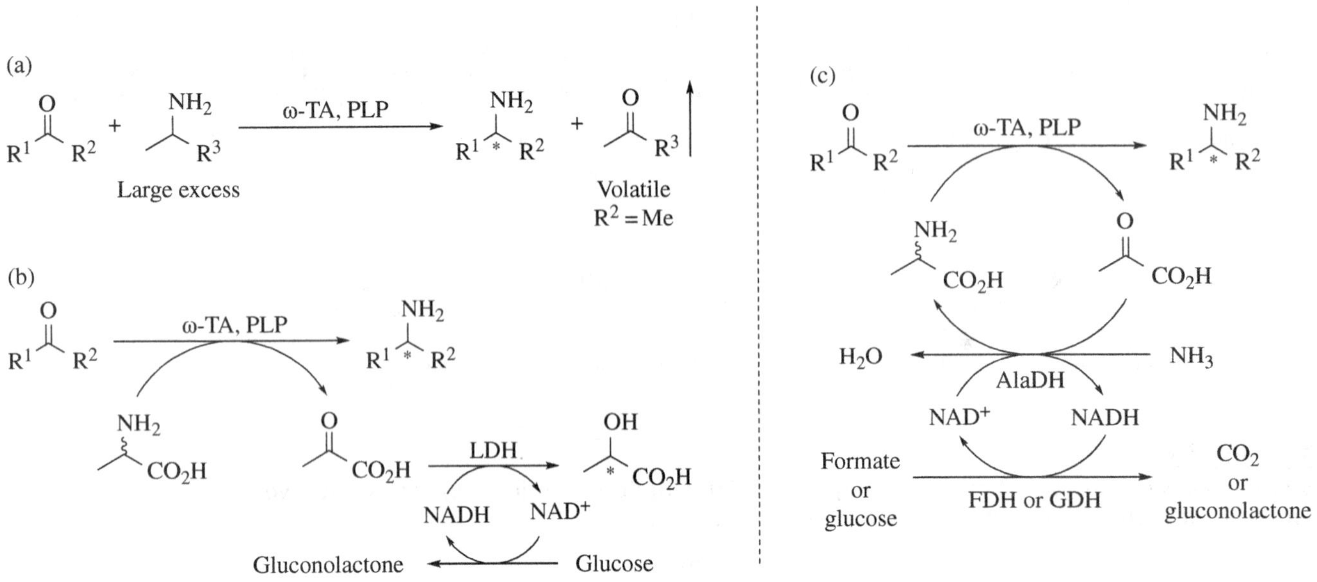

SCHEME 2.3
Overview of selected techniques to shift the equilibrium. (a) Volatile carbonyl coproduct. (b) LDH/GDH multienzyme network. (c) AlaDH in combination with formate or glucose dehydrogenase (FDH/GDH).

Nevertheless, the search for alternative amination systems is still ongoing, aiming to increase the overall efficiency of the methodology and, therefore, making the strategy more attractive for practical applications.

2.2.4 Asymmetric Amination of Linear Ketones

The scope of ω-TAs has been investigated in the last years employing various biocatalysts and amine donors. Most of the examples involve methyl-alkyl or methyl-aryl ketones. The reason for that is in an accepted model ω-TAs possess a small and a large binding pocket; however, the small pocket of ω-TAs is usually unable to accommodate medium or large groups (>C3) [29–31]. By using enzymatic amination, a broad range of aliphatic and aromatic amines have become easily accessible (Table 2.1). Notably, both the (*R*)- and the (*S*)-enantiomers can be synthesized in general with high enantiomeric excesses and reasonable substrate concentration. For instance, the amination of alkyl-methyl ketones (50 mM) proceeds smoothly even for those compounds bearing sensitive ester moieties (Entries 1–6).

In many examples, ω-TAs were overexpressed in *Escherichia coli* and used as freeze-dried catalyst preparation; thus enzyme purification was not necessary [32–34]. Acetophenone derivatives were in general successfully transformed with excellent degrees of enantioselectivity using different enzyme preparations [35, 36]. In a another report, the amination was studied at high substrate concentration (100 mM) and 1 g scale using the Codex® transaminase screening kit [37]. Unexpectedly, the best results were obtained for the sterically hindered *ortho*-derivatives because of the establishment of an intramolecular H-bond driving the reaction to completion (Entries 7–8). Additionally, a series of valuable heterocyclic amines bearing six- or five-membered rings were prepared (Entries 9–12). Methyl-alkyl ketones containing large alkyl or heteroalkyl groups were also converted using the crude overexpressed enzymes as catalyst. The system provided the desired amines with excellent conversion and stereoselectivity (Entries 13–16).

TABLE 2.1 Enzymatic Amination of Linear Ketones using ω-TAs and Amine Donors

Entry	Amine	ω-TA[a]	Donor	Conv. (%)	ee (%)	Reference
1	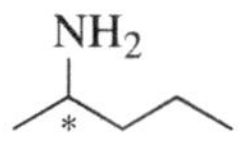	CV, BM	L-Alanine	>99	>99 (*S*)	[32]
2		AT	D-Alanine	>99	>99 (*R*)	[33]
3	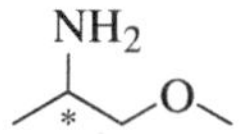	BM	L-Alanine	97	>99 (*S*)	[32]
4		ArR, AT, HN	D-Alanine	>99	>99 (*R*)	[33]
5	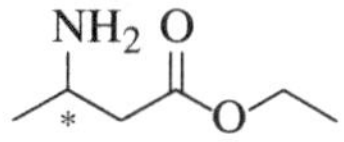	PD, VF	L-Alanine	>99	>99 (*S*)	[34]
6		ArR, AT, HN	D-Alanine	>99	>99 (*R*)	[33]
7	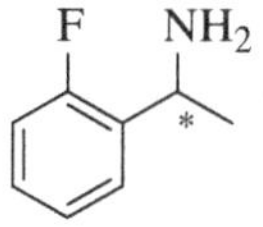	TA-P1-A06	2-PrNH_2	94	>99 (*S*)	[35]
8		ATA-025	2-PrNH_2	95	>99 (*R*)	[35]
9	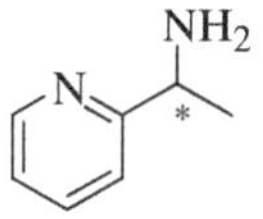	TA-P1-A06	2-PrNH_2	92	>99 (*S*)	[35]
10		ATA-025	2-PrNH_2	93	>99 (*R*)	[35]
11	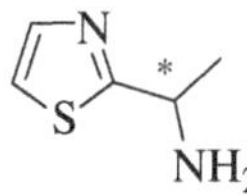	TA-P1-A06	2-PrNH_2	79	>99 (*S*)	[35]
12		ATA-025	2-PrNH_2	83	>99 (*R*)	[35]
13	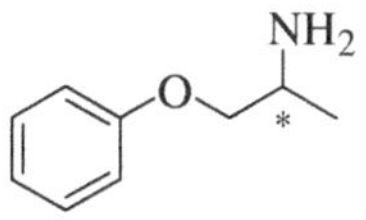	PD, PF	L-Alanine	>99	>99 (*S*)	[34]
14		ArR, AT, HN	D-Alanine	>99	>99 (*R*)	[33]
15	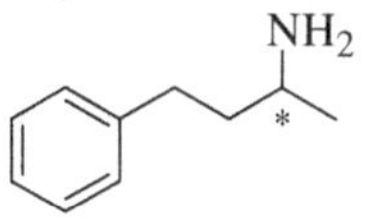	PF	L-Alanine	99	94 (*S*)	[34]
16		AT	D-Alanine	96	>99 (*R*)	[33]

[a] Origin of the ω-TAs: ArR, (*R*)-selective *Arthrobacter* sp.; AT, *Aspergillus terreus*; BM, *Bacillus megaterium*; CV, *Chromobacterium violaceum*; HN, *Hyphomonas neptunium*; TA-P1-A06, ATA-025, ATA-033, ω-TAs from the Codex TA screening kit; VF, *Vibrio fluvialis*.

2.2.5 Asymmetric Amination of Cyclic Ketones

Carbocyclic amines are often key starting materials for the preparation of natural products and are consequently valuable targets for the development of new drugs. For instance, the 1-aminotetralin skeleton can be found in several pharmaceuticals such as sertraline [38], norsertraline [39], or rotigotine [40]. However, the preparation of such amines is often challenging because of harsh reaction conditions (i.e., up to 100 bar of hydrogen pressure) [5]. ω-TAs have been demonstrated to be a mild alternative for the preparation of enantiomerically pure carbocyclic amines (Table 2.2). For instance, diastereomerically pure nonnatural AAs derived from 2-(3-aminocyclohexyl) acetic acid were obtained via a biocatalytic cascade employing ω-TAs in the final step (Entries 1–4) [41]. By the choice of the ω-TA, the *cis*-diastereomer and the *trans*-diastereomer were accessible in an excellent yield and an enantiopure form. In another approach, combining spontaneous racemization of an α-chiral keto ester and amination resulted in a straightforward route to α,β-chiral β-amino esters (Entry 5) [42]. In general, poor diastereoselectivities were observed; however, outstanding results were obtained for cyclic derivatives such as (1*S*,2*S*)-ethyl 2-aminocyclopentanecarboxylate. Additional studies regarding substrate scope revealed that a series of α- or β-tetralone and chromanone derivatives can be successfully aminated using various (*S*)- and (*R*)-selective ω-TAs (Entries 6–11) [43]. For most cases, both enantiomers were accessible in enantiopure form as also demonstrated on a preparative scale (100 mg).

TABLE 2.2 Asymmetric Reductive Amination of Prochiral Cyclic Ketones using ω-TAs and Different Amine Donors

Entry	Amine	ω-TA[a]	Donor	Conv. (%)	ee (%)	Reference
1	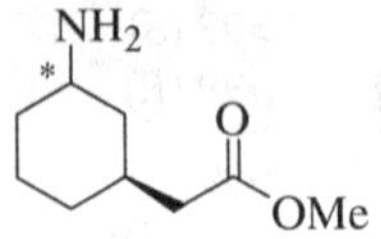	VF	L-Alanine	>99	99 (1*S*,5*S*)	[41]
2		ArRmut11	2-PrNH_2	96	>99 (1*S*,5*R*)	[41]
3	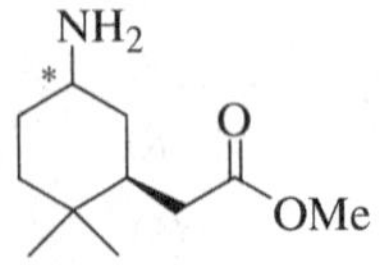	VF	L-Alanine	>99	99 (1*S*,5*S*)	[41]
4		ArRmut11	2-PrNH_2	>99	>99 (1*S*,5*R*)	[41]
5	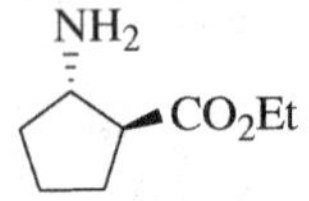	TA-P1-G05	2-PrNH_2	58	>99 (1*S*,2*S*)	[42]
6	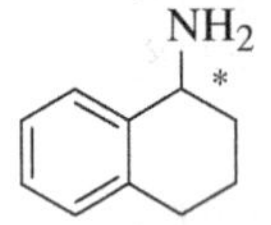	VF	L-Alanine	22	98 (*S*)	[43]
7		BV	D-Alanine	51	>99 (*R*)	[44]
8	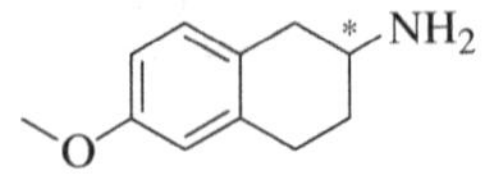	ArS	L-Alanine	90	>99 (*S*)	[43]
9		ArRmut11	2-PrNH_2	>99	>99 (*R*)	[43]
10	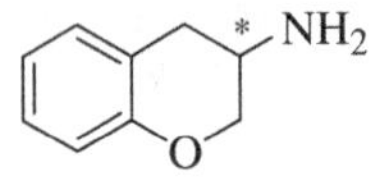	ArS	L-Alanine	>99	>99 (*S*)	[43]
11		HN	D-Alanine	22	>99 (*R*)	[43]

[a] Origin of ω-TAs: ArS, *Arthrobacter citreus*; BM, *Bacillus megaterium*; BV, *Burkholderia vietnamiensis*; HN, *Hyphomonas neptunium*; PF, *Pseudomonas fluorescens*; TA-P1-G05, Codex TA screening kit; VF, *Vibrio fluvialis*.

2.2.6 Application in the Synthesis of Pharmaceutically Active Ingredients

In recent years, different research groups have explored the synthetic potential of ω-TAs for the preparation of active pharmaceutical ingredients (APIs). The broad substrate scope has allowed the preparation of various APIs with an amino moiety including aromatic, cyclic, and also multicyclic systems such as the cholinesterase inhibitor rivastigmine [45], the antidiabetic sitagliptin [46], the antiallergic ramatroban [47], or the antiarrhythmic vernakalant [48]. The first example dealing with the use of ω-TAs in the asymmetric synthesis of pharmaceutical precursors was developed for the chemoenzymatic route of the anticholinergic agent (*S*)-rivastigmine (Scheme 2.4) [49–51]. The drug was prepared in a linear three-step sequence where an ω-TA either from *Paracoccus denitrificans* or *Vibrio fluvialis* was employed in the asymmetric key step (45 mM concentration, 100 mg scale) using L-alanine as donor. The enantiopure (*S*)-amine was obtained in perfect optical purity (>99% ee) and transformed further to afford the enantiopure drug.

However, in some cases, the amination step involving ω-TAs was rather challenging requiring protein engineering for the development of new enzyme variants. For instance, originally there was no enzyme available to transform the bulky–bulky ketone substrate required for the preparation of the antidiabetic drug (*R*)-sitagliptin (Scheme 2.5) [23]. For another API, all known ω-TAs led to the wrong diastereomer required for the preparation of the antiarrhythmic agent (1*R*,2*R*)-vernakalant [52]. In both cases, the most promising ω-TA candidate (commercial enzymes ATA-117 for sitagliptin and ATA-013 for vernakalant) was subjected to a combination of *in silico* design and directed evolution. Enzyme engineering led

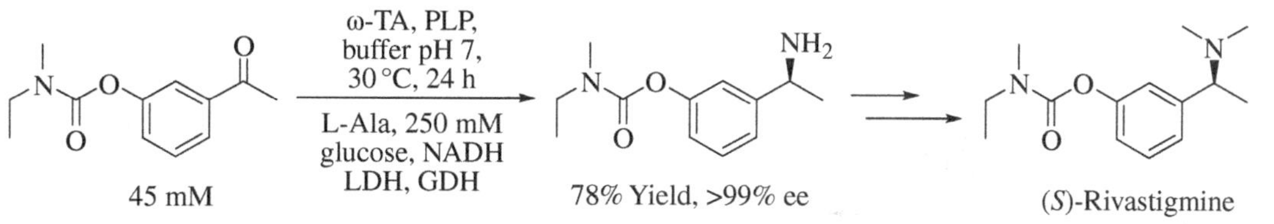

SCHEME 2.4
Three-step chemoenzymatic synthesis of the anticholinergic drug (*S*)-rivastigmine.

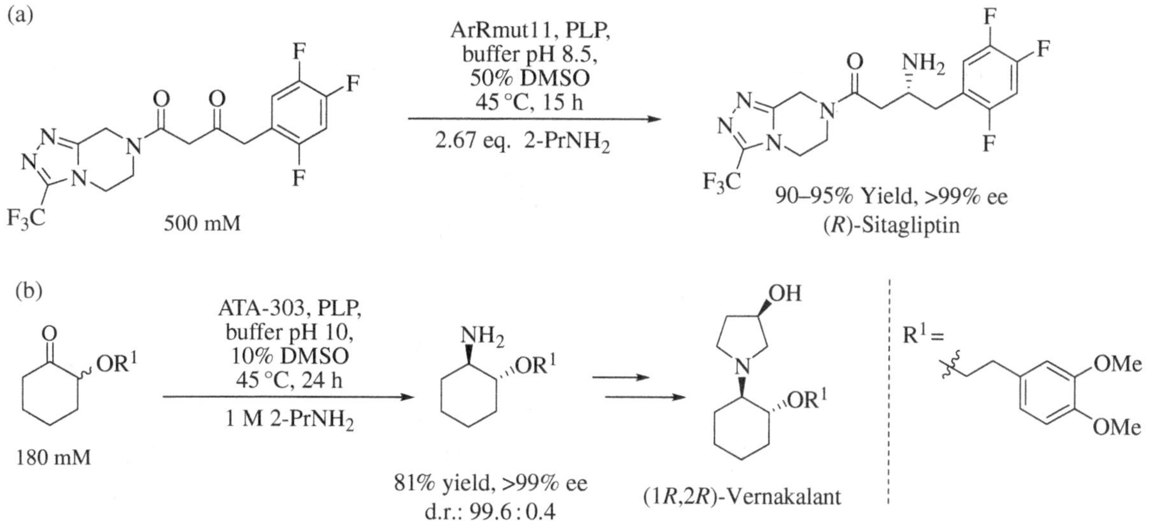

SCHEME 2.5
Biocatalytic routes for the preparation of the antidiabetic drug sitagliptin (a) and the antiarrhythmic agent vernakalant (b) using evolved ω-TAs.

finally to improved variants being able to either accommodate the second large substrate moiety (ArRmut11, sitagliptin) or give reverse diastereoselectivity (ATA-303, vernakalant). The obtained improved variants showed higher activity and increased tolerance to organic solvents and elevated pHs. Notably, both products were isolated in excellent yield and enantiopure form after a short reaction time using 2-propylamine as amine donor.

For other substrates, a careful selection of the amine donor may be necessary to avoid the formation of side products and provide the target compound in high yield. For instance, although 2-propylamine is regarded as an ideal donor, it was responsible for the formation of various side products in the synthesis of (*R*)-2,3,4,9-tetrahydro-1*H*-carbazol-3-amine, a key intermediate for the synthesis of the antiallergic drug ramatroban [53]. The side products were avoided using a sterically more demanding amine donor such as (*R*)-1-phenylethylamine (Scheme 2.6), which afforded the desired (*R*)-isomer in excellent isolated yield on a preparative scale (0.5 g).

Biocatalytic amination was also successfully applied for the synthesis of potentially biologically relevant 17-α-steroids, which have been recently identified as potent sulfatase inhibitors with outstanding possibilities in the treatment of breast cancer [54]. From a broad panel of ω-TAs, the steroid analogs were only accepted by the ω-TA variant evolved for sitagliptin (ArRmut11), highlighting the acceptance of this variant for sterically challenging substrates (Scheme 2.7) [55]. The transamination occurred with excellent diastereoselectivity isolating the α-isomers in high yields (83–89%) on a preparative scale replacing a three-step chemical synthesis.

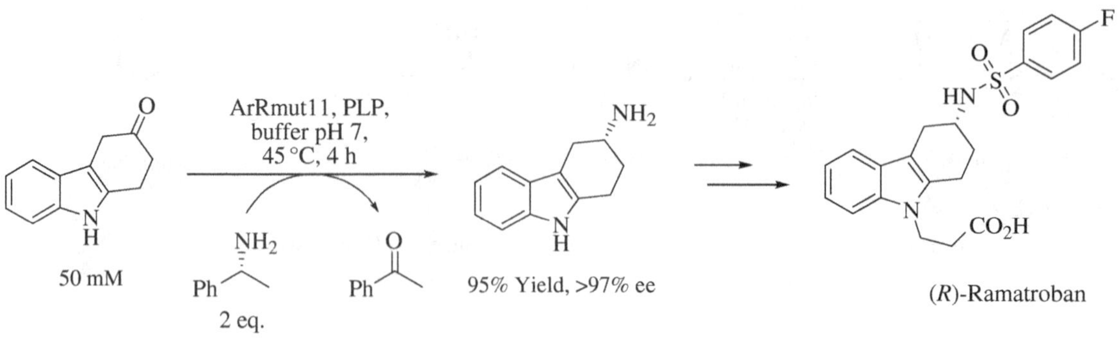

SCHEME 2.6

Reductive amination of the ramatroban ketone precursor using sterically demanding (*R*)-1-phenylethylamine as amine donor.

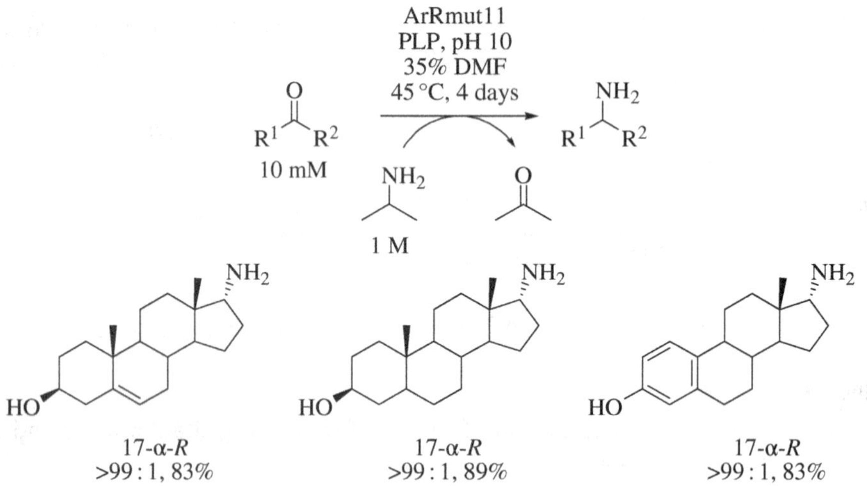

SCHEME 2.7

ω-TA-catalyzed biocatalytic route for the synthesis of 17-α-aminosteroids.

2.2.7 Amination of Ketones in Organic Solvents

Synthetic transformations involving ω-TAs are usually performed in aqueous media under neutral or slightly basic pHs, therefore requiring basification and extraction work-up to recover the neutral amine. As an alternative, synthetic transformation in pure organic solvent would be highly desirable to optimize work-up steps and improve the economy of the process. A recent study suggested that ω-TAs need to be engineered and immobilized before use in pure organic solvents [37]. In contrast, cell-free freeze-dried enzyme preparations [56] were proven to be active in organic solvents such as *tert*-butyl methyl ether (TBME) or ethyl acetate (EtOAc) by carefully tuning the water activity. Maximum activity values were found in TBME (water activity, 0.6) and increased thermal stability of the proteins was observed compared to experiments in buffer. Even more interesting is that all nine enzymes tested were found to accept 2-propylamine as amine donor although some of them failed in buffer solution. Using this reaction setup, a set of aliphatic and aromatic amines was successfully produced observing also a significant increase in the reaction rate for selected substrates compared to experiments in buffer (up to 17-fold); notably, the stereoselectivity was not affected (Scheme 2.8). After the reaction, the catalyst can be recovered from the reaction by filtration and reused several times without losing activity. Just three equivalents of 2-propylamine were necessary for the amination of β-ketoesters and β-oxygenated ketones (50 mM) in high conversion. A possible explanation might be that these products are thermodynamically more stable due to the formation of an intramolecular H-bond between the oxygen atom and the amine.

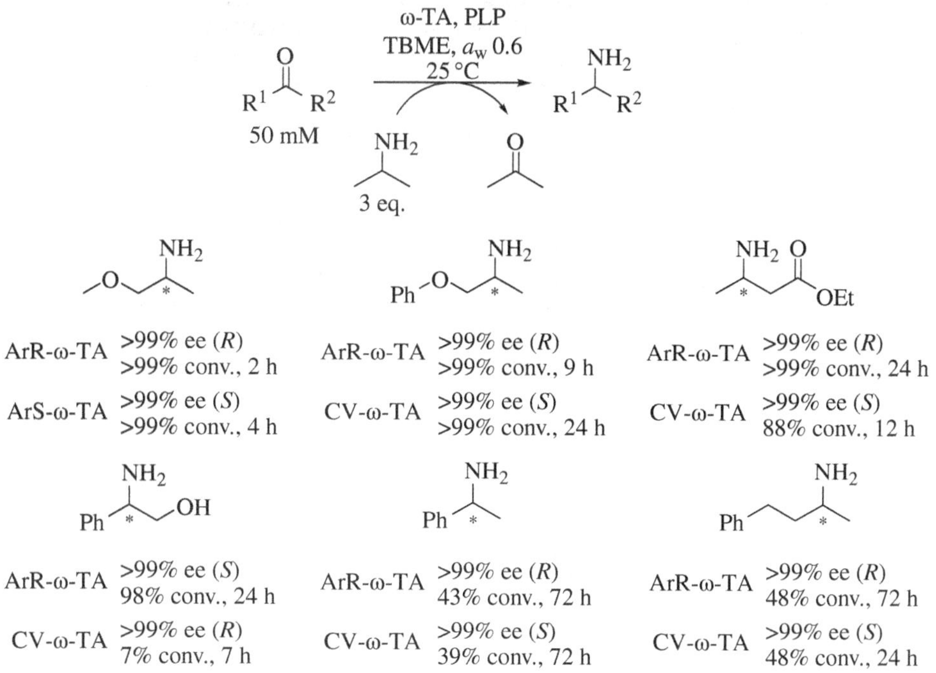

SCHEME 2.8
ω-TA-catalyzed amination of ketones in TBME at a defined water activity of $a_w = 0.6$.

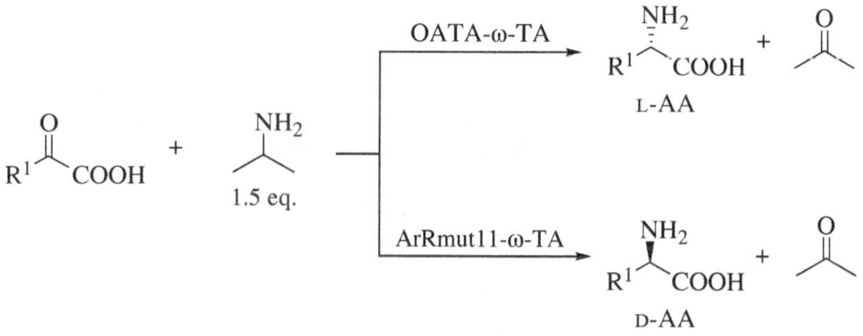

SCHEME 2.9
Asymmetric amination of α-keto acids using 2-propylamine as amine source.

2.2.8 Asymmetric Amination of Keto Acids: Synthesis of Nonnatural Amino Acids

Nonnatural AAs are important precursors for the preparation of bioactive compounds. For instance, D-phenylalanine is a precursor of the antidiabetic drug nateglinide [57], while L-homoalanine is a building block for the antiepileptic drug levetiracetam [58]. Biocatalytic approaches for the transfer of the amino group between α-AAs are usually based on reactions catalyzed by α-TAs; for this reaction, the equilibrium constant is close to unity requiring additional reagents or enzymes to shift it toward product formation [59]. In contrast, the ω-TA-catalyzed amination of α-keto acids using aliphatic amines such as 2-propylamine as amine donor is thermodynamically downhill affording high conversions with reduced amounts of the amine donor (1.5 eq.) (Scheme 2.9) [60]. Initial optimization experiments revealed that the (S)-selective *Ochrobactrum anthropi* ω-TA (OATA) and (R)-selective ArRmut11 were the most active catalysts for the asymmetric synthesis of AAs.

As expected, OATA was only able to convert α-keto acids with (R<C2) because of steric limitations of the small binding pocket (Table 2.3, Entries 1 and 3). On the other hand, ArRmut11 developed for the amination of bulky–bulky ketones efficiently converted α-keto acids with linear groups although it failed for branched groups (Entries 2, 4–7). Preparative-scale experiments (15 mmol) were successfully performed for the synthesis of the pharmacologically relevant L- and D-homoalanine.

TABLE 2.3 Synthesis of Nonnatural Amino Acids through ω-TA-Catalyzed Amination of Keto Acids

Entry	R^1	ω-TA	Conv. (%)	Yield (%)[a]	ee (%)
1	Et	OATA	99	78	>99 (*S*)
2	Et	ArRmut11	99	78	>99 (*R*)
3	Me	OATA	79	n.r.	>99 (*S*)
4	Me	ArRmut11	98	n.r.	>99 (*R*)
5	CH_2OH	ArRmut11	99	n.r.	>99 (*R*)
6	CH_2F	ArRmut11	86	n.r.	>99 (*R*)
7	$(CH_2)_5CH_3$	ArRmut11	99	n.r.	>99 (*R*)

[a] n.r., not reported.

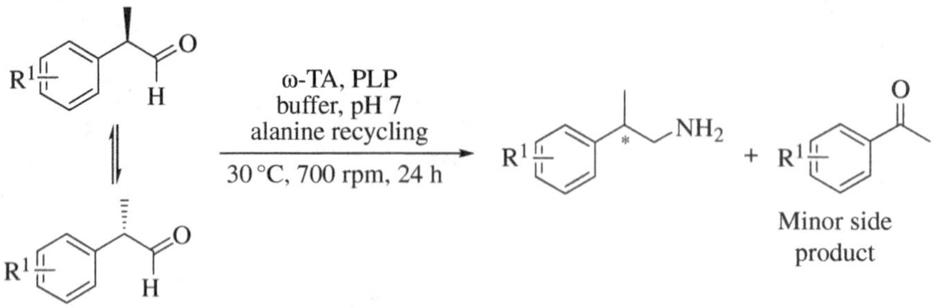

SCHEME 2.10

Synthesis of β-chiral 2-phenylpropanamines through dynamic kinetic resolution of 2-phenylpropanal derivatives.

TABLE 2.4 Amination of Racemic 2-Phenylpropanal Derivatives by DKR on a Preparative Scale

Entry	R^1	Amine donor	ω-TA	Yield (%)	ee (%)
1	*p*-Me	L-Alanine	PP1	62	75 (*R*)
2	*p*-OMe	L-Alanine	PP1	86	92 (*R*)
3	*m*-Me	L-Alanine	PP1	56	80 (*R*)
4	*m*-OMe	L-Alanine	PP1	82	50 (*R*)
5	*o*-Me	L-Alanine	PP1	75	89 (*R*)
6	*o*-OMe	D-Alanine	HN	50	99 (*S*)

2.2.9 Amination of Aldehydes

Although aldehydes are in general more reactive than ketones or α-keto acids, the ω-TA-catalyzed amination of aldehydes has been less frequently explored; most of the examples about aldehydes concern the kinetic characterization of new enzymes rather than preparative synthetic applications. The transformation of 2-phenyl propanal derivatives to the corresponding β-chiral amines has been recently investigated using ω-TAs (Scheme 2.10) [61].

Therefore, the α-substituted aldehydes racemized spontaneously due to the high acidity of the benzylic proton enabling a dynamic kinetic resolution (DKR). Interestingly, the observed enantiopreference of the enzymes for these substrates could not be correlated with the enantiopreference previously observed for methyl ketones with these enzymes. The combination of substrate and medium engineering increased the enantioselectivity of the process obtaining the corresponding amines with up to 99% ee (Table 2.4). In most cases, it was possible to obtain both, the (*R*)- and (*S*)-enantiomers, by the use of stereocomplementary ω-TAs whereby *ortho*-substituted substrates were transformed best with minimized formation of side products (Entries 5–6). Aminations were also performed on preparative conditions (100 mg) obtaining the corresponding amines in high enantiomeric excesses and isolated yields.

2.2.10 Cascade Reactions Involving ω-TAs

Biocatalytic reactions are generally performed under very mild reaction conditions, most of the time at room temperature and physiological pH [9]. For that reason, enzymes are usually compatible with each other, opening the possibility of combining several reactions in the same reaction vessel. Such biocatalytic cascades offer many practical advantages—that is, minimized work-up steps, shortened reaction times, improved atom economy, and easier reaction setups since, for example, the manipulation of unstable intermediates is avoided. In the last years, enzymatic cascades have reached a remarkable level of complexity allowing the one-pot preparation of highly functionalized molecules from bulk chemicals [62–66]. In addition, thermodynamically uphill reactions can be easily driven to completion if the products are removed by a favorable consecutive reaction. In this section, the application in organic synthesis of successful cascades involving asymmetric amination of ketones will be summarized including reaction sequences initiated not only by ω-TAs but also by multienzyme systems [67].

2.2.11 Cascades Initiated by ω-TAs: Synthesis of Chiral Heterocycles

Nitrogen-containing heterocycles including pipecolic acid derivatives, alkaloids, or penicillins have attracted a lot of attention in the last years, not only because of their therapeutic uses but also due to their applications in organocatalysis. For that reason, a significant number of chemical and biocatalytic approaches have been developed in the last years for the preparation of these scaffolds [68]. Here, examples are given where heterocyclic molecules have been synthesized via carbonyl amination followed by spontaneous ring closure. The methodology usually offers high levels of efficiency for the amination step since the cyclization step drives the reaction to the side of the products without requiring high excesses of the amine donor. In this context, amination–lactamization cascades have been successfully used for the production of six-membered heterocyclic compounds (Scheme 2.11). For instance, the ω-TA-catalyzed amination of ethyl 4-acetylbutyrate (50 g/l) with 2-propylamine (1 M) affords spontaneously the corresponding six-membered lactam (Scheme 2.11a) [69]. By selection of the transaminase, both enantiomers were accessible and isolated in enantiopure form and high yield. Using a similar approach, the dual orexin antagonist MK-6096 containing the α-methylpiperidine core was synthesized on a kilogram scale [70]. The concept has also been applied for the synthesis of biologically active γ-aminobutyric acid (Scheme 2.11b) [71]. In this example, the ω-TA-catalyzed DKR of the aromatic aldehyde ester precursor affords the corresponding amino ester, which spontaneously cyclizes to 4-phenylpyrrolidin-2-one. Even though the lactam was obtained in moderate optical purity (68% ee), this contribution represented the first DKR of an α-chiral aldehyde. Using the amination–lactamization cascade, various chemoenzymatic routes have been reported for the preparation of niraparib (Scheme 2.11c) [72], an inhibitor of the poly(ADP-ribose)polymerase (PARP) [73]. Preliminary DKR experiments employing ATA-301 and 2-propylamine afforded the desired (*R*)-lactam in high isolated yield and enantiopure form on a preparative scale (35 g), although an inert atmosphere was essential to avoid the oxidative degradation of the aldehyde. Alternatively, transamination of more stable aldehyde surrogates such as bisulfite adducts or lactol derivatives has also been performed providing improvements in terms of robustness and reliability, which are essential for the development of industrial-scale applications.

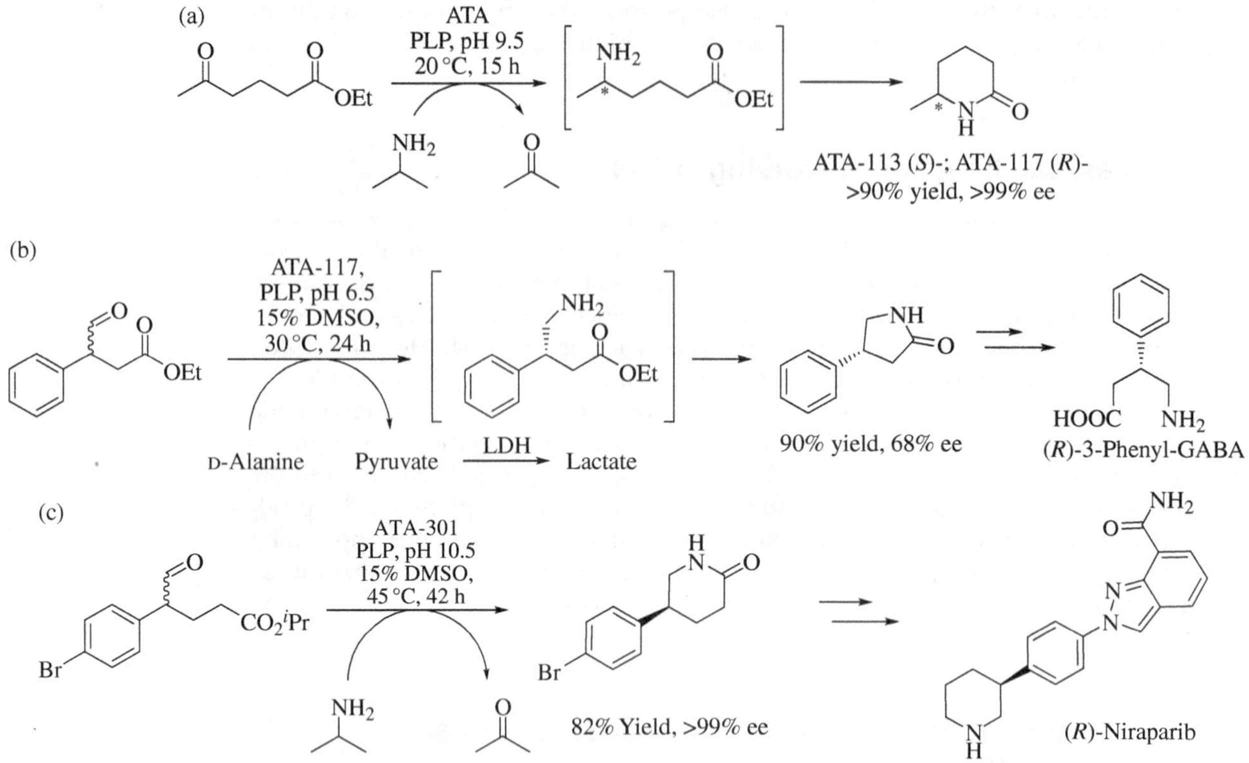

SCHEME 2.11

Amination–lactamization cascade for the preparation of valuable lactams. (a) 6-Methyl-piperidin-2-one, (b) 4-phenylpyrrolidin-2-one, and (c) synthesis of niraparib precursor.

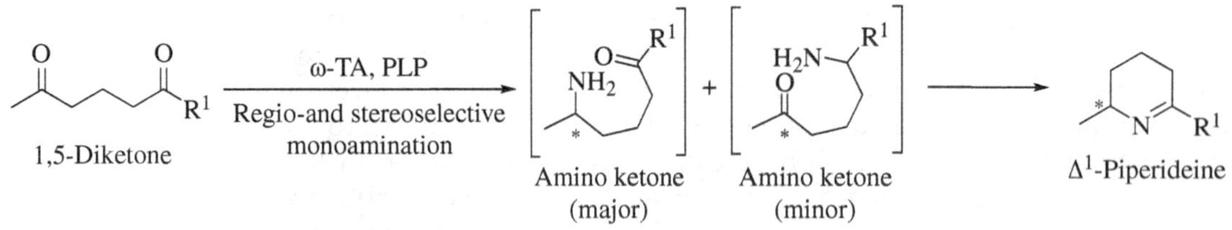

SCHEME 2.12

Regio- and enantioselective monoamination of 1,5-diketones using ω-TAs.

Combining regio- and enantioselective amination of 1,5-diketones provided the shortest route to date for the preparation of 2,6-disubstituted piperidines, a group of bioactive molecules present in a vast number of natural products [74, 75]. Most of the ω-TAs showed exquisite levels of regioselectivity as the amination occurred exclusively at the sterically less demanding ketone (ω-1), which spontaneously cyclized to the corresponding enantiopure cyclic imine denoted as Δ^1-piperideine (Scheme 2.12) [76].

A representative panel of diketones (50 mM) containing cyclic, aromatic, and aliphatic moieties was effectively converted with a perfect stereoselectivity and reactivity (Table 2.5). Simply by the choice of enantiocomplementary ω-TAs, both (*R*)- and (*S*)-Δ^1-piperideines were accessible in enantiopure form.

The practical applicability of the route was shown for the synthesis of the natural alkaloids dihydropinidine and *epi*-dihydropinidine (Scheme 2.13). The monoamination of nonane-2,6-dione (50 mM) using (*R*)- and (*S*)-selective ω-TAs afforded both enantiomers of the imine in enantiopure form with complete regiocontrol. Simple Pd/C-catalyzed hydrogenation exclusively led to the *syn*-isomers. The *anti*-isomers were obtained in high diastereomeric ratio (*anti*/*syn* 84:16) based on a Lewis acid-promoted conformational change during the reduction [77]. A further

TABLE 2.5 Regioselective Monoamination leading to Enantiomerically Pure Δ¹-Piperidines after Spontaneous Cyclization

Entry	R^1	ω-TA[a]	Conv. (%)	ee (%)
1	*n*-Pr	CV-ω-TA	>99	>99 (*S*)
2	*n*-Pr	AT-ω-TA	>99	>99 (*R*)
3	*cyclo*-Pr	VF-ω-TA	>99	>99 (*S*)
4	*cyclo*-Pr	HN-ω-TA	>99	>99 (*R*)
5	*iso*-Pr	ArS-ω-TA	97	>99 (*S*)
6	*iso*-Pr	AT-ω-TA	93	>99 (*R*)
7	*iso*-Bu	BM-ω-TA	>99	>99 (*S*)
8	*iso*-Bu	HN-ω-TA	>99	>99 (*R*)
9	Ph	PF-ω-TA	>99	98 (*S*)
10	Ph	PF-ω-TA	>99	>99 (*R*)

[a] Origin of ω-TAs: ArR, (*R*)-*Arthrobacter* sp.; ArS, *Arthrobacter citreus*; AT, *Aspergillus terreus*; BM, *Bacillus megaterium*; CV, *Chromobacterium violaceum*; HN, *Hyphomonas neptunium*; PF, *Pseudomonas fluorescens*; VF, *Vibrio fluvialis*.

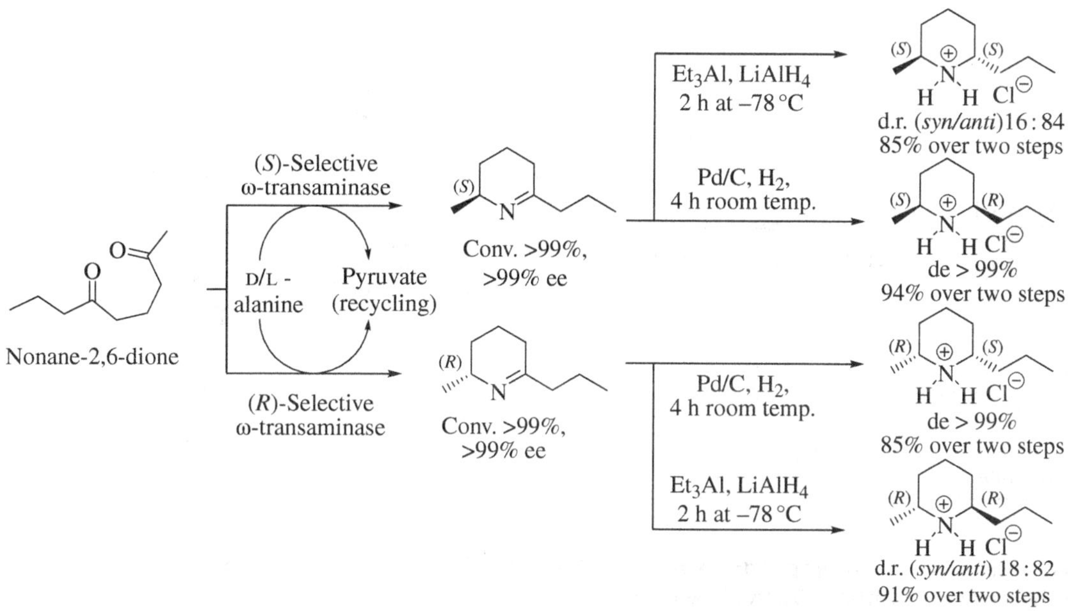

SCHEME 2.13

Chemoenzymatic route for the preparation of the natural alkaloids dihydropinidine (*cis*) and *epi*-dihydropinidine (*trans*).

extension of this concept has been recently described for the preparation of the alkaloid isosolenopsin [78]. In this case, a concise medium engineering in terms of pH, cosolvent, and temperature was essential to achieve a successful amination–cyclization cascade.

The amination–cyclization strategy has also proven successful for the multi-gram enantioselective synthesis of the dual orexin inhibitor suvorexant [79], currently in phase III trials for the treatment of primary insomnia. Starting from the adequate ketone precursor, the enantioselective amination using the ω-TA ArRmut11 with 2-propylamine afforded an unstable amino mesylate (Scheme 2.14) [80]. The intermediate spontaneously underwent cyclization to yield the target 1,4-diazepane ring in enantiopure form without any additional treatment even though the formation of seven-membered cycles is sometimes challenging. A careful optimization of pH and leaving group was essential to avoid the formation of side products in the cyclization step.

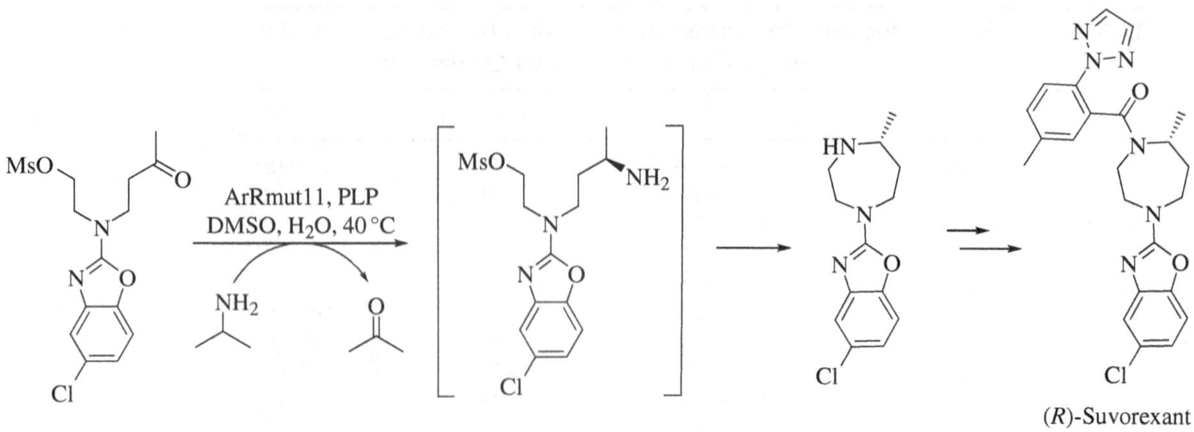

SCHEME 2.14
Tandem enantioselective amination/spontaneous annulation as a key step in the preparation of suvorexant.

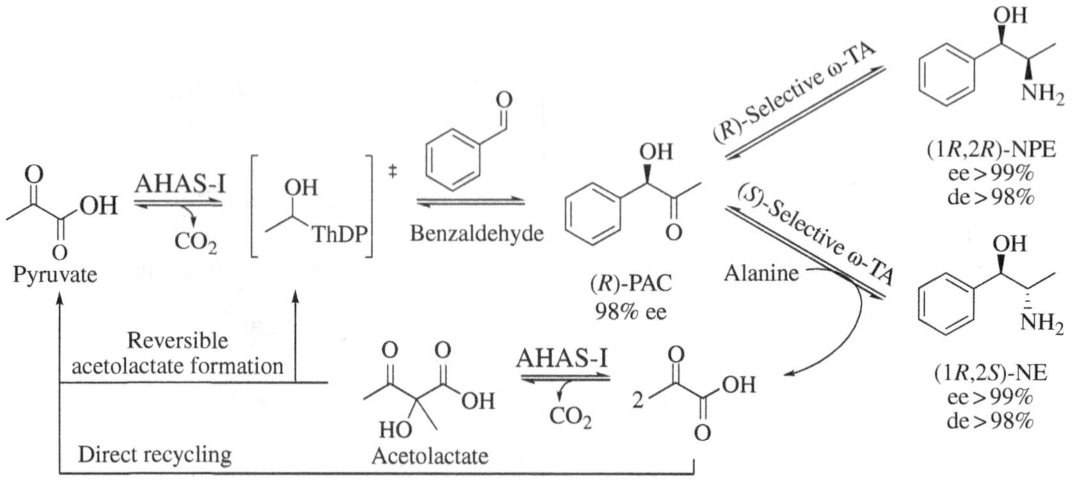

SCHEME 2.15
Two-step cascade for the synthesis of (1*R*,2*R*)-norephedrine and (1*R*,2*S*)-ephedrine combining a thiamine diphosphate-dependent lyase (ThDP) acetohydroxyacid synthase (AHAS-I) with stereocomplementary ω-TAs.

2.2.12 Multienzyme Cascades Involving ω-TA-Catalyzed Amination of Ketones

ω-TAs have been successfully combined with different types of other biocatalysts such as transketolases [81], hydrolases [41], α-TAs [59], etc. for the production of valuable compounds with impressive levels of stereoselectivity. For instance, the combination of a lyase and different ω-TAs has been employed for the preparation of biologically relevant (1*R*,2*R*)-norpseudoephedrine (NPE) or (1*R*,2*S*)-norephedrine (NE) using two enzymes in one pot from inexpensive starting materials: in the first step, the (*R*)-selective thiamine diphosphate (ThDP)-dependant lyase, acetohydroxy-acid synthase I (AHAS-I), affords (*R*)-phenylcarbinol from pyruvate and benzaldehyde (Scheme 2.15) [82]. Next, the intermediate keto alcohol is directly converted to (1*R*,2*R*)-NPE and (1*R*,2*S*)-NE in high stereoisomeric purity and high conversions (80–96%) using two enantiocomplementary ω-TAs. Further optimization allows recycling of the pyruvate produced in the second step as starting material for the AHAS-I reaction, improving the global efficiency of the process.

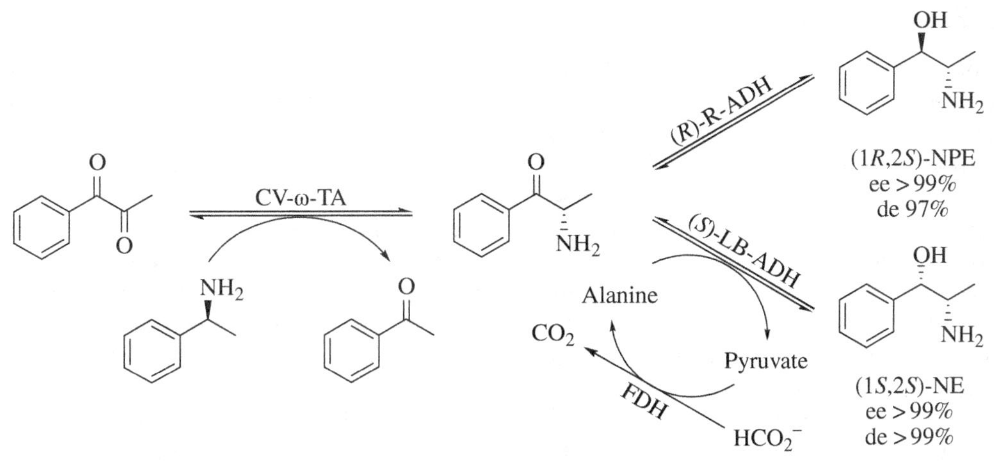

SCHEME 2.16

Synthesis of (1*S*,2*S*)-norephedrine and (1*S*,2*R*)-ephedrine combining an (*S*)-selective ω-TA with (*R*)- or (*S*)-selective ADH.

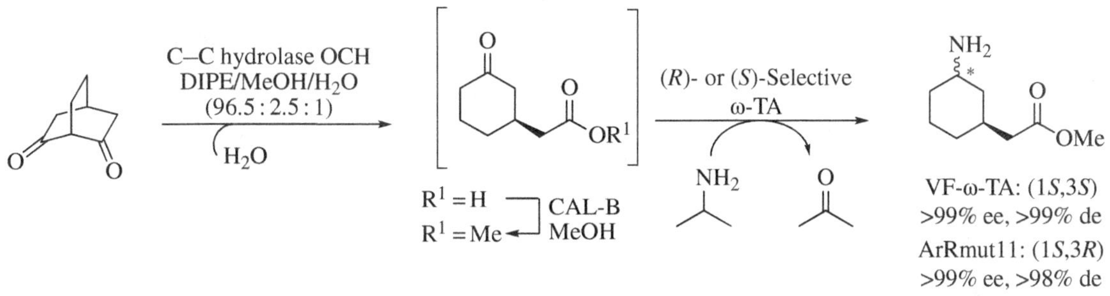

SCHEME 2.17

Diastereoselective preparation of 3-substituted cyclohexylamine derivatives using two hydrolases and one ω-TA.

However, this approach fails for the synthesis of the complementary (1*S*,2*S*)-NPE and (1*S*,2*R*)-NE since no enzyme is known to provide (*S*)-PAC in high optical purity. As an alternative, an ω-TA/alcohol dehydrogenase (ADH) reaction sequence has been investigated in a recent report. The reaction sequence comprises an (*S*)-selective ω-TA (CV ω-TA) with an (*R*)-selective ADH from *Ralstonia* sp. (R-ADH) or the (*S*)-selective ADH from *Lactobacillus brevis* (LB-ADH) (Scheme 2.16) [83]. The cascade has to be performed in a sequential mode due to the divergent reaction conditions required for each step. Besides, deactivation of the ω-TA before the ADH-catalyzed step was mandatory to avoid the formation of 1-phenylpropane-1,2-diol. At 10 mM substrate concentration, the system yielded (1*S*,2*S*)-NPE and (1*S*,2*R*)-NE in moderate to high conversions and perfect optical purity when pure enzymes were used as catalysts.

Hydrolases were also successfully coupled with ω-TAs for the production of 3-cyclohexylamine derivatives from prochiral bicyclic diketones (Scheme 2.17) [41]. The reaction sequence was based on three biocatalytic steps—namely, stereoselective hydrolysis of a C—C bond using a β-diketone hydrolase (6-oxocamphor hydrolase (OCH) from *Rhodococcus* sp.), followed by *Candida antarctica* lipase B (CAL-B)-catalyzed esterification, and finally asymmetric amination by stereocomplementary (*R*)-selective ArRmut11 or (*S*)-selective VF-ω-TA. The first two reactions were performed simultaneously in an organic solvent (diisopropyl ether : H_2O : MeOH 97.5/2.5/1). The amination step could also be conducted in an organic solvent after removing the hydrolases by filtration, avoiding the change of the reaction media.

2.2.13 Deracemization of Primary Amines

The asymmetric amination of ketones is by far the most preferred approach for the preparation of chiral amines using ω-TAs. However, alternative methodologies may be considered if the carbonyl precursor is unstable or the synthesis of the racemic amine is easier to provide better results in economic and/or yield terms. Using racemic amines deracemization strategies allow the preparation of the desired amines in enantiopure form and a theoretical 100% yield [84–86]. This can be achieved by the combination of two stereocomplementary ω-TAs. In the first step, the enantioselective deamination of the racemic amine affords enantiopure untouched amine (50%) and the corresponding ketone (50%). In the second step, an enantiocomplementary ω-TA catalyzes the asymmetric amination of the ketone, leading to the optical pure amine in 100% theoretical yield (Scheme 2.18).

The major challenge of the concept is to avoid the interference between the two enantiocomplementary ω-TAs, which could decrease the final ee of the amine. For that purpose, different alternatives have been employed. Successful deracemization of a group of relevant amines was originally described in a one-pot two-step setup using, as catalysts, (*R*) and (*S*)-selective ω-TAs and stoichiometric amounts of pyruvate for the deamination step (Table 2.6) [87]. Target amines were recovered in high yields and excellent optical purities whereby the interference was prevented by deactivating the first enzyme before the second step by heat treatment. Thus, the recycling of the ω-TA is not possible.

The amount of pyruvate required for the deamination step could be drastically reduced by recycling the coproduct of the deamination step (D- or L-alanine) back to

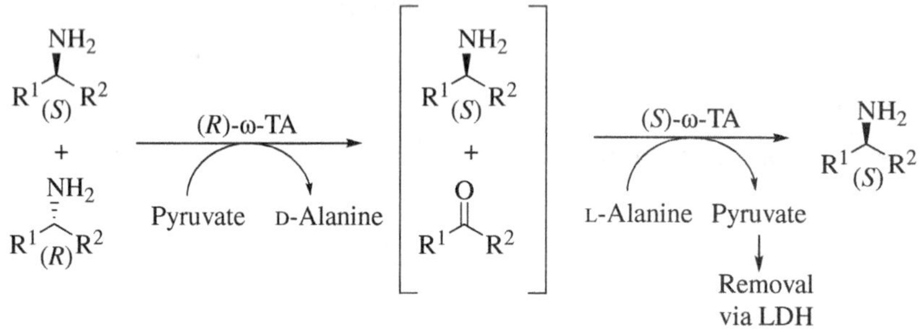

SCHEME 2.18
Deracemization of amines using two enantiocomplementary ω-TAs.

TABLE 2.6 Two-Step Deracemization of Pharmacologically Relevant Amines

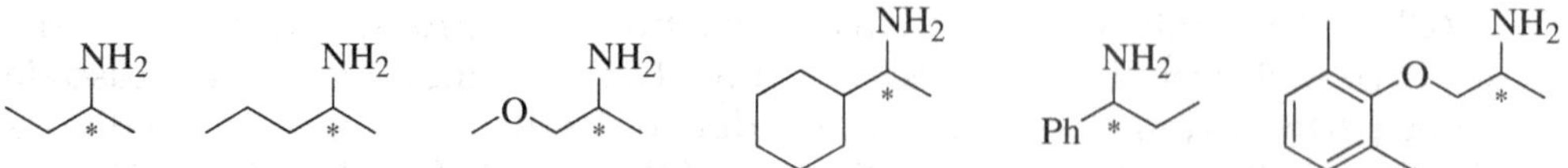

A B C D E F

Entry	Amine	First-ω-TA	Second-ω-TA	Conv. (%)	ee (%)
1	**A**	ATA-117	ATA-113	>99	>99 (*S*)
2	**A**	ATA-114	ATA-117	>99	>99 (*R*)
3	**B**	ATA-117	ATA-113	>99	>99 (*S*)
4	**B**	ATA-114	ATA-117	>99	>99 (*R*)
5	**C**	ATA-117	ATA-113	>99	>99 (*S*)
6	**C**	ATA-114	ATA-117	>99	96 (*R*)
7	**D**	ATA-117	ATA-113	62	>99 (*S*)
8	**D**	ATA-114	ATA-117	88	>99 (*R*)
9	**E**	ATA-117	ATA-113	82	>99 (*S*)
10	**E**	ATA-114	ATA-117	72	>99 (*R*)
11	**F**	ATA-117	ATA-113	98	>99 (*S*)
12	**F**	ATA-114	ATA-117	97	>99 (*R*)

pyruvate using an AAO [88]. Using this improved setup and by simply selecting the order of the ω-TAs, both enantiomers of the antiarrhythmic drug mexiletine were synthesized in an excellent isolated yield and enantiopure form (Scheme 2.19) [89].

The economy of the process was significantly improved by immobilizing the first ω-TAs since the catalyst can be easily removed by filtration after the first step and can be reused several times without losing the activity [90]. (*S*)-Mexiletine was synthesized on a preparative scale in 95% yield and enantiopure form, demonstrating the practical applicability of the methodology.

In a recent report, deracemization protocols have been improved by running both deamination and amination steps simultaneously [91]. The success of the methodology was based on the proper selection of a smart amine acceptor for the (*S*)-selective catalyst but not accepted by the (*R*)-selective ω-TA. α-Ketoglutaric acid was found to be the most suitable candidate (Scheme 2.20).

By combining the (*S*)-selective ω-TA from *Polaromonas* sp. (PS-ω-TA) and the (*R*)-selective TA from *Neosartorya fischeri* (NY-ω-TA) or *Mycobacterium vanbaalenii* (MW-ω-TA), a panel of (*R*)-1-phenylethylamine derivatives (10 mM) was synthesized in high yields and enantiopure form (Table 2.7).

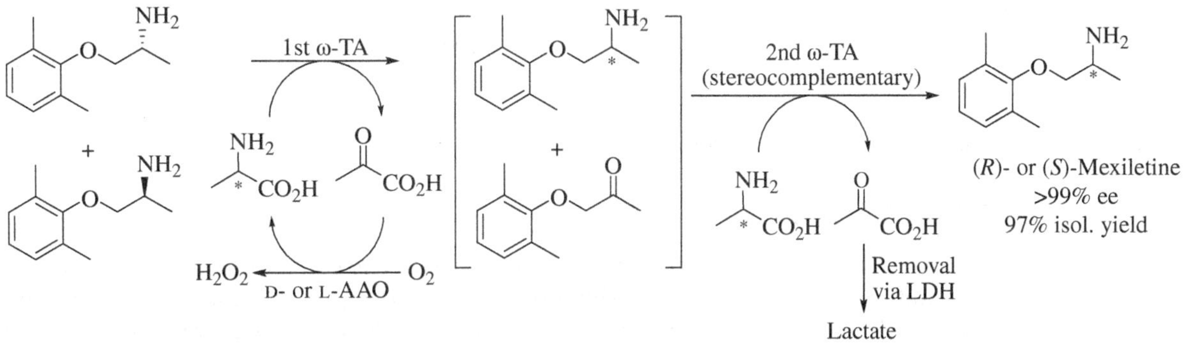

SCHEME 2.19

Deracemization of mexiletine. (*R*)-isomer: 1st ω-TA (ATA-117), 2nd ω-TA (ATA-113). (*R*)-isomer: 1st ω-TA (ATA-113), 2nd ω-TA (ATA-117).

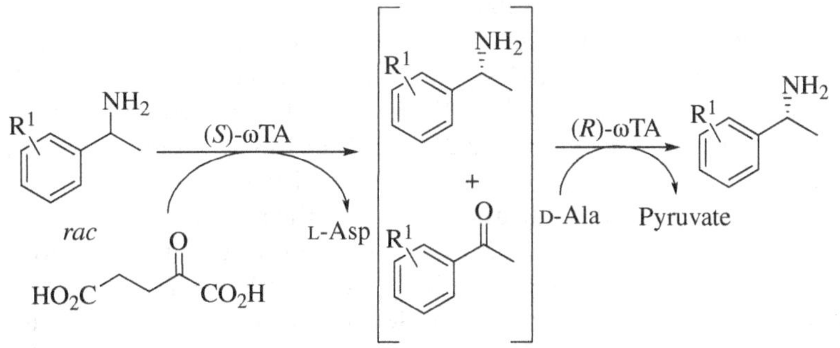

SCHEME 2.20

Simultaneous deracemization of aromatic amines combining two stereocomplementary ω-TAs using an enzyme-specific amine acceptor.

TABLE 2.7 One-Pot Simultaneous Deracemization of Amines Combining (*S*)- and (*R*)-Selective ω-TAs

Entry	R^1	ω-TAs	Conv. (%)	ee (%)
1	4-OH	PS/MW	>99	>99 (*R*)
2	4-F	PS/MW	82	>99 (*R*)
3	4-Me	PS/MW	>99	>99 (*R*)
4	4-Br	PS/NF	97	>99 (*R*)
5	3-Me	PS/MW	98	>99 (*R*)
6	3-CN	PS/MW	>99	96 (*R*)
7	3,4-di-F	PS/MW	98	>99 (*R*)
8	3,5-di-F	PS/NF	94	>99 (*R*)

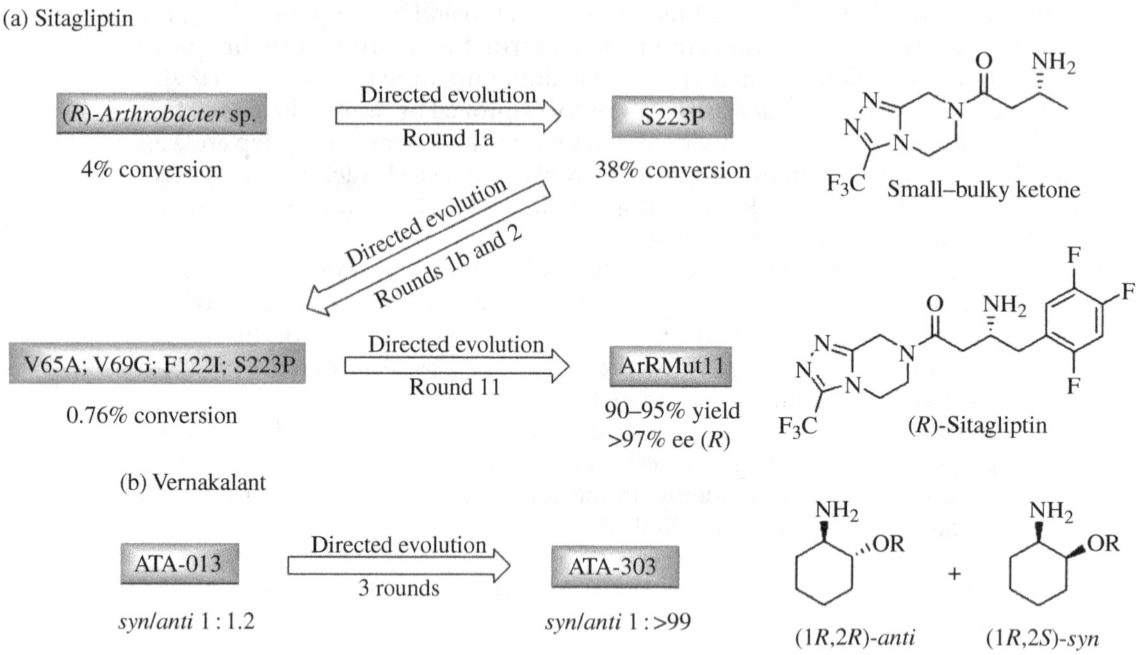

SCHEME 2.21
Extending the scope of ω-TAs by means of rational protein design.

2.2.14 Perspective

In this section, ω-TAs were presented as an alternative to conventional methods for the preparation of primary amines showing excellent degrees of regio-, diastereo-, and enantioselectivity for most of the cases, even in pure organic solvents. The scope of these enzymes has been extended through directed evolution by rational protein design, especially in the last few years when crystal structures became available. For instance, using a structural homology model and modeling the substrate into the active site of ATA-117, (R)-*Arthrobacter* sp., this enzyme was successfully engineered after 11 rounds of directed evolution (27 mutations in total), providing an improved variant for bulky–bulky substrates like the sitagliptin ketone (Scheme 2.21a) [23]. Related to this example, a combination of *in silico* design and directed evolution was used by Merck to invert the diastereoselectivity of ATA-013, required for the preparation of the antiarrhythmic drug vernakalant (Scheme 2.21b) [52]. In this case, the gene sequences for both wild-type and the mutant were not published yet. This information may have been included in the original publication to ensure the reproducibility of the work.

On the other hand, engineering of (S)-selective ω-TAs was found to be more challenging. For instance, the small binding pocket of VF-ω-TA was enlarged for the synthesis of (3S,5R)-ethyl 3-amino-5-methyloctanoate, a precursor of imagabalin [92], leading to a 60-fold activity improvement for the transformation of the bulky–bulky aminoester ($k_{cat} = 0.02\,s^{-1}$).

Although the currently available enzymes provide access to a broad range of enantiopure amines, there are still white spots on the landscape of substrate spectra. For instance, the development of new variants for the production of not only (S)-bulky–bulky amines but also for (R)-bulky–bulky amines containing branched groups would be highly desirable.

2.3 AMINE OXIDASES

The oxidation of chiral amines represents a fundamental key step in various physiological processes, being important, for example, for detoxification, cell growth, and signaling. AOs are the enzymes responsible for this particular transformation. These

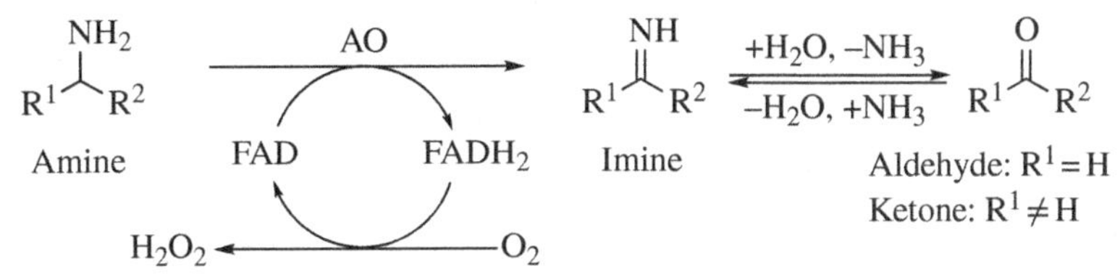

SCHEME 2.22
Simplified reaction scheme for the oxidation of an amine to the corresponding imine through FAD-containing amine oxidases (AOs).

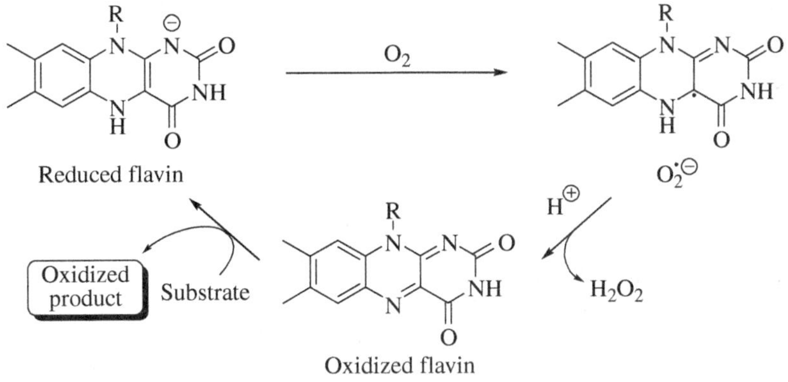

SCHEME 2.23
Redox cycle of the cofactor for flavin-containing enzymes.

enzymes are involved in metabolic pathways of AAs and in the biosynthesis of complex natural products like alkaloids. AOs are members of the oxidoreductase family and can be found in plants, fungi, mammals, and bacteria. Based on the cofactor involved, AOs are classified into type I and II AOs [93]. Type I enzymes require copper and pyrroloquinoline (PQQ) or topaquinone (TPQ) as cofactors [94]; however, as the formed imine remains covalently bound to the protein, no practical applicability is given yet [95]. Type II AOs are flavin dependent wherein the formed imine is released after the oxidation, rendering them suited for biotechnological applications [96]. Prominent examples of flavin adenine dinucleotide (FAD)-dependent enzymes comprise L-amino acid oxidases (L-AAOs; E.C.1.4.3.2), D-amino acid oxidases (D-AAOs; E.C.1.4.3.3), and monoamine oxidases (MAOs; E.C.1.4.3.4) [97]. The enzyme-catalyzed redox reaction employs molecular oxygen as oxidant, which becomes simultaneously reduced to hydrogen peroxide (H_2O_2); the C—N bond is thereby oxidized to C=N (imine), which might be hydrolyzed to the corresponding ketone or the respective aldehyde depending on the reaction conditions employed. The cofactor FAD serves as a shuttle for two electrons between the oxygen and the amine (Scheme 2.22) [98].

The catalytic cycle of oxidases employing FAD as cofactor consists of two half reactions: in a first step, the substrate is oxidized by a two-electron transfer resulting in a reduced flavin (hydroquinone) together with the oxidized product. Regeneration of the cofactor by molecular oxygen occurs in the subsequent oxidative half reaction, generating H_2O_2 and the oxidized flavin, thereby closing the catalytic cycle (Scheme 2.23) [98].

2.3.1 Amino Acid Oxidases

AAOs catalyze the enantioselective oxidation of natural and nonnatural α-AAs to the corresponding α-imino acids. In case a racemate is used as starting material, an oxidative kinetic resolution occurs; thus, in the ideal case, only one enantiomer gets transformed [99]. Spontaneous hydrolysis affords the α-keto acid along with the nonconverted stereoisomer of the AA (Scheme 2.24) [100].

For example, the racemic lactone of 4-hydroxyisoleucine, which was envisioned to be resolved by an AAO in order to afford the bioactive (2*S*,3*R*,4*S*)-enantiomer as the natural product, was found to possess antidiabetic properties [101]. However, the commercially available D-AAO originating from hog kidney failed to give the desired

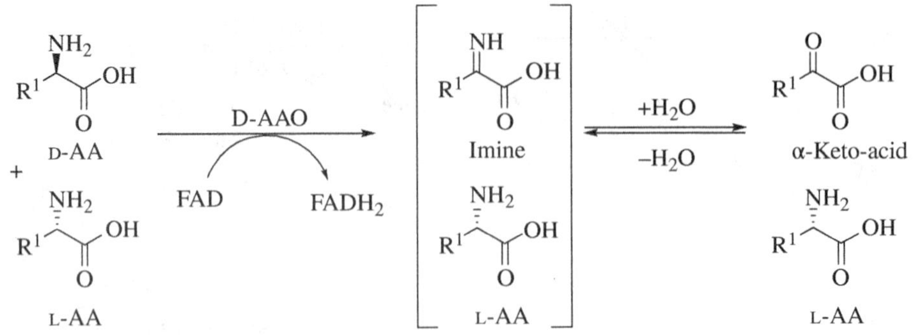

SCHEME 2.24

General reaction scheme for the oxidative kinetic resolution of a racemic amino acid (AA) by a D-AAO; AA, amino acid.

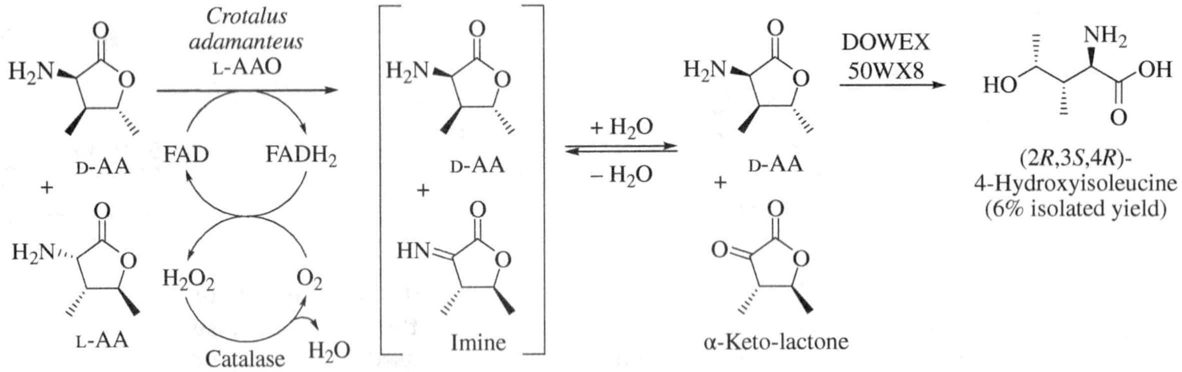

SCHEME 2.25

Oxidative kinetic resolution of a racemic mixture of cyclic 4-hydroxyisoleucine by an L-AAO originating from *Crotalus adamanteus.*

product in reasonable conversion, even after 3 days. The L-selective AAO originating from *Crotalus adamanteus* on the other hand was able to accept the L-enantiomer affording the open-chain form with (2*R*,3*S*,4*R*)-configuration (Scheme 2.25); the isolated yield was only low (6%), but the process demonstrates the high stereopreference of these enzymes in general [102].

Even though the catalytic (enzymatic) kinetic resolution [103, 104] is in general a powerful method for the separation of enantiomers, the major drawback lies in the limitation of 50% maximum yield from the outset [99]. In order to gain more than 50% product yield, alternative techniques based on the asymmetric transformation of a prochiral substrate, DKR [105, 106] or deracemization, were established successfully [84, 86, 107–110]. The latter mentioned technique is either based on the combination of two enantiocomplementary enzymes or by coupling a stereoselective oxidation with a nonstereoselective reduction. This concept is rather powerful as theoretically only seven catalytic cycles are necessary to achieve a single enantiomer in >99% yield [111].

In the 1970s, Hafner and Weller reported the successful deracemization of simple AAs like alanine and leucine by oxidation with a D-AAO originating from porcine kidney in combination with sodium borohydride ($NaBH_4$) as reducing agent (Scheme 2.26) [112]. Although the intention of this study was to prove the enzyme-mediated imine intermediate, the principle of enzymatic stereoinversion and deracemization was established. The concept has been adapted in the following by various research groups and was shown to be feasible for other compounds like proline, pipecolic acid [113, 114], and diverse other α-AAs [115].

A significant improvement regarding the overall efficiency was achieved several years later by changing the nonselective reducing reagent to sodium cyanoborohydride ($NaCNBH_3$) [116]. The major problem associated with the usage of $NaBH_4$ is its tendency to decompose under neutral conditions as well as its high reactivity in water and other protic solvents: up to 500 equivalents were initially necessary for

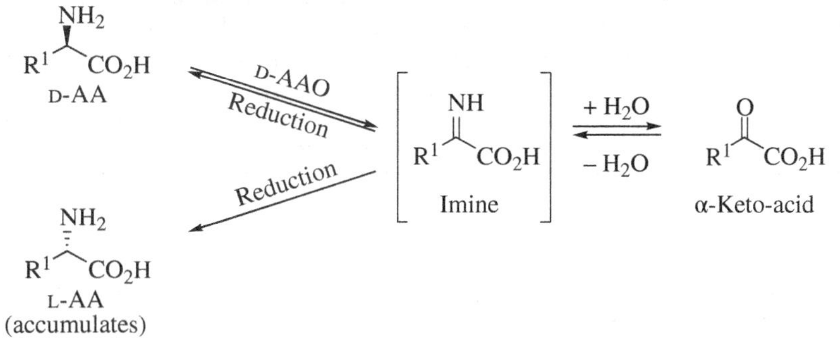

SCHEME 2.26

Deracemization by combining the enantioselective oxidation employing a D-AAO with a nonselective reducing agent (e.g., $NaBH_4$, $NaCNBH_3$, $NH_3{\cdot}BH_3$, etc.).

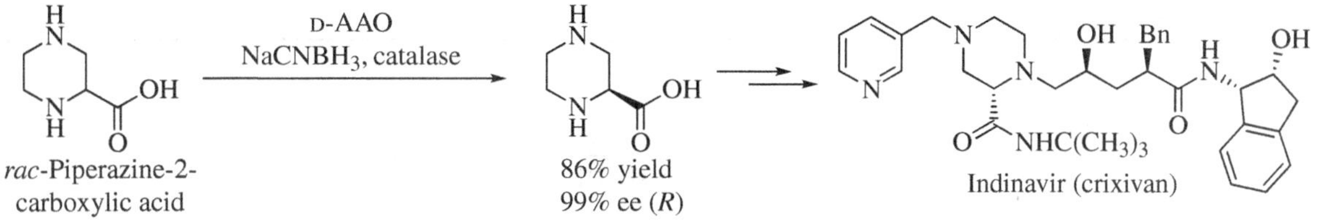

SCHEME 2.27

Deracemization of D/L-piperazine-2-carboxylic acid by D-AAO in combination with sodium cyanoborohydride ($NaCNBH_3$) as nonselective reducing agent.

TABLE 2.8 Scope of the Amino Acid Deracemization Employing the D-AAO Originating from Porcine Kidney with Nonselective Reducing Agents

Entry	Residue (R)	Reducing Agent	Yield (%)	ee (%)
1	Ph	$NaCNBH_3$	75	>99 (L)
2	Bn	$NaCNBH_3$	82	>99 (L)
3	Indolyl-CH_2	$NaCNBH_3$	76	>99 (L)
4	HS-CH_2	$NaBH_4$	77	>99 (L)
5	Et	$NaBH_4$	87	>99 (L)
6	*iso*-Bu	$NaBH_4$	90	>99 (L)
7	Cyclopentyl	$NaCNBH_3$	79	>99 (L)

complete deracemization of the AAs [112]. $NaCNBH_3$, on the other hand, is a milder and water-stable hydride source, which allowed the reducing agent to be diminished to only three equivalents. Even though this reagent is less reactive in comparison with $NaBH_4$, a highly efficient stereoinversion/deracemization procedure can be achieved. For example, the deracemization of racemic piperazine-2-carboxylic acid yielded the optically pure (*R*)-enantiomer, a key building block for the HIV-protease inhibitor indinavir, in remarkable 86% yield (Scheme 2.27) [116].

Further studies regarding the substrate scope revealed that a series of cyclic, alicyclic, and acyclic AAs and derivatives can be successfully converted to the pure L-enantiomers in high yield (Table 2.8). However, in case of the acyclic AAs (Entries 4–6), $NaBH_4$ gave significant better results due to the insufficient stability of the corresponding imines, which led to a competing hydrolysis reaction toward the analogues of α-keto acids.

Even though this concept has been proven to be effective, the toxicity of $NaCNBH_3$ led to the search for alternative reagents whereby amine–boranes turned out to be competent substitutes [117]. They are stable in water at neutral or basic pH and soluble and unreactive toward a wide range of protic and aprotic solvents. Deracemization of several natural and nonnatural AAs with an L-amino acid deaminase (L-AAD)

TABLE 2.9 Selected Examples of the Deracemization of Natural and Nonnatural L-Amino Acids with L-AAD Originating from *Proteus myxofaciens* in Combination with Ammonia Borane Complex

Entry	Product	Reducing Agent	Yield (%)	ee (%)
1	Norvaline	$NH_3{\cdot}BH_3$	81	>99 (D)
2	Norleucine	$NH_3{\cdot}BH_3$	86	>99 (D)
3	Tryptophan	$NH_3{\cdot}BH_3$	82	>99 (D)
4	Phenylalanine	$NH_3{\cdot}BH_3$	82	>99 (D)
5	Allylglycine	$NH_3{\cdot}BH_3$	79	>99 (D)
6	Methionine	$NH_3{\cdot}BH_3$	90	>99 (D)
7	Cyclopentylglycine	$NH_3{\cdot}BH_3$	87	>99 (D)
8	Tyrosine	$NH_3{\cdot}BH_3$	79	>99 (D)

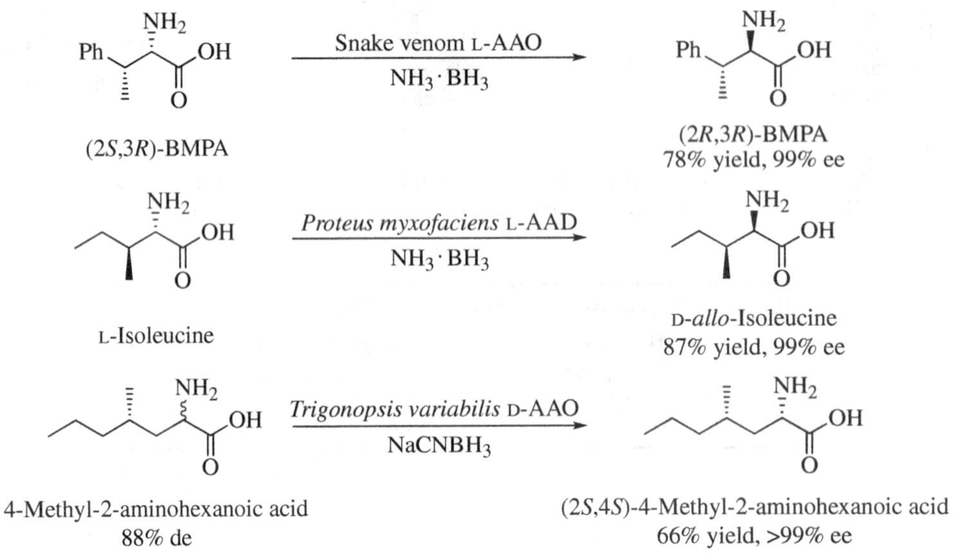

SCHEME 2.28 Deracemization of β- and γ-substituted α-amino acid derivatives by complementary AAOs and L-AADs in combination with different reducing reagents.

from *Proteus myxofaciens* [118] was conducted successfully to yield the optically pure D-AAs in high conversion (Table 2.9). Notably, even sensitive compounds with a terminal double bond were effectively transformed without the observation of side reactions like hydroboration (Entry 5).

An extension of this method with respect to a second stereocenter providing pure diastereomers was also demonstrated afterward: this allows access not only to β-substituted but also γ-substituted α-AA derivatives in good yield and excellent stereoselectivity under mild reaction conditions [119]. For example, enantiomerically pure β-methyl-phenylalanine (BMPA) and L-leucine were effectively epimerized to afford the optically pure products in 78 and 87% yields, respectively (Scheme 2.28). The deracemization of 4-methyl-3-aminohexanoic acid with a methyl group in γ-position proceeded also smoothly in the presence of D-AAO from *Trigonopsis variabilis* in combination with $NaCNBH_3$.

2.3.2 Cascade Reactions Involving AAOs

AAOs were also combined with other biocatalysts in various cascade reactions [66, 120]. A couple of processes involving AAOs were established providing enantiomerically pure building blocks in the form of a multienzyme network. For example, D-methionine was effectively converted to L-methionine using four different enzymes in one pot: oxidation of the D-amino acid to the corresponding α-keto acid was achieved with D-AAO originating from *Arthrobacter protophormiae*,

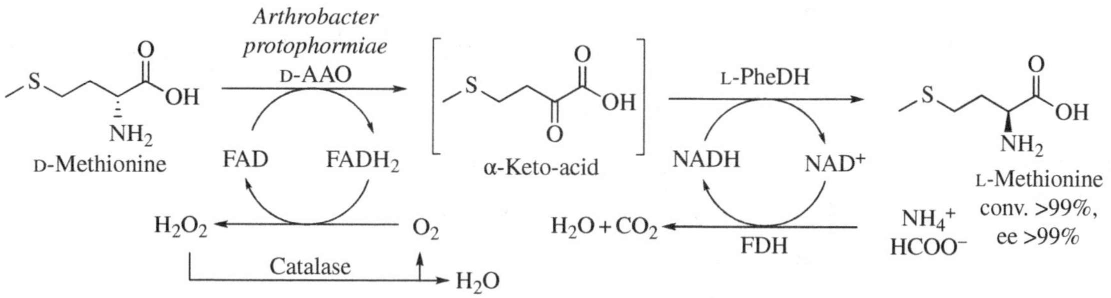

SCHEME 2.29

Stereoinversion of D- into L-methionine by a multienzyme network, combining a D-AAO with L-phenylalanine dehydrogenase (L-PheDH), catalase, and an FDH in one pot.

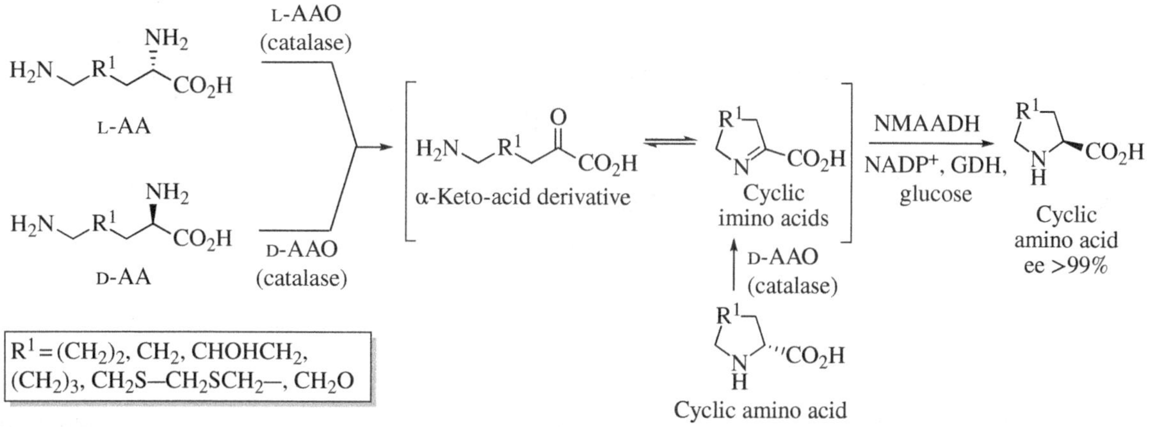

SCHEME 2.30

Enzymatic cascade providing access to cyclic amino acid (AA) derivatives by the combination of AAO with *N*-methyl amino acid dehydrogenase (NMAADH) starting either from a linear or cyclic AA.

and the intermediate α-keto acid was aminated by L-phenylalanine dehydrogenase (L-PheDH) [121]. Catalase was added for the decomposition of the formed H_2O_2 and an FDH for recycling of the cofactor NADH (Scheme 2.29).

Another one-pot multienzyme network to access enantiomerically pure cyclic amino acids (AAs) started either from the enantiopure acyclic- or the racemic cyclic AA derivative (Scheme 2.30) [122]. Starting from the linear precursor, oxidation of the enantiopure diamino acid by L- or D-AAO affords the α-keto acid cyclizing spontaneously to the corresponding imino-acid (Δ^1-piperideine-2-carboxylic acid). The intermediate is not isolated but reduced immediately to the cyclic AA by a *N*-methyl-L-amino acid dehydrogenase (NMAADH) originating from *Pseudomonas putida* [123]. The feasibility of the overall transformation has been demonstrated on a preparative scale for designated compounds affording products in yields between 35% (R = CH_2O) and 70% (R = —CH_2SCH_2—). Although the yields were only moderate at its best, the compatibility of various enzymes and cofactors involved in this cascade could be shown, and further optimization studies are expected to provide a highly efficient process for cyclic AA preparation.

AAOs have also been coupled with other enzymes, like amino transferases, in order to achieve an alternate deracemization process. For example, a racemic α-AA with a bulky heterobiaryl residue was converted to the (*S*)-enantiomer by combination of an oxidative kinetic resolution performed by the D-AAO from *Trigonopsis variabilis* and an ω-TA from *Burkholderia* sp. (Scheme 2.31). A conversion of 85% after 22 h led to 72% isolated product yield with an excellent stereoselectivity (>99.5% ee). Notably, the reaction was initially performed on a liter scale but was then scaled up to convert 1 kg of substrate [124].

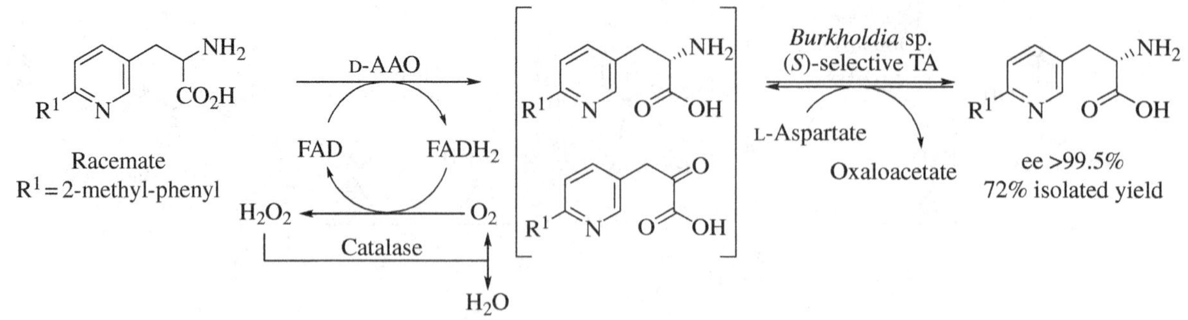

SCHEME 2.31

Enzymatic deracemization by combination of a D-AAO with an ω-TA.

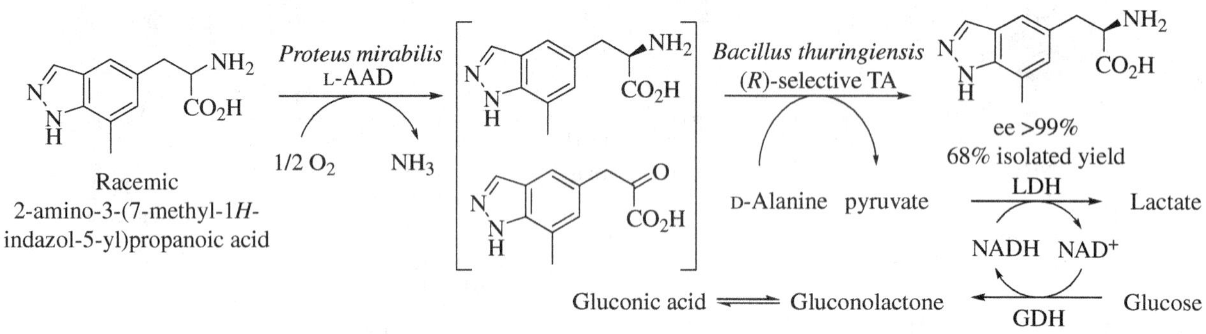

SCHEME 2.32

Multienzyme network for the deracemization of 2-amino-3-(7-methyl-1-*H*-indazol-5-yl)propanoic acid by the combination of a L-AAD with an ω-TA. GDH, glucose dehydrogenase; LDH, lactate dehydrogenase.

SCHEME 2.33

Multienzyme cascade for the conversion of L-amino acids to α-hydroxy acids employing an L-amino acid deaminase (L-AAD) and isocaproate reductases (Hic).

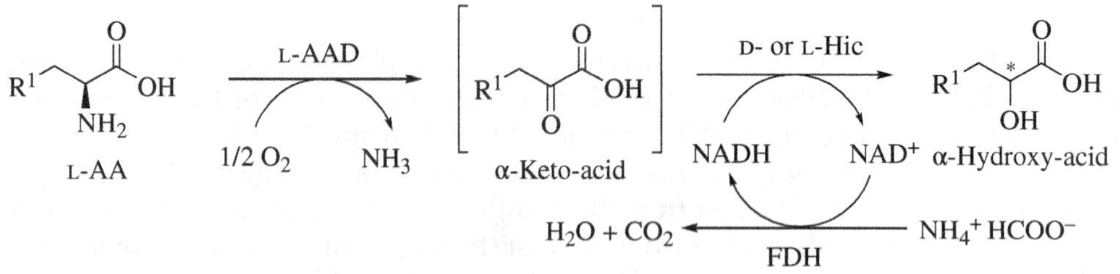

A further extension of this concept was elaborated for a key intermediate needed for the synthesis of diverse drug candidates: 2-amino-3-(7-methyl-1*H*-indazol-5-yl) propanoic acid. An L-AAD was employed instead of an L-AAO for the oxidative kinetic resolution (Scheme 2.32). In order to ensure an efficient asymmetric reduction with the transaminase, LDH and GDH were also incorporated in the network: while an LDH reduced pyruvate to lactate in order to avoid inhibition of the transaminase, GDH enabled an efficient NAD^+ cofactor recycling. Optimization of both steps individually allowed finally running the whole cascade in one pot to afford the optically pure product (>99% ee) with a good isolated yield of 68% [125].

In a another report, the natural and nonnatural α-AAs were also transformed to the corresponding α-hydroxy acids through a multienzyme network: this sequence comprises an L-AAD with complementary isocaproate reductases originating from *Lactobacillus casei* (L-Hic) [126] and *Lactobacillus confusus* (D-Hic) [127], respectively, and an FDH for cofactor recycling. The reaction proceeded via formal biocatalytic retention or inversion of absolute configuration allowing both α-hydroxy acid enantiomers to be accessed simply by the choice of the reductase employed (Scheme 2.33) [128].

TABLE 2.10 Scope of the Reaction Cascade to Afford α-Hydroxyacids through the Combination of an AAD with Complementary Isocaproate Reductases (Hics)

Entry	R^1	Conc. (mM)	HicDH	Conv. (%)[a]	ee (%)
1	Ph	200	L-Hic	>99 (78)	>99 (*S*)
2	Ph	200	D-Hic	>99 (79)	>99 (*R*)
3	3-Indoyl	50	L-Hic	>99 (83)	>99 (*S*)
4	3-Indoyl	50	D-Hic	>99 (85)	>99 (*R*)
5	*iso*-Propyl	100	L-Hic	>99 (81)	>99 (*S*)
6	*iso*-Propyl	100	D-Hic	>99 (85)	>99 (*R*)
7	*n*-Propyl	100	L-Hic	>99 (79)	>99 (*S*)
8	*n*-Propyl	100	D-Hic	>99 (82)	>99 (*R*)
9	CH_3-S-CH_2	200	L-Hic	>99 (77)	>99 (*S*)
10	CH_3-S-CH_2	200	D-Hic	>99 (73)	>99 (*R*)

[a] The values in brackets refer to the isolated yield.

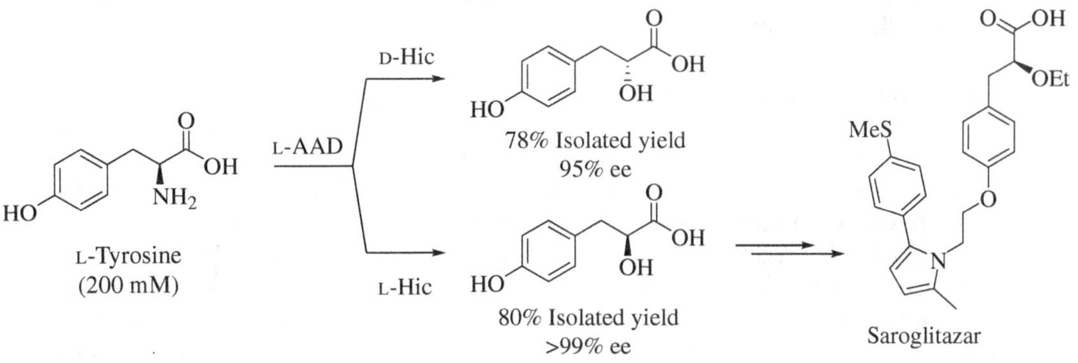

SCHEME 2.34
Synthetic applicability of the cascade reaction comprising an L-amino acid deaminase (L-AAD) and stereocomplementary isocaproate reductases (Hics).

Initially each step was individually investigated and optimized with respect to the oxygen and substrate concentration. The subsequent coupling of both reactions proceeded smoothly, even allowing the substrate concentration to increase up to 200 mM with R = Ph in the presence of one bar of oxygen pressure. It is worth mentioning that no significant amounts of α-keto acid accumulated (<5%) as analyzed by following the reaction by an NMR study demonstrating the excellent interlinkage of both reaction steps.

Representative α-AAs with different functionalities were successfully transformed proving the general feasibility of this cascade, even for sensitive compounds like tyrosine; with an exception of R = 3-indoyl (Entries 3 and 4), very high substrate concentrations (>50 mM) were well accepted to yield the corresponding hydroxy acids in excellent conversion, isolated yield, and enantioselectivity (Table 2.10).

The synthetic applicability was also demonstrated on a preparative scale converting L-tyrosine to the pharmaceutically relevant analogue (*S*)-2-hydroxy-3-(4-hydroxyphenyl)propanoic acid in 80% yield (>99% ee). The product serves as a key building block for the treatment of diabetes II drug saroglitazar (Scheme 2.34).

2.3.3 Monoamine Oxidases

Although AAOs accept broad substrate spectrum with respect to the AA side chain, they are restricted to the carboxylic acid motif limiting their application in the enantioselective C—N bond oxidation. Hence, a wide range of structurally diverse

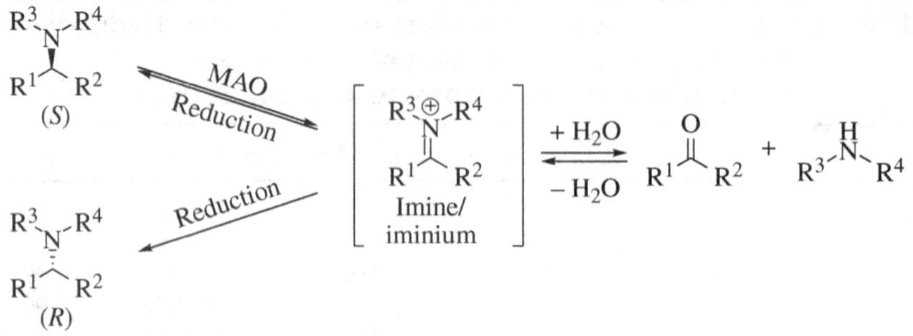

SCHEME 2.35

Deracemization of chiral primary, secondary, and tertiary (*S*)-amines by MAOs in combination with a nonselective reducing agent.

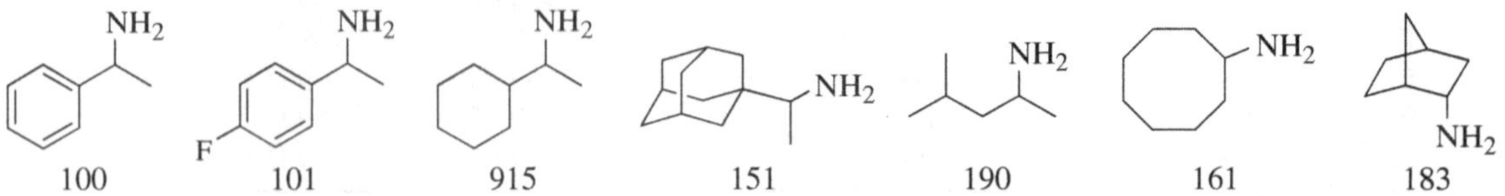

FIGURE 2.1

Selected examples of primary amines being substrates for the Asn336Ser variant of the MAO-N D1 from *Aspergillus niger*. The numbers indicate the activity of the variant relative to 1-phenylethylamine.

amines remain inaccessible with these enzymes [100]. On the other hand, MAOs do not depend on such additional structural motives and are therefore able to oxidize a broad scope of amines. The deracemization of amines via enantioselective oxidation proceeds thereby through an equivalent reaction as described for the AAOs as a combination with nonstereoselective reduction (Scheme 2.35) [96–98].

In particular, recombinant variants of the monoamine oxidase N (MAO-N) originating from *Aspergillus niger* have been established over the past years as excellent biocatalysts for the deracemization and stereoinversion of diverse primary, secondary, and even tertiary amines by Turner and coworkers [18]. A series of tailored variants were created by an intensive study and successive optimization of each catalyst by the combination of directed evolution, rational protein design, and novel techniques of high-throughput screenings [95, 129].

The development of the catalysts started with the finding that the native wild-type MAO-N was highly active in the oxidation of terminal aliphatic amines such as *n*-amyl- and 1-butylamine but also, at a low rate, toward benzylamine [130–132]. Moreover, this initial enzyme displayed a clear enantiopreference for the (*R*)-enantiomer. In order to broaden the substrate specificity, the enzyme was subjected to several rounds of directed evolution using a combination of random mutagenesis and a solid-phase screening method employing 1-phenylethylamine as substrate probe. Out of ~50 000 mutants, sequencing of the best variant revealed a single point mutation in the AA sequence (Asn336Ser), which led to a 47-fold higher activity and sixfold higher enantioselectivity in comparison with the wild-type enzyme [133]. After a subsequent optimization with respect to the protein expression on the genetic level (addition of one more mutation: Met348Leu), exploration of the substrate scope revealed several compounds reacting even faster than 1-phenylethylamine itself (Figure 2.1) [134].

A further round of directed evolution led to beneficial mutations providing a novel variant (MAO-N D3; Asn336Ser, Arg259Lys, Arg260Lys, Met348Lys, Ile246Met) with higher activity toward a wide range of chiral secondary amines than the original parent. After initial small-scale reactions, the practical applicability of the MAO-N D3 variant was demonstrated for the deracemization of 1-methyl-tetrahydroisoquinoline (MTQ, Scheme 2.36a) and 2-phenylpyrrolidine (PPD, Scheme 2.36b) on a preparative scale. The enzyme was additionally immobilized on Eupergit C, and the reactions were conducted in combination with ammonia borane ($NH_3{\cdot}BH_3$) as reducing agent. Both products were obtained in excellent yield and optical purity after sufficient reaction time [135].

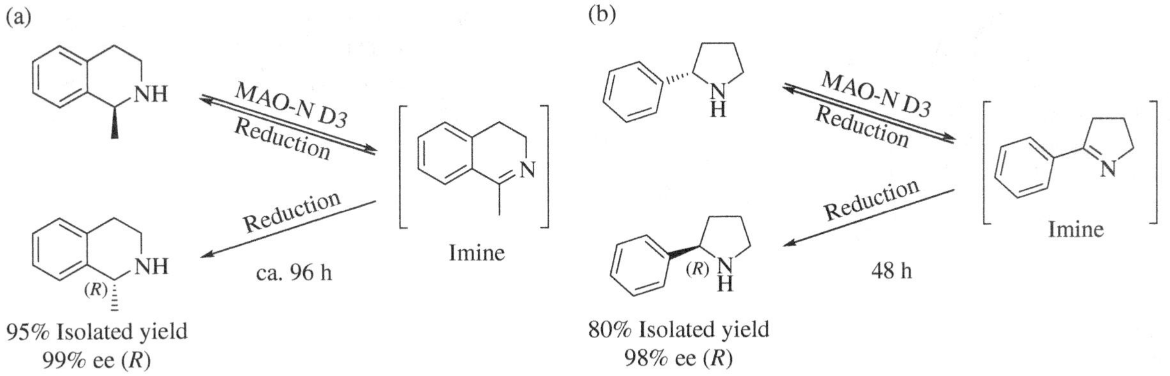

SCHEME 2.36

Deracemizations of chiral secondary amines employing the variant MAO-N D3 in combination with ammonia borane complex (a) 1-methyl-tetrahydroisoquinoline (20 mM) (b) 2-phenylpyrrolidine (100 mM).

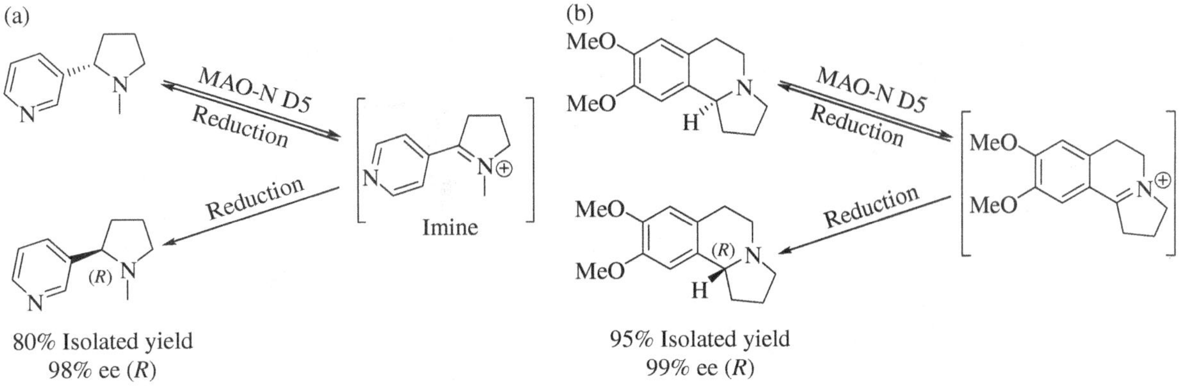

SCHEME 2.37

Stereoinversion of (*S*)-nicotine (a) and deracemization of *rac*-cripsine A (b) on a preparative scale employing the MAO-N D5 variant with a nonselective reduction agent ($NH_3 \cdot BH_3$).

Maintaining the strategy of rational protein design, also tertiary amines have become accessible after introducing further important mutations (MAO-N D5; N336S/I246M/M348L/T384N/D385S); the variant so obtained was successfully used in the stereoinversion of (*S*)-nicotine [136] and in the deracemization of racemic crispine A (Scheme 2.37) [137, 138], a natural alkaloid showing cytotoxic activities against HeLa human cancer cell lines [139]. Notably, as the amines are tertiary, both reactions proceed via the achiral iminium-ion intermediate affording (*R*)-nicotine in 80% and (*R*)-crispine in 95% yield (≥98% ee).

The MAO-N D5 variant was also employed in the desymmetrization of *N*-unprotected *meso*-pyrrolidines yielding the corresponding Δ^1-pyrrolines; these compounds are useful intermediate respective templates for the synthesis of highly substituted proline analogues through a *trans*-selective nucleophilic addition of designated nucleophiles like cyanide, followed by an acid-mediated hydrolysis (Scheme 2.38) [140].

Investigations regarding the scope revealed good activities for a wide range of symmetrical 3,4-*cis*-disubstituted pyrrolidine amines, whereas the rate of oxidation was found to correlate with the bulkiness and lipophilicity as judged by a colorimetric assay (Figure 2.2). Studying the stereochemical outcome of the reactions, the high stereoselectivity already observed for the previous variants MAO-N D1 and MAO-N D3 was retained; hence, most 3,4-disubstituted Δ^1-pyrrolines were recovered in ee's ≥98%.

The practical utility was illustrated hereafter through the desymmetrization of a bicyclic *meso*-pyrrolidine, which was readily converted to the corresponding AA by

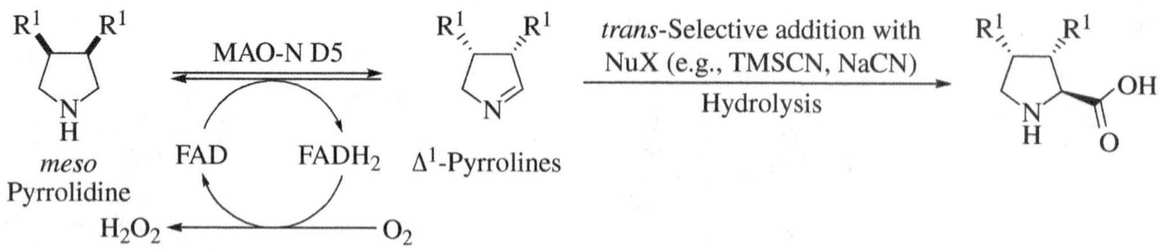

SCHEME 2.38

General reaction scheme for the MAO-N D5-catalyzed desymmetrization of achiral 3,4-disubstituted *meso*-pyrrolidines and a possible application.

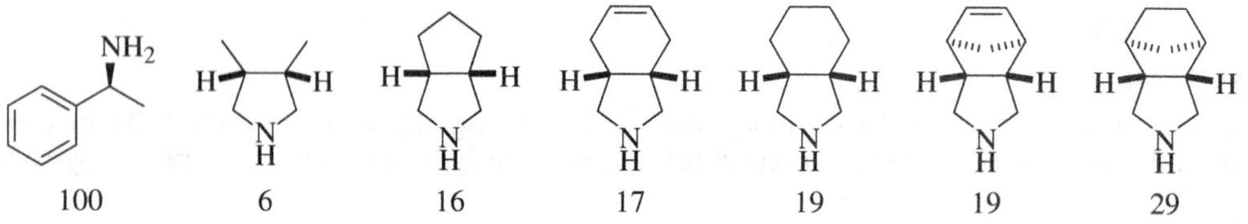

FIGURE 2.2

Selected examples of secondary 3,4-disubstituted pyrrolidines being substrates for the MAO-N D5 variant from *Aspergillus niger*. The numbers indicate the activity relative to 1-phenylethylamine.

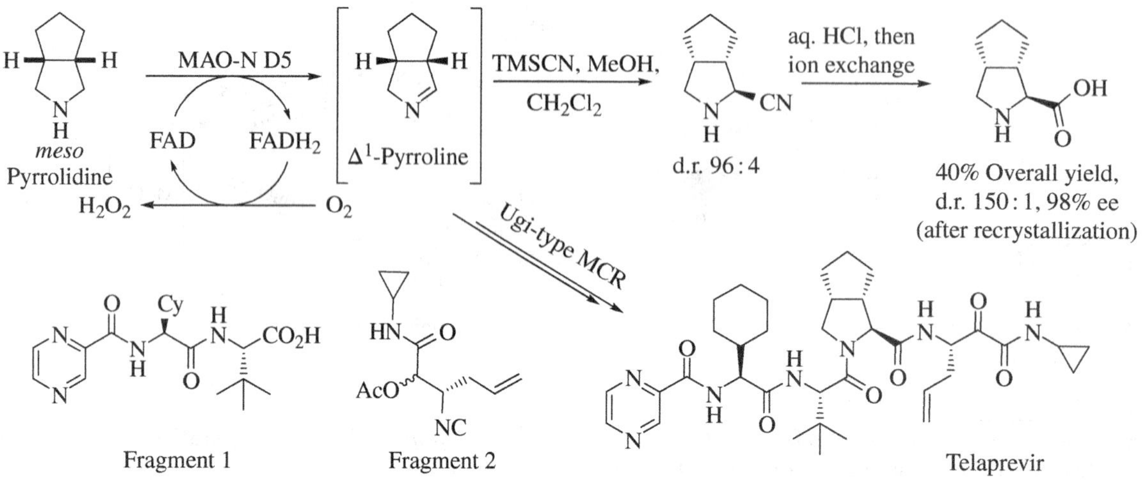

SCHEME 2.39

Stereoselective synthesis of a bicyclic AA through enzymatic desymmetrization of *meso*-pyrrolidine and application of the intermediate as key building block in telaprevir synthesis.

the addition of trimethylsilyl cyanide (TMSCN) to the formed imine and subsequent nitrile hydrolysis. Even though the d.r. was found to be strongly dependent on the reaction conditions employed, the product was obtained with excellent stereoselectivity (optimized conditions): d.r. 95 : 5 (94% ee) and 51% yield over three steps. The optical purity was improved by a single recrystallization step (Scheme 2.38).

A further extension of the MAO-N-catalyzed desymmetrization of bicyclic *meso*-pyrrolidines was also reported in which the formed imine was coupled with the Ugi-type multicomponent reaction (MCR) to access highly substituted prolyl peptides and derivates thereof. The so-obtained products are applicable in medicinal chemistry as well as organocatalysis as proline analogues [141]. Subsequently, this combination (desymmetrization and Ugi-type MCR) [142] has also led to the development of an efficient synthesis of the peptide-based hepatitis C protease inhibitor telaprevir through coupling of the Δ^1-pyrroline with the fragments 1 and 2 (Scheme 2.39) [143].

The latter strategy has been adapted by Merck and Codexis to manufacture a key intermediate of the related peptidomimetic protease inhibitor boceprevir [144]: a variant with improved solubility and thermal stability (up to 50 °C) displaying an eightfold higher activity as the wild type was created employing family shuffling of the MAOs originating from *A. niger* with *Aspergillus oryzae*, random mutagenesis (via epPCR), and homology model-inspired approaches. A highly efficient enzymatic

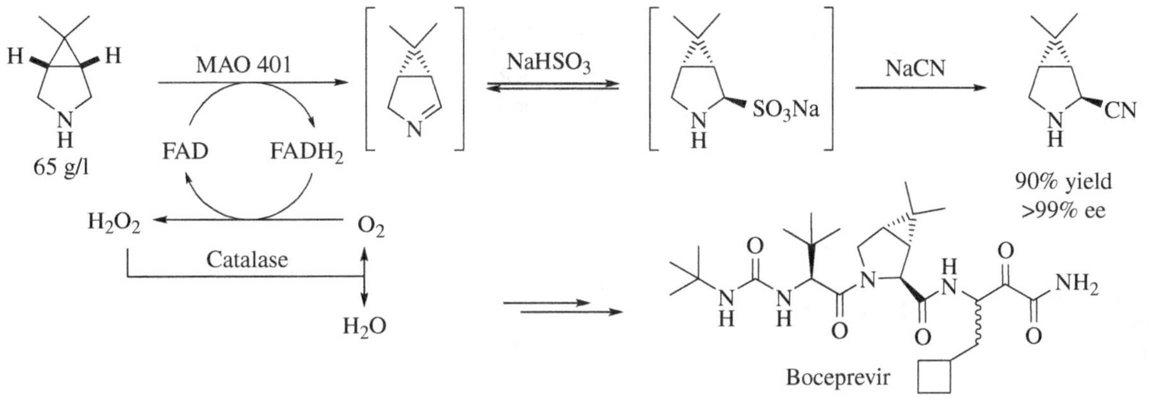

SCHEME 2.40

Merck's approach for the stereoselective synthesis of the bicyclic building block of boceprevir.

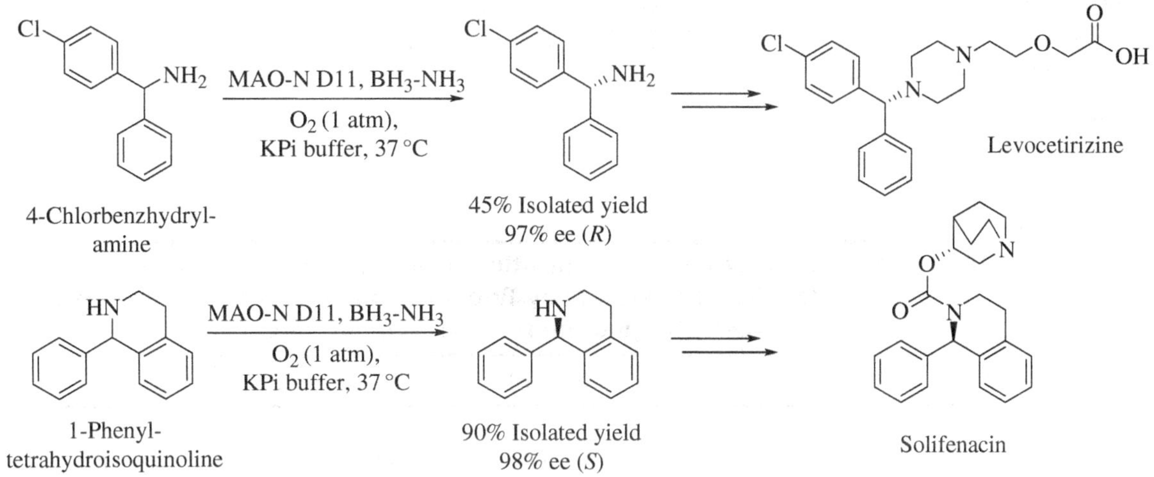

SCHEME 2.41

Deracemization of important building blocks for the pharmaceutical drugs levocetirizine and solifenacin employing MAO-N D11.

method has been established for the asymmetric oxidation on an industrial scale after addressing certain limitations such as the low solubility of the formed Δ^1-pyrrolidine, inhibition of the catalase by cyanide, and epimerization of the cyanide intermediate in an aqueous solution (Scheme 2.40).

The crystal structure of the MAO-N D5 variant has been solved (pdb:2vvm), leading to valuable insights for the binding modes as well as regarding the substrate accessibility [145]. This information has been accommodated in further rational protein design approaches providing novel variants (D9–D11) with a more spacious small active site cavity being active toward sterically demanding substrates. For example, 4-chloro-benzhydrylamine (4-CBHA) and 1-phenyltetrahydroisoquinoline (PTIQ) (Scheme 2.41) were efficiently deracemized with the MAO-N D11 variant to yield the optically pure (≥97% ee) precursors for the pharmaceutical drugs levocetirizine (a nonsedative histamine antagonist) and solifenacin (for treating contraction of overactive bladder) [146]. Even though the isolated yield of 4-CBHA was only moderate (45%) due to competing hydrolysis of the formed imine, the effective discrimination of such similar residues by the enzyme is remarkable.

Another highly sophisticated application represents the oxidative Pictet–Spengler cascade for the synthesis of the natural alkaloid (*R*)-harmicine catalyzed by the MAO-N D9 mutant: enzymatic oxidation of the tertiary amine results in a reactive

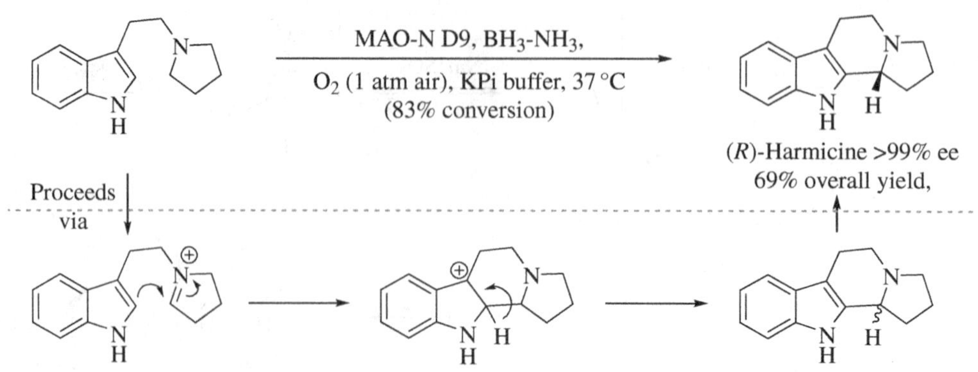

SCHEME 2.42

Formal oxidative Pictet–Spengler cascade catalyzed by the MAO-N D9 variant of *Aspergillus niger*.

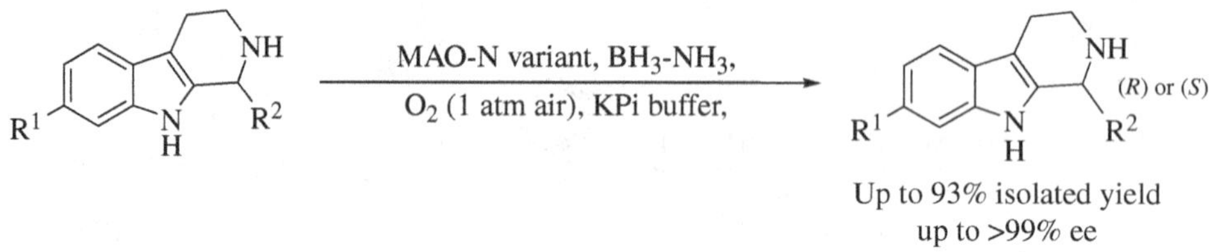

SCHEME 2.43

General reaction scheme for the deracemization of 1-substituted tetrahydro-β-carbolines (TBHCs) by the evolved monoamine oxidase N (MAO-N) in combination with a nonselective reducing agent.

TABLE 2.11 Scope of the Deracemization of Tetrahydro-β-Carbolines (THBCs, Scheme 3.22) by Two Variants of the Monoamine Oxidase N (MAO-N) Originating from *Aspergillus niger*

Entry	R^1	R^2	Time (h)	ee (%) MAO-N D9	ee (%) MAO-N D11
1	OMe	Me	24	99 (*R*)	n.r.
2	H	Me	24	>99 (*R*)	99 (*R*)
3	H	Et	84	36 (*R*)	25 (*R*)
4	H	*iso*-Propyl	72	40 (*S*)	33 (*S*)
5	H	Cyclopropyl	48	86 (n.d.)	35 (n.d.)
6	H	*n*-Propyl	48	99 (*S*)	>99 (*S*)
7	H	*n*-Butyl	48	96 (*S*)	99 (*S*)
8	H	*sec*-Butyl	48	58 (*S*)	>99 (*S*)
9	H	*tert*-Butyl	48	62 (*S*)	96 (*S*)
10	H	Cyclopentyl	48	85 (*S*)	96 (*S*)
11	H	Cyclohexyl	48	80 (*S*)	97 (*S*)
12	H	Phenyl	72	30 (*S*)	92 (*S*)

n.d., not determined; n.r., not reported.

iminium-ion, which underwent a nonselective cyclization to form the racemic *rac*-harmicine after rearomatization. Subsequent deracemization by the enzyme yields the (*R*)-enantiomer in >99% ee and an excellent 69% overall yield (Scheme 2.42). The formal one-pot oxidative asymmetric Pictet–Spengler cascade represents a conceptually new approach to such challenging compounds, highlighting simultaneously the enormous potential for further applications.

MAO-N D11 and MAO-N D9 were also found to be effective in the oxidation of tetrahydro-β-carbolines (THBCs) and related heterocycles. Several 1-substituted aliphatic and alicyclic THBCs were deracemized in good yields and high enantioselectivity by the proper choice of the catalyst employed (Scheme 2.43 and Table 2.11) [146, 147].

This method has opened a straightforward approach for this structural motive widespread in nature, as demonstrated for the alkaloids leptaflorin (R^1=OMe,

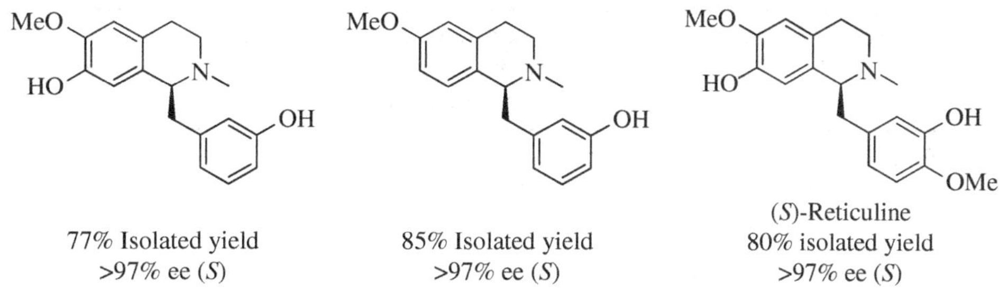

FIGURE 2.3
Selected results of the MAO-N-catalyzed desymmetrization of benzylisoquinoline alkaloids.

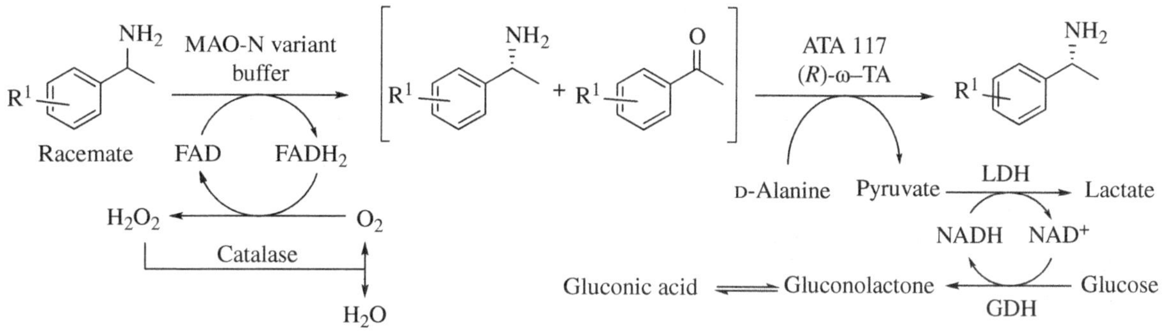

SCHEME 2.44
The reaction scheme for the deracemization of (substituted) 1-phenylethyl amine derivatives by the combination of a monoamine oxidase (MAO) with an ω-TA; ATA (amino transferase) denotes the commercially available ω-TAs from Codexis.

R^2=Me, Table 3.3, Entry 1) and eleganine (R^1=H, R^2=Me, Entry 2). Both products were obtained in enantiomerically pure form (≥99% ee) within reasonable time (24h) on a preparative scale.

Interestingly, depending on the nature of the C-1 substituent (R^2), a switch in stereopreference was observed: while in case of small residues like methyl and ethyl the (*S*)-enantiomer was oxidized, the opposite stereopreference is obtained with increasing size and lipophilicity (Table 2.11, Entries 4–12). Several docking simulations with the MAO-N D11 variant (pdb:3ZDN) [146] were performed furnishing a mechanistic rationalization based on the productive binding modes, which offers at the same time a platform for further directed evolution strategies.

Benzylisoquinoline alkaloids were transformed by the enzyme (MAO-N D11 variant) again with (*R*)-enantiopreference although not always with high enantioselectivity [148]. Protein–ligand docking simulations confirmed the productive binding modes, which were already found for the THBCs [147]. Preparative-scale deracemization reactions with the most active substrates allowed the (*S*)-configured products to be recovered in good yield and with excellent optical purity (Figure 2.3).

2.3.4 Cascade Reactions Involving Monoamine Oxidases

Similar to AAOs, MAO can also be employed either in the stereoinversion starting from an enantiopure amine or in the deracemization starting from the racemate (*vide supra*) [18]. Both approaches have been proven to be remarkably efficient for chiral amine synthesis, in particular for cyclic secondary and tertiary structures. Conversely, for acyclic imine intermediates, a competing hydrolysis reaction toward the carbonyl compound is often observed, thus leading to a diminished product yield. This spontaneous hydrolysis was recently exploited for the deracemization cascade reaction combining an (*S*)-selective MAO with an (*R*)-selective ω-TA (Scheme 2.44) [149].

Starting from a racemic mixture of 1-phenylethyl amine derivatives, the MAO-N-mediated oxidative kinetic resolution of the (*S*)-enantiomer affords the ketone (after hydrolysis of the imine) besides the nonconverted (*R*)-enantiomer of the amine.

TABLE 2.12 Scope of the Deracemization Cascade Combining a MAO with an ω-TA to Afford Enantiomerically Pure 1-Phenylethylamine Derivatives

Entry	R^1	Conv. (%)	ee (%)
1	—	99	>99 (*R*)
2	4-Cl	87	>99 (*R*)
3	4-Br	91	>99 (*R*)
4	4-I	90	>99 (*R*)
5	4-NO_2	99	>99 (*R*)
6	3-Cl	90	>99 (*R*)
7	3-Br	90	>99 (*R*)
8	2-F	81	>99 (*R*)

SCHEME 2.45

Enzymatic *N*-dealkylation of alkyl-benzylamine derivatives by the concurrent oxidation–reduction cascade combining a MAO with an ω-TA.

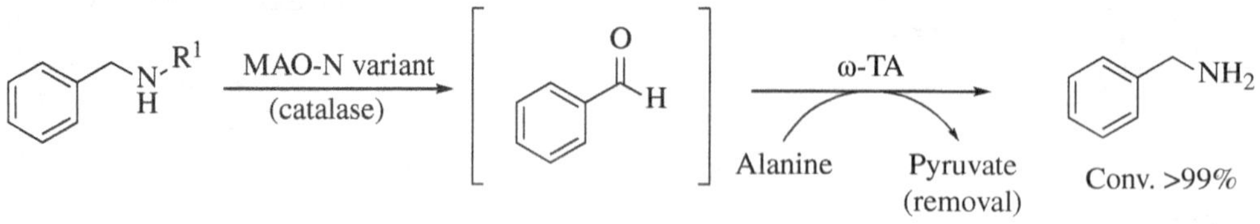

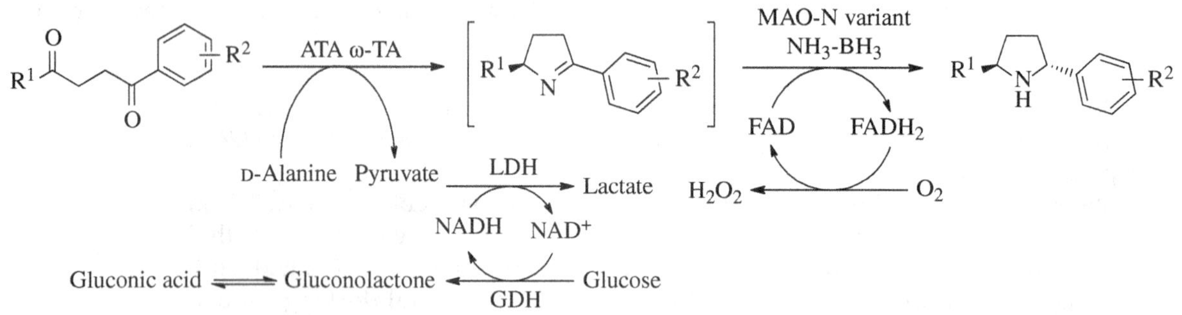

SCHEME 2.46

One-pot cascade to yield 2,5-disubstituted pyrrolidines. ATA (amino transferase) denotes the commercially available ω-TAs from the Codex screening kit.

Subsequent asymmetric reductive amination catalyzed by the ω-TA afforded the optically pure products in excellent conversions between 81 and 99% (Scheme 2.44 and Table 2.12). This simultaneous oxidation–reduction cascade comprises in total five concurrent enzymatic transformations without affecting the conversion or stereochemical outcome observed for each individual step.

The biocatalytic cascade was also demonstrated for the oxidative *N*-dealkylation of various secondary benzylamine derivatives, which usually requires harsh conditions (Scheme 2.45). All investigated substrates (R^1 = Me, Et, 2-Pr) were transformed to completion (conv. >99%) giving the unprotected amine under mild reaction conditions. Notably, the already existing toolbox of tailored MAO-Ns implies that this method is not limited to such simple aryl–alkyl systems but represents a general method for *N*-deprotection.

Another cascade exploiting the MAO-N-mediated oxidation and an ω-TA-catalyzed asymmetric amination has been established to access diastereomerically pure 2,5-disubstituted pyrrolidines, an important scaffold found in diverse pharmaceuticals and natural products. The key steps involve a regioselective mono-amination of various substituted 1,4-diketones, followed by a MAO-N-catalyzed desymmetrization of the intermediate formed Δ^1-pyrroline (Scheme 2.46) [150].

TABLE 2.13 Selected Results of the Regioselective Monoamination–Cyclization–Desymmetrization Multienzyme Cascade

Entry	R^1	R^2	ω-TA	MAO-N	Conv. (%)	de (%)
1	Me	—	ATA-113	D5	>99	>99 (2*S*,5*R*)
2	Me	4-Me	ATA-113	D9	>99	>99 (2*S*,5*R*)
3	Me	4-F	ATA-113	D9	>99	>99 (2*S*,5*R*)
4	Me	4-OMe	ATA-113	D9	>99	>99 (2*S*,5*R*)

ATA (amino transferase) denotes the commercially available ω-TAs from the Codex screening kit.

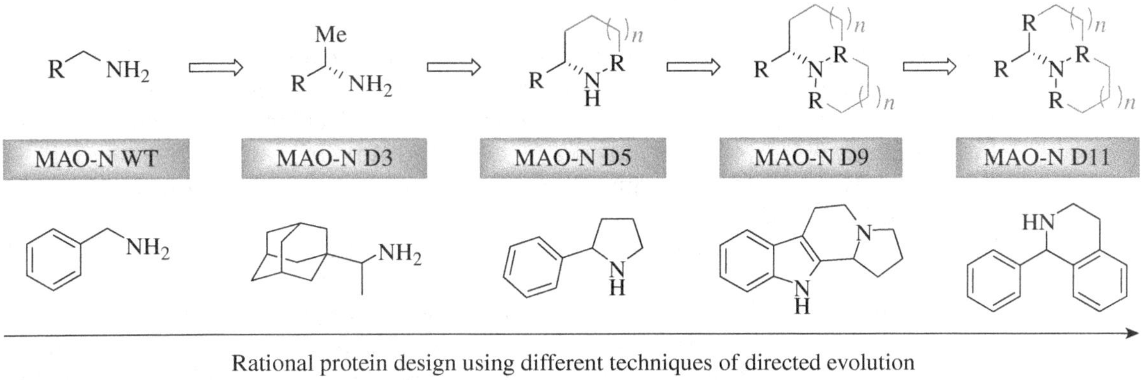

FIGURE 2.4

Consecutive optimization of MAO-N variants originating from *Aspergillus niger* to broaden the substrate scope toward primary, secondary, and tertiary amines.

After suitable reaction conditions were established, the entire cascade gave access to the products with excellent conversion and diastereoselectivity (Table 2.13). Overall, this system demonstrates not only the compatibility of all enzymes employed but also the exquisite regioselectivity of the ω-TAs found previously for related compounds [76–78].

2.3.5 Perspective

Recent advances in rational protein design have enabled the catalyst to be tailored to fit the needs of the process conditions by altering the substrate specificity or other molecular characteristics like stability and activity [14, 151]. Within this context, the evolution of the wild-type MAO from *A. niger* represents so far an unique example of how a catalyst can be adapted to various types of substrates; thus, starting from the wild-type MAO, a set of variants was developed whereby each one is specific for the transformation of either primary, secondary, or tertiary amines (Figure 2.4) [18].

However, all variants display in general a stereopreference for the (*S*)-enantiomer; there are only a few exceptions in which the opposite (*R*)-enantiomer is observed, for which mechanistic rationalizations are provided by productive/unproductive-binding modes within the protein. There is also a demand to provide access to the complementary stereoisomer, in particular by the pharmaceutical industry. The approach might be either to modify the existing enzymes or to identify novel wild-type enzymes [17]. For example, a highly (*R*)-selective AO has been developed just recently by a mutagenesis strategy based on combinatorial active-site testing (CASTing): the enzyme 6-hydroxy-D-nicotine oxidase (6-HDNO) originating from *Arthrobacter nicotinovorans* has been engineered to yield a novel double mutant suitable for the deracemization of a variety of secondary and tertiary amines, yielding the corresponding

(*S*)-enantiomer [152]. Alternatively, the D-AAO from the pig kidney has been engineered losing its requirement of a carboxylic acid in α-position: this has facilitated to change the substrate scope from α-AAs to primary and very few secondary amines [153]. Iterative saturation mutagenesis of cyclohexylamine oxidase (CHAO) has led to beneficial mutations, expanding the toolbox of biocatalyst for the production of chiral amines furthermore [154]. Another promising perspective relies in the combination of evolved biocatalysts with artificial (metallo)enzymes: this allows exploiting attractive features of both catalyst types [155], opening new capabilities as shown recently by the combination of various oxidases with an artificial transfer hydrogenase [156].

2.4 BERBERINE BRIDGE ENZYMES

BBEs (EC 1.21.3.3) are biocovalent flavoenzymes occurring in alkaloid-producing plants, particularly within the *Papaveraceae* (poppy) and *Fumariaceae* (fumitory) families [157]. They catalyze a unique oxidative intramolecular C—C bond formation by bridging a phenol moiety to an *N*-methyl group. In nature, (*S*)-reticuline, a 1-benzyl-1,2,3,4-tetrahydroisoquinoline derivative, is converted to the berbine (*S*)-scoulerine by the BBE at the expense of molecular oxygen (Scheme 2.47).

The BBE from *Eschscholzia californica* (California poppy) is the best-characterized member of this enzyme class due to the availability of an efficient heterologous expression system using *Pichia pastoris* [158, 159]. Moreover, its biochemical properties, structure, and reaction mechanism were thoroughly investigated [160–162]. In addition, the potential of BBE was investigated for biocatalytic applications such as the kinetic resolution of different racemic nonnatural benzylisoquinolines [16, 163, 164].

Table 2.14 gives an overview of the different racemic substrates that have been converted by the BBE.

These results demonstrate that BBE accepts a variety of different substrates bearing substituents on both aromatic rings with good conversion and excellent enantioselectivity in most cases. The studies showed that the hydroxyl group in position 3′ of the phenol moiety is essential for a substrate to be converted. A crucial factor for the successful application of BBEs was the use of cosolvents to allow efficient transformation of the barely water-soluble substrates. Optimization studies showed that BBE tolerated various organic solvents even at remarkable high cosolvent concentrations. Best conditions were found when using a biphasic system with 70% v/v toluene at pH9 and 40 °C, allowing the conversion of substrate concentrations of up to 20 g/l [168].

These optimized conditions were used to perform the preparative transformation on a 0.5 g scale, demonstrating the applicability of the BBE-catalyzed kinetic resolution of different benzylisoquinoline and berbine alkaloids. An impressive example employing the described enzymatic oxidative intramolecular C—C bond formation by the BBE can be found in the first asymmetric total synthesis of the natural product (*S*)-scoulerine, a sedative and muscle-relaxing agent, which was obtained in 7.4% yield over nine linear steps [166].

The BBE-catalyzed kinetic resolution gives access to the berbine products and the remaining benzylisoquinolines, both in optically pure form. Moreover, the reaction

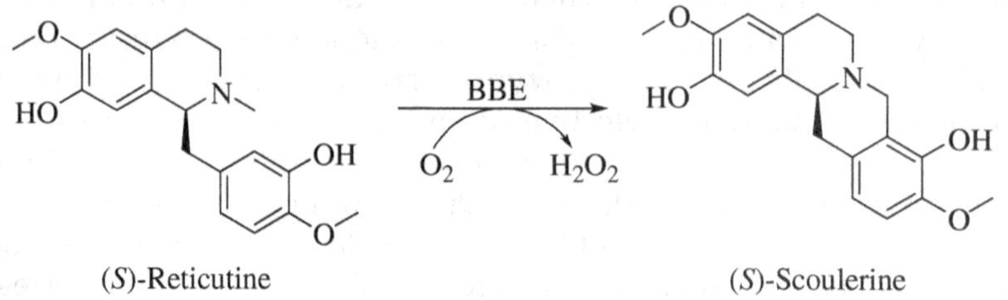

SCHEME 2.47

Natural substrate and reaction catalyzed by the berberine bridge enzyme (BBE).

proceeds with a high selectivity toward the formation of the 9-hydroxyberbines, while the regioisomeric product (11-hydroxyberbines) is in general only formed as the minor isomer.

The regiochemistry of the cyclization was addressed in more detail in a further study, showing that in particular the substitution pattern of the substrate influences the ratio of **B**/**C** (Table 2.14, Entries 11, 14–16). A complete switch in regioselectivity could be achieved by substrate engineering (Entry 16), facilitating in addition a route to (*S*)-11-hydroxyberbines and (*R*)-benzylisoquinolines. Moreover, even though the regioselectivity of the reaction was changed, the enantioselectivity was still perfect (>97% ee) on a preparative scale [165].

To overcome the limitation of a kinetic resolution, different approaches to racemize the substrate of the BBE reaction were investigated to set up a DKR; however, racemization proved to be problematic for the desired compounds. Although various racemization catalysts were tested, none of them acted exclusively

TABLE 2.14 Scope of BBE for the Oxidative Kinetic Resolution 1-Benzylisoquinolines

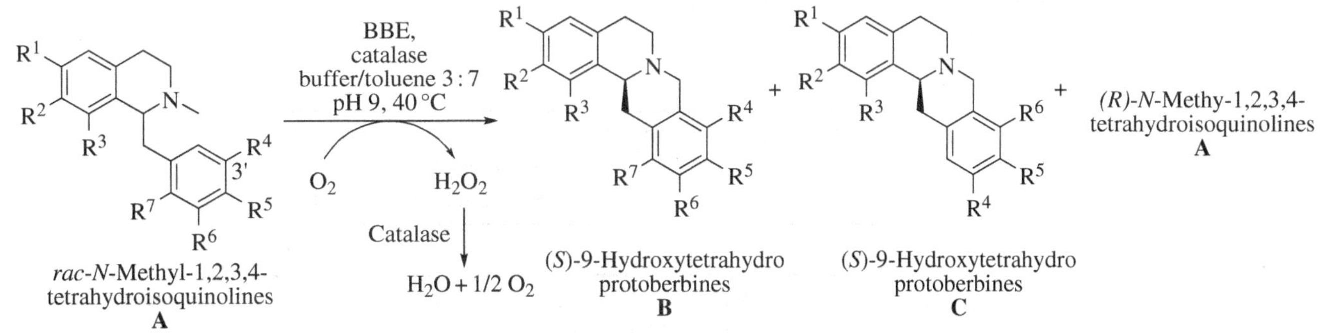

a: R^1 = OMe, R^2 = OH, R^3 = H, R^4 = OH, R^5 = OMe, R^6 = R^7 = H
b: R^1 = R^2 = OMe, R^3 = H, R^4 = OH, R^5 = R^6 = R^7 = H
c: R^1 = R^2 = OCH_2O, R^3 = H, R^4 = OH, R^5 = R^6 = R^7 = H
d: R^1 = OMe, R^2 = R^3 = H, R^4 = OH, R^5 = R^6 = R^7 = H
e: R^1 = R^2 = R^3 = OMe, R^4 = OH, R^5 = R^6 = R^7 = H
f: R^1 = OMe, R^2 = OH, R^3 = H, R^4 = OH, R^5 = R^6 = R^7 = H
g: R^1 = R^2 = R^3 = H, R^4 = OH, R^5 = R^6 = R^7 = H
h: R^1 = OH, R^2 = OMe, R^3 = H, R^4 = OH, R^5 = R^6 = R^7 = H
i: R^1 = OH, R^2 = OMe, R^3 = H, R^4 = OH, R^5 = OMe, R^6 = R^7 = H
j: R^1 = R^2 = OMe, R^3 = H, R^4 = OH, R^5 = H, R^6 = OH, R^7 = H
k: R^1 = R^2 = OMe, R^3 = H, R^4 = OH, R^5 = R^6 = H, R^7 = F
l: R^1 = R^2 = OMe, R^3 = R^4 = R^5 = H, R^6 = OH, R^7 = Cl
m: R^1 = R^2 = OMe, R^3 = R^4 = R^5 = H, R^6 = OH, R^7 = CH_3
n: R^1 = R^2 = OMe, R^3 = R^4 = R^5 = H, R^6 = OH, R^7 = F
o: R^1 = OMe, R^2 = OH, R^3 = R^4 = R^5 = H, R^6 = OH, R^7 = F
p: R^1 = OH, R^2 = OMe, R^3 = R^4 = R^5 = H, R^6 = OH, R^7 = F

Entry	Substrate	Conv. (%)	(*S*)-**B** (%)	ee **B** (%)	(*R*)-**A** (%)	ee **A** (%)	Ratio **B**:**C**	References
1	**a**	50	47	>97	37	>97	>99:<1[a]	[165, 166]
2	**b**	50	42	>97	50	>97	84:16	[167]
3	**c**	50	31	>97	46	>97	82:18	[167]
4	**d**	50	36	>97	36	>97	90:10	[167]
5	**e**	50	39	>97	47	>97	80:20	[167]
6	**f**	50	22	>97	49	>97	97:3[a]	[165, 166]
7	**g**	50	46	>97	49	>97	86:14[a]	[165, 166]
8	**h**	48	n.d.	n.d.	n.d.	n.d.	70:30	[165]
9	**i**	53	n.d.	n.d.	n.d.	n.d.	98:2	[165]
10	**j**	14	n.d.	n.d.	n.d.	n.d.	n.a.	[165]
11	**k**	50	49	>97	49	>97	>99:<1	[165]
12	**l**	<1	n.d.	n.d.	n.d.	n.d.	n.d.	[165]
13	**m**	<1	n.d.	n.d.	n.d.	n.d.	n.d.	[165]
14	**n**	48	43	>97	43	>97	<1:>99	[165]
15	**o**	49	32	>97	32	>97	<1:>99	[165]
16	**p**	49	42	>97	42	>97	<1:>99	[165]

n.a., not applicable; n.d., not determined.
[a] Ratios of 2:3 were determined in a second study at slightly different conversions (52% for Entry 6 and 53% for Entry 7) [165].

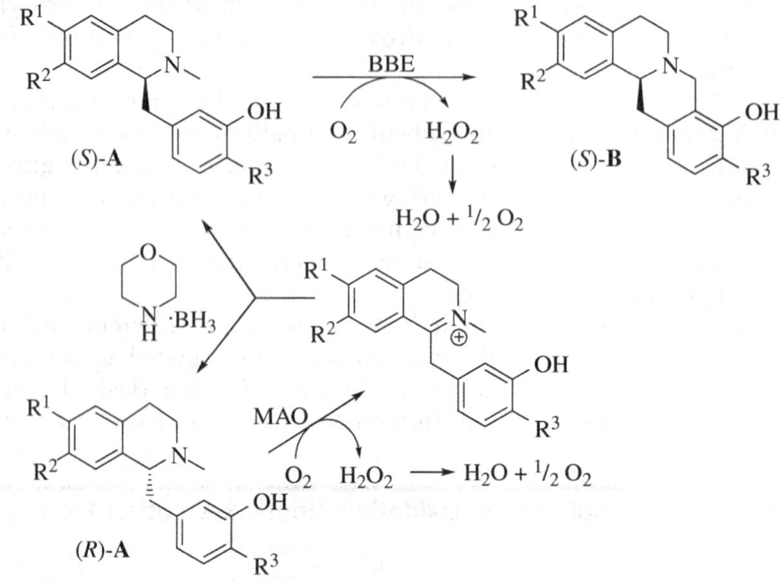

SCHEME 2.48

Deracemization of benzylisoquinolines *rac*-**A** to berbines (*S*)-**B**, employing a monoamine oxidase (MAO), berberine bridge enzyme (BBE), and borane.

TABLE 2.15 Stepwise Deracemization of Benzylisoquinolines to Berbines using MAO, Borane in the First Step and BBE in the Second Step (Scheme 4.2)

Entry	R^1	R^2	R^3	Time (h)	(*S*)-**B** (%)	ee **B** (%)
1	OMe	OH	H	24 + 48	93	>97
2	OMe	OH	OMe	24 + 48	97	>97
3	OMe	H	H	24 + 72	79[a]	>97

[a] Four percent of the regioisomer formed.

on the benzylisoquinoline substrates, leaving the berbine products untouched. Therefore, an alternative strategy combining the concept of stereoinversion, catalyzed by a MAO/reducing agent, with the kinetic resolution catalyzed by the BBE, was investigated (Scheme 2.48).

Only one variant of MAO from *A. niger* (MAO-N variant D11) enabled the stereoinversion; other variants of MAOs did not show activity for the challenging BBE substrates. However, high conversions and excellent ee's were obtained in a one-pot two-step approach (Table 2.15).

Subsequently, both reactions were also performed simultaneously in one pot. After initial experiments, the cascade was performed on a preparative scale [146]. For both compounds (Entries 1 and 2), excellent conversions were obtained: 92 and 98%, respectively. Moreover, the reaction products could be isolated in 88 and 80% yield in optically pure form (>97% ee) employing this chemoenzymatic cascade reaction [169].

2.5 CONCLUSIONS

In the last years, the development of novel and improved synthetic routes for the preparation of chiral amines has attracted the attention of many synthetic researchers. In this context, ω-TAs, AOs, and BBEs provide a wealth of opportunities and alternatives for the preparation of enantio-enriched amines displaying, in most cases, excellent regio-, diastereo-, and stereoselectivity, which has been summarized in this chapter. By combining molecular modeling and directed evolution techniques, new enzyme variants have been created not only with an altered substrate specificity but also with improved activities and stabilities. In the future, the development of new variants and further alternative enzyme reactions will expand the toolbox for the preparation of chiral amines.

REFERENCES

[1] T. C. Nugent and S. M. Marinova, *Synthesis* **45**, 153–166 (2013).
[2] T. C. Nugent, *Chiral Amine Synthesis: Methods, Developments and Applications*, Wiley-VCH, Weinheim, 2010.
[3] J. D. Duan and P. F. Li, *Catal. Sci. Technol.* **4**, 311–320 (2014).
[4] M. Breuer, K. Ditrich, T. Habicher, B. Hauer, M. Kesseler, R. Sturmer, and T. Zelinski, *Angew. Chem. Int. Ed.* **43**, 788–824 (2004).
[5] T. C. Nugent and M. El-Shazly, *Adv. Synth. Catal.* **352**, 753–819 (2010).
[6] A. Bartoszewicz, N. Ahlsten, and B. Martin-Matute, *Chem. Eur. J.* **19**, 7274–7302 (2013).
[7] M. Rueping, J. Dufour, and F. R. Schoepke, *Green Chem.* **13**, 1084–1105 (2011).
[8] P. J. Dunn, *Chem. Soc. Rev.* **41**, 1452–1461 (2012).
[9] K. Faber, *Biotransformations in Organic Synthesis*, Springer-Verlag, Heidelberg, 2011.
[10] P. T. Anastas and T. C. Williamson, *Green Chemistry: Designing Chemistry for the Environment*, Oxford University Press, Oxford, 1998.
[11] E. García-Junceda, *Multi-Step Enzyme Catalysis: Biotransformations and Chemoenzymatic Synthesis*, Wiley-VCH, Weinheim, 2008.
[12] R. C. Simon, F. G. Mutti, and W. Kroutil, *Drug Discov. Today Technol.* **10**, e37–e44 (2013).
[13] R. N. Patel, *ACS Catal.* **1**, 1056–1074 (2011).
[14] U. T. Bornscheuer, G. W. Huisman, R. J. Kazlauskas, S. Lutz, J. C. Moore, and K. Robins, *Nature* **485**, 185–194 (2012).
[15] E. Busto, V. Gotor-Fernandez, and V. Gotor, *Chem. Rev.* **111**, 3998–4035 (2011).
[16] W. Kroutil, E. M. Fischereder, C. S. Fuchs, H. Lechner, F. G. Mutti, D. Pressnitz, A. Rajagopalan, J. H. Sattler, R. C. Simon, and E. Siirola, *Org. Process Res. Dev.* **17**, 751–759 (2013).
[17] H. Kohls, F. Steffen-Munsberg, and M. Höhne, *Curr. Opin. Chem. Biol.* **19**, 180–192 (2014).
[18] D. Ghislieri and N. J. Turner, *Top. Catal.* **57**, 284–300 (2014).
[19] D. Koszelewski, K. Tauber, K. Faber, and W. Kroutil, *Trends Biotechnol.* **28**, 324–332 (2010).
[20] H. Yun, B. Y. Hwang, J. H. Lee, and B. G. Kim, *Appl. Environ. Microbiol.* **71**, 4220–4224 (2005).
[21] G. W. Matcham and A. R. S. Bowen, *Chim. Oggi.* **14**, 20–24 (1996).
[22] M. D. Truppo, J. D. Rozzell, J. C. Moore, and N. J. Turner, *Org. Biomol. Chem.* **7**, 395–398 (2009).
[23] C. K. Savile, J. M. Janey, E. C. Mundorff, J. C. Moore, S. Tam, W. R. Jarvis, J. C. Colbeck, A. Krebber, F. J. Fleitz, J. Brands, P. N. Devine, G. W. Huisman, and G. J. Hughes, *Science* **329**, 305–309 (2010).
[24] A. Iwasaki, Y. Yamada, Y. Ikenaka, and J. Hasegawa, *Biotechnol. Lett.* **25**, 1843–1846 (2003).
[25] D. Koszelewski, I. Lavandera, D. Clay, D. Rozzell, and W. Kroutil, *Adv. Synth. Catal.* **350**, 2761–2766 (2008).
[26] D. Koszeleivski, I. Lavandera, D. Clay, G. M. Guebitz, D. Rozzell, and W. Kroutil, *Angew. Chem. Int. Ed.* **47**, 9337–9340 (2008).
[27] E. S. Park, M. S. Malik, J. Y. Dong, and J. S. Shin, *ChemCatChem.* **5**, 1734–1738 (2013).
[28] K. Fesko, K. Steiner, R. Breinbauer, H. Schwab, M. Schurmann, and G. A. Strohmeier, *J. Mol. Catal. B: Enzym.* **96**, 103–110 (2013).
[29] E. S. Park, M. Kim, and J. S. Shin, *Appl. Microbiol. Biotechnol.* **93**, 2425–2435 (2012).
[30] T. Sehl, R. C. Simon, H. C. Hailes, J. M. Ward, U. Schell, M. Pohl, and D. Rother, *J. Biotechnol.* **159**, 188–194 (2012).
[31] J. S. Shin and B. G. Kim, *J. Org. Chem.* **67**, 2848–2853 (2002).
[32] D. Koszelewski, M. Goritzer, D. Clay, B. Seisser, and W. Kroutil, *ChemCatChem.* **2**, 73–77 (2010).
[33] F. G. Mutti, C. S. Fuchs, D. Pressnitz, J. H. Sattler, and W. Kroutil, *Adv. Synth. Catal.* **353**, 3227–3233 (2011).
[34] F. G. Mutti, C. S. Fuchs, D. Pressnitz, N. G. Turrini, J. H. Sattler, A. Lerchner, A. Skerra, and W. Kroutil, *Eur. J. Org. Chem.* 1003–1007 (2012).
[35] B. Wang, H. Land, and P. Berglund, *Chem. Commun.* **49**, 161–163 (2013).
[36] M. D. Truppo, H. Strotman, and G. Hughes, *ChemCatChem.* **4**, 1071–1074 (2012).

[37] C. E. Paul, M. Rodriguez-Mata, E. Busto, I. Lavandera, V. Gotor-Fernández, V. Gotor, S. García-Cerrada, J. Mendiola, O. de Frutos, and I. Collado, *Org. Process Res. Dev.* **18**, 788–792 (2014).
[38] M. F. Flament, R. M. Lane, R. Zhu, and Z. Ying, *Int. Clin. Psychopharmacol.* **14**, 259–275 (1999).
[39] Z. X. Han, S. G. Koenig, H. Zhao, X. P. Su, S. P. Singh, and R. P. Bakale, *Org. Process Res. Dev.* **11**, 726–730 (2007).
[40] R. L. Watts, J. Jankovic, C. Waters, A. Rajput, B. Boroojerdi, and J. Rao, *Neurology* **68**, 272–276 (2007).
[41] E. Siirola, F. G. Mutti, B. Grischek, S. F. Hoefler, W. M. F. Fabian, G. Grogan, and W. Kroutil, *Adv. Synth. Catal.* **355**, 1703–1708 (2013).
[42] A. Cuetos, I. Lavandera, and V. Gotor, *Chem. Commun.* **49**, 10688–10690 (2013).
[43] D. Pressnitz, C. S. Fuchs, J. H. Sattler, T. Knaus, P. Macheroux, F. G. Mutti, and W. Kroutil, *ACS Catal.* **3**, 555–559 (2013).
[44] J. J. Jiang, X. Chen, J. H. Feng, Q. Q. Wu, and D. M. Zhu, *J. Mol. Catal. B: Enzym.* **100**, 32–39 (2014).
[45] C. M. Spencer and S. Noble, *Drugs Aging* **13**, 391–411 (1998).
[46] D. M. Kendall, R. M. Cuddihy, and R. M. Bergenstal, *Eur. J. Intern. Med.* **20**, S3229–S3339 (2009).
[47] U. Rosentreter, H. Boeshagen, F. Seuter, E. Perzborn, and V. B. Fiedler, *Arneimittelforschung* **39**, 1519–1521 (1990).
[48] E. Vizzardi, F. Salghetti, I. Bonadei, S. Gelsomino, R. Lorusso, A. D. Aloia, and A. Curnis, *Cardiovasc. Ther.* **31**, E55–E62 (2013).
[49] F. Cabirol, A. Gohel, S. Ho-Oh, D. Smith, and B. Wong, WO2011/159910 A159912 (December 22, 2011).
[50] M. Fuchs, D. Koszelewski, K. Tauber, W. Kroutil, and K. Faber, *Chem. Commun.* **46**, 5500–5502 (2010).
[51] M. Fuchs, D. Koszelewski, K. Tauber, J. Sattler, W. Banko, A. K. Holzer, M. Pickl, W. Kroutil, and K. Faber, *Tetrahedron* **68**, 7691–7694 (2012).
[52] J. Limanto, E. R. Ashley, J. Yin, G. L. Beutner, B. T. Grau, A. M. Kassim, M. M. Kim, A. Kaplars, Z. Liu, H. R. Strotman, and M. D. Truppo, *Org. Lett.* **16**, 2716–2719 (2014).
[53] E. Busto, R. C. Simon, B. Grischek, V. Gotor-Fernández, and W. Kroutil, *Adv. Synth. Catal.* **356**, 1937–1942 (2014).
[54] Y. A. Mostafa and S. D. Taylor, *Bioorg. Med. Chem.* **20**, 1535–1544 (2012).
[55] N. Richter, R. C. Simon, W. Kroutil, J. M. Ward, and H. C. Hailes, *Chem. Commun.* **50**, 6098–6100 (2014).
[56] F. G. Mutti and W. Kroutil, *Adv. Synth. Catal.* **354**, 3409–3413 (2012).
[57] G. Maftin and R. E. Pratley, *Ther. Adv. Endocrinol. Metab.* **1**, 5–14 (2010).
[58] M. Sasa, *J. Pharmacol. Sci.* **100**, 487–494 (2006).
[59] E. S. Park, J. Y. Dong, and J. S. Shin, *ChemCatChem.* **5**, 3538–3542 (2013).
[60] E. S. Park, J. Y. Dong, and J. S. Shin, *Org. Biomol. Chem.* **11**, 6929–6933 (2013).
[61] C. S. Fuchs, M. Hollauf, M. Meissner, R. C. Simon, T. Besset, J. N. H. Reek, W. Riethorst, F. Zepeck, and W. Kroutil, *Adv. Synth. Catal.*, **356**, 2257–2265, (2014).
[62] A. Bruggink, R. Schoevaart, and T. Kieboom, *Org. Process Res. Dev.* **7**, 622–640 (2003).
[63] F. Lopez-Gallego and C. Schmidt-Dannert, *Curr. Opin. Chem. Biol.* **14**, 174–183 (2010).
[64] S. F. Mayer, W. Kroutil, and F. Kurt, *Chem. Soc. Rev.* **30**, 332–339 (2001).
[65] I. Oroz-Guinea and E. Garcia-Junceda, *Curr. Opin. Chem. Biol.* **17**, 1039–1039 (2013).
[66] V. Resch, J. H. Schrittwieser, E. Siirola, and W. Kroutil, *Curr. Opin. Biotechnol.* **22**, 793–799 (2011).
[67] R. C. Simon, N. Richter, E. Busto, and W. Kroutil, *ACS Catal.* **4**, 129–143 (2014).
[68] J. Alvarez-Builla, J. J. Vaquero, and J. Barluenga, *Modern Heterocyclic Chemistry*, Wiley-VCH, Weinheim, 2011.
[69] M. D. Truppo, J. D. Rozzell, and N. J. Turner, *Org. Process Res. Dev.* **14**, 234–237 (2010).
[70] M. Girardin, S. G. Ouellet, D. Gauvreau, J. C. Moore, G. Hughes, P. N. Devine, P. D. O'shea, and L. C. Campeau, *Org. Process Res. Dev.* **17**, 61–68 (2013).
[71] D. Koszelewski, D. Clay, K. Faber, and W. Kroutil, *J. Mol. Catal. B: Enzym.* **60**, 191–194 (2009).
[72] C. K. Chung, P. G. Bulger, B. Kosjek, K. M. Belyk, N. Rivera, M. E. Scott, G. R. Humphrey, J. Limanto, D. C. Bachert, and K. M. Emerson, *Org. Process Res. Dev.* **18**, 215–227 (2014).
[73] P. Jones, S. Altamura, J. Boueres, F. Ferrigno, M. Fonsi, C. Giomini, S. Lamartina, E. Monteagudo, J. M. Ontoria, M. V. Orsale, M. C. Palumbi, S. Pesci, G. Roscilli, R. Scarpelli, C. Schultz-Fademrecht, C. Toniatti, and M. Rowley, *J. Med. Chem.* **52**, 7170–7185 (2009).
[74] R. W. Bates and K. Sa-Ei, *Tetrahedron* **58**, 5957–5978 (2002).

[75] M. G. P. Buffat, *Tetrahedron* **60**, 1701–1729 (2004).
[76] R. C. Simon, B. Grischek, F. Zepeck, A. Steinreiber, F. Belaj, and W. Kroutil, *Angew. Chem. Int. Ed.* **51**, 6713–6716 (2012).
[77] R. C. Simon, F. Zepeck, and W. Kroutil, *Chem. Eur. J.* **19**, 2859–2865 (2013).
[78] R. C. Simon, C. S. Fuchs, H. Lechner, F. Zepeck, and W. Kroutil, *Eur. J. Org. Chem.*, 3397–3402 (2013).
[79] C. D. Cox, M. J. Breslin, D. B. Whitman, J. D. Schreier, G. B. McGaughey, M. J. Bogusky, A. J. Roecker, S. P. Mercer, R. A. Bednar, W. Lemaire, J. G. Bruno, D. R. Reiss, C. M. Harrell, K. L. Murphy, S. L. Garson, S. M. Doran, T. Prueksaritanont, W. B. Anderson, C. Y. Tang, S. Roller, T. D. Cabalu, D. H. Cui, G. D. Hartman, S. D. Young, K. S. Koblan, C. J. Winrow, J. J. Renger, and P. J. Coleman, *J. Med. Chem.* **53**, 5320–5332 (2010).
[80] I. K. Mangion, B. D. Sherry, J. J. Yin, and F. J. Fleitz, *Org. Lett.* **14**, 3458–3461 (2014).
[81] M. E. B. Smith, B. H. Chen, E. G. Hibbert, U. Kaulmann, K. Smithies, J. L. Galman, F. Baganz, P. A. Dalby, H. C. Hailes, G. J. Lye, J. M. Ward, J. M. Woodley, and M. Micheletti, *Org. Process Res. Dev.* **14**, 99–107 (2010).
[82] T. Sehl, H. C. Hailes, J. M. Ward, R. Wardenga, E. von Lieres, H. Offermann, R. Westphal, M. Pohl, and D. Rother, *Angew. Chem. Int. Ed.* **52**, 6772–6775 (2013).
[83] T. Sehl, H. C. Hailes, J. M. Ward, U. Menyes, M. Pohl, and D. Rother, *Green Chem.*, **16**, 3341–3348, (2014).
[84] C. C. Gruber, I. Lavandera, K. Faber, and W. Kroutil, *Adv. Synth. Catal.* **348**, 1789–1805 (2006).
[85] A. N. Parvulescu, P. A. Jacobs, and D. E. De Vos, *Chem. Eur. J.* **13**, 2034–2043 (2007).
[86] N. J. Turner, *Curr. Opin. Chem. Biol.* **14**, 115–121 (2010).
[87] D. Koszelewski, D. Clay, D. Rozzell, and W. Kroutil, *Eur. J. Org. Chem.*, 2289–2292 (2009).
[88] M. D. Truppo, N. J. Turner, and J. D. Rozzell, *Chem. Commun.*, 2127–2129 (2009).
[89] D. Koszelewski, D. Pressnitz, D. Clay, and W. Kroutil, *Org. Lett.* **11**, 4810–4812 (2009).
[90] D. Koszelewski, N. Muller, J. H. Schrittwieser, K. Faber, and W. Kroutil, *J. Mol. Catal. B: Enzym.* **63**, 39–44 (2010).
[91] G. Shin, S. Mathew, M. Shon, B. G. Kim, and H. Yun, *Chem. Commun.* **49**, 8629–8631 (2013).
[92] K. S. Midelfort, R. Kumar, S. Han, M. J. Karmilowicz, K. McConnell, D. K. Gehlhaar, A. Mistry, J. S. Chang, M. Anderson, A. Villalobos, J. Minshull, S. Govindarajan, and J. W. Wong, *Protein Eng. Des. Sel.* **26**, 25–33 (2013).
[93] P. L. Dostert, M. S. Benedetti, and K. F. Tipton, *Med. Res. Rev.* **9**, 45–89 (1989).
[94] B. J. Brazeau, B. J. Johnson, and C. M. Wilmot, *Arch. Biochem. Biophys.* **428**, 22–31 (2004).
[95] M. Alexeeva, R. Carr, and N. J. Turner, *Org. Biomol. Chem.* **1**, 4133–4137 (2003).
[96] P. F. Fitzpatrick, *Arch. Biochem. Biophys.* **493**, 13–25 (2010).
[97] V. Joosten and W. J. H. van Berkel, *Curr. Opin. Chem. Biol.* **11**, 195–202 (2007).
[98] W. P. Dijkman, G. de Gonzalo, A. Mattevi, and M. W. Fraaije, *Appl. Microbiol. Biotechnol.* **97**, 5177–5188 (2013).
[99] C. S. Chen, Y. Fujimoto, G. Girdaukas, and C. J. Sih, *J. Am. Chem. Soc.* **104**, 7294–7299 (1982).
[100] N. J. Turner, *Chem. Rev.* **111**, 4073–4087 (2011).
[101] C. Broca, M. Manteghetti, R. Gross, Y. Baissac, M. Jacob, P. Petit, Y. Sauvaire, and G. Ribes, *Eur. J. Pharmacol.* **390**, 339–345 (2000).
[102] V. Rolland-Fulcrand, M. Rolland, M. L. Roumestant, and J. Martinez, *Eur. J. Org. Chem.* 873–877 (2004).
[103] H. Pellissier, *Adv. Synth. Catal.* **353**, 1613–1666 (2011).
[104] E. Vedejs and M. Jure, *Angew. Chem. Int. Ed.* **44**, 3974–4001 (2005).
[105] B. Martin-Matute and J. E. Backvall, *Curr. Opin. Chem. Biol.*, **11**, 226–232 (2007).
[106] H. Pellissier, *Tetrahedron* **67**, 3769–3802 (2011).
[107] M. Rachwalski, N. Vermue, and F. P. J. T. Rutjes, *Chem. Soc. Rev.* **42**, 9268–9282 (2013).
[108] N. J. Turner, *Curr. Opin. Chem. Biol.* **8**, 114–119 (2004).
[109] C. V. Voss, C. C. Gruber, K. Faber, T. Knaus, P. Macheroux, and W. Kroutil, *J. Am. Chem. Soc.* **130**, 13969–13972 (2008).
[110] C. V. Voss, C. C. Gruber, and W. Kroutil, *Synlett* 991–998 (2010).
[111] K. Faber, *Chem. Eur. J.* **7**, 5004–5010 (2001).
[112] E. W. Hafner and D. Wellner, *Proc. Natl. Acad. Sci. U. S. A.* **68**, 987–991 (1971).
[113] J. W. Huh, K. Yokoigawa, N. Esaki, and K. Soda, *J. Ferment. Bioeng.* **74**, 189–190 (1992).
[114] K. Soda, T. Oikawa, and K. Yokoigawa, *J. Mol. Catal. B: Enzym.* **11**, 149–153 (2001).
[115] S. Servi, D. Tessaro, and G. Pedrocchi-Fantoni, *Coord. Chem. Rev.* **252**, 715–726 (2008).
[116] T. M. Beard and N. J. Turner, *Chem. Commun.* 246–247 (2002).

[117] F. R. Alexandre, D. P. Pantaleone, P. P. Taylor, I. G. Fotheringham, D. J. Ager, and N. J. Turner, *Tetrahedron Lett.* **43**, 707–710 (2002).
[118] D. P. Pantaleone, A. M. Geller, and P. P. Taylor, *J. Mol. Catal. B: Enzym.* **11**, 795–803 (2001).
[119] A. Enright, F. R. Alexandre, G. Roff, I. G. Fotheringham, M. J. Dawson, and N. J. Turner, *Chem. Commun.* 2636–2637 (2003).
[120] E. Ricca, B. Brucher, and J. H. Schrittwieser, *Adv. Synth. Catal.* **353**, 2239–2262 (2011).
[121] Z. Findrik and D. Vasic-Racki, *Biotechnol. Bioeng.* **98**, 956–967 (2007).
[122] M. Yasuda, M. Ueda, H. Muramatsu, H. Mihara, and N. Esaki, *Tetrahedron: Asymmetry.* **17**, 1775–1779 (2006).
[123] H. Mihara, H. Muramatsu, R. Kakutani, M. Yasuda, M. Ueda, T. Kurihara, and N. Esaki, *FEBS J.* **272**, 1117–1123 (2005).
[124] R. N. Patel, Y. Chen, S. L. Goldberg, R. L. Hanson, I. Gill, A. Goswami, T. P. Tully, and W. L. Parker, WO 2007/112299 A112292 (2007).
[125] R. L. Hanson, B. L. Davis, S. L. Goldberg, R. M. Johnston, W. L. Parker, T. P. Tully, M. A. Montana, and R. N. Patel, *Org. Process Res. Dev.* **12**, 1119–1129 (2008).
[126] H. Schutte, W. Hummel, and M. R. Kula, *Appl. Microbiol. Biotechnol.* **19**, 167–176 (1984).
[127] W. Hummel, H. Schutte, and M. R. Kula, *Appl. Microbiol. Biotechnol.* **21**, 7–15 (1985).
[128] E. Busto, N. Richter, B. Grischek, and W. Kroutil, *Chem. Eur. J.* **20**, 11225–11228 (2014).
[129] N. J. Turner, I. G. Fotheringham, and R. Speight, *Innov. Pharm. Technol.* **4**, 114–122 (2004).
[130] S. O. Sablin, V. Yankovskaya, S. Bernard, C. N. Cronin, and T. P. Singer, *Eur. J. Biochem.* **253**, 270–279 (1998).
[131] B. Schilling and K. Lerch, *Biochim. Biophys. Acta* **1243**, 529–537 (1995).
[132] B. Schilling and K. Lerch, *Mol. Gen. Genet.* **247**, 430–438 (1995).
[133] M. Alexeeva, A. Enright, M. J. Dawson, M. Mahmoudian, and N. J. Turner, *Angew. Chem. Int. Ed.* **41**, 3177–3180 (2002).
[134] R. Carr, M. Alexeeva, A. Enright, T. S. C. Eve, M. J. Dawson, and N. J. Turner, *Angew. Chem. Int. Ed.* **42**, 4807–4810 (2003).
[135] R. Carr, M. Alexeeva, M. J. Dawson, V. Gotor-Fernandez, C. E. Humphrey, and N. J. Turner, *ChemBioChem.* **6**, 637–639 (2005).
[136] C. J. Dunsmore, R. Carr, T. Fleming, and N. J. Turner, *J. Am. Chem. Soc.* **128**, 2224–2225 (2006).
[137] K. R. Bailey, A. J. Ellis, R. Reiss, T. J. Snape, and N. J. Turner, *Chem. Commun.* 3640–3642 (2007).
[138] I. Rowles, K. J. Malone, L. L. Etchells, S. C. Willies, and N. J. Turner, *ChemCatChem.* **4**, 1259–1261 (2012).
[139] Q. Y. Zhang, G. Z. Tu, Y. Y. Zhao, and T. M. Cheng, *Tetrahedron* **58**, 6795–6798 (2002).
[140] V. Kohler, K. R. Bailey, A. Znabet, J. Raftery, M. Helliwell, and N. J. Turner, *Angew. Chem. Int. Ed.* **49**, 2182–2184 (2010).
[141] A. Znabet, E. Ruijter, F. J. J. de Kanter, V. Kohler, M. Helliwell, N. J. Turner, and R. V. A. Orru, *Angew. Chem. Int. Ed.* **49**, 5289–5292 (2010).
[142] A. Znabet, J. Zonneveld, E. Janssen, F. J. J. De Kanter, M. Helliwell, N. J. Turner, E. Ruijter, and R. V. A. Orru, *Chem. Commun.* **46**, 7706–7708 (2010).
[143] A. Znabet, M. M. Polak, E. Janssen, F. J. J. de Kanter, N. J. Turner, R. V. A. Orru, and E. Ruijter, *Chem. Commun.*, **46**, 7918–7920 (2010).
[144] T. Li, J. Liang, A. Ambrogelly, T. Brennan, G. Gloor, G. Huisman, J. Lalonde, A. Lekhal, B. Mijts, S. Muley, L. Newman, M. Tobin, G. Wong, A. Zaks, and X. Y. Zhang, *J. Am. Chem. Soc.* **134**, 6467–6472 (2012).
[145] K. E. Atkin, R. Reiss, V. Koehler, K. R. Bailey, S. Hart, J. P. Turkenburg, N. J. Turner, A. M. Brzozowski, and G. Grogan, *J. Mol. Biol.* **384**, 1218–1231 (2008).
[146] D. Ghislieri, A. P. Green, M. Pontini, S. C. Willies, I. Rowles, A. Frank, G. Grogan, and N. J. Turner, *J. Am. Chem. Soc.* **135**, 10863–10869 (2013).
[147] D. Ghislieri, D. Houghton, A. P. Green, S. C. Willies, and N. J. Turner, *ACS Catal.* **3**, 2869–2872 (2013).
[148] J. H. Schrittwieser, B. Groenendaal, S. C. Willies, D. Ghislieri, I. Rowles, V. Resch, J. H. Sattler, E.-M. Fischereder, B. Grischek, W.-D. Lienhart, N. J. Turner, and W. Kroutil, *Catal. Sci. Technol.* **4**, 3657–3664 (2014).
[149] E. O'Reilly, C. Iglesias, and N. J. Turner, *ChemCatChem.* **6**, 992–995 (2014).
[150] E. O'Reilly, C. Iglesias, D. Ghislieri, J. Hopwood, J. L. Galman, R. C. Lloyd, and N. J. Turner, *Angew. Chem. Int. Ed.* **53**, 2447–2450 (2014).
[151] N. J. Turner, *Nat. Chem. Biol.* **5**, 568–574 (2009).
[152] R. S. Heath, M. Pontini, B. Bechi, and N. J. Turner, *ChemCatChem.* **6**, 996–1002 (2014).
[153] K. Yasukawa, S. Nakano, and Y. Asano, *Angew. Chem. Int. Ed.* **53**, 4428–4431 (2014).

[154] G. Y. Li, J. Ren, P. Y. Yao, Y. T. Duan, H. L. Zhang, Q. Q. Wu, J. H. Feng, P. C. K. Lau, and D. M. Zhu, *ACS Catal.* **4**, 903–908 (2014).
[155] A. J. Carnell, *ChemCatChem.* **6**, 958–960 (2014).
[156] V. Kohler, Y. M. Wilson, M. Durrenberger, D. Ghislieri, E. Churakova, T. Quinto, L. Knorr, D. Haussinger, F. Hollmann, N. J. Turner, and T. R. Ward, *Nat. Chem.* **5**, 93–99 (2013).
[157] S. Wallner, C. Dully, B. Daniel, and P. Macheroux, *Handbook of Flavoproteins*, Walter de Gruyter, Berlin/Boston, 2012.
[158] H. Dittrich and T. M. Kutchan, *Proc. Natl. Acad. Sci. U. S. A.* **88**, 9969–9973 (1991).
[159] A. Winkler, F. Hartner, T. M. Kutchan, A. Glieder, and P. Macheroux, *J. Biol. Chem.* **281**, 21276–21285 (2006).
[160] H. M. Gaweska, K. M. Roberts, and P. F. Fitzpatrick, *Biochemistry* **51**, 7342–7347 (2012).
[161] A. Winkler, A. Lyskowski, S. Riedl, M. Puhl, T. M. Kutchan, P. Macheroux, and K. Gruber, *Nat. Chem. Biol.* **4**, 739–741 (2008).
[162] A. Winkler, K. Motz, S. Riedl, M. Puhl, P. Macheroux, and K. Gruber, *J. Biol. Chem.* **284**, 19993–20001 (2009).
[163] M. Müller, *Adv. Synth. Catal.* **354**, 3161–3174 (2012).
[164] J. H. Schrittwieser and V. Resch, *RSC Adv.* **3**, 17602–17632 (2013).
[165] V. Resch, H. Lechner, J. H. Schrittwieser, S. Wallner, K. Gruber, P. Macheroux, and W. Kroutil, *Chem. Eur. J.* **18**, 13173–13179 (2012).
[166] J. H. Schrittwieser, V. Resch, S. Wallner, W. D. Lienhart, J. H. Sattler, J. Resch, P. Macheroux, and W. Kroutil, *J. Org. Chem.* **76**, 6703–6714 (2011).
[167] J. H. Schrittwieser, V. Resch, J. H. Sattler, W. D. Lienhart, K. Durchschein, A. Winkler, K. Gruber, P. Macheroux, and W. Kroutil, *Angew. Chem. Int. Ed.* **50**, 1068–1071 (2011).
[168] V. Resch, J. H. Schrittwieser, S. Wallner, P. Macheroux, and W. Kroutil, *Adv. Synth. Catal.* **353**, 2377–2383 (2011).
[169] J. H. Schrittwieser, B. Groenendaal, V. Resch, D. Ghislieri, S. Wallner, E.-M. Fischereder, E. Fuchs, B. Grischek, J. H. Sattler, P. Macheroux, N. J. Turner, and W. Kroutil, *Angew. Chem. Int. Ed.* **53**, 3731–3734 (2014).

CHAPTER 3

Decarboxylation and Racemization of Unnatural Compounds using Artificial Enzymes Derived from Arylmalonate Decarboxylase

Kenji Miyamoto

Department of Biosciences and Informatics, Keio University, Yokohama, Japan

3.1 INTRODUCTION

Decarboxylation is one of the most common processes in natural metabolism. All decarboxylases [EC 4.1.1.-] cleave a substrate carboxylic group with or without the requirement of an enzymatic cofactor. There are three known decarboxylase types: (i) thiamine diphosphate (ThDP)-dependent decarboxylases, (ii) pyridoxal phosphate (PLP)-dependent decarboxylases, and (iii) cofactor-independent decarboxylases (Figure 3.1) [1–4]. Cofactor-independent decarboxylases are specific for activated substrates.

Decarboxylation of malonic acid derivatives is a well studied process in the biosynthesis of biomolecules such as long-chain fatty acids and polyketides. A decarboxylase that exhibits enantioselectivity for substituted malonates would be useful for producing optically active carboxylic acids. In fact, malonyl-CoA decarboxylase does catalyze an enantioselective decarboxylation (Figure 3.2) [5], but malonyl-CoA is an unsuitable precursor for optically active substances. Instead, we focused on the prochiral-activated compound arylmalonate, an intermediate of malonic ester synthesis, to develop a method for enantioselective decarboxylation. Malonates are stable at room temperature but readily decompose to arylacetate and CO_2 at high temperatures. This suggests that the decarboxylation of arylmalonate may occur naturally if arylmalonate acts as a substrate for a decarboxylase.

The profens (α-arylpropionates) are an important family of nonsteroidal anti-inflammatory drugs (NSAIDs) [6]. However, only the (*S*)-profen shows anti-inflammatory activity (Figure 3.3).

Various biocatalytic processes for the synthesis of optically pure (*S*)-profens have been proposed [7], but profens are still often used in their racemic form because a α-arylpropionyl-CoA epimerase in human cells interconverts the enantiomers.

Green Biocatalysis, First Edition. Edited by Ramesh N. Patel.

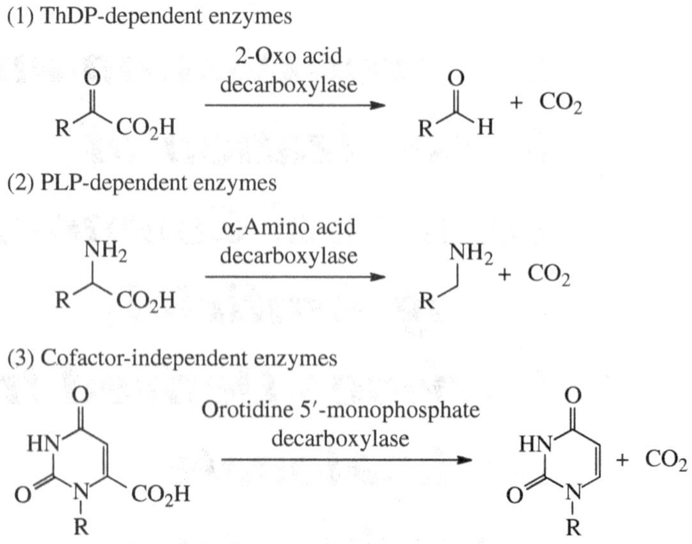

FIGURE 3.1
Representative decarboxylases.

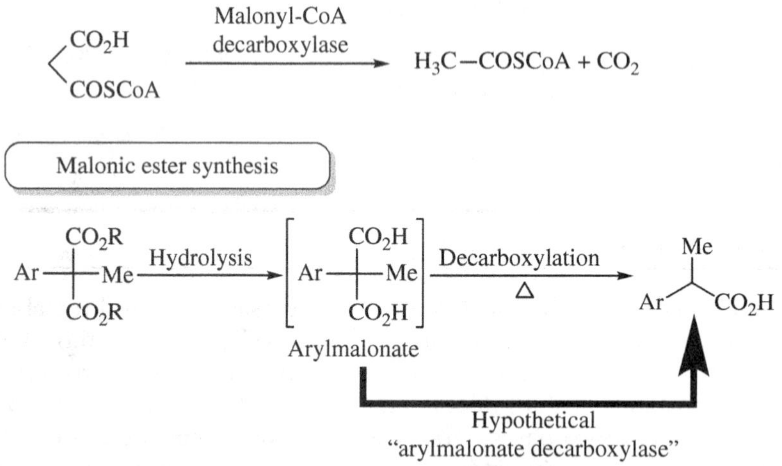

FIGURE 3.2
Malonyl-CoA decarboxylase and hypothetical decarboxylase.

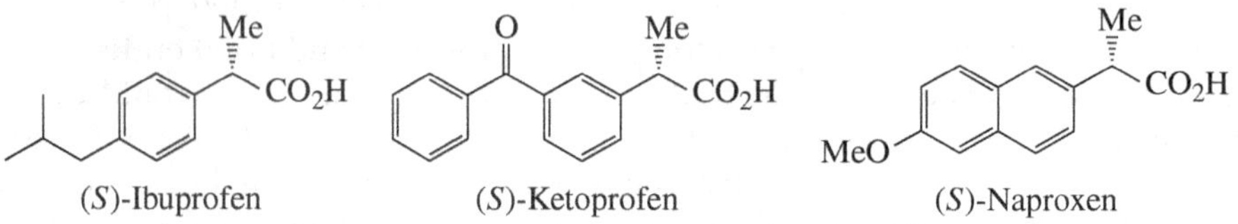

FIGURE 3.3
Representative compounds of profen.

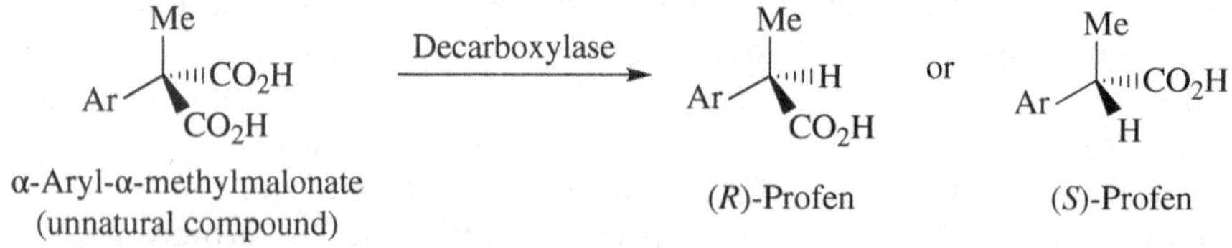

FIGURE 3.4
Asymmetric synthesis of profens using a decarboxylase.

Nonetheless, the inactive (*R*)-enantiomers may have unwanted side effects. Thus, the development of efficient approaches for the synthesis of optically pure profens remains an important challenge in medicinal chemistry. In this chapter, we describe a method for the stereoselective synthesis of profens by asymmetric decarboxylation of unnatural α-aryl-α-methylmalonates using artificial enzymes derived from a bacterial α-aryl-α-methylmalonate decarboxylase (Figure 3.4) [8].

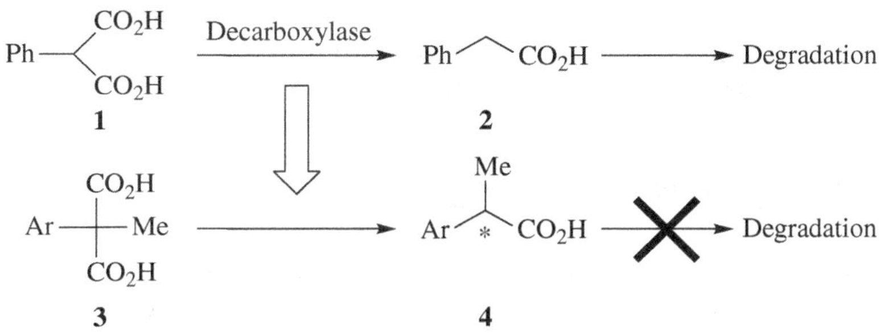

FIGURE 3.5
Screening strategy of a decarboxylase.

3.2 DISCOVERY OF A BACTERIAL α-ARYL-α-METHYLMALONATE DECARBOXYLASE

Soil microorganisms were screened for the capacity to decarboxylate α-aryl-α-methylmalonates (**3**) (Figure 3.5) [9]. An inorganic medium containing phenylmalonate (**1**) as the sole carbon source was used because we speculated that the first step of phenylmalonate metabolism would be decarboxylation to yield phenylacetate (**2**). Accordingly, microorganisms that can grow on this medium are expected to have decarboxylase activity to phenylmalonate. Enzymes that catalyze decarboxylation of phenylmalonate (**1**) are also expected to be active for α-aryl-α-methylmalonates (**3**). If the presence of a methyl group easily inhibits subsequent metabolic degradation, α-arylpropionates (**4**) would accumulate. Various soil samples were tested, and we selected a bacterium identified as *Alcaligenes bronchisepticus* KU1201. We succeeded in obtaining optically active α-arylpropionate (**4**) starting from α-aryl-α-methylmalonate (**3**). The active enzyme was inducible by phenylmalonate (**1**).

3.3 PURIFICATION AND CHARACTERIZATION OF THE DECARBOXYLASE (AMDASE)

We expected that detailed studies of the isolated enzyme would reveal a new mechanism of decarboxylation. *A. bronchisepticus* was cultivated in a medium containing phenylmalonate as the enzyme inducer and carbon source. The decarboxylase was purified according to a standard protocol [10], resulting in a ~300-fold purification to 377 U/mg protein. Sodium dodecyl sulfate polyacrylamide gel electrophoresis (SDS-PAGE) and high-performance liquid chromatography (HPLC) analysis showed that this enzyme was a monomer of 24 000. As the rate of phenylmalonate decarboxylation (**1**) was faster than α-methyl derivative decarboxylation (**3**), we named the enzyme arylmalonate decarboxylase (AMDase) [10].

To clarify the characteristics of AMDase, the effects of additives were examined. The addition of ATP and coenzyme A (CoA) to the enzyme reaction mixture did not enhance the rate of decarboxylation. In the case of malonyl-CoA decarboxylase, ATP and substrate form a mixed anhydride, which in turn reacts with CoA to form a thiol ester of the substrate. In the case of AMDase, however, neither ATP nor CoA had any effect, so this mechanism is unlikely. It is well established that avidin is a potent inhibitor of biotin–enzyme complex formation [11, 12]. In this case, addition of avidin had no influence on decarboxylase activity, indicating that AMDase is not a biotin-dependent decarboxylase. Thus, the cofactor requirements of AMDase are entirely different from known analogous enzymes, such as malonyl-CoA decarboxylases.

In contrast, a strong inhibitory effect of sulfhydryl reagents, including $HgCl_2$, HgCl, $AgNO_3$, iodoacetate, and *p*-chloromercuribenzoate, was found. Thus, AMDase is a thiol decarboxylase. The activity of the enzyme was not lost upon incubation with several divalent metal cations and serine inhibitors. It was concluded that AMDase is an unusual enzyme that does not contain metal ions or coenzymes usually required for decarboxylases and transcarboxylases.

3.4 CLONING OF THE AMDASE GENE

We next cloned and overexpressed the AMDase gene using the direct expression method [13]. Genomic DNA isolated from *A. bronchisepticus* was digested by PstI and cloned into the PstI site of pUC19. *Escherichia coli* DH5a-MCR were transformed by the plasmids and genomic libraries were constructed. The transformants were screened by the change in color of plates containing bromothymol blue (BTB) and phenylmalonate due to the pH change around the active colony resulting from decarboxylation (Figure 3.6).

One in ~700 transformants exhibited AMDase activity. The active plasmid (pAMD100) contained an insert of about 2.8 kbp and was sequenced. The gene consisted of 720 bp, indicating an enzyme of 240 amino acids. A PstI–HindIII (1.2 kbp) fragment was subcloned into pUC19 to generate the pAMD101 plasmid. The enzyme activity in a cell-free extract of *E. coli* JM109/pAMD101 was elevated to 37 250 U/L culture broths. It was calculated that the enzyme comprised over 25% of the total extractable cellular protein. The enzyme produced by the *E. coli* transformant was purified to homogeneity and shown to be identical to that described by SDS-PAGE and HPLC. Both enzymes had the same enzymological properties and N-terminal amino acid sequences. Since these initial experiments, many genes with homology to the AMDase gene have been reported, forming an AMDase family.

3.5 STEREOCHEMICAL COURSE OF AMDASE-CATALYZED DECARBOXYLATION

Subsequent studies examined whether AMDase discriminates between two prochiral carboxyl groups. To this end, we prepared both enantiomers of α-methyl-α-phenylmalonate (**6**) with ^{13}C in one of the two carboxyl groups (Figure 3.7) [14]. Starting from ^{13}C-phenylacetate (^{13}C-**2**), both enantiomers of chiral ^{13}C-containing α-methyl-α-phenylmalonate (**6**) were prepared. The absolute configuration of the chiral substrate was clearly determined by the optical rotation of the resolved hydroxyl acid (**5**).

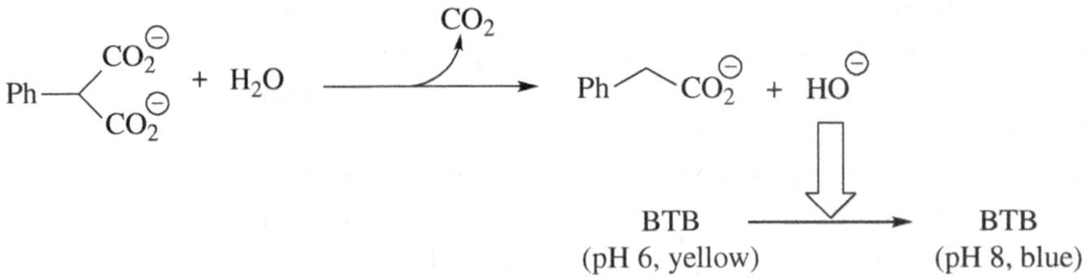

FIGURE 3.6
Visualization of decarboxylation activity.

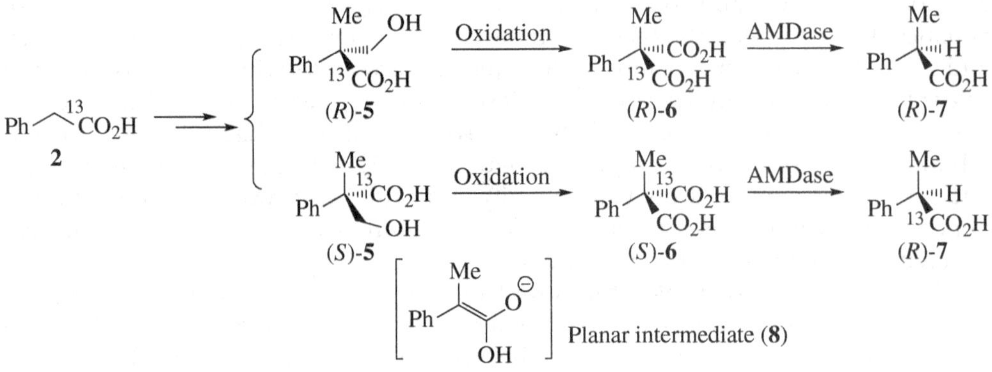

FIGURE 3.7
Stereochemistry of AMDase-catalyzed decarboxylation.

The result of enzymatic decarboxylation was very clear. While (*S*)-**6** resulted in a ^{13}C-containing product (**7**), (*R*)-**6** yielded a product with no more ^{13}C than occurs naturally. Apparently, the reaction proceeds with a net inversion of configuration. Thus, the presence of a planar intermediate (**8**) can be postulated. Enantioface-differentiating protonation to compound **8** will yield the optically active final product (**7**).

3.6 DIRECTED EVOLUTION OF AMDase TO AN ARTIFICIAL PROFEN RACEMASE

The efficient synthesis of profens is of great importance to pharmaceutical chemistry. The (*S*)-enantiomer exhibits much greater bioactivity, so many kinetic resolution and asymmetric synthesis processes have been proposed for stereospecific synthesis [7, 15]. Methods for the preparation of optically pure profens from racemic starting substrates, such as separation by chromatography, crystallization, or kinetic resolution, have achieved only 50% of the maximum yield of the desired (*S*)-enantiomer. This limitation has been overcome by recycling of the undesired (*R*)-enantiomer coupled with racemization. By repetition of the process, it is possible to achieve (*S*)-product yields >50% (Figure 3.8). Sustainable racemization methods are also of great potential value because chemical racemization methods often require extreme conditions [7]. Although enzymatic racemization would be a desirable method, no "profen racemase" has been discovered.

We recently succeeded in determining the three-dimensional (3D) structure of AMDase by X-ray crystallography [16]. AMDase is structurally similar to cofactor-independent amino acid racemases [17, 18]. Considering the similarities of ligand binding to racemases, we propose a binding mode for a model substrate of AMDase, α-methyl-α-phenylmalonate (**6**). In this model, decarboxylation is triggered by destabilization of the pro-(*R*)-carboxylic group in the hydrophobic pocket. Finally, Cys188 protonates the planar intermediate (**8**) to yield optically pure (*R*)-products (**7**).

AMDase has some homology and structural similarity to glutamate racemase, which has two cysteine residues in the active site (Figure 3.9) [19–24]. Thus, we designed a G74C variant of AMDase by site-directed mutagenesis expressing two active-site Cys residues. Although decarboxylation and racemization are different reactions, the key planar intermediate is the same type of species (**8**). Prior to this study, there was no report of a racemase that could act on unnatural compounds such as profens.

The G74C variant did indeed yield the racemic product (**7**) regardless of the configuration of the starting materials [25], while wild-type AMDase has no racemase activity (Table 3.1). Thus, introduction of a single mutation is sufficient to change the decarboxylase to an artificial "profen racemase." We speculated that the presence of two active-site Cys residues enables the two-base mechanism to work similarly to glutamate racemase (Figure 3.9). However, the G74C variant still showed low decarboxylase

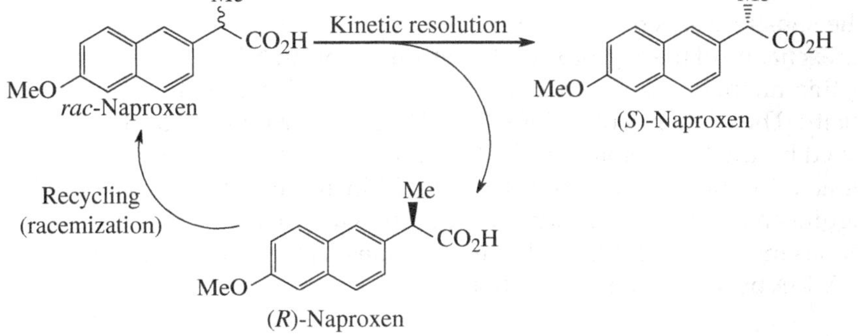

FIGURE 3.8
Kinetic resolution of naproxen and recycling.

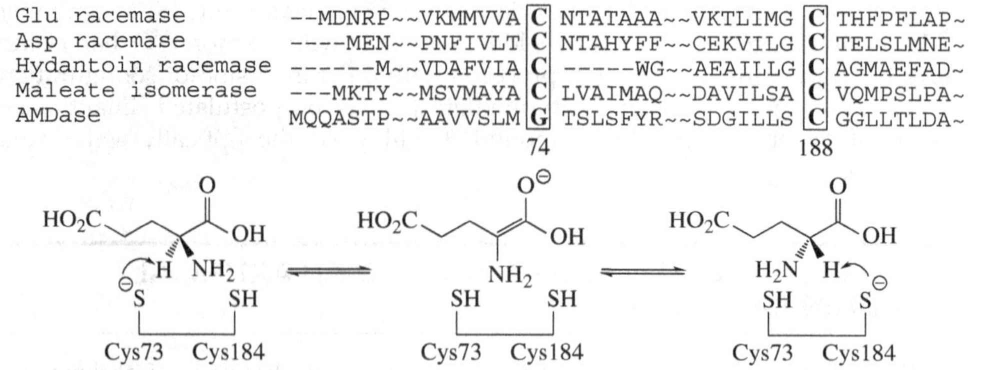

FIGURE 3.9

Homology and reaction mechanism of glutamate racemase.

TABLE 3.1 Reaction Specificity of AMDase and its Variant in Decarboxylation and Racemization

	Decarboxylation			Racemization		
	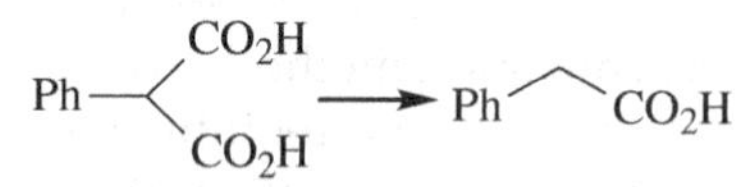			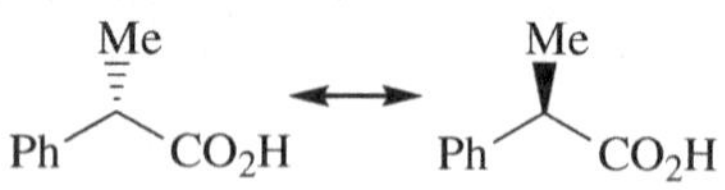		
Enzyme	K_m (mM)	k_{cat} (S^{-1})	Relative Activity (%)	K_m (mM)	k_{cat} (S^{-1})	Relative Activity (%)
Wild type	13.9	35.3	97 700	No reaction		0
G74C	10.4	0.27	100	30.2	0.60	100
G74C/V43A	11.4	0.022	8	29.7	1.1	185

activity compared to wild-type AMDase. Next, we examined the substrate scope of the G74C variant (Table 3.2). In general, good substrates for decarboxylation were also good substrates for racemization. The similarity of substrate specificity between decarboxylation and racemization indicates that both reactions involve the same catalytic machinery in wild-type and G74C AMDase [26]. The small hydrophobic pocket may have contributed to the destabilization of the carboxyl group for the decarboxylation reaction. Molecular dynamics studies of AMDase G74C indicated that Val43 is situated very close to the substrate methyl group (**7**). To increase the racemase activity, we generated several variants at position 43 by site-specific mutagenesis. The G74C/V43A variant exhibited an approximate twofold increase in racemase activity (Table 3.1) but only 8% residual activity in the decarboxylation reaction. This represents a 20-fold shift from the native decarboxylation reaction to more promiscuous racemization. From these mutational analyses, we confirmed the important role of the hydrophobic pocket in decarboxylation and racemization.

The substrate spectrum of these AMDase variants was determined by monitoring the exchange of the α-proton of deuterium using an NMR spectroscopy (Table 3.2). Using this method, we found that G74C and G74C/V43A differ in their substrate specificity. The higher activity of G74C/V43A toward compounds **9** and **10** can be explained by the larger space in the hydrophobic pocket. The V43A mutant showed higher activity toward naproxen. The 30-fold increase in activity toward ketoprofen is a significant change, as it indicates that the variant with a modified hydrophobic pocket has more space for these large substrates. We then continued the screening of G74C/V43A by saturating hydrophobic residues.

TABLE 3.2 Racemization Activity of AMDase Variants

Run	Substrate	Relative Activity (%) G74C	Relative Activity (%) G74C/V43A	Activity Increase (Fold)
1	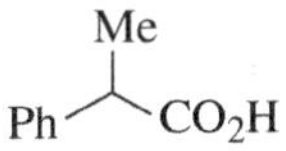 **6**	100	183	1.8
2	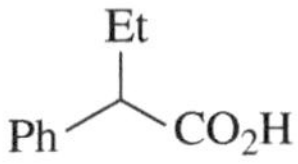 **9**	8	40	5
3	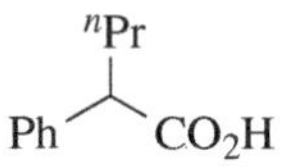 **10**	0	0.7	—
4	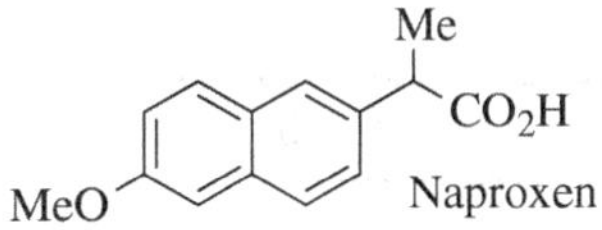 Naproxen	767	2100	2.7
5	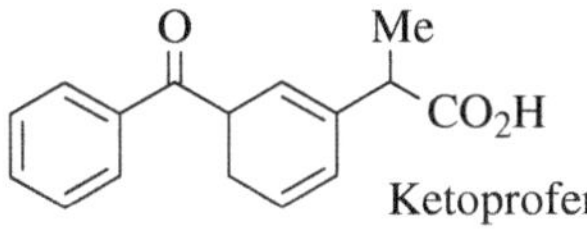 Ketoprofen	1.7	50	30

3.7 INVERSION OF ENANTIOSELECTIVITY DRAMATICALLY IMPROVES CATALYTIC ACTIVITY

Profens are an important group of NSAIDs. The biological activity of these drugs resides exclusively in the (*S*)-enantiomer, so considerable effort has been invested in developing efficient routes for their preparation. For instance, (*S*)-naproxen has been prepared via recrystallization of diastereomeric mixtures. The carboxylesterase-catalyzed kinetic resolution of (*R*/*S*)-naproxen methyl ester achieves excellent optical purity of the product. Nevertheless, the AMDase-catalyzed asymmetrization of prochiral α-aryl-α-methylmalonates gives rise to a 100% theoretical yield of profens, a clear improvement from kinetic resolution (50%). Unfortunately, wild-type AMDase produces only the undesirable (*R*)-enantiomers.

The reaction catalyzed by AMDase proceeds through enantioselective decarboxylation of the substrate and enantioface-selective protonation of the planar enolate intermediate by Cys188 (Figure 3.10).

On the basis of the structural similarity of AMDase to cofactor-independent glutamate racemases (about 30% homology), we designed an AMDase variant, G74C/C188S, which yields the desired (*S*)-enantiomers of **7** with good enantioselectivity (94% ee) [27, 28]. In the active site of the double mutant, Cys74 donates a proton toward the *re*-face of the enolate intermediate, which leads to inverse enantioface selectivity. However, this variant shows very low catalytic activity compared to wild-type AMDase (Table 3.3, Run 1). To improve the enzymatic activity, we conducted random mutagenesis experiments using the mutator *E. coli* strain XL1-Red. Among these mutants, variant S36N exhibited a tenfold increase in activity. Serine 36 is situated on the surface of the protein, so its effect on catalytic activity is difficult to elucidate. Despite this considerable improvement, the activity remained too low for

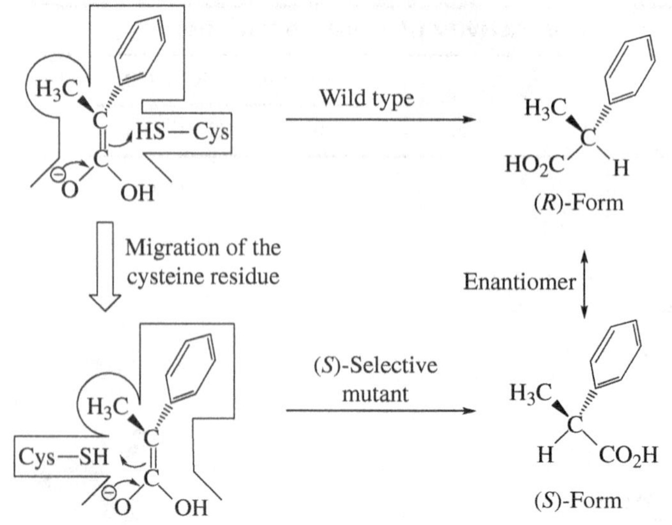

FIGURE 3.10

Rational design strategy for inversion of selectivity.

TABLE 3.3 Specific Activity of Wild-Type AMDase and (*S*)-Selective Variants

Run	Variant	Specific Activity (U/mg)	
		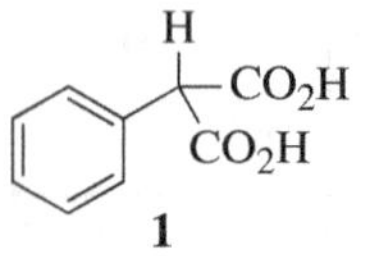 **1**	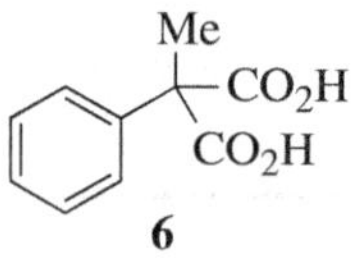 **6**
1	G74C/C188S	0.040	0.0015
2	G74C/C188G	0.10	0.010
3	G74C/C188G/Y48F	0.82	0.014
4	G74C/C188G/M159L	6.2	1.3
5	G74C/C188G/Y48F/M159L	24	0.63
6	Wild type	550	13

preparative purposes. Therefore, we attempted to improve the activity of an (*S*)-selective AMDase variant, but this task proved to be a notoriously difficult problem of protein engineering.

Recently, the catalytic mechanism of AMDase was suggested based on 3D structures. The AMDase active site is composed of a so-called oxyanion hole (Thr75, Ser76, Tyr126, Gly189), an aryl-binding pocket (Pro14, Pro15, Gly189, Gly190), a hydrophobic pocket (Leu40, Val43, Tyr48, Val156, Met159), and a proton donor (Cys188 in wild-type AMDase or Cys74 in variant G74C) (Figure 3.11).

At first, the pro-*S* carboxylate of a substrate is stabilized by the oxyanion hole. Next, the decarboxylation reaction is triggered by destabilization of the pro-*R* carboxylate through an interaction with the hydrophobic pocket. To examine this catalytic mechanism further, molecular docking was used to position phenylmalonate (**1**) in the active site of a structural model of the G74C/C188S mutant. Leucine40, Val43, Tyr48, Leu77, Val156, and Met159 were positioned close to the pro-*R* carboxylate such that they were likely to form a hydrophobic pocket. Intriguingly, both Cys74 and Ser188 potentially form hydrogen bonds with the oxygen atoms of the pro-*S* carboxylate. The activity of the G74C and C188S mutants was much lower than that of the native enzyme. In experiments using the wild-type and the G74C/C188S mutant in deuterium oxide, no kinetic isotope effect was observed. This finding confirms that the rate-limiting step of the reaction catalyzed by AMDase is decarboxylation, not protonation, and is reflected by the G74C/C188S mutant activity decrease.

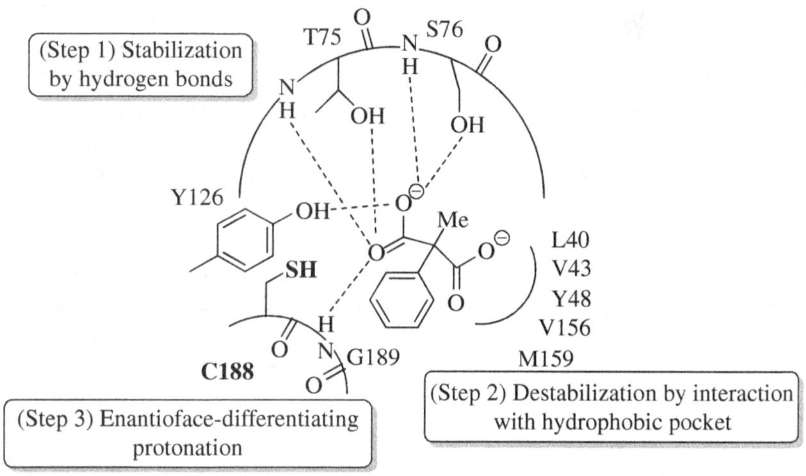

FIGURE 3.11 Structure of the active site of AMDase and reaction mechanism.

We speculated that Cys74 and Ser188 undergo unfavorable interactions with the substrate, which may disturb the stabilization of the pro-*S* carboxylate and the destabilization of the pro-*R* carboxylate.

From this observation, we decided to improve the enzyme activity by applying iterative saturation mutagenesis based on structural information [29]. First, the unfavorable effect of Ser188 was corrected by mutagenesis at this position. Earlier results show that modification of the hydrophobic pocket can strongly increase catalytic activity, which can be explained by an improved destabilization of pro-*R* carboxylate. We speculated that by focusing on residues of the hydrophobic pocket, we could increase the catalytic activity of an (*S*)-selective variant. We followed a strategy of three generations of mutagenesis and screening. The first screening generations focused on Ser188, while the second and third focused on residues Leu40, Val43, Tyr48, Leu77, Val156, and Met159 in the hydrophobic pocket.

Directed evolution was performed as follows. The gene encoding the *A. bronchisepticus* AMDase carrying a C-terminal poly-His-tag was subjected to saturation mutagenesis of targeted amino acids. Site-directed mutagenesis with primers bearing an NNK-degenerating codon was used to construct libraries containing all 20 possible amino acids at the desired position. Variants were screened for phenylmalonate decarboxylation activity by the pH shift of the reaction solution as visualized by the pH indicator BTB (Figure 3.6). In each case, a set of 112 transformants was screened to assure that each set would statistically provide full coverage of all 20 amino acid exchanges.

The first screen showed that the G74C/C188G mutant has 5.6-fold higher activity than the G74C/C188S mutant (Table 3.3, Run 2). The second screening generation identified the triple mutant G74C/M159L/C188G with a 210-fold increase in activity (Table 3.3, Run 4) and Y48F/G74C/C188G with a 23-fold increase in activity (Table 3.3, Run 3) compared to the G74C/C188S mutant. We continued the third screening generation from G74C/M159L/C188G by individually saturating Leu40, Val43, Tyr48, Leu77, and Val156. The quadruple mutant Y48F/G74C/M159L/C188G exhibited a 920-fold activity increase relative to the G74C/C188S mutant (Table 3.3, Run 5).

Next, we analyzed the substrate specificity of these variants (Table 3.3). The G74C/C188G/M159L variant had the highest specific activity toward α-aryl-α-methylmalonates (Run 4), 220-fold higher toward phenylmalonate (**1**) than the G74C/C188S variant, yielding (*S*)-naproxen of excellent optical purity (>99% ee).

The G74C/C188G mutant showed higher activity than the G74C/C188S mutant, indicating that the hydroxyl group of Ser188 or steric hindrance of this residue decreases AMDase activity. This variant had comparatively low enantioselectivity, however, suggesting that Ser188 also interacts with the enolate intermediate. The M159L mutation is predicted to cause a steric effect at the α-position of the substrate, thus altering its binding mode. Moreover, exchanging Tyr48 for Phe should lower the polarity of the hydrophobic pocket. Both events should further destabilize the pro-*R* carboxylate. Surprisingly, the Y48F mutation increased AMDase activity toward compound **1** but decreased the activity toward α-methyl-α-phenylmalonates (**6**). The crystal structure suggested that Tyr48 may form hydrogen bonds with the hydroxyl group of Ser76 in the oxyanion hole. The Y48F mutation conceivably changes the orientation of Ser76, resulting in unfavorable substrate binding for decarboxylation.

Recently, we tried to further improve the activity of G74C/M159L/C188G by optimizing activity toward α-aryl-α-methylmalonates. We continued the third screening generation from G74C/M159L/C188G by saturating hydrophobic residues. Consequently, we have successfully increased the activity of an artificial (*S*)-selective AMDase variant. This most recent variant shows 9500-fold greater activity than the G74C/C188S mutant [30].

3.8 FUTURE PROSPECTS

Introduction of a single cysteine residue at G74 converts AMDase to an unnatural "profen racemase" (Figure 3.12). Moreover, we were able to generate a G74C/V43A variant with a 20-fold shift toward the racemization reaction based on reduced decarboxylase activity and a twofold increase in racemization activity. The decarboxylase activity of the G74C/V43A variant disappeared almost completely compared to wild-type AMDase. Consequently, we have succeeded in the production of an optimal unnatural "profen racemase." We are now trying to increase the catalytic activity and extend the substrate scope of "profen racemase" using a structure-based rational design approach and high-throughput screening.

In addition, we have succeeded in inversion of the enantioselectivity of AMDase using a rational design approach. Although the G74C/C188S mutant produced high yields of the (*S*)-enantiomer, the decarboxylation activity was much lower than wild-type AMDase. We have successfully improved the activity of an artificial (*S*)-selective AMDase variant using directed evolution.

Further advancement of biocatalysis will require the use of directed evolution to bridge the functional gap between wild-type and desired biocatalyst properties. These studies underscore the power of directed evolution to create artificial enzymes derived from wild-type enzymes with the desired catalytic activity. Directed evolution techniques will continue to fulfill the promise of biocatalysis for industrial applications.

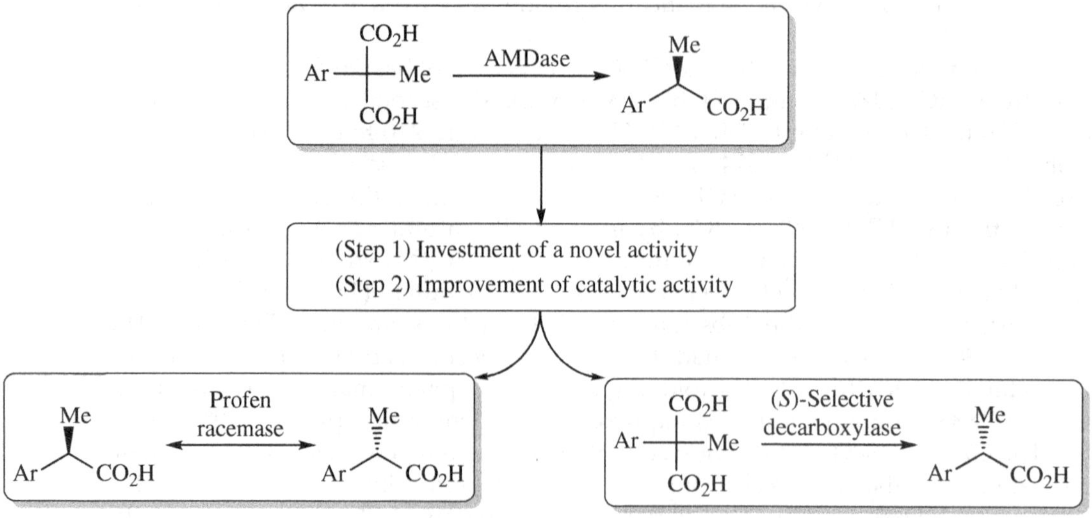

FIGURE 3.12

Directed evolution to create artificial enzymes.

REFERENCES

[1] H. C. Hailes, D. Rother, M. Müller, R. Westphal, J. M. Ward, J. Pleiss, C. Vogel and M. Pohl, "Engineering stereoselectivity of ThDP-dependent enzymes", *FEBS Journal* **280**, 6374–6394 (2013).

[2] F. Jordan and H. Patel, "Catalysis in enzymatic decarboxylations: comparison of selected cofactor-dependent and cofactor-independent examples", *ACS Catalysis* **3**, 1601–1617 (2013).

[3] W. F. Kellett, E. Brunk, B. J. Desai, A. A. Fedorov, S. C. Almo, J. A. Gerlt, U. Rothlisberger and N. G. J. Richards, "Computational, structural, and kinetic evidence that *Vibrio vulnificus* FrsA is not a cofactor-independent pyruvate decarboxylase", *Biochemistry* **52**, 1842–1844 (2013).

[4] T. Li, L. Huoa, C. Pulley and A. Liu, "Decarboxylation mechanisms in biological system", *Bioorganic Chemistry* **43**, 2–14 (2012).

[5] Y. S. Kim and P. E. Kolattukudy, "Stereospecificity of malonyl-CoA decarboxylase, acetyl-CoA carboxylase, and fatty acid synthetase from the uropygial gland of goose", *Journal Biological Chemistry* **255**, 686–689 (1980).

[6] T. Y. Shen, "Perspectives in nonsteroidal anti-inflammatory agents", *Angewandte Chemie International Edition* **11**, 460–472 (1972).

[7] R. Kourist, P. D. Maria and K. Miyamoto, "Biocatalytic strategies for the asymmetric synthesis of profens—recent trends and developments", *Green Chemistry* **13**, 2607–2618 (2011).

[8] K. Miyamoto and H. Ohta, "Enzyme-mediated asymmetric decarboxylation of disubstituted malonic acids", *Journal of the American Chemical Society* **112**, 4077–4078 (1990).

[9] K. Miyamoto and H. Ohta, "Asymmetric decarboxylation of disubstituted malonic acid by *Alcaligenes bronchisepticus* KU1201", *Biocatalysis* **5**, 49–60 (1991).

[10] K. Miyamoto and H. Ohta, "Purification and properties of a novel arylmalonate decarboxylase from *Alcaligenes bronchisepticus* KU1201", *European Journal of Biochemistry* **210**, 475–481 (1992).

[11] J. H. Galivan and S. H. G. Allen, "Methylmalonyl-CoA decarboxylase: partial purification and enzymatic properties", *Archives of Biochemistry and Biophysics* **126**, 838–847 (1968).

[12] A. Hoffmann, W. Hilpert and P. Dimroth, "The carboxyltransferase activity of the sodium-ion-translocating methylmalonyl-CoA decarboxylase of *Veillonella alcalescens*", *European Journal of Biochemistry* **179**, 645–650 (1989).

[13] K. Miyamoto and H. Ohta, "Cloning and heterologous expression of a novel arylmalonate decarboxylase gene from *Alcaligenes bronchisepticus* KU1201", *Applied Microbiology and Biotechnology* **38**, 234–238 (1992).

[14] K. Miyamoto, S. Tsuchiya and H. Ohta, "Stereochemistry of enzyme-catalyzed decarboxylation of a-methyl-a-phenylmalonic acids", *Journal of the American Chemical Society* **114**, 6256–6257 (1992).

[15] R. Wada, T. Kumon R. Kourist, H. Ohta, D. Uemura, S. Yoshida and K. Miyamoto, "Thermally driven asymmetric domino reaction catalyzed by a thermostable esterase and its variants", *Tetrahedron Letters* **54**, 1921–1923 (2013).

[16] R. Obata and M. Nakasako, "Structural basis for inverting the enantioselectivity of arylmalonate decarboxylase revealed by the structural analysis of the Gly74Cys/Cys188Ser mutant in the liganded form", *Biochemistry* **49**, 1963–1969 (2010).

[17] K. A. Gallo and J. R. Knowles, "Purification, cloning, and cofactor independence of glutamate racemase from *Lactobacillus*", *Biochemistry* **32**, 3981–3990 (1993).

[18] M. Yohda, H. Okada and H. Kumagai, "Molecular cloning and nucleotide sequencing of the aspartate racemase gene from lactic acid bacteria *Streptococcus thermophilus*", *Biochimica et Biophysica Acta* **1089**, 234–240 (1991).

[19] K. Watabe, T. Ishikawa, Y. Mukohara and H. Nakamura, "Identification and sequencing of a gene encoding a hydantoin racemase from the native plasmid of *Pseudomonas* sp. strain NS671", *Journal of Bacteriology* **174**, 3461–3466 (1992).

[20] K. Hatakeyama, Y. Asai, Y. Uchida, M. Kobayashi, M. Terasawa and H. Yukawa, "Gene cloning and characterization of maleate cis–trans isomerase from *Alcaligenes faecalis*", *Biochemical and Biophysical Research Communications* **239**, 74–79 (1997).

[21] S. Glavas and M. E. Tanner, "Catalytic acid/base residues of glutamate racemase", *Biochemistry* **38**, 4106–4113 (1999).

[22] K. A. Gallo, M. E. Tanner and J. R. Knowles, "Mechanism of the reaction catalyzed by glutamate racemase", *Biochemistry* **32**, 3991–3997 (1993).

[23] M. E. Tanner, K. A. Gallo and J. R. Knowles, "Isotope effects and the identification of catalytic residues in the reaction catalyzed by glutamate racemase", *Biochemistry* **32**, 3998–4006 (1993).

[24] K. Y. Hwang, C.-S. Cho, S. S. Kim, H.-C. Sung, Y. G. Yu and Y. Cho, "Structure and mechanism of glutamate racemase from *Aquifex pyrophilus*", *Nature Structural Biology* **6**, 422–426 (1999).

[25] Y. Terao, K. Miyamoto and H. Ohta, "Introduction of single mutation changes arylmalonate decarboxylase to racemase", *Chemical Communications* 3600–3602 (2006).

[26] R. Kourist, Y. Miyauchi, D. Uemura and K. Miyamoto, "Engineering the promiscuous racemase activity of an arylmalonate decarboxylase", *Chemistry—A European Journal* **17**, 557–563 (2011).

[27] Y. Ijima, K. Matoishi, Y. Terao, N. Doi, H. Yanagawa and H. Ohta, "Inversion of enantioselectivity of asymmetric biocatalytic decarboxylation by site-directed mutagenesis based on the reaction mechanism", *Chemical Communications* 877–879 (2005).

[28] Y. Terao, Y. Ijima, K. Miyamoto and H. Ohta, "Inversion of enantioselectivity of arylmalonate decarboxylase via site-directed mutation based on the proposed reaction mechanism", *Journal of Molecular Catalysis B: Enzymatic* **45**, 15–20 (2007).

[29] Y. Miyauchi, R. Kourist, D. Uemura and K. Miyamoto, "Dramatically improved catalytic activity of an artificial (*S*)-selective arylmalonate decarboxylase by structure-guided directed evolution", *Chemical Communications* **47**, 7503–7505 (2011).

[30] S. Yoshida, J. Enoki, R. Kourist and K. Miyamoto, "Engineered hydrophobic pocket of (*S*)-selective arylmalonate decarboxylase variant by simultaneous saturation mutagenesis to improve catalytic performance", *Bioscience, Biotechnology, and Biochemistry* **79**, 1965–1971 (2015).

CHAPTER 4

Green Processes for the Synthesis of Chiral Intermediates for the Development of Drugs

Ramesh N. Patel

SLRP Associates, Consultation in Biotechnology, Bridgewater, NJ, USA

4.1 INTRODUCTION

In the development of pharmaceuticals and agrochemicals, the synthesis of single enantiomers of intermediates has become increasingly important since the undesired enantiomer is biologically inactive and even toxic in nature. The advantages of biocatalysis over chemical synthesis are that enzyme-catalyzed reactions are highly enantio-, regio-, and chemoselective. Such high selectivity is extremely desirable in pharmaceutical synthesis as it offers several advantages such as minimizing side reactions and not requiring protection and deprotection steps. They can be carried out under mild conditions such as ambient temperature and atmospheric pressure and in aqueous systems. Biocatalysis generally enables more sustainable routes to key intermediates and active pharmaceutical ingredients (APIs), effectively reducing level of waste. Technological advances have established biocatalysis as an environmentally-friendly and economical alternative to traditional chemical synthesis. Chapter 1 provides more information on considering bioprocesses as green technology. Directed evolution of biocatalysts under process conditions has led to increased enzyme activity, selectivity, and stability of biocatalysts [1–8]. Biocatalysts are now developed as per process requirements, and the immobilization and reusability of biocatalysts have lead to efficient and economical bioprocesses.

A number of review articles [9–20] have described the use of enzymes in organic synthesis. This chapter provides many examples of the industrial use of enzymes for the synthesis of key intermediates for the development of pharmaceuticals.

4.2 SAXAGLIPTIN: ENZYMATIC SYNTHESIS OF (*S*)-*N*-BOC-3-HYDROXYADAMANTYLGLYCINE

Dipeptidyl peptidase 4 (DPP-4) is a ubiquitous proline-specific serine protease responsible for the rapid inactivation of glucagon-like peptide 1 (GLP-1). To alleviate the inactivation of GLP-1, inhibitors of DPP-4 are being evaluated for their ability to provide improved control of blood glucose for diabetics and treatment for type 2 diabetes [21]. Saxagliptin **1** [22–24] (Figure 4.1), a DPP-4 inhibitor developed by Bristol-Myers Squibb (BMS), requires (*S*)-*N*-boc-3-hydroxyadamantylglycine **2** as a key intermediate.

Green Biocatalysis, First Edition. Edited by Ramesh N. Patel.

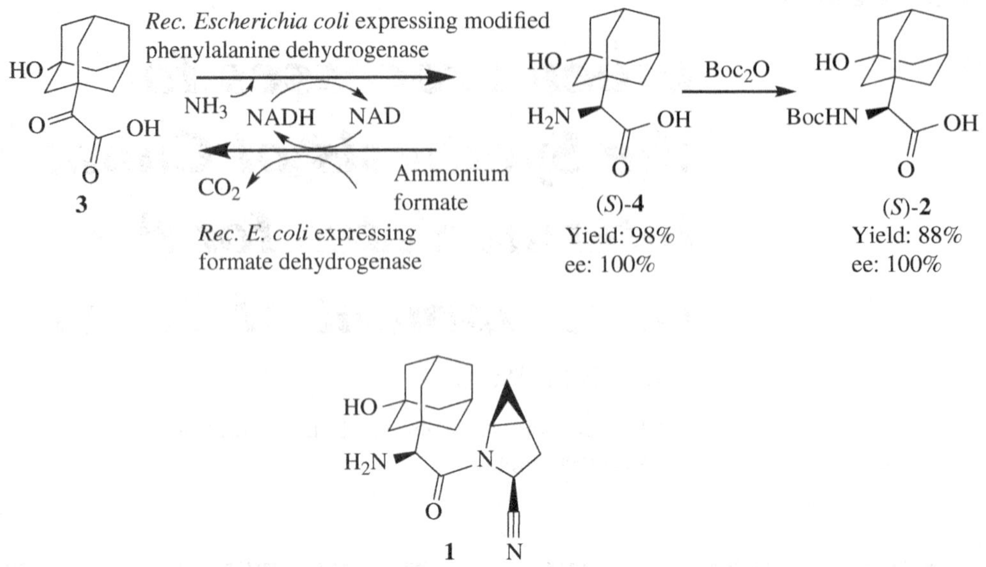

FIGURE 4.1
Saxagliptin: enzymatic preparation of (*S*)-*N*-boc-3-hydroxyadamantylglycine.

A process for converting keto acid **3** to the corresponding amino acid **4** using (*S*)-amino acid dehydrogenases was developed using a modified form of a recombinant phenylalanine dehydrogenase (PDH) cloned from *Thermoactinomyces intermedius* and expressed in *Pichia pastoris* or *Escherichia coli*. The modified PDH contained two amino acid changes at the C-terminus along with a 12-amino acid extension at the C-terminus and was highly active toward keto acid **3** [24]. NAD^+ produced during the reaction was recycled to NADH using formate dehydrogenase (FDH). The production of multikilogram batches was originally carried out with extracts of *P. pastoris* overexpressing the modified PDH from *T. intermedius* and endogenous FDH. The reductive amination process was further scaled up using a preparation of the two enzymes expressed in single recombinant *E. coli* at 100 g/l substrate input to prepare multikilogram batches of amino acid **4**, which was directly protected as its boc derivative without isolation to afford intermediate **2**. Yields of **4** before isolation were close to 98% with 100% ee, and yields of **2** were 88% after isolation with 100% ee [24].

4.3 SITAGLIPTIN: ENZYMATIC SYNTHESIS OF CHIRAL AMINE

Sitagliptin **5** (Januvia, Figure 4.2) is an oral antidiabetic drug of the DPP-4 inhibitor class developed and is marketed by Merck & Co. Sitagliptin is used either alone or in combination with other oral antihyperglycemic agents (such as metformin or a thiazolidinedione) for the treatment of type-2 diabetes.

The chemical synthesis of sitagliptin [25, 26] involved an asymmetric hydrogenation of an enamine at high pressure using a rhodium-based chiral catalyst [27]. The chemical process suffers from inadequate stereoselectivity and a product stream contaminated with rhodium, necessitating additional purification steps at the expense of yield to upgrade both enantiomeric excess and chemical purity.

To develop a biocatalytic process, screening of commercially available transaminases by Merck and Codexis provided no enzyme with detectable activity for amination of the prositagliptin ketone **6** [28]. They therefore applied a combination of *in silico* design and directed evolution in an effort to confer such an enzyme.

By using a transaminase [29–32] scaffold and various protein engineering technologies, an enzymatic process was developed for sitagliptin manufacturing. Starting from an enzyme that had the catalytic machinery to perform the desired chemistry but lacked any activity toward the prositagliptin ketone **6**, by using a substrate walking, modeling, and mutation approach they created a transaminase with marginal activity

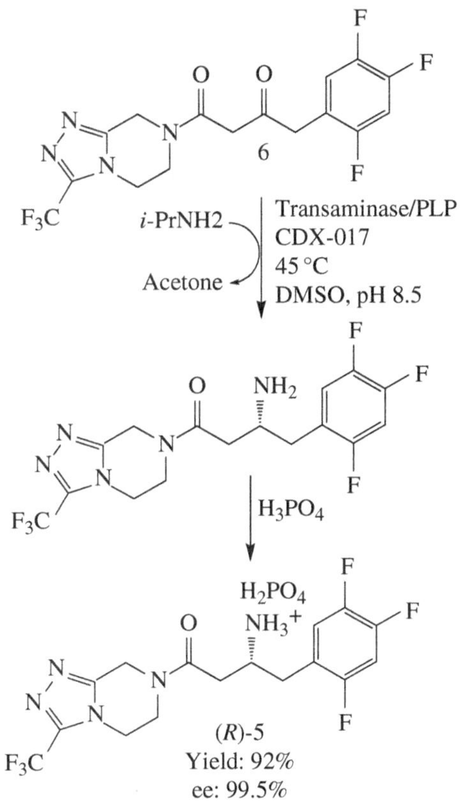

FIGURE 4.2
Sitagliptin: enzymatic synthesis of (*R*)-amine by transamination.

for the synthesis of the chiral amine. This variant was then further engineered via a number of rounds of directed evolution under the desired process conditions for practical application in a manufacturing setting. The substrate concentration was increased from 2 to 100 g/l, the isopropylamine concentration from 0.5 to 1 M, the cosolvent from 5 to 50% DMSO, the pH from 7.5 to 8.5, and the temperature from 22 to 45 °C, ultimately leading to a catalyst that met the required process targets. Under optimal conditions [28], the best variant converted 200 g/l prositagliptin ketone **6** to sitagliptin **5** in >99.95% ee and 92% yield by using 6 g/l enzyme.

In comparison with the rhodium-catalyzed process, the biocatalytic process provides sitagliptin with a 10–13% increase in overall yield, a 53% increase in productivity, a 19% reduction in total waste, the elimination of all heavy metals, and a reduction in total manufacturing cost; the enzymatic reaction is run in multipurpose vessels, avoiding the need for specialized high-pressure hydrogenation equipment [28].

4.4 VANLEV: ENZYMATIC SYNTHESIS OF (*S*)-6-HYDROXYNORLEUCINE

Vanlev **7** (Figure 4.3) is an antihypertensive drug, which acts by inhibiting both angiotensin-converting enzyme (ACE) and neutral endopeptidase enzyme (NEP) [33]. (*S*)-6-Hydroxynorleucine **8** is a key intermediate in the synthesis of Vanlev. The synthesis and complete conversion of 2-keto-6-hydroxyhexanoic acid **9** to (*S*)-6-hydroxynorleucine **8** was developed by enzymatic reductive amination using beef liver glutamate dehydrogenase [34]. The reaction requires ammonia and NADH. NAD^+ produced during the reaction was recycled to NADH by the oxidation of glucose to gluconic acid using glucose dehydrogenase (GDH) from *Bacillus megaterium*. The reaction was completed in about 3 h at 100 g/l substrate input with a reaction yield of 92% and an ee of 99.8% for (*S*)-6-hydroxynorleucine.

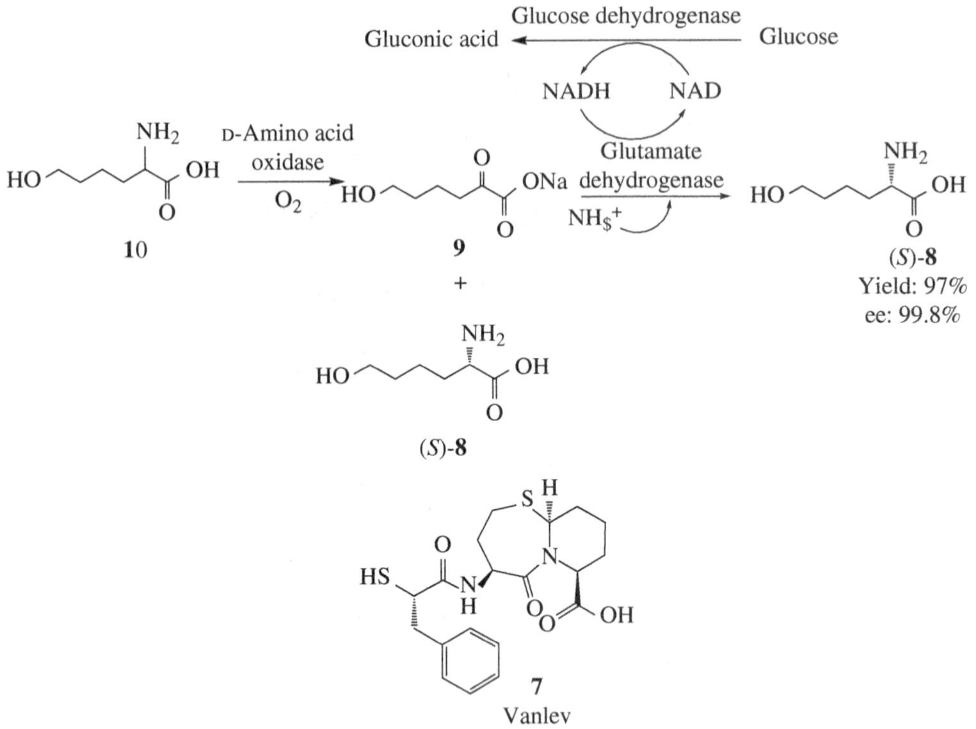

FIGURE 4.3
Vanlev: enzymatic synthesis of (*S*)-6-hydroxynorleucine.

In a second, more convenient process, the keto acid was prepared by the treatment of racemic 6-hydroxynorleucine **10** (produced by the hydrolysis of 5-(4-hydroxybutyl) hydantoin) with (*R*)-amino acid oxidase (Figure 4.3). After ee of the unreacted (*S*)-6-hydroxynorleucine had reached 99.8%, the reductive amination procedure was used to produce (*S*)-6-hydroxynorleucine at 97% yield with 99.8% ee from racemic 6-hydroxynorleucine at 100 g/l substrate concentration [34]. The (*S*)-6-hydroxynorleucine prepared by the enzymatic process was converted chemically to Vanlev **7** [35].

4.5 VANLEV: ENZYMATIC SYNTHESIS OF ALLYSINE ETHYLENE ACETAL

(*S*)-2-Amino-5-(1,3-dioxolan-2-yl)-pentanoic acid [(*S*)-allysine ethylene acetal] (**11**, Figure 4.4) is one of three building blocks used in an alternative synthesis of Vanlev **7** [33]. Enzymatic process for the synthesis of **11** was developed by reductive amination of keto acid acetal **12** using PDH from *T. intermedius* [36]. The reaction required ammonia and NADH; NAD^+ produced during the reaction was recycled to NADH by the oxidation of formate to CO_2 using FDH. *Thermoactinomyces intermedius* PDH was cloned and expressed in *E. coli*, and recombinant culture was used as a source of PDH. *Pichia pastoris* [36] grown on methanol was used as a source of FDH. Expression of *T. intermedius* PDH in *P. pastoris*, inducible by methanol, allowed generation of both enzymes (PDH and FDH) in a single fermentation process. This process was scaled up to prepare a total of 197 kg of **11** with 91 M% yield and >98% ee [36]. (*S*)-allysine ethylene acetal prepared by enzymatic process was converted to Vanlev **7** [35].

4.6 VANLEV: ENZYMATIC SYNTHESIS OF THIAZEPINE

[4*S*-(4a,7a,10ab)]-1-Octahydro-5-oxo-4-[[(phenylmethoxy)carbonyl]amino]-7*H*-pyrido-[2,1-β] [1,3]thiazepine-7-carboxylic acid **13** (Figure 4.5) is a key intermediate for the synthesis of Vanlev **7** [33]. An enzymatic process for the oxidation of ε-amino group of

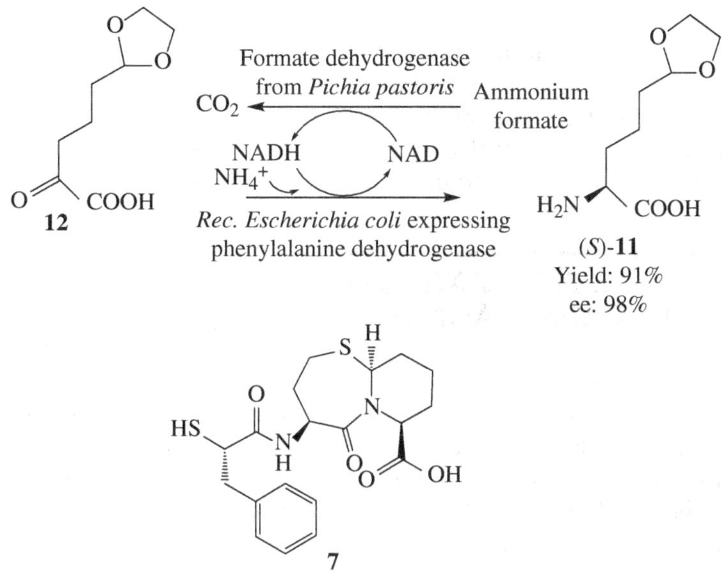

FIGURE 4.4
Vanlev: enzymatic synthesis of allysine ethylene acetal.

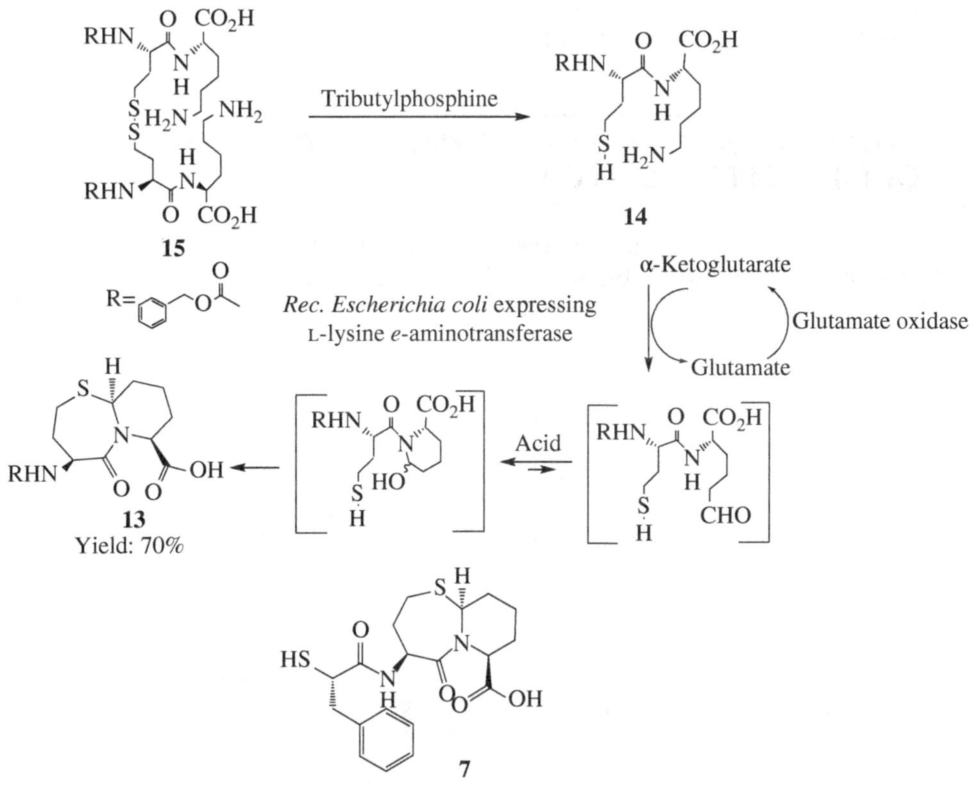

FIGURE 4.5
Vanlev: enzymatic synthesis of thiazepine.

(*S*)-lysine in the thiol **14** generated *in situ* from disulfide N2-[*N*[[(phenylmethoxy) carbonyl] L-homocysteinyl] L-lysine]-1,1-disulfide **15** was developed to produce compound **13** using L-lysine ε-aminotransferase (LAT) from *Sphingomonas paucimobilis* SC 16113 [37] cloned and overexpressed in *E. coli* [38, 39]. The aminotransferase reaction required α-ketoglutarate as the amine acceptor. The glutamate formed during this reaction was recycled back to α-ketoglutarate by glutamate oxidase (GOX) from *Streptomyces noursei* SC 6007.

The biotransformation of compound **15** to compound **13** was carried out using LAT from *E. coli* GI724[pal781-LAT] in the presence of α-ketoglutarate and dithiothreitol or tributylphosphine and glutamate oxidase. Reaction yields of 65–67 M%

were obtained. To reduce the cost of producing two enzymes, the transamination reactions were carried out in the absence of GOX and with higher levels of α-ketoglutarate. The reaction yield in the absence of GOX averaged only about 33–35 M%; however, the reaction yield increased to 70 M%, by increasing the α-ketoglutarate level to 40 mg/ml (a tenfold increase in concentration) and conducting the reaction at 40 °C, equivalent to that in the presence of GOX [37].

4.7 TIGEMONAM: ENZYMATIC SYNTHESIS OF (*S*)-β-HYDROXYVALINE

(*S*)-β-Hydroxyvaline **16** (Figure 4.6) is a key chiral intermediate required for the total synthesis of orally active monobactam antibiotic, Tigemonam **17** [40]. The synthesis of (*S*)-β-hydroxyvaline **16** from α-keto-β-hydroxyisovalerate **18** was developed by enzymatic reductive amination using leucine dehydrogenase from *Bacillus sphaericus* ATCC 4525 [41]. The NADH required for this reaction was regenerated by either FDH from *Candida boidinii* or GDH from *B. megaterium* using NAD^+ as cofactor in catalytic amount. Substrate **18** was generated either from α-keto-β-bromoisovalerate or its ethyl esters by hydrolysis with sodium hydroxide *in situ*. In this process, an overall reaction yield of 98% and an ee of 99.8% were obtained for the L-β-hydroxyvaline **16**.

4.8 AUTOIMMUNE DISEASES: ENZYMATIC SYNTHESIS OF (*S*)-NEOPENTYLGLYCINE

The enantioselective synthesis of (*S*)-neopentylglycine **19** (Figure 4.7) was developed by Groeger *et al.* [42]. Recombinant whole cells containing leucine dehydrogenase and FDH were used in the reductive amination of the corresponding α-keto acid **20**.

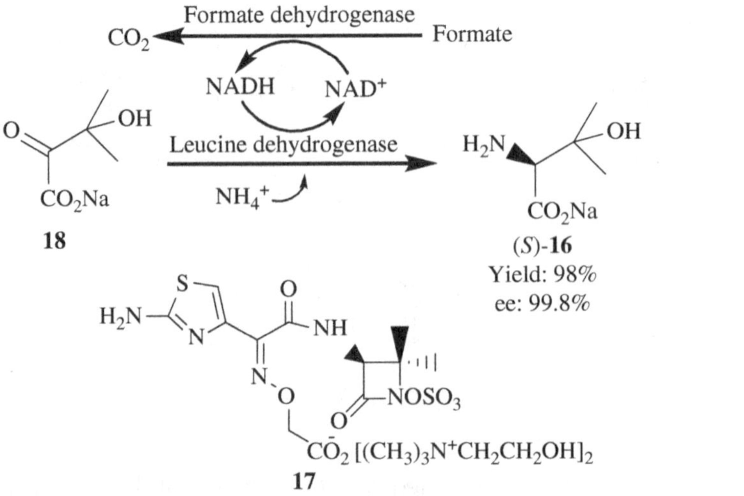

FIGURE 4.6

Tigemonam: enzymatic synthesis of (*S*)-β-hydroxyvaline.

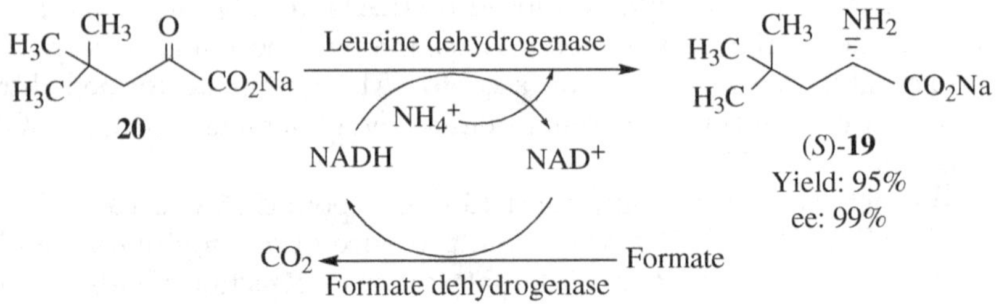

FIGURE 4.7

Autoimmune diseases: enzymatic synthesis of (*S*)-neopentylglycine.

The desired (*S*)-neopentylglycine was obtained with >95% conversion and >99% ee at substrate concentrations of up to 88 g/l. Spiroheterocyclic compounds (morpholine-4-carboxylic acid amides and neopentylglycine derivatives and their analogs) are reversible inhibitors of cysteine proteases such as cathepsin S and, therefore, useful in treating a variety of autoimmune diseases [43].

4.9 ATAZANAVIR: ENZYMATIC SYNTHESIS OF (*S*)-TERTIARY LEUCINE

Atazanavir **21** is an acyclic aza-peptidomimetic and potent human immunodeficiency virus (HIV) protease inhibitor [44, 45]. (*S*)-Tertiary leucine **22** (Figure 4.8) is a key intermediate required for the synthesis of atazanavir along with other drugs containing peptides such as boceprevir and telaprevir [46, 47].

Leucine dehydrogenase and an FDH from *C. boidinii* were used to reduce 2-ketocarboxylic acids enzymatically while regenerating the NADH cofactor *in situ*, and this constitutes an industrially established method for preparing optically active (*S*)-α-amino acids particularly (*S*)-*tert*-leucine, which is produced on the ton scale using a membrane reactor. The method is described in detail in the literature [48, 49]. However, this method required isolated enzymes in purified form.

Soda *et al.* [50] described the use of a whole-cell catalyst for the enzymatic synthesis of (*S*)- and (*R*)-amino acids from α-keto acids with *E. coli* cells that overexpress heterologous genes. L-amino acids were produced with a thermostable (*S*)-amino acid dehydrogenase and an FDH from α-keto acids and ammonium formate. No exogenous cofactor was required as the intracellular pool was sufficient. This group constructed plasmids containing—in addition to the FDH gene—genes encoding a variety of amino acid dehydrogenases such as leucine, alanine, and phenylalanine. These recombinant cells were used to produce L-leucine, L-valine, L-norvaline, L-methionine, L-phenylalanine, and L-tyrosine with high chemical yields (>80%) and high optical purities (up to 100% ee). The preparation of various (*R*)-amino acids from their corresponding α-keto acids was also examined with recombinant *E. coli* cells containing a plasmid encoding four heterologous genes: a thermostable D-amino acid aminotransferase, alanine racemase, L-alanine dehydrogenase, and

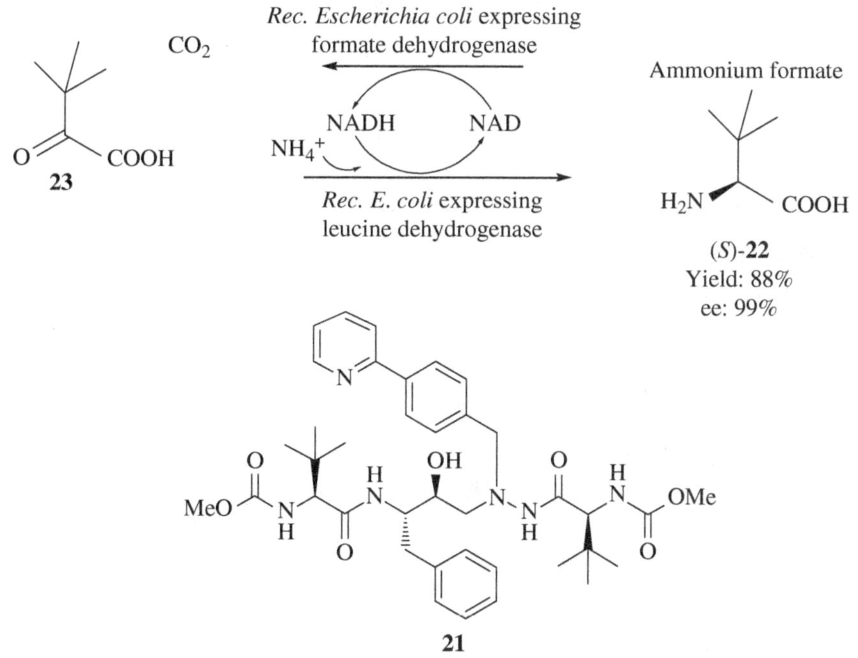

FIGURE 4.8
Atazanavir: enzymatic synthesis of (*S*)-*tert*-leucine.

an FDH. This publication points out taking advantage of the intracellular pool of NAD^+ in *E. coli*; however, the final concentration of product was restricted to about 0.35–0.38 M.

Recently, Groger *et al.* [51] developed a process for preparing (*S*)-tertiary leucine **22** by reacting the corresponding keto acid **23** with an ammonium ion donor in the presence of a whole cell of *E. coli* expressing amino acid dehydrogenase and cofactor-regenerating enzyme FDH. The substrate addition was metered such that the stationary concentration of 2-ketocarboxylic acid remains <500 mM and the external addition of cofactor, based on the total input of substrate, corresponds to <0.0001 equivalents. In this process, a reaction yield of 84% and ee of 99% were obtained for (*S*)-tertiary leucine at 130 g/l substrate input.

4.10 THROMBIN INHIBITOR (INOGATRAN): SYNTHESIS OF (*R*)-CYCLOHEXYLALANINE

(*R*)-Amino acids are increasingly becoming important building blocks in the production of pharmaceuticals and fine chemicals [52, 53]. Using rational and random mutagenesis, Rozzell and Novick [54] have created the broad substrate range highly stereoselective (*R*)-amino acid dehydrogenase. This new enzyme is capable of producing (*R*)-amino acids via the reductive amination of the corresponding 2-keto acid with ammonia. This biocatalyst was the result of three rounds of mutagenesis and screening performed on the enzyme *meso*-diaminopimelate (*R*)-dehydrogenase from *Corynebacterium glutamicum*. The first round targeted the active site of the wild-type enzyme and produced mutants that were no longer strictly dependent on the native substrate. The second and third rounds produced mutants that had an increased substrate range including straight- and branched-aliphatic amino acids and aromatic amino acids. The very high selectivity toward the (*R*)-enantiomer (95 to >99% ee) was shown to be preserved three rounds of mutagenesis and screening [55]. The synthesis of (*R*)-cyclohexylalanine **24** (Figure 4.9) was developed by reductive amination of cyclohexylpyruvate **25** to

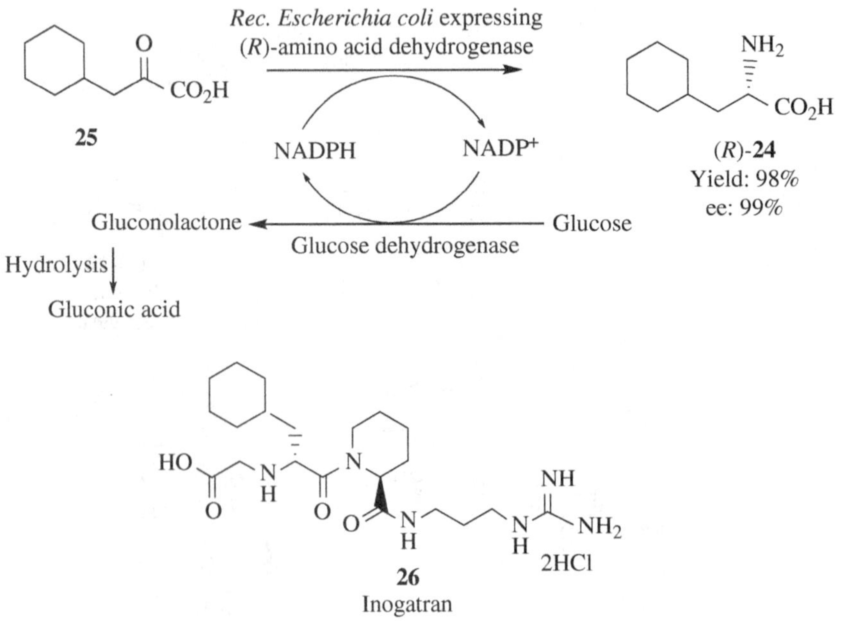

FIGURE 4.9

Thrombin inhibitor (ino-gatran): synthesis of (*R*)-cyclohexylalanine.

yield (*R*)-**24** with a 98% yield and >99% ee. (*R*)-**24** is a potential chiral intermediate for the synthesis of thrombin inhibitor inogatran **26** [56].

4.11 GAMMA SECRETASE INHIBITOR: ENZYMATIC SYNTHESIS OF (*R*)-5,5,5-TRIFLUORONORVALINE

Amyloid-β peptides (Aβ) are a major component of the plaques that are found in the brains of Alzheimer's patients and have been proposed to play a causative role in the disease. These peptides are produced from amyloid precursor protein by β-secretase and γ-secretase to generate Aβ forms, with Aβ42 most closely associated with Alzheimer's. BMS-708163 **27**, a γ-secretase inhibitor (Figure 4.10), causes a significant decrease of Aβ40 levels formed by secretases and developed for the treatment of Alzheimer's disease by Bristol-Myers Squibb (BMS) [57, 58]. (*R*)-5,5,5-Trifluoronorvaline **28** is a key intermediate for the synthesis of compound **27**.

(*R*)-5,5,5-Trifluoronorvaline **28** was prepared from the corresponding keto acid **29** using a commercially available D-amino acid dehydrogenase (biocatalytics) for reductive amination and GDH for cofactor NADPH recycling [59, 60]. This amino acid was also prepared using a D-amino acid transaminase with alanine as the amino donor, but the transamination reaction also requires lactate dehydrogenase (LDH), NAD, formate, and an FDH to remove pyruvate byproduct in order to bring the reaction to completion. An effective proprietary (*R*)-amino acid dehydrogenase was constructed by the modification of the D-diaminopimelic acid dehydrogenase gene from *B. sphaericus*. The cofactor-regenerating enzyme GDH gene was also cloned from *Gluconobacter oxidans*. Both genes were expressed in the same strain of *E. coli*, together with the glutamate dehydrogenase gene, which was inactivated in the expression strain to eliminate background production of the (*S*)-amino acid to improve the ee of the product to 100%. The amino acid could be isolated or converted without isolation to a *p*-chlorophenylsulfonamide carboxamide **30** intermediate needed for the synthetic route to the γ-secretase inhibitor development. The reaction was carried out at 50 g/l keto acid **29** and 382 mg $NADP^+$ in the presence of extract of recombinant *E. coli* containing 1250 units of (*R*)-amino acid dehydrogenase and 7500 units of GDH. After 22 h reaction, the amino acid **28** was produced with an 88.5% yield and 98.9% ee. Extracts from this strain were used for the production of intermediate **28** on a 50 kg scale [59, 60].

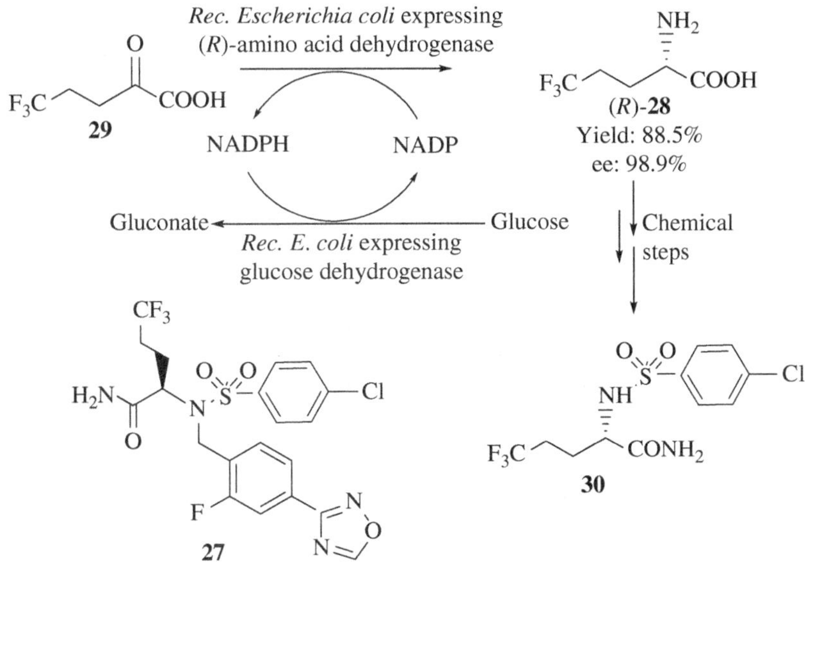

FIGURE 4.10
Gamma secretase inhibitor: enzymatic synthesis (*R*)-5,5,5-trifluoronorvaline.

4.12 NK1/NK2 DUAL ANTAGONISTS: ENZYMATIC DESYMMETRIZATION OF DIETHYL 3-[3′,4′-DICHLOROPHENYL] GLUTARATE

Tachykinins are biologically active neuropeptide hormones that are widely distributed throughout the nervous system and implicated in a variety of biological processes such as pain transmission, inflammation, vasodilatation, and secretion [61]. The nonpeptide NK-receptor antagonists are potentially useful in the treatment of a variety of chronic diseases including asthma, bronchospasm, arthritis, and migraine [62].

The structure–activity relationship of several nonpeptide NK1/NK2 antagonists has led to the discovery of a new class of oxime-based NK1/NK2 dual antagonists [63, 64] such as compound **31** (Figure 4.11). The biological activity of **31** resides mainly in the (*R*,*R*)-diastereomer. An enzymatic process for desymmetrizing the prochiral diethyl 3-[3′,4′-dichlorophenyl] glutarate **32** to the corresponding (*S*)-monoester **33** was developed. Commercially available enzymes were screened for the hydrolysis of **32**. Several enzymes were identified capable of hydrolyzing **32**. Out of 11 candidates with pro-*S* selectivity, lipase B from *Candida antarctica*, Chirazyme L-2, was selected for further development [63].

The reaction was carried out using immobilized lipase B at 100 g/l substrate input and 20 g/l immobilized enzyme input. A reaction yield of 97% and an ee of 99% were obtained for the desired (*S*)-monoester **33** [63]. The process was scaled up to produce 200 kg of product with an 80% overall isolated yield.

DNA family shuffling was used to create a chimeric lipase B protein with improved activity toward **32**. Three homologous lipases from *C. antarctica* ATCC 32657, *Hyphozyma* sp. CBS 648.91, and *Cryptococcus tsukubaensis* ATCC 24555 were cloned and shuffled to generate a diverse gene library. Using a high-throughput screening assay, a chimeric lipase B protein having 20-fold higher activity toward the substrate was identified. In addition, the stability characteristics of several highly active chimeric proteins were also improved as a result of family shuffling [64]. The thermostability of the lipase B from *C. antarctica* was also improved by directed evolution. Two mutants, 23G5 and 195F1, were generated with over a 20-fold increase in half-life at 70 °C compared with the wild-type enzyme [65].

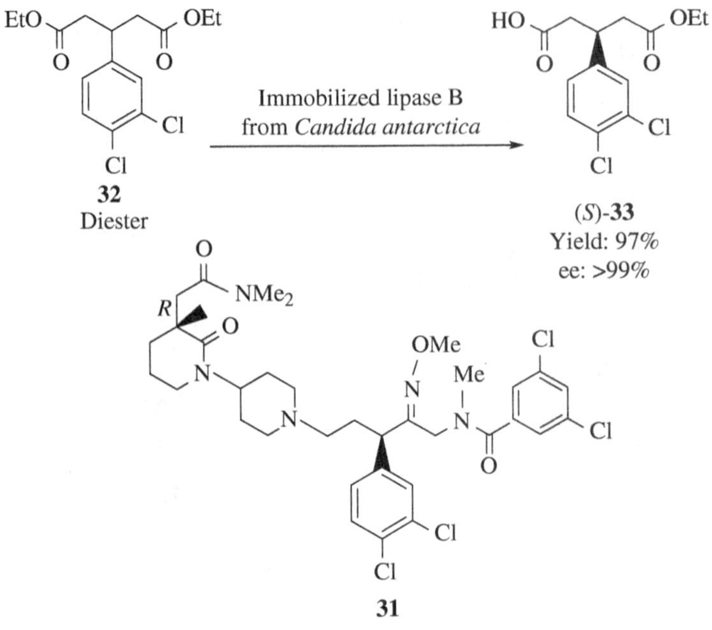

FIGURE 4.11

NK1/NK2 dual antagonists: enzymatic desymmetrization of diethyl 3-[3′,4′-dichlorophenyl] glutarate.

4.13 PREGABALIN: ENZYMATIC SYNTHESIS OF ETHYL (*S*)-3-CYANO-5-METHYLHEXANOATE

Pregabalin **34** (Figure 4.12) is a lipophilic γ-aminobutyric acid (GABA) analog that was developed for the treatment of several central nervous system (CNS) disorders including epilepsy, anxiety, and social phobia [66, 67]. In June 2007, pregabalin was approved by the US Food and Drug Administration, specifically for the treatment of fibromyalgia.

Several routes were examined in detail [68, 69]. The first-generation manufacturing resolution process and asymmetric hydrogenation process were potentially elegant manufacturing processes, but they were surpassed in terms of cost-effectiveness and environmental performance by an enzymatic route [70].

The enzymatic route was developed from an existing, relatively inexpensive racemic precursor **35** and used a Lipolase to generate an enantiopure precursor (*S*)-**36**. This strategy also required a procedure to recycle unused *R*-**37** enantiomer as well as chemistry to convert (*S*)-**36** to pregabalin (Figure 4.12). Lipolase is commercially produced by Novozymes using a submerged fermentation of a genetically modified *Aspergillus oryzae* strain. At high substrate input (>1 M), the reaction was inhibited. Adding divalent ions such as calcium and zinc to the reaction significantly suppressed the inhibition, possibly by forming a complex that remained suspended in the emulsion. The effect of calcium acetate in the reaction medium at higher concentrations of **35** (3 M or 765 g/l) gave conversion values ranging from 42 to 48% after 24 h [70].

Under optimized conditions, the enzymatic process was tested for robustness in multiple runs at a 10 kg scale. Three pilot runs at a 900 kg (1600 l reactor) scale as well as manufacturing trials at 3.5*t* (8000 l reactor) demonstrated the consistently high performance of this enzymatic reaction. A heat-promoted decarboxylation of (*S*)-**36** efficiently generated ethyl ester (*S*)-**38**, a known precursor of pregabalin **1** [70].

This new route dramatically improved process efficiency compared to the first-generation route by setting the stereocenter early in the synthesis and enabling the facile racemization and reuse of (*R*)-**37**. The chemoenzymatic process also reduced organic solvent usage resulting in a mostly aqueous process. Compared to the first-generation manufacturing process, the new process resulted in higher yields of pregabalin and reductions of waste streams.

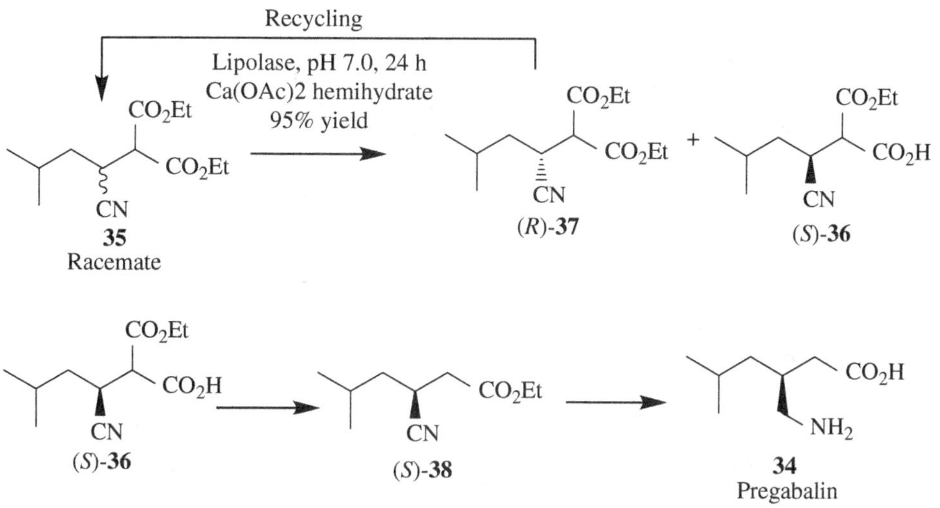

FIGURE 4.12
Pregabalin: enzymatic synthesis of ethyl (*S*)-3-cyano-5-methylhexanoate.

4.14 CHEMOKINE RECEPTOR MODULATOR: ENZYMATIC SYNTHESIS OF (1*S*,2*R*)-2-(METHOXYCARBONYL)CYCLOHEX-4-ENE-1-CARBOXYLIC ACID

The chiral monoester, (1*S*,2*R*)-2-(methoxycarbonyl)cyclohex-4-ene-1-carboxylic acid **39** (Figure 4.13) is a key intermediate for the synthesis of a potential drug candidate **40** for the modulation of chemokine receptor 2 (CCR2) activity useful in treatment of rheumatoid arthritis [71, 72]. The quinine-catalyzed alcoholysis of the anhydride **41** provided kilogram quantities of the (1*S*,2*R*)-monoester **39** with 90.8% ee initially at Bristol-Myers Squibb. To improve the enantioselectivity of **39**, alternative enzymatic processes were evaluated [73]. Screening of various enzymes was carried out to perform hydrolysis of dimethyl ester **42**. After evaluating yield and optical purity of the desired product, reaction rate, and cost of enzymes, the immobilized lipase from *C. antarctica* (Novozym 435) was selected for process development. Under optimized reaction conditions on a 50ml scale (57.2g) of dimethyl ester **42** afforded (1*S*,2*R*)-monoester **39** with a 96% yield and >99.9% ee after 24h reaction. Two reparative batches were carried out at 1.73kg input of dimethyl ester **42** to afford 3.4kg of 1*S*,2*R*-monoester **39** with a 98.1–99.8% yield, 98.3–99.2% HPLC purity, and ee of >99.9% [73].

4.15 ENZYMATIC SYNTHESIS OF (3*S*,5*R*)-3-(AMINOMETHYL)-5-METHYLOCTANOIC ACID

In recent years there has been a great deal of interest in α2δ-ligands, following the discovery of gabapentin **43** and pregabalin **34** (Figure 4.14), which bind with very high affinity to the α2δ-protein, an auxiliary subunit of voltage-gated calcium channels (VGCCs). Both gabapentin and pregabalin have been developed for the treatment of a number of conditions, such as generalized anxiety disorder, insomnia, fibromyalgia, epilepsy, neuropathic pain, anxiety, and depression [74, 75].

A group at Pfizer was interested in developing efficient syntheses to support the development of the α2δ-ligand, (3*S*,5*R*)-3-(aminomethyl)-5-methyloctanoic acid **44** (Figure 4.14), a lipophilic GABA analogue under development for the treatment of interstitial cystitis [76].

Three synthetic approaches, suitable for the large-scale manufacture of the α2δ-ligand, (3*S*,5*R*)-3-(aminomethyl)-5-methyloctanoic acid **44**, have been developed [67, 77, 78]. The selected seven-step manufacturing process was optimized and used to prepare 20kg of API. From this process it was concluded that the medicinal chemistry route, although suitable for the preparation of material to support early toxicological and clinical studies, would not be a suitable long-term manufacturing process

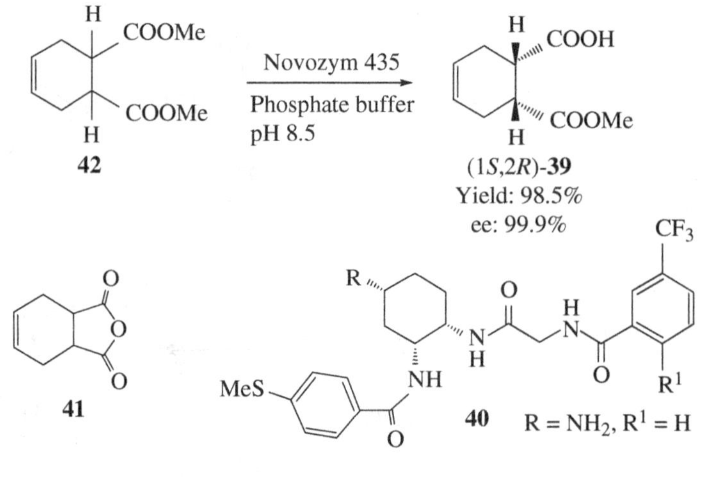

FIGURE 4.13
Chemokine receptor modulator: enzymatic synthesis of (1*S*,2*R*)-2-(methoxycarbonyl) cyclohex-4-ene-1-carboxylic acid.

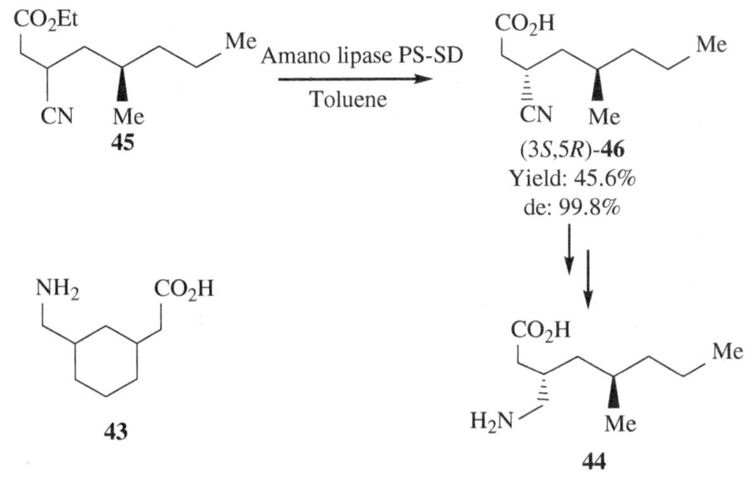

FIGURE 4.14
Enzymatic synthesis of (3*S*,5*R*)-3-(aminomethyl)-5-methyloctanoic acid.

without an extensive modification. The important step of the synthesis includes the use of 4,4,4-trimethoxybutyronitrile, as an efficient four-carbon amino acid equivalent. A highly selective kinetic resolution process was developed for diastereoselective hydrolysis of a cyanoester intermediate **45** using Amano Lipase PS-SD. Extensive process optimization of the route starting from (*R*)-2-methylpentanol led to significant improvements through telescoping to produce 1 kg of API [79].

In enzymatic diastereoselective hydrolysis process, a reactor was charged with water (486 l) and sodium bicarbonate (17.3 kg), and the mixture was stirred until a solution was formed. Lipase PS-SD (10.8 kg, <23 000 U/g) was charged, and the mixture was stirred until a solution was formed. The solution was transferred to a vessel containing substrate **45** in toluene, and the reaction was carried out at 45 °C for 48 h. From the reaction mixture, product **46** was isolated as a TBME solution (202.4 kg) containing 24.8 kg of product with a 45.6% yield [79].

4.16 ATORVASTATIN (LIPITOR): ENZYMATIC DESYMMETRIZATION OF 3-HYDROXYGLUTARONITRILE

Today, HMG-CoA reductase inhibitors (statins) have worldwide sales of approximately $20 billion, led by Atorvastatin **47** and Crestor **48** (Rosuvastatin) (Figure 4.15) The synthetic statins share the chiral 3, 5-dihydroxy acid side chain, which is essential for activity and represents the synthetic challenge for the preparation of these drugs [80].

By screening genomic libraries prepared from environmental samples collected around the globe, Diversa (now Veranium) Corporation discovered over 200 unique nitrilases that allowed mild and selective hydrolysis of the prochiral substrate 3-hydroxyglutaronitrile (3-HGN) **49** to afford (*R*)-4-cyano-3-hydroxybutyric acid **50**, a precursor to Lipitor. Through gene site saturation mutagenesis (GSSM), a mutagenesis technique that effects the combinatorial saturation of each amino acid in the protein to each of the other 19 proteinogenic amino acids, combined with a novel high-throughput mass spectroscopy assay [81–84], a number of improved variants were identified. Transformation and expression of this collection of genes in *E. coli* furnished a comprehensive library of single-site enzyme mutants, which were screened in an attempt to identify mutant enzymes with improved activity and enantioselectivity. A number of improved variants were identified, the best of which is the Ala190His mutant which yields product with 98.5% ee at 3 M substrate loading and a volumetric productivity of 619 g/l/d [45]. Using the best mutant nitrilase enzyme, an efficient process for preparing the Lipitor intermediate (*R*)-**50** was developed [81, 82].

Subsequently, Dowpharma and Chirotech Technology Ltd. developed a three-stage process that started by reacting low-cost Epichlorohydrin **51** (Figure 4.15) with

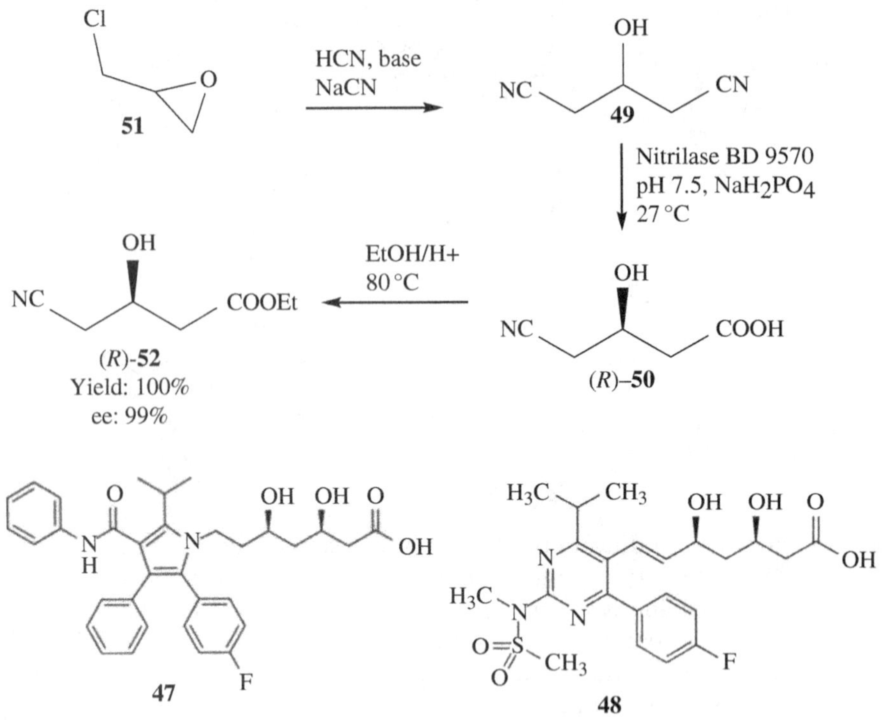

FIGURE 4.15
Atorvastatin (Lipitor): enzymatic desymmetrization of 3-hydroxy-glutaronitrile.

cyanide to give 3-HGN **49**. The second stage of the process utilizes a nitrilase-catalyzed desymmetrization of the 3-HGN **49**. The nitrilase reaction was optimized to work at 3M (330 g/l) substrate concentration, pH 7.5, and 27 °C. Under these conditions, with an enzyme loading of 6 wt%, 100% conversion, and 99% ee, product **50** was obtained in 16 h. This product was then esterified to give the target compound, ethyl (*R*)-4-cyano-3-hydroxybutyrate **52** [83].

4.17 ANTICANCER DRUGS: ENZYMATIC SYNTHESIS OF TAXANE SIDE CHAIN

Paclitaxel (Taxol® **53**, Figure 4.16), a complex, polycyclic diterpene, exhibits a unique mode of action on microtubule proteins responsible for the formation of the spindle during cell division and known to inhibit the depolymerization process of microtubulin. Paclitaxel is approved by the FDA for the treatment of ovarian cancer and metastatic breast cancers. (*R*)-**54** side chain was required for the preparation of the Paclitaxel **53** [85–88] by a semisynthetic process. The enantioselective enzymatic hydrolysis of racemic acetate *cis*-3-(acetyloxy)-4-phenyl-2-azetidinone **55** to the corresponding (*S*)-alcohol **56** and the unreacted desired (*R*)-acetate **57** was demonstrated [89] using lipase PS-30 from *Pseudomonas cepacia* (Amano Enzyme Co.) and BMS lipase (extracellular lipase derived from the fermentation of *Pseudomonas* sp. SC 13856). Reaction yields of >48% (theoretical maximum yield 50%) with ee of >99.5% were obtained for the (*R*)-acetate.

BMS lipase and lipase PS-30 were immobilized on Accurel polypropylene (PP), and the immobilized lipases were used in the hydrolytic reaction. Immobilized enzymes were reused over 10 cycles without the loss of enzyme activity, productivity, or the ee of product **57**. This process was scaled up to 250 l (2.5 kg substrate input) using immobilized lipase PS-30. From the reaction batch, *R*-acetate **57** was isolated with a 45 M% yield (theoretical maximum yield 50%) and 99.5% ee. The (*R*)-acetate was chemically converted to (*R*)-alcohol **54**. The C-13 Paclitaxel side-chain synthon (*R*)-**54** produced by the resolution process was coupled to baccatin III **58** after protection and deprotection to prepare Paclitaxel by a semisynthetic process [87, 89].

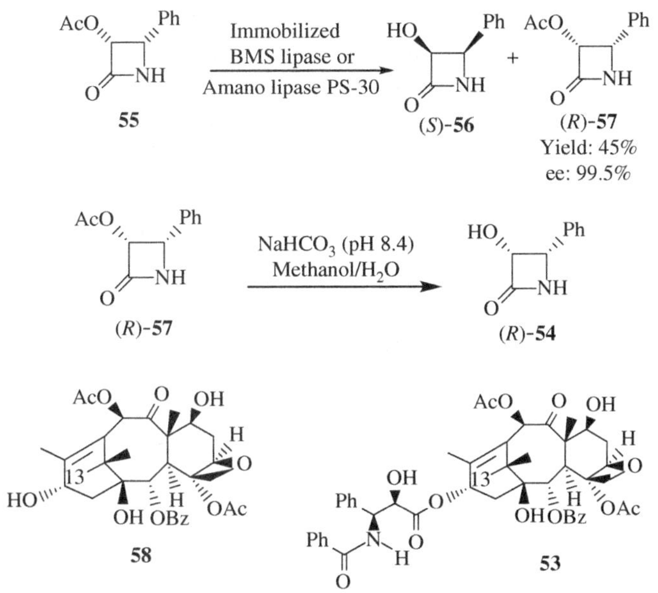

FIGURE 4.16
Anticancer drugs: enzymatic synthesis of taxane side chain.

4.18 ANTIDIABETIC AND CNS DRUGS: ENZYMATIC HYDROLYSIS OF DIMETHYL BICYCLO[2.2.1]HEPTANE-1,4-DICARBOXYLATE

Monoester 4-(methoxycarbonyl)bicyclo[2.2.1]heptane-1-carboxylic acid **59** (Figure 4.17) is a building block of many potential therapeutic candidates for inhibitors of 11-β-hydroxysteroid dehydrogenase type 1 enzyme and their use in the treatment of non-insulin-dependent type 2 diabetes, insulin resistance, obesity, lipid disorders, metabolic syndrome, and CNS disorders. It is also required for the synthesis of 5-hydroxytryptamine receptor agonists, useful for the treatment of anxiety disorders and schizophrenia [90, 91].

A biocatalytic process for the hydrolysis of dimethyl bicyclo[2.2.1]heptane-1,4-dicarboxylate **60** to the corresponding monoester 4-(methoxycarbonyl)bicyclo[2.2.1] heptane-1-carboxylic acid **59** was developed using lipases from *C. antarctica* and *Burkholderia cepacia* [92]. About 100 kg of monoester **59** was prepared with a 78% yield by hydrolysis of diester with a commercially available lipase from *B. cepacia*. A more efficient enzymatic process was developed for the hydrolysis of **60** that gave the monoester **59** with an 82% yield using significantly lower amounts of the commercially available immobilized lipase B from *C. antarctica*. The commercially available immobilized lipase B from *C. antarctica* and porcine liver esterase were also used efficiently for the hydrolysis of dimethyl cyclobut-1-ene-1,2-dicarboxylate **61** (Figure 4.17) to the corresponding monoester **62** with yields of 78% and 87%, respectively [92].

4.19 CLOPIDOGREL (PLAVIX): ENZYMATIC PREPARATION OF 2-CHLOROMANDELIC ACID ESTERS

Clopidogrel bisulfate **63** (Plavix, Figure 4.18) is an orally active inhibitor of platelet aggregation marketed as an antithrombotic agent. Clopidogrel works by helping to prevent harmful blood clots, and studies have shown that clopidogrel is more effective in blocking platelet aggregation than aspirin and ticlopidine even at a much lower dosage [93–95]. Methyl (*R*)-2-chloromandelate **65** is a key intermediate for the synthesis of clopidogrel.

An efficient process for resolving methyl 2-chloromandelate (*R*,*S*)-**64** was developed using a lipase-mediated transesterification. Among 11 hydrolytic enzymes examined, *C. antarctica* lipase A (CAL-A) showed the highest enantioselectivity and

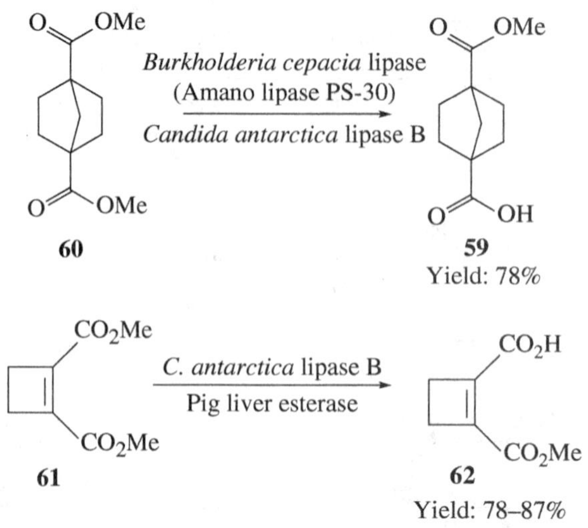

FIGURE 4.17

Antidiabetic and CNS drugs: enzymatic hydrolysis of dimethyl bicyclo[2.2.1]heptane-1, 4-dicarboxylate.

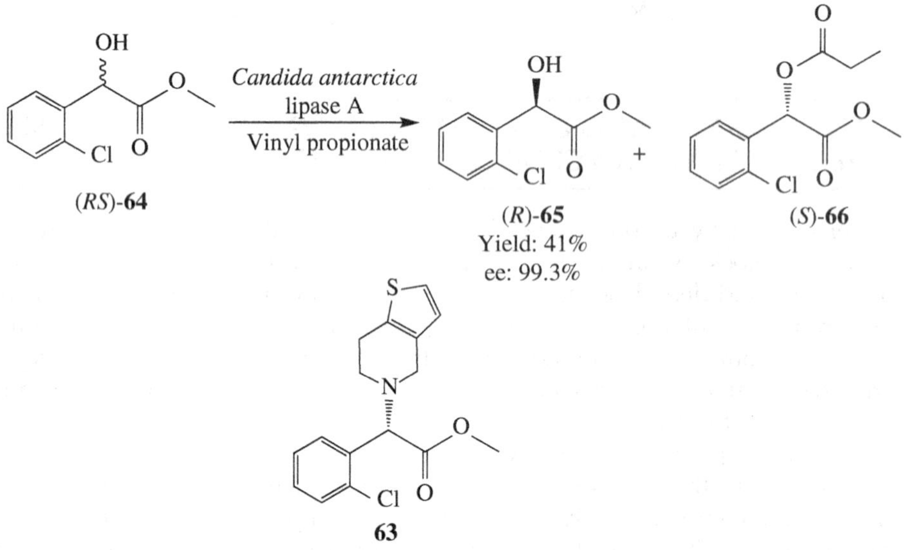

FIGURE 4.18

Clopidogrel (Plavix): enzymatic preparation of 2-chloromandelic acid esters.

reaction rate toward methyl (*S*)-2-chloromandelate, yielding (*S*)-**66** and methyl (*R*)-2-chloromandelate **65**. Moreover, (*R*)-**65** was obtained in an enantiomerically pure form (99% ee) and with a 41% yield through the lipase-mediated resolution under solvent-free conditions [96]. On a preparative-scale reaction, immobilized CAL-A (4.8 g) was added to a solution of (*R*,*S*)-**64** (500 g) and vinyl propionate (500 g). This gave (*R*)-**65** as a colorless oil (205 g, 41% yield); 99.3% ee. CAL-A maintained its catalytic activity during 13 cycles of repeated use without a significant decrease in enantioselectivity, indicating that the method is economical and easy to scale up for commercial production of methyl (*R*)-2-chloromandelate [96].

4.20 ANTIVIRAL DRUG: REGIOSELECTIVE ENZYMATIC ACYLATION OF RIBAVIRIN

Ribavirin **67** (Figure 4.19) is an antiviral agent used in combination with α-2β interferon to treat hepatitis C [97, 98]. Although this therapy is effective against the hepatitis C virus, it has several side effects [99]. To improve the pharmacokinetic profile and reduce side effects, a ribavirin prodrug was considered for development by Schering-Plough. The alanine ester of ribavirin **68** showed improved bioavailability and reduced side effects in preclinical evaluation. The synthesis of **68** required the

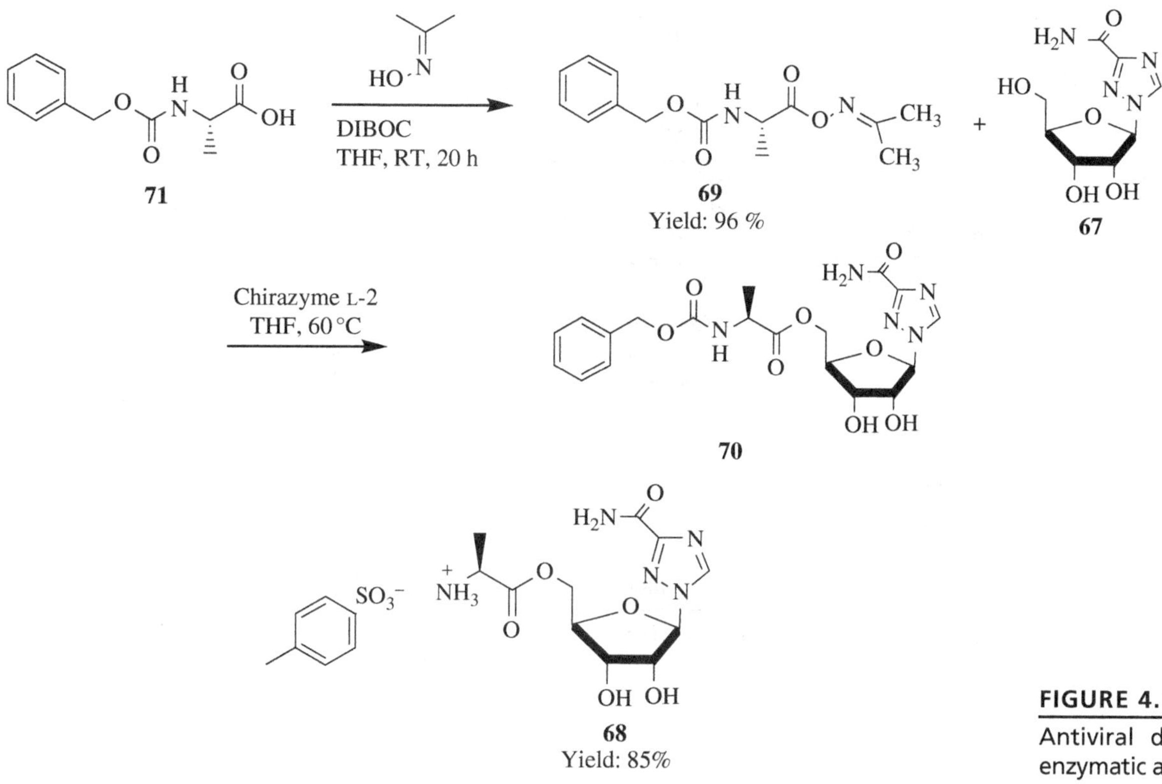

FIGURE 4.19
Antiviral drug: regioselective enzymatic acylation of ribavirin.

acylation of unprotected ribavirin. The chemical acylation gave a mixture of mono-, di-, and triacylated products. An enzymatic process was developed for the regioselective acylation of ribavirin with the oxime ester of L-carbobenzyloxy-alanine **69** to give the desired **70** using *C. antarctica* lipase B (CAL-B) (Novozym 435 or Chirazyme L-2). Chemical deprotection of **70** gave **68**. On a preparative scale, the coupling of **71** with acetone oxime in the presence of di-*t*-butyl dicarbonate in tetrahydrofuran (THF) was carried out giving **69** with a >96% yield. At the end of the reaction, the reaction mixture was diluted threefold with THF, ribavirin was added, and the acylation was initiated by the addition of the Novozym 435. After 24h at 60°C, product **68** was isolated with a 85% yield [100].

4.21 ANTICHOLESTEROL DRUG: ENZYMATIC ACYLATION OF ALCOHOL

The (5*R*,3*R*)-alcohol **72**, [4-[4α,6ß(*E*)]]-6-[4, 4-bis(4-fluorophenyl)-3-(1-methyl-1*H*-tetrazol-5-yl)-1,3-butadienyl]tetrahydro-4-hydroxy-2*H*-pyren-2-one (Figure 4.20) is a potential new anticholesterol drug, which acts by inhibition of HMG-CoA reductase [101].

Using an enzymatic diastereoselective acetylation process, the (5*R*,3*R*)-alcohol **72** was prepared from racemic **73** [102]. Lipase PS-30 efficiently catalyzed the acetylation of racemic **73** (50g/l) to yield the (5*R*,3*S*)-acetate **74** and the unreacted, desired (5*R*,3*R*)-alcohol **72**. A reaction yield of 46M% and an ee of 98% were obtained for (5*R*,3*R*)-alcohol **72** when the reaction was conducted in methyl ethyl ketone in the presence of isopropenyl acetate as an acyl donor. The enzymatic process was scaled up to a 640l preparative batch using immobilized lipase PS-30 (lipase PS-30 immobilized on Accurel PP). Product (5*R*,3*R*)-alcohol **72** was isolated with a 35M% overall

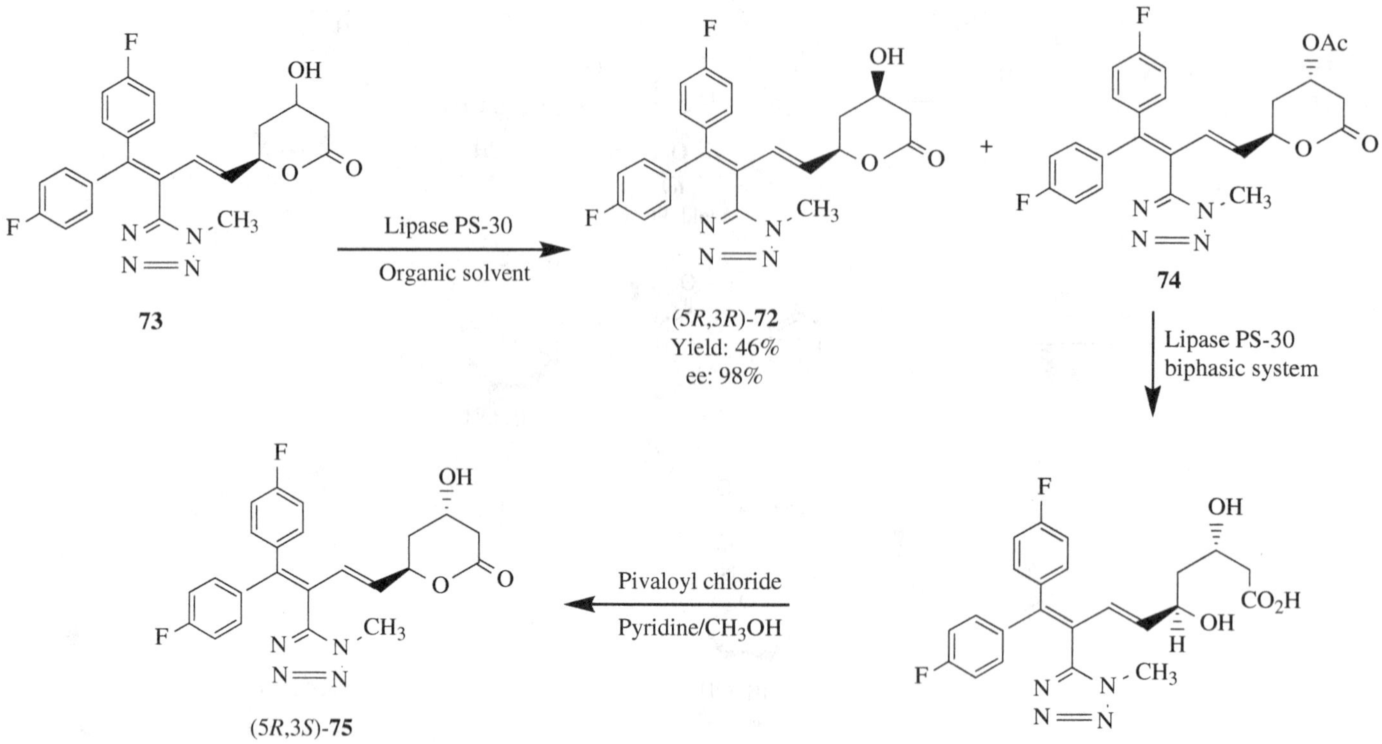

FIGURE 4.20
Anticholesterol drug: enzymatic acylation of alcohol.

yield with 98% ee and 99.5% chemical purity. The (5*R*,3*S*)-acetate **74** produced by this process was hydrolyzed by lipase PS-30 in a biphasic system to prepare the corresponding (5*R*,3*S*)-alcohol **75** [102]. Thus both enantiomers of alcohol **73** were obtained.

4.22 SAXAGLIPTIN: ENZYMATIC SYNTHESIS OF (5*S*)-4,5-DIHYDRO-1*H*-PYRROLE-1,5 DICARBOXYLIC ACID, 1-(1,1-DIMETHYLETHYL)-5-ETHYL ESTER

The synthesis of DPP-4 inhibitor, Saxagliptin **1** (Figure 4.21) required key intermediate (5*S*)-5-aminocarbonyl-4,5-dihydro-1*H*-pyrrole-1-carboxylic acid,1-(1,1-dimethyl ethyl)-ester **76** [22–24]. Direct chemical ammonolysis resulted in unacceptable levels of amide racemization and side-product formation. Milder, two-step hydrolysis–condensation protocols using coupling agents such as 4-(4,6-dimethoxy-1,3,5-triazin-2-yl)-4-methyl-morpholinium chloride (DMT-MM) were compromised by reduced overall yields [103].

An enzymatic process was developed using the CAL-B-mediated ammonolysis of (5*S*)-4,5-dihydro-1*H*-pyrrole-1,5-dicarboxylic acid, 1-(1,1-dimethylethyl)-5-ethyl ester **77** with ammonium carbamate to furnish **76** without racemization and with very low levels of side-product formation [104].

Initial enzymatic process provided amide with yields of 69%, together with 21% of side products (by HPLC). The inclusion of various additives was investigated to solve potential inhibitory phenomena, shifting the equilibrium toward amide synthesis and reducing side-product formation. Drying agents such as calcium chloride gave a significant improvement (79% amide and 13% side-products). Calcium chloride is known to complex alcohols as well as act as a desiccant, and its presumed binding of ethanol released during the course of amide formation served to mitigate any deleterious effects of this alcohol on CAL-B catalysis. A dramatic increase in amide yield (95%) was achieved by including Ascarite, at 200 g/l in the reaction headspace. Ascarite likely

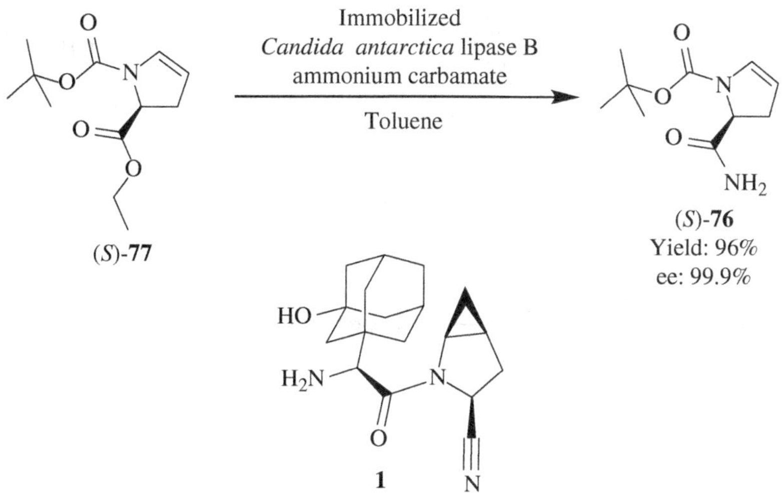

FIGURE 4.21
Saxagliptin: enzymatic synthesis of (5*S*)-4,5-dihydro-1*H*-pyrrole-1, 5-dicarboxylic acid, 1-(1,1-dimethylethyl)-5-ethyl ester.

adsorbed and thereby removed the CO_2 byproduct from decomposition of ammonium carbamate. A further increase in yield of 96–98% was attained via the combined use of 100 g/l of calcium chloride and 200 g/l of Ascarite. A preparative-scale reaction with the process ester feed was used. Ester **77** (220 g/l) was reacted with 90 g/l of ammonium carbamate, 33 g/l of CAL-B, 110 g/l calcium chloride, and 216 g/l of Ascarite (in the headspace). Complete conversion of ester was achieved after 3 days at 50 °C, with the formation of 96% (182 g/l) of amide **76** and 4% of side products, and after workup, 98% potency amide of >99.9% ee was isolated with an 81% yield [104].

4.23 MONTELUKAST: SYNTHESIS OF INTERMEDIATE FOR LTD4 ANTAGONISTS

After the slow-reacting substance of anaphylaxis (SRS-A) and its relation to the leukotrienes (LTC4, LTD4, and LTE4) and asthma were discovered, the search for leukotriene antagonists, useful for the treatment of asthma, has been intensive. Merck identified montelukast **78** (Figure 4.22) as an effective drug for the treatment of asthma [105, 106]. The synthetic route for producing montelukast required the stereoselective reduction of ketone **79** to the (*S*)-alcohol **80** [105–108]. The original strategy for reducing **80** required stoichiometric amounts of the chiral reducing agent (−)-chlorodiisopino camphenylborane [(−)-DIP-chloride]. While (−)-DIP-chloride is selective and avoids the side reactions, it is also corrosive and moisture sensitive, causing burns if it is allowed to contact the skin. The reaction must be carried out at −20 to −25 °C to achieve the best stereoselectivity. The quench and extractive workup generate large volumes of waste solvent for disposal [107].

An enzymatic alternative for reducing ketone **79** was created by Codexis. A ketoreductase (KRED) was developed by directed evolution using high-throughput screens that mimicked the actual process conditions. Beneficial mutations obtained during each round were recombined and new mutations were introduced, guided by ProSAR. The productivity of the final enzyme was improved 2000-fold and stability was also substantially increased [109, 110].

The enzymatic process was carried out with a slurry of (*E*)-methyl 2-(3-(3-(2-(7-chloroquinolin-2-yl)vinyl)phenyl)-3-oxopropyl)benzoate **79** (230 kg) in a mixture of isopropyl alcohol (5 vol), toluene (1 vol), and triethanolamine buffer of pH 8.0 (3 vol). KRED CDX-026 (9.2 kg) and the cofactor NADP-Na (0.23 kg) were added to the reaction mixture, and the reaction was carried out at 40–45 °C under stirring for 40–45 h. From the reaction mixture, the crude (*S*)-2-[3-2[7-chloro-2-quinolinyl]ethenyl]phenyl-3-hydroxypropyl] benzoic acid methyl ester **80** was obtained, which was purified by recrystallization to obtain pure product **80** as monohydrate 3 (233 kg, 97.2% yield, >99.9% ee) [109, 110].

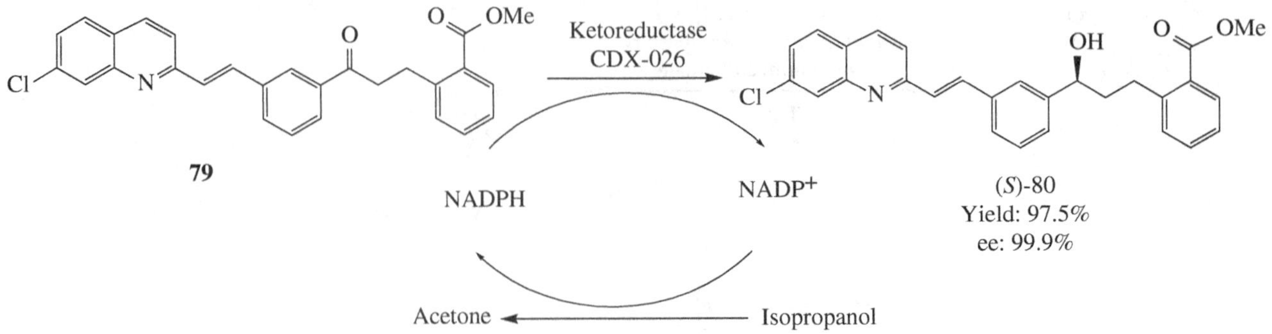

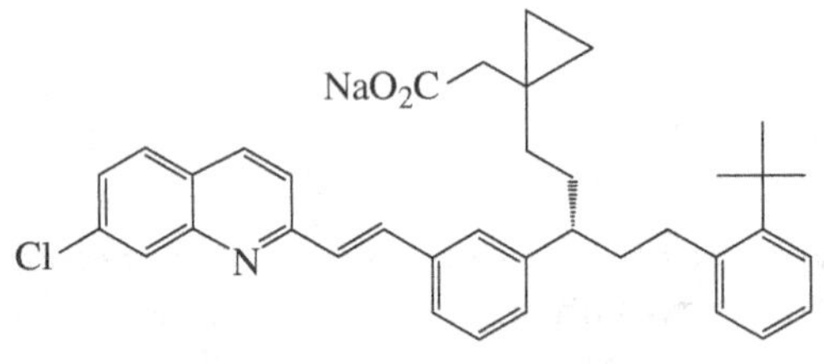

78

FIGURE 4.22
Montelukast: synthesis of intermediate for LTD 4 antagonists.

4.24 ATAZANAVIR: ENZYMATIC SYNTHESIS OF (1*S*,2*R*)-[3-CHLORO-2-HYDROXY-1 (PHENYLMETHYL) PROPYL]-CARBAMIC ACID,1,1-DIMETHYL-ETHYL ESTER

Atazanavir **21** (Figure 4.23) is an acyclic aza-peptidomimetic, a potent HIV protease inhibitor [44, 45] approved by the FDA for the treatment of acquired immunodeficiency syndrome (AIDS). An enzymatic process was developed to prepare (1*S*,2*R*)-[3-chloro-2-hydroxy-1-(phenylmethyl) propyl]carbamic acid, 1,1-dimethylethyl ester **81**, a key chiral intermediate in the synthesis of atazanavir. The diastereoselective reduction of (1*S*)-[3-chloro-2-oxo-1-(phenylmethyl)propyl] carbamic acid, 1,1-dimethylethyl ester **82** was carried out using *Rhodococcus erythropolis* to afford **81** with a >90% yield and with a diastereomeric purity of >98% and an ee of 99.4% [111].

A single-stage fermentation–bioreduction process using cells of *R. erythropolis* SC 13845 gave **81** with a 95% yield with a diastereomeric purity of 98.2% and an ee of 99.4% at a substrate input of 10 g/l. The reduction process was further improved by generating mutants and selecting the desired mutant for converting **82** to (1*S*,2*R*)-**81** at substrate input of 60 g/l. Subsequently, (1*S*,2*R*)-**81** was converted to epoxide **83** and used in the synthesis of atazanavir [111, 112]. Chemical reduction of chloroketone **82** using $NaBH_4$ gave the undesired chlorohydrin diastereomer [112].

Recently, Codexis developed a KRED, which catalyzed the reduction of **82** to (1*S*,2*R*)-**81**. Wild-type recombinant KRED gave only 5% conversion and 80% enantiopurity after 24 h at substrate input of 20 g/l and enzyme input of 5 g/l. The enzyme was evolved based on a ProSAR-generated library and screening of the entire library of mutants. A highly stereoselective and diastereoselective reduction process was developed with isopropanol in water as the solvent and $NADP^+$ as the cofactor by a custom-evolved KRED. The evolved enzyme operated at 200 g/l substrate, 1 g/l enzyme, and 0.01 g/l NADP inputs, with 99.9% conversion and >99% enantiomeric purity in 24 h reaction time [113].

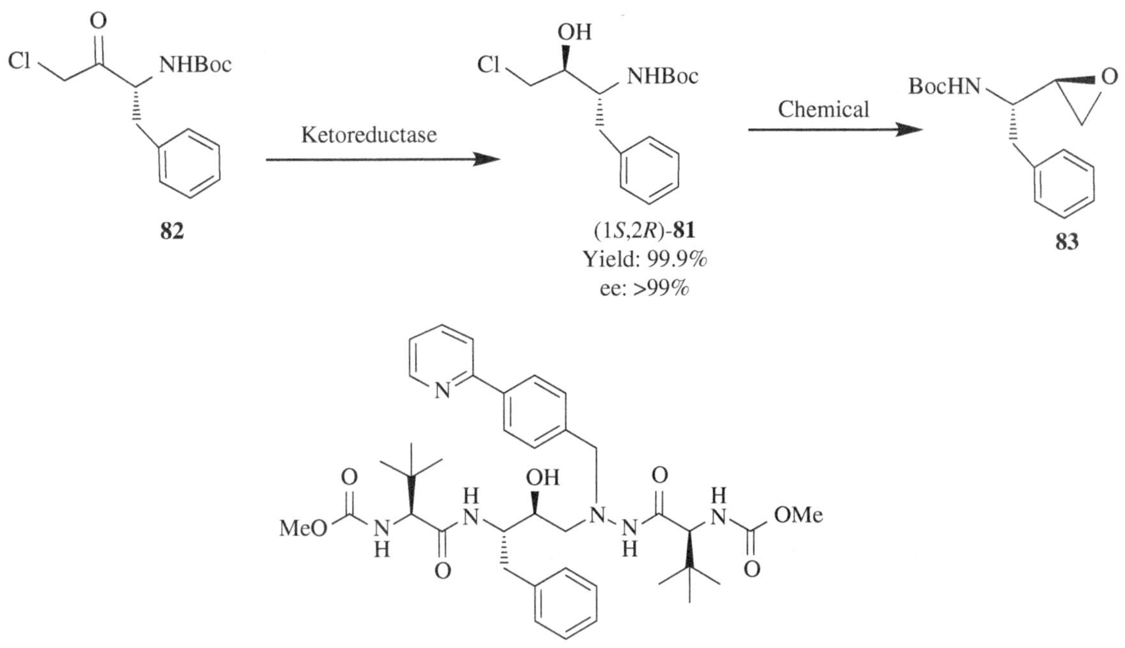

FIGURE 4.23
Atazanavir: enzymatic synthesis of (1*S*,2*R*)-[3-chloro-2-hydroxy-1(phenylmethyl) propyl]-carbamic acid,1,1-dimethylethyl ester.

4.25 ATORVASTATIN: ENZYMATIC SYNTHESIS OF (*R*)-4-CYANO-3-HYDROXYBUTYRATE

An enzymatic process for the preparation of ethyl (*R*)-4-cyano-3-hydroxybutyric acid **84** (Figure 4.24), a key intermediate for the synthesis of Atorvastatin **47** [114] was developed by Codexis [115]. The synthesis of ethyl (*S*)-4-chloro-3-hydroxybutyric acid **85** was carried out by KRED-catalyzed conversion of 4-chloro-3-ketobutyric acid derivative **86** [116]. The genes encoding halohydrin dehydrogenase (HHDH), KRED, GDH, and FDH were separately cloned into *E. coli* BL21. Through several generations of DNA shuffling, GDH activity was improved by a factor of 13 and KRED activity by a factor of 7. The enantioselectivity of the improved KRED remained >99.5%. With the improved enzymes, the reaction was completed in 8 h with substrate input of 160 g/l, KRED loading to 0.57 g/l, and GDH loading to 0.38 g/l.

Similarly after many rounds of DNA shuffling, with screening in the presence of higher concentrations of product, product inhibition was overcome and the HHDH activity was increased >2500-fold compared to the wild-type enzyme.

The biocatalytic reaction was carried out to produce ethyl (*S*)-4-chloro-3-hydroxybutyrate **85**; the reaction mixture contained 570 ml of 100 mM triethanolamine, D-glucose (298 g), KRED (854 mg), and GDH (578 mg) at pH 7.0 and 25 °C. Then, Na-NADP (98 mg) was added followed by butyl acetate (370 ml) and substrate **86** (240 g). The reaction was completed in 8 h to obtain 232.61 g (96%) of **85** with ee of >99.5%. Subsequently for conversion of ethyl (*S*)-4-chloro-3-hydroxybutyrate **85** to ethyl (*R*)-4-cyano-3-hydroxybutyrate **84** in a 1 l jacketed airtight reactor charged with 400 ml 500 mM NaCN. HHDH was charged as an aqueous solution (1.05 g) and the reaction mixture was heated to 40 °C at pH 7.3 and 70 g of **85** was added with a syringe over 10 min. Product **84** was recovered 61.6 g (93%) with chemical purity of 99.5% and >99.5% ee after 50 h reaction time [115].

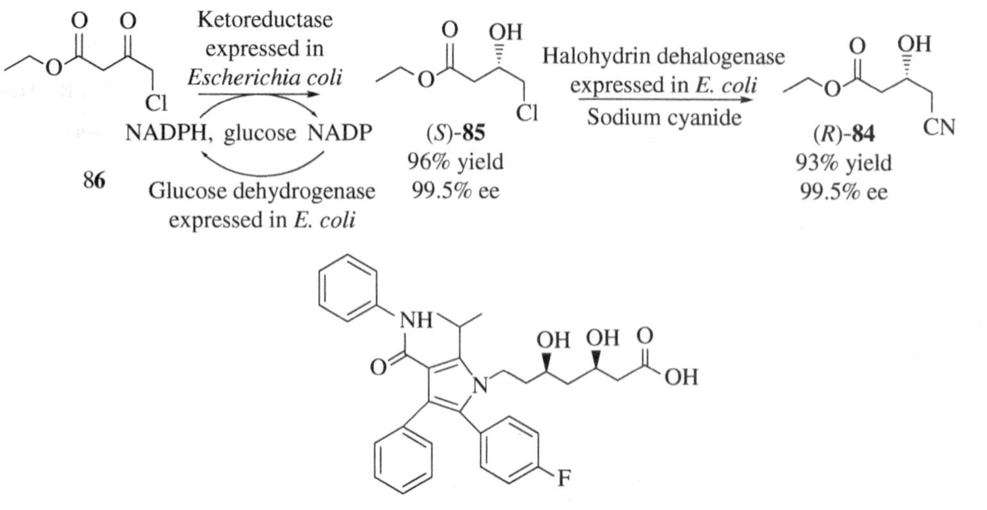

47

FIGURE 4.24
Atorvastatin: enzymatic synthesis of (*R*)-4-cyano-3-hydroxybutyrate.

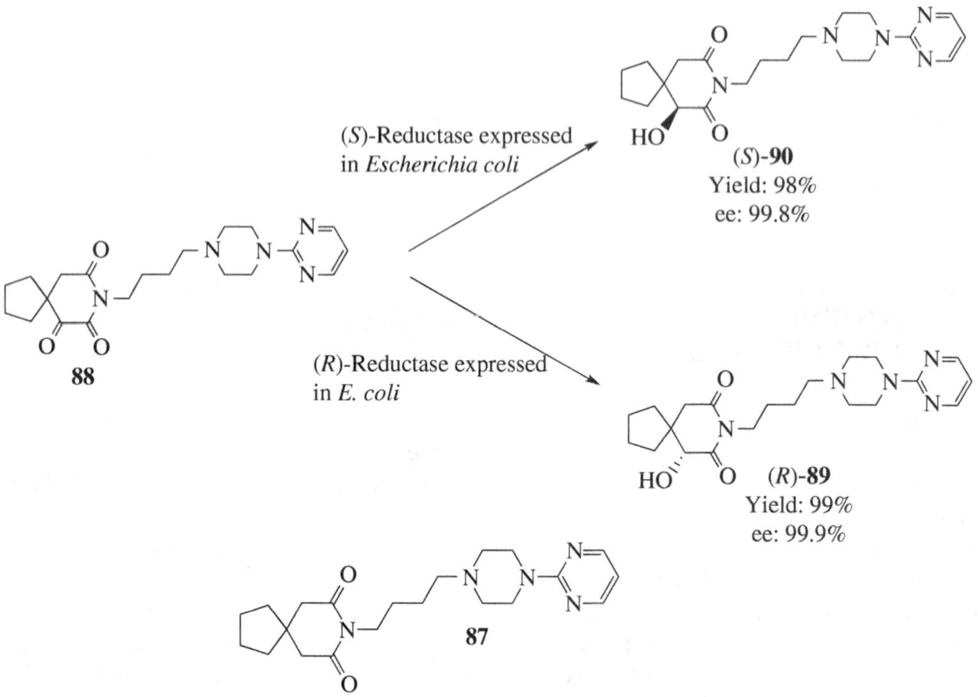

FIGURE 4.25
Antianxiety drug: enzymatic synthesis of 6-hydroxybuspirone.

4.26 ANTIANXIETY DRUG: ENZYMATIC SYNTHESIS OF 6-HYDROXYBUSPIRONE

Buspirone (BuSpar®, **87**, Figure 4.25) binds to the serotonin 5HT1A receptor and is meant for the treatment of anxiety and depression [117–119]. Buspirone is extensively converted to various hydroxylated metabolites a few hours after dosing [120]. A major metabolite, 6-hydroxybuspirone was present at much higher concentrations in human blood than buspirone itself. For the development of 6-hydroxybuspirone as a potential antianxiety drug, the preparation and evaluation of the two enantiomers were of interest. An enantioselective enzymatic process was developed for the reduction of 6-oxobuspirone **88** to (*R*)-6-hydroxybuspirone **89** or (*S*)-6-hydroxybuspirone **90**. About 150 microbial cultures were screened for the reduction of **88**. *Rhizopus stolonifer* SC 13898, *Neurospora crassa* SC 13816, *Mucor racemosus* SC 16198, and *Pseudomonas putida* SC 13817 gave >50% reaction yields and >95% ee of (*S*)-6-hydroxybuspirone. The yeast strains *Hansenula polymorpha* SC 13845 and *Candida maltosa* SC 16112 gave (*R*)-6-hydroxybuspirone with a >60% reaction yield and >97% ee [121]. The NADPH-dependent (*R*)-reductase (RHBR) from *H. polymorpha* SC 13845 was purified to homogeneity, its amino acid sequences were determined, and the corresponding gene was expressed in *E. coli*. To regenerate the NADPH required for reduction, the glucose-6-phosphate dehydrogenase gene from *Saccharomyces cerevisiae* was coexpressed in the same *E. coli* strain. Recombinant *E. coli* coexpressing both enzymes catalyzed the conversion of 6-ketobuspirone to (*R*)-6-hydroxybuspirone **89** with a 99% yield and 99.9% ee at 50 g/l substrate input [122].

(*S*)-Reductase (SHBR) from *P. putida* SC 16269 was also purified to homogeneity, its amino acid sequences were determined, and the corresponding gene was expressed in *E. coli*. To regenerate the cofactor NADH required for reduction, the NAD^+-dependent FDH gene from *P. pastoris* was coexpressed in the same *E. coli* strain. Recombinant *E. coli* coexpressing both enzymes was used to catalyze the conversion of 6-ketobuspirone to (*S*)-6-hydroxybuspirone **90**, in >98% yield and >99.8% ee at 50 g/l substrate input [122].

4.27 PROTEASE INHIBITOR: ENZYMATIC SYNTHESIS OF (*R*)-3-(4-FLUOROPHENYL)-2-HYDROXY PROPIONIC ACID

Human rhinoviruses (HRVs) are the predominant cause of the common cold; however, they can also be associated with more serious illnesses, specifically in individuals with underlying respiratory disorders. Rupintrivir (formerly AG7088) is an irreversible inhibitor of 3C protease that was discovered by using structure-based drug design and was formulated for intranasal delivery in human clinical trials. (*R*)-3-(4-fluorophenyl)-2-hydroxy propionic acid **91** (Figure 4.26) is a key chiral building block for the synthesis of Rupintrivir, a rhinovirus protease inhibitor **92** [123, 124]. An enzymatic reduction process was developed using a membrane reactor for the conversion of keto acid salt **93** to the corresponding *R*-hydroxy acid **91** [125]. D-LDH from *Leuconostoc mesenteroides* (Sigma-Aldrich) and FDH from *C. boidinii* (Julich Fine Chemicals) were used for reduction process. The cofactor itself is oxidized to NAD^+ in the process, which is regenerated by an FDH. To scale up and make the process economically feasible, both D-LDH and FDH were recycled. A continuous membrane

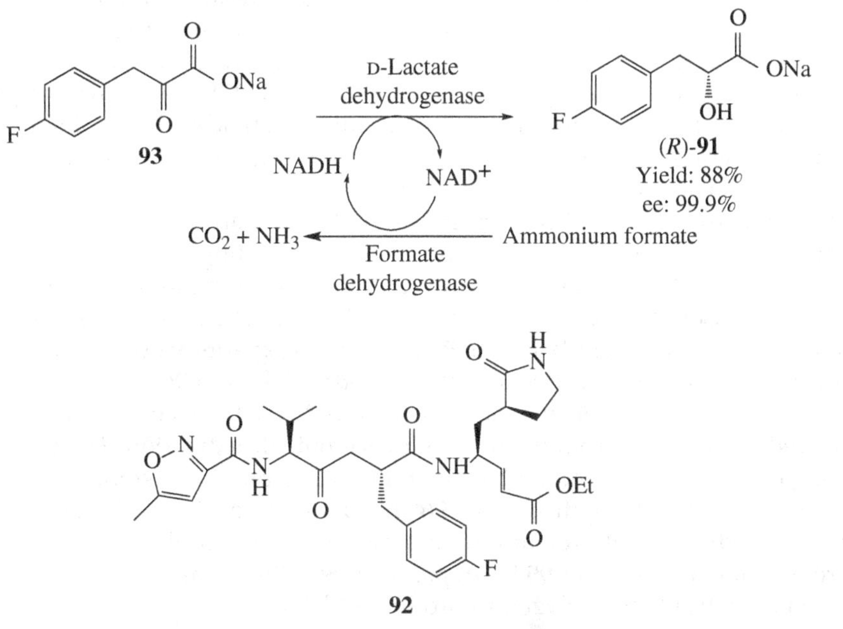

FIGURE 4.26

Protease inhibitor: enzymatic synthesis of (*R*)-3-(4-fluorophenyl)-2-hydroxy propionic acid.

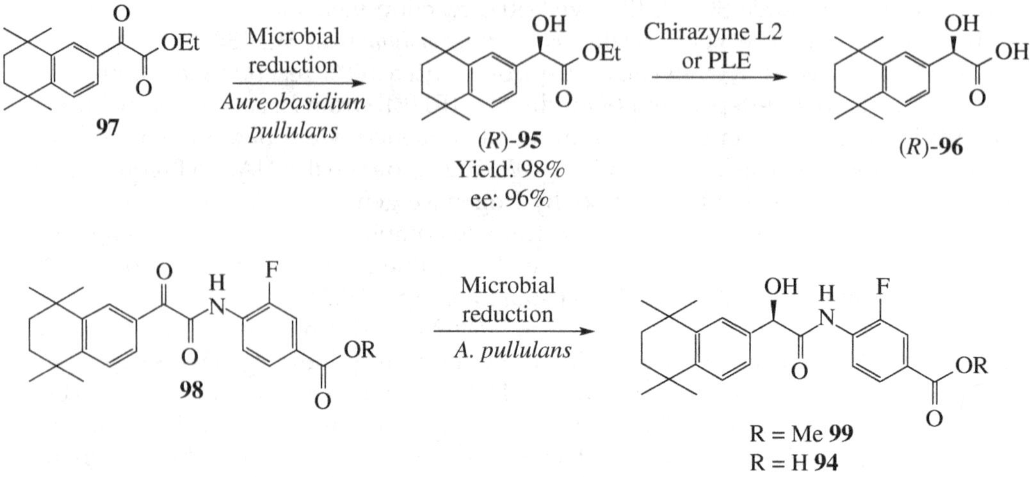

FIGURE 4.27

Dermatological and anticancer drugs: enzymatic synthesis of 2-(*R*)-hydroxy-2-(1′,2′,3′,4′-tetrahydro-1′,1′,4′,4′-tetramethyl-6′-naphthalenyl)acetate.

reactor equipped with an ultrafiltration membrane unit (10000MW cut off in Millipore Pellicon Module) allowed the recycling of both D-LDH and FDH. The reaction was carried out in a 22l reactor equipped with agitator and gas diffuser. The reaction mixture contained EDTA (3.35g), mercaptoethanol (1.41g), ammonium formate (908g), sterile water (18.0l), and keto acid salt **93** (800g). At the end of the reaction, 972g of product **91** was isolated with a yield of 88% and UV purity >90% with ee of >99.9%. A total of 14.5kg of **91** was prepared with a productivity of about 560g/l/day with overall 72% yields [125].

4.28 DERMATOLOGICAL AND ANTICANCER DRUGS: ENZYMATIC SYNTHESIS OF 2-(*R*)-HYDROXY-2-(1′,2′,3′,4′-TETRAHYDRO-1′,1′,4′,4′-TETRAMETHYL-6′-NAPHTHALENYL) ACETATE

Retinoic acid and its natural and synthetic analogs (retinoids) have been shown to affect cellular growth and differentiation and are promising drugs for the treatment of cancers [126, 127]. A few retinoids are already in clinical use for the treatment of dermatological diseases such as acne and psoriasis [127]. (*R*)-3-Fluoro-4-[[hydroxy-(5,6,7,8-tetrahydro-5,5,8,8-tetramethyl-2-naphthalenyl)-acetyl]amino]benzoic acid **94** (Figure 4.27) is a retinoic acid receptor gamma-specific agonist potentially useful as a dermatological and anticancer drug [128].

(*R*)-Hydroxy-2-(1′,2′,3′,4′-tetrahydro-1′,1′,4′,4′-tetramethyl-6′-naphthalenyl) acetate, ethyl ester **95** (Figure 4.27) and the corresponding acid **96** were prepared as intermediates in the synthesis of the retinoic acid receptor gamma-specific agonist **94** [129]. Enantioselective reduction of ethyl 2-oxo-2-(1′,2′,3′,4′-tetrahydro-1′,1′,4′,4′-tetramethyl-6-naphthalenyl) acetate **97** to (*R*)-hydroxyester **95** was carried out using *Aureobasidium pullulans* SC 13849 to afford a 98% yield and 96% ee. At the end of the reaction, hydroxyester **95** was adsorbed onto XAD-16 resin and, after filtration, recovered in 94% yield from the resin with acetonitrile extraction. The recovered (*R*)-hydroxyester **95** was treated with Chirazyme L-2 or pig-liver esterase to convert it to the corresponding (*R*)-hydroxyacid **96** in quantitative yield. The enantioselective microbial reduction of ketoamide methyl ester **98** to the corresponding (*R*)-hydroxyamide methyl ester **99** by *A. pullulans* SC 13849 was also demonstrated, and it was enzymatically hydrolyzed to product **94** [129].

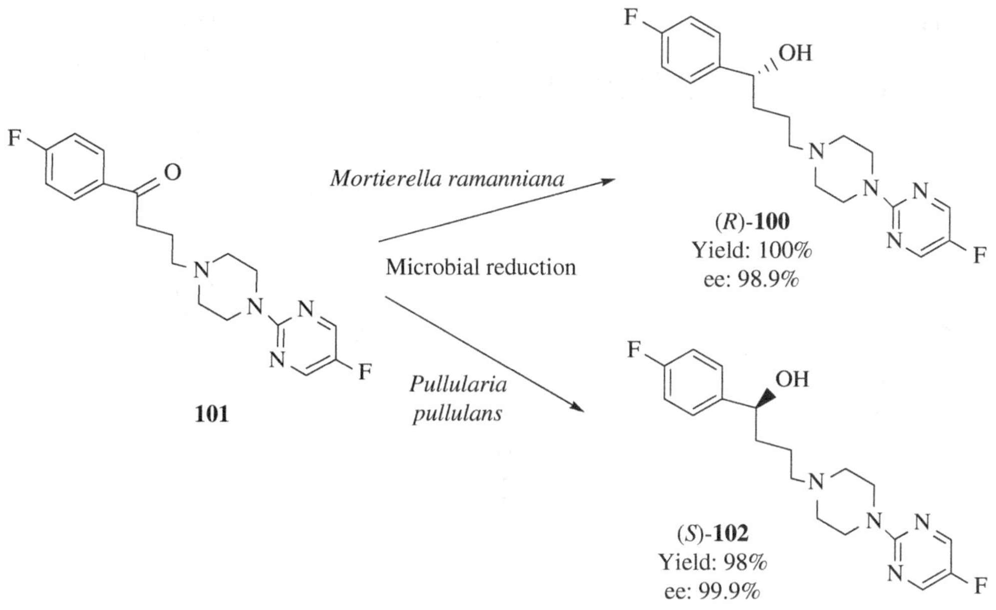

FIGURE 4.28
Antipsychotic drug: enzymatic reduction of 1-(4-fluorophenyl) 4-[4-(5-fluoro-2-pyrimidinyl) 1-piperazinyl]-1-butanone.

4.29 ANTIPSYCHOTIC DRUG: ENZYMATIC REDUCTION OF 1-(4-FLUOROPHENYL)4-[4-(5-FLUORO-2-PYRIMIDINYL) 1-PIPERAZINYL]-1-BUTANONE

The sigma receptor system in the brain and endocrine tissue has been the target for the development of a new class of antipsychotic drugs [55, 130]. Compound (*R*)-**100** (Figure 4.28) is a sigma ligand and has a high affinity for sigma binding sites and antipsychotic efficacy. An enantioselective microbial reduction process was developed for the conversion of ketone **101** to both enantiomers of alcohols [131]. Among microorganisms screened for reduction of 1-(4-fluorophenyl)4-[4-(5-fluoro-2-pyrimidinyl)1-piperazinyl]-1butanone **101**, *Mortierella ramanniana* ATCC 38191 was identified as predominantly reducing the compound **101** to (*R*)-**100**, while *Pullularia pullulans* ATCC 16623 was identified as predominantly reducing compound **101** to (*S*)-**102** with 98% yield and 99.9% ee. A single-stage fermentation/biotransformation process was developed. Cells of *M. ramanniana* were grown in a 20 l fermentor, and after a 40 h growth period, the biotransformation process was initiated by the addition of 40 g ketone **101** and 400 g glucose. The biotransformation process was completed in 24 h with a reaction yield of 100% and an ee of 98.9% for (*R*)-**100**. At the end of the biotransformation process, cells were removed by filtration and product was recovered from the filtrate with an overall yield of 80% [131].

4.30 CHOLESTEROL-LOWERING AGENTS: ENZYMATIC SYNTHESIS OF (3*S*,5*R*)-DIHYDROXY-6-(BENZYLOXY) HEXANOIC ACID, ETHYL ESTER

Compound (3*R*,5*S*)-**103** (Figure 4.29) is a key chiral intermediate for the chemical synthesis of compound **104**, Atorvastatin **47**, and Rosuvastatin **48**, all being anticholesterol drugs that act by inhibiting HMG-CoA reductase [114, 132].

An enzymatic process was developed for the reduction of a diketone, 3,5-dioxo-6-(benzyloxy) hexanoic acid, ethyl ester **105** to (3*R*,5*S*)-dihydroxy-6-(benzyloxy) hexanoic acid, ethyl ester **103** (Figure 4.29) by cells of *Acinetobacter calcoaceticus* SC 13876 [133, 134]. Both the *syn*-4 and *anti*-8 dihydroxy esters were formed in the ratio of about 76:24 after 24 h at 10 g/l of substrate **105** input. There was no significant peak due to a

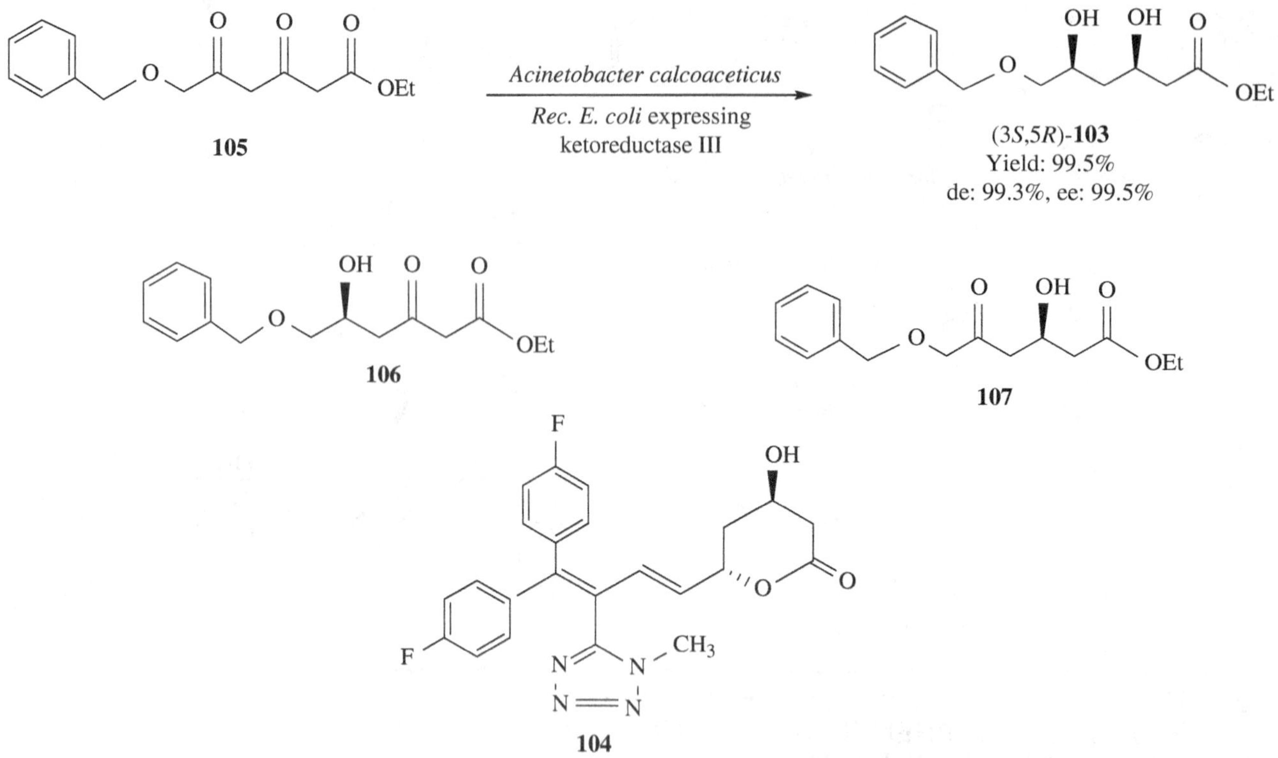

FIGURE 4.29
Cholesterol-lowering agents: enzymatic synthesis of (3*S*,5*R*)-dihydroxy-6-(benzyloxy)hexanoic acid, ethyl ester.

monohydroxy ester. Chiral HPLC determined that the desired *syn*-(3*R*,5*S*)-**103** was the major product with 99.4% ee. Almost complete (>95%) conversion of the diketoester **105** to dihydroxy ester **103** was seen in 24h [133, 134].

A mixture of ethyl 3-keto-5-hydroxy **106** (major) and 5-keto-3-hydroxy **107** (minor) was obtained from partial microbial reduction of **104**. These two mixtures were subjected to microbial reduction by *Acinetobacter* sp. SC 13874 cells for 6h. The results indicated that the second reduction of the monohydroxy compound by SC 13874 cells was quite enantioselective. Reduction of the 3-keto-5-hydroxy **106** provided predominantly the (3*R*)-hydroxy, while reduction of the 3-hydroxy-5-keto ester **107** provided predominantly the (5*S*)-hydroxy compound [134].

Three different KREDs were purified to homogeneity from cell extracts of *A. calcoaceticus* SC 13876, and their biochemical properties were compared. Reductase I only catalyzes the reduction of diketoester **105** to its monohydroxy products whereas reductase II catalyzes the formation of dihydroxy products from monohydroxy substrates. A third reductase (reductase III) was identified, which catalyzes the reduction of diketoester **105** directly to *syn*-(3*R*,5*S*)-dihydroxy ester **103** [134]. Reductase III was cloned and expressed in *E. coli* and *rec E. coli* reduced the diketoester **105** to *syn*-(3*R*,5*S*)-dihydroxy ester **103** with a 99.3% yield, 100% ee, and 99.8% de [116].

4.31 ANTIMIGRAINE DRUGS: ENZYMATIC SYNTHESIS OF (*R*)-2-AMINO-3-(7-METHYL-1*H*-INDAZOL-5-YL) PROPANOIC ACID

The (*R*)-amino acid, (*R*)-2-amino-3-(7-methyl-1*H*-indazol-5-yl)propanoic acid (*R*)-**108** (Figure 4.30), is a key intermediate for the synthesis of calcitonin gene-related peptide (CGRP) receptors **109** antagonist [135], potentially useful for the treatment of migraine and other maladies [136, 137].

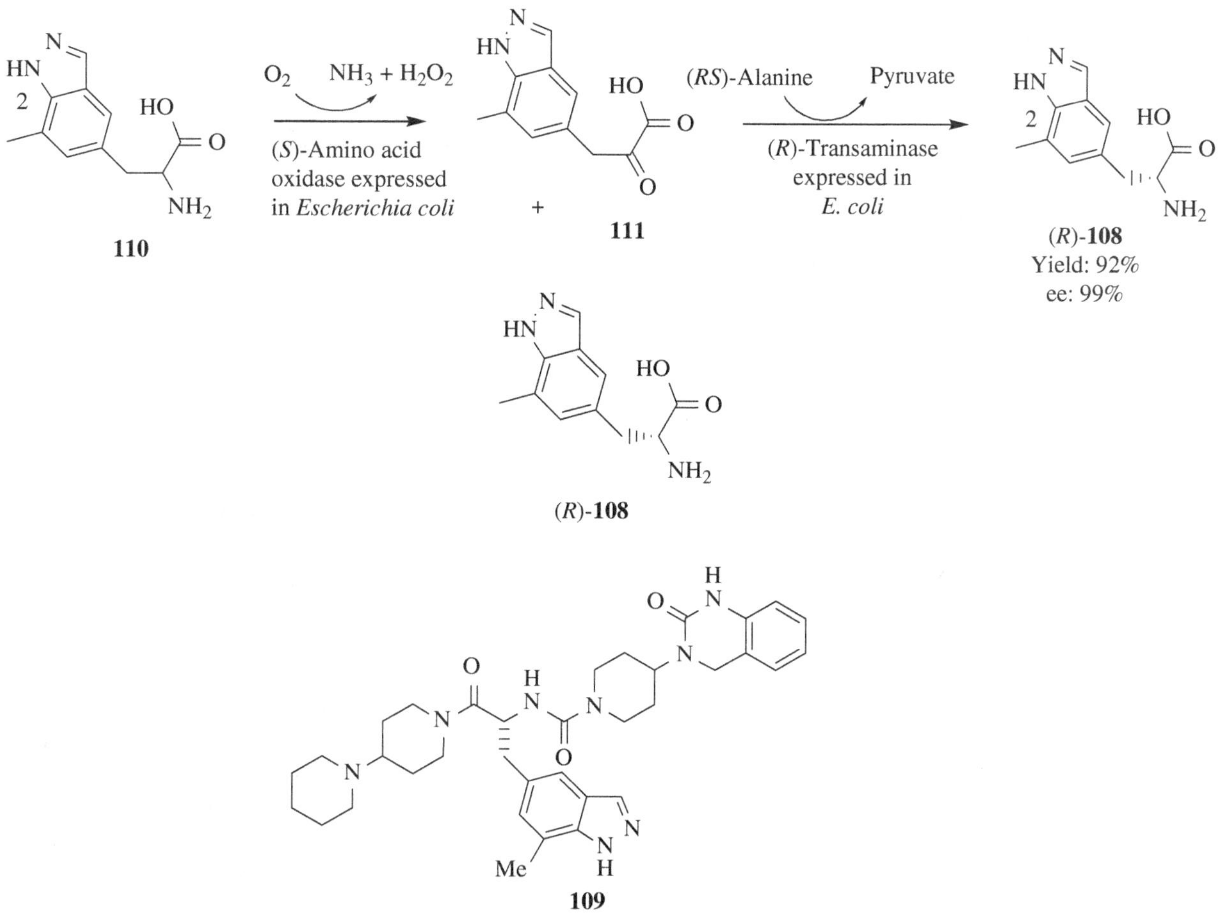

FIGURE 4.30
Antimigraine drugs: enzymatic synthesis of (*R*)-2-amino-3-(7-methyl-1*H*-indazol-5-yl)propanoic acid.

An enzymatic process was developed for the preparation of (*R*)-amino acid **108** from racemic amino acid **110** using (*S*)-amino acid oxidase from *Proteus mirabilis* cloned and expressed in *E. coli* in combination with a commercially available (*R*)-transaminase and (*R*)-alanine as the amino donor [138]. The process was optimized at the 500 ml scale using racemic **110** (20 g), D-alanine (40 g), pyridoxal phosphate (PLP) monohydrate (2.65 mg), 100 g of *E. coli* frozen cell paste containing the cloned (*S*)-amino acid deaminase (27 U/g cells) from *P. mirabilis*, and (*R*)-transaminase (Biocatalytics co., 200 mg). An average 85% solution yield with 96.5% ee was obtained for **108**. This process was scaled up in a pilot plant to prepare 1.79 kg **108** in 99.6% ee from 2.9 kg of racemic substrate **110** [138].

(*R*)-**108** was also prepared with a 79% isolated yield with >99% ee from the corresponding keto acid **111** using the (*R*)-transaminase with racemic alanine as the amino donor. An (*R*)-transaminase was identified and purified from a soil organism *Bacillus thuringiensis* and cloned and expressed in *E. coli*. The recombinant (*R*)-transaminase was very effective for preparing (*R*)-**108** and gave a nearly complete conversion of **111** to (*R*)-**108**. Reaction was carried out at 60 g/l keto acid to afford (*R*)-**108** with a 95% in-process yield and 99.5% ee [138].

4.32 ANTIDIABETIC DRUG (GLP-1 MIMICS): ENZYMATIC SYNTHESIS OF (*S*)-AMINO-3-[3-{6-(2-METHYLPHENYL)} PYRIDYL]-PROPIONIC ACID

(*S*)-Amino-3-[3-{6-(2-methylphenyl)} pyridyl]-propionic acid **112** (Figure 4.31) is a key intermediate required for the synthesis of GLP-1 mimics (GLP-1 receptor modulators), potentially useful for the treatment of type-2 diabetes [139, 140].

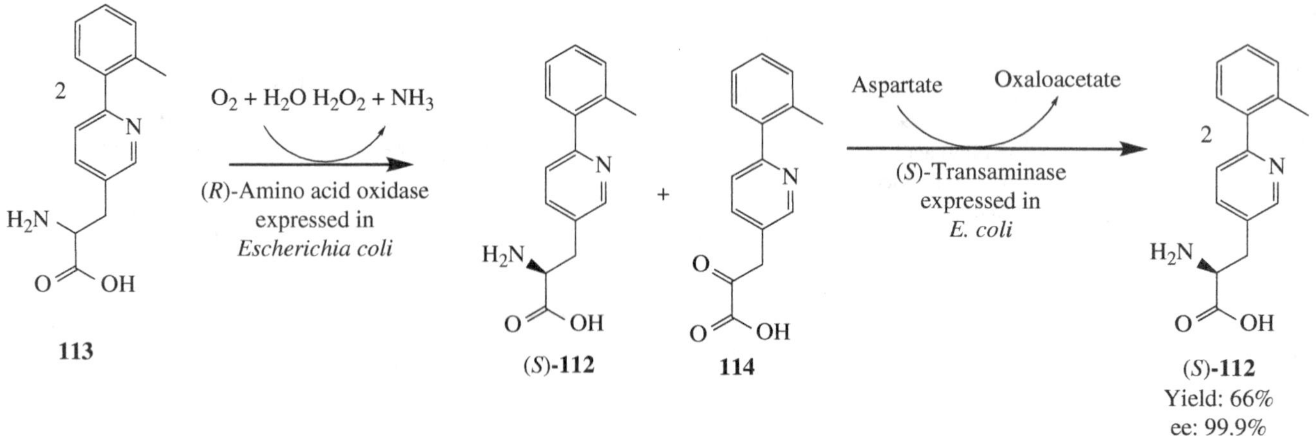

FIGURE 4.31
Antidiabetic drug (GLP-1 mimics): enzymatic synthesis of (*S*)-amino-3-[3-{6-(2-methylphenyl)} pyridyl]-propionic acid.

An enzymatic deracemization process was developed for the preparation of (*S*)-amino-3-[3-{6-(2-methylphenyl)}pyridyl]-propionic acid **112** using a combination of two enzymes: (*R*)-amino acid oxidase from *Trigonopsis variabilis*, cloned and expressed in *E. coli*, and an (*S*)-aminotransferase from soil bacterium *Burkholderia* sp. cloned and expressed in *E. coli* [141]. Racemic amino acid **113** was used as a substrate, and (*S*)-aspartate was used as the amino donor. This process was scaled up to a 70 l scale. Cell extracts from *rec E. coli* SC 16541 cells (12 kg) containing (*S*)-aminotransferase and *rec E. coli* SC 16544 cells (2.4 kg) containing (*R*)-amino acid oxidase were isolated in 46 l of 0.1 M phosphate buffer (pH 8.0) containing 5 μM PLP. A 100 l reactor containing 20 l of water was supplied with aspartate monohydrate (1.67 kg), ascorbic acid (0.33 kg), racemic amino acid **113** monosulfate monohydrate (1 kg, potency 60.7%), and propylene glycol (3.33 kg). The solution was adjusted to pH 7.0. Subsequently, catalase (0.133 kg), PLP (167 mg), clarified cell extract (40 l), and SAG-5693 antifoam (133 g) were added to the reactor, and the reaction was carried out at 30 °C, 250 rpm, and pH 7.5. The bioconversion of **113** to a mixture of (*S*)-**112** and keto acid **114** was completed after 3 h. The conversion of keto acid **114** to product (*S*)-**112** was completed in 25 h. (*S*)-**112** was isolated from the reaction mixture in 66% yield (580 g) with 99.9% ee [141].

In an alternative process, the enzymatic dynamic resolution of racemic amino acid **113** was also demonstrated by combining amino acid oxidase with chemical reduction. (*R*)-selective oxidation with Celite-immobilized (*R*)-amino acid oxidase from *T. variabilis* expressed in *E. coli* in combination with chemical imine reduction with borane–ammonia provided (*S*)-**112** with a 75% process yield and 100% ee [141].

4.33 EPHEDRINE: SYNTHESIS OF (*R*)-PHENYLACETYLCARBINOL

Stereoselective carbon–carbon bond-forming reactions are among the most useful synthetic methods in asymmetric synthesis as they allow the simultaneous creation of up to two adjacent stereocenters. Acyloin formation mediated by thiamine diphosphate-dependent decarboxylase, yeast pyruvate decarboxylase, bacterial benzoylformate decarboxylase, and phenylpyruvate decarboxylase has been reported [142–147].

The industrial process based on yeast pyruvate decarboxylase for the production of (*R*)-phenylacetylcarbinol **115** (Figure 4.32), a precursor to (−)-ephedrine, is still used on an industrial scale [148]. The optimization of a continuous enzymatic reaction yielding (*R*)-phenylacetylcarbinol (*R*)-PAC **115** was evaluated using three different pyruvate decarboxylases (PDC): PDC from *S. cerevisiae*, PDC from *Zymomonas mobilis*, and a potent mutant of the latter, PDC*W392M*, using acetaldehyde and benzaldehyde as substrates. Among the PDCs evaluated, the mutant enzyme was the most active

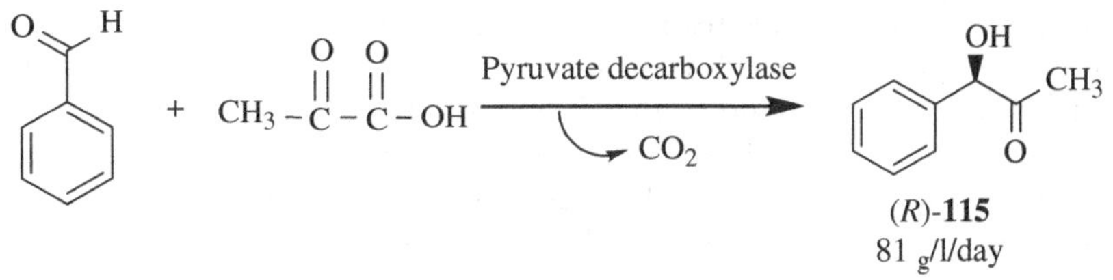

FIGURE 4.32
Ephedrine: synthesis of (*R*)-phenyl-acetyl carbinol.

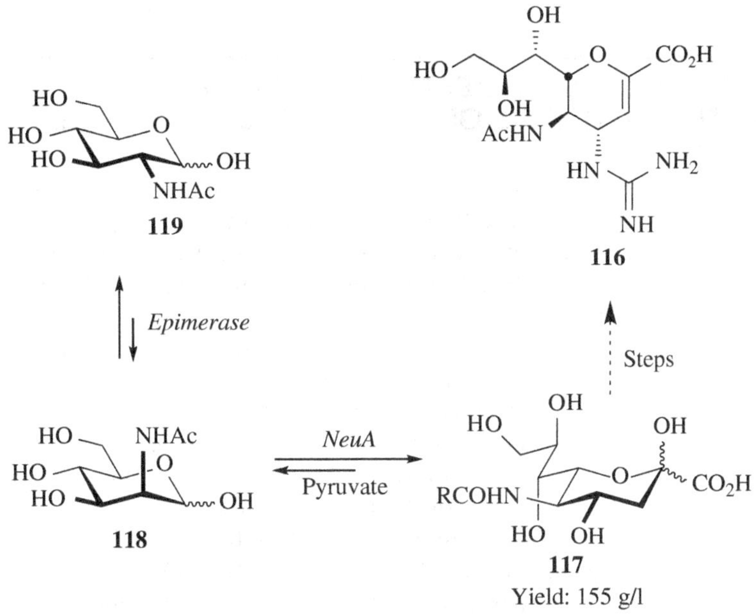

FIGURE 4.33
Zanamivir: enzymatic synthesis of *N*-acetylneuraminic acid.

and most stable. The reaction conditions were optimized and the carboligation was performed using a continuous reaction system and feeding both substrates in equimolar concentration. Initial studies using a continuously operated enzyme-membrane reactor gave (*R*)-PAC with a space–time yield of 81 g/l/day using a substrate concentration of 50 mM of both aldehydes. The yield was increased by a cascadation of enzyme-membrane reactors. The new strategy allows the synthesis of (*R*)-PAC from cheap substrates [149].

4.34 ZANAMIVIR: ENZYMATIC SYNTHESIS OF *N*-ACETYLNEURAMINIC ACID

Zanamivir **116** (Figure 4.33) is a potent and selective inhibitor of influenza virus sialidase (neuraminidase) and has been approved by the FDA for the treatment of type A and B influenza [150]. *N*-Acetylneuraminic acid **117** is a key intermediate required for the synthesis of Zanamivir. The synthesis of *N*-acetylneuraminic acid using the aldolase either from *E. coli* or *Clostridium perfringens* has been reported by batch process from the *N*-acetyl-D-mannosamine **118** and pyruvate using free or immobilized *N*-acetylneuraminic acid aldolase [38, 39]. In order to drive the equilibrium toward *N*-acetylneuraminic acid, the pyruvate was generally used in large excess, making the downstream processing rather difficult.

An elegant continuous process for *N*-acetylneuraminic acid synthesis was developed by introducing the *N*-acetylneuraminic acid-2-epimerase for epimerization of *N*-acetyl glucosamine **119** and integrating the epimerization with synthesis in an enzyme-membrane reactor [151]. Although this process has an excellent space–time yield, the epimerase enzyme is not freely available and the product stream is very dilute and chromatography is required for purification, making scale-up difficult.

Subsequently, the enzyme from *E. coli* has been overexpressed in an inducible system at very high expression levels, and homogenized bacterial cells were directly used to immobilize the enzyme from crude extracts onto Eupergit C beads without any clarification, and the immobilized enzyme was used for the synthesis of *N*-acetylneuraminic acid **117** at 155 g/l concentration. At the manufacturing scale, the same batch of enzymes was reused in >2000 cycles in batch column reactors, without any significant loss of activity, to produce multi-tonne quantities of *N*-acetylneuraminic acid [152].

4.35 EPIVIR: ENZYMATIC DEAMINATION PROCESS FOR THE SYNTHESIS OF (2′*R-CIS*)-2′-DEOXY-3-THIACYTIDINE

(2′*R-cis*)-2′-Deoxy-3-thiacytidine (3TC, Epivir, **120** Figure 4.34) has been approved by the FDA and is marketed for the treatment of HIV. Epivir is a potent and selective inhibitor of the reverse transcriptase, that catalyzes the conversion of the HIV RNA to a double stranded DNA copy. Epivir is also active against hepatitis B virus (HBV) and is sold as lamivudine [153].

The antiviral activity of nucleoside analogs mainly resides in the "natural" ß-D-isomer. However the "unnatural" ß-L-(−)-**120** isomer (Epivir) is substantially less cytotoxic and very effective than its corresponding "natural" ß-D-(+)-isomer [154].

One of the initial routes considered for the preparation of isomers was the enzymatic resolution using 5′-nucleotidase and alkaline phosphatase that would allow access to both enantiomers, racemic compound. The chemically synthesized monophosphate derivative (±)-**121** was resolved using 5′-nucleotidase from *Crotalus atrox* venom, and the resulting mixture was separated by chromatography and purified on silica gel to give (+)-**123** with a 90% ee. Hydrolysis of the remaining monophosphate (−)-**122** with alkaline phosphatase from *E. coli* afforded Epivir [155].

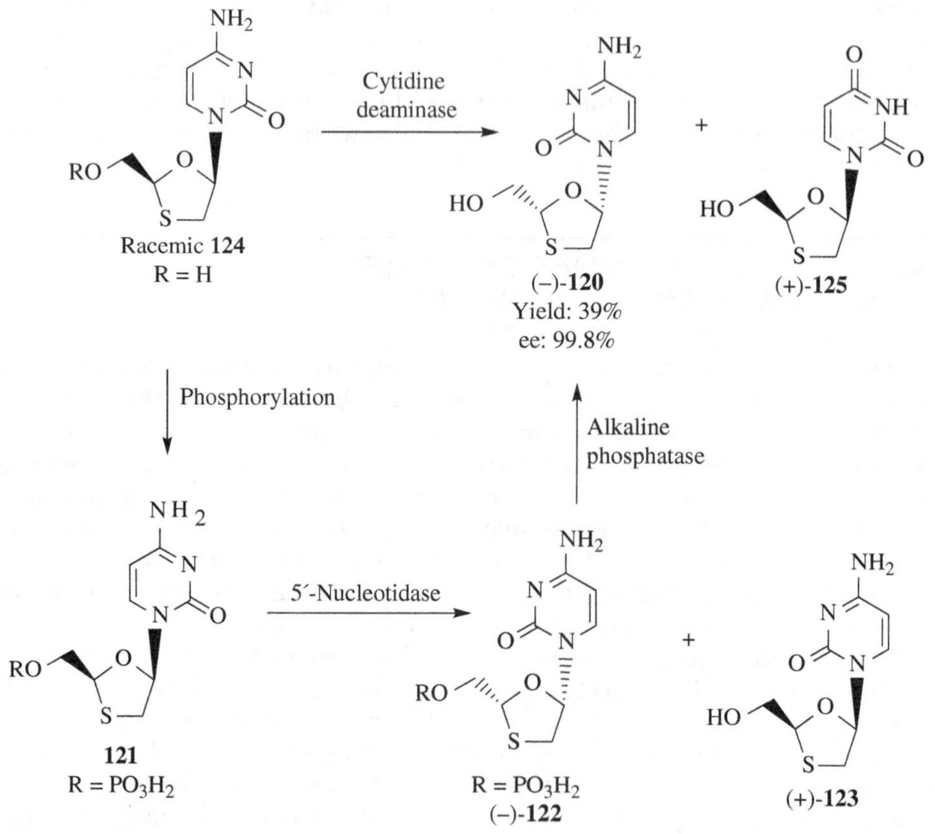

FIGURE 4.34

Epivir: enzymatic deamination process for the synthesis of (2′*R-cis*)-2′-deoxy-3-thiacytidine.

To produce much larger quantities of Epivir, a scaleable end-stage resolution, via enantioselective deamination of (±)-**124** with cytidine deaminase was developed to afford (−)-**120** (Epivir) and (+)-**125**. To use *E. coli* as a source of enzyme, the cloning and overexpression of enzyme, fermentation and immobilization of cytidine deaminase for reuse, and the development of an efficient isolation process were developed for large-scale manufacturing of Epivir [156]. About 1.15 kg of highly pure Epivir was recovered from each 3 kg batch in 97% purity and 99.8% ee. Using this approach, 20 kg of optically pure Epivir was isolated. This process was used on a manufacturing scale to prepare tonnes of Epivir, using immobilized cytidine deaminase from the recombinant strain. The same batch of enzyme was used for at least 15 cycles [156, 157].

4.36 HMG-CoA REDUCTASE INHIBITORS: ALDOLASE-CATALYZED SYNTHESIS OF CHIRAL LACTOL

The chiral lactol **126** is a key intermediate required for the synthesis of various HMG-CoA reductase inhibitors such as atorvastatin and rosuvastatin.

Wong and coworkers reported one-pot aldol reactions catalyzed by a deoxyribose-5-phosphate aldolase (DERA), in which 2 eq of acetaldehyde was added in sequence to two-carbon aldehyde acceptors to afford six-membered lactol derivatives **126** (Figure 4.35). The DERA-catalyzed reaction is an equilibrium process; the intermediate four-carbon adduct **127** is reversibly formed under the reaction conditions. The second condensation between this intermediate and a second equivalent of acetaldehyde drives the equilibrium favorably due to the stability of the cyclized lactol product **126**. DERA has been expressed in *E. coli* [158, 159].

To improve the commercial feasibility of the process, Diversa (now Veranium) Corporation developed DERA that was an improvement over the enzyme from *E. coli* with respect to high tolerance to substrate concentrations and reduced catalyst load. They created large genomic libraries by extracting DNA directly from environmental samples and screened through various high-throughput methods. A volumetric productivity improvement of about 400-fold was obtained with the enzyme, which resulted in a process a that has been run on up to a 100 g scale in a single batch at a rate of 30.6 g/l/h of lactol **126**. The catalyst load was improved tenfold, from 20 to 2.0 wt% DERA. The two stereogenic centers are set by DERA with an ee of >99.9% and a diastereomeric excess of 96.6% of lactol **126** [74]. The conversion of lactol product to the lactone was replaced by using aqueous sodium hypochlorite and acetic acid. Lactone **128** was obtained as a crystalline product, providing a readily scalable process for purification by recrystallization [160].

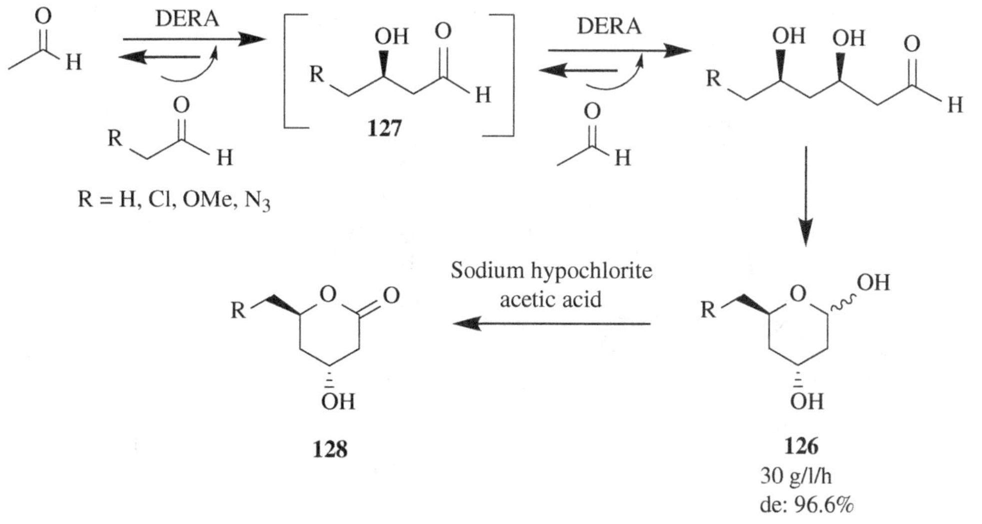

FIGURE 4.35
HMG-CoA reductase inhibitors: aldolase-catalyzed synthesis of chiral lactol.

DSM has also developed a process for the synthesis of atorvastatin chiral intermediate using DERA. *E. coli* gene coding for DERA was cloned and overexpressed. They created random genetic diversity using error-prone PCR to generate over 10 000 DERA variants. Using directed evolution and high-throughput screening technology, they identified several improved variants toward increased tolerance to chloroacetaldehyde substrate and process productivity. By combining the most beneficial mutations a tenfold improved variant was obtained compared to wild-type DERA with regard to (3*R*,5*S*)-6-chloro-2,4,6-trideoxy hexapyranoside synthesis, under industrially relevant conditions. This process has been commercialized [161].

4.37 BOCEPREVIR: OXIDATION OF 6,6-DIMETHYL-3-AZABICYCLO[3.1.0]HEXANE BY MONOAMINE OXIDASE

Boceprevir **129** (Victrelis, Figure 4.36) is developed for treating chronic hepatitis C infection. It is recommended as an addition to the current standard regimen of PEG-interferon and ribavirin in treating patients infected with hepatitis C virus (HCV). Boceprevir is a peptidomimetic protease inhibitor with four moieties, P1–P3 and a Cap, where P1 is a racemic β-aminoamide, P2 is a chiral dimethylcyclopropylproline analog, P3 is (*S*)-*tert*-leucine, and Cap is a *tert*-butylcarbamoyl group [162, 163].

The chemical method for the synthesis of P2 moiety suffers from at least 50% yield loss in the resolution process [164]. Biocatalytic desymmetrization of substituted pyrrolidines to give the desired enantiomer is an attractive approach to access proline analogs.

Turner's group demonstrated that a variety of 3,4-substituted *meso*-pyrrolidines can be enantioselectively oxidized by the monoamine oxidase (MAO) mutant. The Δ^1-pyrroline products were converted to amino acids through the corresponding nitriles. In another application, the chiral Δ^1-pyrrolines were directly used to prepare prolyl peptides in an Ugi reaction. This reaction has been applied to an efficient synthesis of Telaprevir. The directed evolution of MAO from *Aspergillus niger* provided variant with a broadened substrate specificity and high enantioselectivity for several α-methyl amines. In further enzyme evolution, an improved variant was identified for secondary amines [165, 166].

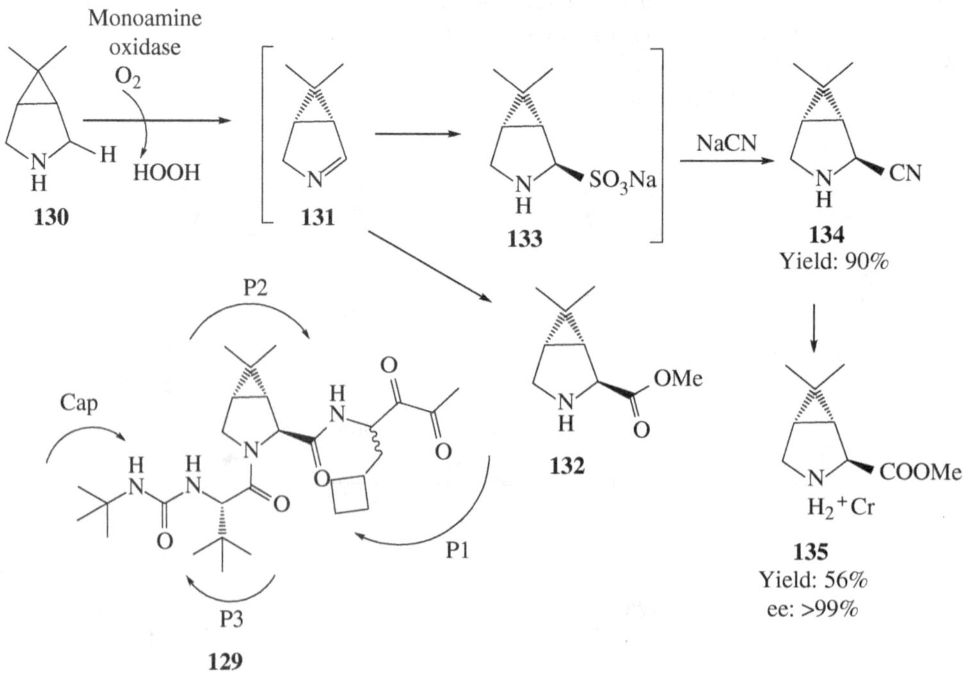

FIGURE 4.36

Boceprevir: oxidation of 6,6-dimethyl-3-azabicyclo[3.1.0] hexane by monoamine oxidase.

Codexis and Schering-Plough jointly further developed a chemoenzymatic asymmetric synthesis of P2 by enzymatic oxidative desymmetrization of the prochiral amine using MAO from *A. niger* and *A. oryzae* acting on compound **130**. Product **131** was converted to compound **132** and found to be the desired enantiomer with an ee of >99%. The enzyme optimization was carried out using the *A. niger* enzyme (MAON). Initial mutant libraries were generated via random mutagenesis using error-prone PCR as well as via a homology-modeling approach. Screening of these round 1 libraries gave the variant MAON156 with 2.4-fold higher activity than that of the wild-type enzyme. MAON156 was selected as the backbone for the next round of evolution, and a parallel strategy was performed by family shuffling of MAON156 with the homolog from *A. oryzae*. Improved variants were identified from the second round evolution, and the best variant was used to identify variants with tolerance to substrate and product inhibition and high thermo stability. This final hybrid variant MAON 401 was 2.8-fold more active and was stable upon storage [167].

The oxidation of **130** by oxygen under pressure was developed at lab scale. Under optimal conditions, the substrate/bisulfite mixture was added to a solution of **130** (3.9 g/l), MAON401 and catalase over 20 h. Substrate **130** (65 g/l) was converted to sulfonate **133** with a small amount of **131** (<10%). The enzyme reaction stream was telescoped for cyanation to afford only *trans*-nitrile **134** in 90% yields from **130**. Subsequently the nitrile was transformed to the methyl ester and the product was converted to the free base **132**. In the final step the free base was crystallized to afford **135** in 56% yield and >99% ee. The conditions of the procedure were successfully applied to pilot plant scale. Compared with the resolution method in the enzymatic process the product yield was increased by 150%, raw material use was reduced by 59.8%, consumption of water was reduced by 60.7%, and the overall process waste was reduced by 63.1% [167].

4.38 CRIXIVAN: ENZYMATIC SYNTHESIS OF INDANDIOLS

Hydroxylation reactions are catalyzed by oxygenases. A great deal of our understanding of the structure, function, and substrate specificity of aromatic hydrocarbon dioxygenases has come from studies of toluene dioxygenase (TDO) and naphthalene dioxygenase (NDO). To date, more than 100 Rieske non-heme iron oxygenases have been identified based on biological activity or nucleotide sequence identity, and about 300 arene *cis*-dihydrodiols have been obtained using TDO, which have been used extensively as chiral precursors in organic synthesis [168–173].

cis-(1*S*,2*R*)-Indandiol **135** or *trans*-(1*R*,2*R*)-indandiol **136** (Figure 4.37) are potential precursors to *cis*-(1*S*,2*R*)-1-aminoindan-2-ol **137**, a chiral synthon for Crixivan (indinavir) **138**, an HIV protease inhibitor [174]. Enrichment and isolation of microbial cultures yielded two strains, *Rhodococcus* sp. B 264-1 (MB 5655) and I-24 (MA 7205), capable of oxidizing indene **139** to *cis*-(1*S*,2*R*)-indandiol and *trans*-(1*R*,2*R*)-indandiol, respectively. Isolate MB 5655 was found to contain a TDO, while isolate MA 7205 was found to harbor both toluene and naphthalene dioxygenases that catalyze the earlier mentioned biotransformation. When scaled up in a 14 l fermentor, MB5655 produced up to 2.0 g/l of *cis*-(1*S*,2*R*)-indandiol **135** with >99% ee. *Rhodococcus* sp. MA 7205 cultivated under similar conditions produced up to 1.4 g/l of *trans*-(1*R*,2*R*)-indandiol **136** with >98% ee. Process development studies yielded titers >4.0 g/l of *trans*-(1*R*,2*R*)-indandiol [174].

A metabolic engineering approach [175] and directed evolution techniques [176] were evaluated to avoid side reactions, block degradative pathways, and enhance the desired reaction (conversion of indene to *cis*-amino indanol **137** or *cis*-indanediol). Multiparameter flow cytometry was used to assess indene toxicity, and it was shown that concentrations up to 0.25 g/l of indene (0.037 g indene per gram dry cell wt.) in batch bioconversions did not influence reaction rate. Using this information, a single-phase

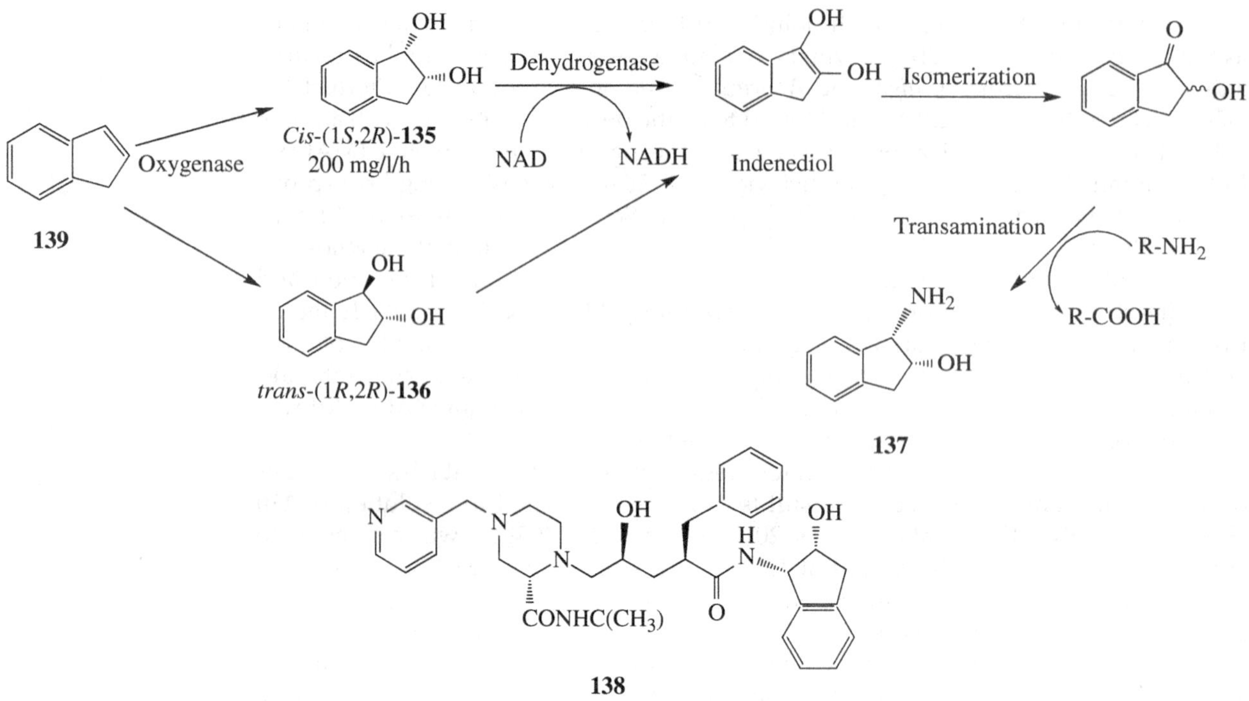

FIGURE 4.37
Crixivan: enzymatic synthesis of indandiols.

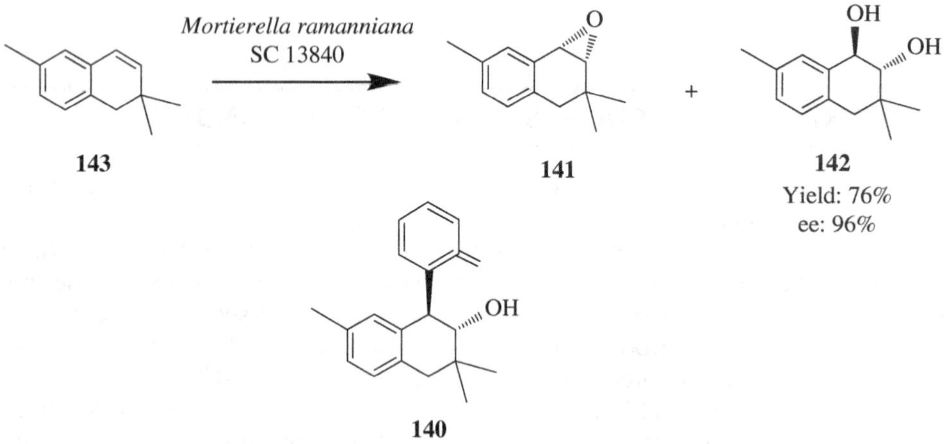

FIGURE 4.38
Potassium channel opener: preparation of chiral epoxide and *trans*-diol.

indene-fed batch bioconversion was carried out by indene supply at a rate of 0.1 g/l/h. *cis*-(1*S*,2*R*)-Indandiol **135** production rates were enhanced up to 200 mg/l/h by a combination of suitable indene feeding rates in the stationary phase and by operating with a high biomass concentration [177].

4.39 POTASSIUM CHANNEL OPENER: PREPARATION OF CHIRAL EPOXIDE AND *TRANS*-DIOL

Potassium (K) channels play a critical role in the basic electrical and mechanical function of a wide variety of tissues, including smooth and cardiac muscle [178]. Highly specific compounds that either open or block K channels have been developed [179]. The synthesis and antihypertensive activity of K-channel openers based on monosubstituted *trans*-4-amino-3,4-dihydro-2,2-dimethyl-2*H*-1-benzopyran-3-ol **140** (Figure 4.38) has been demonstrated [180]. Chiral epoxide **141** and *trans*-diol **142** are potential intermediates for the synthesis of **140**. Microbial process for oxygenation of 6-cyano-2,2-dimethyl-2*H*-1-benzopyran **143** to the corresponding chiral epoxide **141** and *trans*-diol **142** was demonstrated [181]. *Mortierella ramanniana* SC 13840 and *Corynebacterium* sp. SC 13876 gave 67.5 M% and 32 M% yield and 96% ee and 89% ee, respectively, for *trans*-diol **142**. *Corynebacterium* sp. SC 13876 also gave chiral epoxide **141** in 17 M% yields and 88% ee.

A single-stage process (fermentation/bio-oxidation) for the biotransformation of **143** was developed using *M. ramanniana* SC 13840 in a 25 l fermentor to afford *trans*-diol **142** in 61 M% yield and 92.5% ee. In a two-stage process using a cell suspension (10% w/v, wet cells) of *M. ramanniana* SC 13840, the *trans*-diol **142** was obtained in 76 M% yield and 96% ee when the reaction was carried out in a 5 l fermentor. Glucose was supplied to regenerate NADH required for this reaction. From the reaction mixture, *trans*-diol **142** was isolated with a 65 M% overall yield with 97% ee and 98% chemical purity [181].

4.40 EPOTHILONES (ANTICANCER DRUGS): EPOTHILONE B AND EPOTHILONE F

The clinical success of Paclitaxel has stimulated research into compounds with similar modes of action with antineoplastic efficacy while minimizing its less desirable properties, such as water insolubility, difficult synthesis, and emerging resistance. The epothilones are a novel class of natural product cytotoxic compounds derived from the fermentation of the *Sorangium cellulosum*, which are non-taxane microtubule-stabilizing compounds that trigger apoptosis [182–189]. The natural product epothilone B **144** (Figure 4.39) has demonstrated broad-spectrum antitumor activity

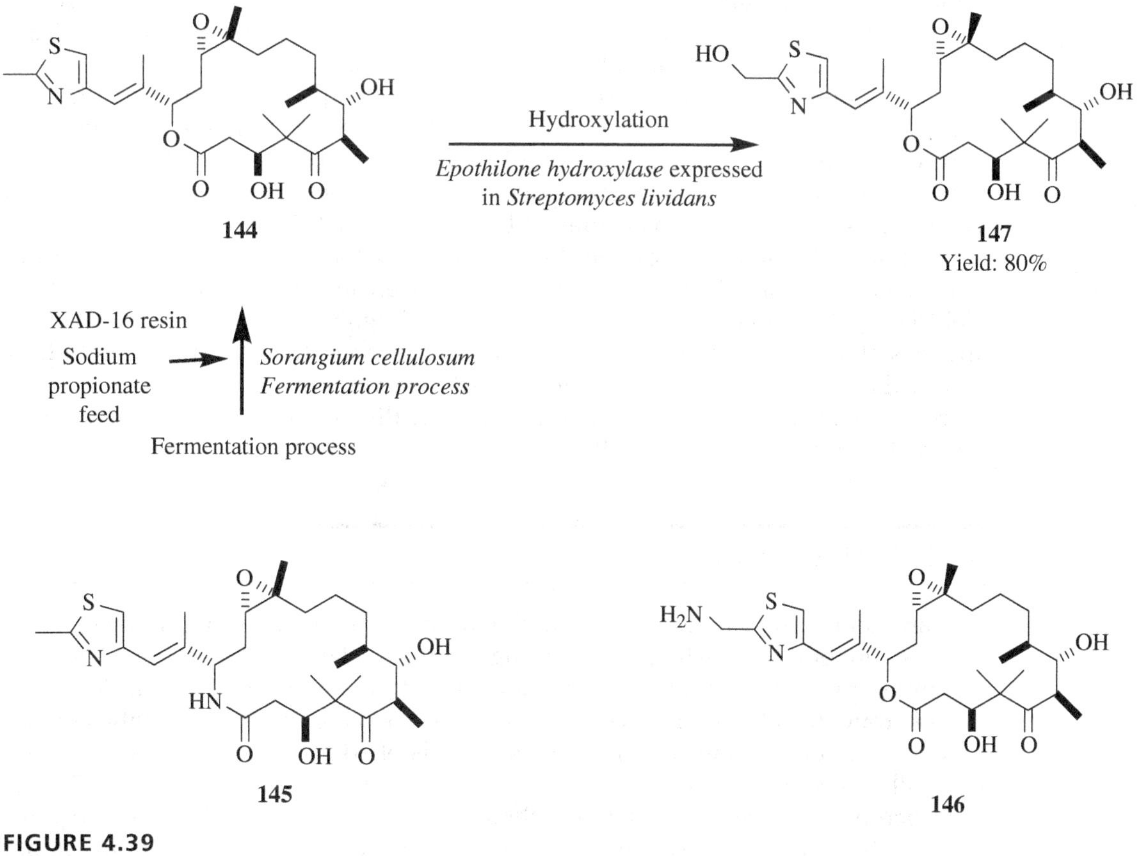

FIGURE 4.39
Epothilones (anticancer drugs): epothilone B and epothilone F.

in vitro and *in vivo*, including tumors with Paclitaxel resistance. Ixabepilone **145** (BMS-247550), a novel epothilone derivative, was developed by BMS has been approved by the US FDA for the treatment of metastatic or locally advanced breast cancer [182].

A fermentation process was developed for the production of epothilone B **144**, and the titer of epothilone B was increased by a continuous feed of sodium propionate during fermentation. The inclusion of XAD-16 resin during fermentation to adsorb epothilone B and to carry out volume reduction made the recovery of product very simple [182].

Another epothilone derivative BMS-310705 **146** developed by BMS is a semisynthetic analog of epothilone B. A microbial hydroxylation process was developed for converting epothilone B **144** to epothilone F **147** by *Amycolatopsis orientalis* SC 15847. Epothilone F is a key intermediate for the synthesis of an anticancer drug **146**. A bioconversion yield of 37–47% was obtained when the process was scaled up to 100–250 l. The epothilone B hydroxylase along with the ferredoxin gene has been cloned and expressed in *Streptomyces rimosus* from *A. orientalis* SC 15847. Variants of this cloned enzyme have been used in the hydroxylation of epothilone B, providing 80% yields of epothilone F [188, 189].

4.41 β-ADRENERGIC BLOCKING AGENTS: SYNTHESIS OF INTERMEDIATES FOR PROPRANOLOL AND DENOPAMINE

Deracemization of a number of pharmaceutically valuable building blocks has been carried out by biocatalytic processes. They include epoxides, alcohols, amines, and acids [12, 14, 190–192]. Dynamic kinetic resolution (DKR) involves the combination of an enantioselective transformation with an *in situ* racemization process so that, in principle, both enantiomers of the starting material can be converted to the product with a high yield and ee. The racemization step can be catalyzed either enzymatically by racemases or nonenzymatically by transition metals.

Propanolol **148** (Figure 4.40) belongs to the group of β-adrenergic blocking agents of the general structure $ArOCH_2CH(OH)CH_2NHR$, where Ar is aryl and R is alkyl. These compounds are potentially useful for the treatment of hypertension [193]. β-Adrenergic receptor blocking activity mainly resides in the (*S*)-enantiomers. The synthesis of (*S*)-**148** was achieved by DKR of (±)-**149** using Novozym 435 in toluene at 80 °C and *p*-chlorophenyl acetate as acyl donor in the presence of ruthenium complex **150**. (*R*)-Acetate **151** was produced in >99% ee and 86% isolated yield [193]. The enzyme was recycled and used again for another cycle without any loss of activity. Using the same procedure, racemic **152** was subjected to DKR to afford the (*R*)-acetate **153** with 99% ee, 92% conversion, and 84% isolated yield, a precursor of (*R*)-denopamine **154**, a potent orally active β_1 receptor agonist for the treatment of hypertension [194].

4.42 CONCLUSION

Demand for chiral molecules is very high mainly due to their use in pharmaceuticals but also in other industries such as agricultural, flavor, and aroma chemicals. Examples presented in this chapter are only from a few selected articles that demonstrate the interest of process scientists for the use of biotransformations in the preparation of chiral molecules for the synthesis of APIs. Biocatalytic processes are generally green processes.

Microbes and enzymes derived therefrom are highly enantio-, chemo-, and regioselective across a diverse range of reactions under mild conditions of pH, temperature, and pressure. Availability of "off-the-shelf" enzymes has been increased.

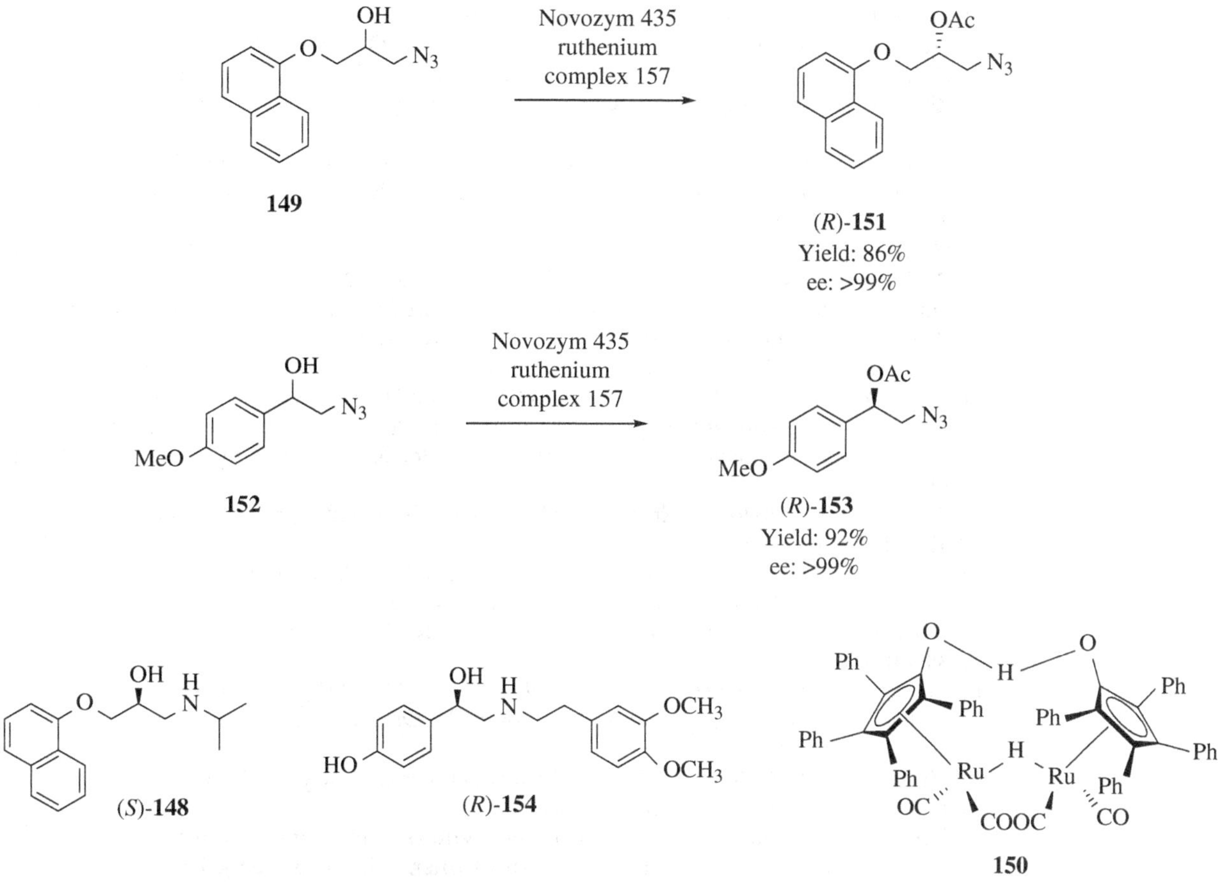

FIGURE 4.40
β-Adrenergic blocking agents: synthesis of intermediates for propranolol and denopamine.

In addition to hydrolytic enzymes such as lipases, proteases, and esterases, now more enzymes, such as KREDs, amino acid dehydrogenases, transaminases, nitrilases, acylases, amino acid dehydrogenases, and amidases are available.

Since the mid-2000s progress in fermentation technology, protein chemistry, molecular cloning, enzyme immobilization, random and site-directed mutagenesis, and directed evolution of biocatalysts has opened up unlimited access to a variety of enzymes and microbial cultures as tools in organic synthesis. Currently biocatalysts are evolved under suitable process conditions to meet process requirement with high activity toward the desired substrate.

REFERENCES

[1] Oliver, M., Voigt, C.A., Arnold, F.H. Enzyme engineering by directed evolution. In *Enzyme Catalysis in Organic Synthesis*, 2nd edition (eds Drauz, K. and Waldmann, H.), 2002, Wiley-VCH, Weinheim, Germany, 95–138.

[2] Kazlauskas, R.J. *Current Opinion in Chemical Biology*, 2005, **9**(2), 195–201.

[3] Schmidt, M., Bauman, M., Henke, E., Konarzycka-Bessler, M., Bornscheuer, U.T. *Methods in Enzymology (Protein Engineering)*, 2004, **388**, 199–207.

[4] Bornscheuer, U.T., Huisman, G.W., Kazlauskas, R.J., Lutz, S., Moore, J.C., Robins, K. *Nature*, 2012, **485**, 185–194.

[5] Huisman, G.W., Lalonde, J.J. Enzyme evolution for chemical process applications. In *Biocatalysis in the Pharmaceutical and Biotechnology Industries* (ed. Patel, R.N.), 2007, CRC Press, Boca Raton, FL, 717–742.

[6] (a) Reetz, M.T. Directed evolution of enzymes. In *Enzyme Catalysis in Organic Synthesis*, 3rd edition (eds Drauz, K., Gröger, H., and May, O.), 2012, Wiley–VCH, Weinheim, Germany, 214–223; (b) Turner, N.J., O'Reilly, E. *Nature Chemical Biology*, 2013, **9**, 285–343.
[7] Zhao, H., Chockalingom, K., Chen, Z. *Current Opinion in Biotechnology*, 2002, **13**(2), 104–110.
[8] Reetz, M.T., Torre, C., Eipper, A., Lohmer, R., Hermes, M., Brunner, B., Maichele, A., Bocola, M., Arand, M., Cronin, A., Genzel, Y., Archelas, A., Furstoss, R. *Organic Letters*, 2004, **6**(2), 177–180.
[9] Hans-Peter Meyer, H.-P., Eichhorn, E., Hanlon, S., Lutz, S., Schürmann, M., Wohlgemuth, R., Coppolecchia, R. *Catalysis Science & Technology*, 2013, **3**, 29–40.
[10] DiCosimo, R. Nitrilases and nitrile hydratases. In *Biocatalysis in the Pharmaceutical and Biotechnology Industries* (ed. Patel, R.N.), 2007, CRC Press, Boca Raton, FL, 1–26.
[11] Patel, R.N. *ACS Catalysis*, 2011, **1**, 1056–1074.
[12] Simeo, Y., Kroutil, W., Faber, K. Biocatalytic deracemization: dynamic resolution, stereoinversion, enantioconvergent processes and cyclic deracemization. In *Biocatalysis in the Pharmaceutical and Biotechnology Industries* (ed. Patel, R.N.), 2007, CRC Press, Boca Raton, FL, 27–51.
[13] Simons, C., Hanefeld, U., Arends, I., Maschmeyer, T., Sheldon, R. *Topics in Catalysis*, 2006, **40**(1–4), 35–44.
[14] Turner, N.J. *Current Opinion in Chemical Biology*, 2004, **8**(2), 114–119.
[15] Hall, M., Stueckler, C., Ehammer, H., Pointner, E., Oberdorfer, G., Gruber, K., Hauer, B., Stuermer, R., Kroutil, W., Macheroux, P., Faber, K. *Advanced Synthesis & Catalysis*, 2008, **350**(3), 411–418.
[16] Reetz, M.T. *Journal of the American Chemical Society*, 2013, **135**, 12480–12496.
[17] Wells, A.S., Finch, G.L., Michaels, P.C., Wong, J.W. *Organic Process Research & Development*, 2012, **16**, 1986–1993.
[18] Huisman, G.W., Collier, S.J. *Current Opinion in Chemical Biology*, 2013, **17**, 284–292.
[19] Patel, R.N. *Biomolecules*, 2013, **3**, 741–777.
[20] Patel, R.N. Biocatalytic key steps in semisynthesis and total synthesis. In *Science of Synthesis* (eds Faber, K., Fessner, W.-D., and Turner, N.J.), 2015, Georg Thieme Verlag KG, Stuttgart, Germany/New York, 403–441.
[21] Sinclair, E.M., Drucker, D.J. *Current Opinion in Endocrinology Diabetes*, 2005, **12**, 146–151.
[22] Augeri, D.J., Robl, J.A., Betebenner, D.A., Magnin, D.R., Khanna, A., Robertson, J.G., Wang, A., Simpkins, L.M., Taunk, P., Huang, Q., Han, S.-P., Abboa-Offei, B., Cap, M., Xin, L., Tao, L., Tozzo, E., Welzel, G.E., Egan, D.M., Marcinkeviciene, J., Chang, S.Y., Biller, S.A., Kirby, M.S., Parker, R.A., Hamann, L.G. *Journal of Medicinal Chemistry*, 2005, **248**, 5025–5037.
[23] Vu, T.C., Brzozowski, D.B., Fox, R., Godfrey, J.D. Jr., Hanson, R.L., Kolotuchin, S.V., Mazzullo, J.A. Jr., Patel, R.N., Wang, J., Wong, K., Yu, J., Zhu, J., Magnin, R.D., Augeri, D.J., Hamann, L.G. PCT International Application, 2004, 101 pp. WO 2004052850.
[24] Hanson, R.L., Goldberg, S.L., Brzozowski, D.B., Tully, T.P., Cazzulino, D., Parker, W.L., Lyngberg, O.K., Vu, T.C., Wong, M.K., Patel, R.N. *Advanced Synthesis & Catalysis*, 2007, **349**(8+9), 1369–1378.
[25] Kim, D., Wang, L., Beconi, M., Eiermann, G.J., Fisher, M.H., He, H., Hickey, G.J., Kowalchick, J.E., Bleating, B., Lyons, K., Marsilio, F., McCann, M.E., Patel, R.A., Petrov, A., Scapin, G., Patel, S.B., Roy, R.S., Wu, J.K., Wyvratt, M.J., Zhang, B.B., Zhu, L., Thornberry, N.A., Weber, A.E. *Journal of Medicinal Chemistry*, 2005, **48**, 141–151.
[26] Hansen, K.B., Hsiao, Y., Xu, F., Rivera, N., Clausen, A., Kubryk, M., Krska, S., Rosner, T., Simmons, B., Balsells, J., Ikemoto, N., Sun, Y., Spindler, F., Malan, C., Grabowski, E.J.J., Armstrong, J.D. *Journal of the American Chemical Society*, 2009, **131**, 8798–8804.
[27] Johnson, N.B., Lennon, I.C., Moran, P.H., Ramsden, J.A. *Accounts of Chemical Research*, 2007, **40**, 1291–1297.
[28] Savile, C.K., Janey, J.M., Mundorff, E.C., Moore, J.C., Tam, S., Jarvis, W.R., Colbeck, J.C., Krebber, A., Fleitz, F.J., Brands, J., Devine, P.N., Huisman, G.W., Hughes, G.J. *Science*, 2010, **329**, 305–309.
[29] Shin, J.-S., Kim, B.-G. *Journal of Organic Chemistry*, 2002, **67**, 2848–2853.
[30] Cho, B.-K., Park, H.-Y., Seo, J.-H., Kim, J., Kang, T.-J., Lee, B.-S., Kim, B.-G. *Biotechnology and Bioengineering*, 2008, **99**, 275–284.
[31] Höhne, M., Kühl, S., Robins, K., Bornscheuer, U.T. *ChemBioChem*, 2008, **9**, 363–365.
[32] Truppo, M.D., Rozzell, J.D., Moore, J.C., Turner, N.J. *Organic & Biomolecular Chemistry*, 2009, **7**, 395–398.

[33] Robl, J., Sun, C., Stevenson, J., Ryono, D., Simpkins, L., Cimarusti, M., Dejneka, T., Slusarchyk, W., Chao, S., Stratton, L., Misra, R., Bednarz, M., Asaad, M., Cheung, H., Aboa-Offei, B., Smith, P., Mathers, P., Fox, M., Schaeffer, T., Seymour, A., Trippodo, N. *Journal of Medicinal Chemistry*, 1997, **40**, 1570–1577.

[34] Hanson, R.L., Schwinden, M.D., Banerjee, A., Brzozowski, D.B., Chen, B.-C. Patel, B.P., McNamee, C.G., Kodersha, G.A., Kronenthal, D.R., Patel, R.N., Szarka, L.J. *Bioorganic & Medicinal Chemistry*, 1999, **7**(10), 2247–2252.

[35] Patel, R.N. *Biomolecular Engineering*, 2001, **17**, 167–182.

[36] Hanson, R.L., Howell, J., LaPorte, T., Donovan, M., Cazzulino, D., Zannella, V., Montana, M., Nanduri, V., Schwarz, S., Eiring, R., Durand, S., Wasylyk, J., Parker, L., Liu, M., Okuniewicz, F., Chen, B.-C., Harris, J., Natalie, K., Ramig, K., Swaminathan, S., Rosso, V., Pack, S., Lotz, B., Bernot, P., Rusowicz, A., Lust, D., Tse, K., Venit, J., Szarka, L., Patel, R.N. *Enzyme and Microbial Technology*, 2000, **26**, 348–358.

[37] Patel, R.N., Banerjee, A., Nanduri, V., Goldberg, S., Johnston, R., Hanson, R., McNamee, C., Brzozowski, D., Tully, T., Ko, R., LaPorte, T., Cazzulino, D., Swaminathan, S., Parker, L., Venit, J. *Enzyme and Microbial Technology*, 2000, **27**, 376–389.

[38] Auge, C., David, S., Gautheron, C. *Tetrahedron Letters*, 1984, **25**, 4663–4664.

[39] Kim, M.J., Hennen, W.J., Sweers, H.M., Wong, C.H. *Journal of the American Chemical Society*, 1988, **110**, 6481–6486.

[40] Gordon, E.M., Ondetti, M.A., Pluscec, J., Cimarusti, C.M., Bonner, D.P., Sykes, R.B. *Journal of the American Chemical Society*, 1982, **104**, 6053–6060.

[41] Hanson, R.L., Singh, J., Kissick, T.P., Patel, RN., Szarka, L., Mueller, R. *Bioorganic Chemistry*, 1990, **18**(2), 116–130.

[42] Groeger, H., May, O., Werner, H., Menzel, A., Altenbuchner, J. *Organic Process Research & Development*, 2006, **10**(3), 666–669.

[43] Emmanuel, M.J., Frye, L.L., Hickey, E.R., Liu, W., Morwick, T.M., Spero, D.M., Sun, S., Thomson, D.S., Ward, Y.D., Young, E.R.R. PCT International Application, 2001, **361** pp. WO 2001019816.

[44] Bold, G., Faessler, A., Capraro, H.-G., Cozens, R., Klimkait, T., Lazdins, J., Mestan, J., Poncioni, B., Roesel, J., Stover, D., Tintelnot-Blomley, M., Acemoglu, F., Beck, W., Boss, E., Eschbach, M., Huerlimann, T., Masso, E., Roussel, S., Ucci-Stoll, K., Wyss, D., Lang, M. *Journal of Medicinal Chemistry*, 1998, **41**(8), 3387–3401.

[45] Robinson, B.S., Riccardi, K.A., Gong, Y.F., Guo, Q., Stock, D.A., Blair, W.S., Terry, B.J., Deminie, C.A., Djang, F., Colonno, R.J., Lin, P.-F. *Antimicrobial Agents and Chemotherapy*, 2000, **44**(8), 2093–2099.

[46] Degertekin, B., Lok, A.S. *Current Opinion in Gastroenterology*, 2008, **24**(3), 306–311.

[47] Njoroge, F.G., Chen, K.X., Shih, N.Y., Piwinski, J.J. *Accounts of Chemical Research*, 2008, **41**(1), 50–59.

[48] Kragl, U., Nasic-Racki, D., Wandrey, C. *Bioprocess Engineering*, 1996, **14**, 291–297.

[49] Krix, A.S.G., Bommarius, A.S., Kottenhahn, K., Schwarm, M., Kula, M.R. *Journal of Biotechnology*, 1997, **53**, 29–39.

[50] Galkin, A., Kulakova, L., Yoshimura, T., Soda, K., Esaki, N. *Applied and Environmental Microbiology*, 1997, **63**, 4651–4656.

[51] Menzel, A., Werner, H., Altenbuchner, J., Gröger, H. *Engineering in Life Sciences*, 2004, **4**(6), 573–576.

[52] Straathof, A.J.J., Panke, S.A. *Current Opinion in Biotechnology*, 2002, **13**(6), 548–556.

[53] Spero, D.M., Kapadia, S.R. *Journal of Organic Chemistry*, 1996, **61**(21), 7398–7401.

[54] Vedha-Peters, K., Gunawardana, M., Rozzell, J.D., Novick, S.J. *Journal of the American Chemical Society*, 2006, **128**, 10923–10929.

[55] Ferris, C.D., Hirsch, D.J., Brooks, B.P., Snyder, S.H. *Journal of Neurochemistry*, 1991, **57**(3), 729–737.

[56] Gustafsson, D., Elg, M., Lenfors, S., Boerjesson, I., Teger-Nilsso, A.-C. *Blood Coagulation & Fibrinolysis*, 1996, **7**(1), 69–79.

[57] Gillman, K.W., Starrett, J.E. Jr., Parker, M.F., Xie, K., Bronson, J.J., Marcin, L.R., McElhone, K.E., Bergstrom, C.P., Mate, R.A., Williams, R., Meredith, J.E., Burton, C.R., Barten, D.M., Toyn, J.H., Roberts, S.B., Lentz, K.A., Houston, J.G., Zaczek, R., Albright, C.F., Decicco, C.P., Macor, J.E., Olson, R.E. *ACS Medicinal Chemistry Letters*, 2010, **1**, 120.

[58] Ojima, I., Koji, K., Nakahashi, K. *Journal of Organic Chemistry*, 1989, **54**, 4511.

[59] Goswami, A., Goldberg, S.L., Johnston, R.M., Hanson, R.L., Parker, W.L. PCT International Application, 2012. WO 2012106579.

[60] Hanson, R.L., Johnston, R.M., Goldberg, S.L., Parker, W.L., Goswami, A. *Organic Process Research & Development*, 2013, **17**, 693–700.

[61] Veronesi, B., Carter, J.D., Devlin, R.B., Simon, S.A., Oortgiesen, M. *Neuropeptides (Edinburgh)*, 1999, **33**(6), 447–456.

[62] Reichard, G.A., Ball, Z.T., Aslanian, R., Anthes, J.C., Shih, N.-Y., Piwinski, J.J. *Bioorganic & Medicinal Chemistry Letters*, 2000, **10**(20), 2329–2332.

[63] Homann, M.J., Vail, R., Morgan, B., Sabesan, V., Levy, C., Dodds, D.R., Zaks, A. *Advanced Synthesis & Catalysis*, 2001, **343**(6+7), 744–749.

[64] Suen, W.-C., Zhang, N., Xiao, L., Madison, V., Zaks, A. *Protein Engineering, Design & Selection*, 2004, **17**(2), 133–140.

[65] Zhang, N., Suen, W.-C., Windsor, W., Xiao, L., Madison, V., Zaks, A. *Protein Engineering*, 2003, **16**(8), 599–605.

[66] Selak, I. *Current Opinion in Investigational Drugs*, 2001, **2**(6), 828–834.

[67] Hoekstra, M.S., Sobieray, D.M., Schwindt, M.A., Mulhern, T.A., Grote, T.M. Huckabee, B.K., Hendrickson, V.S., Franklin, L.C., Granger, E.J., Karrick, G.L. *Organic Process Research & Development*, 1997, **1**, 26–38.

[68] Burk, M.J., de Koning, P.D., Grote, T.M., Hoekstra, M.S., Hoge, G., Jennings, R.A., Kissel, W.S., Le, T.V., Lennon, I.C., Mulhern, T.A., Ramsden, J.A., Wade, R.A. *Journal of Organic Chemistry*, 2003, **68**, 5731–5734.

[69] Hoge, G., Wu, H.-P., Kissel, W.S., Pflum, D.A., Greene, D.J., Bao, J. *Journal of the American Chemical Society*, 2004, **126**, 5966–5967.

[70] Martinez, C.A., Hu, S., Dumond, Y., Tao, J., Kelleher, P., Liam Tully, L. *Organic Process Research & Development*, 2008, **12**(3), 392–398.

[71] Cherney, R.J., Brogan, J.B., Mo, R., Lo, Y.C., Yang, G., Miller, P.B., Scherle, P.A., Molino, B.F., Carter, P.H., Decicco, C.P. *Bioorganic & Medicinal Chemistry Letters*, 2009, **19**, 597–601.

[72] Carter, P. *Expert Opinion on Therapeutic Patents*, 2013, **23**(5), 549–568.

[73] Goswami, A., Kissick, T.P. *Organic Process Research & Development*, 2009, **13**(3), 483–488.

[74] Silverman, R.B., Andruszkiewicz, R., Nanavati, S.M., Taylor, C.P., Vartanian, M.G. *Journal of Medicinal Chemistry*, 1991, **34**, 2295–2298.

[75] Dooley, D.J., Taylor, C.P., Donevan, S., Feltner, D. *Trends in Pharmacological Sciences*, 2007, **28**, 75–82.

[76] Aiyer, L.P., Brown, P.K., Gajda, P.A., Mastrell, C.E. PCT International Application, 2009. WO 2009136375 A1.

[77] Evans, D.A., Ennis, M.D., Mathre, D.J. *Journal of the American Chemical Society*, 1982, **104**, 1737–1739.

[78] Gooding, O.W., Voladri, R., Bautista, A., Hopkins, T., Huisman, G., Jenne, S., Ma, S., Mundorff, E.C., Savile, M.M., Truesdell, S.J., Wong, J.W. *Organic Process Research & Development*, 2010, **14**, 119–126.

[79] Murtagh, L., Dunne, C., Gabellone, G., Panesar, N.J., Field, S., Reeder, L.M., Saenz, J., Smith, G.P., Kissick, K., Martinez, C., John, G. Van Alsten, J.G., Evans, M.C., Franklin, L.C., Nanninga, T., Wong, J. *Organic Process Research & Development*, 2011, **15**, 1315–1327.

[80] DeSantis, G., Zhu, Z., Greenberg, W.A., Wong, K., Chaplin, J., Hanson, S.R., Farwell, B., Nicholson, L.W., Rand, C.L., Weiner, D.P., Robertson, D.E., Burk, M.J. *Journal of the American Chemical Society*, 2002, **124**, 9024–9025.

[81] De Santis, G., Wong, K., Farwell, B., Chapman, K., Zhu, Z., Tomlinson, G., Huang, H., Tan, X., Bibbs, L., Chen, P., Kretz, K., Burk, N.J. *Journal of the American Chemical Society*, 2003, **125**(38), 11476–11477.

[82] Reetz, M.T., Becker, M.H., Klein, H.-W., Stockigt, D. *Angewandte Chemie, International Edition*, 1999, **38**, 1758–1761.

[83] Guo, J., Wu, J., Siuzdak, G., Finn, M.G. *Angewandte Chemie, International Edition*, 1999, **38**, 1755–1758.

[84] Bergeron, S., Chaplin, D.A., Edwards, J.H., Ellis, B.S.W., Hill, C.L., Holt-Tiffin, K., Knight, J.R., Mahoney, T., Osborne, A.P., Ruecroft, G. *Organic Process Research & Development*, 2006, **10**, 661–665.

[85] Suffness, M., Wall, M.E. Discovery and development of taxol. In *Taxol: Science and Application* (ed. Suffness, M.), 1995, CRC Press, Boca Raton, FL, 3–25.

[86] Holton, R., Biediger, R., Joatman, P. Semisynthesis of taxol and taxotere. In *Taxol: Science and Application* (ed. Suffness, M.), 1995, CRC Press, Boca Raton, FL, 97–123.

[87] Patel, R.N. *Annual Review of Microbiology*, 1995, **98**, 361–395.

[88] Baloglu, E., Kingston, D.G. *Journal of Natural Products*, 1999, **62**(7), 1068–1071.

[89] (a) Patel, R.N., Banerjee, A., Ko, R.Y. Howell, J., Li, W.-S., Comezoglu, F., Partyka, R., Szarka, L. *Biotechnology and Applied Biochemistry*, 1994, **20**(1), 23–33; (b) Al Hussainy, R., Verbeek, J., van der Born, D., Braker, A.H., Leyson, J.E., Knol, R.J., Booij, J., Herscheid, J.D.M. *Journal of Medicinal Chemistry*, 2010, **54**, 3480–3491.
[90] Peddi, S., Patel, M.V., Rohde, J.J. US Patent Application, October 21, 2010. US 20100267738 A1.
[91] Bennett, B.L., Elsner, J., Erdman, P., Hilgraf, R., Lebrun, L.A., Mccarrick, M., Moghaddam, M.F., Nagy, M.A., Norris, S., Paisner, D.A., Sloss, M., Romanow, W.J., Satoh, Y., Tikhe, J., Yoon, W.H., Delgrado, M. PCT International Application, October 26, 2012. WO 2012145569 A1.
[92] Guo, Z., Wong, M.K.Y., Hickey, M.R., Patel, B.P., Qian, X., Goswami, A. *Organic Process Research & Development*, 2014, **18**, 774–780.
[93] Gurbel, P.A., O'Connor, C.M., Cummings, C.C., Serebruany, V.L. *Pharmacological Research*, 1999, **40**(2), 107–111.
[94] Coukell, A.J., Markham, A. *Drugs*, 1997, **54**, 745–751.
[95] Wang, L., Shen, J., Tang, Y., Chen, Y., Wen Wang, W., Cai, Z., Du, Z. *Organic Process Research & Development*, 2007, **11**(3), 487–489.
[96] Uhm, K.-N., Lee, S.-J., Kim, H.-K., Kang, H.-Y., Lee, Y.J. *Journal of Molecular Catalysis B: Enzymatic*, 2007, **45**, 34–38.
[97] Pianko, S., McHutchison, J.G. *Journal of Gastroenterology and Hepatology*, 2000, **15**(6), 581–586.
[98] Ferenci, P., Brunner, H., Nachbaur, K., Datz, C., Gschwantler, M., Hofer, H., Stauber, R., Hackl, F., Jessner, W., Rosenbeiger, M., Petra, M.-S., Hegenbarth, K., Gangl, A., Vogel, W. *Hepatology*, 2001, **34**(5), 1006–1011.
[99] Bonkovsky, H.L., Stefancyk, D., McNeal, K., Banner, B.F., Liu, Q., Zucker, G.M., Israel, J., Stagias, J., Colker, J. *Digestive Diseases and Sciences*, 2001, **46**(10), 2051–2059.
[100] Tamarez, M., Morgan, B., Wong, G.S.K., Tong, W., Bennett, F., Lovey, R., McCormick, J.L., Zaks, A. *Organic Process Research & Development*, 2003, **7**(6), 951–953.
[101] Sit, S.Y., Parker, R.A., Motoc, I., Han, W., Balasubramanian, N., Catt, J.D., Brown, P.J., Harte, W.E., Thompson, M.D., Wright, J.J. *Journal of Medicinal Chemistry*, 1990, **33**(11), 2982–2999.
[102] Patel, R.N., McNamee, C.M., Szarka, L.J. *Applied Microbiology and Biotechnology*, 1992, **38**, 56–60.
[103] Kunishima, M., Kawachi, C., Hioki, K., Terao, S., Tani, S. *Tetrahedron*, 2001, **57**(8), 1551–1557.
[104] Gill, I., Patel, R.N. *Bioorganic & Medicinal Chemistry Letters*, 2006, **16**(3), 705–709.
[105]. King, A.O., Corley, E.G., Anderson, R.K., Larsen, R.D., Verhoeven, T.R., Reider, P.J., Xiang, Y.B., Belley, M., Leblanc, Y., Labelle, M., Prasit, P., Zamboni, R.J. *Journal of Organic Chemistry*, 1993, **58**, 3731–3735.
[106] Shinkai, I., King, A.O., Larsen, R.D. *Pure and Applied Chemistry*, 1994, **66**(7), 1551–1556.
[107] Zhao, M., King, A.O., Larsen, R.D., Verhoeven, T.R., Reider, P.J. *Tetrahedron Letters*, 1997, **36**, 2641–2644.
[108] Shafiee, A., Motamedi, H., King, A. *Applied Microbiology and Biotechnology*, 1998, **49**, 709–717.
[109] Liang, J., Borup, B., Mitchell, V., Mundorff, E., Lalonde, J., Huisman, G.W. PCT International Application, 2009. WO 2009042984 A1.
[110] Liang, J., Lalonde, J., Borup, B., Mitchell, V., Mundorff, E., Trinh, N., Kochrekar, D.A., Nair, C.R., Pai, G.G. *Organic Process Research & Development*, 2010, **14**, 193–198.
[111] Patel, R.N., Chu, L., Mueller, R.H. *Tetrahedron: Asymmetry*, 2003, **14**(20), 3105–3109.
[112] Xu, Z., Singh, J., Schwinden, M.D., Zheng, B., Kissick, T.P., Patel, B., Humora, M.J., Quiroz, F., Dong, L., Hsieh, D.-M., Heikes, J.E., Pudipeddi, M., Lindrud, M.D., Srivastava, S.K., Kronenthal, D.R., Mueller, R.H. *Organic Process Research & Development*, 2002, **6**(3), 323–328.
[113] Bong, Y.K., Vogel, M., Collier, S.J., Mitchell, V., Mavinahalli, J. PCT International Application, 2011. WO 2011005527.
[114] Roth, B.D. *Progress in Medicinal Chemistry*, 2002, **40**, 1–22.
[115] Ma, S.K., Gruber, J., Davis, C., Newman, L., Gray, D., Wang, A., Grate, J., Huisman, G.W., Sheldon, R.A. *Green Chemistry*, 2012, **12**, 81–86.
[116] Goldberg, S., Guo, Z., Chen, S., Goswami, A., Patel, R.N. *Enzyme and Microbial Technology*, 2008, **43**, 544–549.
[117] Jajoo, H., Mayol, R., LaBudde, J., Blair, I. *Drug Metabolism and Disposition*, 1989, **17**(6), 634–640.

[118] Mayol, R. US Patent Application, 2000. US 6150365.
[119] Yevich, J., New, J., Lobeck, W., Dextraze, P., Bernstein, E., Taylor, D., Yocca, F., Eison, M., Temple, D. Jr. *Journal of Medicinal Chemistry*, 1992, **35**(24), 4516–4525.
[120] Yevich, J., Mayol, R., Li, J., Yocca, F. US Patent Application, 2003. US 2003022899.
[121] Patel, R., Chu, L., Nanduri, V., Jianqing, L., Kotnis, A., Parker, W., Liu, M., Mueller, R. *Tetrahedron: Asymmetry*, 2005, **16**(16), 2778–2783.
[122] Goldberg, S., Nanduri, V., Chu, L., Johnston, R., Patel, R. *Enzyme and Microbial Technology*, 2006, **39**(7), 1441–1450.
[123] Zalman, L.S., Brothers, M.A., Dragovich, P.S., Zhou, R., Prins, T.J., Worland, S.T., Patick, A.K. *Antimicrobial Agents and Chemotherapy*, 2000, **44**, 1236–1241.
[124] Dragovich, P.S., Prins, T.J., Zhou, R., Webber, S.E., Marakovits, J.T., Fuhrman, S.A., Patick, A.K., Matthews, D.A., Lee, C.A., Ford, C.E., Burke, B.J., Rejto, P.A., Hendrickson, T.F., Tuntland, T., Brown, E.L., Meador, J.W. III, Ferre, R.A., Harr, J.E.V., Kosa, M.B., Worland, S.T. *Journal of Medicinal Chemistry*, 1999, **42**, 1213–1224.
[125] Tao, J., McGee, K. *Organic Process Research & Development*, 2002, **6**, 520–524.
[126] Kagechika, H., Kawachi, E., Hashimoto, Y., Shudo, K., Himi, T. *Journal of Medicinal Chemistry*, 1989, **32**(12), 2583–2588.
[127] Morriss-Kay, G.M. Retinoids in mammalian embryonic development. In *Retinoids: Their Physiological Function and Therapeutic Potential* (ed. Sherbet, G.V.), 1997, JAI Press Inc., Greenwich, CT, 79–93.
[128] Moon, R.C., Mehta, R.G. *Advances in Experimental Medicine and Biology*, 1986, **206**, 399–411.
[129] Patel, R.N., Chu, L., Chidambaram, R., Zhu, J., Kant, J. *Tetrahedron: Asymmetry*, 2002, **13**(4), 349–355.
[130] Junien, J.L., Leonard, B.E. *Clinical Neuropharmacology*, 1989, **12**(5), 353–374.
[131] Patel, R.N., Banerjee, A., Liu, M., Hanson, R.L., Ko, R., Howell, J.M., Szarka, L.J. *Biotechnology and Applied Biochemistry*, 1993, **17**(2), 139–153.
[132] Balasubramanian, N., Brown, P.J., Catt, J.D., Han, W.T., Parker, R.A., Sit, S.Y., Wright, J.J. *A potent, tissue-selective, synthetic inhibitor of HMG-CoA reductase. Journal of Medicinal Chemistry*, 1989, **32**(9), 2038–2041.
[133] Patel, R.N., Banerjee, A., McNamee, C.G., Brzozowski, D.B., Hanson, R.L., Szarka, L.J. *Enzyme and Microbial Technology*, 1993, **15**(12), 1014–1021.
[134] Guo, Z., Chen, Y., Goswami, A., Hanson, R.L., Patel, R.N. *Tetrahedron: Asymmetry*, 2006, **17**(1), 1589–1602.
[135] Chaturvedula, P.V., Chen, L., Civiello, R., Degnan, A.P., Dubowchik, G.M., Han, X., Jiang, J.J., Macor, J.E., Poindexter, G.S., Tora, G.O., Luo, G. US Patent Application, 2007. US 20070149503 A1.
[136] Han, X., Civiello, R.L., Conway, C.M., Cook, D.A., Davis, C.D., Macci, R., Pin, S.S., Ren, S.X., Schartman, R., Signor, L.J., Thalody, G., Widmann, K.A., Xu, C., Chaturvedula, P.V., Macor, J.E., Dubowchik, G.M. *Bioorganic & Medicinal Chemistry Letters*, 2012, **22**(14), 4723–4727.
[137] Chaturvedula, P.V., Mercer, S.E., Pin, S.S., Thalody, G., Xu, C., Conway, C.M., Keavy, D., Signor, L., Cantor, G.H., Mathias, N., Moench, P., Denton, R., Macci, R., Schartman, R., Whiterock, V., Davis, C., Macor, J.E., Dubowchik, G.M. *Bioorganic & Medicinal Chemistry Letters*, 2013, **23**(11), 3157–3161.
[138] Hanson, R.L., Davis, B.L., Goldberg, S.L., Johnston, R.M., Parker, W.L., Tully, T.P., Montana, M.A., Patel, R.N. *Organic Process Research & Development*, 2008, **12**(6), 1119–1129.
[139] Haque, T.S., Ewing, W.R., Mapelli, C., Lee, V.G., Sulsky, R.B., Riexinger, D.J., Martinez, R.L., Zhu, Y.Z. PCT International Application, 2007. WO 2007082264.
[140] Mapelli, C., Natarajan, S.I., Meyer, J.-P., Bastod, M.M., Bernatowicz, M.S., Lee, V.G., Pluscec, J., Riexinger, D.J., Sieber-McMaster, E.S., Constantine, K.L., Smith-Monroy, C.A., Golla, R., Ma, Z., Longhi, D.A. Shi, D., Xin, L., Taylor, J.R., Koplowitz, B., Chi, C.L., Khanna, A., Robinson, G.W., Seethala, S., Antal-Zimanyi, H.A., Stoffel, R.H., Han, S., Whaley, J.M., Huang, C.S., Krupinski, J., Ewing, W.R. *Journal of Medicinal Chemistry*, 2009, **52**(23), 7788–7799.
[141] Chen, Y., Goldberg, S.L., Hanson, R.L., Parker, W.L., Gill, I., Tully, T.P., Montana, M., Goswami, A., Patel, R.N. *Organic Process Research & Development*, 2011, **15**(1), 241–248.
[142] Pohl, M., Lingen, B., Muller, M. *Chemistry—A European Journal*, 2002, **8**(23), 5288–5295.
[143] Ward, O.P., Singh, A. Recent developments in enzymatic acyloin condensations. In *Biocatalysis in the Pharmaceutical and Biotechnology Industries* (ed. Patel, R.N.), 2007, CRC Press, Boca Raton, FL, 605–621.

[144] Ohta, H., Sugai, T. Enzyme-mediated decarboxylation reactions in organic synthesis. In *Stereoselective Biocatalysis* (ed. Patel, R.N.), 2000, Marcel Dekker, New York, 487–526.
[145] Pohl, M., Liese, A. Industrial processes using lyases for C–C, C–N, and C–O-bond formation. In *Biocatalysis in the Pharmaceutical and Biotechnology Industries* (ed. Patel, R.N.), 2007, CRC Press, Boca Raton, FL, 661–676.
[146] Resch, V., Schrittwieser, J.H., Siirola, E., Kroutil, W. *Current Opinion in Biotechnology*, 2011, **22**(6), 793–799.
[147] Sprenger, G.A., Pohl, M. *Journal of Molecular Catalysis B: Enzymatic*, 1999, **6**(3), 145–159.
[148] Goetz, G., Iwan, P., Hauer, B., Breuer, M., Pohl, M. *Biotechnology and Bioengineering*, 2001, **74**(4), 317–325.
[149] Lingen, B., Pohl, M., Demir, A.S., Liese, A., Mueller, M. *Oxidative Stress and Disease*, 2004, **11**(Thiamine), 113–129.
[150] Moscona, A. *The New England Journal of Medicine*, 2005, **353**(13), 1363–1373.
[151] Kragl, U., Gygax, D., Ghisalba, O., Wandrey, C. *Angewandte Chemie, International Edition*, 1991, **103**(7), 854–855.
[152] Mahmoudian, M., Noble, D., Drake, C.S., Middleton, R.F., Montgomery, D.S., Piercey, J.E., Ramlakhan, D., Todd, M., Dawson, M.J. *Enzyme and Microbial Technology*, 1997, **20**(5), 393–400.
[153] Fauci, A.S., Masur, H., Gelmann, E.P., Markham, P.D., Hahn, B.H., Lane, H.C. *Annals of Internal Medicine*, 1985, **102**(6), 800–813.
[154] Coates, J.A.V., Cammack, N., Jenkinson, H.J., Helen, J., Mutton, I.M., Pearson, B.A., Storer, R., Cameron, J.M., Penn, C.R. *Antimicrobial Agents and Chemotherapy*, 1992, **36**, 202–205.
[155] Storer, R., Clemens, I.R., Lamont, B., Noble, S.A., Williamson, C., Belleau, B. *Nucleosides & Nucleotides*, 1993, **12**, 225–236.
[156] Mahmoudian, M., Baines, B.S., Drake, C.S., Hale, R.S., Jones, P., Piercey, J.E., Montgomery, D.S., Purvis, I.J., Storer, R. *Enzyme and Microbial Technology*, 1993, **15**, 749–755.
[157] Mahmoudian, M., Dawson, M.J. Chemoenzymic production of the anti-viral agent Epivir (3TC). In *Biotechnology of Antibiotics*, 2nd edition (ed. Strohl, W.R.), 1997, Marcel Dekker Inc., New York, 753–777.
[158] Gijsen, H.J.M., Wong, C.-H. *Journal of the American Chemical Society*, 1994, **116**, 8422–8423.
[159] Wong, C.-H., Garcia-Junceda, E., Chen, L., Blanco, O., Gijsen, H.J.M., Steensma, D.H. *Journal of the American Chemical Society*, 1995, **117**, 3333–3339.
[160] Greenberg, W.A., Varvak, A., Hanson, S.R., Wong, K., Huang, H., Chen, P., Burk, M.J. *Proceedings of the National Academy of Sciences of the United States of America*, 2004, **101**(16), 5788–5793.
[161] Wolberg, M., Hilker, I., Luiten, R., Wubbolts, M., Mink, D. *Biotechnology Journal*, 2006, **1**(5), 537–548.
[162] Chen, K.X., Njoroge, F.G. *Progress in Medicinal Chemistry*, 2010, **49**, 1–36.
[163] Chen, K.X., Nair, L., Vibulbhan, B., Yang, W., Arasappan, A., Bogen, S.L., Venkatraman, S., Bennett, F., Pan, W., Blackman, M.L., Padilla, A.I., Prongay, A., Cheng, K.-C., Tong, X., Shih, N.-Y., Njoroge, G.F. *Journal of Medicinal Chemistry*, 2009, **52**, 1370–1379.
[164] Zhang, R., Mamai, A., Madalengoitia, J.S. *Journal of Organic Chemistry*, 1999, **64**, 547–555.
[165] Koehler, V., Bailey, K.R., Znabet, A., Raftery, J., Helliwell, M., Turner, N.J. *Angewandte Chemie, International Edition*, 2010, **49**, 2182–2184.
[166] Znabet, A., Ruijter, E., de Kanter, F.J.J., Koehler, V., Helliwell, M., Turner, N.J., Orru, R.V.A. *Angewandte Chemie, International Edition*, 2010, **49**, 5289–5292.
[167] Li, T., Liang, J., Ambrogelly, A., Brennan, B., Gloor, G., Huisman, G., Lalonde, J., Lekhal, A., Mijts, B., Muley, S., Newman, L., Tobin, M. Wong, G., Zaks, A., Zhang, X. *Journal of the American Chemical Society*, 2012, **134**, 6467–6472.
[168] Gibson, D.T., Parales, R.E. *Current Opinion in Biotechnology*, 2000, **11**, 236–243.
[169] Parales, R.E., Resnick, S.M. Applications of aromatic hydrocarbon dioxygenases. In *Biocatalysis in the Pharmaceutical and Biotechnology Industries* (ed. Patel, R.N.), 2007, CRC Press, Boca Raton, FL, 299–332.
[170] Louise, C., Nolan, A.E., O'Connor, K.E. *Biotechnology Letters*, 2008, **30**, 1879–1891.
[171] Karlsson, A., Parales, J.V., Parales, R.E., Gibson, D.T., Eklund, H., Ramaswamy, S. *Science* (Washington, DC), 2003, **299**(5609), 1039–1042.
[172] Boyd, D.R., Sharma, N.D., Bowers, N.I., Brannigan, I.N., Groocock, M.R., Malone, J.F., McConville, G., Allen, C.C.R. *Advanced Synthesis & Catalysis*, 2005, **347**(7+8), 1081–1089.

[173] Boyd, D.R., Bugg, T.D.H. *Organic & Biomolecular Chemistry*, 2006, **4**(2), 181–192.
[174] Chartrain, M., Jackey, B., Taylor, C., Sandfor, V., Gbewonyo, K., Lister, L., Dimichele, L., Hirsch, C., Heimbuch, B., Maxwell, C., Pascoe, D., Buckland, B., Greasham, R. *Journal of Fermentation Technology*, 1998, **86**, 550–558.
[175] Chartrain, M., Salmon, P.M., Robinson, D.K., Buckland, B.C. *Current Opinion in Biotechnology*, 2000, **11**, 209–214.
[176] Zhang, N., Stewart, B.G., Moore, J.C., Greasham, R.L., Robinson, D.K., Buckland, B., Lee, C. *Metabolic Engineering*, 2000, **2**(4), 339–348.
[177] Amanullah, A., Hewitt, C.J., Nienow, A.W., Lee, C., Chartrain, M., Buckland, B.C., Drew, S.W., Woodley, J.M. *Biotechnology and Bioengineering*, 2003, **81**(4), 405–420.
[178] Edwards, G., Weston, A.H. *Pharmacology & Therapeutics*, 1990, **48**(2), 237–258.
[179] Robertson, D.W., Steinberg, M.I. *Annual Reports in Medicinal Chemistry*, 1989, **24**, 91–100.
[180] Ashwood, V.A., Buckingham, R.E., Cassidy, F., Evans, J.M., Far, E.A., Hamilton, T.C., Nash, D.J., Stemp, G., Willcocks, K. *Journal of Medicinal Chemistry*, 1986, **29**(11), 2194–2201.
[181] Patel, R.N., Banerjee, A., Davis, B., Howell, J., McNamee, C., Brzozowaski, D., North, J., Kronenthal, D., Szarka, L. *Bioorganic & Medicinal Chemistry*, 1994, **2**(6), 535–542.
[182] Benigni, D., Stankavage, R., Chiang, S.-J., Hou, H., Eagan, B., Gu, D., Hou, D., Mintzmyer, L., Tully, T.P., Davis, B.L., Hargro, I., Mascari, M., Galvin, G., Stein, G., McConlogue, C.W., Comezoglu, F.T. PCT International Application, 2004. WO 2004026254 A1.
[183] Goodin, S., Kane, M.P., Rubin, E.H. *Journal of Clinical Oncology*, 2004, **22**(10), 2015–2025.
[184] Nicolaou, K.C., Roschangar, F., Vourloumis, D. *Angewandte Chemie, International Edition*, 1998, **37**(15), 2014–2045.
[185] Altmann, K.-H. *Organic & Biomolecular Chemistry*, 2004, **2**(15), 2137–2152.
[186] Lin, N., Brakora, K., Seiden, M. *Current Opinion in Investigational Drugs*, 2003, **4**(6), 746–756.
[187] Low, J.A., Wedam, S.B., Lee, J.J., Berman, A.W., Brufsky, A., Yang, S.X., Poruchynsky, M.S., Steinberg, S.M., Mannan, N., Fojo, T., Swain, S.M. *Journal of Clinical Oncology*, 2005, **23**(12), 2726–2734.
[188] Basch, J.D., Chiang, S.-J., Liu, S.W., Nayeem, A., Sun, Y.-L. PCT International Application, 2004. WO 2004078978 A1.
[189] Basch, J., Chiang, S.J. *Journal of Industrial Microbiology and Biotechnology*, 2007, **34**(2), 171–176.
[190] Kim, M.-J., Ahn, Y., Park, J. Dynamic kinetic resolution and asymmetric transformations by enzyme-metal combination. In *Biocatalysis in the Pharmaceutical and Biotechnology Industries* (ed. Patel, R.N.), 2007, CRC Press, Boca Raton, FL, 249–272.
[191] Martin-Matute, B., Baeckvall, J.-E. *Current Opinion in Chemical Biology*, 2007, **11**(2), 226–232.
[192] Azerad, R., Buisson, D. *Current Opinion in Biotechnology*, 2000, **11**(6), 565–571.
[193] Ishide, T. *Current Medical Research and Opinion*, 2002, **18**(7), 404–413.
[194] Pamies, O., Baeckvall, J.-E. *Journal of Organic Chemistry*, 2001, **66**(11), 4022–4025.

CHAPTER 5

Dynamic Kinetic Resolution of Alcohols, Amines, and Amino Acids

Jusuk Lee, Yoon Kyung Choi, Jaiwook Park, and Mahn-Joo Kim

Department of Chemistry, Pohang University of Science and Technology, Pohang, South Korea

5.1 INTRODUCTION

Dynamic kinetic resolution (DKR) provides a powerful methodology for the complete transformation of racemates into single enantiomers [1–3]. In DKR, one enantiomer is selectively converted to the product with the simultaneous racemization of the other enantiomer, thus affording a high yield approaching 100%. Since the mid-2000s, several groups including ours have explored intensively a new approach coupling an enzymatic kinetic resolution with a metal-catalyzed racemization for DKR (Scheme 5.1) [4, 5]. As a result, several useful procedures have been developed for the DKR of secondary alcohols, primary amines, and amino acids. This chapter provides a review of these DKR procedures with detailed examples for each of them.

5.1.1 Kinetic and Dynamic Kinetic Resolution

The classical enzymatic kinetic resolution of a racemic substrate provides two enantiomers as the product and unreacted substrate. In this case, the theoretical maximum yield for one enantiomer is 50%. To improve the yield, the opposite enantiomer is racemized and then recycled for the second enzymatic kinetic resolution. The racemization–recycling process should be repeated at least three times to achieve more than a 90% yield. In the metalloenzymatic DKR, the enzymatic kinetic resolution occurs simultaneously with the metal-catalyzed racemization, so all the substrates can be converted to the products to provide a near 100% yield in a single operation. Each DKR process, however, gives only single enantiomeric products. Accordingly, two stereocomplementary enzymes are needed for the synthesis of a pair of enantiomeric products via DKR.

5.1.2 Enzymes as the Resolution Catalysts for DKR

Hydrolases such as lipases and proteases have been employed as the resolution catalysts for DKR. They include *Candida antarctica* lipase A (CAL-A), *C. antarctica* lipase B (CAL-B), *Burkholderia (formerly Pseudomonas) cepacia* lipase (BCL), *Pseudomonas*

Green Biocatalysis, First Edition. Edited by Ramesh N. Patel.

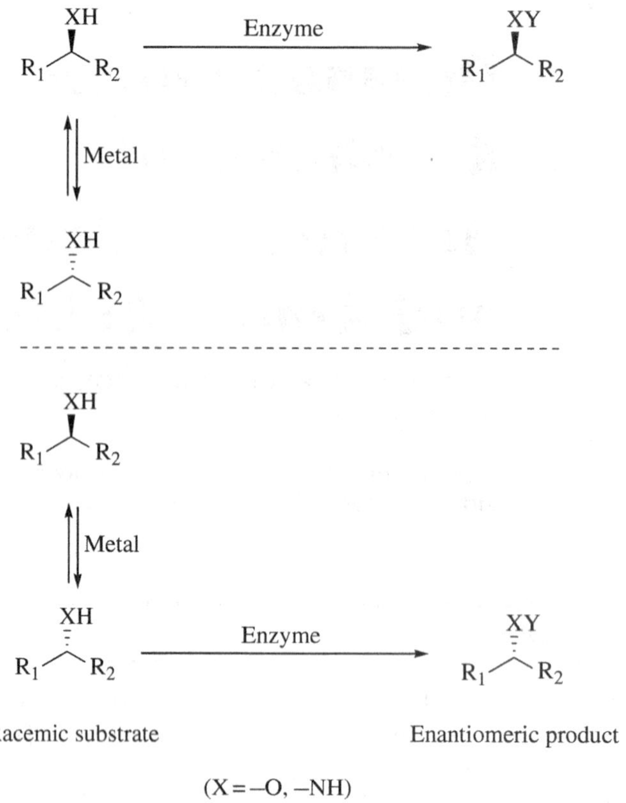

SCHEME 5.1
Metalloenzymatic dynamic kinetic resolution.

stutzeri lipase (PSL), *Candida rugosa* lipase (CRL), and subtilisin Carlsberg (SC) from *Bacillus licheniformis*. Among them, CAL-B (trade name, Novozym 435; immobilized on polyacrylamide resin) has been most frequently employed owing to its high activity and excellent thermostability. It has good stability, even at 100 °C. However, it has rather narrow substrate specificity. Recently, CAL-B was genetically engineered to show broader specificity [6]. BCL (trade name, lipase PS) accepts a wider range of substrates compared to CAL-B and has good thermostability. It is stable up to 60 °C. Commercially available BCL, however, has low activity. We recently developed a highly active BCL, which was coated with an ionic surfactant via lyophilization [7]. The ionic surfactant-coated BCL (ISCBCL) was more active than CAL-B (Novozym 435) and showed good performance in the DKR of secondary alcohols. CAL-A, PSL (trade name, lipase TL), and SC are relatively unstable and thus should be used at room temperature.

5.1.3 The Enantioselectivity of Enzymes in DKR

C. antarctica lipase B and BCL follow the Kazlauskas rule [8] for enantioselectivity in the transesterification of secondary alcohols: (i) the enantiomer shown in Figure 5.1 reacts more rapidly than the other if two substituents at the hydroxymethine center of secondary alcohol are different in size and (ii) the enantioselectivity increases with increasing difference in size between two substituents.

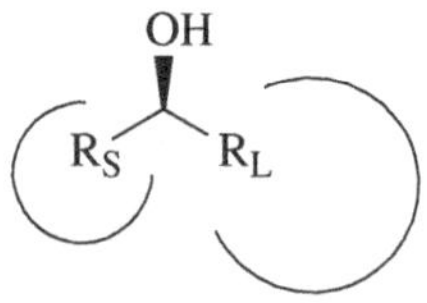

FIGURE 5.1
The enantiomer shown reacts more rapidly than the other in the lipase-catalyzed transesterification. R_L, large substituent; R_S, small substituent.

According to the Kazlauskas rule, CAL-B and BCL normally provide (*R*)-products in the DKR of simple secondary alcohols such as 1-phenyl-1-alkanols (Scheme 5.2a). On the other hand, SC is stereocomplementary [8] to CAL-B and BCL. It provides (*S*)-products in the DKR of simple secondary alcohols (Scheme 5.2b). CAL-A and PSL are somehow different from CAL-B and BCL in substrate specificity and stereospecificity. They can accept sterically more demanding secondary alcohols such as 1,2-diarylethanols, which are poorly reactive with CAL-B and BCL. Interestingly, they are stereocomplementary toward these substrates: CAL-A accepts

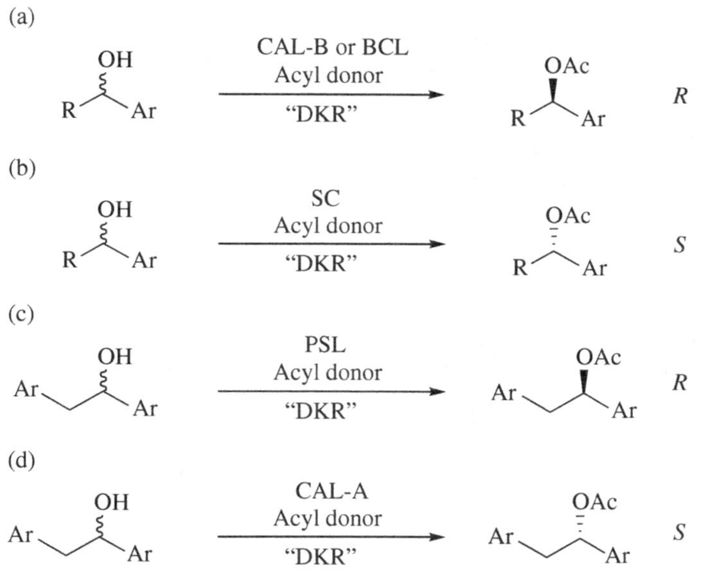

SCHEME 5.2 The enantioselectivity in the DKR of secondary alcohols with CAL-A, CAL-B, BCL, PSL, and SC.

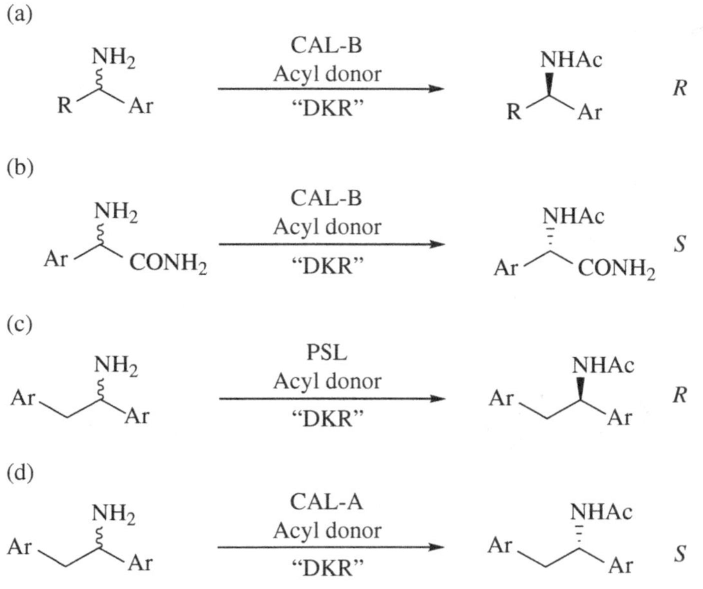

SCHEME 5.3 The enantioselectivity in the DKR of primary amines and amino acids with some lipases.

(*S*)-enantiomer but PSL does (*R*)-enantiomer (Scheme 5.2c and d). The lipases display the same enantiopreference toward primary amines and amino acids, which are isosteric with secondary alcohols, as shown in Scheme 5.3.

5.1.4 Metal (Complexes) as the Racemization Catalysts for DKR

Various transition metal complexes (Rh, Ru, and V) have been employed as the racemization catalysts for the DKR of secondary alcohols. Among them, ruthenium (Ru)-based racemization catalysts are described in Chart 5.1. Dimeric Ru complex **1** (Shvo's catalyst) displays satisfactory activity at high temperature (80 °C or higher). So it should be used with a thermostable enzyme such as Novozym 435 for successful DKR. Monomeric Ru complexes **5–10** are active at room temperature so that they can be coupled with thermally less stable enzymes. Ru complexes **7** and **9** are particularly useful because they have good stability and high activity under air. Their polymer-bound counterparts **8** and **10** are reusable several times with no significant loss in activity. Dimeric Ru complex **11** is active under light, where it dissociates into two monomeric species displaying good racemization activity at room temperature.

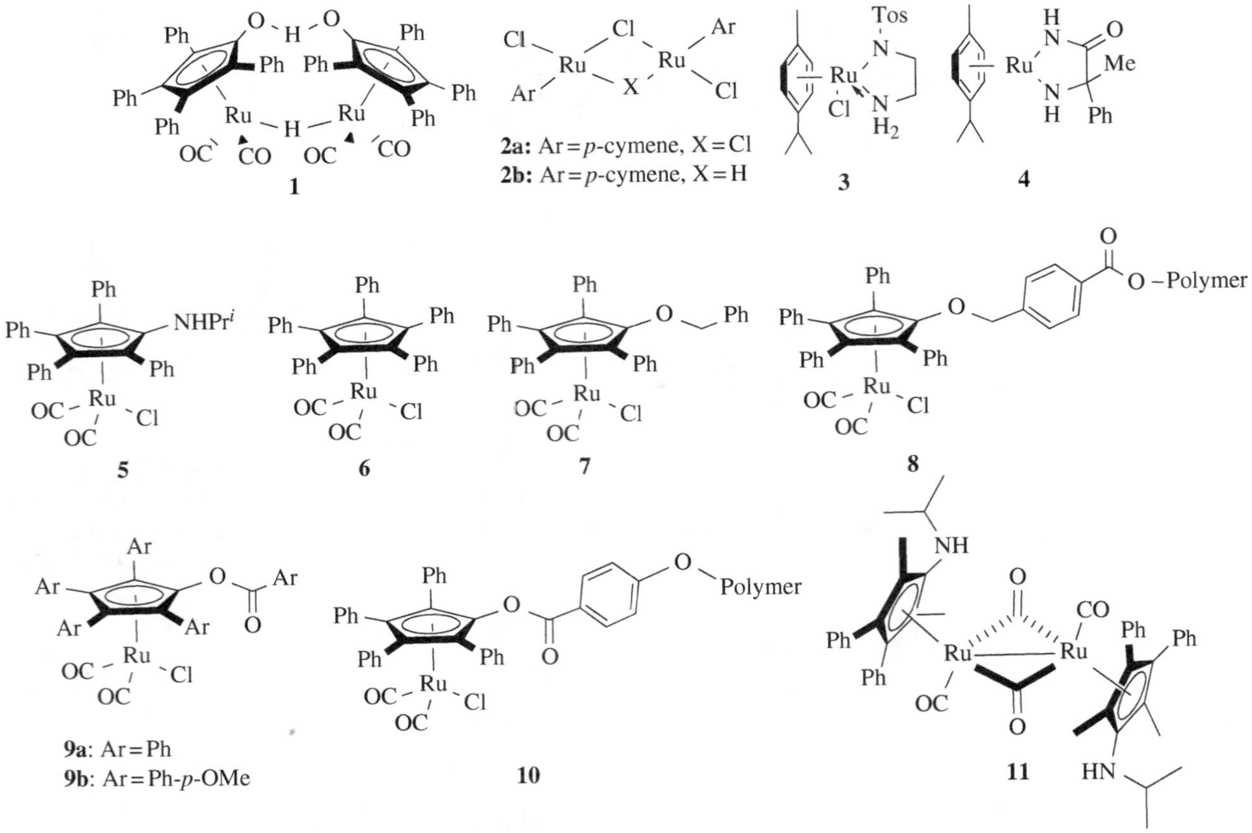

CHART 5.1
Ruthenium-based racemization catalysts for the DKR of secondary alcohols.

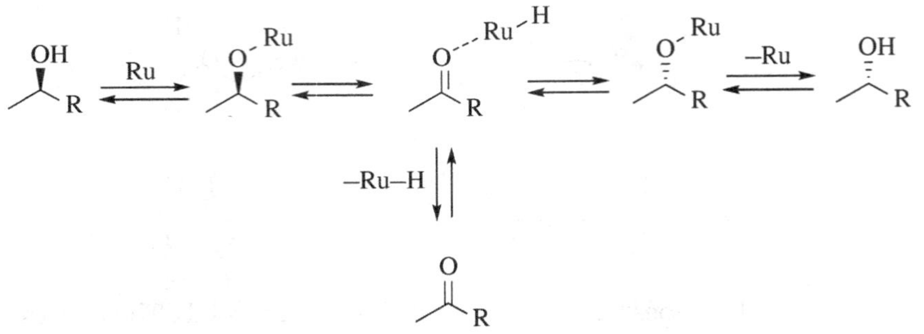

SCHEME 5.4
Ru-catalyzed racemization of secondary alcohol.

A simplified mechanism for the Ru-catalyzed racemization of secondary alcohol is described in Scheme 5.4. The racemization takes place via ketone. Ru catalysts **5–10** normally catalyze the racemization without releasing free ketone and thus provide the higher yields in DKR.

On the other hand, the racemization of amines is more difficult compared to that of alcohols. Several metal systems based on palladium (Pd), Ru, nickel (Ni), cobalt (Co), and Ir have been employed as the racemization catalysts. Pd-based catalysts include Pd/C, Pd/$BaSO_4$, and Pd/AlO(OH). They are readily available but require higher temperatures for satisfactory racemization. So they should be coupled with thermostable enzymes such as Novozym 435 for the successful DKR. A possible mechanism for the Pd-catalyzed racemization of amine is described in Scheme 5.5. The racemization occurs via reversible dehydrogenation/hydrogenation steps including an imine intermediate. The imine intermediate can react with starting material to afford a secondary amine as the byproduct. The deamination of substrate and byproduct are also possible at elevated temperature. In case the

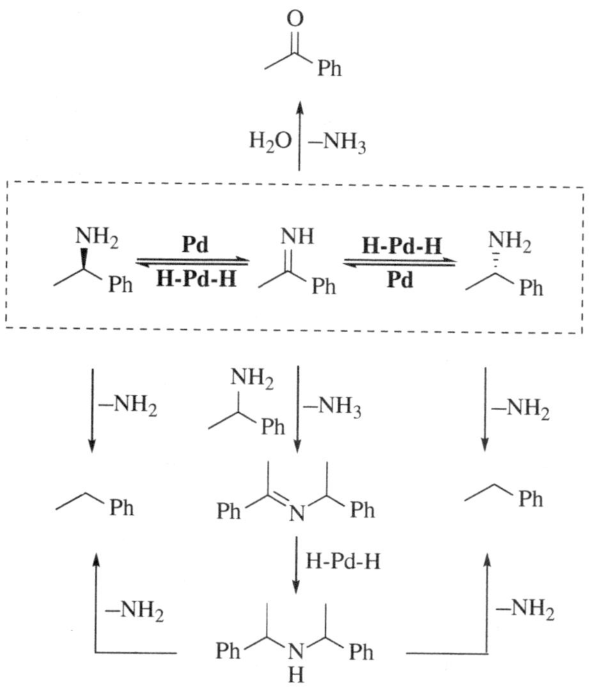

SCHEME 5.5
Pd-catalyzed racemization and possible side reactions.

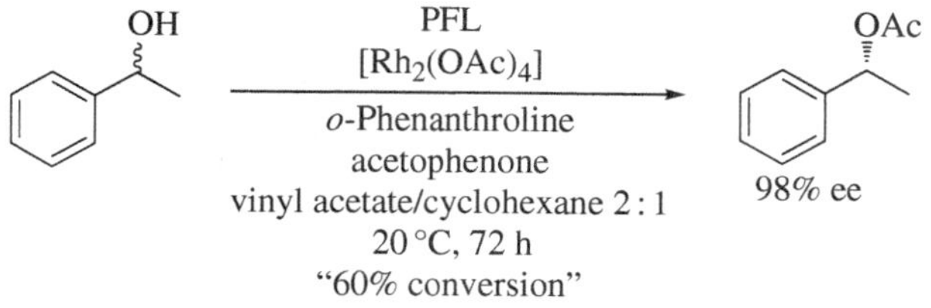

SCHEME 5.6
DKR of 1-phenylethanol by the combination of a *Pseudomonas fluorescens* lipase and a Rh complex.

enzyme employed for DKR contains a significant amount of water, the hydrolysis of imine intermediate takes place during the DKR process to yield the corresponding ketone as the by-product. Owing to these side reactions, the DKR of amines often provides lower yields.

5.2 DYNAMIC KINETIC RESOLUTION OF SECONDARY ALCOHOLS

The Williams group reported the DKR of 1-phenylethanol as the first metalloenzymatic DKR [9]. This DKR employed *Pseudomonas fluorescens* lipase (PFL) as the resolution enzyme, a rhodium complex as the racemization catalyst, and vinyl acetate as the acyl donor in cyclohexane, but the reaction required a stoichiometric amount of acetophenone for hydrogen transfer and stopped at 60% conversion (Scheme 5.6).

The Bäckvall group reported a more practical DKR procedure using CAL-B (Novozym 435) as the resolution enzyme and the Shvo's catalyst **1** as the racemization catalyst (Scheme 5.7) [10]. In this DKR, *p*-chlorophenyl acetate (PCPA) was the best acyl donor, and several secondary alcohols were successfully transformed to the corresponding acetates with high enantiomeric excesses and good yields (Chart 5.2). The DKRs of diols [11] and β-azidoalcohols [12] were also successfully accomplished with CAL-B and **1** in the presence of PCPA (Charts 5.3 and 5.4).

For the DKR of hydroxy esters [13] and β-haloalcohols [14], the Bäckvall group used *Pseudomonas cepacia* lipase (PCL) immobilized on ceramic (Lipase PS-C) instead

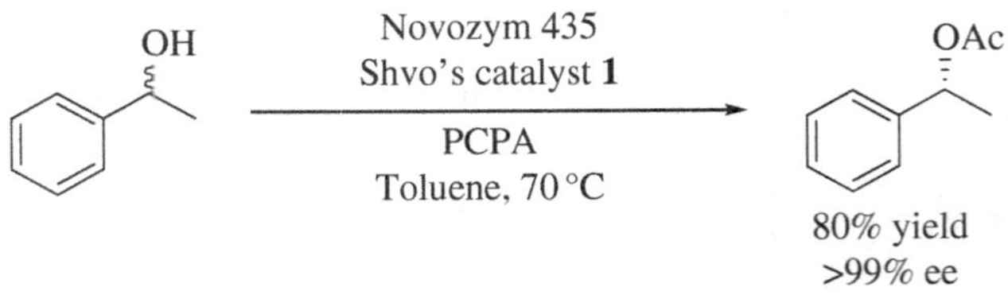

SCHEME 5.7

DKR of 1-phenylethanol by the combination of Novozym 435 and Shvo's catalyst.

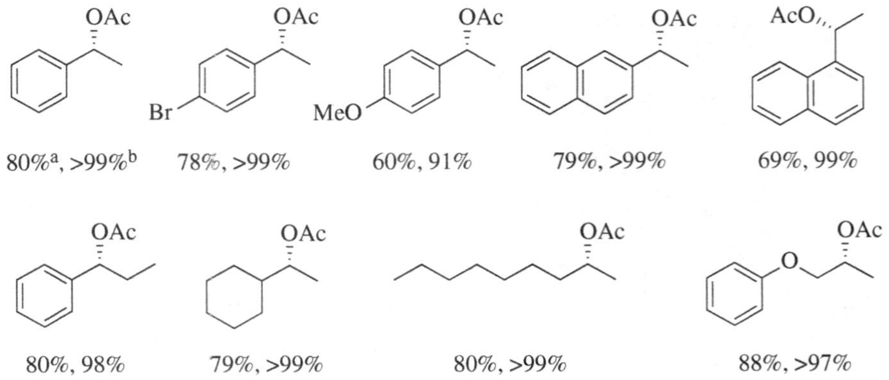

[a]Yield, [b]enantiopurity

CHART 5.2

Products from the DKR of simple secondary alcohols.

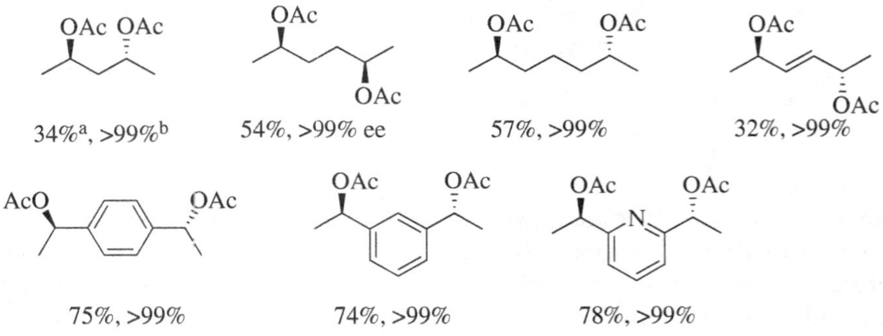

[a]Yield, [b]enantiopurity

CHART 5.3

Products from the DKR of diols.

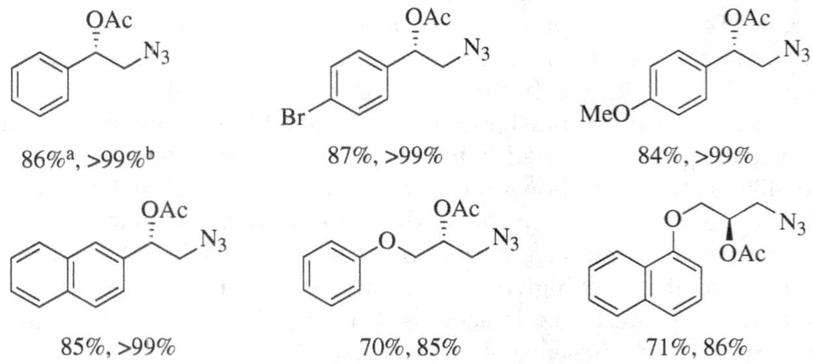

[a]Yield, [b]enantiopurity

CHART 5.4

Products from the DKR of β-azido alcohols.

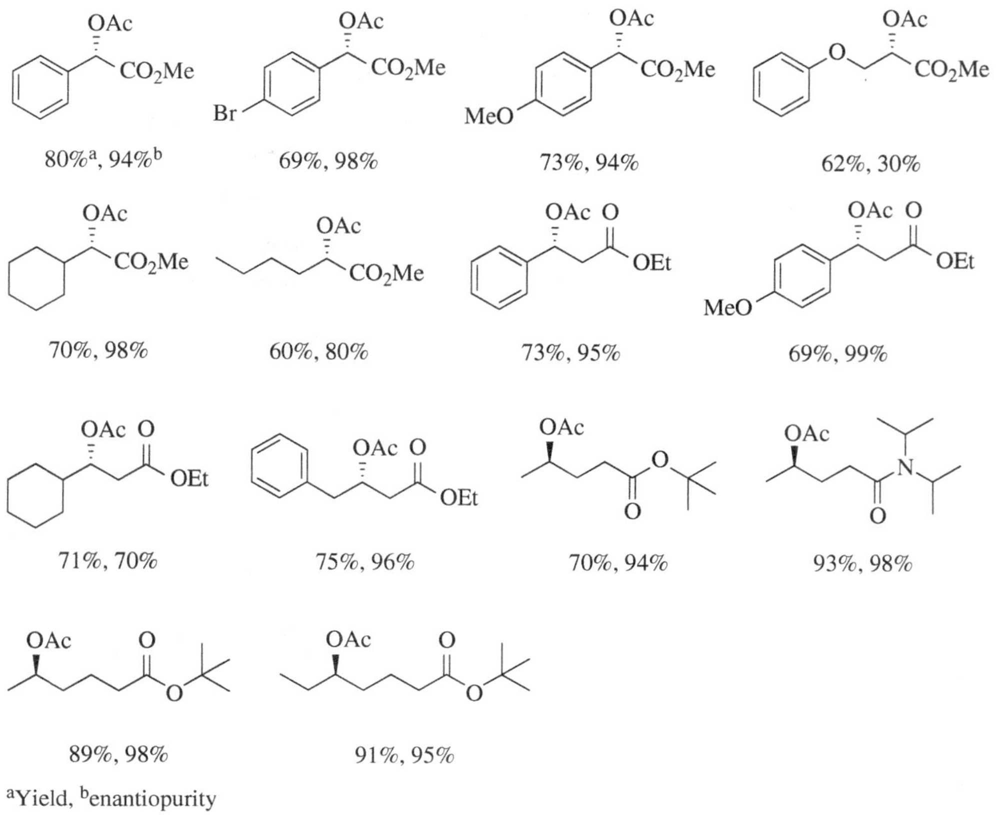

CHART 5.5 Products from the DKR of hydroxy esters.

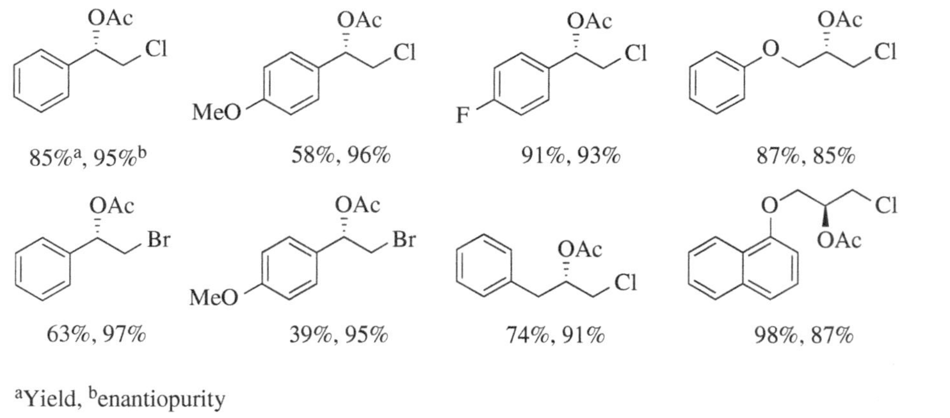

CHART 5.6 Products from the DKR of β-haloalcohols.

of CAL-B as the resolution enzyme because the former displayed better performance toward these substrates. The products from the DKR of hydroxy esters and β-haloalcohols are described in Charts 5.5 and 5.6, respectively.

We reported the efficient DKR of 1,2- and 1,3-diols protected with a trityl group using PCL immobilized on earth (brand name, lipase PS-D) and Shvo's catalyst (Scheme 5.8) [15]. In this DKR, the trityl group introduced for the protection of primary alcohol served as the steric auxiliary to induce the high enantioselectivity of enzyme. The DKR reactions thus proceeded with high enantioselectivity to give the products of excellent enantiopurity with good yields. The DKR of α- and β-hydroxyl aldehydes protected with a sterically bulky group also proceeded with high enantioselectivity.

We applied the DKR procedure to the asymmetric reductive acylation of ketones and the asymmetric hydrogenation of enol acetates [16]. In the asymmetric reductive

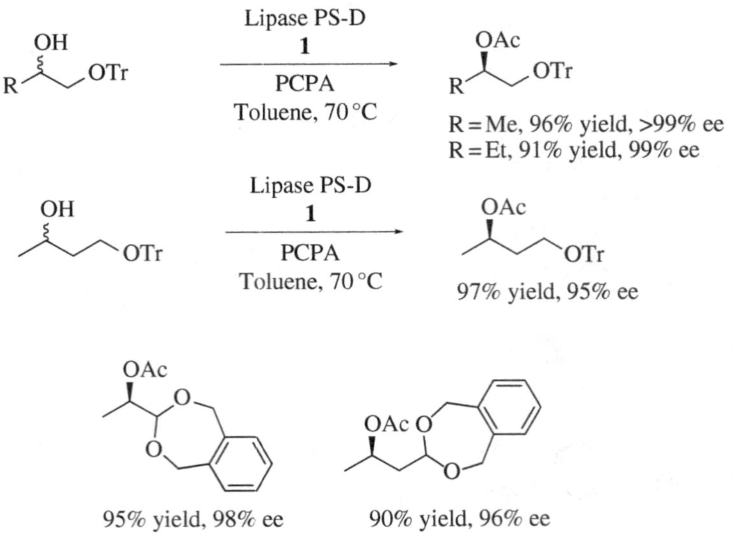

SCHEME 5.8
DKR of substrates protected with a bulky group.

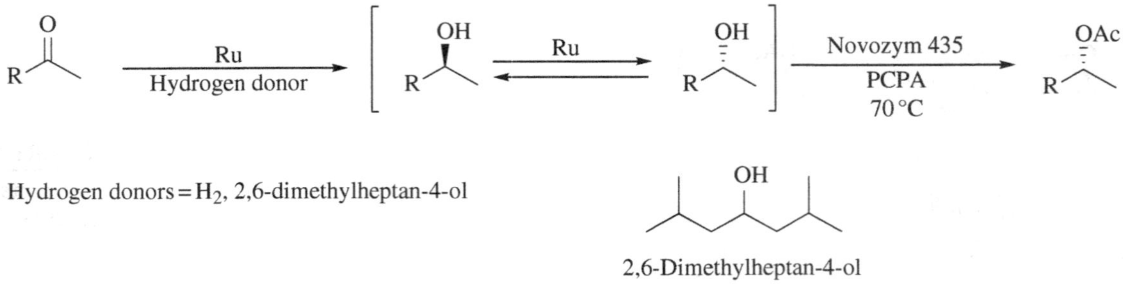

SCHEME 5.9
Asymmetric reductive acylation of ketones based on DKR.

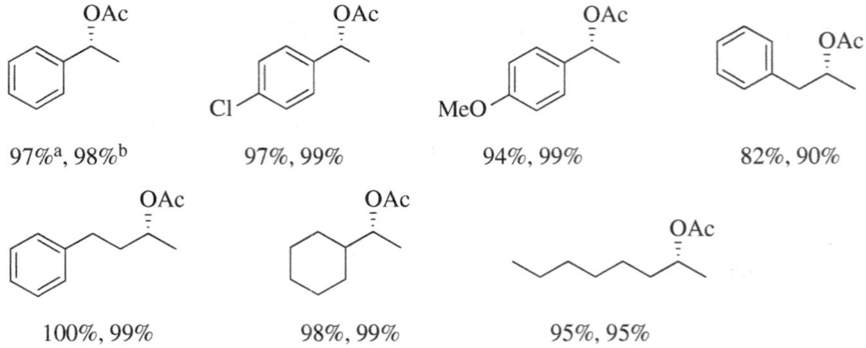

[a]Yield, [b]enantiopurity

CHART 5.7
Products from the asymmetric reductive acylation of ketones.

acylation of ketones, three different reactions take place in tandem. Ketones are first hydrogenated to secondary alcohols, which then undergo racemization and resolution for DKR (Scheme 5.9 and Chart 5.7). Here, the same Ru catalyst catalyzes both hydrogenation and racemization. This asymmetric transformation was achieved with Novozym 435, Shvo's catalyst, hydrogen molecule or 2,6-dimethylheptan-4-ol as the hydrogen source, and PNPA as the acyl donor at 70 °C. It is noted that 2,6-dimethylheptan-4-ol is not acetylated during the DKR reaction because Novozym 435 normally does not accept secondary alcohols with two sterically bulky substituents at the hydroxymethine center.

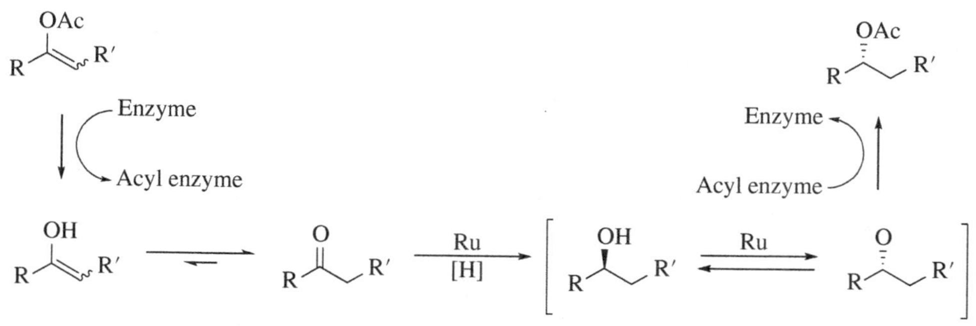

SCHEME 5.10
Five sequential reactions in the asymmetric hydrogenation of enol acetates.

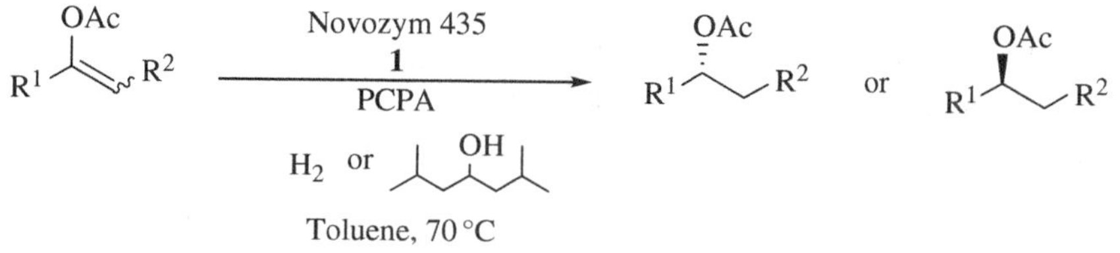

SCHEME 5.11
Asymmetric hydrogenation of enol acetates based on the DKR.

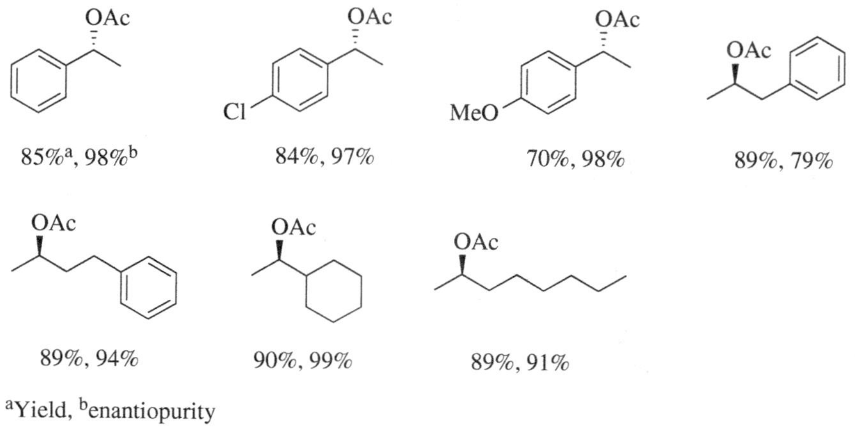

CHART 5.8
Products from the asymmetric hydrogenation of enol acetates.

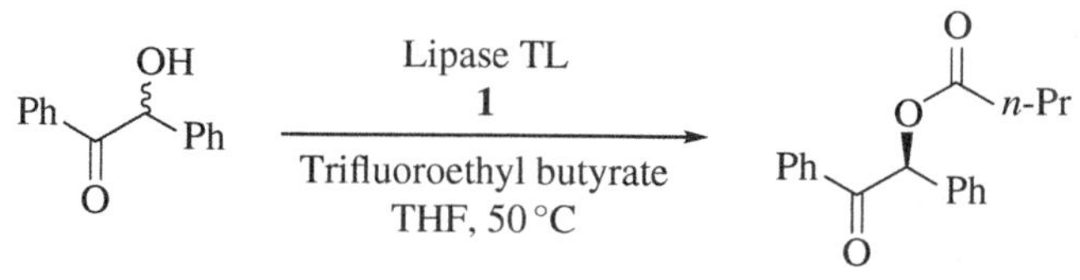

SCHEME 5.12
DKR of benzoins by the combination of PSL and Shvo's catalyst.

In the asymmetric reduction of enol acetates, five different reactions take place sequentially: deacetylation of enol acetate, keto–enol isomerization, hydrogenation of ketone, and finally racemization and resolution of alcohol for DKR (Scheme 5.10). Here, the enol acetate acts as both the precursor of ketone and the acyl donor. The overall transformation was performed with Novozym 435 and Shvo's catalyst in the presence of hydrogen molecule or 2,6-dimethylheptan-4-ol as the hydrogen donor to provide the products of high enantiopurity with good yields in most cases (Scheme 5.11 and Chart 5.8).

Hoyos *et al.* achieved the DKR of benzoins using *P. stutzeri* lipase (PSL; trade name, lipase TL) and **1** in the presence of trifluoroethyl butyrate as the acyl donor (Scheme 5.12 and Chart 5.9) [17]. It is noted that benzoins are poorly reactive with Novozym 435 and lipase PS because they have two sterically demanding substituents at the hydroxymethine center.

We reported the first DKR of allylic alcohols using cymene–Ru complex **2** as the racemization catalyst (Scheme 5.13 and Chart 5.10) [18]. In this DKR, lipase PS-C

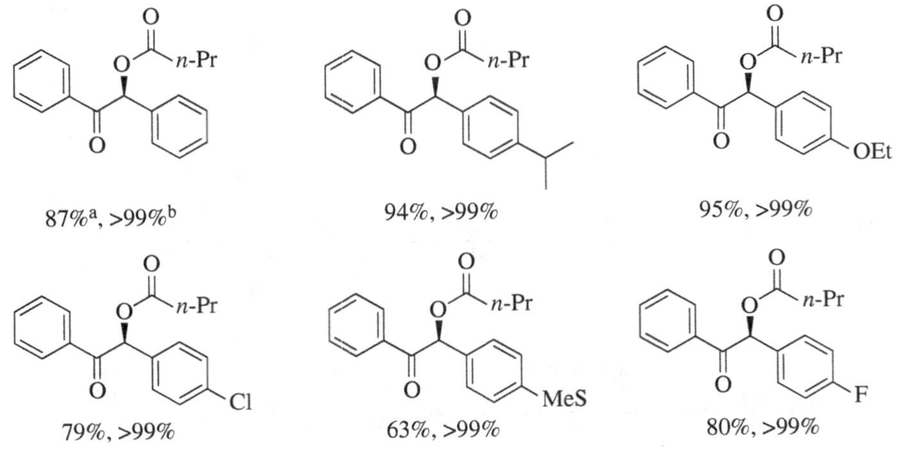

aYield, benantiopurity

CHART 5.9
Products from the DKR of benzoins.

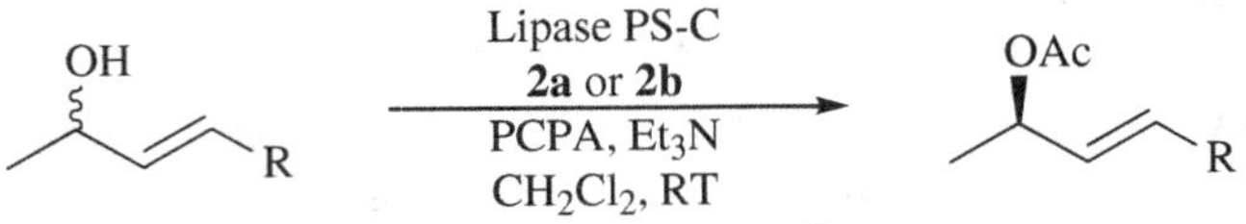

SCHEME 5.13
DKR of allylic alcohols with lipase PS-C and **2**.

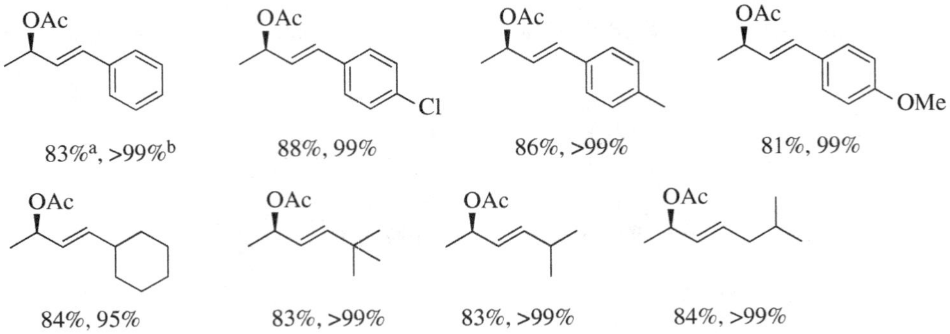

aYield, benantiopurity

CHART 5.10
Products from the DKR of allylic alcohols with lipase PS-C and **2**.

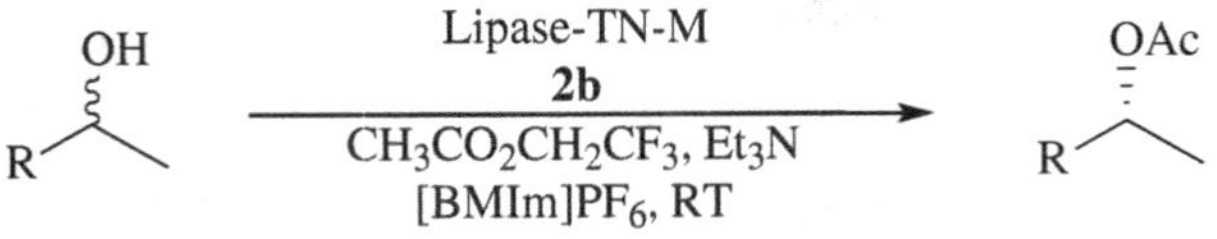

SCHEME 5.14
DKR of secondary alcohols in ionic liquid.

(lipase PS immobilized on celite) was employed as the resolution enzyme with PCPA as the acyl donor. Later, the Bäckvall group reported the DKR of cyclic allylic alcohols using lipase PS-IM and **1** [19].

We found that the cymene–Ru complex **2b** displayed good activity as the racemization catalyst in ionic liquids such as [EMIm]BF_4 ([EMIm]=1-ethyl-3-methylimidazolium) and [BMIm]PF_6 ([BMIm]=1-butyl-3-methylimidazolium) at room temperature. Thus the first DKR in ionic liquid was achieved using lipase PS-C, **2b**, and trifluoroethyl acetate as the acyl donor at room temperature (Scheme 5.14 and Chart 5.11) [20]. The products were readily removed by simple extraction with ether, and the remaining ionic liquid containing enzyme and Ru catalyst was reused without further treatment for the second run. This easy recovery and recycling of reaction medium and catalysts is a big advantage of DKR in ionic liquid.

Ru catalysts **3** and **4** were introduced by Sheldon *et al.* and Meijer *et al.* as the racemization catalyst for the DKR of alcohols. The DKR of 1-phenylethanol with **3**

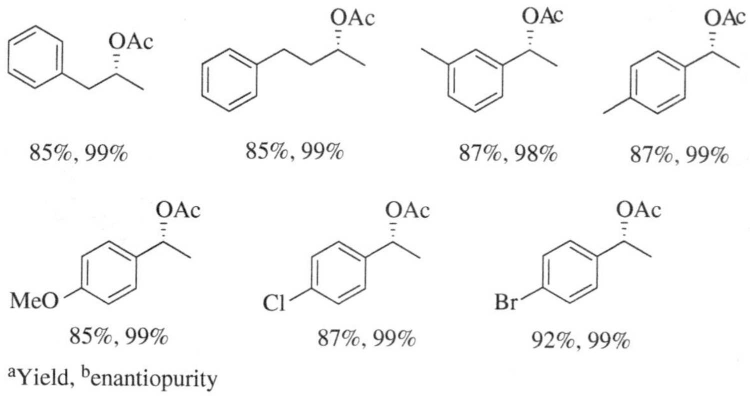

CHART 5.11
Products from the DKR of secondary alcohols in ionic liquid.

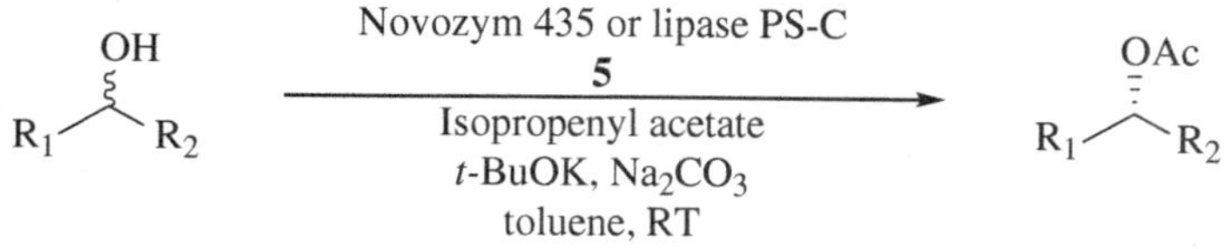

SCHEME 5.15
DKR of secondary alcohols using **5** as the racemization catalyst.

and CAL-B gave chiral acetate with 76% yield and 99% ee [21]. Catalyst **4** was employed in an iterative DKR to produce enantiomeric oligoesters [22].

Most of the DKR processes described previously employ **1** or **2** as the racemization catalyst. As mentioned in the "Introduction" section, these catalysts are active at elevated temperature and thus should be coupled with thermally stable enzymes for successful DKR. In particular, **1** displayed satisfactory activity at 80 °C and thus was coupled with the highly thermostable Novozym 435. Furthermore, it was not compatible with readily available acyl donors such as vinyl and isopropenyl acetates but with PCPA, which was difficult to remove from the reaction mixture.

In 2002, we reported that monomeric Ru catalyst **5** had a good racemization activity at room temperature and excellent compatibility with isopropenyl acetate [23]. We thus accomplished the first DKR of secondary alcohols, at room temperature by combining **5** with Novozym 435 or lipase PS-C in the presence of isopropenyl acetate (Scheme 5.15). A wide range of secondary alcohols including simple alcohols, allylic alcohols, alkynyl alcohols, diols, hydroxyl esters, and chlorohydrins were transformed to their acetates with good yields and excellent enantiomeric excesses in the DKR using **5** (Chart 5.12) [24].

The good activity of Ru catalyst **5** at room temperature allowed the use of a thermally fragile protease (SC) in the DKR. The first (*S*)-selective DKR of secondary alcohols was achieved with **5** and SC at room temperature (Scheme 5.16 and Chart 5.13) [25]. It is noted that commercial SC was poorly active in organic solvent and thus treated with a surfactant, Brij 56, before use to improve its activity in organic solvent. The treated SC displayed a 4000-fold enhanced activity.

Later, the Bäckvall group reported that the Ru catalyst **6**, an analogue of **5**, also displayed a good racemization activity at room temperature [26]. It was compatible with isopropenyl acetate. Dynamic kinetic resolution with Novozym 435 and **6** in the presence of isopropenyl acetate gave good results at room temperature (Chart 5.14). The catalyst **6** was also combined with lipase PS-C [27] in other DKR applications.

It was also possible to perform the (*S*)-selective DKR by combining catalyst **6** with SC [28] or a mutant (CAL-B W104A) of CAL-B [29, 30] (Scheme 5.17). The Ema group combined **6** with a mutant of BCL for the DKR of benzyl alcohols with a long alkyl substituent at the α-position (Scheme 5.18 and Chart 5.15) [31].

The Ru catalysts **1–6** were air sensitive, so they were difficult to reuse. We found that the Ru catalyst **7** was air stable and its polymer-attached form **8** was reusable several times without losing any significant activity [32]. Later, we found another

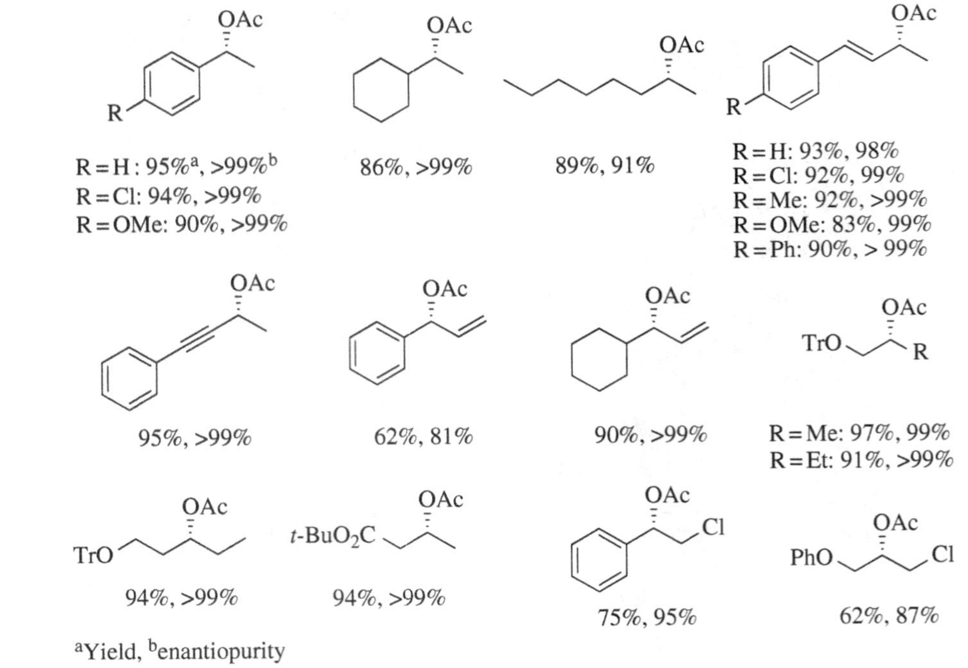

CHART 5.12

Products from the DKR of secondary alcohols using **5** as the racemization catalyst.

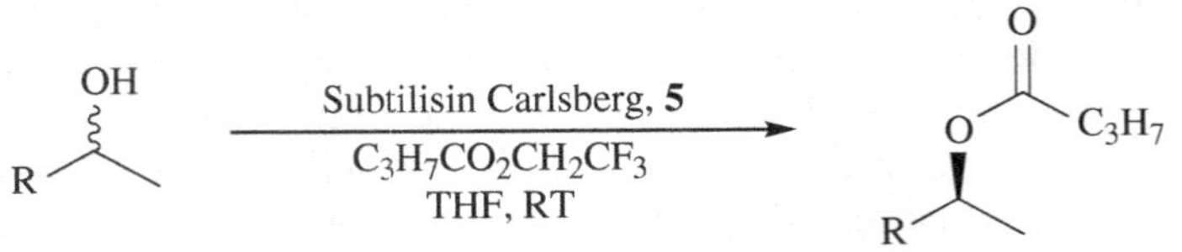

SCHEME 5.16

(*S*)-Selective DKR of secondary alcohols with subtilisin Carlsberg and **5**.

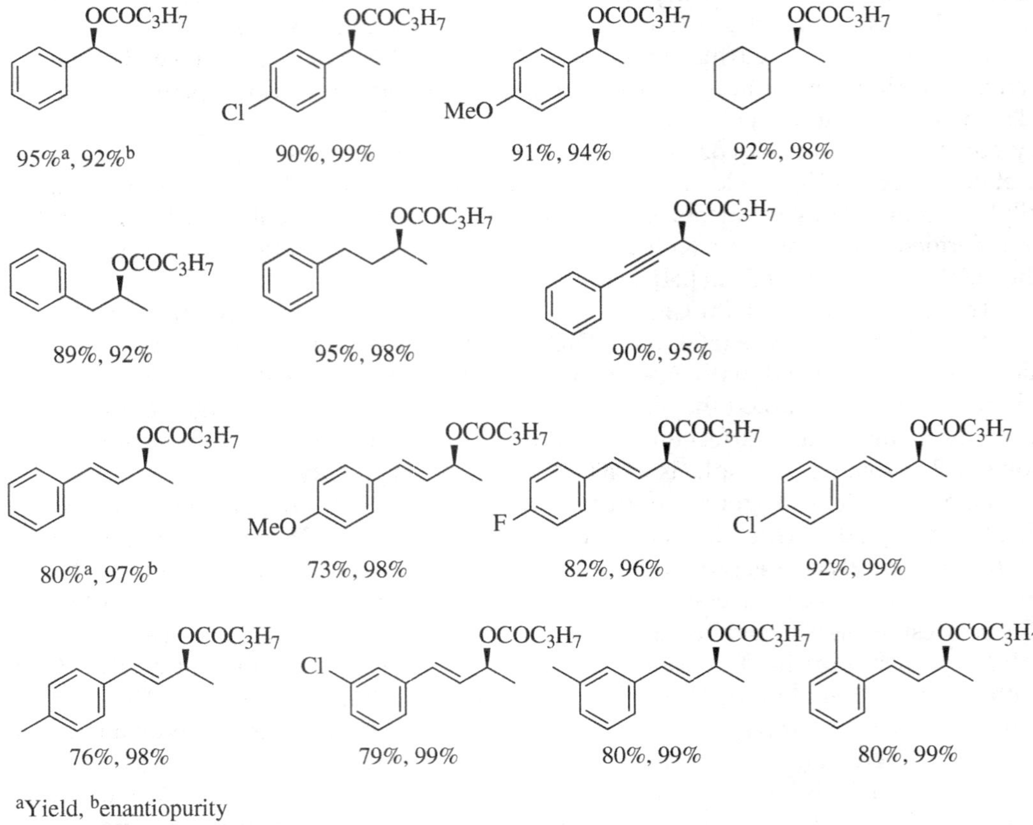

CHART 5.13

Products from the (*S*)-selective DKR of secondary alcohols with subtilisin Carlsberg and **5**.

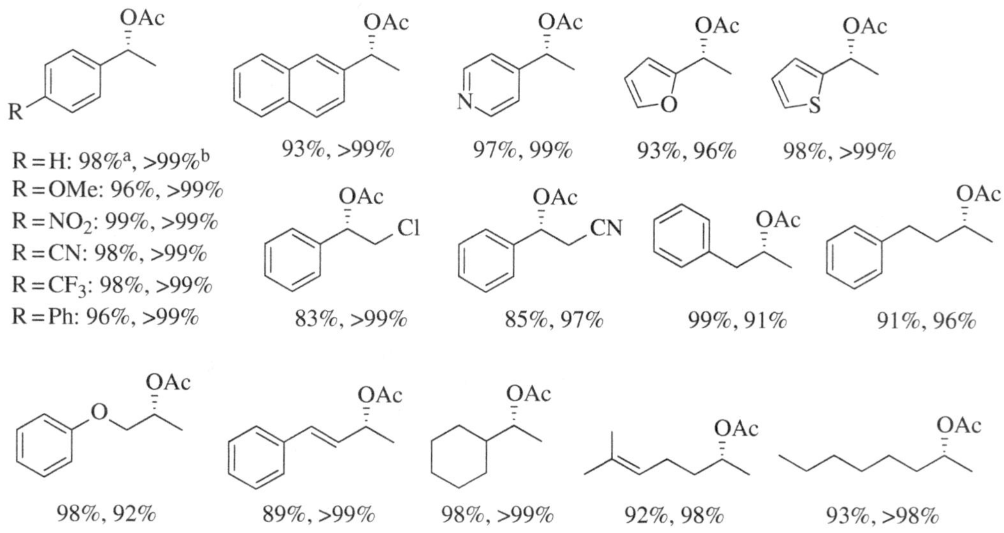

[a]Yield, [b]enantiopurity

CHART 5.14

Products from the DKR of secondary alcohols using **6** and Novozym 435.

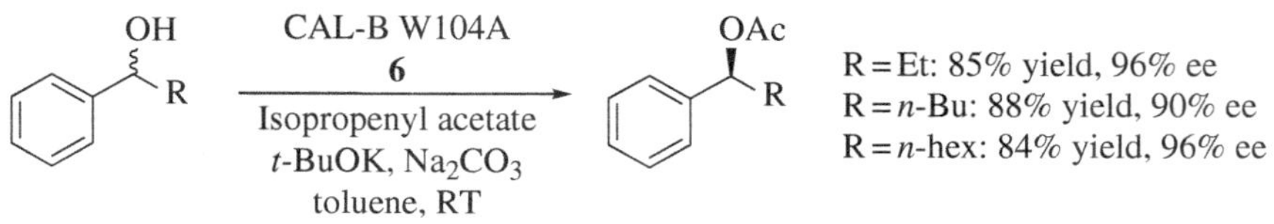

SCHEME 5.17

(*S*)-Selective DKR of 1-phenyl-1-alkanols with CAL-B W104A and **6**.

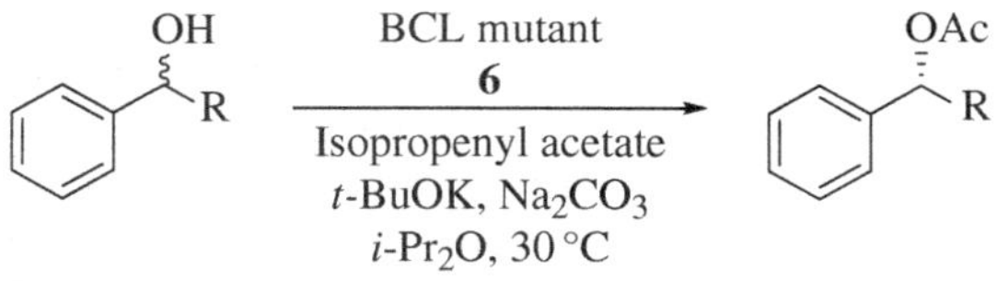

SCHEME 5.18

DKR of 1-phenyl-1-alkanols with a BCL mutant and **6**.

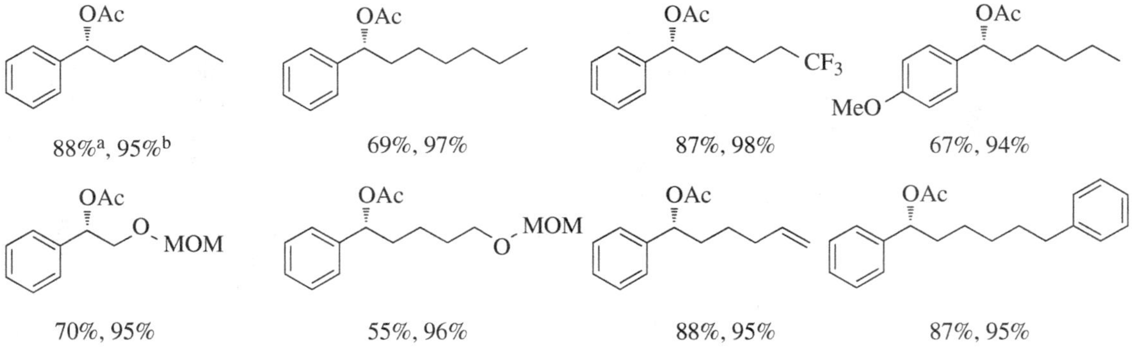

[a]Yield, [b]enantiopurity

CHART 5.15

Products from the DKR of 1-phenyl-1-alkanols with a BCL mutant and **6**.

air-stable catalyst **9a**, which was more practical to synthesize compared to **7** [33]. Its polymer-bound form **10** was also reusable several times without losing any significant activity [34]. As an application of air-stable **9a**, we performed the DKR of 1,2-diarylethanols. In this DKR, PSL was used as its enzyme partner, which was highly (*R*)-enantioselective ($E > 200$) (Scheme 5.19 and Chart 5.16) [33]. It is noted that

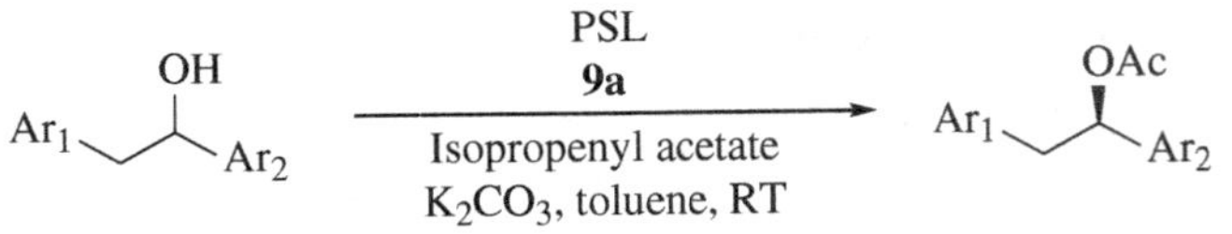

SCHEME 5.19

DKR of 1,2-diarylethanols with PSL and **9a**.

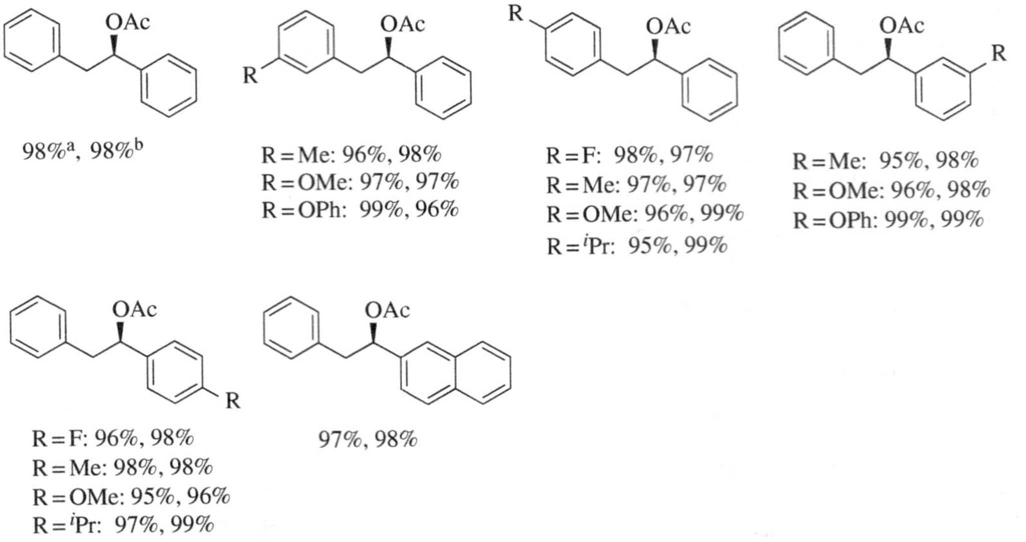

CHART 5.16

Products from the DKR of 1,2-diarylethanols with PSL and **9a**.

[a]Yield, [b]enantiopurity

Novozym 435 and lipase PS-C are inapplicable to the DKR of 1,2-diarylethanols because they are poorly active toward these substrates.

Recently, we found that **9b** was 14 times more active than **9a** in the racemization of 1-phenylethanol [35]. This Ru catalyst was coupled with a highly active lipase preparation (ionic surfactant-coated *B. cepacia* lipase, ISCBCL) [7] to achieve the fastest DKR at room temperature (Scheme 5.20). The DKR of 1-phenylethanol was complete in 1 h and provided 95% isolated yield and 99% ee. A wide range of secondary alcohols including alkyl aryl carbinols, γ-chlorohydrins, homoallylic alcohols, propargyl alcohols, boron-containing alcohols, and diarylmethanols, which had been previously unexplored as the substrates for DKR, were subject to the DKR using **9b** and ISCBCL [36]. All the γ-chlorohydrins and homoallylic alcohols were transformed into acetates of high enantiopurity with high yields (Charts 5.17 and 5.18). In case of alkyl aryl carbinols, high enantiopurity (99% ee) was obtained if a bulky substituent (isopropyl or *t*-butyl) was present at the *p*-position of an aryl group (Chart 5.19). α-Substituted propargyl alcohols carrying a bulky trimethylsilyl (TMS) or *t*-butyl group at the terminal carbon were also converted to the products of high enantiopurity (92–99% ee) (Chart 5.20). Here, it is noteworthy that the stereochemistry of the products from the TMS or *t*-butyl-substituted α-arylpropargyl alcohols (Chart 5.20b and c) was opposite to that of those from α-arylalkanols and α-arylallyl alcohols (Charts 5.17–5.19). This implies that ISCBCL recognizes the TMS or *t*-butyl-substituted acetylenic group as the bulkier of two substituents at the hydroxymethine center. In case of alkyl aryl carbinols with a bulky boronic substituent on the benzene ring, the enantiopurity of product was high in case the alkyl chain was small (Me and Et) (Chart 5.21). The enantioenriched boron-containing products are useful as the precursors for the synthesis of more complex compounds via the Suzuki coupling. Diarylmethanols with two bulky substituents at the hydroxymethine center were the most difficult to transform among the substrates tested. Satisfactory yields and good enantiopurities were obtained in only a few cases (Chart 5.22).

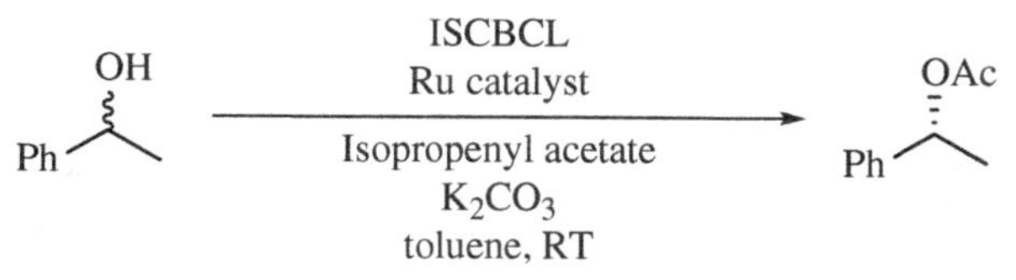

SCHEME 5.20
DKR of 1-phenylethanol with **9b** and ISCBCL.

Ru Catalyst	Time (h)	Yield (%)	ee (%)
5	1	75	99
6	1	85	99
9b	1	95	99

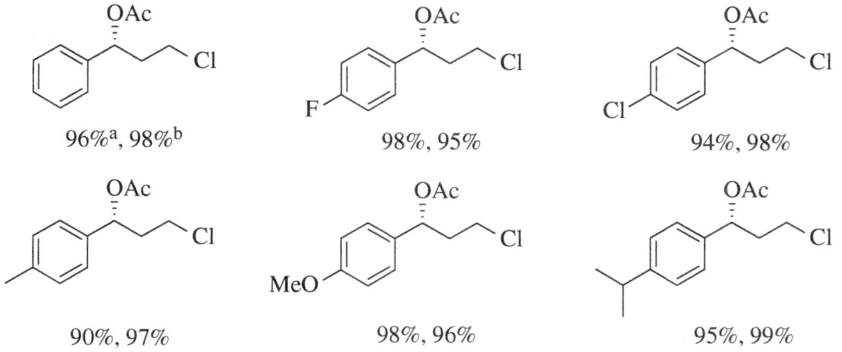

[a]Yield, [b]enantiopurity

CHART 5.17
Products from the DKR of γ-chlorohydrins with **9b** and ISCBCL.

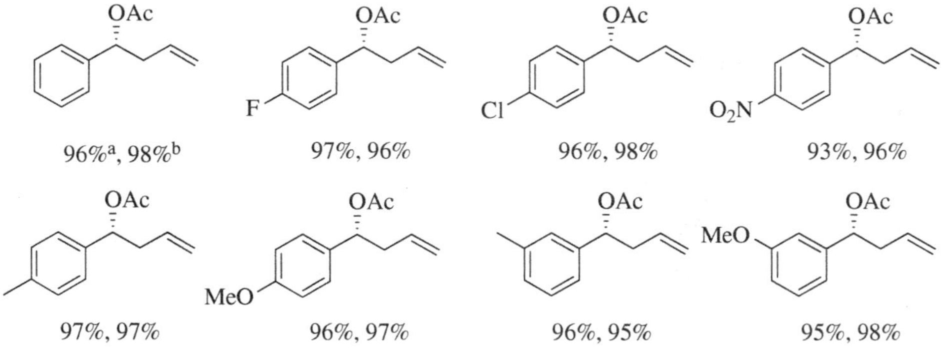

[a]Yield, [b]enantiopurity

CHART 5.18
Products from the DKR of homoallylic alcohols with **5** and ISCBCL.

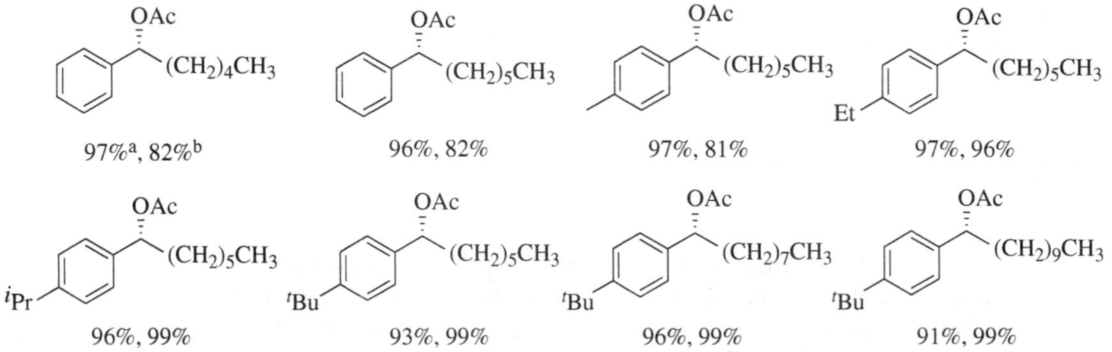

[a]Yield, [b]enantiopurity

CHART 5.19
Products from the DKR of alkyl aryl carbinols with **9b** and ISCBCL.

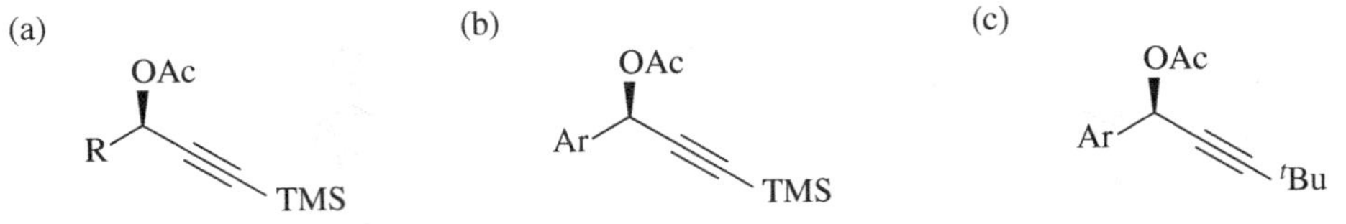

(a)
R = Me: 92% yield, 99% ee
R = Et: 92% yield, 99% ee
R = *n*-Pr: 93% yield, 98% ee
R = *n*-Bu: 94% yield, 97% ee
R = *n*-Pen: 97% yield, 92% ee
R = *n*-Hex: 97% yield, 92% ee

(b)
Ar = Ph: 94% yield, >99% ee
Ar = Ph-*p*-F: 91% yield, 96% ee
Ar = Ph-*p*-Cl: 90% yield, 96% ee
Ar = Ph-*p*-Br: 91% yield, 96% ee
Ar = Ph-*p*-Me: 94% yield, >99% ee
Ar = Ph-*p*-OMe: 90% yield, 98% ee
Ar = Ph-*m*-OMe: 91% yield, 98% ee

(c)
Ar = Ph: 97% yield, 99% ee
Ar = Ph-*p*-F: 97% yield, 95% ee
Ar = Ph-*p*-Cl: 96% yield, 95% ee
Ar = Ph-*p*-Me: 96% yield, 95% ee

CHART 5.20
Products from the DKR of propargyl alcohols with **5** or **9b** and ISCBCL.

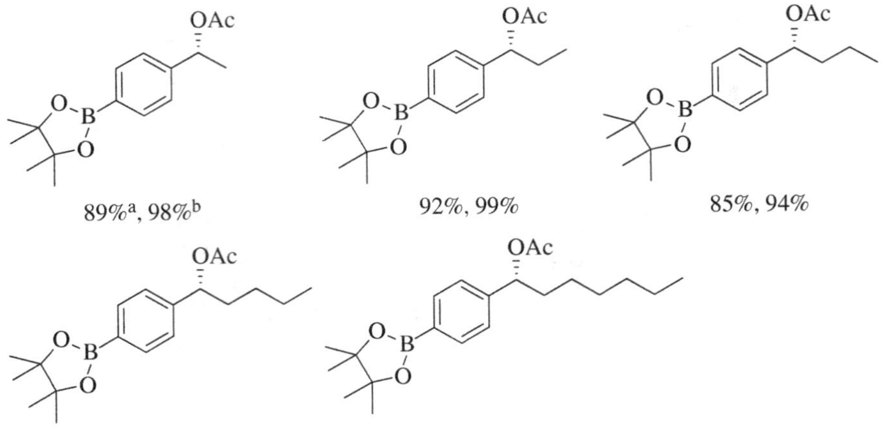

89%[a], 98%[b] | 92%, 99% | 85%, 94%

87%, 94% | 88%, 94%

[a]Yield, [b]enantiopurity

CHART 5.21
Products from the DKR of boron-containing alcohols with **9b** and ISCBCL.

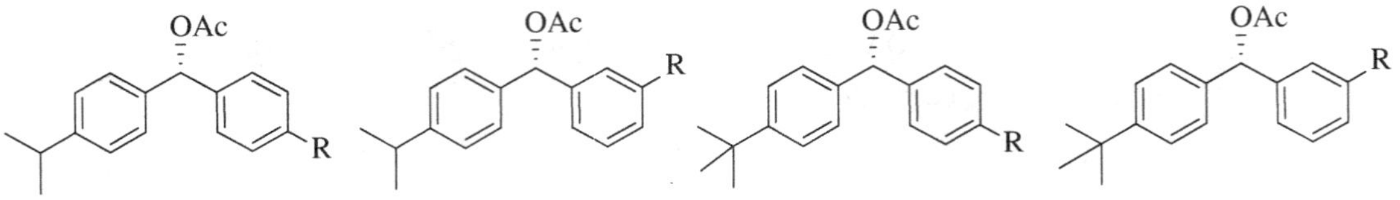

R = H, 86%[a], 91%[b]
R = F, 57%, 91%
R = Cl, 61%, 85%
R = OMe, 19%, 61%
R = NO_2, 75%, 75%

R = F, 65%, 79%
R = Me, 46%, 83%

R = H, 87%, 95%
R = F, 76%, 95%
R = Cl, 55%, 94%
R = OMe, 42%, 87%
R = NO_2, 76%, 71%

R = F, 69%, 86%
R = Me, 68%, 93%

[a]Yield, [b]enantiopurity

CHART 5.22
Products from the DKR of diarylmethanols with **9b** and ISCBCL.

It was found that the Ru catalyst **11** displayed a good racemization activity under household fluorescent light [37]. Its activity was comparable to those of **5** and **6**. The DKR of simple secondary alcohols with **11** and Novozym 435 gave products of excellent enantiopurity with high yields (Scheme 5.21).

Other metal systems have also been employed as racemization catalysts for the DKR of alcohols. Akai *et al.* reported the use of oxovanadium compound ($VO(OSiPh_3)_3$) as a racemization catalyst in the DKR of allylic alcohols [38, 39]. In this

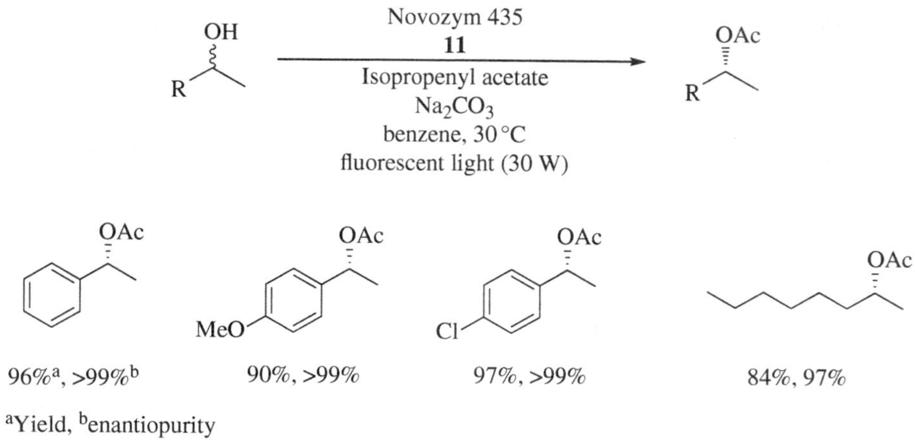

SCHEME 5.21

DKR of secondary alcohols with ruthenium catalyst **11** under light.

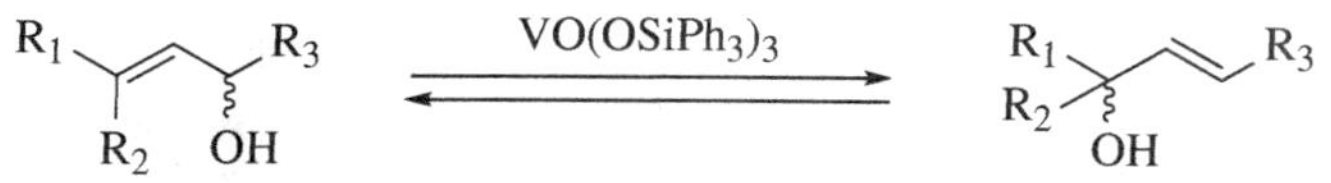

SCHEME 5.22

Vanadium-catalyzed racemization of allylic alcohols.

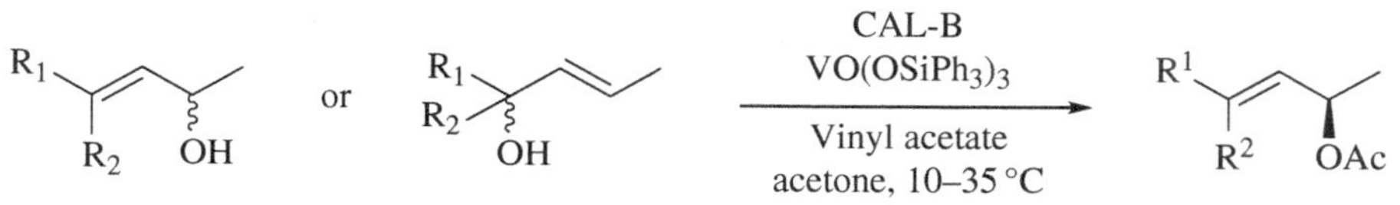

SCHEME 5.23

DKR of allylic alcohols with $VO(OSiPh_3)_3$ and CAL-B.

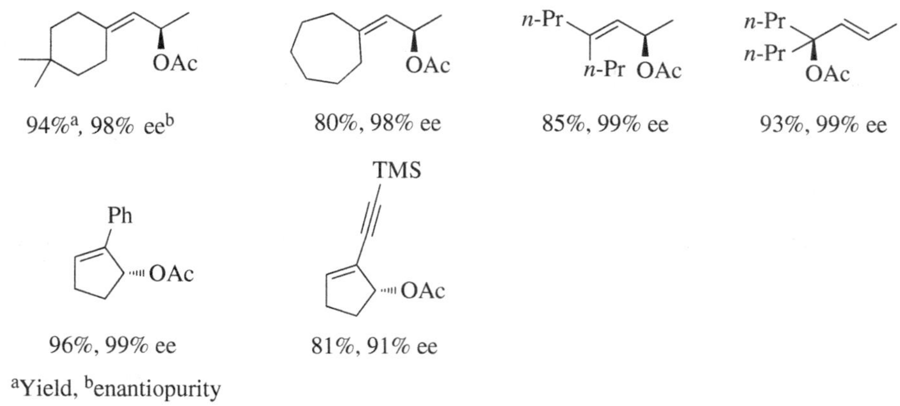

CHART 5.23

Products from the DKR of allylic alcohols with $VO(OSiPh_3)_3$ and CAL-B.

procedure, the vanadium complex catalyzed the reversible 1,3-migration of OH group (Scheme 5.22). The catalyst was compatible with several lipases such as Novozym 435, lipase AK, and lipase PS-D. It also worked well with vinyl acetate, which was inapplicable with the Ru catalysts. In addition, it showed good stability in the presence of oxygen and moisture. The DKR reactions with $VO(OSiPh_3)_3$ and CAL-B provided high yields and excellent enantiopurities (Scheme 5.23 and Chart 5.23). Later, the same group reported the use of a practical oxovanadium catalyst (V-MPS) coimmobilized with CAL-B inside mesoporous silica (MPS) in the DKR (Chart 5.24) [67].

Berkessel and coworkers reported the DKR of secondary alcohols based on the Meerwein–Ponndorf–Verley–Oppenauer (MPVO) reaction (Scheme 5.24) [40]. Trimethylaluminum ($AlMe_3$) was employed for the *in situ* generation of aluminum

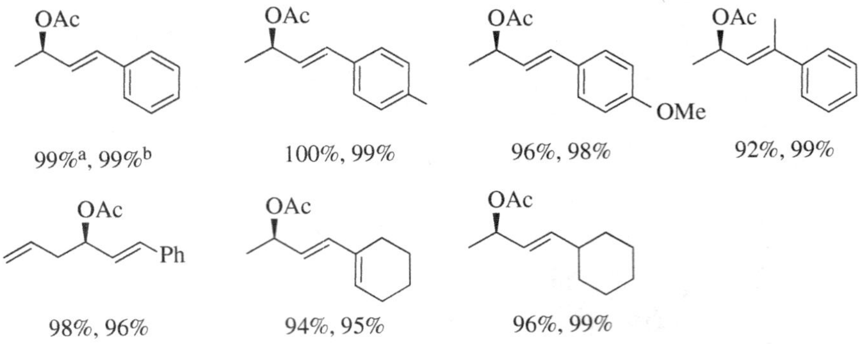

aYield, benantiopurity

CHART 5.24

Products from the DKR of allylic alcohols with immobilized $VO(OSiPh_3)_3$ and CAL-B.

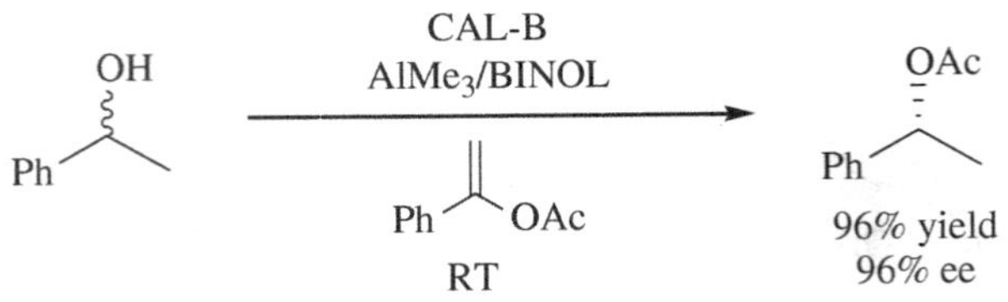

SCHEME 5.24

DKR of secondary alcohols catalyzed by the combination of lipase and $AlMe_3$.

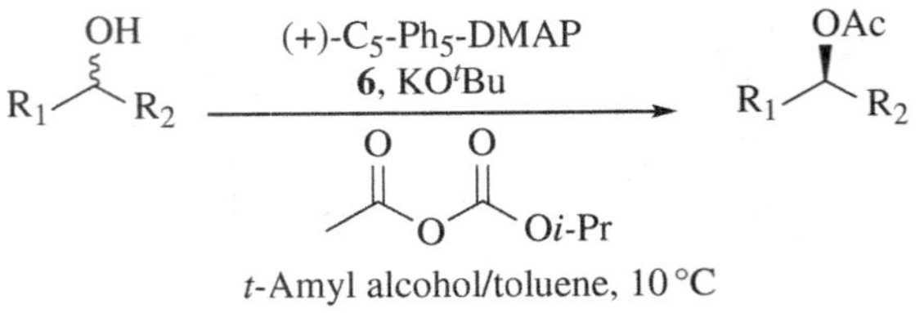

SCHEME 5.25

Nonenzymatic DKR of secondary alcohols.

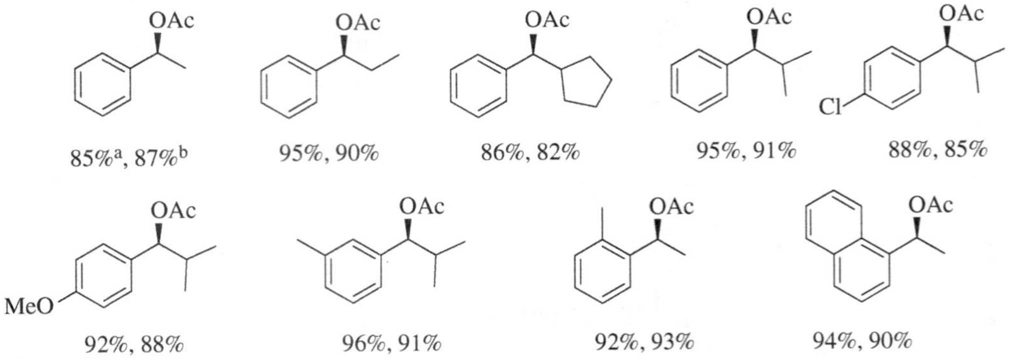

aYield, benantiopurity

CHART 5.25

Products from the nonenzymatic DKR of secondary alcohols.

alkoxide acting as the racemization catalyst. Dynamic kinetic resolution with $AlMe_3$, however, has a limitation that a customized enol ester must be used as the acyl donor for each substrate.

The Fu group has reported the first nonenzymatic DKR of secondary alcohols [41]. In this DKR, a planar-chiral DMAP derivative ((+)-C_5-Ph_5-DMAP) as the resolution catalyst was coupled with a Ru complex **6** as the racemization catalyst in the presence of an acyl carbonate (Scheme 5.25). The DKR of simple secondary alcohols provided good yields but lower enantiopurities compared to the enzymatic DKRs. It is noteworthy that the DKR of sterically more demanding substrates carrying a branched side chain (isopropyl or cyclopentyl) also provided similarly good results (Chart 5.25).

5.3 DYNAMIC KINETIC RESOLUTION OF AMINES AND AMINO ACIDS

The Reetz group reported the first DKR of amine, which was achieved with 1-phenylethylamine as the substrate, CAL-B as the resolution catalyst, Pd/C as the racemization catalyst, and ethyl acetate as the acyl donor in triethylamine (Scheme 5.26) [42]. The DKR required a long time (8 days) and provided a modest yield (64%) although the enantiopurity of product was high (99% ee). Later, we reported the DKR of several amines starting from ketoximes [43]. In this case, amines were generated *in situ* from ketoximes by Pd/C-catalyzed hydrogenation and then directly subjected to the DKR with CAL-B and Pd/C (Scheme 5.27). It is noted that a tertiary amine ($^{i}Pr_2NEt$) was added to suppress the reductive elimination of amino group. The DKR provided good yields with high enantiomeric excesses (Chart 5.26).

The Jacobs group reported the DKR of amines using Pd/$BaSO_4$ as the racemization catalyst (Scheme 5.28) [44, 45]. The DKR was carried out with CAL-B and Pd/$BaSO_4$ under molecular hydrogen and provided excellent enantiopurities and good yields (Chart 5.27).

For the more efficient amine DKR, we developed a novel type of Pd nanocatalyst, Pd/AlO(OH), in which Pd nanoparticles were entrapped into AlO(OH) matrix via sol–gel process [46]. The racemization of enantiopure 1-phenylethylamine with 1 mol% of Pd/AlO(OH) at 70 °C was completed within 24 h. The DKR of various amines with Pd/AlO(OH) and CAL-B provided high yields and excellent enantiomeric excesses [47] (Scheme 5.29 and Chart 5.28). The Pd nanocatalyst was readily recyclable together with CAL-B. They were reused ten times without losing any significant activity. The use of ethyl acetate as the acyl donor required a relatively large amount of enzymes containing a substantial amount of water, which could cause the hydrolysis of imine intermediate. Accordingly, it was necessary to add molecular sieves for trapping water. The use of more active ethyl methoxyacetate alleviated this

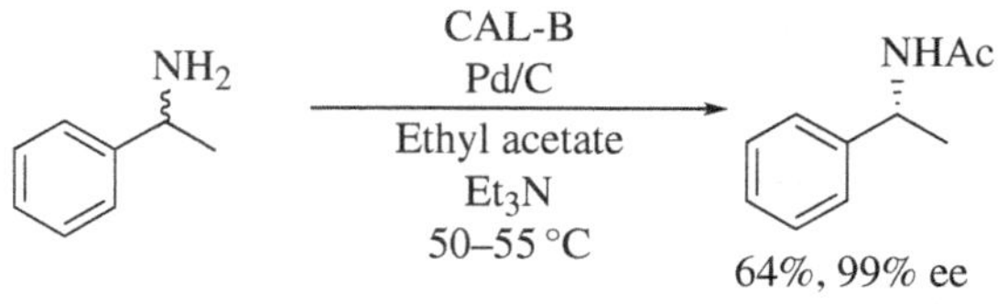

SCHEME 5.26
DKR of 1-phenylethylamine with Pd/C and CAL-B.

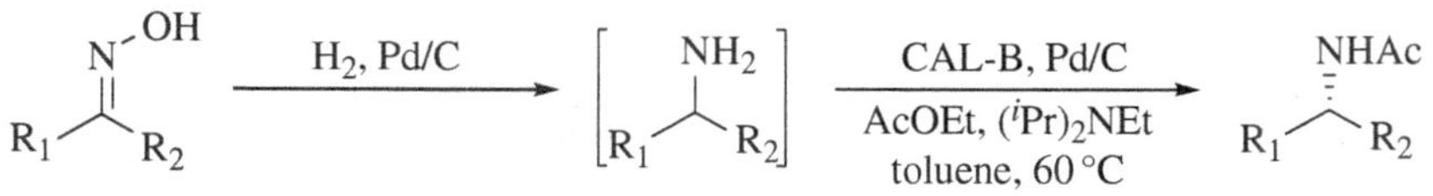

SCHEME 5.27
Asymmetric transformation of ketoximes to chiral amides via the DKR of amines.

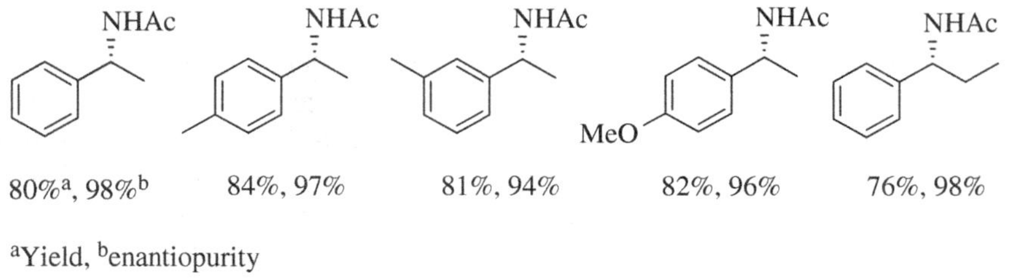

CHART 5.26
Products from the asymmetric transformation of ketoximes to chiral amides via the DKR of amines.

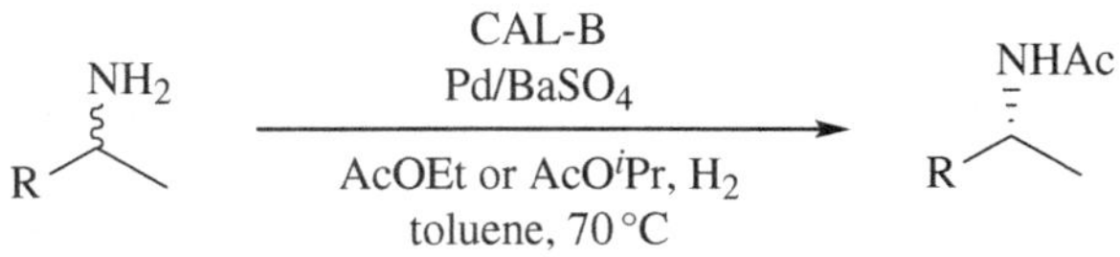

SCHEME 5.28
DKR of benzylic amines with CAL-B and Pd/$BaSO_4$.

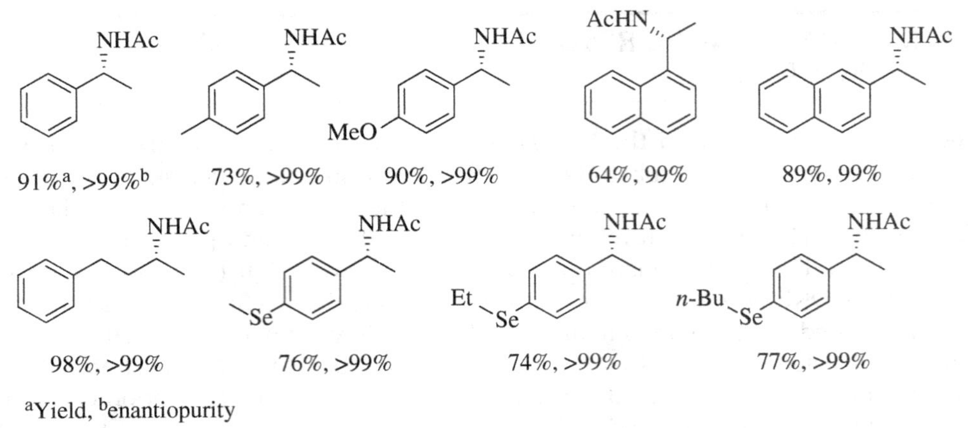

CHART 5.27

Products from the DKR of benzylic amines with CAL-B and Pd/$BaSO_4$.

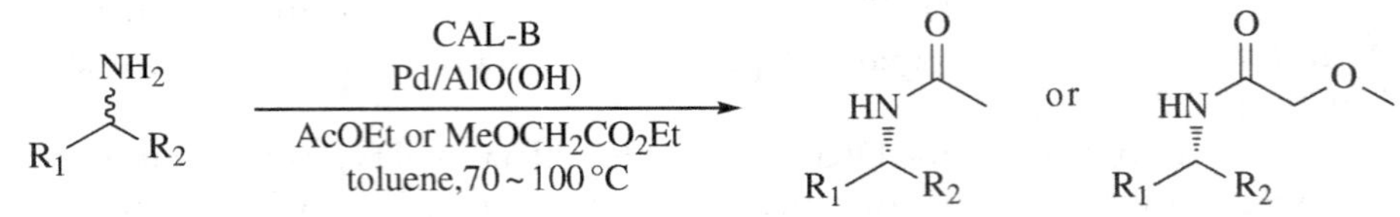

SCHEME 5.29

DKR of benzylic amines with Pd/AlO(OH) and CAL-B.

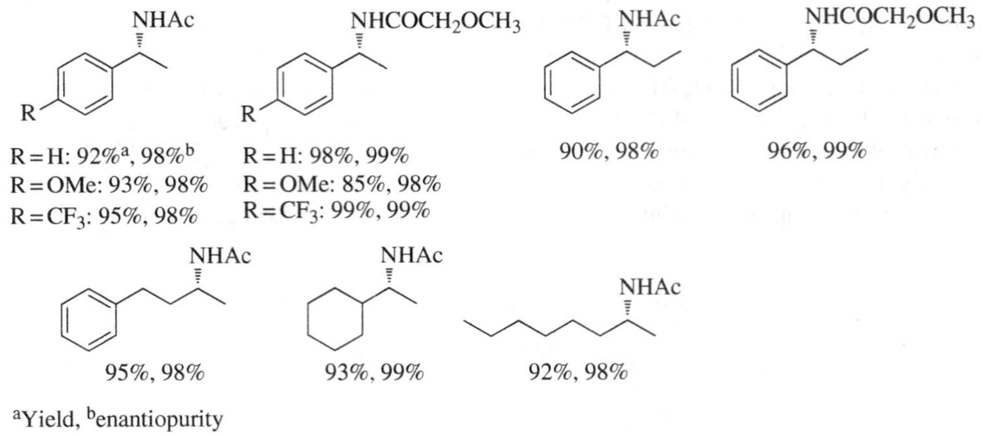

CHART 5.28

Products from the DKR of benzylic amines with Pd/AlO(OH) and CAL-B.

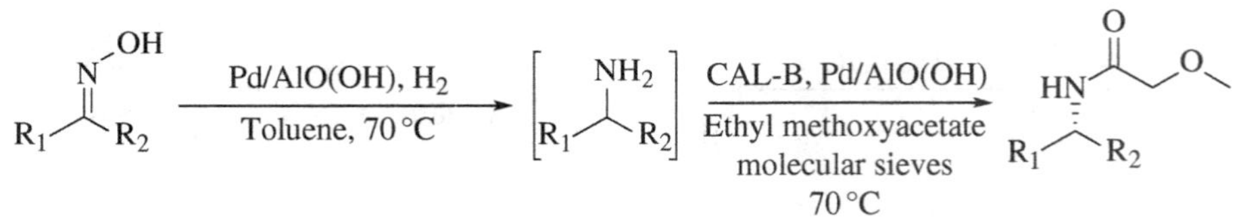

SCHEME 5.30

Asymmetric reductive acetylation of benzylic ketoximes with CAL-B and Pd/AlO(OH).

problem and thus led to higher yields. The Pd/AlO(OH) was also effective as the amine racemization catalyst for the asymmetric reductive acylation of benzylic ketoximes (Scheme 5.30) [48]. The use of Pd/AlO(OH) reduced the reaction time and enhanced the yield relative to the previous case using Pd/C (Chart 5.29). The racemization activity of Pd/AlO(OH) was enhanced by reducing the sizes of Pd nanoparticles to achieve a fast DKR of 1-phenylethylamine (6h at 70 °C) [49].

The Wu and Yang group reported a different type of Pd catalyst (Pd/layered double-hydroxide-dodecyl sulfate anion) which showed a good racemization activity at 55 °C [50]. The Li group showed that the use of Pd nanoparticles modified with various alkalic salts improved the selectivity toward the amine racemization [51].

The Bäckvall group introduced a Ru-based racemization catalyst **12** for the DKR of primary amines [52]. The dimeric Ru catalyst showed good performance in the

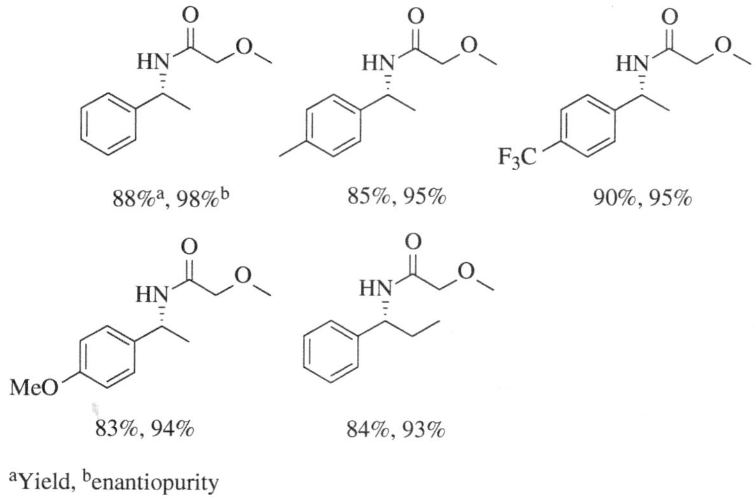

CHART 5.29

Products from the asymmetric reductive acetylation of benzylic ketoximes with CAL-B and Pd/AlO(OH).

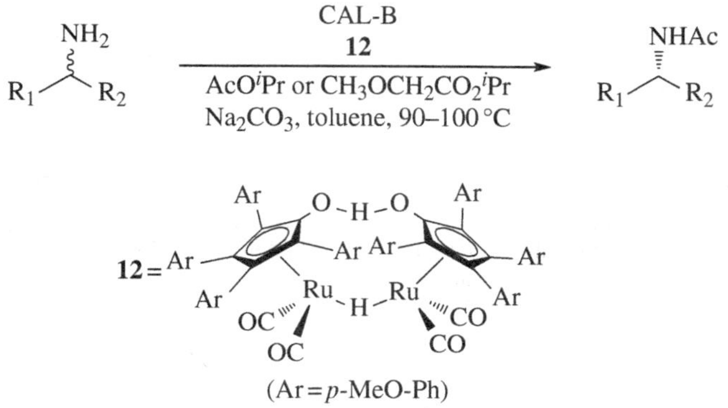

SCHEME 5.31

DKR of primary amines with CAL-B and ruthenium complex **12**.

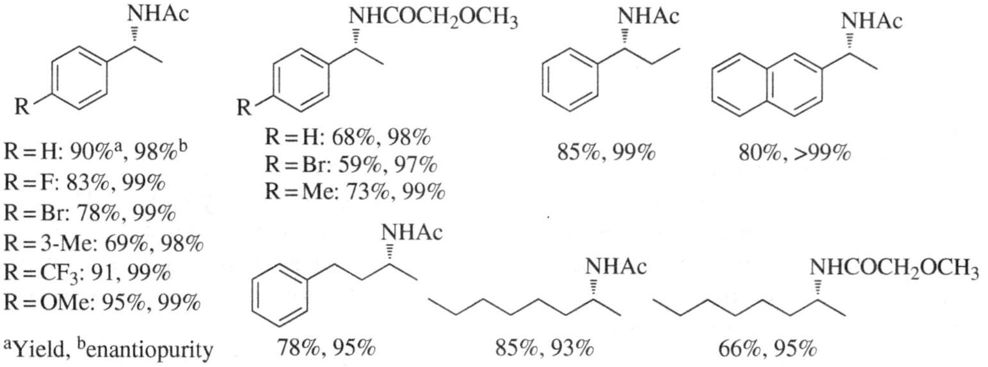

CHART 5.30

Products from the DKR of primary amines with CAL-B and ruthenium complex **12**.

DKR of both benzylic and aliphatic amines [53] (Scheme 5.31 and Chart 5.30). Isopropyl acetate and 2-methoxyacetate were employed as the acyl donors. Dynamic kinetic resolution with **12** proceeded more rapidly in the presence of 2,4-dimethyl-3-pentanol as hydrogen donor at 100 °C [54].

The previous DKR procedures employing alkyl acetates as the acyl donors had a limitation that the deacylation of products to obtain free amines required a harsh conditions. This limitation was avoided by using dibenzyl carbonate as the acyl donor (Scheme 5.32 and Chart 5.31). The benzyloxycarbonyl group of the product was readily removed by Pd-catalyzed hydrogenolysis [55].

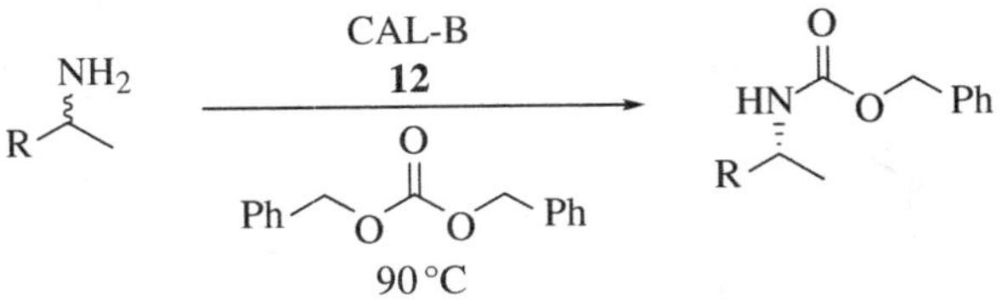

SCHEME 5.32

DKR of primary amines using dibenzyl carbonate as the acyl donor.

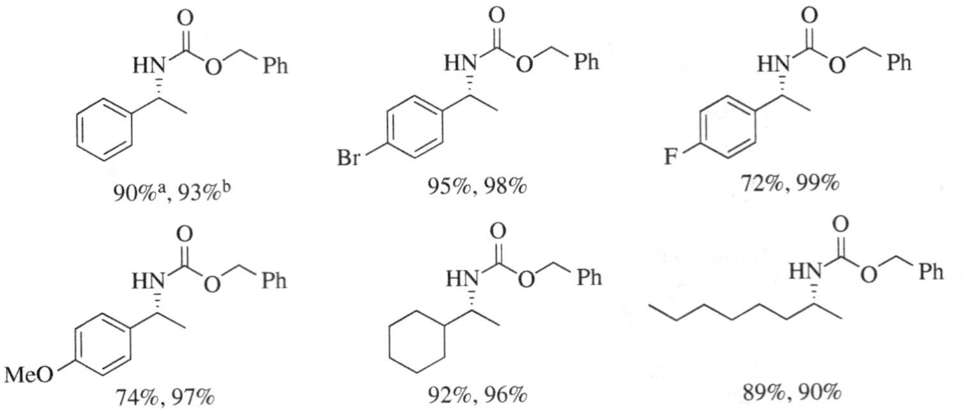

aYield, benantiopurity

CHART 5.31

Products from the DKR of primary amines using dibenzyl carbonate as the acyl donor.

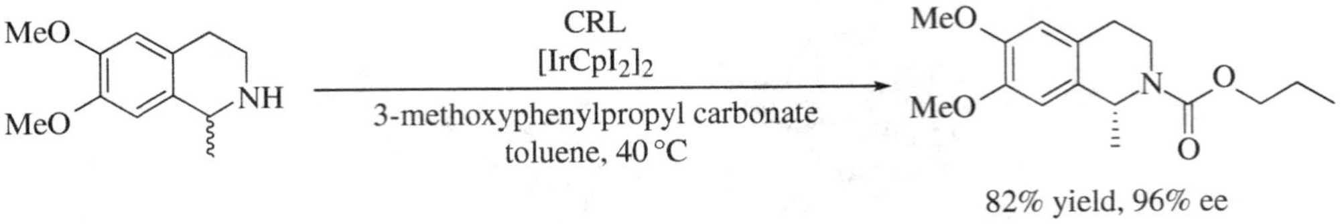

SCHEME 5.33

DKR of a secondary amine with $[IrCpI_2]_2$ and lipase.

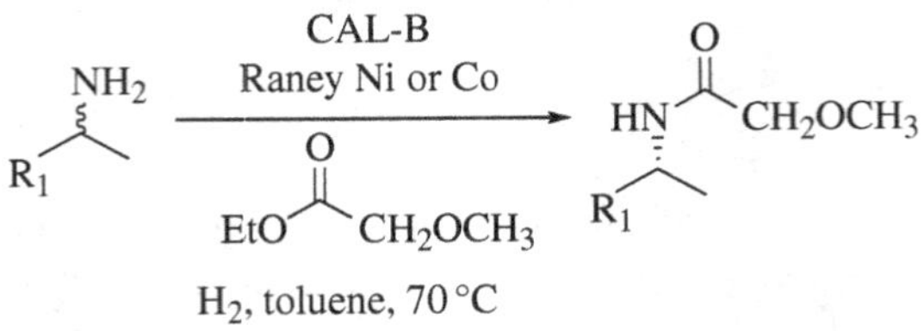

SCHEME 5.34

DKR of primary amines with Raney Ni or Co.

As described earlier, Pd and Ru catalysts were useful for the DKR of primary amines. However, they, were not effective for the DKR of secondary amines. The Page group introduced an iridium complex, $[IrCpI_2]_2$, as the racemization catalyst for the DKR of secondary amines [56]. They accomplished the DKR of an isoquinoline-based secondary amine with $[IrCpI_2]_2$, *C. rugosa* lipase (CRL), and 3-methoxyphenyl propyl carbonate as the acyl donor (Scheme 5.33).

Raney Ni and Co were also tested as the racemization catalysts for the DKR of amines [57]. Raney Ni displayed better activity and selectivity (Scheme 5.34). Specifically, Raney Ni displayed good performance in the DKR of aliphatic amines (Chart 5.32). The DKR of 1-phenylethylamine with Raney Ni did not proceed to completion even after 5 days, whereas the DKR of 2-hexylamine was complete within 4 days to provide 98% yield and 97% ee. On the other hand, the DKR of 1-phenylethylamine was performed in a two-pot process for higher yield, in which the reaction medium was alternatingly shuttled between two separate vessels containing the racemization catalysts and the enzymes, respectively. After four cycles, the yield reached 94% yield with 91% ee.

The Ramstrom group reported a metal-free procedure for the DKR of α-aminonitriles (Scheme 5.35) [58]. The Ramstrom group found that native lipase catalyzed both

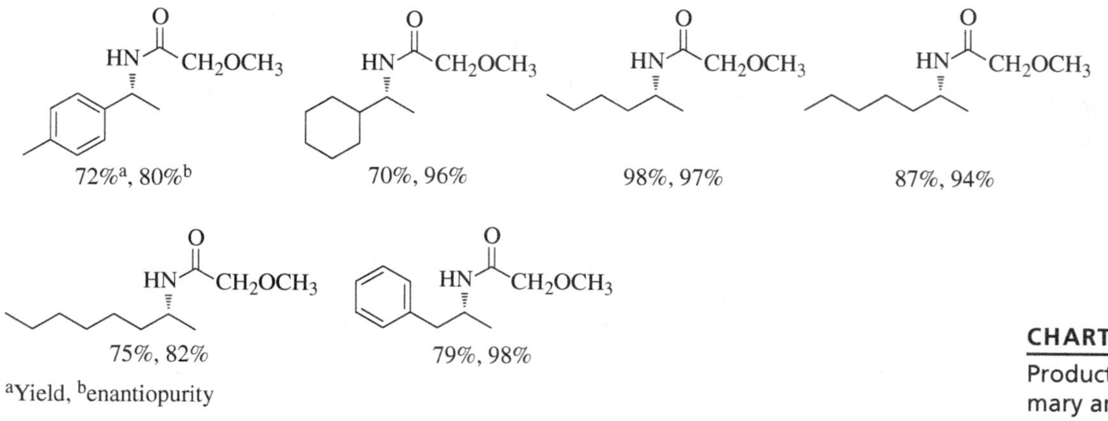

CHART 5.32

Products from the DKR of primary amines with Raney Ni.

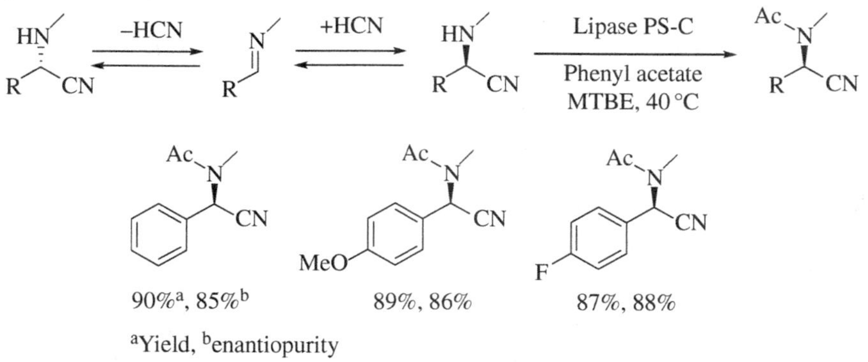

SCHEME 5.35

DKR of *N*-methyl α-aminonitriles in which the racemization took place via the lipase-catalyzed retro-Strecker and Strecker reaction.

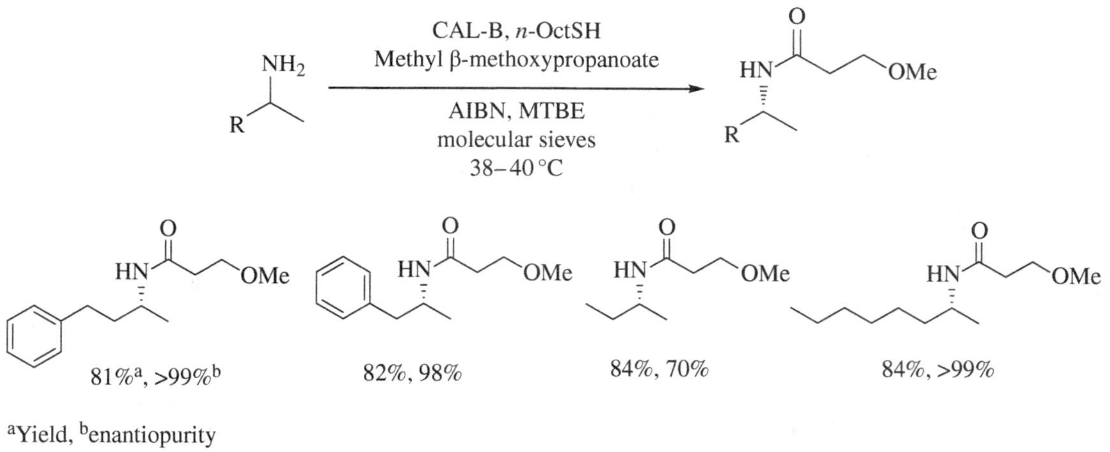

SCHEME 5.36

DKR of primary amines with CAL-B and a radical racemizing agent.

racemization and acylation of *N*-substituted α-aminonitriles. It was proposed that the racemization occurred via the retro-Strecker and Strecker reaction at the active site of enzyme. This DKR procedure provides a convenient route to *N*-methyl-α-amino acids and derivatives. Poulhes *et al.* reported an interesting procedure employing radical racemization for the DKR of nonbenzylic amines (Scheme 5.36) [59]. The DKR was performed with CAL-B, octanethiol, and AIBN in the presence of methyl β-methoxypropanoate as the acyl donor at 38–40 °C, in which the racemization of amine was catalyzed by octanethiol radical (which was generated by the reaction of octanethiol and AIBN).

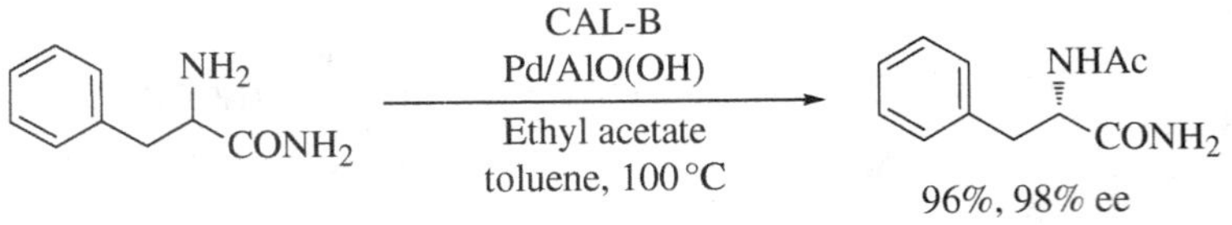

SCHEME 5.37
DKR of *dl*-phenylalanine amide with CAL-B and Pd/AlO(OH).

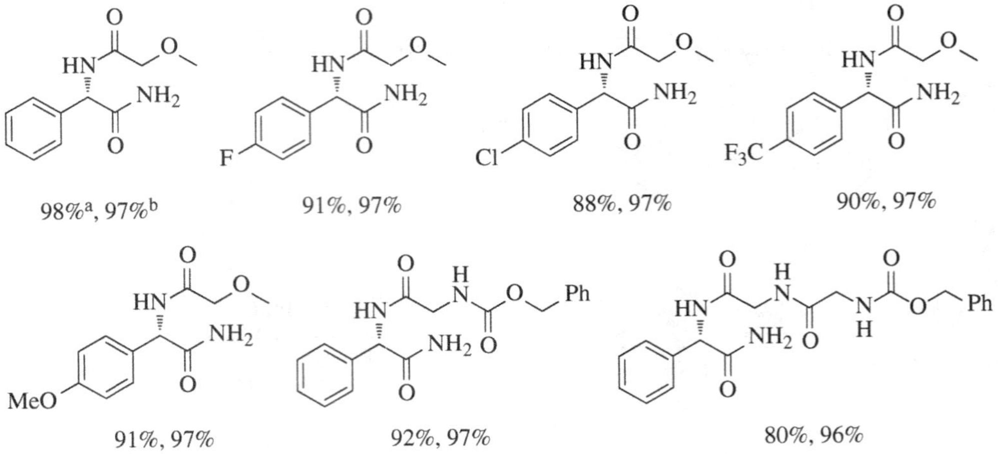

[a]Yield, [b]enantiopurity

CHART 5.33
Products from the DKR of phenylglycine amides with CAL-B and Pd/AlO(OH).

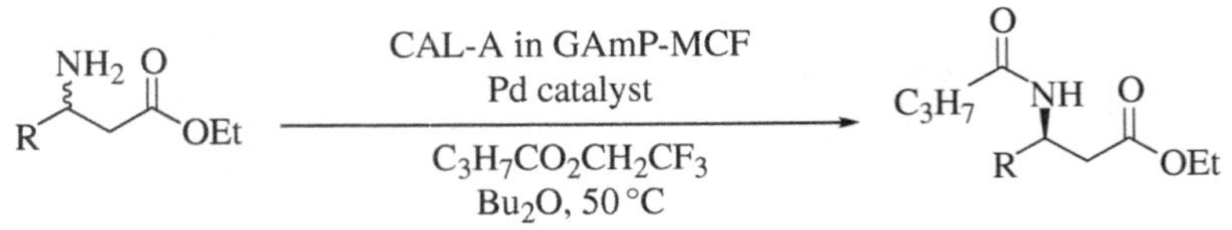

SCHEME 5.38
DKR of amino acid ester with Pd catalyst and MCF-immobilized CAL-A.

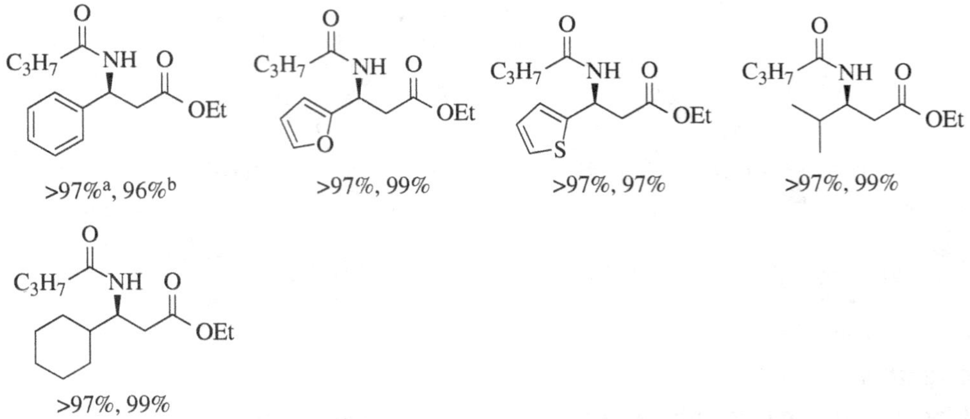

[a]Yield, [b]enantiopurity

CHART 5.34
Products from the DKR of amino acid ester with Pd/AlO(OH) and MCF-immobilized CAL-A.

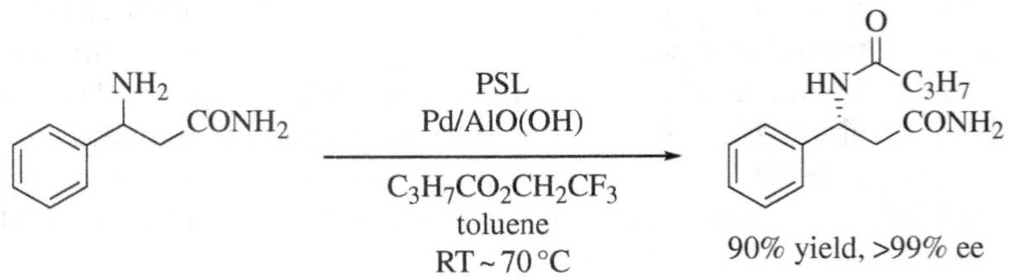

SCHEME 5.39
DKR of β-amino acid with PSL and Pd/AlO(OH).

The first DKR of α-amino acids was achieved using their amides as the substrate, Novozym 435 as the resolution enzyme, Pd/AlO(OH) as the racemization catalyst, and ethyl acetate as the acyl donor [46]. The DKR of phenylalanine amide was successful at 100 °C (Scheme 5.37). In contrast to this, the DKR of phenylglycine amide and simple derivatives was achieved at lower temperature (60 °C) [60]. On the other hand, the use of *N*-cbz-glycine and *N*-cbz-gly-glycine esters as the acyl donors provided enantioenriched dipeptides and tripeptides (Chart 5.33).

The Bäckvall group reported the DKR of β-amino acids using CAL-A and Ru or Pd catalyst (Scheme 5.38) [61]. CAL-A was immobilized in the functionalized mesocellular form (MCF) to improve its stability and enantioselectivity. The DKR with Pd catalyst and the MCF-immobilized CAL-A provided (*S*)-products, with better yields and higher enantiopurities (Chart 5.34). Our group explored the DKR of β-amino acids for preparing the products of opposite configuration with PSL and Pd/AlO(OH) (Scheme 5.39) [62]. As PSL was not thermally stable, the PSL-catalyzed resolution and Pd-catalyzed racemization were performed alternatively at two different temperatures (RT and 70 °C) to give (*R*)-product with 90% yield and >99% ee (Scheme 5.39).

5.4 APPLICATIONS OF DYNAMIC KINETIC RESOLUTION

Several chiral drugs were synthesized as examples of the synthetic applications of metalloenzymatic DKR. The Bäckvall group reported the synthesis of (*S*)-Me-imidacloprid (**15**), an insect neurotoxin, including the DKR of secondary alcohol as the key step [63]. In this synthesis, to obtain enantiopure secondary alcohol (*R*)-**13** as a key intermediate, the DKR of its racemic form was carried out with CAL-B (Novozym 435), Ru catalyst **6**, and isopropenyl acetate as the acyl donor followed by hydrolysis. Then (*R*)-**13** was converted to the target (98% ee) via two more steps in the overall 32% yield (Scheme 5.40).

The same group reported the formal synthesis of sertraline (**19**), an antidepressant, based on the DKR of diol [64]. In this synthesis, the DKR of *dl/meso*-diol **16** was carried out with CAL-B (Novozym 435) and **6** in the presence of isopropenyl acetate, followed by enzymatic alcoholysis, to obtain single enantiomeric diol (*R*,*R*)-**16**, which was then converted to the key intermediate (**18**) of sertraline via four steps (Scheme 5.41).

The Bäckvall group also reported the enantioselective synthesis of duloxetine (**23**), a drug for the treatment of major depressive disorder and generalized anxiety disorder, including the DKR of β-hydroxynitrile as the key step (Scheme 5.42) [65]. The DKR of racemic β-hydroxynitrile (*rac*-**20**) was performed with CAL-B (Novozym 435), **6**, and isopropenyl acetate to obtain an enantioenriched acetate (*R*)-**21** (98% ee), which was converted to (*R*)-duloxetine (96% ee) via a key intermediate (*R*)-**22**. The target molecule was synthesized in a 37% overall yield from the starting material.

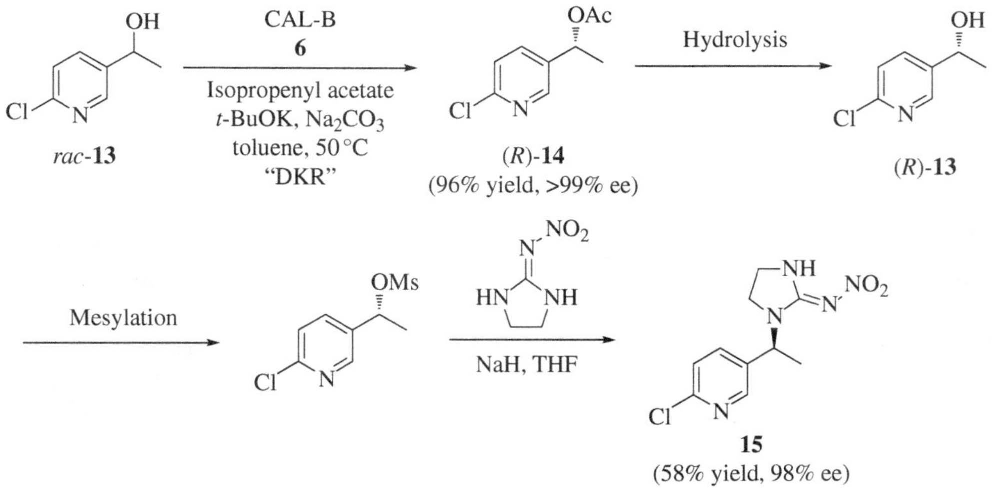

SCHEME 5.40
Synthesis of (*S*)-Me-imidacloprid.

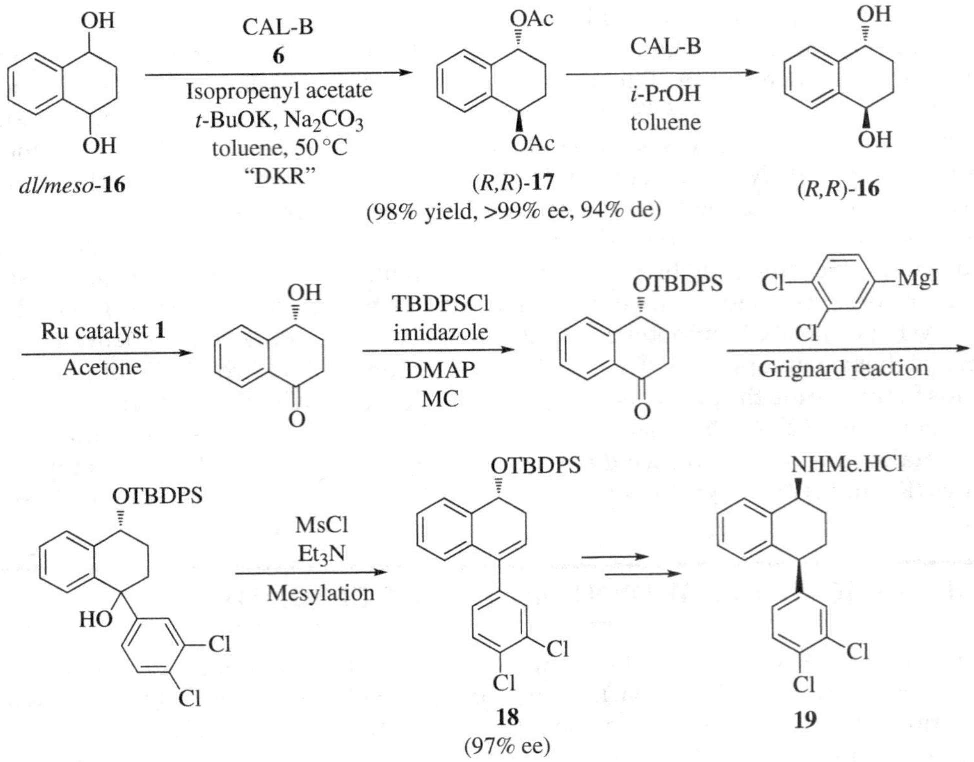

SCHEME 5.41
Formal synthesis of sertraline.

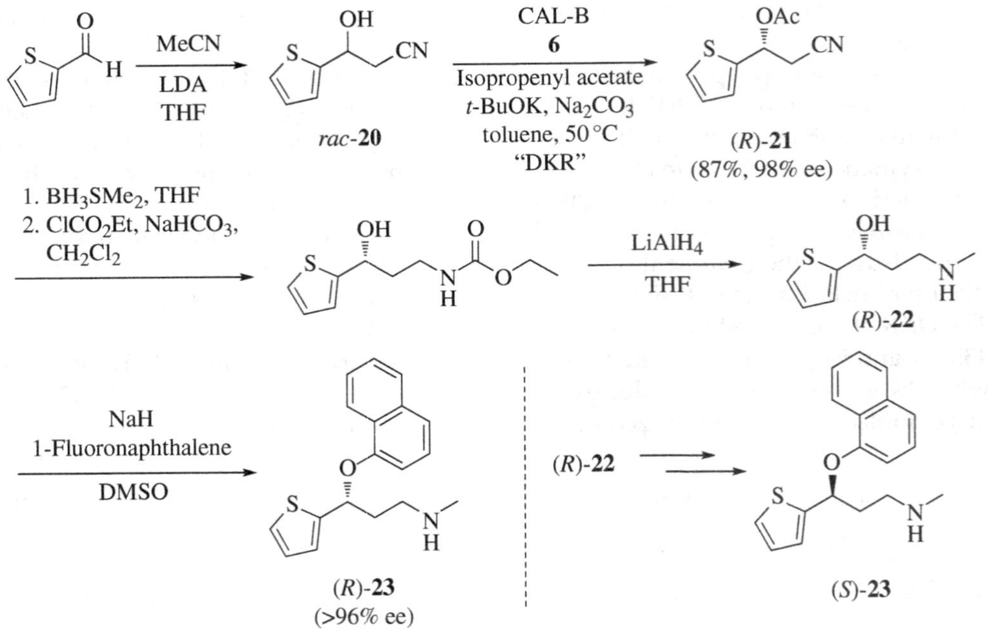

SCHEME 5.42
Synthesis of duloxetine.

Our group achieved the synthesis of (*S*)-rivastigmine (**26**), a drug for the treatment of Alzheimer's and Parkinson's diseases, employing the DKR of secondary alcohol as the key step (Scheme 5.43) [34]. In this synthesis, chiral alcohol (*R*)-**24** as a key intermediate was obtained from its racemate via the DKR catalyzed by CAL-B (Novozym 435) and a polymer-bound Ru catalyst **10**. The conversion of (*R*)-**24** to the target molecule (*S*)-**26** (97% ee) was accomplished via two steps. The overall yield was 57% from the starting material.

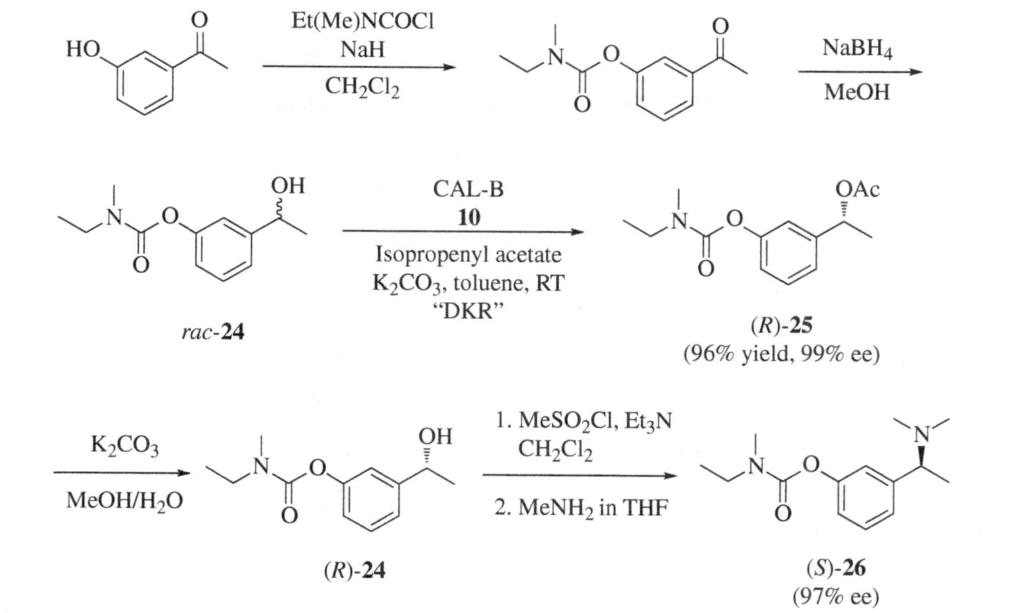

SCHEME 5.43
Synthesis of (*S*)-rivastigmine.

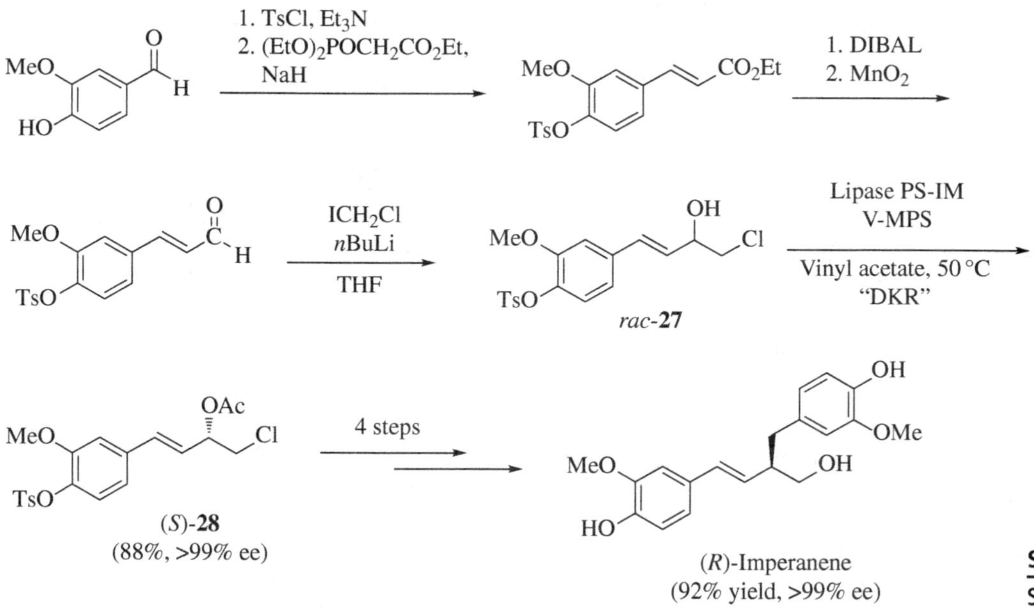

SCHEME 5.44
Synthesis of (*R*)-imperanene.

The Akai group synthesized (*R*)-imperanene by employing the DKR of allylic alcohol as the key step. In this synthesis, racemic alcohol intermediate *rac*-**27** was converted to its enantiomeric acetate (*S*)-**28** via the DKR, which was performed using lipase PS-IM and oxovanadium catalyst immobilized inside mesoporous silica (V-MPS). The target molecule of >99% ee was then obtained via four steps (Scheme 5.44) [66].

The Bäckvall group applied the amine DKR methodology to the synthesis of norsertraline [67] (Scheme 5.45), which is an antidepressant of the selective serotonin reuptake inhibitor (SSRI) class. In the first step, the DKR of racemic amine *rac*-**29** was performed with CAL-B (Novozym 435) and dimeric Ru catalyst **12** to obtain the enantiomeric amide (*R*)-**30** (99% ee). The (*R*)-amide was converted via several chemical steps to the target (99% ee) with an overall 28% yield.

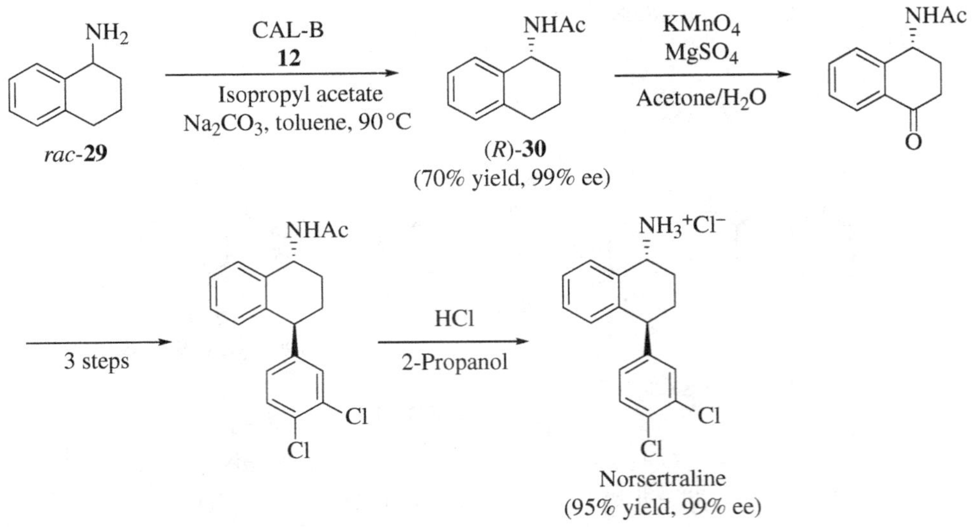

SCHEME 5.45
Synthesis of norsertraline.

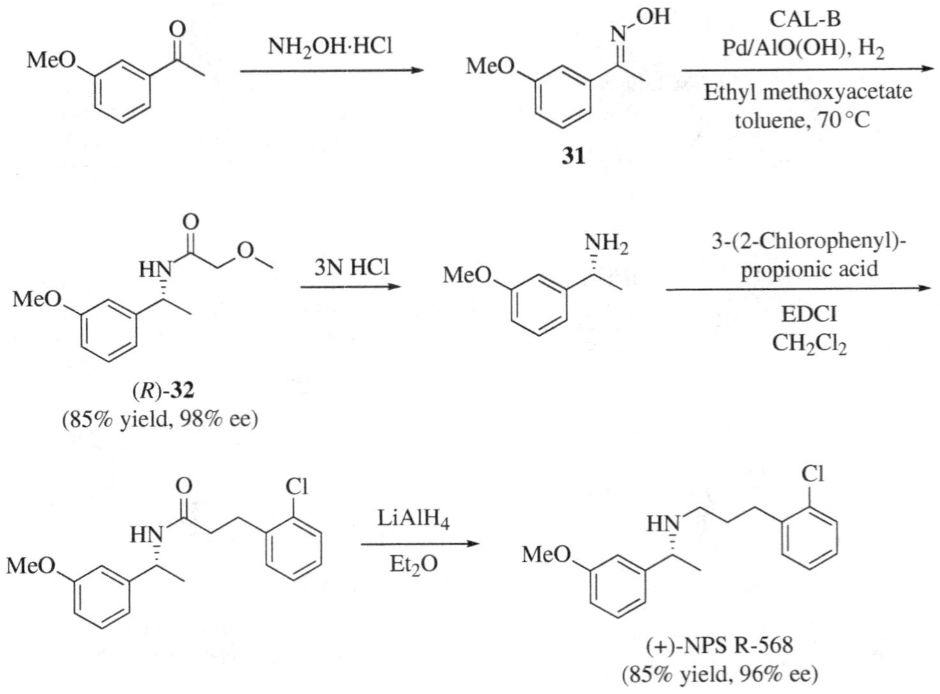

SCHEME 5.46
Synthesis of calcimimetics (+)-NPS R-568.

Our group applied the asymmetric reductive acetylation of ketoxime including the amine DKR step to the synthesis of calcimimetics (+)-NPS R-568 (Scheme 5.46) [68]. In the second step, the asymmetric reductive acylation of ketoxime **31** was carried out with CAL-B (Novozym 435) and Pd/AlO(OH) under molecular hydrogen to obtain the enantiomeric amide (*R*)-**32** (98% ee). The target molecule (96% ee) was prepared from (*R*)-**32** via five steps. The overall yield of calcimimetics (+)-NPS R-568 was 63% from the starting material.

Nanda *et al.* applied an alcohol DKR methodology to the total synthesis of a natural product, stagonolide-C, which exhibited antibacterial and antifungal activities (Scheme 5.47) [69]. In this synthesis, the DKRs of two monoprotected diols *rac*-**33** and *rac*-**36** were performed with CAL-B (Novozym 435) and Ru catalyst **5**, followed by

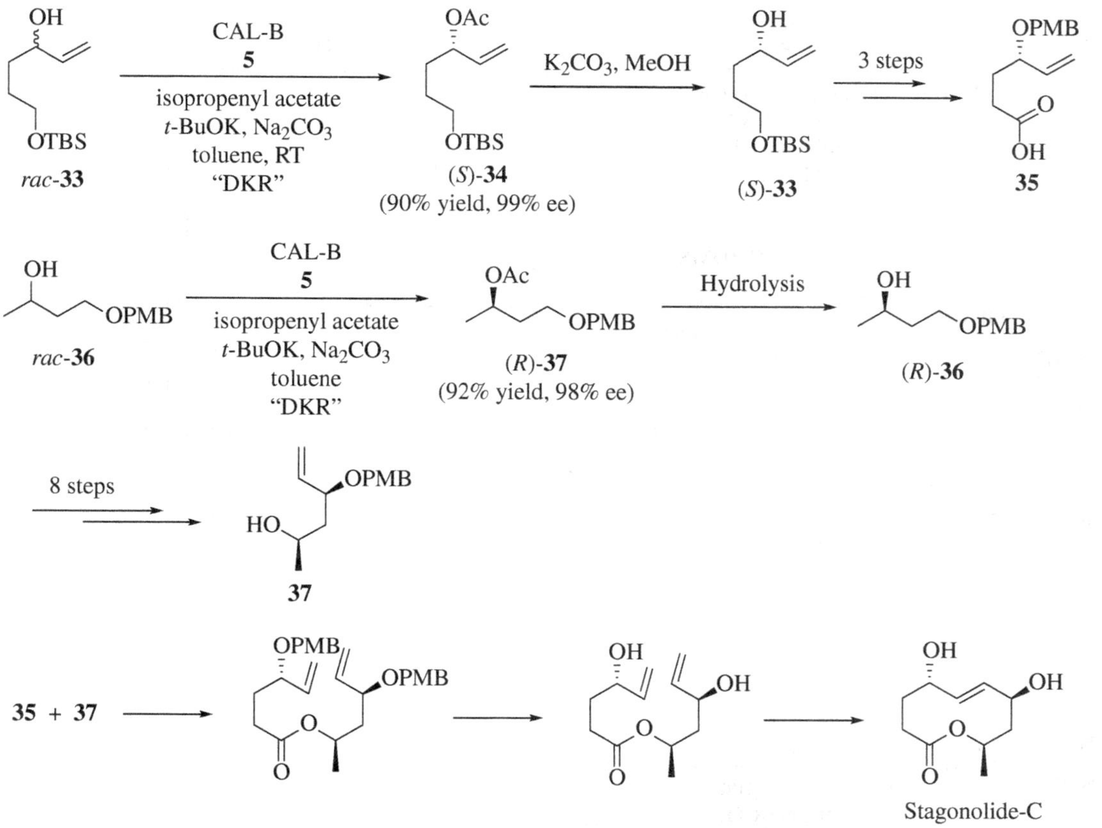

SCHEME 5.47
Total synthesis of stagonolide-C.

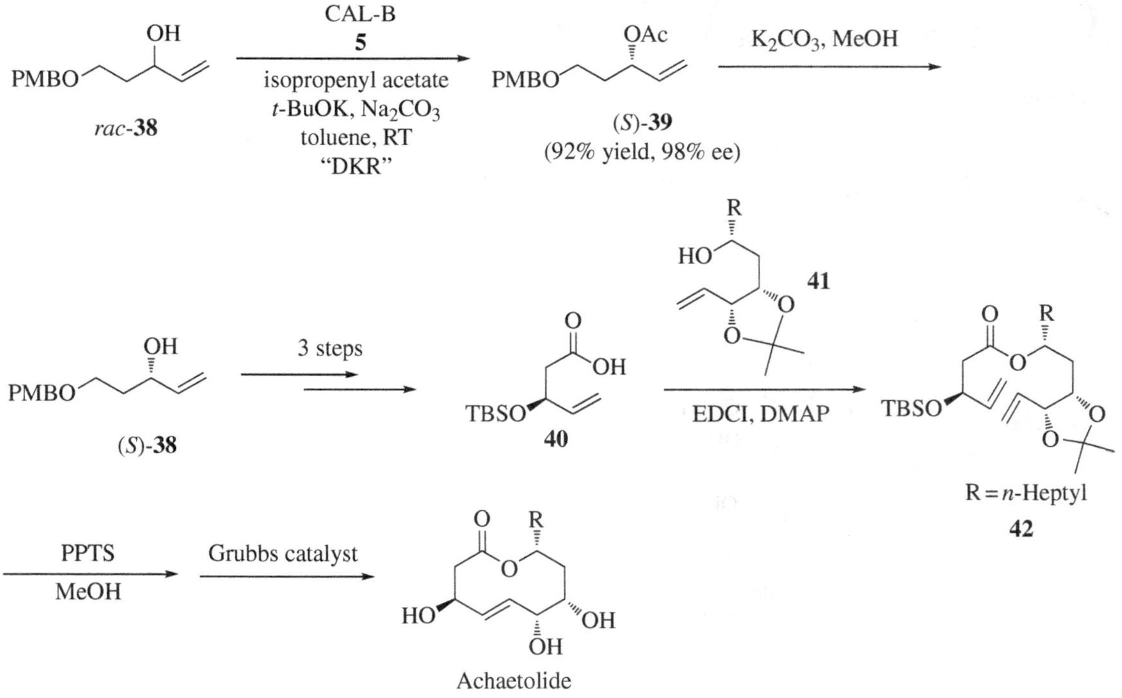

SCHEME 5.48
Total synthesis of achaetolide.

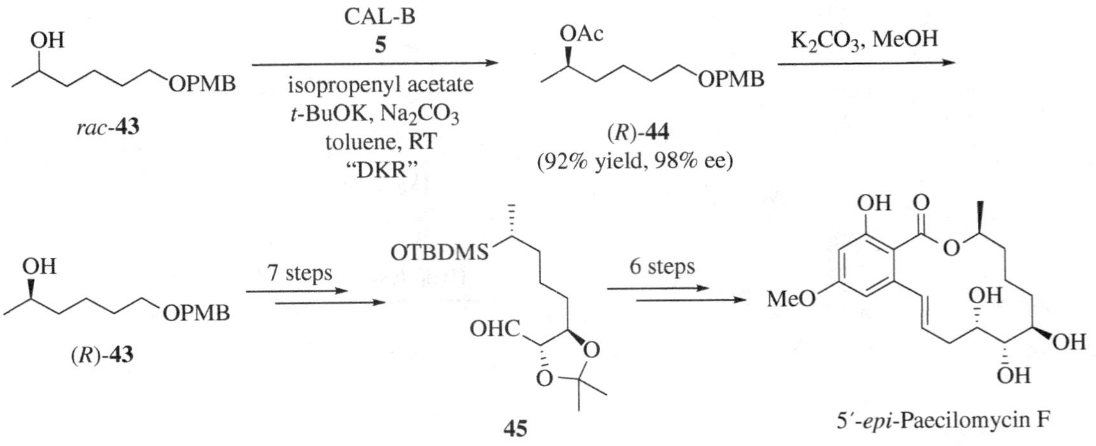

SCHEME 5.49

Total synthesis of 5′-*epi*-paecilomycin-F.

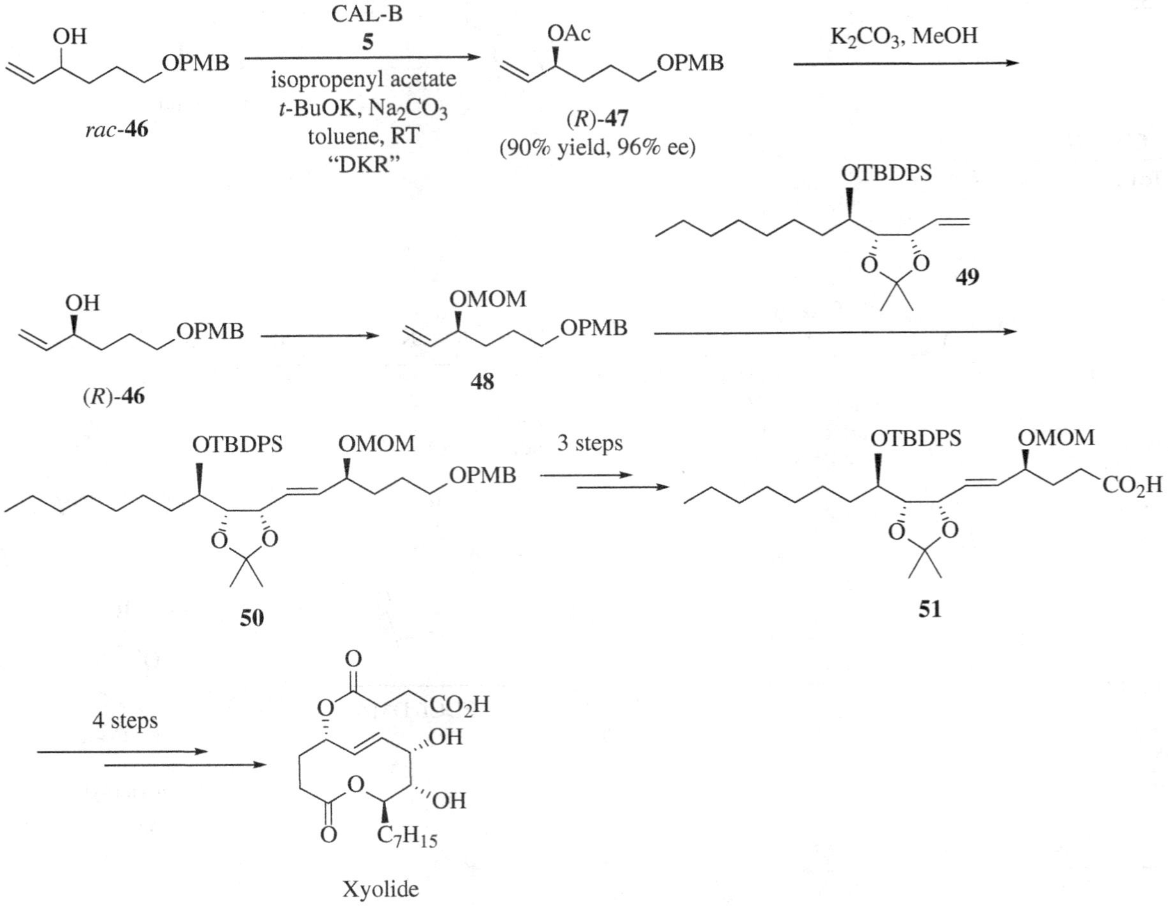

SCHEME 5.50

Total synthesis of xyolide.

hydrolysis to obtain their enantiomers (*S*)-**33** and (*R*)-**36**, respectively. The two enantiomers were converted via several chemical steps to two key intermediates **35** and **37**, which in turn were coupled by esterification and cyclized by ring-closing metathesis to provide the target molecule.

The same group accomplished the first synthesis of another ten membered macrolide, achaetolide, including a DKR step (Scheme 5.48) [70]. In this synthesis, the DKR of racemic alcohol *rac*-**38** was performed with CAL-B (Novozym 435) and Ru catalyst **5** to obtain (*S*)-**38** via its acetate (*S*)-**39** (98% ee). The enantioenriched alcohol (*S*)-**38** was then converted via three steps to a key intermediate **40**, which in turn was coupled with **41** to yield an ester intermediate **42**. The target molecule was synthesized via two steps, deprotection and cyclization, from **42**.

The DKR of secondary alcohol employing CAL-B (Novozym 435) and Ru catalyst **5**, which had been successfully applied to the total syntheses of stagonolide-C and achaetolide, also was applicable to the total syntheses of 5-*epi*-paecilomycin F [71] (Scheme 5.49) and xyolide [72] (Scheme 5.50).

5.5 SUMMARY

The coupling of enzyme-catalyzed resolution with metal-catalyzed racemization constitutes a powerful DKR methodology for the synthesis of enantioenriched alcohols, amines, and amino acids. In many cases, the metalloenzymatic DKRs provide high yields and excellent enantiopurities, both approaching 100%, and thus provide useful alternatives to the chemical catalytic asymmetric reactions employing transition metals (complexes) or organocatalysts. The wider applications of a metalloenzymatic DKR method, however, are often limited by the low activity, narrow substrate specificity, or modest enantioselectivity of the enzyme employed. The low activities of metal-based catalysts, particularly in the racemization of amines and amino acids, also limit the wider applications of DKR. It is expected that further efforts to overcome these limitations with the developments of new enzyme–metal combinations will make the metalloenzymatic DKR more attractive as a tool for asymmetric synthesis in the future.

APPENDIX: LIST OF ABBREVIATIONS

Techniques

de	Diastereomeric excess
DKR	Dynamic kinetic resolution
ee	Enantiomeric excess
KR	Kinetic resolution

Reagents, solvents, etc.

AIBN	Azobisisobutyronitrile
Cy	Cyclohexyl
Ms	Methanesulfonyl
MsCl	Methanesulfonyl chloride
MTBE	Methyl *tert*-butyl ether
TBDPS	t-Butyldiphenylsilyl
TBDPSCl	t-Butyldiphenylsilyl chloride
Tr	Trityl (triphenylmethyl)
Ts	Tosyl

REFERENCES

[1] Ward, R. S. *Tetrahedron: Asymmetry* 1995, **6**, 1475.

[2] Ebbers, E. J.; Ariaans, G. J. A.; Houbiers, J. P. M.; Bruggink, A.; Zwanenburg, B. *Tetrahedron* 1997, **53**, 9417.

[3] (a) Pellissier, H. *Tetrahedron* 2003, **59**, 8291; (b) Pellissier, H. *Tetrahedron* 2008, **64**, 1563.

[4] (a) Kim, M.-J.; Ahn, Y.; Park, J. *Curr. Opin. Biotechnol.* 2002, **13**, 578; (b) Pamies, O.; Bäckvall, J. E. *Chem. Rev.* 2003, **103**, 3247; (c) Kim, M.-J.; Ahn, Y.; Park, J. *Bull. Korean Chem. Soc.* 2005, **26**, 515; (d) Ahn, Y.; Ko, S.-B.; Kim, M.-J.; Park, J. *Coord. Chem. Rev.* 2008, **252**, 647; (e) Lee, J. H.; Han, K.; Kim, M.-J.; Park, J. *Eur. J. Org. Chem.* 2010, 999; (f) Kim, Y.; Park, J.; Kim, M.-J. *ChemCatChem* 2011, **3**, 271.

[5] (a) Pamies, O.; Bäckvall, J. E. *Curr. Opin. Biotechnol.* 2003, **14**, 407; (b) Pamies, O.; Bäckvall, J. E. *Trends Biotechnol.* 2004, **22**, 130; (c) Martin-Matute, B.; Bäckvall, J. E. *Curr. Opin. Chem. Biol.* 2007, **11**, 226.

[6] (a) Engstrom, K.; Vallin, M.; Syren, P. O.; Hult, K.; Bäckvall, J. E. *Org. Biomol. Chem.* 2011, **9**, 81; (b) Ema, T.; Nakano, Y.; Yoshida, D.; Kamata, S.; Sakai, T. *Org. Biomol. Chem.* 2012, **10**, 6299.

[7] Kim, H.; Choi, Y. K.; Lee, J.; Lee, E.; Park, J.; Kim, M.-J. *Angew. Chem. Int. Ed.* 2011, **50**, 10944.

[8] (a) Kazlauskas, R. J.; Weissfloch, A. N. E.; Rappaport, A. T.; Cuccia, L. A. *J. Org. Chem.* 1991, **56**, 2656; (b) Kazlauskas, R. J.; Weissfloch, A. N. E. *J. Mol. Catal. B: Enzym.* 1997, **3**, 65.

[9] Dinh, P. M.; Howarth, J. A.; Hudnott, A. R.; Williams, J. M. J.; Harries, W. *Tetrahedron Lett.* 1996, **37**, 7623.

[10] (a) Larsson, A. L. E.; Persson, B. A.; Bäckvall, J. E. *Angew. Chem. Int. Ed.* 1997, **36**, 1211; (b) Persson, B. A.; Larsson, A. L. E.; Le Ray, M.; Bäckvall, J. E. *J. Am. Chem. Soc.* 1999, **121**, 1645.

[11] Persson, B. A.; Huerta, F. F.; Bäckvall, J. E. *J. Org. Chem.* 1999, **64**, 5237.

[12] Pamies, O.; Bäckvall, J. E. *J. Org. Chem.* 2001, **66**, 4022.

[13] (a) Huerta, F. F.; Laxmi, Y. R. S.; Bäckvall, J. E. *Org. Lett.* 2000, **2**, 1037; (b) Huerta, F. F.; Bäckvall, J. E. *Org. Lett.* 2001, **3**, 1209; (c) Pamies, O.; Bäckvall, J. E. *J. Org. Chem.* 2002, **67**, 1261; (d) Runmo, A.-B. L.; Pamies, O.; Faber, K.; Bäckvall, J. E. *Tetrahedron Lett.* 2002, **43**, 2983.

[14] Pamise, O.; Bäckvall, J. E. *J. Org. Chem.* 2002, **67**, 9006.

[15] Kim, M.-J.; Choi, Y. K.; Choi, M. Y.; Kim, M. J.; Park, J. *J. Org. Chem.* 2001, **66**, 4736.

[16] Jung, H. M.; Koh, J. K.; Kim, M.-J.; Park, J. *Org. Lett.* 2000, **2**, 409.

[17] (a) Hoyos, P.; Fernandez, M.; Sinisterra, J. V.; Alcantara, A. R. *J. Org. Chem.* 2006, **71**, 7632; (b) Hoyos, P.; Quezada, M. A.; Sinisterra, J. V.; Alcantara, A. R. *J. Mol. Catal. B: Enzym.* 2011, **72**, 20; (c) Hoyos, P.; Pace, V.; Sinisterra, J. V.; Alcantara, A. R. *Tetrahedron* 2011, **67**, 7321.

[18] Lee, D.; Huh, E. A.; Kim, M. J.; Jung, J. Y.; Kim, M.-J. *Org. Lett.* 2000, **2**, 2377.

[19] Lihammar, R.; Millet, R.; Bäckvall, J. E. *J. Org. Chem.* 2013, **78**, 12114

[20] Kim, M.-J.; Kim, H. M.; Kim, D.; Ahn, Y.; Park, J. *Green Chem.* 2004, **6**, 471.

[21] Dijksman, A.; Elzinga, J. M.; Li, Y.-X.; Arends, I. W. C. E.; Sheldon, R. A. *Tetrahedron: Asymmetry* 2002, **13**, 879.

[22] (a) van As, B. A. C.; van Buijtenen, J.; Heise, A.; Broxterman, Q. B.; Verzijl, G. K. M.; Palmans, A. R. A.; Meijer, E. W. *J. Am. Chem. Soc.* 2005, **127**, 9964; (b) Van Buijtenen, J.; van As, B. A. C.; Meuldijk, J.; Palmans, A. R. A.; Vekemans, J. A. J. M.; Hulshof, L. A.; Meijer, E. W. *Chem. Commun.* 2006, 3169.

[23] Choi, J. H.; Kim, Y. H.; Nam, S. H.; Shin, S. T.; Kim, M.-J.; Park, J. *Angew. Chem. Int. Ed.* 2002, **41**, 2373.

[24] Choi, J. H.; Choi, Y. K.; Kim, Y. H.; Park, E. S.; Kim, E. J.; Kim, M.-J.; Park, J. *J. Org. Chem.* 2004, **69**, 1972.

[25] (a) Kim, M.-J.; Chung, Y. I.; Choi, Y. K.; Lee, H. K.; Kim, D.; Park, J. *J. Am. Chem. Soc.* 2003, **125**, 11494; (b) Kim, M.-J.; Lee, H.; Park, J. *Bull. Korean Chem. Soc.* 2007, **28**, 2096.

[26] (a) Martin-Matute, B.; Edin, M.; Bogar, K.; Kaynak, F. B.; Bäckvall, J. E. *J. Am. Chem. Soc.* 2005, **127**, 8817; (b) Martin-Matute, B.; Edin, M.; Bogar, K.; Bäckvall, J. E. *Angew. Chem. Int. Ed.* 2004, **43**, 6535.

[27] (a) Martin-Matute, B.; Edin, M.; Bäckvall, J. E. *Chem. Eur. J.* 2006, **12**, 6053; (b) Fransson, A.-B. L.; Xu, Y.; Leijondahl, K.; Bäckvall, J. E. *J. Org. Chem.* 2006, **71**, 6309; (c) Olofsson, B.; Bogar, K.; Fransson, A.-B. L.; Bäckvall, J. E. *J. Org. Chem.* 2006, **71**, 8256; (d) Edin, M.; Martin-Matute, B.; Bäckvall, J. E. *Tetrahedron: Asymmetry* 2006, **17**, 708.

[28] Boren, L.; Martin-Matute, B.; Xu, Y.; Cordova, A.; Bäckvall, J. E. *Chem. Eur. J.* 2006, **12**, 225.
[29] (a) Magnusson, A. O.; Takwa, M.; Hamberg, A.; Hult, K. *Angew. Chem. Int. Ed.* 2005, **44**, 4582; (b) Magnusson, A. O.; Rotticci-Mulder, J. C.; Santagostino, A.; Hult, K. *ChemBioChem* 2005, **6**, 1051.
[30] Engstrom, K.; Vallin, M.; Syren, P. O.; Hult, K.; Bäckvall, J. E. *Org. Biomol. Chem.* 2011, **9**, 81.
[31] Ema, T.; Nakano, Y.; Yoshida, D.; Kamata, S.; Sakai, T. *Org. Biomol. Chem.* 2012, **10**, 6299.
[32] Kim, N.; Ko, S.-B.; Kwon, M. S.; Kim, M.-J.; Park, J. *Org. Lett.* 2005, **7**, 4523.
[33] Kim, M.-J.; Choi, Y. K.; Kim, S.; Kim, D.; Han, K.; Ko, S.-B.; Park, J. *Org. Lett.* 2008, **10**, 1295.
[34] Han, K.; Kim, C.; Park, J.; Kim, M.-J. *J. Org. Chem.* 2010, **75**, 3105.
[35] Lee, J. H.; Kim, N.; Kim, M.-J.; Park, J. *ChemCatChem* 2011, **3**, 354.
[36] Kim, C.; Lee, J.; Cho, J.; Oh, Y.; Choi, Y. K.; Choi, E.; Park, J.; Kim, M.-J. *J. Org. Chem.* 2013, **78**, 2571.
[37] Do, Y.; Hwang, I.-C.; Kim, M.-J.; Park, J. *J. Org. Chem.* 2010, **75**, 5740.
[38] Akai, S.; Tanimoto, K.; Kanao, Y.; Egi, M.; Yamamoto, T.; Kita, Y. *Angew. Chem. Int. Ed.* 2006, **45**, 2592.
[39] Akai, S.; Hanada, R.; Fujiwara, N.; Kita, Y.; Egi, M. *Org. Lett.* 2010, **12**, 4900.
[40] Berkessel, A.; Sebastian-Ibarz, M. L.; Müller, T. N. *Angew. Chem. Int. Ed.* 2006, **45**, 6567.
[41] (a) Lee, S. Y.; Murphy, J. M.; Ukai, A.; Fu, G. C. *J. Am. Chem. Soc.* 2012, **134**, 15149; (b) Diaz-Alvarez, A. E.; Mesas-Sanchez, D.; Diner, P. *Angew. Chem. Int. Ed.* 2013, **52**, 502.
[42] Reetz, M. T.; Schimossek, K. *Chimia* 1996, **50**, 668.
[43] Choi, Y. K.; Kim, M.-J.; Ahn, Y. *Org. Lett.* 2001, **3**, 4099.
[44] (a) Parvulescu, A.; De Vos, D.; Jacobs, P. *Chem. Commun.* 2005, 5307; (b) Parvulescu, A. N.; Jacobs, P. A.; De Vos, D. E. *Chem. Eur. J.* 2007, **13**, 2034.
[45] Andrade, L. H.; Silva, A. V.; Pderozo, E. C. *Tetrahedron Lett.* 2009, **50**, 4331.
[46] (a) Kwon, M. S.; Kim, N.; Park, C. M.; Lee, J. S.; Kang, K. Y.; Park, J. *Org. Lett.* 2005, **7**, 1077; (b) Kwon, M. S.; Kim, N.; Seo, S. H.; Park, I. S.; Cheedrala, R. K.; Park, J. *Angew. Chem. Int. Ed.* 2005, **44**, 6913; (c) Kwon, M. S.; Park, I. S.; Jang, J. S.; Lee, J. S.; Park, J. *Org. Lett.* 2007, **9**, 3417.
[47] Kim, M.-J.; Kim, W.-H.; Han, K.; Choi, Y. K.; Park, J. *Org. Lett.* 2007, **9**, 1157.
[48] Han, K.; Park, J.; Kim, M.-J. *J. Org. Chem.* 2008, **73**, 4302.
[49] Kim, Y.; Park, J.; Kim, M.-J. *Tetrahedron Lett.* 2010, **51**, 5581.
[50] Xu,G.; Dai, X.; Fu, S.; Wu, J.; Yang, L. *Tetrahedron Lett.* 2014, **55**, 397.
[51] Jin, Q.; Jia, G.; Zhang, Y.; Li, C. *Catal. Sci. Technol.* 2014, **4**, 464.
[52] Pàmies, O.; Éll, A. H.; Samec, J. S. M.; Hermanns, N.; Bäckvall, J. E. *Tetrahedron Lett.* 2002, **43**, 4699.
[53] Paetzold, J.; Bäckvall, J. E. *J. Am. Chem. Soc.* 2005, **127**, 17620.
[54] Veld, M. A. J.; Hult, K.; Palmans, A. R. A.; Meijer, E. W. *Eur. J. Org. Chem.* 2007, 5416.
[55] Hoben, C. E.; Kanupp, L.; Bäckvall, J. E. *Tetrahedron Lett.* 2008, **49**, 977.
[56] Stirling, M.; Blacker, J.; Page, M. I. *Tetrahedron Lett.* 2007, **48**, 1247.
[57] Parvulescu, A. N.; Jacobs, P. A.; De Vos, D. E. *Adv. Synth. Catal.* 2008, **350**, 113.
[58] Vongvilai, P.; Linder, M.; Sakulsombat, M.; Humble, M. S.; Berglund, P.; Brinck, T.; Ramstrom, O. *Angew. Chem. Int. Ed.* 2011, **50**, 6592.
[59] Poulhes, F.; Vanthuyne, N.; Betrand, M. P.; Gastaldi, S.; Gil, G. *J. Org. Chem.* 2011, **76**, 7281.
[60] Choi, Y. K.; Kim, Y.; Han, K.; Park, J.; Kim, M.-J. *J. Org. Chem.* 2009, **74**, 9543.
[61] (a) Shakeri, M.; Engström, K.; Sandström, A. G.; Bäckvall, J. E. *ChemCatChem* 2010, **2**, 534; (b) Engstrom, K.; Shakeri, M.; Bäckvall, J. E. *Eur. J. Org. Chem.* 2011, 1827.
[62] Choi, E.; Kim, Y.; Ahn, Y.; Park, J.; Kim, M.-J. *Tetrahedron: Asymmetry* 2013, **24**, 1449.
[63] Krumlinde, P.; Bogár, K.; Bäckvall, J. E. *J. Org. Chem.* 2009, **74**, 7407.
[64] Krumlinde, P.; Bogár, K.; Bäckvall, J. E. *Chem. Eur. J.* 2010, **16**, 4031.
[65] Traff, A.; Lihanmar, R.; Bäckvall, J. E. *J. Org. Chem.* 2011, **76**, 3917.
[66] Egi, M.; Sugiyama, K.; Saneto, M.; Hanada, R.; Kato, K.; Akai, S. *Angew. Chem. Int. Ed.* 2013, **52**, 3654.
[67] Thalén, L. K.; Zhao, D.; Sortais, J.-B.; Paetzold, J.; Hoben, C.; Bäckvall, J. E. *Chem. Eur. J.* 2009, **15**, 3403.
[68] Han, K.; Kim, Y.; Park, J.; Kim, M.-J. *Tetrahedron Lett.* 2010, **51**, 3536.
[69] Jana, N.; Mahapatra, T.; Nanda, S. *Tetrahedron: Asymmetry* 2009, **20**, 2622.
[70] Das, T.; Bhuniya, R.; Nanda, S. *Tetrahedron: Asymmetry* 2010, **21**, 2206.
[71] Jana, N.; Nanda, S. *Tetrahedron: Asymmetry* 2012, **23**, 802.
[72] Rej, R. K.; Jana, A.; Nanda, S. *Tetrahedron* 2014, **70**, 2634.

CHAPTER 6

Recent Developments in Flavin-Based Catalysis: Enzymatic Sulfoxidation

Patricia B. Brondani[1], Marco W. Fraaije[1], and Gonzalo de Gonzalo[2]

[1] Groningen Biomolecular Sciences and Biotechnology Institute, University of Groningen, Groningen, the Netherlands

[2] Departamento de Química Orgánica, Universidad de Sevilla, Sevilla, Spain

6.1 INTRODUCTION

Chiral sulfoxides are compounds that have attracted much interest in synthetic chemistry as building blocks or end products. They have been extensively employed in asymmetric synthesis as chiral auxiliaries and intermediates [1–3]. Their application as chiral synthons is mainly indebted to the high asymmetric induction achieved by the chiral sulfinyl group. Chiral sulfoxides that induce optical activity are very versatile, participating in several organic reactions, and can be easily removed by employing mild reaction conditions. The latter represents an advantage for the preparation of optically active compounds. Optically active sulfoxides have also been employed as catalysts in enantioselective organocatalytic processes [4].

The sulfinyl moiety is present in several compounds of biological interest. For example, one of the world's most sold pharmaceuticals, esomeprazole, is an optically active sulfoxide. The (*S*)-enantiomer of this sulfoxide is used to treat gastric or duodenal ulcers. Other examples of bioactive chiral sulfoxides with interesting properties are modafinil (a vigilance-promoting drug), aprikalim (a selective potassium channel opener), or oxisurane (an immunosuppressor).

Several approaches have been described for the preparation of optically active sulfoxides [5–7]. The three main routes to obtain these compounds are as follows: (i) the asymmetric sulfoxidation of prochiral sulfides, (ii) nucleophilic substitution using a chiral sulfur precursor, and (iii) the kinetic resolution of racemic sulfoxides. The first of these methods involves the use of various oxidants and catalysts and has been the most extensively employed. There are many examples in the scientific literature and reviews are available on this approach. In recent years, much attention has been focused on the synthesis of organic sulfoxides by employing conditions compatible with the green chemistry procedures [8–10]. For this reason, mild oxidants such as molecular oxygen or hydrogen peroxide are considered in combination with novel catalysts in order to develop a mild and environmentally friendly process.

Green Biocatalysis, First Edition. Edited by Ramesh N. Patel.

Biocatalysis, which is the use of biological catalysts (including whole cells, enzymatic preparations, or purified enzymes) for performing chemical reactions, fits perfectly into this trend [11]. By using biocatalysts, it is feasible to develop green chemical processes for the synthesis of optically active sulfoxides. Example studies in which enzymes have been used as biocatalysts for the synthesis of (enantiopure) sulfoxides have appeared since the mid-1990s. Often, the key biocatalyst for these processes required flavin-containing enzyme.

Flavins are a group of natural enzyme cofactors with interesting redox and photochemical properties that can participate in a wide set of reactions [12]. The most common flavin cofactors are flavin mononucleotide (FMN) or flavin adenine dinucleotide (FAD). Enzymes harboring one of these cofactors are called flavoenzymes. A large number of flavoenzymes have been extensively studied for their structural and mechanistic properties, and they are gaining momentum in industrial biocatalytic applications [13, 14]. Flavoenzymes have evolved to become powerful oxidative biocatalysts: they can catalyze not only simple alcohol oxidations but they were also found to be efficient in catalyzing, for example, oxidative C—C bond formation and enantioselective sulfoxidations.

Flavoprotein oxidases and oxygenases are of special interest because they are typically highly enantio- and regioselective and can be used for oxyfunctionalizations. Flavoprotein oxidases convert their substrates with the concomitant reduction of molecular oxygen to hydrogen peroxide [15]. Flavoprotein monooxygenases introduce a single atom of molecular oxygen into the substrate at the expense of molecular oxygen and a reduced nicotinamide coenzyme [16, 17].

6.2 ENZYMATIC SULFOXIDATION CATALYZED BY FLAVOPROTEIN OXIDASES

Flavoprotein oxidases do not catalyze directly the oxidation of prochiral sulfides, but these biocatalysts have been used in combination with other oxidative enzymes to perform asymmetric sulfoxidations. Thus, flavoprotein oxidases are able to generate hydrogen peroxide *in situ* as a byproduct, which will be used as an oxidant by, for example, peroxidases or peroxygenases. This cascade methodology enhances the operational stability of peroxidases when compared with the one-pot addition of hydrogen peroxide. The direct addition of peroxides often leads to rapid inactivation of the employed enzyme [18].

By using this approach of *in situ* hydrogen peroxide generation, a set of aryl methyl sulfoxides were synthesized on preparative scale with excellent yields (82–97%) and good to high optical purities (73–97%). Hydrogen peroxide was generated by the oxidation of alanine catalyzed by the D-amino oxidase from *Trigonopsis variabilis* (*Tv*DAO), while the sulfoxidation was performed by an immobilized fungal peroxidase from *Coprinus cinereus* (CiP) [19].

In order to avoid the use of alanine, an expensive reactant, thioanisole, was selectively oxidized to (*S*)-methyl phenyl sulfoxide in a 1.0 g scale with a 72% yield and 75% ee after 24 h by using the alcohol oxidase from *Pichia pastoris*. This flavoprotein oxidase was able to oxidize methanol to formaldehyde, generating hydrogen peroxide, which is employed by CiP [20].

CiP was also used in the synthesis of heteroaryl methyl sulfoxides combined with glucose oxidase as a hydrogen peroxide source, as shown in Scheme 6.1 [21]. Optically active heteroaryl alkyl sulfoxides are interesting compounds in organic chemistry, as they present a chelating center that can be used in asymmetric synthesis. Moderate to good results were achieved depending on the substrate structure. The best results were obtained for the oxidation of sulfides with electron-rich heterocycles.

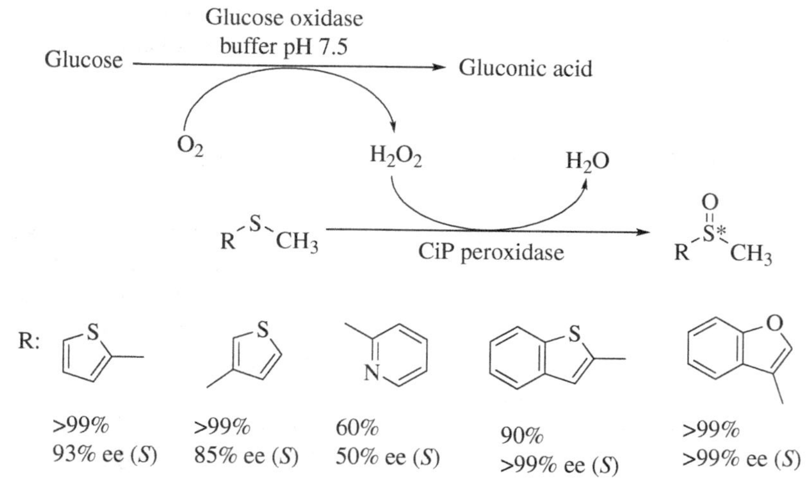

SCHEME 6.1

Asymmetric sulfoxidation of heteroaryl methyl sulfides employing the enzymatic cascade system, glucose oxidase, and CiP peroxidase.

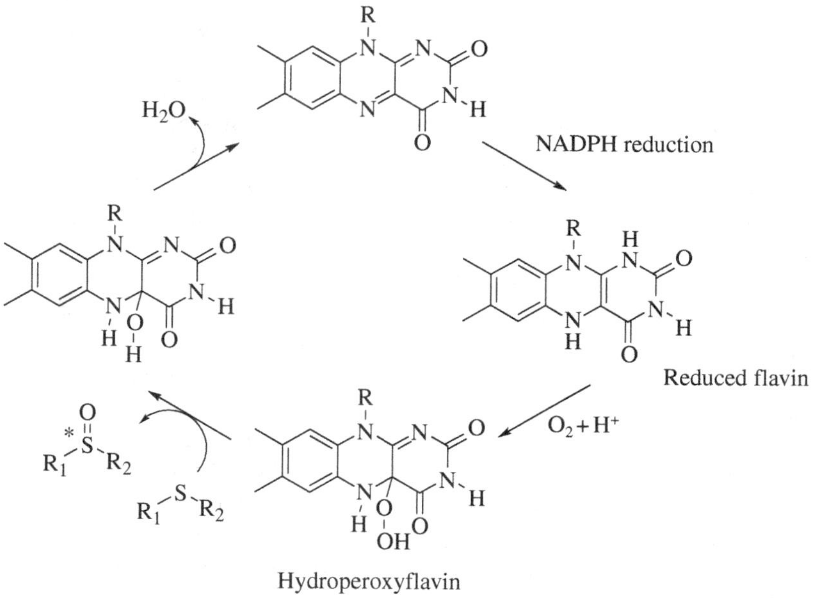

SCHEME 6.2

Mechanism of the enzymatic sulfoxidation of prochiral sulfides catalyzed by flavoprotein monooxygenases.

6.3 USE OF FLAVOPROTEIN MONOOXYGENASES FOR THE SYNTHESIS OF CHIRAL SULFOXIDES

Flavoprotein monooxygenases participate in a wide variety of metabolic processes both in prokaryotic and eukaryotic cells and are the main group of flavin-containing enzymes able to perform sulfoxidation reactions. In order to provide the two electrons necessary to catalyze this transformation, flavin-containing monooxygenases (FMOs) employ NADH or NADPH as the electron donor. In their catalytic cycle, the flavin cofactor is reduced by the nicotinamide coenzyme, after which it reacts with molecular oxygen to form a hydroperoxyflavin intermediate. This reactive flavin intermediate is responsible for the insertion of one of the oxygen atoms into the sulfide (see Scheme 6.2). Many flavoprotein monooxygenases have been characterized, and in 2006 these enzymes were classified into six distinct classes, depending on their structural and mechanistic properties. Two of these classes (classes A and B) are especially attractive for biocatalytic processes [22]. Flavoprotein monooxygenases from classes A and B are single-component enzymes containing a tightly bound FAD flavin cofactor. Representatives of these two flavoprotein monooxygenase classes have been applied for various sulfoxidation reactions in recent years.

6.3.1 Sulfoxidations Catalyzed by Baeyer–Villiger Monooxygenases

Among the flavoprotein monooxygenases, Baeyer–Villiger monooxygenases (BVMOs) are the most commonly used in sulfoxidation reactions. These flavin-containing enzymes are NAD(P)H dependent and catalyze Baeyer–Villiger oxidations, boron oxidations, as well as the oxidations of other heteroatoms such as sulfur, selenium, nitrogen, phosphorus, and iodine [23–25]. BVMOs can be divided in two types, Type I and Type II, but almost all BVMOs that have been applied as biocatalysts belong to the group of Type I BVMOs. These enzymes are FAD-containing biocatalysts, which require the NADPH as an electron source. Only recently, a new group of sequence-related flavoprotein monooxygenases has been identified for which indifference in nicotinamide coenzyme usage is observed. Whether members of this newly identified group of microbial monooxygenases are useful biocatalysts has still to be established.

Cyclohexanone monooxygenase from *Acinetobacter calcoaceticus* NCIB 9871 ($CHMO_{Acineto}$) is the most studied and applied BVMO. It has been demonstrated that this bacterial enzyme is an extremely versatile biocatalyst with an exceptionally broad substrate acceptance. $CHMO_{Acineto}$ has been employed in hundreds of oxidation procedures [26]. In fact, $CHMO_{Acineto}$ is a well known catalyst in the synthesis of (asymmetric) sulfoxides, being able to oxidize many sulfur-containing compounds such as thiosulfinates, sulfites, and 1,3-dithioacetals.

$CHMO_{Acineto}$ has been used in the diastereoselective oxidation of different β-hydroxy sulfides to the corresponding chiral β-hydroxy sulfoxides. Chiral β-hydroxy sulfoxides represent interesting compounds used as chiral auxiliaries in asymmetric synthesis, asymmetric ligands, or as building blocks for the synthesis of cyclic sulfides, benzoxathiepines, allylic alcohols, or leukotrienes [27]. The sulfoxidation of these substrates is a kinetic resolution, in which both the sulfide and the sulfoxide can be obtained in chiral form. Oxidation of the cyclohexyl derivative (Table 6.1, entry 2) by a semipurified preparation of $CHMO_{Acineto}$ in the presence of the enzymatic regeneration system D-glucose-6-phosphate/glucose-6-phosphate dehydrogenase (G6P/G6PDH) from *Leuconostoc mesenteroides* occurred with a high enantioselectivity (*E*). After 5 h of conversion, the major diastereoisomer of the β-hydroxy sulfoxide was recovered with an excellent optical purity. The biooxidation of the cyclopentyl and cycloheptyl analogues also showed a good diastereoselectivity, but both substrate and product were achieved with low to moderate *E* (entries 1 and 3). When the hydrogen atom in the *para*-position was replaced by a methoxy group (entry 4), a further decrease in the diastereoselectivity was observed, while the presence of bulkier substituents did not afford the formation of any sulfoxide.

In 2005, the recognition of the protein sequence motif common for the known BVMOs allowed the identification of new biocatalysts. Since then, the list of known and recombinant BVMOs has steadily grown. A particularly interesting novel BVMO has been identified in the thermophilic actinomycetes *Thermobifida fusca*. The respective enzyme can be easily expressed in *Escherichia coli* and was found to convert phenylacetone, hence its name: phenylacetone monooxygenase (PAMO) [28]. PAMO is thermostable and tolerates many organic solvents. Several crystal structures of PAMO have been solved, and its catalytic mechanism has been elucidated. The available structural information makes this protein an ideal candidate for protein engineering projects, while its thermostability and tolerance toward organic solvents are attractive for practical applications. However, PAMO has a somewhat limited substrate scope by only preferring relatively simple aromatic compounds. Initial experiments using purified PAMO and G6P/G6PDH as a coenzyme regeneration system showed the best (*S*)- or (*R*)-selectivities (ee ≥ 80%) in the oxidation of benzyl or phenylethyl

TABLE 6.1 Asymmetric Synthesis of β-Hydroxy Sulfoxides Catalyzed by $CHMO_{Acineto}$

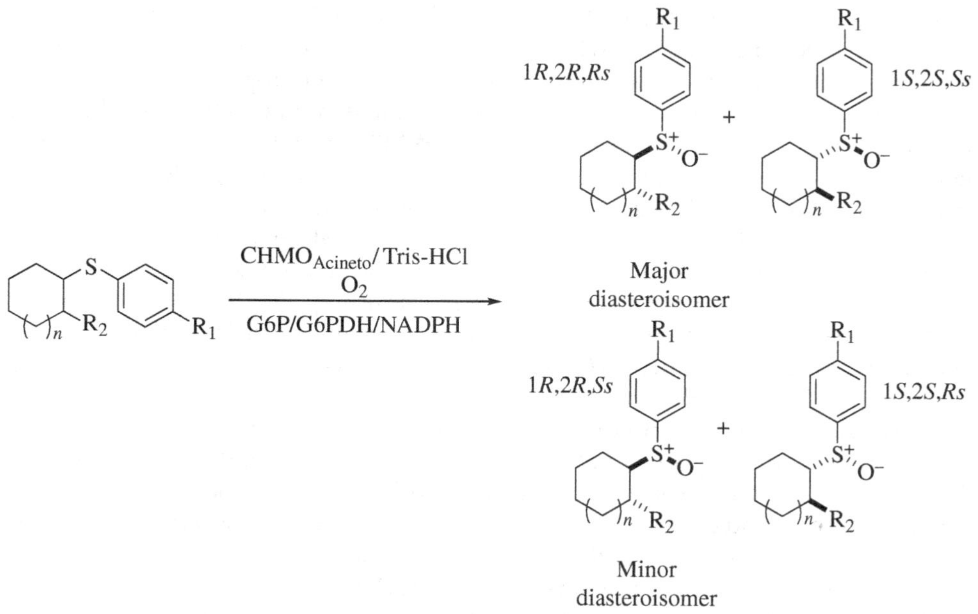

Entry	Sulfide	*t* (h)	*c* (%)	ee Sulfide (%)	Diastereomeric Ratio	ee_{major} (%)	ee_{minor} (%)	E^a
1	$n=0$, $R^1=H$, $R^2=OH$	24	97	69	83:17	53	91	1.6
2	$n=1$, $R^1=H$, $R^2=OH$	5	47	87	99:1	≥98	≥95	>200
3	$n=2$, $R^1=H$, $R^2=OH$	3	78	79	82:18	45	78	3.3
4	$n=1$, $R^1=OMe$, $R^2=OH$	48	95	13	3 peaks in 8:3:1 ratio[b]			1.1

[a] Enantiomeric ratio.
[b] Configuration not assignable.

alkyl sulfides, respectively, presenting a small alkyl group [29]. Bulkier groups resulted in decreased selectivity, while the oxidation of phenyl sulfides offers a wide range of enantioselectivities and conversions depending on the substituents. Phenylacetone monooxygenase was not a good biocatalyst for the preparation of alkyl heteroaryl sulfoxides. In general low activities and/or selectivities were achieved, with the exception of (*R*)-methyl 2-methylfurfuryl sulfoxide, which was recovered with almost complete conversion and 81% ee. Besides being primarily active on aromatic substrates, PAMO can also convert nonaromatic substrates [30]. Thus, (*S*)-cyclohexyl methyl sulfoxide was obtained with complete conversion and moderate optical purity (ee = 35%). Oxidation of butyl alkyl sulfides led to the corresponding (*S*)-sulfoxides with moderate conversions (40–34%) and optical purities around 50–65% for the methyl and ethyl derivatives, respectively. However, longer alkyl chains are not accepted by the enzyme.

One interesting property of PAMO is that the selectivity of thioanisole oxidation can be modulated depending on the pH of the reaction medium employed in the biooxidation [31]. Thus, by increasing the pH, an increase in the enantiomeric excess of (*R*)-methyl phenyl sulfoxide was obtained, ranging from 10% at pH 6.0 to 45% at pH 10.0. The same result was observed in the preparation of (*S*)-cyclohexyl methyl sulfoxide, with the highest selectivity at pH 10.5 (43%). This effect may be explained by the different protonation state of the hydroperoxyflavin responsible for the oxidative attack at different pHs.

BVMO-catalyzed sulfoxidations have been performed mainly in aqueous media, due to the low stability of $CHMO_{Acineto}$. Yet, PAMO was shown to be very robust, and it was found that it can catalyze sulfoxidations even in the presence of 30% $v\,v^{-1}$ of different cosolvents. Especially cosolvent with a hydrophilic character were well tolerated [32]. The use of these nonconventional reaction media has a negative effect on the enzymatic activity (lower conversions) and stability but, surprisingly, the addition of short alkyl chain alcohols (MeOH, EtOH) resulted in an increase in the PAMO selectivity when oxidizing alkyl phenyl sulfides. By using a cosolvent it was even possible to reverse the enantiopreference in the synthesis of phenyl propyl sulfoxide.

PAMO has been covalently fused to a thermostable mutant of the phosphite dehydrogenase (PTDH) from *Pseudomonas stutzeri* WM88. This bacterial dehydrogenase regenerates the NADPH cofactor by oxidizing phosphite to phosphate. An improved fusion protein (CRE2–PAMO) was expressed and purified in order to create a self-sufficient BVMO. CRE2–PAMO does not require an external cofactor regenerating system. In comparison, other enzymes require additional NADPH regenerating systems, such as D-glucose dehydrogenase from *Bacillus* sp. (GDH), G6PDH, and the PTDH. The oxidation of thioanisole to (*R*)-methyl phenyl sulfoxide led to the highest conversion when using glucose/GDH as secondary enzyme system (95% after 24h). However, the (*R*)-sulfoxide was recovered with a slightly lower enantiomeric excess (35% ee) when compared with the optical purities achieved for the rest of other tested cofactor regeneration systems (40% ee and conversions around 60–80%) [33]. This effect has been previously shown and can be explained by an allosteric effect on the enzyme by the second enzyme system [34]. No differences were observed between the fused protein (CRE2–PAMO) and the uncoupled system.

The main drawback for the application of PAMO as a biocatalyst is its narrow substrate acceptance. This biocatalyst only accepts relatively small aromatic substrates. Enzyme engineering attempts have focused on expanding the substrate acceptance of PAMO by introducing mutations close to the active site [35, 36]. A recent attempt has been performed by comparing PAMO structure with cyclopentanone monooxygenase from *Comamonas* sp. strain NCIMB 9872 ($CPMO_{Coma}$). The latter bacterial BVMO is able to oxidize a great number of ketones and sulfides. It also displays significant sequence homology with PAMO (41% sequence identity). Analysis of the PAMO structure and its comparison with $CPMO_{Coma}$ has revealed that 15 amino acids in the PAMO active site can be altered in order to modify the substrate specificity [37]. Thus, 30 single and multiple PAMO mutants were expressed in *E. coli* in which the specific mutation was inspired by the residues that are present in the CPMO sequence. Initial activity screenings revealed seven mutants with altered biocatalytic properties. These mutant enzymes were purified, characterized, and employed in combination with PTDH in the sulfoxidation of several aromatic sulfides. Most of the mutants led to an increase in the selectivity of thioanisole oxidation, yielding (*R*)-methyl phenyl sulfoxide with up to 75% enantiomeric excess. Intriguingly, the A442G PAMO was able to reverse the enantiopreference, affording the *S* enantiomer, albeit with low selectivity. Conversions ranged from 24 to 78% after 24h. None of the mutants were able to improve the results of the wild-type enzyme in the synthesis of (*S*)-benzyl methyl sulfoxide, but for most of the mutants good conversions and high optical purities were observed. More interesting results are shown in the biooxidation of bulky sulfides. While these sulfides are poorly accepted by the wild-type PAMO, they were converted more efficiently by some of the mutants. Thus, the fivefold mutant S441A/A442G/S444C/M446G/L447P was able to catalyze the preparation of (*R*)-benzyl phenyl sulfoxide with good conversion and excellent optical purity after 48h. Furthermore, a few mutants oxidized methyl 2-naphthyl sulfide with higher conversions and enantioselectivities higher than those of the wild-type enzyme. (*S*)-Sulfoxide was obtained for all the enzymes, achieving a 56% ee when the fivefold mutant was employed, while the use of A442G afforded 36% of the (*R*)-sulfoxide with 28% ee. A good result was also obtained in the presence of V54I PAMO: (*S*)-methyl 2-naphthyl sulfoxide was recovered with good conversion (c=51%) and a 45% ee.

Another way of altering the substrate specificity of PAMO was shown in 2012 by creating some chimeric biocatalysts [38]. Thus, using structure-inspired subdomain exchange, PAMO was blend with $CHMO_{Acineto}$ or steroid monooxygenase (STMO) from *Rhodococcus rhodochrous*. For this, 106 amino acids corresponding to the

TABLE 6.2 Enantioselective Oxidation of Prochiral Aromatic Sulfides Using Chimeric BVMOs

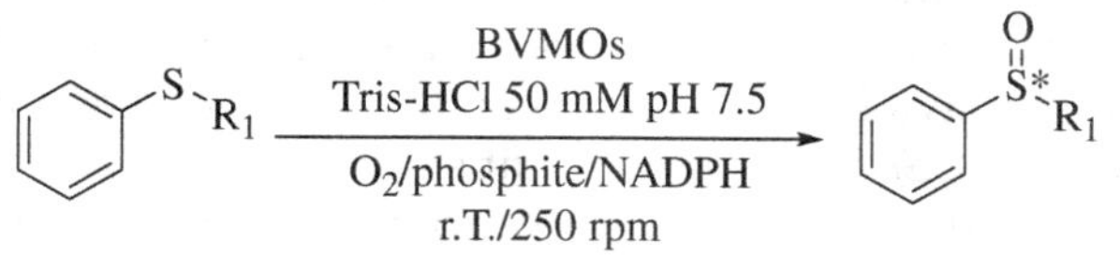

Entry	BVMO	R[1]	Time (h)	*c* (%)	ee (%)
1	PAMO	Me	24	93	16 (*R*)
2	PACHMO	Me	24	45	60 (*R*)
3	PASTMO	Me	24	79	62 (*S*)
4	PAMEMO1	Me	24	82	60 (*S*)
5	PAMO	Bn	48	7	36 (*R*)
6	PACHMO	Bn	48	14	70 (*R*)
7	PAMEMO1	Bn	48	39	56 (*S*)

C-terminal part of PAMO were replaced by the respective subdomains of other BVMOs in order to create the novel biocatalysts PACHMO and PASTMO. Another chimeric enzyme (PAMEMO1) was also prepared by blending PAMO with a putative BVMO obtained from the metagenome. The three chimeric BVMOs were successfully expressed in *E. coli* as fused biocatalysts with a thermostable mutant of PTDH from *P. stutzeri*. The created chimeric BVMOs exhibited some of the thermostability of PAMO being more stable when compared with $CHMO_{Acineto}$ and STMO. Sulfoxidations catalyzed by the three chimeric BVMOs resulted in novel and interesting results. Regarding the oxidation of thioanisole at pH7.5, PACHMO increased the *R* selectivity of PAMO (entry 2, Table 6.2). After 24h, PASTMO afforded (*S*)-methyl phenyl sulfoxide with 62% ee, improving the *S* selectivity of wild-type STMO, as shown in entry 3. PAMEMO1 also led to the (*S*)-enantiomer with 60% ee and 80% conversion (entry 4). The bulky substrate benzyl phenyl sulfide was also tested, as this sulfide was not oxidized by both STMO and $CHMO_{Acineto}$ and was poorly converted by PAMO (7% conversion into the (*R*)-sulfoxide with 36% ee after 48h). PASTMO did not convert this substrate, while oxidation catalyzed by PACHMO (entry 6) resulted in 14% of (*R*)-benzyl phenyl sulfoxide with good selectivity (70% ee). Intriguingly, the oxidation catalyzed by PAMEMO1 resulted in the most effective conversion (39%), generating the opposite *S* enantiomer with a 56% ee.

PAMO has been coimmobilized with the NADPH recycling secondary enzymatic system (G6PDH) onto several polyphosphazene $\{NP[O_2C_{12}H_{8-x}(NH_2)_x]\}_n$ solid supports [39]. The biocatalysts were linked to the polymer through a glutaraldehyde connector. By this, it was possible to obtain a set of synthetic "self-sufficient" systems with best results for the polyphosphazene with $x=0.5$. This novel immobilized biocatalyst was able to oxidize thioanisole and benzyl methyl sulfide with no significant alteration in *E*. Unfortunately, due to the low stability of the immobilized monooxygenase, the preparation lost most of its activity after a few cycles.

The M446G PAMO mutant was obtained from the comparison between PAMO and $CPMO_{Coma}$ structures [40]. The substitution of the methionine at the 446 position of the enzyme active site by a smaller residue led to a novel biocatalyst with interesting properties when oxidizing prochiral sulfides. Thus, isolated M446G showed an excellent selectivity in the synthesis of some alkyl phenyl sulfoxides, such as (*R*)-methyl phenyl, (*R*)-methyl 4-methylphenyl, and (*R*)-ethyl phenyl, which were obtained with optical purities higher than 90%. Even in the synthesis of ethyl phenyl sulfoxide, there is a reversal of the enantiopreference by using the mutant BVMO. Both (*R*)-benzyl phenyl sulfoxide (ee=97%) and (*S*)-methyl 2-naphthyl sulfoxide (ee=78%) are good substrates for M446G PAMO. Reversals in selectivities in the sulfoxidation of benzyl or phenylethyl sulfides are lower than those achieved for the wild-type PAMO. M446G catalyzed the oxidation of methyl 3-pyridyl sulfide to the

(*R*)-sulfoxide with good conversion and high optical purity (ee = 88%) [30]. This enzyme was also active on nonaromatic substrates. It could be used to prepare (*S*)-cyclohexyl methyl sulfoxide with a 67% conversion and 49% ee using the corresponding sulfide. The PAMO mutant also oxidized *n*-butyl alkyl sulfides but with modest results in terms of activity (45–39% conversion for the methyl and ethyl derivatives) and selectivity (53–36% ee). It was found that sulfoxidation selectivity of cyclohexyl methyl sulfide catalyzed by M446G PAMO also depends on the pH of the medium. The optical purity of the (*R*)-sulfoxide increased with the pH until raising a maximum value at pH 9.0. Another interesting property of M446G PAMO is the effect of some polar cosolvents in the reaction media [41]. When the sulfoxidations were carried out in Tris/HCl pH 9.0 with 5% v v^{-1} MeOH, an increase in the conversions was measured in all the biotransformations. Moreover, cyclohexyl propyl sulfide was not a substrate for the enzyme in aqueous medium, but in presence of a cosolvent such as methanol conversion was observed yielding 11% (*S*)-cyclohexyl propyl sulfoxide with a 21% ee.

4-Hydroxyacetophenone monooxygenase (HAPMO) from *Pseudomonas fluorescens* ACB is another BVMO cloned and expressed in *E. coli*, demonstrated extensive application in biocatalysis [42]. Initial experiments showed that this enzyme, coupled with G6P/G6PDH, was able to oxidize methyl phenyl sulfides to the (*S*)-sulfoxides as, for example, (*S*)-methyl phenyl sulfoxide (96% conversion, ≥99% ee) [43]. Other phenyl sulfides containing longer alkyl chains were also oxidized with good conversions and good to excellent enantioselectivities to the corresponding (*S*)-sulfoxides, while the oxidation of compounds presenting the sulfur atom further from the aromatic ring led to a decrease in the conversion and a lower selectivity, but still with moderate to good values.

HAPMO has shown to be a good biocatalyst for the preparation of alkyl heteroaryl sulfoxides (Scheme 6.3) [30]. Both methyl 2- and 3-pyridyl sulfides were converted to the enantiopure (*S*)-sulfoxides with excellent conversions, while enantiopure (*R*)-methyl 4-pyridyl sulfoxide was achieved with 63% conversion after 24 h. Enantiopure (*S*)-ethyl 2-pyridyl sulfoxide was also recovered with complete conversion, while the propyl and allyl analogues are obtained with 99% ee but low conversions. HAPMO was also able to oxidize selectively 2-methylfurfuryl and thiophenyl methyl sulfides with conversions around 70%, yielding the (*R*)-enantiomer for the oxygenated heterocycle, while the (*S*)-enantiomer was obtained for the thiophene derivative. Regarding nonaromatic sulfides, HAPMO showed complete conversion and selectivity in the preparation of (*S*)-cyclohexyl methyl sulfoxide and its ethylated analogue. Cyclic sulfides such as tetrahydrothiophene and tetrahydro-2*H*-thiopyran were oxidized by HAPMO, leading to high amounts (close to 80%) of the final sulfoxides. No starting sulfides were detected, as a moderate amount of sulfone was formed. (*S*)-*n*-Butyl methyl and (*S*)-*n*-butyl ethyl sulfoxides were achieved in enantiopure form and with complete conversion after 24 h. Oxidation of a linear sulfide containing a longer alkyl chain such as methyl *n*-octyl sulphide gave 75% of enantiopure (*S*)-methyl *n*-octyl sulfoxide, observing the formation of 12% of the sulfone.

Both PAMO and HAPMO are not only able to perform the desymmetrization of prochiral sulfides but they can also catalyze the kinetic resolution of racemic sulfoxides by oxidizing selectively one of the sulfoxide enantiomers to the sulfone, leaving

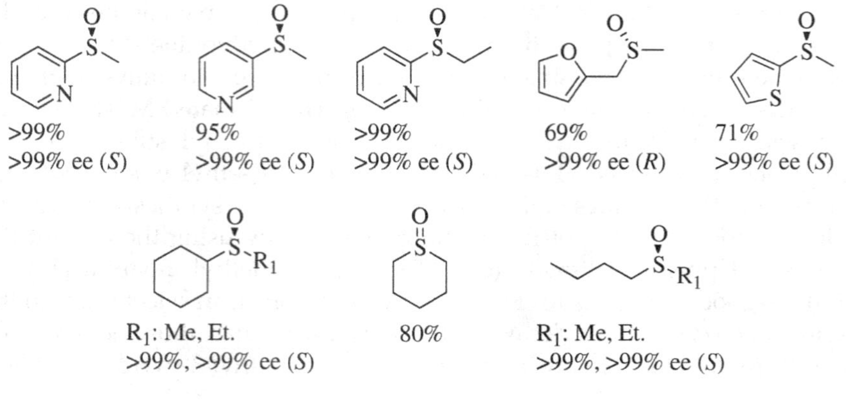

SCHEME 6.3

Optically active alkyl heteroaryl and dialkyl sulfoxides obtained by HAPMO-catalyzed sulfoxidation.

TABLE 6.3 Synthesis of Optically Active Sulfoxides by Kinetic Resolution of Racemic Sulfoxides Employing BVMOs

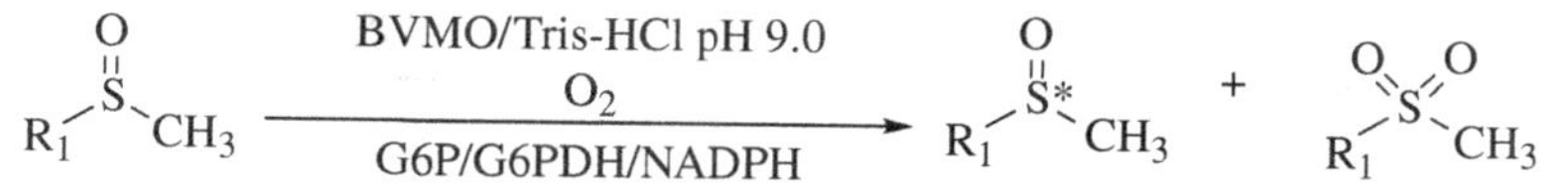

Entry	BVMO	R¹	Time (h)	*c* (%)	ee (%)
1	PAMO	Bn	4	39	60 (*S*)
2	PAMO	$PhCH_2CH_2$	8	51	95 (*R*)
3	HAPMO	2-Pyridyl	0.5	60	71 (*S*)
4	HAPMO	Cy	5	60	82 (*S*)
5	HAPMO	Octyl	4.5	55	99 (*S*)

the other sulfoxide enantioenriched, as shown in Table 6.3 [29]. Regarding PAMO, the kinetic resolution of alkyl phenyl sulfoxides occurred with low to moderate *E*. In contrast, alkyl benzyl or alkyl phenylethyl sulfoxides are resolved with excellent selectivity (entries 1 and 2). Tests performed using HAPMO in the resolution of alkyl aryl sulfoxides showed that this oxidation occurred, but with no selectivity [43]. Yet, HAPMO is able to oxidize alkyl pyridyl sulfoxides, for example, (±)-methyl 2-pyridyl sulfoxide oxidation occurred with 60% conversion after 0.5 h, leading to the (*S*)-sulfoxide with a good enantiomeric excess [30]. When the nitrogen atom was located further away from the sulfoxide moiety, reactions were still fast but the enzyme selectivity was much lower, giving enantiomeric excesses close to 20%. As expected, HAPMO-catalyzed oxidation of nonaromatic sulfoxides occurred with lower activity as compared to that of heteroaromatic analogues. Incubations of 5 h are required to reach conversions of around 60% for (±)-cyclohexyl methyl sulfoxide and (±)-methyl *n*-octyl sulfoxide, but with relatively higher *E*. The unreacted sulfoxides could be recovered with 82% and 99% ee, respectively (entries 4 and 5).

Aspergillus has been widely used to perform several biotransformations due to its metabolic activity. Nine strains of *Aspergillus* displaying BVMO activity have been tested recently in the sulfoxidation of cyclohexyl methyl sulfide and thioanisole into their sulfoxides [44]. For almost all the microorganisms, best results were achieved in the oxidation of the cyclohexyl methyl sulfide, choosing this compound for a further optimization regarding the reaction medium, the type of the cells, the substrate concentration, and the presence of organic cosolvents. It was found that *Aspergillus japonicus* is a promising biocatalyst as it enabled the recovery of enantiopure (*R*)-cyclohexyl methyl sulfoxide with complete conversion after 5 days when using growing cells and substrate concentrations between 1 and 10 mM in the presence of 0.5% $v\,v^{-1}$ of isopropanol. Recently, a BVMO isolated from an *Aspergillus* strain (*Aspergillus fumigatus* Af293, $BVMO_{Af1}$) was cloned and successfully overexpressed when fused to the cofactor regenerating enzyme PTDH [45]. The fused biocatalyst showed a good stability toward temperature, pH, and organic cosolvents. $BVMO_{Af1}$ was able to oxidize thioanisole and ethyl benzyl sulfide to the enantiopure (*S*)-sulfoxides. After 3 h, (*S*)-methyl phenyl sulfoxide was achieved with excellent conversion, but the benzyl sulfoxide was only recovered in a low yield (25%) and the sulfone was the main oxidation product (75%). When the oxidation was stopped after 1 h, conversion was decreased to 56%, but still a high amount of sulfone was achieved (37%). These studies show that *Aspergillus* contain a variety of BVMOs that await biocatalytic exploration.

A new BVMO from the bacterium *Dietzia* sp. D5 (BVMO4) has been characterized and employed in different oxidative processes, including the sulfoxidation of prochiral organic sulfides [46]. BVMO4 was able to oxidize thioanisole with moderate selectivity (ee = 50%) but also tolerates small substituents in *para*- or *meta*-position in order to achieve the (*R*)-sulfoxides with high selectivity in cases of the *p*-methyl and *p*-fluoro derivatives. It is very interesting to notice that BVMO4 increased significantly its stereospecificity in the oxidation of ethyl phenyl sulfide (≥99% for the

TABLE 6.4 Oxidation of Prochiral Aromatic Sulfides Catalyzed by Cells of *Rhodococcus* sp.

Entry	Substrate	Yield (%)	ee (%)
1	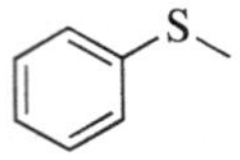	44	99 (*S*)
2	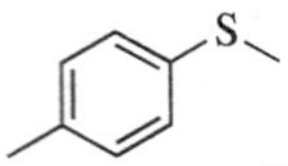	86	97 (*S*)
3	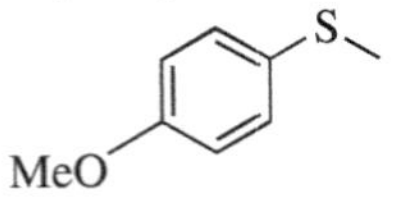	61	38 (*R*)
4	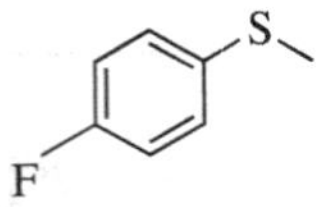	70	92 (*S*)
5	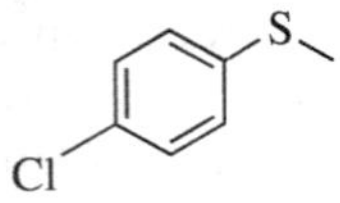	60	99 (*S*)
6	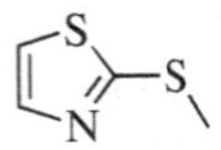	20	90 (–)[a]
7	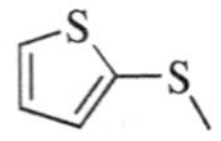	14	71 (*S*)

[a] Absolute configuration not determined.

(*R*)-enantiomer), when compared with thioanisole, while the reaction rate was comparable. BVMO4 was not able to oxidize *O*-methoxy or nitro-substituted methyl phenyl sulfides, as well as heteroaromatic sulfides or *tert*-butyl methyl sulfide.

Resting cells of the microorganism *Rhodococcus* sp. (strain ECU0066) have been used in the asymmetric oxidation of several aryl alkyl and heteroaryl alkyl sulfides, as shown in Table 6.4 [47]. Oxidation of thioanisole led to the formation of enantiopure (*S*)-methyl phenyl sulfoxide in a two-step procedure, in which a first oxidation to (*S*)-sulfoxide with 80% ee was produced, followed by the oxidation of (*R*)-enantiomer of the sulfoxide to the achiral sulfone. To confirm this mechanism racemic methyl phenyl sulfoxide was selectively oxidized by the microorganism affording the *S* enantiomer with 99% ee, while the *R* enantiomer led to the sulfone. *Rhodococcus* sp. cells have been also employed to oxidize other thioanisole analogues. In the case of *para* substitutions, high yields and almost enantiopure (*S*)-sulfoxides were recovered except for the methoxy derivative, for which the *R* enantiomer was formed with moderate enantiomeric excess. The selectivity of the transformation does not depend on the electronic properties of the *para* substituents. Sulfides with a heterocycle (entries 6 and 7) are not good substrates for the bacterial cells as low yields were obtained and an unknown byproduct appeared, though the stereoselectivity was very high.

The bacterium *Rhodococcus jostii* RHA1 is a promising source of biocatalysts. Its predicted proteome contains an unusually high variety of oxidative enzymes [48]. Genome mining analyses showed that in this microorganism more than 20 putative BVMOs are present. In 2012, after an optimization procedure, 22 of these BVMOs were cloned and expressed in *E. coli* cells [49]. The biocatalysts were employed as cell extracts in Baeyer–Villiger oxidations and sulfoxidations, using phosphite/PTDH as a secondary enzymatic system to regenerate NADPH. Some of the BVMOs were able to oxidize thioanisole and benzyl ethyl sulfide. The best results were obtained with two of these enzymes, BVMO3 and BVMO24, for which it was possible to obtain (*S*)- and (*R*)-methyl phenyl sulfoxide, respectively, with high conversions and optical purities. BVMO3 led to 60% of (*R*)-benzyl ethyl sulfoxide with 75% ee, while BVMO24 afforded the enantiopure (*S*)-sulfoxide with excellent conversion.

6.3.2 Oxidative Processes Employing Styrene Monooxygenases

Two-component flavoprotein monooxygenases are emerging biocatalysts that generally consist of a monooxygenase and a reductase component [50]. Styrene monooxygenases (SMOs; EC 1.14.13) are two-component NADH-dependent flavoenzymes known to catalyze the enantioselective oxidation of the vinyl side chain of styrene to form styrene epoxide. SMOs can also perform sulfoxidations. *E. coli* BL21 cells expressing SMO from *Pseudomonas putida* CA-3 have been used enantioselective sulfoxidations [51]. The wild-type enzyme and an engineered SMO (SMOeng R3-11) were able to catalyze the oxidation of thioanisole and other substituted aryl methyl sulfides with low to good selectivities depending on the substrate structure. (*S*)-Methyl 2-methylphenyl sulfoxide was recovered with 84% ee, while the (*S*)-4-chlorophenyl derivative was obtained with only 10% ee. For other sulfides tested, (*R*)-enantiomeric sulfoxides with modest ee values were obtained. The two SMO enzymes are also able to catalyze the oxidation of benzo[*b*]thiophene and its 2-methyl analogue (Scheme 6.4), but the main drawback for this procedure is that the final sulfoxides quickly racemized at the employed reaction conditions.

The screening of a metagenomic library derived from loam soil afforded a novel two-component SMO, which has been heterologously expressed in *E. coli* [52]. This biocatalytic system was able to perform selective sulfoxidations of alkyl aryl sulfides, providing the recovered (*R*)-sulfoxides with good to high selectivities and moderate conversions after 16h. Ethyl phenyl sulfide was the best substrate for this enzyme, as the final sulfoxide was obtained with 92% ee. It therefore represents an interesting SMO alternative.

6.3.3 Enzymatic Sulfoxidations Catalyzed by Flavin-Containing Monooxygenases

Flavin-containing monooxygenases (FMOs) are a class of flavin monooxygenases that contain tightly bound FAD and require NADPH as coenzyme. These enzymes catalyze the monooxygenation of different heteroatoms, with the natural role of participating in the detoxification of drugs and xenobiotics [22, 53]. Most of the FMOs are membrane associated, which makes it difficult to obtain them for further application. The human proteome contains five FMO isoforms, FMO1–5. FMO3 seems to be the dominant enzyme in the human body.

Some examples have been shown in which highly interesting sulfoxides have been obtained using eukaryotic FMOs, as shown in Scheme 6.5. For instance, albendazole and fenbendazole sulfoxides, benzimidazole derivatives with anthelmintic activity, were obtained from the corresponding sulfides using liver, lung, and intestinal microsomes from sheep and cattle, which contained FMOs as oxidative biocatalyst [54]. Thus, the FMO-mediated liver sulfoxidation of albendazole led to enantiopure sulfoxide while the oxidation of fenbendazole occurred with 65–79% ee. Several mutants of the FMO1 found in humans have been applied as biocatalysts in the oxidation of fenthion, an organothiophosphate insecticide, with excellent selectivity for the (+)-sulfoxide [55]. Sulindac, a nonsteroidal anti-inflammatory drug (NSAID), was obtained by oxidation of its precursor sulfide employing human liver or kidney microsomes and cDNA-expressed microsomes [56]. This oxidation was shown to be catalyzed by FMOs obtaining the (*R*)-sulfoxide with up to 97% ee.

In 2003, the first FMO from bacterial origin was reported from bacterium *Methylophaga* sp. strain SK1 (mFMO) [57]. This soluble enzyme can be easily expressed in *E. coli*, which increases the biocatalytic potential FMOs as biocatalysts. Recently,

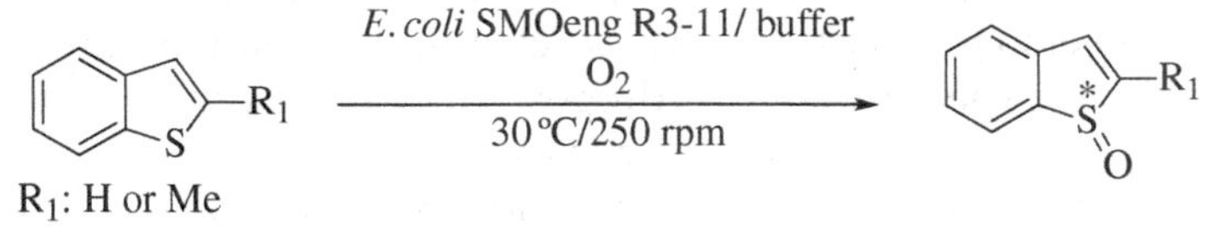

SCHEME 6.4 Oxidation of benzo[*b*]thiophene and 2-methyl benzo[*b*]thiophene employing *E. coli* cells expressing an SMO from *Pseudomonas putida* CA-3.

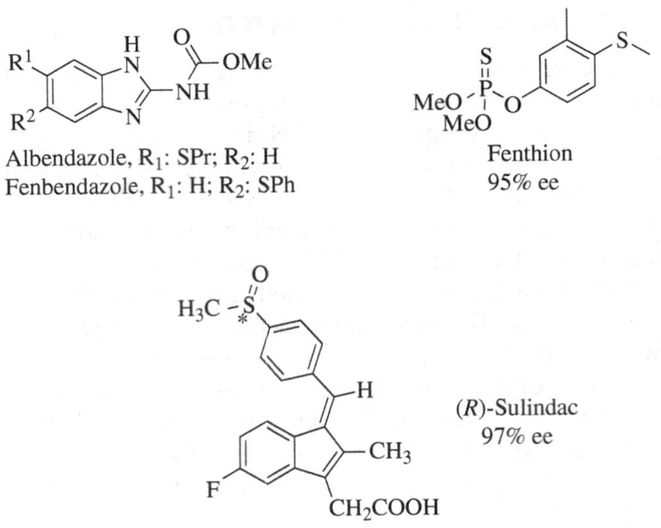

SCHEME 6.5
Some examples of sulfoxides with biological interest obtained employing FMOs.

mFMO has been fused to PTDH from *P. stutzeri* (PTDH-FMO) [58]. PTDH-mFMO enables the preparation of different alkyl aryl, alkyl heteroaryl, and dialkyl sulfoxides with good results in general, as shown in Scheme 6.6. The oxidation of thioanisole and analogues occurred with good conversion and optical purity, especially for those derivatives containing a *p*-Cl or a *p*-Me moiety. mFMO is also able to synthesize other (*S*)-alkyl phenyl sulfoxides with high optical purities, while the preparation of (*R*)-butyl alkyl sulfoxides occurred with high conversion and good selectivity. The presence of 5% $v\,v^{-1}$ hexane as cosolvent has an activating effect in mFMO-catalyzed oxidation of thioanisole resulting in an increase in both the conversion and the ee of the formed (*S*)-methyl phenyl sulfoxide.

Eight FMOs from *R. jostii* RHA1 have been recently cloned and expressed in soluble form in order to explore them as biocatalysts [59]. These eight enzymes do not present all typical properties of the previously known FMOs, so they have been classified as Type II FMOs. When used in the biooxidation of thioanisole, the final sulfoxide was recovered with low to good conversions and low to moderate enantioselectivities. Both enantiomers could be obtained depending on the biocatalyst employed. Recently it was discovered that a gene from the marine bacterium *Stenotrophomonas maltophilia* also encodes an FMO (SMFMO) [60]. This biocatalyst was able to employ either NADH or NADPH as an electron source and catalyzes both the Baeyer–Villiger oxidation of carbonyl compounds and the sulfoxidation of prochiral sulfides. SMFMO converted a wide set of aromatic sulfides into the (*R*)-sulfoxides, with the exception of methyl 4-nitrophenyl sulfide, which is not a substrate for the biocatalyst. Processes performed in the presence of NADH are considerably faster than when using its phosphorylated analog, while the enzymatic selectivity is dependent on the substrate, there being some slight differences depending on the coenzyme tested. (*R*)-4-Chlorophenyl methyl sulfoxide can be obtained with an optical purity higher than 80%. Interestingly, some of the FMOs from *R. jostii* RHA1 also showed an NADPH/NADH indifference, which makes these newly discovered Type II FMOs an interesting class of biocatalyst [59].

6.4 ASYMMETRIC SULFOXIDATION USING FLAVINS AS CATALYSTS

Flavin catalysts have provided a synthetic methodology for the organocatalytic oxidation of sulfides, simulating to a certain extent the enzymatic processes catalyzed by flavin-containing enzymes [10, 61]. These chemical oxidations are formally performed by a reactive species, flavin hydroperoxide, which can be generated by two different ways: (i) the direct addition of hydrogen peroxide to the flavin or (ii) through flavin reduction and subsequent incorporation of molecular oxygen. The flavin

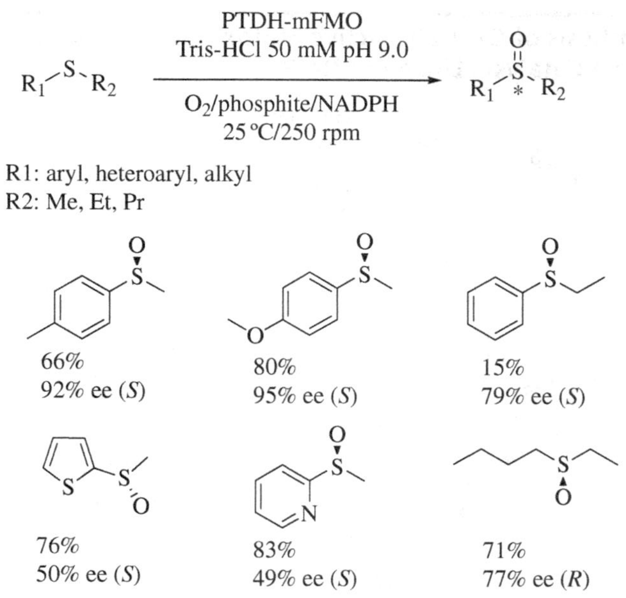

SCHEME 6.6 Oxidation of prochiral sulfides catalyzed by the fused biocatalyst PTDH-mFMO.

hydroperoxide transfers one of its oxygen atoms to the substrate, while the other atom is eliminated as water in a dehydration process to restart the catalytic cycle. By modifying the flavin structure, it is possible to obtain different reactivities in order to use flavins as organocatalysts in sulfoxidation procedures.

Most of the flavin-catalyzed oxidations have been performed to obtain the racemic sulfoxides from the starting sulfides. However, in recent years, some efforts have been made to carry out asymmetric organocatalyzed sulfoxidations by creating novel systems mimicking flavoenzymes. Thus, a set of β-cyclodextrins have been linked to different neutral N^5-ethylflavins in order to create chiral flavin catalysts [62, 63]. These systems have been employed in the oxidation of thioanisole and its aryl analogues in aqueous medium in the presence of hydrogen peroxide. Depending on the substrate structure, different results were achieved. Thus, sulfoxides containing electron-donating groups, such as *p*-hydroxy or *p*-methoxy, can be obtained with excellent conversions and good enantioselectivities, while sulfides with electron-withdrawing groups are oxidized with moderate selectivity. The catalyst load can be reduced to 0.2 mol% in the oxidation of methyl *p*-methylphenyl sulfide. This methodology was extended by preparing four novel catalysts using the alloxazine flavin connected to the primary face of both α- and β-cyclodextrins via linkers of variable length. *n*-Alkyl methyl sulfides are oxidized with excellent conversions and moderate to good enantioselectivities by both α- and β-cyclodextrins attached to the flavin by the longest linker. Those substrates containing bulkier substituents such as benzyl, phenyl, cyclohexyl, or *tert*-butyl are oxidized with higher conversions by the flavin-cyclodextrin conjugates, as these present a bigger cavity for the sulfide accommodation. Oxidation of *tert*-butyl methyl sulfide led to the formation of the (−)-sulfoxide with complete conversion and 91% ee after 1 h, while both (−)-cyclohexyl and (+)-phenyl methyl sulfoxide also showed high selectivity.

A different approach for the synthesis of optically active sulfoxide has been developed recently, creating artificial and self-sufficient oxidative biocatalysts by incorporating a set of nonnatural flavins to flavin-binding proteins. Riboflavin-binding protein (RfBP) from *Gallus gallus* does not present catalytic activity but is able to bind a wide set of modified flavins [64]. Taking into account this ability, apo-RfBP was reconstituted with different N^5-ethyl flavins that can form stable 4a-peroxyflavins in the presence of hydrogen peroxide [65]. The unnatural enzymes created were shown to be able to catalyze peroxide-driven sulfoxidations. Such catalytic activity, peroxygenase activity, has been never found in natural flavoenzymes. The five synthesized artificial biocatalysts have been employed in the asymmetric sulfoxidation of different alkyl aryl or dialkyl sulfides using hydrogen peroxide as an oxidant (Table 6.5). Conversions were performed at 4 °C

TABLE 6.5 Synthesis of Optically Active Sulfoxides Employing Novel Oxidative Flavin Catalysts Linked to RfBP

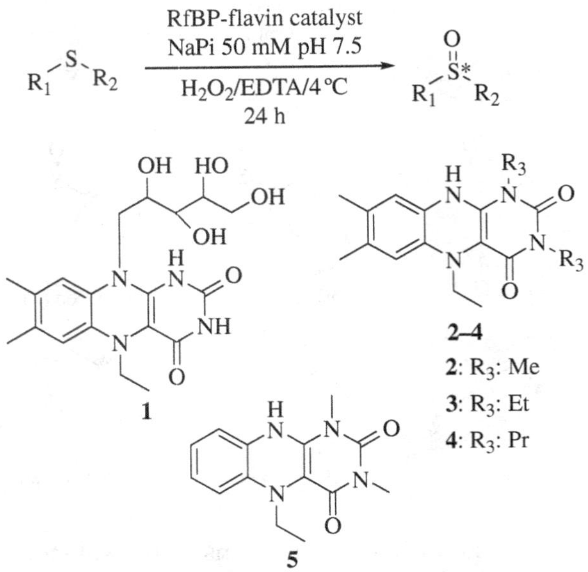

Entry	Flavin	R_1	R_2	*c* (%)	ee (%)
1	**1**	*p*-Tol	Me	44	30 (*R*)
2	**2**	*p*-Tol	Me	45	25 (*S*)
3	**2**	Ph	Et	44	15 (*S*)
4	**3**	Ph	Me	48	25 (*S*)
5	**5**	Bn	Me	60	36 (*R*)
6	**5**	Bu	Et	53	33 (*R*)

in the presence of EDTA in order to minimize the noncatalytic sulfoxidation while using a 1% $w\,w^{-1}$ of catalyst. All the generated artificial biocatalysts were active in the oxidation processes. Moderate conversions and low to moderate optical purities in the formation of (*R*)-sulfoxides were obtained when the modified riboflavin **1** was employed. N^5-ethylated flavins presenting methyl (**2**), ethyl (**3**), or propyl (**4**) groups in N^1 and N^3 were also tested in the catalyzed sulfoxidations, obtaining low to moderate conversions (up to 50%) and enantiomeric excesses (16–25%) in the formation of the (*S*)-sulfoxides. The use of a demethylated analog (**6**) of lumiflavin **2** afforded the (*R*)-sulfoxides, but again, the conversions and selectivities obtained were low. The best result obtained was the formation of (*R*)-benzyl methyl sulfoxide with 60% conversion and 36% ee after 24 h. The fact that the catalyst selectivity is strongly dependent on the type of flavin cofactor employed is a very attractive feature as it allows the preparation of both enantiomers of the product by changing the type of flavin cofactor. This modification of the selectivity by altering the flavin structure can be explained by docking studies, revealing that the flavins can be bound to the RfBP in two different ways that will determine the enantiopreference of the oxidation. Another approach to improve the catalytic performance of these unnatural biocatalysts is to employ directed evolution, which has been shown to be very effective in changing the catalytic properties of enzymes.

6.5 SUMMARY AND OUTLOOK

Since the mid-2000s, biocatalysis has been demonstrated to be a very powerful tool for the preparation of optically active sulfoxides using mild conditions. Among all the biocatalysts employed for the preparation of chiral sulfoxides, in particular flavoenzymes have proven their efficiency. Flavoprotein oxidases have been employed as *in situ* hydrogen peroxide generators to support sulfoxidations catalyzed by other peroxide-dependent oxidative enzymes. Flavoprotein monooxygenases are able to perform selective sulfoxidations using molecular oxygen and a reduced nicotinamide cofactor. Well established examples are BVMOs, SMOs, and FMOs. These biocatalysts have been widely exploited in chemo- and enantioselective sulfoxidations. Recent advances in molecular biology have allowed a large variety of different flavoprotein monooxygenases to be obtained. Furthermore, enzyme engineering approaches have been shown to be successful in generating more selective and/or more robust biocatalysts. With these developments it has become relatively easy to develop a dedicated flavin-based biocatalyst for the synthesis of a specific sulfoxide.

REFERENCES

[1] M. C. Carreño, G. Hernández-Torres, M. Rigaborda, and A. Urbano, *Chem. Commun.*, 6129–6144 (2009).
[2] H. Pellissier, *Tetrahedron* **62**, 5559–5601 (2006).
[3] R. Bentley, *Chem. Soc. Rev.* **34**, 609–624 (2005).
[4] I. Fernández, V. Valdivia, M. P. Leal, and N. Khiar, *Org. Lett.* **9**, 2215–2218 (2007).
[5] G. O'Mahony, A. Ford, and A. R. Maguire, *J. Sulfur Chem.* **34**, 301–341 (2013).
[6] E. Wojaczynska and J. Wojaczynski, *Chem. Rev.* **110**, 4303–4356 (2010).
[7] J. Legros, J. R. Dehli, and C. Bolm, *Adv. Synth. Catal.* **347**, 19–31 (2007).
[8] H. L. Holland, *Nat. Prod. Rep.* **18**, 171–181 (2001).
[9] R. D. Schmid and V. Urlacher, eds. *Modern Biooxidation: Enzymes, Reactions and Applications*, Wiley-VCH, Weinheim, 2007.
[10] G. de Gonzalo and M. W. Fraaije, *ChemCatChem* **5**, 403–415 (2012).
[11] R. N. Patel, *ACS Catal.* **1**, 1056–1074 (2011).
[12] V. Massey, *Biochem. Soc. Trans.* **28**, 283–296 (2000).
[13] C. T. Walsh and T. A. Wencewicz, *Nat. Prod. Rep.* **30**, 175–200 (2013).
[14] V. Joosten and W. J. H. van Berkel, *Curr. Opin. Chem. Biol.* **11**, 195–202 (2007).
[15] W. P. Dijkman, G. de Gonzalo, A. Mattevi, and M. W. Fraaije, *Appl. Microbiol. Biotechnol.* **97**, 5177–5188 (2013).
[16] D. E. Torres Pazmiño, M. Winkler, A. Glieder, and M. W. Fraaije, *J. Biotechnol.* **146**, 9–24 (2010).
[17] M. M. Huijbers, S. Montersino, A. H. Westphal, D. Tischler, and W. J. H. van Berkel, *Arch. Biochem. Biophys.* **544**, 2–17 (2014).
[18] F. van de Velde, N. D. Lourenço, M. Bakker, F. van Rantwijk, and R. A. Sheldon, *Biotechnol. Bioeng.* **69**, 286–291 (2000).
[19] K. Okrasa, A. Falcimaigne, E. Guibé-Jampelb, and M. Therisod, *Tetrahedron: Asymmetry* **13**, 519–522 (2002).
[20] F. Pezzoti and M. Therisod, *Tetrahedron: Asymmetry* **18**, 701–704 (2007).
[21] F. Pezzoti, K. Okrasa, and M. Therisod, *Tetrahedron: Asymmetry* **16**, 2681–2683 (2005).
[22] W. J. H. van Berkel, N. M. Kamerbeek, and M. W. Fraaije, *J. Biotechnol.* **124**, 670–689 (2006).
[23] K. Balke, M. Kadow, H. Mallin, S. Saβ, and U. T. Bornscheuer, *Org. Biomol. Chem.* **10**, 6249–6265 (2012).
[24] H. Leisch, K. Morley, and P. C. K. Lau, *Chem. Rev.* **111**, 4165–4222 (2011).
[25] G. de Gonzalo and M. W. Fraaije, *ChemBioChem* **11**, 2208–2231 (2010).
[26] J. D. Stewart, *Curr. Org. Chem.* **2**, 195–216 (1998).
[27] S. Colonna, V. Pironti, F. Zambianchi, G. Ottolina, N. Gaggero, and G. Celentano, *Eur. J. Org. Chem.*, 363–368 (2007).
[28] M. W. Fraaije, J. Wu, D. P. H. M. Heuts, E. W. van Hellemond, J. H. Lutje Spelberg, and D. B. Janssen, *Appl. Microbiol. Biotechnol.* **66**, 393–400 (2005).
[29] G. de Gonzalo, D. E. Torres Pazmiño, G. Ottolina, M. W. Fraaije, and G. Carrea, *Tetrahedron: Asymmetry* **16**, 3077–3083 (2005).
[30] A. Rioz-Martínez, G. de Gonzalo, D. E. Torres Pazmiño, M. W. Fraaije, and V. Gotor, *Eur. J. Org. Chem.*, 6409–6416 (2010).
[31] F. Zambianchi, M. W. Fraaije, G. Carrea, G. de Gonzalo, C. Rodríguez, V. Gotor, and G. Ottolina, *Adv. Synth. Catal.* **349**, 1327–1331 (2007).
[32] G. de Gonzalo, G. Ottolina, F. Zambianchi, M. W. Fraaije, and G. Carrea, *J. Mol. Catal. B: Enzym.* **39**, 91–97 (2006).

[33] C. Rodríguez, G. de Gonzalo, and V. Gotor, *J. Mol. Catal. B: Enzym.* **74**, 138–143 (2012).
[34] M. Hall, C. Stueckler, H. Ehammer, E. Pointer, G. Oberdorfer, K. Gruber, B. Hauer, R. Stuermer, W. Kroutil, P. Macheroux, and K. Faber, *Adv. Synth. Catal.* **350**, 411–418 (2008).
[35] M. Bocola, F. Schulz, F. Leca, A. Vogel, M. W. Fraaije, and M. T. Reetz, *Adv. Synth. Catal.* **347**, 979–986 (2005).
[36] H. M. Dudek, D. E. Torres Pazmiño, C. Rodríguez, G. de Gonzalo, V. Gotor, and M. W. Fraaije, *Appl. Microbiol. Biotechnol.* **88**, 1135–1143 (2010).
[37] H. M. Dudek, G. de Gonzalo, D. E. Torres Pazmiño, P. Stepniak, L. W. Wyrwicz, L. Rychlewski, and M. W. Fraije, *Appl. Environ. Microbiol.* **77**, 5730–5738 (2011).
[38] H. L. van Beek, G. de Gonzalo, and M. W. Fraaije, *Chem. Commun.* **48**, 3288–3290 (2012).
[39] A. Cuetos, A. Rioz-Martínez, M. L. Valenzuela, I. Lavandera, G. de Gonzalo, G. A. Carriedo, and V. Gotor, *J. Mol. Catal. B: Enzym.* **74**, 178–183 (2013).
[40] D. E. Torres Pazmiño, R. Snajdrova, D. V. Rial, M. D. Mihovilovic, and M. W. Fraaije, *Adv. Synth. Catal.* **349**, 1361–1368 (2007).
[41] G. de Gonzalo, C. Rodríguez, A. Rioz Martínez, and V. Gotor, *Enzyme Microb. Technol.* **50**, 43–49 (2012).
[42] N. M. Kamerbeek, M. J. H. Moonen, J. G. M. van der Ven, W. J. H. van Berkel, M. W. Fraaije, and D. B. Janssen, *Eur. J. Biochem.* **268**, 2547–2557 (2001).
[43] G. de Gonzalo, D. E. Torres Pazmiño, G. Ottolina, M. W. Fraaije, and G. Carrea, *Tetrahedron: Asymmetry* **17**, 130–135 (2006).
[44] M. L. Mascotti, A. A. Orden, F. R. Bisogno, G. de Gonzalo, and M. Kurina-Sanz, *J. Mol. Catal. B: Enzym.* **82**, 32–36 (2012).
[45] M. L. Mascotti, M. J. Ayub, H. Dudek, M. K. Sanz, and M. W. Fraaije, *AMB Express* **3**, 33 (2013).
[46] S. Bisagni, B. Summers, S. Kara, R. Hatti-Kaul, G. Grogan, G. Mamo, and F. Hollmann, *Top. Catal.* **57**, 366–375 (2014).
[47] A.-T. Li, J.-D. Zhang, J.-H. Xu, W.-Y. Lu, and G.-Q. Lin, *Appl. Environ. Microbiol.* **75**, 551–556 (2009).
[48] C. Szolkowy, L. D. Eltis, N. C. Bruce, and G. Grogan, *ChemBioChem* **10**, 1208–1217 (2009).
[49] A. Riebel, H. M. Dudek, G. de Gonzalo, P. Stepniak, L. Rychlewski, and M. W. Fraaije, *Appl. Microbiol. Biotechnol.* **95**, 1479–1489 (2012).
[50] S. Montersino, D. Tischler, G. T. Gassner, and W. J. H. van Berkel, *Adv. Synth. Catal.* **353**, 2301–2319 (2011).
[51] J. Nikodinovic-Runic, L. Coulombel, D. Francuski, N. D. Sharma, D. R. Boyd, R. O. M. Ferrall, and K. E. O'Connor, *Appl. Microbiol. Biotechnol.* **97**, 4849–4858 (2013).
[52] E. W. van Hellemond, D. B. Janssen, and M. W. Fraaije, *Appl. Environ. Microbiol.* **73**, 5832–5839 (2007).
[53] D. M. Ziegler, *Drug. Metab. Rev.* **34**, 503–511 (2005).
[54] G. Virkel, A. Lifschitz, J. Sallovitz, A. Pis, and C. Lanusse, *Drug Metab. Dispos.* **32**, 536–544 (2004).
[55] B. Furnes and D. Schlenk, *Toxicol. Sci.* **78**, 196–203 (2004).
[56] M. A. Hamman, B. D. Haehner-Daniels, S. A. Wrighton, A. E. Rettie, and S. D. Hall, *Biochem. Pharmacol.* **60**, 7–17 (2000).
[57] H. S. Choi, J. K. Kim, E. H. Cho, Y. C. Kim, J. I. Kim, and S. W. Kim, *Biochem. Biophys. Res. Commun.* **306**, 930–936 (2003).
[58] A. Rioz-Martínez, M. Kopacz, G. de Gonzalo, D. E. Torres Pazmiño, V. Gotor, and M. W. Fraaije, *Org. Biomol. Chem.* **9**, 1337–1341 (2011).
[59] A. Riebel, G. de Gonzalo, and M. W. Fraaije, *J. Mol. Catal. B: Enzym.* **88**, 20–25 (2013).
[60] C. N. Jensen, J. Cartwright, J. Ward, S. Hart, J. P. Turkenburg, S. T. Ali, M. J. Allen, and G. Grogan, *ChemBioChem* **13**, 872–878 (2012).
[61] F. G. Gelalcha, *Chem. Rev.* **107**, 3338–3361 (2007).
[62] V. Mojr, V. Herzig, M. Buděšinský, R. Cibulka, and T. Kraus, *Chem. Commun.* **46**, 7599–7601 (2010).
[63] V. Mojr, M. Buděšinský, R. Cibulka, and T. Kraus, *Org. Biomol. Chem.* **9**, 7318–7326 (2011).
[64] F. Müller and W. J. H. van Berkel, Methods to reversibly resolve flavoproteins into the constituents apoprotein and prosthetic group, in F. Müller, ed. *Chemistry and Biochemistry of Flavoenzymes*, CRC Press, Boca Raton, vol. 1, 1991, p. 261.
[65] G. de Gonzalo, C. Smit, J. Lin, A. J. Minnaard, and M. W. Fraaije, *Chem. Commun.* **47**, 11050–11052 (2011).

CHAPTER 7

Development of Chemoenzymatic Processes: An Industrial Perspective

Rajesh Kumar, Carlos Martinez, Van Martin, and John Wong

Biocatalysis Center of Emphasis, Chemical R&D, PharmaTherapeutics Pharmaceutical Science, Pfizer Worldwide R&D, Groton, CT, USA

7.1 INTRODUCTION

Biotransformations have been used by humankind for several thousand years. The classic examples are the conversion of ethanol to acetic acid by *Acetobacter* or ethanol production by sugar-based fermentations [1, 2]. However, biocatalysis technology has only emerged as a powerful tool since the mid-2000s or so and has rapidly become an established method in organic chemical synthesis, as evident from the large number of publications in this area [3–5]. This exponential growth is driven by a combination of factors, which include the following:

1. Recent advances in biotechnology, also described as the "third wave of biocatalysis" by Bornscheuer *et al.* [6], have made available numerous whole-genome sequences leading to a dramatic increase in the number of potentially available enzymes.
2. In addition, the ability to modify enzymes via enzyme engineering to meet process targets, such as activity, selectivity, substrate specificity, enzyme stability, and loadings, has made enzymes more suitable for industrial manufacturing [7].
3. Advances in recombinant protein expression and fermentation technology have enabled rapid access to industrial quantities of custom enzymes for manufacturing.
4. Successful application of biocatalysis at the manufacturing scale without the need for specialized equipment or plants has generated confidence in biocatalysis as a robust, cost-effective manufacturing technology [3, 5]. In particular, the use of highly active lyophilized or liquid recombinant enzymes has simplified plant operations, and these are now treated as standard chemical processes.
5. Biocatalytic processes are often intrinsically "green" and therefore contribute to waste reduction. Many biocatalytic reactions are performed in aqueous media under mild temperature and pressure using enzymes that are sustainably produced from renewable raw materials [8].
6. Biocatalysis can generate intellectual property (IP) or preserve freedom to operate (FTO) by enabling novel routes to active pharmaceutical ingredients (APIs).

Green Biocatalysis, First Edition. Edited by Ramesh N. Patel.

These attributes have resulted in numerous applications, especially in the food and chemical industries, where high reaction selectivity on complex substrates is critical [4, 9]. Recent advances have now broadened the application of enzyme catalysis to include the synthesis of high-value pharmaceutical intermediates and APIs, further adding to the value of biocatalysis technology [5, 10]. Biocatalysis and biotransformation have a long history at Pfizer, going back to the early 1900s, with the production of citric acid and penicillin by fermentation [11]. However, the application of biocatalysis for the synthesis of pharmaceutical intermediates and APIs started in the mid-1990s, and significant progress has been made over two decades. This includes the development of chemoenzymatic processes for several Pfizer drugs including the blockbusters, Lyrica™ [12] and Lipitor™.

Development of a commercial manufacturing route for an API requires a multidisciplinary team comprised of many skillsets, including organic chemistry, analytical chemistry, reaction engineering, process safety, and others. When biocatalysis is involved, additional skillsets may be required, such as molecular biology, microbiology, and bioinformatics, especially when a custom enzyme is needed. At Pfizer, the development of a chemoenzymatic route to an API is a truly multidisciplinary effort combining the resources of the Biocatalysis Center of Emphasis, which includes organic chemistry, microbiology, and molecular biology, with those from groups in Chemical R&D and Pfizer Global Supply (PGS) that focus on chemical process development and implementation of manufacturing routes. This chapter will give an overview of the chemical process development involving biocatalysis with examples from the Pfizer portfolio.

7.2 SYNTHETIC ROUTE DESIGN AND INTEGRATION OF BIOCATALYSIS

Synthetic strategies for pharmaceutical intermediates and APIs have so far shown limited incorporation of the diverse pool of biocatalysts that are currently available. This is possibly due to a lack of awareness on the part of synthetic chemists as to biocatalysis and its applications. An article [13] underscoring the need for better integration of biocatalysis, has proposed developing guidelines and rules for "biocatalytic retrosynthesis" and including this in training and education for organic chemists. As shown in the Figure 7.1, synthetic route design is the critical first step for incorporating biocatalysis into a synthetic route. This requires close interaction between biocatalysis experts and synthetic chemists and should start at the brainstorming stage to design a new route or enable a current route. The scope for biocatalysis integration in a synthetic route will be highly dependent on the stage

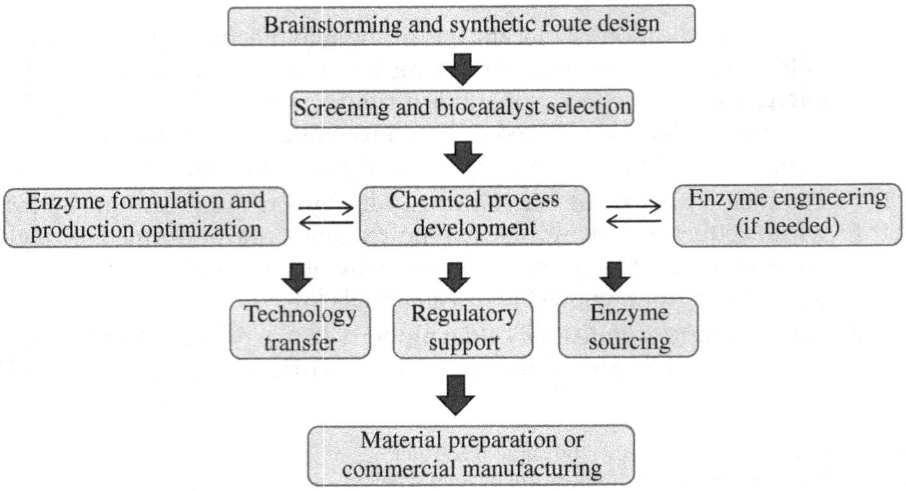

FIGURE 7.1
Generic work flow for developing chemoenzymatic processes.

of program, scope of improvement, and challenges in the existing route. The goal of this interaction and the bringing in of a biocatalytic step should be value driven. This could be simply providing a solution for a current challenging chemistry step or developing an altogether new route.

Imagabalin is an example from the Pfizer portfolio that illustrates the integration of biocatalysis into synthetic route design. Imagabalin was a late-stage candidate with high projected volumes and high cost of goods (COGs), which drove process improvement efforts. The enabling route synthesis (Scheme 7.1) was highly efficient- and was successfully used to produce material on large scale, for clinical supply [14]. However, to overcome the high COGs, two new routes were proposed, including a biocatalytic route (Scheme 7.1) projected to reduce COGs by 50–55%.

The enzymatic transformation for stereoselective direct conversion of keto ester **2** to amino ester **3** using a transaminase was attractive, but considered risky as commercial application of transaminases for API synthesis was unprecedented at the time (since then several new reports have appeared on the successful use of transaminases) [15] and few transaminases were commercially available. However, these risks were considered acceptable in view of the perceived value of the biocatalytic route, and subsequent efforts resulted in partial optimization of a transaminase [16] prior to development being halted for the project.

Another example, from the Pfizer portfolio, of successful incorporation of biocatalysis to make functional group interconversions (FGIs) more efficient in API synthesis is in a program for a smoothened (SMO) receptor inhibitor [17]. Introduction of a transaminase-catalyzed stereoselective reductive amination of a 4-piperidone **4** with concurrent dynamic kinetic resolution (DKR) gave amine **5** (Scheme 7.2), resulting in the highly efficient incorporation of two chiral centers in a single step.

Developments of enzymatic routes for the side chain of various drugs in the statin family, such as Lipitor from Pfizer and Crestor from Astra Zeneca, are very elegant applications of biocatalysis. Early on, several chemoenzymatic routes were developed to synthesize the side chain for statins, and these were based on replacing a chemical catalysis step with a biocatalytic step for the desired FGIs [18]. The substitution of stereoselective chemical reduction of keto ester to diol under cryogenic

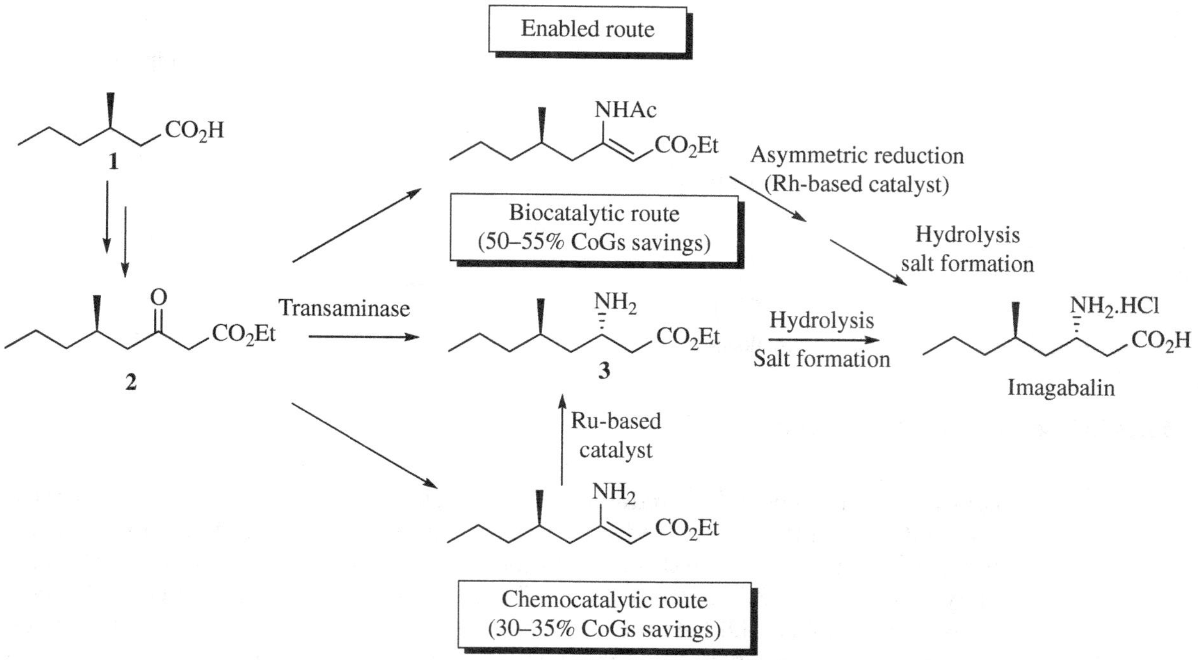

SCHEME 7.1
Synthetic and chemoenzymatic routes to imagabalin.

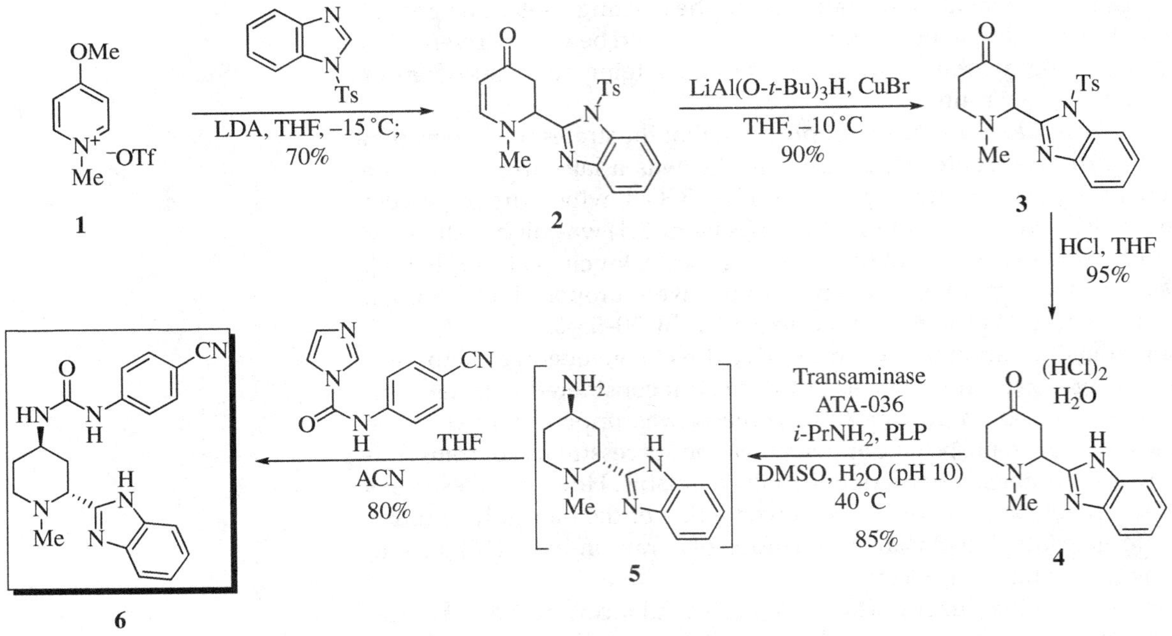

SCHEME 7.2

Chemoenzymatic synthesis of an SMO inhibitor.

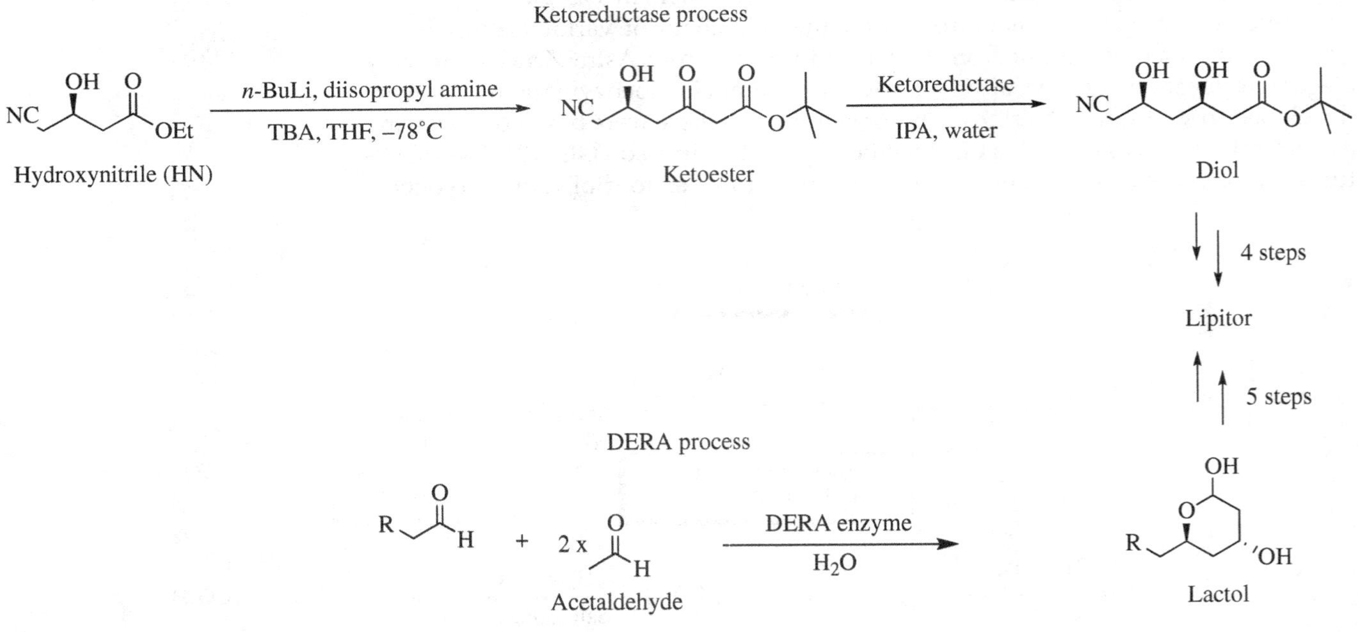

SCHEME 7.3

Chemoenzymatic routes for the synthesis of atorvastatin.

conditions with biocatalytic reduction using a ketoreductase (Scheme 7.3; top part) at ambient temperature provided an efficient way to make the side chain. Later on more efficient routes were designed based on C–C bond formation, using aldolase enzymes (Scheme 7.3; bottom part). These approaches utilized 2-deoxyribose-5-phosphate aldolase (DERA) for the asymmetric synthesis of a lactol intermediate, starting from inexpensive materials [19]. This resulted in the formation of two stereocenters in one step, which would be difficult to accomplish as efficiently by conventional means.

7.3 SCREENING AND BIOCATALYST SELECTION

Access to diverse enzyme libraries for various classes of enzymes with broad substrate scope is critical for finding an enzyme hit for a desired chemical transformation. The biocatalytic toolbox has been considerably expanded since the mid-2000s or so, as novel enzyme activities and chemoenzymatic reactions are discovered [20]. This is being fueled by the advent of recombinant DNA technology, which has enabled the cloning and functional expression of target enzymes, making a significantly larger number of enzymes available for each enzyme class. Table 7.1 shows a list of chemical transformations and types of enzymes that could be used to perform them. These enzymes can be obtained from vendors or by cloning and expression from genes of interest, commonly available in protein data banks. Several companies offer various enzyme screening kits, including kits where enzymes are arranged in a 96-well format and are ready for screening. These enzyme kits facilitate rapid feasibility evaluation with a small amount of substrate (25–100 mg) [21]. Having a good analytical assay, with the possibility of running 96-well plates both for screening and analysis, could provide a streamlined and faster access to data and enzyme hits.

7.4 CHEMOENZYMATIC PROCESS DEVELOPMENT

Identifying an enzyme hit for a desired reaction is a big leap forward in developing a chemoenzymatic process. The next step is to access an appropriate enzyme that fits the current process requirements, which vary depending upon the stage of the program (i.e., discovery, early development, late-stage development, or manufacturing).

7.4.1 Reaction Engineering versus Enzyme Engineering

The development of a process is generally determined by a set of target parameters and drivers, which are defined by the stage of development of the target molecule, the specific nature of the process, and whether it involves a single chemical step or a

TABLE 7.1 Commonly Used Biocatalytic Functional Group Interconversions

Chemical Transformations	Enzymes
Ester/amide hydrolysis or synthesis	Hydrolases (lipases, esterases, amidases and proteases)
Chiral alcohol synthesis (secondary alcohols)	Alcohol dehydrogenases or ketoreductases
Chiral amine synthesis	Transaminases (1° amine), amine dehydrogenases (1° and 2° amine), imine reductases (2° and 3° amine), and monoamine oxidases (cyclic and 1° amine)
Asymmetric C═C bond reduction (for activated alkenes conjugated with activating groups such as $-NO_2$, –CHO, –CO–, –COOH, –COOR)	Ene/enoate reductases
Selective nitrile hydrolysis (to acids and amides)	Nitrilases and nitrile hydratases
Asymmetric alcohol oxidations	Alcohol dehydrogenases, oxidases, and peroxidases
Chiral cyanohydrin synthesis	Hydroxy nitrile lyase
Aromatic hydroxylation	Monooxygenases, dioxygenases
Asymmetric aldol additions	Aldolases
Asymmetric epoxide hydrolysis	Epoxide hydrolases

TABLE 7.2 Common Drivers and Targets for Defining a Commercially Viable Manufacturing Process

Drivers (considerations/requirements)	• Cost • Timeline • Quality • Capacity/equipment • Regulatory strategy (RSM/API)
Targets (tunable parameters)	• Substrate and enzyme loads • Solvent (buffer solutions) • Temperature • pH • Biocatalyst formulation • Space time yield • Selectivity

series of reactions (Table 7.2). The main drivers that influence the development of a process are costs, time lines, and quality requirements/specifications (chemical and stereochemical purity) of the isolated product. Cost itself could be defined simply as COGs (raw materials) or more realistically as total cost, which also includes all direct and indirect operating costs related to the physical activities of producing the target compound. This typically includes costs associated with materials, processing, utilities, and waste treatment. The time available to develop a given process defines how much reaction or catalyst optimization can be done and how far a given target parameter such as substrate or enzyme load can be pushed. Time available for development is closely linked to the stage of development of a particular compound and whether the material is part of a regulatory synthesis or will ultimately be prepared at a vendor site. Other relevant drivers, such as the equipment available and capacity, may define the type of reactor used (batch, continuous, or semicontinuous) as well as the preferred isolation scheme. This driver typically becomes a large impediment in terms of chemical reaction engineering (CRE) since the pharma industry typically manufactures products with relatively low volumes and short commercial life spans, hence the dominant presence of processes run in batch reactors [22].

One of the most critical factors in the successful development of an enzymatic manufacturing process is the throughput or volumetric productivity of the enzymatic step, which is determined mainly by the concentration of substrate and the reaction rate. Identifying an efficient biocatalyst (or at least one that can be easily engineered) in the initial screen is the key to achieving the desired volumetric productivity downstream. This is why it is important to ensure that screening is conducted on a comprehensive library of enzymes to ensure that the best possible candidate is chosen for further optimization. In order to achieve the desired process targets, the level of optimization required varies depending on how good the starting enzyme is, both in terms of activity and operational stability under process conditions. Therefore, most of the work done in enzymatic process development centers around the enzyme itself and how to maintain high reaction rates at substrate loads that are high enough to enable an economically viable process, typically ≥50 g/L. When developing conditions for optimal operational stability, the thermodynamic stability of the enzyme is usually not the most important factor, since most industrially relevant enzymatic reactions are run at <50 °C. Rather, it is the enzyme's kinetic stability that is more relevant, and this is readily measurable by monitoring the activity under process conditions [23].

Enzymatic reaction engineering (or optimization) is the systematic alteration of reaction components and/or conditions, such as solvent, additive, temperature, pH, enzyme formulation, substrate, etc., to access desired activity. Optimization can be performed in a linear fashion or using design of experiment (DOE) approaches [24]. In terms of the form of biocatalyst used, the main options are whole cells, stabilized lysate, and lyophilized enzyme. Alternative conditions that confer stability to enzymes, such as immobilization, presence of additives, and cross-linking, are routinely employed to maximize the throughput of an enzymatic process [25]. Most immobilized biocatalysts are prepared from crude enzyme preparations, although some cross-linked enzyme crystals and aggregates may require semipure enzyme samples.

In recent, intensive efforts have been directed toward the improvement of enzyme activity, selectivity, and stability via protein engineering. A multitude of methods are now available for the selection of key residues for protein engineering [26], thus enabling the rapid screening of hundreds of variants as opposed to the thousands (or tens of thousands) required when gene shuffling or random mutagenesis methodologies are used. Achieving a successful throughput target requires an enzyme with not only the desired activity and selectivity but also with high stability. The improvement of enzyme robustness by means of site-directed mutagenesis has been reported for several decades as a means to increase enzyme thermodynamic stability. More recently, the ability of an enzyme to maintain its activity under process conditions (kinetic or process stability) has been the focus of engineering programs for enzymes used in industry [6]. The study reported by Xie and collaborators [27] illustrates the use of saturation mutagenesis of active site residues situated on flexible segments of the protein. The increase in rigidity of regions of the active site proved to improve the kinetic stability. This and other studies support the idea that optimizing the correct conformation of those active site residues playing a key role in enzyme catalysis is a key to achieving robust enzyme variants, with kinetic stability in a broad range of process conditions.

Even with the most robust biocatalysts, what is often encountered at high substrate loads (>1 M concentrations) is the issue of enzyme inhibition or deactivation. Improving enzyme activity and stability, as well as overcoming substrate and product inhibition, is an essential requirement in ensuring a process with high volumetric productivity. The use of reactor configurations that minimize enzyme inhibition has been widely reported and adopted in industry [28]. More recently, the use of flow chemistry technology has shown promise in the formation of hazardous and unstable products where controlled formation is a preference. It remains to be seen if similar approaches can be easily implemented for enzyme-mediated reactions. One of the main impediments is typically the integration of the hardware such as pumps, valves, analytical equipment, and the heating/cooling zone in labs and manufacturing plants dominated by the use of batch reactors [29].

7.4.2 Product Isolation

The types of challenges found in the recovery of products from both whole cell and isolated enzyme biocatalytic reactions can vary depending on the type of reactor (stirred tank or membrane reactors), the form of the enzyme (immobilized or free form), the reaction media (aqueous or organic), and the amount of biocatalyst used. In general, reactions performed in organic media offer easier product recovery as the biocatalyst can be removed by simple filtration followed by solvent evaporation. Recovering products from biocatalytic reactions performed in aqueous media or in the presence of cosolvents can be more challenging. In reactions carried out in stirred tank reactors, product recovery is often performed using filtration (via crystallization of product) or organic solvent extraction technologies. This is due to the simplicity associated with those techniques and the fact that most organic chemistry labs and plants are well equipped to deal with them. Immobilized enzymes can greatly simplify product recovery as the biocatalyst can be removed by a filtration step.

The extractive recovery of products and/or remaining substrates from aqueous or aqueous/organic reaction systems is often complicated by the tendency of enzymes (and other components found in enzyme preparations) and whole cells to form emulsions when in contact with aqueous–organic interfaces. Formation of stable emulsions that hamper product recovery is one of the challenges encountered when applying biocatalysis to organic synthesis. Continuous centrifugation can be used to separate phases from stable emulsions, but it is not always available in manufacturing facilities and therefore could require capital investment.

7.4.3 Scale-Up of Enzymatic Processes

This section will discuss selected topics pertaining to scale-up and transfer of enzymatic processes for small-molecule API manufacture from laboratory to pilot plant and full commercial scale.

Equipment Requirements

Pharmaceutical pilot plant and commercial manufacturing facilities are typically batch manufacturing facilities, and in general these serve well for enzymatic processes. In terms of specialized equipment needs, pH monitoring and control is very useful for enzymatic reactions and is not always available in chemical synthesis facilities but can easily be added. Continuous centrifugation can also be useful for workup of enzymatic reactions as some processes tend to form stable emulsions. Enzymes typically require refrigerated (4–8 °C) or freezer (−20 °C) storage depending on the formulation, and this should be considered when determining requirements for manufacturing facilities (and for enzyme formulation). Depending on the activity assay specification for the enzyme, scale-up facilities may need analytical instrumentation such as a GC, HPLC, or spectrophotometers to perform assays to verify the activity of an enzyme. Gel electrophoresis equipment may also be useful for monitoring residual protein levels in an API.

Sourcing of Biocatalysts and Biocatalyst-Derived Materials

Many of the issues that should be considered in the sourcing of biocatalysts and biocatalyst-derived materials are much the same as for traditional reagents and chemistry-derived materials. Some of these are cost, quality, reliability of the vendor, control and security of intellectual property (IP), freedom to operate (FTO), the vendor's business model, avoidance of being single sourced, etc. Some of the perceived differences and issues are often due to the fact that biocatalysts, and sourcing aspects associated with them, are still an emerging or developing area.

Identity of the Enzyme and Being Single Sourced

There are some special considerations involved with sourcing of biocatalysts. The first of these is that in many cases the biocatalyst is proprietary or a trade secret and its exact identity is not disclosed to the customer. It is common for vendors, through the use of rigorous material transfer agreements (MTAs) and the like, to greatly restrict the knowledge of the enzyme, its structure, or its amino acid sequence from the customer. This situation may result in reliance on a sole supplier for a biocatalyst and is in contrast to traditional chemical reagents, where typically the exact molecular structure of the reagent being used is known, and except for proprietary catalysts, there will often be multiple existing sources of the reagent. The typical remedy for this situation is to screen enzymes from multiple sources to identify alternative suppliers. However, in cases where the biocatalyst is proprietary and only a single supplier is available, having more than one manufacturing site for the biocatalyst can help mitigate potential supply risks.

Need for Multiple Vendors/CMOs to Execute a Complete Process

Sourcing of a biocatalyst-derived material (starting material, intermediate, RSM) may require capabilities and expertise: (i) to prepare, isolate, and formulate an enzyme, (ii) to run the biocatalysis step, (iii) to handle the downstream processing necessary to remove the enzyme and other potential enzyme-related impurities (i.e., cell debris, residual enzyme, protein fragments, etc.), and (iv) to handle all of the usual traditional organic chemistry reactions and processing operations. Although this is changing, as more vendors develop the expertise, capability, and confidence to handle both biocatalysis-based chemistry and traditional organic chemistry, currently there are not a lot of vendors/CMOs that offer services that incorporate all four steps. If the reaction involves an enzyme that is commercially available, steps 2–4 can often be handled at one location.

7.4.4 Enzyme Supply Scenarios

The various issues that arise when sourcing enzymes and biocatalytic processes can be organized by examining two major factors. These are (i) whether the enzyme is commercially available and (ii) whether, and how, the customer has the FTO. The combination of these factors leads to five typical scenarios.

1. Enzyme Commercially Available; Customer has Freedom to Operate
   This is the simplest situation and the one most like sourcing a traditional chemical reagent or catalyst. In this case, the enzyme is already commercially available, likely from multiple sources, and the enzyme has no IP encumbrances. Typically, especially for larger, well-established vendors, production processes are already in place, and usually being single sourced is not an issue.
2. Enzyme Not Commercially Available; Customer has Freedom to Operate
   In this situation, the enzyme is likely one that was obtained through cloning and expression of a gene obtained through research or the public domain (literature, public databases, expired or lapsed patents, etc.). The major issue here is that a production process must be developed for the enzyme and this can add to the time required to establish a supply. The plus side to this situation is that, because of the FTO situation, multiple vendors can be developed by the customer, if necessary. It may also be useful to have a strategy for maintaining the FTO for the enzyme in question as it is possible that others may independently discover the same enzyme or a close variant and potentially block the FTO by obtaining a patent.
3. Enzyme Not Commercially Available; Enzyme is Proprietary to the Customer
   This situation occurs when the customer has used enzyme engineering, either in-house or with a CRO/CMO, to develop an enzyme for its own specific process. The major issues here are twofold. First, as in the preceding scenario, a production process must be developed, either by the customer or the vendor, and the vendor must develop sufficient knowledge, experience, and process understanding to be able to provide the enzyme reliably in the amounts and with the quality that the customer needs. If the enzyme is sufficiently different from related enzymes, for which expression conditions are already known, this may require significant development time. The second issue is one of IP management. Because of the level of sophistication of modern molecular biology techniques, the identification and duplication of an engineered enzyme (either from the enzyme itself or from the genes used to produce it) is extremely easy. Extra care must be taken to choose a vendor that is trusted or that is located in a country with strong IP protections in place.

4. Enzyme Commercially Available; Enzyme is Proprietary to the Vendor
   In this situation, the vendor becomes responsible for all aspects of enzyme manufacture and formulation. If the enzyme of interest is one that is already being manufactured at very large scales for other industries, as is the case for many commodity enzymes (e.g., for commercial detergents), these can often be purchased at very low cost. Enzymes offered solely for biocatalysis applications are usually more expensive. The main issue in both cases is that there is usually only a single supplier for the enzyme.
5. Enzyme Not Commercially Available; Enzyme is Proprietary to the Vendor
   This situation, at least on the surface, is similar to the preceding scenario, except that the enzyme is not commercially available. There are many situations in which a customer may find an enzyme that will work well for a process, but the vendor or IP owner may not want to commercialize the enzyme. This may simply be due to a case of insufficient demand to justify custom manufacture of an enzyme, or it can occur when a company changes its business focus or is bought by another company with no desire to manufacture the enzyme. If the lack of commercial availability is not due to technical reasons, then a supply of enzyme could be developed by negotiation with the IP owner and may involve a third party if the IP owner does not have production capability. This scenario will likely result in reliance on a single supplier unless a suitable agreement can be reached with the IP owner. If a production process has not been developed for the enzyme, then this can add to lead time to establish a supply and increase risk that a suitable supply may not be developed.

7.4.5 Manufacture of APIs using Enzymes: Quality and Safety Aspects

Drug quality and patient safety are of paramount importance in the manufacture of APIs, and in regard to these topics, the use of enzymes in API manufacture brings up special considerations that have been discussed in the literature [30]. While existing guidance relating to API purity from major organizations (Food and Drug Administration, European Medicines Agency, International Conference on Harmonization) deals clearly with the quality of starting materials, solvents, and reagents, it does not explicitly address the use of enzymes. This lack of specific guidance can result in ambiguity concerning requirements for filing enzyme-based processes. This is an important topic since the increasing availability of enzymes, particularly engineered enzymes, should lead to more successful applications of enzymes for API manufacture and the subsequent need for a clear path for filing enzymatic processes. In this section, the application of biocatalysis for API manufacture will be discussed from the perspectives of quality and safety and will highlight selected topics from the paper by Wells *et al.* [30].

Biocatalyst Source and Quality

Since the key input material in an enzymatic process is the enzyme itself, the source and quality of the enzyme are obviously important factors to consider in assessing potential risks to drug quality and patient safety. Enzymes can be obtained from a variety of sources, including plants, animals, and microorganisms. However, for pharmaceutical applications, enzymes produced in recombinant microorganisms offer several distinct advantages. These advantages include production efficiency, since microbial fermentation can be used to produce enzymes in a cost-effective manner at very large scale, ability to access engineered enzymes, and from a quality and safety perspective, avoiding potential risks associated with using enzymes from natural sources. A good case in point is pig liver esterase (PLE), which is naturally sourced from pig livers. PLE is a useful enzyme for hydrolytic kinetic resolution of esters, but its use in pharmaceutical applications is hampered by potential risks associated with animal viruses [31]. Recombinant PLE produced by microbial fermentation eliminates the issue of using animal-derived materials in pharmaceutical applications and could provide greater selectivity compared to PLE sourced from pig liver since individual isoforms of PLE are separately produced by recombinant technology [32]. Recombinant technology could also mitigate potential risks associated with using enzymes from native microbial strains, such as the production of toxins by some bacterial and fungal strains, by using host strains with a long history of safe use [33]. Thus bacterial strains derived from *Escherichia coli* K12, yeast strains such as *Pichia pastoris* and *Saccharomyces cerevisiae*, and fungal strains such as *Aspergillus oryzae* are suitable for enzyme production as they are nontoxic, are nonpathogenic, and have a long history of safe use for the production of industrial enzymes, including those used in food production.

Enzymes sourced from vendors should come with a certificate of analysis (CoA) that provides information regarding specifications. Typical specifications on a CoA include physical form (solid, liquid, or suspension), appearance (color), and, most importantly, a specific activity measurement (U/ml or U/mg protein), which ideally would be measured on the substrate of interest. A BSE/TSE statement is a standard requirement for material used in c-GMP manufacture of pharmaceuticals and certainly applies in the case of enzymes.

Potential Enzyme-Related Impurities and Purge Strategies

A consideration of potential impurities that may be associated with small-molecule APIs manufactured using enzymes should include the enzyme used to catalyze the transformation, other biomolecules (proteins, polynucleotides, endotoxins, cell-wall debris) from the source material (native or recombinant microorganism, plant, or animal tissues), and materials used to formulate the enzyme (stabilizers, preservatives). In general, biomolecules are highly water soluble and likely to be degraded and purged under conditions that are typically used in chemical processing such as high temperatures, pH extremes, and organic solvents. Although not strictly applicable to small-molecule APIs, but relevant to the current topic, FDA and EMEA guidance on fermentation products and semisynthetics states that protein contamination is a minimal concern after processing steps such as solvent extractions, washings, and crystallizations [5]. Therefore, a risk assessment for potential enzyme-related impurities should take into account the number and type of processing steps after the enzymatic step, with the highest theoretical concern for processes in which an enzyme is used in the final step of the process. Testing should be carried out to demonstrate purge of the enzyme, and a variety of techniques are available for protein determination such as colorimetric assays (Bradford, Lowry), mass spectral techniques, gel electrophoresis, and digestion and amino acid analysis. As to acceptable levels of protein impurities, Wells *et al.* [1] proposed that proteins should be treated as normal organic impurities under ICH guidelines with a 0.1% level for identification and qualification. This was based on a risk assessment that concluded that (i) the large majority of industrial enzymes did not cause any systemic toxicity and (ii) orally ingested enzymes are considered intrinsically safe as they are typically found in fresh and processed foods and are degraded into peptides and amino acids along with other dietary proteins [34].

The main adverse events reported for industrial enzymes were occupational asthma and allergy, and irritation of skin, eye, and mucosal membranes related to enzymes developed for the detergent industry.

DNA from the host cell used to produce the enzyme is a potential impurity, but the risk of orally ingested DNA is considered to be very low as humans typically ingest a significant amount of DNA in their diet, and dietary DNA has not been associated with any toxicity. Sensitive methods for the detection of DNA are available, such as Threshold™ Total DNA assay that could be used to detect potential residual DNA in API.

Endotoxins are also potential impurities that could result from enzyme production using *E. coli* or other bacterial strains and are a potential concern for inhaled or injectable drugs. Specific guidelines for acceptable endotoxin limits are available, as are appropriate assay methods [35].

Tiered Risk Assessment Approach

Based on their comprehensive review and analysis of the literature, Wells *et al.* [30] proposed a tiered risk-assessment approach in which the route of drug administration and purity of the biocatalyst were two major factors to consider. Tier 1 represents the lowest level of concern and is to be applied for cases in which the drug is administered orally and when the enzyme used for API manufacture is relatively pure and produced by a precedented host. In Tier 1 cases, it was recommended that the guidance contained in ICH Q3A could be followed. If either the drug in question is to be administered nonorally or the enzyme to be used is not relatively pure or produced by a precedented host, then a Tier 2 approach is triggered, in which control strategies that go beyond ICH Q3A are considered. A hypothetical Tier 2 case could be one where an inhaled drug is manufactured using an enzyme. Tier 3 cases represent the highest level of potential concern and would be triggered when risk assessment of a Tier 2 case indicates a need for additional risk management approaches. Extending the reasoning presented in the paper, a hypothetical Tier 3 case could be required for an inhaled drug manufactured using an enzyme that has been associated with occupational asthma.

7.5 CONCLUSIONS

Integration of biocatalysis in synthetic routes for the development of pharmaceutical intermediates and APIs holds great promise, as is evident by the implementation of an increasing number of chemoenzymatic processes for commercial manufacturing. This trend is expected to grow further as new capabilities and technologies will be developed. This in turn will lead to an increase in the development of chemoenzymatic routes to APIs, which is a truly multidisciplinary effort and includes organic chemistry, microbiology, molecular biology, and process engineering. As emphasized, there are several critical components to develop a successful chemoenzymatic process, and they need to be addressed based on the development stage of the compound. This chapter has provided an outline of various components from industrial perspective that could be used for discussions and guidance.

REFERENCES

[1] Park, Y. S.; Toda, K.; Fukaya, M.; Okumura, H.; Kawamura, Y., Production of a high concentration acetic acid by Acetobacter aceti using a repeated fed-batch culture with cell recycling. *Applied Microbiology and Biotechnology* 1991, **35**(2), 149–153.

[2] Lin, Y.; Tanaka, S., Ethanol fermentation from biomass resources: Current state and prospects. *Applied Microbiology and Biotechnology* 2006, **69**(6), 627–642.

[3] Schmid, A.; Dordick, J.; Hauer, B.; Kiener, A.; Wubbolts, M.; Witholt, B., Industrial biocatalysis today and tomorrow. *Nature* 2001, **409**(6817), 258–268.

[4] Huisman, G. W.; Gray, D., Towards novel processes for the fine-chemical and pharmaceutical industries. *Current Opinion in Biotechnology* 2002, **13**(4), 352–358.

[5] Pollard, D. J.; Woodley, J. M., Biocatalysis for pharmaceutical intermediates: The future is now. *Trends in Biotechnology* 2007, **25**(2), 66–73.

[6] Bornscheuer, U.; Huisman, G.; Kazlauskas, R.; Lutz, S.; Moore, J.; Robins, K., Engineering the third wave of biocatalysis. *Nature* 2012, **485**(7397), 185–194.
[7] Woodley, J. M., Protein engineering of enzymes for process applications. *Current Opinion in Chemical Biology* 2013, **17**(2), 310–316.
[8] Leresche, J. E.; Meyer, H.-P., Chemocatalysis and biocatalysis (biotransformation): Some thoughts of a chemist and of a biotechnologist. *Organic Process Research and Development* 2006, **10**(3), 572–580.
[9] Ikunaka, M., Biocatalysis from the perspective of an industrial practitioner: Let a biocatalyst do a job that no chemocatalyst can. *Catalysis Today* 2004, **96**(3), 93–102.
[10] Lorenz, P.; Eck, J., Screening for novel industrial biocatalysts. *Engineering in Life Sciences* 2004, **4**(6), 501–504.
[11] Lombardino, J. G., A brief history of Pfizer Central Research. *Bulletin for the History of Chemistry* 2000, **25**(1), 10.
[12] Martinez, C. A.; Hu, S.; Dumond, Y.; Tao, J.; Kelleher, P.; Tully, L., Development of a chemoenzymatic manufacturing process for pregabalin. *Organic Process Research and Development* 2008, **12**(3), 392–398.
[13] Turner, N. J.; O'Reilly, E., Biocatalytic retrosynthesis. *Nature Chemical Biology* 2013, **9**(5), 285–288.
[14] Birch, M.; Challenger, S.; Crochard, J.-P.; Fradet, D.; Jackman, H.; Luan, A.; Madigan, E.; Mathew, J. S.; McDowall, N.; Meldrum, K., A Simplified process for the manufacture of imagabalin hydrochloride (PD-0332334), an α2δ-ligand for the treatment of generalised anxiety disorder. *Organic Process Research and Development* 2011, **15**(6), 1358–1364.
[15] Savile, C. K.; Janey, J. M.; Mundorff, E. C.; Moore, J. C.; Tam, S.; Jarvis, W. R.; Colbeck, J. C.; Krebber, A.; Fleitz, F. J.; Brands, J., Biocatalytic asymmetric synthesis of chiral amines from ketones applied to sitagliptin manufacture. *Science* 2010, **329**(5989), 305–309.
[16] Midelfort, K. S.; Kumar, R.; Han, S.; Karmilowicz, M. J.; McConnell, K.; Gehlhaar, D. K.; Mistry, A.; Chang, J. S.; Anderson, M.; Villalobos, A., Redesigning and characterizing the substrate specificity and activity of Vibrio fluvialis aminotransferase for the synthesis of imagabalin. *Protein Engineering Design and Selection* 2013, **26**(1), 25–33.
[17] Peng, Z.; Wong, J. W.; Hansen, E. C.; Puchlopek-Dermenci, A. L.; Clarke, H. J., Development of a concise, asymmetric synthesis of a smoothened receptor (SMO) inhibitor: Enzymatic transamination of a 4-piperidinone with dynamic kinetic resolution. *Organic Letters* 2014, **16**(3), 860–863.
[18] Panke, S.; Held, M.; Wubbolts, M., Trends and innovations in industrial biocatalysis for the production of fine chemicals. *Current Opinion in Biotechnology* 2004, **15**(4), 272–279.
[19] Müller, M., Chemoenzymatic synthesis of building blocks for statin side chains. *Angewandte Chemie International Edition* 2005, **44**(3), 362–365.
[20] Drepper, T.; Eggert, T.; Hummel, W.; Leggewie, C.; Pohl, M.; Rosenau, F.; Wilhelm, S.; Jaeger, K. E., Novel biocatalysts for white biotechnology. *Biotechnology Journal* 2006, **1**(7–8), 777–786.
[21] Yazbeck, D. R.; Tao, J.; Martinez, C. A.; Kline, B. J.; Hu, S., Automated enzyme screening methods for the preparation of enantiopure pharmaceutical intermediates. *Advanced Synthesis and Catalysis* 2003, **345**(4), 524–532.
[22] Levenspiel, O., Chemical reaction engineering. *Industrial and Engineering Chemistry Research* 1999, **38**(11), 4140–4143.
[23] Polizzi, K. M.; Bommarius, A. S.; Broering, J. M.; Chaparro-Riggers, J. F., Stability of biocatalysts. *Current Opinion in Chemical Biology* 2007, **11**(2), 220–225.
[24] (a) Kaul, P.; Asano, Y., Strategies for discovery and improvement of enzyme function: State of the art and opportunities. *Microbial Biotechnology* 2012, **5**(1), 18–33; (b) Martins, A. B.; Friedrich, J. L.; Rodrigues, R. C.; Garcia-Galan, C.; Fernandez-Lafuente, R.; Ayub, M. A., Optimized butyl butyrate synthesis catalyzed by Thermomyces lanuginosus lipase. *Biotechnology Progress* 2013, **29**(6), 1416–1421; (c) Sontakke, J. B.; Yadav, G. D., Kinetic modeling and statistical optimization of lipase catalyzed enantioselective resolution of (R, S)-2-pentanol. *Industrial & Engineering Chemistry Research* 2011, **50**(23), 12975–12983.
[25] Illanes, A., Stability of biocatalysts. *Electronic Journal of Biotechnology* 1999, **2**, 1–9.
[26] (a) Verma, R.; Schwaneberg, U.; Roccatano, D., Computer-aided protein directed evolution: A review of web servers, databases and other computational tools for protein engineering. *Computational and Structural Biotechnology Journal* 2012, **2**, 1–12; (b) Höhne, M.; Schätzle, S.; Jochens, H.; Robins, K.; Bornscheuer, U. T., Rational assignment of key motifs for function guides in silico enzyme identification. *Nature Chemical Biology* 2010, **6**(11), 807–813.

[27] (a) Xie, Y.; An, J.; Yang, G.; Wu, G.; Zhang, Y.; Cui, L.; Feng, Y., Enhanced enzyme kinetic stability by increasing rigidity within the active site. *Journal of Biological Chemistry* 2014, **289**(11), 7994–8006; (b) Eijsink, V. G.; Bjørk, A.; Gåseidnes, S.; Sirevåg, R.; Synstad, B.; van den Burg, B.; Vriend, G., Rational engineering of enzyme stability. *Journal of Biotechnology* 2004, **113**(1), 105–120; (c) Tsou, C.-L., Inactivation precedes overall molecular conformation changes during enzyme denaturation. *Biochimica et Biophysica Acta (BBA)-Protein Structure and Molecular Enzymology* 1995, **1253**(2), 151–162; (d) Tsou, C.-L., Conformation flexibility of enzyme active sites. *Science* 1993, **262**(5132), 380–381.

[28] (a) Liese, A.; Kragl, U.; Kierkels, H.; Schulze, B., Membrane reactor development for the kinetic resolution of ethyl 2-hydroxy-4-phenylbutyrate. *Enzyme and Microbial Technology* 2002, **30**(5), 673–681; (b) Liese, A.; Karutz, M.; Kamphuis, J.; Wandrey, C.; Kragl, U., Enzymatic resolution of 1-phenyl-1, 2-ethanediol by enantioselective oxidation: Overcoming product inhibition by continuous extraction. *Biotechnology and Bioengineering* 1996, **51**(5), 544–550; (c) Rao, N.; Lutz, S.; Seelbach, K.; Liese, A.; Liese, A.; Seelbach, K.; Wandrey, C., *Industrial Biotransformations*. Liese, A., Seelbach, K., and Wandrey, C., Eds. Weinheim/New York: Wiley-VCH, 2006, 115–145.

[29] (a) Hessel, V., Novel process windows–gates to maximizing process intensification via flow chemistry. *Chemical Engineering and Technology* 2009, **32**(11), 1641–1641; (b) Krühne, U.; Heintz, S.; Ringborg, R.; Rosinha, I. P.; Tufvesson, P.; Gernaey, K. V.; Woodley, J. M., Biocatalytic process development using microfluidic miniaturized systems. *Green Processing and Synthesis* 2014, **3**(1), 23–31.

[30] Wells, A. S.; Finch, G. L.; Michels, P. C.; Wong, J. W., Use of enzymes in the manufacture of active pharmaceutical ingredients—a science and safety-based approach to ensure patient safety and drug quality. *Organic Process Research and Development* 2012, **16**(12), 1986–1993.

[31] (a) Gaede, B. J.; Nardelli, C. A., Viral clearance issues associated with the use of an animal-derived enzyme in the synthesis of emtricitabine. *Organic Process Research and Development* 2005, **9**(1), 23–29; (b) Gaede, B. J.; Nardelli, C. A., Viral clearance issues associated with the use of an animal-derived enzyme in the synthesis of emtricitabine. *Organic Process Research and Development* 2004, **9**(1), 23–29.

[32] Hummel, A.; Brüsehaber, E.; Böttcher, D.; Trauthwein, H.; Doderer, K.; Bornscheuer, U. T., Isoenzymes of pig-liver esterase reveal striking differences in enantioselectivities. *Angewandte Chemie International Edition* 2007, **46**(44), 8492–8494.

[33] Pariza, M. W.; Johnson, E. A., Evaluating the safety of microbial enzyme preparations used in food processing: Update for a new century. *Regulatory Toxicology and Pharmacology* 2001, **33**(2), 173–186.

[34] Olempska-Beer, Z. S.; Merker, R. I.; Ditto, M. D.; DiNovi, M. J., Food-processing enzymes from recombinant microorganisms: A review. *Regulatory Toxicology and Pharmacology* 2006, **45**(2), 144–158.

[35] Magalhães, P. O.; Lopes, A. M.; Mazzola, P. G.; Rangel-Yagui, C.; Penna, T.; Pessoa Jr, A., Methods of endotoxin removal from biological preparations: A review. *Journal of Pharmacy and Pharmaceutical Sciences* 2007, **10**(3), 388–404.

CHAPTER 8

Epoxide Hydrolases and their Application in Organic Synthesis

Alain Archelas[1], Gilles Iacazio[1], and Michael Kotik[2]

[1] Institut des Sciences Moléculaires de Marseille, Université d'Aix-Marseille, Marseille, France

[2] Institute of Microbiology, Academy of Sciences of the Czech Republic, Prague, Czech Republic

8.1 INTRODUCTION

Organic chemists have become interested in enzymes as catalysts due to their high efficiencies and specificities. Moreover, recent progress in molecular biology and enzyme-related research areas enabled and simplified the production and purification of recombinant enzymes in large quantities and their engineering toward tailor-made biocatalysts using straightforward mutagenesis and screening techniques. This is also true for epoxide hydrolases (EHs), as evidenced by the many published research papers about the synthetic applications of naturally occurring or engineered EHs.

EHs catalyze the opening of oxirane rings, generating a vicinal diol as the final product (Figure 8.1). From a synthetic chemistry point of view, the most valuable EHs are those with either (i) high enantioselectivities or (ii) a combination of low enantiopreference with a high level of enantioconvergence. While the former generate enantiopure epoxides with a maximum yield of 50% in a kinetic resolution process, the latter produce ideally an enantiopure diol product with a theoretical yield of 100%. Thus, EHs provide convenient access to enantiopure epoxides or diols from racemic epoxides. Furthermore, the enantiopure diols can then often be chemically transformed back to the corresponding epoxides with no effect on the enantiomeric excess (ee). High values of enantioselectivity or enantioconvergence are a consequence of particular enzyme–substrate interactions, which can be modulated by specific reaction conditions or through the exchange of specific amino acids of the biocatalyst. The final stereochemical outcome of an EH-catalyzed reaction depends solely on the regioselectivity coefficients, which determine the absolute configuration and the ee of the diol product when the reaction reaches 100% conversion. During the reaction, the oxirane ring of each enantiomer is often attacked at either carbon atom, resulting in a mixture of diol enantiomers (Figure 8.2). Unfortunately, several authors used the product-derived *E*-value, E_p (calculated from *c* and ee_p using Sih's equation; see Ref. [1]) for characterizing the stereochemistry of the diol formation of an EH-mediated hydrolysis reaction [2, 3]. This is wrong and misleading for epoxide-opening reactions since there are two possible positions of attack for each oxirane enantiomer. The ideal enantioconvergent EH leads to deracemization of a racemic mixture of an epoxide and exhibits reversed regioselectivity for either substrate enantiomer, that is,

Green Biocatalysis, First Edition. Edited by Ramesh N. Patel.

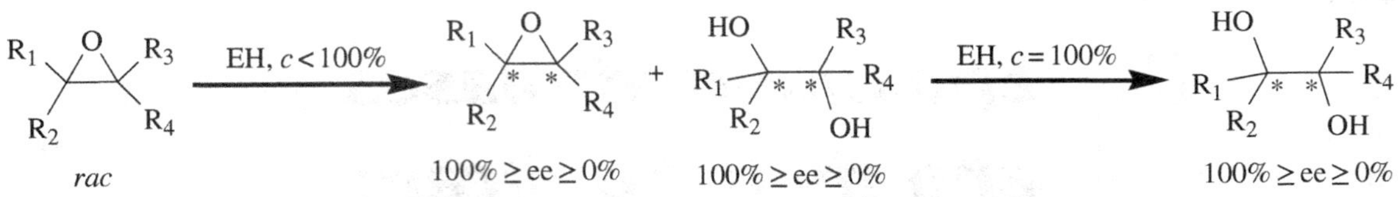

FIGURE 8.1

EH-catalyzed hydrolysis of a racemic epoxide, resulting in the formation of a vicinal diol with an enantiomeric excess of $100\% \geq ee_p \geq 0\%$ at complete conversion ($c=100\%$) of the substrate. Depending on the EH-substrate interactions and consequently its substrate-related enantiomeric ratio or *E*-value (which characterizes the ability of the enzyme to discriminate between the two competing substrate enantiomers [1]), enantiopure epoxide are obtained at a specific degree of conversion within the range of $50\% < c < 100\%$. The kinetics of highly enantioselective EHs is characterized by a rapid hydrolysis of the preferred epoxide enantiomer, followed by a much slower hydrolysis of the remaining epoxide.

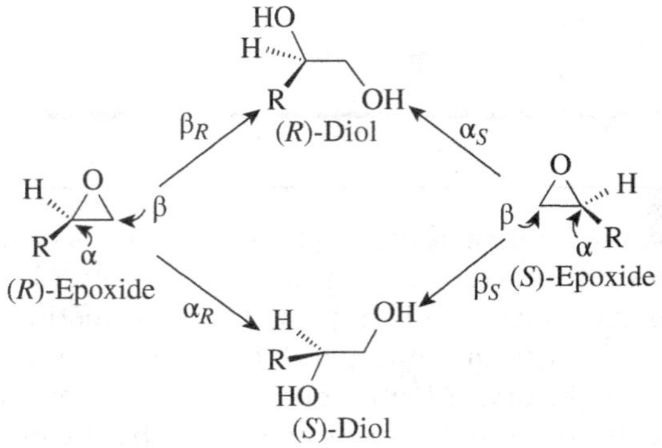

FIGURE 8.2

Attack at the oxirane ring can occur at either the terminal carbon atom, which results in a diol product with retained configuration (β-attack), or the carbon atom with the substituent R, resulting in a diol with an inverted configuration (α-attack). The percentages of epoxide molecules following a particular reaction pathway are represented by the corresponding regioselectivity coefficients; the following relationships between the four regioselectivity coefficients are valid: $\alpha_R + \beta_R = 100\%$, and $\alpha_S + \beta_S = 100\%$.

$\alpha_S = 100\%$ and $\alpha_R = 0\%$, or $\alpha_S = 0\%$ and $\alpha_R = 100\%$, affording an enantiomerically pure diol in a theoretical yield of 100%. However, enantioconvergence levels above 90% (i.e., $ee_p > 90\%$ at complete conversion of the substrate) are quite rare with wild-type EHs. EHs can also be used to deracemize *meso*-epoxides, and in this case, the ee of the formed diol is kept constant during the entire reaction. The resulting ee is again fixed by the four regioselectivity coefficients. In the ideal case, the diol product is obtained in 100% ee and 100% yield.

Enantiopure epoxides and diols are chiral intermediates or building blocks, which can be used for the synthesis of biologically active compounds such as antibiotics [4], antifungal drugs [5], antiandrogens [6], receptor agonists and antagonists [7–9], 11-heterosteroids [10], HIV protease inhibitors [11, 12], nonsteroidal anti-inflammatory drugs [13, 14], and other natural products [15–19]. An EH-triggered cascade reaction enabled the asymmetric synthesis of two antitumor agents, which occur naturally in *Panax ginseng* [20]. Further, substituted tetrahydrofurans, whose structural motifs are frequently found in acetogenins and polyether antibiotics, were also accessed through EH-mediated cascade reactions [21].

The scope of this chapter is to provide a comprehensive overview of the most interesting applications of EHs for the generation of enantiopure epoxides or diols, which are valuable chiral building blocks and key intermediates for the synthesis of various natural products or pharmacologically active compounds.

8.2 SOURCES AND REACTION MECHANISM OF EHs

8.2.1 Sources of EHs

EHs have been found in many prokaryotic and eukaryotic organisms, including bacteria, fungi, yeast, plants, insects, fish, and mammals (Figure 8.3). The physiological role of EHs appears to be manifold: they are involved in the detoxification of potentially harmful, naturally occurring, or anthropogenic epoxides [23]; in lipid metabolism in plants and animals [24, 25]; and in the metabolism of juvenile hormones in insects [25]. Recently, a new role of EHs in the biosynthesis of two antibiotics has been established in two *Streptomyces* strains [26, 27]. The substrates of EHs are structurally very diverse, representing a broad range of metabolites and xenobiotics. The substrate specificity of individual EHs appears to be diverse as well, being in many cases broad, but occasionally limited to a few available epoxidic compounds [28–30]. Sources of novel EHs are not limited to known (micro)organisms. Metagenomic or environmental DNA (eDNA), that is, the total microbial DNA of a microcosm, such as a small soil or groundwater sample, can serve as a source of novel EHs without the need to isolate and cultivate the microorganisms. PCR-based amplification of EH gene fragments in conjunction with genome-walking techniques [31] has been used to retrieve entire genes encoding α/β-hydrolase fold EHs directly from the metagenomic DNA [29, 32]. Moreover, the activity

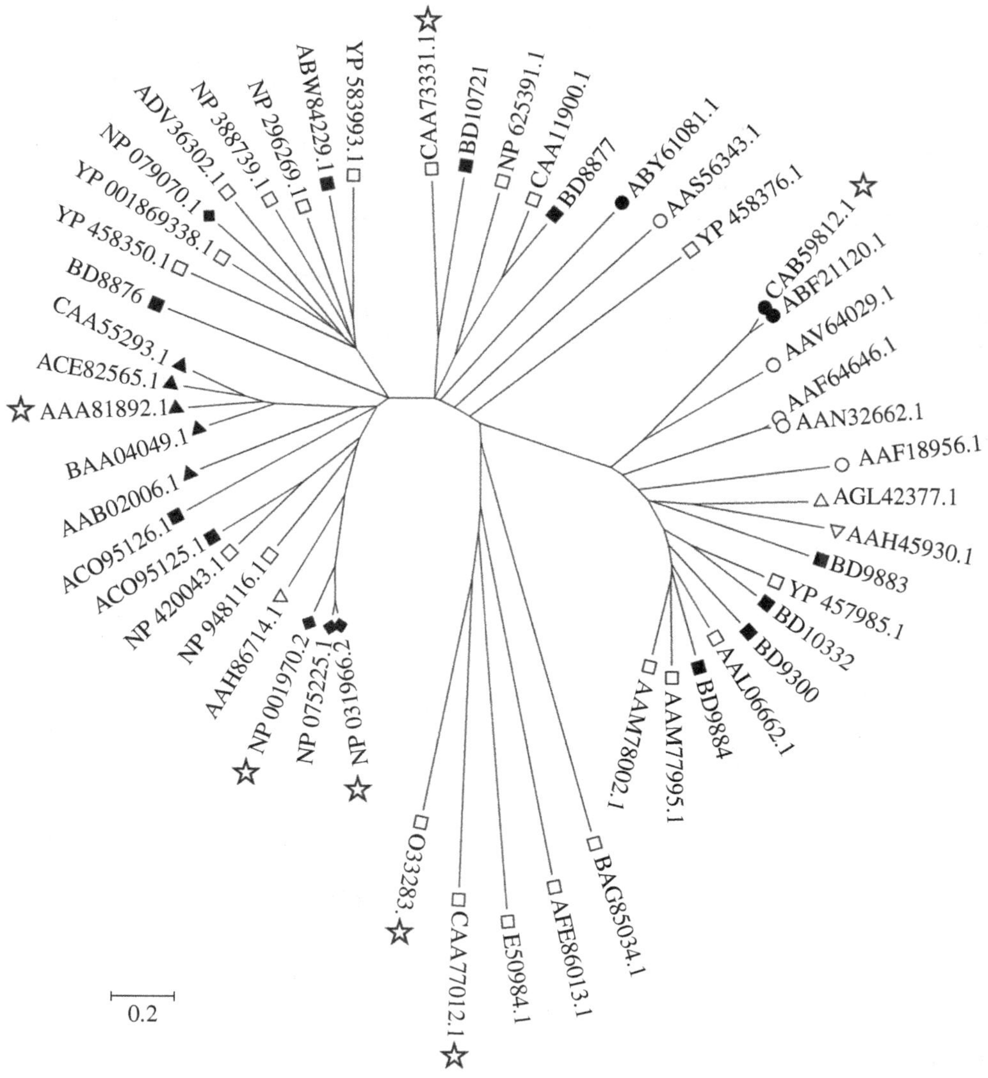

FIGURE 8.3

Phylogenetic relationships among protein sequences encoding EHs with confirmed activities. Each sequence is represented by its GenBank accession number; for sequences and further data regarding EHs with codes starting with BD, see Zhao *et al.* [22]. The data set includes sequences of mammalian EHs (♦), plant EHs (▲), fish EHs (▽), insect EHs (Δ), yeast EHs (○), fungal EHs (●), bacterial EHs (□), and eDNA-derived EHs (■). A star represents an EH with a determined X-ray protein structure: human EH (NP_001970), murine EH (NP_031966), potato EH (AAA81892), a bacterial EH from *Agrobacterium radiobacter* AD1 (CAA73331), a fungal EH from *Aspergillus niger* LCP 521 (CAB59812), and two bacterial EHs, which are not members of the α/β-hydrolase fold superfamily (CAA77012 and O33283). The bootstrap consensus tree was inferred from 1000 replicates. The evolutionary distances were computed using the Poisson correction method. The bar represents 0.2 amino acid substitutions per site.

screening of recombinant clones containing fragments of eDNA and hybridization to EH-specific target sequences led to the discovery of novel eDNA-derived EHs with considerable potential for biotransformations [22].

8.2.2 Heterologous Expression of EHs

Most EH expression systems were based on *Escherichia coli* as a host; however, heterologous expression of EHs was also established in mammalian cells [33]; the baculovirus system with *Spodoptera frugiperda* and *Trichoplusia ni* insect cell lines [34]; in the yeast strains *Pichia pastoris, Saccharomyces cerevisiae,* and *Yarrowia lipolytica* [35–37]; and in *Aspergillus niger* NW219 (originally published as *A. niger* NW171; [38]). The latter heterologous host offered the possibility of using a low-cost culture medium with inexpensive corn steep liquor as the main component. Further, *E. coli* RE3 as the recombinant host enabled EH production in a minimal growth medium with inexpensive sucrose as the sole carbon source [39]. Coexpression of molecular chaperones together with the optimization of culture conditions resulted in lower levels of inclusion bodies in the recombinant strain *E. coli* BL21(DE3) when overexpressing the EH from *Rhodotorula glutinis* [40].

8.2.3 Reaction Mechanisms of EHs

A large fraction of EHs belongs to the α/β-hydrolase fold superfamily, which contains—besides EHs—other structurally related hydrolytic enzymes with a characteristic arrangement of α-helices and β-sheets: esterases, haloalkane dehalogenases, lipases, amidases, and some more [41, 42]. It appears that all EHs of the α/β-hydrolase fold superfamily share a common three-step reaction mechanism, which involves the action of the active site-located catalytic triad (Asp-His-Glu/Asp), two tyrosine residues, and a water molecule (Figure 8.4). In the first step, the carboxylic acid of aspartate attacks an oxirane carbon of the bound epoxide substrate, resulting in a transiently formed ester intermediate. Two conserved tyrosines assist in this step of catalysis, polarizing the epoxide ring by hydrogen bonding with the oxirane oxygen. The second step of the reaction mechanism, which is often rate limiting, is characterized by the hydrolysis of the ester intermediate, catalyzed by an activated water molecule; the amino acid pair His-Glu/Asp of the catalytic triad is responsible for this activation. In the third and final step, the formed diol product is released, leaving behind the restored catalytic triad of the enzyme.

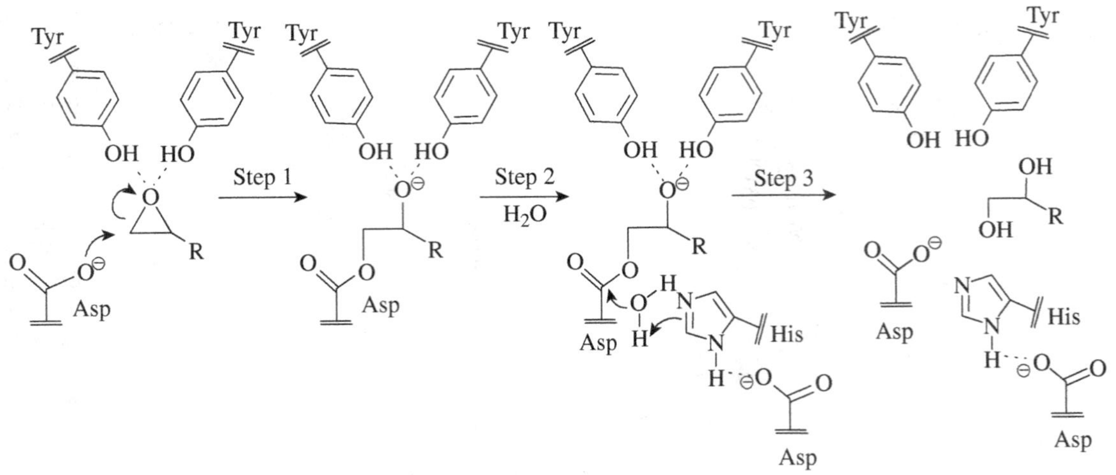

FIGURE 8.4
Proposed catalytic mechanism of EHs, which are members of the α/β-hydrolase fold superfamily.

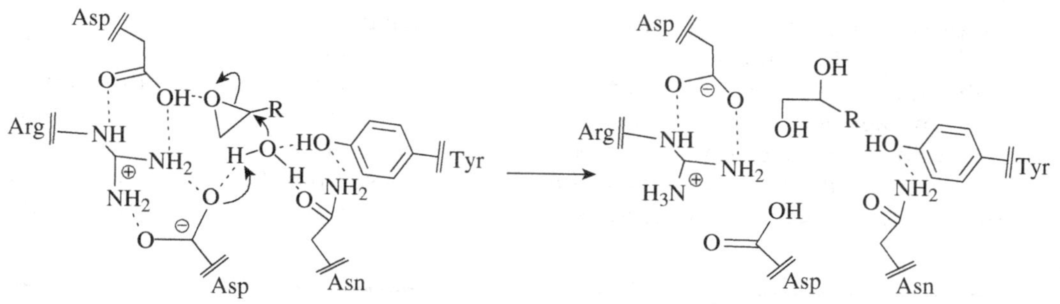

FIGURE 8.5

Proposed catalytic mechanism of the limonene-1,2-epoxide hydrolase from *Rhodococcus erythropolis* and the EH from *Mycobacterium tuberculosis*. The catalytic water molecule is activated by the formation of hydrogen bonds involving aspartic acid, asparagine, and tyrosine residues. At the same time, polarization and activation of the oxirane ring are achieved by hydrogen bonding with the oxirane oxygen involving another aspartic acid residue.

Some EHs are not members of the α/β-hydrolase fold superfamily, as shown for the limonene-1,2-epoxide hydrolase (EH) from *Rhodococcus erythropolis* [43] and the EH from *Mycobacterium tuberculosis* [44]. These two enzymes, which are dimers, have similar overall structures, each subunit consisting of a curved six-stranded β-sheet and four helices. The active site is located in a deep pocket with an Asp-Arg-Asp catalytic triad at its bottom. In contrast to the catalytic mechanism of α/β-hydrolase fold EHs mentioned already, a single-step push–pull mechanism has been proposed, which includes the activation of a water molecule by hydrogen bonding, resulting in a nucleophilic attack at the epoxide ring; at the same time, the epoxide is polarized and thereby activated by making available a proton to the oxirane oxygen (acid catalysis) (Figure 8.5).

8.3 DIRECTED EVOLUTION AND GENETIC ENGINEERING OF EHs

Nature is a very rich source of EHs (Figure 8.3), each enzyme having its own specific substrate range. Nevertheless, the perfect EH, which satisfies all the needs of a chemist for a given application—high chiral selectivity, low product inhibition, acceptable activity, and process stability—is rarely found. The generation of improved EH variants by genetic engineering and directed evolution approaches is the solution to the increasing demand for specifically tailored biocatalysts. Mutations in residues that are involved in catalysis or in residues that are flanking the nucleophilic aspartate of the catalytic triad can have a strong influence on the enantioselectivity and EH activity. For instance, replacing the acidic charge relay residue Glu with Asp in the rat microsomal EH led to an increase in V_{max} by more than 20-fold for the substrates styrene oxide (SO) and 9,10-epoxystearic acid [45]. In another example, one of the two oxirane-polarizing tyrosines was exchanged by phenylalanine in the active site of the EH from *Agrobacterium radiobacter* AD1, which resulted in a substantial increase in *E*-values for racemic SO and its *para*-nitro and *para*-chloro derivatives. However, the enzyme variants were significantly less active when compared to the wild-type EH [46]. More recently, the nucleophile-flanking residue Phe-108 was replaced in the same enzyme by one of the remaining 19 amino acids using saturation mutagenesis. Activity measurements of the resulting enzymes revealed an increase in activity and/or enantioselectivity for a number of enzyme–substrate variant pairs [47].

Due to the complexity of biocatalyst engineering and its limitations when using site-directed mutagenesis of a few mutational sites, *in vitro* enzyme evolution has emerged as an additional powerful tool for improving enzymes [48]. Using this approach, higher values of enantioselectivity have been achieved for a number of EH-catalyzed reactions (Table 8.1). Three main techniques have been used for the laboratory evolution of EHs: error-prone PCR, DNA shuffling, and iterative saturation mutagenesis. A critical and thorough comparison of these methods has not been done until recently when Reetz and coworkers compared the efficiencies of various evolutionary methods for improving a lipase from *Pseudomonas aeruginosa* in the stereoselective hydrolytic kinetic resolution of an ester [53]. Their results

TABLE 8.1 Selected EH Variants Evolved toward Higher Enantioselectivities

Source of EH	Racemic Substrate	*E*-values[a] (Abs. Conf. Remaining Epoxide)	No. of Mutations Introduced	Technique of Evolution	Screening Method	Reference and Remarks
Aspergillus niger LCP 521	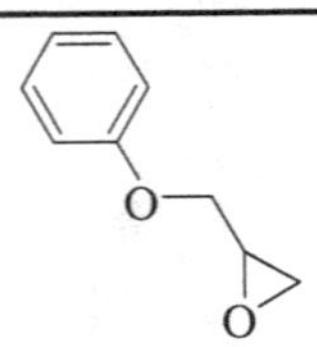	4.6→115 (*R*)	9	Iterative saturation mutagenesis	Two-step screening: (1) growth on epoxide-containing agar plates and (2) MS-based system	Reetz *et al.* [49]; NNK codon randomization
		5→160 (*R*)	6 Plus insert of 13 residues[b]		Chiral HPLC-based test	Reetz and Zheng [50]; NDT codon randomization; use of chaperones
	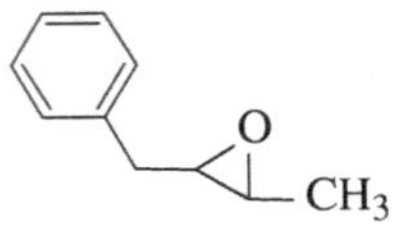	–[c]→>200 (2*R*,3*R*)	3		Cell-based adrenaline test	Reetz *et al.* [51]; NDT codon randomization
Agrobacterium radiobacter AD1	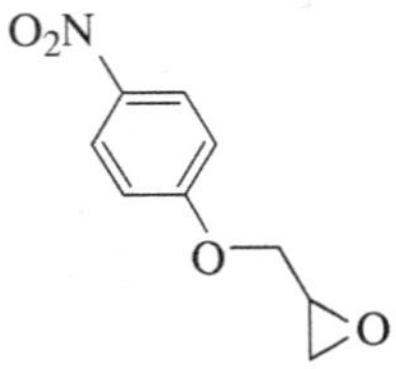	3.4→32 (*S*)	4	Error-prone PCR and DNA shuffling	Three-step screening: (1) activity screening using safranin O indicator plates, (2) progress curves with whole cells, and (3) cell-free extracts	van Loo *et al.* [52]; significant reduction in activity for the mutant EHs

[a] The change in *E*-value from wild-type or initial variant to final mutant enzyme is indicated.
[b] A stretch of 13 residues was accidentally incorporated between sites 318 and 319 during directed evolution. This insertion had a significant enantioselectivity-enhancing effect.
[c] There is virtually no activity of the wild-type enzyme toward this substrate.

TABLE 8.2 Selected EH Variants Evolved toward Higher Enantioconvergence using Iterative Saturation Mutagenesis (Including Desymmetrization of *meso*-Epoxides)

Source of EH	Substrate	ee_p (%)[a] (Abs. Conf. of Main Diol)	No. of Mutations Introduced	Screening Method	Reference and Remarks
Aspergillus niger M200	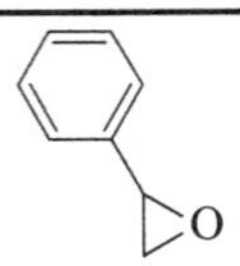	3→70 (*R*)	9	Two-step screening: (1) activity screening using 4-(4-nitrobenzyl) pyridine and (2) chiral GC	Kotik *et al.* [55]; NNK codon randomization
Rhodococcus erythropolis DCL 14	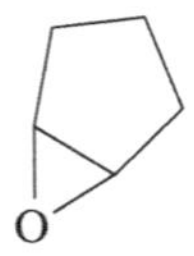	14→93 (*S*,*S*) 14→80 (*R*,*R*)	4 5	Two-step screening: (1) cell-based adrenaline test and (2) chiral GC	Zheng and Reetz [56]; NDT codon randomization

[a] The change in ee_p-value from wild-type to final mutant enzyme is indicated.

suggest a significantly higher efficiency of iterative saturation mutagenesis over error-prone PCR, saturation mutagenesis at hot spots, and/or DNA shuffling in generating highly enantioselective mutants. In each round of iterative saturation mutagenesis, randomization at two or more sites is performed simultaneously, exploiting the possible occurrence of cooperative effects operating between the randomized sites [54]. Usually, these randomization sites are selected in the vicinity of the active site, a minimal requirement of this technique being a sound structural model of the enzyme. Improvement in enantioconvergence of EHs was also

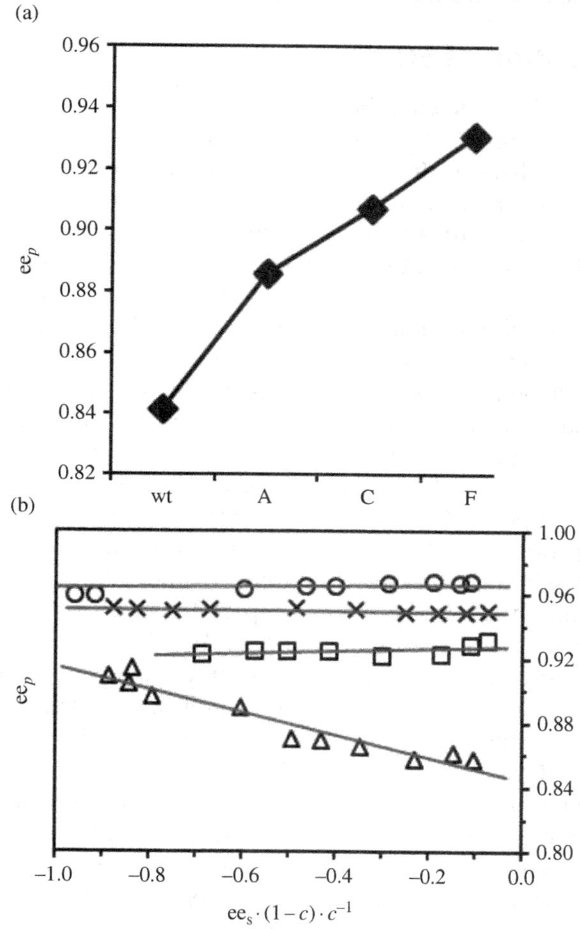

FIGURE 8.6

(a) Directed evolution of the metagenome-derived EH Kau2 toward an enantioconvergent biocatalyst. Five amino acid substitutions were sufficient to increase the degree of enantioconvergence from 84% ee_p for the wild-type EH (wt) to 93% ee_p for the final evolved variant (F), enabling the production of the *R*-diol with an ee value of 93% at 28 °C in a complete conversion of racemic *para*-chlorostyrene oxide. A and C represent intermediate variants of the *in vitro* evolution process. Higher ee_p- values of up to 97% were determined in enantioconvergent reactions using lower temperatures [57]. (b) Determination of the regioselectivity coefficients α_S and α_R of the EH-catalyzed reaction using the wild-type and final evolved a Kau2 variant with racemic *para*-chlorostyrene oxide as the substrate. Plotting ee_p against $ee_s \cdot (1-c) \cdot c^{-1}$ enabled the determination of α_R and subsequently α_S as described in Kotik *et al.* [32], using the intercept *q* and the slope *m* from the linear regression; for the calculations, nonabsolute values for ee_p and ee_s were used (see Moussou *et al.* [58]). The following relationships are valid: $\alpha_R+\beta_R=1$, and $\alpha_S+\beta_S=1$. Wild-type EH at 28 °C (Δ): $\alpha_R=0.12$, $\alpha_S=0.95$; final evolved variant at 28 °C (□): $\alpha_R=0.03$, $\alpha_S=0.96$; at 18 °C (x): $\alpha_R=0.03$, $\alpha_S=0.98$; at 8 °C (O): $\alpha_R=0.02$, $\alpha_S=0.98$.

reported using iterative saturation mutagenesis (Table 8.2; Figure 8.6). For example, higher levels of enantioconvergence toward racemic *para*-chlorostyrene oxide were achieved in only three productive mutagenesis rounds (Figure 8.6a), using a manual two-step screen based on a colorimetric activity assay followed by a chiral GC test for ee_p determination. The wild-type enzyme and the evolved EHs were analyzed for all four regioselectivity coefficients (see Figure 8.2), using the racemic epoxide as a substrate and a chiral GC column for separating both diol enantiomers (Figure 8.6b).

8.4 IMMOBILIZED EHs AND REACTIONS IN NONAQUEOUS MEDIA

8.4.1 Immobilization of EHs

Immobilization can help to increase the operational stability of a biocatalyst, which is a particularly important aspect of preparative and industrial-scale biotransformations. The following immobilization techniques have been reported for EHs: carrier-free approaches, adsorption or covalent attachment of the enzyme to a support, whole-cell encapsulation and entrapment, and immobilized metal ion affinity binding via genetically engineered His tags. The latter technology can be used for both one-step extraction from cellular crude extracts and facile immobilization of the recombinant EH.

Often, enzyme enrichment using recombinant protein production and/or purification precedes the immobilization to increase the enzyme load on the support. For instance, a polyethylene glycol-based aqueous two-phase system was established to extract a recombinant metagenome-derived EH from the clarified crude cell lysate. This enriched EH preparation was then used directly for immobilization onto oxirane-activated supports. The highest per-gram activity of the immobilized EH was achieved with Eupergit C 250L, reaching 44 μmol/min (gram of dry support)$^{-1}$ for *trans*-1-phenylpropene oxide at 28 °C. The immobilized EH exhibited an increase in thermal stability when compared to the free enzyme [39].

Carrier-free immobilized EHs were prepared using the cross-linked enzyme aggregate (CLEA) technology, as exemplified by the CLEA formation of a mixture of two EHs from mung beans. Compared to the free enzymes, the CLEAs exhibited significantly shorter reaction times and higher enantioconvergence for the hydrolysis of SO in a biphasic *n*-hexane buffer reaction system, which ensured low levels of nonenzymatic hydrolysis of the substrate [59].

Immobilization reactions based on biocatalyst adsorption or covalent binding onto supports were described for various EHs; the most interesting results are summarized in the following paragraphs. A clarified crude extract of recombinant *E. coli* biomass was used for the immobilization of the *Solanum tuberosum* EH onto glyoxyl agarose in the presence of dextran. This covalently immobilized enzyme preparation exhibited a marked increase in thermostability by a factor of 300 at 60 °C and nearly no change in the regioselectivity coefficients and *E*-values for the hydrolysis of racemic SO [60].

Two EHs, one from *S. tuberosum* and the other from *A. niger* LCP 521, were separately immobilized onto DEAE-cellulose by adsorption and used sequentially for the transformation of racemic *para*-chlorostyrene oxide into the corresponding (*R*)-diol with an enantiopurity of 89%. In total, five cycles were performed, each immobilized biocatalyst being recovered after each cycle and reused in the following biotransformation reaction [61].

A magnetically separable biocatalyst was created by immobilizing the thermally unstable EH from *Mugil cephalus* within the cavities of magnetite-containing mesoporous silica. First, the purified enzyme was adsorbed onto the carrier material and then cross-linked within the pores of the carrier with glutaraldehyde, creating a so-called nanoscale enzyme reactor. The immobilized biocatalyst exhibited significantly higher stability at 30 °C compared to the free enzyme. Repeated hydrolytic kinetic resolution experiments were performed with racemic SO as the substrate, the EH-containing nanoparticles being recovered and reused in a total of seven cycles [62].

The results described in the following five reports were all obtained with a partially purified recombinant EH from *A. niger* LCP 521. Yildirim and coworkers covalently immobilized the EH onto modified Florisil and Eupergit C supports, leading to biocatalyst preparations with improved thermostabilities and a two-fold higher *E*-value for the hydrolysis of racemic SO when compared to the free enzyme [63].

Petri and coworkers used an epoxide-derivatized silica gel for the covalent immobilization of the EH, resulting in an immobilized enzyme preparation with enhanced stability in a reaction mixture containing dimethylsulfoxide, which was needed as a cosolvent for efficient solubilization of the substrate *para*-nitrostyrene oxide. Nine cycles of repeated batch reactions were performed without a significant loss of biocatalyst activity [64].

Immobilization onto DEAE-cellulose by ionic adsorption resulted in an EH preparation with significantly reduced half-life of enzyme activity at all investigated reaction temperatures. Nevertheless, the immobilized EH was successfully recycled in a repeated biphasic batch process, kinetically resolving in five cycles a total of 3 g of racemic *para*-chlorostyrene oxide at a concentration of 2 M at 4 °C [65]. The same substrate at a concentration of 4 mM and the same EH-containing support were used in a series of hydrolytic kinetic resolutions in heptane, which was equilibrated at a water activity of 0.9. A decrease in *E*-value but an increase in operational stability during biocatalyst recycling were determined, when compared to the free EH [66].

An enhancement in *E*-value for the hydrolysis of racemic SO and *para*-chlorostyrene oxide was observed for the EH mentioned already when immobilized onto an ethylene diamine-modified epoxy-activated support. A repeated batch experiment with 12 cycles of hydrolytic kinetic resolution of racemic SO was performed using this immobilized biocatalyst preparation without significant loss in enzyme activity [67].

Whole cells of the bacterium *Nocardia tartaricans* with hydrolytic activity toward *cis*-epoxysuccinate were encapsulated in a carrier material consisting of sodium alginate, cellulose sulfate, and poly(methylene-*co*-guanidine) for the production of L-(+)-tartrate. The encapsulation led to improved performance of the cells in terms of EH activity and storage stability when compared to the cells that were entrapped in calcium pectate gel beads [68].

Entrapment of *Rhodosporidium toruloides* cells in calcium alginate led to an immobilized EH-containing biocatalyst for the enantioselective kinetic resolution of racemic 1,2-epoxyoctane. The entrapped cells exhibited a much lower loss in enzyme activity during storage at room temperature in the absence of substrate [69].

Recently, several immobilization methods based on the His-tag technology [70] were reported for EHs. A magnetically separable Fe_3O_4–silica-based composite material with NiO nanoparticles attached to its surface was produced, which enabled the immobilization of a His-tagged EH from *R. glutinis*. The immobilized biocatalyst was reused in repeated batch experiments for the hydrolytic kinetic resolution of racemic SO using a magnet for separating the biocatalyst from the reaction mixture [71]. Wang and coworkers synthesized magnetic iron oxide-based nanoparticles with Ni-nitrilotriacetic acid functions on their surface. The His-tagged EH from *S. tuberosum* was selectively immobilized on these nanoparticles, and the resulting biocatalyst was reused in repeated hydrolytic kinetic resolutions with racemic *para*-chlorostyrene oxide as the substrate. The nanoparticle-based biocatalyst retained 80% of its productivity after eight cycles of kinetic resolution [72]. Co(II)-resorcinol-grafted silica beads were prepared for the one-step extraction–immobilization of His-tagged enzymes. Immobilized *S. tuberosum* EH was prepared in this way and tested in water-saturated cyclohexane for the formation of hydrobenzoin from *trans*-stilbene oxide [73].

8.4.2 EH-Catalyzed Reactions in Organic Solvent- or Ionic Liquid-Containing Media

The use of nonaqueous reaction media for EH-based biotransformations can be advantageous: higher solubilities and stabilities of the relatively hydrophobic epoxidic substrates can be achieved, and in some cases, higher enzyme thermostabilities have been reported. Stirred two-phase systems can ensure an optimal reaction environment for the biocatalyst in the aqueous phase and at the same time a high concentration of the epoxidic substrate with a low nonenzymatic hydrolysis rate in the organic phase. For instance, the hydrolytic kinetic resolution of phenyl glycidyl ether (PGE) was achieved in a stirred water–isooctane two-phase system using whole cells of *Bacillus megaterium* ECU1001 [74].

An increase in *E*-value in an *n*-hexane-containing two-phase reaction system was determined for the EH-mediated enantioselective hydrolysis of SO using a freeze-dried cell-free extract from *Sphingomonas* sp. [75].

The tendency of epoxides to undergo spontaneous nonenzymatic hydrolysis in water was also reduced in reaction media composed of ionic liquids, as shown by Chiappe and coworkers. Under these reaction conditions with crude or purified EHs from cress or mouse as the biocatalyst, the reaction rates and stereoselectivities for the hydrolysis of *trans*-β-methylstyrene oxide were found to be—with a few exceptions—generally comparable to those of buffer-based reactions [76]. The beneficial effects of ionic liquids on EH-mediated asymmetric hydrolysis reactions, that is, reducing the nonenzymatic hydrolysis rate and increasing the solubility of the epoxidic substrate, were confirmed in more recent reports. Chiappe and coworkers tested a number of ionic liquids and established optimal reaction conditions for the kinetic resolution of racemic 2-*tert*-butyloxirane using recombinant soluble and microsomal EHs from mouse, cress, and rat. It is worth noting that several of the tested ionic liquids were found to be incompatible with this epoxidic substrate, presumably due to a nonenzymatic reaction between the epoxide and the anion of the ionic liquid compound [2]. Further, Chen and coworkers studied the asymmetric hydrolysis of SO in two-phase systems using an enriched EH-containing preparation from mung beans. The biphasic systems were composed of either hydrophobic ionic liquid—buffer or *n*-hexane—buffer in the presence of small amounts of hydrophilic ionic liquids. Using optimized reaction conditions, the two EHs in mung bean converted the racemic substrate into (*R*)-1-phenyl-1,2-ethanediol with an ee value of up to 97% and a product yield of 49% [77, 78].

Compared to an aqueous medium, an increased thermostability in heptane for an enriched and freeze-dried EH preparation from *A. niger* LCP 521 was reported. An initial water activity (a_w) of 0.9 for the heptane reaction medium was determined to be optimal in terms of enzyme activity and *E*-value. On the other hand, the use of heptane greatly increased the apparent K_m value of the enzyme for *para*-chlorostyrene oxide by a factor of 460, thereby reducing the catalytic efficiency of the enzyme [79].

A high substrate concentration in an aqueous reaction mixture can also lead to biphasic systems with the substrate forming a second phase. Such two-phase systems were established for a number of EH-based reactions [5, 80, 81].

8.5 MONOFUNCTIONAL EPOXIDES AS CHIRAL BUILDING BLOCKS FOR THE SYNTHESIS OF BIOLOGICALLY ACTIVE COMPOUNDS

Monofunctional epoxides are the simplest EH substrates and some of them, such as SO and derivatives thereof, are routinely used as standard substrates for EH activity measurements. Even within such a simple class of compounds, it is possible to distinguish between, for example, mono-, di-, and trisubstituted epoxides; aromatic and nonaromatic epoxides; and *meso*-epoxides. The following reflects such a subdivision of epoxides and may assist in finding the right enzyme for as yet not-studied epoxides just by structure comparison with already described substrates of EHs.

8.5.1 Monosubstituted Aromatic Epoxides

Styrene Oxide

Both enantiomers of SO and phenyl-1,2-ethanediol are common chiral aromatic molecular building blocks for the synthesis of pharmaceuticals and other specialty chemicals. In recent decades a substantial amount of work has been dedicated to the enantioselective biohydrolysis of SO, the reason probably being the commercial availability of SO and the corresponding diol in both racemic and enantiopure forms. In 1993, Furstoss and coworkers described the first preparative access to both enantiomers of SO by enantioselective hydrolysis of the racemate using cells of two fungal strains that were enantiocomplementary, that is, enantioselectively hydrolyzed either of the two SO enantiomers [82, 83]. In addition, it was observed that the hydrolysis of racemic SO using cells of *A. niger* LCP 521 proceeded with the retention of configuration at the chiral center, generating the (*R*)-diol and leaving behind the (*S*)-epoxide as the residual compound. On the other hand, hydrolysis of SO using cells of *Beauveria sulfurescens* ATCC 7159 resulted in the formation of the (*R*)-diol from the (*S*)-epoxide by inversion of configuration, leaving behind the unreacted (*R*)-epoxide. These findings were used for a biohydrolysis reaction in the presence of both molds, resulting in the first enantioconvergent bienzymatic process for the production of (*R*)-phenyl-1,2-ethanediol in high yield (92%) and with an ee as high as 89% (Figure 8.7). More recently, this bienzymatic enantioconvergent strategy was applied for the preparation of the same (*R*)-diol compound in a higher ee, using different combinations of EHs. For example, by mixing two purified EHs, a wild-type EH from *S. tuberosum* and an evolved EH from *A. radiobacter* AD1, SO was rapidly converted to the corresponding (*R*)-diol with 98% ee and 100% yield [84]. However, this process had to be carried out at a low substrate concentration of 5 mM due to product inhibition of the *S. tuberosum* EH. Moreover, a mixture of recombinant whole cells harboring the EH-encoding genes from *A. niger* LK and *Caulobacter crescentus* was also used in a preparative-scale batch reaction for the enantioconvergent hydrolysis of 1.2 g of racemic SO at a concentration of 33 mM; 1.3 g of (*R*)-phenyl-1,2-ethanediol with an enantiopurity of 91% was obtained with an overall yield of 95%. Substrate concentrations exceeding 50 mM could not be used due to product inhibition of the bacterial EH [85]. A similar approach was described later in which the *A. niger* LK EH was replaced by an EH mutant from *M. cephalus*, a marine fish [86]. After optimization of the reaction conditions, (*R*)-phenyl-1,2-ethanediol was obtained in 90% ee and 95% yield from 50 mM racemic SO. Shen and coworkers isolated two bacterial EHs (SgcF and NcsF2), which are involved in the biosynthesis of enediynes, antitumor antibiotics produced by *Streptomyces globisporus* and *Streptomyces carzinostaticus*. Using SO as a substrate mimic, SgcF and NcsF2 were shown to be enantiocomplementary, leading to the formation of (*R*)-phenyl-1,2-ethanediol with a 99% ee and 87% yield from racemic SO (14 mM) when used together

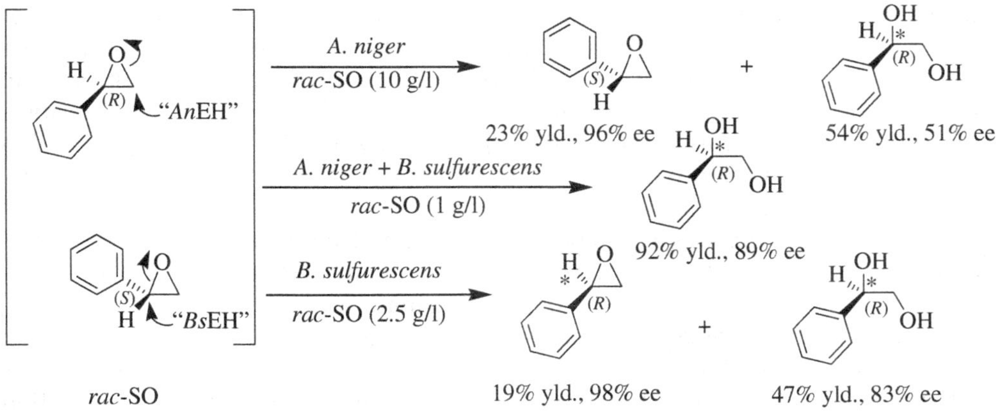

FIGURE 8.7 Biohydrolytic kinetic resolution of racemic styrene oxide (*rac*-SO) catalyzed by *Aspergillus niger* LCP 521 (*An*EH) and/or *Beauveria sulfurescens* ATCC 7159 (*Bs*EH). Using both fungi together led to an enantioconvergent production of (*R*)-phenyl-1,2-ethanediol.

in the reaction mixture [27]. An interesting enantioconvergent process with whole cells of *Aspergillus tubingensis* TF1 was recently described by Duarah *et al.* [87]. (*R*)-Phenyl-1,2-ethanediol was isolated from the reaction mixture in 97% ee after 45 min, reaching >99% conversion of racemic SO (8.75 mM). Although a single enzyme was claimed to be responsible for this process, the possibility of two enantiocomplementary EHs with opposite regioselectivity being present in the microorganism cannot be ruled out from the described data.

Although a great number of wild-type EHs was found to kinetically resolve racemic SO, to the best of our knowledge, no EH showed a very high enantioselectivity level. Nevertheless, several kinetic resolutions at high SO concentrations have been described since the mid-2000s using EHs from yeast, bacteria, and plants. For example, (*S*)-SO with 98% ee was obtained in 41% yield from racemic SO at a very high concentration of 1.8 M using a *P. pastoris* strain overexpressing the EH from *R. glutinis*. Such a high substrate concentration in the reaction mixture called for optimized reaction conditions, that is, the reaction was taking place at 4 °C in the presence of 40% (v/v) Tween 20 and 5% (v/v) glycerol [88]. Biomass of *Achromobacter* sp. MTCC 5605, which was isolated from a petroleum-contaminated sludge sample, enabled the hydrolytic kinetic resolution of racemic SO at a high concentration of 0.5 M using a biphasic reaction system composed of isooctane and buffer. Under these conditions, an enantiomeric ratio of 64 was determined; the remaining (*S*)-SO (42% yield) and the formed (*R*)-phenyl-1,2-ethanediol were isolated in >99% and 65% ee, respectively [89]. It is worth mentioning that an improvement of the enantioselectivity was obtained after covalent immobilization of the multimeric EH from *A. niger* LCP 521 onto Eupergit C, which was partially modified with ethylene diamine (Eupergit C/EDA), resulting in an *E*-value of 56 instead of 25 [67].

Chlorostyrene Oxide

(*R*)-*para*- and (*R*)-*meta*-chlorostyrene oxides are important building blocks for the synthesis of various biologically active molecules. Indeed, these compounds are, for example, essential chiral intermediates for the production of eliprodil [90], an effective NMDA receptor antagonist, and various β-3-adrenergic receptor agonists such as SR 58611A or AJ-9677 [91]. Numerous biohydrolytic kinetic resolutions have been described in the literature with the purpose of preparing these chiral synthons in enantiopure form. However, it is well known that one of the general drawbacks of including a resolution step in a chemical synthesis is its intrinsic 50% yield limitation. This is the reason why various enantioconvergent processes have been elaborated with the aim of approaching the ideal situation of "100% yield and 100% ee".

As far as *para*-chlorostyrene oxide (*p*-ClSO) is concerned, enzymatic extracts of overexpressed EHs from *A. niger* LCP 521 (*An*EH) and *S. tuberosum* (*St*EH) were shown to resolve this racemic epoxide efficiently (*E*-values of 100 at 0 °C). The enantiocomplementarity of these two enantioselective EHs enabled the preparation of both enantiomers of *p*-ClSO, the (*R*)-*p*ClSO and (*S*)-*p*ClSO being preferentially hydrolyzed by the *An*EH and the *St*EH, respectively. The absolute configuration of the formed diol was determined to be *R* in both cases. Preparative-scale resolutions were performed at very high substrate concentrations of 306 g/l and 30.6 g/l using, respectively, *An*EH and *St*EH as the biocatalysts [92]. Similar results were also obtained in a repeated batch reaction with both enzymes immobilized onto DEAE-cellulose by ionic adsorption [65].

Based on the complementary enantio- and regioselectivities of these two EHs, an enantioconvergent production of (*R*)-*para*-chlorophenyl-1,2-ethanediol from *rac*-*p*-ClSO was established using a sequential bienzymatic strategy. Thus, a preparative-scale experiment was carried out at a substrate concentration of 30.6 g/l using first the *St*EH followed by the *An*EH. The (*R*)-diol was obtained with an overall yield as high as 93% and 96% ee (Figure 8.8) [92]. Unfortunately, the substrate concentration

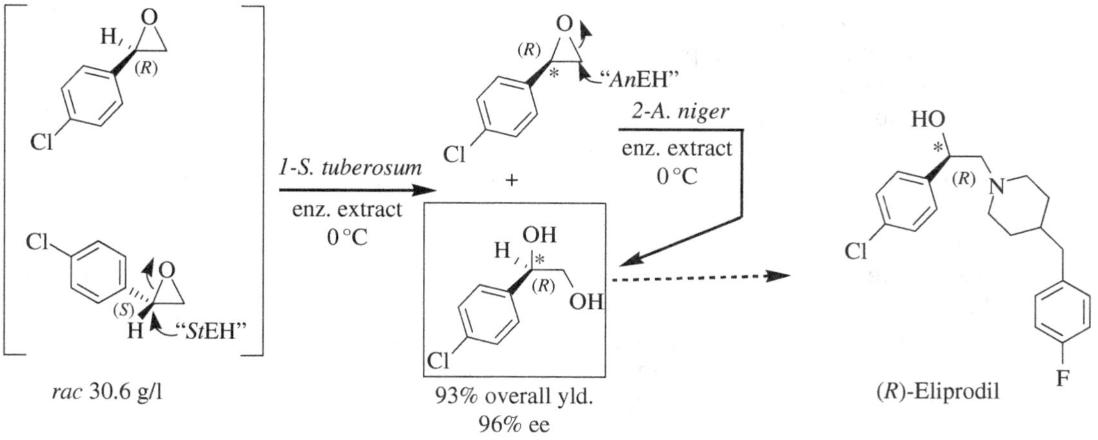

FIGURE 8.8

Enantioconvergent biohydrolytic transformation of *para*-chlorostyrene oxide using a bienzymatic process. The sequential use of *Solanum tuberosum* and *Aspergillus niger* EHs as biocatalysts led to the formation of enantiopure (*R*)-*para*-chlorophenyl-1,2-diol, a chiral building block for the synthesis of (*R*)-eliprodil.

had to be decreased by a factor of 10 in this bienzymatic process, compared to a single-enzyme resolution process with *An*EH, to diminish the inhibitory effect of the formed diol on the *St*EH activity. Later, a repeated batch experiment using these two EHs, separately immobilized onto DEAE-cellulose, was also described [61]. More recently, other teams reported a similar strategy using two EHs. A sequential bienzymatic hydrolysis of *p*-ClSO was described by Min and Lee [86], using a heterologously expressed EH from *C. crescentus* and an EH mutant from *M. cephalus*. The combined use of whole cells overexpressing these two EHs enabled the production of (*R*)-*para*-chlorophenyl-1,2-ethanediol with 92% enantiopurity and 71% yield, starting from 17 g/l of *rac*-*p*-ClSO. In a previous paper, the same authors described an enantioconvergent process that was based on a monoenzymatic approach using the EH from *C. crescentus* [93]. This EH was shown to have opposite enantiopreference and regioselectivity toward either enantiomer of *rac*-*p*-ClSO, which led to the almost exclusive formation of the (*R*)-diol. With the enantioselectivity being not too high (*E*-value = 30), a preparative-scale biohydrolysis at a substrate concentration of 16.8 g/l resulted in the formation of (*R*)-*para*-chlorophenyl-1,2-ethanediol with 98% ee and 78% overall yield. Kotik *et al.* have described the first example of regioselectivity engineering in EHs by directed evolution starting from a nonenantioconvergent enzyme [55]. The substrate binding cavity of the EH from *A. niger* M200 was redesigned to generate an enantioconvergent biocatalyst by guiding the point of nucleophilic attack to the benzylic oxirane position of the bound (*S*)-enantiomer. After nine amino-acid exchanges, the final enzyme variant transformed racemic *p*-ClSO to the (*R*)-diol with an ee of 70.5%. These authors reported in the same article a sequential bienzymatic reaction using the wild-type EH from *A. niger* M200 and its evolved variant, resulting in the formation of the (*R*)-diol with an ee value of 88%. More recently, Kotik and coworkers reported that an EH (named Kau2), whose gene was isolated from a biofilter-derived metagenome, exhibited an opposite regioselectivity for the two enantiomers of *p*-ClSO, which enabled them to obtain (*R*)-*para*-chlorophenyl-1,2-ethanediol in 84% ee at 100% conversion [32]. Later, an improvement in the enantioconvergence of Kau2 was achieved using iterative saturation mutagenesis (Figure 8.6), leading to a higher enantiomeric excess of 94% for the diol product at complete conversion of the racemic *p*-ClSO substrate [57].

An enantioconvergent preparative-scale production of (*R*)-*meta*-chlorophenyl-1,2-ethanediol was described by Monterde *et al.* using the EH from *S. tuberosum*. The enzyme exhibited a low enantioselectivity (*E*-value of 6) in conjunction with an opposite regioselectivity for each enantiomer of *m*-ClSO, which are ideal conditions for an enantioconvergent process. Starting from *rac*-*m*-ClSO, nine cycles of a repeated batch experiment in a stirred reactor at 10 g/l of substrate concentration furnished the (*R*)-diol with an ee value of 97% and an 88% overall yield [8]. Very recently, Li and coworkers have studied the biohydrolysis of numerous epoxides (including *meso* compounds) using the EH from *Sphingomonas* sp. HXN-200 overexpressed in *E. coli* (*Sp*EH) [94]. This EH with an *E*-value of 41 was shown to be more enantioselective than any other known EH for the hydrolysis of *rac*-*m*-ClSO. Interestingly, *Sp*EH reacted preferentially with the (*R*)-epoxide, forming the (*R*)-diol and leaving behind the (*S*)-*m*-ClSO, which is a useful chiral building block for the preparation of an IGF-1R kinase inhibitor [95]. A gram-scale kinetic resolution of *rac*-*m*-ClSO was performed in a two-phase system (buffer/*n*-hexane) at 15 g/l of substrate with resting *Sp*EH-containing cells. Enantiopure (*S*)-*m*-ClSO was obtained in 37.9% yield.

Nitrostyrene Oxide

(*R*)-*para*-Nitrostyrene oxide (*p*-NSO) is the key chiral synthon for the synthesis of nifenalol, a compound showing β-blocking activity and used in the treatment of hypertensive diseases [96]. This epoxide, which is mostly insoluble in water, has been resolved with good enantioselectivity by fungal, bacterial, and yeast EHs (Table 8.3). The first examples were described with whole cells of the fungi *A. niger* LCP 521 and *B. sulfurescens* ATCC7159 [104]. One year later, a cell-free extract from *A. niger* LCP 521 was used as a biocatalyst [97]. To improve the solubility of *p*-NSO in the reaction mixture 20% of a miscible solvent (DMSO or DMF) were added. Under these experimental conditions, the hydrolysis was relatively enantioselective (*E*-value = 48) and could be performed at a high substrate concentration of 54 g/l (330 mM) to produce (*S*)-*p*-NSO (99% ee; 49% yield). Later, high enantioselectivity with an *E*-value of >100 at 10 °C was also determined with another EH present in the strain *A. niger* M200, which was isolated from industrial biofilters [100, 105]. Starting from the wild-type EH from *A. radiobacter* AD1 with an *E*-value of 65 [98], a substantial improvement of the enantiopreference was obtained by error-prone PCR and DNA shuffling, reaching an *E*-value >200 [52, 99]. Interestingly, a very high enantioselectivity (*E*-value >200) together with a reversed (*S*)-enantiopreference was observed for the EH from

TABLE 8.3 Selected Kinetic Resolutions of *rac*-*p*-Nitrostyrene Oxide

Source of EH	Notes	*E*-value	Substrate Conc. (mM)	Abs. Conf. Residual Epoxide/Formed Diol	References
Aspergillus niger LCP 521	Enzymatic extract	48	330	*S*/*R*	[97]
Agrobacterium radiobacter AD1	Purified enzyme	65	3	*S*/*R*	[98]
A. radiobacter AD1, variant Y215F		>200	3	*S*/*R*	[99]
A. radiobacter AD1, variant S7	Error-prone PCR and DNA shuffling; cell-free extract or purified EH	>200	2	*S*/*R*	[52]
A. niger GBCF 79 (recombinant strain)	Enzymatic extract, EH immobilized onto silica gel, cosolvent: 20% DMSO	85	4.5	*S*/*R*	[64]
A. niger M200	Purified enzyme	>100	8	*S*/*R*	[100]
Kau2 (metagenome)	Crude extract, EH overexpressed in *Escherichia coli*	80	3.5	*R*/*R*	[32]
Kau8 (metagenome)		65	3.5	*S*/*R*	
Oxy-4	Whole cells, EH overexpressed in *Yarrowia lipolytica*	>100	500	*S*/*R*	[101, 102]
Oxy-10		>100	500	*R*/*R*	
Bacillus megaterium ECU1001	Crude extract, EH overexpressed in *E. coli*	>200	20	*R*/–	[103]

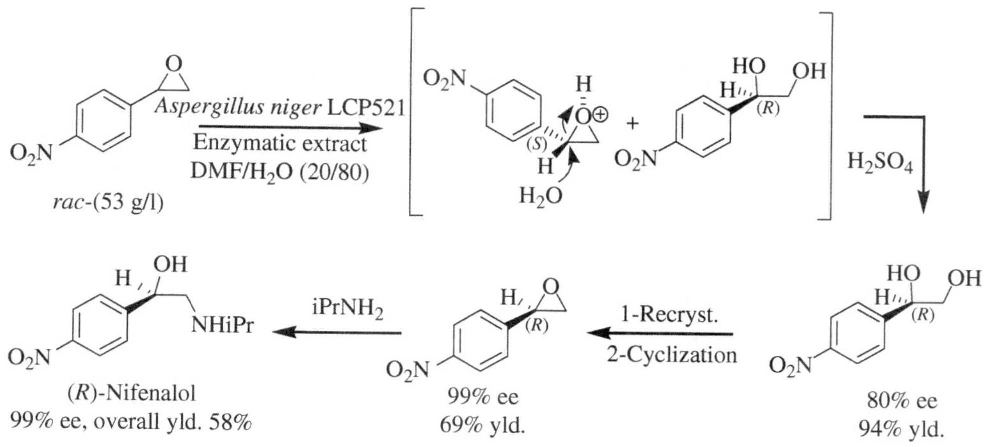

FIGURE 8.9 Enantioconvergent synthesis of the β-blocker (*R*)-nifenalol using a combined chemoenzymatic approach.

B. megaterium ECU1001, retaining the useful (*R*)-*p*-NSO for the direct synthesis of (*R*)-nifenalol [103]. More recently, EHs exhibiting opposite enantiopreference toward the two antipodes of *p*-NSO were also described. Two of them, Kau2 and Kau8, were isolated from biofilter-derived metagenomes [32] and two others, Oxy-4 and Oxy-10, from yeast [102]. It should be noticed that the remaining epoxides of Kau2- and Oxy-10-mediated kinetic resolutions were determined to be (*R*)-*p*-NSO, which is the correct enantiomer for the synthesis of (*R*)-nifenalol.

A one-pot chemoenzymatic enantioconvergent process for the production of (*R*)-*p*-NSO was described by Furstoss *et al.* in 1997 [9]. The strategy was based on the sequential hydrolysis of the (*R*)-epoxide using an enzymatic extract of *An*EH, followed by an acid-catalyzed hydrolysis of the remaining (*S*)-epoxide (Figure 8.9). A 330 mM (54 g/l) solution of *p*-NSO was hydrolyzed within 6 h to furnish the (*S*)-epoxide in 49% yield and 99% ee. Then, the controlled acid hydrolysis of the reaction mixture yielded the (*R*)-diol (80% ee) as a result of steric inversion upon acid hydrolysis of the unreacted (*S*)-epoxide. After recrystallization, the (*R*)-diol could be easily recycled to the epoxide and transformed into (*R*)-nifenalol. Interestingly, Botes *et al.* [101, 102] described the enantioconvergent production of the (*R*)-diol in one pot at a high substrate concentration using the combined action of the Oxy-4 and Oxy-10 biocatalysts (bienzymatic process).

Trifluomethylstyrene Oxide

Enantiocontrolled synthesis of fluorinated organic compounds has gained tremendous impetus in recent years because it is well known that the presence of fluorine atoms in a molecule can have dramatic effects on its biological activity [106]. In this context, the kinetic resolution of a specific trifluoro-methyl-substituted aromatic epoxide family was studied using a recombinant EH from *A. niger* LCP 521, and the productivity of the biotransformation process was evaluated [107]. A two liquid–liquid phase methodology with an appropriate cosolvent (isooctane, 10–35% (v/v)) and optimized operational conditions led to a very efficient and cost-effective resolution process. The best results (high *E*-value, high TON and TOFF) were obtained in the case of *para*-substituted CF_3, OCF_3, and SCF_3 derivatives. For example, resolution of (4-trifluoromethoxyphenyl)-oxirane could be performed at 250 g/l, resulting in the residual (*S*)-epoxide and the formed (*R*)-diol in good yields and very high enantiomeric excess (Figure 8.10).

Pyridyl Oxirane

Enantiopure 2-, 3-, and 4-pyridyloxiranes are key building blocks for the synthesis of several biologically active compounds, such as β-adrenergic receptor agonists or anti-obesity drugs [108–110]. Up to now, none of these products could be obtained in a

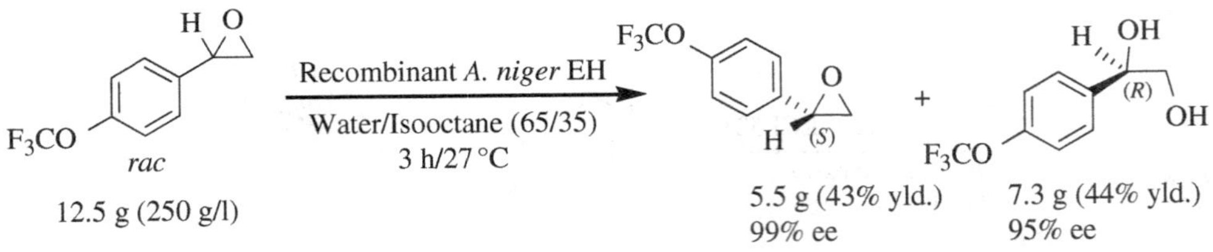

FIGURE 8.10

Preparative-scale synthesis of enantiopure (4-trifluoromethoxyphenyl)-oxirane using a partially purified recombinant EH from *Aspergillus niger* LCP 521 as a biocatalyst.

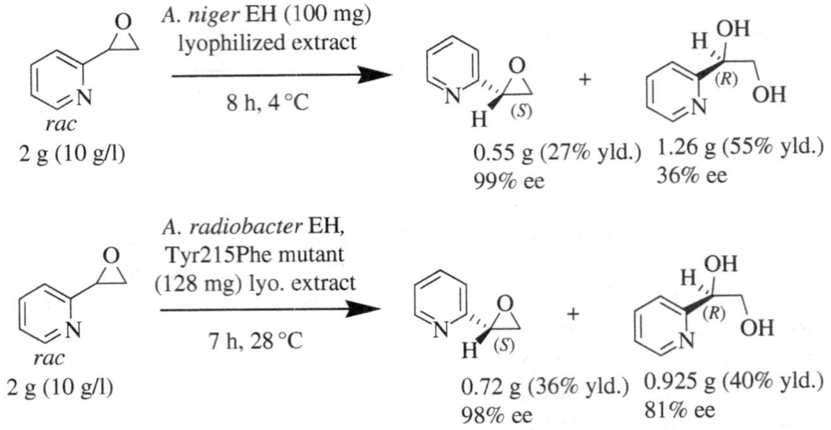

FIGURE 8.11

Preparative kinetic resolution of 2-pyridyloxirane with EHs from *Aspergillus niger* and *Agrobacterium radiobacter* AD1 (mutant Tyr215Phe).

satisfactory enantiopure form using the most effective heavy metal-containing catalysts (Jacobsen epoxidation, Jacobsen HKR, or Sharpless dihydroxylation). Interestingly, the recombinant EH from *A. niger* LCP 521 exhibited a rather high enantioselectivity toward all three substrates with *E*-values of 96, 27, and 47, respectively, hydrolyzing preferentially the (*R*)-enantiomer and thus enabling the recovery of the slowly reacting (*S*)-epoxide. Unfortunately, it was shown that the *E*-value decreased with increasing substrate concentration. Nevertheless, the preparative-scale synthesis of each pyridyloxirane could be performed at about 10 g/l substrate concentration, which enabled these three target compounds to be obtained in nearly enantiopure form [111]. In the same year, it was shown that a Tyr215Phe mutation in the EH from *A. radiobacter* AD1 resulted in an enzyme variant that could efficiently resolve 2-pyridyloxirane (*E*-value = 55) at a substrate concentration as high as 127 mM (15.5 g/l) (Figure 8.11) [112].

8.5.2 Disubstituted Aromatic Epoxides

p-*Bromo-*, p-*Isobutyl-*, *and* p-*Trifluoromethyl*-alpha-*Methyl Styrene Oxide*

To overcome the problem of low solubility of aromatic epoxides in the water phase, Furstoss and collaborators studied, in 1998, enzymatic resolutions at high substrate concentration without adding organic solvents [80]. They showed that at a high concentration of 80 g/l of *para*-bromo-α-methyl styrene oxide, a biphasic system was formed with the epoxide constituting one phase by itself. This enabled a good kinetic resolution of the aromatic epoxide using an EH-containing extract from *A. niger* LCP 521 as a biocatalyst. Under these experimental conditions, the residual epoxide was found to be of (*S*) configuration, whereas the formed product was the corresponding (*R*)-diol. Surprisingly, the use of this procedure led to a dramatic enhancement in the enantioselectivity with the *E*-value increasing from 20 at low substrate concentration (1.7 g/l) to 260 at 80 g/l (Figure 8.12).

One year later, the same authors studied the biohydrolysis of seven differently substituted α-methylstyrene oxide derivatives, including the *para*-bromo-α-methyl styrene oxide, using ten different EHs [14]. The best results were obtained with the

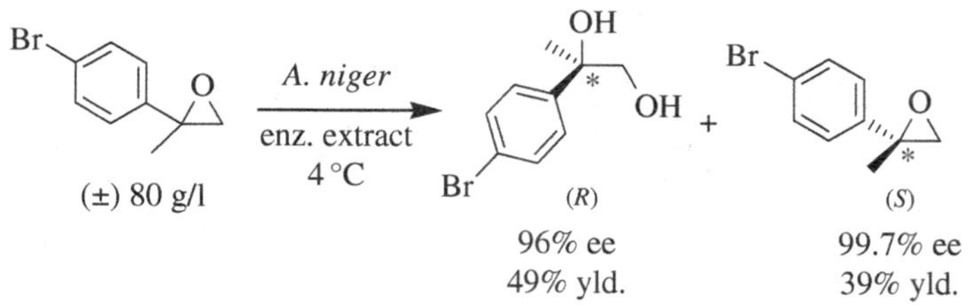

FIGURE 8.12

Kinetic hydrolytic resolution at high substrate concentration of *para*-bromo-α-methyl styrene oxide using an enzymatic extract of *Aspergillus niger* LCP 521.

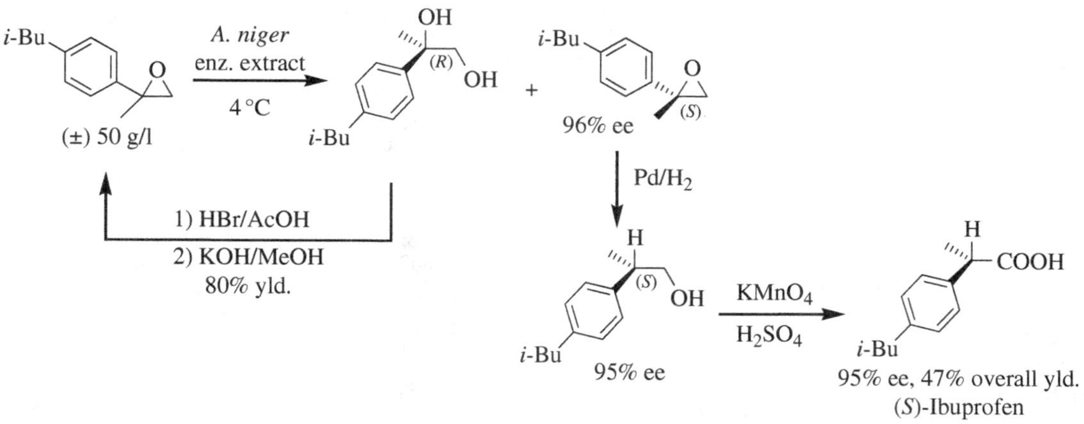

FIGURE 8.13

Preparative-scale resolution of *rac*-4-isobutyl-α-methylstyrene oxide using an enzymatic extract of *Aspergillus niger* LCP 521. A four-step enantioconvergent procedure enabled the synthesis of (*S*)-ibuprofen.

EH from *A. niger* LCP 521; however, the *E*-values were relatively moderate. A four-step synthesis of (*S*)-ibuprofen, a nonsteroidal anti-inflammatory drug, was performed to illustrate the synthetic potential of EHs. The strategy was to achieve the enantioselective hydrolysis of *rac*-4-isobutyl-α-methylstyrene oxide using the EH from *A. niger*, which has been shown to hydrolyze specifically the undesired (*R*)-enantiomer and to further transform the enantiopure residual (*S*)-epoxide into (*S*)-ibuprofen using classical chemical synthesis. As in the case of the *para*-bromo derivative, the biohydrolysis was performed at a high substrate concentration of 50 g/l, leading to a biphasic process, and at a low reaction temperature of 4 °C to enhance enzyme stability and decrease the spontaneous hydrolysis of the substrate. Following this strategy, the overall yield of (*S*)-ibuprofen was only 27%. Recycling of the formed diol via chemical racemization substantially improved the process yield. Indeed, treatment of the formed diol with HBr/AcOH and subsequent cyclization of the bromohydrin intermediate under basic conditions afforded racemic 4-isobutyl-α-methylstyrene oxide in 80% yield, which could thus be resubmitted to the enzymatic resolution step. Under these conditions, the overall yield increased from 27 to 47% (Figure 8.13).

Ten years later, Furstoss's group confirmed the high kinetic resolving power of the EH from *A. niger* LCP 521 by showing that the enzyme could also be used for the efficient resolution of *para*-trifluoromethyl-α-methyl styrene oxide [107]. Indeed, the preparative-scale biohydrolysis at 100 g/l could be performed in a short reaction time in a biphasic reaction medium containing water–organic solvent. Isooctane (25% v/v) was added to the reactor to obtain a good substrate emulsion, leading to an optimal transfer of the substrate from the organic phase to the water phase. Under these experimental conditions, a loading of 25 g of racemic epoxide resulted in the formation of 10.5 g (42% yield) of (*S*)-epoxide (99.7% ee) and 13.4 g (78.5% yield) of (*R*)-diol (78.5% ee) (Figure 8.14).

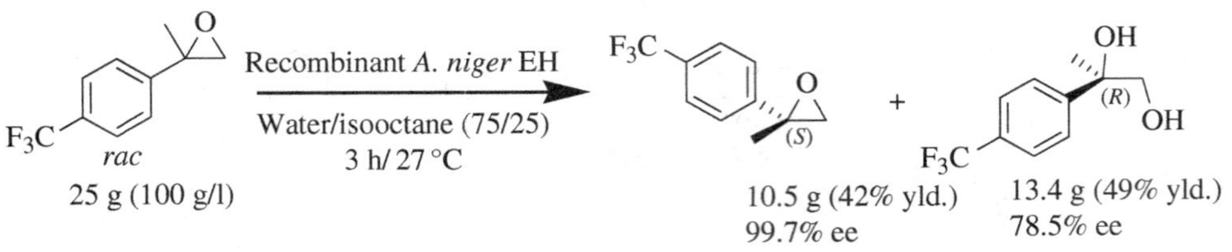

FIGURE 8.14

Preparative-scale synthesis of enantiopure *para*-trifluoromethyl-α-methyl styrene oxide using a partially purified recombinant EH from *Aspergillus niger* LCP 521 as a biocatalyst.

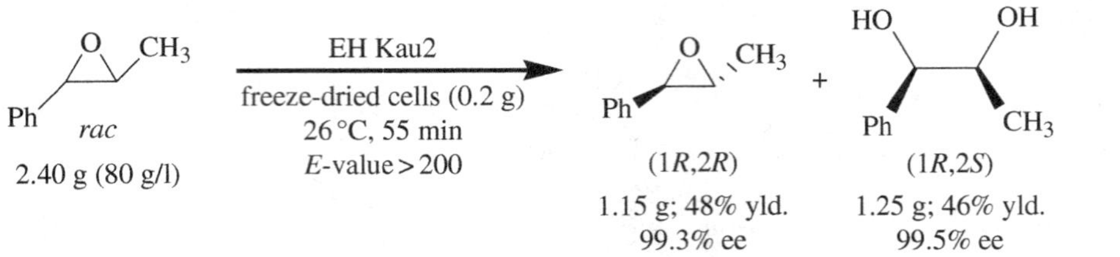

FIGURE 8.15

Preparative-scale kinetic resolution of racemic *trans*-β-methylstyrene oxide using the EH Kau2 at high substrate concentration.

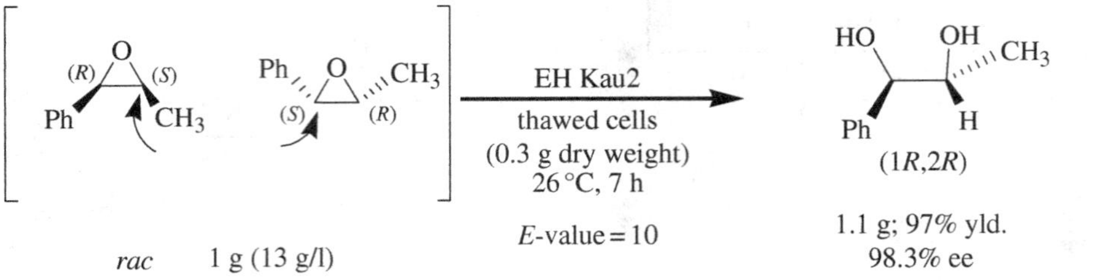

FIGURE 8.16

Preparative enantioconvergent biohydrolysis of racemic *cis*-β-methylstyrene oxide using the EH Kau2.

cis- and trans-β-Methylstyrene Oxide

trans-β-Methylstyrene oxide in an enantiopure form has been used as a building block for the synthesis of a potential cocaine abuse therapeutic agent and an anti-obesity drug [113, 114]. Very satisfactory resolutions of *trans*-β-methylstyrene oxide were achieved with a metagenome-derived EH (termed Kau2) [32] and EHs from fungi, for example, *B. sulfurescens* [115] and yeast (*R. glutinis* CIMW147 [116], *R. glutinis* UOFS Y-0123 [117], and *Rotoruloides mucoides* UOFSY-0471 [118]). A comparison of all the described results leads to the conclusion that the hydrolysis of *trans*-β-methylstyrene oxide proceeded with a similar enantioselectivity and stereochemistry. Indeed, in all cases, the (1*S*,2*S*)-epoxide was preferentially hydrolyzed to the (1*R*,2*S*)-*erythro*-diol, indicating that the enzymatic attack occurred at the benzylic position. The best result was achieved in a preparative-scale reaction at 80 g/l of substrate concentration, using freeze-dried *E. coli* RE3 cells harboring the plasmid pSEKau2 for expression of the EH Kau2. Both (1*R*,2*R*)-epoxide and the corresponding (1*R*,2*S*)-diol were isolated in high enantiomeric excess (>99%) and good yield (>45%), corresponding to a very high enantioselectivity (*E*-value >200) (Figure 8.15).

In contrast to these results, Chiappe *et al.* [76] reported that the reaction catalyzed by the EH from cress was partially stereoconvergent, furnishing the corresponding (1*S*,2*R*)-*erythro*-1-phenylpropane-1,2-diol as the main product. Using pure enantiomers as starting epoxides, it was determined that the formal nucleophilic attack of water occurred exclusively at C1 in the case of (1*R*,2*R*)-epoxide, while the same enzyme was practically nonselective for the same carbon atom of its antipode.

As far as *cis*-β-methylstyrene oxide is concerned, very few EHs with the ability to hydrolyze this compound have been described. Three similar and interesting enantioconvergent processes were described using fungal EHs from *B. sulfurescens* ATCC7159 [115], *Aspergillus terreus* [58], and the metagenome-derived EH Kau2 [32]. Indeed, it was observed that the two antipodes were hydrolyzed with a low enantioselectivity (*E*-value = 10) and an opposite regioselectivity, leading to the formation of the same (1*R*,2*R*)-diol in almost optically pure form and nearly quantitative yield (Figure 8.16). It should be noticed that the modification of the stereochemistry (*trans* to *cis*) of the starting epoxide resulted in a significant decrease of the reaction rate.

Indene Oxide

Chiral (1*S*,2*R*)-indene oxide is a valuable precursor for the synthesis of the side chain of the HIV protease inhibitor MK 639, which was developed by Merck Research Laboratories. Several teams searched for EHs capable of catalyzing the kinetic resolution of racemic indene oxide. Selected strains of bacteria [119], yeast [116], and fungi [11, 12] were shown to be useful for the preparation of the two antipodes of indene oxide (Figure 8.17). However, low enantioselectivity of the implied EHs and instability of indene oxide in water resulted in low yields when attempting to reach 100% ee of the residual epoxide.

Recently, kinetic resolution performed at high substrate concentration was described in a patent by Botes *et al.* [118] using a recombinant yeast EH. Indeed, a preparative-scale biohydrolysis at 264 g/l (2 M) substrate concentration was performed using whole cells of a recombinant *Y. lipolytica* strain expressing the EH from the yeast *Rhodosporidium paludigenum* NCYC 3179. In this process, crystalline indene oxide powder (26.4 g) was directly added to 100 ml of phosphate buffer, which contained thawed whole cells (13.5 g wet weight). After 150 min at 25 °C, the resolution process was complete and 8.5 g (32% yield) of residual enantiopure (1*R*,2*S*)-indene oxide were isolated. Unfortunately, the recovered (1*R*,2*S*)-indene oxide exhibited the wrong stereochemistry for the synthesis of indinavir.

8.5.3 Nonaromatic Epoxides

Monosubstituted Alkyl Epoxides

Highly enantioselective EHs toward monoaliphatic terminal oxiranes were detected in some species of specific yeast genera such as *Rhodotorula* sp. and *Rhodosporidium* sp. [120, 121]. Although enantiomeric distinction of these highly flexible molecules is believed to be a difficult task for the enzyme, high *E*-values of >100 were reported in several cases. All enantioselective EHs reacted preferentially with the (*R*)-epoxide, forming (*R*)-diols and leaving behind the (*S*)-epoxide. Using whole cells of *Rhodotorula araucariae* CBS 6031 or *R. toruloides* CBS 0349 it was possible to perform a preparative-scale hydrolysis of racemic 1,2-epoxyoctane at a 500 mM substrate concentration (Figure 8.18).

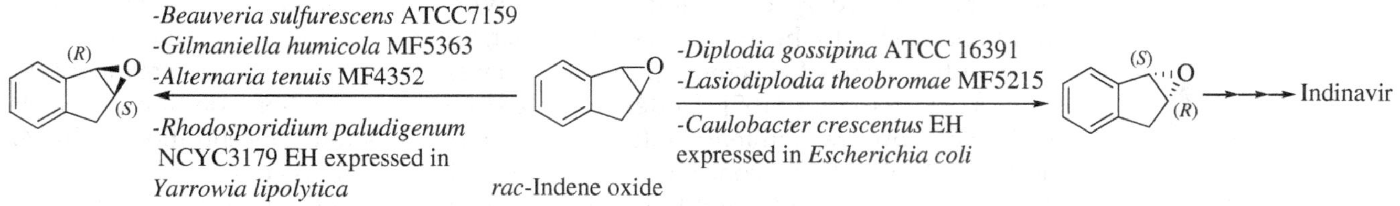

FIGURE 8.17

Biohydrolytic kinetic resolution of racemic indene oxide. Access to (1*S*,2*R*)- or (1*R*,2*S*)-indene oxide as a function of the biocatalyst used.

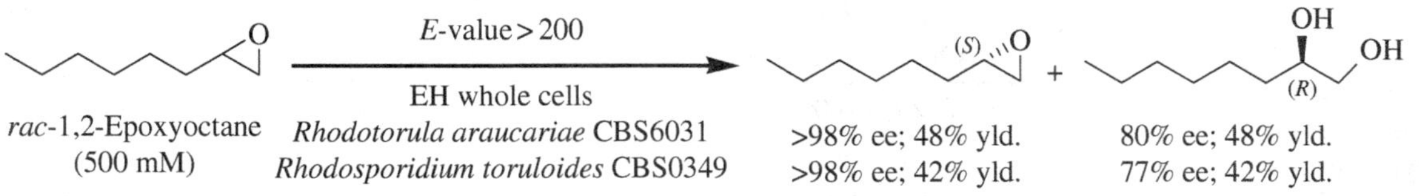

FIGURE 8.18

Kinetic resolution of 1,2-epoxyoctane at high substrate concentration with two yeast EHs.

Glycidyl Ether and Derivatives Thereof

Biohydrolytic kinetic resolutions of alkyl and aryl glycidyl ethers, which are important building blocks for the production of various bioactive compounds, have been extensively investigated using EHs from bacteria, yeast, and filamentous fungi. Concerning PGE and various derivatives thereof, interesting results were obtained with the EH from *B. megaterium* ECU1001. In contrast to the majority of other EHs, this enzyme exhibited an unusual (*R*)-enantioselectivity for PGE, retaining the useful (*S*)-PGE for the synthesis of β-blocker compounds; it also exhibited the highest enantioselectivity ($E=58$) among all known wild-type EHs [103]. In addition, it was observed that introducing a methyl substituent at the phenyl ring of PGE had a pronounced influence on the enantioselectivity of the hydrolysis reaction. As a general trend, the *E*-value increased as the substituent was shifted from the *para* ($E=11$) to the *meta* position ($E=19$), with the *ortho* position exhibiting the highest *E*-value of more than 200. It appears that wild-type EHs are not very efficient catalysts for the preparation of (*R*)-PGE; the best results were obtained with the EHs from *A. radiobacter* (*E*-value = 12) [122], *Trichosporon loubierii* (*E*-value = 20) [123], and a *Rhodobacterales* species (*E*-value = 38) [124]. However, as described at the beginning of this chapter, molecular engineering techniques can be used to improve enzymatic properties such as enantioselectivity or enantioconvergence. For example, starting from the wild-type EH from *A. niger* LCP 521, which catalyzes the kinetic resolution of PGE with quite low enantioselectivity (*E*-value = 4.6), Reetz and collaborators generated by directed evolution the highly enantioselective variant LW202, which exhibited an *E*-value of 115 [125]. Interestingly, in all cases involving other monosubstituted epoxides such as substituted glycidyl ethers or alkyl and aromatic epoxides, a substantial increase in *E*-value of the evolved EH compared to the wild-type EH was observed as well, leading to Reetz's conclusion: "The traditional credo in directed evolution, 'You get what you screen for', can be extended by the corollary 'You may get more than what you originally screened for'." Two years later, Reetz and collaborators managed to improve the expression efficiency and enantioselectivity of the LW202 EH variant by laboratory evolution again [50]. The strategy was to focus first on expression and then improve the enantioselectivity. The expression of the generated mutant EH222 was 50 times higher than that of the wild-type *An*EH, and a very high enantiomeric ratio (*E*-value = 160) in favor of the (*S*)-diol was detected (Figure 8.19).

Kotik *et al.* screened 270 microbial isolates from biofilters and petroleum-contaminated bioremediation sites for enantioselective EHs using *tert*-butyl glycidyl ether, benzyl glycidyl ether, and allyl glycidyl ether as substrates. The best results were obtained with the most substituted ether, *tert*-glycidyl ether, for which a moderate enantioselective EH activity (*E*-value = 30) was found in a fungal isolate identified as a *A. niger* species [105]. Under optimized biotransformation conditions, which included a low reaction temperature of 5 °C, a low substrate concentration of 5 mM, and a high biocatalyst concentration, the enantiomeric ratio could be increased to 100, and enantiopure (*R*)-*tert*-butyl glycidyl ether and (*S*)-3-butoxy-1,2-propanediol were isolated as residual epoxide and formed diol.

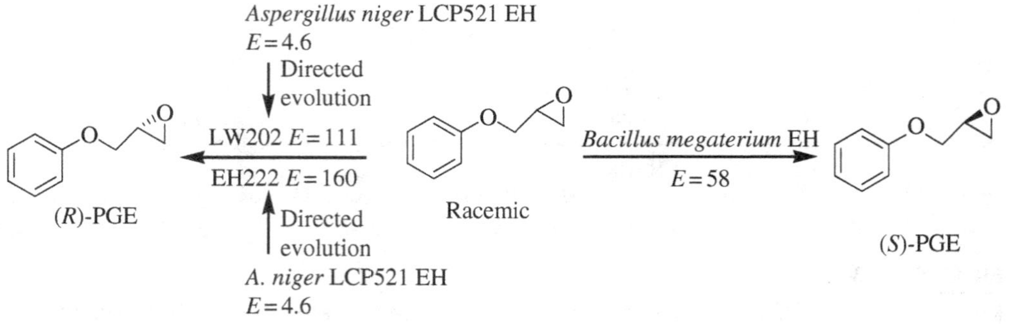

FIGURE 8.19
Biohydrolytic kinetic resolutions of racemic-PGE. Use of enantiocomplementary wild-type and evolved mutant EHs for the preparation of two antipodes of PGE.

Disubstituted Alkyl Epoxides

gem-*Disubstituted Oxiranes*

gem-Disubstituted oxiranes bearing linear alkyl, alkenyl, alkynyl, or benzyl substituents were enantioselectively hydrolyzed by EHs of the following bacterial genera: *Rhodococcus*, *Mycobacterium*, *Nocardia*, *Methylobacterium*, and *Arthrobacter*. Only EHs reacting with the (*S*)-enantiomer have been found so far, forming the (*S*)-diol (Table 8.4). Using these results, Faber and coworkers developed chemoenzymatic approaches for the deracemization of several *gem*-disubstituted oxiranes (Table 8.4, entries 1, 5, 9, 10). Enantioselective biohydrolysis of the (*S*)-epoxide proceeded with the retention of configuration, resulting in the formation of the corresponding (*S*)-1,2-diol. In a subsequent step, acid-catalyzed hydrolysis of the residual (*R*)-epoxide took place exclusively at the substituted oxirane atom with complete inversion of configuration, yielding the same (*S*)-1,2-diol. The combination of both reaction steps resulted in the generation of the (*S*)-1,2-diol in enantiopure form and in almost quantitative yield (Figure 8.20). Valuable illustrations of these enantioconvergent processes are the synthesis of natural compounds such as frontalin (Table 8.4, entry 5), fridamycin (Table 8.4, entry 9), and mevalonolactone (Table 8.4, entry 10), which will be described in detail in Chapter 7.

The kinetic resolution of a range of methyl-substituted 1-oxaspiro[2,5]octanes was investigated by Weijers *et al.* in 2005 using a yeast EH from *R. glutinis* [130]. It was observed that the positioning of substituents close to the spiroepoxide carbon atom resulted in a decreased reaction rate but increased enantioselectivity. The best

TABLE 8.4 Selected Kinetic Resolutions of Some *gem*-Disubstituted Alkyl Epoxides

Entry	Substrate	Source of EH	*E*-value	Subs. Conc. (g/l)	Abs. Conf.[a]	References
1		*Rhodococcus ruber* DSM 43338 *Nocardia* H8, TB1, and EH1	>200	20	*R/S*	[126, 127]
2		*Rhodococcus* sp. NCIMB 11216	126	2.5	*R/S*	[128]
3		*Rhodococcus* sp. NCIMB 11216 *Arthrobacter* sp. DSM 312	>200 172	5	*R/S*	[129]
4		*Rhodococcus* sp. NCIMB 11216	>200	2.5	*R/S*	[128]
5	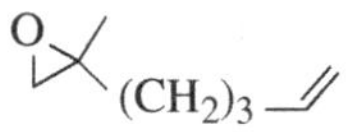	*Nocardia* EH1	>200	20	*R/S*	[126]
6	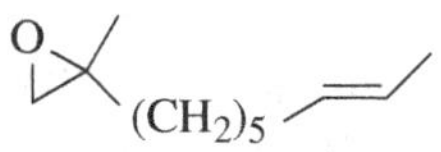	*Rhodococcus* sp. NCIMB 11216	125	5	*R/S*	[129]
7	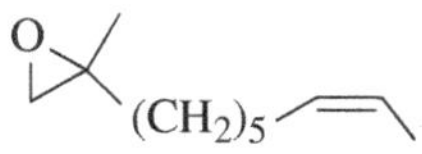	*Rhodococcus* sp. NCIMB 11216	142	5	*R/S*	[129]
8	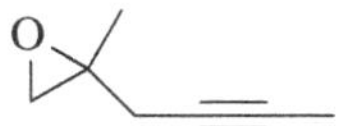	*Rhodococcus* sp. NCIMB 11216	>200	5	*R/S*	[129]
9		*Methylobacterium* sp. FCC 031	>200	10	*R/S*	[4]
10		*Nocardia* EH1	123	40	*R/S*	[18]

[a] Absolute configuration of the remaining epoxide and the formed diol, respectively.

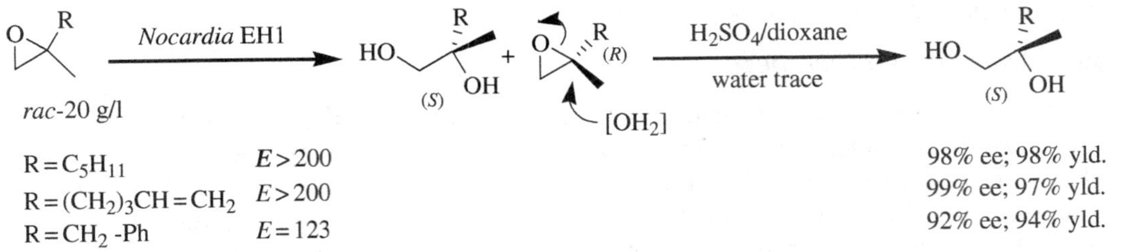

FIGURE 8.20

Chemoenzymatic deracemization of *rac-gem*-disubstituted oxiranes.

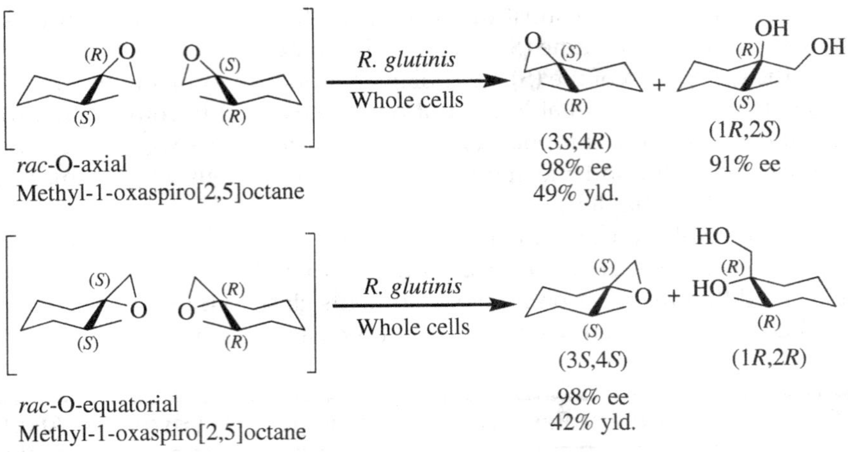

FIGURE 8.21

Kinetic resolution of O-axial or O-equatorial 4-methyl-1-oxaspiro [2,5]octane by *Rhodotorula glutinis* EH.

enantioselectivity with an *E*-value >100 was obtained for O-axial or O-equatorial 4-methyl-1-oxaspiro[2,5]octane (Figure 8.21).

The first deracemization of a *trans*-divinyl spiroepoxide, a strategic key building block of 11-heterosteroids, was described in 2007 using two enantiocomplementary microbial EHs as biocatalysts [10]. One enzyme was the partially purified recombinant EH from *A. niger* LCP 521 (*An*EH); the other enzyme, the so-called "Limonene EH" (LEH), was isolated from the bacterium *R. erythropolis*. The residual (*R*,*R*)-spiroepoxide (from the *An*EH-mediated reaction) and the residual (*S*,*S*)-spiroepoxide (using LEH) were isolated in nearly enantiopure forms (99% ee). However, because of the moderate *E*-values of around 20 in both kinetic resolutions, the respective reaction yields did not exceed 26%. A process-improving strategy enabled transformation of the formed enantiomerically enriched diols of opposite absolute configuration, that is, (*S*,*S*)-spirodiol and (*R*,*R*)-spirodiol, back to the corresponding epoxides, which were then submitted to a second enzymatic resolution cycle using the enantiocomplementary enzyme (i.e., the LEH for (*S*,*S*)-spiroepoxide or the *An*EH for (*R*,*R*)-spiroepoxide). In conclusion, both enantiomers of the substrate could be obtained in high enantiomeric purity (99%) and a reasonable yield of 50% (Figure 8.22).

cis- *and* trans-*Disubstituted Epoxides*

2,3-Disubstituted aliphatic oxiranes have been reported to be hydrolyzed by EHs from fungi [131], yeast, and bacteria. The most interesting results were observed with yeast and bacterial EHs. As far as kinetic resolution is concerned, it was shown by Weijers [116] that *R. glutinis* catalyzed the enantioselective hydrolysis of *cis*-2,3- and *trans*-2,3-epoxypentane, resulting in residual (2*R*)-epoxides with yields that approached the theoretical maximum of 50%. More interestingly, biocatalytic transformations of racemic 2,3-disubstituted oxiranes to vicinal diols with high ees at

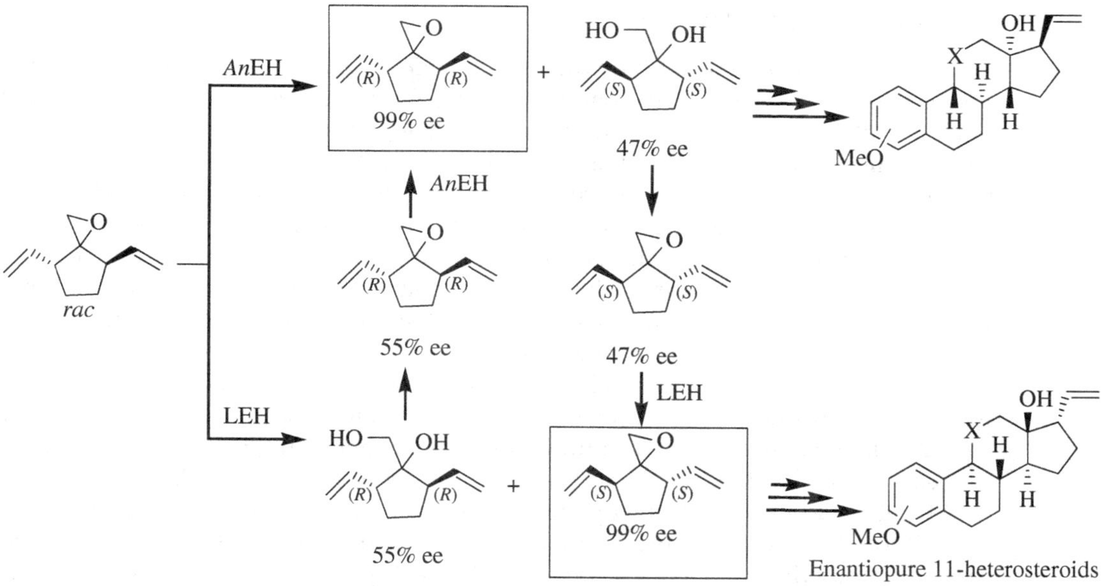

FIGURE 8.22

Preparative hydrolytic resolution of *trans*-divinyl spiroepoxide using *An*EH and LEH as biocatalysts, enabling the synthesis of enantiopure 11-heterosteroids.

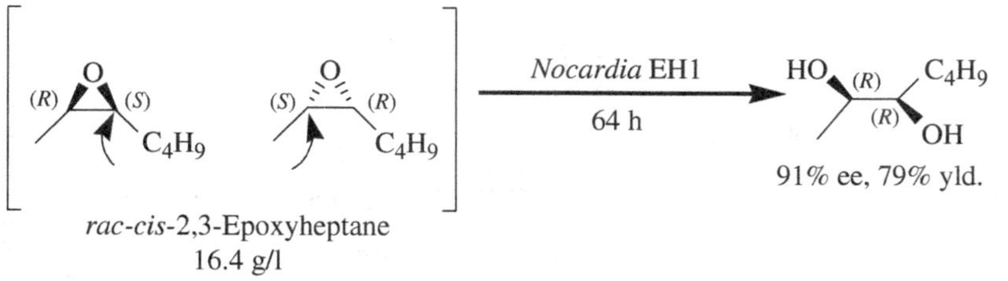

FIGURE 8.23

Deracemization of *rac-cis*-2, 3-epoxyheptane via enantio-convergent biohydrolysis using *Nocardia* EH1, leading to the formation of an (*R*,*R*)-diol.

complete conversion were obtained using bacterial EHs. Satisfactory results were described for the first time by Faber's group after a screening of 18 strains. Lyophilized biomass of *Nocardia* EH1 proved to be the best biocatalyst, leading to almost deracemization of racemic *cis*-2,3-epoxy-heptane at a concentration of 16 g/l, thus producing the corresponding (*R*,*R*)-diol in 79% overall yield and 91% ee in gram-scale amounts (Figure 8.23) [132]. It was shown by ^{18}O labeling that the hydrolysis of both enantiomers occurred with opposite regioselectivity via an attack at the (*S*)-configured oxirane carbon atom with concomitant inversion of the configuration for both enantiomers. It is worth mentioning that, although the *trans*-isomers were easily hydrolyzed by all the tested strains, no enantioconvergent hydrolysis occurred. Based on these results, the same group developed several preparative syntheses of natural products such as antitumor agents (two stereoisomers of panaxytriol), (+)-pestalotin, and a constituent of Jamaican rum (see Chapter 7).

Recently, Reetz and coworkers reported on a laboratory evolution experiment with the *A. niger* LCP 521 EH. Several EH variants were isolated that accepted *trans*-2-benzyl-3-methyloxirane as a substrate, in contrast to the wild-type EH, which was found to be inactive. Two variants exhibited high enantioselectivity (*E*-value >200) toward this epoxide [51]. Later, using the same substrate, enantioconvergence was detected with one of these selected mutants, that is, enantiopure (2*R*,3*S*)-diol at a high conversion ratio of 92% was formed. Directed evolution with this mutant identified second-generation mutants showing higher reaction rates while maintaining the enantioconvergence [53].

Trisubstituted Epoxides

One of the first examples of biohydrolytic kinetic resolution of trisubstituted epoxides was the synthesis of both enantiomers of Bower's compound, which is a potent analogue of the insect juvenile hormone [133]. The enantioselective biohydrolysis of the racemic Bower's compound using whole cells of *A. niger* LCP 521 as a biocatalyst resulted in the formation of the corresponding (6*S*)-diol in 48% yield and 70% ee (Figure 8.24). The remaining (6*S*)-epoxide (96% ee) was isolated in 36% yield, and starting from this pure enantiomer, its antipode was easily prepared by chemical means in two steps. Interestingly, it was shown that the (6*R*)-enantiomer was ten times more active against the yellow meal worm *Tenebrio molitor* than its antipode.

In 1996, Archer *et al.* reported a highly enantioselective chemoenzymatic resolution of *rac*-1-methyl-1,2-epoxycyclohexane using whole cells of *Corynebacterium* C12, giving rise to the (1*R*,2*S*)-epoxide in >99% ee (30% yield) and the (1*S*,2*S*)-diol in 92% ee (42% yield) [134]. Further, an additional efficient chemoenzymatic deracemization process was run in tandem using the EH from *Corynebacterium* C12 and perchloric acid for the acid-catalyzed ring opening of the residual epoxide (Figure 8.25).

Although enantioconvergent biohydrolysis of trisubstituted epoxides seemed rather unlikely due to their steric bulkiness, Faber's group succeeded in finding several bacterial strains that could perform such a reaction. The biohydrolysis of three racemic trialkyl epoxides at 5–6 g/l using lyophilized cells of *Rhodococcus* and *Mycobacterium* sp. resulted in the formation of the corresponding (*R*)-diols with enantiopurities exceeding 80% ee (Figure 8.26) [135]. The same group used the EH from *Rhodococcus ruber* for the preparative biohydrolysis of two trisubstituted epoxides

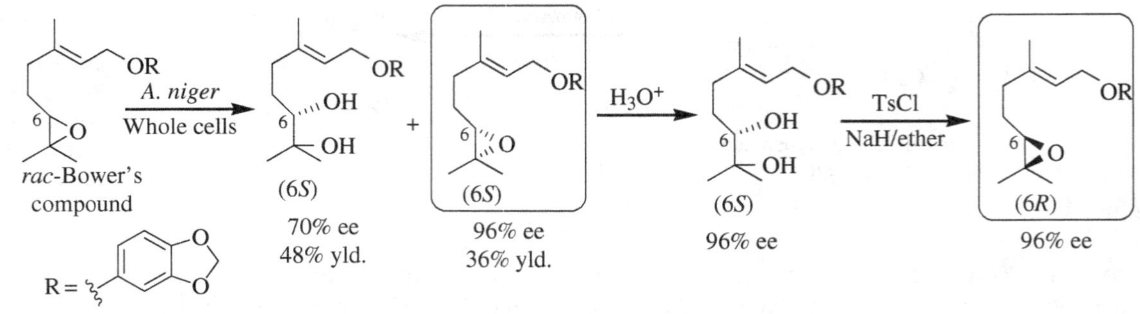

FIGURE 8.24
Synthesis of both enantiomers of Bower's compound using whole cells of *Aspergillus niger* LCP 521 as a biocatalyst.

FIGURE 8.25
Chemoenzymatic resolution and deracemization of *rac*-1-methyl-1,2-epoxycyclohexane using whole cells of *Corynebacterium* C12.

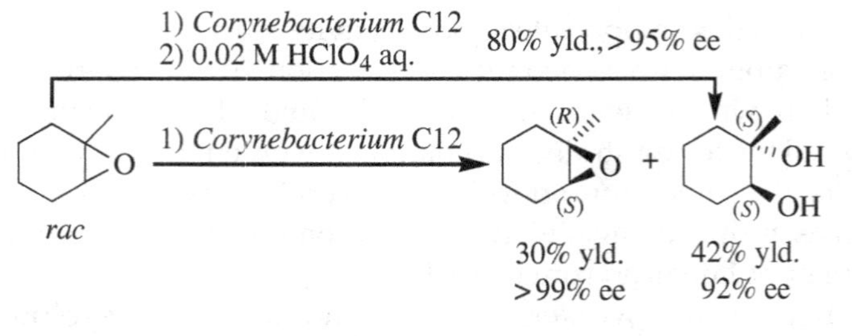

FIGURE 8.26
Deracemization of three *rac*-trialkyl oxiranes via enantioconvergent biohydrolysis using whole cells of *Mycobacterium paraffinicum* and *Rhodococcus ruber*.

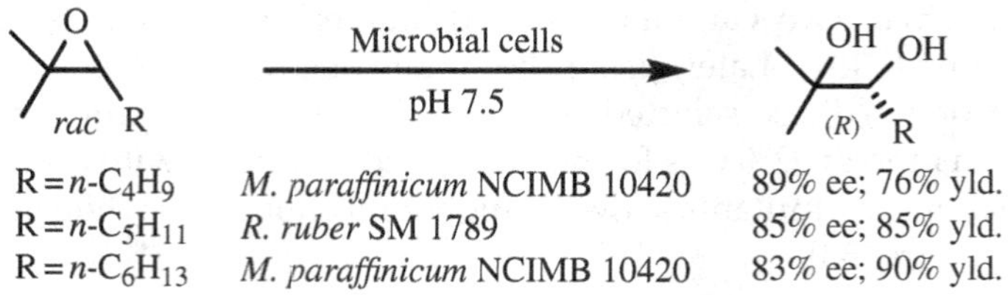

bearing an olefinic side chain with one or two double bonds. The formed corresponding diols enabled them to prepare natural compounds such as (*R*)-myrcenediol and a beer aroma constituent (see Chapter 7).

8.5.4 *meso*-Epoxides

Interestingly, desymmetrization of *meso*-epoxides can produce optically enriched vicinal (*R*,*R*)- or (*S*,*S*)-diols in 100% theoretical yield through a stereoselective attack at only one carbon atom of the oxirane. It is worth mentioning that the enantiopurity of the formed diol does not change with reaction time or conversion ratio. Desymmetrization of *meso*-epoxides by microbial biotransformations are scarce. For example, EHs from the yeast *R. glutinis* ATCC 201718 [116] and from the bacterium *Sphingomonas* sp. HXN-200 [136] have been used to stereoselectively hydrolyze cyclohexene oxide to the corresponding (*R*,*R*)-diols with enantiomeric excesses of 90% and 87%, respectively. It should be also mentioned that Botes and coworkers described in a patent various yeast strains that were able to produce optically active vicinal diols from *meso*-epoxides [137]. Recombinant *Y. lipolytica* cells expressing exogenous yeast EHs enabled them to obtain (*R*,*R*)-vicinal diols in high enantiomeric excess starting from *meso*-epoxides, such as *cis*-epoxybutane and cyclopentene oxide.

An interesting desymmetrization of a *meso*-bis-epoxide was recently described by Faber and coworkers [21]. EHs from various sources (bacteria, fungi, and plants) were found to catalyze the transformation of 6,7:9,10-bis(epoxy)pentadecane by hydrolysis/cyclization cascades, leading to different tetrahydrofuran derivatives with excellent de and ee values (Figure 8.27). The reaction pathway was initiated by EH-catalyzed hydrolysis of an oxirane moiety, followed by spontaneous ring closure of the epoxy diol intermediate. Based on these results, the authors suggested that the formation of tetrahydrofuran moieties found in numerous natural products such as acetogenins proceeds through a nucleophilic cascade mechanism starting from bis-epoxide without the involvement of a "cyclase."

Besides classical screening the following strategies were used for the discovery or generation of EHs capable of desymmetrization of *meso*-epoxides. The first was based on high-throughput screening of DNA libraries, which were generated from environmental samples. Using this approach, 50 novel microbial EHs have been discovered, and among these, 11 were able to desymmetrize not only cyclic *meso*-epoxides (cyclopentyl and cyclohexyl epoxides) but also bulky internal epoxides such as *cis*-stilbene oxide with various substituents including dipyridyl analogues (Figure 8.28) [22]. The second strategy comprised a screening of genomic databases for the presence of EH-encoding genes [30]. Five recombinant EHs were found to be active toward *meso*-epoxides such as cyclohexene oxide and *cis*-2,3-epoxybutane. Finally, Reetz and coworkers applied an iterative saturation mutagenesis strategy to the LEH from

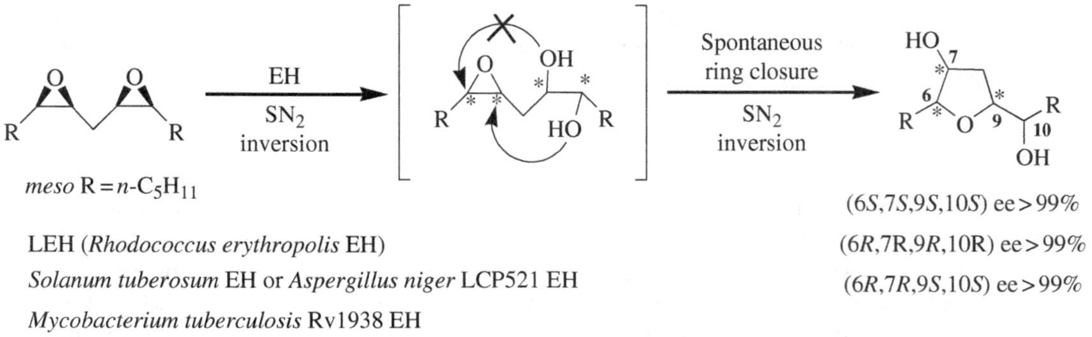

FIGURE 8.27

Stereochemical courses of EH-catalyzed hydrolysis/cyclization cascades of 6,7:9,10-bis (epoxy) pentadecane to yield tetrahydrofurane derivatives.

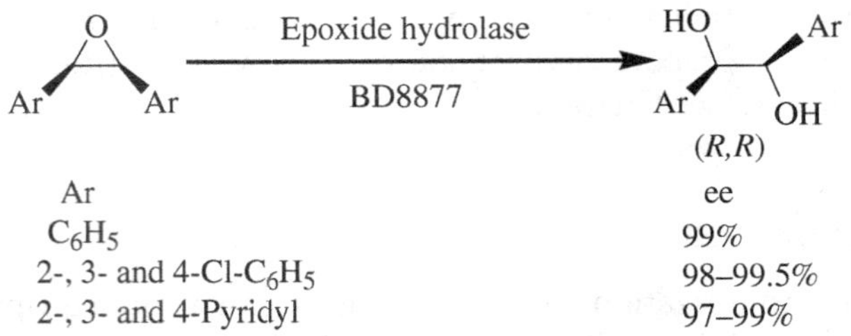

FIGURE 8.28

Desymmetrization of aryl *meso*-epoxides with BD887 obtained from DNA libraries of environmental samples.

R. erythropolis DCL 14 and obtained three LEH variants (H150, H173, H178) that catalyzed the desymmetrization of cyclic *meso*-epoxides and *cis*-1,2-homodisubstituted *meso*-epoxides with stereoselective formation of either the (*R*,*R*)-diol or the (*S*,*S*)-diol on an optional basis [56].

Li and coworkers have described the gram-scale preparative desymmetrization of cyclohexene oxide, cyclopentene oxide, and *N*-benzyloxycarbonyl-3,4-epoxypyrrolidine using the EH from *Sphingomonas* sp. HXN-200 expressed in the recombinant host *E. coli* (SpEH) [94]. Desymmetrization of 10 g of cyclohexene oxide (500 mM substrate concentration) with resting cells of *E. coli* (SpEH) (10 g cdw/l) afforded 10.3 g (89% isolated yield) of (1*R*,2*R*)-1,2-cyclohexanediol in 86% ee. Desymmetrization of the two other *meso*-epoxides (200 mM substrate concentration) afforded (1*R*,2*R*)-1,2-cyclopentanediol in 87% ee and 70.4% isolated yield and (3*R*,4*R*)-*N*-benzyloxycarbonyl-3,4-dihydroxypyrrolidine in 93% ee and 94.1% isolated yield, respectively.

8.6 PREPARATION OF VALUABLE CHIRAL BUILDING BLOCKS FOR THE SYNTHESIS OF BIOLOGICALLY ACTIVE COMPOUNDS STARTING FROM BIFUNCTIONAL EPOXIDES

8.6.1 Halogenated Epoxides

Alkyl Chloroepoxide

Epichlorohydrin

Bifunctional small molecules are particularly useful synthons that give access to various valuable biologically active products. A typical example is epichlorohydrin, which has been used [138] for the synthesis of—among others—the nutritional supplement L-carnitine; the β-adrenergic blocking agent (*S*)-atenolol; (+)-trehazolin, which is a potent inhibitor of glycosidases; and the beetle pheromone (*S*)-Ipsenol. Both (*S*)-epichlorohydrin and (*R*)-epichlorohydrin have been obtained in optically or nearly optically pure form through the kinetic resolution of the racemate at high to very high substrate concentrations using the purified EH from *A. radiobacter* overexpressed in *E. coli* [139] (Figure 8.29) or the purified EH from *Agromyces mediolanus* [140] (Figure 8.30). In the former case, the highest concentrations of racemic epichlorohydrin called for a biphasic system with 40% cyclohexane.

Aromatic Chloroepoxide

Furstoss and collaborators have shown that the partially purified recombinant EH from *A. niger* LCP 521 efficiently catalyzes the kinetic resolution of 1-chloro-2-(2,4-difluorophenyl)-2,3-epoxypropane at a very high substrate concentration of 500 g/l using a biphasic process [5]. The unreacted (*S*)-chloro-epoxide and the formed (*R*)-chloro-diol were obtained in nearly enantiopure form and nearly quantitative yield. Due to the fact that the formed (*R*)-chloro-diol was easily chemically transformed into the (*S*)-chloro-epoxide, an enantioconvergent process could be set up. Using the difference in chemical reactivity between the oxirane ring and the chlorine

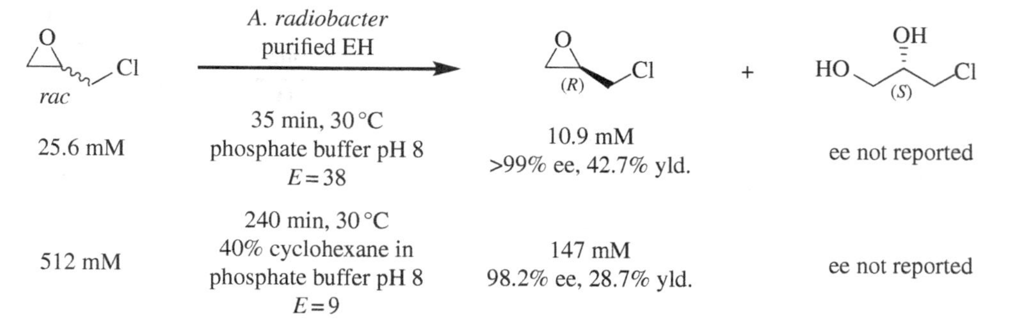

FIGURE 8.29

Kinetic resolution of racemic epichlorohydrin using purified EH from *Agrobacterium radiobacter*, leading to the formation of enantiomerically pure (*R*)-epichlorohydrin.

FIGURE 8.30

Kinetic resolution of racemic epichlorohydrin using the purified EH from *Agromyces mediolanus*, leading to the formation of enantiomerically pure (*S*)-epichlorohydrin.

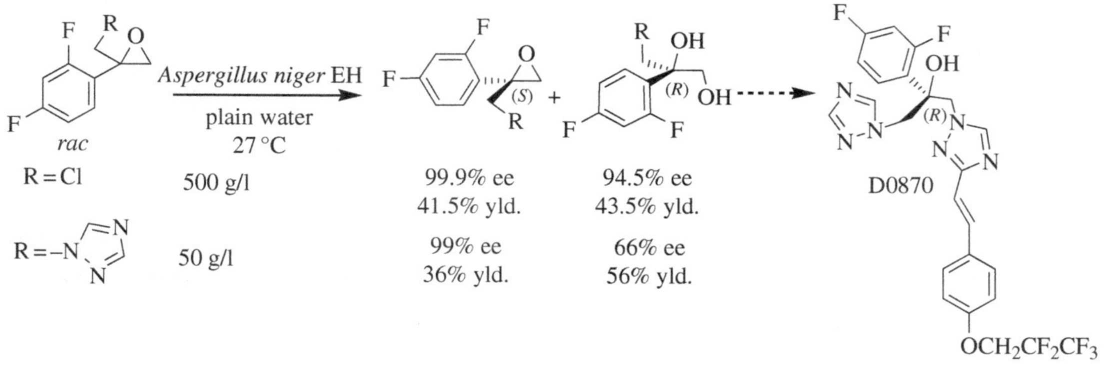

FIGURE 8.31

Preparative hydrolytic resolution of *rac*-1-chloro- and 1-triazole-2(2,4-difluorophenyl)-2,3-epoxypropane using *An*EH as a biocatalyst. The latter compound is a useful building block for the synthesis of enantiopure D0870.

substituent, it was possible to use this enantiopure chloro-epoxide as a building block for the synthesis of D0870, a triazole drug derivative known to display an efficient activity against human fungal infections (Figure 8.31). Such an enantioselective hydrolysis was also performed with the corresponding epoxytriazole compound at 50 g/l (N. Monfort, A. Archelas, and R. Furstoss, unpublished results, 2004).

Bromo Epoxide

Yarrowia lipolytica has been established as a heterologous host for EH overexpression in the company Oxyrane Ltd. [102]. The origin of some EHs that were used in various biotransformation reactions was released in different patents [35, 118, 141]. The following biotransformation reaction was performed with the *Y. lipolytica* Oxy-9 strain, which overexpressed an EH whose origin was not released. Thus, racemic 4-bromo-1,2-epoxybutane (60 g) was kinetically resolved with 5 g of wet cells in 40 ml of 0.1 M phosphate buffer at pH 7.5 [102]. The reaction was stopped after 90 min, and after extraction and purification (S)-4-bromo-1,2-epoxybutane was obtained in 26% yield and 98.6% ee (Figure 8.32). The residual epoxide was then easily transformed

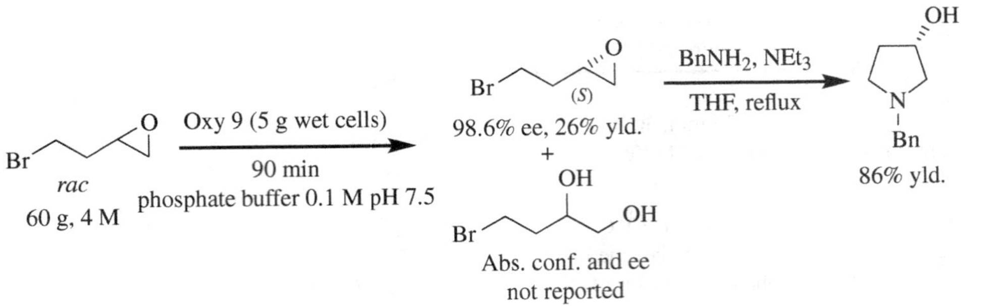

FIGURE 8.32
Kinetic resolution of racemic 4-bromo-1,2-epoxybutane using purified Oxy-9 EH and its use in the synthesis of (*S*)-*N*-benzyl-3-hydroxypyrrolidine.

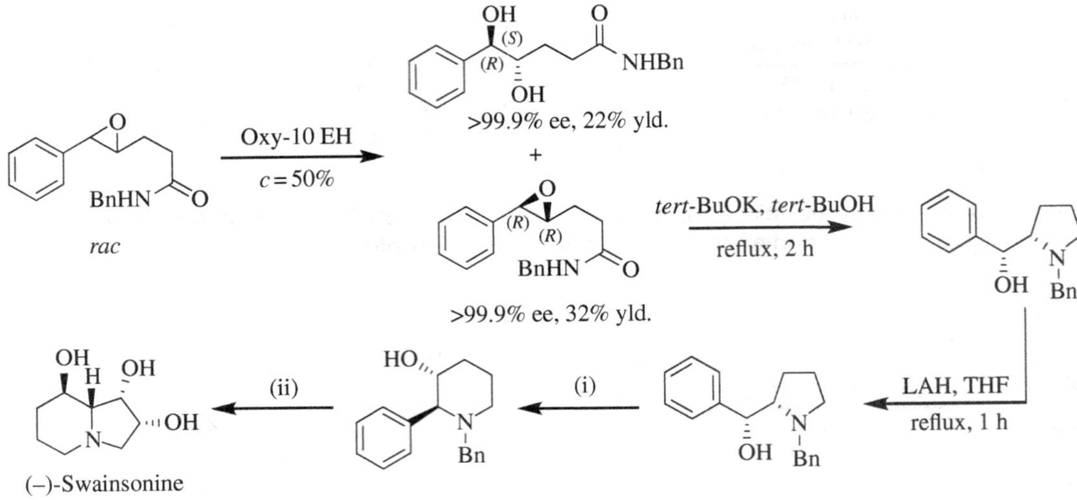

FIGURE 8.33
Kinetic resolution of racemic epoxyamide using purified Oxy-10 EH and its application to the synthesis of (*S*)-*N*-benzyl-3-hydroxypyrrolidine en route to (−)-swainsonine. (i) See Calvez *et al.* [143]; (ii) see Haddad and Larchevêque [144], and Ferreira *et al.* [145].

into a hydroxy pyrrolidine derivative, which was used as a chiral synthon for the synthesis of the calcium antagonist barnidipine [142].

8.6.2 Epoxyamide

Using another EH overexpressed in *Y. lipolytica* (Oxy-10), the kinetic resolution of an epoxyamide was performed at 50% conversion, resulting in the optically pure residual (*R,R*)-epoxide and the formed *threo*-diol with yields of 32% and 22%, respectively [102] (Figure 8.33). The remaining (*R,R*)-epoxide is of synthetic interest, because it is used for the synthesis of the α-mannosidase inhibitor (−)-swainsonine.

8.6.3 Protected Epoxy Alcohols

With yet another overexpressed EH in *Y. lipolytica* (Oxy-3) (*S*)-4-acetoxy-buten-1,2-oxide was obtained in 20% yield and 97.8% ee starting from the racemate [102]. The (*S*)-epoxide can easily be converted to (*S*)-3-hydroxytetrahydrofuran (Figure 8.34), a compound used for the synthesis of HIV protease inhibitors amprenavir and fosamprenavir [147].

Another protected epoxy-alcohol, *rac*-2-methylglycidyl benzyl ether, has been the subject of numerous investigations. Two types of EHs proved to be useful: those from various *Rhodococcus* strains [148–151] and the one from *Bacillus subtilis* [6, 152].

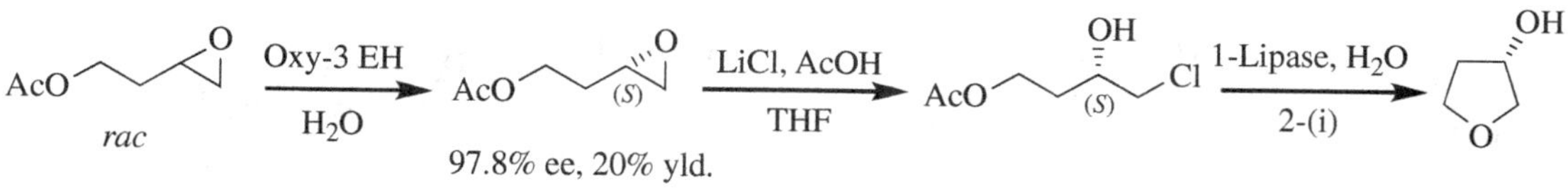

FIGURE 8.34

Chemoenzymatic access to (*S*)-3-hydroxytetrahydrofuran. (i) See Yuasa and Tsuruta [146]; 79% yield

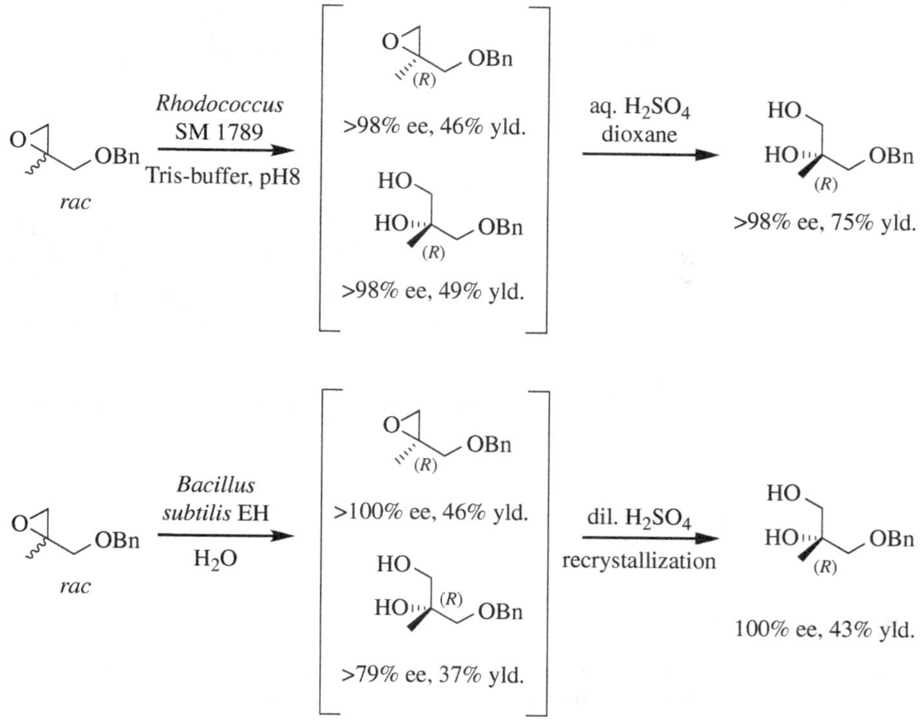

FIGURE 8.35

Enantioconvergent chemoenzymatic access to (*R*)-3-benzyloxy-2-methylpropane-1,2-diol. See the note and Figure 8.36 concerning the correct determination of the absolute configuration of epoxides.

Both types of EHs exhibited the same enantiopreference, generating the (*R*)-diol and leaving behind the unreacted (*R*)-epoxide (see note below), and were used for the enantioconvergent access to the highly enantio-enriched (*R*)-diol (Figure 8.35). Enantiopure (*R*)-3-benzyloxy-2-methylpropane-1,2-diol was the starting material for the synthesis of (*R*)-bicalutamide, a synthetic antiandrogen [6]. The same (*R*)-diol was used as the starting material for the bio-assisted synthesis of the intermediate (*R*)-3-hydroxy-3-methyl-5-hexanoic acid *p*-methoxybenzyl ester, which was then used in the synthesis of taurospongin A [153], a natural product inhibiting DNA polymerase β and HIV reverse transcriptase.

Note: several authors mentioned in their publications that the biohydrolysis led to the formation of a diol with an inverted absolute configuration when compared to the configuration of the reacting epoxide, the reason being a switch in the Cahn–Ingold–Prelog (CIP) priority and not an inversion at the stereogenic center. This conclusion is wrong according to the CIP rule applied to chiral centers in a ring [154, 155]. The application of this rule ("Each cyclic ligand is detached at the distal end, where it doubles back upon the chiral center; the chiral center is placed as a phantom atom (with no further ligands) at the end of the chain so generated"; see [154]) leads to the conclusion that the absolute configuration at the stereogenic centers of the epoxides should be reversed in these publications [6, 148–152] (Figure 8.36). In these cases, the (*R*)-diol is formed by the retention of the configuration of the (*R*)-epoxide, the opening of the oxirane ring thus occurring at the terminal carbon atom.

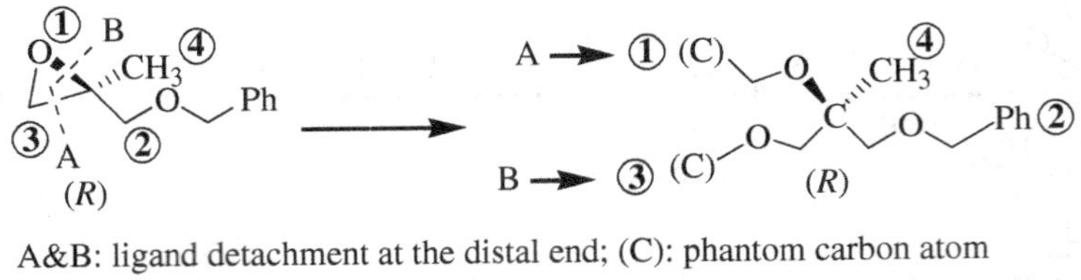

FIGURE 8.36
Application of the CIP rule for the determination of the absolute configuration of an epoxide.

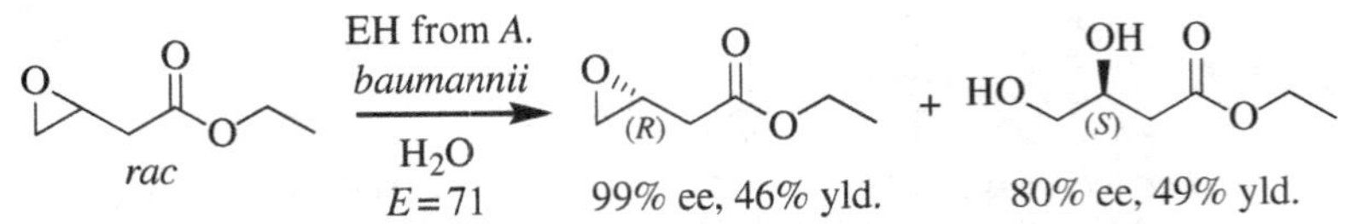

FIGURE 8.37
Kinetic resolution of *rac*-ethyl-3, 4-epoxybutyrate by *Acinetobacter baumannii*.

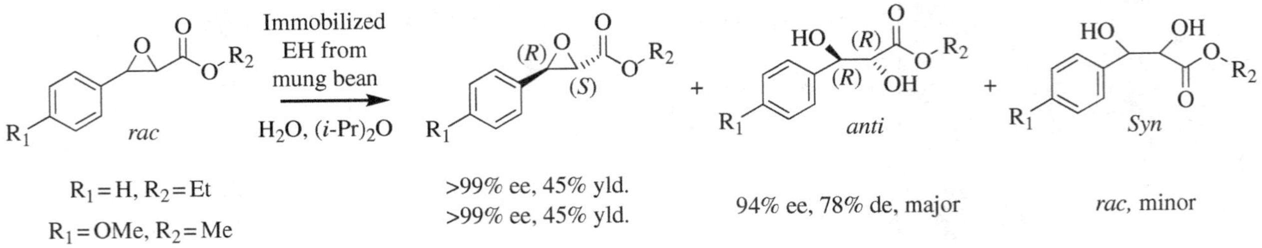

FIGURE 8.38
Kinetic resolution of two *trans*-(±)-3-phenyl glycidates by immobilized mung bean EH.

8.6.4 Epoxy Ester

As a result of a screening for microorganisms with high EH activities using enrichment cultures with alkenes as the sole carbon source, Choi *et al.* isolated a bacterial strain that was identified as *Acinetobacter baumannii* and showed high selectivity in the kinetic resolution of ethyl-3,4-epoxybutyrate [156]. Using the wet cells of *A. baumannii* ethyl-3,4-epoxybutyrate was kinetically resolved at a concentration of 60 mM, affording, after 2 h of reaction, (*R*)-ethyl-3,4-epoxybutyrate (ee > 99%) in 46% yield and (*S*)-ethyl-3,4-dihydroxybutyrate (ee = 80%) in 49% yield (Figure 8.37). The unreacted (*R*)-epoxide is a valuable intermediate that can be used in the synthesis of (*R*)-GABOB, (*R*)-carnitine, the anticancer agent lobatamide C, and the statin Lipitor®.

Other bifunctional epoxy esters such as some *trans*-(±)-3-phenyl glycidates have been studied as substrates for EH-mediated reactions [157] in order to obtain the useful chiral synthons ethyl (2*S*,3*R*)-*trans*-3-phenylglycidate and methyl (2*S*,3*R*)-*trans*-3-(4-methoxyphenyl)glycidate, which can be used to gain access to the taxol side chain and Diltiazem, respectively (Figure 8.38). The EH from the mung bean was immobilized in a gelatin gel for this kinetic resolution, and diisopropyl ether was used as an immiscible organic cosolvent to minimize spontaneous hydrolysis of the substrates. In both cases, the (2*S*,3*R*)-glycidate ester was obtained in 45% yield (90% of the theoretical value) and very high enantiomeric purity (>99%).

8.6.5 Epoxy Aldehyde

Glycidyl acetal derivatives, that is, C3 chiral building blocks bearing one stereogenic center and two different and chemically differentiable functions (such as protected aldehyde and epoxide) located on a short carbon skeleton, are of particular interest due their high chemical versatility. Five enantiopure glycidyl acetal derivatives were prepared by using the partially purified recombinant EH from *A. niger* LCP 521 as a biocatalyst [81]. All the epoxides of (*R*) absolute configuration were prepared in high enantiomeric excess (ee > 99%), whereas the formed diols were of (*S*) absolute configuration and showed moderate to excellent ees (45–97%) (Figure 8.39). The *E*-values were shown to be modest to excellent, depending on the structure of the acetal moiety. The best results were obtained when the protecting group was a cyclic acetal (*E*-value of 126) and a diisopropyl acetal (*E*-value >200). As a proof of principle a 50 g scale resolution of glycidaldehyde 2,2-dimethyltrimethylene acetal was performed at 200 g/l, leading to 22.8 g of residual (*R*)-epoxide (ee > 98%) and 25.4 g of the formed (*S*)-diol (ee = 92%). It is worthwhile to note that this enzymatic transformation was performed as a biphasic process using solely demineralized water as a solvent.

With the aim of exploring the metabolism of fructose, Bolte and coworkers synthesized 4-deoxy-D-fructose 6-phosphate in four steps, which included two enzymatic reactions [158] starting from racemic 1,1-diethoxy-3,4-epoxybutane. In the first step, optically pure (3*S*)-1,1-diethoxy-3,4-epoxybutane was obtained using an EH-catalyzed kinetic resolution. After a screening of several fungal strains, the best results were obtained with the *A. niger* LCP 521 EH. Although the enantioselectivity was only moderate (*E*-value of 15), a large-scale resolution with 15 g of substrate at 50 g/l enabled generation of 4.5 g of the (*S*)-residual epoxide in high enantiomeric excess (98% ee). Opening of the epoxide with inorganic phosphate, followed by deprotection of the acetal moiety in acidic conditions, led to the formation of enantiopure (*S*)-2-hydroxy 4-oxobutyl 1-phosphate. In the last step, the transketolase-catalyzed reaction of this enantiomer with L-erythrulose enabled the stereochemical control of the second asymmetric center of the formed enantiopure 4-deoxy-D-fructose 6-phosphate. Recycling of the NADH cofactor, required for shifting the equilibrium of the transketolase-mediated reaction toward the formation of 4-deoxy-D-fructose 6-phosphate by removing the formed glycolaldehyde, was achieved with formate dehydrogenase (Figure 8.40).

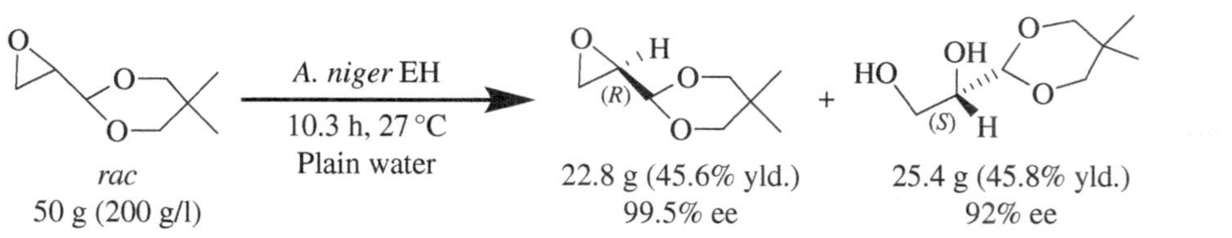

FIGURE 8.39

Gram-scale and high substrate concentration preparation of enantiopure glycidaldehyde 2,2-dimethyltrimethylene acetal using partially purified recombinant EH from *Aspergillus niger* LCP 521 as a biocatalyst.

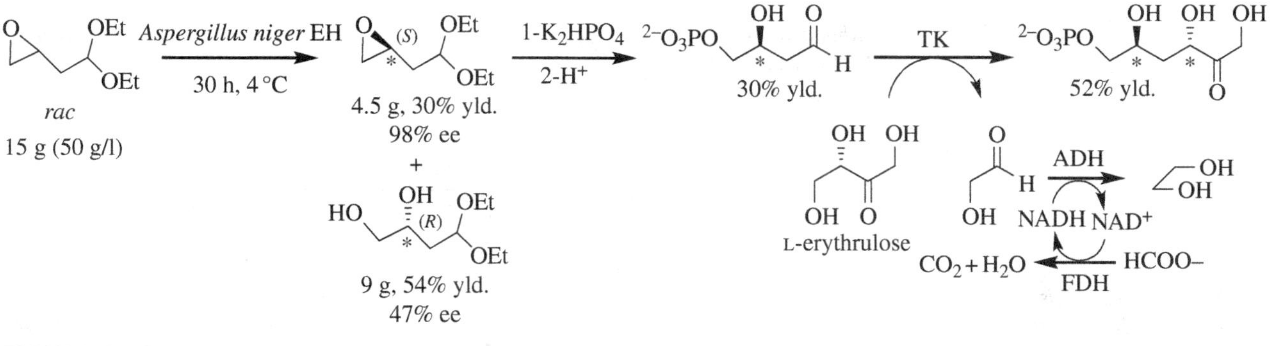

FIGURE 8.40

Synthesis of enantiopure 4-deoxy-D-fructose 6-phosphate using two enzymatic steps.

8.7 APPLICATION TO NATURAL PRODUCT SYNTHESIS

8.7.1 Disparlure

Several syntheses of natural products have been achieved with EHs as the key steps to introduce chirality. One of the first was the synthesis of the biologically active enantiomer of disparlure, the sex pheromone of the moth *Lymantria dispar*, which causes severe damage to trees in some parts of the world [159]. The synthesis was based on the kinetic resolution, mediated by a *Pseudomonas* strain, of *rac*-9,10-epoxy-15-methyl hexadecanoic acid, a precursor of the pheromone. The residual epoxide was then chemically converted in a Kolbe reaction into (+)-disparlure in 95% ee (Figure 8.41).

8.7.2 Linalool

Enantiopure *trans*- and *cis*-linalool oxides, which are constituents of several plants and fruits, are among the main aroma components of oolong and black teas. These oxides were prepared from (3*RS*,6*R*)-2,3-epoxylinalyl acetate via a chemoenzymatic route [160]. The key step was the separation of the diastereomeric mixture of the starting compound using *Rhodococcus* sp. NCIMB 11216, yielding diol and remaining epoxide in excellent diastereomeric excess (de>98%). Chemical transformations followed, which gave both linalool oxide isomers on a preparative scale in excellent diastereomeric and enantiomeric purities. From a mechanistic point of view, it was shown by means of ^{18}O labeling that the enzyme-catalyzed reaction proceeded in an enantioconvergent fashion, leading to the (3*S*,6*R*)-diol at 100% conversion (Figure 8.42).

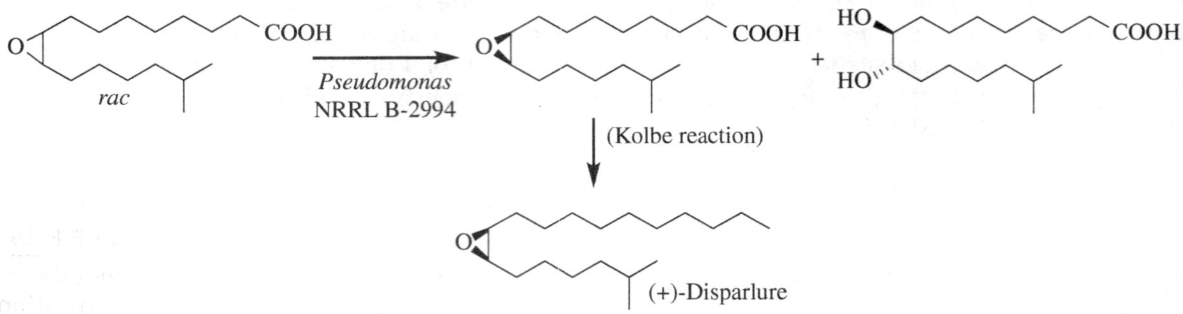

FIGURE 8.41
Chemoenzymatic synthesis of disparlure.

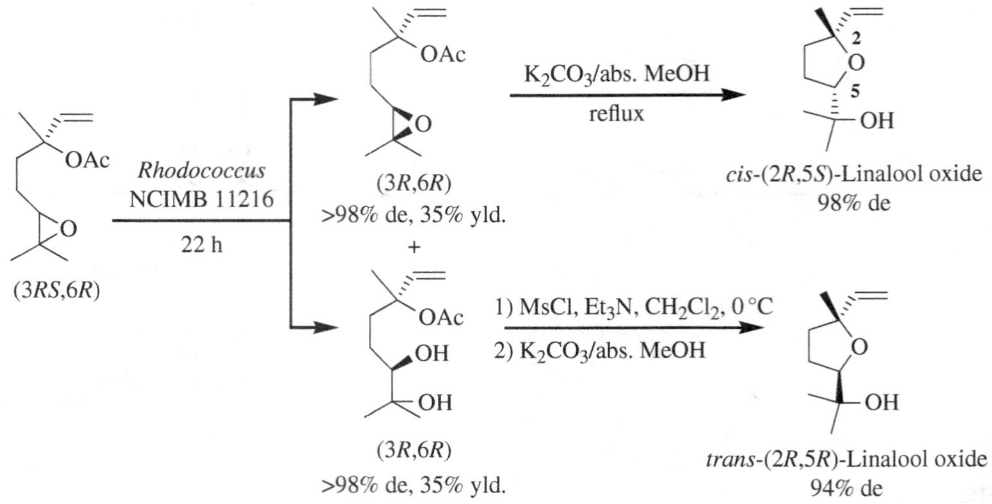

FIGURE 8.42
Chemoenzymatic synthesis of *cis*- and *trans*-linalool oxide using bacterial EH.

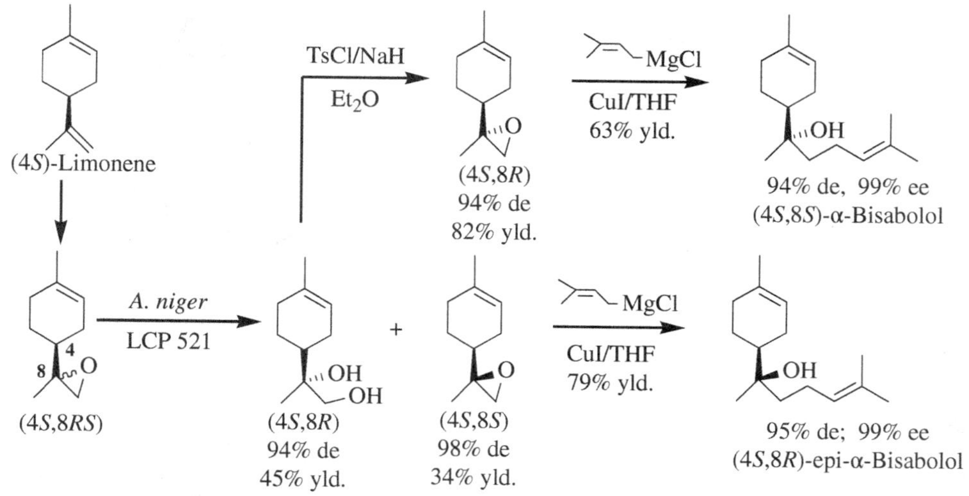

FIGURE 8.43 Diastereoselective biohydrolysis of (4*S*,8*RS*)-limonene oxide using whole cells of the strain *Aspergillus niger* LCP 521 as a biocatalyst, leading to the synthesis of α-bisabolols.

8.7.3 Bisabolol

Using the fungal strain *A. niger* LCP 521, Furstoss and collaborators described the synthesis of the four stereoisomers of bisabolol, one of them, (−)-(4*S*,8*S*)-α-bisabolol, being of industrial value for the cosmetic industry [13]. The approach was based on the diastereoselective biohydrolysis of a mixture of 8,9-epoxylimonene diastereomers, obtained from (*S*)- or (*R*)-limonene, implying whole cells of *A. niger* LCP 521. The biohydrolysis of the racemic mixture of (4*S*,8*RS*)-epoxides resulted in the unreacted (4*S*,8*S*)-epoxide and the (4*S*,8*R*)-diol (Figure 8.43). The former was chemically converted into (4*S*,8*R*)-epi-α-bisabolol, and the diol was cyclized to give the corresponding (4*S*,8*R*)-epoxide, which led to the synthesis of (4*S*,8*S*)-bisabolol. These two bisabolol stereoisomers were isolated in high enantiomeric (99% ee) and diastereomeric excess (94% de).

8.7.4 Frontalin

In 1997, Faber and coworkers described the synthesis of (*S*)-(−)-frontalin with 94% ee, an important aggregation pheromone of pine beetles of the *Dendroctonus* family. The synthesis was performed in five steps (but with rather low overall yield) via a chemoenzymatic route, which implied an enzymatic kinetic resolution of an epoxide [16]. In the key step racemic 2-methyl-2-(pent-4-en-1-yl)oxirane was enzymatically resolved with an *E*-value of 39 using freeze-dried whole cells of *Rhodococcus equi* IFO 3730. The reaction was stopped at 18% conversion, enabling the isolation of the formed (*S*)-diol in 94% ee. The latter compound was chemically transformed via Wacker oxidation and subsequent ketalization in a one-pot reaction into the pheromone (*S*)-(−)-frontalin (Figure 8.44). An improvement of the synthesis was reported in a review 3 years later [161]. The improvement was based on the fact that racemic 2-methyl-2-(pent-4-en-1-yl)oxirane could be more efficiently resolved with an *E*-value >200 using the *Nocardia* EH1 biocatalyst instead of the *Rhodococcus* EH [162]. A subsequent chemical hydrolysis of the remaining (*R*)-epoxide using sulfuric acid in dioxane with a trace amount of water afforded (*S*)-2-methyl-hept-6-ene-1,2-diol in 97% yield and 99% ee. Therefore, (*S*)-frontalin was obtained in higher yield and ee value using the same route.

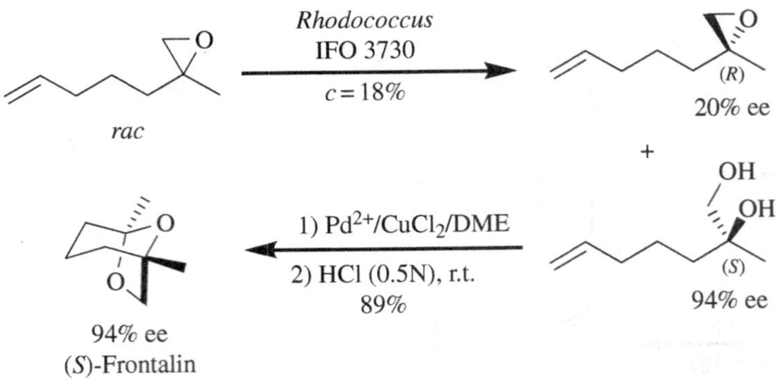

FIGURE 8.44
Key steps for the chemoenzymatic synthesis of (*S*)-frontalin using whole cells of *Rhodococcus* IFO 3730.

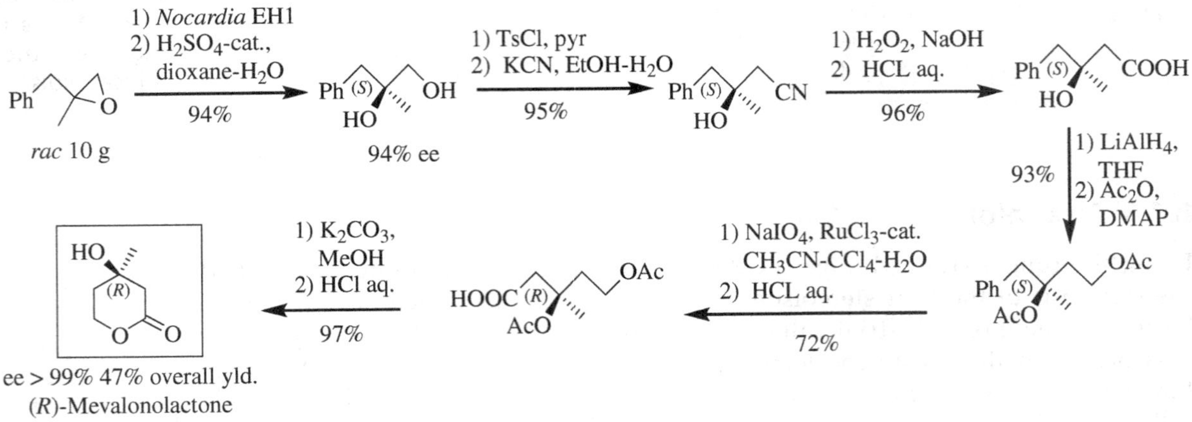

FIGURE 8.45
Asymmetric chemoenzymatic total synthesis of (*R*)-mevalonolactone.

8.7.5 Mevalonolactone

(*R*)-Mevalonolactone, a naturally occurring intermediate in a broad spectrum of cellular processes and their regulation, was synthesized in eight steps in 55% overall yield and >99% optical purity in an enantioconvergent chemoenzymatic process (Figure 8.45) [18]. In the key step, 10 g of 2-benzyl-2-methyl oxirane were deracemized using lyophilized cells of *Nocardia* EH1, followed by aqueous sulfuric acid treatment in dioxane. In this way, the substrate could be resolved with high enantioselectivity (*E*-value = 123); the subsequent chemical hydrolysis of the remaining (*R*)-epoxide resulted in the (*S*)-diol in 94% chemical yield and 94% ee, thus establishing an efficient enantioconvergent process. The limiting step in this total synthesis was the ruthenium tetroxide-based oxidation of the aryl moiety to form the carboxylic acid. Saponification followed by acidic lactonization, resulting in (*R*)-mevalonolactone in 47% overall yield and 94% ee.

8.7.6 Myrcenediol and Beer Aroma

Enantioconvergent preparative-scale production of (*R*)-myrcenediol, a plant constituent isolated from the roots of *Bidens graveolens* and from flowers of *Tanacetum annuum*, was accomplished in one step from *rac*-6,7-epoxy-7-methyl-3-methylene-1-octene using lyophilized cells of *R. ruber* DSM 44540 [19], affording (*R*)-myrcenediol in 92% yield and 83% ee (Figure 8.46).

Chemoenzymatic asymmetric total synthesis of (*S*)-7,7-dimethyl-6,8-dioxabicyclo[3.2.1]octane, which is a volatile contributor to the aroma of beer, was also accomplished using lyophilized cells of *R. ruber* DSM 44540 in an enantioconvergent enzymatic step. Inversion at the stereogenic center of (*R*)-2-methyl-7-octene-2,3-diol,

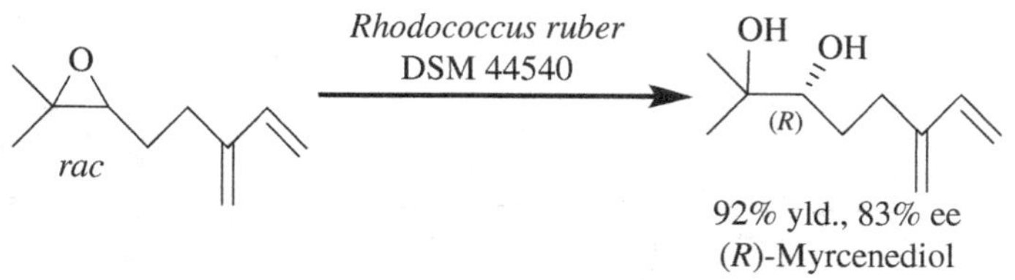

FIGURE 8.46

Synthesis of (*R*)-myrcenediol via an enantioconvergent biohydrolysis using *Rhodococcus ruber* DSM 44540 EH as a biocatalyst.

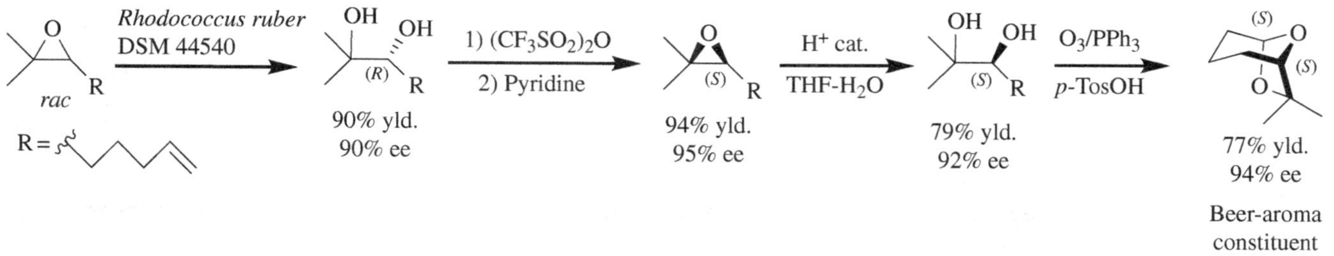

FIGURE 8.47

Enantioconvergent biohydrolytic access to a volatile contributor of beer aroma.

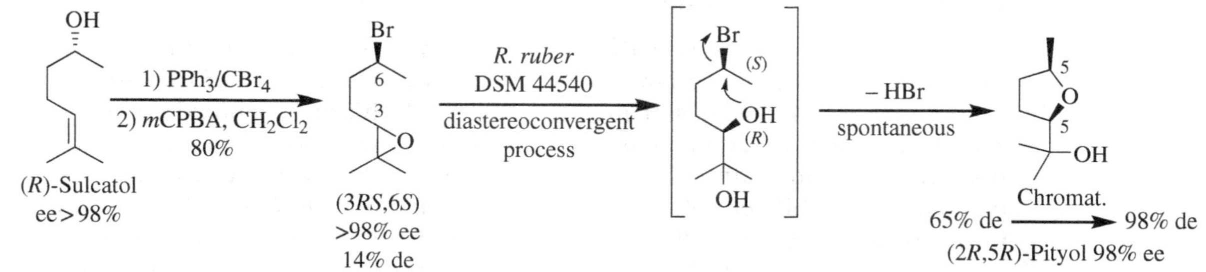

FIGURE 8.48

Synthesis of the pheromone (2*R*,5*R*)-pityol via a diastereoconvergent biohydrolysis using *Rhodococcus ruber* DSM 44540 EH as a biocatalyst.

which was obtained from the biohydrolysis of the corresponding racemic epoxide, was achieved via an epoxide closing–reopening sequence, which proceeded with an inversion and retention of absolute configuration. Oxidation of the (*S*)-diol by ozonolysis, followed by acid-catalyzed ring closure in a one-pot reaction, finally gave (*S*)-7,7-dimethyl -6,8-dioxabicyclo[3.2.1]octane in 77% yield and 94% ee (Figure 8.47).

8.7.7 Pityol

Using lyophilized cells of *R. ruber* DSM 44540, Faber and coworkers described the synthesis of two enantiomerically pure diastereoisomers of the bark beetle pheromone pityol [163]. Their approach was based on a diastereoconvergent biohydrolysis of a mixture of (3*RS*,6*S*)- or (3*RS*,6*R*)-6-bromo-2-methyl-2-heptene oxide diastereomers, which were obtained, respectively, from (*R*)- and (*S*)-sulcatol after bromination and epoxidation of the double bond. As exemplified in the succeeding text, the (3*RS*,6*R*)-6-bromo-2-methyl-2-heptene oxide mixture of diastereomers afforded upon EH catalysis the corresponding (3*R*,6*S*)-bromo-diol, which was formed as the sole intermediate. Due to the presence of a bromine atom in the molecule, this intermediate underwent spontaneous ring closure affording (2*R*,5*R*)-pityol in 54% yield and 98% ee after separation of the formed minor (2*S*,5*R*) diastereomer (12% yield) (Figure 8.48). Both enantiomers of sulcatol were prepared in enantiopure form via *Candida antarctica* lipase B-catalyzed kinetic resolution via an acyl transfer or ester hydrolysis.

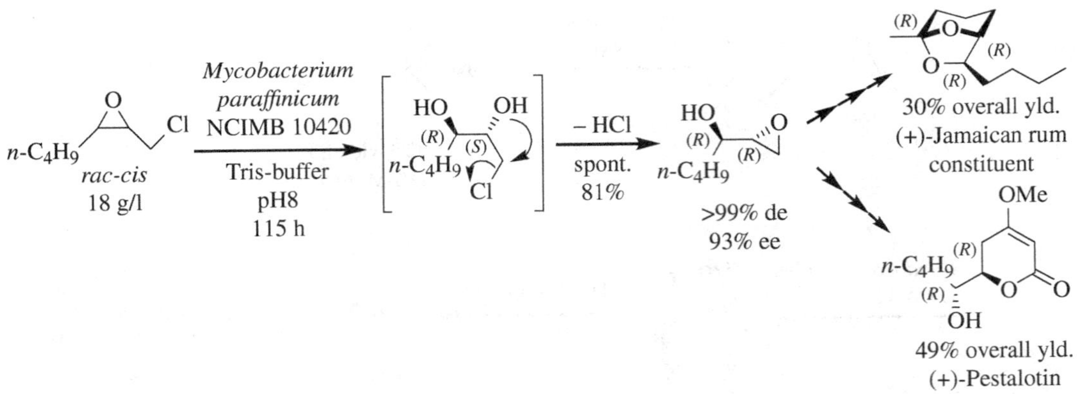

FIGURE 8.49

Enantioconvergent enzyme-triggered cascade reaction of *rac-cis*-1-chloro-2,3-epoxyheptane as a key step for the total synthesis of a (+)-Jamaican rum constituent and (+)-pestalotin.

8.7.8 Pestalotin: Jamaican Rum Constituent

In 2002 Faber and coworkers described a chemoenzymatic route to two natural products, a constituent of Jamaican rum and the (+)-antipode of the gibberellin synergist (−)-pestalotin, starting from *rac-cis*-1-chloro-2,3-epoxyheptane [17]. The strategy was based on an enzyme-triggered cascade reaction. First, an enantioconvergent biocatalytic hydrolysis of *rac-cis*-1-chloro-2,3-epoxyheptane in the presence of lyophilized cells of *Mycobacterium paraffinicum* NCIMB 10420 afforded the corresponding *threo*-(2*S*,3*R*)-diol as an intermediate. Due to the presence of a chlorine atom in the molecule, this intermediate underwent spontaneous ring closure to yield (2*R*,3*R*)-1,2-epoxy-3-heptanol in 81% yield, 99% de, and 93% ee as the sole product. This almost enantiopure epoxy alcohol was used as a building block for the total synthesis of (+)-exo-7-butyl-5-methyl-6,8-dioxabicyclo [3.2.1]octane, which is a constituent of Jamaican rum, and (+)-pestalotin (Figure 8.49).

8.7.9 Panaxytriol

A total asymmetric synthesis of two enantiopure diastereoisomeric components of *Panax ginseng* showing antitumor activity was accomplished by connecting two chiral building blocks, which were obtained via two distinct enantioconvergent enzymatic processes implying EH and lipase as biocatalysts [20]. The first building block was obtained via an enantioconvergent biocatalytic hydrolysis of *rac-cis*-1-chloro-2,3-epoxydecane, using lyophilized cells of *Streptomyces sp.* FCC008 as a biocatalyst. The corresponding *threo*-(2*S*,3*R*)-diol was formed as an intermediate that underwent acidic spontaneous ring closure to yield (2*R*,3*R*)-1,2-epoxy-3-decanol in 92% yield and 95% ee as the sole product. For the second building block, a *C. antarctica* lipase B-catalyzed kinetic resolution via acyl transfer or ester hydrolysis furnished separately both enantiomers of hept-1-ene-4,6-diyn-3-ol. In addition, both kinetic resolutions were linked to an *in situ* inversion of configuration to avoid the occurrence of the unwanted enantiomer. Thus, (*R*)- and (*S*)-hept-1-ene-4,6-diyn-3-ol were obtained in >98% ee from the racemate in 82% and 81% yields, respectively. The last step of the synthesis of (3*R*,9*R*,10*R*)- and (3*S*,9*R*,10*R*)-panaxytriol constituted the protection of the free hydroxyl of the first building block, leading to (2*R*,3*R*)-*tert*-butyldimethylsilyloxy-1,2-epoxy-3-decane and its chemical coupling with (*S*)- or (*R*)-hept-1-ene-4,6-diyn-3-ol (Figure 8.50).

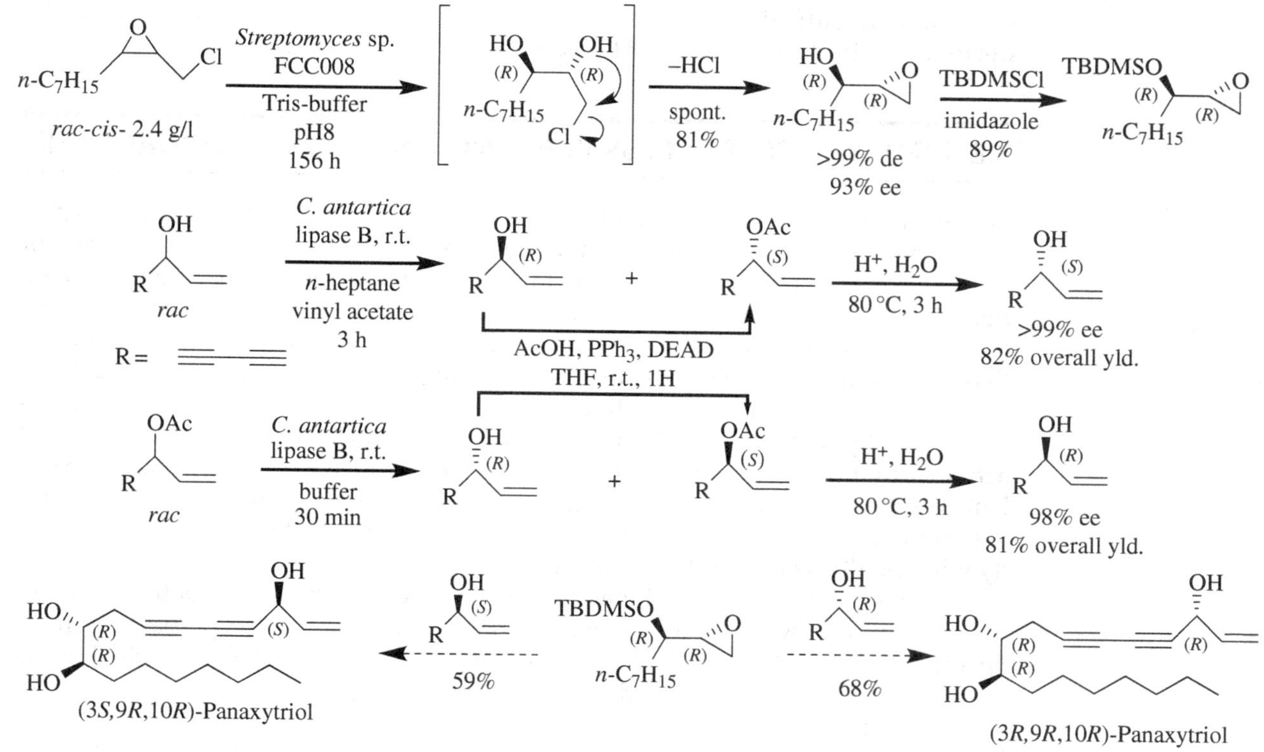

FIGURE 8.50

Synthesis of (3*R*,9*R*,10*R*)- and (3*S*,9*R*,10*R*)-panaxytriol via two distinct enantioconvergent enzymatic processes implying EH from *Streptomyces* sp. FCC008 and *Candida antarctica* lipase B as biocatalysts.

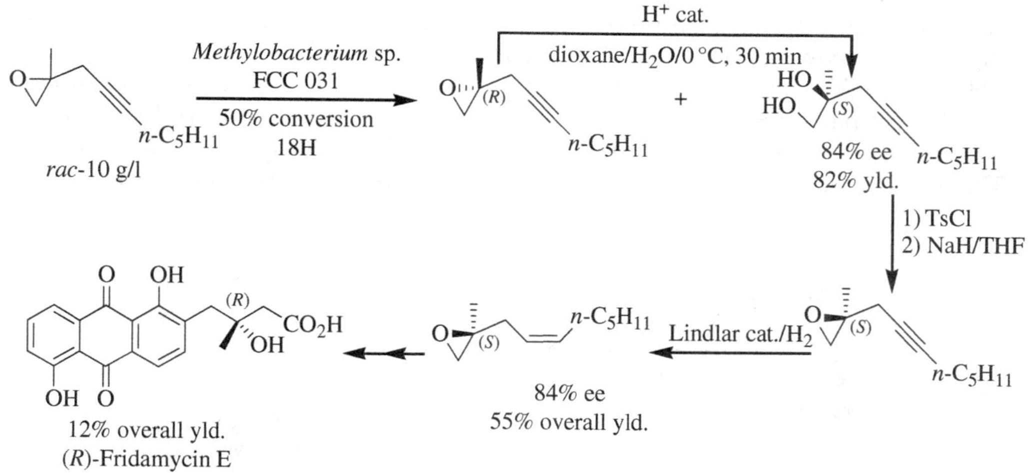

FIGURE 8.51

Asymmetric total synthesis of (*R*)-fridamycin E.

8.7.10 Fridamycin E

A total synthesis of the antibiotic (*R*)-fridamycin E was accomplished using as a key chiral building block, the (*S*)-diol product obtained through EH-catalyzed kinetic resolution of 2-methyl-2-(oct-2-yn-1-yl)oxirane, followed by stereoinversion of the remaining (*R*)-epoxide by subsequent chemical hydrolysis (Figure 8.51). The biocatalytic kinetic resolution (*E*-value = 66) was performed with lyophilized cells of *Methylobacterium* sp. FCC 031. Chemical hydrolysis of the remaining epoxide with sulfuric acid in dioxane and a trace amount of water provided the desired diol building block in 84% ee and 82% overall yield [4].

8.8 BIENZYMATIC PROCESS IMPLYING ONE EPOXIDE HYDROLASE

Enzymes work within biosynthetic or biodegradation pathways, the product of one enzyme being the substrate of the following. It is thus tempting to use two or more enzymes in a biomimetic fashion for multistep transformations of specific compounds; this is especially true for one-pot transformations, thereby simplifying the experimental procedures. Only a few examples of multienzyme processes involving EHs have been described to date. In one such example, *Sphingomonas* sp. HXN-200 was used to conduct the enantioselective *trans*-specific dihydroxylation of *N*-substituted 1,2,5,6-tetrahydropyridines (Figure 8.52) owing to both monooxygenase and EH activities found in this strain [164]. The formed *trans*-diol was obtained with an excellent ee and in good yield, suggesting that either the EH was enantioconvergent for the intermediate epoxide or the monooxygenase exhibited a high degree of selectivity with the EH being highly regioselective.

A few years later [165], a biphasic reaction was established using both a recombinant *E. coli* strain with styrene monooxygenase activity and an EH preparation from *Sphingomonas* sp. HXN-200, generating highly enantiomerically enriched diols (ee > 99%) from various styrene derivatives (Figure 8.53).

The monooxygenase was found to be particularly stereoselective, leading to the formation of (*S*)-SO in very high ee (>99%). The EH from *Sphingomonas* sp. attacked as a highly regioselective biocatalyst the less hindered carbon atom of the epoxide functionality, resulting in the formation of the corresponding (*S*)-diol.

In a continuous effort to develop a widely applicable biocatalytic system that would permit both *S*- and *R*-specific dihydroxylation of monosubstituted aryl olefins and *cis*- and *trans*-specific dihydroxylation of α,β-disubstituted aryl olefins through cascade reactions, Li's group reported the engineering of an *E. coli* strain expressing both styrene monooxygenase and either one of the two regiocomplementary EHs from *Sphingomonas* sp. HXN-200 or *S. tuberosum* [166]. The two *E. coli* strains, which were named SSP1 and SST1, were first tested as resting cells for the biotransformation of 15 variously substituted styrene derivatives and styrene itself (Table 8.5). The reaction was conducted for 8 h at a 20 mM substrate concentration in a biphasic

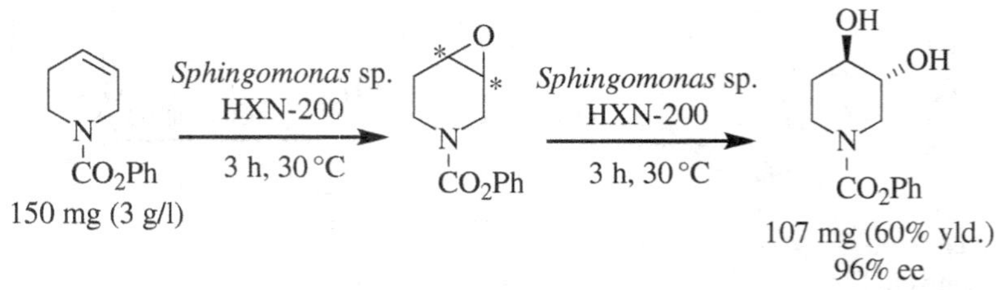

FIGURE 8.52

trans-Specific dihydroxylation of *N*-substituted 1,2,5,6-tetrahydropyridines using *Sphingomonas* sp. HXN-200 exhibiting monooxygenase and EH activities.

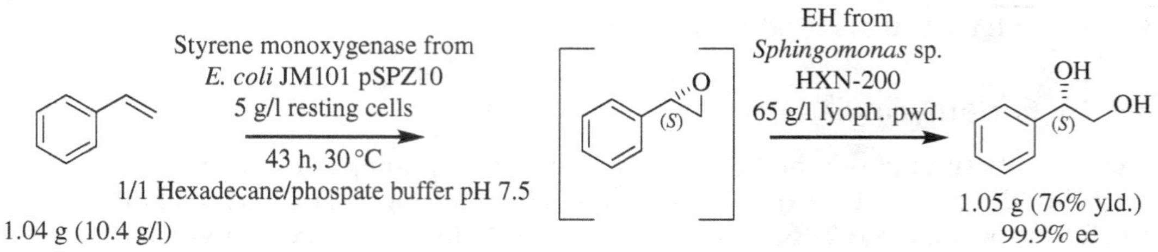

FIGURE 8.53

trans-Specific dihydroxylation of styrene using *Escherichia coli* JM101 (pSPZ10) with styrene monooxygenase activity and the EH from *Sphingomonas* sp. HXN-200 in a biphasic system.

TABLE 8.5 Enantioselective *S*- and *R*-Specific Dihydroxylation of Styrene Oxide and Some of Its Monosubstituted Derivatives

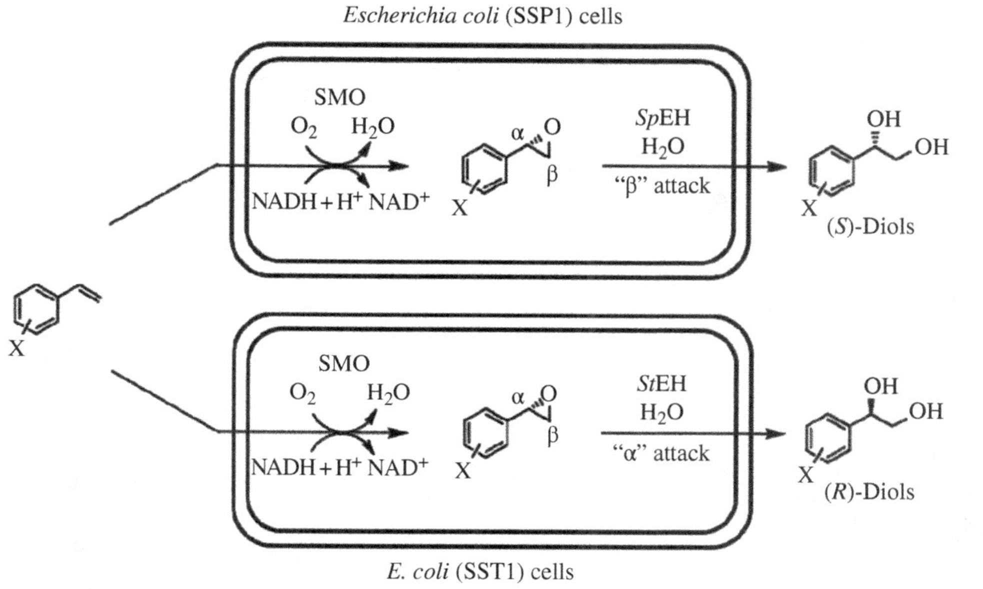

Cells	Anal. Yield %	Diol Abs. Conf./ee %	Anal. Yield %	Diol Abs. Conf./ee %	Anal. Yield %	Diol Abs. Conf./ee %	Anal. Yield %	Diol Abs. Conf./ee %
	X=H		**X=*p*-F**		**X=*p*-Cl**		**X=*p*-Br**	
SSP1	92	(*S*)/98.1	88	(*S*)/97.9	67	(*S*)/97.8	34	(*S*)/97.7
SST1	93	(*R*)/95.5	90	(*R*)/95.2	89	(*R*)/95.6	86	(*R*)/94.4
	X=*p*-CH_3		**X=*p*-OCH_3**		**X=*p*-CF_3**		**X=*m*-F**	
SSP1	86	(*S*)/93.9	67	(*S*)/83.2	25	(*S*)/97.5	>99	(*S*)/98.4
SST1	85	(*R*)/87.7	65	(*R*)/85.4	19	(*R*)/87.7	>99	(*R*)/94.2
	X=*m*-Cl		**X=*m*-Br**		**X=*m*-CH_3**		**X=*m*-OCH_3**	
SSP1	95	(*S*)/97.5	67	(*S*)/97.5	91	(*S*)/93.1	96	(*S*)/97.6
SST1	95	(*R*)/95.8	86	(*R*)/84.2	92	(*R*)/98.2	>99	(*R*)/87.3
	X=*m*-CF_3		**X=*o*-F**		**X=*o*-Cl**		**X=*o*-CH_3**	
SSP1	46	(*S*)/97.6	94	(*S*)/98.6	34	(*S*)/92.2	34	(*S*)/65.7
SST1	13	(*S*)/74.0	89	(*R*)/68.1	10	(*R*)/36.9	15	(*R*)/89.9

Adapted from Wu *et al.* [166].

(1 : 1) phosphate buffer/*n*-hexadecane reaction medium with 10 g cdw/l of either the SSP1 or SST1 biomass.

The analytical yields were mostly excellent with ees of the formed diols reaching values of up to 98.6%. The lowest yields and/or the lowest ees were obtained for styrene derivatives bearing electron-withdrawing substituents or substituents in *ortho*-positions. The use of either the SSP1 or SST1 cells enabled the preferential formation of the (*S*)-diol enantiomer and (*R*)-diol enantiomer, respectively. Using the same experimental conditions as described already, the substrate range was then extended to some di- and trisubstituted or cyclic aryl olefins. Once again, high yields and high selectivities were generally determined, however, with some exceptions (Figure 8.54).

In order to demonstrate the usefulness of the developed process as a synthetic chemistry tool, the substrate and the biomass concentrations were increased to 50 mM and 20 g cdw/l, respectively; the performance of the biotransformation set up was tested with some of the best substrates using either the SSP1 or SST1 cells. Very good isolated yields as well as high selectivities were determined, enabling the

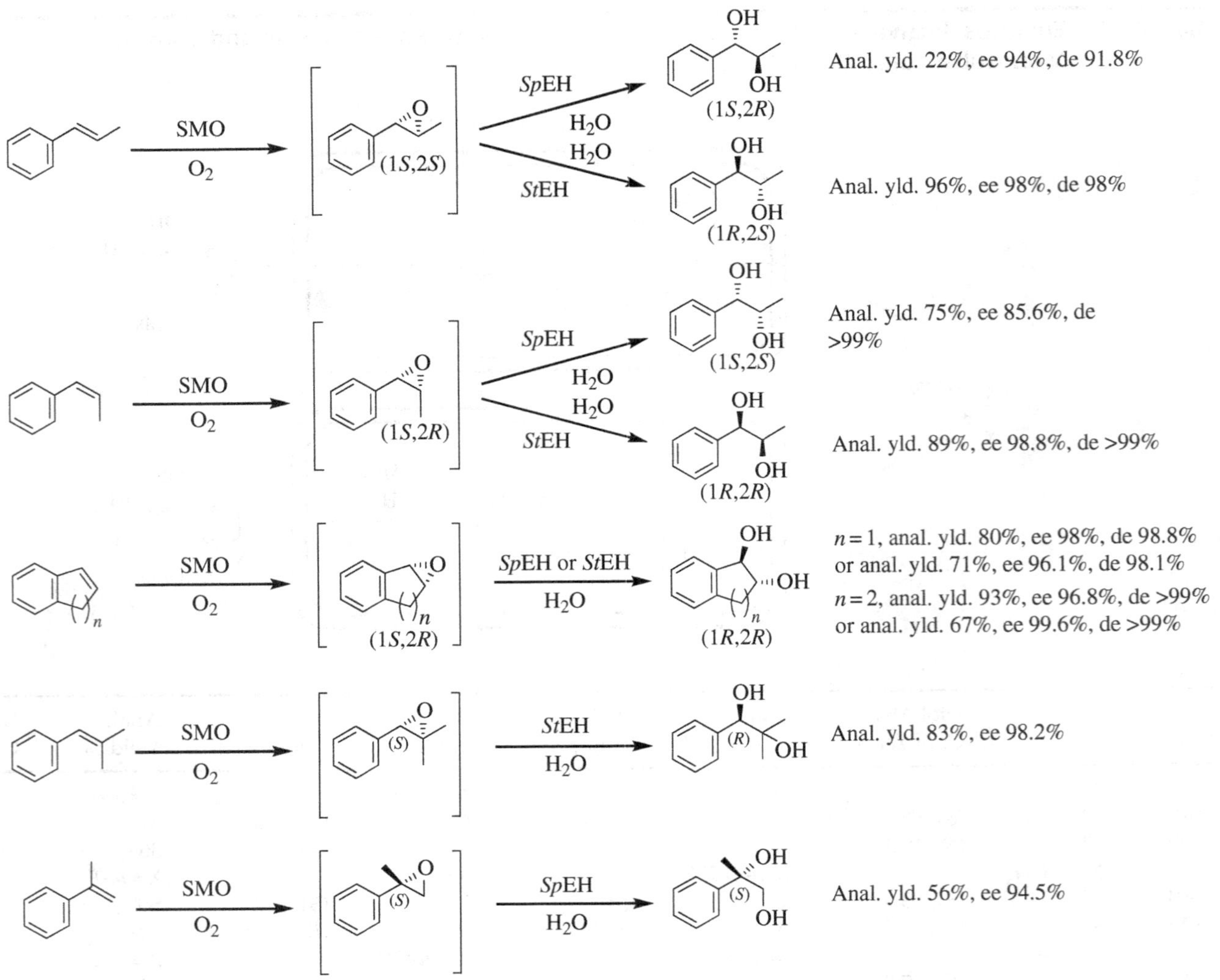

FIGURE 8.54

Enantioselective *cis-*, *trans-*, *R-*, or *S-*specific dihydroxylation of some di-, tri-, or cyclic-aryl olefins.

isolation of about 300 mg of each diol. Finally, the dihydroxylation of styrene was conducted with growing SST1 cells in a bioreactor. In this case, 120 mM (16.6 g/l) of (*R*)-1-phenyl-1,2-ethanediol was obtained in 96.2% ee with a volumetric productivity of 3.3 g/l/h, styrene being added continuously over a period of 5 h.

The same group also developed a one-pot bienzymatic sequential process for obtaining access to *trans*-diols from cyclic olefins using lipase-mediated epoxidation of olefins in the presence of H_2O_2 and lauric acid in conjunction with the EH-catalyzed hydrolysis of the formed epoxide. The method was developed with cyclohexene and *N*-benzyloxycarbonyl 3-pyrroline as substrates. The first chemoenzymatic reaction, that is, the lipase-mediated epoxidation, was conducted in acetonitrile, which was compatible with the next biocatalyst suspended in phosphate buffer, that is, whole cells from *Sphingomonas* sp. HXN-200 bearing an EH activity. The initial substrate concentration was 0.5 M, that of H_2O_2 was 1.25 M, and that of lauric acid either 0.25 M with cyclohexene as a substrate or 1.25 M with the pyrroline derivative as a substrate. 25 or 35 g/l of Novozym 435 lipase and 24 or 15 g/l of *Sphingomonas* sp. HXN-200 cells were used to obtain 9.5 mM of 1*R*,2*R*-*trans*-cyclohexane diol (84% ee) after 72 h or 9.4 mM of 3*R*,4*R*-*trans*-N-benzyloxycarbonyl 3-pyrroline diol (93% ee) after 168 h (Figure 8.55).

Another example of an EH-based bienzymatic transformation was described which involved a haloalcohol dehalogenase and 1,3-dichloro-2-propanol as the

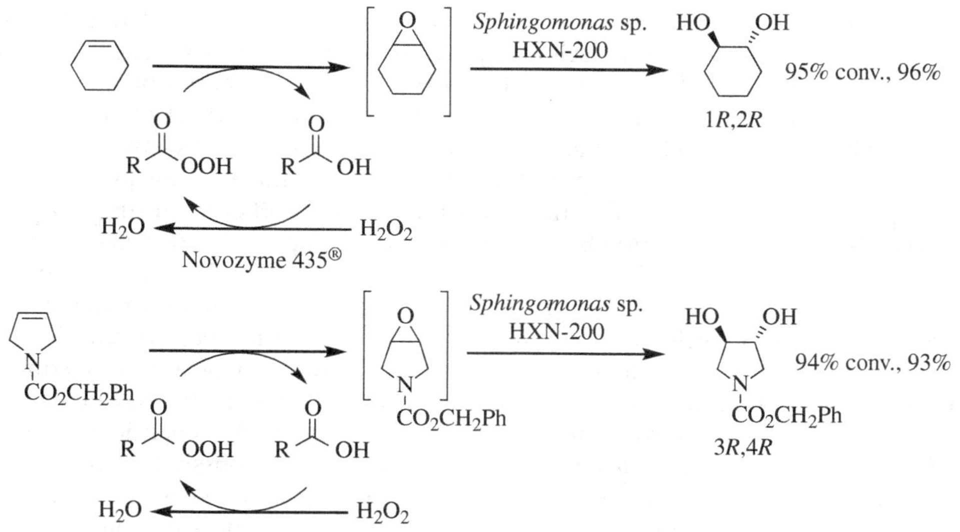

FIGURE 8.55

Asymmetric *trans*-dihydroxylation of cyclohexene and *N*-benzyloxycarbonyl 3-pyrroline.

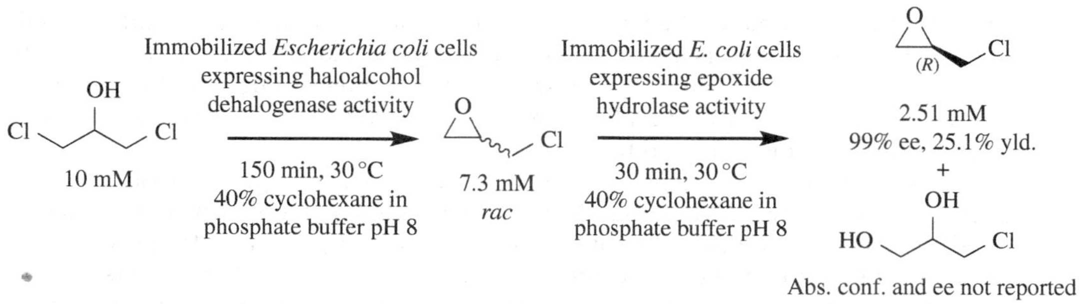

FIGURE 8.56

Access to (*R*)-epichlorohydrin using two successive biotransformation reactions in a bienzymatic system involving haloalcohol dehalogenase and EH.

substrate [167]. The two enzymes were expressed in two different *E. coli* strains. The cells were then immobilized onto perlite and placed in two distinct bubble reactor columns, the final reaction medium containing 40% cyclohexane. The first column exhibiting haloalcohol dehalogenase activity was fed with 10 mM 1,3-dichloro-2-propanol, and after 150 min the reaction reached an equilibrium, generating racemic epichlorohydrin (7.3 mM, 73% yield). The second column with EH activity was then filled with the reaction medium of the first column (without the cells), and after 30 min, 2.51 mM of essentially enantiopure (*R*)-epichlorohydrin (99% ee, 25% yield) was obtained (Figure 8.56). In summary, the EH enabled the kinetic resolution of the racemic epichlorohydrin, which was formed via the enzymatic dehalogenation of the starting compound 1,3-dichloro-2-propanol.

From those examples, it becomes apparent that enzymes that are involved in the synthesis of epoxides are particularly well suited to be used in conjunction with EHs for the generation of optically active diols or epoxides. As these compounds are highly attractive chiral synthons, one can expect that, in the future, more and more bi (or tri) enzymatic processes will be developed with EHs as the key components.

8.9 CONCLUSIONS

For more than 20 years EHs have proved to be outstanding biocatalysts in a large number of applications in fine chemistry. This success is largely due to the fact that EHs are robust enzymes, which do not need any cofactors for catalysis. Furthermore, although these enzymes catalyze a hydrolytic reaction, they are tolerant to both water-miscible and water-immiscible organic solvents. From a practical point of view, this makes possible to increase the used substrate concentration, and consequently, in a lot of cases, it helps alleviate one of the main drawbacks in using EHs, that is, substrate and product inhibition. Indeed, the presence of a water-immiscible solvent limits the substrate and product concentration in the aqueous phase, which consequently limits enzyme inhibition to a large extent. Another interesting property of EHs is their relatively large substrate spectrum as exemplified in this chapter. Even trisubstituted epoxides could act as substrates for some EHs. Numerous types of biocatalytic reactions have been performed with EHs; this should arouse the interest of many organic chemists in these catalysts in the near future. Besides the classical kinetic resolution of a racemate with its intrinsic limitation of a 50% maximum theoretical yield in both residual epoxide and formed diol, EHs can also be used in stereoselective desymmetrization of *meso*-epoxides or in the enantioconvergent hydrolysis of racemic epoxides, both processes offering the possibility to generate enantiopure diol products in theoretically 10% yield. The latter case is a characteristic of EHs and is the result of the intrinsic capacity of some of these enzymes to react with either of the two oxirane carbon atoms in an enantiomer-dependent fashion, resulting in the same diol enantiomer at complete conversion of the racemic substrate. Such a situation is also encountered in classical kinetic resolutions when the remaining epoxide is chemically transformed (e.g., under acidic conditions) into the same enantiomer of the enzymatically formed diol product or when two enantiocomplementary enzymes are used together.

The development of molecular biology tools has had a great impact on many aspects of biocatalysis. Numerous chemistry labs are now familiar with gene cloning and protein overexpression techniques, which offer the possibility to obtain, as yet, uncharacterized enzymes. Screening complex samples of high biodiversity for a specific activity, in particular, can uncover enzymes with novel properties in terms of activity, substrate range, and stereo-, regio-, and chemoselectivity. As a consequence, recombinant hosts such as *E. coli* or yeast overexpressing EHs are now routinely used in biotransformation reactions as a powerful alternative to naturally occurring EH-containing microorganisms. The high attainable level of overexpression enables extremely high substrate concentrations to be used, in conjunction with significantly shorter reaction times. Consequently, the time-consuming preparation of purified or partially purified enzymes is no longer a prerequisite for high enzymatic activities. Site-directed mutagenesis and laboratory evolution (error-prone PCR, DNA shuffling, and iterative saturation mutagenesis) further offer the possibility to modify existing enzymes in order to improve one or more enzyme characteristics. Such tailor-made catalysts can now be developed to fulfill specific needs of the chemical industry. Another emerging area is the use of multienzyme-based transformations to perform several consecutive reactions in one pot. Combining epoxide-generating enzymes with EHs in a single process or host can considerably reduce the complexity of biotransformation processes. In conclusion, EHs are extremely useful enzymes in organic synthesis due to their ease of use, robustness, ubiquity, and applicability, and will probably continue to be a field of intense research in the near future, from both an applied and a fundamental point of view.

REFERENCES

[1] Chen, C.-S., Fujimoto, Y., Girdaukas, G. and Sih, C.J. (1982) Quantitative analyses of biochemical kinetic resolutions of enantiomers. *J. Am. Chem. Soc.*, **104**, 7294–7299.

[2] Chiappe, C., Leandri, E., Hammock, B. and Morisseau, C. (2007) Effect of ionic liquids on epoxide hydrolase-catalyzed synthesis of chiral 1,2-diols. *Green Chem.*, **9**, 162–168.

[3] Rui, L., Cao, L., Chen, W., Reardon, K.F. and Wood, T.K. (2005) Protein engineering of epoxide hydrolase from *Agrobacterium radiobacter* AD1 for enhanced activity and enantioselective production of (*R*)-1-phenylethane-1,2-diol. *Appl. Environ. Microbiol.*, **71**, 3995–4003.

[4] Ueberbacher, B.J., Osprian, I., Mayer, S.F. and Faber, K. (2005) A chemoenzymatic, enantio-convergent, asymmetric total synthesis of (*R*)-Fridamycin E. *Eur. J. Org. Chem.*, 1266–1270.

[5] Monfort, N., Archelas, A. and Furstoss, R. (2004) Enzymatic transformations. Part 55: Highly productive epoxide hydrolase catalysed resolution of an azole antifungal key synthon. *Tetrahedron*, **60**, 601–605.

[6] Fujino, A., Asano, M., Yamaguchi, H., Shirasaka, N., Sakoda, A., Ikunaka, M., Obata, R., Nishiyama, S. and Sugai, T. (2007) *Bacillus subtilis* epoxide hydrolase-catalyzed preparation of enantiopure 2-methylpropane-1,2,3-triol monobenzyl ether and its application to expeditious synthesis of (*R*)-bicalutamide. *Tetrahedron Lett.*, **48**, 979–983.

[7] Genzel, Y., Archelas, A., Broxterman, Q.B., Schulze, B. and Furstoss, R. (2002) Microbiological transformations. 50. Selection of epoxide hydrolases for enzymatic resolution of 2-, 3-, or 4-pyridyloxirane. *J. Mol. Catal. B: Enzym.*, **16**, 217–222.

[8] Monterde, M.I., Lombard, M., Archelas, A., Cronin, A., Arand, M. and Furstoss, R. (2004) Enzymatic transformations. Part 58: Enantioconvergent biohydrolysis of styrene oxide derivatives catalysed by the *Solanum tuberosum* epoxide hydrolase. *Tetrahedron: Asymmetry*, **15**, 2801–2805.

[9] Pedragosa-Moreau, S., Morisseau, C., Baratti, J., Zylber, J., Archelas, A. and Furstoss, R. (1997) Microbiological transformations. 37. An enantioconvergent synthesis of the *beta*-blocker ®-Nifénalol using a combined chemoenzymatic approach. *Tetrahedron*, **53**, 9707–9714.

[10] Bottalla, A.L., Ibrahim-Ouali, M., Santelli, M., Furstoss, R. and Archelas, A. (2007) Epoxide hydrolase-catalyzed kinetic resolution of a spiroepoxide, a key building block of various 11-heterosteroids. *Adv. Synth. Catal.*, **349**, 1102–1110.

[11] Pedragosa-Moreau, S., Archelas, A. and Furstoss, R. (1996) Microbiological transformations. 31. Synthesis of enantiopure epoxides and vicinal diols using fungal epoxide hydrolase mediated hydrolysis. *Tetrahedron Lett.*, **37**, 3319–3322.

[12] Zhang, J., Reddy, J., Roberge, C., Senanayake, C., Greasham, R. and Chartrain, M. (1995) Chiral bio-resolution of racemic indene oxide by fungal epoxide hydrolases. *J. Ferment. Bioeng.*, **80**, 244–246.

[13] Chen, X.-J., Archelas, A. and Furstoss, R. (1993) Microbiological transformations. 27. The first examples for preparative-scale enantioselective or diastereoselective epoxide hydrolases using microorganisms. An unequivocal access to all four bisabolol stereoisomers. *J. Org. Chem.*, **58**, 5528–5532.

[14] Cleij, M., Archelas, A. and Furstoss, R. (1999) Microbial transformations. 43. Epoxide hydrolase as tools for the synthesis of enantiopure alpha-methylstyrene oxides: a new and efficient synthesis of (*S*)-Ibuprofen. *J. Org. Chem.*, **64**, 5029–5035.

[15] Edegger, K., Mayer, S.F., Steinreiber, A. and Faber, K. (2004) Chemo-enzymatic enantio-convergent asymmetric synthesis of (*R*)-(+)-Marmin. *Tetrahedron*, **60**, 583–588.

[16] Kroutil, W., Osprian, I., Mischitz, M. and Faber, K. (1997b) Chemoenzymatic synthesis of (*S*)-(−)-frontalin using bacterial epoxide hydrolases. *Synthesis*, 156–158.

[17] Mayer, S.F., Steinreiber, A., Goriup, M., Saf, R. and Faber, K. (2002) Chemoenzymatic asymmetric total syntheses of a constituent of Jamaican rum and of (+)-Pestalotin using an enantioconvergent enzyme-triggered cascade reaction. *Tetrahedron: Asymmetry*, **13**, 523–528.

[18] Orru, R.V.A., Osprian, I., Kroutil, W. and Faber, K. (1998) An efficient large-scale synthesis of (*R*)-(−)-mevalonolactone using simple biological and chemical catalysts. *Synthesis*, 1259–1263.

[19] Steinreiber, A., Mayer, S.F. and Faber, K. (2001) Asymmetric total synthesis of a beer-aroma constituent based on enantioconvergent biocatalytic hydrolysis of trisubstituted epoxides. *Synthesis*, **13**, 2035–2039.

[20] Mayer, S.F., Steinreiber, A., Orru, R.V.A. and Faber, K. (2002) Chemoenzymatic asymmetric total synthesis of antitumor agents (3*R*,9*R*,10*R*)- and (3*S*,9*R*,10*R*)-Panaxytriol and (*R*)- and (*S*)-Falcarinol from *Panax ginseng* using an enantioconvergent enzyme-triggered cascade reaction. *J. Org. Chem.*, **67**, 9115–9121.

[21] Ueberbacher, B.T., Oberdorfer, G., Gruber, K. and Faber, K. (2009) Epoxide-hydrolase-initiated hydrolysis/rearrangement cascade of a methylene-interrupted bis-epoxide yields chiral THF moieties without involvement of a "cyclase". *ChemBioChem*, **10**, 1697–1704.

[22] Zhao, L., Han, B., Huang, Z., Miller, M., Huang, H., Malashock, D.S., Zhu, Z., Milan, A., Robertson, D.E., Weiner, D.P. and Burk, M.J. (2004) Epoxide hydrolase-catalyzed enantioselective synthesis of chiral 1,2-diols via desymmetrization of *meso*-epoxides. *J. Am. Chem. Soc.*, **126**, 11156–11157.

[23] Decker, M., Arand, M. and Cronin, A. (2009) Mammalian epoxide hydrolases in xenobiotic metabolism and signalling. *Arch. Toxicol.*, **83**, 297–318.
[24] Morisseau, C. (2013) Role of epoxide hydrolases in lipid metabolism. *Biochimie*, **95**, 91–95.
[25] Newman, J.W., Morisseau, C. and Hammock, B.D. (2005) Epoxide hydrolases: their roles and interactions with lipid metabolism. *Prog. Lipid Res.*, **44**, 1–51.
[26] Lin, S., Horsman, G.P., Chen, Y., Li, W. and Shen, B. (2009) Characterization of the SgcF epoxide hydrolase supporting an (*R*)-vicinal diol intermediate for enediyne antitumor antibiotic C-1027 biosynthesis. *J. Am. Chem. Soc.*, **131**, 16410–16417.
[27] Lin, S., Horsman, G.P. and Shen, B. (2010) Characterization of the epoxide hydrolase NcsF2 from the neocarzinostatin biosynthetic gene cluster. *Org. Lett.*, **12**, 3816–3819.
[28] Elfström, L.T. and Widersten, M. (2005) The *Saccharomyces cerevisiae* ORF YNR064c protein has characteristics of an "orphaned" epoxide hydrolase. *Biochim. Biophys. Acta*, **1748**, 213–221.
[29] Kotik, M., Štěpánek, V., Marešová, H., Kyslik, P. and Archelas, A. (2009) Environmental DNA as a source of a novel epoxide hydrolase reacting with aliphatic terminal epoxides. *J. Mol. Catal. B: Enzym.*, **56**, 288–293.
[30] van Loo, B., Kingma, J., Arand, M., Wubbolts, M.G. and Janssen, D.B. (2006) Diversity and biocatalytic potential of epoxide hydrolases identified by genome analysis. *Appl. Environ. Microbiol.*, **72**, 2905–2917.
[31] Kotik, M. (2009) Novel genes retrieved from environmental DNA by polymerase chain reaction: current genome-walking techniques for future metagenome applications. *J. Biotechnol.*, **144**, 75–82.
[32] Kotik, M., Štěpánek, V., Grulich, M., Kyslik, P. and Archelas, A. (2010) Access to enantiopure aromatic epoxides and diols using epoxide hydrolases derived from total biofilter DNA. *J. Mol. Catal. B: Enzym.*, **65**, 41–48.
[33] Grant, D.F., Storms, D.H. and Hammock, B.D. (1993) Molecular cloning and expression of murine liver epoxide hydrolase. *J. Biol. Chem.*, **268**, 17628–17633.
[34] Kamita, S.G., Oshita, G.H., Wang, P., Morisseau, C., Hammock, B.D., Nandety, R.S. and Falk, B.W. (2013) Characterization of Hovi-mEH1, a microsomal epoxide hydrolase from the glassy-winged sharpshooter *Homalodisca vitripennis*. *Arch. Insect Biochem. Physiol.*, **83**, 171–179.
[35] Botes, A.L., Labuschagne, M., Roth, R., Mitra, R.K., Lotter, J., Lalloo, R., Ramduth, D., Rohitlall, N., Simpson, C. and Van Zyl, P. (2008) Recombinant yeasts for synthesizing epoxide hydrolases. U.S. Patent 0171359 A1 (Oxyrane (UK) Ltd, July 17, 2008).
[36] Kim, H.S., Lee, S.J. and Lee, E.Y. (2006) Development and characterization of recombinant whole-cell biocatalysts expressing epoxide hydrolase from *Rhodotorula glutinis* for enantioselective resolution of racemic epoxides. *J. Mol. Catal. B: Enzym.*, **43**, 2–8.
[37] Labuschagne, M. and Albertyn, J. (2007) Cloning of an epoxide hydrolase-encoding gene from *Rhodotorula mucilaginosa* and functional expression in *Yarrowia lipolytica*. *Yeast*, **24**, 69–78.
[38] Naundorf, A., Melzer, G., Archelas, A., Furstoss, R. and Wohlgemuth, R. (2009) Influence of pH on the expression of a recombinant epoxide hydrolase in *Aspergillus niger*. *Biotechnol. J.*, **4**, 756–765.
[39] Grulich, M., Maršálek, J., Kyslik, P., Štěpánek, V. and Kotik, M. (2011) Production, enrichment and immobilization of a metagenome-derived epoxide hydrolase. *Process Biochem.*, **46**, 526–532.
[40] Visser, H., de Oliveira Villela Filho, M., Liese, A., Weijers, C.A.G.M. and Verdoes, J.C. (2003) Construction and characterisation of a genetically engineered *Escherichia coli* strain for the epoxide hydrolase-catalysed kinetic resolution of epoxides. *Biocatal. Biotransformation*, **21**, 33–40.
[41] Heikinheimo, P., Goldman, A., Jeffries, C. and Ollis, D.L. (1999) Of barn owls and bankers: a lush variety of α/β hydrolases. *Structure*, **7**, R141–R146.
[42] Lenfant, N., Hotelier, T., Velluet, E., Bourne, Y., Marchot, P. and Chatonnet, A. (2013) ESTHER, the database of the α/β-hydrolase fold superfamily of proteins: tools to explore diversity of functions. *Nucleic Acids Res.*, **41**, D423–D429.
[43] Arand, M., Hallberg, B.M., Zou, J., Bergfors, T., Oesch, F., van der Werf, M.J., de Bont, J.A.M., Jones, T.A. and Mowbray, S.L. (2003) Structure of *Rhodococcus erythropolis* limonene-1,2-epoxide hydrolase reveals a novel active site. *EMBO J.*, **22**, 2583–2592.
[44] Johansson, P., Unge, T., Cronin, A., Arand, M., Bergfors, T., Jones, T.A. and Mowbray, S.L. (2005) Structure of an atypical epoxide hydrolase from *Mycobacterium tuberculosis* gives insights into its function. *J. Mol. Biol.*, **351**, 1048–1056.

[45] Arand, M., Müller, F., Mecky, A., Hinz, W., Urban, P., Pompon, D., Kellner, R. and Oesch, F. (1999) Catalytic triad of microsomal epoxide hydrolase: replacement of Glu^{404} with Asp leads to a strongly increased turnover rate. *Biochem. J.*, **337**, 37–43.

[46] Rink, R., Lutje Spelberg, J.H., Pieters, R.J., Kingma, J., Nardini, M., Kellogg, R.M., Dijkstra, B.W. and Janssen, D.B. (1999) Mutation of tyrosine residues involved in the alkylation half reaction of epoxide hydrolase from *Agrobacterium radiobacter* AD1 results in improved enantioselectivity. *J. Am. Chem. Soc.*, **121**, 7417–7418.

[47] van Loo, B., Kingma, J., Heyman, G., Wittenaar, A., Lutje Spelberg, J.H., Sonke, T. and Janssen, D.B. (2009) Improved enantioselective conversion of styrene epoxides and *meso*-epoxides through epoxide hydrolases with a mutated nucleophile-flanking residue. *Enzyme Microb. Technol.*, **44**, 145–153.

[48] Jochens, H., Hesseler, M., Stiba, K., Padhi, S.K., Kazlauskas, R.J. and Bornscheuer, U.T. (2011) Protein engineering of α/β-hydrolase fold enzymes. *ChemBioChem*, **12**, 1508–1517.

[49] Reetz, M.T., Wang, L.-W. and Bocola, M. (2006) Directed evolution of enantioselective enzymes: iterative cycles of CASTing for probing protein-sequence space. *Angew. Chem. Int. Ed.*, **45**, 1236–1241.

[50] Reetz, M.T. and Zheng, H. (2011) Manipulating the expression rate and enantioselectivity of an epoxide hydrolase by using directed evolution. *ChemBioChem*, **12**, 1529–1535.

[51] Reetz, M.T., Kahakeaw, D. and Lohmer, R. (2008) Addressing the numbers problem in directed evolution. *ChemBioChem*, **9**, 1797–1804.

[52] van Loo, B., Spelberg, J.H.L., Kingma, J., Sonke, T., Wubbolts, M.G. and Janssen, D.B. (2004) Directed evolution of epoxide hydrolase from *A. radiobacter* toward higher enantioselectivity by error-prone PCR and DNA shuffling. *Chem. Biol.*, **11**, 981–990.

[53] Reetz, M.T., Prasad, S., Carballeira, J.D., Gumulya, Y. and Bocola, M. (2010) Iterative saturation mutagenesis accelerates laboratory evolution of enzyme stereoselectivity: rigorous comparison with traditional methods. *J. Am. Chem. Soc.*, **132**, 9144–9152.

[54] Mildvan, A.S. (2004) Inverse thinking about double mutants of enzymes. *Biochemistry*, **43**, 14517–14520.

[55] Kotik, M., Archelas, A., Faměrova, V., Oubrechtova, P. and Křen, V. (2011) Laboratory evolution of an epoxide hydrolase—towards an enantioconvergent biocatalyst. *J. Biotechnol.*, **156**, 1–10.

[56] Zheng, H. and Reetz, M.T. (2010) Manipulating the stereoselectivity of limonene epoxide hydrolase by directed evolution based on iterative saturation mutagenesis. *J. Am. Chem. Soc.*, **132**, 15744–15751.

[57] Kotik, M., Zhao, W., Iacazio, G. and Archelas, A. (2013) Directed evolution of metagenome-derived epoxide hydrolase for improved enantioselectivity and enantioconvergence. *J. Mol. Catal. B: Enzym.*, **91**, 44–51.

[58] Moussou, P., Archelas, A., Baratti, J. and Furstoss, R. (1998) Microbiological transformations. Part 39: Determination of the regioselectivity occurring during oxirane ring opening by epoxide hydrolases: a theoretical analysis and a new method for its determination. *Tetrahedron: Asymmetry*, **9**, 1539–1547.

[59] Sheldon, R.A., van Pelt, S., Kanbak-Aksu, S., Rasmussen, J. and Janssen, M.H.A. (2013) Cross-linked enzyme aggregates (CLEAs) in organic synthesis. *Aldrichim. Acta*, **46**, 81–93.

[60] Mateo, C., Fernandez-Lafuente, R., Archelas, A., Guisan, J.M. and Furstoss, R. (2007) Preparation of a very stable immobilized *Solanum tuberosum* epoxide hydrolase. *Tetrahedron: Asymmetry*, **18**, 1233–1238.

[61] Karboune, S., Archelas, A., Furstoss, R. and Baratti, J. (2005) Immobilization of the *Solanum tuberosum* epoxide hydrolase and its application in an enantioconvergent process. *Biocatal. Biotransformation*, **23**, 397–405.

[62] Kim, Y.H., Lee, I., Choi, S.H., Lee, O.K., Shim, J., Lee, J., Kim, J. and Lee, E.Y. (2013) Enhanced stability and reusability of marine epoxide hydrolase using ship-in-a-bottle approach with magnetically-separable mesoporous silica. *J. Mol. Catal. B: Enzym.*, **89**, 48–51.

[63] Yildirim, D., Tükel, S.S., Alptekin, Ö. and Alagöz, D. (2013) Immobilized *Aspergillus niger* epoxide hydrolases: cost-effective biocatalysts for the preparation of enantiopure styrene oxide, propylene oxide and epichlorohydrin. *J. Mol. Catal. B: Enzym.*, **88**, 84–90.

[64] Petri, A., Marconcini, P. and Salvadori, P. (2005) Efficient immobilization of epoxide hydrolase onto silica gel and use in the enantioselective hydrolysis of racemic *para*-nitrostyrene oxide. *J. Mol. Catal. B: Enzym.*, **32**, 219–224.

[65] Karboune, S., Archelas, A., Furstoss, R. and Baratti, J. (2005) Immobilization of epoxide hydrolase from *Aspergillus niger* onto DEAE-cellulose: enzymatic properties and application for the enantioselective resolution of a racemic epoxide. *J. Mol. Catal. B: Enzym.*, **32**, 175–183.

[66] Karboune, S., Archelas, A. and Baratti, J.C. (2010) Free and immobilized *Aspergillus niger* epoxide hydrolase-catalyzed hydrolytic kinetic resolution of racemic *p*-chlorostyrene oxide in a neat organic solvent medium. *Process Biochem.*, **45**, 210–216.

[67] Mateo, C., Archelas, A., Fernandez-Lafuente, R., Guisan, J.M. and Furstoss, R. (2003) Enzymatic transformations. Immobilized *A. niger* epoxide hydrolase as a novel biocatalytic tool for repeated-batch hydrolytic kinetic resolution of epoxides. *Org. Biomol. Chem.*, **1**, 2739–2743.

[68] Bučko, M., Vikartovská, A., Lacík, I., Kolláriková, G., Gemeiner, P., Pätoprstý, V. and Brygin, M. (2005) Immobilization of a whole-cell epoxide-hydrolyzing biocatalyst in sodium alginate–cellulose sulfate–poly(methylene-co-guanidine) capsules using a controlled encapsulation process. *Enzyme Microb. Technol.*, **36**, 118–126.

[69] Maritz, J., Krieg, H.M., Yeates, C.A., Botes, A.L. and Breytenbach, J.C. (2003) Calcium alginate entrapment of the yeast *Rhodosporidium toruloides* for the kinetic resolution of 1,2-epoxyoctane. *Biotechnol. Lett.*, **25**, 1775–1781.

[70] Hochuli, E., Bannwarth, W., Döbeli, H., Gentz, R. and Stüber, D. (1988) Genetic approach to facilitate purification of recombinant proteins with a novel metal chelate adsorbent. *Nat. Biotechnol.*, **6**, 1321–1325.

[71] Lee, K.S., Woo, M.H., Kim, H.S., Lee, E.Y. and Lee, I.S. (2009) Synthesis of hybrid Fe_3O_4–silica–NiO superstructures and their application as magnetically separable high-performance biocatalysts. *Chem. Commun.*, 3780–3782.

[72] Wang, W., Wang, D.I.C. and Li, Z. (2011) Facile fabrication of recyclable and active nanobiocatalyst: purification and immobilization of enzyme in one pot with Ni-NTA functionalized magnetic nanoparticles. *Chem. Commun.*, **47**, 8115–8117.

[73] Cassimjee, K.E., Kourist, R., Lindberg, D., Larsen, M.W., Hong Thanh, N., Widersten, M., Bornscheuer, U.T. and Berglund, P. (2011) One-step enzyme extraction and immobilization for biocatalysis applications. *Biotechnol. J.*, **6**, 463–469.

[74] Gong, P.-F. and Xu, J.-H. (2005) Bio-resolution of a chiral epoxide using whole cells of *Bacillus megaterium* ECU1001 in a biphasic system. *Enzyme Microb. Technol.*, **36**, 252–257.

[75] Liu, Z., Michel, J., Wang, Z., Witholt, B. and Li, Z. (2006) Enantioselective hydrolysis of styrene oxide with the epoxide hydrolase of *Sphingomonas* sp. HXN-200. *Tetrahedron: Asymmetry*, **17**, 47–52.

[76] Chiappe, C., Leandri, E., Lucchesi, S., Pieraccini, D., Hammock, B.D. and Morisseau, C. (2004) Biocatalysis in ionic liquids: the stereoconvergent hydrolysis of *trans*-β-methylstyrene oxide catalyzed by soluble epoxide hydrolase. *J. Mol. Catal. B: Enzym.*, **27**, 243–248.

[77] Chen, W.-J., Lou, W.-Y. and Zong, M.-H. (2012) Efficient asymmetric hydrolysis of styrene oxide catalyzed by mung bean epoxide hydrolases in ionic liquid-based biphasic systems. *Bioresour. Technol.*, **115**, 58–62.

[78] Chen, W.-J., Lou, W.-Y., Yu, C.-Y., Wu, H., Zong, M.-H. and Smith, T.J. (2012) Use of hydrophilic ionic liquids in a two-phase system to improve mung bean epoxide hydrolases-mediated asymmetric hydrolysis of styrene oxide. *J. Biotechnol.*, **162**, 183–190.

[79] Karboune, S., Archelas, A. and Baratti, J. (2006) Properties of epoxide hydrolase from *Aspergillus niger* for the hydrolytic kinetic resolution of epoxides in pure organic media. *Enzyme Microb. Technol.*, **39**, 318–324.

[80] Cleij, M., Archelas, A. and Furstoss, R. (1998) Microbiological transformations. Part 42: A two-liquid-phase preparative scale process for an epoxide hydrolase catalysed resolution of *para*-bromo-α-methyl styrene oxide. Occurrence of a surprising enantioselectivity enhancement. *Tetrahedron: Asymmetry*, **9**, 1839–1842.

[81] Doumèche, B., Archelas, A. and Furstoss, R. (2006) Enzymatic transformations. 62. Preparative scale synthesis of enantiopure glycidylacetals using an *A. niger* epoxide hydrolase catalysed kinetic resolution. *Adv. Synth. Catal.*, **348**, 1948–1957.

[82] Pedragosa-Moreau, S., Archelas, A. and Furstoss, R. (1993) Microbiological transformations. 28. Enantiocomplementary epoxide hydrolyses as a preparative access to both enantiomers of styrene oxide. *J. Org. Chem.*, **58**, 5533–5536.

[83] Pedragosa-Moreau, S., Archelas, A. and Furstoss, R. (1995) Preparative access to both enantiomers of styrene oxide by hydrolysis of the racemate using *Aspergillus niger* or *Beauveria sulfurescens*. In *Preparative Biotransformation*, Vol. **1**, John Wiley & Sons, Ltd, Chichester, 18.1–18.7.

[84] Cao, L., Lee, J., Chen, W. and Wood, T.K. (2006) Enantioconvergent production of (*R*)-1-phenyl-1,2-ethanediol from styrene oxide by combining the *Solanum tuberosum* and an evolved *Agrobacterium radiobacter* AD1 epoxide hydrolases. *Biotechnol. Bioeng.*, **94**, 522–529.

[85] Hwang, S., Choi, C.Y. and Lee, E.Y. (2008) One-pot biotransformation of racemic styrene oxide into (*R*)-1,2-phenylethanediol by two recombinant microbial epoxide hydrolases. *Biotechnol. Bioprocess Eng.*, **13**, 453–457.
[86] Min, J.Y. and Lee, E.Y. (2012) Biosynthesis of (*R*)-1,2-phenylethanediol and (*R*)-4-chloro-1,2-phenylethanediol by using two recombinant cells expressing enantiocomplementary epoxide hydrolases. *J. Ind. Eng.*, **18**, 160–164.
[87] Duarah, A., Goswami, A., Bora, T.C., Talukdar, M. and Gogoi, B.K. (2013) Enantioconvergent biohydrolysis of racemic styrene oxide to *R*-phenyl-1,2-ethanediol by a newly isolated filamentous fungus *Aspergillus tubingensis* TF1. *Appl. Biochem. Biotechnol.*, **170**, 1965–1973.
[88] Yoo, S.S., Park, S. and Lee, E.Y. (2008) Enantioselective resolution of racemic styrene oxide at high concentration using recombinant *Pichia pastoris* expressing epoxide hydrolase of *Rhodotorula glutinis* in the presence of surfactant and glycerol. *Biotechnol. Lett.*, **30**, 1807–1810.
[89] Kamal, A., Khanna, R., Kumar, C.G., Shaik, A.B. and Kumar, M.S. (2013) A novel bacterial strain of *Achromobacter* sp. MTCC 5605 and a highly enantioselective epoxide hydrolase isolated therefrom. WO Patent 030851 A1 (Council of Scientific & Industrial Research, March 7, 2013).
[90] Pabel, J., Hofner, G. and Wanner, K.T. (2000) Synthesis and resolution of racemic eliprodil and evaluation of the enantiomers of eliprodil as NMDA receptor antagonist. *Bioorg. Med. Chem. Lett.*, **10**, 1377–1380.
[91] Harada, H., Hirokawa, Y., Suzuki, K., Hiyama, Y., Oue, M., Kawashima, H., Yoshida, N., Furutani, Y. and Kato, S. (2003) Novel and potent human and rat β3-Adrenergic receptor agonists containing substituted 3-indolylalkylamines. *Bioorg. Med. Chem. Lett.*, **13**, 1301–1305.
[92] Manoj, K.M., Archelas, A., Baratti, J. and Furstoss, R. (2001) Microbiological transformations. Part 45. A green chemistry preparative scale synthesis of enantiopure building blocks of Eliprodil: elaboration of a high substrate concentration epoxide hydrolase-catalyzed hydrolytic kinetic resolution process. *Tetrahedron*, **57**, 695–701.
[93] Hwang, S., Choi, C.Y. and Lee, E.Y. (2008) Enantioconvergent bioconversion of *p*-chlorostyrene oxide to (*R*)-*p*-chlorophenyl-1,2-ethandiol by the bacterial epoxide hydrolase of *Caulobacter crescentus*. *Biotechnol. Lett.*, **30**, 1219–1225.
[94] Wu, S., Li, A., Chin, Y.S. and Li, Z. (2013) Enantioselective hydrolysis of racemic and *meso*-epoxides with recombinant *Escherichia coli* expressing epoxide hydrolase from *Sphingomonas* sp. HXN-200: preparation of epoxides and vicinal diols in high ee and high concentration. *ACS Catal.*, **3**, 752–759.
[95] Wittman, M., Carboni, J., Attar, R., Balasubramanian, B., Balimane, P., Brassil, P., Beaulieu, F., Chang, C., Clarke, W., Dell, J., Eummer, J., Frennesson, D., Gottardis, M., Greer, A., Hansel, S., Hurlburt, W., Jacobson, B., Krishnananthan, S., Lee, F.Y., Li, A., Lin, T.-A., Liu, P., Ouellet, C., Sang, X., Saulnier, M.G., Stoffan, K., Sun, Y., Velaparthi, U., Wong, H., Zang, Z., Zimmermann, K., Zoeckler, M. and Vyas, D. (2005) Discovery of a 1*H*-(benzoimidazol-2-yl)-1*H*-pyridin-2-one (BMS-536924) inhibitor of insulin-like growth factor I receptor kinase with in vivo antitumor activity. *J. Med. Chem.*, **48**, 5639–5643.
[96] Murmann, W., Rumore, G. and Gamba, A. (1967) Pharmacological properties of 1-(4′-nitrophenyl)-2-isopropylamino-ethanol (INPEA), a new beta-adrenergic receptor antagonist. V. Effects of the optical isomers D(minus) and L(plus) INPEA on heart rate, oxygen consumption and body temperature and on the cardiac and metabolic effects of adrenaline and noradrenaline in urethane-anesthetized rats. *Boll. Chim. Farm.*, **106**, 251–268.
[97] Morisseau, C., Nellaiah, H., Archelas, A., Furstoss, R. and Baratti, J.C. (1997) Asymmetric hydrolysis of racemic *para*-nitrostyrene oxide using an epoxide hydrolase preparation from *Aspergillus niger*. *Enzyme Microb. Technol.*, **20**, 446–452.
[98] Spelberg, J.H.L., Rink, R., Archelas, A., Furstoss, R. and Janssen, D. (2002) Biocatalytic potential of the epoxide hydrolase from *Agrobacterium radiobacter* AD1 and a mutant with enhanced enantioselectivity. *Adv. Synth. Catal.*, **344**, 980–985.
[99] Rink, R., Kingma, J., Spelberg, H.L. and Janssen, D. (2000) Tyrosine residues serve as proton donor in the catalytic mechanism of epoxide hydrolase from *Agrobacterium radiobacter*. *Biochemistry*, **39**, 5600–5613.
[100] Kotik, M. and Kyslik, P. (2006) Purification and characterisation of a novel enantioselective epoxide hydrolase from *Aspergillus niger* M200. *Biochim. Biophys. Acta*, **1760**, 245–252.
[101] Botes, A.L., Mitra, R. and Plenaar, D. (2007) Epoxide Hydrolases: Process Applications. *Innov. Pharm. Technol.*, 90–92.

[102] Pienaar, D.P., Mitra, R.K., van Deventer, T.I. and Botes, A.L. (2008) Synthesis of a variety of optically active hydroxylated heterocyclic compounds using epoxide hydrolase technology. *Tetrahedron Lett.*, **49**, 6752–6755.
[103] Zhao, J., Chu, Y.-Y., Li, A.-T., Ju, X., Kong, X.-D., Pan, J., Tang, Y. and Xu, J.-H. (2011) An unusual (*R*)-selective epoxide hydrolase with high activity for facile preparation of enantiopure glycidyl ethers. *Adv. Synth. Catal.*, **353**, 1510–1518.
[104] Pedragosa-Moreau, S., Morisseau, C., Zylber, J., Archelas, A., Baratti, J. and Furstoss, R. (1996) Microbiological transformations. 33. Fungal epoxide hydrolases applied to the synthesis of enantiopure *para*-substituted styrene oxides. A mechanistic approach. *J. Org. Chem.*, **61**, 7402–7407.
[105] Kotik, M., Brichac, J. and Kyslik, P. (2005) Novel microbial epoxide hydrolases for biohydrolysis of glycidyl derivatives. *J. Biotechnol.*, **120**, 364–375.
[106] Soloshonok, V.A. (1999) *Enantiocontrolled Synthesis of Fluoro-Organic Compounds: Stereochemical Challenges and Biomedicinal Targets*, John Wiley & Sons, Ltd, Chichester.
[107] Deregnaucourt, J., Archelas, A., Barbirato, F., Paris, J.-M. and Furstoss, R. (2007) Enzymatic transformations. 63. High-concentration two liquid-liquid phase *Aspergillus niger* epoxide hydrolase-catalysed resolution: application to trifluoromethyl-substituted aromatic epoxides. *Adv. Synth. Catal.*, **349**, 1405–1417.
[108] Devries, K.M., Dow, R.L. and Wright, S.W. (1998) Procédé de preparation de pyridines substituées. WO Patent 021184 A1 (May 22, 1998).
[109] Fisher, M.H., Naylor, E.M., Ok, D., Weber, A.E., Shih, T. and Ok, H. (1996) Substituted sulfonamides as selective β3 agonists for the treatment of diabetes and obesity. U.S. Patent 5561142 A (Merck & Co, Inc., October 1, 1996).
[110] Mathvink, R.J., Barritta, A.M., Candelore, M.R., Cascieri, M.A., Deng, L., Tota, L., Strader, C.D., Wyvratt, M.J., Fisher, M.H. and Weber, A.E. (1999) Potent selective human β-3 adrenergic receptor agonists containing substituted indoline-5-sulfonamide pharmacophore. *Bioorg. Med. Chem. Lett.*, **9**, 1869–1874.
[111] Genzel, Y., Archelas, A., Broxterman, Q.B., Schulze, B. and Furstoss, R. (2001) Microbiological transformations. 47. A step toward a green chemistry preparation of enantiopure (*S*)-2-, -3-, and -4-pyridyloxirane via an epoxide hydrolase catalysed kinetic resolution. *J. Org. Chem.*, **66**, 538–543.
[112] Genzel, Y., Archelas, A., Lutje, J.F.H., Janssen, D.B. and Furstoss, R. (2001) Microbiological transformations. Part 48: Enantioselective biohydrolysis of 2-,3- and 4-pyridyloxirane at high substrate concentration using the *Agrobacterium radiobacter* AD1 epoxide hydrolase and its Tyr215Phe mutant. *Tetrahedron*, **57**, 2775–2779.
[113] Hsin, L.-W., Prisinzano, T., Wilkerson, C.R., Dersch, C.M., Horel, R., Jacobson, A.E., Rothman, R.B. and Rice, K.C. (2003) Synthesis and dopamine transporter affinity of chiral 1-[2-[bis(4-fluorophenyl)methoxy]ethyl]-4-(2-hydroxypropyl)piperazines as potential cocaine abuse therapeutic agents. *Bioorg. Med. Chem. Lett.*, **13**, 553–556.
[114] Lin, L.S., Lanza, T.J. Jr., Jewell, J.P., Liu, P., Shah, S.K., Qi, H., Tong, X., Wang, J., Xu, S.S., Fong, T.M., Shen, C.-P., Lao, J., Xiao, J.C., Shearman, L.P., Stribling, D.S., Rosko, K., Strack, A., Marsh, D.J., Feng, Y., Kumar, S., Sumuel, K., Yin, W., Van der Ploeg, L.H.T., Goulet, M.T. and Hagmann, W.K. (2006) Discovery of *N*-[(1*S*,2*S*)-3-(4-chlorophenyl)-2-(3-cyanophenyl)-1-methylpropyl]-2-methyl-2-{[5-(trifluoro-methyl)pyridin-2-yl]oxy}propanamide (MK-0364), a novel, acyclic cannabinoid-1 receptor inverse agonist for the treatment of obesity. *J. Med. Chem.*, **49**, 7584–7587.
[115] Pedragosa-Moreau, S., Archelas, A. and Furstoss, R. (1996) Microbial transformations. 32. Use of epoxide hydrolase mediated biohydrolysis as a way to enantiopure epoxides and vicinal diols: application to substituted styrene oxide derivatives. *Tetrahedron*, **52**, 4593–4606.
[116] Weijers, C.A.G.M. (1997) Enantioselective hydrolysis of aryl, alicyclic and aliphatic epoxides by *Rhodotorula glutinis*. *Tetrahedron: Asymmetry*, **8**, 639–647.
[117] Lotter, J., Botes, A.L., van Dyk, M.S. and Breytenbach, J.C. (2004) Hydrolytic kinetic resolution of the enantiomers of the structural isomers *trans*-1-phenylpropene oxide and (2,3-epoxypropyl)benzene by yeast epoxide hydrolase. *Biotechnol. Lett.*, **15**, 1197–1200.
[118] Botes, A.L., Labuschagne, M., Roth, R., Mitra, R.K., Lotter, J., Abrahams, N., Simpson, C. and Van der Westhuizen, C. (2007) Methods for obtaining optically active epoxides and diols from 2,3-disubstituted and 2,3-trisubstituted epoxides. WO Patent 069079 A2 (CSIR, June 21, 2007).
[119] Hwang, S., Hyun, H., Lee, B., Park, Y., Lee, E.Y. and Choi, C. (2006) Purification and characterization of a recombinant *Caulobacter crescentus* epoxide hydrolase. *Biotechnol. Bioprocess Eng.*, **11**, 282–287.

[120] Botes, A.L., Steenkamp, J.A., Weijers, C.A.G.M. and van Dyk, M.S. (1998) Biocatalytic resolution of 1,2-epoxyoctane using resting cells of different yeast strains with novel epoxide hydrolase activities. *Biotechnol. Lett.*, **20**, 421–426.

[121] Botes, A.L., Weijers, C.A.G.M., Botes, P.J. and van Dyk, M.S. (1999) Enantioselectivities of yeast epoxide hydrolases for 1,2-epoxides. *Tetrahedron: Asymmetry*, **10**, 3327–3336.

[122] Spelberg, J.H.L., Rink, R., Kellogg, R.M. and Janssen, D.B. (1998) Enantioselectivity of a recombinant epoxide hydrolase from *Agrobacterium radiobacter*. *Tetrahedron: Asymmetry*, **9**, 459–466.

[123] Xu, Y., Xu, J.-H., Pan, J. and Tang, Y.-F. (2004) Biocatalytic resolution of glycidyl aryl ethers by *Trichosporon loubierii*: cell/substrate ratio influences the optical purity of (*R*)-epoxides. *Biotechnol. Lett.*, **26**, 1217–1221.

[124] Woo, J.-H., Kang, J.-H., Hwang, H.-O., Cho, J.-C., Kim, S.-J. and Kang, S.G. (2010) Biocatalytic resolution of glycidyl phenyl ether using a novel epoxide hydrolase from a marine bacterium, *Rhodobacterales bacterium* HTCC2654. *J. Biosci. Bioeng.*, **109**, 539–544.

[125] Reetz, M.T., Bocola, M., Wang, L.-W., Sanchis, J., Cronin, A., Arand, M., Zou, J., Archelas, A., Bottalla, A.-L., Naworyta, A. and Mowbray, S.L. (2009) Directed evolution of an enantioselective epoxide hydrolase: uncovering the source of enantioselectivity at each evolutionary stage. *J. Am. Chem. Soc.*, **131**, 7334–7343.

[126] Orru, R.V.A., Mayer, S.F., Kroutil, W. and Faber, K. (1998) Chemoenzymatic deracemization of (±)-2,2-disubstituted oxiranes. *Tetrahedron*, **54**, 859–874.

[127] Osprian, I., Kroutil, W., Mischitz, M. and Faber, K. (1997) Biocatalytic resolution of 2-methyl-2-(aryl)alkyloxiranes using novel bacterial epoxide hydrolases. *Tetrahedron: Asymmetry*, **18**, 65–71.

[128] Wandel, U., Mischitz, M., Kroutil, W. and Faber, K. (1995) Highly selective asymmetric hydrolysis of 2,2-disubstituted epoxides using lyophilized cells of *Rhodococcus* sp. NCIMB 11216. *J. Chem. Soc. Perkin Trans.* **1**, 735–736.

[129] Osprian, I., Stampfer, W. and Faber, K. (2000) Selectivity enhancement of epoxide hydrolase catalyzed resolution of 2,2-disubstituted oxiranes by substrate modification. *J. Chem. Soc. Perkin Trans.* **1**, 3779–3785.

[130] Weijers, C.A.G.M., Meeuwse, P., Herpers, R.L.J.M., Franssen, M.C.R. and Sudhoelter, E.J.R. (2005) Stereoselectivity and substrate specificity in the kinetic resolution of methyl-substituted 1-oxaspiro[2.5]octanes by *Rhodotorula glutinis* epoxide hydrolase. *J. Org. Chem.*, **70**, 6639–6646.

[131] Moussou, P., Archelas, A. and Furstoss, R. (1998) Microbiological transformations. 40. Use of fungal epoxide hydrolases for the synthesis of enantiopure alkyl epoxides. *Tetrahedron*, **54**, 1563–1572.

[132] Kroutil, W., Mischitz, M. and Faber, K. (1997) Deracemisation of (±)-2,3-disubstituted oxiranes via biocatalytic hydrolysis using bacterial epoxide hydrolases: kinetics of an enantioconvergent process. *J. Chem. Soc. Perkin Trans.* **1**, 3629–3636.

[133] Archelas, A., Delbecque, J.P. and Furstoss, R. (1993) Microbiological transformations. 30. Enantioselective hydrolysis of racemic epoxides: the synthesis of enantiopure insect juvenile hormone analogs (Bower's compound). *Tetrahedron: Asymmetry*, **4**, 2445–2446.

[134] Archer, I.V.J., Leak, D.J. and Widdowson, D.A. (1996) Chemoenzymatic resolution and deracemisation of (±)-1-methyl-1,2-epoxycyclohexane: the synthesis of (1-*S*,2-*S*)-1-methylcyclohexane-1,2-diol. *Tetrahedron Lett.*, **37**, 8819–8822.

[135] Steinreiber, A., Mayer, S.F., Saf, R. and Faber, K. (2001) Biocatalytic asymmetric enantioconvergent hydrolysis of trisubstituted oxiranes. *Tetrahedron: Asymmetry*, **12**, 1519–1528.

[136] Chang, D., Wang, Z., Heringa, M.F., Wirthner, R., Witholt, B. and Li, Z. (2003) Highly enantioselective hydrolysis of alicyclic *meso*-epoxides with a bacterial epoxide hydrolase from *Sphingomonas* sp. HXN-200: simple synthesis of alicyclic vicinal *trans*-diols. *Chem. Commun.*, 960–961.

[137] Botes, A.L., Lotter, J. and Labuschagne, M. (2007) Methods of obtaining optically active epoxides and vicinal diols from *meso*-epoxides. U.S. Patent 0275448 A1 (CSIR, November 29, 2007).

[138] Kasai, N., Suzuki, T. and Furukawa, Y. (1998) Chiral C3 epoxides and halohydrins: their preparation and synthetic application. *J. Mol. Catal. B: Enzym.*, **4**, 237–252.

[139] Jin, H.-X., Liu, Z.-Q., Hu, Z.-C. and Zheng, Y.-G. (2013) Biosynthesis of (*R*)-epichlorohydrin at high substrate concentration by kinetic resolution of racemic epichlorohydrin with a recombinant epoxide hydrolase. *Eng. Life Sci.*, **13**, 385–392.

[140] Xue, F., Liu, Z.-Q., Zou, S.-P., Wan, N.-W., Zhu, W.-Y., Zhu, Q. and Zheng, Y.-G. (2014) A novel enantioselective epoxide hydrolase from *Agromyces mediolanus* ZJB120203: cloning, characterization and application. *Process Biochem.*, **49**, 409–417.

[141] Botes, A.L., Lotter, J., Labuschagne, M. and Mitra, R.K. (2005) Methods for the preparation of optically active epoxides and vicinal diols from styrene epoxides using enantioselective epoxide hydrolases derived from yeasts. WO Patent 100569 A2 (CSIR, October 27, 2005).

[142] Li, Z., Feiten, H.-J., Chang, D., Duetz, W.A., van Beilen, J.B. and Witholt, B. (2001) Preparation of (*R*)- and (*S*)-*N*-protected 3-hydroxypyrrolidines by hydroxylation with *Sphingomonas* sp. HXN-200, a highly active, region- and stereoselective, and easy to handle biocatalyst. *J. Org. Chem.*, **66**, 8424–8430.

[143] Calvez, O., Chiaroni, A. and Langlois, N. (1998) Enantioselective synthesis of 2,3-disubstituted piperidines from (*S*)-methylpyroglutamate. *Tetrahedron Lett.*, **39**, 9447–9450.

[144] Haddad, M. and Larchevêque, M. (2001) An alternative stereoselective synthesis of *trans*-(2*R*,3*R*)-3-hydroxypipecolic acid. *Tetrahedron Lett.*, **42**, 5223–5225.

[145] Ferreira, F., Greck, C. and Genêt, J.P. (1997) Stereocontrolled syntheses of *trans*-3-hydroxypipecolic acids and application to (–)-Swainsonine. *Bull. Soc. Chim. Fr.*, **134**, 615–621.

[146] Yuasa, Y. and Tsuruta, H. (1997) Practical syntheses of (*S*)-4-hydroxytetrahydrofuran-2-one, (*S*)-3-hydroxytetrahydrofuran and their (*R*)-enantiomers. *Liebigs Ann. Recl.*, **9**, 1877–1879.

[147] Honda, Y., Katayama, S., Kojima, M., Suzuki, T., Kishibata, N. and Izawa, K. (2004) New approaches to the industrial synthesis of HIV protease inhibitors. *Org. Biomol. Chem.*, **2**, 2061–2070.

[148] Fuchs, M., Simeo, Y., Ueberbacher, B.T., Mautner, B., Netscher, T. and Faber, K. (2009) Enantiocomplementary chemoenzymatic asymmetric synthesis of (*R*)- and (*S*)-chromanemethanol. *Eur. J. Org. Chem.*, **6**, 833–840.

[149] Hellstrom, H., Steinreiber, A., Mayer, S.F. and Faber, K. (2001) Bacterial epoxide hydrolase-catalyzed resolution of a 2,2-disubstituted oxirane: optimization and upscaling. *Biotechnol. Lett.*, **23**, 169–173.

[150] Simeó, Y. and Faber, K. (2006) Selectivity enhancement of enantio- and stereo-complementary epoxide hydrolases and chemo-enzymatic deracemization of (±)-2-methylglycidyl benzyl ether. *Tetrahedron: Asymmetry*, **17**, 402–409.

[151] Steinreiber, A., Hellstrom, H., Mayer, S.F., Orru, R.V.A. and Faber, K. (2001) Chemoenzymatic enantioconvergent synthesis of C4-building blocks containing a fully substituted chiral carbon center using bacterial epoxide hydrolases. *Synlett*, 111–113.

[152] Shimizu, K.I., Sakamoto, M., Hamada, M., Higashi, T., Sugai, T. and Shoji, M. (2010) The scope and limitation of the regio- and enantioselective hydrolysis of aliphatic epoxides using *Bacillus subtilis* epoxide hydrolase, and exploration toward chirally differentiated tris(hydroxymethyl)methanol. *Tetrahedron: Asymmetry*, **21**, 2043–2049.

[153] Fujino, A. and Sugai, T. (2008) Chemoenzymatic approach to enantiomerically pure (*R*)-3-hydroxy-3-methyl-4-pentenoic acid ester and its application to a formal total synthesis of taurospongin A. *Adv. Synth. Catal.*, **350**, 1712–1716.

[154] Eliel, E.L. (1985) The R/S system. A method for assignment and some recent modifications. *J. Chem. Educ.*, **62**, 223–224.

[155] Prelog, V. and Helmchem, G. (1982) Basic principles of the CIP-system and proposals for a revision. *Angew. Chem. Int. Ed. Engl.*, **21**, 567–583.

[156] Choi, W.J., Puah, S.M., Tan, L.L. and Ng, S.S. (2008) Production of (*R*)-ethyl-3,4-epoxybutyrate by newly isolated *Acinetobacter baumannii* containing epoxide hydrolase. *Appl. Microbiol. Biotechnol.*, **79**, 61–67.

[157] Devi, A.V., Lahari, C., Swarnalatha, L. and Fadnavis, N.W. (2008) Gelozymes in organic synthesis. Part IV: resolution of glycidate esters with crude mung bean (*Phaseolus radiatus*) epoxide hydrolase immobilized in gelatin matrix. *Tetrahedron: Asymmetry*, **19**, 1139–1144.

[158] Guérard, C., Alphand, V., Archelas, A., Demuynck, C., Hecquet, L., Furstoss, R. and Bolte, J. (1999) Transketolase-mediated synthesis of 4-deoxy-D-fructose 6-phosphate by epoxide hydrolase-catalysed resolution of 1,1-diethoxy-3,4-epoxybutane. *Eur. J. Org. Chem.*, 3399–3402.

[159] Otto, P.P.J.H.L., Stein, F. and Van der Willigen, C.A. (1988) A bio-organic synthesis of (+)-disparlure, sex pheromone of *Lymantria dispar* (gypsy moth). *Agric. Ecosyst. Environ.*, **21**, 121–123.

[160] Mischitz, M. and Faber, K. (1996) Chemo-enzymatic synthesis of (2*R*,5*S*)- and (2*R*,5*R*)-5-(1-hydroxy-1-methylethyl)-2-vinyl-tetrahydrofuran ("linalool oxide"): preparative application of a highly selective bacterial epoxide hydrolase. *Synlett*, 978–980.

[161] Orru, R.V.A., Archelas, A., Furstoss, R. and Faber, K. (1999) Epoxide hydrolases and their synthetic applications. *Adv. Biochem. Eng.*, **63**, 145–167.
[162] Orru, R.V.A., Kroutil, W. and Faber, K. (1997) Deracemization of (±)-2,2-disubstituted epoxides via enantioconvergent chemoenzymatic hydrolysis using *Nocardia* EH1 epoxide hydrolase and sulfuric acid. *Tetrahedron Lett.*, **38**, 1753–1754.
[163] Steinreiber, A., Edegger, K., Mayer, S.F. and Faber, K. (2001) Enantio- and diastereo-convergent synthesis of (2*R*,5*R*)- and (2*R*,5*S*)-Pityol through enzyme-triggered ring closure. *Tetrahedron: Asymmetry*, **12**, 2067–2071.
[164] Chang, D., Heringa, M.F., Witholt, B. and Li, Z. (2003) Enantioselective trans dihydroxylation of nonactivated C–C double bounds of aliphatic heterocycles with *Sphingomonas* sp. HXN-200. *J. Org. Chem.*, **68**, 8599–8606.
[165] Xu, Y., Jia, X., Panke, S. and Li, Z. (2009) Asymmetric dihydroxylation of aryl olefins by sequential enantioselective epoxidation and regioselective hydrolysis with tandem biocatalysts. *Chem. Commun.*, 1481–1483.
[166] Wu, S., Chen, Y., Xu, Y., Li, A., Xu, Q., Glieder, A. and Li, Z. (2014) Enantioselective *trans*-dihydroxylation of aryl olefins by cascade biocatalysis with recombinant *Escherichia coli* coexpressing monooxygenase and epoxide hydrolase. *ACS Catal.*, **4**, 409–420.
[167] Jin, H.-X., Liu, Z.-Q., Hu, Z.-C. and Zheng, Y.-G. (2013) Production of (*R*)-epichlorohydrin from 1,3-dichloro-2-propanol by two-step biocatalysis using haloalcohol dehalogenase and epoxide hydrolase in two-phase system. *Biochem. Eng. J.*, **74**, 1–7.

CHAPTER 9

Enantioselective Acylation of Alcohol and Amine Reactions in Organic Synthesis

Vicente Gotor-Fernández and Vicente Gotor

Departamento de Química Orgánica e Inorgánica, Instituto Universitario de Biotecnología de Asturias, Universidad de Oviedo, Oviedo, Spain

9.1 INTRODUCTION

Nowadays, the use of biotransformations for synthetic purposes plays nowadays an important role in synthetic chemistry [1, 2]. Enzymes have shown a high level of reactivity and very high selectivity, catalyzing chemo-, regio-, and stereoselective transformations under mild reaction conditions [3]. Thus, their use sometimes assists current methods for accessing chiral chemicals [4–6]. This fact has led to the incorporation of enzymatic catalysis into the synthetic organic expert's toolbox [7, 8]. Additionally, the advances in protein engineering techniques broaden the number of enzymes and possibilities to find an adequate hit for selected transformations [9]. This is especially relevant when the transfer of knowledge arrives to industrial biotechnology, and fortunately many companies have taken advantage of biotransformations because of their low environmental impact in terms of energy requirement, raw material used, and low amount of waster production [10–14].

9.1.1 General Considerations for Hydrolase-Catalyzed Reactions

Hydrolases (EC 3), without any doubt are one of the most common enzymes in the development of asymmetric transformations [15]. In particular, esterases, lipases, and proteases have allowed the development of efficient transformations for the production of different families of compounds in enantiomerically pure form such as alcohols, amines, amides, carboxylic acids, carbonates, and esters [16, 17]. These facts have led to their successful use in some representative applications for the industrial sector [18]. Their high tolerances for nonnatural reaction conditions, particularly highlighted in their activity in organic media, make hydrolases a favorite class of enzymes for organic chemists. Moreover, their availability from a large number of commercial sources at reasonable prices, lack of cofactor dependency, and broad substrate specificity facilitate their acceptance in the scientific community.

This chapter deals with the use of hydrolases for enantioselective transformations of alcohols and amines, describing acylation reactions as synthetic challenges. Initially, general comments regarding the action and use of hydrolases in different media will be discussed. Special attention will be paid to the selection of reactants

Green Biocatalysis, First Edition. Edited by Ramesh N. Patel.

and conditions for the development of synthetically useful transformations, leading to the desired final products in excellent yields and with high optical purity. In this manner we will try to offer guidelines for acylation reactions for both new and expert researchers in the field. The acylation of alcohols and amines will be extensively examined later. For this reason, a variety of strategies will be discussed such as classical kinetic resolution (KR) and dynamic kinetic resolution (DKR) of racemic alcohols and amines, and also the desymmetrization of *meso*-diols. Relevant examples of the production of pharmaceuticals by chemoenzymatic methods have been selected, hydrolases being involved in the successful asymmetrization of an adequate precursor or the final drug.

9.1.2 Serine Hydrolase Mechanism for the Acylation of Alcohols and Amines

Hydrolases possess a common feature for its mechanistic action. Their active site is built around three key amino acid residues, which are histidine (His), serine (Ser), and aspartic acid (Asp) or alternatively glutamic acid (Glu). This site, namely, the catalytic triad, forms a charge-relay network to polarize and activate the nucleophile, allowing the serine hydrolase mechanism to start [19]. This review covers the acylation of alcohols and amines using mainly esters; so Figure 9.1 depicts the mechanism for these types of reactions is depicted.

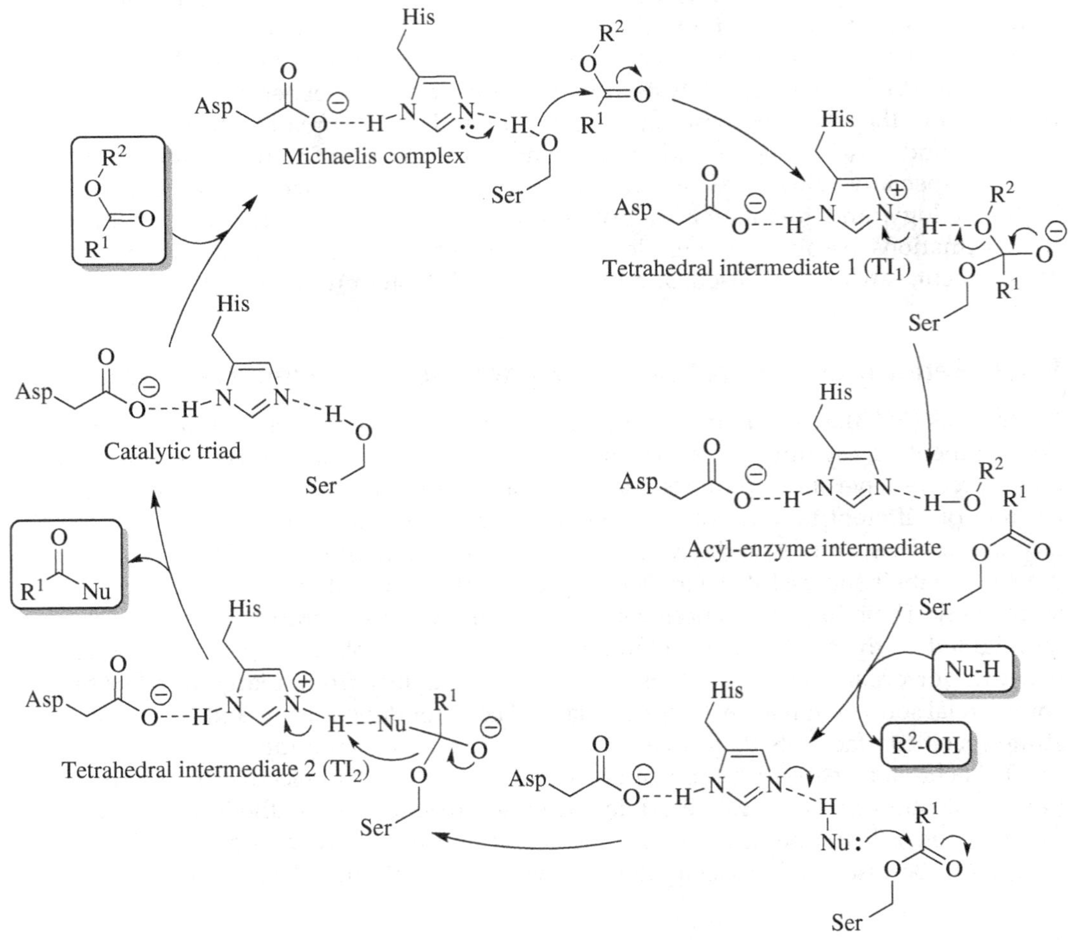

FIGURE 9.1
Mechanism for hydrolase-catalyzed acylation reactions of different nucleophiles such as alcohols or amines.

Initially, the ester (R^1COOR^2) is directed to the hydrolase active site, forming a noncovalent Michaelis complex and then allowing the nucleophilic attack of the serine hydroxyl group, whose pK_a has been reduced by the histidine residue acting in an orchestral way, on the ester carbonyl group. The acyl moiety of the ester therefore becomes covalently linked to the hydrolase, forming a tetrahedral intermediate 1 (TI_1). Next, electronic and atomic reorganization takes place, forming the acyl-enzyme intermediate by releasing an alcohol (R^2—OH). Then, a nucleophile (Nu—H) attacks the carbonyl group of the acyl-enzyme intermediate, leading to the formation of the tetrahedral intermediate 2 (TI_2). After a new rearrangement, the desired ester of amide product (R^1—CO—Nu) is released, and the free active enzyme starts a new cycle. Throughout the process the positively charged oxyanion hole stabilizes the charge density of both tetrahedral intermediates through the side chains of various residues.

9.1.3 Use of Organic Solvents for Hydrolase-Catalyzed Acylation Reactions

Although their natural function is the hydrolysis of carboxyl, peptide, or amide groups among others, hydrolases can also accelerate the reverse reactions, favoring synthesis over hydrolysis. Importantly, their high levels of catalytic activity and stability in organic solvents [20–22] and neoteric solvents [23–26] have led to the full recognition of hydrolases for synthetic transformations. After the formation of the active acyl-enzyme complex, they can catalyze, apart from hydrolysis, a broad number of reactions, such as aminolysis, ammonolysis, hydrazinolysis, perhydrolysis, thiolysis, and transesterification reactions (Figure 9.2).

The use of enzymes in organic media offers many advantages, which have been fully exploited since the discovery of the high activity in this type of solvents: (i) development of transformations that are impossible to conduct in water due to the formation of side products (decomposition, elimination, hydrolysis, polymerization, racemization, etc.), (ii) the better solubility of substrates and products, (iii) the easy

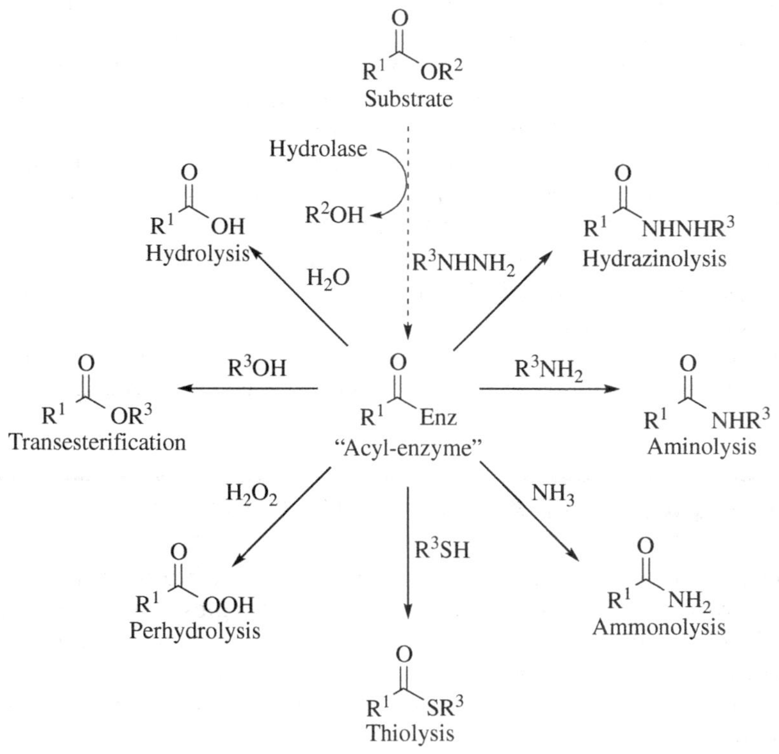

FIGURE 9.2
Most significant transformations catalyzed by hydrolases.

removal of the solvent by distillation in comparison with the limitations of water because of its high boiling point and heat vaporization, and (iv) high isolated yields of the resulting products [27]. Overall, the use of organic solvents is very highly recommended but is worth mentioning that a minimum of water is usually needed (<1%) in order to avoid a rigid catalyst that immediately results in a quick deactivation [28].

The development of immobilization techniques has contributed greatly to the success of hydrolases for synthetic purposes, as they can improve enzyme stability toward pH and temperature changes, leading to an active enzyme conformation [29–34]. Importantly, the use of supported enzymes allows the easy recovery of the biocatalyst at the end of the process through a simple filtration or centrifugation step, facilitating the purification of the final products and leading generally to high isolated yields.

From the hydrolases toolbox, probably lipases have been the most demanding catalysts for synthetic application. Their natural function involves the hydrolysis of triacylglycerol ester bonds, compounds that are poorly soluble in water. Thus, the reaction usually occurs in an organic–aqueous interface. This phenomenon involving the conformational change of the selected lipase is called interfacial activation [35], and it provides an inherent affinity for hydrophobic media to the enzyme.

Different reaction media can be considered for biotransformations depending on the miscibility and relative ratio of the organic solvent with water as described by Doukyu and Ogino [36]: (i) a water and water-miscible organic solvent system, the organic solvent acting just as a cosolvent; (ii) a water and water-immiscible organic solvent system for the development of reactions in a biphasic system; and (iii) a nearly anhydrous organic solvent, which will be the main one used for the examples described in this chapter. Some of the most representative solvents are tetrahydrofuran (THF), methyl *tert*-butyl ether (MTBE), and toluene. These are hydrophobic solvents with a high value of logP [37]. The hydrolysis and the reverse acylation reaction of alcohols and amines are in competition. The equilibrium can be shifted toward synthesis by using dried solvents under an inert atmosphere. In some cases the use of molecular sieves (MS) has been also employed, which act as water reservoir to preserve the enzymatic activity and can be easily removed from the reaction medium through a filtration process at the end of the reactions.

The development of catalytic methods at a large scale is well established nowadays in the pharmaceutical industry [38, 39]. Biocatalysis has emerged as a powerful technology for the selective modification of final targets or adequate precursors through a variety of reactions [40]. Hydrolases, in particular, have reacted with excellent levels of selectivity for the production of optically active alcohol, esters, amines, amides, carbonates, and carboxylate with also good conversion values [41]. Importantly, the introduction of chirality or stereodiscrimination for the selective formation of one single enantiomer has been achieved by the use of hydrolytic enzymes, which is extremely important because of the different biological profiles of enantiomers [42]. Recent examples of successful hydrolase-catalyzed acylation reactions will be described in this chapter, considering alcohols and amines as substrates. Examples will be highlighted focusing on the substrate structure, type of acylating agent and racemization agent, scale-up of the reactions, and the usefulness of catalyst recycling.

9.2 ENANTIOSELECTIVE ACYLATION OF ALCOHOLS

Hydrolases and mainly lipases have appeared as valuable biocatalysts for the development of asymmetric transformations. Several strategies have been carried out involving the classical KR and DKR of racemic alcohols and the desymmetrization of *meso*- and prochiral diols. Taking into account the reversibility of this type of process in complementary hydrolysis pathway and adequate conditions must be established to favor synthetic acylation reactions.

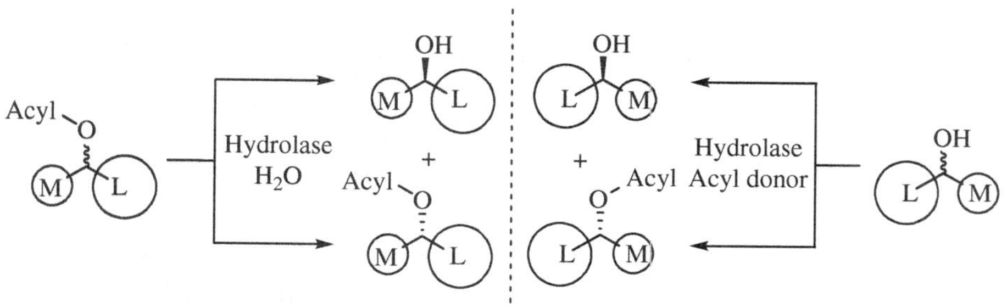

FIGURE 9.3 Complementarity of acylation and hydrolysis reactions using hydrolases as biocatalysts.

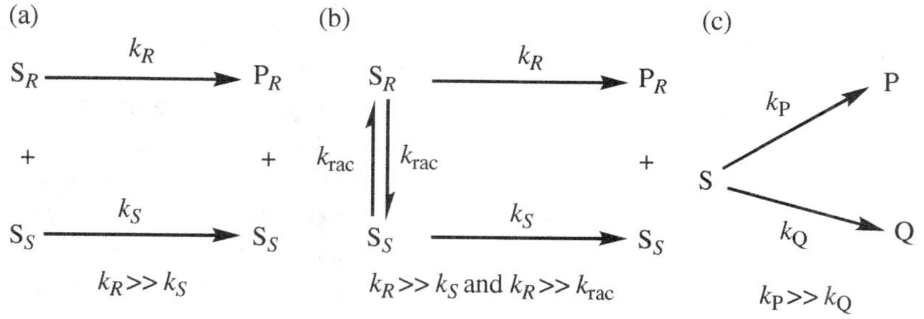

FIGURE 9.4 General representation of classical kinetic resolutions (a), dynamic kinetic resolutions (b), and desymmetrization processes (c).

The symmetry of both processes has been represented in Figure 9.3 according to the Kazlauskas rule for secondary alcohols [43], which established empirically that lipases favors the hydrolysis of (*R*)-esters to form (*R*)-alcohols, maintaining unaltered the (*S*)-esters. On the other hand the formation of (*R*)-esters and (*S*)-alcohols is preferred in acylation reactions. This rule is directly linked to the size and Cahn–Ingold–Prelog preferences of the substituents.

The best stereoselectivities usually start from secondary alcohols, in comparison with most hindered tertiary alcohols, or if the transformation must occur far from the stereogenic center in primary alcohols. Examples of all these types of reactions will be described in detail next.

A classical KR is defined as the selective modification of one enantiomer from a starting racemate, the transformation into products occurring at different rate levels (Figure 9.4a). This strategy allows the recovery of two different products with opposite stereochemistries, which is important in pharmaceutical chemistry to test the different biological profiles of enantiomers. However at the same time, some limitations are inherent to KRs as the theoretical yield is limited to a 50%, so for that reason DKRs offer the advantage of reaching a theoretical 100% for a single enantiomer [44, 45]. In this case, a racemization agent is involved in the process (Figure 9.4b), which may be compatible with the biocatalyst action. Alternatively, desymmetrization processes are able to introduce chirality in a target *meso*- or prochiral compound instead of using a racemate as initial material (Figure 9.4c) [46].

9.2.1 Classical Kinetic Resolution of Racemic Alcohols

The establishment of suitable conditions for an irreversible equilibrium is of dramatic importance for the development of acylative KRs. To circumvent the problem of a possible undesired hydrolytic reverse reaction, the selection of an adequate acyl donor is crucial in order to drive the reaction to completion. For the acylation of alcohols, an acyl donor must be selected from a variety of esters, anhydrides, and carboxylic acids. Some revisions have been published in this field giving outstanding selection guides [47, 48]. As mentioned before, the use of dry solvents is important to avoid hydrolytic side reactions, and also in some cases the use of MS has been described.

A key aspect resides in the selection of a proper acyl donor, which activates the hydrolase forming the acyl-enzyme complex. In order to avoid the reversible

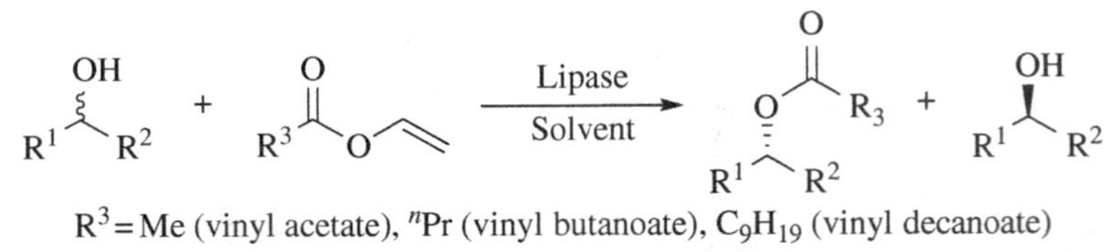

FIGURE 9.5

General representation for the kinetic resolution of racemic secondary alcohols.

hydrolytic process that drives the reaction to a complete conversion, the use of activated acyl donors is required for the resolution of secondary alcohols. The use of these reagents has been extensively reviewed in the literature [47, 48], enol and oxime esters being the most common resolving agents [49]. Both esters shift the equilibrium toward the formation of acylated products due to the formation of acetaldehyde or a weak nucleophile in the reaction medium, respectively.

Probably vinyl acetate (VinOAc) (Figure 9.5) is nowadays the most common reagent for the esterification of secondary alcohols, displaying significant advantages in comparison with other acyl donors such as its high reactivity, commercial availability, and low boiling point, which allow an easy purification of the final products. VinOAc enables the acylation of the target alcohol, forming acetaldehyde, which allows the development of a practically irreversible reaction because of its high volatility and nonnucleophilic character based on the fast tautomerization of the enol to the keto form. Similarly other vinyl esters such as vinyl butanoate and vinyl decanoate also serve as good acyl donor for acylative processes.

For the resolution of a secondary alcohol, forming an optically active alcohol enantiomer and the corresponding ester antipode, purification is usually carried out by silica gel chromatography because of the chemical nature of both final products, and adequate derivatization can be carried out, facilitating the isolation of the final products by filtration or distillation [50]. Some elegant and more sophisticated acyl donors have been described for the KRs of secondary alcohols, which allow an easy purification of the final products and the removal of the acyl donor by simple extraction protocols. These include the use of cyclic anhydrides such as succinic anhydride [51, 52] or vinyl esters with amino acid tags [53].

Some reviews have described elegant examples for the enantioselective acylation of racemic alcohols [14–18]. Here, we have selected some reactions of considerable importance due to different factors attending to the substrate structure, the possibility to scale up the biotransformations, and the enzyme recycling.

1-Phenylethanol is usually the model substrate for the enzymatic study of commercially and newly developed hydrolases, so the use of this substrate and its derivatives has been extensively described in the literature [54–60]. In addition, the development of continuous flow *Pseudomonas cepacia lipase* (PSL)-catalyzed enantioselective transesterification has also been possible in microreactors [54]. The substitution in the aromatic ring has a great influence in the KR of iodophenylethanols catalyzed by *Candida antarctica* lipase type B (CAL-B) using VinOAc in hexane at 32 °C and 130 rpm, obtaining 50% conversion values for the *meta*- and *para*-monosubstituted alcohols, while just a 25% conversion is achieved for the *ortho*-iodo-1-phenylethanol although with excellent selectivity (Table 9.1, entries 1–3) [55]. This methodology is also compatible with the presence of heteroatom-containing functionalities in the substrate structure as described by Andrade and Barcellos for the KR of boron-containing racemic alcohols using CAL-B or PSL in combination with 3 equivalents of VinOAc in a mixture of THF and hexane at 30 °C [56], a strategy also effective in the resolution of other organic selenides and tellurides [57]. Similarly, the resolution of racemic allylic alcohols was reported using two equivalents of VinOAc and PS-30 lipase from *P. cepacia* in the presence of MS (Table 9.1, entries 4–8) [58]. The benefits of immobilizing PSL on supermagnetic nanoparticles were demonstrated in the resolution of 2-bromo-1-phenylethanol, a poor substrate for the commercially available PSL ($E < 30$), although the conversion was far from 50% (Table 9.1, entry 9) [59]. In some cases vinyl butanoate can be used instead of VinOAc to improve the selectivity of the process, as occurs in the KR of fluorinated propargyl alcohols catalyzed by CAL-B in hexane at 60 °C [60].

TABLE 9.1 Lipase-Catalyzed Kinetic Resolution of Racemic Alcohols with Vinyl Acetate in Organic Solvent

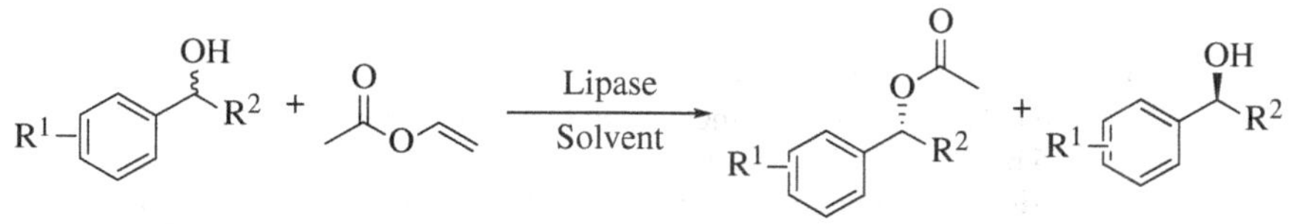

Entry	R^1	R^2	Lipase	*t* (h)	ee_p (%)	ee_S (%)	*c* (%)	*E*	Reference
1	2-I	CH_3	CAL-B	24	>98	24	25	>200	[55][a]
2	3-I	CH_3	CAL-B	2	>98	>98	50	>200	[55][a]
3	4-I	CH_3	CAL-B	2	>98	>98	50	>200	[55][a]
4	H	$CH{=}CH_2$	PSL	48	96	63	46	97	[58][b]
5	4-Br	$CH{=}CH_2$	PSL	48	>99	22	30	>200	[58][b]
6	3-Br	$CH{=}CH_2$	PSL	48	97	16	23	93	[58][b]
7	4-F	$CH{=}CH_2$	PSL	50	>99	38	40	>200	[58][b]
8	$4\text{-}NO_2$	$CH{=}CH_2$	PSL	50	97	92	50	>200	[58][b]
9	H	CH_2Br	PSL	50	97	>99	34	>200	[59][c]

[a] Hexane was used as solvent at 32 °C and 130 rpm.
[b] Toluene was used as solvent at room temperature and magnetic stirring.
[c] Toluene was used as solvent at 52 °C and 800 rpm.

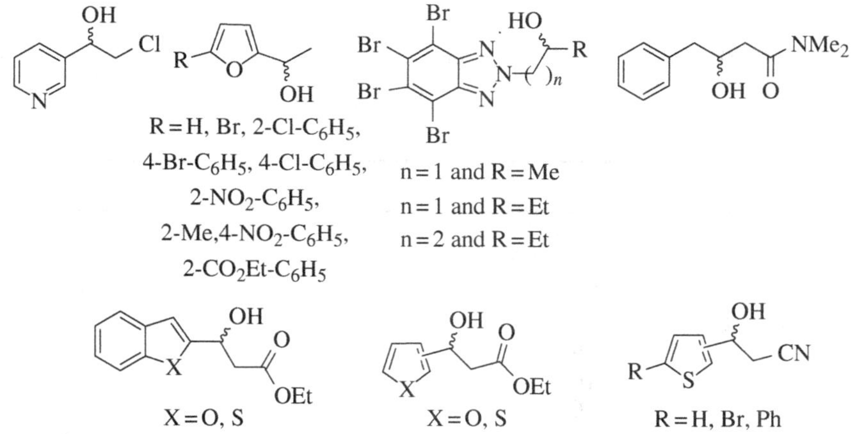

FIGURE 9.6 Structure of racemic heteroaromatic alcohols and hydroxyamino acids suitable for enantioselective lipase-catalyzed KR through acylation procedures.

The asymmetric synthesis of heterocyclic structures is a challenging task, and hydrolase-catalyzed processes have allowed the formation of interesting alcohols and acetates in enantioenriched form through acylation reactions (Figure 9.6). For instance, 2-chloro-1-(pyridin-3-yl)ethanol was enantioselectively acylated using a excess of VinOAc (8.5 equivalents) in hexane at 37 °C after a long reaction time (6–7 days) [61]. Irimie and coworkers reported the gram-scale KR of seven furan-based alcohols using two equivalents of VinOAc and CAL-B, yielding the 1-(furan-2-yl)ethanols and their corresponding esters with high selectivity and short reaction times using diisopropyl ether (DIPE, $^iPr_2O$) as solvent [62]. More hindered hydroxyalkyl 4,5,6,7-tetrabromo-1*H*-benzotriazoles have also been recognized as suitable substrates for the lipase Amano AK from *Pseudomonas fluorescens* in MTBE at 23 °C and 200 rpm using VinOAc, but also other vinyl esters such as isopropenyl acetate (IPA) (^{i}PrOAc), vinyl butyrate, or vinyl decanoate [63].

Hydroxyamino acid derivatives are interesting compounds with multiple applications as synthetic building blocks. Lipases have effectively acylated by different substrates such as γ-hydroxy amides using AK lipase and IPA [64], 3-(hetero)aryl-3-hydroxypropanoates with a variety of lipases and vinyl esters with short or long chains, such as vinyl decanoate [65–67], and sulfur heterocyclic β-hydroxynitriles with PSL and VinOAc [68].

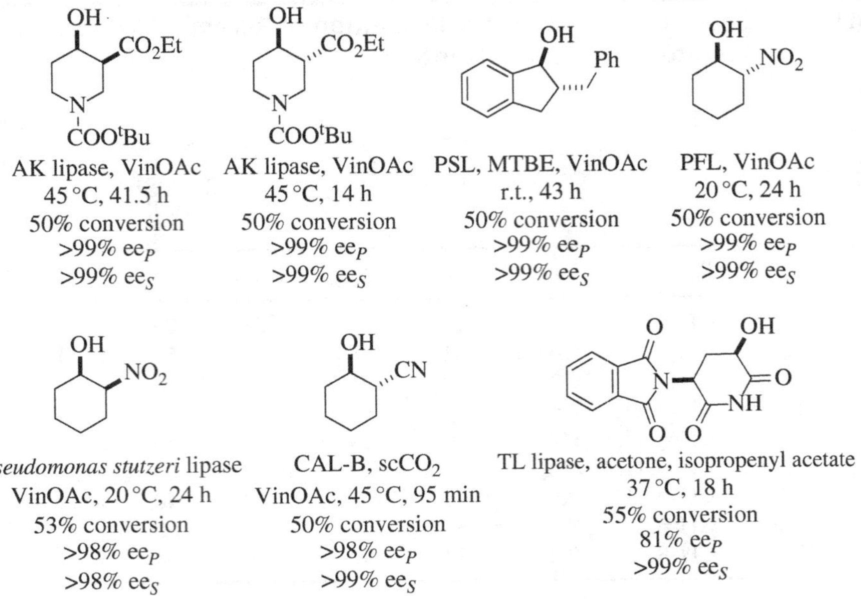

FIGURE 9.7

Structure of racemic alcohol bearing two stereogenic structures suitable for enantioselective acylation using lipases and vinyl esters.

TABLE 9.2 Asymmetric Acetylation of α-Allenols Using Vinyl Butanoate and PPL at Room Temperature in DIPE

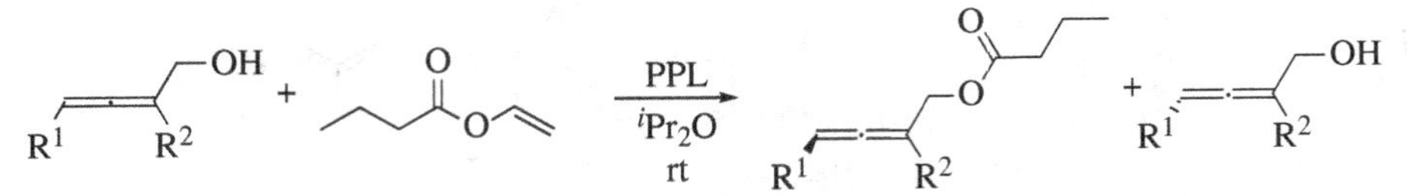

Entry	R^1	R^2	t (h)	ee_P (%)	ee_S (%)	c (%)	E
1	Ph	Me	24	98	67	42	>200
2	Ph	Et	24	97	61	39	127
3	2-Me-C_6H_5	Me	30	94	21	19	40
4	3-Me-C_6H_5	Me	42	97	74	43	176
5	4-Me-C_6H_5	Me	24	95	82	46	102
6	4-Cl-C_6H_5	Me	28	97	96	49	>200
7	4-CF_3-C_6H_5	Me	30	99	23	10	>200
8	2-Naphthyl	Me	31	98	58	37	>200

The selective acylation of hydroxyl functionalities of compounds bearing two stereogenic centers can lead to the formation of single diastereomers in enantiomerically pure form, as occurs in the acylation of different cyclic alcohols such as *cis*- and *trans*-1-(*tert*-butoxycarbonyl)-4-hydroxypiperidine-3-carboxylate [69], *trans*-2-benzylindan-1-ol [70], *trans*-2-nitrocyclohexanol [71], *trans*-2-hydroxycyclohexanecarbonitriles, and *cis*-5′-hydroxythalidomide (Figure 9.7), in the latter case using supercritical carbon dioxide as solvent [72].

The resolution of racemic alcohols in which the hydroxyl residue is not at the stereogenic center is of particular interest and can be done through so-called remote resolutions [73]. The best results have been found when the stereocenter is in β-position with respect to the alcohol function, while when moving the stereocenter to a further position the processes result in a loss of chiral recognition. In this context, Bäckvall and Deska have reported the KR of axially chiral primary allenic alcohols, which occurred with high enantioselectivities by using a crude preparation of porcine pancreatic lipase (PPL) in combination with five equivalents of vinyl butanoate using DIPE at room temperature (Table 9.2) [74]. In some cases a quaternary stereocenter contains a hydroxymethyl residue, and the remote resolution is achieved

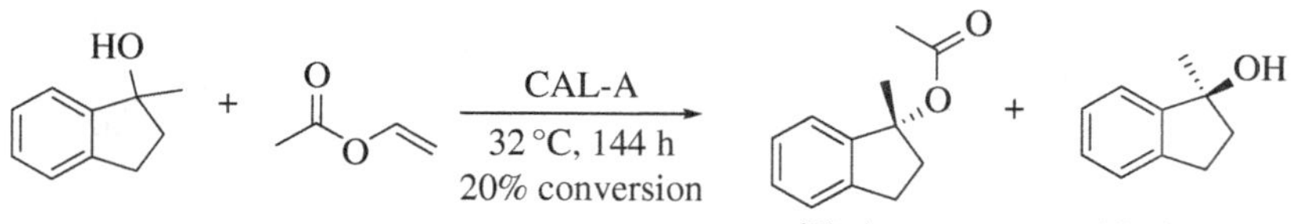

FIGURE 9.8
Kinetic resolution of 1-methyl-1,2,3,4-tetrahydronaphthalen-1-ol using CAL-A and vinyl acetate.

by lipase-catalyzed acylation of the alcohol. For instance, Nanda and coworkers described the resolution of α-tetralones using eight equivalents of VinOAc and PSL in DIPE with good selectivity and a variety of substitutions bearing both the aromatic and the aliphatic ring [75, 76]. Similarly, the resolution of α-hydroxymethylcycloalkanones has been successfully achieved using PSL in pentane or AK lipase in toluene with VinOAc as acyl donor [77].

Tertiary alcohols are interesting building blocks for a variety of pharmaceuticals, but only limited examples regarding their hydrolase-catalyzed enantioselective acylation have been currently reported. In these examples *Candida antarctica* lipase type A (CAL-A) [78] has been employed in combination with VinOAc as an acyl donor. The first example was reported by Bornscheuer and coworkers for the resolution of 2-phenylbut-3-yn-2-ol with low to high selectivities using CAL-A ($E<90$) with conversions up to 25% [79]. Using this lipase for the resolution of aromatic ring-fused cyclic tertiary alcohols such as 1-methyl-1,2,3,4-tetrahydronaphthalen-1-ol and 1-methyl-2,3-dihydro-1*H*-inden-1-ol, despite using a large excess of VinOAc, only low conversion was obtained, although for 1-methyl-1,2,3,4-tetrahydronaphthalen-1-ol there was excellent selectivity (Figure 9.8) [80]. Trying to overcome this problem, complementary hydrolytic processes have been attempted using biocatalysts from metagenomic resources, although moderate selectivity was usually achieved in the reaction with the corresponding racemic acetates [81–83].

Finally, the resolution of amino alcohols has been considered; the nucleophilicity of the amino group leads to its preferred acylation compared to hydroxyl functionality. For that reason, the selective protection of the amine functionality is often required for successful resolution through the alcohol moiety, avoiding undesired migration reactions, which lead to a decrease in the optical activity of the final products. For example, in the first attempt of lipase-catalyzed resolution of 2-amino-1-alkanols through transesterification in ethyl acetate (EtOAc), *N*-protection was needed to avoid the nonstereospecific acylation of the amino group [84]. In this case steapsin and pancreatin, two mammalian lipases, showed good stereospecificities (>95% ee); it is also worth noting that their activity rose when adsorbed onto Celite. Kamal and Rao reported the resolution of 2-propanol amines achieving the stereoselective acetylation using 1.3 equivalents of trichloroacetic anhydride and PPL as biocatalyst in 1,4-dioxane at high temperatures (70–85 °C), leading to moderate to high enantiomeric excess of the final products [85]. The resolution of 2-amino-1-phenylethanol containing the amino functionality protected with carbamate functionalities was efficiently resolved using PSL and 2,2,2-trifluoroethyl butanoate [86]. Similarly, the resolution of *trans*-2-aminocyclohexanol and *trans*-2-aminocyclopentanol with VinOAc and PSL in 1,4-dioxane proceeded with excellent selectivities when the amino group was protected with the benzyloxycarbonyl group (Cbz), leading to the (1*R*,2*R*)-amino esters in near enantiopure form (>98% ee) and the remaining (1*S*,2*S*)-substrate with 49% ee for the cyclohexanol and >99% ee for the cyclopentanol derivative after 6 and 5 days, respectively (Figure 9.9) [87]. Similarly good results were obtained in the resolution of the homologues *cis*- and *trans*-*N*-Cbz-3-aminocyclohexanols in the reaction catalyzed by CAL-B with 3 equivalents of VinOAc at 30 °C in THF, leading after 3 h to the amino esters in >97% ee (Figure 9.9) [88]. More recently, Brocklehurst and coworkers have developed this resolution at the 660 g substrate scale using CAL-B, THF, and 2.8 equivalents of EtOAc in a higher substrate concentration (4 M), achieving a 52% conversion after 4.5 days at room temperature [89]. Similarly,

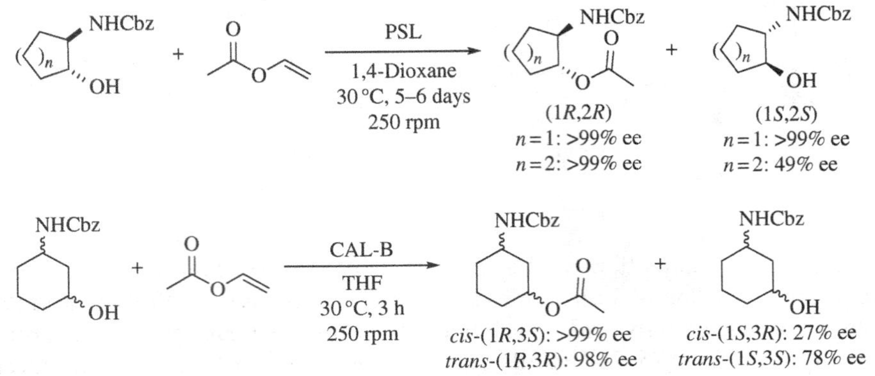

FIGURE 9.9

Lipase-catalyzed acetylation of racemic *trans*-2-aminocycloalcanols, *trans*-3-aminocycloalcanol, and *cis*-3-aminocycloalcanol using lipases and ethyl acetate.

the same strategy has allowed the resolution of *cis*- and *trans*-3-aminoindan-1-ols using CAL-B, MTBE, and EtOAc for the final synthesis of enantiopure orthogonally protected *cis*- and *trans*-indane-1,3-diamines [90].

9.2.2 Dynamic Kinetic Resolution of Racemic Alcohols

The development of multistep-catalyzed processes has allowed the design of synthetic strategies to overcome the limitations of traditional individual steps such as classical KRs. This is possible through biochemical DKR, in which the lability of a stereogenic center is used in an efficient KR [91], but alternatively chemoenzymatic processes using metal catalysis as racemization agent have been widely used [92–96]. In this case a racemization agent that allows the constant presence of the racemic starting substrate (generally an alcohol or an amine) is used in combination with a suitable hydrolase. For this purpose, the compatibility of both catalysts in the reaction medium is required, and additional issues must be taken into account such as [97, 98] (i) the development of an efficient KR, the reaction of one enantiomer being extremely favored in comparison with the slower antipode; (ii) an efficient and fast racemization of the slower enantiomer, which must be extremely faster than the favored biotransformation; and (iii) the avoidance of the product racemization that will depend on its chemical properties.

Satisfyingly a broad number of hydrolase-catalyzed DKR processes have been reported, mainly using lipases as described in previous comprehensive bibliographic revisions [99–103]. Prior to the description of selected examples for the DKR resolution of racemic alcohols via hydrolase-catalyzed acylation, three issues must be highlighted and explained in detail. These are (i) the use of an adequate acyl donor, (ii) the role of the solvent in the reaction, and (iii) the nature of the racemization agent and its compatibility with the hydrolase.

Previously, the use of activated esters such as enol esters, especially VinOAc, has been shown for the classical KR of alcohols. Nevertheless, since the leaving acetaldehyde can react in the DKR with the racemization catalyst, other acyl donors must be considered, like IPA (^{i}PrOAc) or *p*-chlorophenyl acetate (PCPA), the structures of which are shown in Figure 9.10. IPA is a very interesting acyl donor as its byproduct in acylation reactions is acetone, which is a less reactive carbonyl compound that usually does not caused biocatalyst deactivation, its lower atom economy being the only drawback for synthetic purposes.

The solvent is also a critical parameter as it must be compatible with both catalysts avoiding their deactivation and must be inert with the acyl donor. The requirement of an initial activation of metal catalyst for the racemization step has led to the common selection of THF and toluene as ideal solvents, the latter having the advantage of its higher boiling point where a high temperature is required in the system although temperatures over 100 °C are not recommendable due to enzyme denaturation.

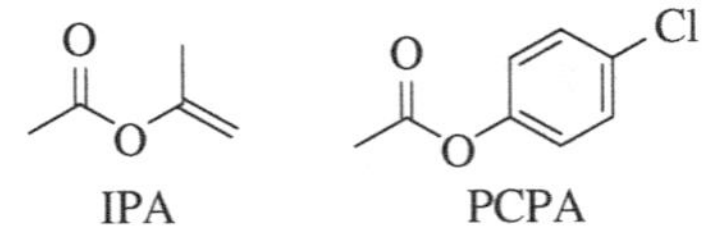

FIGURE 9.10

Structures of suitable acyl donors in hydrolase-catalyzed dynamic kinetic resolution of racemic alcohols.

Other green solvents, such as 2-methyltetrahydrofuran [104] or ionic liquids [105] also have additional advantages for safety reasons.

Since the first DKR of secondary alcohols described by Williams and coworkers in 1996 combining $Rh_2(OAc)_4$, VinOAc, and the *Pseudomonas fluorescens* lipase (PFL) for the asymmetrization of 1-phenylethanol [106], the development of chemoenzymatic DKR using transition metal complexes such as aluminum [107], iridium [108], palladium [109, 110], rhodium [111], vanadium [112, 113], and mainly ruthenium catalysts has been well reported, but materials such as hydrophobic zeolites [114] ionic resins [115–121] or silica-supported reagents [122, 123] can be also used as racemization agents. Alternatively, in some cases the reaction conditions allowed a fast racemization through concomitant reactions such as ring-opening reactions in activated positions [124, 125].

The use of ruthenium catalysts are without any doubt the most common approach for the racemization of the alcohol in a hydrolase–metal combo DKR. The racemization step occurs via hydrogen transfer involving metal hydrides as intermediates, and usually requires the use of an external base as cocatalyst, which can affect enzyme activity and the formation of side products. From this toolbox, Bäckvall and coworkers reported for the first time the use of a dimeric ruthenium catalyst (2 mol%), compatible with CAL-B in the enantioselective acetylation of 1-phenylethanol at 70 °C (Figure 9.11). This ruthenium catalyst was not fully compatible with conventional acyl donors for KR of alcohols such as VinOAc or IPA because of the formation of acetaldehyde or acetone, respectively, which inactivates the metal catalyst. PCPA (three equivalents) was found as an ideal acyl donor, yielding (*R*)-α-methylbenzyl acetate in enantiopure form and 92% isolated yield using *tert*-BuOH as solvent at 70 °C under argon. The presence of one equivalent of acetophenone was required as cocatalyst to promote the ruthenium-catalyzed hydrogen transfer conditions. An optimization of the reaction conditions led to the DKR of a wide panel of secondary alcohols after 24–72 h using identical conditions but employing toluene as solvent and without the addition of acetophenone in the reaction medium [126]. This strategy has also been applied satisfactorily to a vast family of substrates such as allylic alcohols [127], α-hydroxy esters [128], β-hydroxy esters [129, 130], δ-hydroxy esters [131], β-hydroxyalkyl sulfones [132], β-hydroxynitriles [133, 134], and γ-hydroxy amides [135, 136].

On addition other ruthenium catalysts have also been employed, including the use of an indenyl-ruthenium complex in combination with PSL and three equivalents of PCPA in the presence of Et_3N and a catalytic amount of molecular oxygen, which led to the recovery of optically enriched acetates in good yields [137], and dichloro(*p*-cymene)ruthenium(II) dimer has been reported as an ideal catalyst for the synthesis of allylic acetates [138]. Another racemization agent commonly employed in combination with hydrolases such as wild-type or evolved lipases and subtilisin is the chlorodicarbonyl(1,2,3,4,5-pentaphenylcyclopentadienyl)ruthenium(II), which has been used for DKR of racemic aliphatic and aromatic alcohols, chlorohydrins, diols, amino alcohols, and azido alcohols (Figure 9.12) [139–145]. Remarkably, an aminocyclopentadienyl ruthenium chloride complex in combination with an isopropenyl ester and a hydrolase has allowed the DKR of a wide variety of racemic alcohols under mild

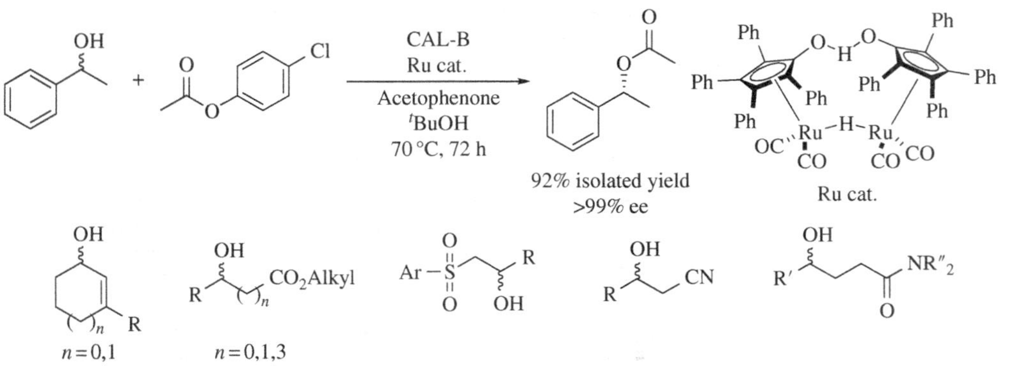

FIGURE 9.11
DKR of racemic 1-phenylethanol and other families of alcohols using a ruthenium dimeric catalyst as racemization agent.

reaction conditions [146, 147]. In addition, the compatibility of this metal catalyst with subtilisin has allowed the synthesis of the opposite enantiomers to those obtained with lipases; these are (*S*)-butanamides using 2,2,2-trifluoroethyl butyrate as acyl donor (Figure 9.12) [148].

9.2.3 Desymmetrization of Diols

Enantioselective hydrolase-catalyzed desymmetrization is a powerful strategy for the selective modification of polyfunctionalized molecules [46]. Particularly acylation procedures have been fully exploited, the acylation of 1,3-propanediol being one of the most prolific alternatives found in the literature. Usually, these studies have not been only focused on the acylation reaction but also on the hydrolytic procedure in order to obtain both monoacylated enantiomers through complementary strategies. The desymmetrization of *N*-Boc-serinol was achieved through its selective monoacetylation with VinOAc as acylating agent and solvent using PPL as biocatalyst after 2 h at 30 °C, yielding the (*R*)-monoacetate in enantiopure form and 69% isolated yield (Table 9.3, entry 1) [149]. This has been exploited to prepare enantiomerically pure Evans auxiliary 4-hydroxymethyl-1,3-oxazolidin-2-one. Similarly, Choi and Borch reported the desymmetrization of *N*-protected serinols for the synthesis of both enantiomers of protected and unprotected 2-hydroxymethylaziridines (Table 9.3, entries 2 and 3) [150].

The fragrance industry has taken advantage of the use of enantioselective desymmetrization processes; for instance, the selective acylation of 2-hydroxymethyl-3-(4-isopropylphenyl)-1-propanol with vinyl butanoate led to the corresponding enantiopure monoester in 97% yield (Figure 9.13), which is a precursor of cyclamen aldehyde, an important component for obtaining special blossom notes in perfume compositions [151]. In a 5 g scale the successful desymmetrization of 2-(2′,2′-dimethoxyethyl) propane-1,3-diol has been reported using PFL and 1.3 equivalents of VinOAc in

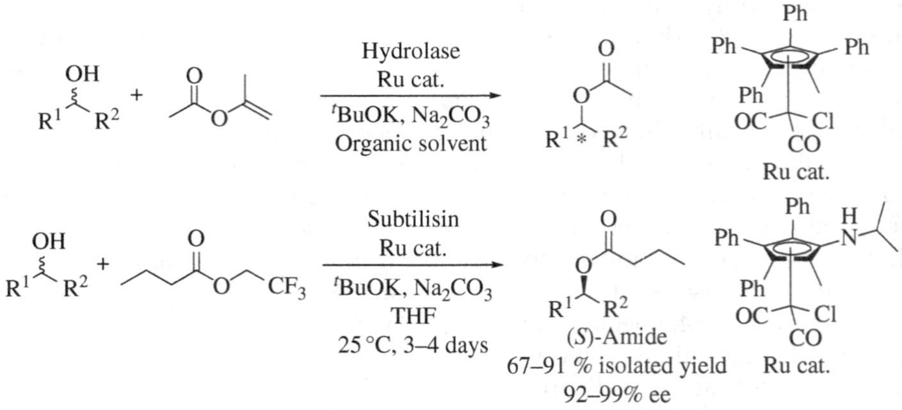

FIGURE 9.12 DKR of racemic alcohols using the different ruthenium complex in combination with lipases or subtilisin.

TABLE 9.3 Desymmetrization of *N*-Protected Serinols Using PPL and Vinyl Acetate

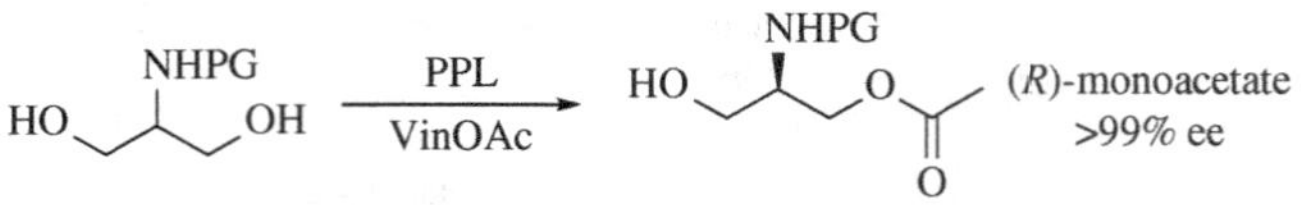

Entry	PG	T	*t* (h)	Isolated Yield (%)	Reference
1	Boc	30 °C	2	69	[149]
2	Fmoc	rt	8	91	[150]
3	Cbz	rt	3	86	[150]

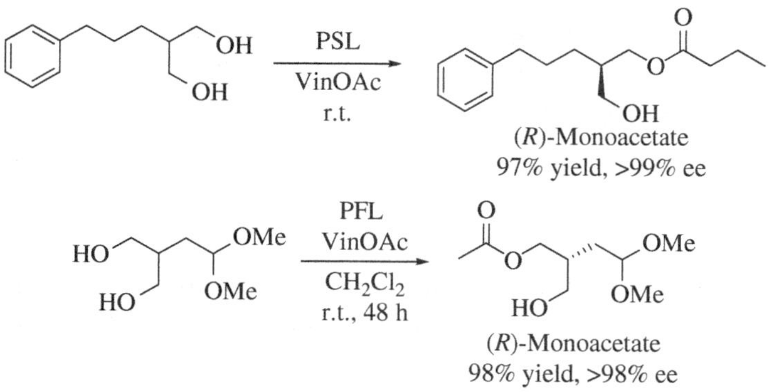

FIGURE 9.13

Desymmetrization of 2-substituted propane-1,3-diols through acetylation reactions using vinyl acetate and a lipase.

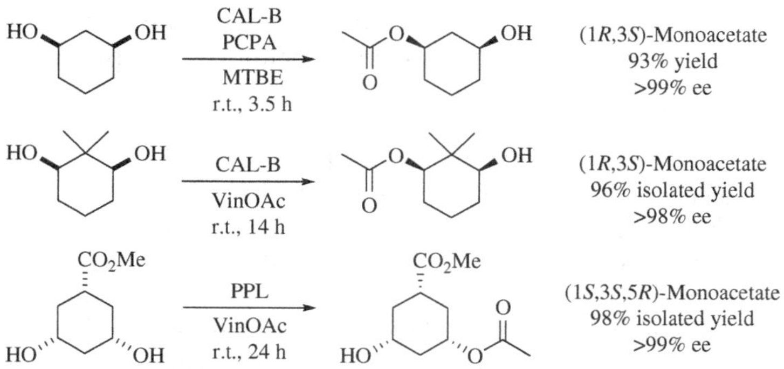

FIGURE 9.14

Lipase-catalyzed desymmetrization of cyclohexanediols.

dichloromethane, yielding (*R*)-monoacetate with excellent results, a compound that served as an ideal precursor for the jaborandi alkaloid (+)-pilocarpine [152].

The desymmetrization of cycloalkane polyols has allowed the synthesis of valuable enantioenriched monoacetates by means of hydrolase-catalyzed transformations. For instance, the efficient desymmetrization of *cis*-1,3-cyclohexanediol to obtain enantiopure (1*S*,3*R*)-3-acetoxy-1-cyclohexanol was reached via CAL-B-catalyzed acetylation using three equivalents of PCPA in 93% yield after 3.5h at room temperature in MTBE (Figure 9.14) [153]. Similarly, *meso*-2,2-dimethyl-1,3-cyclohexanediol was acetylated with VinOAc as an acyl donor and solvent in the presence of CAL-B for the selective production of (1*R*,3*S*)-3-hydroxy-2,2-dimethylcyclohexyl acetate in a 1.5 g scale reaction (Figure 9.14) [154]. *cis,cis*-3,5-dDihydroxy-1-(methoxycarbonyl)cyclohexane has efficiently acylated PPL-catalyzed reaction on a 400g scale with VinOAc as a solvent and acyl donor, yielding after 24h at room temperature the (1*S*,3*S*,5*R*)-monoacetate in enantiopure form with 98% isolated yield (Figure 9.14) [155]. Following this strategy, also efficient acylation of other more structurally complex families of compounds has been achieved, such as 1,3,5-cyclohexanetriols or [156] tetrahydrooxepins [157].

Finally, it is worthy mentioning other families of diols that are less known but have also been selectively desymmetrized using lipase acylation protocols. Hammel and Deska reported the acetylation of prochiral tetrasubstituted allenic diols, yielding highly enantioenriched axially chiral allenyl monoesters (68–99% ee) with good yields (59–90%), after their reaction with five equivalents of vinyl butanoate in 1,4-dioxane at 40 °C using PFL as biocatalyst [158]. Other prochiral diols bearing a heteroatom such as boron [159] or sulfur [160], have also been studied, leading usually to modest yields or selectivities.

9.2.4 Selected Examples of Acylation Reaction with Interest for the Pharmaceutical Industry

The synthesis of enantiopure drugs has taken advantage of the development of biocatalyzed reactions [161], especially of asymmetric hydrolase-catalyzed processes [41, 162]. In this context, many different biocatalysts have been used to achieve

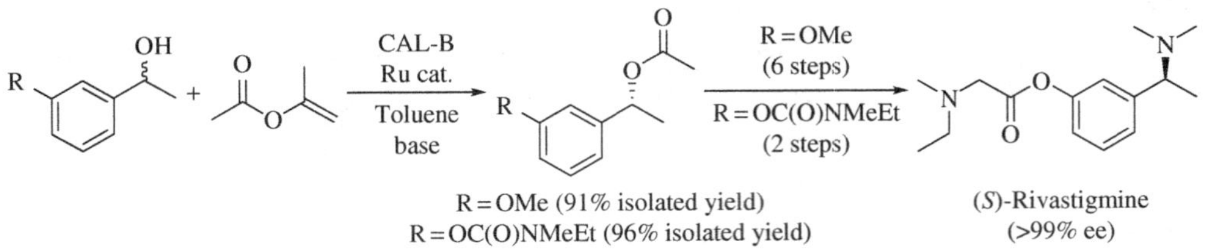

FIGURE 9.15
DKR of racemic 3-substituted (1-phenyl)ethanols using CAL-B and isopropenyl acetate in combination with a ruthenium catalyst.

a successful selective modification of a final drug, or most commonly, of an adequate intermediate to follow the synthetic strategy toward the formation of the chiral drug later.

We have made a brief selection of synthetically useful transformations for the production of a valuable pharmaceutical based on the action of the enzyme in terms of enantiomeric excess and conversion, or alternatively because of the complexity of the molecule and the impossibility of obtaining better results through other current enzymatic or nonenzymatic processes.

Rivastigmine is an acetylcholinesterase inhibitor of the carbamate type, the (*S*)-enantiomer having applications in mild to moderate dementia caused by Parkinson's disease as the commercialized tartrate salt called *Exelon*. Two independent DKRs based on lipase-catalyzed acetylation combined with metal catalysis have been described so far (Figure 9.15). Firstly, racemic 1-(3-methoxyphenyl) ethanol was enantioselective acylated with CAL-B and 1.5 equivalents of IPA using chlorodicarbonyl(1,2,3,4,5-pentaphenylcyclopentadienyl)ruthenium(II) as racemization agent, yielding the corresponding enantiopure (*R*)-acetate in 91% isolated yield after 24h at 50 °C in toluene [163]. Alternatively, at room temperature, the acetylation of 3-(1-hydroxyethyl)phenyl ethyl(methyl)carbamate was successfully achieved in 96% isolated yield using a ruthenium polymer-bound catalyst that allowed the recycling of both the metal and the enzyme, which did not suffer significant loss of activity and selectivity until the fourth recycling use, improving their action again when one additional equivalent of K_2CO_3 was added [164]. (*S*)-Clopidogrel is an antithrombotic agent commercialized with the brand name *Plavix*, which has been obtained through the resolution of racemic methyl-2-chloromandelate using CAL-A in a solvent-free transesterification reaction. Thus, the methyl (*R*)-2-chloromandelate was obtained in an enantiomerically pure form and 41% yield in a 500g substrate scale after 18h at 30 °C [165]. Significantly, although the enantioselectivity was not very high (E = 34.7), CAL-A maintained its catalytic activity for 13 cycles without a significant decrease in the activity and stereoselectivity values.

Dichotomine A and dichotomide II are β-carboline alkaloids that possess different biological profiles including antiallergic drugs. Their synthesis has been possible after enantioselective acetylation of a key alcohol intermediate, namely, methyl 1-(1-hydroxyethyl)pyrido[3,4-*b*]indole-3-carboxylate [166]. The QLM lipase in combination with VinOAc in MTBE led to the formation of the (*S*)-(−)-alcohol and the (*R*)-(+)-acetate both in 96% ee after 72h at 32 °C (Figure 9.16). Another indole structure is ramatroban, which is commercialized in Japan with the trade name Baynas, showing remarkable applications for the treatment of asthma, rhinitis, and coronary diseases, with its activity mainly residing in the (*R*)-enantiomer. CAL-B has allowed the selective acylation of racemic 2,3,49-tetrahydro-1*H*-carbaz-3-ol using three equivalents of VinOAc at 30 °C in THF for 8h, yielding in enantiopure form both the (*R*)-acetate and the (*S*)-alcohol (Figure 9.16) [167].

Fluoxetine, tomoxetine, and nisoxetine are three well-known antidepressants for the treatment of psychiatric disorders and metabolic problems (Figure 9.17). All of them possess a 3-aryloxy-3-phenylpropylamine core, their enantiomers having very different biological responses, so asymmetric synthetic strategies are required for the access to these valuable drugs. In this context, β-hydroxynitriles and in particular 3-hydroxy-3-phenylpropanenitrile seem to be an adequate precursor, so the lipase-catalyzed acylation has been optimized using six equivalents of VinOAc, PSL, and

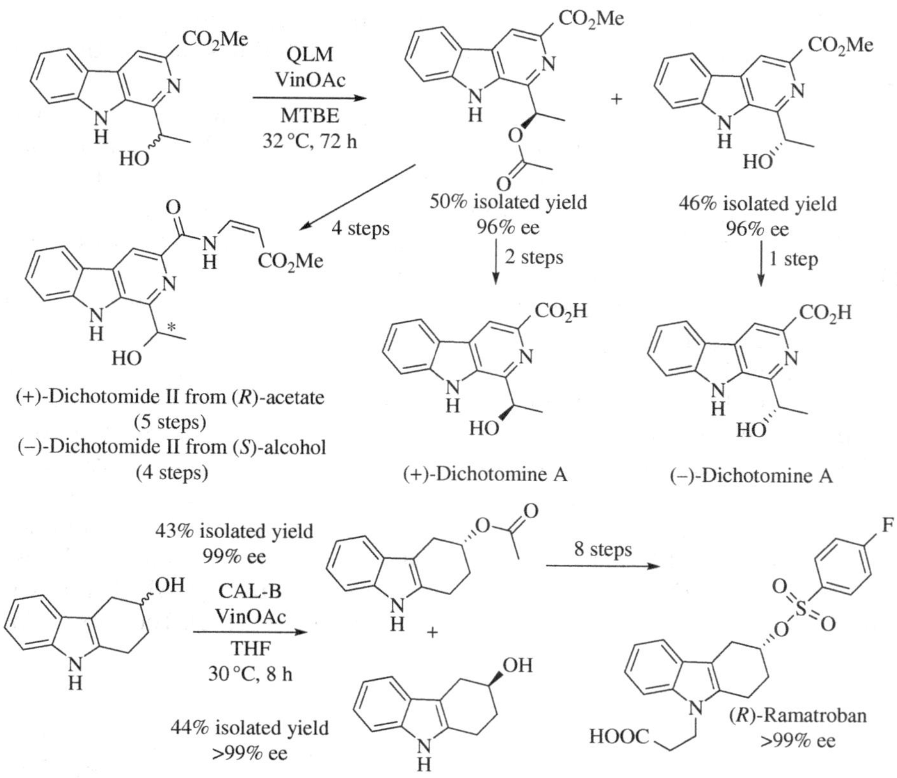

FIGURE 9.16 Lipase-catalyzed acetylation of indole derivatives for the asymmetric synthesis of β-carboline alkaloids and ramatroban.

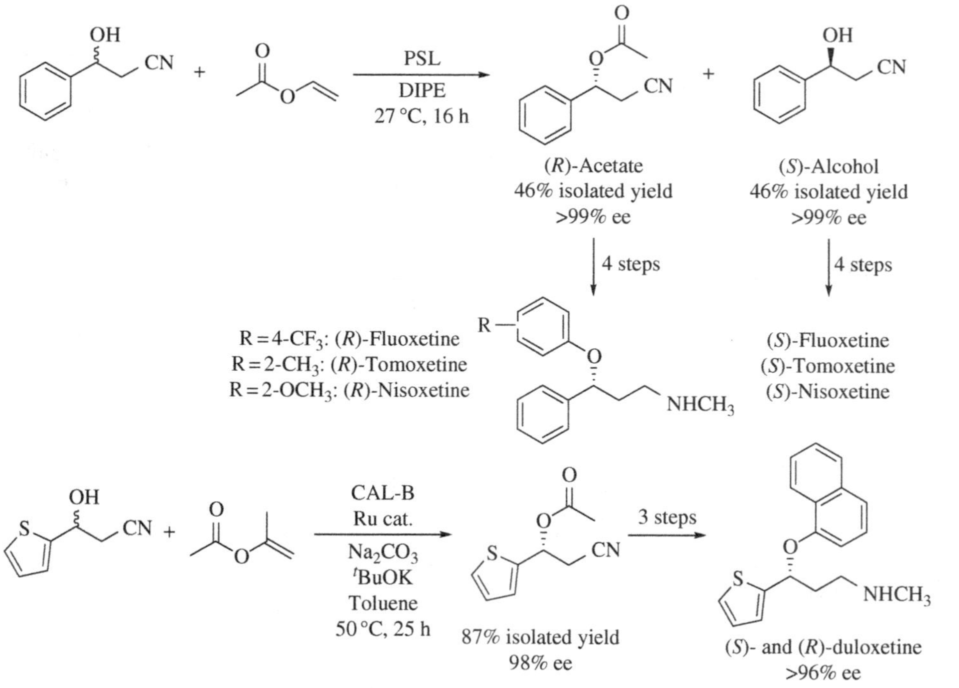

FIGURE 9.17 Lipase-catalyzed acetylation of 3-hydroxy-3-substituted propanenitriles for the global synthesis of fluoxetine, tomoxetine, nisoxetine, and duloxetine enantiomers.

DIPE as solvent [168]. The access to both (*R*)- and (*S*)-enantiomers was elegantly developed by the enzyme-catalyzed KR. Similarly, the lipase-catalyzed acetylation of 3-hydroxy-3-(2-thienyl)propanenitrile using PSL and later chemical modifications led to the synthesis of duloxetine enantiomers, which are also antidepressant agents [169]. The DKR of the β-hydroxynitrile precursor of duloxetine reported by Bäckvall and coworkers has allowed the improvement of the overall yield for the synthesis of these drug enantiomers (Figure 9.17) [170].

Bufuralol is a potent β-adrenergic receptor antagonist with applications in the treatment of hypertension, the (*S*)-enantiomer being 100 times more active than its counterpart. The enantioselective acylation of racemic 2-bromo-1-(7-ethylbenzofuran-2-yl)ethanol has provided access to the (*R*)-alcohol and different (*S*)-esters depending on the acyl donor (Figure 9.18). Thus, CAL-B has catalyzed the formation of both alcohol and ester in enantiopure form using 2 equivalents of vinyl dodecanoate in toluene after 16 h at room temperature in a 1 g scale reaction [171]. The DKR of 2-chloro-1-(7-ethylbenzofuran-2-yl)ethanol was effectively achieved in this case using the combination of PSL, IPA, and chlorodicarbonyl(1,2,3,4,5-pentaphenylcyclopentadienyl) ruthenium(II) at 40 °C for 24 h, yielding the enantiopure (*R*)-acetate in 96% yield [172].

The use of long-chain vinyl esters such as vinyl stearate has also allowed the multigram KR of 1-chloro-3-[4-(2-methoxyethyl)phenoxy]2-propanol, which is an intermediate in the chemoenzymatic synthesis of (2*S*)-2-[[(2*R*)-2-hydroxy-3-[4-(2-methoxyethyl)phenoxy]propyl]amino]-1-butanol, a potent antihypertensive agent (Figure 9.16) [173]. CAL-B showed an excellent discrimination capability in a mixture composed of hexane and acetone (95:5) at 27 °C during 24 h, giving a 35% yield of the (*S*)-ester using only 0.5 equivalents of the acyl donor. (*S*)-Propranolol is also widely used as antihypertensive drug, and its chemoenzymatic synthesis was achieved through the DKR of a β-azido alcohol precursor using CAL-B and PCPA in combination with a dimeric ruthenium catalyst (Figure 9.19) [174]. The corresponding

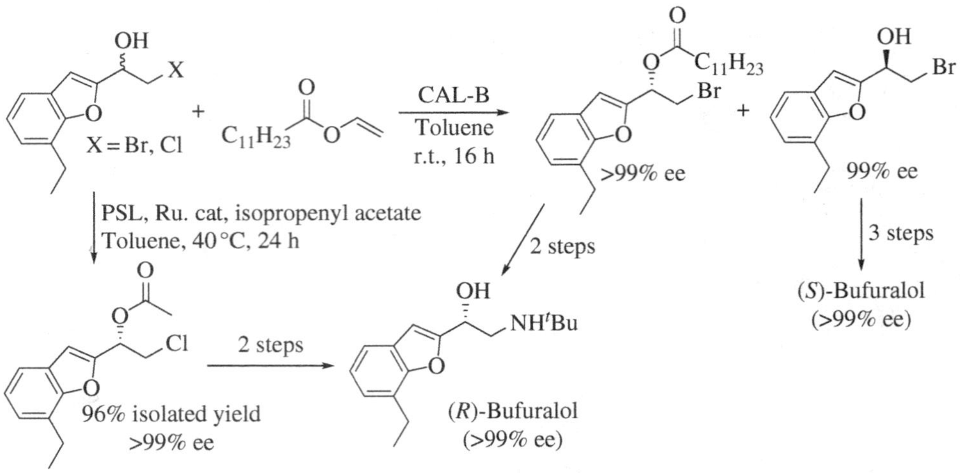

FIGURE 9.18

Lipase-catalyzed KR and DKR of 2-halo-1-(7-ethylbenzofuran-2-yl) ethanols for the chemoenzymatic asymmetric synthesis of bufuralol enantiomers.

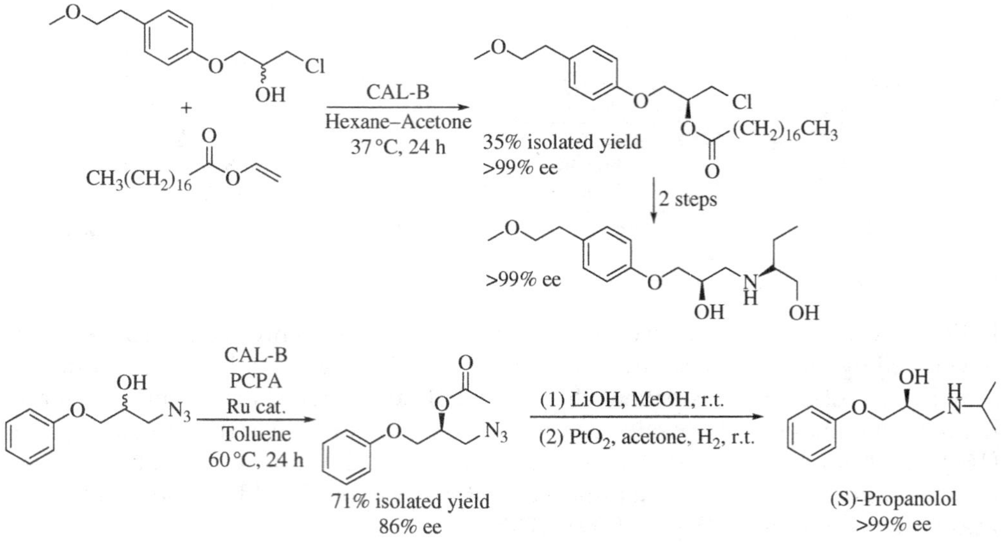

FIGURE 9.19

Chemoenzymatic synthesis of two antihypertensive agents: (2*R*,2′*S*)-β-hydroxyhomometoprolol and (*S*)-propranolol through CAL-B-catalyzed acylation reactions.

(*S*)-acetate was obtained in 86% ee and 71% yield, improving the optical purity of the final products after its chemical synthesis and final recrystallization purification.

In some cases, the lipase-catalyzed acylation provides moderate selectivity, which can be improved by including a second KR; this is more tedious, but the development of a double KR can be used in order to obtain both substrate and product in high optical purity as performed with PPL and VinOAc in the acetylation of 2-arylpropan-1-ols (Figure 9.20) [175]. For instance, when the formation of the optically enriched (*S*)-acetate reached around 60% conversion, the alcohol and the acetate were separated by flash chromatography, and then the enantioenriched acetate was submitted to a chemical hydrolysis process and another lipase-catalyzed resolution to obtain the desired (*S*)-acetate in enantiomerically pure form. Thus, the corresponding (*R*)-2-(2-methoxy-4-methylphenyl)propan-1-ol has served as adequate intermediate in the synthesis of the phenolic sesquiterpene (*S*)-turmenorol B, which is a strong antioxidant and lipoxygenase inhibitor. Similarly, the KR of other substituted 2-aryl-propan-1-ols can lead to the formation of interesting sesquiterpenes such as (*R*)-curcumene, (*R*)-curcuphenol, (*R*)-xanthorrhizol, and (*R*)-curcuhydroquinone.

SCH 51048 is a THF-based antifungal agent, which can be synthesized through the lipase-catalyzed desymmetrization of an adequate homoallylic diol. CAL-B have allowed the selective synthesis of the enantiopure (*S*)-monoacetate in 71% isolated yield at a 30 kg batch reaction using two equivalents of VinOAc in acetonitrile (MeCN) at 0 °C after 6 h (Figure 9.21) [176]. Complete conversion of the diol was observed, yielding byproduct diacetate in around 30%, which was separated by column chromatography. The desymmetrization of other 1,3-propanediol subunits has been

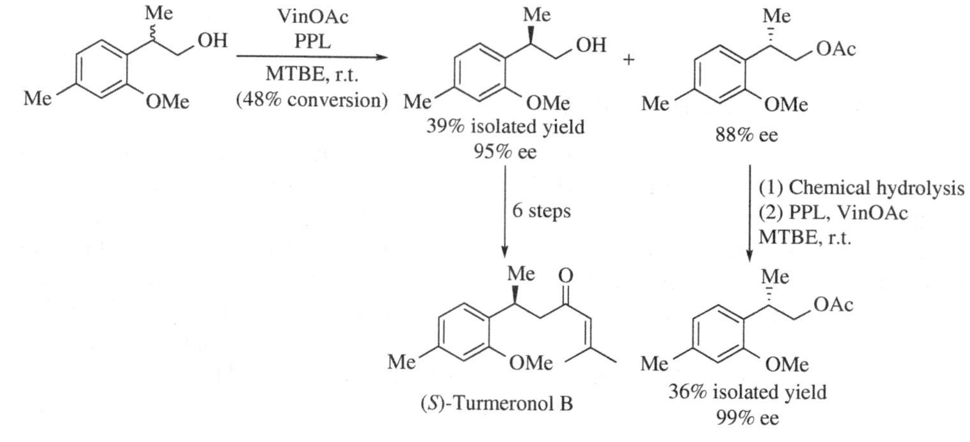

FIGURE 9.20
Double kinetic resolution of 2-phenylpropan-1-ol with vinyl acetate using PPL as biocatalyst for the synthesis of (*S*)-turmenorol B.

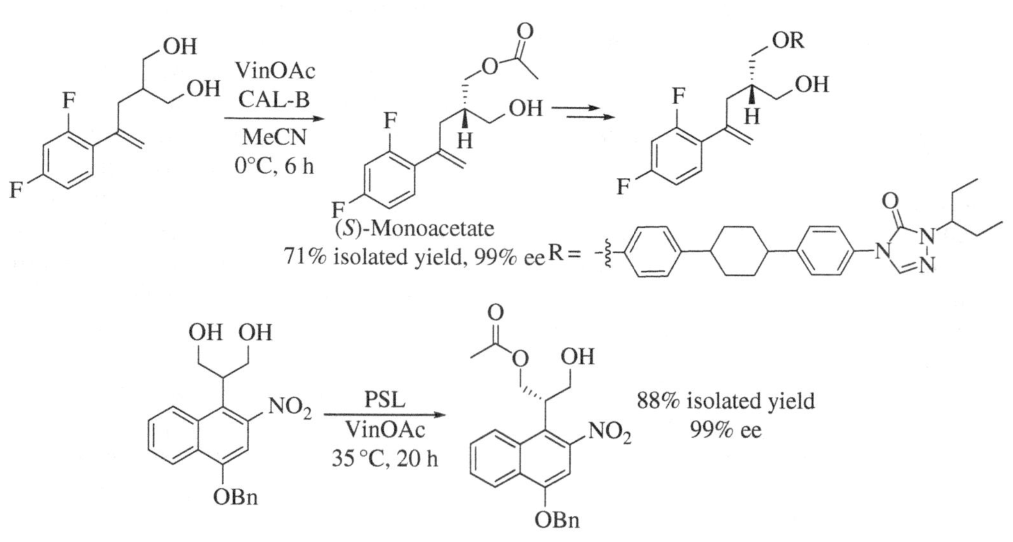

FIGURE 9.21
Lipase-catalyzed desymmetrization of prochiral diol intermediates for the synthesis of antifungal and antitumor antibiotics.

successfully carried out with applications in the synthesis of drugs or interesting synthetic precursors; for instance the asymmetric synthesis of 1,2,9,9a-tetrahydrocyclopropa[c]benzo[e]indol-4-one has been achieved, the key step consisting in the desymmetrization of a prochiral diol using PSL and VinOAc as solvent and acyl donor, leading to the desired (*S*)-monoacetate in 88% isolated yield and 99% ee (Figure 9.21) [177].

9.3 ACYLATION OF AMINES

The search for optimum methods and conditions in the production of chiral amines is nowadays considered as a key issue for the industrial sector [178]. Traditionally, hydrolases have played a major role in this area [179, 180], although other classes of enzymes such as amine dehydrogenases, monoamine oxidases, phenyl aminomutases, and transaminases have opened a myriad of possibilities in this field [181, 182].

In this context, van Rantwijk and Sheldon reported in the last decade some remarkable examples of the potential of serine hydrolase for the enantioselective acylation of amines [183]. Here, more recent reactions will be discussed starting with KR of primary and secondary racemic amines and a few examples of the desymmetrization of diamines, a reaction that has scarcely been exploited. Acylation reactions are quite selective transformations for the formation of the desired final optically active amines and amides; however, in some cases, a sister reaction such as the alkoxycarbonylation process using a carbonate instead of an ester will be mentioned in order to present some advantages, especially for the resolution of secondary cyclic amines.

9.3.1 Kinetic Resolution of Racemic Amines

The enzymatic KR is based on the reaction between the most favored enantiomer of a racemic amine and a suitable acyl donor in an organic solvent. The acylation of amines can also be considered as an aminolysis process, the examples for the KR on primary amines being more common in the literature than the resolution of secondary amines. From the hydrolases toolbox, lipases seem to be the most suitable enzymes due to the fact that they scarcely ever hydrolyze an amide bond [184, 185], and, therefore, the reaction is practically irreversible. CAL-B has usually shown the best activity and selectivity for the resolution of racemic amines [186, 187].

A key aspect is the selection of a good acyl donor, which may be reactive enough to allow a fast selective acylation but avoiding a chemical reaction that would lead to a decrease in the enantiomeric excess of the resulting amide. Thus, irreversible acyl donors such as vinyl or isopropenyl esters, widely employed in the KR of alcohols as stated before, are not appropriate when reacting with amines since they generate acetaldehyde and acetone, respectively, leading to the formation of imines as by products. Nonactivated esters such as EtOAc or alkyl methoxyacetates have been widely employed not only as acyl donors but also as solvents (Figure 9.22). These donors also possess the advantage of being liquids and are commercially available, so their use as both acyl donors and solvents is also possible when low kinetic rates are observed. On the contrary, their use in large amounts can lead in some cases to uncatalyzed chemical reactions, so the use of 2–3 equivalents of the nonactivated esters in combination with dry solvents as MBTE and THF is a common practice.

The isolation of the amine and amide products is not limited to column chromatography, and their purification can be efficiently carried out after simple extraction procedures involving the formation of the corresponding amine salt that remains in the aqueous phase, while the amide can be extracted with an organic solvent as depicted in Figure 9.22.

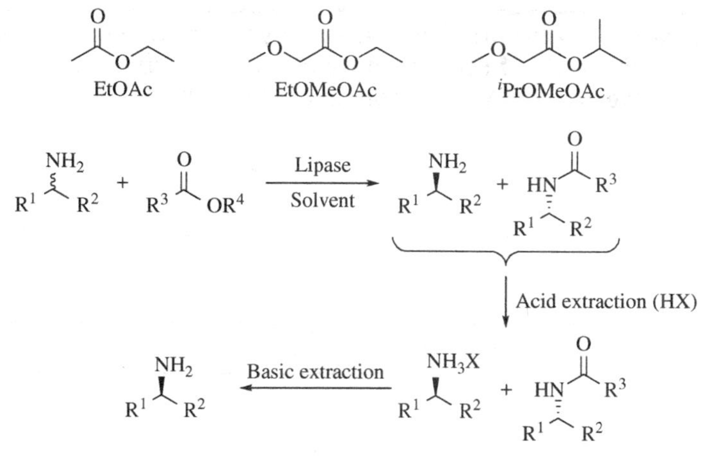

FIGURE 9.22 Structures of common non-activated esters for the kinetic resolution of racemic primary amines through acylation reactions and schematic representation of the isolation and purification procedure.

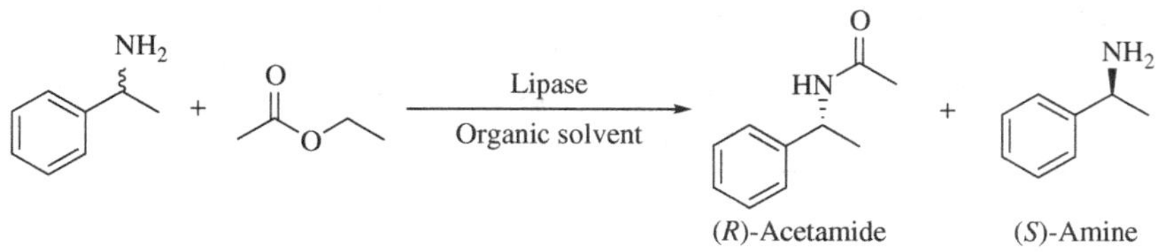

FIGURE 9.23 Lipase-catalyzed kinetic resolution of α-methylbenzylamine using lipases in organic solvents.

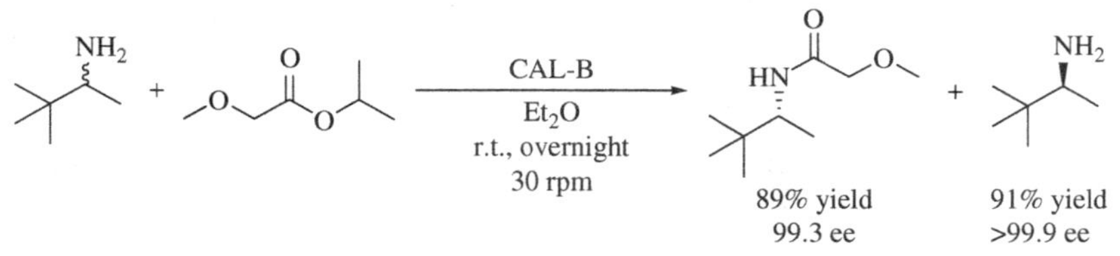

FIGURE 9.24 Lipase-catalyzed kinetic resolution of 3,3-dimethyl-2-butanamine using isopropyl methoxyacetate in diethyl ether.

The resolution of α-methylbenzylamine also, called 1-phenylethylamine, has received considerable attention as occurs with 1-phenylethanol in the screening of appropriate biocatalysts for secondary alcohol resolution (Figure 9.23). Souza and coworkers have developed an efficient method for the resolution of α-methylbenzylamine in a continuous flow system using EtOAc and short residence times (40 min) [188]. Similarly, the combination of EtOAc and CAL-B has also served for the KR of boron-containing 1-phenylethylamine derivatives [189].

The use of another acyl donor such as isopropyl methoxyacetate has allowed the resolution of 1-phenylethlamine-substituted derivatives under free solvent conditions in the presence of CAL-B [190], this also being effective for aliphatic amines containing long side chains, and naphthyl and pyridine derivatives using the same lipase and diethyl ether (Et_2O) as solvent (Figure 9.24) [191].

Ethyl methoxyacetate (EtOMeOAc) has also been employed successfully for the resolution of racemic amines such as 1-aryl-2-fluoroethylamines [192], 1-phenylbut-3-en-1-amine [193], or 1-arylallylamines [194]. The use of this acyl donor improves the results for sterically hindered aminoalkylpyridines with respect to the ones obtained with EtOAc in some cases (Table 9.4) [195]. Significantly, the kinetic and stereoselectivity values of the reactions dramatically decrease with bulkier alkyl substituents.

The resolution of primary amines with long-chain esters and their corresponding carboxylic acids has been less explored in comparison with the use of EtOAc and alkyl methoxyacetates. However, lauric acid ($C_{11}H_{23}CO_2H$) has served as an ideal acyl donor for the CAL-B-catalyzed resolution of aliphatic and aromatic amines at 80 °C using heptane as solvent (Figure 9.25) [196]. The use of carboxylic acids led to a marked acceleration of the reaction rates compared to their ethyl ester counterparts.

TABLE 9.4 Kinetic Resolution of Aminoalkylpyridines Using Lipases and 5 Equivalents of Ethyl Acetate (EtOAc) or Ethyl Methoxyacetate (EtOMeOAc) as Acyl Donors in MTBE at 30 °C

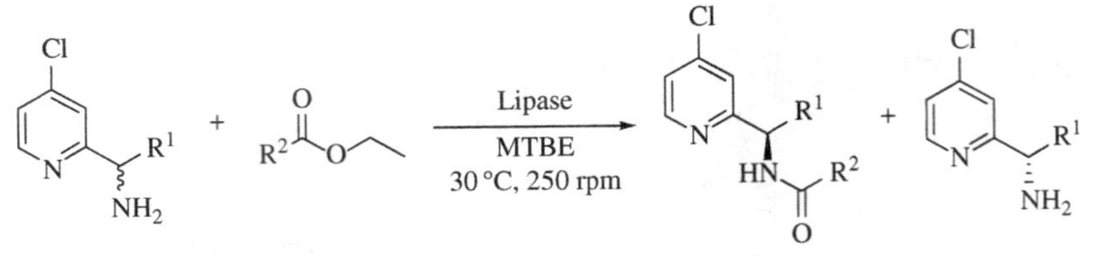

Entry	R	Enzyme	Acyl Donor	*t* (h)	ee_p (%)	ee_S (%)	*c* (%)	*E*
1	Me	CAL-B	EtOAc	4	>99	>99	50	>200
2	Me	PSL	EtOAc	32	>99	>99	50	>200
3	Et	CAL-B	EtOAc	31.5	>99	>99	50	>200
4	Et	PSL	EtOAc	48	>99	95	49	>200
5	*n*-Pr	CAL-B	EtOAc	72	74	10	12	7
6	*n*-Pr	PSL	EtOAc	72	96	29	23	68
7	*n*-Pr	PSL	EtOMeOAc	48	93	97	51	114
8	*n*-Bu	PSL	EtOMeOAc	48	92	87	49	70

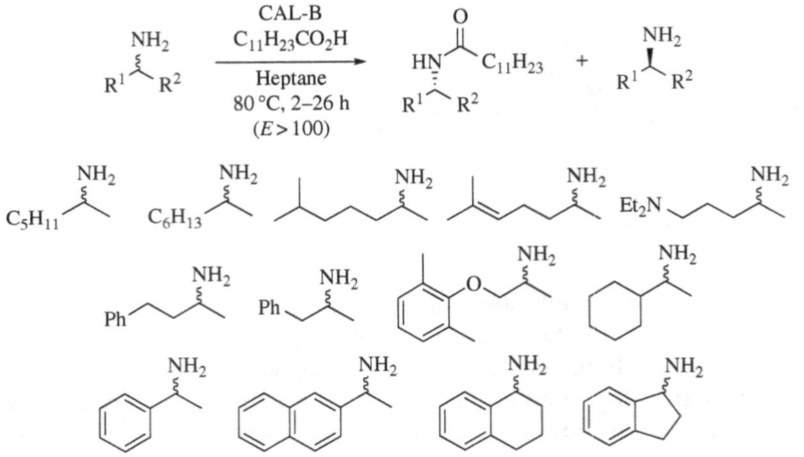

FIGURE 9.25

Lipase-catalyzed kinetic resolution of primary amines using lauric acid as acyl donor.

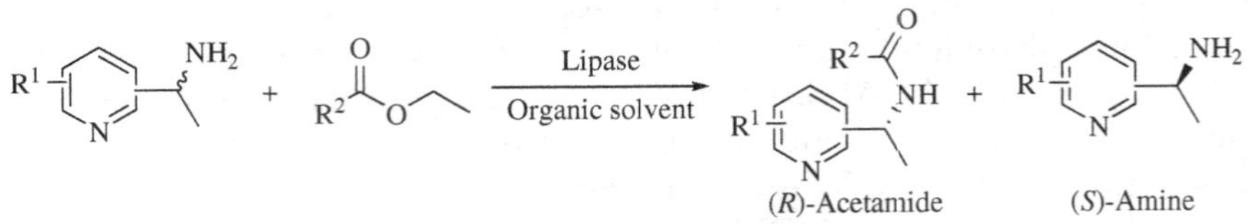

FIGURE 9.26

Enzymatic kinetic resolution of pyridine derivatives (R^2 = Me for EtOAc as acyl donor).

Similarly, *N*-acyl glycinates like *N*-octanoylglycine trifluoroethyl ester [197] and *N*-octanoyldimethylglycine trifluoroethyl ester [198] have been successfully employed as acyl donors in subtilisin-catalyzed KR of aliphatic and benzylic amines. Based on the opposite selectivity displayed by subtilisin and lipases, in these cases the (*S*)-amides and (*R*)-amines were obtained, both with excellent stereodiscrimination values.

The resolution of optically active 1-(heteroaryl)ethanamines has received considerable attention as they are good substrates for lipase-catalyzed resolution using EtOAc as an acyl donor in combination with an organic solvent or as both solvent and acylating agent [199–201], obtaining the corresponding amines and amides with high enantiomeric excesses (Figure 9.26).

Structurally more complex 1-(heteroaryl)ethanamines [202, 203] and 1-(2 phenylthiazol-4-yl)ethanamines [204] have also been efficiently acylated using CAL-B and different acyl donors such as EtOAc, isopropyl butanoate, or ethyl butyrate in organic solvents, probing the efficiency of the use of lipases for the asymmetric preparation of enantiopure amines and amides (Figure 9.27).

The KR of racemic amines bearing more than one stereocenter is also challenging, and lipases have acted with a high level of stereodiscrimination in combination with nonactivated esters. For instance, the CAL-B-catalyzed resolution of *trans*-2-phenylcyclopentanamine with EtOAc proceeds with an excellent enantioselectivity, while the reaction with *cis*-2-phenylcyclopentanamine led to poorer results ($E = 16$ and 28% conversion) [205]. Nevertheless, this problem has been overcome using other acyl donors such as racemic *cis*-2-phenylcyclopentyl methoxyacetate (Figure 9.28). This strategy can be also successfully used for the resolution of *trans*-2-phenylcyclohexanamine [206], *trans*-*N*-substituted-cyclopentane-1,2-diamines [207, 208], *trans*-*N*-substituted-cyclohexane-1,2-diamines, [209] and *cis*-cyclohex-4-ene-1,2-diamine [210].

The KR of secondary cyclic amines through hydrolase-catalyzed acylation reactions has scarcely been reported. One of the main drawbacks of these reactions is the hydrolysis side reactions of the obtained amides, which lead to a decrease in the enantioselectivity of the process [211]. To circumvent this problem, alkyl carbonates are usually employed as alkoxycarbonylation agents. These reactions are irreversible, as the resulting carbamates are not adequate substrates for the serine-type hydrolases. Unfortunately, this methodology cannot be applied to primary amines because a significant background reaction is observed without enzyme [212]. Breen reported the use of a symmetrical carbonates combining allyl alcohol and a phenol residue, such as allyl 3-methoxyphenylcarbonate for the KR of 1-methyl-tetrahydroisoquinoline using a cross-linked enzyme crystal of *Candida rugosa* lipase (Figure 9.29) [211], achieving excellent selectivity values in comparison

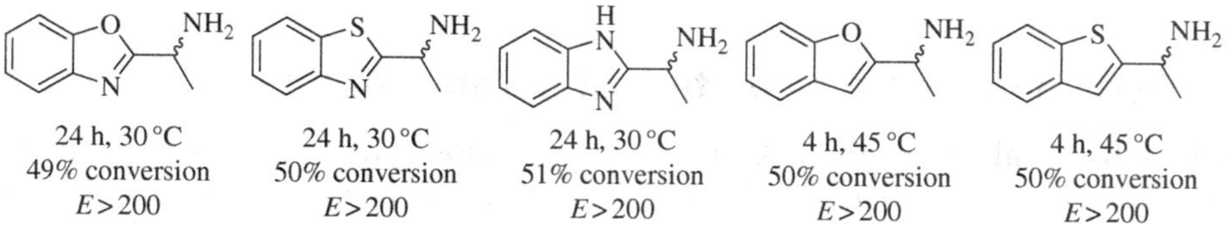

FIGURE 9.27 Structures of racemic 1-(heteroaryl) ethanamines and representative activity and stereoselectivity values for their kinetic resolution.

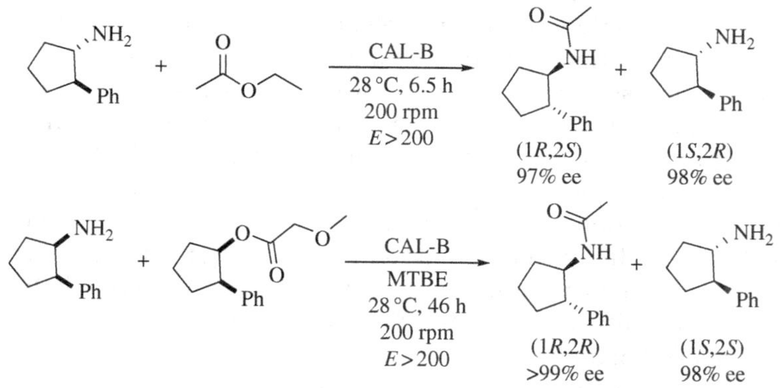

FIGURE 9.28 Lipase-catalyzed acetylation of racemic *trans*- and *cis*-2-phenylcyclopentanamine.

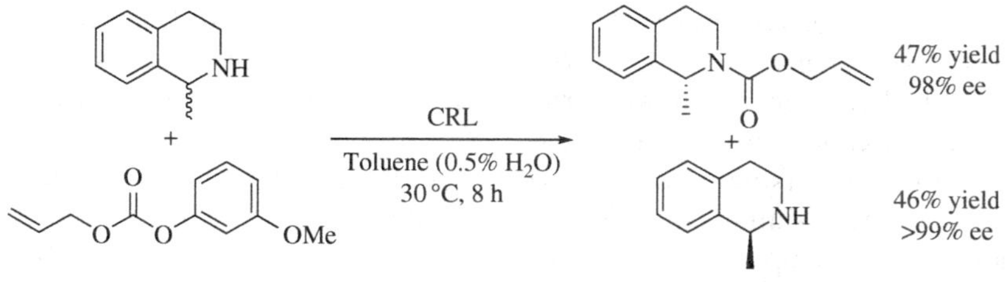

FIGURE 9.29 Lipase-catalyzed kinetic resolution of 1-methyl-tetrahydroisoquinoline with allyl 3-methoxyphenyl carbonate using CRL as biocatalyst.

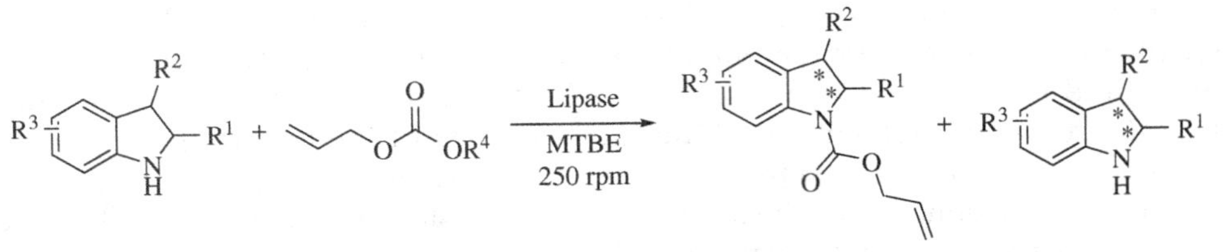

FIGURE 9.30
Lipase-catalyzed kinetic resolution of substituted indolines.

with the recovery of an 8% of racemic amide when using EtOAc as acyl donor. In a similar approach but using CAL-A and 3-methoxyphenyl allyl carbonate, the KR of the alkaloid 1-methyl-6,7-dimethoxy-1,2,3,4-tetrahydroisoquinoline was achieved obtaining the (*R*)-carbonate and the (*S*)-amine both in 98% with 50% conversion after 72 h at 40 °C in toluene [213].

The same strategy has been employed successfully for the KR of mono-2- or mono-3-substituted indolines [214] and 2,3-disubstituted indolines [215] using CAL-A or CAL-B (Figure 9.30). Similarly, cyclic α-amino esters derived from the 2,3-dihydroindole and octahydroindole cores have also been resolved successfully using allyl carbonates and CAL-A, attaining excellent selectivity and conversion values [216]. Although less explored, complementary hydrolytic processes have also been reported, as, for instance, Ostaszewski and coworkers used "homemade" animal liver acetone powders for the KR of 2-acetyl-4-phenyl-1,4-dihydro-2*H*-isoquinolin-3-one [217].

Lipases can also perform the KR of conformational isomers, processes that have been useful in the synthesis of a key intermediate of the farnesyl protein transferase inhibitor, SCH 6636 [218]. The use of the lipase Toyobo LIP-300 and 3 equivalents of trifluoroethyl isobutyrate for a 2.3 g of substrate *N*-acylation reaction allowed the separation of the piperidine enantiomers, which exist due to atropisomerism about an exocyclic double bond, obtaining after 26 h at room temperature the substrate and product in around 97% ee after 50% conversion.

9.3.2 Dynamic Kinetic Resolution of Racemic Amines

Like the DKR of alcohols, the DKR of amines provides an efficient access to enantiopure compounds in theoretically 100% yield. In this case amides are the final products obtained after the combination of an adequate acyl donor and a racemization agent [96, 99–101, 219]. Amine racemization usually requires higher temperatures and/or strongly basic media in comparison with alcohol racemization [96] and must be compatible with preserving the enzyme activity.

The racemization strategies for the DKR of amines can occur mainly through two different pathways. On one hand, the amine racemization can proceed via metal-catalyzed formation of an achiral imine intermediate, which is subsequently hydrogenated in a nonselective fashion to obtain the racemic amine [220]. Alternatively racemization can be achieved by means of the reversible homolytic abstraction of the α-hydrogen atom of the amine, allowing the interconversion of the amine enantiomers via a carbon-centered α-amino radical intermediate. Owing to its high thermal stability, even at 90–100 °C, CAL-B has been the most common hydrolase in the DKR of amines.

Next, the development of chemoenzymatic DKR for primary and secondary amines is discussed, starting from the first DKR of a primary amine, in this case 1-phenylethylamine. This example was described by Reetz and Schimossek using four equivalents of EtOAc as acyl donors, palladium on charcoal as racemization catalyst, CAL-B as biocatalyst, and triethylamine (Et_3N) as solvent, recovering the corresponding (*R*)-acetamide in 64% yield after 8 days at 55 °C [221]. The use of palladium supported by alkaline earth salts ($BaSO_4$, $CaCO_3$, or $BaCO_3$) allows the reduction of the reaction time in similar conditions, yielding the enantiopure (*R*)-*N*-(1-phenylethyl) acetamide and other optically active benzylamines in good yields using Pd/$BaSO_4$

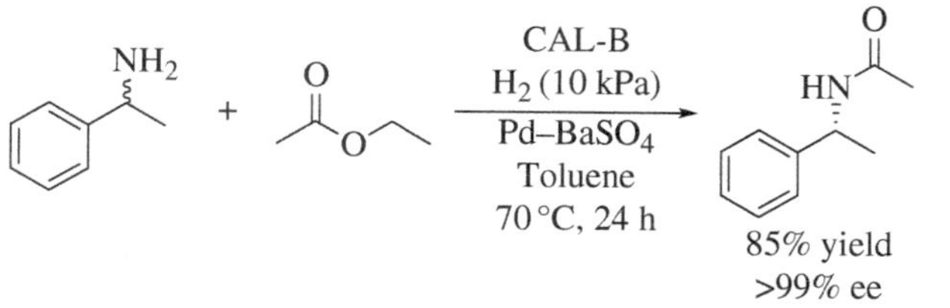

FIGURE 9.31
DKR of 1-phenylethylamine using CAL-B and palladium supported on $BaSO_4$.

TABLE 9.5 DKR of Phenylglycine Amide Derivatives Using 5 mol% of Pd/AlO(OH), CAL-B, and 2 Equivalents of Ethyl Methoxyacetate in Toluene at 60 °C for 3 Days

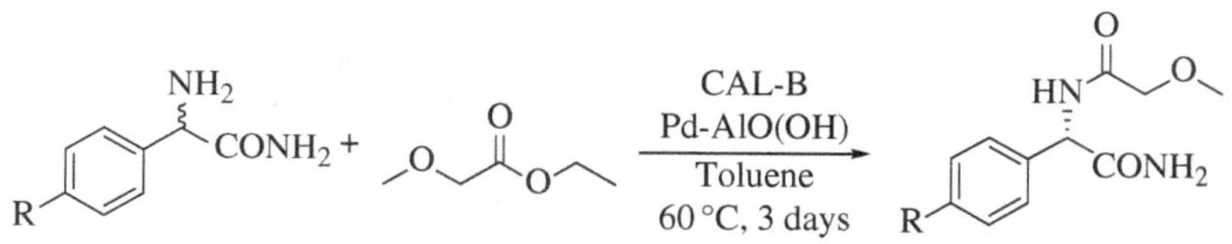

Entry	R	Yield (%)	ee_p (%)
1	F	91	97
2	Cl	88	97
3	OCH_3	91	97
4	CF_3	90	97

(Figure 9.31) [222–225] or Pd/K_2CO_3 with multiple incipient wetness impregnation [226]. Selenium-containing chiral amines have also been used efficiently in DKR processes combining EtOAc, CAL-B, and Pd/$BaSO_4$ in the presence of hydrogen, yielding the corresponding enantiopure amides with up to 87% conversion [227]. On the other hand, the use of palladium supported by aluminum oxide (Pd/Al_2O_3) in combination with CAL-B and isopropyl acetate led to the DKR of α-trifluoromethylated amines; the enantioselectivity increased (>90% ee) when the size difference between trifluoromethyl and substituents connected with the stereocenter of the amines also increased [228].

The use of palladium nanocatalyst entrapped in aluminum hydroxide [Pd/AlO(OH)] has been successfully employed for the DKR of benzyl and aliphatic primary amines at 70 °C [229] or even lower temperatures (40–50 °C) [230], obtaining the corresponding acetamides and methoxyacetamides with good yields and high optical purities. The use of this system was also extended to the DKR of racemic amino acid amides, achieving the production of optically active amino acid derivatives with excellent enantiomeric excess (Table 9.5) [231]. However it is not only CAL-B that is able to perform elegant DKR process of amino acid derivatives using the Pd/AlO(OH) as racemization agent. For instance, Bäckvall and coworkers also reported the use of CAL-A immobilized on different mesoporous materials for the DKR of aromatic, heteroaromatic, and aliphatic β-amino esters using the Pd/AlO(OH) and two equivalents of 2,2,2-trifluoroethyl butyrate in dibutyl ether (Bu_2O) at 50 °C [232]. Similarly, the DKR of aromatic β-amino acid amides has been reported using *Pseudomonas stutzeri* lipase in combination with two equivalents of trifluoroethyl butanoate [233].

Recently, the DKR of arylamines has been successfully achieved with Pd/layered double hydroxide-dodecyl sulfate anion as racemization agent, a catalyst that was effectively reused more than 30 times without loss of activity and stereodiscrimination [234]. Its high catalytic efficiency allows the reaction to be performed at 55 °C, opening the possibility of working with less thermostable enzymes. In this case, 4-chlorophenyl valerate was used as acyl donor. The search for a novel palladium catalyst

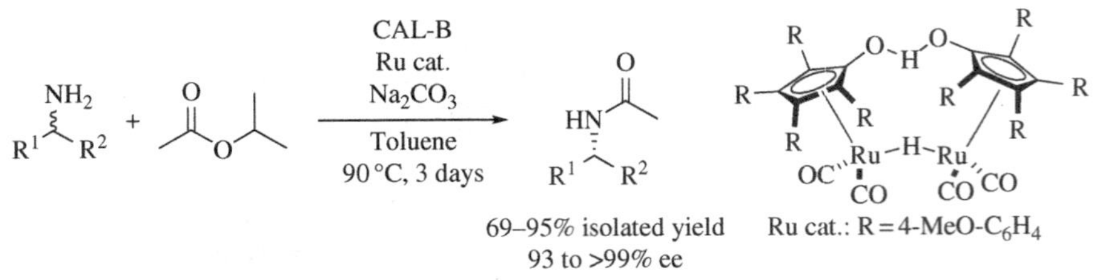

FIGURE 9.32

DKR of aliphatic and benzylic amines using CAL-B, isopropyl acetate, and a dimeric ruthenium complex.

is still in progress, and, for instance, the compatibility of palladium nanoparticles on siliceous amino-functionalized mesocellular foam with CAL-B or PSL has been demonstrated recently for the DKR of benzylic amines at 50 and 70 °C [235]. Finally, the use of a multifunctional hybrid catalyst containing a lipase (CAL-B) and palladium nanoparticles allows the DKR of 1-phenylethylamine in an orchestral fashion, opening a myriad of possibilities for the development of new DKR with racemic amines [236, 237].

Homogeneous ruthenium-containing catalysts can also be used for the racemization of primary amines in DKR processes. Monomeric ruthenium catalysts that need strong base activation failed in DKR processes, while dimeric compounds of Shvo's catalyst type are more effective and selective as they act under more neutral conditions, and just minimum amounts are required (i.e., 4 mol%) [238]. Sodium carbonate should be added to the reaction mixture since any trace of acid, originating either from the enzyme support, the acyl donor, or the enzyme itself, interferes with the ruthenium catalyst. Regarding the acyl donor, isopropyl acetate is a good choice as acylating agent because, apart from leading to high enantioselectivities, it releases 2-propanol, which acts as an additional hydrogen donor that can lower the amount of by-products. This effect can also be achieved by adding 0.5–1.0 equivalents of 2,4-dimethyl-3-pentanol as an additive [239], this also being compatible with the use of isopropyl methoxyacetate as acyl donor [240]. Thus, aliphatic and aromatic amines have been efficiently acylated, leading to the corresponding (*R*)-acetamides in high to quantitative conversions and excellent enantiomeric excess after 3 days at 90 °C in toluene (Figure 9.21) [238]. The commercially available Shvo's catalyst (Ru cat. with R = Ph in Figure 9.32) can be also used in similar conditions for the DKR of 1-aryl- and 1-heteroarylpropan-2-amines, leading to the (*R*)-methoxyacetamides in high enantiomeric excess [241]. In addition, other lipases such as CAL-A have been fully compatible with ruthenium catalyst for DKR processes, for instance, catalyzing the acylation of ethyl 3-amino-3-phenylpropanoate with two equivalents of trifluoroethyl butyrate in dibutyl ether at 90 °C for two days, obtaining the corresponding amide in 85% yield and 89% ee [242].

Bäckvall and coworkers have also developed a practical method for the chemoenzymatic DKR of primary amines using dibenzyl carbonate as acyl donor, combining the use of CAL-B and the ruthenium complex mentioned above in toluene at 90 °C for the production of enantioenriched (*R*)-carbamates (60–95% yield, 90–99% ee; Table 9.6) [243]. The main advantage of this method is that the benzyloxycarbonyl group (Cbz) can be easily removed by hydrogenolytic cleavage without any loss of the carbamate optical purity (compound in entry 1 of Table 9.6) [244].

As an alternative to palladium and ruthenium catalysts as racemization agents, De Vos and coworkers reported the use of Raney nickel for the DKR of aliphatic amines in toluene and a low pressure of hydrogen (0.01–0.02 MPa) at 70 °C. Its combination with CAL-B and EtOAcMeOAc led to the desired (*R*)-methoxyacetamides in 64–98% conversion and 80–98% ee [245]. The production of optically active amines has been also possible using ketoximes instead of racemic amines as starting materials, and using metal catalysis for the racemization step. Thus, Kim and coworkers reported the asymmetric reductive acylation a system including the use of palladium on charcoal for the reduction/racemization sequence in combination with CAL-B, 2 equivalents of EtOAc, three equivalents of diisopropylethylamine, and 1 H_2 atm in toluene at

TABLE 9.6 DKR of Primary Amines with 2.5 Equivalents of Dibenzyl Carbonate Catalyzed by CAL-B and 4 mol% of a Ruthenium Catalyst in the Presence of Na_2CO_3

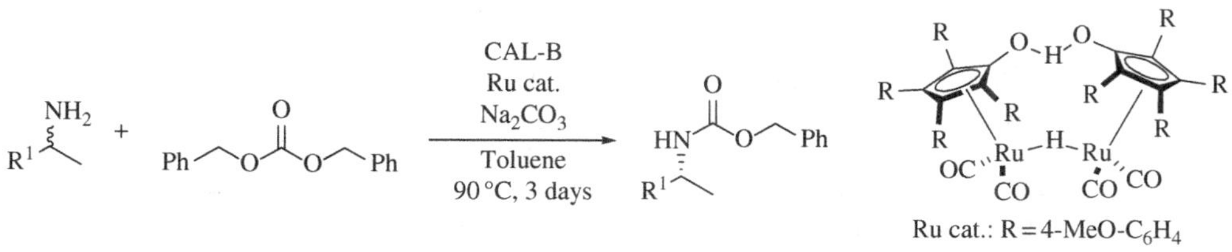

Entry	R	Yield (%)	ee_p (%)
1	C_6H_5	90	93
2	4-Br-C_6H_5	95	98
3	4-F-C_6H_5	72	99
4	4-MeO-C_6H_5	74	97
5	Cyclohexyl	92	96
6	Heptyl	89	90
7	*iso*-Propyl	60	99

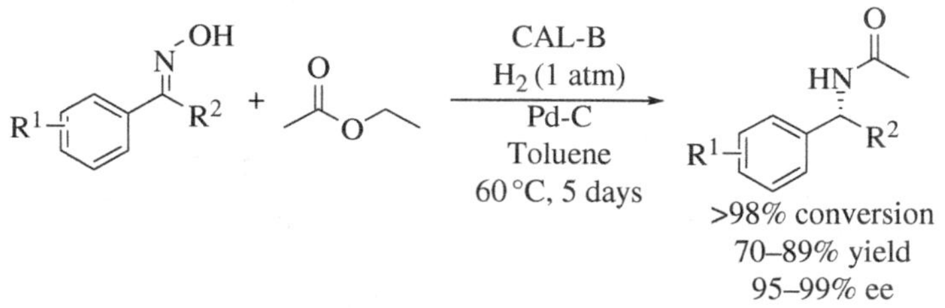

FIGURE 9.33 DKR of ketoximes for the production of benzylic amines using CAL-B as biocatalyst.

60 °C for 5 days (Figure 9.33) [246]. Excellent results were also found using the combination of CAL-B, EtOAcMeOAc, and Pd supported on AlO(OH) in the presence of hydrogen, yielding eight methoxyacetamides in good yields (83–92%) and high optical purity (93–98% ee) [247].

Racemization of nonactivated amines often requires the use of harsh reaction conditions for the activation of metal catalysis, which sometimes is not compatible with the presence of additional functional groups. Trying to solve this problem, alkylsulfanyl radicals have been able to facilitate the racemization of amines via reversible hydrogen abstraction at the chiral center in an α-position to the nitrogen nucleus. The association between CAL-B for the lipase-catalyzed resolution and *in situ* racemization mediated with the thiyl radical has allowed the DKR of nonbenzylic amines, although at a high temperature (80 °C) [248]. The DKR of nonbenzylic amines was achieved using 1.5 equivalents of ethyl laurate as acyl donor and 1.2 equivalents of *N,N*-diethyl-2-sulfanylpropionamide [$Et_2NOCCH(SH)CH_3$] for the racemization in the presence of azobisisobutyronitrile (AIBN), leading to the (*R*)-amides in 47–81% yield and 86 to >99% ee after 8 h in heptane at 80 °C (Table 9.7).

Due to the fact that the synthesis of (*S*)-amides by means of hydrolase-catalyzed processes can only be achieved by using proteases, which have less thermal stability, the same authors performed the racemization at a temperature that does not exceed 30 °C. With that purpose sulfanyl radicals were generated via photolysis of 2,2,2-trifluoroethanethiol (CF_3CH_2SH) under photochemical irradiation in the presence of AIBN using alkaline protease as biocatalyst and *N*-octanoyl-L-alanine trifluoroethyl ester as acyl donor in a mixture of *tert*-BuOH and THF as solvent for the development of the DKR, obtaining the (*S*)-amides in good yields and diastereomeric excess (Figure 9.34) [249]. A similar photochemical procedure has been applied to the

TABLE 9.7 DKR of Nonbenzylic Amines Using CAL-B, Ethyl Laurate, and *N,N*-Diethyl-2-sulfanylpropionamide as Racemization Agent

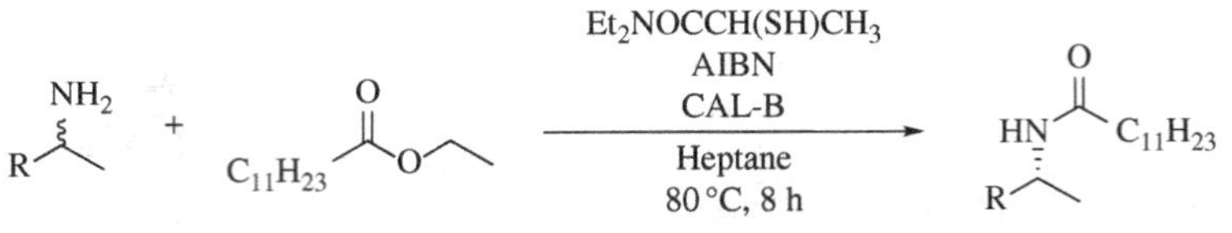

Entry	R	Yield (%)	ee_p (%)
1	$Ph(CH_2)_2$	70	99
2	$Me(CH_2)_5$	81	>99
3	*tert*-BuOCOCH$_2$	47	92
4	$Me_2C{=}CH(CH_2)_2$	68	94
5	CH_3CH_2	57	86

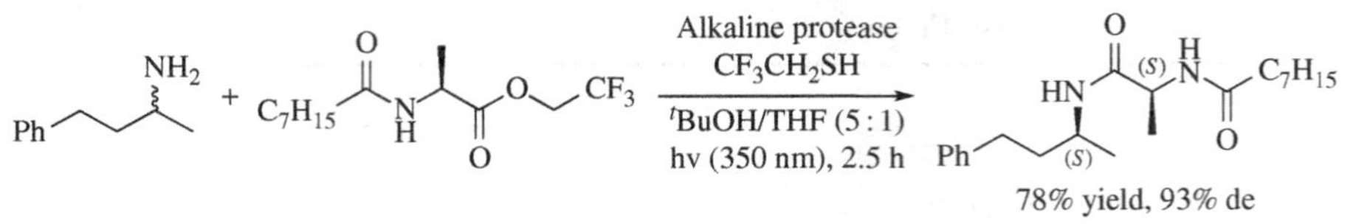

FIGURE 9.34
DKR of 4-phenylbutan-2-amine involving sulfanyl radical-induced racemization via photochemical irradiation with CAL-B as biocatalyst.

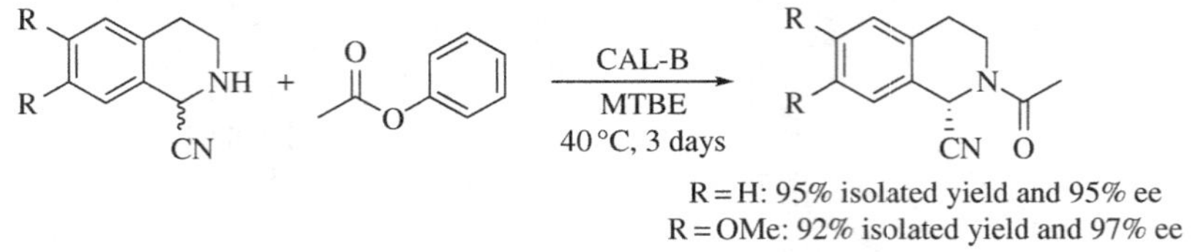

FIGURE 9.35
DKR of 1-cyano-1,2,3,4-tetrahydroisoquinolines using CAL-B and phenyl acetate.

synthesis of (*R*)-amides using CAL-B as biocatalyst, octanethiol as radical racemization agent, and methyl-β-methoxypropionate as acyl donor at 38–40 °C in MTBE [250].

Finally, spontaneous racemization can be achieved depending on the primary amine structure under study, a fact that facilitates the development of mild conditions for the DKR reaction through acylation processes. A series of examples has been described in the literature as the CAL-B-catalyzed acylation of 8-amino-5,6,7,8-tetrahydroquinoline via the spontaneous formation of the corresponding ketone that follows a condensation/hydrolysis sequence with the remaining (*S*)-amine for the formation of an enamine prior to the recovery of the racemic amine [251]. The process carried out with EtOAc as acyl donor and toluene as solvent at 50 °C led to the recovery of the (*R*)-acetamide in enantiopure form and 78% yield after 48 h. The influence of alkoxycarbonyl groups in the racemization of *cis*-*N*-(alkoxycarbonyl)cyclopentane-1,2-diamines has been demonstrated when studying their KR process by means of CAL-B-catalyzed acetylation with phenyl acetate in the presence of Et_3N at 50 °C [252]. A simple KR was found for the Boc derivative, while a DKR proceeded with Cbz, allyloxycarbonyl, and ethoxycarbonyl derivatives, the racemization occurring due to the *N,N'*-intramolecular migration of the alkoxycarbonyl group.

As with the KR processes, the DKR of secondary amines has been explored less than the reactions with primary racemic amines. Ramström and coworkers have reported the acylation of 1-cyano-1,2,3,4-tetrahydroisoquinoline and 6,7-dimethoxy-1,2,3,4-tetrahydroisoquinoline, discovering CAL-B as the best lipase for the acetylation reaction using three equivalents of phenyl acetate as acyl donor (Figure 9.35). The corresponding (*S*)-amides were obtained in very high enantiomeric excess and quantitative conversion after three days at 40 °C in MTBE [253].

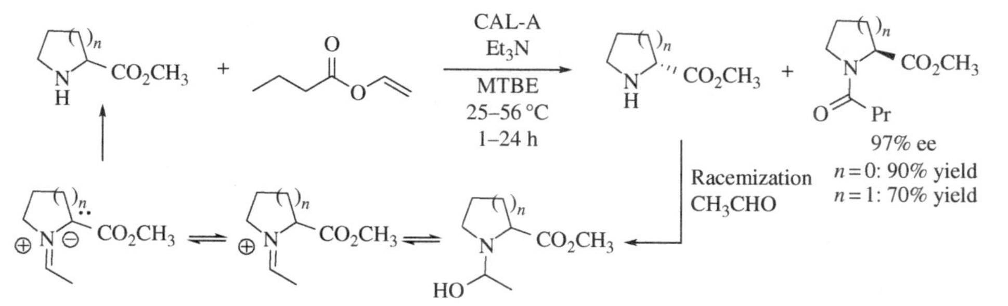

FIGURE 9.36 Mechanism of the DKR for proline and pipecolic acid methyl esters using CAL-A and vinyl butanoate.

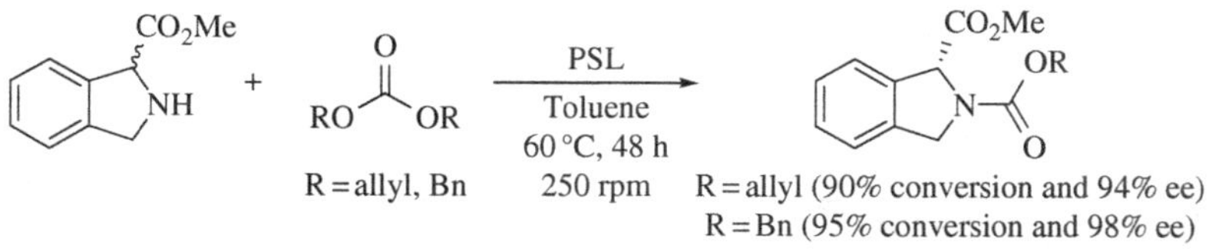

FIGURE 9.37 DKR of 1,3-dihydro-2*H*-isoindole-1-carboxylic acid methyl ester using carbonates and PSL in toluene.

The development of asymmetric transformation for the resolution of cyclic secondary amines has also been reported using alkoxycarbonylation reactions instead of acylation processes because of the high stability of the so-obtained carbamates toward undesired hydrolytic side reactions. For example, Page and coworkers described the combination of a lipase with an air-stable metal catalyst such as pentamethylcyclopentadienyliridium(III) iodide dimer ($[IrCp^*I_2]_2$), which is responsible for the racemization step in mild reaction conditions [254]. Thus, 1-methyl-1,2,3,4-tetrahydroisoquinoline was reacted with 3-methoxyphenyl propyl carbonate for 23 h at 40 °C in toluene, yielding the corresponding (*R*)-carbamate in 90% conversion and 96% ee. The reaction was scaled up to 3 g of substrate, and the product was recovered with the same enantiopurity and 82% isolated yield [255].

Amino esters are also good substrates for DKR using hydrolases. Kanerva and coworkers developed a DKR of proline and pipecolic acid methyl esters using CAL-A as biocatalyst and vinyl butanoate as acyl donor [256]. After acylation of the preferred (*S*)-amino ester enantiomer, acetaldehyde is released from the acylating agent promoting the racemization of the remaining (*R*)-amino ester through the formation of an imine intermediate in the presence of Et_3N (Figure 9.36). Thus, (*S*)-butanamides in 97% ee were isolated in high to excellent yields.

The presence of labile proton at chiral center positions greatly favors the racemization step under mild reaction conditions, in some cases without the addition of an acid–base catalyst, a metal, or an oxidation agent. In this context, the racemization of 1,3-dihydro-2*H*-isoindole-1-carboxylic acid methyl ester was possible through an equilibrium consisting in a deprotonation–protonation process through an achiral-stabilized enolate intermediate [257]. High temperatures led to a significant increase in the conversion rate for the reaction between the racemic amino esters with diallyl or dibenzyl carbonates catalyzed by PSL (Figure 9.37). The corresponding (*R*)-carbamates were obtained with high optical purity and conversion levels using both alkoxycarbonylation agents.

9.3.3 Selected Examples of Acylation Reactions with Interest for the Pharmaceutical Industry

A selection of KR and DKR of amines is presented from a myriad of examples described in the literature. In this section we have focused on the selection of biotransformations with a variety of amine cores including primary and secondary amines and amino alcohols, leading to the recovery of the final products in excellent enantiomeric excess through KR and DKR reactions. The possibility of scaling up the

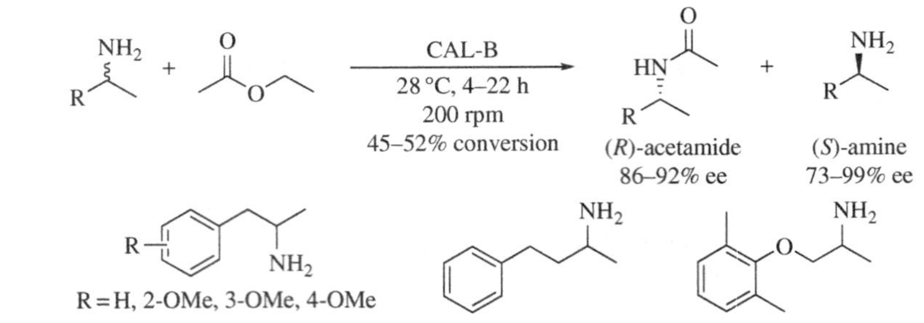

FIGURE 9.38 CAL-B-catalyzed stereoselective acetylation of amphetamine derivatives.

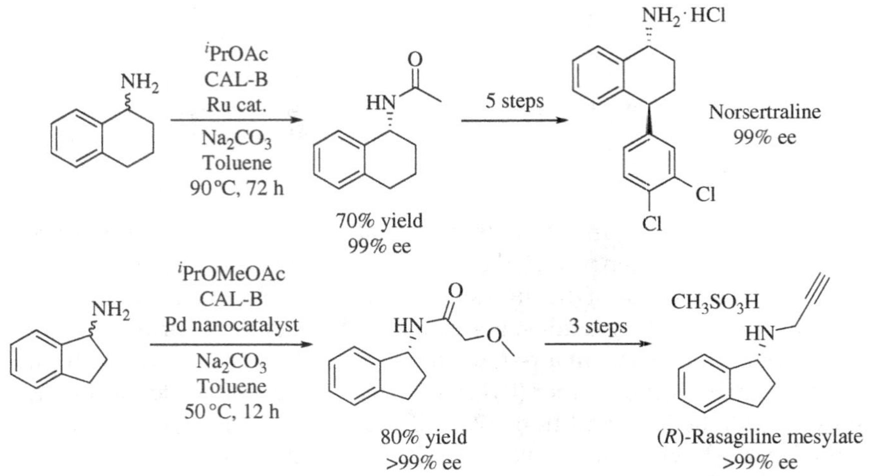

FIGURE 9.39 DKR of racemic amines for the synthesis of norsertraline and (*R*)-rasagiline mesylate.

process with satisfactorily synthetic purposes has been highlighted in representative examples.

The KR of pharmacologically active β-substituted isopropylamines has been studied, based on the importance of these structures. For instance, (*S*)-amphetamine (1-phenylpropan-2-amine) is a stimulant and hyperthermic agent, (*R*)-4-methoxyamphetamine is an important building block of the bronchodilator (*R*,*R*)-formoterol, (*R*)-4-phenylbutan-2-amine is a precursor of the antihypertensive dilevalol, and (*R*)-mexiletine [1-(2,6-dimethylphenoxy)propan-2-amine] is an antiarrhythmic agent. Racemic amines from the amphetamine family, 4-phenylbutan-2-amine, and mexiletine have been stereoselectively acylated using CAL-B and EtOAc at 28 °C, obtaining the corresponding (*S*)-amines and the (*R*)-acetamides in conversions close to 50% yield and with excellent enantioselectivities (Figure 9.38) [258]. The optical purity of the final products was improved by a simple recrystallization of hexane/chloroform mixtures.

Norsertraline is the chief metabolite of sertraline, which is marketed by Pfizer as Zoloft as a selective serotonin reuptake inhibitor for the treatment of depression. Its synthesis has been possible through the DKR of commercially available 1,2,3,4-tetrahydro-1-naphthylamine using CAL-B and isopropyl acetate in combination with a ruthenium dimeric catalyst for amine racemization (see structure in Table 9.6) [244]. The corresponding (*R*)-acetamide was obtained in a 70% yield and 99% ee after 72 h at 90 °C (Figure 9.39). Rasagiline is an irreversible and selective monoamine oxidase inhibitor used in the treatment of Parkinson's disease in the form of (*R*)-rasagiline mesylate. Its synthesis has been recently reported, identifying the DKR of 2,3-dihydro-1-indanamine as a key step [259]. This chemoenzymatic step was performed on a 73 g scale at 200 g/l, using, in this case, a palladium nanocatalyst as a racemization agent in combination with CAL-B and isopropyl methoxyacetate, obtaining the desired amide in enantiopure form after a recrystallization purification (Figure 9.39).

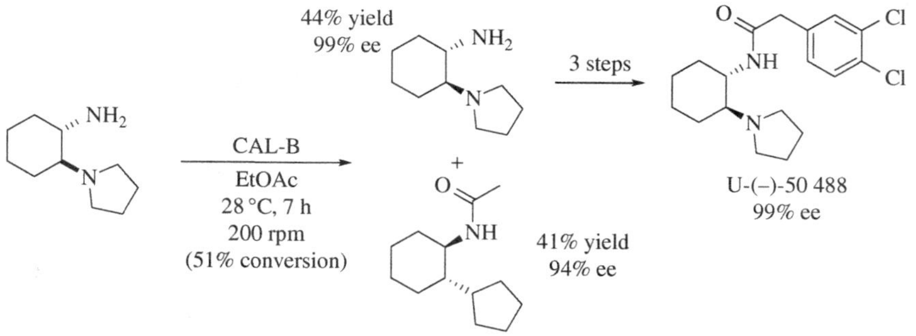

FIGURE 9.40

Chemoenzymatic synthesis of the analgesic U-(−)-50488.

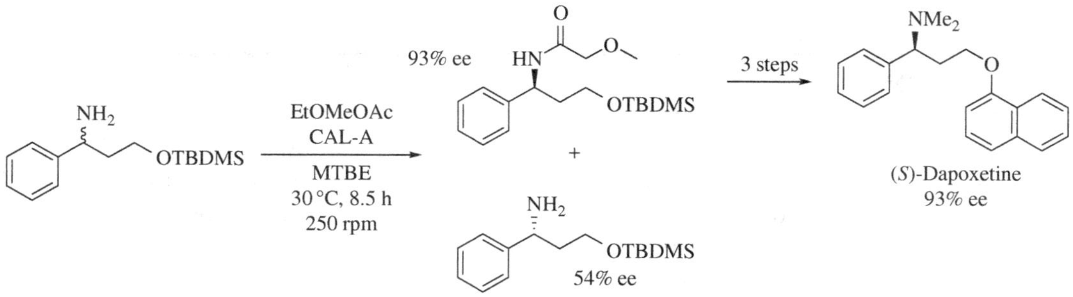

FIGURE 9.41

Chemoenzymatic synthesis of (*S*)-dapoxetine through enantioselective acetylation of an *O*-protected amino alcohol precursor with CAL-A.

The *trans*-cyclohexano-1,2-diamine derivatives are an important class of nitrogenated compounds due to their use as ligands in asymmetric organocatalysis and chiral building blocks. In fact, this core is present in the analgesic U-(−)-50488 possessing the (1*S*,2*S*)-configuration (Figure 9.40). Its chemoenzymatic preparation was possible through the KR of racemic *trans*-2-(pyrrolidin-1-yl)cyclohexanamine using EtOAc as both acyl donor and solvent and CAL-B as biocatalyst, yielding the desired (1*S*,2*S*)-amine in 99% ee and 44% yield [260].

(*S*)-Dapoxetine is a potent serotonin reuptake inhibitor for treating depression and different disorders such as bulimia, anxiety, or premature ejaculation. The biocatalyzed KR resolution of an *N*-protected amino alcohol precursor leads to low enantioselectivities as the modification occurs in a remote position with respect to the chiral center, so the best solution for the KR of 3-amino-3-phenylpropan-1-ol is the development of an *O*-protection and later the resolution through the free amino group (Figure 9.30) [261]. In this case, the best results were achieved when the oxygen atom was protected as a *tert*-butyldimethylsilyl (TBDMS) ether and using EtOMeOAc as an acyl donor in the presence of CAL-A as a biocatalyst, leading to the desired amide in 93% ee (Figure 9.41). Improvements for this route have been achieved by means of lipase-catalyzed hydrolysis toward amino ester intermediate in order to reached (*S*)-dapoxetine in enantiomerically pure form [262].

NPS R-568 is a potent calcimimetic for the treatment of primary and secondary hyperparathyroidism, the activity of the (*R*)-enantiomer being 10–100 times more active than its corresponding counterpart. A ketoxime directly obtained from 3-methoxyacetophenone was subjected to an asymmetric reductive acylation process using Pd-AlO(OH), CAL-B, and 1.5 equivalents of EtOMeOAc in a hydrogen atmosphere, isolating the (*R*)-methoxyacetamide in 85% isolated yield and 98% ee after 3 days at 70 °C (Figure 9.42) [263]. Interestingly, the recyclability of the system was demonstrated for four cycles without any loss of the catalyst activities.

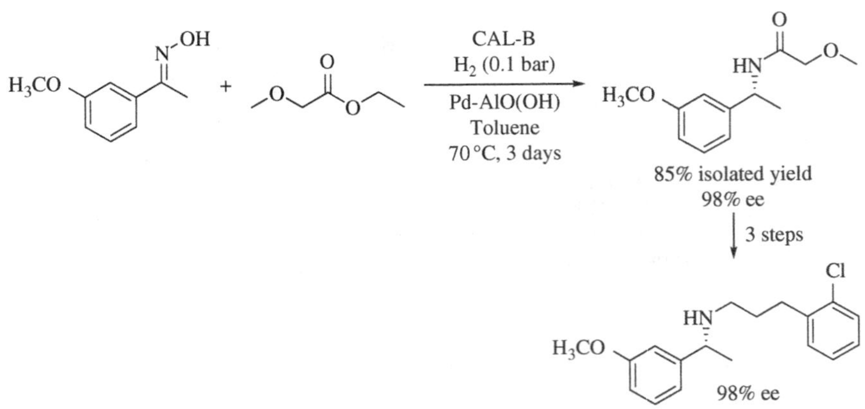

FIGURE 9.42
Synthesis of (+)-NPS R-568 via asymmetric reductive acylation of a ketoxime in the presence of CAL-B as biocatalyst and Pd/AlO(OH) as racemization agent.

9.4 CONCLUSIONS

The application of enzymes has gain importance in recent decades for the development of efficient transformations with a high level of selectivity, having the possibility for combined use with chemical catalysis that leads to the production of highly pure chemicals, and sometimes in multistep one-pot reactions [264, 265]. Although a new generation of biocatalysts is emerging and attracting considerable attention [266–268], the use of hydrolases still remains as a pillar for applied synthetic transformations because of their simple handling and use, commercial availability, lack of cofactor dependency, and the possibility to apply them in aqueous, organic, and neoteric solvents. The search for efficient asymmetric transformations has largely contributed to the identification of hydrolase-catalyzed nonhydrolytic processes and their implementation at an industrial level. Apart from the possibilities of natural hydrolases, the application of continuous flow reactions [269, 270], immobilization techniques, site-specific chemical modification [271], fermentation engineering [272], and directed evolution of proteins [273–276] has allowed the improvement of their catalytic activity, stability, and substrate specificity toward nonnatural substrates. Thus, the development of a more robust hydrolase-catalyzed asymmetric method will continue, having importance for the industrial biotechnology and chemical sector in future years.

REFERENCES

[1] Faber, K. ed. *Biotransformations in Organic Chemistry. A Textbook*, 6th Ed. Springer, Heidelberg: 2011.
[2] Clouthier, C. M.; Pelletier, J. N. *Chem. Soc. Rev.* 2012, **41**, 1585–1605.
[3] Milner, S. E.; Maguire, A. R. *Arkivoc* 2012, **i**, 321–382.
[4] Gotor, V.; Alfonso, I.; García-Urdiales, E. ed. *Asymmetric Organic Synthesis with Enzymes*, Willey-VCH, Weinheim: 2008.
[5] Tao, J.; Kazlauskas, R. J. ed. *Biocatalysis for Green Chemistry and Chemical Process Development*, John Wiley & Sons, Inc., Hoboken: 2011.
[6] Wells, A.; Meyer, H.-P. *ChemCatChem* 2014, **6**, 918–920.
[7] Hudlicky, T.; Reed, J. W. *Chem. Soc. Rev.* 2009, **38**, 3117–3132.
[8] Reetz, M. T. *J. Am. Chem. Soc.* 2013, **135**, 12480–12496.
[9] Behrens, G. A.; Hummel, A.; Padhi, S. K.; Schätzle, S.; Bornscheuer, U. T. *Adv. Synth. Catal.* 2011, **353**, 2191–2215.
[10] Sanchez, S.; Demain, A. L. *Org. Process Res. Dev.* 2011, **15**, 224–230.
[11] Wohlgemuth, R. *Curr. Opin. Biotechnol.* 2010, **21**, 713–724.
[12] Nestl, B. M.; Nebel, B. A.; Hauer, B. *Curr. Opin. Chem. Biol.* 2011, **15**, 187–193.
[13] Gröger, H.; Asano, Y.; Bornscheuer, U. T.; Ogawa, J. *Chem. Asian J.* 2012, **7**, 1138–1153.
[14] Solano, D. M.; Hoyos, P.; Hernáiz, M. J.; Alcántra, A. R.; Sánchez-Montero, J. M. *Bioresour. Technol.* 2012, **115**, 196–207.

[15] Bornscheuer, U. T.; Kazlauskas, R. J. ed. *Hydrolases in Organic Synthesis*, 2nd Ed. Wiley-VCH, Weinheim: 2006.
[16] Ghanem, A.; Aboul-Enein, H. Y. *Tetrahedron: Asymmetry* 2004, **15**, 3331–3351.
[17] Ghanem, A. *Tetrahedron* 2007, **63**, 1721–1754.
[18] Meyer, H.-P.; Eichhorn, E.; Hanlon, S.; Lütz, S.; Schürman, M.; Wohlgemuth, R.; Coppolecchia, R. *Catal. Sci. Technol.* 2013, **3**, 29–40.
[19] Fersht, A. ed. *Structure and Mechanism in Protein Science: A Guide to Enzyme Catalysis and Protein Folding*, Freeman, New York: 1999.
[20] Klibanov, A. M. *Trends Biochem. Sci.* 1989, **14**, 141–144.
[21] Carrea, G.; Ottolina, G.; Riva, S. *Trends Biotechnol.* 1995, **13**, 63–70.
[22] Carrea, G.; Riva, S. ed. *Organic Synthesis with Enzymes in Non-Aqueous Media*, Wiley-VCH, Weinheim: 2008.
[23] Cantone, S.; Hanefeld, U.; Basso, A. *Green Chem.* 2007, **9**, 954–971.
[24] Fan, Y.; Qian, J. *J. Mol. Catal. B: Enzym.* 2010, **66**, 1–7.
[25] Lozano, P. *Green Chem.* 2010, **12**, 555–569.
[26] Hernáiz, M. J.; Alcántara, A. R.; García, A. I.; Sinisterra, J. V. *Chem. Eur. J.* 2010, **16**, 9422–9437.
[27] Kirchner, G.; Scollar, M. P.; Klibanov, A. M. *J. Am. Chem. Soc.* 1985, **107**, 7072–7076.
[28] Klibanov, A. M. *Nature* 2001, **409**, 241–246.
[29] Palomo, J. M. *Curr. Org. Synth.* 2009, **6**, 1–14.
[30] Tran, D. N.; Balkus, Jr. K. J. *ACS Catal.* 2011, **1**, 956–968.
[31] Garcia-Galan, C.; Berenguer-Murcia, Á.; Fernandez-Lafuente, R.; Rodrigues, R. C. *Adv. Synth. Catal.* 2011, **353**, 2885–2904.
[32] Sheldon, R. A. *Org. Process Res. Dev.* 2011, **15**, 213–223.
[33] Adlercreutz, P. *Chem. Soc. Rev.* 2013, **42**, 6406–6436.
[34] Stepankova, V.; Bidmanova, S.; Koudekalova, T.; Prokop, Z.; Chaloupkova, R.; Damborsky, J. *ACS Catal.* 2013, **3**, 2823–2836.
[35] Verger, R. *Trends Biotechnol.* 1997, **15**, 32–38.
[36] Doukyu, N.; Ogino, H. *Biochem. Eng. J.* 2010, **48**, 270–282.
[37] Laane, C.; Boeren, S.; Vos, K.; Veeger, C. *Biotechnol. Bioeng.* 1987, **30**, 81–87.
[38] Busacca, C. A.; Fandrick, D. R.; Song, J. J.; Senanayake, C. H. *Adv. Synth. Catal.* 2011, **353**, 1825–1864.
[39] Magano, J.; Dunetz, J. R. *Chem. Rev.* 2011, **111**, 2177–2250.
[40] Patel, R. N. *ACS Catal.* 2011, **1**, 1056–1074.
[41] Gotor-Fernández, V.; Brieva, R.; Gotor, V. *J. Mol. Catal. B: Enzym.* 2006, **40**, 111–120.
[42] Kasprzyk-Hordern, B. *Chem. Soc. Rev.* 2010, **39**, 4466–4503.
[43] Kazlauskas, R. J.; Weissfloch, A. N. E.; Rappaport, A. T.; Cuccia, L. A. *J. Org. Chem.* 1991, **56**, 2656–2665.
[44] Pellissier, H. *Tetrahedron* 2011, **67**, 3769–3802.
[45] Pellissier, H. *Adv. Synth. Catal.* 2011, **353**, 659–676.
[46] García-Urdiales, E.; Alfonso, I.; Gotor, V. *Chem. Rev.* 2011, **111**, PR110–PR180.
[47] Faber, K.; Riva, S. *Synthesis* 1992, 895–910.
[48] Hanefeld, U. *Org. Biomol. Chem.* 2013, **1**, 2405–2415.
[49] Paravidino, M.; Hanefeld, U. *Green Chem.* 2011, **13**, 2651–2657.
[50] Maywald, M.; Pfaltz, A. *Synthesis* 2009, 3654–3660.
[51] Goswami, A.; Howell, J. M.; Hua, E. Y.; Mirfakhrae, K. D.; Sourmeillant, M. C.; Swaminathan, S.; Qian, X.; Quiroz, F. A.; Vu, T. C. Wang, X.; Zheng, B.; Kronenthal, D. R.; Patel, R. N. *Org. Process Res. Dev.* 2001, **5**, 415–420.
[52] Bouzemi, N.; Debbeche, H.; Aribi-Zouiou, L.; Fiaud, J.-C. *Tetrahedron Lett.* 2004, **45**, 627–630.
[53] Brossat, M.; Moody, T. S.; de Nanteuil, F.; Taylor, S. J. C.; Vaughan, F. *Org. Process Res. Dev.* 2009, **13**, 706–709.
[54] Kataoka, S.; Takeuchi, Y.; Harada, A.; Yamada, M.; Endo, A. *Green Chem.* 2010, **12**, 331–337.
[55] Rocha, L. C.; Rosset, I. G.; Luiz, R. F.; Raminelli, C.; Porto, A. L. M. *Tetrahedron: Asymmetry* 2010, **21**, 926–929.
[56] Andrade, L. H.; Barcellos, T. *Org. Lett.* 2009, **11**, 3052–3055.
[57] Comasseto, J. V.; Gariani, R. A. *Tetrahedron* 2009, **65**, 8447–8459.
[58] Chen, P.; Xiang, P. *Tetrahedron Lett.* 2011, **52**, 5758–5760.
[59] Andrade, L. H.; Rebelo, L. P.; Netto, C. G. C. M.; Toma, H. E. *J. Mol. Catal. B: Enzym.* 2010, **66**, 55–62.

[60] Ko, S.-J.; Lim, J. Y.; Jeon, N. Y.; Won, K.; Ha, D.-C.; Kim, B. T.; Lee, H. *Tetrahedron: Asymmetry* 2009, **20**, 1109–1114.

[61] Perrone, M. G.; Santandrea, E.; Giorgio, E.; Bleve, L.; Scilimati, A.; Tortorellam, P. *Bioorg. Med. Chem.* 2006, **14**, 1207–1214.

[62] Hara, P.; Turcu, M. C.; Sundell, R.; Toşa, M.; Paizs, C.; Irimie, F.-D. *Tetrahedron: Asymmetry* 2013, **24**, 142–150.

[63] Wawro, A. M.; Wielechowska, M.; Bretner, M. *J. Mol. Catal. B: Enzym.* 2013, **87**, 44–50.

[64] Ketterer, C.; Wünsch, B. *Eur. J. Org. Chem.* 2012, 2428–2444.

[65] Brem, J.; Paizs, C.; Toşa, M. I.; Vass, E.; Irimie, F. D. *Tetrahedron: Asymmetry* 2009, **20**, 489–496.

[66] Brem, J.; Liljeblad, A.; Paizs, C.; Toşa, M. I.; Irimie, F. D.; Kanerva, L. T. *Tetrahedron: Asymmetry* 2011, **22**, 315–322.

[67] Brem, J.; Naghi, M.; Toşa, M.-I.; Boros, Z.; Poppe, L.; Irimie, F.-D.; Paizs, C. *Tetrahedron: Asymmetry* 2011, **22**, 1672–1679.

[68] Turcu, M. C.; Perkiö, P.; Kanerva, L. T. *Arkivoc* 2009, **iii**, 251–263.

[69] Solymár, M.; Forró, E.; Fülöp, F. *Tetrahedron: Asymmetry* 2004, **15**, 3281–3287.

[70] Pinedo-Revilla, C.; Aleu, J.; Benito, M. G.; Collado, I. G. *Org. Biomol. Chem.* 2010, **8**, 3784–3789.

[71] Milner, S. E.; Brossat, M.; Moody, T. S.; Elcoate, C. J.; Lawrence, S. E.; Maguire, A. R. *Tetrahedron: Asymmetry* 2010, **21**, 1011–1016.

[72] Utczás, M.; Székely, E.; Forró, E.; Szóllósy, Á.; Fülöp, F.; Simándi, B. *Tetrahedron Lett.* 2011, **52**, 3916–3918.

[73] Blasco, M. A.; Gröger, H. *Bioorg. Med. Chem.* 2014, **22**, 5539–5546.

[74] Deska, J.; Bäckvall, J.-E. *Org. Biomol. Chem.* 2009, **7**, 3379–3381.

[75] Mahpatra, T.; Nanda, N. J. S. *Tetrahedron: Asymmetry* 2008, **19**, 1224–1232.

[76] Bhuniya, R.; Nanda, S. *Org. Biomol. Chem.* 2012, **10**, 536–547.

[77] Guerrab, Z.; Schweiger, S.; Daou, B.; Ahmar, M.; Cazes, B. *Tetrahedron: Asymmetry* 2010, **21**, 1752–1757.

[78] Domínguez de María, P.; Carboni-Oerlemans, C.; Tuin, B.; Bargeman, G.; van der Meer, A.; van Gemert, R. *J. Mol. Catal. B: Enzym.* 2005, **37**, 36–46.

[79] Krihna, S. H.; Persson, M.; Bornscheuer, U. T. *Tetrahedron: Asymmetry* 2002, **13**, 2693–2696.

[80] Özdemirhan, D.; Sezer, S.; Sönmez, Y. *Tetrahedron: Asymmetry* 2008, **19**, 2717–2720.

[81] Kourist, R.; Krishna, S. H.; Patel, J. S.; Bartnek, F.; Hitchman, T. S.; Weiner, D. P.; Bornscheuer, U. T. *Org. Biomol. Chem.* 2007, **5**, 3310–3313.

[82] Wiggers, M.; Holt, J.; Kourist, R.; Bartsch, S.; Arends, I. W. C. E.; Minnaard, A. J.; Bornscheuer, U. T.; Hanefeld, U. *J. Mol. Catal. B: Enzym.* 2009, **60**, 82–86.

[83] Nguyen, G.-S.; Thompson, M. L.; Grogan, G.; Bornscheuer, U. T.; Kourist, R. *J. Mol. Catal. B: Enzym.* 2011, **70**, 88–94.

[84] Francalanci, F.; Cesti, P.; Cabri, W.; Bianchi, D.; Martinengo, T.; Foà, M. *J. Org. Chem.* 1987, **52**, 5079–5082.

[85] Kamal, A.; Rao, M. V. *Tetrahedron: Asymmetry* 1991, **2**, 751–754.

[86] Kanerva, L. T.; Rahiala, K.; Vänttinen, E. *J. Chem. Soc. Perkin Trans.* **1** 1992, 1759–1762.

[87] Maestro, A.; Astorga, C.; Gotor, V. *Tetrahedron: Asymmetry* 1997, **8**, 3153–3159.

[88] Levy, L. M.; de Gonzalo, G.; Gotor, V. *Tetrahedron: Asymmetry* 2004, **15**, 2051–2056.

[89] Brocklehurst, C. E.; Laumen, K.; La Veccia, L.; Shaw, D.; Vögtle, M. *Org. Process Res. Dev.* 2011, **15**, 294–300.

[90] López-García, M.; Alfonso, I.; Gotor, V. *Chem. Eur. J.* 2004, **10**, 3006-3014.

[91] Schnell, B.; Faber, K.; Kroutil, W. *Adv. Synth. Catal.* 2003, **345**, 653–666.

[92] Huerta, F. F.; Minidis, A. B. E.; Bäckvall, J.-E. *Chem. Soc. Rev.* 2001, **30**, 321–331.

[93] Kim, M.-J.; Ahn, Y.; Park, J. *Curr. Opin. Biotechnol.* 2002, **13**, 578–587.

[94] Bäckvall, J.-E.; Pàmies, O. *Chem. Rev.* 2003, **103**, 3247–3261.

[95] Martín-Matute, B.; Bäckvall, J.-E. *Curr. Opin. Chem. Biol.* 2007, **11**, 226–232.

[96] Ahn, Y.; Ko, S.-B.; Ki, M.-J.; Park, J. *Coord. Chem. Rev.* 2008, **252**, 647–658.

[97] Pàmies, O.; Bäckvall, J.-E. *Curr. Opin. Biotechnol.* 2003, **14**, 407–413.

[98] Pàmies, O.; Bäckvall, J.-E. *Trends Biotechnol.* 2004, **22**, 130–135.

[99] Lee, J. H.; Han, K.; Kim, M.-J.; Park, J. *Eur. J. Org. Chem.* 2010, 999–1015.

[100] Hoyos, P.; Pace, V.; Alcántara, A. *Adv. Synth. Catal.* 2012, **354**, 2585–2611.

[101] Marcos, R.; Martín-Matute, B. *Isr. J. Chem.* 2012, **52**, 639–652.

[102] Ahmed, M.; Kelly, T.; Ghanem, A. *Tetrahedron* 2012, **68**, 6781–6802.

[103] Akai, S. *Chem. Lett.* 2014, **43**, 746–754.

[104] Hoyos, P.; Quezada, M. A.; Sinisterra, J. V.; Alcántara, A. R. *J. Mol. Catal. B: Enzym.* 2011, **72**, 20–24.
[105] Kim, M.-J.; Kim, H. M.; Kim, D.; Ahn, Y.; Park, J. *Green Chem.* 2004, **6**, 471–474.
[106] Dinh, P. M.; Howarth, J. A.; Hudnott, A. R.; Williams, M. J. Harris, W. *Tetrahedron Lett.* 1996, **37**, 7623–7626.
[107] Berkessel, A.; Sebastian-Ibarz, M. L.; Müller, T. N. *Angew. Chem. Int. Ed.* 2006, **45**, 6567–6570.
[108] Sato, Y.;Kayaki, Y.; Ikariya, T. *Chem. Commun.* 2012, **48**, 3635–3637.
[109] Choi, Y. K.; Suh, J. H.; Lee, D.; Lim, I. T.; Jung, J. Y.; Kim, M.-J. *J. Org. Chem.* 1999, **64**, 8423–8424.
[110] Deska, J.; Ochoa, C. P.; Bäckvall, J.-E. *Chem. Eur. J.* 2010, **16**, 4447–4451.
[111] Marr, A. C.; Pollock, C. L.; Saunders, G. C. *Organometallics* 2007, **26**, 3283–3285.
[112] Akai, S.; Tanimoto, K.; Kanao, Y.; Egi, M.; Yamamoto, T.; Kita, Y. *Angew. Chem. Int. Ed.* 2006, **45**, 2592–2595.
[113] Akai, S.; Hanada, R.; Fujiwara, N.; Kita, Y.; Egi, M. *Org. Lett.* 2010, **12**, 4900–4903.
[114] Zhu, Y.; Fow, K.-L.; Chuah, G.-K.; Jaenicke, S. *Chem. Eur. J.* 2007, **13**, 541–547.
[115] Inagaki, M.; Hiratake, J.; Nishioka, T.; Oda, J. *J. Org. Chem.* 1992, **57**, 5643–5649.
[116] Paizs, C.; Toşa, M.; Majdik, C.; Tähtinen, P.; Irimie, F. D.; Kanerva, L. T. *Tetrahedron: Asymmetry* 2003, **14**, 619–627.
[117] Paizs, C.; Tähtinen, P.; Lundlell, K.; Poppe, L.; Irimie, F.-D.; Kanerva, L. T. *Tetrahedron: Asymmetry* 2003, **14**, 1895–1904.
[118] Paizs, C.; Tähtinen, P.; Toşa, M.; Majdik, C.; Irimie, F.-D.; Kanerva, L. T. *Tetrahedron* 2004, **60**, 10533–10540.
[119] Ödman, P.; Wessjohann, L. A.; Bornscheuer, U. T. *J. Org. Chem.* 2005, **70**, 9551–9555.
[120] Cheng, Y.; Xu, G.; Wu, J.; Zhang, C.; Yang, L. *Tetrahedron Lett.* 2010, **51**, 2366–2369.
[121] Xu, G.; Wang, L.; Chen, Y.; Cheng, Y.; Wu, J.; Yang, L. *Tetrahedron Lett.* 2013, **54**, 5026–5030.
[122] Brand, S.; Jones, M. F.; Rayner, C. M. *Tetrahedron Lett.* 1995, **36**, 8493–8496.
[123] Sakai, T.; Wang, K.; Ema, T. *Tetrahedron* 2008, **64**, 2178–2183.
[124] Sharfuddin, M.; Narumi, A.; Iwai, Y.; Miyazawa, K.; Yamada, S.; Kakuchi, T.; Kaga, H. *Tetrahedron: Asymmetry* 2003, **14**, 1581–1585.
[125] Nunes, J. P. M.; Veiros, L. F.; Vaz, P. D.; Afonso, C. A. M.; Caddick, S. *Tetrahedron* 2011, **67**, 2779–2787.
[126] Persson, B. A.; Larsson, A. L. E.; Le Ray, M. M; Bäckvall, J.-E. *J. Am. Chem. Soc.* 1999, **121**, 1645–1650.
[127] Lilhammar, R.; Millet, R.; Bäckvall, J.-E. *J. Org. Chem.* 2013, **78**, 12114–12120.
[128] Huerta, F. F.; Laxmi, Y. R. S.; Bäckvall, J.-E. *Org. Lett.* 2000, **2**, 1037–1040.
[129] Kim, M.-J.; Choi, Y. K.; Choi, M. Y.; Kim, M. J.; Park, J. *J. Org. Chem.* 2001, **66**, 4736–4738.
[130] Huerta, F. F.; Bäckvall, J.-E. *Org. Lett.* 2001, **3**, 1209–1212.
[131] Pàmies, O.; Bäckvall, J.-E. *J. Org. Chem.* 2002, **67**, 1261–1265.
[132] Kiełbasiński, P.; Rachwalski, M.; Mikołajczyk, M.; Moelands, M. A. H.; Zwanenburg, B.; Rutjes, F. P. J. T. *Tetrahedron: Asymmetry* 2005, **16**, 2157–2160.
[133] Pàmies, O.; Bäckvall, J.-E. *Adv. Synth. Catal.* 2001, **343**, 726–731.
[134] Pàmies, O.; Bäckvall, J.-E. *Adv. Synth. Catal.* 2002, **344**, 947–952.
[135] Runmo, A-B. L.; Pàmies, O.; Faber, K.; Bäckvall, J.-E. *Tetrahedron Lett.* 2002, **43**, 2983–2986.
[136] Fransson, A.-B. L.; Borén, L.; Pàmies, O.; Bäckvall, J.-E. *J. Org. Chem.* 2005, **70**, 2582–2587.
[137] Koh, J. H.; Jung, H. M.; Kim, M.-J.; Park, J. *Tetrahedron Lett.* 1999, **40**, 6281–6284.
[138] Lee, D.; Huh, E. A.; Kim, M.-J.; Jung, H. M.; Koh, J. H.; Park, J. *Org. Lett.* 2000, **2**, 2377–2379.
[139] Martín-Matute, B.; Edin, M.; Bogar, K.; Bäckvall, J.-E. *Angew. Chem. Int. Ed.* 2004, **43**, 6535–6539.
[140] Martín-Matute, B.; Edin, M.; Bogar, K.; Kaynak, F. B.; Bäckvall, J.-E. *J. Am. Chem. Soc.* 2005, **127**, 8817–8825.
[141] Borén, L.; Martín-Matute, B.; Xu, Y.; Córdova, A.; Bäckvall, J.-E. *Chem. Eur. J.* 2006, **12**, 225–232.
[142] Träff, A.; Bogár, K.; Warner, M.; Bäckvall, J.-E. *Org. Lett.* 2008, **10**, 4807–4810.
[143] Lihammar, R.; Nillet, R.; Bäckvall, J.-E. *Adv. Synth. Catal.* 2011, **353**, 2321–2327.
[144] Engström, K.; Vallin, M.; Syrén, P.-O.; Hult, K.; Bäckvall, J.-E. *Org. Biomol. Chem.* 2011, **9**, 81–82.
[145] Warner, M. C.; Nagendiran, A.; Bógar, K.; Bäckvall, J.-E. *Org. Lett.* 2012, **14**, 5094–5097.
[146] Choi, J. H.; Kim, Y. H.; Nam, S. H.; Shin, S. T.; Kim, M.-J.; Park, J. *Angew. Chem. Int. Ed.* 2002, **41**, 2373–2376.
[147] Choi, J. H.; Choi, Y. K.; Kim, Y. H.; Park, E. S.; Kim, E. J.; Kim, M.-J.; Park, J. *J. Org. Chem.* 2004, **69**, 1972–1977.

[148] Kim, M.-J.; Chung, Y. I.; Choi, Y. K.; Lee, H. K.; Kim, D.; Park, J. *J. Am. Chem. Soc.* 2003, **125**, 11494–11495.

[149] Neri, C.; Williams, J. M. J. *Adv. Synth. Catal.* 2003, **345**, 835–848.

[150] Choi, J. Y.; Borch, R. *Org. Lett.* 2007, **9**, 215–218.

[151] Kawasaki, M.; Goto, M.; Hu, D.; Toyooka, N.; Kometani, T. *J. Mol. Catal. B: Enzym.* 2013, **96**, 27–33.

[152] Davies, S. G.; Roberts, P. M.; Stephenson, P. T.; Storr, H. R.; Thomson, J. E. *Tetrahedron* 2009, **65**, 8283–8296.

[153] Fransson, A.-B. L.; Xu, Y.; Leijondahl, K.; Bäckvall, J.-E. *J. Org. Chem.* 2006, **71**, 6309–6316.

[154] Chênevert, R.; Lévesque, C.; Morin, P. *J. Org. Chem.* 2008, **73**, 9501–9503.

[155] Zhao, Y.; Wu, Y.; De Clercq, P.; Vandewalle, M.; Maillos, P.; Pascal, J.-C. *Tetrahedron: Asymmetry* 2000, **11**, 3887–3900.

[156] Wirz, B.; Iding, H.; Hilpert, H. *Tetrahedron: Asymmetry* 2000, **11**, 4171–4178.

[157] Oishi, T.; Maruyama, M.; Shoji, M.; Maeda, K. Kumahara, N.; Tanaka, S.-I.; Hirama, M. *Tetrahedron* 1999, **55**, 7471–7498.

[158] Hammel, M.; Deska, J. *Synthesis* 2012, **44**, 3789–3796.

[159] Wiktelius, D.; Johansson, M. J.; Luthman, K.; Kann, N. *Org. Lett.* 2005, **7**, 4491–4494.

[160] Rachwalski, M.; Kwiatkowska, M.; Drabowicz, J.; Kłos, M.; Wieczorek, W. M.; Szyrej, M.; Sieroń, L.; Kiełbasiński, P. *Tetrahedron: Asymmetry* 2008, **19**, 2096–2101.

[161] Patel, R. N. *Biomolecules* 2013, **3**, 741–777.

[162] Patel, R. N. ed. *Biocatalysis in the Pharmaceutical and Biotechnology Industries*, CRC Press, Boca Raton: 2006, 203–248.

[163] Mangas-Sánchez, J.; Rodríguez-Mata, M.; Busto, E.; Gotor-Fernández, V.; Gotor, V. *J. Org. Chem.* 2009, **74**, 5304–5310.

[164] Han, K.; Kim, C.; Park, J.; Kim, M.-J. *J. Org. Chem.* 2010, **75**, 3105–3108.

[165] Uhm, K.-N.; Lee, S.-J.; Kim, H.-k.; Kang, H.-Y.; Lee, Y. *J. Mol. Catal. B: Enzym.* 2007, **45**, 34–38.

[166] Tagawa, S.; Choshi, T.; Okamoto, A.; Nishiyama, T.; Watanabe, S.; Hatae, N.; Ishikura, M.; Hibino, S. *Eur. J. Org. Chem.* 2013, **77**, 1805–1810.

[167] Busto, E.; Gotor-Fernández, V.; Gotor, V. *J. Org. Chem.* 2012, **77**, 4842–4848.

[168] Kamal, A.; Khanna, G. B. R.; Ramu, R. *Tetrahedron: Asymmetry* 2002, **13**, 2039–2051.

[169] Kamal, A.; Khanna, G. B. R.; Ramu, R.; Krishnaji, T. *Tetrahedron Lett.* 2003, **44**, 4783–4787.

[170] Träff, A.; Lihammar, R.; Bäckvall, J.-E. *J. Org. Chem.* 2011, **76**, 3917–3921.

[171] Nagy, B.; Dima, N.; Paizs, C.; Brem, J.; Irimie, F. D.; Toşa, M. I. *Tetrahedron: Asymmetry* 2014, **25**, 1316–1322.

[172] Johnston, E. V.; Bogár, K.; Bäckvall, J.-E. *J. Org. Chem.* 2010, **75**, 4596–4599.

[173] Regla, I.; Luviano-Jardón, A.; Demare, P.; Hong, E.; Torres-Gavilán, A.; López-Munguía, A.; Castillo, E. *Tetrahedron: Asymmetry* 2008, **19**, 2439–2442.

[174] Pàmies, O.; Bäckvall, J.-E. *J. Org. Chem.* 2001, **66**, 4022–4025.

[175] Serra, S. *Tetrahedron: Asymmetry* 2011, **22**, 619–628.

[176] Saksena, A. K.; Girijavallabhan, V. M.; Lovey, R. G.; Pike, R. E.; Wang, H. *Tetrahedron Lett.* 1995, **36**, 1787–1790.

[177] Kastrinsky, D. B.; Boger, D. L. *J. Org. Chem.* 2004, **69**, 2284–2289.

[178] Breuer, M.; Ditrich, K.; Habicher, T.; Hauer, B.; Keβeler, M.; Stürmer, R.; Zelinsky, T. *Angew. Chem. Int. Ed.* 2004, **43**, 788–824.

[179] Gotor, V. *Bioorg. Med. Chem.* 1999, **7**, 2189–2197.

[180] Busto, E.; Gotor-Fernández, V.; Gotor, V. *Chem. Rev.* 2011, **11**, 3998–4035.

[181] Höhne, M.; Bornscheuer, U. T. *ChemCatChem* 2009, **1**, 42–51.

[182] Ghislieri, D.; Turner, N. J. *Top. Catal.* 2014, **57**, 284–300.

[183] van Rantwijk, F.; Sheldon, R. A. *Tetrahedron* 2004, **60**, 501–519.

[184] Liljeblad, A.; Kallio, P.; Vainio, M.; Niemi, J.; Kanerva, L. T. *Org. Biomol. Chem.* 2010, **8**, 886–895.

[185] Busto, E.; Gotor-Fernández, V.; Gotor, V. *Chem. Soc. Rev.* 2010, **39**, 4504–4523.

[186] Gotor-Fernández, V.; Gotor, V. *Curr. Org. Chem.* 2006, **10**, 1125–1143.

[187] Gotor-Fernández, V.; Busto, E.; Gotor, V. *Adv. Synth. Catal.* 2006, **348**, 797–812.

[188] de Miranda, A. S.; Miranda, L. S. M.; de Souza, R. O. M. A. *Org. Biomol. Chem.* 2013, **11**, 3332–3336.

[189] Andrade, L. H.; Barcellos, T.; Santiago, C. G. *Tetrahedron: Asymmetry* 2010, **21**, 2419–2424.

[190] Päiviö, M.; Perkiö, P.; Kanerva, L. T. *Tetrahedron: Asymmetry* 2012, **23**, 230–236.

[191] Ditrich, K. *Synthesis* 2008, 2283–2287.

[192] Thvedt, T. H. K.; Fuglseth, E.; Sundby, E.; Hoff, B. H. *Tetrahedron* 2010, **66**, 6733–6743.

[193] Hietanen, A.; Saloranta, T.; Rosenberg, S.; Laitinen, E.; Leino, R.; Kanerva, L. T. *Eur. J. Org. Chem.* 2010, 909–919.
[194] Knežević, A.; Landek, G.; Dokli, I.; Vinković, V. *Tetrahedron: Asymmetry* 2011, **22**, 936–941.
[195] Torre, O.; Busto, E.; Gotor-Fernández, V.; Gotor, V. *Adv. Synth. Catal.* 2007, **349**, 1481–1488.
[196] Nechab, M.; Azzi, N.; Vanthuyne, N.; Bertrand, M. P.; Gastaldi, S.; Gil, G. *J. Org. Chem.* 2007, **72**, 6918–6923.
[197] Nechab, M.; El Blidi, L.; Vanthuyne, N.; Gastaldi, S.; Bertrand, M. P.; Gil, G. *Org. Biomol. Chem.* 2008, **6**, 3917–3920.
[198] Queyroy, S.; Vanthuyne, N.; Gastaldi, S.; Bertrand, M. P.; Gil, G. *Adv. Synth. Catal.* 2012, **354**, 1759–1764.
[199] Iglesias, L. E.; Sánchez, V. M.; Rebolledo, F.; Gotor, V. *Tetrahedron: Asymmetry* 1997, **8**, 2675–2677.
[200] Skupinsa, K. A.; McEachern, E. J.; Baird, I. R.; Skerlj, R. T.; Bridger, G. J. *J. Org. Chem.* 2003, **68**, 3546–3551.
[201] Sigmund, A. E.; DiCosimo, R. *Tetrahedron: Asymmetry* 2004, **15**, 2797–2799.
[202] Alatorre-Santamaría, S.; Gotor-Fernández, V.; Gotor, V. *Eur. J. Org. Chem.* 2009, 2533–2538.
[203] Brem, J.; Bencze, L.-C.; Liljeblad, A.; Turcu, M. C.; Paizs, C.; Irimie, F.-D.; Kanerva, L. T. *Eur. J. Org. Chem.* 2012, 3288–3294.
[204] Radu, A.; Moisă, M. E.; Toşa, M. I.; Dima, N.; Zaharia, V.; Irimie, F. D. *J. Mol. Catal. B: Enzym.* 2014, **107**, 114–119.
[205] González-Sabín, J.; Gotor, V.; Rebolledo, F. *Tetrahedron: Asymmetry* 2004, **15**, 481–488.
[206] González-Sabín, J.; Gotor, V.; Rebolledo, F. *Tetrahedron: Asymmetry* 2005, **16**, 3070–3076.
[207] González-Sabín, J.; Gotor, V.; Rebolledo, F. *J. Org. Chem.* 2007, **72**, 1309–1314.
[208] Peña, C.; González-Sabín, J.; Rebolledo, F.; Gotor, V. *Tetrahedron: Asymmetry* 2008, **19**, 751–755.
[209] Quijada, F. J.; González-Sabín, J.; Rebolledo, F.; Gotor, V. *Tetrahedron* 2009, **65**, 8028–8034.
[210] Quijada, F. J.; Rebolledo, F.; Gotor, V. *Tetrahedron* 2012, **68**, 7670–7674.
[211] Breen, G. F. *Tetrahedron: Asymmetry* 2004, **15**, 1427–1430.
[212] Orsat, B.; Alper, P. B.; Moree, W.; Mak, C.-P.; Wong, C.-H. *J. Am. Chem. Soc.* 1996, **118**, 712–713.
[213] Ding, W.; Li, M.; Dai, R.; Deng, Y. *Tetrahedron: Asymmetry* 2012, **23**, 1376–1379.
[214] Gotor-Fernández, V.; Fernández-Torres, P.; Gotor, V. *Tetrahedron: Asymmetry* 2006, **17**, 2558–2564.
[215] López-Iglesias, M.; Busto, E.; Gotor, V.; Gotor-Fernández, V. *J. Org. Chem.* 2012, **77**, 8049–8055.
[216] Alatorre-Santamaría, S.; Rodríguez-Mata, M.; Gotor-Fernández, V.; de Mattos, M. C.; Sayago, F. J.; Jiménez, A. I.; Cativiela, C.; Gotor, V. *Tetrahedron: Asymmetry* 2008, **19**, 1714–1719.
[217] Koszelewski, D.; Cwiklak, M.; Ostazewski, R. *Tetrahedron: Asymmetry* 2012, **23**, 1256–1261.
[218] Morgan, B.; Zaks, A.; Dodds, D. R.; Liu, J.; Jain, R.; Megati, S.; Njoroge, F. G.; Girijavallabhan, V. M. *J. Org. Chem.* 2000, **65**, 5451–5459.
[219] Turner, N. J. *Curr. Opin. Chem. Biol.* 2010, **14**, 115–121.
[220] Kim, Y.; Park, J.; Kim, M.-J. *ChemCatChem* 2011, **3**, 271–277.
[221] Reetz, M. T.; Schimossek, K. *Chimia* 1996, **50**, 668–669.
[222] Parvulescu, A.; De Vos, D.; Jacobs, P. *Chem. Commun.* 2005, 5307–5309.
[223] Parvulescu, A. N.; Jacobs, P. A.; De Vos, D. E. *Chem. Eur. J.* 2007, **13**, 2034–2043.
[224] Parvulescu, A.; Jacobs, P.; De Vos, D. *Appl. Catal. A: Gen.* 2009, **368**, 9–16.
[225] Parvulescu, A.; Van der Eycken, E.; Jacobs, P.; De Vos, D. E. *J. Catal.* 2008, **255**, 206–212.
[226] Jin, Q.; Jia, G.; Zhang, Y.; Li, C. *Catal. Sci. Technol.* 2014, **4**, 464–471.
[227] Andrade, L. H.; Silva, A. V.; Pedrozo, E. C. *Tetrahedron Lett.* 2009, **50**, 4331–4334.
[228] Cheng, G.; Xia, B.; Wu, Q.; Lin, X. *RSC Adv.* 2013, **3**, 9820–9828.
[229] Kim, M.-J.; Kim, W.-H.; Han, K.; Choi, Y. K.; Park, J. *Org. Lett.* 2007, **9**, 1157–1159.
[230] Kim, Y.; Park, J.; Kim, M.-J. *Tetrahedron Lett.* 2010, **51**, 5581–5584.
[231] Choi, Y. K.; Kim, Y.; Han, K.; Park, J.; Kim, M.-J. *J. Org. Chem.* 2009, **74**, 9543–9545.
[232] Engström, K.; Shakeri, M.; Bäckvall, J.-E. *Eur. J. Org. Chem.* 2011, 1827–1830.
[233] Choi, E.; Kim, Y.; Ahn, Y.; Park, J.; Kim, M.-J. *Tetrahedron: Asymmetry* 2013, **24**, 1449–1452.
[234] Xu, G.; Dai, X.; Fu, S.; Wu, J.; Yang, L. *Tetrahedron Lett.* 2014, **55**, 397–402.
[235] Gustafson, K. P. J.; Lihammar, R.; Verho, O.; Engström, K.; Bäckvall, J.-E. *J. Org. Chem.* 2014, **79**, 3747–3751.
[236] Engström, K.; Johnston, E. V.; Verho, O.; Gustafson, K. P. J.; Shakeri, M.; Tai, C.-W.; Bäckvall, J.-E. *Angew. Chem. Int. Ed.* 2013, **52**, 14006–14010.
[237] Filice, M.; Marciello, M.; Morales, M. P.; Palomo, J. M. *Chem. Commun.* 2013, **49**, 6876–6878.

[238] Paetzold, J.; Bäckvall, J.-E. *J. Am. Chem. Soc.* 2005, **127**, 17620–17621.
[239] Pàmies, O.; Éll, A. H.; Samec, J. S. M.; Hermanns, N.; Bäckvall, J.-E. *Tetrahedron Lett.* 2002, **43**, 4699–4702.
[240] Veld, M. A. J.; Hult, K.; Palmans, A. R. A.; Meijer, E. W. *Eur. J. Org. Chem.* 2007, 5416–5421.
[241] Rodríguez-Mata, M.; Gotor-Fernández, V.; González-Sabín, J.; Rebolledo, F.; Gotor, V. *Org. Biomol. Chem.* 2011, **9**, 2274–2278.
[242] Shakeri, M.; Engström, K.; Sandström, A. G.; Bäckvall, J.-E. *ChemCatChem* 2010, **2**, 534–538.
[243] Hoben, C. E.; Kanupp, L.; Bäckvall, J.-E. *Tetrahedron Lett.* 2008, **49**, 977–979.
[244] Thalén, L. K.; Zhao, D.; Sortais, J.-B.; Paetzold, J.; Hoben, C.; Bäckvall, J.-E. *Chem. Eur. J.* 2009, **15**, 3403–3410.
[245] Parvulescu, A. N.; Jacobs, P. A.; De Vos, D. E. *Adv. Synth. Catal.* 2008, **350**, 113–121.
[246] Choi, Y. K.; Kim, M. J.; Ahn, Y.; Kim, M.-J. *Org. Lett.* 2001, **3**, 4099–4101.
[247] Han, K.; Park, J.; Kim, M.-J. *J. Org. Chem.* 2008, **73**, 4302–4304.
[248] Gastaldi, S.; Escoubet, S.; Vanthuyne, N.; Gil, G.; Bertrand, M. P. *Org. Lett.* 2007, **9**, 837–839.
[249] El Blidi, L.; Vanthuyne, N.; Siri, D.; Gastaldi, S.; Bertrand, M. P.; Gil, G. *Org. Biomol. Chem.* 2010, **8**, 4165–4168.
[250] Poulhès, F.; Vanthuyne, N.; Bertrand, M. P.; Gastaldi, S.; Gil, G. *J. Org. Chem.* 2011, **76**, 7281–7286.
[251] Crawford, J. B.; Skerlj, R. T.; Bridger, G. J. *J. Org. Chem.* 2007, **72**, 669–671.
[252] Quijada, F. J.; Gotor, V.; Rebolledo, F. *Org. Lett.* 2010, **12**, 3602–3605.
[253] Sakulsombat, M.; Vongvilai, P.; Ramström, O. *Chem. Eur. J.* 2004, **20**, 11322–11325.
[254] Blacker, A. J.; Stirling, M. J.; Page, M. I. *Org. Process Res. Dev.* 2007, **11**, 642–648.
[255] Stirling, M.; Blacker, J.; Page, M. I. *Tetrahedron Lett.* 2007, **48**, 1247–1250.
[256] Liljeblad, A.; Kiviniemi, A.; Kanerva, L. T. *Tetrahedron* 2004, **60**, 671–677.
[257] Morán-Ramallal, R.; Gotor-Fernández, V.; Laborda, P.; Sayago, F. J.; Cativiela, C.; Gotor, V. *Org. Lett.* 2012, **14**, 1696–1699.
[258] González-Sabín, J.; Gotor, V.; Rebolledo, F. *Tetrahedron: Asymmetry* 2002, **13**, 1315–1320.
[259] Ma, G.; Xu, Z.; Zhang, P.; Liu, J.; Hao, X.; Ouyang, J.; Liang, P.; You, S.; Jia, X. *Org. Process Res. Dev.* 2014, **18**, 1169–1174.
[260] González-Sabín, J.; Gotor, V.; Rebolledo, F. *Chem. Eur. J.* 2004, **10**, 5788–5794.
[261] Torre, O.; Gotor-Fernández, V.; Gotor, V. *Tetrahedron: Asymmetry* 2006, **17**, 860–866.
[262] Rodríguez-Mata, M.; García-Urdiales, E.; Gotor-Fernández, V.; Gotor, V. *Adv. Synth. Catal.* 2010, **352**, 395–406.
[263] Han, K.; Kim, Y.; Park, J.; Kim, M.-J. *Tetrahedron Lett.* 2010, **51**, 3536–3537.
[264] Drauz, K.; Gröger, H.; May, O. ed. *Enzyme Catalysis in Organic Synthesis*, Wiley-VCH, Weinheim: 2012.
[265] Denard, C. A.; Hartwig, J. F.; Zhao, H. *ACS Catal.* 2013, **3**, 2856–2864.
[266] Bornscheuer, U. T.; Huisman, G. W.; Kazlauskas, R. J.; Lutz, S.; Moore, J. C.; Robins, K. *Nature* 2012, **485**, 185–194.
[267] Nestl, B. M.; Hammer, S. C.; Nebel, B. A.; Hauer, B. *Angew. Chem. Int. Ed.* 2014, **53**, 3070–3095.
[268] Yang, G.; Ding, Y. *Bioorg. Med. Chem.* 2014, **22**, 5604–5612.
[269] Itabaiana Jr., I.; Miranda, L. S. M.; de Souza, R. O. M. S. *J. Mol. Catal. B: Enzym.* 2013, **85–86**, 1–9.
[270] Boros, Z.; Falus, P.; Márkus, M.; Weiser, D.; Oláh, M.; Hornyánszky, G.; Nagy, J.; Poppe, L. *J. Mol. Catal. B: Enzym.* 2013, **85–86**, 119–125.
[271] Marciello, M.; Filicie, M.; Palomo, J. M. *Catal. Sci. Technol.* 2012, **2**, 1531–1543.
[272] Shu, Z.-Y.; Juang, H.; Lin, R. F.; Jiang, Y.-M.; Lin, L.; Huang, J.-Z. *J. Mol. Catal. B: Enzym.* 2010, **62**, 1–8.
[273] Bommarius, A. S.; Blum, J. K.; Abrahamson, M. J. *Curr. Opin. Chem. Biol.* 2011, **15**, 194–200.
[274] Wang, M.; Si, T.; Zhao, H. *Bioresour. Technol.* 2012, **115**, 117–125.
[275] Reetz, M. T. *Tetrahedron* 2012, **68**, 7530–7548.
[276] Bornscheuer, U. T. *Synlett* 2013, **24**, 150–156.

CHAPTER 10

Recent Advances in Enzyme-Catalyzed Aldol Addition Reactions

Pere Clapés

Department of Biological Chemistry and Molecular Modeling, Instituto de Química Avanzada de Cataluña, IQAC-CSIC, Barcelona, Spain

10.1 INTRODUCTION

Aldol addition reaction is a well-known and powerful synthetic tool for the formation of carbon–carbon bonds, which enables the concomitant functionalization and creation of new stereogenic centers [1–6]. A challenging goal of this methodology is therefore the control of both the relative and absolute stereochemistry of the newly formed stereogenic centers. There are two main strategies for the aldol reactions: directed and direct. In the directed aldol reactions, the enolate nucleophile intermediate (e.g., silyl or metal) is preformed, and the asymmetric induction is conducted by a covalent incorporation of a chiral auxiliary or promoted by catalytic quantities of a chiral species (e.g., the Mukaiyama reaction) [3–5, 7]. In the direct strategy, the reactive nucleophile species and the stereochemical outcome of the reaction are controlled by catalysis (e.g., metal complexes and organocatalysis). Among the catalytic methods, direct aldol additions mediated by aldolases are finding increasing acceptance in chemical research and in the production of asymmetric compounds due to their catalytic efficacy and high stereoselectivity [8–17]. Moreover, biocatalytic aldol reactions offer a preferable tool to perform asymmetric carbon–carbon bond formation in a sustainable and environmentally benign fashion in terms of reaction conditions and atom economy. Considerable advantages are evident for biologically active compounds that are usually polyfunctional and often water soluble, such as amino acids or carbohydrates and analogues, for example, iminocyclitols. Aldolases catalyze reversibly the aldol addition of a nucleophile donor component (i.e., aldehyde or ketone) to an electrophile acceptor (i.e., aldehyde or ketone), through an enolate or enamine generated at the active site of the aldolase. Concomitant to functionalization, the formation of one or two sterogenic centers occurred, whose absolute and relative stereochemistry is, in most cases, strictly controlled by the enzyme. Aldol adducts are usually typified as β-hydroxycarbonyl compounds: a structural element that is frequently found in the framework of complex molecules, either naturally occurring or synthetic [13]. Aldolases can be classified according to their donor specificity (Figure 10.1) into (i) pyruvate and related analogues, (ii) dihydroxyacetone phosphate (DHAP)-, (iii) DHA- and other unphosphorylated analogues, (iv) glycine/alanine-, and (v) acetaldehyde-dependent aldolases.

Green Biocatalysis, First Edition. Edited by Ramesh N. Patel.

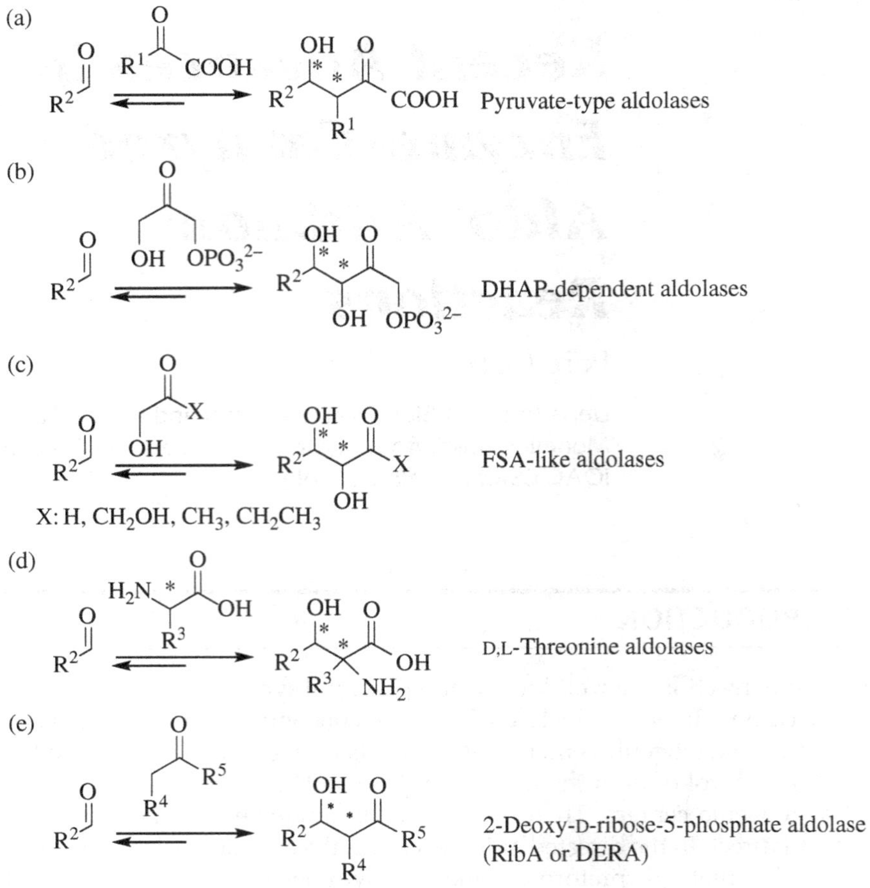

FIGURE 10.1
Classification of aldolases according to their donor selectivity: (a) Pyruvate aldolases, (b) dihydroxyacetone phosphate (DHAP)-dependent aldolases, (c) DHA- and other unphosphorylated analogues or DHA utilizing aldolases, (d) glycine/alanine aldolases, and (e) acetaldehyde-dependent aldolases.

Mechanistically, aldolases promote the reactive nucleophile species formation by the abstraction of the α-proton of the donor substrate, remaining bound at the enzyme's active site, protected from the solvent molecules, and shielding one of its enantiotopical faces to secure correct diastereofacial discrimination. Therefore, the absolute stereochemistry of the stereogenic center generated from the donor component is usually conserved throughout the reaction regardless of the acceptor substrate [18–22]. Aldolases have a stringent specificity for the donor substrate, while they are much more flexible for the acceptor electrophiles, tolerating a broad structural variety. The consequence of this relaxed selectivity for the acceptor is that the facial enantiotopic discrimination is, in some instances, compromised by the structure of the acceptor obtaining mixtures of enantiomers (i.e., when no stereogenic center is generated from the donor) or diasteroisomers when two stereogenic centers are produced [19, 23, 24]. Considering the type of the reactive nucleophilic species generated, the aldolases can be further divided into Class I and Class II (Figure 10.2). In the Class I, an enamine is produced by covalent binding to a conserved lysine residue, whereas in the Class II, an enediol is formed by chelating coordination to a transition metal cation (mostly Zn^{2+}), which acts as a Lewis acid promotor [14, 18, 20].

Aldolases constitute attractive tools in the asymmetric construction of molecular frameworks as well as in the synthesis of chiral bioactive compounds, such as carbohydrates, amino acids, and their analogues. Indeed, molecular complexity through enzymatic C—C connection can be built up under mild conditions, without a need for iterative steps of protection and deprotection of sensitive or reactive functional groups, increasing the atom economy of the transformation. Moreover, the same or different aldolase types can be applied independently in consecutive

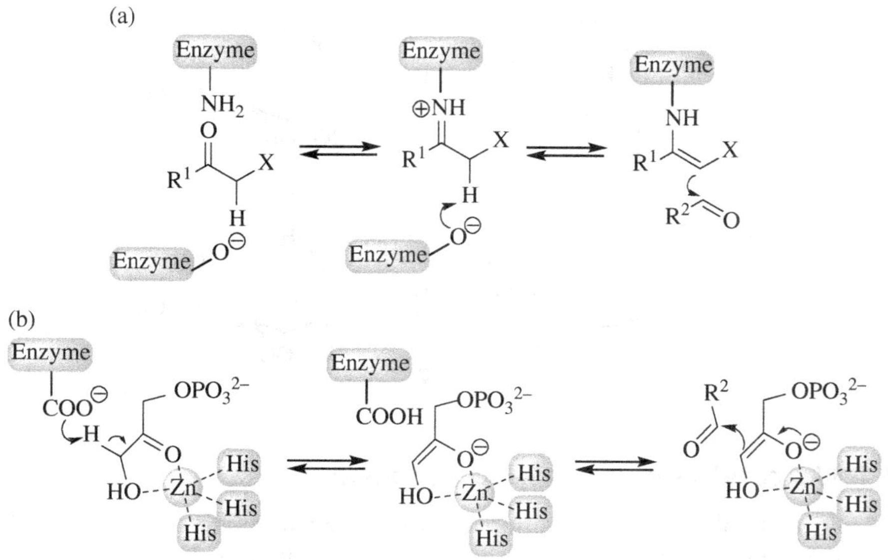

FIGURE 10.2

Mechanistic classification of aldolases. (a) Class I, an enamine is produced by covalent binding to a conserved lysine residue, (b) Class II, an enolate is formed by chelating coordination to a transition metal cation (mostly Zn^{2+}), which acts as a Lewis acid promotor.

cascade aldol additions in which the aldol adduct product of one would be the acceptor of the next [25–27]. Furthermore, the application of diverse stereocomplementary aldolases with the same donor selectivity may allow access to collections of products with functional and configurational diversity [24]. The application of cascade reactions to the asymmetric construction of complex natural products, such as the ones that can be envisaged with aldolases, has undeniable benefits including the atom economy of the process, economies in time, labor, resource management, and waste generation avoiding the tedious work-up and purification of the intermediates [28].

The discovery of novel aldolases in nature and the redesign of the active sites of the known ones either by directed evolution or by structure-guided approaches has dramatically expanded the field. This is fueled by the implementation of novel screening and selection strategies from the pool of enzymes in nature and protein engineering technique [29–31]. Ultimately, the knowledge of protein structure is extremely useful for gaining information about the role of the amino acid residues involved in the catalysis and substrate binding to propose further specific modifications. Progress toward this goal is growing, for example, the application of extremely intense X-ray flashes that allow the elucidation of the structure of a protein without the need for crystallization [32–34]. Furthermore, new mechanistic insights may help develop novel and better aldolases and even *de novo* design of enzyme-like proteins with unconventional catalytic activities [35, 36].

Since the mid-1990s, an increasing number of applications of aldolases in stereoselective synthesis have been reported. For a more comprehensive coverage of the general topic, we refer the readers to previous reviews [11, 15, 37–44].

10.2 PYRUVATE-DEPENDENT ALDOLASES

Pyruvate-dependent aldolases reversibly catalyze the aldol addition of pyruvate or analogues to aldehydes yielding γ-hydroxy-α-oxoacids (Figure 10.1). They exist as Class I aldolases (i.e., Schiff base/enamine formation) and Class II (i.e., metal cofactor and enolate formation) aldolases (Figure 10.2) [45, 46]. Class II pyruvate aldolases contain a Mg^{2+}, Mn^{2+}, or Co^{2+} divalent metal cation in octahedral coordination, which stabilizes the nucleophile (i.e., pyruvate anion) in the active site [45, 47].

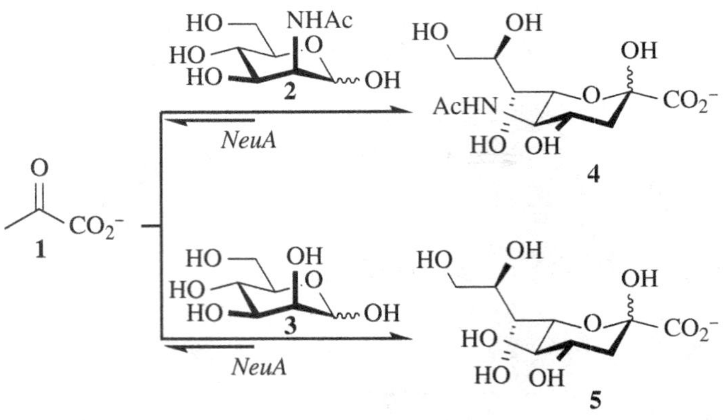

SCHEME 10.1

N-Acetylneuraminic acid aldolase (NeuA) catalyzed synthesis of sialic acid (**4**) and deoxy-D-*glycero*-D-*galacto*-2-nonulosonic acid (KDN, **5**) through aldol addition of pyruvate to *N*-acetyl-D-mannosamine (**2**) and D-manose (**3**).

10.2.1 *N*-Acetylneuraminic Acid Aldolase

Class I pyruvate aldolases are the most widely used in organic synthesis [10, 11]. Among them, the *N*-acetylneuraminic acid aldolase or sialic acid aldolase (NeuA, EC 4.1.3.3) has been subjected to numerous studies and a significant number of synthetic applications have emerged, for example, NeuA was the first aldolase used to produce a precursor to the antiviral drug Zanamivir at the multiton scale [48]. NeuA catalyzes the reversible *si*-face attack of pyruvate (**1**) to the carbonyl group of *N*-acetyl-D-mannosamine (**2**) furnishing sialic acid (**4**) with the installation of one new stereogenic center (Scheme 10.1).

Sialic acids are nine-carbon α-keto aldonic acids and exist in a great variety of derivatives on a common structural scaffold [49, 50]. They play important roles in biological, pathological, and immunological processes [51]. NeuA has the ability to catalyze the aldol addition of pyruvate to a variety of aldohexose sugars configurationally related to *N*-acetyl-D-mannosamine (D-ManNAc), that is, its natural substrate and D-mannose derivatives (e.g., 3-deoxy-D-*glycero*-D-*galacto*-2-nonulosonic acid, KDN derivatives) as well as other sugar derivatives larger or equal to pentoses [52–56] that lead to important construction of a large library of natural and designed sialic acid derivatives (Figure 10.3): for example, modifications at C5/C9 [9, 11, 57]; *N*-acetyl moiety replacement by other groups [58, 59], for example, *N*-Cbz **6** [60, 61], *N*-(hexinoyl) **7** [62], *N*-(*O*-Bn-glycolyl) **8** [63], amino acid conjugates **9** [64], and phenyl group **10** [58]; **11**, C9-bearing ether/ester moieties [65–68], for example, Boc-glycyl **12** [69], phenyl **13**, and 2-(perfluorohexyl)ethoxymethyl-tag **14** [70]; regiospecific modifications of Neu5Ac and KDN by OAc, OMe, H, F, or N_3 replacements at C9, C8, C7, and C5 (e.g., **15–24**) [71–78].

NeuA also tolerates a structural variety of disaccharides (e.g., **25**) carrying a reducing D-Man or D-ManNAc constituting a general and efficient methodology for producing disaccharides containing sialic acid at the reducing end (e.g., **26–31**) (Figure 10.4) [79, 80]. Sterically hindered 1,4-linked disaccharides and derivatives with sugar units coupled to mannosamine by an *N*-glycolyl linker are tolerated as substrates as well with uncompromised stereochemical fidelity.

The access to 1,6-linked disaccharides using NeuA catalysis envisaged the synthesis of sialoconjugates, such as α2,3- and α2,6-linked sialylated trisaccharides and tetrasaccharides, with one-pot, three-step cascade reactions using a combination of NeuA, CMP-sialic acid synthetase (CSS) and a sialyltransferase (SiaT) [79–81]. By coupling the aldol addition reaction with the practically irreversible enzymatic nucleotide activation and sialyl transfer, the thermodynamically disfavored aldol addition of pyruvate catalyzed by NeuA can be overcome, and none of the synthetic intermediates needs to be isolated and purified, adding to the overall efficiency of the procedure. One example of this approach was the preparation of α2–6- (**32**) and α2–3- (**33**) linked sialyl *p*-nitrophenyl β-galactoside (GalβpNP) (Scheme 10.2) [73, 82]. Other siaoloconjugates prepared using this procedures have been recently reported [75, 76, 83, 84].

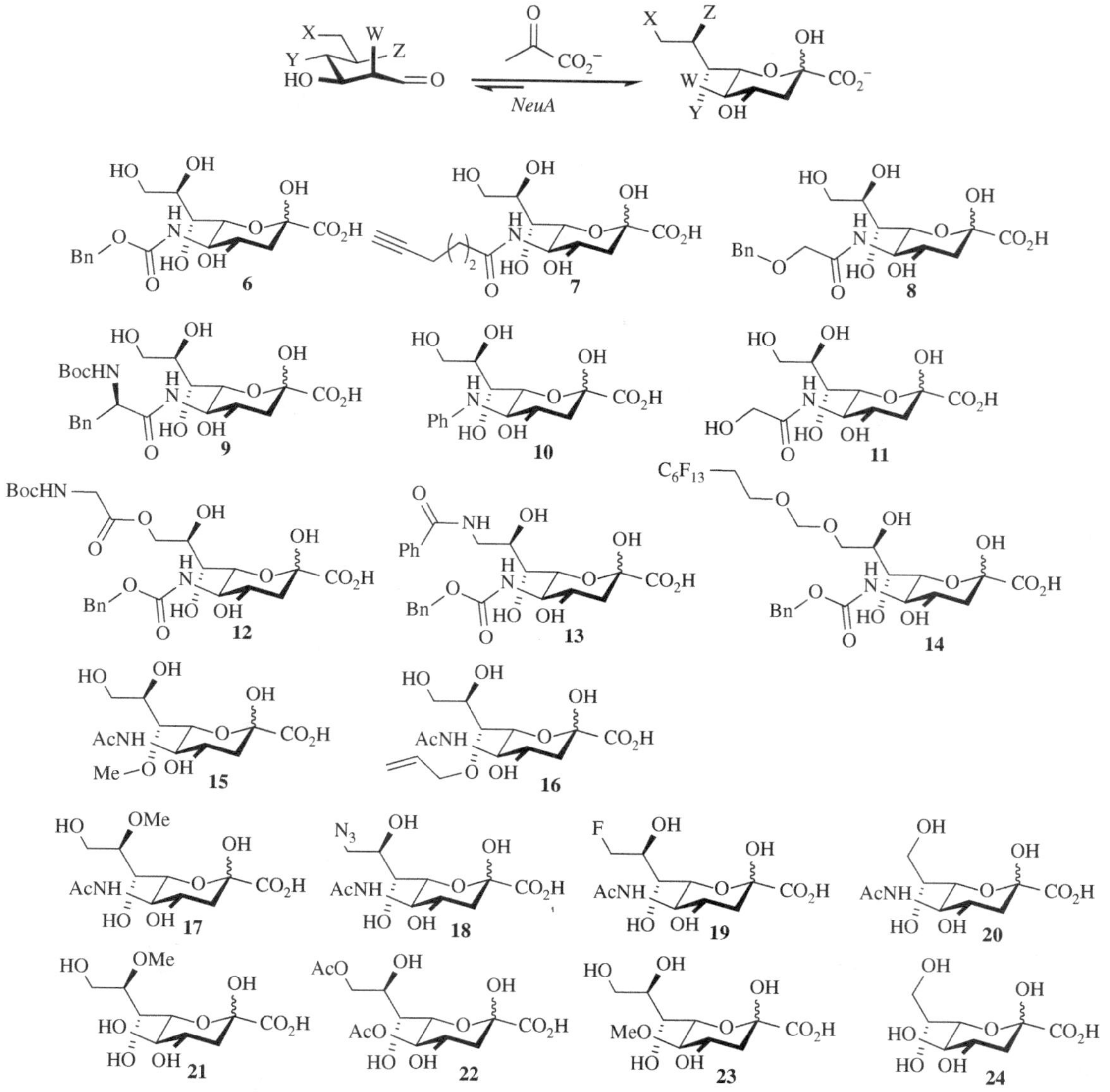

FIGURE 10.3

Examples of structurally diverse sialic acid, D-ManNAc and D-mannose derivatives synthesized using NeuA.

10.2.2 Other Pyruvate-Dependent Aldolases

A pyruvate aldolase from *Pseudomonas taetrolens* catalyzes the aldol addition of pyruvate to indole-pyruvic acid (**34**), a ketone [85]. This is an interesting example of an aldolase that catalyzes an aldol addition to a ketone acceptor. A variant of this aldolase was used in the stereoselective synthesis of a precursor (**35**) of monatin (**36**), whose 2*R*,4*R* stereoisomer is 2700-fold sweeter than sucrose (Scheme 10.3) [86].

4-Hydroxy-4-methyl-2-oxoglutarate/4-carboxy-4-hydroxy-2-oxoadipate (HMG/CHA) aldolase (EC 4.1.3.17) from *Pseudomonas putida* is a Class II pyruvate-dependent aldolase that catalyzes reversibly the homo-aldol addition of pyruvate (HMG aldolase) and the addition of pyruvate to oxaloacetate (**38**) (CHA aldolase), the last step of the bacterial protocatechuate 4,5-cleavage pathway (Scheme 10.4) [87].

Remarkably, HMG/CHA aldolase, as 4-hydroxy-2-keto-4-methyl glutarate aldolase (EC 4.1.3.17) [88, 89], utilizes a ketone (i.e., **1** or **38**) as acceptor, an activity that has not been exploited synthetically. Insights into the active site and catalytic mechanism are reported by Wang *et al.*, which will facilitate efforts for its development as a potential biocatalyst [87].

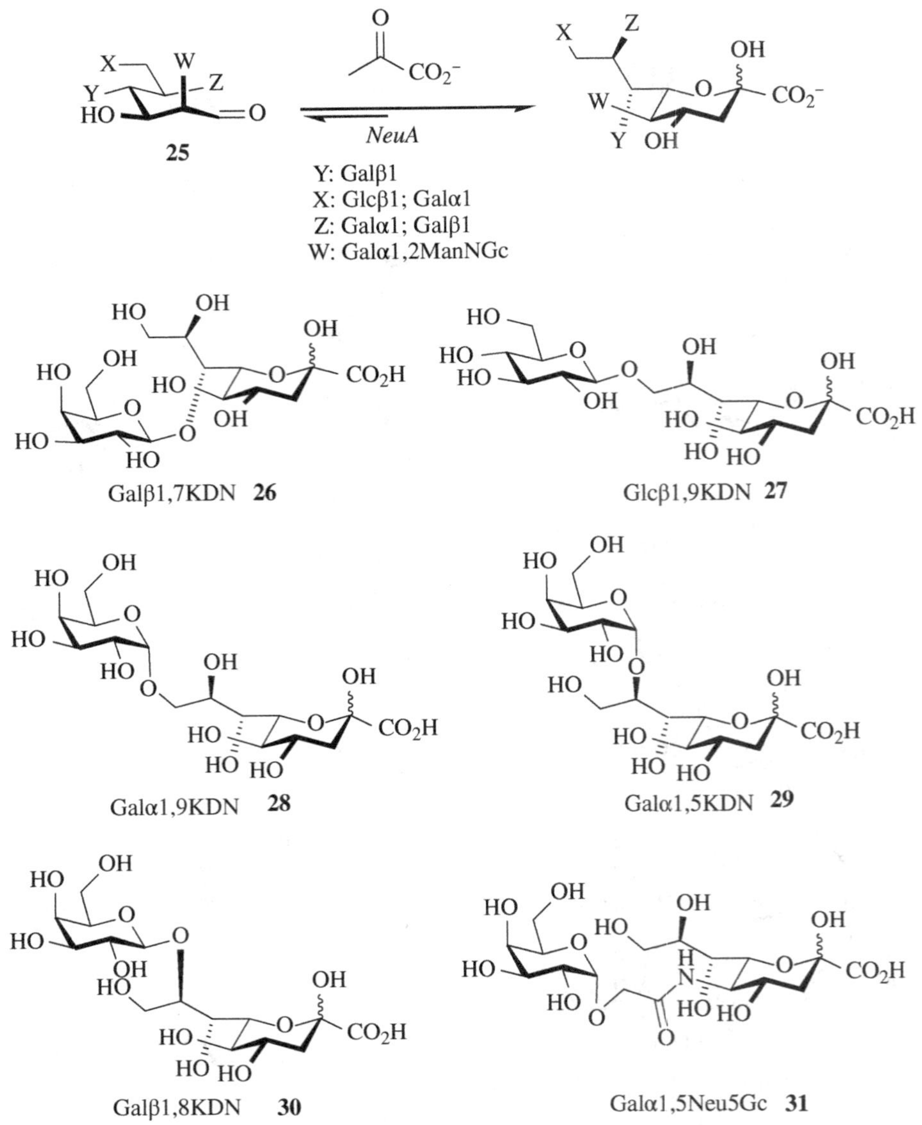

FIGURE 10.4

NeuA catalyzed synthesis of disaccharides containing sialic acid at the reducing end (e.g., **26–31**) by means of aldol addition of pyruvate to disaccharides bearing a mannose derivative at the reducing end.

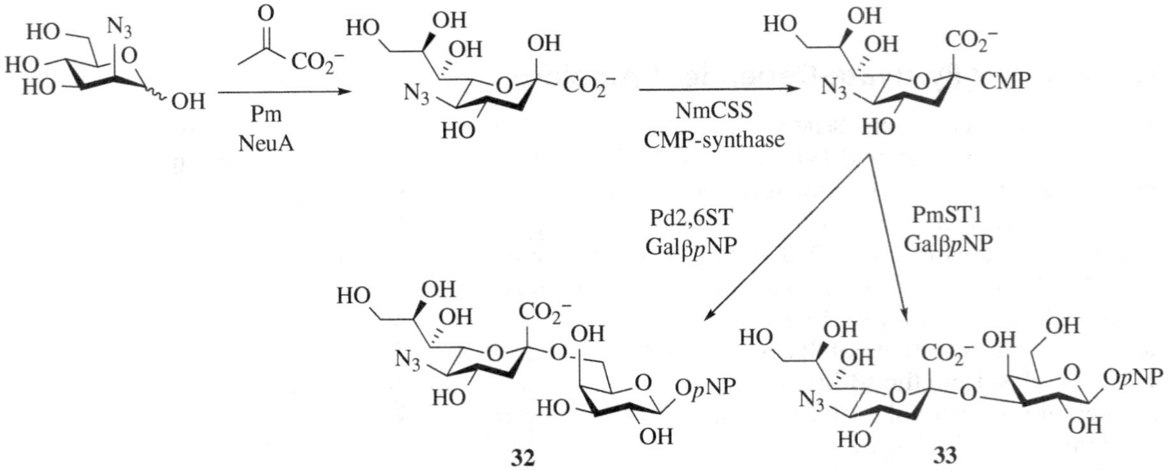

SCHEME 10.2

Example of the one-pot, three-enzyme synthesis of Siaα2–6GalbpNP (**32**) and Siaα2–3GalβpNP (**33**). Abbreviations: Pm: NeuA aldolase from, *Pasteurella multocida*; NmCSS CMP-synthetase: CMP-sialic acid synthetase from *Neisseria meningitidis*; PmST1: *P. multocida* sialyltransferase; Pd2,6ST: α2–6-sialyltransferase from *Photobacterium damsela*; GalβpNP *p*-nitrophenyl β-galactoside.

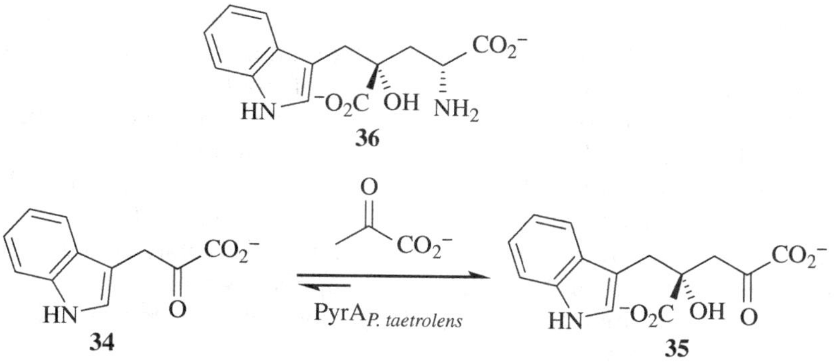

SCHEME 10.3

Enzymatic synthesis of a precursor of monatin (**36**) using pyruvate-dependent aldolases from *Pseudomonas taetrolens* (PyrA$_{P. taetrolens}$).

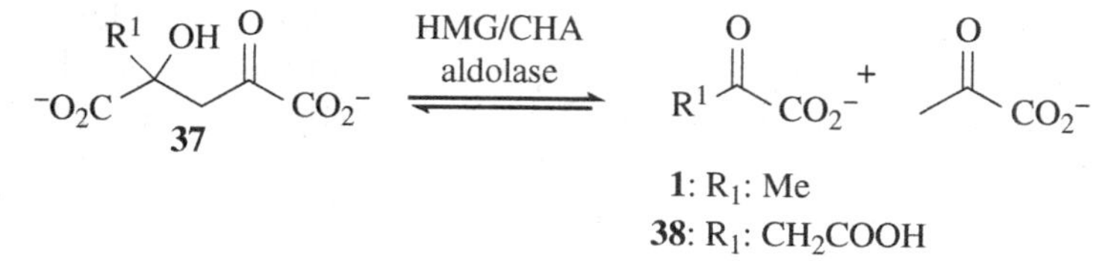

SCHEME 10.4

4-Hydroxy-4-methyl-2-oxoglutarate/4-carboxy-4-hydroxy-2-oxoadipate (HMG/CHA) aldolase catalyzed reversibly the homo-aldol addition of pyruvate (**1**) (HMG aldolase) and the addition of **1–38** (CHA aldolase).

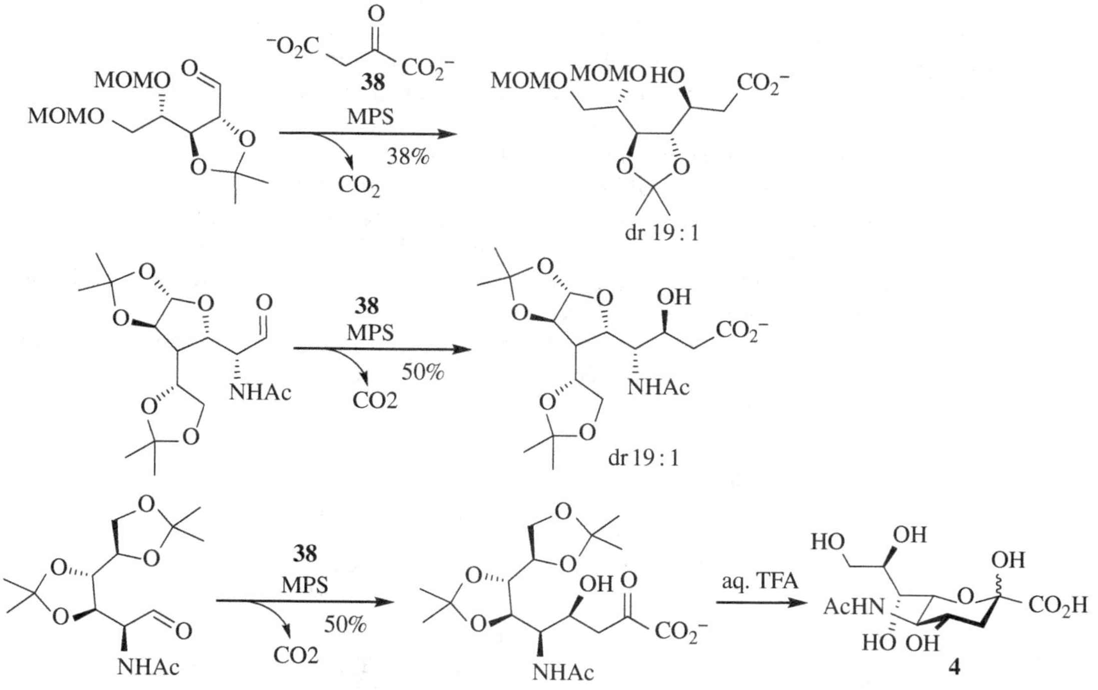

SCHEME 10.5

Protected polyoxygenated acceptor aldehyde substrates for macrophomate synthase (MPS) that rendered 19 : 1 *S* : *R* stereochemical outcome, and new chemoenzymatic synthesis of sialic acid (**4**).

Macrophomate synthase (MPS) from *Macrophoma commelinae* is able to catalyze the aldol addition of pyruvate to a variety of aldehyde acceptors [90]. The enzyme utilizes oxaloacetate (**38**) as donor substrate that, after decarboxylation at the enzyme active site, yields the enzyme–pyruvate enolate complex intermediate. The MPS showed a wide tolerance for structurally varied aldehyde acceptors, but it exhibited low enantioselectivity with small aldehydes [90]. Interestingly, bulky acceptors showed excellent diastereoselectivity, this being a remarkable strategy for the preparation of masked 3-deoxysugars from protected hydroxyaldehydes (Scheme 10.5) [91]. This tactic was used successfully for a new chemoenzymatic synthesis of sialic acid (**4**) (Scheme 10.5) [92]. The large size of the binding pocket of MPS adjacent to the magnesium-bound pyruvate allows the allocation of a broad variety of substrate structures, but only the bulkier ones are able to be stereoselectively approached to the nucleophile rendering the *S*-configured stereocenter.

During an investigation on the biosynthesis of peptidyl nucleoside antibiotics, namely nikkomycins, two enzymes acting in concert were discovered to be involved: *san*M and *san*N [46, 93]. *San*M is a Class II aldolase (i.e., requires Mn^{2+} for its activity) that catalyzes the addition of 2-ketobutyrate (**40**) to picolinaldehyde (**39**), to furnish ketoacid (**41**) with excellent stereoselectivity (Scheme 10.6). Like *san*M, 4-hydroxy-3-methyl-2-keto-pentanoate aldolase from *Arthrobacter simplex* AKU 626 catalyzes the addition of **40** to acetaldehyde (Scheme 10.6). This aldolase was applied for the synthesis of 4-hydroxyisoleucine (**42**), an insulinotropic agent, using a one-pot, two-step cascade enzymatic aldol and transamination reactions [94, 95]. It is noteworthy that *san*M and the aldolase from *A. simplex* AKU 626 rendered inverse stereochemistry at C3/C4, being enantiocomplementary pyruvate aldolases. Most importantly, both aldolases yielded two new stereogenic centers, thus opening up new opportunities in biocatalysis, because this is different from the situation with shorter C3 donors used by the pyruvate aldolases (i.e., pyruvate or phosphoenolpyruvate) where only one stereocenter is formed.

Class II pyruvate-dependent 4-hydroxy-2-oxopentanoate aldolases (EC 4.1.2.-), BphI and HpaI, catalyze the retroaldol reactions of 4-hydroxy-2-oxopentanoate (**43**) to pyruvate (**1**) and acetaldehyde and 4-hydroxy-2-oxo-1,7-heptanedioate (**44**) to **1** and succinic semialdehyde (**45**) (Scheme 10.7) [45, 96]. BphI is selective for (*S*)-**43** stereoisomer, whereas HpaI accepts both enantiomers [97].

The catalytic properties of BphI and HpaI as aldol addition catalysts showed that, 2-ketobutyrate (Table 10.1, **46**, $R^2=CH_3$) can also be tolerated as donor substrate by HpaI, although at a lower rate, whereas BphI was only able to accept pyruvate [98]. As pointed out before, the use of **46** ($R^2=CH_3$) as a nucleophile results in the formation of two new stereocenters.

Concerning the acceptor selectivity, HpaI tolerates linear and branched aliphatic aldehydes as well as hydroxyaldehydes using pyruvate or 2-ketobutyrate (Table 10.1, **46**, $R^2=H$ or CH_3) as donors. BphI preferentially accepts linear aliphatic aldehydes, acetaldehyde (Table 10.1, entry 1) and propionaldehyde (Table 10.1, entry 3) being the best acceptors in terms of k_{cat}/K_m. Concerning the stereochemistry, the aldol addition of pyruvate to acetaldehyde catalyzed by BphI produced exclusively the 4*S* stereoisomer. This is particularly promising for other additions catalyzed with this enzyme. It is also remarkable that BphI tolerates simple aliphatic aldehydes as acceptors, a

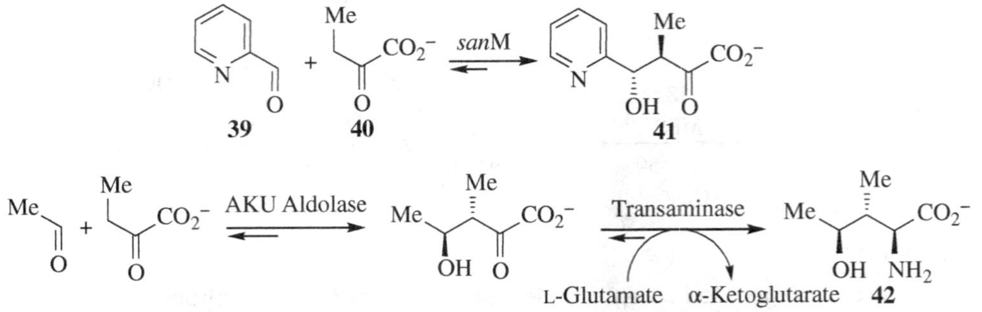

SCHEME 10.6

SanM and AKU aldolase catalyzed the aldol addition of 2-ketobutyrate (**40**) to picolinaldehyde (**39**) and acetaldehyde, respectively.

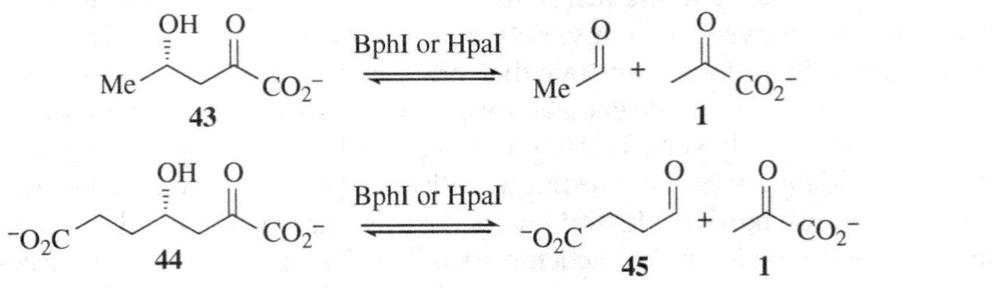

SCHEME 10.7

Retro aldol reactions of 4-hydroxy-2-oxopentanoate (**43**) to pyruvate (**1**) and acetaldehyde and 4-hydroxy-2-oxo-1,7-heptanedioate (**44**) to **1** and succinic semialdehyde (**45**) catalyzed by 4-hydroxy-2-oxopentanoate aldolases, BphI and HpaI.

TABLE 10.1 Acceptor/Donor Tolerance of BphI and HpaI Aldolases

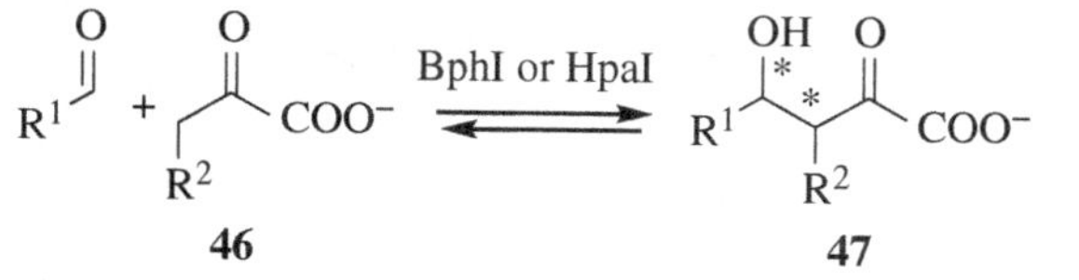

Entry	R^1	R^2	HpaI	BphI
1	CH_3	H	++	+++
2	CH_3	CH_3	+	–[a]
3	CH_2CH_3	H	+++	+++
4	$CH_2CH_2CH_3$	H	+++	++
5	$CH_2CH_2CH_2CH_3$	H	+++	+
6	$CH_2CH_2(CH_3)_2$	H	+++	+
7	CH_2OH	H	++	++
8	*rac*-$CHOHCH_2OH$	H	+	–[a]
9	CH_2CH_2COOH	H	+++	–[a]

*The stereochemistry of the aldol addition not determined.
+++ $k_{cat}/K_m > 10^4$; ++ $5 \cdot 10^3 > k_{cat}/K_m > 3 \cdot 10^3$, + $k_{cat}/K_m < 10^3\,M^{-1}s^{-1}$.
[a] No product was detected.

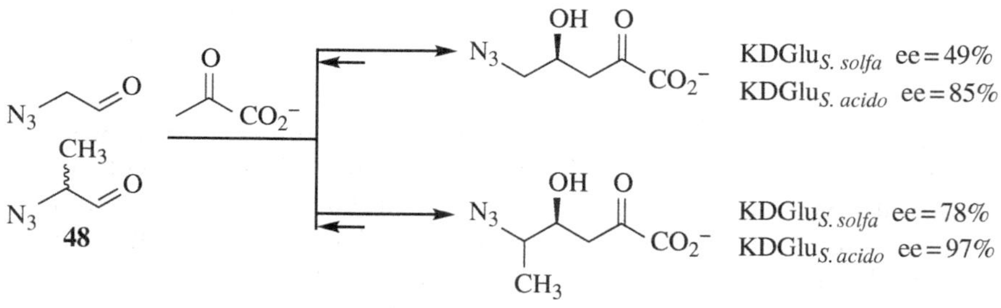

SCHEME 10.8 2-Keto-3-deoxygluconate (KDGlu) aldolase from *Sulfolobus solfataricus* (KDGlu$_{S.\ solfa}$) and from *Sulfolobus acidocaldarius* (KDGlu$_{S.\ acido}$) catalyzed aldol addition of pyruvate to azido aldehydes **48**.

property not found in NeuA, 2-keto-3-deoxy-6-phospho-D-gluconate (GlcA) and 2-keto-3-deoxy-6-phospho-D-galactonate (GalA) wild-type aldolases [99–107].

On the other hand, HpaI furnished a racemic mixture of 4*S*:4*R* aldol adducts limiting their potential synthetic applications. The authors reasoned that the BphI active site is smaller than that of HpaI and, therefore, the steric constraints prevent alternative conformations of the acceptor aldehyde sustaining the C4 stereochemistry.

2-Keto-3-deoxygluconate aldolase from hyperthermophile *Sulfolobus solfataricus* (KDGlu$_{S.\ solfa}$) is a Class I pyruvate aldolase with broad acceptor specificity, which includes nonphosphorylated aldehydes, but also has a poor facial stereocontrol [108, 109]. This was partially overcome by modifying the substrate to a more structurally rigid analogue [108]. This is a similar case to that of MPS, discussed previously in which the structure of substrate may compromise the stereochemistry at C4 (vide supra) [90, 91].

KDGlu$_{S.\ solfa}$ and another one from the hyperthermophile *Sulfolobus acidocaldarius* (KDGlu$_{S.\ acido}$) were used as catalyst for the addition of pyruvate to azido-substituted aldehydes (**48**) to prepare nitrogen heterocycles (Scheme 10.8). The KDGlu$_{S.\ acido}$ gave higher stereoselectivity than KDGlu$_{S.\ solfa}$ and this appeared to be related to the A128 residue in the *S. solfataricus* one which is S195 in that from *S. acidocaldarius* [110].

10.2.3 Structure-Guided Pyruvate Aldolase Modification

Structure-guided site-directed mutagenesis offers a great potential for the rapid preparation of mutants with new/improved activity when the structure of the enzyme and its catalytic mechanism are known [29, 111–114].

To overcome the poor stereoselectivity of KDGlu$_{S.\ solfa}$, a pair of stereocomplementary KDGlu$_{S.\ solfa}$ variants were obtained for the aldol addition of pyruvate to

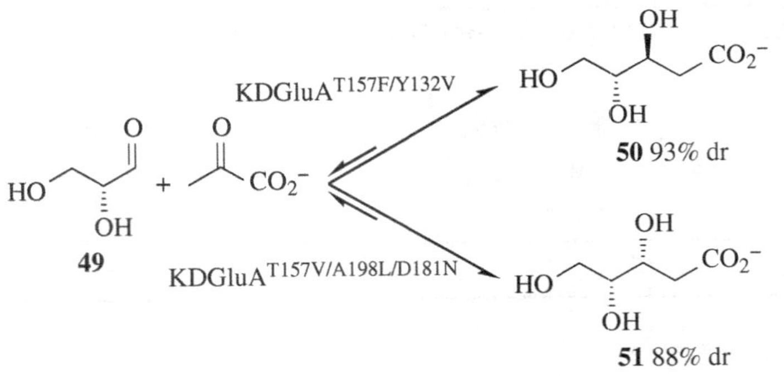

SCHEME 10.9

2-Keto-3-deoxygluconate aldolase (KDGluA) variants T157F/Y132V (KDGluA$^{T157F/Y132V}$) and T157V/A198L/D181N (KDGluA$^{T157V/A198L/D181N}$) from the hyperthermophile *Sulfolobus solfataricus* catalyzed the stereoselective aldol addition of pyruvate to D-glyceraldehyde (**49**).

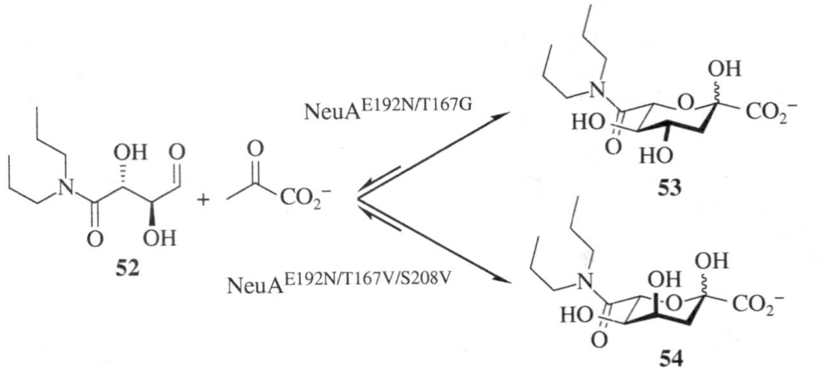

SCHEME 10.10

Stereodivergent synthesis of 4*S*- and 4*R*-configured 6-dialkylaminocarbonyl *N*-acetylneuraminic acid derivatives (**53** and **54**), catalyzed by NeuA E192N/T167G (NeuA$^{E192N/T167G}$) and NeuA E192N/T167V/S208V (NeuA$^{E192N/T167V/S208V}$) enantiocomplementary variants.

D-glyceraldehyde (**49**, Scheme 10.9) by using the structure-guided approach [115]. Amino acids T157 and T132 were identified as key residues for controlling the stereoselectivity. Saturation mutagenesis on T157 and a further substitution of Y132 by valine, intended to destroy the interaction of Y132 with the C5—OH group of **51**, gave the double mutant T157F/Y132V that furnished **50** in a 93% dr (Scheme 10.9). For the synthesis of the other diasteroisomer (**51**) the mutation, T157G in addition to A198L/D181Q, intended to maximize the hydrophobic interactions between the Ala198 side chain and the C1—C3 hydrophobic face of the D-glyceraldehyde backbone, improved the stereoselectivity of **51** up to 88% dr (Scheme 10.9).

The same goal was previously achieved on *N*-acetylneuraminic acid aldolase from *Escherichia coli* by combination of error-prone PCR and a structure-guided approach of saturation and site-directed mutagenesis [116]. This resulted in two variants E192N/T167G and E192N/T167V/S208V, which were ≈50-fold selective toward the retroaldol reaction of 4*S*- and 4*R*-configured 6-dialkylaminocarbonyl *N*-acetylneuraminic acid derivatives (**53** and **54**), respectively. Both variants catalyzed the aldol addition of pyruvate to (2*R*,3*S*)-2,3-dihydroxy-4-oxo-*N*,*N*-dipropylbutyramide (**52**) with complementary C4 stereochemistry and more than 98% diastereoselectivity (Scheme 10.10).

A third example of stereochemical modification was performed with the previously mentioned 4-hydroxy-2-oxopentanoate aldolase, BphI [97]. Double variants L87N/Y290F and L87W/Y290F were constructed and demonstrated to exclusively utilize (*R*)-4-hydroxy-2-oxopentanoate (*ent*-**44**, Scheme 10.7) as the substrate, while the wild-type only tolerates the enantiomer (**44**).

10.3 DIHYDROXYACETONE PHOSPHATE (DHAP)-DEPENDENT ALDOLASES, D-FRUCTOSE-6-PHOSPHATE ALDOLASE (FSA) AND TRANSALDOLASES

10.3.1 DHAP-Dependent Aldolases

DHAP-dependent aldolases constitute a set of four stereocomplementary enzymes that catalyze stereoselectively the reversible aldol addition reaction of DHAP to a large variety of aldehyde acceptors. Among them, D-fructose-1,6-phosphate aldolase

(FruA) from different sources, L-rhamnulose-1-phosphate aldolase (RhuA) and L-fuculose-1-phosphate aldolase (FucA) from *E. coli* have been extensively investigated and applied in the synthesis of carbohydrates and their derivatives such as deoxysugars, iminocyclitols and aminocyclitols [44]. An exception was D-tagatose-1,6-bisphosphate aldolase (TagA), whose lack of stereoselection constitutes a major limitation of this aldolase, which frustrates the attempts to obtain the four possible stereoisomers from a single aldehyde, one of the hallmarks of enzymatic aldol additions [117, 118].

A general drawback of DHAP-dependent aldolases is their strict selectivity toward DHAP, which is chemically unstable, particularly under alkaline conditions, and expensive [44]. Moreover, DHAP decomposition generates inorganic phosphate and methyl glyoxal, both of them inhibitors of the aldolase [10]. For this reason, numerous methods were developed for the chemical or enzymatic *in situ* preparation of DHAP, including cascade enzymatic reactions [15, 17, 41, 119, 120]. As a recent example, a multienzyme system using a dihydroxyacetone kinase (DHAK) from *Citrobacter freundii*, and acetate kinase (AK) for adenosine triphosphate (ATP) regeneration, has been proved to generate DHAP *in situ* efficiently (Scheme 10.11). The one-pot, two-step reaction was completed by the addition of the aldehyde acceptor and aldol biocatalyst namely FruA from rabbit muscle (RAMA), RhuA or FucA from *E. coli* [121]. In the same line of thinking, a bifunctional fusion protein between the DHAK from *C. freundii* and FruA from *Staphylococcus carnosus* has been reported, which catalyzes the phosphorylation of DHA and its subsequent aldol addition in one-pot with 20-fold improved kinetics as compared to the use of separate enzymes (Scheme 10.11) [120, 122].

A novel method was also reported for the generation of DHAP from DHA, based on the use of the acid phosphatase from *Shigella flexneri* (PhoN-Sf) using pyrophosphate (PPi) as phosphate donor (Scheme 10.12) [123, 124]. In the reaction, PhoN-Sf can also use the phosphorylated aldol adduct as phosphate donor to phosphorylate DHA and release the unphosphorylated aldol adduct (Scheme 10.12). A directed evolution approach was devised to obtain a variant PhoN-Sf V178L, which produced 25% more DHAP than the wild type at pH 6.0 in less than half the time, and it is far more effective in the cascade aldol reaction [125].

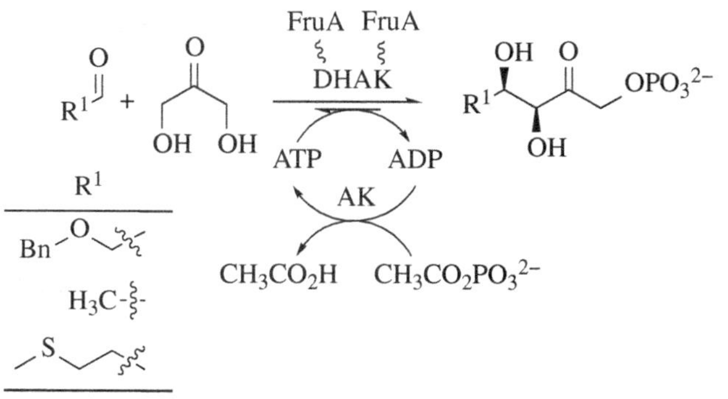

SCHEME 10.11

Fusion protein between the dihydroxyacetone kinase (DHAK) from *Citrobacter freundii* and FruA from *Staphylococcus carnosus* catalyzed aldol addition of DHA to aldehydes to render phosphorylated products.

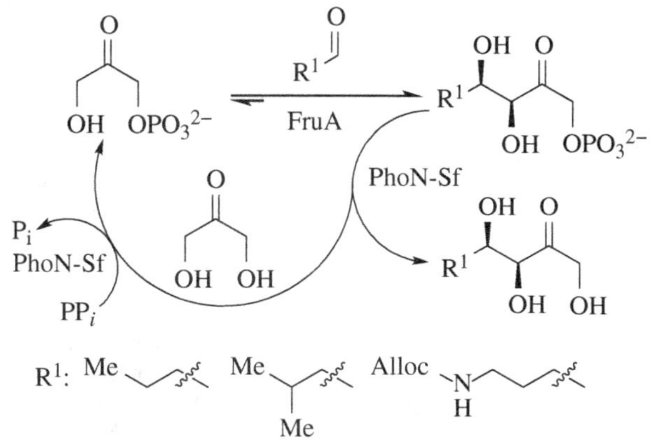

SCHEME 10.12

Generation of DHAP from DHA catalyzed by the acid phosphatase from *Shigella flexneri* (PhoN-Sf) using pyrophosphate (PPi) as primary phosphate donor, followed by the ensuing aldol addition reaction of DHAP to selected aldehyde examples and the *in situ* utilization of the corresponding phosphorylated aldol adduct as phosphate donor for the PhoN-Sf.

Although multienzymatic methods for DHAP generation render, in general, good results, the direct use of inexpensive nonphosphorylated DHA as nucleophile is the preferred choice, especially when the phosphate group of the product must be removed in a separate reaction. In this sense, a significant activity of RhuA wild type in aldol additions using DHA as donor, although with high K_M values (~1.1 M) was found [113]. To improve the activity of wild-type RhuA toward DHA, the main residues interacting with the phosphate moiety [126], N29, N32, S75, T115, and S116, were replaced by aspartate [113], intending to establish new polar contacts that may stabilize bound DHA [104, 111, 127]. After suitable screening, $RhuA^{N29D}$ proved to be the most active mutant for the aldol addition of DHA [113].

Besides protein engineering, it was uncovered that RhuA wild type can improve the aldol addition of DHA to aldehyde between 35- and 100-fold when the reactions were carried out in the presence of borate [113, 128]. Moreover, retroaldol rates for some aldol adducts in the presence of borate were low or negligible as compared with the aldol ones, making the process irreversible [128, 129]. However, the tolerance to unphosphorylated DHA by of a Class II DHAP-dependent aldolase appears to be an exclusive property of RhuA, since the stereocomplementary FucA had no detectable activity with DHA either with or without borate added [113].

10.3.2 Iminocyclitol, Pipecolic Acids, Homoiminocyclitols, and Aminocyclitol Synthesis

The chemo-enzymatic access to iminocyclitols mediated by aldolases consists of two steps: the aldol addition of DHAP to a synthetic equivalent of an aminoaldehyde, followed by the intramolecular reductive amination (Scheme 10.13) [130, 131]. As a synthetic equivalent of amino aldehyde, *N*-benzyloxycarbonyl (Cbz)-protected aminoaldehyde derivatives have been widely used [23, 132, 133]. An advantage is that *N*-protected aminoaldehydes can be easily obtained from the wide structural variety of readily available optically pure α- or β-amino acids or alcohols and their derivatives [134, 135], and that both Cbz deprotection and the intramolecular reductive amination can be performed catalytically as a one-pot, two-step synthesis.

Using this synthetic strategy (Scheme 10.13), a structural variety of pyrrolidine- and pyrrolizidine-type iminocyclitols can be obtained (Scheme 10.14) [24, 112, 136].

RhuA tolerated both linear and branched C-α-substituted *N*-Cbz-aminoaldehydes (**55**; R^1: ethyl. propyl, butyl, isopropyl, Scheme 10.14), including the conformationally restricted (*R*)- and (*S*)-*N*-Cbz-prolinal derivatives (**56**, Scheme 10.15) [24]. RhuA was fully stereoselective (>98%) for the linear and branched C-α alkyl-substituted (*S*)-*N*-Cbz-aminoaldehydes including both (*R*)- and (*S*)-*N*-Cbz-prolinal rendering the *syn* (3*R*,4*S*) configured aldol adducts.[1] FucA from *E. coli* catalysis gave full control of the newly formed stereogenic centers, furnishing always the *anti* (3*R*,4*R*) configured aldol adducts (i.e., *anti* (3*R*,4*R*):*syn* (3*R*,4*S*) diastereomeric ratios >98:2 for the (*R*)-*N*-Cbz-aminoaldehydes and >90:10 for the (*S*)-counterparts). However, FucA accepted preferentially C-α linear alkane substitutions [24]. FruA from rabbit muscle (RAMA) was the most restrictive on aldehyde selectivity and only tolerates the unbranched *N*-Cbz-aminoaldehydes, for example,

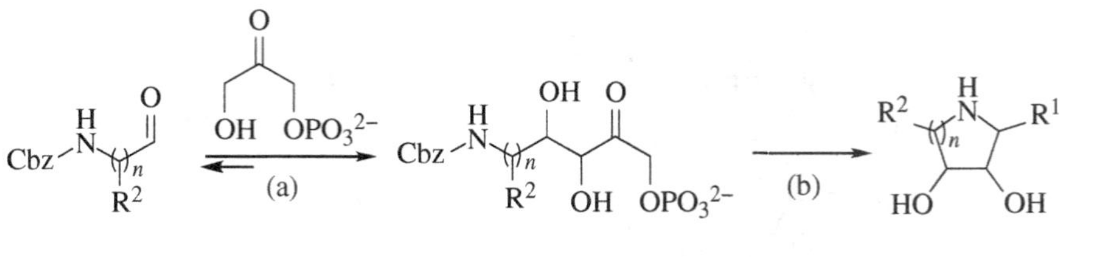

SCHEME 10.13
Two-step chemo-enzymatic synthesis of iminocyclitols: (a) aldolase-catalyzed aldol addition and phosphate hydrolysis and (b) one-pot Cbz removal and reductive amination.

[1] Please note that in the original article [136] the incorrect (1*S*)-configuration was assigned to the polyhydroxylated pyrrolizidines. Based on this assignment, the aldol reaction was thought to proceed with low diastereoselectivity. The correct stereochemical assignment was unequivocally established in a later study [112]. Therefore, RhuA was fully stereoselective in the aldol addition of DHAP to *N*-Cbz-prolinal derivatives.

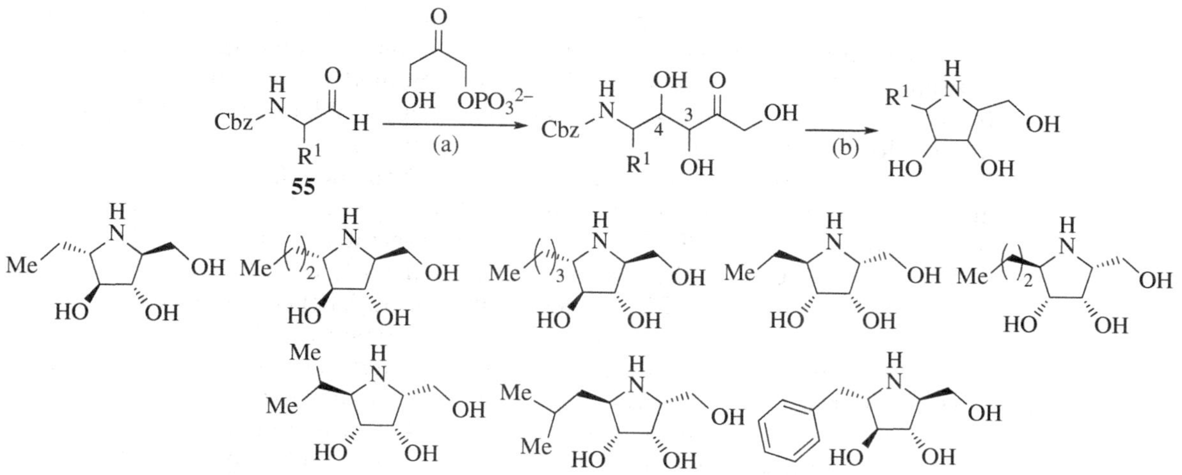

SCHEME 10.14

Preparation of diverse pyrrolidine type iminocyclitols mediated by DHAP-dependent aldolases, L-rhamnulose-1-phosphate aldolase (RhuA) and L-fuculose-1-phosphate aldolase (FucA) from *Escherichia coli*. (a) DHAP-dependent aldolase, namely and acid phosphatase to remove the phosphate group, and (b) one-pot Cbz removal and reductive amination.

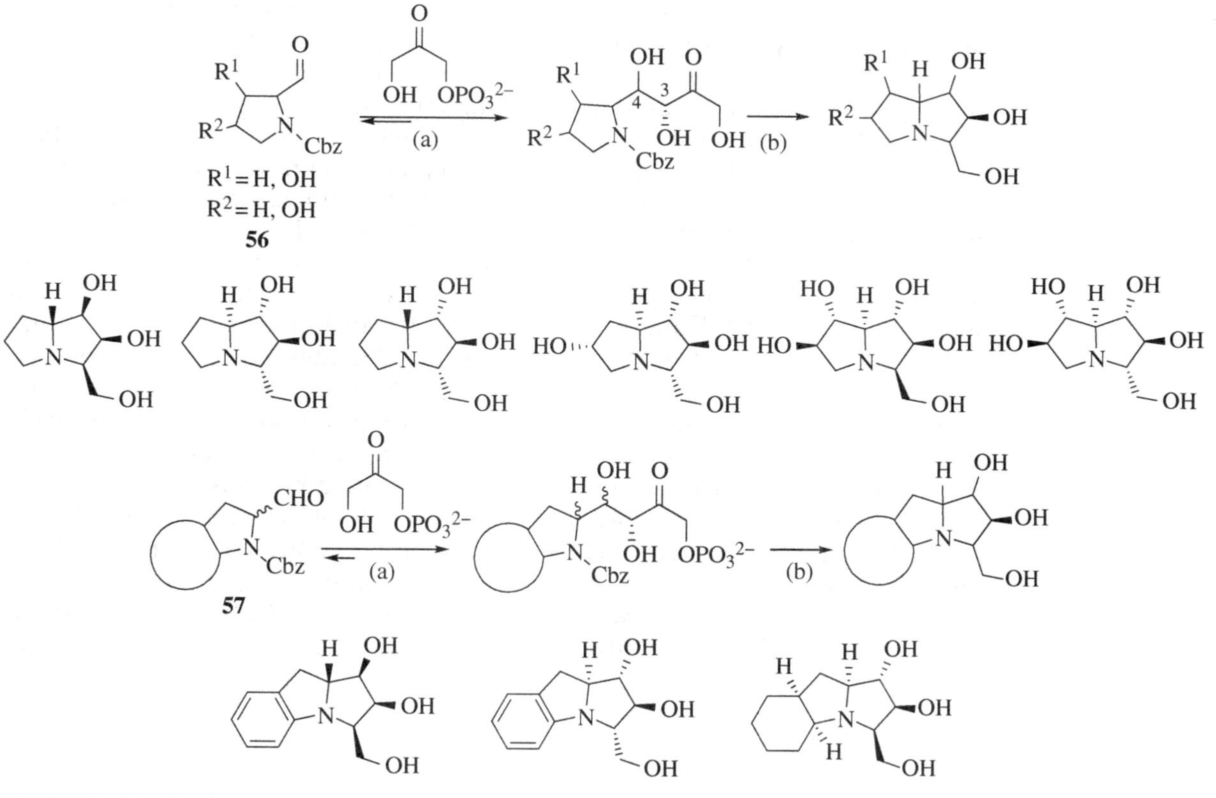

SCHEME 10.15

Preparation of diverse pyrrolizidine type iminocyclitols mediated by DHAP-dependent aldolases, L-rhamnulose-1-phosphate aldolase (RhuA) and L-fuculose-1-phosphate aldolase (FucA) from *Escherichia coli* variant F131A. (a) DHAP-dependent aldolase, namely, RhuA and/or FucA F131A and acid phosphatase to remove the phosphate group and (b) one-pot Cbz removal and reductive amination.

N-Cbz-3-aminopropanal and *N*-Cbz-2-aminoethanal [132], being the linear or branched C-α substituted aminoaldehydes including the prolinal derivatives no substrates. The reductive amination was normally conducted with H_2 in the presence of Pd/C. This reaction proceeded in many instances with high selectivity, but for some structures, mixtures of diastereoisomers were obtained depending on the aldol adduct [23, 24, 132].

The limitation of wild-type FucA toward the acceptor substrate was overcome by a structure-guided redesign of the active site using a site-directed mutagenesis approach [112]. The goal was to alleviate constrained space in the active site and thus facilitate the accommodation of bulky acceptor aldehydes, including conformationally restricted cyclic derivatives. Among the variants designed and assayed, FucA F131A, was the most active, tolerating branched C-α-substituted *N*-Cbz-aminoaldehydes including the (*R*)- and (*S*)-*N*-Cbz-prolinal, 3- and 4-hydroxyprolinal (**56**, Scheme 10.15), *N*-Cbz-indoline-and *N*-Cbz-octahydroindole-2-carbaldehydes derivatives (**57**, Scheme 10.15) with remarkable activity [112, 137, 138]. Concerning the stereochemistry, FucA F131A yielded the *anti* (3*R*,4*R*) configured aldol adduct for the (*R*)-*N*-Cbz-aminoaldehydes in high diastereoselectivity (>2:98 *syn* (3*R*,4*S*):*anti* (3*R*,4*R*) ratio). Interestingly, the (*S*)-prolinal and hydroxyprolinal derivatives rendered the *syn* (3*R*,4*S*) stereoselectivity, while for the rest of (*S*)-cyclic acceptors gave mixtures of *syn* (3*R*,4*S*):*anti* (3*R*,4*R*) adduct depending on the aldehyde [137]. Molecular models of the aldolases with the substrates were built up to gain insight in the acceptor binding mode that led to this distinct stereochemical outcome [112].

RhuA wild type was synthetically useful in the preparation of new indolizidine- and quinolizidine-type iminocyclitols (Scheme 10.16) [139]. The aldol addition of DHAP to (*R*) and (*S*)-*N*-Cbz-piperidin-2-carbaldehyde (**58**, $n=1$) and rac-*N*-Cbz-2-(piperidin-4-yl)acetaldehyde (Scheme 10.16, **58**, $n=2$) was the key reaction. In this case, FucA F131A rendered only meager yields that were not satisfactory from a preparative point of view. The RhuA-catalyzed aldol addition of DHAP to (*R*-**58**, $n=1$) gave a 2:3 *syn* (3*R*,4*S*):*anti* (3*R*,4*R*) diasteromeric aldol mixture, which was regarded as a useful opportunity to expand the configurational diversity of the compounds. Moreover, RhuA did not show any kinetic discrimitation when using rac-*N*-Cbz-2-(piperidin-4-yl)acetaldehyde (**58**, $n=2$) as acceptor, yielding also aldol adducts as *syn*:*anti* mixtures. The mixtures of indolizidines and quinolizidines generated were isolated and purified by cation exchange chromatography [139].

The synthesis of iminocyclitols was also accomplished using aldol additions of the unphosphorylated analogs of DHAP, dihydroxyacetone (DHA). As mentioned before, RhuA^{N29D} and RhuA wild type/borate buffers proved their utility with examples reported in the literature [113, 128, 129]. High conversions, for example 90–99%, were accomplished with RhuA/borate buffer, which are comparable to those achieved under different optimized conditions using DHAP donors. Importantly, the full equivalence of the stereochemical outcome with the additions of DHAP indicated the unbiased orientation of DHA in the active site of RhuA catalyst. Remarkably, the additions of DHA-borate to (*R*)-*N*-Cbz-aminoaldehydes furnished exclusively the *anti* (3*R*,4*R*) configured adducts, whereas the (*S*)-*N*-Cbz-aminoaldehydes always yielded the *syn* (3*R*,4*S*) adducts. This high stereoselectivity toward the *R* enantiomers of *N*-Cbz-aminoaldehydes at 25 °C contrasted with the different *syn*/*anti* mixtures of aldol adduct obtained using DHAP [24].

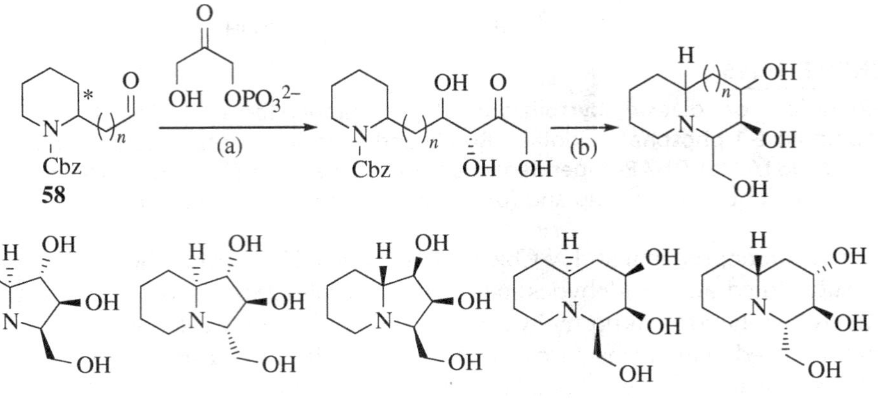

SCHEME 10.16

RhuA wild-type catalyzed aldol additions reactions of DHAP to *N*-Cbz piperidine carbaldehyde derivatives (**58**) for the synthesis of indolizidine and quinolizidine iminocyclitols. (a) RhuA wild-type catalyst and dephosphorylation by acid phosphatase and (b) one-pot Cbz removal and reductive amination.

10.3.3 Synthesis of Polyhydroxylated Pipecolic Acids and Homoiminocyclitols

By a combination of glycine-dependent aldolases and (DHAP)-dependent aldolases, a new strategy was reported for stereodiverse synthesis of polyhydroxypipecolic acid analogues (**60**) and homoiminocyclitols (**61**, Scheme 10.17) [140]. The strategy started from simple achiral materials (i.e., glycine and dimethoxyacetaldehyde (**59**)) and consisted of two key enzymatic aldol addition steps: first, the addition of glycine to dimethoxyacetaldehyde (**59**) catalyzed by L- and D-glycine aldolases, and second, the addition of DHAP to the aldol adduct from the first aldol addition using FucA F131A and RhuA as catalysts (or D-fructose-6-phophate aldolase and variants, see section 10.3.4) followed by catalytic reductive amination. The strategy creates up to five new stereogenic centers in three steps, four of them being controlled in two enzymatic reactions (Scheme 10.17). Moreover, it allowed the installation of diverse functionalities in the molecules. This was possible by taking the advantage of using aldolases in a multistep approach by virtue of their stereocomplementarity, stereoselectivity, and broad substrate tolerance.

10.3.4 Aminocyclitol Synthesis

The aminocyclitol synthesis mediated by DHAP-dependent aldolases consists of a double aldol reaction, the first one enzymatically controlled by aldolases and the second one a spontaneous intramolecular nitro-aldol reaction (i.e., the Henry reaction) (Scheme 10.18). The latter makes use of the electrophilic carbonyl unit introduced by the first aldol addition on the nucleophilic character of carbon bonded to a nitro group installed from the acceptor (e.g., **62**). This twofold C—C bond-forming reaction cascade was shown to deliver, after nitro group reduction, aminocyclitol analogs of valiolamine (**63**), of interest as inhibitors of intestinal glycosidases [141].

10.3.5 DHA-Utilizing Enzymes

The recent discovery and synthetic developments of D-fructose-6-phosphate aldolase (FSA) from *E. coli*, and the structurally related transaldolase B variant F178Y (TalB F178Y) has given a completely new dimension to the field of enzymatic aldol addition reactions [114, 142–150]. FSA accepts unphosphorylated DHA donor and analogs such as hydroxyacetone (HA) and hydroxybutanone (HB) with unprecedented high activity. Furthermore, FSA accepts glycolaldehyde (GO) as a donor, which is rather unprecedented for DHAP-aldolases as they act only on ketoses with the exception of 2-deoxy-D-ribose 5-phosphate aldolase (DERA) (see later). The use of unphosphorylated substrates is a great advantage from the synthetic point of view, especially when the phosphate group must be removed from the final product. Most importantly, the stereochemical outcome of the reactions is remarkably high and often uncompromised by the structure of both donor and acceptor. Apart from the relative relaxed donor flexibility, FSA possesses a broad acceptor substrate tolerance, which makes FSA, TalB F178Y, and their variants extremely useful tools in the preparation of iminocyclitol, aminocyclitol, and deoxysugars among other carbohydrate derivatives.

Redesign of the FSA and TalB F178Y active site was envisaged to overcome the problems concerning the tolerance toward both donor and acceptor substrates [114, 151]. Early studies showed that FSA A129S variant (Figure 10.5), designed to resemble the donor binding site of TalB [152], exhibited improved tolerance toward DHA (DHA > HA, HB > GO) as compared with the wild type (HA > HB, GO > DHA) [114]. In the opposite direction, the poor tolerance of TalB F178Y toward HA could be improved by substituting the hydrophilic S176 for the hydrophobic Ala (i.e., the double mutant TalB F178Y/A176S), the equivalent A129 in FSA [150].

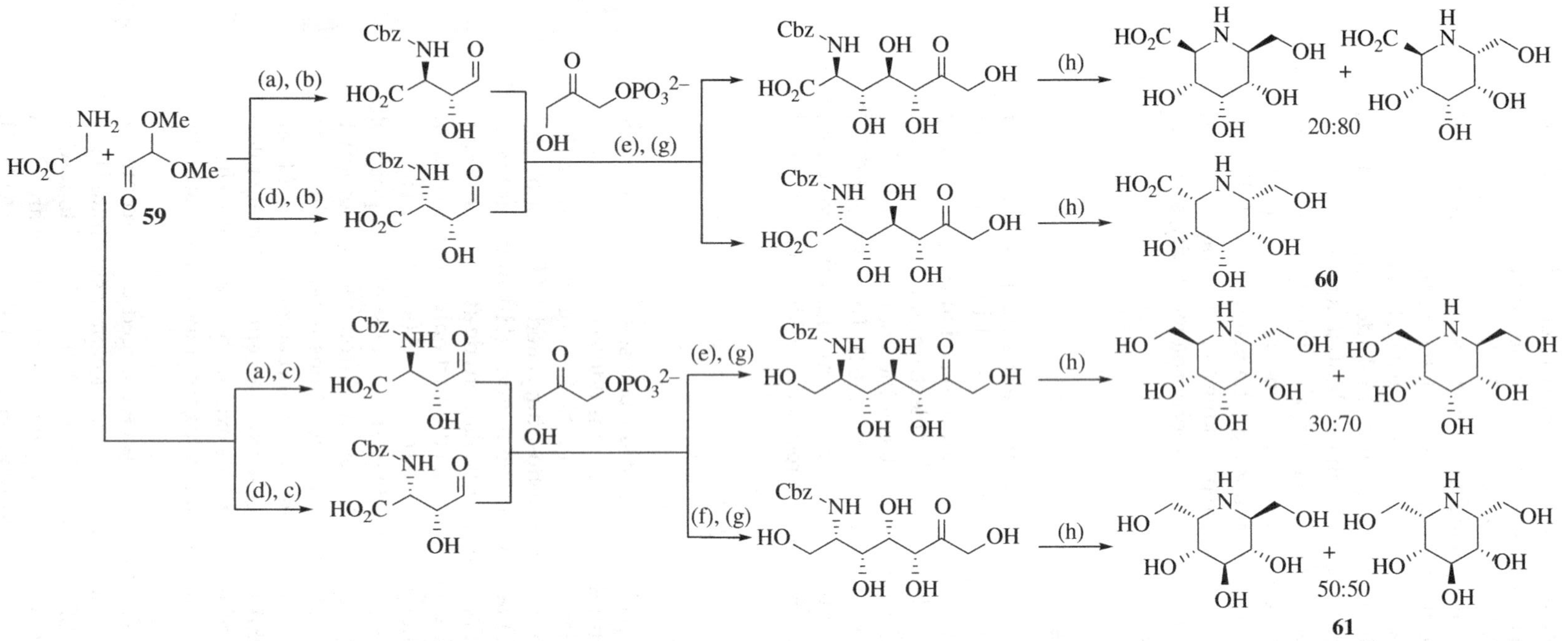

SCHEME 10.17

Example of multistep chemo-enzymatic synthesis of pipecolic acid derivatives (**60**) and homoiminocyclitols (**61**) by two consecutive enzymatic aldol addition reactions: (a) L-serine hydroxy-methyltransferase from *Streptococcus thermophilus* ($LSHMT_{sth}$); (b) Cbz-OSu; CH_3CN/aqueous HCl; (c) Cbz-OSu; $MeOH/SO_2Cl_2$; $CaCl_2$, $NaBH_4$, EtOH/THF; CH_3CN/aqueous HCl; (d) D-threonine aldolase from *Achromobacter xylosoxidans* ($DThrA_{axy}$); (e) FucA F131A; (f) RhuA wild-type; (g) acid phosphatase from potato type II; and (h) H_2 Pd/C.

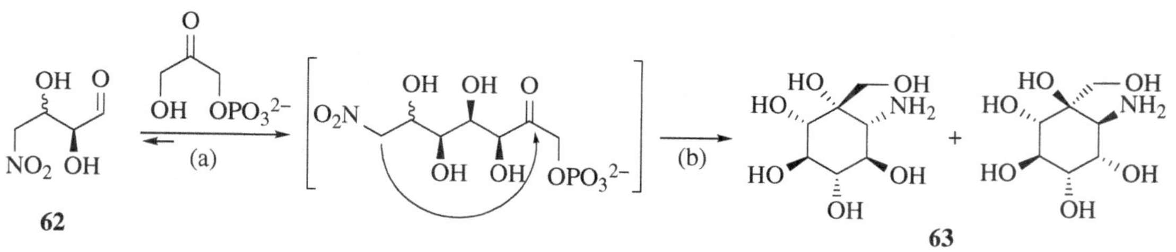

SCHEME 10.18

Aminocyclitol (e.g., **63**) synthesis from nitroaldehydes (e.g., **62**). Aldol addition of DHAP to nitroaldehyde **62** catalyzed by DHAP-dependent aldolases followed by an ensuing spontaneous intramolecular Henry reaction. (a) DHAP-dependent aldolase and (b) Dephosphorylation by acid phosphatase and subsequent catalytic reduction of the nitro group.

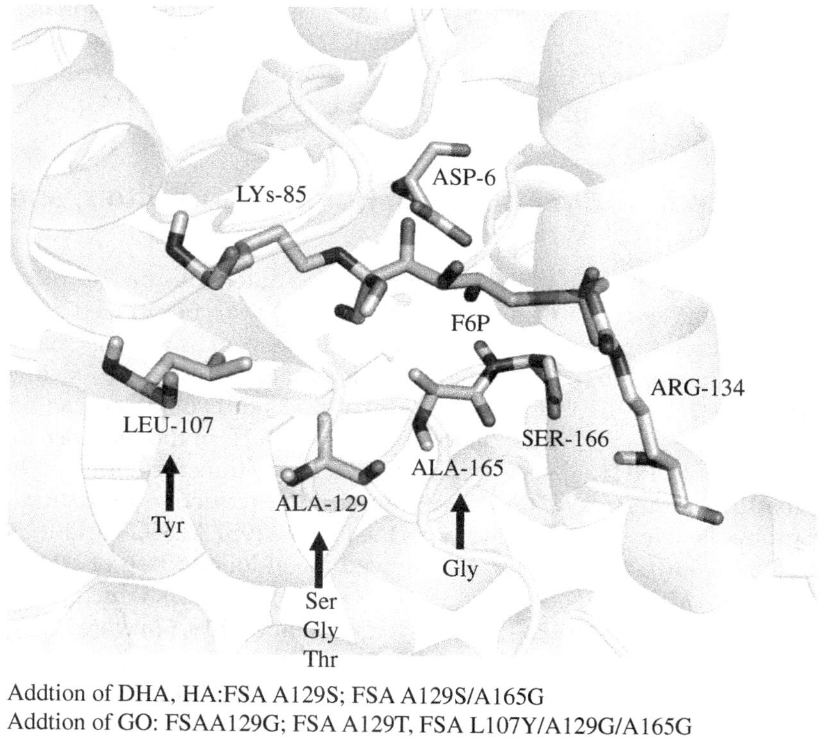

FIGURE 10.5

Modeled active site structure of wild-type FSA in complex with D-fructose-6-phosphate (F6P) enamine intermediate. Position and changes in the amino acid residues to alter the donor and acceptor substrate tolerance.

A structure-guided approach of saturation mutagenesis, site-directed mutagenesis, and computational modeling was applied to construct a set of FSA variants that improved the catalytic efficiency toward the homoaldol addition to GO up to 1800-fold [153]. The combination of mutations in the positions of L107, A129, and A165 provided a toolbox of FSA variants, for example, A129G, A129T, and L107Y/A129G/A165G that expanded the synthetic possibilities toward the preparation of aldose-like carbohydrate compounds through cross-aldol addition of GO (Figure 10.5).

On the acceptor site, the reported FSA A165G was found to enhance the activity with previously tolerated acceptors and made possible the reaction with, for example, alpha-substituted *N*-Cbz-amino aldehydes [151]. Combination of two mutations lead to FSA A129S/A165G variant with an activity between 5- and >900-fold higher than that of wild type in aldol additions of DHA and HA toward *N*-Cbz-aminoaldehyde derivatives (Figure 10.5). Computational models confirm that A165G mutation generates the required space to allocate the C-α methyl group without clashing with the protein residues, and suggests a potentially activating polar interaction of residue S129 with the substrate–enzyme intermediate [151]. The combination of A129G and A165G mutations to expand the tolerance for the acceptor substrate in GO additions was detrimental for the activity, since the double A129G/A165G variants do not directly introduce any additional interaction with the substrates but rather generate additional empty space in the active-site cavity. It could be speculated that this empty space could be either filled with water molecules that could interact with the substrates modifying their reactivity, or that they provide larger flexibility to the protein, allowing some residues to adopt different conformations and modulate the enzymatic reaction [153]. The introduction of a tyrosine at position 107, that is, FSA L107Y/A129G/A165G, greatly improves the activity and substrate tolerance of the aldol additions of GO. From the molecular models, it appears that the hemiaminal hydroxyl group acts as hydrogen bond donor to the phenol group of Y107, which at the same time is hydrogen-bonded to the backbone carbonyl of L163. This interaction may increase the basicity of the hemiaminal hydroxyl group, facilitating its protonation and dehydration to yield the corresponding Schiff base [153]. The aldehyde-binding site of TalB F178Y was also engineered to improve its efficiency toward unphosphorylated substrates [148]. Three positions corresponding to the putative phosphate-binding site of the acceptor, that is, R181, S226, and R228, were targeted with saturation mutagenesis, and the screening provided the TalB F178Y/R181E, with enhanced tolerance toward GO and D- or L-glyceraldehyde as acceptor substrates [148].

10.3.6 Iminocyclitol, Pipecolic Acid, Homoiminocyclitols, and Aminocyclitol Synthesis

Synthetic equivalents of aminoaldehydes such azido-aldehydes or *N*-protected aminoaldehyde derivatives (e.g., **64** and **65**, Scheme 10.19) can be utilized as acceptors [145, 146, 154]. Importantly, unposphorylated DHA as well as analogs, namely HA, HB, and GO can be used for broadening the structural diversity of the existing iminocyclitols (Scheme 10.19). In addition, the aldol adducts thus obtained can be treated directly, without the need to be dephosphorylated, with H_2 in the presence of a metal catalyst, for example, Pd/C to perform the deprotection/transformation of the amino equivalent into an amino group and the intramolecular-reductive amination with the carbonyl group. Besides, the FSA wild type, the FSA A129S/A165G variant improved the rates of aldol adduct formation and allowed the aldol addition of DHA and HA to (*S*)-, (*R*)-*N*-Cbz-alaninal (Scheme 10.19, **64**, R^1 = (*S*)-, (*R*)-CH_3), which was not possible using wild-type FSA. Nitro aldehydes (e.g., Scheme 10.19, **66**) were also accepted by FSA A129S rendering the corresponding nitrocyclitols, which can be transformed into aminocyclitols by catalytic reduction of the NO_2 group [114, 155].

Similarly to the strategy using (DHAP)-dependent aldolases (see Sections 10.3.1 and 10.3.2), wild-type FSA and variants can also be used for stereodiverse synthesis of polyhydroxypipecolic acid analogs, (e.g., **67**) and homoiminocyclitols (e.g., **68**) (Scheme 10.20). Aldol additions of DHA, HA, and GO catalyzed by FSA variants gave aldol adducts **69** in good to excellent yields (68–90 %). The reactions were fully stereoselective regardless of the donor substrate always furnishing the corresponding *syn* (2*S*,3*R*) configured aldol adducts. FSA A129S or FSA A129/A165G were the best for the addition of DHA; wild-type FSA for HA, while FSA A129G was the catalyst of choice for the addition of GO.

10.3.7 Carbohydrates, Deoxysugars, and Sugar Phosphate Synthesis

Diverse aldehydes have been utilized as acceptors in combination with DHA, GO, HA, and HB donors to yield a plethora of structurally diverse carbohydrate, deoxysugars, and sugar phosphate analogues as well as polyhydroxylated molecules with exquisite stereoselectivity (Figure 10.6) [114, 146–148, 150, 156–160].

Simple aldehydes, for example, formaldehyde, acetaldehyde, propanal, butanal, rendered aldol adduct (i.e., **70–75**) with exquisite stereoselectivity, a fact which is

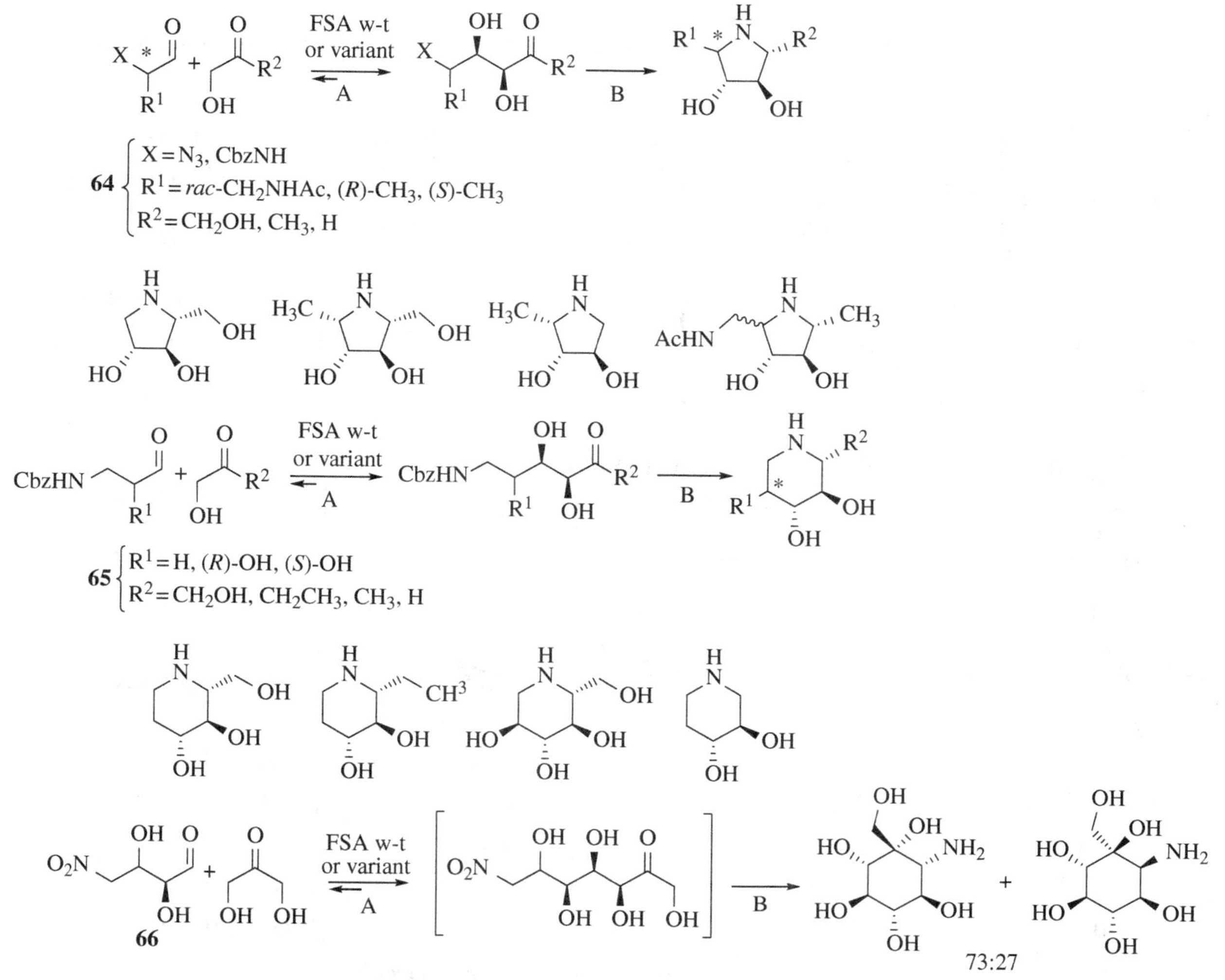

SCHEME 10.19

Examples of the chemo enzymatic synthesis of iminocyclitols, pipecolic acid, and aminocyclitols mediated by wild-type FSA and variants. (a) Aldol addition of DHA, HA, HB, or GO catalyzed by wild-type FSA and variants and (b) reductive amination with H_2 catalyzed by Pd/C.

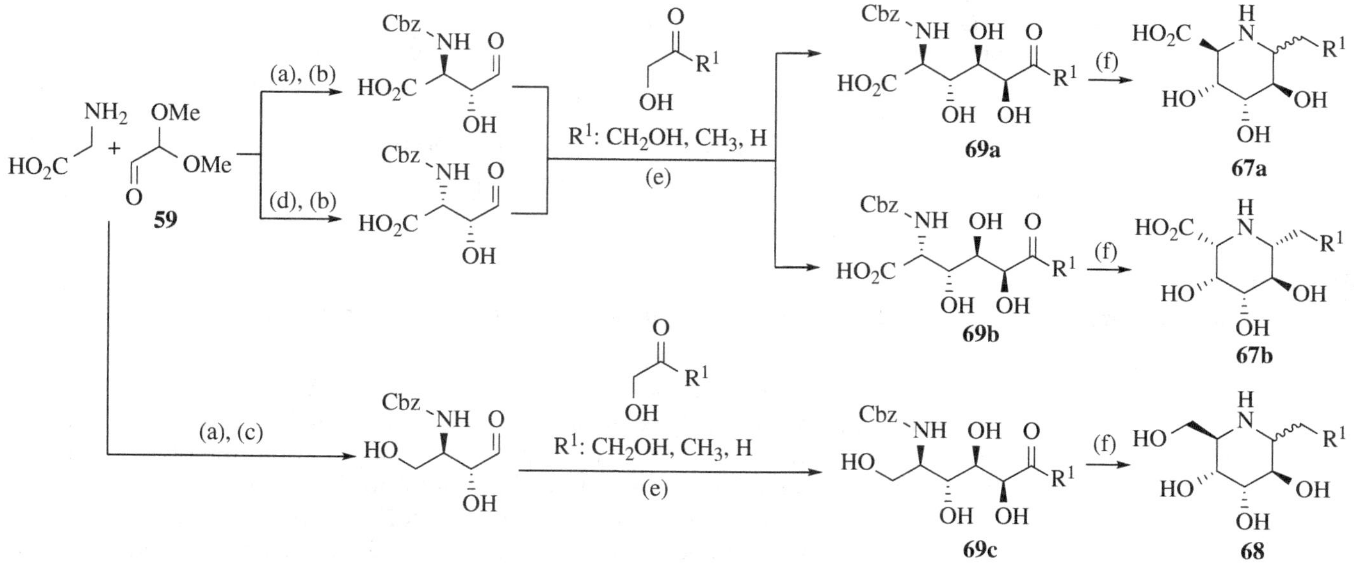

SCHEME 10.20

Multistep chemo-enzymatic synthesis of polyhydroxylated pipecolic acid (**67**) derivatives and homoiminocyclitols (**68**) by two consecutive aldol addition reactions. (a) L-serine hydroxy-methyltransferase from *Streptococcus thermophilus* (LSHMT$_{sth}$); (b) Cbz-OSu; CH_3CN/aqueous HCl; (c) Cbz-OSu; MeOH/SO_2Cl_2; $CaCl_2$, $NaBH_4$, EtOH/THF; CH_3CN/aqueous HCl; (d) D-threonine aldolase from *Achromobacter xylosoxidans* (DThrA$_{axy}$); (e) FSA A129S/A165G for $R^1 = CH_2OH$, FSA wild type for $R^1 = CH_3$, FSA A129G for $R^1 = H$; and (f) H_2, Pd/C.

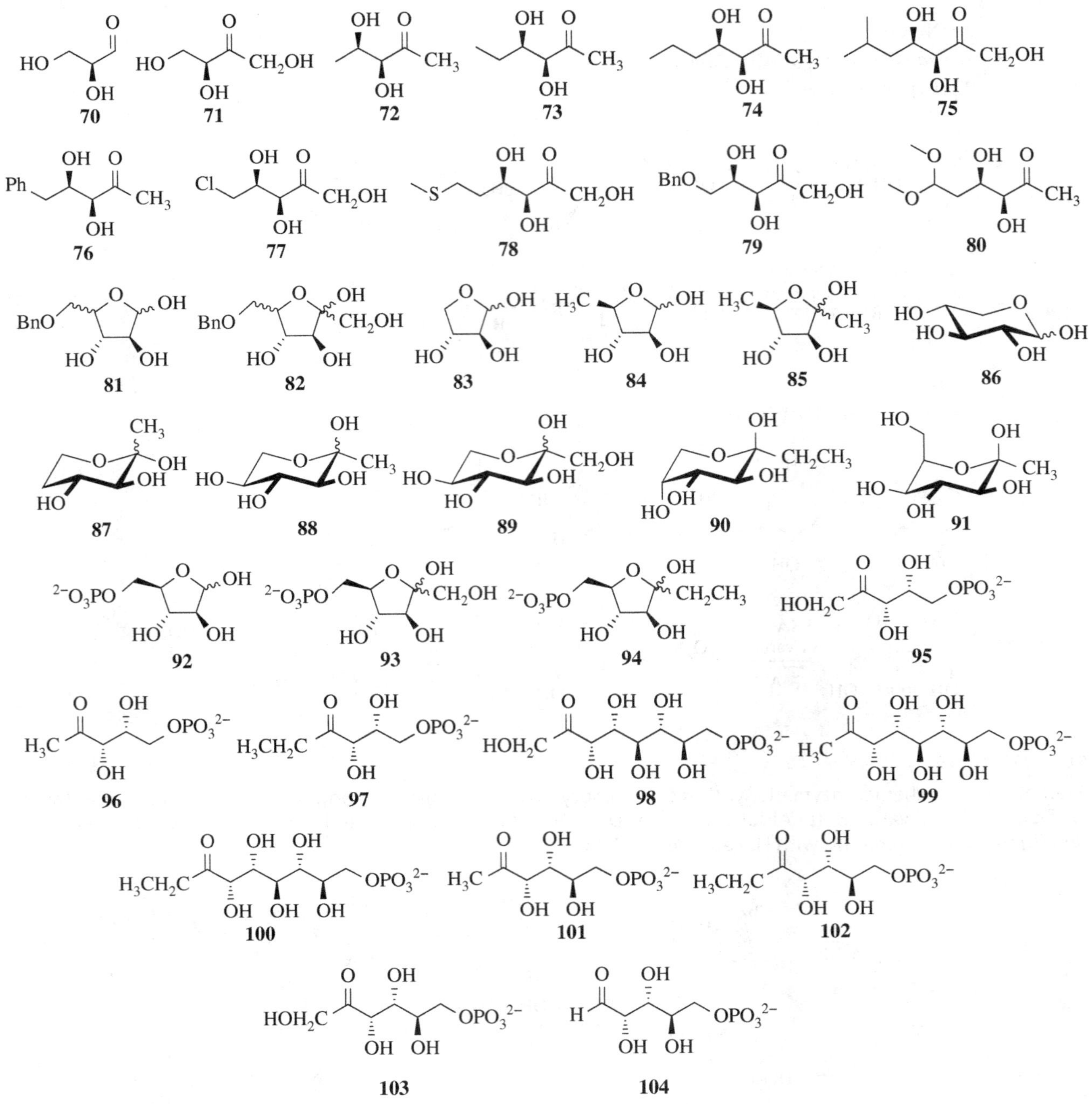

FIGURE 10.6

Examples of carbohydrates, deoxysugars and sugar phosphate compound accessible by FSA catalysis.

unattainable for many aldolases. The use of different FSA variant was fundamental toward the synthesis of polyoxygenated molecules. This can be exemplified in the case of GO, which can act as donor or acceptor depending on both the other competing nucleophiles and the FSA variant. Indeed, the aldol addition of DHA to GO catalyzed by FSA wild-type furnished exclusively D-threose (**83**) from the homo-aldol addition of GO acting as donor and acceptor (Scheme 10.21) [147]. The same reaction using FSA A129S gave D-xylulose (**86**) (Scheme 10.21) [114, 150]. Furthermore, 1-deoxy-D-xylulose (**105**) can be prepared from the aldol addition of HA to GO using FSA wild type [146], which it is expected according to the donor preference (Scheme 10.21) [147].

In another remarkable example, D-threose (**83**) was tolerated as acceptor when HA was the donor, obtaining the 1-deoxy-D-ido-hept-2-ulose (**98**), whereas no product

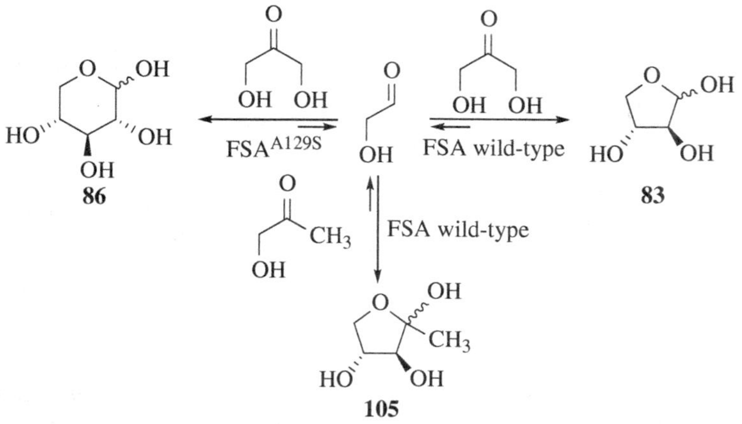

SCHEME 10.21

Wild-type FSA and FSA A129S variant catalyzed synthesis of D-threose, D-xylulose, and 1-deoxy-D-xylulose.

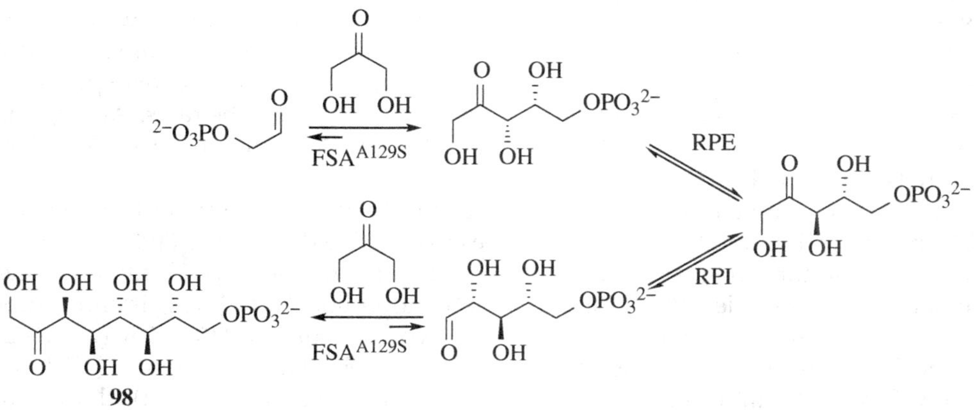

SCHEME 10.22

One-pot synthesis of D-glycero-D-altro-2-octulose 8-phosphate (**98**). Abbreviations: RPE: ribulose-5-phosphate epimerase; RPI: ribose-5-phosphate isomerase.

was detected either with DHA or GO [146]. Phosphorylated aldehydes such as, GO phosphate and D-ribose-5-phosphate were good acceptors for FSA A129S catalyzing the addition of DHA, HA, and HB, furnishing terminally phosphorylated C5 to C9 D-ketoses such as D-xylulose-5-phosphate (**95**), 1-deoxy-D-xylulose-5-phosphate (**96**), (2*R*,3*S*)-2,3-dihydroxy-4-oxohexyl phosphate (**97**), D-glycero-D-*altro*-octulose-8-phosphate (**98**), 1-deoxy-D-glycero-D-*altro*-octulose-8-phosphate (**99**), and (2*R*,3*R*,4*R*,5*R*,6*S*)-2,3,4,5,6-pentahydroxy-7-oxononyl phosphate (**100**) [160]. D-Glyceraldehyde-3-phosphate was demonstrated to be a very good acceptor for FSA, which after the addition of selected donors furnished 1-deoxy-D-fructose-6-phosphate (**101**), 1,2-dideoxy-D-*arabino*-hept-3-ulose 7-phosphate (**102**), D-fructose-6-phosphate (**103**), and D-arabinose 5-phosphate (**104**) [158, 159]. The highly sensitive D-glyceraldehyde-3-phosphate was generated *in situ* using a multienzymatic cascade reaction from DHA using DHAK and triosephosphate isomerase (TPI) or, alternatively, from D-fructose-1,6-bisphosphate using D-fructose-1,6-bisphosphate aldolase and TPI [158, 159]. Interestingly, D-glycero-D-altro-2-octulose 8-phosphate (**98**) can also be produced in one pot cascade of four enzymatic reactions involving FSA A129S, ribulose-5-phosphate epimerase (RPE) and ribose-5-phosphate isomerase (RPI) (Scheme 10.22) [160].

10.4 THREONINE ALDOLASES

The pyridoxal-5′-phosphate-dependent threonine aldolases (ThrA; EC 4.1.2.5) and serine hydroxymethyltransferase (SHMT; EC 2.1.2.1) catalyze the aldol addition of glycine to aldehydes with the formation of two new stereogenic centers [15, 44, 161–163]. Hence, four possible products can be formally obtained from a single aldehyde, depending on the specificity of the threonine aldolase. All glycine aldolases assayed

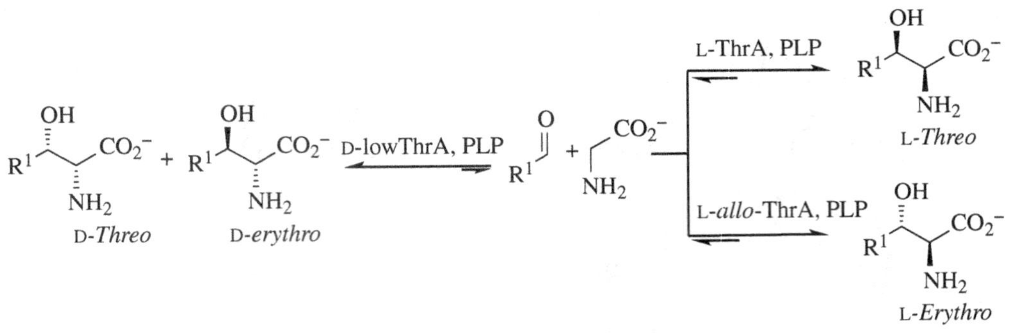

SCHEME 10.23

Set of accessible threonine aldolases for adol addition of glycine to aldehydes. Abbreviations: PLP = Pyridoxal-5′-phosphate.

up to now exquisitely control the configuration of sterogenic center generated from the nucleophile, that is, (2*S*) or (2*R*) for the L- or D-threonine-type aldolases, respectively (Scheme 10.23). With respect to the stereogenic center generated by the aldehyde acceptor, the L-specific threonine aldolases can further be divided into three sub-types: L-threonine aldolase (LThrA), which preferentially cleaves L-*threo* adducts; L-*allo*-threonine aldolase (L-*allo*ThrA), which cleaves L-*erythro*; and L-low specificity threonine aldolase, which accepts both L-*threo* and L-*erythro* substrates. Among the D-series, only D-low-specificity threonine aldolase (DTA) could so far be found in nature [161, 162, 164, 165].

β-Hydroxy-amino acids (Figure 10.7) are multifunctional compounds with valuable interest as intermediates for the synthesis of statine derivatives (**106**) [166–168], protease inhibitors [169], antivirals [170, 171], peptide mimetics [172], idulonic acid mimetics, for example, 3*R*,5*R*-dihydroxy-L-homoproline (**111**) [173], immunosuppressive lipid mycestericin D (**112**) [174], 3,4-dihydroxyprolines (**113**) [175], (2*S*,3*R*)-2-amino-3-hydroxybutyrolactone, precursor of monobactam antibiotics [176], or L-*threo*-3-[4-(methylthio)phenylserine] precursor of thiamphenicol (**114**), florfenicol (**115**) [177], sialyl Lewis x mimetics (**117**) [178], β-hydroxyornithine (**109**), a relevant building block for the β-lactamase inhibitor, clavulanic acid, and the antibiotic and anticancer acivicin [179], surveyed in previous reviews [41, 57].

The synthesis of some of these precursors can be accomplished via aldol addition of *N*-Cbz-protected aminoaldehydes and glycine catalyzed by an SHMT from *Streptococcus thermophilus* (LSHMT$_{sth}$) and L-ThrA from *E. coli* (e.g., Scheme 10.24) [180]. Both biocatalyst turned out to be stereocomplementary when reactions were conducted at 4 and 25 °C just by varying the reaction temperature (Scheme 10.24) [180].

The possibility of using other nucleophiles rather than glycine has opened new possibilities for the synthesis of β-hydroxy-α,α-dialkyl-α-amino acids. α,α-Dialkyl-α-amino acids and β-hydroxy-substituted derivatives are important as building blocks of pharmaceuticals, as enzyme inhibitors, and as conformational modifiers of physiologically active peptides [181–186]. Moreover, they can be found as constituents of biologically active molecules such as myriocin (**121**), sphingofungines E (**122**) and F (**123**), and lactacystin (**124**) (Figure 10.8). These compounds are not naturally occurring, and their synthesis is still a challenging task, comprising several steps and with difficult control of stereoselectivity [181, 184, 187–189].

Screening for novel biocatalyst also afforded some interesting synthetic approaches to this challenging goal. A PLP-dependent α-methylserine aldolase from different microorganisms [190, 191] was uncovered to catalyze the addition of L-alanine and L-2-aminobutyric acid to formaldehyde (Scheme 10.25), which resulted in the stereoselective formation of a quaternary C-center, rather unprecedented for naturally occurring aldolase. The enzyme from *Ralstonia* sp. Strain AJ110405 was applied in a whole-cell catalyzed stereospecific synthesis of α-methyl-L-serine [190].

Another screening study uncovered a natural L-*allo*-threonine aldolase from *Aeromonas jandaei* (L-*allo*ThrA$_{A. jandaei}$) and a D-threonine aldolase from *Pseudomonas* sp. (D-ThrA$_{P. sp}$) that catalyze stereoselectively the retroaldol reaction of β-hydroxy-α-methylthreonine, an β-hydroxy-α,α-dialkyl-α-amino acid, to produce lactaldehyde

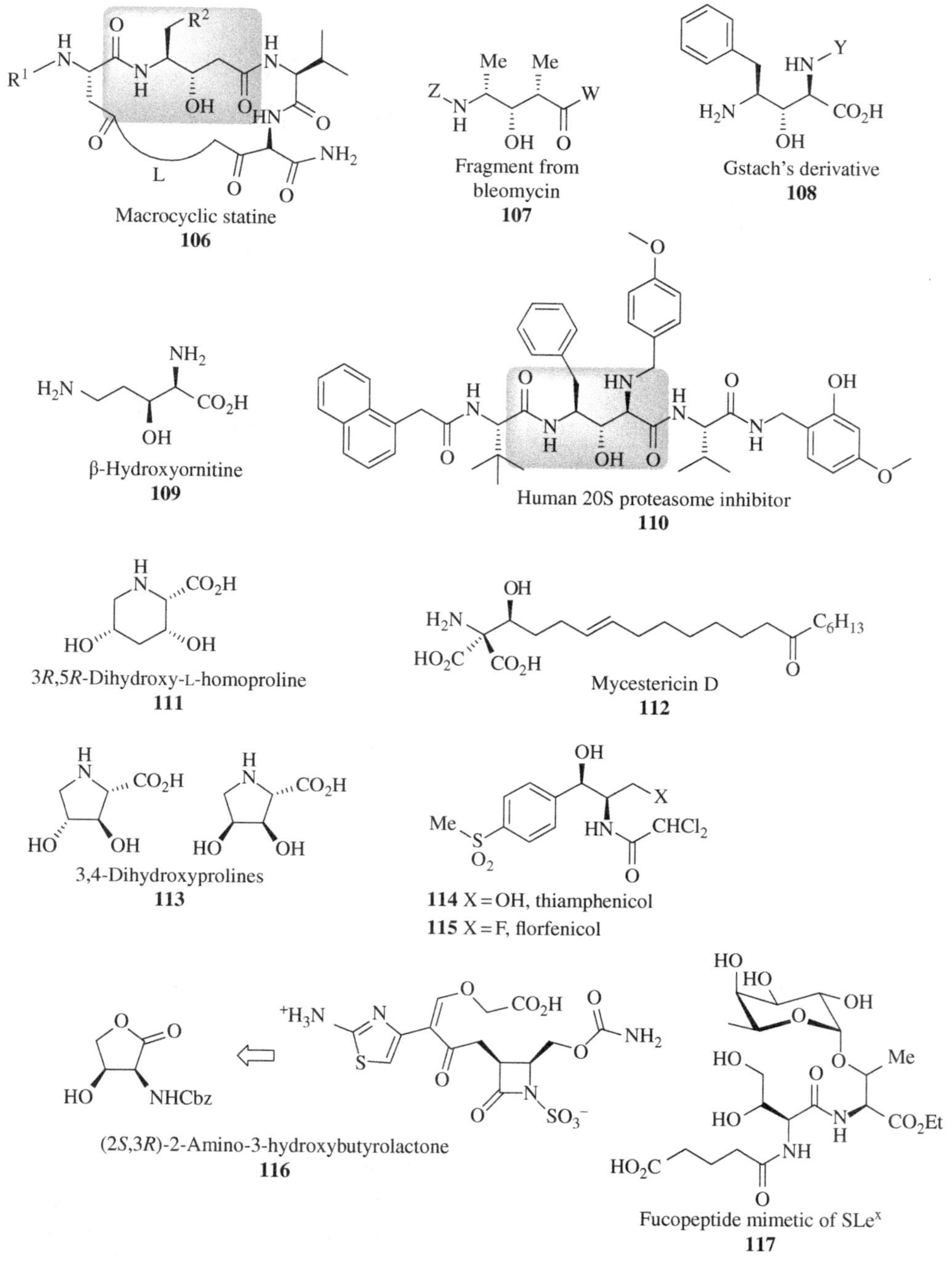

FIGURE 10.7

Examples of β-hydroxy-amino acids as components of products with biological or therapeutical interest.

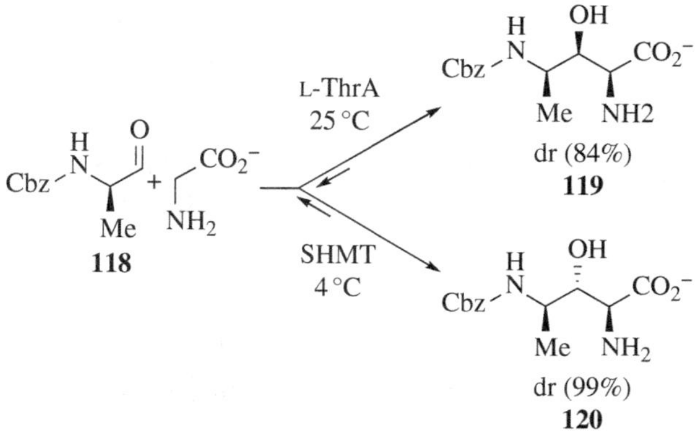

SCHEME 10.24

Example of SHMT from *Streptococcus thermophilus* (LSHMT$_{sth}$) and L-ThrA from *Escherichia coli* catalyze synthesis of β-Hydroxy-α,ω-diamino acid derivatives (e.g., (2*S*,3*R*,4*R*)-(**119**) and (2*S*,3*S*,4*R*)-2-amino-4-(benzyloxycarbonylamino)-3-hydroxypentanoic acid, (**120**)) via aldol addition of glycine to Cbz-protected aminoaldehyde (**118**).

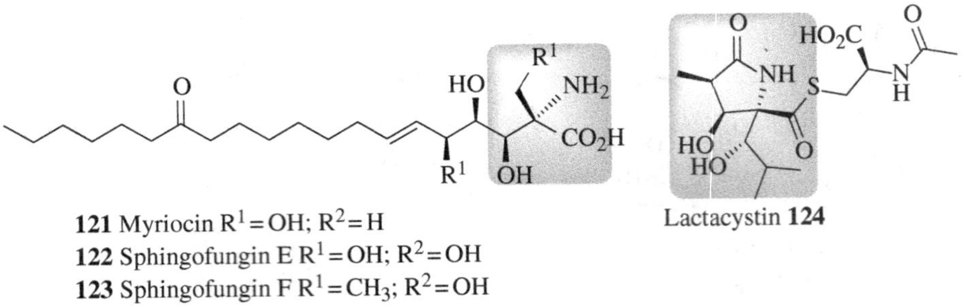

FIGURE 10.8
Structures of myriocin (**121**), sphingofungines E (**122**) and F (**123**), and lactacystin (**124**) that incorporate hydroxy-α,α-dialkyl-α-amino acid moieties.

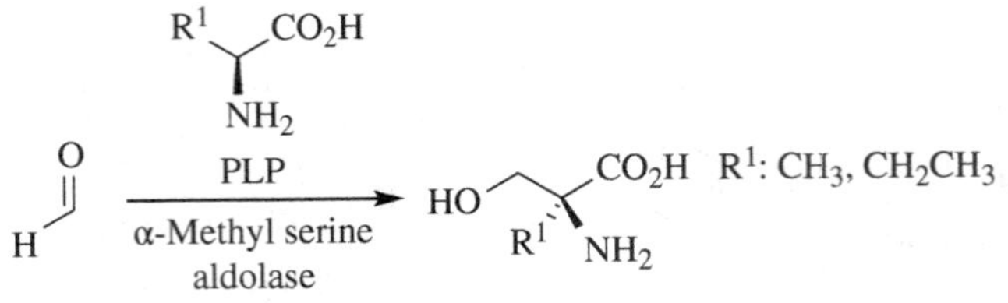

SCHEME 10.25
PLP-dependent α-methylserine aldolase catalyzed the addition of L-alanine and L-2-aminobutyric acid to formaldehyde.

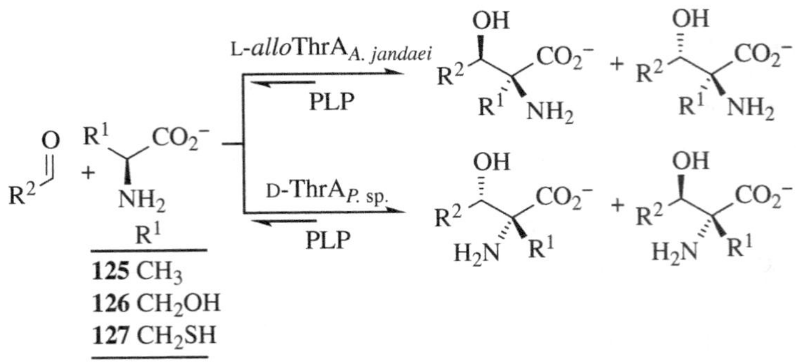

SCHEME 10.26
L-*allo*-Threonine aldolase from *Aeromonas jandaei* (L-*allo*-ThrA$_{A.\ jandaei}$) and D-threonine aldolase from *Pseudomonas* sp. (D-ThrA$_{P.\ sp.}$) catalyzed synthesis of β-hydroxy-α,α-dialkyl-α-amino acids.

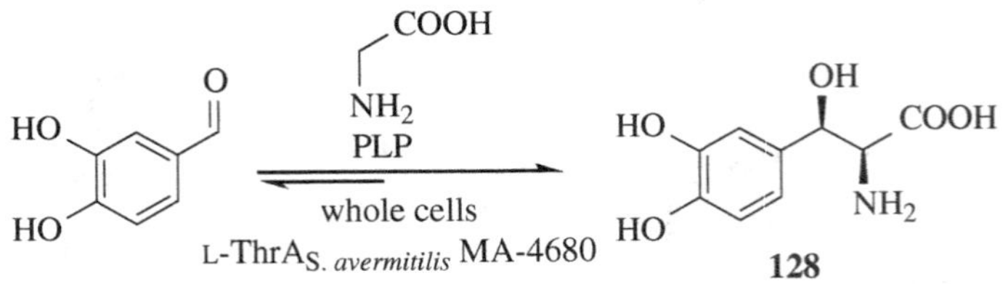

SCHEME 10.27
Synthesis of L-threo-3-(3,4-dihydroxyphenyl)serine (**128**) by means of L-ThrA from *Streptomyces avermitilis* MA-4680 in a whole-cell high-density bioreactor.

and D-alanine [192]. In the aldol direction, D-alanine (**125**), D-serine (**126**) and, to a lesser extent, D-cysteine (**127**) (but not the L enantiomers) were tolerated as donor substrates exhibiting broad acceptor selectivity (Scheme 10.26). Conversions varied between 6 and 84% with moderate to low stereoselectivity, a general drawback found in many known threonine aldolase.

In an industrial-scale application, L-*threo*-3-(3,4-dihydroxyphenyl)serine (**128**), a drug for treatment of Parkinson's disease, was prepared using a L-ThrA from *Streptomyces avermitilis* MA-4680 as catalyst in a whole-cell high-density bioreactor (Scheme 10.27). The method allows productivities around 8 g l^{-1} of L-*threo*-**128** [193].

The limitations imposed by the thermodynamic relations due to the reversible nature of the aldol reactions, may be overcome by introducing an ensuing complementary enzymatic reaction in a cascade fashion. In this way, it is possible to improve the stereochemical outcome of threonine aldolases by selectively transforming one of the diastereosiomers *in situ* by the action of another enzyme. Toward this end, utilization of L-ThrA in tandem with L-tyrosine decarboxylase (LTyrDC, PLP-dependent) to produce (*R*)-2-amino-1-phenylethanol in 89% isolated yield represents a significant example (Scheme 10.28) [194].

One important application of glycine-dependent aldolases is for the kinetic resolution of β-hydroxy-α-amino acids [162]. A recent example is the kinetic resolution of β-phenylserine (**130**), β-(nitrophenyl) serine (**131**) and β-(methylsulfonylphenyl) serine (**132**) catalyzed by immobilized *E. coli* cells overexpressing $SHMT_{E. coli}$ (93–97% conversion was obtained with ca 98% ee) (Scheme 10.29). The immobilized cells where reused up to ten times exhibiting an excellent operational stability and an average conversion of 60%.

10.4.1 2-Deoxy-D-Ribose 5-Phosphate Aldolase

2-Deoxy-D-ribose 5-phosphate aldolase (DERA) catalyzes *in vivo* the reversible aldol addition of acetaldehyde to D-glyceraldehyde-3-phosphate to furnish 2-deoxy-D-ribose 5-phosphate (Scheme 10.30).

As with to FSA and GO, the most significant feature of DERA is the ability to catalyze self- and cross-aldol additions of acetaldehyde. Therefore, the first aldol addition furnishes another aldehyde that can be used as acceptor by DERA, or in combination with other aldolases, for cascade aldol reactions (Scheme 10.31) [25–27, 195].

This sequential aldol reactions has attracted much attention for the synthesis of a chiral precursor for the side chain of the statin drugs (Figure 10.9).

The target raldol reaction is the tandem addition of two equivalents of acetaldehyde to one equivalent of chloroacetaldehyde to furnish (3*R*,5*S*)-6-chloro-2,4,6-trideoxyhexose, which after a follow-up chemistry gave the statin chiral building blocks (Scheme 10.32) [196, 197].

One limitation of this methodology is that DERA is inactivated by high aldehyde concentrations of chloroacetaldehyde. Therefore, industrial applications needed the development of an engineered DERA variant, robust enough to ensure an efficient transformation process [198]. The DSM team applied directed evolution and

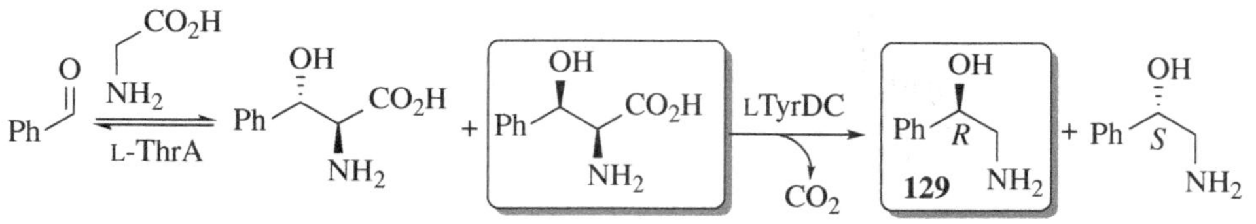

SCHEME 10.28

L-ThrA in tandem with L-tyrosine decarboxylase (LTyrDC, PLP-dependent)-catalyzed synthesis of (*R*)-2-amino-1-phenylethanol (**129**).

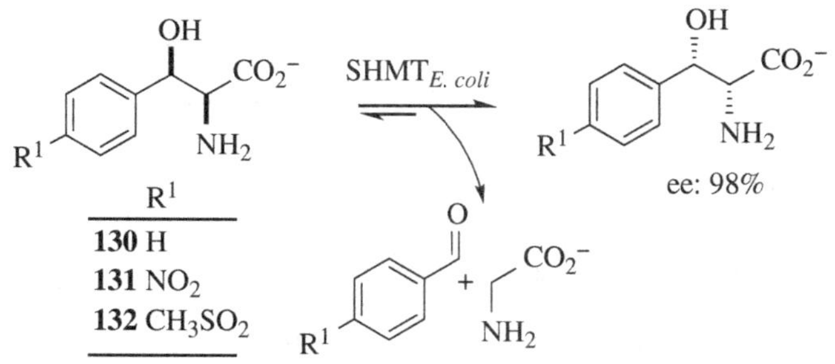

SCHEME 10.29

Immobilized *E. coli* cells overexpressing serine hydroxymethyl transferase ($SHMT_{E. coli}$) catalyzed the kinetic resolution of **130**, **131**, and **132**.

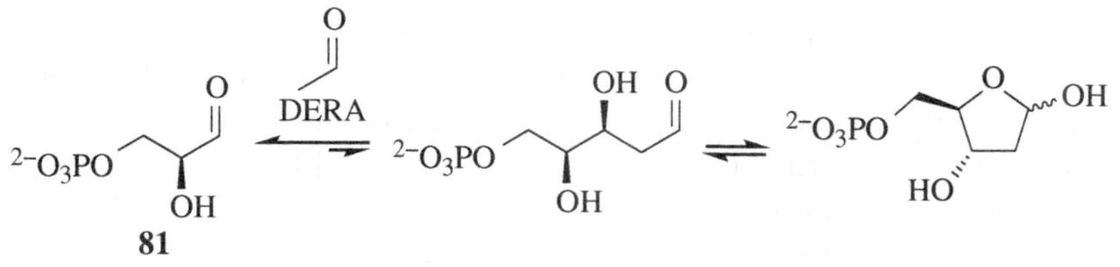

SCHEME 10.30

RibA catalyzed the reversible aldol addition of acetaldehyde to D-glyceraldehyde-3-phosphate.

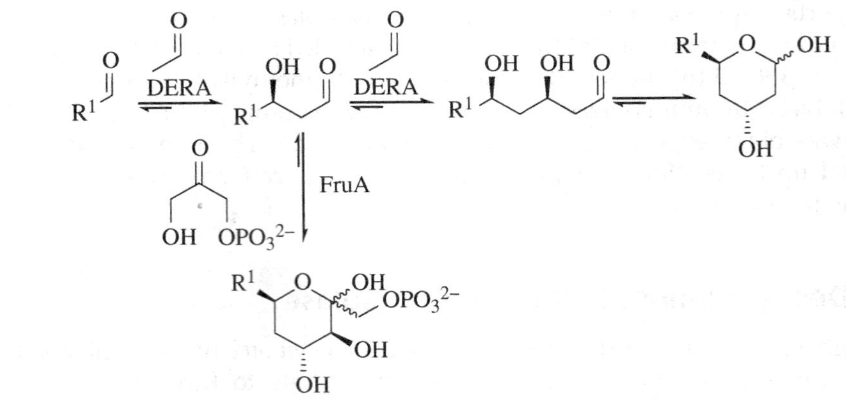

SCHEME 10.31
One-pot double addition of acetaldehyde to an aldehyde catalyzed by DERA and two consecutive aldol additions of acetaldehyde and dihydroxyacetone phosphate (DHAP) catalyzed by a tandem of DERA and FruA catalysts.

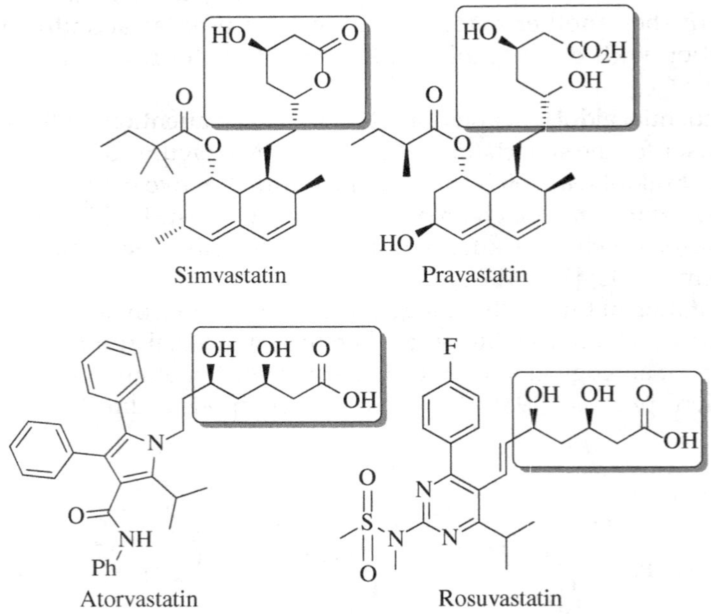

FIGURE 10.9
Structures of statin drugs.

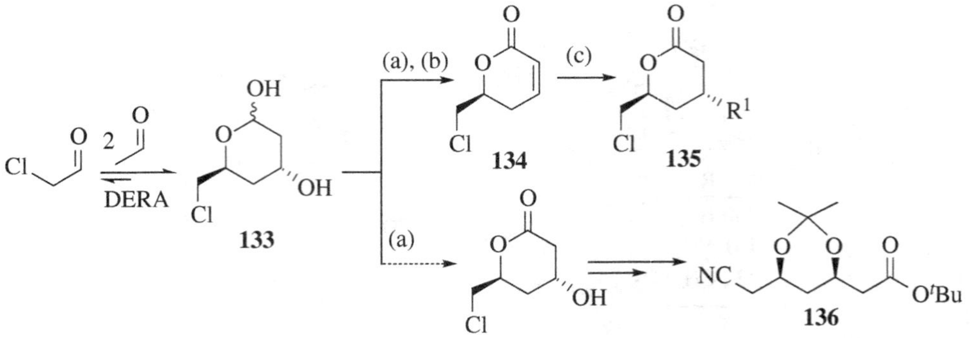

SCHEME 10.32
Double addition of acetaldehyde to chloroacetaldehyde and follow-up chemistry for the synthesis of the statin chiral building blocks. Follow-up chemistry (a) Br_2, H_2O, pH 5–6, (b) cat. TsOH, toluene, Δ, and (c) $R^1 = Nu$-H.

site-directed mutagenesis to increase DERA resistance to chloroacetaldehyde, as well as its volumetric productivity. This resulted in a DERA variant with about tenfold greater productivity than the wild-type enzyme at high aldehyde concentrations. In a subsequent study, they further exploit the synthetic utility of the lactone pyranoid obtained by RibA-mediated catalysis [196]. They converted **133** to the α,β-unsaturated D-lactone **134** (Scheme 10.32). The α,β-unsaturated D-lactone is a characteristic structural motif in naturally occurring α-pyrones (5,6-dihydropyran-2-ones), a large family of physiologically active secondary metabolites. Besides the α,β-unsaturated D-lactone is of particular interest because the activated double bond C2—C3 leads to a highly diastereoselective addition of various nucleophiles at the electrophilic C3 position. The products of this reaction are versatile pyranoid building blocks **135** for the synthesis of pharmaceuticals and natural products [196].

In view of such applications, DERA enzymes from hyperthermophilic *Pyrobaculum aerophilum* and *Thermotoga maritima*, were reported [199]. Confirming expectations that enzymes from hyperthermophiles may also show improved stability against inactivating agents, the enzymes indeed efficiently performed tandem aldol additions at low temperature with much higher stability than wild-type DERA from *E. coli.*

10.5 ALDOL TYPE REACTIONS CATALYZED BY NON-ALDOLASES

Synthetic peptide dendrimers, catalytic antibodies, RNA catalysts, peptide foldamers as well as other native or modified enzymes with completely different functions were discovered to catalyze carbon–carbon bond formation [15]. 4-Oxalocrotonate tautomerase (4-OT) catalyzes *in vivo* the conversion of 2-hydroxy-2,4-hexadienedioate (**136**) to 2-oxo-3-hexenedioate (**137**) (Scheme 10.33a), and it belongs to the catabolic pathway for aromatic hydrocarbons in *P. putida* mt-2 [200]. This enzyme carries a catalytic amino-terminal proline, which could act as catalyst in the same fashion as the proline mediated by organocatalytic reactions. Initial studies demonstrate that this enzyme was able to catalyze aldol condensations of acetaldehyde to a variety of electrophiles **138** (Scheme 10.33b) [200]. This enzyme was also examined as a potential catalyst for carbon–carbon bond forming Michael-type reactions of acetaldehyde to nitroolefins **139** (Scheme 10.33c) [201, 202].

Wild-type or engineered proteases and lipases were also found to catalyze aldol additions. First, examples were reported about the self-aldol addition of α,β-unsaturated aliphatic aldehydes (**140**) catalyzed by *Pseudozyma antarctica* lipase B (PalB) (formerly *Candida antarctica* CALB lipase) S105A and S105G variants (Scheme 10.34) [203, 204]. The PalB S105A variant was also reported to catalyze aldol

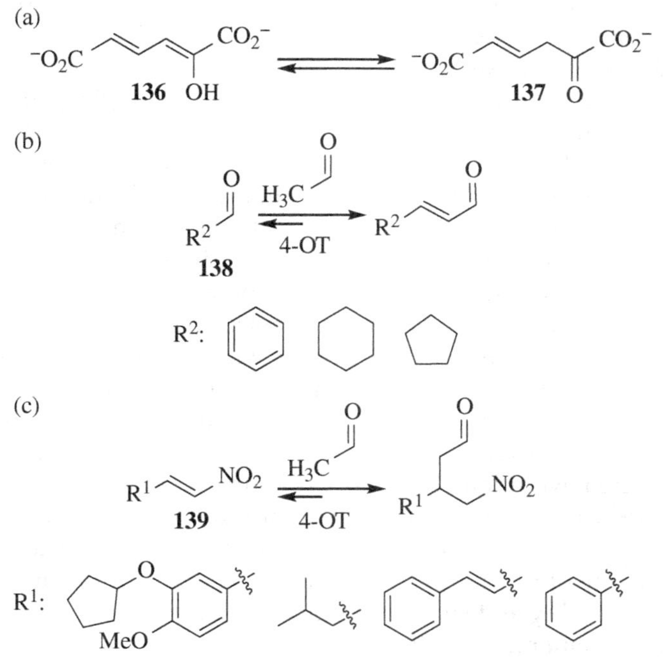

SCHEME 10.33
Oxalocrotonate tautomerase (4-OT), catalyzed reactions: (a) conversion of 2-hydroxy-2,4-hexadienedioate (**136**) to 2-oxo-3-hexenedioate (**137**) *in vivo* reaction, (b) aldol condensations of acetaldehyde to aldehydes **138**, and (c) Michael-type reactions of acetaldehyde to nitroolefins **139**.

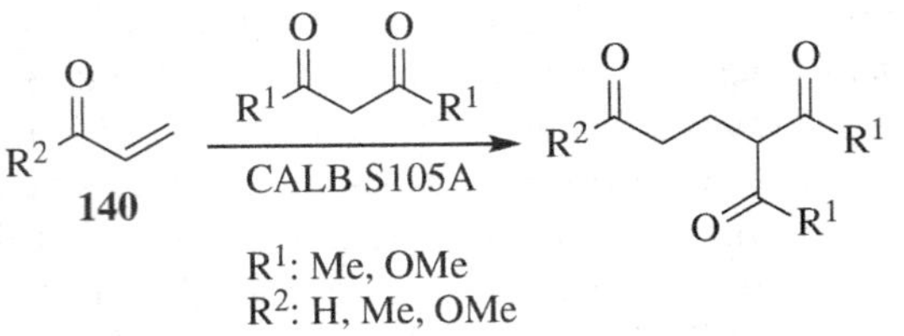

SCHEME 10.34

Homo-aldol addition of α,β-unsaturated aliphatic aldehydes (**140**) catalyzed by *Pseudozyma antarctica* lipase B (PalB) S105A variant.

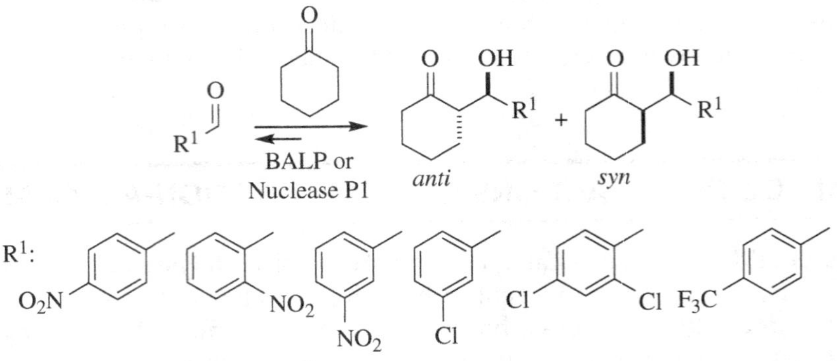

SCHEME 10.35

Aldol addition of cyclohexanone to aromatic aldehydes catalyzed by alkaline protease from *Bacillus licheniformis* (BALP) and Nuclease P1 from *Penicillium citrinum*.

addition of 1,3-dicarbonyl compounds to α,β-unsaturated carbonyl compounds (Scheme 10.34) [205, 206].

Moreover, an alkaline protease from *Bacillus licheniformis* (BALP) [207] and Nuclease P1 (EC 3.1.30.1) from *Penicillium citrinum* [208] were also found to be catalysts for carbon–carbon bond formation by aldol addition of cyclohexanone to aromatic aldehydes. In the examples (Scheme 10.35), BALP rendered yields between 81 and 91%, *anti*:*syn* 58:42 and 95:5, and ee 36 and 70%, whereas Nuclease P1 rendered yields between 21 and 28%, *anti*:*syn* 87:13 and 94:6, and ee 89 and 92%.

10.6 COMPUTATIONAL *DE NOVO* ENZYME DESIGN

De novo computational design of protein catalysts with tailored activities is a promising research area fostered by the number of new algorithms, potential energy functions as well as computer potency that allows more sophisticated and precise protein design methods [209–212]. Regarding aldolases, a de novo designed protein with retroaldolase activity was reported as a pioneering example [35]. The de novo construction of this retroaldolase consisted of first to define one or more potential catalytic mechanisms for the desired reaction and, second, to construct active sites with key amino acid residues in positions compatible with the multiple transition states (TS) and intermediates of the reaction pathway, for example, carbinolamine formation and water elimination, carbon–carbon bond cleavage, and release of bound products. Then, the protein scaffolds that can accommodate the designed TS ensemble were identified. In this way, a protein with retroaldol activity was obtained with acceleration rate up to four orders of magnitude with respect to uncatalyzed reaction [35]. However, there is still a gap between naturally evolved enzymes with catalytic efficiencies of 10^5–$10^9\,M^{-1}s^{-1}$ and the designed protein catalyst with efficiencies typically 1–100$\,M^{-1}s^{-1}$. The original activity was subsequently increased more than a 1000-fold by directed evolution [213, 214]. Structural and mechanistic studies reveal that the engineered catalytic apparatus is dramatically changed during the optimization process. Structures of the initial *in silico* design, a mechanistically promiscuous intermediate and one of the most evolved variants, highlight the importance of factors such as loop mobility and supporting functional groups in the emergence of the optimized catalytic center [214–217].

10.7 CONCLUSIONS AND PERSPECTIVES

Aldolases are an important class of biocatalyst for carbon–carbon bond formation, useful in the synthesis of naturally occurring compounds and analogs and with the unique characteristic that they are capable of building up new polyfunctional molecular frameworks through consecutive or cascade reactions.

The scope and limitations of aldolases were stated in a plethora of reports surveyed in the literature and in this chapter. Advances in protein engineering techniques and screening of novel aldol activities in nature can offer promising perspectives for the application of aldolases with broad substrate and stereochemical scope. Aldolase redesign by protein engineering is mainly focused on the modification of substrate selectivity and control of the stereochemical outcome of the reactions. The combination of rational and random protein engineering is essential for future biocatalyst development and particularly for aldolases. Properties of native enzymes such as catalytic efficiency, substrate tolerance, stereoselectivity, and stability are often to be modified to fulfill the requirements for synthetic substrates and/or reaction conditions. In addition, detailed studies on metabolic pathways will lead to the discovery of new aldolases opening up opportunities to broaden the scope of application and filling the toolbox with alternative carboligation biocatalysts.

The discovery of the precise mechanism of the enzymatic aldol addition process in combination with the elucidation of enzyme structure is also of paramount importance. In this way, it becomes more practicable to elucidate the precise transition state(s) involved in the catalytic process, and this is of utmost significance in modifying rationally existing enzymes or de novo design new activities.

The exploitation of FSA-like aldolases represents a qualitative progress in aldolase-catalyzed synthesis as these enzymes accept diverse donors with the hitherto unique requirement of the hydroxymethylcarbonyl moiety. A systematic study of potential donor substrates with innovative functionalities and structures is still necessary to assess the potential of these catalysts. Additionally, further mutagenic works on the active site may provide FSA-like aldolases with broader structural tolerance for both donor and acceptor substrates, which would be of paramount utility in organic synthesis. A significant limitation of the FSA-like aldolases is the lack of a set of stereocomplementary enzymes, which has not been found in nature yet.

REFERENCES

[1] Evans, D. A., Nelson, J. V., and Taber, T. R., Stereoselective aldol condensations. *Top. Stereochem.* 1982, **13**, 1–115.

[2] Mukherjee, S., Yang, J. W., Hoffmann, S., and List, B., Asymmetric enamine catalysis. *Chem. Rev.* 2007, **107** (12), 5471–5569.

[3] Mahrwald, R., *Modern Aldol Reactions*, Vol. 2: Metal Catalysis. Wiley-VCH: Weinheim, 2004.

[4] Mahrwald, R., *Modern Aldol Reactions*, Vol. 1: Enolates, Organocatalysis, Biocatalysis and Natural Product Synthesis. Wiley-VCH: Weinheim, 2004.

[5] Mahrwald, R., *Modern Methods in Stereoselective Aldol Reactions*. Wiley-VCH: Weinheim, 2013; Vol. 1.

[6] Markert, M. and Mahrwald, R., Total syntheses of carbohydrates: Organocatalyzed aldol additions of dihydroxyacetone. *Chem. Eur. J.* 2008, **14** (1), 40–48.

[7] Palomo, C., Oiarbide, M., and Garcia, J. M., Current progress in the asymmetric aldol addition reaction. *Chem. Soc. Rev.* 2004, **33** (2), 65–75.

[8] Wong, C.-H., Engineering enzymes for chemoenzymatic synthesis. Part I: Practical routes to aza-sugars and complex carbohydrates. *Trends Biotechnol.* 1992, **10** (10), 337–341.

[9] Gijsen, H. J. M., Qiao, L., Fitz, W., and Wong, C.-H., Recent advances in the chemoenzymic synthesis of carbohydrates and carbohydrate mimetics. *Chem. Rev.* 1996, **96**, 443–473.

[10] Fessner, W.-D. and Walter, C., Enzymic C—C bond formation in asymmetric synthesis. *Top. Curr. Chem.* 1997, **184** (Bioorganic Chemistry), 97–194.

[11] Fessner, W.-D., Enzyme mediated C—C bond formation. *Curr. Opin. Chem. Biol.* 1998, **2** (1), 85–97.

[12] Fessner, W.-D., Enzymatic asymmetric synthesis using aldolases. In *Stereoselective Biocatalalysis*, Patel, R. N., Ed. Marcel Dekker Inc.: New York, NY, 2000; pp 239, 265.

[13] Fessner, W.-D. and Helaine, V., Biocatalytic synthesis of hydroxylated natural products using aldolases and related enzymes. *Curr. Opin. Biotechnol.* 2001, **12** (6), 574–586.

[14] Fessner, W.-D., Biocatalytic C—C bond formation in asymmetric synthesis. In *Asymmetric Synthesis with Chemical and Biological Methods*, Enders, D. and Jaeger, K.-E., Eds. Wiley-VCH Verlag GmbH & Co KGaA: Weinheim, 2007; pp 351–375.

[15] Clapés, P., Fessner, W.-D., Sprenger, G. A., and Samland, A. K., Recent progress in stereoselective synthesis with aldolases. *Curr. Opin. Chem. Biol.* 2010, **14** (2), 154–167.

[16] Clapés, P. and Joglar, J., Enzyme-catalyzed aldol additions. In *Modern Methods in Stereoselective Aldol Reactions*, Mahrwald, R., Ed. John Wiley & Sons, Ltd: Chichester, 2013; pp 475–528.

[17] Clapés, P. and Garrabou, X., Current trends in asymmetric synthesis with aldolases. *Adv. Synth. Catal.* 2011, **353** (13), 2263–2283.

[18] Gefflaut, T., Blonski, C., Perie, J., and Willson, M., Class I aldolases: Substrate specificity, mechanism, inhibitors and structural aspects. *Prog. Biophys. Mol. Biol.* 1995, **63** (3), 301–340.

[19] Fessner, W.-D., Sinerius, G., Schneider, A., Dreyer, M., Schulz, G. E., Badia, J., and Aguilar, J., Enzymes in organic synthesis. Part 1. Diastereoselective, enzymatic aldol addition with L-rhamnulose- and L-fuculose-1-phosphate aldolases from *E. coli*. *Angew. Chem. Int. Ed.* 1991, **30** (5), 555–558.

[20] Fessner, W.-D., Schneider, A., Held, H., Sinerius, G., Walter, C., Hixon, M., and Schloss, J. V., Enzymes in organic synthesis. 12. The mechanism of class II, metal-dependent aldolases. *Angew. Chem. Int. Ed.* 1996, **35** (19), 2219–2221.

[21] Dalby, A., Dauter, Z., and Littlechild, J. A., Crystal structure of human muscle aldolase complexed with fructose 1,6-bisphosphate: Mechanistic implications. *Protein Sci.* 1999, **8** (2), 291–297.

[22] Hall, D. R., Leonard, G. A., Reed, C. D., Watt, C. I., Berry, A., and Hunter, W. N., The crystal structure of *Escherichia coli* class II fructose-1,6-bisphosphate aldolase in complex with phosphoglycolohydroxamate reveals details of mechanism and specificity. *J. Mol. Biol.* 1999, **287** (2), 383–394.

[23] Espelt, L., Bujons, J., Parella, T., Calveras, J., Joglar, J., Delgado, A., and Clapés, P., Aldol additions of dihydroxyacetone phosphate to N-Cbz-amino aldehydes catalyzed by L-fuculose-1-phosphate aldolase in emulsion systems: Inversion of stereoselectivity as a function of the acceptor aldehyde. *Chem. Eur. J.* 2005, **11** (5), 1392–1401.

[24] Calveras, J., Egido-Gabás, M., Gómez, L., Casas, J., Parella, T., Joglar, J., Bujons, J., and Clapés, P., Dihydroxyacetone phosphate aldolase-catalyzed synthesis of structurally diverse polyhydroxylated pyrrolidine derivatives and evaluation of their glycosidase inhibitory properties. *Chem. Eur. J.* 2009, **15**, 7310–7328.

[25] Gijsen, H. J. M. and Wong, C.-H., Unprecedented asymmetric aldol reactions with three aldehyde substrates catalyzed by 2-deoxyribose-5-phosphate aldolase. *J. Am. Chem. Soc.* 1994, **116** (18), 8422–8423.

[26] Gijsen, H. J. M. and Wong, C.-H., Sequential one-pot aldol reactions catalyzed by 2-deoxyribose-5-phosphate aldolase and fructose-1,6-diphosphate aldolase. *J. Am. Chem. Soc.* 1995, **117** (10), 2947–2948.

[27] Gijsen, H. J. M. and Wong, C.-H., Sequential three- and four-substrate aldol reactions catalyzed by aldolases. *J. Am. Chem. Soc.* 1995, **117** (29), 7585–7591.

[28] Nicolaou, K. C., Edmonds, D. J., and Bulger, P. G., Cascade reactions in total synthesis. *Angew. Chem. Int. Ed.* 2006, **45** (43), 7134–7186.

[29] Toscano, M. D., Woycechowsky, K. J., and Hilvert, D., Minimalist active-site redesign: Teaching old enzymes new tricks. *Angew. Chem. Int. Ed.* 2007, **46** (18), 3212–3236.

[30] Jäckel, C. and Hilvert, D., Biocatalysts by evolution. *Curr. Opin. Biotechnol.* 2010, **21** (6), 753–759.

[31] Giger, L., Toscano, M. D., Bouzon, M., Marlière, P., and Hilvert, D., A novel genetic selection system for PLP-dependent threonine aldolases. *Tetrahedron* 2012, **68** (37), 7549–7557.

[32] Ihee, H., Wulff, M., Kim, J., and Adachi, S.-I., Ultrafast X-ray scattering: Structural dynamics from diatomic to protein molecules. *Int. Rev. Phys. Chem.* 2010, **29** (3), 453–520.

[33] Saldin, D. K., Shneerson, V. L., and Spence, J. C. H., X-ray structure solution without crystallization. *Abstracts, 38th Great Lakes Regional Meeting of the American Chemical Society*, Chicago, IL, May 13–16, 2009, GLRM-085.

[34] Shneerson, V. L., Ourmazd, A., and Saldin, D. K., Crystallography without crystals. I. The common-line method for assembling a three-dimensional diffraction volume from single-particle scattering. *Acta Crystallogr., Sect. A: Found. Crystallogr.* 2008, **A64** (2), 303–315.

[35] Jiang, L., Althoff, E. A., Clemente, F. R., Doyle, L., Rothlisberger, D., Zanghellini, A., Gallaher, J. L., Betker, J. L., Tanaka, F., Barbas, C. F., Hilvert, D., Houk, K. N., Stoddard, B. L., and Baker, D., *De novo* computational design of retro-aldol enzymes. *Science* 2008, **319** (5868), 1387–1391.

[36] Siegel, J. B., Zanghellini, A., Lovick, H. M., Kiss, G., Lambert, A. R., St. Clair, J. L., Gallaher, J. L., Hilvert, D., Gelb, M. H., Stoddard, B. L., Houk, K. N., Michael, F. E., and Baker, D., Computational design of an enzyme catalyst for a stereoselective bimolecular Diels–Alder reaction. *Science* 2010, **329** (5989), 309–313.

[37] Petersen, M., Zannetti, M. T., and Fessner, W.-D., Tandem asymmetric C—C bond formations by enzyme catalysis. *Top. Curr. Chem.* 1997, **186** (Glycoscience Synthesis of Oligosaccharides and Glycoconjugates), 87–117.

[38] Machajewski, T. D. and Wong, C.-H., The catalytic asymmetric aldol reaction. *Angew. Chem. Int. Ed.* 2000, **39** (8), 1353–1374.

[39] Samland, A. K. and Sprenger, G. A., Microbial aldolases as C—C bonding enzymes—unknown treasures and new developments. *Appl. Microbiol. Biotechnol.* 2006, **71** (3), 253–264.

[40] Fessner, W.-D. and Jennewein, S., Biotechnological applications of aldolases. In *Biocatalysis in the Pharmaceutical Biotechnological Industries*, Patel, R. N., Ed. Marcel Dekker: New York, 2007; pp 363–400.

[41] Fessner, W.-D., Aldolases: Enzymes for making and breaking C—C bonds. In *Asymmetric Organic Synthesis with Enzymes*, Gotor, V., Alfonso, I., and Garcia-Urdiales, E., Eds. Wiley-VCH Verlag GmbH & Co. KGaA: Weinheim, 2008; pp 275–318.

[42] Dean, S. M., Greenberg, W. A., and Wong, C.-H., Recent advances in aldolase-catalyzed asymmetric synthesis. *Adv. Synth. Catal.* 2007, **349** (8–9), 1308–1320.

[43] Bolt, A., Berry, A., and Nelson, A., Directed evolution of aldolases for exploitation in synthetic organic chemistry. *Arch. Biochem. Biophys.* 2008, **474** (2), 318–330.

[44] Clapés, P. and Fessner, W.-D., Enzymatic direct aldol additions. In *Science of Synthesis. Stereoselective Synthesis 2. Stereoselective Reactions of Carbonyl and Imino Groups*, Molander, G. A., Ed. Georg Thieme Verlag KG: Stuttgart (Germany), 2011; Vol. 2, pp 677–734.

[45] Wang, W. and Seah, S. Y. K., Purification and biochemical characterization of a pyruvate-specific class II aldolase, HpaI. *Biochemistry* 2005, **44** (27), 9447–9455.

[46] Ling, H., Wang, G., Tian, Y., Liu, G., and Tan, H., SanM catalyzes the formation of 4-pyridyl-2-oxo-4-hydroxyisovalerate in nikkomycin biosynthesis by interacting with SanN. *Biochem. Biophys. Res. Commun.* 2007, **361** (1), 196–201.

[47] Coincon, M., Wang, W., Sygusch, J., and Seah, S. Y. K., Crystal structure of reaction intermediates in pyruvate class II aldolase: Substrate cleavage, enolate stabilization, and substrate specificity. *J. Biol. Chem.* 2012, **287** (43), 36208–36221.

[48] Blayer, S., Woodley, J. M., Lilly, M. D., and Dawson, M. J., Characterization of the chemoenzymic synthesis of *N*-acetyl-D-neuraminic acid (Neu5Ac). *Biotechnol. Prog.* 1996, **12** (6), 758–763.

[49] Varki, A., Diversity in the sialic acids. *Glycobiology* 1992, **2**, 25–40.

[50] Varki, A., Glycan-based interactions involving vertebrate sialic-acid-recognizing proteins. *Nature (London, U. K.)* 2007, **446** (7139), 1023–1029.

[51] Chen, X. and Varki, A., Advances in the biology and chemistry of sialic acids. *ACS Chem. Biol.* 2009, **5** (2), 163–176.

[52] Kim, M. J., Hennen, W. J., Sweers, H. M., and Wong, C.-H., Enzymes in carbohydrate synthesis: *N*-acetylneuraminic acid aldolase catalyzed reactions and preparation of *N*-acetyl-2-deoxy-D-neuraminic acid derivatives. *J. Am. Chem. Soc.* 1988, **110** (19), 6481–6486.

[53] Auge, C., David, S., Gautheron, C., Malleron, A., and Cavaye, B., Preparation of six naturally occurring sialic acids with immobilized acylneuraminate pyruvate lyase. *New J. Chem.* 1988, **12** (8–9), 733–744.

[54] Kragl, U., Gygax, D., Ghisalba, O., and Wandrey, C., Enzymatic, two-step synthesis of N-acetylneuraminic acid in a enzyme-membrane reactor. *Angew. Chem. Int. Ed.* 1991, **30** (7), 827–828.

[55] Fitz, W., Schwark, J.-R., and Wong, C.-H., Aldotetroses and C(3)-modified aldohexoses as substrates for *N*-acetylneuraminic acid aldolase: A model for the explanation of the normal and the inversed stereoselectivity. *J. Org. Chem.* 1995, **60** (12), 3663–3670.

[56] Lin, C. H., Sugai, T., Halcomb, R. L., Ichikawa, Y., and Wong, C.-H., Unusual stereoselectivity in sialic acid aldolase-catalyzed aldol condensations: Synthesis of both enantiomers of high-carbon monosaccharides. *J. Am. Chem. Soc.* 1992, **114** (26), 10138–10145.

[57] Fessner, W.-D., Enzyme-catalyzed aldol additions. In *Modern Aldol Reactions. Enolates, Organocatalysis, Biocatalysis and Natural Product Synthesis*, Rainer, M., Ed. Wiley-VCH Verlag GmbH & Co. KgaA: Weinheim, 2004; Vol. **1**, pp 201–272.

[58] Augé, C., Bouxom, B., Cavayé, B., and Gautheron, C., Scope and limitations of the aldol condensation catalyzed by immobilized acylneuraminate pyruvate lyase. *Tetrahedron Lett.* 1989, **30** (17), 2217–2220.

[59] Liu, J. L. C., Shen, G. J., Ichikawa, Y., Rutan, J. F., Zapata, G., Vann, W. F., and Wong, C.-H., Overproduction of CMP-sialic acid synthetase for organic synthesis. *J. Am. Chem. Soc.* 1992, **114** (10), 3901–3910.

[60] Zhou, P., Salleh, H. M., and Honek, J. F., Facile chemoenzymic synthesis of 3-(hydroxymethyl)-6-epicastanospermine. *J. Org. Chem.* 1993, **58** (1), 264–266.

[61] Sparks, M. A., Williams, K. W., Lukacs, C., Schrell, A., Priebe, G., Spaltenstein, A., and Whitesides, G. M., Synthesis of potential inhibitors of hemagglutination by influenza virus: Chemoenzymic preparation of N-5 analogs of *N*-acetylneuraminic acid. *Tetrahedron* 1993, **49** (1), 1–12.

[62] He, N., Yi, D., and Fessner, W.-D., Flexibility of substrate binding of cytosine-5′-monophosphate-N-acetylneuraminate synthetase (CMP-sialate synthetase) from Neisseria meningitidis: An enabling catalyst for the synthesis of neo-sialoconjugates. *Adv. Synth. Catal.* 2011, **353** (13), 2384–2398.

[63] Yi, D., He, N., Kickstein, M., Metzner, J., Weiß, M., Berry, A., and Fessner, W.-D., Engineering of a cytidine 5′-monophosphate-sialic acid synthetase for improved tolerance to functional sialic acids. *Adv. Synth. Catal.* 2013, **355** (18), 3597–3612.

[64] Lin, C.-C., Lin, C.-H., and Wong, C.-H., Sialic acid aldolase-catalyzed condensation of pyruvate and N-substituted mannosamine: A useful method for the synthesis of N-substituted sialic acids. *Tetrahedron Lett.* 1997, **38** (15), 2649–2652.

[65] Murakami, M., Ikeda, K., and Achiwa, K., Chemo-enzymic synthesis of neuraminic acid analogs structurally varied at C-5 and C-9 as potential inhibitors of the sialidase from influenza virus. *Carbohydr. Res.* 1996, **280** (1), 101–110.

[66] Kong, D. C. M. and Von Itzstein, M., The chemoenzymic synthesis of 9-substituted 3,9-dideoxy-D-glycero-D-galacto-2-nonulosonic acids. *Carbohydr. Res.* 1998, **305** (3–4), 323–329.

[67] Kiefel, M. J., Wilson, J. C., Bennett, S., Gredley, M., and von Itzstein, M., Synthesis and evaluation of C-9 modified N-acetylneuraminic acid derivatives as substrates for N-acetylneuraminic acid aldolase. *Bioorg. Med. Chem.* 2000, **8** (3), 657–664.

[68] Calveras, J., Nagai, Y., Sultana, I., Ueda, Y., Higashi, T., Shoji, M., and Sugai, T., New chemo-enzymatic route toward N-acetylneuraminic acid derivatives with alkyl groups at C-7 hydroxyl group. *Tetrahedron* 2010, **66** (24), 4284–4291.

[69] Fitz, W. and Wong, C.-H., Combined use of subtilisin and N-acetylneuraminic acid aldolase for the synthesis of a fluorescent sialic acid. *J. Org. Chem.* 1994, **59** (26), 8279–8280.

[70] Ikeda, K., Mori, H., and Sato, M., Preparation of a fluorous protecting group and its application to the chemoenzymatic synthesis of sialidase inhibitor. *Chem. Commun.* 2006, (29), 3093–3094.

[71] Hartlieb, S., Guenzel, A., Gerardy-Schahn, R., Muenster-Kuehnel, A. K., Kirschning, A., and Draeger, G., Chemoenzymatic synthesis of CMP-*N*-acetyl-7-fluoro-7-deoxyneuraminic acid. *Carbohydr. Res.* 2008, **343** (12), 2075–2082.

[72] Li, Y. H., Yu, H., Cao, H. Z., Lau, K., Muthana, S., Tiwari, V. K., Son, B., and Chen, X., *Pasteurella multocida* sialic acid aldolase: A promising biocatalyst. *Appl. Microbiol. Biotechnol.* 2008, **79** (6), 963–970.

[73] Cao, H., Li, Y., Lau, K., Muthana, S., Yu, H., Cheng, J., Chokhawala, H. A., Sugiarto, G., Zhang, L., and Chen, X., Sialidase substrate specificity studies using chemoenzymatically synthesized sialosides containing C5-modified sialic acids. *Org. Biomol. Chem.* 2009, **7** (24), 5137–5145.

[74] Yu, H., Cao, H., Tiwari, V. K., Li, Y., and Chen, X., Chemoenzymatic synthesis of C8-modified sialic acids and related alpha2-3- and alpha2-6-linked sialosides. *Bioorg. Med. Chem. Lett.* 2011, **21** (17), 5037–5040.

[75] Khedri, Z., Muthana, M. M., Li, Y., Muthana, S. M., Yu, H., Cao, H., and Chen, X., Probe sialidase substrate specificity using chemo-enzymatically synthesized sialosides containing C9-modified sialic acid. *Chem. Commun.* 2012, **48** (27), 3357–3359.

[76] Yu, H., Cheng, J., Ding, L., Khedri, Z., Chen, Y., Chin, S., Lau, K., Tiwari, V. K., and Chen, X., Chemoenzymatic synthesis of GD3 oligosaccharides and other disialyl glycans containing natural and non-natural sialic acids. *J. Am. Chem. Soc.* 2009, **131** (51), 18467–18477.

[77] Song, X., Yu, H., Chen, X., Lasanajak, Y., Tappert, M. M., Air, G. M., Tiwari, V. K., Cao, H., Chokhawala, H. A., Zheng, H., Cummings, R. D., and Smith, D. F., A sialylated glycan microarray reveals novel interactions of modified sialic acids with proteins and viruses. *J. Biol. Chem.* 2011, **286** (36), 31610–31622.

[78] Li, Y., Yu, H., Cao, H., Muthana, S., and Chen, X., Pasteurella multocida CMP-sialic acid synthetase and mutants of Neisseria meningitidis CMP-sialic acid synthetase with improved substrate promiscuity. *Appl. Microbiol. Biotechnol.* 2012, **93** (6), 2411–2423.

[79] Yu, H. and Chen, X., Aldolase-catalyzed synthesis of β-D-Galp-(1–9)-D-KDN: A novel acceptor for sialyltransferases. *Org. Lett.* 2006, **8** (11), 2393–2396.

[80] Huang, S., Yu, H., and Chen, X., Disaccharides as sialic acid aldolase substrates: Synthesis of disaccharides containing a sialic acid at the reducing end. *Angew. Chem. Int. Ed.* 2007, **46** (13), 2249–2253.

[81] Yu, H., Huang, S., Chokhawala, H., Sun, M., Zheng, H., and Chen, X., Highly efficient chemoenzymatic synthesis of naturally occurring and non-natural α-2,6-linked sialosides: A P. damsela α-2,6-sialyltransferase with extremely flexible donor-substrate specificity. *Angew. Chem. Int. Ed.* 2006, **45** (24), 3938–3944.

[82] Chokhawala, H. A., Yu, H., and Chen, X., High-throughput substrate specificity studies of sialidases by using chemoenzymatically synthesized sialoside libraries. *ChemBioChem* 2007, **8** (2), 194–201.

[83] Ding, L., Yu, H., Lau, K., Li, Y., Muthana, S., Wang, J., and Chen, X., Efficient chemoenzymic synthesis of sialyl Tn-antigens and derivatives. *Chem. Commun.* 2011, **47** (30), 8691–8693.

[84] Sugiarto, G., Lau, K., Qu, J., Li, Y., Lim, S., Mu, S., Ames, J. B., Fisher, A. J., and Chen, X., A sialyltransferase mutant with decreased donor hydrolysis and reduced sialidase activities for directly sialylating Lewisx. *ACS Chem. Biol.* 2012, **7** (7), 1232–1240.

[85] Sugiyama, M., Watanabe, K., Funakoshi, N., Amino, Y., Kawahara, S., and Takemoto, T. Process for production of glutamic acid derivatives. 2002-JP12852, 2003056026, WO 2003/056026, 20021209., 2003, Ajinomoto Co., Inc.

[86] Sugiyama, M., Kashiwagi, T., Mori, K., and Suzuki, E. Mutagenesis and crystal structure of Pseudomonas taetrolens aldolase and the use of the enzyme for production of optically activity IHOG. 2004-75256, 2005261231, US 2005/0244939 Al, 20040316., 2005, Ajinomoto Co., Inc.

[87] Wang, W., Mazurkewich, S., Kimber, M. S., and Seah, S. Y. K., Structural and kinetic characterization of 4-hydroxy-4-methyl-2-oxoglutarate/4-carboxy-4-hydroxy-2-oxoadipate aldolase, a protocatechuate degradation enzyme evolutionarily convergent with the HpaI and DmpG pyruvate aldolases. *J. Biol. Chem.* 2010, **285** (47), 36608–36615.

[88] Moehler, H., Decker, K., and Wood, W. A., Structure of 2-keto-3-deoxy-6-phosphogluconate aldolase. IV. Structural features revealed by treatment with urea and Ellman's reagent. *Arch. Biochem. Biophys.* 1972, **151** (1), 251–260.

[89] Wood, W. A., *2-Keto-3-Deoxy-6-Phosphogluconic and Related Aldolases*. Academic Press: New York, NY, 1972; Vol. 7.

[90] Serafimov, J. M., Gillingham, D., Kuster, S., and Hilvert, D., The putative diels-alderase macrophomate synthase is an efficient aldolase. *J. Am. Chem. Soc.* 2008, **130** (25), 7798–7799.

[91] Gillingham, D. G., Stallforth, P., Adibekian, A., Seeberger, P. H., and Hilvert, D., Chemoenzymatic synthesis of differentially protected 3-deoxysugars. *Nat. Chem.* 2010, **2** (2), 102–105.

[92] Stallforth, P., Matthies, S., Adibekian, A., Gillingham, D. G., Hilvert, D., and Seeberger, P. H., De novo chemoenzymatic synthesis of sialic acid. *Chem. Commun.* 2012, **48** (98), 11987–11989.

[93] Moon, M. and Van Lanen, S. G., Characterization of a dual specificity aryl acid adenylation enzyme with dual function in nikkomycin biosynthesis. *Biopolymers* 2010, **93** (9), 791–801.

[94] Ogawa, J., Yamanaka, H., Mano, J., Doi, Y., Horinouchi, N., Kodera, T., Nio, N., Smirnov, S. V., Samsonova, N. N., Kozlov, Y. I., and Shimizu, S., Synthesis of 4-hydroxyisoleucine by the aldolase-transaminase coupling reaction and basic characterization of the aldolase from Arthrobacter simplex AKU 626. *Biosci., Biotechnol., Biochem.* 2007, **71** (7), 1607–1615.

[95] Smirnov, S. V., Samsonova, N. N., Novikova, A. E., Matrosov, N. G., Rushkevich, N. Y., Kodera, T., Ogawa, J., Yamanaka, H., and Shimizu, S., A novel strategy for enzymatic synthesis of 4-hydroxyisoleucine: Identification of an enzyme possessing HMKP (4-hydroxy-3-methyl-2-keto-pentanoate) aldolase activity. *FEMS Microbiol. Lett.* 2007, **273** (1), 70–77.

[96] Baker, P., Pan, D., Carere, J., Rossi, A., Wang, W., and Seah, S. Y. K., Characterization of an aldolase–dehydrogenase complex that exhibits substrate channeling in the polychlorinated biphenyls degradation pathway. *Biochemistry* 2009, **48** (27), 6551–6558.

[97] Baker, P. and Seah, S. Y. K., Rational design of stereoselectivity in the class II pyruvate aldolase BphI. *J. Am. Chem. Soc.* 2012, **134** (1), 507–513.

[98] Wang, W. J., Baker, P., and Seah, S. Y. K., Comparison of two metal-dependent pyruvate aldolases related by convergent evolution: Substrate specificity, kinetic mechanism, and substrate channeling. *Biochemistry* 2010, **49** (17), 3774–3782.

[99] Shelton, M. C., Cotterill, I. C., Novak, S. T. A., Poonawala, R. M., Sudarshan, S., and Toone, E. J., 2-Keto-3-deoxy-6-phosphogluconate aldolases as catalysts for stereocontrolled carbon–carbon bond formation. *J. Am. Chem. Soc.* 1996, **118** (9), 2117–2125.

[100] Floyd, N. C., Liebster, M. H., and Turner, N. J., A simple strategy for obtaining both enantiomers from an aldolase reaction: Preparation of L- and D-4-hydroxy-2-ketoglutarate. *J. Chem. Soc. Perkin Trans. 1* 1992 (9), 1085–1086.

[101] Cotterill, I. C., Shelton, M. C., Machemer, D. E. W., Henderson, D. P., and Toone, E. J., Effect of phosphorylation on the reaction rate of unnatural electrophiles with 2-keto-3-deoxy-6-phosphogluconate aldolase. *J. Chem. Soc. Perkin Trans.* 1998, **1** (7), 1335–1342.

[102] Henderson, D. P., Cotterill, I. C., Shelton, M. C., and Toone, E. J., 2-Keto-3-deoxy-6-phosphogalactonate aldolase as a catalyst for stereocontrolled carbon–carbon bond formation. *J. Org. Chem.* 1998, **63** (4), 906–907.

[103] Allen, S. T., Heintzelman, G. R., and Toone, E. J., Pyruvate aldolases as reagents for stereospecific aldol condensation. *J. Org. Chem.* 1992, **57** (2), 426–427.

[104] Cheriyan, M., Toone, E. J., and Fierke, C. A., Mutagenesis of the phosphate-binding pocket of KDPG aldolase enhances selectivity for hydrophobic substrates. *Protein Sci.* 2007, **16** (11), 2368–2377.

[105] Walters, M. J., Srikannathasan, V., McEwan, A. R., Naismith, J. H., Fierke, C. A., and Toone, E. J., Characterization and crystal structure of *Escherichia coli* KDPGal aldolase. *Bioorg. Med. Chem.* 2008, **16** (2), 710–720.

[106] Fong, S., Machajewski, T. D., Mak, C. C., and Wong, C.-H., Directed evolution of D-2-keto-3-deoxy-6-phosphogluconate aldolase to new variants for the efficient synthesis of D- and L-sugars. *Chem. Biol.* 2000, **7** (11), 873–883.

[107] Wymer, N., Buchanan, L. V., Henderson, D., Mehta, N., Botting, C. H., Pocivavsek, L., Fierke, C. A., Toone, E. J., and Naismith, J. H., Directed evolution of a new catalytic site in 2-keto-3- deoxy-6-phosphogluconate aldolase from *Escherichia coli*. *Structure* 2001, **9** (1), 1–9.

[108] Lamble, H. J., Heyer, N. I., Bull, S. D., Hough, D. W., and Danson, M. J., Metabolic pathway promiscuity in the archaeon Sulfolobus solfataricus revealed by studies on glucose dehydrogenase and 2-keto-3-deoxygluconate aldolase. *J. Biol. Chem.* 2003, **278** (36), 34066–34072.

[109] Lamble, H. J., Royer, S. F., Hough, D. W., Danson, M. J., Taylor, G. L., and Bull, S. D., A thermostable aldolase for the synthesis of 3-deoxy-2-ulosonic acids. *Adv. Synth. Catal.* 2007, **349** (6), 817–821.

[110] Schurink, M., Wolterink-van Loo, S., van der Oost, J., Sonke, T., and Franssen, M. C. R., Substrate specificity and stereoselectivity of two sulfolobus 2-keto-3-deoxygluconate aldolases towards azido- substituted aldehydes. *ChemCatChem* 2014, **6** (4), 1073–1081.

[111] DeSantis, G., Liu, J. J., Clark, D. P., Heine, A., Wilson, I. A., and Wong, C.-H., Structure-based mutagenesis approaches toward expanding the substrate specificity of D-2-deoxyribose-5-phosphate aldolase. *Bioorg. Med. Chem.* 2003, **11** (1), 43–52.

[112] Garrabou, X., Gomez, L., Joglar, J., Gil, S., Parella, T., Bujons, J., and Clapés, P., Structure-guided minimalist redesign of L-fuculose-1-phosphate aldolase active site. Expedient synthesis of novel polyhydroxylated pyrrolizidines and their inhibitory properties against glycosidases and intestinal disaccharidases *Chem. Eur. J.* 2010, **16**, 10691–10706.

[113] Garrabou, X., Joglar, J., Parella, T., Bujons, J., and Clapés, P., Redesign of the phosphate binding site of L-rhamnulose-1-phosphate aldolase towards a dihydroxyacetone dependent aldolase. *Adv. Synth. Catal.* 2011, **353** (1), 89–99.

[114] Castillo, J. A., Guérard-Hélaine, C., Gutiérrez, M., Garrabou, X., Sancelme, M., Schürmann, M., Inoue, T., Hélaine, V., Charmantray, F., Gefflaut, T., Hecquet, L., Joglar, J., Clapés, P., Sprenger, G. A., and Lemaire, M., A mutant D-fructose-6-phosphate aldolase (Ala129Ser) with improved affinity towards dihydroxyacetone for the synthesis of polyhydroxylated compounds. *Adv. Synth. Catal.* 2010, **352** (6), 1039–1046.

[115] Royer, S. F., Haslett, L., Crennell, S. J., Hough, D. W., Danson, M. J., and Bull, S. D., Structurally informed site-directed mutagenesis of a stereochemically promiscuous aldolase to afford stereochemically complementary biocatalysts. *J. Am. Chem. Soc.* 2010, **132** (33), 11753–11758.

[116] Williams, G. J., Woodhall, T., Farnsworth, L. M., Nelson, A., and Berry, A., Creation of a pair of stereochemically complementary biocatalysts. *J. Am. Chem. Soc.* 2006, **128** (50), 16238–16247.

[117] Fessner, W.-D. and Eyrisch, O., Enzymes in organic synthesis. 2. One-pot synthesis of tagatose 1,6-bisphosphate via diastereoselective enzymic aldol addition. *Angew. Chem. Int. Ed.* 1992, **31** (1), 56–58.

[118] Hall, D. R., Bond, C. S., Leonard, G. A., Watt, I., Berry, A., and Hunter, W. N., Structure of tagatose-1,6-bisphosphate aldolase—Insight into chiral discrimination, mechanism, and specificity of class II aldolases. *J. Biol. Chem.* 2002, **277** (24), 22018–22024.

[119] Schümperli, M., Pellaux, R., and Panke, S., Chemical and enzymatic routes to dihydroxyacetone phosphate. *Appl. Microbiol. Biotechnol.* 2007, **75** (1), 33–45.

[120] Iturrate, L., Sánchez-Moreno, I., Oroz-Guinea, I., Pérez-Gil, J., and García-Junceda, E., Preparation and characterization of a bifunctional aldolase/kinase enzyme: A more efficient biocatalyst for C—C bond formation. *Chem. Eur. J.* 2010, **16** (13), 4018–4030.

[121] Sánchez-Moreno, I., Iturrate, L., Doyaguez, E. G., Martínez, J. A., Fernández-Mayoralas, A., and García-Junceda, E., Activated alpha,beta-unsaturated aldehydes as substrate of dihydroxyacetone phosphate (DHAP)-dependent aldolases in the context of a multienzyme system. *Adv. Synth. Catal.* 2009, **351** (17), 2967–2975.

[122] Iturrate, L., Sánchez-Moreno, I., Doyaguez, E. G., and García-Junceda, E., Substrate channelling in an engineered bifunctional aldolase/kinase enzyme confers catalytic advantage for C—C bond formation. *Chem. Commun.* 2009, **13**, 1721–1723.

[123] van Herk, T., Hartog, A. F., Schoemaker, H. E., and Wever, R., Simple enzymatic in situ generation of dihydroxyacetone phosphate and its use in a cascade reaction for the production of carbohydrates: Increased efficiency by phosphate cycling. *J. Org. Chem.* 2006, **71** (16), 6244–6247.

[124] Babich, L., van Hemert, L. J. C., Bury, A., Hartog, A. F., Falcicchio, P., van der Oost, J., van Herk, T., Wever, R., and Rutjes, F. P. J. T., Synthesis of non-natural carbohydrates from glycerol and aldehydes in a one-pot four-enzyme cascade reaction. *Green Chem.* 2011, **13** (10), 2895–2900.

[125] van Herk, T., Hartog, A. F., Babich, L., Schoemaker, H. E., and Wever, R., Improvement of an acid phosphatase/DHAP-dependent aldolase cascade reaction by using directed evolution. *ChemBioChem* 2009, **10** (13), 2230–2235.

[126] Kroemer, M., Merkel, I., and Schulz, G. E., Structure and catalytic mechanism of L-rhamnulose-1-phosphate aldolase. *Biochemistry* 2003, **42** (36), 10560–10568.

[127] Joerger, A. C., Mueller-Dieckmann, C., and Schulz, G. E., Structures of L-fuculose-1-phosphate aldolase mutants outlining motions during catalysis. *J. Mol. Biol.* 2000, **303** (4), 531–543.

[128] Sugiyama, M., Hong, Z. Y., Whalen, L. J., Greenberg, W. A., and Wong, C.-H., Borate as a phosphate ester mimic in aldolase-catalyzed reactions: Practical synthesis of L-fructose and L-iminocyclitols. *Adv. Synth. Catal.* 2006, **348** (18), 2555–2559.

[129] Garrabou, X., Calveras, J., Joglar, J., Parella, T., Bujons, J., and Clapés, P., Highly efficient aldol additions of DHA and DHAP to *N*-Cbz-amino aldehydes catalyzed by L-rhamnulose-1-phosphate and L-fuculose-1-phosphate aldolases in aqueous borate buffer. *Org. Biomol. Chem.* 2011, **9** (24), 8430–8436.

[130] Look, G. C., Fotsch, C. H., and Wong, C.-H., Enzyme-catalyzed organic synthesis: Practical routes to aza sugars and their analogs for use as glycoprocessing inhibitors. *Acc. Chem. Res.* 1993, **26** (4), 182–190.

[131] Whalen, L. J. and Wong, C.-H., Enzymes in organic synthesis: Aldolase-mediated synthesis of iminocyclitols and novel heterocycles. *Aldrichimica Acta* 2006, **39** (3), 63–71.

[132] Espelt, L., Parella, T., Bujons, J., Solans, C., Joglar, J., Delgado, A., and Clapés, P., Stereoselective aldol additions catalyzed by dihydroxyacetone phosphate dependent aldolases in emulsion systems: Preparation and structural characterization of linear and cyclic aminopolyols from aminoaldehydes. *Chem. Eur. J.* 2003, **9** (20), 4887–4899.

[133] Calveras, J., Bujons, J., Parella, T., Crehuet, R., Espelt, L., Joglar, J., and Clapés, P., Influence of N-amino protecting group on aldolase-catalyzed aldol additions of dihydroxyacetone phosphate to amino aldehydes. *Tetrahedron* 2006, **62** (11), 2648–2656.
[134] Jurczak, J. and Golebiowski, A., Optically active N-protected alpha-amino aldehydes in organic synthesis. *Chem. Rev.* 1989, **89** (1), 149–164.
[135] Jurczak, J., Gryko, D., Kobrzycka, E., Gruza, H., and Prokopowicz, P., Effective and mild method for preparation of optically active alpha-amino aldehydes via TEMPO oxidation. *Tetrahedron* 1998, **54** (22), 6051–6064.
[136] Calveras, J., Casas, J., Parella, T., Joglar, J., and Clapés, P., Chemoenzymatic synthesis and inhibitory activities of hyacinthacines A1 and A2 stereoisomers. *Adv. Synth. Catal.* 2007, **349** (10), 1661–1666.
[137] Laborda, P., Sayago, F. J., Cativiela, C., Parella, T., Joglar, J., and Clapés, P., Aldolase-catalyzed synthesis of conformationally constrained iminocyclitols: Preparation of polyhydroxylated benzopyrrolizidines and cyclohexapyrrolizidines. *Org. Lett.* 2014, **16** (5), 1422–1425.
[138] Concia, A. L., Gómez, L., Parella, T., Joglar, J., and Clapés, P., Casuarine stereoisomers from achiral substrates. Chemo-enzymatic synthesis and inhibitory properties. *J. Org. Chem.* 2014, **79** (11), 5386–5389.
[139] Gómez, L., Garrabou, X., Joglar, J., Bujons, J., Parella, T., Vilaplana, C., Cardona, P. J., and Clapés, P., Chemoenzymatic synthesis, structural study and biological activity of novel indolizidine and quinolizidine iminocyclitols. *Org. Biomol. Chem.* 2012, **10** (31), 6309–6321.
[140] Soler, A., Garrabou, X., Hernández, K., Gutiérrez, M. L., Busto, E., Bujons, J., Parella, T., Joglar, J., and Clapés, P., Sequential biocatalytic aldol reactions in multi-step asymmetric synthesis: Pipecolic acid, piperidine and pyrrolidine (homo)iminocyclitol derivatives from achiral building blocks. *Adv. Synth. Catal.* 2014, **356** (14–15), 3007–3024.
[141] El Blidi, L., Assaf, Z., Bres, F. C., Veschambre, H., Thery, V., Bolte, J., and Lemaire, M., Fructose-1,6-bisphosphate aldolase-mediated synthesis of aminocyclitols (analogues of valiolamine) and their evaluation as glycosidase inhibitors. *ChemCatChem* 2009, **1** (4), 463–471.
[142] Schürmann, M., Schürmann, M., and Sprenger, G. A., Fructose 6-phosphate aldolase and 1-deoxy-D-xylulose 5-phosphate synthase from *Escherichia coli* as tools in enzymatic synthesis of 1-deoxysugars. *J. Mol. Catal. B: Enzym.* 2002, **19**, 247–252.
[143] Schürmann, M. and Sprenger, G. A., Fructose-6-phosphate aldolase is a novel class I aldolase from *Escherichia coli* and is related to a novel group of bacterial transaldolases. *J. Biol. Chem.* 2001, **276** (14), 11055–11061.
[144] Castillo, J. A., Calveras, J., Casas, J., Mitjans, M., Vinardell, M. P., Parella, T., Inoue, T., Sprenger, G. A., Joglar, J., and Clapés, P., Fructose-6-phosphate aldolase in organic synthesis: Preparation of D-fagomine, *N*-alkylated derivatives, and preliminary biological assays. *Org. Lett.* 2006, **8** (26), 6067–6070.
[145] Sugiyama, M., Hong, Z., Liang, P. H., Dean, S. M., Whalen, L. J., Greenberg, W. A., and Wong, C.-H., D-Fructose-6-phosphate aldolase-catalyzed-one-pot synthesis of iminocyclitols. *J. Am. Chem. Soc.* 2007, **129** (47), 14811–14817.
[146] Concia, A. L., Lozano, C., Castillo, J. A., Parella, T., Joglar, J., and Clapés, P., D-Fructose-6-phosphate aldolase in organic synthesis: Cascade chemical-enzymatic preparation of sugar-related polyhydroxylated compounds. *Chem. Eur. J.* 2009, **15** (15), 3808–3816.
[147] Garrabou, X., Castillo, J. A., Guérard-Hélaine, C., Parella, T., Joglar, J., Lemaire, M., and Clapés, P., Asymmetric self- and cross-aldol reaction of glycolaldehyde catalyzed by D-fructose-6-phosphate aldolase. *Angew. Chem. Int. Ed.* 2009, **48**, 5521–5525.
[148] Schneider, S., Gutiérrez, M., Sandalova, T., Schneider, G., Clapés, P., Sprenger, G. A., and Samland, A. K., Redesigning the active site of transaldolase TalB from *Escherichia coli*: New variants with improved affinity towards non-phosphorylated substrates. *ChemBioChem* 2010, **11** (5), 681–690.
[149] Schneider, S., Sandalova, T., Schneider, G., Sprenger, G. A., and Samland, A. K., Replacement of a phenylalanine by a tyrosine in the active site confers fructose 6-phosphate aldolase activity to the transaldolase of *Escherichia coli* and human origin. *J. Biol. Chem.* 2008, **283** (44), 30064–30072.
[150] Rale, M., Schneider, S., Sprenger, G. A., Samland, A. K., and Fessner, W.-D., Broadening deoxysugar glycodiversity: Natural and engineered transaldolases unlock a complementary substrate space. *Chem. Eur. J.* 2011, **17**, 2623–2632.

[151] Gutierrez, M., Parella, T., Joglar, J., Bujons, J., and Clapés, P., Structure-guided redesign of D-fructose-6-phosphate aldolase from *E. coli*: Remarkable activity and selectivity towards acceptor substrates by two-point mutation. *Chem. Commun.* 2011, **47** (20), 5762–5764.

[152] Sprenger, G. A., Schürmann, M., Schürmann, M., Johnen, S., Sprenger, G., Sahm, H., Inoue, T., and Schörken, U., C—C-bonding microbial enzymes: Thiamine diphosphate-dependent enzymes and class I aldolases. In *Asymmetric Synthesis with Chemical and Biological Methods*, Enders, D. and Jaeger, K.-E., Eds. Wiley-VCH Verlag GmbH & Co KGaA: Weinheim, 2007; pp 312–326.

[153] Szekrenyi, A., Soler, A., Garrabou, X., Guérard-Hélaine, C., Parella, T., Joglar, J., Lemaire, M., Bujons, J., and Clapés, P., Engineering the donor selectivity of D-fructose-6-phosphate aldolase for biocatalytic asymmetric cross-aldol additions of glycolaldehyde. *Chem. Eur. J.* 2014, **20** (39), 12572–12583.

[154] Castillo, J. A., Parella, T., Inoue, T., Sprenger, G. A., Joglar, J., and Clapés, P., Synthesis of D-fagomine by aldol addition of dihydroxyacetone to *N*-Cbz-3-aminopropanal catalyzed by D-fructose-6-phosphate aldolase. In *Practical Methods for Biocatalysis and Biotransformations*, Sutton, J. W. P., Ed. John Wiley & Sons, Ltd: Chichester, UK, 2009; pp 206–2012.

[155] Camps Bres, F., Guérard-Hélaine, C., Fernandes, C., Castillo, J. A., and Lemaire, M., Fructose-6-phosphate aldolases as versatile biocatalysts for nitrocyclitol syntheses. *Tetrahedron: Asymmetry* 2013, **24** (18), 1075–1081.

[156] Guerard-Helaine, C., Legeret, B., Fernandes, C., Prevot, V., Forano, C., and Lemaire, M., Efficient immobilization of fructose-6-phosphate aldolase in layered double hydroxide: Improved stereoselective synthesis of sugar analogues. *New J. Chem.* 2011, **35**, 776–779.

[157] Simon, G., Eljezi, T., Legeret, B., Charmantray, F., Castillo, J. A., Guérard-Hélaine, C., Lemaire, M., Bouzon, M., Marlière, P., Hélaine, V., and Hecquet, L., Synthesis of specially designed probes to broaden transketolase scope. *ChemCatChem* 2013, **5** (3), 784–795.

[158] Sanchez-Moreno, I., Helaine, V., Poupard, N., Charmantray, F., Legeret, B., Hecquet, L., Garcia-Junceda, E., Wohlgemuth, R., Guerard-Helaine, C., and Lemaire, M., One-pot cascade reactions using fructose-6-phosphate aldolase: Efficient synthesis of D-arabinose 5-phosphate, D-fructose 6-phosphate and analogues. *Adv. Synth. Catal.* 2012, **354** (9), 1725–1730.

[159] Fessner, W. D., Heyl, D., and Rale, M., Multi-enzymatic cascade synthesis of D-fructose 6-phosphate and deoxy analogs as substrates for high-throughput aldolase screening. *Catal. Sci. Technol.* 2012, **2** (8), 1596–1601.

[160] Guerard-Helaine, C., Debacker, M., Clapes, P., Szekrenyi, A., Helaine, V., and Lemaire, M., Efficient biocatalytic processes for highly valuable terminally phosphorylated C5 to C9 D-ketoses. *Green Chem.* 2014, **16** (3), 1109–1113.

[161] Steinreiber, J., Fesko, K., Reisinger, C., Schürmann, M., van Assema, F., Wolberg, M., Mink, D., and Griengl, H., Threonine aldolases—An emerging tool for organic synthesis. *Tetrahedron* 2007, **63** (4), 918–926.

[162] Dückers, N., Baer, K., Simon, S., Gröger, H., and Hummel, W., Threonine aldolases, screening, properties and applications in the synthesis of non-proteinogenic β-hydroxy-α-amino acids. *Appl. Microbiol. Biotechnol.* 2010, **88** (2), 409–424.

[163] Baer, K., Dückers, N., Hummel, W., and Gröger, H., Expanding the application range of aldolases: Novel asymmetric syntheses of α-methylated β-hydroxy α-amino acids and β-amino alcohols. *ChemCatChem* 2010, **2** (8), 939–942.

[164] Fesko, K., Reisinger, C., Steinreiber, J., Weber, H., Schürmann, M., and Griengl, H., Four types of threonine aldolases: Similarities and differences in kinetics/thermodynamics. *J. Mol. Catal. B: Enzym.* 2008, **52–53**, 19–26.

[165] Liu, J. Q., Dairi, T., Itoh, N., Kataoka, M., Shimizu, S., and Yamada, H., Diversity of microbial threonine aldolases and their application. *J. Mol. Catal. B: Enzym.* 2000, **10** (1–3), 107–115.

[166] Schostarez, H. J. and Chrusciel, R. A. Preparation of statine derivatives for the treatment of Alzheimer's disease. WO2003006021, 2003, Elan Pharmaceuticals, Inc., USA; Pharmacia & Upjohn Company.

[167] Farran, D., Toupet, L., Martinez, J., and Dewynter, G., Stereocontrolled synthesis of 2,4-diamino-3-hydroxyacids starting from diketopiperazines: A new route for the preparation of statine analogues. *Org. Lett.* 2007, **9** (23), 4833–4836.

[168] Barazza, A., Götz, M., Cadamuro, S. A., Goettig, P., Willem, M., Steuber, H., Kohler, T., Jestel, A., Reinemer, P., Renner, C., Bode, W., and Moroder, L., Macrocyclic statine-based inhibitors of BACE-1. *ChemBioChem* 2007, **8** (17), 2078–2091.

[169] Furet, P., Imbach, P., Furst, P., Lang, M., Noorani, M., Zimmermann, J., and Garcia-Echeverria, C., Modeling of the binding mode of a non-covalent inhibitor of the 20S proteasome. Application to structure-based analogue design. *Bioorg. Med. Chem. Lett.* 2001, **11** (10), 1321–1324.

[170] Kroemer, R. T., Ettmayer, P., and Hecht, P., 3D-quantitative structure-activity relationships of human immunodeficiency virus type-1 proteinase inhibitors: Comparative molecular field analysis of 2-heterosubstituted statine derivatives-implications for the design of novel inhibitors. *J. Med. Chem.* 1995, **38** (25), 4917–4928.

[171] Billich, A., Charpiot, B., Ettmayer, P., Gstach, H., Lehr, P., and Scholz, D. Preparation of 2,4-diamino-3-hydroxycarboxylic acid-derivative HIV proteinase inhibitors. EP615969, 1994, Sandoz Ltd., Switzerland; Sandoz-Patent-G.m.b.H.; Sandoz-Erfindungen Verwaltungsgesellschaft m.b.H.

[172] Zuo, Z., Luo, X., Zhu, W., Shen, J., Shen, X., Jiang, H., and Chen, K., Molecular docking and 3D-QSAR studies on the binding mechanism of statine-based peptidomimetics with b-secretase. *Bioorg. Med. Chem.* 2005, **13** (6), 2121–2131.

[173] Miura, T. and Kajimoto, T., Application of L-threonine aldolase-catalyzed reaction to the preparation of protected 3*R*,5*R*-dihydroxy-L-homoproline as a mimetic of idulonic acid. *Chirality* 2001, **13** (9), 577–580.

[174] Shibata, K., Shingu, K., Vassilev, V. P., Nishide, K., Fujita, T., Node, M., Kajimoto, T., and Wong, C.-H., Kinetic and thermodynamic control of L-threonine aldolase catalyzed reaction and its application to the synthesis of mycestericin D. *Tetrahedron Lett.* 1996, **37** (16), 2791–2794.

[175] Fujii, M., Miura, T., Kajimoto, T., and Ida, Y., Facile synthesis of 3,4-dihydroxyprolines as an application of the L-threonine aldolase-catalyzed aldol reaction. *Synlett* 2000, **7**, 1046–1048.

[176] Vassilev, V. P., Uchiyama, T., Kajimoto, T., and Wong, C.-H., An efficient chemo-enzymatic synthesis of α-amino-β-hydroxy-γ-butyrolactone. *Tetrahedron Lett.* 1995, **36** (28), 5063–5064.

[177] Liu, J. Q., Odani, M., Dairi, T., Itoh, N., Shimizu, S., and Yamada, H., A new route to L-threo-3-[4-(methylthio)phenylserine], a key intermediate for the synthesis of antibiotics: Recombinant low-specificity D-threonine aldolase-catalyzed stereospecific resolution. *Appl. Microbiol. Biotechnol.* 1999, **51** (5), 586–591.

[178] Wu, S.-H., Shimazaki, M., Lin, C.-C., Qiao, L., Moree, W. J., Weitz-Schmidt, G., and Wong, C.-H., Synthesis of fucopeptides as sialyl Lewisx mimetics. *Angew. Chem. Int. Ed.* 1996, **35** (1), 88–90.

[179] DeMong, D. E. and Williams, R. M., An efficient asymmetric synthesis of (2*S*,3*S*)- and (2*R*,3*R*)-β-hydroxyornithine. *Tetrahedron Lett.* 2001, **42** (2), 183–185.

[180] Gutierrez, M. L., Garrabou, X., Agosta, E., Servi, S., Parella, T., Joglar, J., and Clapés, P., Serine hydroxymethyl transferase from *Streptococcus thermophilus* and L-threonine aldolase from *Escherichia coli* as stereocomplementary biocatalysts for the synthesis of β-hydroxy-β,ω-diamino acid derivatives. *Chem. Eur. J.* 2008, **14** (15), 4647–4656.

[181] Cativiela, C. and Díaz-de-Villegas, M. D., Stereoselective synthesis of quaternary α-amino acids. Part 1: Acyclic compounds. *Tetrahedron: Asymmetry* 1998, **9** (20), 3517–3599.

[182] Cativiela, C. and Díaz-de-Villegas, M. D., Stereoselective synthesis of quaternary α-amino acids. Part 2: Cyclic compounds. *Tetrahedron: Asymmetry* 2000, **11** (3), 645–732.

[183] Vogt, H. and Brase, S., Recent approaches towards the asymmetric synthesis of α,α-disubstituted α-amino acids. *Org. Biomol. Chem.* 2007, **5** (3), 406–430.

[184] Avenoza, A., Cativiela, C., Corzana, F., Peregrina, J. M., and Zurbano, M. M., Asymmetric synthesis of all isomers of α-methyl-β-phenylserine. *Tetrahedron: Asymmetry* 2000, **11** (10), 2195–2204.

[185] Tanaka, M., Design and synthesis of chiral α,α-disubstituted amino acids and conformational study of their oligopeptides. *Chem. Pharm. Bull.* 2007, **55** (3), 349–358.

[186] Ohfune, Y. and Shinada, T., Enantio- and diastereoselective construction of α,α-disubstituted α-amino acids for the synthesis of biologically active compounds. *Eur. J. Org. Chem.* 2005, **2005** (24), 5127–5143.

[187] Rao, A. V. R., Gurjar, M. K., Devi, T. R., and Kumar, K. R., A formal synthesis of a novel immunosuppressant ISP-1: Stereocontrolled Pd(O) catalysed cis-hydroxyamination of carbohydrate derived vinyl epoxide. *Tetrahedron Lett.* 1993, **34** (10), 1653–1656.

[188] Papa, C. and Tomasini, C., Synthesis and ring opening of methyl 2-alkyl-3-(alkyl/aryl)-1-benzoylaziridine-2-carboxylates: Synthesis of polysubstituted amino acids. *Eur. J. Org. Chem.* 2000, **2000** (8), 1569–1576.

[189] Pines, S. H., Karady, S., and Sletzinger, M. Antihypertensive L-isomers of α-alkyl-β-phenylserines and alanines. US3517057A, 1970, Merck and Co., Inc.

[190] Nozaki, H., Kuroda, S., Watanabe, K., and Yokozeki, K., Purification and gene cloning of α-methylserine aldolase from *Ralstonia* sp. strain AJ110405 and application of the enzyme in the synthesis of α-methyl-L-serine. *Appl. Environ. Microbiol.* 2008, **74** (24), 7596–7599.

[191] Nozaki, H., Kuroda, S., Watanabe, K., and Yokozeki, K., Gene cloning of α-methylserine aldolase from *Variovorax paradoxus* and purification and characterization of the recombinant enzyme. *Biosci., Biotechnol., Biochem.* 2008, **72** (10), 2580–2588.

[192] Fesko, K., Uhl, M., Steinreiber, J., Gruber, K., and Griengl, H., Biocatalytic access to α,α-dialkyl-α-amino acids by a mechanism-based approach. *Angew. Chem. Int. Ed.* 2010, **49** (1), 121–124.

[193] Baik, S.-H. and Yoshioka, H., Enhanced synthesis of L-threo-3,4-dihydroxyphenylserine by high-density whole-cell biocatalyst of recombinant L-threonine aldolase from *Streptomyces avermitilis*. *Biotechnol. Lett.* 2009, **31** (3), 443–448.

[194] Steinreiber, J., Schürmann, M., Wolberg, M., van Assema, F., Reisinger, C., Fesko, K., Mink, D., and Griengl, H., Overcoming thermodynamic and kinetic limitations of aldolase-catalyzed reactions by applying multienzymatic dynamic kinetic asymmetric transformations. *Angew. Chem. Int. Ed.* 2007, **46** (10), 1624–1626.

[195] Wong, C.-H., Garcia-Junceda, E., Chen, L., Blanco, O., Gijsen, H. J. M., and Steensma, D. H., Recombinant 2-deoxyribose-5-phosphate aldolase in organic synthesis: Use of sequential two-substrate and three-substrate aldol reactions. *J. Am. Chem. Soc.* 1995, **117** (12), 3333–3339.

[196] Wolberg, M., Dassen, B. H. N., Schürmann, M., Jennewein, S., Wubbolts, M. G., Schoemaker, H. E., and Mink, D., Large-scale synthesis of new pyranoid building blocks based on aldolase-catalysed carbon–carbon bond formation. *Adv. Synth. Catal.* 2008, **350** (11–12), 1751–1759.

[197] Muller, M., Chemoenzymatic synthesis of building blocks for statin side chains. *Angew. Chem. Int. Ed.* 2005, **44** (3), 362–365.

[198] Jennewein, S., Schürmann, M., Wolberg, M., Hilker, I., Luiten, R., Wubbolts, M., and Mink, D., Directed evolution of an industrial biocatalyst: 2-Deoxy-D-ribose 5-phosphate aldolase. *Biotechnol. J.* 2006, **1** (5), 537–548.

[199] Sakuraba, H., Yoneda, K., Yoshihara, K., Satoh, K., Kawakami, R., Uto, Y., Tsuge, H., Takahashi, K., Hori, H., and Ohshima, T., Sequential aldol condensation catalyzed by hyperthermophilic 2-deoxy-D-ribose-5-phosphate aldolase. *Appl. Environ. Microbiol.* 2007, **73** (22), 7427–7434.

[200] Zandvoort, E., Baas, B.-J., Quax, W. J., and Poelarends, G. J., Systematic screening for catalytic promiscuity in 4-oxalocrotonate tautomerase: Enamine formation and aldolase activity. *ChemBioChem* 2011, **12** (4), 602–609.

[201] Zandvoort, E., Geertsema, E. M., Baas, B. J., Quax, W. J., and Poelarends, G. J., Bridging between organocatalysis and biocatalysis: Asymmetric addition of acetaldehyde to ss-nitrostyrenes catalyzed by a promiscuous proline-based tautomerase. *Angew. Chem. Int. Ed.* 2012, **51** (5), 1240–1243.

[202] Geertsema, E. M., Miao, Y., Tepper, P. G., de Haan, P., Zandvoort, E., and Poelarends, G. J., Biocatalytic michael-type additions of acetaldehyde to nitroolefins with the proline-based enzyme 4-oxalocrotonate tautomerase yielding enantioenriched γ-nitroaldehydes. *Chem. Eur. J.* 2013, **19** (43), 14407–14410.

[203] Branneby, C., Calrlqvist, P., Magnusson, A., Hult, K., Brink, T., and Berglund, P., Carbon–carbon bonds by hydrolytic enzymes. *J. Am. Chem. Soc.* 2003, **125**, 874–875.

[204] Branneby, C., Carlqvist, P., Hult, K., Brinck, T., and Berglund, P., Aldol additions with mutant lipase: Analysis by experiments and theoretical calculations. *J. Mol. Catal. B: Enzym.* 2004, **31** (4–6), 123–128.

[205] Svedendahl, M., Hult, K., and Berglund, P., Fast carbon–carbon bond formation by a promiscuous lipase. *J. Am. Chem. Soc.* 2005, **127** (51), 17988–17989.

[206] Svedendahl, M., Jovanovic, B., Fransson, L., and Berglund, P., Suppressed native hydrolytic activity of a lipase to reveal promiscuous Michael addition activity in water. *ChemCatChem* 2009, **1** (2), 252–258.

[207] Li, H.-H., He, Y.-H., and Guan, Z., Protease-catalyzed direct aldol reaction. *Catal. Commun.* 2011, **12** (7), 580–582.

[208] Li, H.-H., He, Y.-H., Yuan, Y., and Guan, Z., Nuclease p1: A new biocatalyst for direct asymmetric aldol reaction under solvent-free conditions. *Green Chem.* 2011, **13** (1), 185–189.

[209] Dahiyat, B. I. and Mayo, S. L., De novo protein design: Fully automated sequence selection. *Science* 1997, **278** (5335), 82–87.

[210] Boas, F. E. and Harbury, P. B., Potential energy functions for protein design. *Curr. Opin. Struct. Biol.* 2007, **17** (2), 199–204.

[211] Rothlisberger, D., Khersonsky, O., Wollacott, A. M., Jiang, L., DeChancie, J., Betker, J., Gallaher, J. L., Althoff, E. A., Zanghellini, A., Dym, O., Albeck, S., Houk, K. N., Tawfik, D. S., and Baker, D., Kemp elimination catalysts by computational enzyme design. *Nature* 2008, **453** (7192), 190–195.

[212] Suárez, M. and Jaramillo, A., Challenges in the computational design of proteins. *J. R. Soc. Interface* 2009, **6** (Suppl 4), S477–S491.

[213] Althoff, E. A., Wang, L., Jiang, L., Giger, L., Lassila, J. K., Wang, Z., Smith, M., Hari, S., Kast, P., Herschlag, D., Hilvert, D., and Baker, D., Robust design and optimization of retroaldol enzymes. *Protein Sci.* 2012, **21** (5), 717–726.

[214] Giger, L., Caner, S., Obexer, R., Kast, P., Baker, D., Ban, N., and Hilvert, D., Evolution of a designed retro-aldolase leads to complete active site remodeling. *Nat. Chem. Biol.* 2013, **9** (8), 494–498.

[215] Lassila, J. K., Baker, D., and Herschlag, D., Origins of catalysis by computationally designed retroaldolase enzymes. *Proc. Natl. Acad. Sci. U. S. A.* 2010, **107** (11), 4937–4942.

[216] Obexer, R., Studer, S., Giger, L., Pinkas, D. M., Grutter, M. G., Baker, D., and Hilvert, D., Active site plasticity of a computationally designed RetroAldolase enzyme. *ChemCatChem* 2014, **6** (4), 1043–1050.

[217] Bjelic, S., Kipnis, Y., Wang, L., Pianowski, Z., Vorobiev, S., Su, M., Seetharaman, J., Xiao, R., Kornhaber, G., Hunt, J. F., Tong, L., Hilvert, D., and Baker, D., Exploration of alternate catalytic mechanisms and optimization strategies for retroaldolase design. *J. Mol. Biol.* 2014, **426** (1), 256–271.

CHAPTER 11

Enzymatic Asymmetric Reduction of Carbonyl Compounds

Tomoko Matsuda[1], Rio Yamanaka[2], and Kaoru Nakamura[3]

[1] Department of Bioengineering, Tokyo Institute of Technology, Yokohama, Japan

[2] Faculty of Pharmaceutical Science, Himeji Dokkyo University, Himeji, Japan

[3] Science Shop, Kobe University, Kobe, Japan

11.1 INTRODUCTION

The development of effective asymmetric synthetic methods using chemical and biological catalysts is essential for drug synthesis as more than half of drug candidate molecules have more than one chiral center [1]. Now, about 10% of the total drug synthesis depends on the biocatalysts. A majority of reports use hydrolysis/esterification due to the high stability and ease of handling of hydrolases, and asymmetric reduction has accounted for the second largest portion. It will be expected to increase as the number of commercially available and easy-to-handle oxidoreductases has been increasing. In this chapter, biocatalytic asymmetric synthesis through reduction of carbonyl compounds and dynamic kinetic resolution is reviewed.

11.2 MECHANISMS

Dehydrogenases and reductases require a coenzyme such as nicotinamide adenine dinucleotide (NADH), nicotinamide adenine dinucleotide phosphate (NADPH), flavin, etc. The reduction of a carbonyl compound with NAD(P)H with auxiliary substance proceeds as follows (Figure 11.1a).

1. Reduced form of coenzyme (NAD(P)H) and substrate (carbonyl compound) bind to an enzyme.
2. The substrate carbonyl group is reduced to product hydroxyl group, while the coenzyme is oxidized to $NAD(P)^+$.
3. The oxidized coenzyme ($NAD(P)^+$) and the product dissociate from the enzyme.
4. An oxidized coenzyme ($NAD(P)^+$) and auxiliary substance (hydroxyl compound) bind to an enzyme.
5. The auxiliary substrate hydroxyl group is oxidize to carbonyl group, while the coenzyme is reduced to NAD(P)H.
6. The reduced coenzyme (NAD(P)H) and auxiliary product dissociate from the enzyme.

Green Biocatalysis, First Edition. Edited by Ramesh N. Patel.

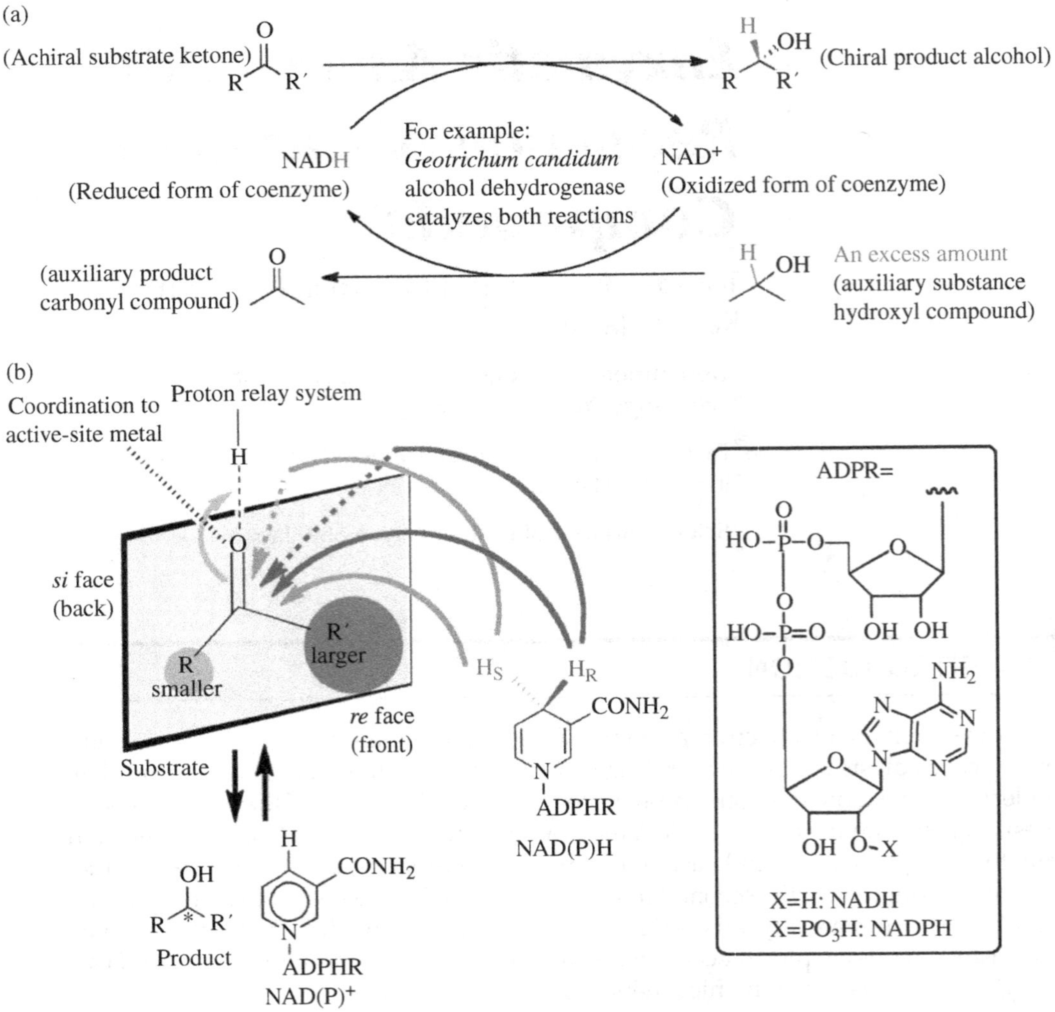

FIGURE 11.1

(a) Reduction of substrate and recycling of NADH using alcohol and (b) hydride transfer from coenzyme, NAD(P)H, to carbonyl compound and proton relay system.

Auxiliary substrate, hydrogen sources for the coenzyme recycling, can be alcohols such as ethanol and 2-propanol, sugars such as glucose, glucose-6-phosphate(G6P), and glucose-6-sulfate; formic acid; amino acids such as glutamic acid, and dihydrogen. Power sources such as electric and light energy can also be used. For example, 2-propanol was used to reduce a ketone to a chiral alcohol by an alcohol dehydrogenase from *Geotrichum candidum* with NAD^+. An excess amount of 2-propanol is used to push the equilibrium to the formation of product alcohols.

Regarding the stereochemistry for NAD(P)H-dependent dehydrogenase, there are four patterns for transfer of a hydride from coenzyme C4 carbon, NAD(P)H, to substrate ketone as shown in Figure 11.1b. The hydride attacks either the *si*-face or *re*-face of the carbonyl group depending of the orientation of the binding of the substrate to the enzyme, which results in the formation of (*R*)- and (*S*)-alcohols, respectively. On the other hand, enzyme transfers either pro-(*R*)-hydride or pro-(*S*)-hydride of the coenzyme depending on the kind of enzyme.

To complete the reduction, a proton is transferred to the carbonyl oxygen through the proton relay system, transferring a proton from outside water. For example, in the horse liver alcohol dehydrogenase-catalyzed reduction, the proton comes from hydroxyl group of Ser48 hydrogen-bonded to 2-hydroxyl group of the NADH ribose, which, in turn, is also hydrogen-bonded to His51 [2]. The role

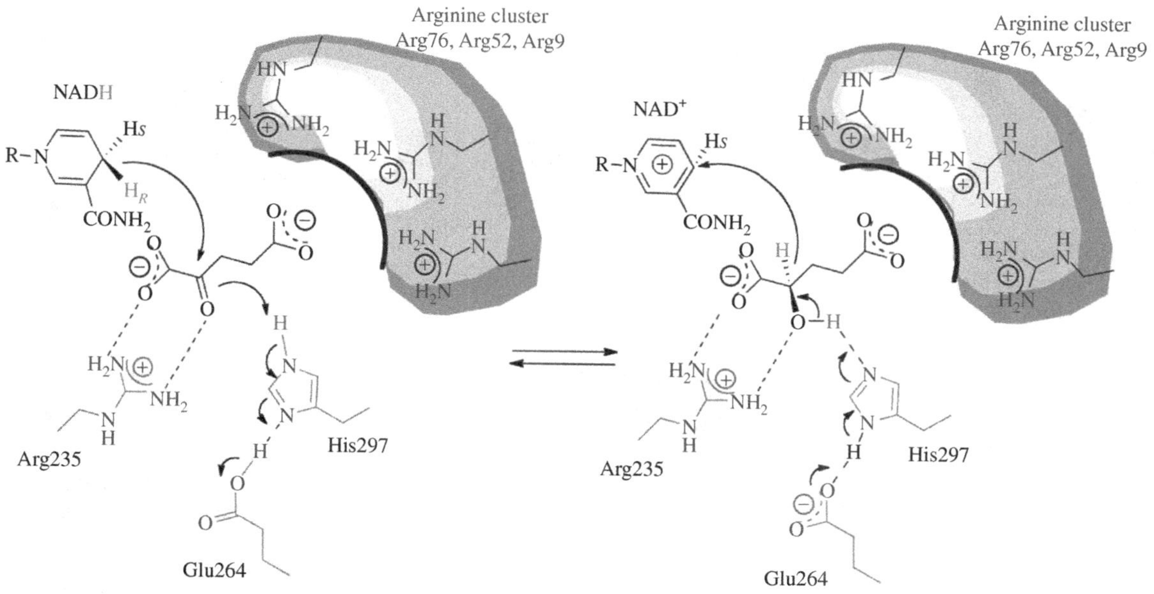

FIGURE 11.2

Transfer of a hydride between substrate and coenzyme and proton relay in active site of an NADH-dependent D-2-hydroxy acid dehydrogenase.

of the active-site zinc is also very important. It is required to coordinate to the carbonyl carbon for catalysis to occur.

Example of the transfer of the pro-(*R*)-hydride of NADH to *si*-face of the substrate is shown using crystal structure of an NADH-dependent enzyme, (*R*)-2-hydroxyglutarate dehydrogenase (HGDH) [3] (Figure 11.2). When HGDH catalyzes the reduction of 2-oxoglutarate, Arg235 interacts with the substrate's α-carboxylate and carbonyl groups, having a dual role in both substrate binding and activation. The γ-carboxylate group can dock at an arginine cluster composed of Arg76, Arg52, and Arg9. The proton relay system built up by Glu264 and His297 permits His297 to act as acid–base catalyst. Thus, the pro-(*R*)-hydrogen of NADH is transferred to the carbonyl group from the *si*-face leading to form (*R*)-2-hydroxyglutarate.

11.3 PREPARATION OF BIOCATALYSTS

Methods to find a unique and useful enzyme and to mutate enzymes are described in this section. Screening sources from which to search for new biocatalysts became drivers. Enzymes from culturable and nonculturable organisms can be used. Biocatalysts from autotrophs as well as heterotrophs are shown.

11.3.1 Screening of Enzymes from Culturable Microorganisms

Screening of enzymes expressed in microorganisms, which are found in the environment and cultivated in laboratory, has the merit of finding novel enzymes because the method does not use the sequence information of already-known enzymes. However, an enzyme cannot be found if the microorganism cannot be cultured under laboratory cultivation conditions, or the microorganism does not express the enzyme under certain laboratory cultivation conditions.

A novel screening method using multi-well plates has been reported [4]. Microbial cultures of about 300 microbes, demonstrating utility in reducing model ketones, were arrayed in multi-well plates, stored until use at −80 °C, and used to rapidly

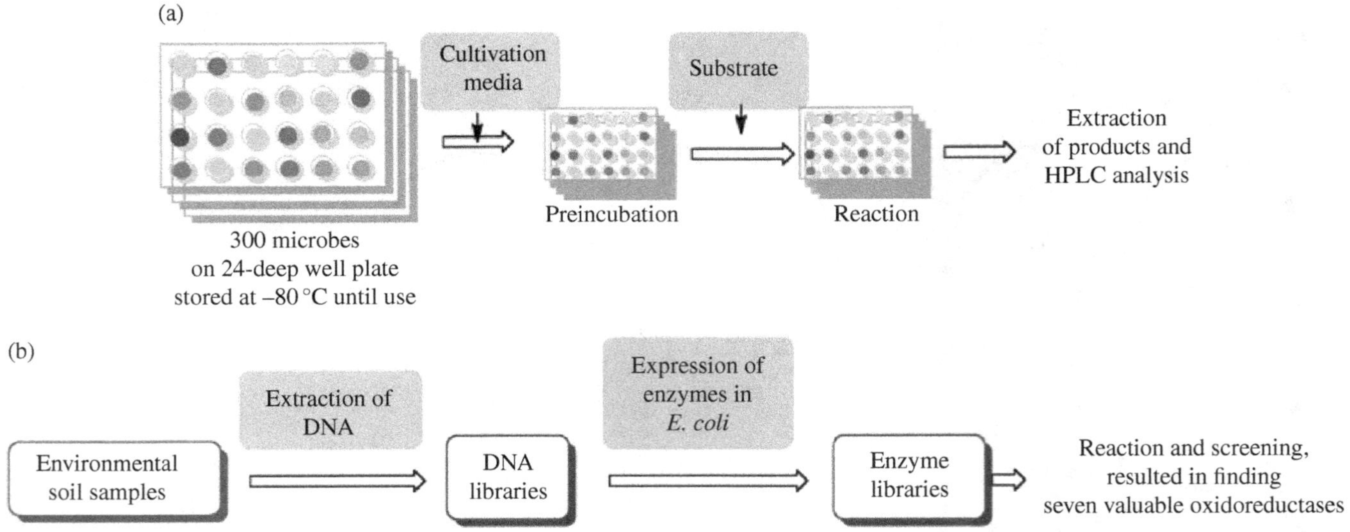

FIGURE 11.3
Screening of biocatalysts (a) from culturable microorganisms using multi-well plates (b) using metagenomes of complex microbial consortia derived from soils.

identify specific organisms capable of producing chiral alcohols used as intermediates for several drug candidates as shown in Figure 11.3a. Approximately 60 cultures were shown to reduce a series of alkyl aryl ketones selectively, providing both the *R* and *S* enantiomers of the corresponding alcohols in 92–99% ee with yields up to 95% at 1–4 g/l.

Another example of screening the culturable microorganism is to find catalysts to convert 2,2,2-trifluoroacetophenone to the corresponding (*S*)-alcohol in the presence of 2-propanol as a hydrogen donor [5]. By screening of styrene-assimilating bacteria (~900 strains) isolated from soil samples, *Leifsonia* sp. strain S749 was found to have an enzyme to reduce 2,2,2-trifluoroacetophenone to (*S*)-1-phenyltrifluoroethanol and also acetophenone to (*R*)-1-phenylethanol both in >99% and 99% ee, respectively.

11.3.2 Screening of Enzymes using Metagenomes

A metagenome, is a collection of DNA extracted directly from environmental samples, can be screened to find an enzyme. This screening method enables enzymes to be found in organisms that cannot be cultured and/or expressed in laboratory conditions. By using this method, oxidoreductase for the formation of carbonyls from short-chain polyols was found [6]. As shown in Figure 11.3b, metagenomic DNA libraries from three different soil samples (meadow, sugar beet field, and cropland) were constructed. These libraries, comprising approximately 167 000 independent clones (~4.05 Gbp of DNA), were screened for the production of carbonyls from short-chain polyols such as 1,2-ethanediol, 2,3-butanediol; a mixture of glycerol and 1,2-propanediol, and 7 oxidoreductase was found.

11.3.3 Screening of Enzymes of Microorganisms of Known Genome Data

The screening of an enzyme from genome database is useful to search for a catalyst for a desired reaction [7]. For example, the enantio- and diastereoselective reduction of ethyl 2-chloro-3-oxoalkanoates were investigated using a reductase library (18 known and putative enzymes) from baker's yeast expressed in *Escherichia coli*

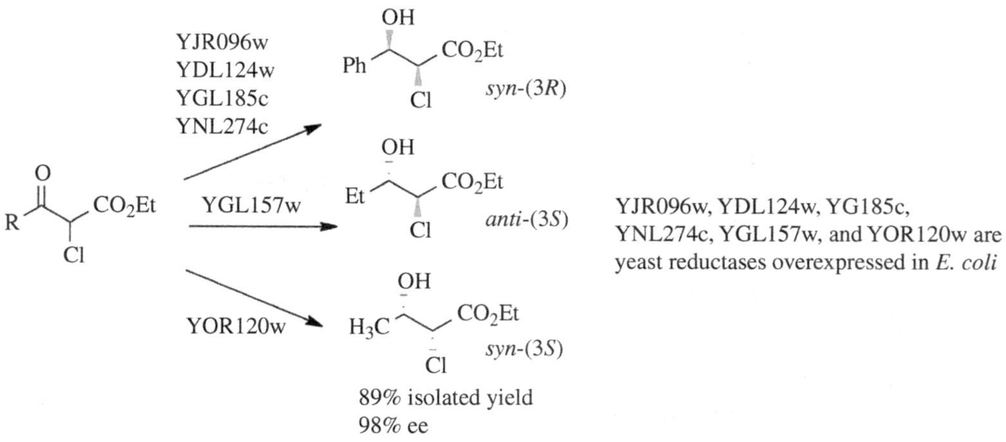

FIGURE 11.4 Screening of reductase library from baker's yeast expressed in *E. coli* for reduction of ethyl 2-chloro-3-oxoalkanoates.

(Figure 11.4). It was possible to produce at least two of the four possible 3-chloro-2-hydroxy ester diastereomers with high optical purities. Four enzymes (YJR096w, YDL124w, YGL185c, and YNL274c) reduced 2-chloro-3-phenylpropanoate to the corresponding *syn*-(3*R*)-hydroxy ester. 2-Chloro-3-oxopentanoate was reduced by YGL157w to the *anti*-(3*S*)-hydroxy ester selectively, and 2-chlorobutanoate was reduced by YOR120w to the *syn*-(3*S*)-product.

11.3.4 Mutation of Enzymes

Improvement of enzyme performance by mutation can be classified into two categories: (i) rational mutation such as point mutation and site-directed saturation mutagenesis (change of 1 amino acid to all 19 naturally occurring amino acids at a specific site within a protein) and (ii) random mutation such as directed evolution (change of amino acids at random to create a large library of mutants proteins followed by the selection of the proteins possessing the desired property; the processes of the random mutation and selection are repeated many times as in natural evolution). When the structure of the enzyme is not known or cannot be predicted, only random mutation can be performed.

Rational Mutation to Widen Substrate Specificity

Rational design in the modification of enzymes by point mutation has been reported [8]. The secondary alcohol dehydrogenase from *Thermoanaerobacter ethanolicus* 39E (*Te*SADH) is highly thermostable and solvent-stable, and it is active on a broad range of substrates. However, *Te*SADH has no detectable activity on (*S*)-1-phenyl-2-propanol, a precursor to major pharmaceuticals containing secondary alcohol groups, but it is highly active on 2-butanol. From the structural model research, tryptophan-110 obstructs the proper fitting of (*S*)-1-phenyl-2-propanol, while the residue does not interfere with the fitting of 2-butanol. Then, tryptophan-110 (W110) was replaced with alanine (A). The W110A mutant could use (*S*)-1-phenyl-2-propanol, (*S*)-4-phenyl-2-butanol, and the corresponding ketones, as substrates and produce (*S*)-4-phenyl-2-butanol from benzylacetone with >99% ee as shown in Figure 11.5a.

Random mutation followed by rational mutation was used to improve the catalytic efficiency of phenylacetaldehyde reductase (PAR). PAR from *Rhodococcus* sp. was mutated to improve the conversion efficiency in high concentrations of substrate and 2-propanol [9]. Here, 2-propanol acts as a solvent as well as a hydrogen donor in coupled cofactor regeneration. First, the PAR library was generated by mutagenic PCR. With only a single selection round of a PAR mutant library followed by a combination of advantageous mutations, PAR was successfully adapted with six amino acid replacements for the conversion of high concentrations of substrate in

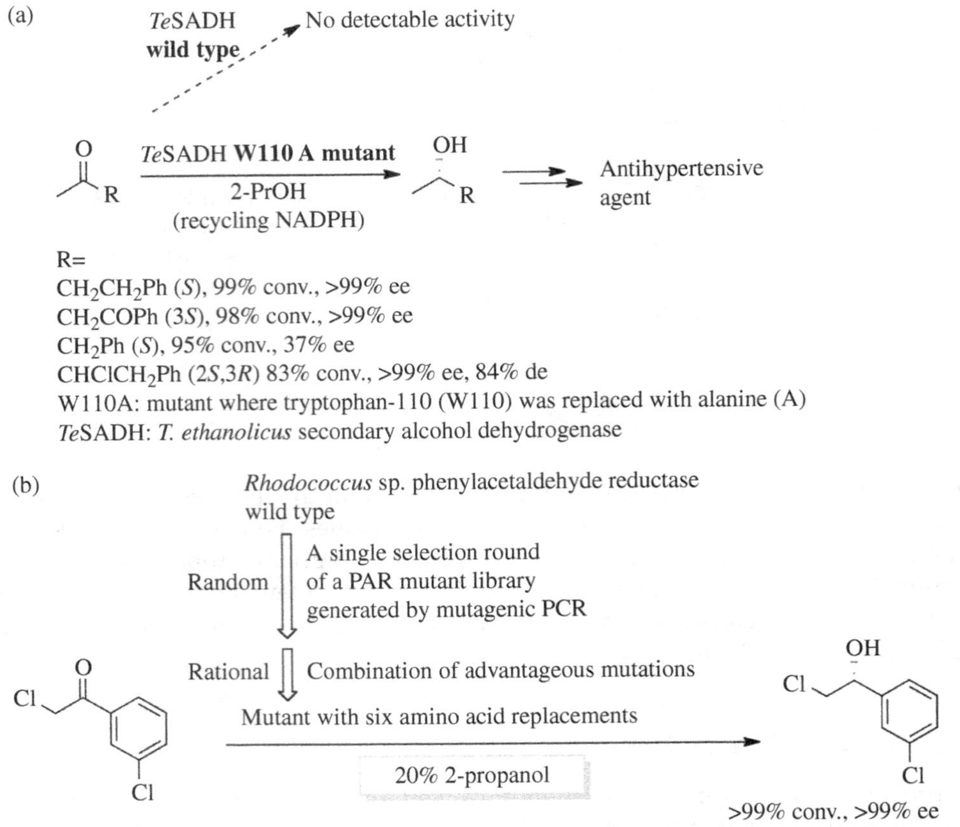

FIGURE 11.5
(a) Rational mutation of *Thermoanaerobacter ethanolicus* 39E secondary alcohol dehydrogenase to widen substrate specificity and (b) mutation to improve catalytic efficiency in concentrated 2-propanol and substrate.

concentrated 2-propanol. *m*-Chlorophenacylchloride was reduced by the mutant in 20% (V/V) 2-propanol to the corresponding (*R*)-alcohol in >99% conversion with >99% ee as shown in Figure 11.5b.

11.3.5 Hyperthermophilic Enzyme as a Biocatalyst

Thermophiles are an important source for biocatalyst due to the thermostability of the enzyme. Here is an interesting example where dehydrogenase shows a high resistance to thermal inactivation. Asymmetric reduction of simple aromatic ketones and keto esters has been investigated by an alcohol dehydrogenase from the hyperthermophilic archaeon *Pyrococcus furiosus* (PFADH) obtained by overexpression in *E. coli* [10]. The half-life at 100 °C was 130 min. The increase in reaction temperature raised the enzyme activity, but exerted no effect on the enantioselectivity. This enzyme also showed a high tolerance to organic solvents such as dimethyl sulfoxide (DMSO), *iso*-propanol, methyl *tert*-butyl ether, and hexane, which is a particularly important and useful feature for the reduction of ketones with a low solubility in aqueous buffers.

11.3.6 Photosynthetic Organism as a Biocatalyst "Photobiocatalyst"

Photosynthetic organisms play important roles in carbon dioxide absorption as well as in the utilization of solar energy on the earth. The additional value on the ecological aspect would make photosynthetic organisms the next generation of a biocatalyst from the standpoint of green chemistry [11–24]. The photosynthetic organism as a biocatalyst can be called "photobiocatalyst." Cyanobacteria, microalgae, plant cell cultures, and germinated plant sprouts have been used as photobiocatalysts so far,

TABLE 11.1 Reactions of Exogenous Substrates by Various "Photobiocatalysts"

Photobiocatalyst	Reaction	References
Cyanobacteria		
Synechococcus PCC7942	 >96% ee	[11–13]
	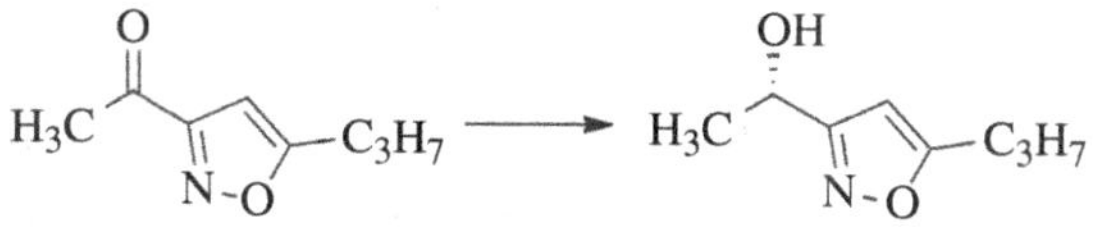	[14]
	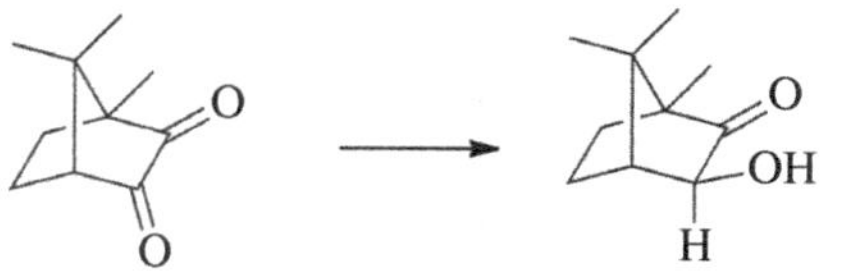	[15]
Synechocystis PCC 6803	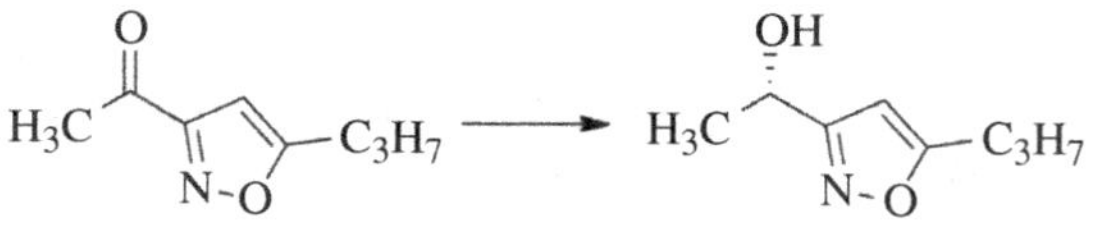	[14]
	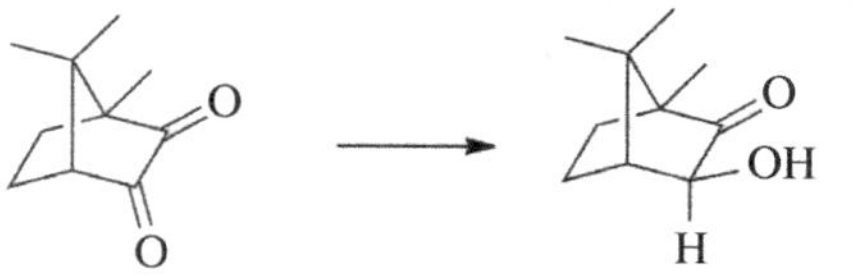	[15]
Microalgae		
Chlorella sp. MK201 (green algae)	 >96% ee	[16]
	 97% ee	[16]
Cyanidioschyzon merolae or *Cyanidium caldarium* (red algae)	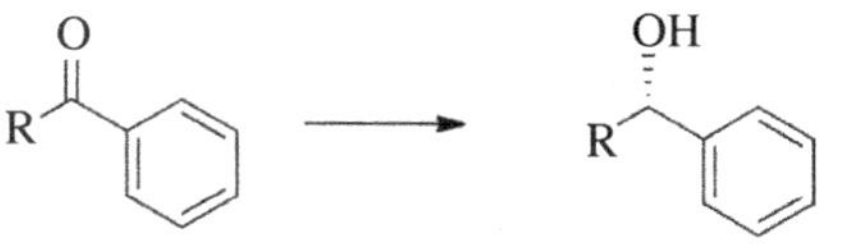 $R = C_nH_{2n+1}$ $n = 1\sim7$ >37~87% ee	[17]
Plants		
Nicotiana tabacum var. Samsun NN (cell cultures)	 98% ee	[18]
Arabidopsis thaliana (germinated sprout)	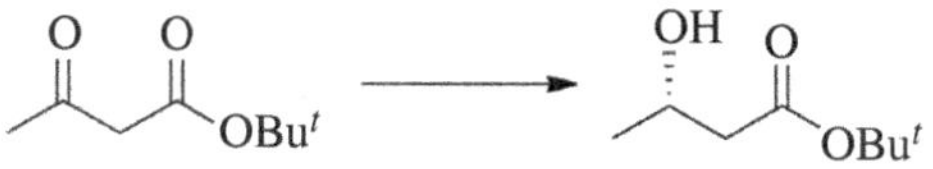 >99% ee	[19]
Raphanus sativus L. (germinated sprout)	 >95% ee	[20]

and there are some notable biotransformations as shown in Table 11.1. Photosynthetic microbes such as cyanobacteria (*Synechococcus* PCC7942 and *Synechocystis* PCC6803) and green algae (*Chlorella* sp. MK201) catalyzed reductions of acetophenone derivatives with high enantioselectivities. Primitive red algae (*Cyanidioschyzon merolae* and

Cyanidium caldarlum) reduced aromatic ketones. Plant cell cultures (*Nicotiana tabacum* var. Samsun NN cells) reduced *t*-butyl acetoacetate to the corresponding (*S*)-alcohol with high optical yield. Radish (*Raphanus sativus* L.) sprout reduced α, α, α-trifluoroacetophenone and *Arabidopsis* seedlings sprout reduced *t*-butyl acetoacetate to the corresponding chiral alcohol in eantiomeric excess.

The Reaction Mechanism in Photobiocatalytic Reaction (Photobiocatalysis)

In photobiocatalytic reactions, the photosynthetic process may be a key regulator. Photosynthetic organisms capture light energy to generate NADPH from $NADP^+$ through photosynthetic electron transfer reactions and CO_2 is converted into sugar, generally using NADPH. The reducing power of NADPH generated through photosynthesis can also be used in the reduction of exogenous substrates. Thus, photobiocatalysts natively provide the cofactor recycling system for the biocatalytic reduction. Accordingly, we can use solar energy directly for bioconversion of artificial substrates as shown in Figure 11.6 [12]. The mechanism of reduction of exogeneously added ketones with *Synechococcus* PCC7942 was investigated in detail using several inhibitors for photosynthetic electron transport system and enzymes in the Calvin cycle. Light intensity and several inhibitors of photosynthetic electron transport significantly affected the reduction of the ketones, which indicates that the reduction strictly depends on photosynthetic activity. Moreover, cofactor (reductant) in the reduction of the exogeneous ketones is NADPH generated photosynthetically, not NADH [21]. The ketoreductase that carries out the reduction of artificial ketones in *Synechococcus* PCC 7942 was identified as 3-ketoacyl-acyl carrier protein reductase and NADPH was used as the coenzyme [12, 21, 22].

The Control of Photobiocatalysis

The reaction mechanism in photobiocatalyses is different from the usual microbial reactions, so photobiocatalyses may need novel methods to control chemical and stereochemical yields. The important factors such as light and CO_2 for photosynthesis can be of great significance for photobiocatalyses. Until now, several methods have been reported as shown in Figure 11.7.

1. **Effect of light intensity (light quantity)**
   The stereochemical course of the asymmetric reduction of ketones by a photosynthetic microbe is largely regulated by light (Figure 11.7a). Thus, the poor enantioselectivities (14% ee) of the reduction of α, α-difluoroacetophenone by a cyanobacterium in the dark was improved by illumination. To confirm the effect of photosynthesis on enantioselectivity, 3-(3,4-dichlorophrnyl)-1, 1-dimethylurea (DCMU), a photosynthetic inhibitor was used to see the decrease of enantioselectivity even under illumination [13].
2. **Effect of light wavelength (light quality)**
   The light wavelength is associated with the efficiency of photosynthesis and can also have an impact on the photobiocatalysis. Asymmetric reduction of 2′, 3′, 4′,

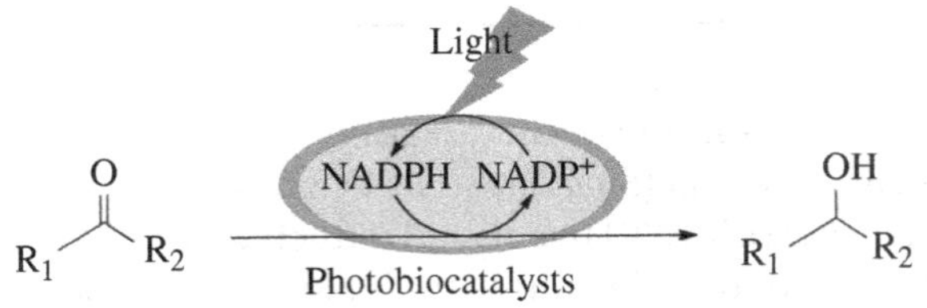

FIGURE 11.6
Reduction of ketones by photobiocatalysts with the aid of light energy.

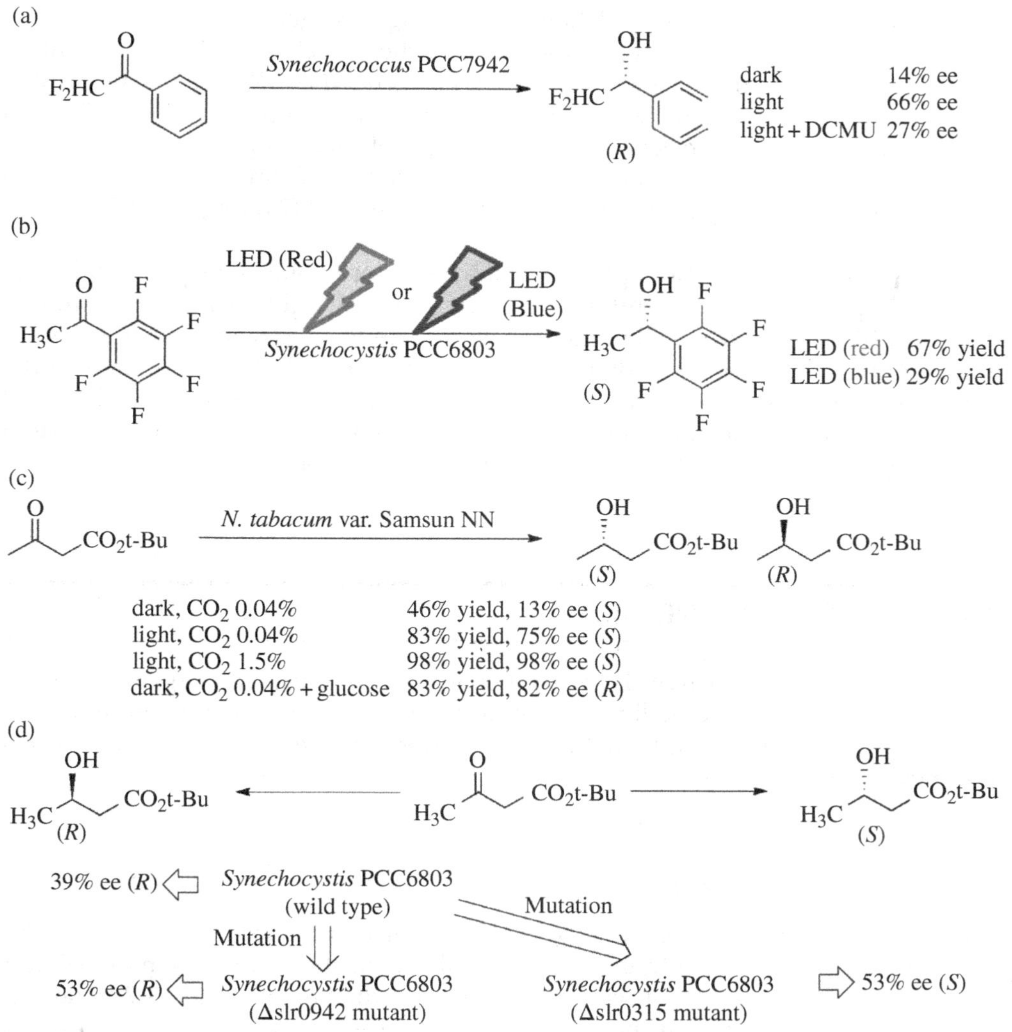

FIGURE 11.7

(a) Effect of light intensity on reduction of α,α-difluoroacetophenone by *Synechococcus* PCC7942; (b) effect of light wave length on reduction of 2′, 3′, 4′, 5′, 6′-pentafluoroacetophenone by *Synechocystis* PCC6803 under LED illumination (620/660 or 470 nm); (c) effect of CO_2 concentrations and light on reduction of *t*-butyl acetoacetate by *Nicotiana tabacum* var. Samsun NN cells; and (d) effect of mutation with deletion of dehydrogenase gene on photobiocatalysis.

5′, 6′-pentafluoroacetophenone by *Synechocystis* PCC6803 smoothly afforded the corresponding (*S*)-alcohol in an excellent optical yield (>99% ee) with the aid of illumination by orange (612 nm) and red (620/660 nm) LED lights that were more than twice as effective as other LEDs such as blue (470 nm) and green (535 nm) lights in the chemical yields (Figure 11.7b) [13].

3. Effect of CO_2 concentrations

Nicotiana tabacum var. Samsun NN cells as a photosynthetic biocatalyst reduced *t*-butyl acetoacetate to the corresponding (*S*)-alcohol with a low chemical yield (46%) and optical yield (13% ee) in the dark (Figure 11.7c). However, the reaction under illumination gave the corresponding (*S*)-alcohol with high chemical yield (83%) and optical yield (75% ee) in the natural atmospheric CO_2 concentrations (0.04%). Furthermore, the high atmospheric CO_2 concentration (1.5%) improved both the chemical yield (up to 98%) and the optical yield (up to 98%) in the reaction.

On the other hand, the antipode, (*R*)-alcohol was obtained in 83% yield with 82% ee when glucose was added to the reaction in the dark with natural atmospheric CO_2 concentrations (0.04%). This suggests that the enzymatic system giving the (*R*)-alcohol is different from the reduction system that affords the (*S*)-alcohol, and the latter system is directly associated with photosynthesis [18].

4. Effect of mutation with deletion of dehydrogenase gene
   The reduction of *t*-butyl acetoacetate by the *Synechocystis* PCC6803 mutant (Δslr0942) with the deletion of slr0942 gene (NADP$^+$-dependent alcohol dehydrogenase) afforded (*R*)-alcohol with high optical yield (53% ee) compared to the reaction with wild type of *Synechocystis* PCC6803 under illumination. On the other hand, the deletion of slr0315 gene (probable oxidoreductase) gave the corresponding (*S*)-alcohol [24] (Figure 11.7d).

As shown in this section, photobiocatalysis affords high selectivity and wide substrate specificity. Photosynthetic organisms can use light energy for the reduction of artificial substrates, the biocatalysis proceeds effectively, and the ratio of the amount of biocatalyst to that of substrate is extremely low compared to the other biocatalytic reactions. Microalgae and plant sprouts can be easily manipulated and have high growth rates. Photosynthetic biocatalysis is an ecological system since photosynthetic organisms absorb sunlight and carbon dioxide and generates oxygen.

11.4 SOLVENT ENGINEERING

Oxidoreductases are active in nonaqueous solvent such as organic solvent, supercritical CO_2, and ionic liquids although immobilization may be necessary to stabilize the enzyme. Examples of the use of oxidoreductases in nonaqueous solvents are shown in this section.

11.4.1 Organic Solvent

Water-miscible and -immiscible organic solvents have been used for the reaction by oxidoreductase. For example, sol-gel-encapsulated alcohol dehydrogenase (W110A mutant of *T. ethanolicus*) was used for asymmetric reduction of 4-phenyl-2-butanone to (*S*)-4-phenyl-2-butanol in hexane [25]. Glycerol is also used for the reduction because it is a nontoxic, biodegradable and recyclable liquid manufactured as a by-product of transesterfication of a triglyceride in the production of natural fatty acid derivatives and bio diesel. The use of glycerol allowed easy separation of the product by simple extraction with diethyl ether. Free and immobilized baker's yeast was successfully employed in the asymmetric reduction of β-keto esters and ketones in glycerol [26]. The merit of using fluorous media is a simple work-up procedure. The reduction of ethyl 2-oxocyclopentanecarboxylate with immobilized baker's yeast (alginate) in perfluorooctane was investigated [27]. The reaction proceeded smoothly and gave the corresponding *syn*-(1*S*,2*R*)-hydroxyester in 98% de with 99% ee.

11.4.2 CO_2

CO_2, an abundantly existing and nontoxic solvent, can be a solvent for biocatalysis instead of organic solvents as the concentration of atmospheric CO_2 has increased since the beginning of the age of industrialization. Figure 11.8 shows examples using supercritical CO_2 for the reduction by alcohol dehydrogenase, immobilized *G. candidum*; high activities and excellent enantioselectivities were observed for the asymmetric reduction of aromatic and cyclic ketones as shown in Figure 11.8a [28a–c] The reaction by horse liver alcohol dehydrogenase (HLADH) using fluorinated NADH is also reported, as shown in Figure 11.8b [28d].

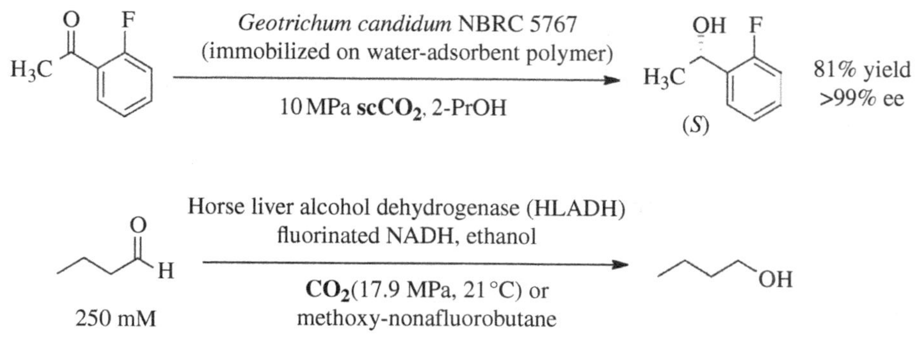

FIGURE 11.8
Biocatalytic reduction in supercritical CO_2.

11.4.3 Ionic Liquid

Ionic liquids, recyclable green solvents with no vapor pressure, have been used for oxidoreductase-catalyzed reactions such as the asymmetric reduction by *G. candidum* in hydrophobic ionic liquid ([bmin]PF6) [29a]. When the cell was immobilized on a water-absorbing polymer containing water, the reaction proceeded smoothly with excellent enantioselectivity, while the reaction without the immobilization on polymer did not proceed. For the reduction of 4-chloroacetophenone by *Lactobacillus kefir* to decrease substrate toxicity toward microbes, the biphsic ionic liquid-water system was investigated [29b]. At 600 mM of substrate concentration, the substrate was highly toxic toward the cells, and the reduction did not proceed in the aqueous system. The addition of 20% BMIM[Tf2N] enabled the reduction to proceed smoothly, giving the (*R*)-alcohol in 92.8% yield with 99.7%. The W110A TeADH catalyzed reaction also takes advantage of the use of ionic liquids [29(c)].

11.5 EXAMPLES FOR BIOCATALYTIC ASYMMETRIC REDUCTIONS

Examples for biocatalytic reduction of carbonyl compounds are described in Refs. [30–99].

11.5.1 Reduction of Ketones

Various ketones have been reduced; the list of the products by the reduction of ketones to (*R*)-alcohols or (*S*)-alcohols and ketoseters to (*R*)-hydroxy esters or (*S*)-hydroxy esters is given in Figure 11.9. Some examples for the synthesis of bioactive compounds are described in detail, as shown in Figure 11.10.

Synthesis of an Intermediate for Arachidonic Acid Metabolites

Useful intermediates for the synthesis of a variety of arachidonic acid metabolites were prepared (Figure 11.10a). (*R*)-5-Hydroxyhept-6-enonoate was obtained from the reduction of ethyl 5-oxo-6-heptenoate by *Thermoanaerobacter* sp. alcohol dehydrogenase (ADH) expressed in *E. coli* [56]. The opposite enantiomer, (*S*)-5-hydroxyhept-6-enonoate, was obtained by using *Lactobacillus brevis* ADH. These chiral alcohols are important intermediates for prostaglandins, leukotrienes, isoprostanes, and atractyligenin.

Synthesis of an Intermediate for Antagonists

Asymmetric reduction of 1-[3,5-bis(trifluoromethyl)phenyl]ethanone was investigated using various enzymes because optically active 1-[3,5-bis(trifluoromethyl) phenyl] ethanols are intermediates for antagonists (Figure 11.10b). The reduction with *L. kefir* gave the corresponding (*R*)-alcohol in >99% ee, an intermediate for tachykinin NK1

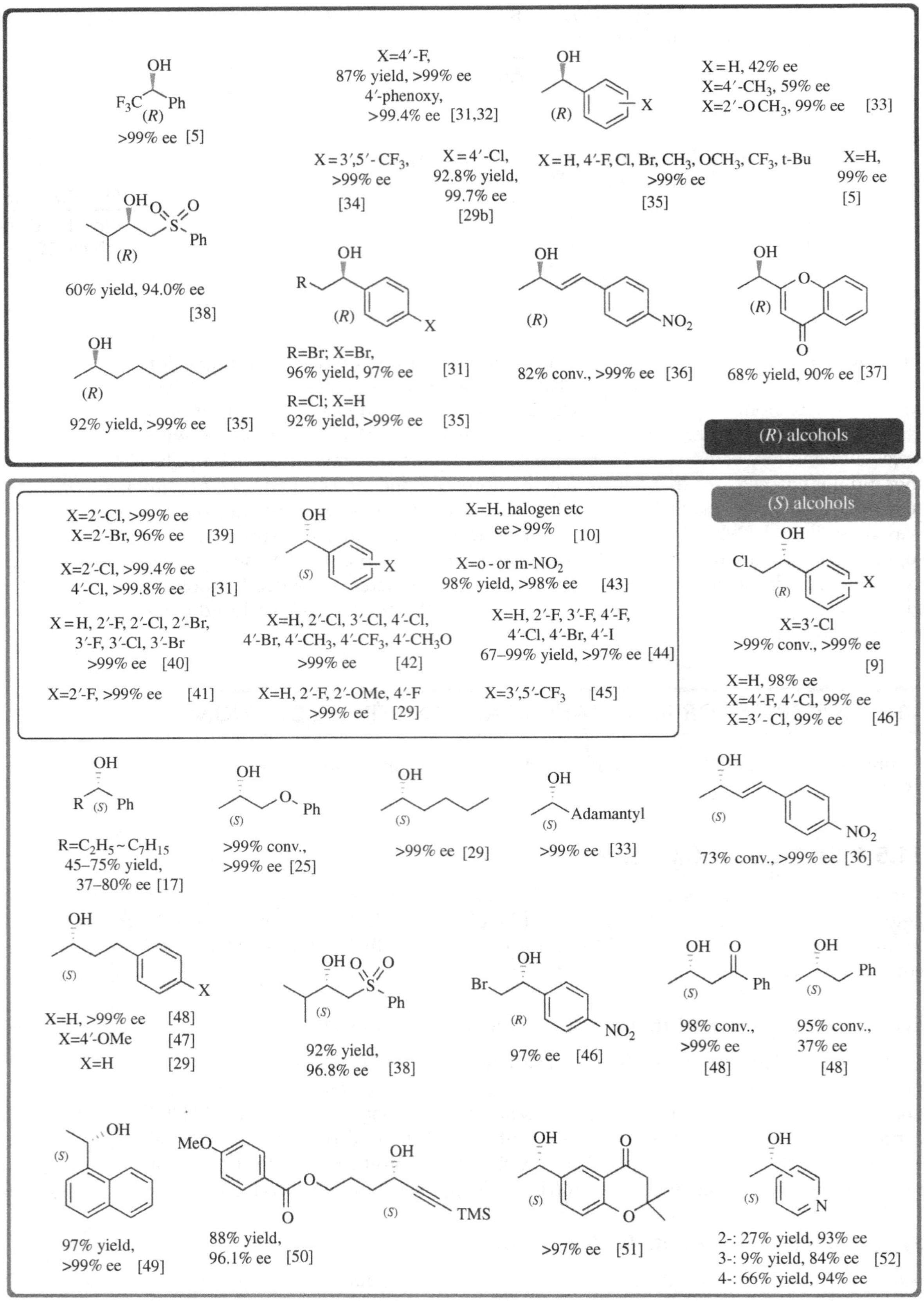

FIGURE 11.9

Products obtained by reduction of ketones and keto esters.

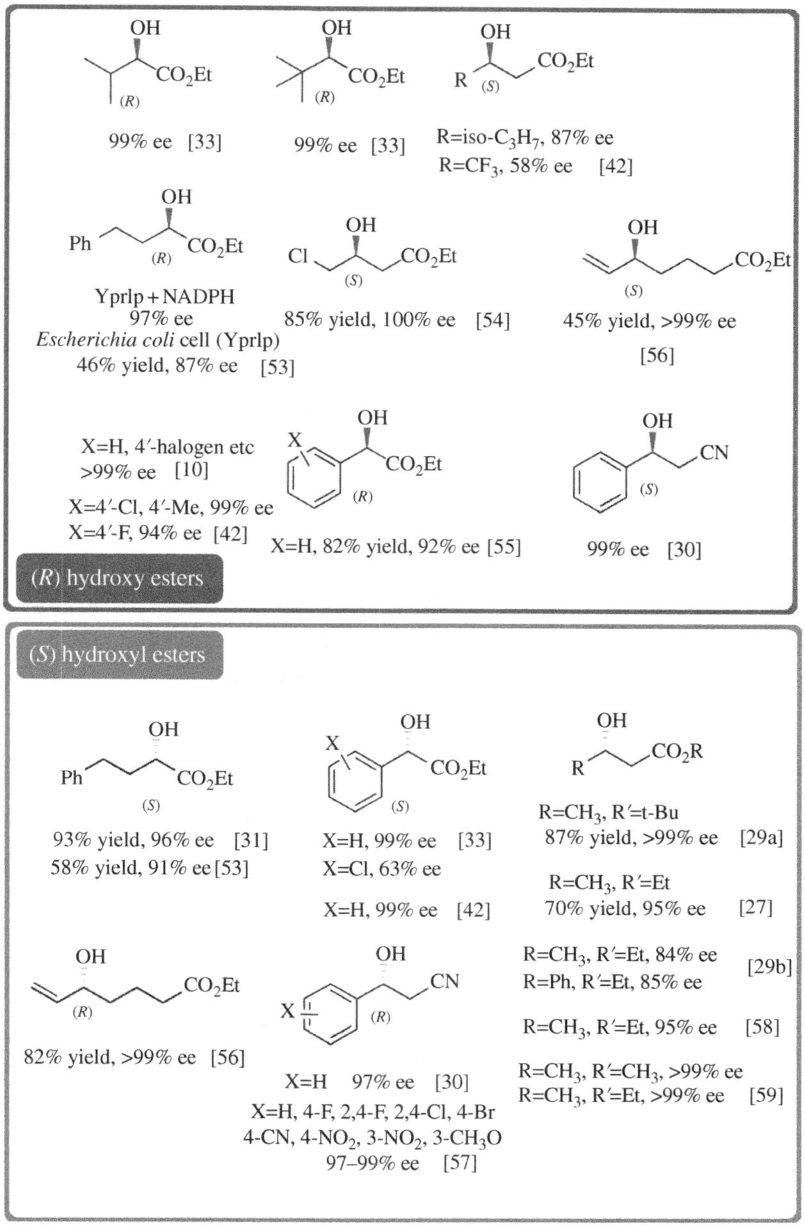

FIGURE 11.9 *(Continued)*

receptor [34]. On the other hand, the reduction by ADH from *Rhodococcus erythropolis* gave the corresponding (S)-alcohol, an intermediate for antagonists currently under clinical evaluation [45]. A pilot-scale reaction in pH6.5 phosphate buffer gave the product (S)-alcohol in 96% yield with 99% ee. The substrate concentrations could be increased to 580 mM with a space–time yield of 260 g/l d.

Synthesis of an Intermediate for Antibacterial Quinolone

For the synthesis of new-generation antibacterial quinolone carboxylic acids, reduction of azaspiroketones by *Phaeocrepsis* sp. JCM 1880 afforded the corresponding (R)-alcohols in high ee as shown in Figure 11.10c [60, 61].

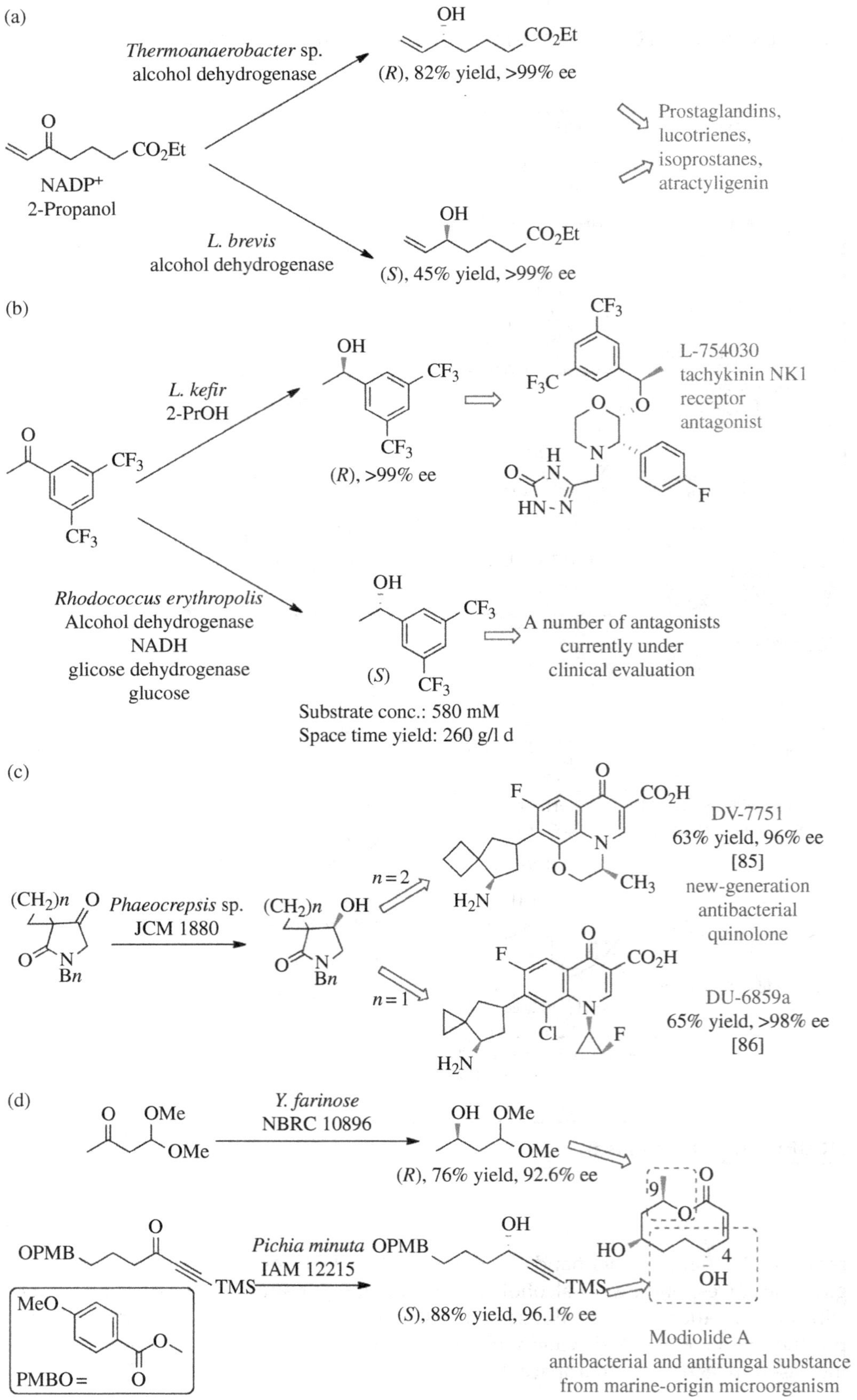

FIGURE 11.10

Synthesis of (a) an intermediate for arachidonic acid metabolites by reduction of ketones by enzymes from *Thermoanaerobacter* sp. and *Lactobacillus brevis*; (b) an intermediates for NK-1 receptor antagonists by *Lactobacillus kefir* and antagonists currently under clinical evaluation by *Rhodococcus erythropolis;* (c) intermediates for antibacterial quinolone by *Phaeocrepsis* sp. JCM 1880; (d) modiolide A using *Pichia minuta* IAM 12215 and *Yamadazyma farinose* NBRC 10896;

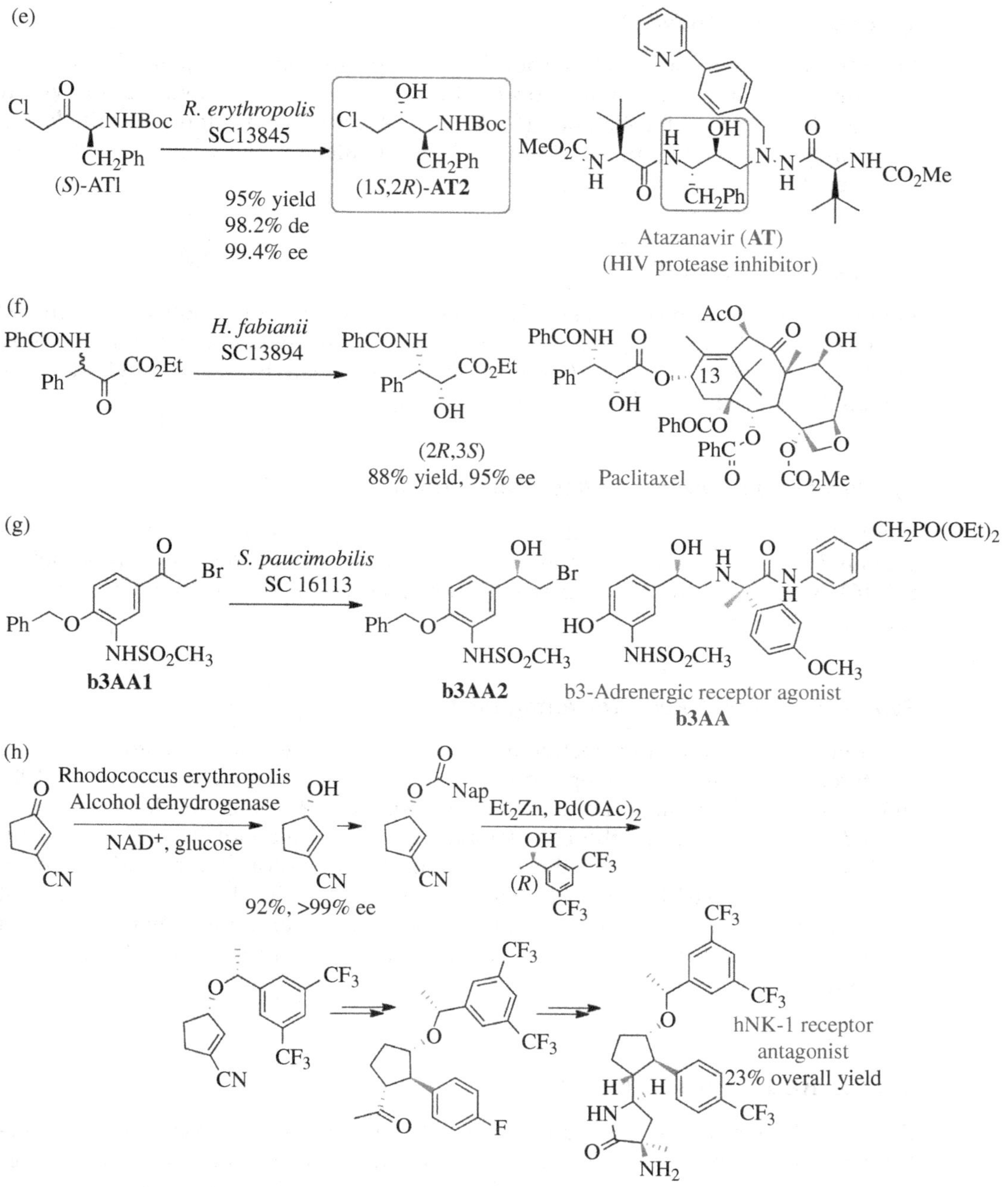

FIGURE 11.10 (*Continued*)

(e) main flame of Atazanavir by reduction with *R. erythropolis* SC 13845; (f) paclitaxel side chain using reduction of aminoketoester with *Hansenula fabianii* SC 13894; (g) intermediate for β3-Adrenergic receptor agonist by the reduction with *Sphingomonas paucimobilis* SC 16113; and (h) an hNK-1 receptor antagonist by the reduction of 3-cyanocyclopentenone with alcohol dehydrogenase from *R. erythropolis*.

Total Synthesis of Modiolide A

Modiolide A, a ten-membered ring lactone from a marine origin with antibacterial and antifungal activities, was synthesized using two microbial reductions as shown in Figure 11.10d. An important chiral building block for constructing the chirality at C-4, (*S*)-6-[(4-methoxybenzyl)oxy]-1-trimethylsilyl-1-hexyn-3-ol, was obtained by asymmetric reduction of the corresponding ketone with *Pichia minuta* IAM 12215 in 96.1% ee. Another chiral center, C-9, was synthesized from the (*R*)-acetal alcohol obtained by asymmetric reduction of the corresponding acetal ketone with *Yamadazyma farinose* NBRC 10896 in 76% yield with 92.6% ee [50].

Synthesis of Atazanavir (HIV Protease Inhibitor)

Main frame of Atazanavir (**AT**), an acyclic aza-peptidomimetic, potent HIV protease inhibitor was prepared using biocatalytic asymmetric reduction. As shown in Figure 11.10e (1*S*)-chloroketone ((*S*)-**AT1**) was reduced diastereoselectively by *R. erythropolis* SC 13845 to provide (1*S*,2*R*)-aminoalcohol (**AT2**) in 98.2% de with 99.4% ee [62].

Synthesis of Paclitaxel

Figure 11.10f shows the synthesis of chiral C-13 paclitaxel side chain, key precursor for the paclitaxel synthetic process by a preparative-scale reduction of aminoketoester by *Hansenula fabianii* SC 13894; (2*R*,3*S*)- *N*-benzoyl-3-phenylisoserine ethyl ester was obtained in 88% yield with 95% ee [63].

Synthesis of β 3-Adrenergic Receptor Agonist

A chiral intermediate for the synthesis of β3-Adrenergic receptor agonist (**β3AA**) was prepared by the reduction of bromoketone (**β3AA1**) with *Sphingomonas paucimobilis* SC 16113 to the corresponding (*R*)-alcohol (**β3AA2**) in 85% yield with 99.5% ee (Figure 11.10g) [64].

Synthesis of hNK-1 Receptor Antagonist

A convergent and enantioselective route to the hNK-1 receptor antagonist having six stereogenic centers in 11 steps and 23% overall yield was reported (Figure 11.10h) [65]. Asymmetric reduction of 3-cyanocyclopentenone by alcohol dehydrogenase from *R. erythropolis* gave the corresponding (*S*)-allyl alcohol in 92% yield with >99% ee. Pd-catalyzed allylic etherification of the corresponding naphthoate ester with 1-(*R*)-3′,5′-bistrifluoromethylphenylethanol gave the desired ether without epimerization. The total synthesis of the hNK1receptor antagonist was completed via the cuprate conjugate addition of 4-fluorophenyl group and alkylation of oxazolidinone.

11.5.2 Reduction of Diketones

Regio- and enantioselective reduction of diketones have been conducted successfully by biocatalysis. Examples for the reduction of diketones to hydroxyl ketones and diols are shown in Figure 11.11. Figure 11.11a shows the preparation of a key intermediate for the synthesis of terpenoids by a baker's yeast-catalyzed reduction of a σ-cyclohexanedione. DMSO (10%) was used to solubilize the substrate [67]. Figure 11.11b shows the reduction of 3,5-dioxo-6-(benzyloxy)hexanoic acid ethyl ester by *Acinetobacter* sp. SC 13874 to the corresponding *syn*-(3*R*,5*S*)-diol, potential intermediates for the synthesis of HMG-CoA reductase inhibitors, in 99.4% ee with 52–74% de depending on substrate concentrations (74% de in 2 g/l and 52% de in 10 g/l) [66]. After the reaction, XAD-16 resin was added to facilitate the recovery process by adsorbing the product.

11.5.3 Dynamic Kinetic Resolution Through Reduction

Dynamic kinetic resolution of racemic substrate proceeds through asymmetric reduction when the substrate racemizes and the product does not under the applied experimental conditions. For example, by dynamic kinetic resolution of

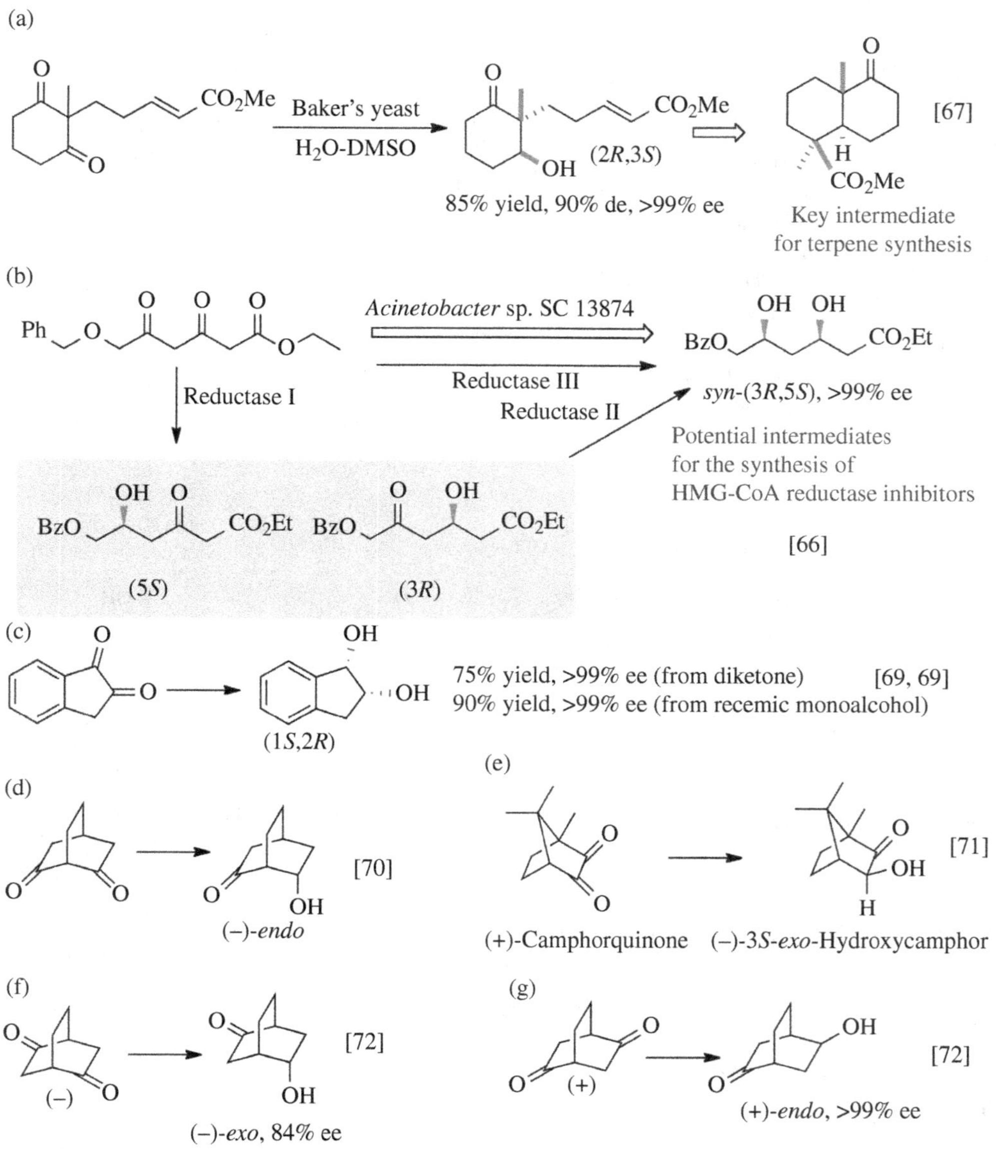

FIGURE 11.11
Reduction of diketones to hydroxyl ketone and diol.

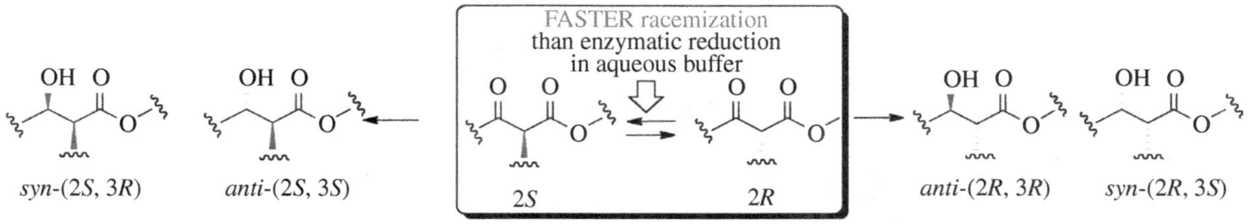

FIGURE 11.12
Possible products of dynamic kinetic resolution of α-methyl β-keto ester through reduction.

α-alkyl β-keto ester, one isomer, out of the four possible products for the unselective reduction, can be selectively synthesized (Figure 11.12). Various examples of biocatalytic kinetic resolution through reduction are summarized in Figure 11.13. Recent development in dynamic kinetic resolution has been reviewed [79].

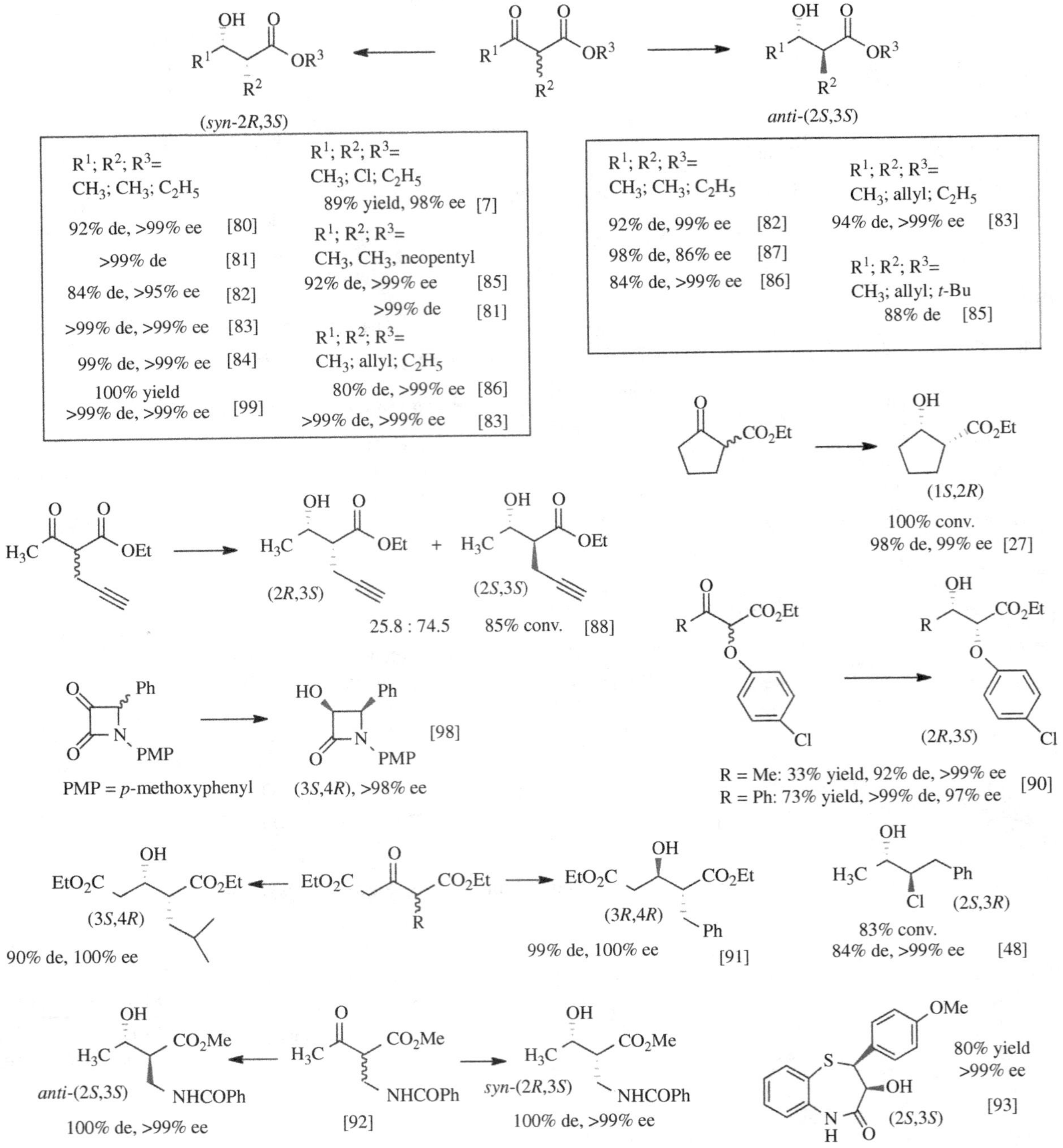

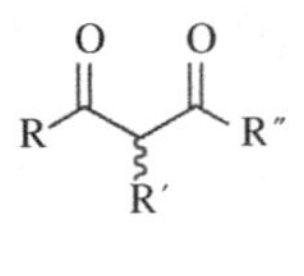	R	R′	R″	Product
	Me	Me	$CH_3CO_2Bu^t$	4*S*, 5*R* 66% yield, 94% de, 99.2% ee [94]
	Me	Me, Et, $-CH_2CHMe_2$, allyl, $-(CH_2)_2CH=CH_2$, $-CH_2CH=CMe_2$	Me	3*R*, 4*S* 100% yield, >99% de [95]
	Me	Et	Me	3*S*, 4*S* 100% yield, >98% de [95]
	Me	Pr	Me	3*S*, 4*S* 100% yield, 90% de [95]
	Me	allyl	Me	3*S*, 4*S* 100% yield, 98% de [95]
	Me	Et	Me	3*S*, 4*R* 100% yield, 90% de [95]
	Me	allyl	Me	3*S*, 4*R* 100% yield, 98% de [95]

FIGURE 11.13

Various examples of dynamic kinetic resolution.

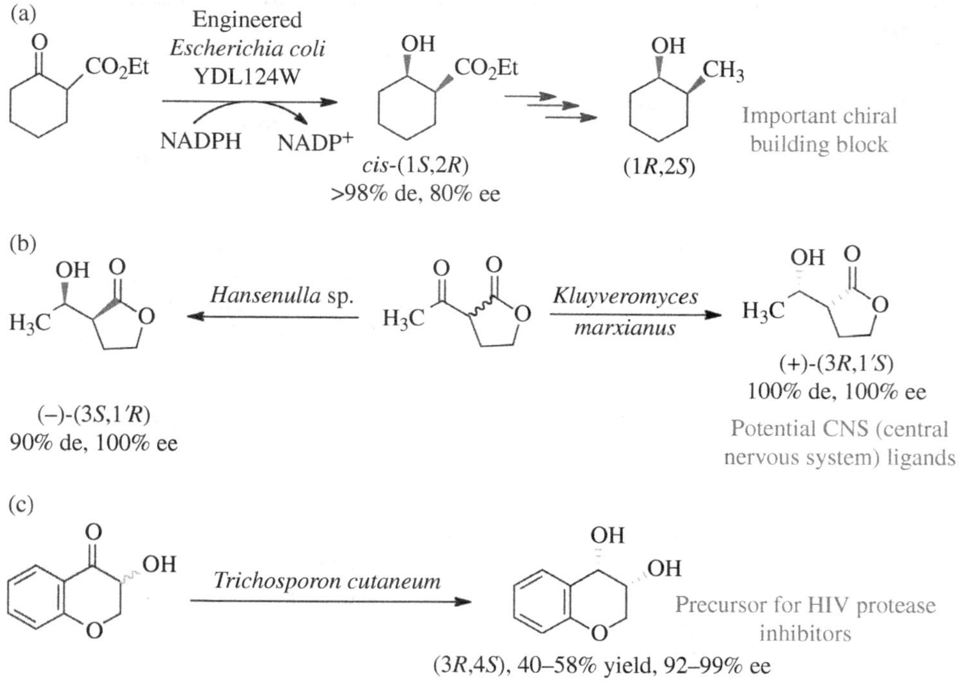

FIGURE 11.14

Synthesis of important compounds by dynamic kinetic resolution through reduction.

Figure 11.14a shows dynamic kinetic resolution of cyclic β-keto esters by enzymes from baker's yeast [96]. Twenty purified dehydrogenases from baker's yeast expressed in *E. coli* were tested for their ability to reduce three homologous cyclic β-keto esters. The majority of dehydrogenases reduced ethyl 2-oxo-cyclopentanecarboxylate, yielding a pair of diastereomeric alcohols with consistent (1*R*)-stereochemistry. *Escherichia coli* cells overexpressing the YDL124w gene were used in a dynamic kinetic resolution of ethyl 2-oxo-cyclohexanecarboxylate to produce the key intermediate in a chemo-enzymatic synthesis of (1*R*,2*S*)-2-methyl-1-cyclohexanol, an important chiral building block. Figure 11.14b shows the dynamic kinetic resolution of α-acetyl-γ-butyrolactone. The reaction with *Kluyveromyces marxianus* gave (+)-(3*R*,1′*S*)-α-1′-hydroxyethyl-γ-butyrolactone in 100% ee and 100% de. The corresponding (–)-(3*S*,1′*R*)-enantiomer was produced with *Hansenula* sp. in 100% ee and 90% de. The products are potential central nervous system (CNS) ligands or as attractive intermediates to new bioactive compounds [89]. Figure 11.14c shows dynamic kinetic resolution of 3-hydroxy-4-chromanone with *Trichosporon cutaneum* CCT 1903 through oxidation of (*S*)-hydroxychromanone and reduction of (*R*)-hydroxychromanone. (3*R*,4*S*)-3,4-Chromanediol, precursor for HIV protease inhibitors, was obtained in 40–58% yield with up to 99% ee [97].

11.6 CONCLUSIONS

Biocatalytic asymmetric reduction is introduced in this chapter. Recent developments in crystallography reveal the detailed mechanism of the reductions. Now, enzyme sources are becoming diverse; enzymes expressed and unexpressed in natural environment can be used. Photosynthetic organisms have been also proved to be effective biocatalysts. Furthermore, to improve the catalysts, performance, extensive research was conducted on mutations as well as investigation of reaction conditions including the use of innovative media such as supercritical CO_2. With these progresses in biocatalysis, industrial applications using oxidoreductase will increase in the future.

REFERENCES

[1] (a) Carey JS, Laffan D, Thomson C, Williams MT. Analysis of the reactions used for the preparation of drug candidate molecules. *Org. Biomol. Chem.* 2006; **4**: 2337–2347; (b) Patel RN. Synthesis of chiral pharmaceutical intermediates by biocatalysis. *Coord. Chem. Rev.* 2008; **252**: 659–701; (c) Nakamura K, Yamanaka R, Matsuda T, Harada T. Recent developments in asymmetric reduction of ketones with biocatalysts. *Tetrahedron Asymmetry* 2003; **14**: 2659–2681; (d) Nakamura K, Matsuda T. Biocatalytic reduction of carbonyl group. *Curr. Org. Chem.* 2006; **10**: 1217–1246; (e) Nakamura K, Matsuda T. Enzymatic reduction reaction. In: Gotor V, Alfonsa I, and Garcia-Urdiales E (Eds.). *Asymmetric Organic Synthesis with Enzyme.* Wiley-VCH, 2008: 193–228; (f) Nakamura K, Matsuda T. Biocatalysis in water. In: Ulf L (Ed.). *Organic Reactions in Water-Principles, Strategies and Methods.* Blackwell, 2007: 301–349.

[2] Eklund H, Ramaswamy S. Three-dimensional structures of MDR alcohol dehydrogenases. *Cell. Mol. Life Sci.* 2008; **65**: 3907–3917.

[3] Martins BM, Macedo-Ribeiro S, Bresser J, Bucke W, Messerschm A. Structural basis for stereo-specific catalysis in NAD+-dependent (*R*)-2-hydroxyglutarate dehydrogenase from *Acidaminococcus fermentans*. *FEBS J.* 2005; **272**: 269–281.

[4] Homann MJ, Vail RB, Previte E, Tamarez M, Morgan B, Dodds DR, Zaks A. Rapid identification of enantioselective ketone reductions using targeted microbial libraries. *Tetrahedron* 2004; **60**: 789–797.

[5] Inoue K, Makino Y, Itoh N. Purification and characterization of a novel alcohol dehydrogenase from *Leifsonia* sp strain S749: A promising biocatalyst for an asymmetric hydrogen transfer bioreduction. *Appl. Environ. Microbiol.* 2005; **71**: 3633–3641.

[6] Knietsch A, Waschkowitz T, Bowien S, Henne A, Daniel R. Metagenomes of complex microbial consortia derived from different soils as sources for novel genes conferring formation of carbonyls from short-chain polyols on *Escherichia coli*. *J. Mol. Microbiol. Biotechnol.* 2003; **5**: 46–56.

[7] Kaluzna IA, Feske BD, Wittayanan W, Ghiviriga I, Stewart JD. Stereoselective biocatalytic reductions of α-chloro-β-keto esters. *J. Org. Chem.* 2005; **70**: 342–345.

[8] Ziegelmann-Fjeld KI, Musa MM, Phillips RS, Zeikus JG, Vieille CA. *Thermoanaerobacter ethanolicus* secondary alcohol dehydrogenase mutant derivative highly active and stereoselective on phenylacetone and benzylacetone. *Protein Eng. Des. Sel.* 2007; **20**: 47–55.

[9] Makino Y, Inoue K, Dairi T, Itoh N. Engineering of phenylacetaldehyde reductase for efficient substrate conversion in concentrated 2-propanol. *Appl. Environ. Microbiol.* 2005; **71**: 4713–4720.

[10] Zhu D, Malik HT, Hua L. Asymmetric ketone reduction by a hyperthermophilic alcohol dehydrogenase. The substrate specificity, enantioselectivity and tolerance of organic solvents. *Tetrahedron Asymmetry* 2006; **17**: 3010–3014.

[11] Nakamura K, Yamanaka R, Tohi K, Hamada H. Cyanobacterium-catalyzed asymmetric reduction of ketones. *Tetrahedron Lett.* 2004; **1**: 6799–6802.

[12] Nakamura K, Yamanaka R. Light mediated cofactor recycling system in biocatalytic asymmetric reduction of ketone. *J. Chem. Soc. Chem. Commun.* 2002; **16**: 1782–1783.

[13] Nakamura K, Yamanaka R. Light-mediated regulation of asymmetric reduction of ketones by a cyanobacterium. *Tetrahedron Asymmetry* 2002; **13**: 2529–2533.

[14] Itoh K, Sakamaki H, Nakamura K, Horiuchi CA. Biocatalytic asymmetric reduction of 3-acetylisoxazoles. *Tetrahedron Asymmetry* 2005; **16**: 1403–1408.

[15] Utsukihara T, Chai W, Kato N, Nakamura K, Horiuchi CA. Reduction of (+)- and (−)-camphorquinones by cyanobacteria. *J. Mol. Catal. B: Enzym.* 2004; **31**: 19–24.

[16] Itoh K, Nakamura K, Aoyama T, Matsuba R, Kakimoto T, Murakami M, Yamanaka R, Muranaka T, Sakamaki H, Takido T. Photobiocatalyzed asymmetric reduction of ketones using *Chlorella* sp. MK201. *Biotechnol. Lett.* 2012; **34**: 2083–2086.

[17] Utsukihara T, Misumi O, Kato N, Kuroiwa T, Horiuchi CA. Reduction of various ketones by red algae. *Tetrahedron Asymmetry* 2006; **17**: 1179–1185.

[18] Kojima H, Okada A, Takeda S, Nakamura K. Effect of carbon dioxide concentrations on asymmetric reduction of ketones with plant-cultured cells. *Tetrahedron Lett.* 2009; **50**: 7079–7081.

[19] Takeda S, Ogata Y, Kojima H, Okada A, Uranishi Y, Nakamura K. Arabidopsis thaliana: A novel biocatalyst for asymmetric reductions. *Plant Biotechnol.* 2011; **28**: 77–82.

[20] Matsuo K, Kawabe S, Tokuda Y, Eguchi T, Yamanaka R, Nakamura K. Asymmetric reduction of ketones with a germinated plant. *Tetrahedron Asymmetry* 2008; **19**: 157–159.
[21] Yamanaka R, Nakamura K, Murakami A. Reduction of exogenous ketones depends upon NADPH generated photosynthetically in cells of he cyanobacterium *Synechococcus* PCC 7942. *AMB Express* 2011; **1**: 24.
[22] Holsch K, Havel J, Haslbeck M, Weuster-Botz D. Identification, cloning, and characterization of a novel ketoreductase from the cyanobacterium *Synechococcus* sp. Strain PCC 7942. *Appl. Environ. Microbiol.* 2008; **74**: 6697–6702.
[23] Itoh K, Nakamura K, Aoyama T, Kakimoto T, Murakami M, Takido T. The influence of wavelength of light on cyanobacterial asymmetric reduction of ketone. *Tetrahedron Lett.* 2014; **55**: 435–437.
[24] Takemura T, Akiyama K, Umeno N, Tamai Y, Ohta H, Nakamura K. Asymmetric reduction of a ketone by knockout mutants of a cyanobacterium. *J. Mol. Catal. B: Enzym.* 2009; **60**: 93–95.
[25] Musa MM, Ziegelmann-Fjeld KI, Vieille C, Zeikus JG, Phillips RS. Xerogel-encapsulated W110A secondary alcohol dehydrogenase from *Thermoanaerobacter ethanolicus* performs asymmetric reduction of hydrophobic ketones in organic solvents. *Angew. Chem. Int. Ed. Engl.* 2007; **46**: 3091–3094.
[26] Wolfson A, Dlugya C, Tavora D, Blumenfelda J, Shotlanda Y. Baker's yeast catalyzed asymmetric reduction in glycerol. *Tetrahedron Asymmetry* 2006; **17**: 2043–2045.
[27] Yajima A, Naka K, Yabuta G. Immobilized baker's yeast reduction in fluorous media. *Tetrahedron Lett.* 2004; **45**: 4577–4579.
[28] (a) Matsuda T, Harada T, Nakamura K. Alcohol dehydrogenase is active in supercritical carbon dioxide. *J. Chem. Soc. Chem. Commun.* 2000, Issue 10: 1367–1368; (b) Matsuda T, Watanabe K, Kamitanaka T, Harada T, Nakamura K. Biocatalytic reduction of ketones by semi-continuous flow process using supercritical carbon dioxide. *J. Chem. Soc. Chem. Commun.* 2003, Issue 10: 1198–1199; (c) Matsuda T, Marukado R, Mukouyama M, Harada T, Nakamura K. Asymmetric reduction of ketones by *Geotrichum candidum*: Immobilization and application to reaction using supercritical carbon dioxide. *Tetrahedron Asymmetry* 2008; **19**: 2272–2275; (d) Panza JL, Russell AJ, Beckman EJ. Synthesis of fluorinated NAD as a soluble coenzyme for enzymatic chemistry in fluorous solvents and carbon dioxide. *Tetrahedron* 2002; **58**: 4091–4104.
[29] (a) Matsuda T, Yamagishi Y, Koguchi S, Iwai N, Kitazume T. An effective method to use ionic liquids as reaction media for asymmetric reduction by *Geotrichum candidum*. *Tetrahedron Lett.* 2006; **47**: 4619–4622; (b) Pfruender H, Amidjojo M, Kragl U, Weuster-Botz D. Efficient whole-cell biotransformation in a biphasic ionic liquid/water system. *Angew. Chem. Int. Ed. Engl.* 2004; **43**: 4529–4531; (c) Musa MM, Ziegelmann-Fjeld KI, Vieille C, Phillips RS. Activity and selectivity of W110A secondary alcohol dehydrogenase from *Thermoanaerobacter ethanolicus* in organic solvents and ionic liquids: Mono-and biphasic media. *Org. Biomol. Chem.* 2008; **6**: 887–892.
[30] Hammond RJ, Poston BW, Ghiviriga I, Feske BD. Biocatalytic synthesis towards both antipodes of 3-hydroxy-3-phenylpropanitrile a precursor to fluoxetine, atomoxetine and nisoxetine. *Tetrahedron Lett.* 2007; **48**: 1217–1219.
[31] Gröger H, Chamouleau F, Orologas N, Rollmann C, Drauz K, Hummel W, Weckbecker A, May O. Enantioselective reduction of ketones with designer cells at high substrate concentrations: Highly efficient access to functionalized optically active alcohols. *Angew. Chem. Int. Ed. Engl.* 2006; **45**: 5677–5681.
[32] Gröger H, Rollmann C, Chamouleau F, Sebastien I, May O, Wienand W, Drauz K. Enantioselective reduction of 4-fluoroacetophenone at high substrate concentration using a tailor-made recombinant whole-cell catalyst. *Adv. Synth. Catal.* 2007; **349**: 709–712.
[33] Zhu D, Yang Y, Buynak JD, Hua L. Stereoselective ketone reduction by a carbonyl reductase from *Sporobolomyces salmonicolor*. Substrate specificity, enantioselectivity and enzyme-substrate docking studies. *Org. Biomol. Chem.* 2006; **4**: 2690–2695.
[34] Gelo-Pujic M, Le Guyader F, Schlama T. Microbial and homogenous asymmetric catalysis in the reduction of 1-[3,5-bis(trifluoromethyl)phenyl]ethanone. *Tetrahedron Asymmetry* 2006; **17**: 2000–2005.
[35] Zhu D, Yang Y, Hua L. Stereoselective enzymatic synthesis of chiral alcohols with the use of a carbonyl reductase from *Candida magnoliae* with *anti*-prelog enantioselectivity. *J. Org. Chem.* 2006; **71**: 4202–4205.
[36] Krausser M, Hummel W, Gröger H. Enantioselective one-pot two-step synthesis of hydrophobic allylic alcohols in aqueous medium through the combination of a Wittig reaction and an enzymatic ketone reduction. *Eur. J. Org. Chem.* 2007: 5175–5179.

[37] Bisel P, Walter L, Nieger M, Hummel W, Müller M. Stereochemical clarification of the enzyme-catalysed reduction of 2-acetylchromen-4-one. *Tetrahedron Asymmetry* 2007; **18**: 1142–1144.

[38] Hiraoka C, Matsuda M, Suzuki Y, Fujieda S, Tomita M, Fuhshuku K, Obata R, Nishiyama S, Sugai T. Screening, substrate specificity and stereoselectivity of yeast strains, which reduce sterically hindered isopropyl ketones. *Tetrahedron Asymmetry* 2006; **17**: 3358–3367.

[39] Kurbanoglu EB, Zilbeyaz K, Kurbanoglu NI, Kilic H. Enantioselective reduction of substituted acetophenones by *Aspergillus niger*. *Tetrahedron Asymmetry* 2007; **18**: 1159–1162.

[40] Kurbanoglu EB, Zilbeyaz K, Kurbanoglu NI, Kilic H. Asymmetric reduction of acetophenone analogues by *Alternaria alternata* using ram horn peptone. *Tetrahedron Asymmetry* 2007; **18**: 2332–2335.

[41] Morikawa S, Nakai T, Yasohara Y, Nanba H, Kizaki T, Hasegawa J. Highly active mutants of carbonyl reductase S1 with inverted coenzyme specificity and production of optically active alcohols. *Biosci. Biotechnol. Biochem.* 2005; **69**: 544–552.

[42] Yang Y, Zhu D, Piegat TJ, Hua L. Enzymatic ketone reduction: Mapping the substrate profile of a short-chain alcohol dehydrogenase (YMR226c) from *Saccharomyces cerevisiae*. *Tetrahedron Asymmetry* 2007; **18**: 1799–1803.

[43] Comasseto JV, Assis LF, Andrade LH, Schoenlein-Crusius IH, Porto ALM. Deracemization of aryl ethanols and reduction of acetophenones by whole fungal cells of *Aspergillus terreus* CCT 4083, *A. terreus* CCT 3320 and *Rhizopus oryzae* CCT 4964. *J. Mol. Catal. B: Enzym.* 2006; **39**: 24–30.

[44] Soni P, Banerjee UC. Enantioselective reduction of acetophenone and its derivatives with a new yeast isolate *Candida tropicalis* PBR-2 MTCC 5158. *Biotechnol. J.* 2006; **1**: 80–85.

[45] Pollard D, Truppo M, Pollard J, Chen C, Moore J. Effective synthesis of (*S*)-3,5-bistrifluoromethylphenyl ethanol by asymmetric enzymatic reduction. *Tetrahedron Asymmetry* 2006; **17**: 554–559.

[46] Yang W, Xu JH, Xie Y, Xu Y, Zhao G, Lin GQ. Asymmetric reduction of ketones by employing *Rhodotorula* sp. AS2.2241 and synthesis of the β-blocker (*R*)-nifenalol. *Tetrahedron Asymmetry* 2006; **17**: 1769–1774.

[47] Erdélyia B, Szabóa A, Seresb G, Birincsika L, Ivanicsc J, Szatzkerd G, Popped L. Stereoselective production of (*S*)-1-aralkyl- and 1-arylethanols by freshly harvested and lyophilized yeast cells. *Tetrahedron Asymmetry* 2006; **17**: 268–274.

[48] Musa MM, Ziegelmann-Fjeld KI, Vieille C, Zeikus JG, Phillips RS. Asymmetric reduction and oxidation of aromatic ketones and alcohols using W110A secondary alcohol dehydrogenase from *Thermoanaerobacter ethanolicus*. *J. Org. Chem.* 2007; **72**: 30–34.

[49] Kamble AL, Soni P, Banerjee UC. Biocatalytic synthesis of *S*(−)-1-(1′-naphthyl) ethanol by a novel isolate of *Candida viswanathii*. *J. Mol. Catal. B: Enzym.* 2005; **35**: 1–6.

[50] Matsuda M, Yamazaki T, Fuhshuku K, Sugai T. First total synthesis of modiolide A, based on the whole-cell yeast-catalyzed asymmetric reduction of a propargyl ketone. *Tetrahedron* 2007; **63**: 8752–8760.

[51] Orden AA, Bisogno FR, Cifuente DA, Giordano OS, Sanz MK. Asymmetric bioreduction of natural xenobiotic diketones by *Brassica napus* hairy roots. *J. Mol. Catal. B: Enzym.* 2006; **42**: 71–77.

[52] Uwai K, Konn N, Kitamura S, Ohta S, Takeshita M. Purification and characterization of rat liver enzyme catalyzing stereoselective reduction of acetylpyridines. *Chirality* 2005; **17**: 494–500.

[53] Kaluzna I, Andrew AA, Bonilla M, Martzen MR, Stewart JD. Enantioselective reductions of ethyl 2-oxo-4-phenylbutyrate by *Saccharomyces cerevisiae* dehydrogenases. *J. Mol. Catal. B: Enzym.* 2002; **17**: 101–105.

[54] Kizaki N, Yasohara Y, Hasegawa J, Wada M, Kataoka M, Shimizu S. Synthesis of optically pure ethyl (*S*)-4-chloro-3-hydroxybutanoate by *Escherichia coli* transformant cells coexpressing the carbonyl reductase and glucose dehydrogenase genes. *Appl. Microbiol. Biotechnol.* 2001; **55**: 590–595.

[55] Milagrea HMS, Milagre CDF, Morana PJS, Santanab MHA, Rodrigues JAR. Reduction of ethyl benzoylformate mediated by *Saccharomyces cerevisiae* entrapped in alginate fibers with double gel layers in a continuously operated reactor. *Enzym. Microb. Technol.* 2005; **37**: 121–125.

[56] Fischer T, Pietruszka J. Efficient synthesis of either enantiomer of ethyl 5-hydroxyhept-6-enoate. *Adv. Synth. Catal.* 2007; **349**: 1533–1536.

[57] Zhu D, Ankati H, Mukherjee C, Yang Y, Biehl ER, Hua L. Asymmetric reduction of β-ketonitriles with a recombinant carbonyl reductase and enzymatic transformation to optically pure beta-hydroxy carboxylic acids. *Org. Lett.* 2007; **9**: 2561–2563.

[58] Fitzpatrick TB, Amrhein N, Macheroux P. Characterization of YqjM, an old yellow enzyme homolog from *Bacillus subtilis* involved in the oxidative stress response. *J. Biol. Chem.* 2003; **278**: 19891–19897.

[59] Surette MA, Sturz AV, Lada RR, Nowak J. Bacterial endophytes in processing carrots (*Daucus carota L. var. sativus*): Their localization, population density, biodiversity and their effects on plant growth. *Plant Soil* 2003; **253**: 381–390.

[60] Miyadera A, Satoh K, Imura A. Efficient synthesis of a key intermediate of DV-7751 via optical resolution or microbial reduction. *Chem. Pharm. Bull.* 2000; **48**: 563–565.

[61] Satoh K, Imura A, Miyadera A, Kanai K, Yukimoto Y. An efficient synthesis of a key intermediate of DU-6859a via asymmetric microbial reduction. *Chem. Pharm. Bull.* 1998; **46**: 587–590.

[62] Patel RN, Chu L, Mueller R. Diastereoselective microbial reduction of (*S*)-[3-chloro-2-oxo-1-(phenylmethyl)propyl]carbamic acid, 1,1-dimethylethyl ester. *Tetrahedron Asymmetry* 2003; **14**: 3105–3109.

[63] Patel RN, Banerjee A, Hawell JM, MCNarnee C, Brozozowski D, Mirfakhrae D, Nanduri V, Thottathil JK, Szarka LJ. Microbial synthesis of (2*R*,3*S*)-(−)-*N*-benzoyl-3-phenyl isoserine ethyl ester-a taxol side-chain synthon. *Tetrahedron Asymmetry* 1993; **4**: 2069–2084.

[64] Palel RN, Banerjee A, Chu L, Brozozowski D, Nanduri V, Szarka L. Microbial synthesis of chiral intermediates for β-3-receptor agonists. *J. Am. Oil Chem. Soc.* 1998; **75**: 1473–1482.

[65] Campos KR, Klapars A, Kohmura Y, Pollard D, Ishibashi H, Kato S, Takezawa A, Waldman JH, Wallace DJ, Chen C, Yasuda N. Enantioselective, biocatalytic reduction of 3-substituted cyclopentenones: Application to the asymmetric synthesis of an hNK-1 receptor antagonist. *Org. Lett.* 2011; **13**: 1004–1007.

[66] Guo Z, Chen Y, Goswami A, Hanson RL, Patel RN. Synthesis of ethyl and t-butyl (3*R*,5*S*)-dihydroxy-6-benzyloxy hexanoates via diastereo- and enantioselective microbial reduction. *Tetrahedron Asymmetry* 2006; **17**: 1589–1602.

[67] Katoh T, Mizumoto S, Fudesaka M, Takeo M, Kajimoto T, Node M. An efficient route for the synthesis of methyl (−)-1,4a-dimethyl-5-oxodecahydronaphthalene-1-carboxylate by using baker's yeast-catalyzed asymmetric reduction. *Tetrahedron Asymmetry* 2006; **17**: 1655–1662.

[68] Conceicao GJA, Moran PJS, Rodrigues JAA. Homochiral (1*S*,2*R*)-1,2-indandiol from asymmetric reduction of 1,2-indanedione by resting cells of the yeast *Trichosporon cutaneum*. *Tetrahedron Asymmetry* 2003; **14**: 2327–2330.

[69] Cazetta T, Lunardi I, Conceicao GJA, Morana PJS, Rodrigues JAR. *Trichosporon cutaneum*-promoted deracemization of (+/−)-2-hydroxindan-l-one: A mechanistic study. *Tetrahedron Asymmetry* 2007; **18**: 2030–2036.

[70] Katz M, Frejd T, Hahn-Hägerdal B, Gorwa-Grauslund MF. Efficient anaerobic whole cell stereoselective bioreduction with recombinant *Saccharomyces cerevisiae*. *Biotechnol. Bioeng.* 2003; **84**: 573–582.

[71] Utsukihara T, Watanabe S, Tomiyama A, Chai W, Horiuchi CA. Stereoselective reduction of ketones by various vegetables. *J. Mol. Catal. B: Enzym.* 2006; **41**: 103–109.

[72] Friberg A, Johanson T, Franzén J, Gorwa-Grauslund MF, Frejd T. Efficient bioreduction of bicyclo[2.2.2]octane-2,5-dione and bicyclo[2.2.2]oct-7-ene-2,5-dione by genetically engineered *Saccharomyces cerevisiae*. *Org. Biomol. Chem.* 2006; **4**: 2304–2312.

[73] Shimoda K, Kubota N, Hirata T, Kondo Y, Hamada H. Stereoselective reduction of 2-butenolides to chiral butanolides by reductases from cultured cells of *Glycine max*. *Tetrahedron Lett.* 2007; **48**: 1345–1347.

[74] Swiderska MA, Stewart JD. Stereoselective enone reductions by *Saccharomyces carlsbergensis* old yellow enzyme. *J. Mol. Catal. B: Enzym.* 2006; **42**: 52–54.

[75] Hirata T, Shimoda K, Gondai T. Asymmetric hydrogenation of the C—C double bond of enones with the reductases from *Nicotiana tabacum*. *Chem. Lett.* 2000; **29**: 850–851.

[76] Shimoda K, Izumi S, Hirata T. A novel reductase participating in the hydrogenation of an exocyclic C—C double bond of enones from *Nicotiana tabacum*. *Bull. Chem. Soc. Jpn.* 2002; **75**: 813–816.

[77] Shimoda K, Kubota N, Hamada H, Kaji M, Hirata T. Asymmetric reduction of enones with *Synechococcus* sp. PCC 7942. *Tetrahedron Asymmetry* 2004; **15**: 1677–1679.

[78] Shimoda K, Kubota N, Hamada H. Asymmetric reduction of α,β-unsaturated carbonyl compounds with reductases from *Nicotiana tabacum*. *Tetrahedron Asymmetry* 2004; **15**: 2443–2446.
[79] Pellissier H. Recent developments in dynamic kinetic resolution. *Tetrahedron* 2008; **64**: 1563–1601.
[80] Nakamura K, Kawai Y, Miyai T, Ohno A. Stereochemical control in diastereoselective reduction with baker's yeast. *Tetrahedron Lett.* 1990; **31**: 3631–3632.
[81] Nakamura K, Kawai Y, Miyai T, Honda S, Nakajima N, Ohno A. Stereochemical control in microbial reduction. 18. Mechanism of stereochemical control in the diastereoselective reduction with baker's yeast. *Bull. Chem. Soc. Jpn.* 1991; **64**: 1467–1470.
[82] Nakamura K, Miyoshi H, Sugiyama T, Hamada H. Diastereo- and enantio-selective reduction of ethyl 2-methyl-3-oxobutanoate by plant cell cultures. *Phytochemistry* 1995; **40**: 1419–1420.
[83] Kaluzna IA, Matsuda T, Sewell AK, Stewart JD. Systematic investigation of *Saccharomyces cerevisiae* enzymes catalyzing carbonyl reductions. *J. Am. Chem. Soc.* 2004; **126**: 12827–12832.
[84] Miya H, Kawada M, Sugiyama Y. Stereoselective reduction of ethyl 2-methyl-3-oxobutanoate by bacteria. *Biosci. Biotechnol. Biochem.* 1996; **60**: 95–98.
[85] Nakamura K, Miyai T, Nagar A, Oka S, Ohno A. Stereochemical control in microbial reduction. 9. Diastereoselective reduction of 2-alkyl-3-oxobutanoate with bakers' yeast. *Bull. Chem. Soc. Jpn.* 1989; **62**: 1179–1187.
[86] Nakamura K, Miyai T, Fukushima K, Kawai Y, Babu BR, Ohno A. Stereochemical control in microbial reduction. XV. Preparation of (2*R*,3*S*)-2-allyl-3-hydroxybutanoate. *Bull. Chem. Soc. Jpn.* 1990; **63**: 1713–1715.
[87] Speicher A, Roeser H, Heisel R. Stereoselective oxidoreductase type bioconversions of exogenous substrates by cell suspension cultures of bryophytes. *J. Mol. Catal. B: Enzym.* 2003; **22**: 71–77.
[88] Rodríguez P, Barton M, Aldabalde V, Onetto S, Panizza P, Menéndez P, Gonzalez D, Rodríguez S. Are endophytic microorganisms involved in the stereoselective reduction of ketones by *Daucus carota* root? *J. Mol. Catal. B: Enzym.* 2007; **49**: 8–11.
[89] Ribeiro JB, Andrade De Sousa LM, da Volta SM, Ramos MDC, Radler de Aquino Neto F, Fraga CAM, Leite SGF, Cordeiro Y, Antunes OAC. Microbial reduction of alpha-acetyl-gamma-butyrolactone. *Tetrahedron Asymmetry* 2006; **17**: 984–988.
[90] Perrone MG, Santandrea E, Scilimati V. Tortorella V, Capitelli F, Bertolasi V. Baker's yeast-mediated reduction of ethyl 2-(4-chlorophenoxy)-3-oxoalkanoates intermediates for potential PPARalpha ligand. *Tetrahedron Asymmetry* 2004; **15**: 3501–3510.
[91] Kambourakis S, Rozzell JD. Ketoreductases in the synthesis of valuable chiral intermediates: Application in the synthesis of α-hydroxy β-amino and β-hydroxy γ-amino acids. *Tetrahedron* 2004; **60**: 663–669.
[92] Shimoda K, Kubota N, Hamada H, Hamada H. Diastereoselective reduction of β-keto carbonyl compounds by cultured plant cells. *Tetrahedron Lett.* 2006; **47**: 1541–1544.
[93] Kometani T, Sakai Y, Matsumae H, Shibatani T, Matsuno R. Production of (2*S*,3*S*)-2,3-dihydro-3-hydroxy-2-(4-methoxyphenyl)-1,5-benzothiazepin-4(5*H*)-one, a key intermediate for diltiazem synthesis, by bakers' yeast-mediated reduction. *J. Ferment. Bioeng.* 1997; **84**: 195–199.
[94] Ji A, Wolberg M, Hummel W, Wandrey C, Müller M. Dynamic kinetic resolution of tert-butyl 4-methyl-3,5-dioxohexanoate through enzymatic reduction. *Chem. Commun.* 2001: 57–58.
[95] Kalaitzakis D, Rozzell JD, Smonou I, Kambourakis S. Synthesis of valuable chiral intermediates by isolated ketoreductases: Application in the synthesis of alpha-alkyl-beta-hydroxy ketones and 1,3-diols. *Adv. Synth. Catal.* 2006; **348**: 1958–1969.
[96] Padhi SK, Kaluzna IA, Buisson D, Azerad R, Stewart JD. Reductions of cyclic β-keto esters by individual *Saccharomyces cerevisiae* dehydrogenases and a chemo-enzymatic route to (1*R*,2*S*)-2-methyl-1-cyclohexanol. *Tetrahedron Asymmetry* 2007; **18**: 2133–2138.
[97] Lunardi I, Conceicao GJA, Moran PJS, Rodrigues JAR. Highly stereoselective preparation of (3*R*,4*S*)-3,4-chromanediol by deracemization of (+/−)-3-hydroxy-4-chromanone by *Trichosporon cutaneum*. *Tetrahedron Asymmetry* 2005; **16**: 2515–2519.
[98] Yang Y, Drolet M, Kayser MM. The dynamic kinetic resolution of 3-oxo-4-phenyl-β-lactam by recombinant *E. coli* overexpressing yeast reductase Ara1p. *Tetrahedron Asymmetry* 2005; **16**: 2748–2753.
[99] Kalaitzakis D, Rozzell JD, Kambourakis S, Smonou I. Highly stereoselective reductions of α-alkyl-1,3-diketones and α-alkyl-β-keto esters catalyzed by isolated NADPH-dependent ketoreductases. *Org. Lett.* 2005; **7**: 4799–4801.

CHAPTER 12

Nitrile-Converting Enzymes and their Synthetic Applications

Ludmila Martínková

Laboratory of Biotransformation, Institute of Microbiology, Academy of Sciences of the Czech Republic, Prague, Czech Republic

12.1 INTRODUCTION

The enzymatic hydrolysis of nitriles proceeds via two different pathways: a two-step pathway catalyzed by nitrile hydratase and amidase, and a single-step pathway catalyzed by nitrilase. Nitrile hydratase and nitrilase belong to different protein superfamilies, and their phylogenetic distribution is also different (nitrilases in plants, fungi, and bacteria but nitrile hydratases almost exclusively in bacteria). Nitrilases (EC 3.5.5.-.) belong to the nitrilase superfamily, that is, non-peptide C—N hydrolases with a Glu-Lys-Cys catalytic triad, and nitrile hydratases (4.2.1.84) are metalloenzymes containing a ferric or a cobalt (III) ion in their catalytic site. Due to their ability to transform an immense number of man-made nitriles, the nitrile-converting enzymes have found widespread use in synthetic organic chemistry. The advantages of enzymatic processes reside in their mild conditions, high yields, and high purities of the products and in their regio- or enantioselectivities.

Previous reviews of this field of research focused mainly on the biocatalytic uses of nitrilases [1], structure and function of nitrilases [2, 3], nitrilases in filamentous fungi [4], stereoselective biotransformations catalyzed by nitrile-converting enzymes [5], and the structure, function, and uses of nitrile hydratases [6]. For a comprehensive review of nitrile-converting enzymes known at the time, the study by Banerjee *et al.* [7] has been helpful. Some of the most recent reviews focused on nitrilases, specifically their sources, properties, and use, [8], and on methodologies for their screening [9].

The aim of this chapter is to summarize the progress achieved in research into nitrile-converting enzymes since approximately the mid-2000s. In this period the spectrum of characterized nitrile-converting enzymes increased due to using new sources (metagenomes, databases) and methods (site-directed mutagenesis, directed evolution, high-throughput screening), and the enzymes were used to develop and optimize processes for fine chemical production. A lot of effort was also put into the development of cascade processes that combined multiple enzymes or enzymatic and chemical steps.

Green Biocatalysis, First Edition. Edited by Ramesh N. Patel.

12.2 SCREENING METHODOLOGY

Modern methodologies used to obtain new nitrile-converting enzymes consisted largely of metagenome screening on one hand and the rational analysis of known nitrilase sequences on the other. In some cases, the enzymes were modified by site-directed mutagenesis or directed evolution of the corresponding genes.

The number of commercial nitrilases has recently increased considerably. Catalysts suitable for the transformation of a specific substrate can be selected from nitrilase sets available from Codexis and Prozomix Limited [10, 11]. A similar set of nitrile hydratases (NHases) is also available from the latter company.

12.2.1 Screening Metagenomic Libraries

Metagenomic nitrilase genes were first exploited by Robertson *et al.* [12]. This study provided a set of 137 new nitrilases, the corresponding genes being largely obtained from clones growing on adiponitrile. This nitrilase library was screened for enzymes suitable for the hydrolysis of industrially important substrates. For instance, the monohydrolysis of prochiral 3-hydroxyglutaronitrile enabled the preparation of enantiopure monoacids of which the *R*-isomer is a precursor of a cholesterol-lowering drug, Lipitor®. A selected nitrilase was further improved by site-directed mutagenesis toward better enantioselectivity for highly concentrated 3-hydroxyglutaronitrile (3M) into (*R*)-4-cyano-3-hydroxybutyric acid (98.5% ee) [13].

Recently, metagenomic library screening provided a few more nitrilases, whose genes were obtained from clones growing on cinnamonitrile or a nitrile mixture. One of the enzymes was selected for the monohydrolysis of 2-methylglutaronitrile into 4-cyanopentanoic acid, a precursor of 1,5-dimethyl-2-piperidone [14]. A different method based on PCR was used to screen metagenomic DNA for new nitrile hydratases and resulted in the isolation of six active enzymes [15].

12.2.2 Database Mining

Some recent studies focused on the exploitation of databases as potential sources of new nitrile-converting enzymes. Two different approaches were used to obtain the gene of interest, that is, either its amplification from the genomic DNA or cDNA of the wild-type strain or its artificial synthesis.

The genes to be expressed were selected according to sequential analysis. Clustering of the sequences and, in some cases, analysis of specific regions enabled arylacetonitrilases acting on mandelonitrile to be predicted [16, 17]. In a similar way, it was possible to predict fungal arylacetonitrilases and aromatic nitrilases [18]. Other putative nitrilases were examined in an analogous way, with the aim of obtaining enzymes suitable for the hydrolysis of iminodiacetonitrile [19].

Recently, the spectrum of known nitrilases was significantly enlarged by a screening of 290 prokaryotic nitrilase genes [20]. Of these, 163 were expressed and 125 protein products proved to exhibit nitrilase activities for at least some of the 25 nitriles tested. For instance, nitrilases from *Sphingomonas wittichii* and *Syntrophobacter fumaroxidans* were optimal catalysts for the synthesis of (*R*)-4-methoxy-4-oxo-3-phenylbutanoic acid and 3-oxocyclopentanecarboxylic acid (building blocks), respectively.

12.2.3 Construction of Enzyme Variants

The structure–activity relationships in nitrilases were probably best described in the nitrilase from *Pseudomonas fluorescens* which was subjected to a number of specific mutations [21–24]. The activities, enantioselectivities, and chemoselectivities (amide

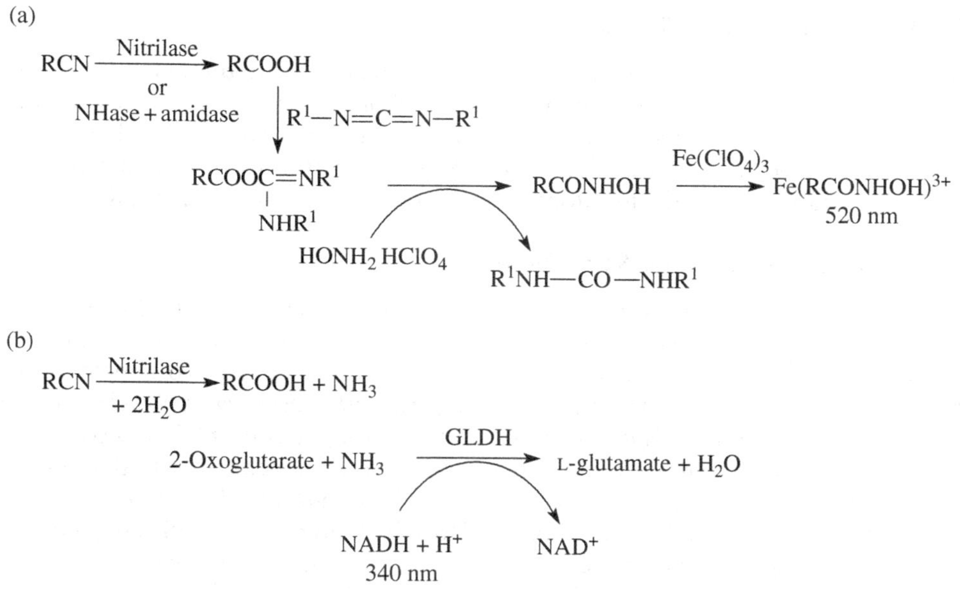

FIGURE 12.1

Principles of coupled nitrilase assays [20, 29]. (a) R1 = cyclohexyl; (b) GLDH = glutamate dehydrogenase.

production) of this enzyme were influenced by point mutations in the neighborhood of the catalytic cysteine, point mutations and deletions in the C-terminal region and by the construction of chimeric variants. Analogous point mutations in the arylacetonitrilases from *Aspergillus niger* and *Neurospora crassa* exhibited similar effects despite low amino acids sequence similarities between the bacterial and fungal enzymes [25].

A different method consisting of directed evolution was used with the nitrilases from *P. fluorescens* [26] and *Alcaligenes faecalis* [27]. The amide production by the former and the operational stability of the latter at a low pH were improved in this way.

The high-throughput screening necessary to evaluate the nitrilase variants resulting from directed evolution may be difficult, as fast spectrophotometric assays can rarely be used for nitrilases [28]. Therefore, coupled assays were used that were based on the colorimetric determination of the Fe(III)–hydroxamate complex [26] or ammonia [27]. A modification of the former method was also developed for nitrilase screening, based on hydroxamic acid production from carboxylic acid in the presence of dicyclohexylcarbodiimide and hydroxylamine perchlorate [29] (Figure 12.1a). An alternative method was developed by adapting the Nessler reaction to microscale conditions [30]. Another coupled method was based on the determination of ammonia using an enzymatic reaction catalyzed by glutamate dehydrogenase in the second step [20]. In this case, the disappearance of NADH during the conversion of α-ketoglutaric acid into glutamic acid was monitored (Figure 12.1b).

12.3 NITRILASES

Microbial nitrilases have been classified into three subtypes with differing substrate specificities—arylacetonitrilases, aromatic nitrilases, and aliphatic nitrilases [3]. The first two types occur in bacteria and fungi, while the last type has only been found in bacteria so far. Plant nitrilases form a distinct group both in terms of their primary structure and substrate specificities. Their preferential substrates are 3-phenylpropionitrile and similar compounds (in NIT1 from *Arabidopsis thaliana* [31] and its homologues) or β-cyano alanine (in NIT4 from the same organism [32] and its homologues). Enzymes with significant identities to plant nitrilases (over 50%) are also found in bacteria [33, 34].

12.3.1 Arylacetonitrilases

Arylacetonitrilases are the most widely used nitrilases due to their enantioselectivities. Most application studies with them focused on the production of (*R*)-mandelic acid—a synthetic intermediate and chiral resolving agent [16, 35–39] (Figure 12.2). A similar procedure was developed for the manufacture of (*R*)-*o*-chloromandelic acid—a building block of Clopidogrel®, a blood anticoagulant agent [40].

Substrate Specificities and Selectivities

Two major groups can be distinguished in the newly obtained arylacetonitrilases which, differed in their enantioselectivities for mandelonitrile (Table 12.1). The nitrilases of the first group are highly selective for *R*-mandelonitrile. These enzymes originate from the *Alcaligenes* genus [39, 47], *Pseudomonas putida* [41], *Burkholderia cenocepacia* [17, 43], and some fungi (*A. niger*, *Nectria haematococca*, and *N. crassa*) [45, 46, 48]. The second group consists of nitrilases that are poorly enantioselective or nonenantioselective for mandelonitrile. These enzymes occur in *P. fluorescens* [42], *Burkholderia xenovorans* [16], and *Bradyrhizobium japonicum* (nitrilase bll6402) [16, 44]. A single eukaryotic enzyme of this type has so far been found in the filamentous fungus *Arthroderma benhamiae* [46]. The nitrilases from *P. fluorescens* and *N. crassa* and primarily some of their artificial variants form significant amounts of (*S*)-mandelamide from racemic mandelonitrile [23, 25, 26, 49]. These new arylacetonitrilases also differ from each other in their substrate specificities, for instance, their relative activities for mandelonitrile compared to phenylacetonitrile (Table 12.1).

The arylacetonitrilases from *A. faecalis* [50] and *B. japonicum* [51] were also able to hydrolyze dinitriles (Figure 12.3). The selectivity of both enzymes was excellent, enabling various aliphatic ω-cyanocarboxylic acids to be obtained in high yields.

α-Hydroxy and α-Amino Acid Production by Wild-Type and Mutant Enzymes

Some arylacetonitrilases were promising for the manufacture of enantiopure (*R*)-mandelic acid due to their high activities and enantioselectivities for mandelonitrile, but high concentrations of the substrate exhibited a negative effect on the conversion rate and product ee. Therefore, various methods were examined to eliminate this

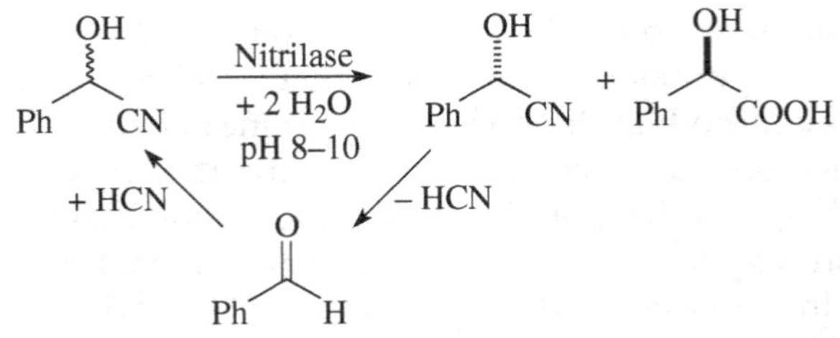

FIGURE 12.2 Production of (*R*)-mandelic acid from racemic mandelonitrile at alkaline pH [16, 35–39].

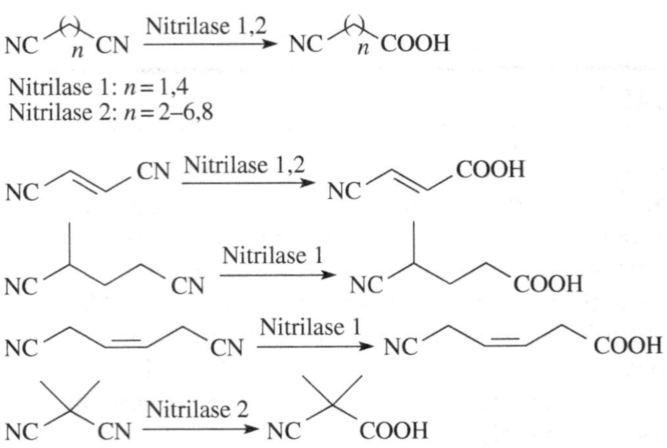

FIGURE 12.3 Hydrolysis of aliphatic dinitriles into cyano carboxylic acids by nitrilases (nitrilase 1 from *Alcaligenes faecalis*, nitrilase 2 from *Bradyrhizobium japonicum*) [50, 51].

TABLE 12.1 Catalytic Properties of Purified Arylacetonitrilases

Enzyme Source	Substrate (U/mg)	Product from (*R*,*S*)-Mandelonitrile: Amide (% in Total Product)	Product from (*R*,*S*)-Mandelonitrile: Product ee (Acid/Amide, %)	References
Alcaligenes sp.[a,b]	Phenylacetonitrile (103) Mandelonitrile (19) 2-Phenylglycinonitrile (17)	n.d.	97 (*R*)/–	[39]
Pseudomonas putida[b,c]	Phenylacetonitrile (11)[d] Indole-3-acetonitrile (10) Mandelonitrile (3.3)	n.d.	99.9 (*R*)/–	[35, 41]
Pseudomonas fluorescens[b]	Phenylacetonitrile (68) 2-Thiopheneacetonitrile (205) 2-Phenylvaleronitrile (230) Mandelonitrile (33)	19	31 (*R*)/85 (*S*)	[42]
Burkholderia cenocepacia[b]	Phenylacetonitrile (3.1) Mandelonitrile (28) 2-Chloromandelonitrile (9.3)	n.d.	98.4 (*R*)/–	[17]
Burkholderia xenovorans[b]	Mandelonitrile (1.5) Adiponitrile (0.51) Glutaronitrile (0.34)	n.d.	0	[16]
Bradyrhizobium japonicum[b]	Mandelonitrile (24) Phenylacetonitrile (5.3)	n.d.	0	[44]
Aspergillus niger[b]	Phenylacetonitrile (11) Mandelonitrile (12)[d]	n.d.	93 (*R*)[e]/–	[25, 45]
Arthroderma benhamiae[b]	Phenylacetonitrile (16) Mandelonitrile (15)	1.6	63 (*R*)/n.a.	[46]
Nectria haematococca[b]	Phenylacetonitrile (64) Mandelonitrile (17)	6.2	89 (*R*)/n.a.	[46]
Neurospora crassa[b]	Phenylacetonitrile (17.5)[d] Mandelonitrile (10)[d]	40	91 (*R*)/97 (*S*)[e]	[25, 45]

n.d., not detected; n.a., not assayed.

[a] Other similar enzymes were characterized in the same genus [3, 37, 47].

[b] Purified from *Escherichia coli*.

[c] Purified from wild-type producing strain.

[d] V_{max}.

[e] Determined with whole cells.

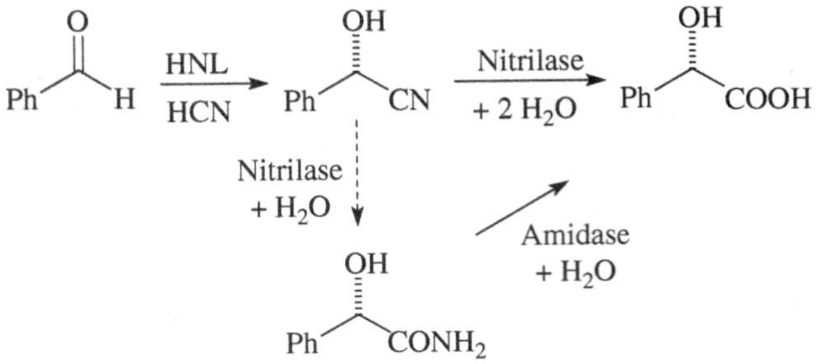

FIGURE 12.4
Synthesis of (*S*)-mandelic acid using *S*-selective hydroxynitrile lyase (HNL) from *Manihot esculenta*, nonselective nitrilase from *Pseudomonas fluorescens* EBC 191, and amidase from *Rhodococcus erythropolis* [55].

drawback, and performing the process in fed-batch mode or addition of an immiscible cosolvent (toluene, ethyl acetate) was found to alleviate the effect of high substrate concentrations [37, 38, 43]. Thus (*R*)-mandelic acid was produced at a concentration of 0.48 M with 98% ee by the nitrilase from *Alcaligenes* sp. [38]. Furthermore, the hydrolysis of 1 M mandelonitrile was achieved with immobilized *Escherichia coli* expressing the nitrilase gene from *B. cenocepacia*, yielding (*R*)-mandelic acid with a 99% yield and 95% ee [43].

In a similar process, an arylacetonitrilase from *Labrenzia aggregata* was used for the manufacture of (*R*)-*o*-chloromandelic acid (0.276 M, 96% ee) [40].

Arylacetonitrilases with low or no enantioselectivity for mandelonitrile may be employed in the manufacture of (*S*)-mandelic acid. This method consists of an enzymatic synthesis of (*S*)-mandelonitrile from benzaldehyde and HCN, followed by nonselective enzymatic hydrolysis of the nitrile at low pH [52, 53] and, optionally, with ionic liquids as cosolvents [54]. Coupling of the third enzyme, amidase enabled the conversion of the by-product (*S*)-mandelamide into (*S*)-mandelic acid [55] (Figure 12.4). Using site-directed mutagenesis, variants of the enzymes from *P. fluorescens* and *N. crassa* were constructed, which produced mandelamide at up to 70–90% of the total product [23, 25, 49].

D-Phenylglycine and analogs were produced from α-amino nitriles in a similar way but the nitrilases were different. Suitable enzymes were selected from the metagenomic nitrilase library, which were able to operate at a pH high enough for substrate racemization [56]. This process was recently also performed with a new nitrilase from *S. wittichii* and optimized using a biphasic organoaqueous system with 1-octanol as the organic solvent [57].

β-Hydroxy Acid Production by a Bioenzymatic Process

The enzyme mentioned above from *B. japonicum* also transformed β-hydroxynitriles into acids but with low enantioselectivities. This problem was circumvented by using optically pure β-hydroxynitriles as substrates, which were prepared by the enantioselective reduction of β-ketonitriles catalyzed by carbonyl reductase [58]. The advantage of the nitrilase-catalyzed step was its mild and stereoretentive conditions. In this way, various substituted 3-aryl-3-hydroxypropanoic acids were prepared in both enantiopure forms (Figure 12.5).

Production of Alicyclic and Heteroalicyclic Amino Acids

Nitrilases suitable for the transformations of alicyclic five- and six-membered γ-amino nitriles were selected from the set of commercial nitrilases available from BioCatalytics Inc. (now Codexis) [59] and the effects of the substrate structure (ring size, protecting group, and *trans* versus *cis* configuration) were studied. *N*-tosylated derivatives of *cis*-3-amino cyclopentane- and cyclohexanecarboxylic acids were obtained with the highest enantiopurities. Production of *N*-substituted pyrrolidine- and piperidinecarboxylic acids was also possible using the same nitrilases, but the enantiopurities of these products were generally low [60].

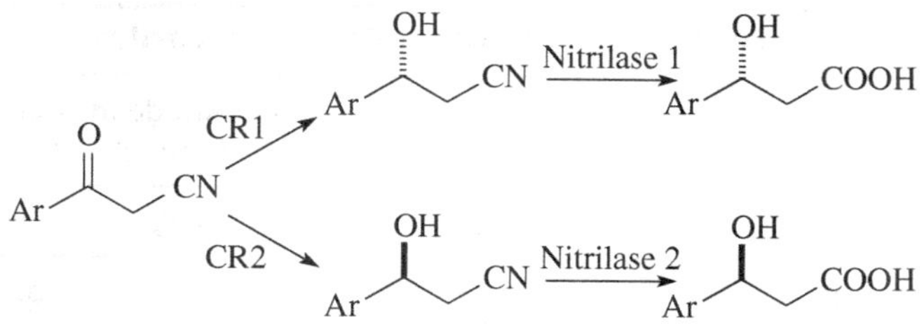

FIGURE 12.5

Synthesis of (*S*)- and (*R*)-3-aryl-3-hydroxypropanoic acids using different carbonyl reductases (CR1, carbonyl reductase from *Candida magnoliae*; CR2, alcohol dehydrogenase from *Saccharomyces cerevisiae*) and different nitrilases (nitrilase 1 from *Synechocystis* sp., nitrilase 2 from *Bradyrhizobium japonicum*) [58].

12.3.2 Aromatic Nitrilases

Aromatic nitrilases were first described in *Rhodococci* in the 1980s [3]. The enzyme from *Rhodococcus rhodochrous* J1 is one of the best characterized enzymes of this type [61]. A similar enzyme with a moderate level of identity (61.6%) was recently described in *P. putida* [62].

Aromatic nitrilases of fungal origin have also been known for decades [3], but their amino acid (AA) sequences were only determined later by MS analysis [63–66]. These sequences were similar to those of enzymes from *Gibberella moniliformis* and *Gibberella intermedia*, which were then produced in *E. coli* [43, 67] and confirmed as aromatic nitrilases.

Production of Heterocyclic Carboxylic Acids

Superior substrates for aromatic nitrilases are 3- and 4-cyanopyridine (Table 12.2), which are industrially important as precursors of nicotinic acid (vitamin B3) and isonicotinic acid, respectively. The latter serves to synthesize isonicotinic acid hydrazide, a tuberculostatic drug. The production of niconitic or isonicotinic acid was examined with various types of catalysts such as immobilized nitrilase [65] or whole cells of *E. coli* producing bacterial or fungal nitrilases [18, 69, 70]. Fungal nitrilases exhibited high specific activities for these substrates, especially for 4-cyanopyridine, but some of them, such as the nitrilase from *A. niger*, may produce high amounts of isonicotinamide as a by-product. However, this effect may be alleviated by adding amidase to convert the amide into carboxylic acid [71].

2-Cyanopyridine (precursor of picolinic acid, intermediate of pharmaceuticals) is transformed by aromatic nitrilases at lower relative rates compared to arylacetonitrilases. Surprisingly, the cyanide hydratase from *A. niger* K10 exhibited the highest specific activity for this substrate among the enzymes examined [18, 72]. This enzyme, as well as other cyanide hydratases (from the *Fusarium* genus), accepted a number of nitrile substrates, albeit at much lower relative rates than HCN [3, 72].

Transformation of Acrylonitrile

The majority of aromatic nitrilases hydrolyze not only (hetero) aromatic but also aliphatic saturated or unsaturated nitriles (Table 12.2). The ability of some of the enzymes to hydrolyze acrylonitrile has an industrial impact, as it enables the synthesis of acrylic acid (a polymer building block), the elimination of acrylonitrile from wastewaters [73], or the construction of acrylonitrile biosensors [74]. The nitrilase from *R. rhodochrous* J1 required activation by ammonium sulfate prior to use for the transformation of acrylonitrile. This activation resided in a rearrangement of the enzyme's quaternary structure, resulting in an increase in molecular weight from

TABLE 12.2 Catalytic Properties of Purified Aromatic Nitrilases

Enzyme Source	Substrate (U/mg)	Amide in Total Product (%)		References
		3-CP	4-CP	
Rhodococcus rhodochrous[a,b]	Benzonitrile (15) 3-Cyanopyridine (11) Acrylonitrile (8.6)	n.d.	n.d.	[61, 68]
Pseudomonas putida[a]	3-Cyanopyridine (54) 4-Cyanopyridine (43) Benzonitrile (38) Acrylonitrile (9.4)	n.d.	n.d.	[62]
Aspergillus niger[a,c]	4-Cyanopyridine (376) Benzonitrile (92) 3-Cyanopyridine (30) Valeronitrile (18) Phenylacetonitrile (10)	8.9	33.8	[63, 64]
Fusarium solani (strain CCF 3635)[b,c]	4-Cyanopyridine (203) Benzonitrile (156) 3-Cyanopyridine (44) Valeronitrile (41)	<3	<3	[65]
F. solani (strain IMI 196840)[b]	4-Cyanopyridine (312) Benzonitrile (144) 3-Cyanopyridine (59)	<2	<2	[66]
Gibberella intermedia[a,d]	Adiponitrile (3.0) 3-Cyanopyridine (2.8) 4-Cyanopyridine (1.9) Benzonitrile (1.2) Valeronitrile (1.3) Acrylonitrile (1.2) Phenylacetonitrile (0.60)	n.a.	n.a.	[67]
Gibberella moniliformis[a]	Benzonitrile (9.7)[e] 4-Cyanopyridine (7.3)[e] 3-Cyanopyridine (4.2)[e] Phenylacetonitrile (1.2)[e]	0	54	[45]

n.a., not assayed; n.d., not detected.
[a] Purified from *Escherichia coli*.
[b] Purified from wild-type strain.
[c] A similar nitrilase was purified from a wild-type strain of *Fusarium oxysporum* f. sp. *melonis* but its AA sequence is unknown.
[d] Identity between nitrilase from *G. moniliformis* and *G. intermedia* is 97%.
[e] V_{max}.

80 (dimer) to 410 kDa (probably a decamer) [68]. In addition, the enzyme subunit underwent a self-catalyzed cleavage of its C-terminal part during long-term storage. This posttranslational modification was associated with an increase in molecular weight to 480 kDa (probably a dodecamer) [61].

12.3.3 Aliphatic Nitrilases

The first enzyme designated as aliphatic nitrilase was from *R. rhodochrous* K22 [3]. Later, a nitrilase from *Acidovorax facilis* was also classified to this group, the similarity of this enzyme to the nitrilase from the K22 strain being relatively high (68%). The enzyme from *A. facilis* was primarily utilized in the hydrolysis of dinitriles such as 2-methylglutaronitrile [75] and iminodiacetonitrile [76]. Other nitrilases of this type, which were specific for fumaronitrile and malononitrile, were then reported in *Synechocystis* sp. [77] and *Pyrococcus abyssi* [78]. The nitrilase from *Synechocystis* sp.

was used for the monohydrolysis of aliphatic α,ω-dinitriles with four to seven carbon atoms, dimethylmalononitrile and fumaronitrile, cyano acids being obtained in good to excellent yields [79]. The identities of these enzymes to each other and to previously characterized aliphatic nitrilases were low (20–30%). The highly thermostable enzyme from *P. abyssi* was the first crystallized nitrilase [80], but its sequence similarity to other characterized nitrilases is low. This makes its utility for homology modeling limited (Table 12.3).

12.3.4 Plant Nitrilases and their Bacterial Homologues

Plant nitrilases and related enzymes in some plant-associated bacteria preferentially utilize arylaliphatic nitriles such as 3-phenylpropionitrile or 3-phenylbutyronitrile. These enzymes also exhibit remarkable activities for aliphatic nitriles (Table 12.4).

The nitrilase NIT1 in *A. thaliana*, which was overproduced in *E. coli*, proved to have a wide synthetic use, its products being aliphatic and arylaliphatic carboxylic acids or amides depending on the structure of the substrate [81]. A similar nitrilase (ZmNIT2) was characterized in *Zea mays* (maize) [82]. This enzyme also

TABLE 12.3 Catalytic Properties of Purified Aliphatic Nitrilases

Enzyme Source	Substrate (U/mg)	Product from Fumaronitrile (% Cyano Acid)	References
Acidovorax facilis[a]	2-Methylglutaronitrile (21)	98	[75]
Synechocystis sp.[a]	Fumaronitrile (109) Indole-3-acetonitrile (1.5) 2-Chloropropionitrile (1.4)[b]	100	[77]
Pyrococcus abyssi[a]	Malononitrile (0.14) Fumaronitrile (n.r.)	100	[78, 80]

n.r., not reported.
[a] Purified from *Escherichia coli*.
[b] V_{max}.

TABLE 12.4 Catalytic Properties of Plant Nitrilases and their Bacterial Homologues

Enzyme Source[a]	Substrate (U/mg)	Amide Production: Substrate (% in Total Product)	References
Arabidopsis thaliana	3-Phenylpropionitrile (10) Octanenitrile (4.0) 4-Phenylbutyronitrile (2.1)	Fumaronitrile (93) 3-Nitroacrylonitrile (95) α-Fluorophenylacetonitrile (85)	[81]
Zea mays	4-Phenylbutyronitrile (2.0)[b] 3-Phenylpropionitrile (0.35)[b] Indole-3-acetonitrile (0.29)[b]	Analogues of 3-hydroxy-3-phenylpropionitrile (63–88)	[82, 83]
Bradyrhizobium japonicum	Hydrocinnamonitrile (10) Heptanenitrile (2.5) Phenylacetonitrile (2.3) 4-Phenylbutyronitrile (2.0)	n.d.	[84]
Syntrophobacter fumaroxidans	3-Phenylpropionitrile (n.r.) *trans*-Cinnamonitrile (n.r.) Glutaronitrile (n.r.) Thiopheneacetonitrile (n.r.)	n.d.	[20]

n.d., not detected; n.r., not reported.
[a] Purified from *Escherichia coli*.
[b] V_{max} (determined in crude extracts).

has a number of potential synthetic applications such as the transformation of β-hydroxynitriles into amides as major products [83]. A bacterial nitrilase (blr3397) similar to plant nitrilases with over 50% identity was characterized in *B. japonicum*, this enzyme differing from the nitrilase mentioned above (bll6402) from the same organism in its substrate specificity. In contrast to the previous enzyme, nitrilase blr3397 hydrolyzed mandelonitrile with low relative rates but its preferential substrates were hydrocinnamonitrile, 2-phenylacetonitrile, 4-phenylbutyronitrile, and some aliphatic nitriles [84]. Another characterized nitrilase of this type was from *S. fumaroxidans* [20]. This enzyme also exhibited a broad substrate specificity, and its synthetic use was mainly demonstrated with 3-oxocyclopentanecarbonitrile as substrate, which was hydrolyzed with a moderate *S*-selectivity [20].

12.4 NITRILE HYDRATASES

Nitrile hydratases (NHases) exhibit broader substrate specificities than nitrilases, and they are more tolerant of sterically demanding substrates. In contrast, NHases mainly exhibit lower enantioselectivities than nitrilases, although biotransformations of a few specific nitriles by NHase proceeded with excellent enantioselectivities [5, 6]. Low NHase enantioselectivity, however, can be compensated for by using enantioselective amidases in the next step. NHases are usually less thermostable than nitrilases, but a few of them are resistant to increased temperatures or organic solvent concentrations [8, 9]. The majority of characterized NHases are Fe or Co type, except for a new NHase with three metal ions (Co, Cu, Zn) reported in *Rhodococcus jostii* [85].

12.4.1 Fe-type Nitrile Hydratase

Fe-type NHases are typically produced by *Rhodococcus erythropolis*, and only a few of them are known in other organisms such as *Bacillus* sp., *Pseudomonas chlororaphis* [7] or *P. putida* [86]. NHases share a high level of amino acid sequence identities, which leads us to expect similar catalytic properties. Apparent differences in the substrate specificities of NHases of different origin may also be due to different assays used. One of the least similar NHases obtained from *P. putida* by genome mining still exhibited remarkable identities to the well-known *Rhodococcal* NHases (about 62 and 57% in the α and β subunit, respectively) [86].

The majority of the synthetic applications of this NHase type were demonstrated with the enzyme from *R. erythropolis* AJ270 [5, 87, 88]. The NHase in this strain shares a high degree of amino acid sequence identity to many other Fe-type NHases. The same holds for the similarities of the enantioselective amidases in this and other strains. Therefore, it is likely that other strains may be also used for these applications.

Using whole cells of this and other strains of *Rhodococci*, a number of applications of NHases and amidases were developed and, in some cases, integrated into multistep routes leading to valuable building blocks (Table 12.5). The products were enantioenriched or enantiopure α-, β-, and γ-substituted carboxylic acids or carboxamides, cyano carboxamides, cyano carboxylic acids and (hetero) cyclic carboxylic acids, and carboxamides. Thus, for instance, both enantiomers of cyclophellitol and *epi*-cyclophelitol (glucosidase inhibitors) were prepared using a cascade of chemical steps and three enzymes—lipase, NHase, and amidase—where stereodiscrimination mainly took place in the lipase-catalyzed reaction [94, 95] (Figure 12.6).

The NHase and amidase were largely nonselective for cyanohydrins and the corresponding 2-hydroxy amides, respectively, but they were suitable for the hydrolysis of enantiopure cyanohydrins prepared from aldehydes and HCN by oxynitrilases [89, 90]. The cascade of NHase and amidase, in which the latter enzyme catalyzed an acyl transfer reaction, was suitable for the preparation of

TABLE 12.5 Synthetic Applications of Fe-type NHases from Various Strains of *Rhodococci*

Enzyme Source	Substrate(s)	Product(s)[a]	Applications	References
Rhodococcus erythropolis NCIMB 11540	Cyanohydrins	2-Hydroxy acids 2-Hydroxy amides	Enantioretentive transformation of enantiopure cyanohydrins	[89–91]
R. erythropolis A4[b]	Aromatic and aliphatic nitriles	Aromatic and aliphatic hydroxamic acids	Bienzymatic synthesis of hydroxamic acids (chelators); NHase screening	[92]
R. erythropolis A4[b]	Heterocyclic dinitriles	Heterocyclic cyano amides and cyano acids	Synthesis of functionalized heterocyclic compounds	[93]
R. erythropolis A4[b]	Cyano cyclitols	Cyclohexenecarboxylic acids	Chemoenzymatic synthesis of cyclophellitol or *epi*-cyclophellitol	[94, 95]
R. erythropolis A4	7-Ferrocenyl heptanenitrile	7-Ferrocenyl heptanamide 7-Ferrocenyl heptanoic acid	Chemoenzymatic synthesis of ferrocene derivatives	[96]
Rhodococcus sp. R-312[c]	2,3-Dihydro-1,4-benzodioxine-2-carbonitriles	(*R*)-2,3-Dihydro-1,4-benzodioxine-2-carboxylic acids	Chemoenzymatic synthesis of building blocks	[97]
Rhodococcus sp. R-312[c] *R. erythropolis* A4[b] *R. erythropolis* NCIMB 11540	β-Amino nitriles	β-Amino amides and carboxylic acids	Chemoenzymatic synthesis of unnatural amino acids and oligopeptides	[98]
R. erythropolis AJ270	3-Aryl-3-hydroxy-2-methylenepropionitriles	(*S*)-3-Aryl-3-hydroxy-2-methylenepropionamides (*R*)-3-Aryl-3-hydroxy-2-methylenepropionic acids	Chemoenzymatic synthesis of Baylis-Hillman adducts	[99]
R. erythropolis AJ270	Cyclopropane, oxirane, aziridine carbonitriles	Enantiopure *trans*- and *cis*-cyclopropanecarboxylic acids and carboxamides, oxirane and aziridine carboxamides	Chemoenzymatic synthesis of various building blocks	[100–104]
R. erythropolis AJ270	(4-Oxo)Azetidine-2-carbonitriles	Enantiopure (4-oxo)azetidine-2-carboxylic acids and carboxamides	Chemoenzymatic synthesis of various building blocks	[105, 106]
R. erythropolis AJ270	2,3-Allenenitriles	(*R*)-2,3-Allenamides (*S*)-2,3-Allenoic acids	Chemoenzymatic synthesis of functionalized heterocyclic compounds	[107]

[a] Amides hydrolyzed by enantioselective amidase in the same strain.
[b] Formerly *Rhodococcus equi*.
[c] Reclassified as *R. erythropolis* [108].

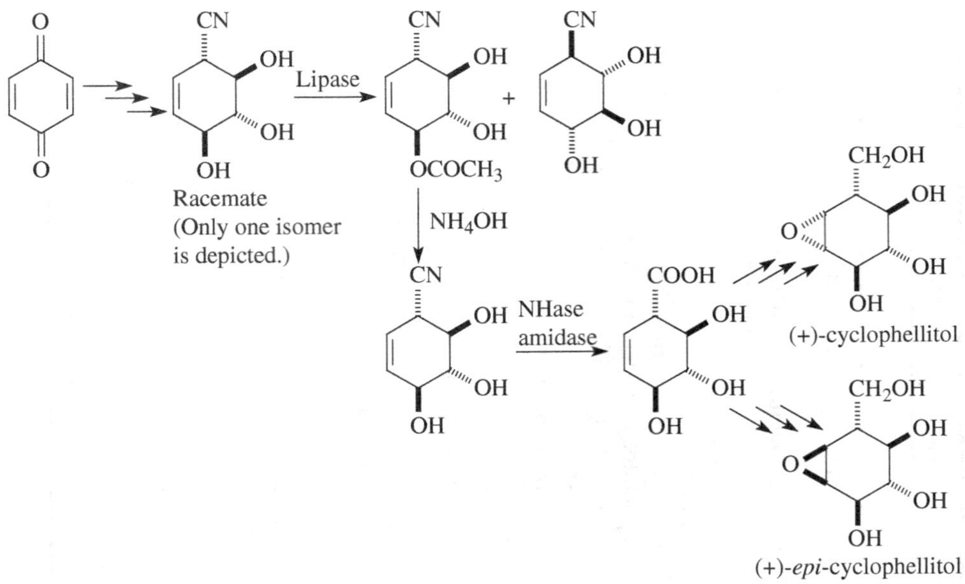

FIGURE 12.6 Synthesis of (+)-cyclophellitol and (+)-*epi*-cyclophellitol from *p*-benzoquinone using lipase from *Rhizomucor miehei*, nitrile hydratase (NHase)/amidase complex from *Rhodococcus erythropolis*, and chemical steps. (−)-Cyclophellitol and (−)-*epi*-cyclophellitol were prepared in an analogous way from the unesterified nitrile isomer [94, 95].

TABLE 12.6 Synthetic Applications of Co-type NHases

Enzyme Source	Substrate(s)	Product(s)	Applications	References
Rhodococcus rhodochrous ATTC BAA-870	4-Aryloxy-3-hydroxybutanenitriles	(*R*)-4-Aryloxy-3-hydroxybutanamides, (*S*)-4-aryloxy-3-hydroxybutanoic acids[a]	Chemoenzymatic synthesis of building blocks	[109]
	3-Amino-3-arylpropanenitriles	(*S*)-3-Amino-3-phenylpropanamides		[110]
R. rhodochrous PA-34[b]	Ferrocene nitrile, ferrocene acetonitrile (6-nitrile hexyl)ferrocene	Amide and acid analogues	Chemoenzymatic synthesis of ferrocene derivatives	[96]
Rhodococcus opacus 71D	α-Aminonitriles	(*R*)- or (*S*)-α-Amino acids[c]	Multienzymatic synthesis of enantiopure α-amino acids	[111, 112]

[a] Amides hydrolyzed by enantioselective amidase in the same strain.
[b] Reclassified as *Rhodococcus ruber* [18].
[c] Amides hydrolyzed by enantioselective amidases from other organism.

hydroxamic acids from nitriles, without requiring intermediate (amide) isolation. This bienzymatic reaction was also applicable for the spectrophotometric screening of NHase activities [91].

12.4.2 Co-type Nitrile Hydratase

Co-type NHases are typical for *R. rhodochrous*, in which two subtypes of this enzyme were found—a low-molecular-weight and a high-molecular-weight NHase consisting of four and 18–20 subunits, respectively. The latter enzyme seems to only occur in this species. A few Co-type NHases were found in *Pseudonocardia thermophila* and *P. putida*; these enzymes have low molecular weights. The high-molecular-weight NHase, which is resistant to heat and cosolvents, has been industrially exploited in bulk chemical (acrylamide, nicotinamide) production, immobilized whole cells of *R. rhodochrous* J1 being used as catalyst [6, 7, 113].

The synthetic utility of this NHase, produced together with an enantioselective amidase, was demonstrated in the synthesis of enantiopure or enantioenriched α- and β-substituted amides and acids (Table 12.6). Thus β-hydroxy amides and acids

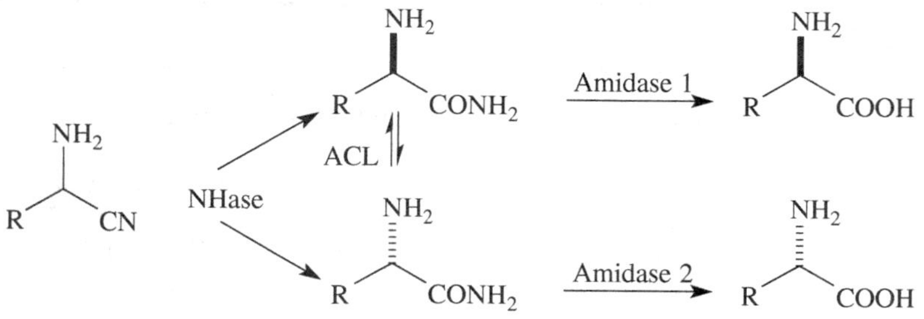

FIGURE 12.7

Synthesis of (*S*)- and (*R*)-α-amino acids using nitrile hydratase (NHase) from *Rhodococcus opacus*, α-amino-ε-caprolactam racemase (ACL) from *Achromobacter obae*, and different amidases (amidase 1, D-amino-peptidase from *Ochrobactrum anthropi*; amidase 2, L-amino acid amidase from *Brevundimonas diminuta*) [111, 112]. *R*=CH_3, CH_3CH_2, $CH_3CH(CH_3)$, $CH_3CH(CH_3)CH_2$.

and β-amino amides were prepared in differing enantiopurities from the corresponding nitriles, their stereochemistry being mainly recognized by the amidase in the same organism, *R. rhodochrous* ATTC BAA-870 [109, 110]. A similar bienzymatic conversion of α-amino nitriles to enantiopure α-amino acids was recently optimized by using a racemase that converted the nonreacted amide enantiomer to the amidase substrate [109, 110, 111, 112] (Figure 12.7).

The Co-type NHase proved to be more suitable for the transformation of ferrocenes compared to the Fe-type NHase, as ferrocene nitrile and ferrocene acetonitrile were only accepted by the former NHase [96]. Differences in the catalytic properties of Co- and Fe-type NHases were also demonstrated by comparing the enantioselectivities of the enzymes for various substituted phenylacetonitriles, the majority of which was transformed with high *E*-values by the former enzyme but with low *E*-values by the latter [114].

12.5 CONCLUSIONS

The use of modern screening methods (metagenomic screening, genome mining) began a new phase in the studies of nitrile-converting enzymes. The knowledge of structure–activity relationships in these enzymes is being continuously refined and enables new enzymes with the required catalytic properties to be obtained. However, only a small proportion of the newly described enzymes are fit for industrial applications. In many cases, improvements need to be made to the enzyme stability, the volumetric and catalyst productivity, and the product purity. To this end, the construction of enzyme variants and high-throughput screening systems are being increasingly used. The need to employ the enzymes in multistep processes is also being emphasized. These approaches are likely to lead to a more intensive exploitation of these enzymes in industrial biotechnology and to provide economically viable and environment-friendly processes for the production of, for example, bioactive compounds and the building blocks of pharmaceuticals.

ACKNOWLEDGEMENTS

Financial support via project P504/11/0394 and internal project RVO61388971 (Institute of Microbiology of the Academy of Sciences of the Czech Republic) is gratefully acknowledged.

REFERENCES

[1] Martínková, L. and Křen, V. (2010). Biotransformations with nitrilases. *Current Opinion in Chemical Biology*, **14**, 130–137.
[2] Thuku, R.N., Brady, D., Benedik, M.J., and Sewell, B.T. (2009). Microbial nitrilases: Versatile, spiral forming, industrial enzymes. *Journal of Applied Microbiology*, **106**, 703–727.
[3] O'Reilly, C. and Turner, P.D. (2003). The nitrilase family of CN hydrolysing enzymes: A comparative study. *Journal of Applied Microbiology*, **95**, 1161–1174.
[4] Martínková, L., Vejvoda, V., Kaplan, O., Kubáč, D., Malandra, A., Cantarella, M., Bezouška, K., and Křen, V. (2009). Fungal nitrilases as biocatalysts: Recent developments. *Biotechnology Advances*, **27**, 661–670.
[5] Wang, M.-X. (2009). Progress of enantioselective nitrile biotransformations in organic synthesis. *CHIMIA International Journal for Chemistry*, **63**, 331–333.
[6] Prasad, S. and Bhalla, T.C. (2010). Nitrile hydratases (NHases): At the interface of academia and industry. *Biotechnology Advances*, **28**, 725–741.
[7] Banerjee, A., Sharma, R., and Banerjee, U.C. (2002). The nitrile-degrading enzymes: Current status and future prospects. *Applied Microbiology and Biotechnology*, **60**, 33–44.
[8] Gong, J.-S., Lu, Z.-M., Li, H., Shi, J.-S., Zhou, Z.-M., and Xu, Z.-H. (2012). Nitrilases in nitrile biocatalysis: Recent progress and forthcoming research. *Microbial Cell Factories*, **11**, 142.
[9] Gong, J.-S., Lu, Z.-M., Li, H., Zhou, Z.-M., Shi, J.-S., Xu, and Z.-H. (2013). Metagenomic technology and genome mining: Emerging areas for exploring novel nitrilases. *Applied Microbiology and Biotechnology*, **97**, 6603–6611.
[10] www.codexis.com (accessed on 12 December, 2015).
[11] www.prozomix.com (accessed on 12 December, 2015).
[12] Robertson, D.E., Chaplin, J.A., DeSantis, G., Podar, M., Madden, M., Chi, E., Richardson, T., Milan, A., Miller, M., Weiner, D.P., Wong, K., McQuaid, J., Farwell, B., Preston, L.A., Tan, X., Snead, M.A., Keller, M., Mathur, E., Kretz, P.L., Burk, M.J., and Short, J.M. (2004). Exploring nitrilase sequence space for enantioselective catalysis. *Applied and Environmental Microbiology*, **70**, 2429–2436.
[13] DeSantis, G., Wong, K., Farwell, B., Chatman, K., Zhu, Z., Tomlinson, G., Huang, H., Tan, X., Bibbs, L., Chen, P., Kretz, K., and Burk, M.J. (2003). Creation of a productive, highly enantioselective nitrilase through gene site saturation mutagenesis (GSSM). *Journal of the American Chemical Society*, **125**, 11476–11477.
[14] Bayer, S., Birkemeyer, C., and Ballschmiter, M. (2011). A nitrilase from a metagenomic library acts regioselectively on aliphatic dinitriles. *Applied and Environmental Microbiology*, **89**, 91–98.
[15] Liebeton, K. and Eck, J. (2004). Identification and expression in *E. coli* of novel nitrile hydratases from the metagenome. *Engineering in Life Sciences*, **4**, 557–562.
[16] Seffernick, J.L., Samanta, S.K., Louie, T.M., Wackett, L.P., and Subramanian, M. (2009). Investigative mining of sequence data for novel enzymes: A case study with nitrilases. *Journal of Biotechnology*, **143**, 17–26.
[17] Wang, H., Sun, H., and Wei, D. (2013). Discovery and characterization of a highly efficient enantioselective mandelonitrile hydrolase from *Burkholderia cenocepacia* J2315 by phylogeny-based enzymatic substrate specificity prediction. *BMC Biotechnology*, **13**, 14.
[18] Kaplan, O., Veselá, A.B., Petříčková, A., Pasquarelli, F., Pičmanová, M., Rinágelová, A., Bhalla, T.C., Pátek, M., and Martínková, L. (2013). A comparative study of nitrilases identified by genome mining. *Molecular Biotechnology*, **54**, 996–1003.
[19] Liu, Z.-Q., Baker, P.J., Cheng, F., Xue, Y.-P., Zheng, Y.G., and Shen, Y.-C. (2013). Screening and improving the recombinant nitrilases and application in biotransformation of iminodiacetonitrile to iminodiacetic acid. *PLoS ONE*, **8**, e67197.
[20] Vergne-Vaxelaire, C., Bordier, F., Fossey, A., Besnard-Gonnet, M., Debard, A., Mariage, A., Pellouin, V., Perret, A., Petit, J.-L., Stam, M., Salanoubat, M., Weissenbach, J., De Berardinis, V., Zaparucha, A. (2013). Nitrilase activity screening on structurally diverse substrates: Providing biocatalytic tools for organic synthesis. *Advanced Synthesis & Catalysis*, **355**, 1763–1779.
[21] Kiziak, C., Klein, J., and Stolz, A. (2007). Influence of different carboxy-terminal mutations on the substrate-, reaction- and enantiospecificity of the arylacetonitrilase from *Pseudomonas fluorescens* EBC191. *Protein Engineering Design and Selection*, **20**, 385–396.
[22] Kiziak, C. and Stolz, A. (2009). Identification of amino acid residues responsible for the enantioselectivity and amide formation capacity of the arylacetonitrilase from *Pseudomonas fluorescens* EBC191. *Applied and Environmental Microbiology*, **75**, 5592–5599.

[23] Sosedov, O., Baum, S., Bürger, S., Matzer, K., Kiziak, C., and Stolz, A. (2010). Construction and application of variants of the *Pseudomonas fluorescens* EBC191 arylacetonitrilase for increased production of acids or amides. *Applied and Environmental Microbiology*, **76**, 3668–3674.

[24] Baum, S., Williamson, D.S., Sewell, T., and Stolz, A. (2012). Conversion of sterically demanding α, α-disubstituted phenylacetonitriles by the arylacetonitrilase from *Pseudomonas fluorescens* EBC191. *Applied and Environmental Microbiology*, **78**, 48–57.

[25] Petříčková, A., Sosedov, O., Baum, S., Stolz, A., and Martínková, L. (2012). Influence of point mutations near the active site on catalytic properties of fungal arylacetonitrilases from *Aspergillus niger* and *Neurospora crassa*. *Journal of Molecular Catalysis B: Enzymatic*, **77**, 74–80.

[26] Sosedov, O. and Stolz, A. (2014). Random mutagenesis of the arylacetonitrilase from *Pseudomonas fluorescens* EBC191 and identification of variants, which form increased amounts of mandeloamide from mandelonitrile. *Applied Microbiology and Biotechnology*, **98**, 1595–1607.

[27] Schreiner, U., Hecher, B., Obrowsky, S., Waich, K., Klempier, N., Steinkellner, G., Gruber, K., Rozzell, J.D., Glieder, A., and Winkler, M. (2010). Directed evolution of *Alcaligenes faecalis* nitrilase. *Enzyme and Microbial Technology*, **47**, 140–146.

[28] Martínková, L., Vejvoda, V., and Křen, V. (2008). Selection and screening for enzymes of nitrile metabolism. *Journal of Biotechnology*, **133**, 318–326.

[29] He, Y.-C., Ma, C.-L., Xu, J.-H., and Zhou, L. (2011). A high-throughput screening strategy for nitrile-hydrolyzing enzymes based on ferric hydroxamate spectrophotometry. *Applied Microbiology and Biotechnology*, **89**, 817–823.

[30] Coady, T.M., Coffey, L.V., O'Reilly, C., Owens, E.B., and Lennon, C.M. (2013). A high throughput screening strategy for the assessment of nitrile-hydrolyzing activity towards the production of enantiopure β-hydroxy acids. *Journal of Molecular Catalysis B: Enzymatic*, **97**, 150–155.

[31] Vorwerk, S., Biernacki, S., Hillebrand, H., Janzik, I., Müller, A., Weiler, E.W., and Piotrowski, M. (2001). Enzymatic characterization of the recombinant *Arabidopsis thaliana* nitrilase subfamily encoded by the *NIT2/NIT1/NIT3*-gene cluster. *Planta*, **212**, 508–516.

[32] Piotrowski, M., Schönfelder, S., and Weiler, E.W. (2001). The *Arabidopsis thaliana* isogene *NIT4* and its orthologs in tobacco encode β-cyano-L-alanine hydratase/nitrilase. *The Journal of Biological Chemistry*, **276**, 2616–2621.

[33] Howden, A.J.M., Harrison, C.J., and Preston, G.M. (2009). A conserved mechanism for nitrile metabolism in bacteria and plants. *The Plant Journal*, **57**, 243–253.

[34] Howden, A.J.M. and Preston, G.M. (2009). Nitrilase enzymes and their role in plant-microbe interactions. *Microbial Biotechnology*, **2**, 441–451.

[35] Banerjee, A., Dubey, S., Kaul, P., Barse, B., Piotrowski, M., and Banerjee, U.C. (2009). Enantioselective nitrilase from *Pseudomonas putida*: Cloning, heterologous expression, and bioreactor studies. *Molecular Biotechnology*, **41**, 35–41.

[36] Detzel, C., Maas, R., and Jose, J. (2011). Autodisplay of nitrilase from *Alcaligenes faecalis* in *E. coli* yields a whole cell biocatalyst for the synthesis of enantiomerically pure (*R*)-mandelic acid. *ChemCatChem*, **3**, 719–725.

[37] Zhang, Z.-J., Xu, J.-H., He, Y.-C., Ouyang, L.-M., Liu, Y.-Y., and Imanaka, T. (2010). Efficient production of (*R*)-(−)-mandelic acid with highly substrate/product tolerant and enantioselective nitrilase of recombinant *Alcaligenes* sp. *Process Biochemistry*, **45**, 887–891.

[38] Zhang, Z.-J., Pan, J., Liu, J.-F., Xu, J.-H., He, Y.-C., and Liu, Y.-Y. (2011). Significant enhancement of (*R*)-mandelic acid production by relieving substrate inhibition of recombinant nitrilase in toluene–water biphasic system. *Journal of Biotechnology*, **152**, 24–29.

[39] Zhang, Z.-J., Xu, J.-H., He, Y.-C., Ouyang, L.-M., and Liu, Y.-Y. (2011). Cloning and biochemical properties of a highly thermostable and enantioselective nitrilase from *Alcaligenes* sp. ECU0401 and its potential for (*R*)-(−)-mandelic acid production. *Bioprocess and Biosystems Engineering*, **34**, 315–322.

[40] Zhang, C.-S., Zhang, Z.-J., Li, C.-X., Yu, H.-L., Zheng, G.-W., and Xu, J.-H. (2012). Efficient production of (*R*)-*o*-chloromandelic acid by deracemization of *o*-chloromandelonitrile with a new nitrilase mined from *Labrenzia aggregata*. *Applied Microbiology and Biotechnology*, **95**, 91–99.

[41] Banerjee, A., Kaul, P., and Banerjee, U.C. (2006). Purification and characterization of an enantioselective arylacetonitrilase from *Pseudomonas putida*. *Archives of Microbiology*, **184**, 407–418.

[42] Kiziak, C., Conradt, D., Stolz, A., Mattes, R., and Klein, J. (2005). Nitrilase from *Pseudomonas fluorescens* EBC191: Cloning and heterologous expression of the gene and biochemical characterization of the recombinant enzyme. *Microbiology*, **151**, 3639–3648.

[43] Ni, K., Wang, H., Zhao, L., Zhang, M., Zhang, S., Ren, Y., and Wei, D. (2013). Efficient production of (*R*)-(−)-mandelic acid in biphasic system by immobilized recombinant *E. coli*. *Journal of Biotechnology*, **167**, 433–440.

[44] Zhu, D., Mukherjee, C., Biehl, E.R., and Hua, L. (2007). Discovery of a mandelonitrile hydrolase from *Bradyrhizobium japonicum* USDA110 by rational genome mining. *Journal of Biotechnology*, **129**, 645–650.

[45] Petříčková, A., Veselá, A.B., Kaplan, O., Kubáč, D., Uhnáková, B., Malandra, A., Felsberg, J., Rinágelová, A., Weyrauch, P., Křen, V., Bezouška, K., and Martínková, L. (2012). Purification and characterization of heterologously expressed nitrilases from filamentous fungi. *Applied Microbiology and Biotechnology*, **93**, 1553–1561.

[46] Veselá, A.B., Petříčková, A., Weyrauch, P., and Martínková, L. (2013). Heterologous expression, purification and characterization of arylacetonitrilases from *Nectria haematococca* and *Arthroderma benhamiae*. *Biocatalysis and Biotransformation*, **31**, 49–56.

[47] Liu, Z.-Q., Dong, L.-Z., Cheng, F., Xue, Y.-P., Wang, Y.-S., Ding, J.-N., Zheng, Y.-G., and Shen, Y.-C. (2011). Gene cloning, expression, and characterization of a nitrilase from *Alcaligenes faecalis* ZJUTB10. *Journal of Agricultural and Food Chemistry*, **59**, 11560–11570.

[48] Petříčková, A., Veselá, A.B., Kaplan, O., Kubáč, D., Uhnáková, B., Malandra, A., Felsberg, J., Rinágelová, A., Weyrauch, P., Křen, V., Bezouška, K., and Martínková, L. (2013). Erratum to: Purification and characterization of heterologously expressed nitrilases from filamentous fungi. *Applied Microbiology and Biotechnology*, **97**, 9263–9264.

[49] Sosedov, O., Matzer, K., Bürger, S., Kiziak, C., Baum, S., Altenbuchner, J., Chmura, A., van Rantwijk, F., and Stolz, A. (2009). Construction of recombinant *Escherichia coli* catalysts which simultaneously express an (*S*)-oxynitrilase and different nitrilase variants for the synthesis of (*S*)-mandelic acid and (*S*)-mandelic amide from benzaldehyde and cyanide. *Advanced Synthesis & Catalysis*, **351**, 1531–1538.

[50] Rey, P., Rossi, J.-C., Taillades, J., Gros, G., and Nore, O. (2004). Hydrolysis of nitriles using an immobilized nitrilase: Applications to the synthesis of methionine hydroxy analogue derivatives. *Journal of Agricultural and Food Chemistry*, **52**, 8155–8162.

[51] Zhu, D., Mukherjee, C., Biehl, E.R., and Hua, L. (2007). Nitrilase-catalyzed selective hydrolysis of dinitriles and green access to the cyanocarboxylic acids of pharmaceutical importance. *Advanced Synthesis & Catalysis*, **349**, 1667–1670.

[52] Rustler, S., Motejadded, H., Altenbuchner, J., and Stolz, A. (2008). Simultaneous expression of an arylacetonitrilase from *Pseudomonas fluorescens* and a (*S*)-oxynitrilase from *Manihot esculenta* in *Pichia pastoris* for the synthesis of (*S*)-mandelic acid. *Applied Microbiology and Biotechnology*, **80**, 87–97.

[53] Mateo, C., Chmura, A., Rustler, S., van Rantwijk, F., Stolz, A., and Sheldon, R.A. (2006). Synthesis of enantiomerically pure (*S*)-mandelic acid using an oxynitrilase–nitrilase bienzymatic cascade: A nitrilase surprisingly shows nitrile hydratase activity. *Tetrahedron: Asymmetry*, **17**, 320–323.

[54] Baum, S., van Rantwijk, F., and Stolz, A. (2012). Application of a recombinant *Escherichia coli* whole-cell catalyst expressing hydroxynitrile lyase and nitrilase activities in ionic liquids for the production of (*S*)-mandelic acid and (*S*)-mandeloamide. *Advanced Synthesis & Catalysis*, **354**, 113–122.

[55] Chmura, A., Rustler, S., Paravidino, M., van Rantwijk, F., Stolz, A., and Sheldon, R.A. (2013). The combi-CLEA approach: Enzymatic cascade synthesis of enantiomerically pure (*S*)-mandelic acid. *Tetrahedron: Asymmetry*, **24**, 1225–1232.

[56] Chaplin, J.A., Levin, M.D., Morgan, B., Farid, N., Li, J., Zhu, Z., McQuaid, J., Nicholson, L.W., Rand, C.A., and Burk, M.J. (2004). Chemoenzymatic approaches to the dynamic kinetic asymmetric synthesis of aromatic amino acids. *Tetrahedron: Asymmetry*, **15**, 2793–2796.

[57] Qiu, J., Su, E., and Wang, W. (2014). High yield synthesis of D-phenylglycine and its derivatives by nitrilase mediated dynamic kinetic resolution in aqueous-1-octanol biphasic system. *Tetrahedron Letters*, **55**, 1448–1451.

[58] Ankati, H., Zhu, D., Yang, Y., Biehl, E.R., Hua, L. (2009) Asymmetric synthesis of both antipodes of β-hydroxy nitriles and b-hydroxy carboxylic acids via enzymatic reduction or sequential reduction/hydrolysis. *Journal of Organic Chemistry*, **74**, 1658–1662.

[59] Winkler, M., Knall, A.C., Kulterer, M.R., and Klempier, N. (2007). Nitrilases catalyze key step to conformationally constrained GABA analogous γ-amino acids in high optical purity. *Journal of Organic Chemistry*, **72**, 7423–7426.

[60] Winkler, M., Meischler, D., and Klempier, N. (2007). Nitrilases-catalyzed enantioselective synthesis of pyrrolidine and piperidinecarboxylic acids. *Advanced Synthesis and Catalysis*, **349**, 1475–1480.

[61] Thuku, R.N., Weber, B.W., Varsani, A., and Sewell, B.T. (2007). Post-translational cleavage of recombinantly expressed nitrilase from *Rhodococcus rhodochrous* J1 yields a stable, active helical form. *FEBS Journal*, **274**, 2099–2108.

[62] Zhu, X.-Y., Gong, J.-S., Li, H., Lu, Z.-M., Zhou, Z.-M., Shi, J.-S., and Xu, Z.-H. (2013). Characterization and functional cloning of an aromatic nitrilase from *Pseudomonas putida* CGMCC3830 with high conversion efficiency toward cyanopyridine. *Journal of Molecular Catalysis B: Enzymatic*, **97**, 175–183.

[63] Kaplan, O., Vejvoda, V., Plíhal, O., Pompach, P., Kavan, D., Bojarová, P., Bezouška, K., Macková, M., Cantarella, M., Jirků, V., Křen, V., and Martínková, L. (2006). Purification and characterization of a nitrilase from *Aspergillus niger* K10. *Applied Microbiology and Biotechnology*, **73**, 567–575.

[64] Kaplan, O., Vejvoda, V., Plíhal, O., Pompach, P., Kavan, D., Bojarová, P., Bezouška, K., Macková, M., Cantarella, M., Jirků, V., Křen, V., and Martínková, L. (2013). Erratum to: Purification and characterization of a nitrilase from *Aspergillus niger* K10 (vol. 73, p. 567, 2006). *Applied Microbiology and Biotechnology*, **97**, 3745–3746.

[65] Vejvoda, V., Kaplan, O., Bezouška, K., Pompach, P., Šulc, M., Cantarella, M., Benada, O., Uhnáková, B., Rinágelová, A., Lutz-Wahl, S., Fischer, L., Křen, V., and Martínková, L. (2008). Purification and characterization of a nitrilase from *Fusarium solani* O1. *Journal of Molecular Catalysis B: Enzymatic*, **50**, 99–106.

[66] Vejvoda, V., Kubáč, D., Davidová, A., Kaplan, O., Šulc, M., Šveda, O., Chaloupková, R., and Martínková, L. (2010). Purification and characterization of nitrilase from *Fusarium solani* IMI196840. *Process Biochemistry*, **45**, 1115–1120.

[67] Gong, J.-S., Li, H., Zhu, X.-Y., Lu, Z.-M., Wu, Y., Shi, J.-S., and Xu, Z.-H. (2012). Fungal His-tagged nitrilase from *Gibberella intermedia*: Gene cloning, heterologous expression and biochemical properties. *PLoS ONE*, **7**, e50622.

[68] Nagasawa, T., Wieser, M., Nakamura, T., Iwahara, H., Yoshida, T., and Gekko, K. (2000). Nitrilase of *Rhodococcus rhodochrous* J1. *European Journal of Biochemistry*, **267**, 138–144.

[69] Luo, H., Fan, L., Chang, Y., Ma, J., Yu, H., and Shen, Z. (2010). Gene cloning, overexpression, and characterization of the nitrilase from *Rhodococcus rhodochrous* tg1-A6 in *E. coli*. *Applied Biochemistry and Biotechnology*, **160**, 393–400.

[70] Pai, O., Banoth, L., Ghosh, S., Chisti, Y., and Banerjee, U.C. (2014). Biotransformation of 3-cyanopyridine to nicotinic acid by free and immobilized cells of recombinant *Escherichia coli*. *Process Biochemistry*, **49**, 655–659.

[71] Vejvoda, V., Kaplan, O., Kubáč, D., Křen, V., and Martínková, L. (2006). Immobilization of fungal nitrilase and bacterial amidase—two enzymes working in accord. *Biocatalysis and Biotransformation*, **24**, 414–418.

[72] Rinágelová, A., Kaplan, O., Veselá, A.B., Chmátal, M., Křenková, A., Plíhal, O., Pasquarelli, F., Cantarella, M., and Martínková, L. (2014). Cyanide hydratase from *Aspergillus niger* K10: Overproduction in *Escherichia coli*, purification, characterization and use in continuous cyanide degradation. *Process Biochemistry*, **49**, 445–450.

[73] Roach, P.C.J., Ramsden, D.K., Hughes, J., and Williams, P. (2004). Biocatalytic scrubbing of gaseous acrylonitrile using *Rhodococcus ruber* immobilized in synthetic silicone polymer (ImmobaSil (TM)) rings. *Biotechnology and Bioengineering*, **85**, 450–455.

[74] Roach, P.C.J., Ramsden, D.K., Hughes, J., and Williams, P. (2003). Development of a conductimetric biosensor using immobilised *Rhodococcus ruber* whole cells for the detection and quantification of acrylonitrile. *Biosensors & Bioelectronics*, **19**, 73–78.

[75] Chauhan, S., Wu, S., Blumerman, S., Fallon, R.D., Gavagan, J.E., DiCosimo, R., and Payne, M.S. (2003). Purification, cloning, sequencing and over-expression in *Escherichia coli* of a regioselective aliphatic nitrilase from *Acidovorax facilis* 72W. *Applied Microbiology and Biotechnology*, **61**, 118–122.

[76] Liu, Z.-Q., Zhou, M., Zhang, X.H., Xu, J.M., Xue, Y.P., and Zheng, Y.G. (2012). Biosynthesis of iminodiacetic acid from iminodiacetonitrile by immobilized recombinant *Escherichia coli* harboring nitrilase. *Journal of Molecular Microbiology and Biotechnology*, **22**, 35–47.

[77] Heinemann, U., Engels, D., Bürger, S., Kiziak, C., Mattes, R., Stolz, A. (2003). Cloning of a nitrilase gene from the cyanobacterium *Synechocystis* sp. strain PCC6803 and heterologous expression and characterization of the encoded protein. *Applied and Environmental Microbiology*, **47**, 672–681.

[78] Müller, P., Egorova, K., Vorgias, C.E., Boutou, E., Trauthwein, H., Verseck, S., and Antranikian, G. (2006). Cloning, overexpression, and characterization of a thermoactive nitrilase from the hyperthermophilic archaeon *Pyrococcus abyssi*. *Protein Expression and Purification*, **47**, 672–681.
[79] Mukherjee, C., Zhu, D., Biehl, E.R., Hua, L. (2006) Exploring the synthetic applicability of a cyanobacterium nitrilase as catalyst for nitrile hydrolysis. *European Journal of Organic Chemistry*, 5238–5242.
[80] Raczynska, J.E., Vorgias, C.E., Antranikian, G., and Rypniewski, W. (2011). Crystallographic analysis of a thermoactive nitrilase. *Journal of Structural Biology*, **173**, 294–302.
[81] Osswald, S., Wajant, H., and Effenberger, F. (2002). Characterization and synthetic applications of recombinant AtNIT1 from *Arabidopsis thaliana*. *European Journal of Biochemistry*, **269**, 680–687.
[82] Park, W.J., Kriechbaumer, V., Müller, A., Piotrowski, M., Meeley, R.B., Gierl, A., and Glawischnig, E. (2003). The nitrilase ZmNIT2 converts indole-3-acetonitrile to indole-3-acetic acid. *Plant Physiology*, **133**, 794–802.
[83] Mukherjee, C., Zhu, D.M., Biehl, E.R., Parmar, R.R., and Hua, L. (2006). Enzymatic nitrile hydrolysis catalyzed by nitrilase ZmNIT2 from maize. An unprecedented β-hydroxy functionality enhanced amide formation. *Tetrahedron*, **62**, 6150–6154.
[84] Zhu, D., Mukherjee, C., Yang, Y., Rios, B.E., Gallagher, D.T., Smith, N.N., Biehl, E.R., and Hua, L. (2008). A new nitrilase from *Bradyrhizobium japonicum* USDA 110. Gene cloning, biochemical characterization and substrate specificity. *Journal of Biotechnology*, **133**, 327–333.
[85] Okamoto, S. and Eltis, L.D. (2007). Purification and characterization of a novel nitrile hydratase from *Rhodococcus* sp. RHA1. *Molecular Microbiology*, **65**, 828–838.
[86] Pei, X.L., Yang, L.R., Xu, G., Wang, Q.Y., and Wu, J.P. (2014). Discovery of a new Fe-type nitrile hydratase efficiently hydrating aliphatic and aromatic nitriles by genome mining. *Journal of Molecular Catalysis B: Enzymatic*, **99**, 26–33.
[87] Wang, M.-X. (2005). Enantioselective biotransformations of nitriles in organic synthesis. *Topics in Catalysis*, **35**, 117–130.
[88] Song, L.Y., Yuan, H.J., Coffey, L., Doran, J., Wang, M.X., Qian, S.J., and O'Reilly, C. (2008). Efficient expression in *E coli* of an enantioselective nitrile hydratase from *Rhodococcus erythropolis*. *Biotechnology Letters*, **30**, 755–762.
[89] Osprian, I., Fechter, M.H., and Griengl, H. (2003). Biocatalytic hydrolysis of cyanohydrins: An efficient approach to enantiopure α-hydroxy carboxylic acids. *Journal of Molecular Catalysis B: Enzymatic*, **24–25**, 89–98.
[90] Reisinger, C., Osprian, I., Glieder, A., Schoemaker, H.E., Griengl, H., and Schwab, H. (2004). Enzymatic hydrolysis of cyanohydrins with recombinant nitrile hydratase and amidase from *Rhodococcus erythropolis*. *Biotechnology Letters*, **26**, 1675–1680.
[91] van Pelt, S., van Rantwijk, F., and Sheldon, R.A. (2009). Synthesis of aliphatic (S)-α-hydroxycarboxylic amides using a one-pot bienzymatic cascade of immobilised oxynitrilase and nitrile hydratase. *Advanced Synthesis & Catalysis*, **351**, 397–404.
[92] Vejvoda, V., Martínková, L., Veselá, A.B., Kaplan, O., Lutz-Wahl, S., Fischer, L., and Uhnáková, B. (2011). Biotransformation of nitriles to hydroxamic acids via a nitrile hydratase–amidase cascade reaction. *Journal of Molecular Catalysis B: Enzymatic*, **71**, 51–55.
[93] Vejvoda, V., Šveda, O., Kaplan, O., Přikrylová, V., Elišáková, V., Himl, M., Kubáč, D., Pelantová, H., Kuzma, M., Křen, V., and Martínková, L. (2007). Biotransformation of heterocyclic dinitriles by *Rhodococcus erythropolis* and fungal nitrilases. *Biotechnology Letters*, **29**, 1119–1124.
[94] D'Antona, N., Morrone, R., Bovicelli, P., Gambera, G., Kubáč, D., and Martínková, L. (2010). A novel chemo-multienzymatic synthesis of bioactive cyclophellitol and *epi*-cyclophellitol in both enantiopure forms. *Tetrahedron: Asymmetry*, **21**, 2448–2454.
[95] D'Antona, N., Nicolosi, G., Morrone, R., Kubáč, D., Kaplan, O., and Martínková, L. (2010). Synthesis of novel cyano-cyclitols and their stereoselective biotransformation catalyzed by *Rhodococcus erythropolis* A4. *Tetrahedron: Asymmetry*, **21**, 695–702.
[96] D'Antona, N., Morrone, R., Nicolosi, G., and Pedotti, S. (2013). Novel enzymatic recognition of the ferrocene framework: Nitrile hydratase/amidase catalyzed cascade biotransformations. *RSC Advances*, **3**, 11456–11458.
[97] Benz, P., Muntwyler, R., and Wohlgemuth, R. (2007). Chemoenzymatic synthesis of chiral carboxylic acids via nitriles. *Journal of Chemical Technology and Biotechnology*, **82**, 1087–1098.

[98] Winkler, M., Martínková, L., Knall, A.C., Krahulec, S., and Klempier, N. (2005). Synthesis and microbial transformation of β-amino nitriles. *Tetrahedron*, **61**, 4249–4260.

[99] Wang, M.-X. and Wu, Y. (2003). Nitrile biotransformations for the synthesis of enantiomerically enriched Baylis-Hillman adducts. *Organic & Biomolecular Chemistry* **1**, 535–540.

[100] Wang, M.-X. and Feng, G.Q. (2003). Nitrile biotransformation for highly enantioselective synthesis of 3-substituted 2,2-dimethylcyclopropanecarboxylic acids and amides. *Journal of Organic Chemistry*, **68**, 621–624.

[101] Wang, M.-X., Feng, G.Q., and Zheng, Q.Y. (2004). Synthesis of high enantiomeric purity *gem*-dihalocyclopropane derivatives from biotransformations of nitriles and amides. *Tetrahedron: Asymmetry*, **21**, 347–354.

[102] Wang, M.-X. and Feng, G.-Q. (2002). A novel approach to enantiopure cyclopropane compounds from biotransformation of nitriles. *New Journal of Chemistry*, **26**, 1575–1783.

[103] Wang, J.-Y., Wang, D.-X.; Pan, J.; Huang, Z.-T.; Wang, M.-X. (2007). Nitrile and amide biotransformations for the synthesis of enantiomerically pure 3-arylaziridine-2-carboxamide derivatives and their stereospecific ring-opening reactions. *Journal of Organic Chemistry* **72**, 9391–9394.

[104] Wang, M.-X., Lin, S.-J., Liu, C.-S., Zheng, Q.-Y., Li, J.-S. (2003). Nitrile biotransformations for highly efficient and enantioselective syntheses of electrophilic oxiranecarboxamides. *Journal of Organic Chemistry* **68**, 4570–4573.

[105] Leng, D.-H., Wang, D.-X., Huang, Z.-T., and Wang, M.-X. (2010). Highly efficient and enantioselective biotransformations of β-lactam carbonitriles and carboxamides and their synthetic applications. *Organic & Biomolecular Chemistry*, **8**, 4736–4743.

[106] Leng, D.-H., Wang, D.-X., Pan, J., Huang, Z.-T., and Wang, M.-X. (2009). Highly efficient and enantioselective biotransformations of racemic azetidine-2-carbonitriles and their synthetic applications. *Journal of Organic Chemistry*, **74**, 6077–6082.

[107] Ao, Y.-F., Wang, D.-X., Zhao, L., and Wang, M.-X. (2014). Biotransformations of racemic 2,3-allenenitriles in biphasic systems: Synthesis and transformations of enantioenriched axially chiral 2,3-allenoic acids and their derivatives. *Journal of Organic Chemistry*, **79**, 3103–3110.

[108] Brandão, P.F.B., Clapp, J.P., and Bull, A.T. (2003). Diversity of nitrile hydratase and amidase enzyme genes in *Rhodococcus erythropolis* recovered from geographically distinct habitats. *Applied and Environmental Microbiology*, **69**, 5754–5766.

[109] Kinfe, H.H., Chhiba, V., Frederick, J., Bode, M.L., Mathiba, K., Steenkamp, P.A., and Brady, D. (2009). Enantioselective hydrolysis of β-hydroxy nitriles using the whole cell biocatalyst *Rhodococcus rhodochrous* ATCC BAA-870. *Journal of Molecular Catalysis B: Enzymatic*, **59**, 231–236.

[110] Chhiba, V., Bode, M.L., Mathiba, K., Kwezi, W., and Brady, D. (2012). Enantioselective biocatalytic hydrolysis of β-aminonitriles to β-amino-amides using *Rhodococcus rhodochrous* ATCC BAA-870. *Journal of Molecular Catalysis B: Enzymatic*, **76**, 68–74.

[111] Yasukawa, K., Hasemi, R., and Asano, Y. (2011). Dynamic kinetic resolution of α-aminonitriles to form chiral α-amino acids. *Advanced Synthesis & Catalysis*, **353**, 2328–2332.

[112] Yasukawa, K. and Asano, Y. (2012). Enzymatic synthesis of chiral phenylalanine derivatives by a dynamic kinetic resolution of corresponding amide and nitrile substrates with a multi-enzyme system. *Advanced Synthesis & Catalysis*, **354**, 3327–3332.

[113] Shaw, N.M., Robins, K.T., Kiener, A. (2003) Lonza. 20 years of biotransformations. *Advanced Synthesis & Catalysis*, **345**, 425–435.

[114] van Pelt, S., Zhang, M., Otten, L.G., Holt, J., Sorokin, D.Y., van Rantwijk, F., Black, G.W., Perry, J.J., and Sheldon, R.A. (2011). Probing the enantioselectivity of a diverse group of purified cobalt-centred nitrile hydratases. *Organic & Biomolecular Chemistry*, **9**, 3011–3019.

CHAPTER 13

Biocatalytic Epoxidation for Green Synthesis

Hui Lin[1], Meng-Yu Xu[1,2], Yan Liu[1,2], and Zhong-Liu Wu[1]

[1] Key Laboratory of Environmental and Applied Microbiology & Environmental Microbiology Key Laboratory of Sichuan Province, Chengdu Institute of Biology, Chinese Academy of Sciences, Chengdu, China

[2] University of the Chinese Academy of Sciences, Beijing, China

13.1 INTRODUCTION

Epoxides are versatile organic intermediates due to the reactive oxirane moiety, which can be opened by various nucleophiles or undergo elimination, reduction, or rearrangements to a multitude of more elaborate intermediates with the retention or inversion of chirality [1, 2]. Enantiopure epoxides, in particular, play an important role in organic synthesis and the pharmaceutical industry [3, 4]. In recent decades, there have been enormous advances in the development of catalysts, including chemical catalysts and biocatalysts, for the production of chiral epoxides [5–10].

Direct epoxidation of alkenes and kinetic resolution of racemic epoxides are the two major approaches for the synthesis of chiral epoxides [5–9]. Although great success has been achieved in the hydrolytic kinetic resolution of epoxides [11–14], asymmetric epoxidation of alkenes remains the most attractive approach due to better atom efficiency of direct incorporation of the C—O functional group [15, 16]. Considerable efforts have been made in this area by synthetic chemists [17–20]. The first breakthrough of direct asymmetric epoxidation was developed by Katsuki and Sharpless. It allowed the synthesis of enantiopure epoxides from allylic alcohols catalyzed with diisopropyl tartrate and titanium tetraisopropoxide using *tert*-butyl hydroperoxide as oxidant [19]. Salen–Mn (III) complexes were later developed by Katsuki and Jacobsen *et al.* for the highly stereoselective epoxidation of *cis*-di- and trisubstituted alkenes [9, 18, 21]. Organocatalysts have also been successfully applied in the asymmetric epoxidation of *trans*-alkenes with excellent stereoselectivity [22].

The microbial epoxidation of alkenes was originally reported in 1963, and the reaction was later found to be stereospecific in 1973 using an alkane monooxygenase [23, 24]. In fact, enzymes involved in epoxide formation exist widely in nature as indicated by the variety of nature products containing the chiral oxirane function [25], which could serve as potential reservoir for asymmetric epoxidation. Established biocatalytic procedures with good stereoselectivity mainly involve the application of monooxygenases or chloroperoxidases (CPOs) using recombinant enzymes or microorganisms producing those enzymes. They have shown remarkable advantages such as biodegradability, highly specificity, and mild reaction conditions. They are particularly

Green Biocatalysis, First Edition. Edited by Ramesh N. Patel.

valuable for those substrates that are poorly accepted by chemical procedures, such as nonfunctional terminal alkenes [3, 5–8]. The main part of this review describes the scope and limitation of several predominant enzymatic approaches for the asymmetric epoxidation of alkenes. Protein engineering for enzymes with better characteristics is also discussed briefly, emphasizing on the screening methods and successful examples for epoxidation reactions.

13.2 ENZYMES FOR ASYMMETRIC EPOXIDATION

13.2.1 Monooxygenases

Monooxygenases catalyze the insertion of an oxygen atom from O_2 into an organic substrate in many metabolic pathways. They utilize cofactors to transfer electrons to molecular oxygen to overcome the spin forbidden of O_2. The majority of monooxygenases are NAD(P)H dependent and require cofactors, such as heme, flavin, copper ion, nonheme iron, or pterin [26]. The type of intermediate formed depends on which cofactor is present in the enzyme. Monooxygenases can catalyze a variety of reactions with chemo-, regio-, and enantioselectivity. Here, those monooxygenases with activities toward alkenes to form chiral epoxides are described in detail.

Heme-Dependent Monooxygenases

Heme-dependent monooxygenases, also referred to as cytochrome P450 monooxygenases (abbreviated as P450 or CYP), are ubiquitous proteins in all domains of life, which can be found in organisms from protists to plants and to human beings. Despite of the low sequence identities, P450s of different origins share a similar protein structure with a highly conserved catalytic core, as well as the same overall catalytic mechanism. P450s acquire electrons from the cofactors NAD(P)H to activate O_2 by the heme prosthetic group [27]. The generally accepted catalytic cycle includes the major events of substrate binding, 1-electron reduction, O_2 binding, a second 1-electron reduction, followed by a series of less defined steps such as protonation, hemolytic scission of the O—O bond to yield an active perferryl FeO species, reaction with substrates, and release of the product [28].

The P450s catalyze a variety of reactions, including hydroxylation, epoxidation, peroxidation, sulfoxidation, dealkylation, deamination, etc., depending on the substrate structures. Many substrates, including alkenes, aromatic hydrocarbons, heterocycles, vinyl halides, ethyl carbamate, vinyl nitrosamines, and aflatoxin B1, have been epoxidized with good stereoselectivities by the membrane-bound mammalian P450s [29].

In the past decades, soluble microbial P450s, which show higher stability, catalytic activity, and availability, have been investigated with respect to the production of fine chemicals. They are typically applied in the form of recombinant *Escherichia coli* whole cells expressing those P450s. However, the performance of native P450s in bioepoxidation is unsatisfactory in terms of activity and stereoselectivity. For example, it is reported that $P450_{cam}$ from *Pseudomonas putida* transfers *cis*-beta-methylstyrene to the (1*S*, 2*R*)-epoxide enantiomer with only 78% ee [30]. In recent decades, with the development of protein engineering technologies, many P450 mutants, which have high reaction activities, coupling efficiency, or stereoselectivities toward unnatural substrates, have been generated from microbial P450s [31–33]. The studies mainly focus on the self-sufficient enzyme $P450_{BM3}$ (CYP102A1) from *Bacillus megaterium*, as well as class I enzymes such as $P450_{cam}$ (CYP101) from *P. putida* and $P450_{pyr}$ from *Sphingomonas* sp. HXN-200.

Nonheme Iron-Dependent Monooxygenases

Nonheme iron-dependent monooxygenases, which use two iron atoms as cofactors, are bacterial multicomponent monooxygenases. They mainly consist of three components, including a monooxygenase, a reductase component, and a small regulatory protein [26]. Some members contain an additional Rieske-type ferredoxin component that facilitates the electron transport between the reductase and monooxygenase components [34].

Nonheme iron-dependent monooxygenases require NAD(P)H as an electron donor to catalyze the initial hydroxylation or epoxidation in the pathways for the degradation of hydrocarbon substrates in the presence of molecular oxygen. The reaction starts with the reduction of Fe(III)–Fe(III) to Fe(II)–Fe(II) by NADH, followed by the oxygen molecular bind to iron. Then, one of these oxygen atoms is inserted into the substrate, resulting in the conversion of alkanes to alcohols and alkenes to oxides [35]. Enzymes of this family that are involved in asymmetric epoxidation are mostly bacterial hydrocarbon hydroxylases, including soluble proteins, such as alkene monooxygenases and toluene monooxygenases, and integral membrane proteins, such as alkane monooxygenases and xylene monooxygenases.

Alkene monooxygenases are isolated from alkene-utilizing bacteria. They are able to discriminate between nonactivated C—H bonds and double bonds and typically catalyze the epoxidation, but not the hydroxylation of alkenes. One of the most extensively studied and versatile alkene-utilizing bacteria for the epoxidation of alkenes is *R. rhodochrous* (formerly *Nocardia corallina, Rhodococcus corallinus*) B-276, which is isolated from soil with propylene as the carbon source [36–40]. *Rhodococcus rhodochrous* B-276 can grow on glucose or sucrose and remains active. The alkene monooxygenase from this strain (referred to as AMO) is a three-component monooxygenase encoded by the *amoABCD* operon [38, 39].

Another well studied alkene monooxygenase is PAMO, which belongs to the four-component alkene/aromatic monooxygenase subdivision [22]. PAMO is isolated from *Xanthobacter autotrophicus* Py2 [41] and exhibits a higher degree of similarity to the isofunctional protein of toluene 4-monooxygenase (T4MO) than to AMO, and its substrate spectra also overlaps with that of aromatic monooxygenases, catalyzing not only the epoxidation of alkenes but also the hydroxylation of benzene, toluene, and phenol [42–44]. The T4MO complex from *Pseudomonas mendocina* KR1 is the first reported four-component nonheme diiron-containing monooxygenase, which consists of four soluble proteins encoded by the *tmo*A-F gene cluster. The hydroxylase component is composed of the *tmo*A, *tmo*B, and *tmo*E gene products [34]. Toluene monooxygenase catalyzes the hydroxylation of the aromatic ring of toluene, forming *ortho-*, *meta-*, and *para*-cresols, and the epoxidation of short-chain alkenes with varied stereoselectivities [45].

Pseudomonas oleovorans is an alkane-utilizing bacterium that produces an alkane monooxygenase (alkane hydroxylase). It catalyzes the omega-hydroxylation of fatty acids and alkanes and the epoxidation of terminal alkenes. The enzyme consists of a soluble NADH–rubredoxin reductase, a soluble rubredoxin, and an integral membrane oxygenase referred to as ω- or alkane hydroxylase (AlkB) encoded by the first cistron of the *alk*BAC operon [46]. It is related to another three-component bacterial membrane-spanning protein, xylene monooxygenase from *P. putida* mt-2 [47] with 25% sequence identity for the hydroxylase component, and a histidine-rich active site and significant homology in both an eight-histidine motif and in the presence of three long hydrophobic domains thought to be membrane-spanning regions [48]. Xylene monooxygenase also catalyzes the hydroxylation of alkanes and the epoxidation of alkenes [47, 49].

Although many nonheme iron-dependent monooxygenases have been purified and cloned from hydrocarbon-utilizing bacteria, the complicated nature of this enzyme family often places a considerable synthetic burden on the host cells during heterologous expression. Therefore, the whole cells of the native strains or recombinant *E. coli*, instead of free enzymes, are often applied in biocatalytic reactions.

Styrene Monooxygenases

Styrene monooxygenases (abbreviated as SMOs) are mostly two-component flavin-dependent monooxygenases composed of a FAD-dependent styrene epoxidase (StyA) and a NADH-dependent flavin reductase (StyB) [50, 51], which requires NADH as the coenzyme. SMOs catalyze the transformation of styrene into (*S*)-styrene oxide in the upper catabolic pathway of styrene degradation [52, 53]. Most two-component SMOs are isolated from the genera of *Pseudomonas* and *Rhodococcus* and share high sequence similarities [50, 51, 54–58]. Besides the two-component SMOs, several self-sufficient one-component SMOs are isolated from *Rhodococcus opacus* [59–61]. SMOs are competitive enzymes for the synthesis of epoxides, owning to its exquisite regio- and enantioselectivities [54, 55, 57, 62].

The crystal structure of StyA from *P. putida* S12 shows that it contains two distinct domains, in which the large cavity opening to the surface forms the FAD binding site and another cavity at the base of this pocket likely represents the styrene binding site [63, 64]. For the epoxidation, StyA binds tightly to the reduced FAD to form the StyA–FAD_{red} complex, which reacts with molecular oxygen to yield the FAD C(4a)-peroxide intermediate, and then styrene reacts with the peroxide intermediate and is transformed into oxide [64], while the other oxygen atom is reduced to water. StyB catalyzes NADH oxidation with two-electron reduction of FAD, the steady-state kinetics for StyB indicate a mechanism of sequential binding of NADH and flavin to StyB, and the NADH oxidation activity does not depend on the presence of StyA [51].

13.2.2 Chloroperoxidases

Chloroperoxidase (CPO) is a heme-thiolate protein secreted by the fungus *Caldariomyces fumago* [65]. CPO utilizes H_2O_2 or *tert*-butyl hydroperoxide (TBHP) as a terminal oxidant instead of molecular oxygen and cofactor NAD(P)H, and is reactive at low pH with an optimum reactivity at pH 2.8 [66]. Although the primary biological function of CPO is to catalyze the hydrogen peroxide-dependent chlorination of electron-rich compounds, CPO also catalyzes halide-independent oxidative reactions characteristic of other heme peroxidases, catalase, and cytochrome P450, such as dehydrogenation and oxygen insertion, including stereoselective epoxidation of alkenes [67, 68], oxidation of sulfides to chiral sulfoxides, and the oxidation of alcohols, aldehydes, and amines [69].

CPO has a hybrid active site pocket that shares similar structure features with peroxidases and cytochrome P450s, but it folds into a novel tertiary structure dominated by eight helical segments. The catalytic base in CPO is glutamic acid rather than histidine as in other peroxidases [70]. The significance of Glu-183 has been demonstrated by mutational studies in which the E183H mutant resulted in 85% loss in chlorination activity [71]. The catalytic cycle starts with the replacement of a water molecule on the distal side of the heme by the incoming hydrogen peroxide. Then Glu-183 abstracts a proton from the hydrogen peroxide, yielding a short-lived peroxo-anion species (compound 0) [72]. Subsequently, the heterolytic cleavage of the O—O bond and releasing of water form an oxyferryl (IV) porphyrin π cation radical referred to as compound I, followed by the oxidation of substrates [73].

13.3 APPLICATION OF BIOEPOXIDATION IN ORGANIC SYNTHESIS

13.3.1 Asymmetric Epoxidation of Aliphatic Alkenes

For the past several decades, several chemocatalysts and biocatalysts have been used for the epoxidation of nonfunctionalized aliphatic alkenes. The Katsuki–Jacobsen epoxidation, which is catalyzed by chiral Mn(III)–salen with NaOCl/PhIO as an oxidant, achieves good yields and high stereoselectivities (84–94% ee) for the epoxidation of *cis*-alkenes [21]. The Shi epoxidation, which is catalyzed by the fructose-derived ketone and oxone, has been successfully used in the epoxidation of *trans*-alkenes, yielding the corresponding oxides with 93–98% ee [74]. However, for nonfunctionalized terminal aliphatic alkenes, chemocatalysts typically display low stereoselectivity [7].

On the other hand, some oxidoreductases accept those substrates well enough to serve as complementary catalysts. Among them, alkene monooxygenases and CPO have been widely used in the epoxidation of aliphatic alkenes including some non-functional terminal alkenes with excellent stereoselectivities. In a lot of cases, alkene/alkane-utilization microorganisms producing uncharacterized enzymes that might belong to the family of alkene/alkane monooxygenase are successfully applied to the epoxidation of aliphatic alkenes with high enantioselectivities. Other enzymes, such as CYPs and SMOs, also show good epoxidation activity toward aliphatic alkenes with high stereoselectivity for certain types of substrates.

Reactions Catalyzed with Nonheme Iron-Dependent Monooxygenases

Microorganisms that use gaseous olefins as a carbon source are widely exist in nature, and they typically produce nonheme iron-dependent alkene/alkane monooxygenases. Although in a lot of cases, the identification of the key enzyme hasn't been attempted; many of those alkene-utilizing bacteria have been used in the asymmetric epoxidation of aliphatic alkenes with high stereoselectivities [26].

Mahmoudian *et al.* have tested 18 organisms that use ethene or propene as the carbon source, including *Aerococcus*, *Alcaligenes*, *Micrococcus*, and *Staphylococcus* spp., in the epoxidation of short-chain aliphatic olefins, and all of them stereospecifically produce (*R*)-1,2-epoxypropane (**3**, 90–96% ee), (*R*)-1,2-epoxybutane (**4**, 90–98% ee), and (2*R*, 3*R*)-2,3-dimethyloxirane (**6**, 64–88% ee) from the corresponding olefins (Scheme 13.1) [75]. Other alkene-utilizing bacteria from the genus of *Nocardia*, *Xanthobacter*, *Mycobacterium*, etc. have also been used in the production of **3**, **4**, and **8** with up to 98% ee (Scheme 13.1) [76, 77].

The alkene-utilizing bacterium *R. rhodochrous* B-276, which produces a three-component alkene monooxygenase, has been extensively studied in the epoxidation of alkenes [38]. *R. rhodochrous* B-276 mainly catalyzes the epoxidation of terminal and subterminal alkenes, including C6–C18 l-alkenes, 2-octene, 2-methyl-l-alkenes, and styrene derivatives, yielding the epoxide **9–18** with >80% ee (up to 97% ee) (Scheme 13.2) [36, 37, 40, 78]. For example, (*R*)-propene oxide (**12**) and (*R*)-1,2-epoxytetradecane (**13**) can be achieved by the whole-cell catalyst from the corresponding alkenes with 86% and 85% ee, respectively. The resting cells of *R. rhodochrous* B-276 are employed in a two-phase system with a nontoxic solvent for C6–C12 1-alkenes to release the product inhibition, while growing cells are applied to longer chain alkenes with relatively low toxicity, allowing product accumulation of 80 g l^{-1} in

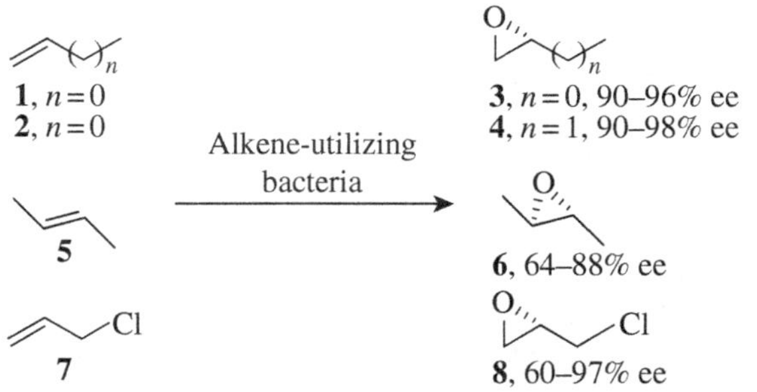

SCHEME 13.1 Asymmetric epoxidation of aliphatic alkenes catalyzed by alkene-utilizing bacteria.

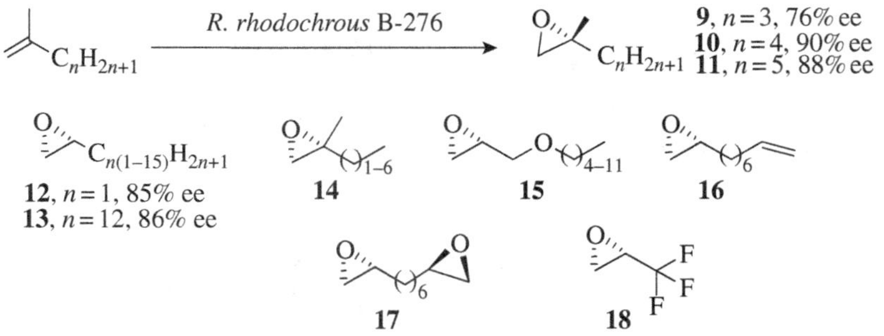

SCHEME 13.2 Asymmetric epoxidation of aliphatic alkenes catalyzed by *Rhodococcus rhodochrous* B-276.

case of 1,2-epoxytetradecane (**13**). Twelve commercial C7–C18 1,2-epoxyalkanes are produced by this biotransformation [24, 36].

The alkane-utilizing bacteria *P. oleovorans*, which contains a three-component alkane monooxygenase, also accepts terminal olefins as substrates for asymmetric epoxidation [79]. For example, it can catalyze the asymmetric epoxidation of 1,7-octadiene to (*R*)-7,8-epoxyoctene and (*R*)-1,2,7,8-diepoxyoctane and also for the production of 1,2-epoxydecane and 1,2-epoxyoctane with moderate stereoselectivity (60–84% ee) [80, 81].

Recombinant *E. coli* cells expressing the toluene monooxygenases from *P. mendocina* KR1, *Ralstonia pickettii* PKO1, *Burkholderia cepacia* G4, *Burkholderia* sp. strain ENVBF1, and *Pseudomonas* sp. strain ENVPC5 were used in the asymmetric epoxidation of butadiene and pentene, resulting in (*S*)-butadiene monoepoxide and (*R*)-pentene epoxide, respectively [45]. Among them, toluene-2-monooxygenase from *B. cepacia* G4 displayed the best stereoselectivity for the production of (*S*)-butadiene monoepoxide and (*R*)-pentene epoxide with 84% and 100% ee, respectively [45].

Reactions Catalyzed with CPOs

The CPO from *C. fumago* is an excellent catalyst for the asymmetric epoxidation of some *cis*-alkenes with TBHP as terminal oxidant but not of *trans*-alkenes. It catalyzes the asymmetric epoxidation of *cis*-disubstituted alkenes with moderate to good stereoselectivities (66–96% ee) (Scheme 13.3) [82]. Among them, *cis*-1-methyl substituted alkenes (C6–C9) appear to be the best substrates for CPO in term of stereoselectivity, yielding the corresponding epoxides with 92–97% ee (Scheme 13.3a) [82, 83]. Functionalized *cis*-2-alkenes, such as those substituted with carboxylic ester or terminal bromide, are also well accepted by CPO, achieving the corresponding epoxides with moderate to high enantiomeric excesses (68–95% ee) (Scheme 13.3b) [84]. In all cases, the chiral carbon atom adjacent to the terminal methyl/ethyl group of the product is of the *R*-configuration, and the length of the carbon chain is essential for the selectivity and efficiency of the reaction.

The CPO can catalyze the bioepoxidation of terminal short-chain aliphatic alkenes as well. The stereoselectivity depends largely on the length of the carbon chain of the substrate, and the 1-methyl group is essential for stereoselectivity and activity. For 1,1-disubstituted terminal alkenes, with one substitute being the methyl group, the reactions yielded epoxides (*R*)-**20** and (*R*)-**22** with 94% and 95% ee, respectively (Scheme 13.4a). However, for similar substrates without methyl substituents, the

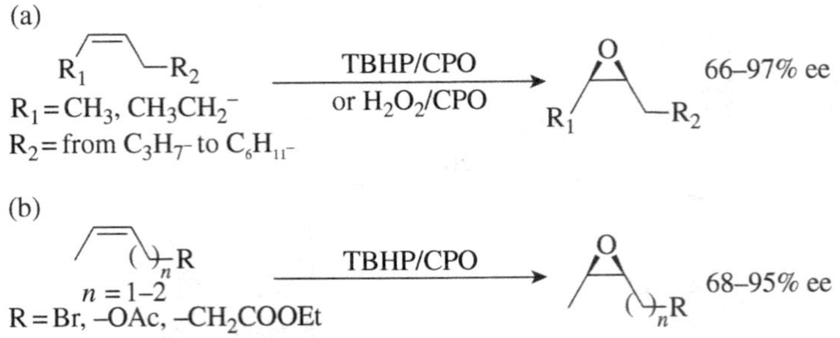

SCHEME 13.3

CPO-catalyzed asymmetric epoxidation of (a) *cis*-disubstituted alkenes and (b) functionalized *cis*-2-alkenes.

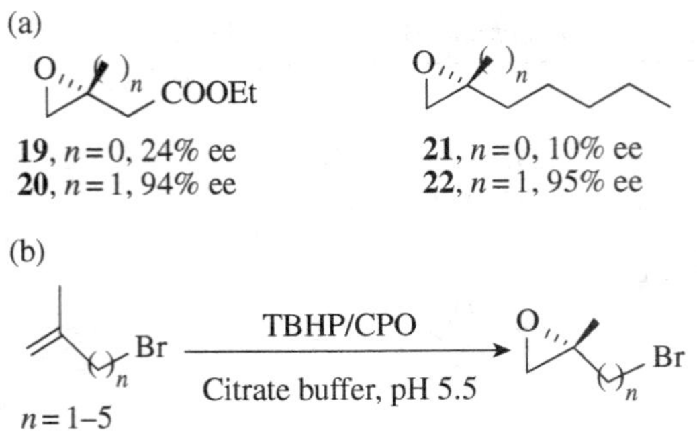

SCHEME 13.4

CPO-catalyzed asymmetric epoxidation of terminal olefins, (a) 1,1-disubstituted terminal alkenes and (b) ω-bromo-2-methylalkenes.

CPO-catalyzed reaction resulted in epoxides (*S*)-**18** and (*S*)-**20** with only 24 and 10% ee (Scheme 13.4a) [85]. For ω-bromo-2-methylalkenes with varied lengths of carbon chain, CPO first showed a progressive increase in selectivity with the increase of the substrate size, with the best result achieved for 5-bromo-2-methyl-1-pentene to yield the (*R*)-epoxide with 95% ee (Scheme 13.4b). Then decreased stereoselectivity was observed toward alkenes possessing additional carbons, such as 6-bromo-2-methyl-1-hexene and 7-bromo-2-methyl-1-heptene, resulting in the epoxides with 87% and 50% ee, respectively [86].

Based on the CPO-catalyzed stereoselective epoxidation of terminal alkenes, Lakner *et al.* designed a concise synthetic route for the preparation of the natural lactone (*R*)-mevalonolactone **27** (Scheme 13.5). The key step of the sequence was the epoxidation of ethyl 3-methyl-3-butenoate catalyze by CPO, which yielded the corresponding (*R*)-**24** in 67% yield and 93% ee. The final product was achieved in 57% overall yield and identical purity (Scheme 13.5) [87].

CPO has also been used in the epoxidation of both conjugated and nonconjugated dienes. Only one double bond was epoxidized to produce monoepoxide with high enantioselectivities (91–97% ee) in all of the reported cases (Scheme 13.6). The selectivity extended to symmetrical dienes, which were epoxidized to monoepoxidess, albeit with moderate stereoselectivity (81% ee) (Scheme 13.6b) [88, 89]. Side reactions have been reported for dienes conjugated to an ester group that leads to aldehyde **35** and **36** (Scheme 13.6c) [89].

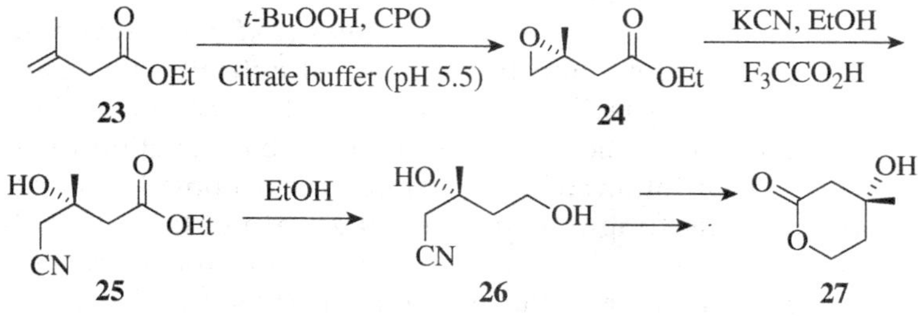

SCHEME 13.5

CPO-catalyzed epoxidation as the key step for the synthesis of (*R*)-mevalonolactone.

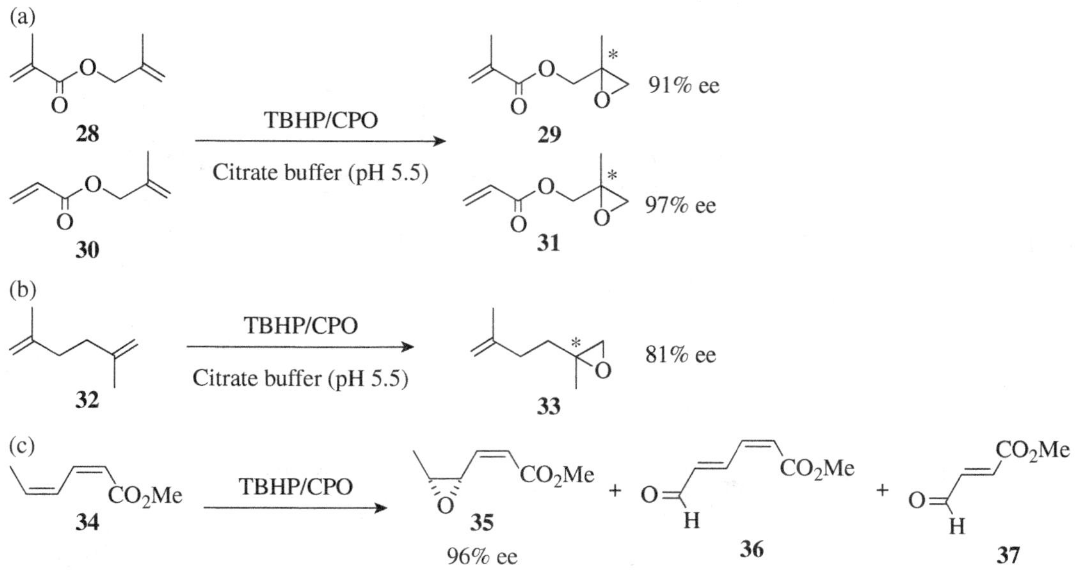

SCHEME 13.6

CPO-catalyzed asymmetric epoxidation of (a) dienes, (b) symmetrical dienes, and (c) conjugated dienes.

Reactions Catalyze with Other Enzymes or Microorganisms

Engineered cytochromes P450$_{BM3}$, designated as SH-44 and RH-47, have been reported to catalyze the stereoselective epoxidation of terminal alkenes to (*S*)- and (*R*)-epoxides, respectively. The variants displayed much improved turnover and epoxide selectivity, producing ~90% epoxide, while the wild type mainly produced the allylic hydroxylation product. However, only moderate enantiomeric excesses were achieved for substrates such as 1-pentene, 1-hexene, 1-heptene, and 1-octene (55–83% ee) [90]. Another engineered mutant of P450$_{BM3}$, designated as G4 (R47L/Y51F/F87A/A328N), was designed for the selective oxidation of amorphadiene (**38**), yielding artemisinic-11*S*, 12-epoxide (**39**) at titers >250 mg l^{-1} in *E. coli*, followed by a high-yielding synthetic chemistry route to dihydroartemisinic acid and onward to artemisinin (**40**) (Scheme 13.7) [91].

Recently, the SMO from *Rhodococcus* sp. ST-5 and ST-10 have been applied to catalyze the epoxidation of 1-hexene and 1-octene to the corresponding (*S*)-epoxides with up to 99% ee [58]. For 1-hexene, the activity was comparable to that acquired with the native substrate styrene [58], but the initial conversion rate was four times lower than styrene [92]. The excellent (*S*)-stereoselectivity is a complementary to the (*R*)-selectivity of majority of the enzymes involved in the asymmetric epoxidation of aliphatic alkenes.

In addition, an unspecific peroxygenase secreted by the fungus *Agrocybe aegerita* was used in the oxidation of 20 different alkenes. Besides epoxidation, hydroxylation also occurred in a lot of cases with peroxygenase as catalyst. The stereoselectivitiesto-ward n-hexene, n-heptene, and n-octene were reported to be 67%, 72%, and 66% ee, respectively, for the (*S*)-enantiomers [93].

As to more complex synthesis design, Archelas *et al.* have reported an elegant four-step chemoenzymatic route for the synthesis of optically pure pityol, a pheromone of the bark beetle *Pityophthorus pityographus*. The method includes two biotransformation steps, which allows the asymmetric bioreduction of the carbonyl group using baker's yeast, followed by the stereoselective epoxidation of (*S*)-urethane **42** using the filamentous fungus *Aspergillus niger* at neutral pH (Scheme 13.8). The presence of the carbamate moiety of **42** is necessary for the bioepoxidation to proceed, as the direct product from the bioreduction, (*S*)-sulcatol **41** (98.5% ee), is not accepted as a substrate by the fungus. The resulting (2*S*, 5*S*)-**43** from the two-step biotransformation undergoes further chemical transformation to yield (2*R*, 5*S*)-**44** in 90% analytical yield with excellent diastereoselectivity and enantioselectivity (100% ee and 98% de) [94].

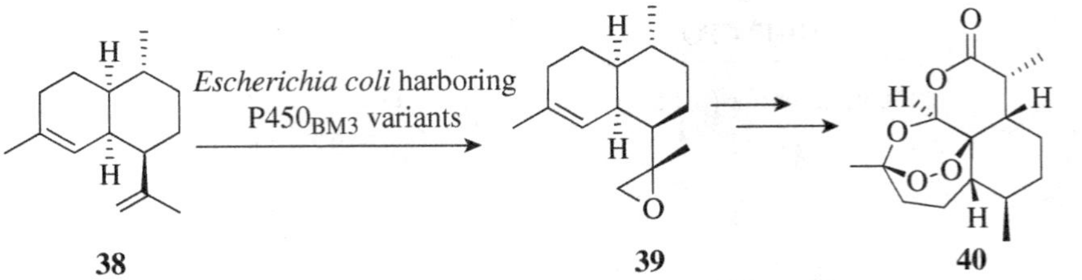

SCHEME 13.7

Cytochrome P450$_{BM3}$-catalyzed epoxidation in the route to artemisinin.

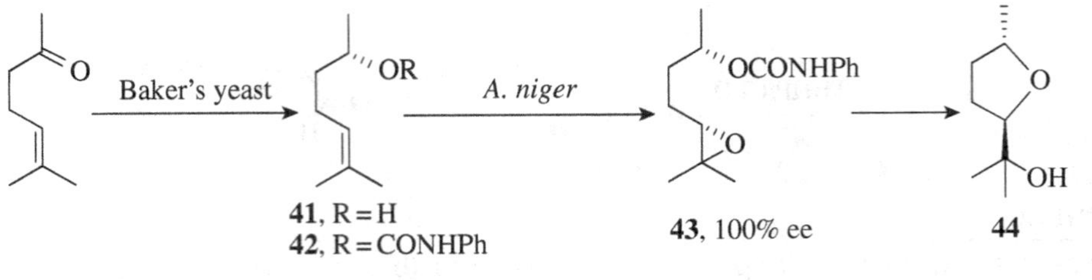

SCHEME 13.8

Asymmetric epoxidation of sulcatol derivative catalyzed by *Aspergillus niger* for the synthesis of (2*R*,5*S*)-pityol.

13.3.2 Asymmetric Epoxidation of Aromatic Alkenes

Asymmetric Epoxidation of Styrene Derivatives and Analogues

The direct epoxidation of terminal aromatic alkenes using classic chemocatalysts, such as Jacobsen's salen transition metal catalysts, suffer from poor stereoselectivity (~36% ee) [95] and/or impractical reaction temperatures of −78 °C (86% ee) [96]. To achieve optically pure styrene oxide, a more universal approach is to use hydrolytic kinetic resolution following unselective epoxidation reactions, which leads to a product with high optical purity but only 50% yield [11].

On the other hand, the biocatalytic epoxidation of styrene and derivatives can be achieved with excellent stereoselectivity using SMOs of various origins (Scheme 13.9). Although isolated SMO has been successfully applied in combination with enzymatic NADH regeneration [92, 97] or reductive electrochemical cofactor regeneration [98–101], the process based on *E. coli* whole cells expressing those SMOs has been proven superior in terms of productivity due to the limited stability of cell-free enzymes, and consequently it has been applied in the majority of reported studies.

Most SMOs can catalyze the conversion of the native substrate, styrene, to (*S*)-styrene epoxide with >99% ee (Scheme 13.9) [7]. In addition, a spectrum of styrene derivatives and analogs with substitution on the benzene ring, or on the double bond, or with charged aromatic core, can be epoxidized with high to excellent selectivity using SMOs (Scheme 13.9) [55, 92, 102]. However, electron-withdrawing and bulky groups significantly impair the activity. For example, methyl cinnamate, containing an electron-withdrawing group, and (*E*)-1,2-diphenylethene, containing a bulky group, were not accepted by SMO [55].

The reactions are routinely carried out in a two-liquid phase system that alleviates the product/substrate inhibition by *in situ* product extraction and substrate supply [103, 104]. In the pilot-scale preparation of enantiopure (*S*)-styrene oxide using the SMO from *Pseudomonas* sp. VLB120 expressed in recombinant *E. coli*, a bis(2-ethylhexyl) phthalate (BEHP)/aqueous biphasic system was applied with the product concentration reaching up to $36.3\,\mathrm{g\,l_{tot}^{-1}}$ and volumetric productivities of $4.19\,\mathrm{g\,l_{tot}^{-1}\,h^{-1}}$. The bioprocess performed best in terms of production costs compared with three chemocatalysis processes [103, 105].

Base on those results, a tandem biocatalysts system was successfully developed using the same SMO and a regioselective epoxide hydrolase (SpEH) for the enantioselective dihydroxylation of aryl olefins to give (*S*)-vicinal diols [106]. By removing epoxide quickly from the aqueous phase through the subsequent hydrolysis, any inhibiting effects are reduced, and the conversion of styrene was accelerated by a factor of 2 (Scheme 13.10a). The strategy was further developed into a cascade biocatalysis via intracellular epoxidation and hydrolysis using recombinant *E. coli* coexpressing SMO and SpEH with expanded substrate spectrum [107]. In addition, the stereocomplementary process was also established by applying an epoxide hydrolase (StEH) with opposite regioselectivity, resulting in the inversion of configuration to produce (*R*)-vicinal diols (Scheme 13.10b) [107].

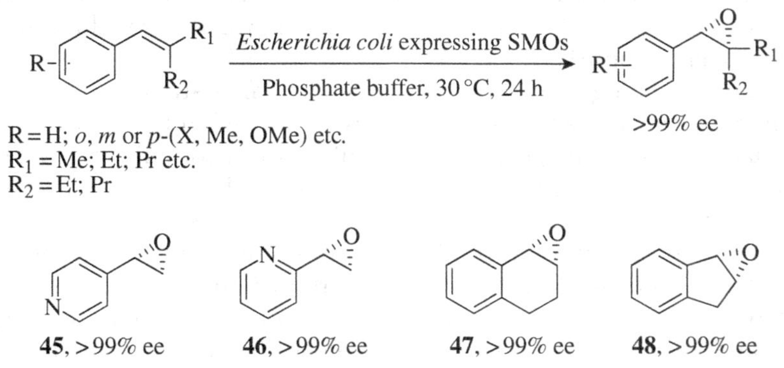

SCHEME 13.9
SMO-catalyzed asymmetric epoxidation of styrene derivatives and analogues.

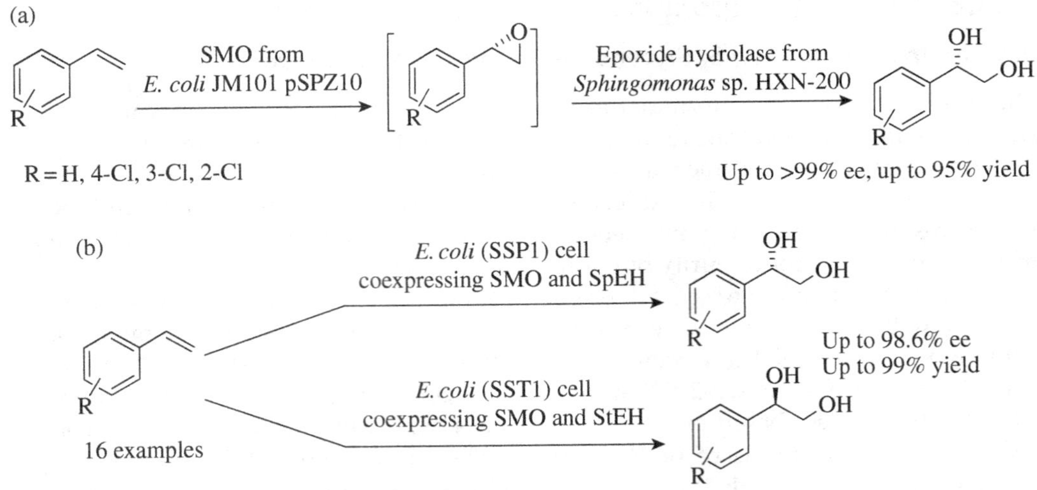

SCHEME 13.10

Enantiomeric dihydroxylation of aryl olefins catalyzed with SMO and epoxide hydrolase in (a) a tandem system or (b) using recombinant *Escherichia coli* coexpressing SMO and epoxide hydrolase.

SCHEME 13.11

CPO-catalyzed asymmetric epoxidation of styrene derivatives.

Besides SMOs, recombinant *E. coli* expressing the xylene monooxygenase from *P. putida* mt-2, which typically oxidizes toluene and xylenes to the corresponding benzyl alcohol derivatives, can as well catalyze the epoxidation of styrene into (*S*)-styrene oxide [49, 108]. However, the substrate spectrum for the enzyme was extremely narrow. Only styrene, 3-chlorostyrene, and 4-chlorostyrene were oxidized without side reactions, yielding epoxides with 92%, 96%, and 37% ee, respectively, while 3-methylstyrene and 4-methylstyrene were primarily oxidized at the methyl moiety rather than the vinyl group [49].

In contrast to SMOs, which exclusively catalyze the formation of (*S*)-epoxides, CPOs are able to yield chiral (*R*)-epoxides from styrene derivatives, which could serve as a stereocomplementary method (Scheme 13.11). However, the substrate spectrum appeared to be very limited [85, 109]. Two *cis*-disubstituted aryl-substituted alkenes, *cis*-α-methylstyrene and 1,2-dihydronaphthalene, were effectively epoxidized in the presence of H_2O_2, and the later underwent spontaneous hydrolytic ring opening to afford the corresponding trans-diol. The enantiomeric excesses of the products were as high as 96% and 97% ee, respectively [83]. Terminal alkenes such as styrene and derivatives with a halide substituent on the benzene ring [109] or an alpha-alkyl substituent [85] could be accepted by CPO, and underwent epoxidation using *t*-BuOOH or H_2O_2 as the oxidant, but the (*R*)-epoxides were achieved with only low to medium enantiopurity [83, 85, 109].

The ethene-oxidizing microorganism *Mycobacterium* strain NBB4 contains an ethylene monooxygenase, which have recently been used in the synthesis of stereocomplementary (*R*)-styrene oxide with 98% ee [110]. The reaction efficiency was enhanced in a biphasic system with styrene as the organic phase, but the enantioselectivity of the reaction decreased and yielded the oxides with 66–86% ee. The authors proposed a possible mechanism that a second, less stereoselective monooxygenase might be induced that also contributed to styrene oxide production.

Cytochrome P450 monooxygenase is another group of enzymes for the preparation of chiral styrene oxide derivatives. However, the stereoselectivities of native

P450s are often low. A recent report describes using chiral carboxylic acids as decoy molecules to enhance the stereoselectivity of a H_2O_2-dependent cytochrome P450$_{SP}\alpha$ in the epoxidation of styrene. The presence of (*R*)-ibuprofen dramatically increased enantiopurity of (*S*)-epoxide to 63 and 88% ee for the wild type and the F288G mutant, respectively [111].

Some engineered P450s display good stereoselectivity in bioepoxidation reactions [30, 90]. For example, residue size at position 87 of P450$_{BM3}$ (CYP102A1) was found to affect its stereoselectivity in epoxidation. The F87G mutant displayed much higher selectivity than the wild type, or F87V and F87A mutants, resulting in (*R*)-styrene oxide and (*R*)-3-chlorostyrene oxide with 92% and 94.6% ee, respectively [112, 113]. Very recently, Li *et al.* have reported the successful bioepoxidation of several *para*-substituted styrenes with excellent (*R*)-selectivity (up to 99.5% ee) and high conversion (82–97%) using the resting cells of *E. coli* expressing a triple mutant of P450$_{pyr}$ (P450$_{pyr}$ TM), which was generated via directed evolution (Scheme 13.12). Interestingly, this enzyme was also found to catalyze the epoxidation of *ortho*-substituted styrenes with high (*S*)-selectivity (up to 97.7% ee) [114]. The same group also reported the discovery of P450$_{tol}$ monooxygenase from *Rhodococcus coprophilus* TC-2, which was coexpressed with the ferredoxin reductase and ferredoxin from *Sphingomonas* sp. HXN-200, and a glucose dehydrogenase in *E. coli* T7. This native P450 showed unique and excellent enantioselectivity, catalyzing the (*R*)-epoxidation of several *ortho*- and *meta*-substituted styrenes (97.5–99.7% ee), as well as the (*S*)-epoxidation of two *para*-substituted styrenes (90% ee) [115].

Asymmetric Epoxidation of Allylic Alcohols

The Sharpless epoxidation serves as the predominant approach for the asymmetric epoxidation of allylic alcohols, which is catalyzed with L-(+)/D-(−)-diethyl tartrate and titanium tetraisopropoxide (Ti(O-*i*-Pr)$_4$) using TBHP as the oxidant [9, 19]. Primary allylic alcohols are typical substrates for Sharpless epoxidation, and the same system could be used to kinetically resolve racemic secondary allylic alcohols as well [116, 117].

As an alternative, SMOs can catalyze the asymmetric epoxidation of primary allylic alcohols bearing a phenyl group. For example, (*Z*)-cinnamyl alcohol was catalyzed by recombinant *E. coli* harboring the SMO from *Pseudomonas* species and yielded the corresponding (2*S*, 3*S*)-epoxide with >99% ee (Scheme 13.13a) [62, 102]. Recombinant *E. coli* coexpressing the SMO from *Rhodococcus* sp. ST-10 and the alcohol dehydrogenase from *Leifsonia* sp. S749 also catalyzed the epoxidation of (*Z*)-cinnamyl alcohol and yielded enantiopure oxide [92].

A recent study demonstrated that SMO can also catalyze the kinetic resolution of secondary phenyl allylic alcohols with excellent stereoselectivity [118]. Using the whole cells of recombinant *E. coli* expressing the SMO from *Pseudomonas* sp. LQ26, the kinetic resolution of racemic 1-phenylprop-2-enol yielded (1*R*, 2*R*)-phenyl glycidol with >99% ee and 98% de, and (*R*)-alcohol was recovered with >99% ee at 50% conversion for 2 h (Scheme 13.13b) [118], which displayed an advantage over previously established chemistry methods, such as the Sharpless epoxidation and

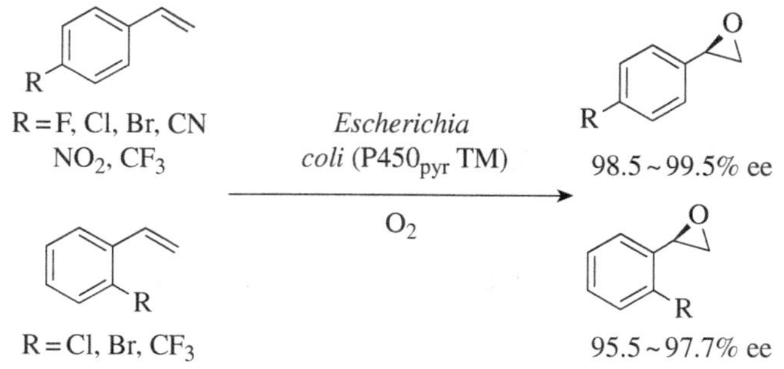

SCHEME 13.12 P450$_{pyr}$ mutant-catalyzed asymmetric epoxidation of styrene derivative.

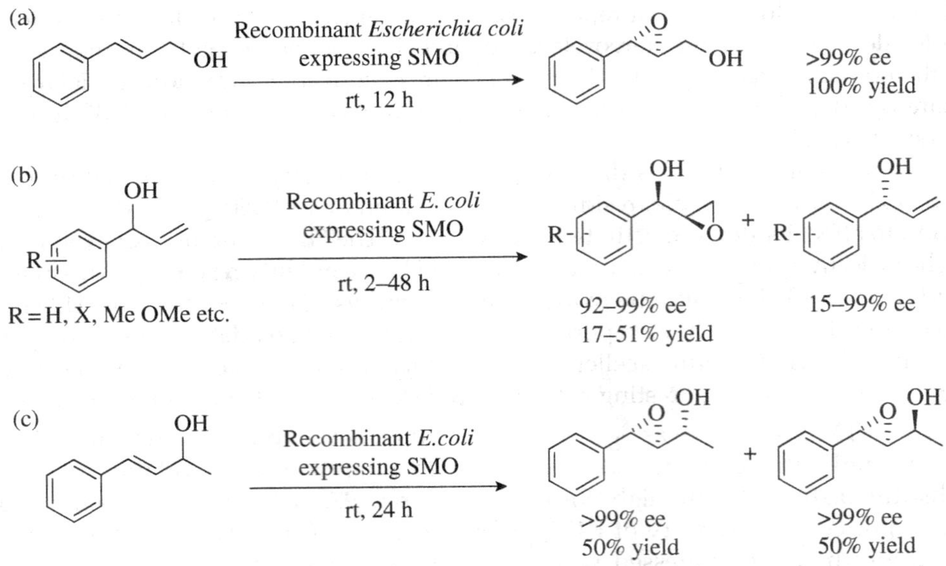

SCHEME 13.13

SMO-catalyzed enantioselective epoxidation of (a) primary allylic alcohols, (b) nonconjugated, and (c) conjugated secondary allylic alcohols.

vanadium-based method, which required up to 12 days for the same substrate to reach 50% conversion, and yielded the epoxides with 90–93% ee [117, 119]. The enzyme can accept a series of aromatic substrates as well as aliphatic allylic alcohols, but high enantio- and diastereoselectivity are only observed for aromatic substrates [118].

Conjugated secondary allylic alcohols, such as (*E*)-4-phenylbut-3-en-2-ol (**51**), are also good substrates for SMO. Epoxide products with excellent stereoselectivity (>99% ee) could be achieved, but the diastereoselectivity was poor with both (*R*)- and (*S*)-**51** epoxidized to **52** and **53**, respectively (Scheme 13.13c) [102].

Asymmetric Epoxidation of Other Aromatic Alkenes

Pseudomonas oleovorans ATCC 29347 harboring an alkane monooxygenase was at first reported to catalyze the asymmetric epoxidation of 1,7-octadiene with moderate stereoselectivity [80]. Further investigation showed that allyl phenyl ether and allyl benzyl ether derivatives could be better substrates in terms of both stereoselectivity and activity, which yielded the corresponding (*S*)-epoxides with up to 99% ee (Scheme 13.14a) [120]. This process has found industrial applications in the syntheses of the active component of the beta-adrenergic receptor blocking agents, metoprolol (**56**) and atenolol (**57**), which were achieved with 95–100% and 97% ee, respectively. The microbial-catalyzed epoxidation could yield over 7 g l^{-1} product within 6-h reactions (Scheme 13.14b) [8, 121].

CPO showed low activity toward allyl phenyl ether, but it catalyzed the epoxidation of 2-methylallyl phenyl ether with moderate stereoselectivity (89% ee), which was better than that achieved for most of the terminal alkenes [85]. The whole cells of *Mycobacterium* M156, which were isolated using propene as the sole carbon source, and contained propene monooxygenase [122], were also reported to catalyze the asymmetric epoxidation of allyl phenyl ether to produce chiral phenyl glycidyl ether in a two-liquid phase system [123].

13.4 PROTEIN ENGINEERING FOR BIOCATALYTIC EPOXIDATION REACTION

Protein engineering, a design or optimization process of proteins, has become a powerful tool in biotechnology to generate engineered biocatalysts to meet the production requirements of pharmaceuticals, biofuels, and other chemicals [124–128].

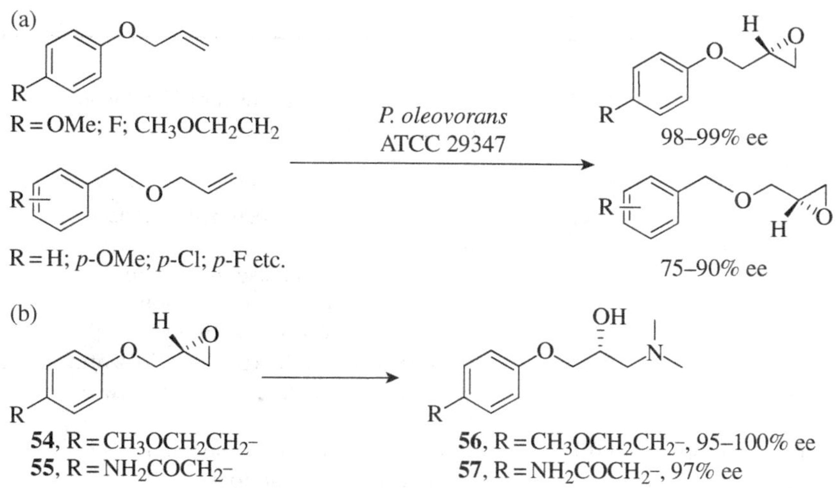

SCHEME 13.14

(a) Asymmetric epoxidation of allyl phenyl ether and allyl benzyl ether derivatives catalyzed by *Pseudomonas oleovorans* and (b) the subsequent synthesis of metoprolor and atenolol.

Taking advantage of the development of efficient screening strategies and elucidation of crystal structures, both directed evolution and rational design approaches have been successfully used to engineer enzymes involved in bioepoxidation for improved activity, stereoselectivity, or protein stability, or modified substrate specificity (Table 13.1).

13.4.1 Screening Methods

Screening strategies involve the characterization of individual clones of a mutant library according to specified criteria. An efficient, sensitive, and robust high-throughput screening system is the prerequisite for protein engineering using the directed evolution approach. Those screening systems can often be adapted to the mining of novel enzymes from natural sources as well. Several screening methods, including indigo-formation-based screening; γ-(4-nitrobenzyl)pyridine, *p*-nitrothiophenolate, and picric acid-based screening, and epoxide-based medium-throughput GC/HPLC assay have been developed for screening epoxidation enzyme mutants.

Indigo-Formation-Based Screening Method

Several types of enzymes that catalyze the epoxidation of alkenes are able to catalyze the oxidative coupling of indole, probably through indole oxide as the intermediate [145] (Scheme 13.15). Indole is naturally formed in typical recombinant *E. coli* cells from tryptophan, and the resulting dimeric product, indigo, is a dark blue pigment which can be visually detected on agar plates. Indigo is insoluble in water and nontoxic to the cells, so that the positive clones could be collected directly. As with most of the strategies based on agar plate, only qualitative results could be achieved, but the primary advantage of this screening method is high throughput and technical simplicity, which makes it suitable for first-stage preliminary screening. This strategy has been employed in the screening of a metagenomic library containing ~65 000 colonies derived from loam soil, which led to the identification of a novel SMO with excellent enantioselectivity in the production of (*S*)-epoxides [146]. The engineering of the SMO from *P. putida* CA-3 also adapted such a strategy (Table 13.1, entry 6) [135].

UV–Visible Spectrophotometer-Based Screening Method

Terminal epoxides react with γ-(4-nitrobenzyl)pyridine (NBP) to form a blue- (or purple-) colored precursor dye, exhibiting a broad absorption spectrum with a maximum at 540–600 nm [90, 147–149]. There is a linear relationship between the concentration of epoxides and absorbance of the blue NBP product, which make it feasible

TABLE 13.1 Examples of Enzymes Involved in Bioepoxidation

Entry	Enzymes	Mutagenesis Methods	Screening	Results	References
Activity					
1	$P450_{BM3}$ from *Bacillus megaterium*	Random mutagenesis/ recombination	N/A[a, b]	Epoxidation rate of mutant 139-3 was enhanced up to 100-fold	[129, 130]
2	$P450_{BM3}$ from *B. megaterium* (parent: variant 9-10A)	Saturation mutagenesis	NBP assay	30-fold increase in TTN for 1-hexene and epoxidation selectivity increased to 90%	[90, 131]
3	$P450_{CAM}$ from *Pseudomonas putida*	Site-directed mutagenesis	N/A[a]	25- and 9-fold increase in styrene oxide formation rate for mutants Y196F and Y96A, respectively	[132]
4	$P450_{CAM}$ from *P. putida*	Site-directed mutagenesis	N/A[a]	NADH coupling increased to 54% and styrene oxide formation rate increase to 70 min^{-1}	[133]
5	$P450_{pyr}$ from *Sphingomonas* sp. HXN-200	Iterative saturation mutagenesis	HPLC[b]	Activities toward *p*-substituted styrene increased up to 3-fold	[33, 134]
6	SMO from *P. putida* CA-3	Random mutagenesis	Indigo formation	8- and 12-fold improvement in styrene oxide and indene oxide formation	[135]
7	SMO from *Pseudomonas* sp. LQ26	Site-directed mutagenesis	N/A[a]	2.6- and 2.3-fold epoxidation activity toward styrene and *trans*-β-methyl styrene, respectively	[136]
8	Chloroperoxidase (CPO) from *Caldariomyces fumago*	Random mutagenesis	UV–vis	8-fold increase in epoxidation activity and 20-fold increase in conversion of *p*-nitrostyrene	[137]
Substrate specificity					
9	$P450_{BM3}$ from *B. megaterium*	Saturation mutagenesis	GC	Mutant accepts amorphadiene as substrate	[91]
10	$P450_{BM3}$ from *B. megaterium* (parent: variant 139-3)	Random mutagenesis	Picric acid assay	Mutant converts large steroid progesterone to 16,17-epoxysterol	[138]
11	SMO from *Pseudomonas* sp. LQ26	Site-directed mutagenesis	N/A[a]	Changed substrate preference from styrene to substituted styrene	[139]
Stereoselectivity					
12	$P450_{BM3}$ from *B. megaterium* (parent: variant 139-3)	Saturation mutagenesis	pNTP assay and GC	Position 184 is important for stereoselectivity	[140]
13	$P450_{BM3}$ from *B. megaterium* (parent: variant 9-10A)	Saturation mutagenesis	GC	Two mutants showed reversed stereoselectivity	[90, 131]
14	$P450_{BM3}$ from *B. megaterium*	Site-directed mutagenesis	N/A[a]	Stereoselectivity for styrene oxide is increased from 0 to 64% ee	[141]
15	$P450_{BM3}$ from *B. megaterium*	Site-directed mutagenesis	N/A[a]	Position 87 is important for stereoselectivity	[113]
16	$P450_{pyr}$ from *Sphingomonas* sp. HXN-200	Iterative saturation mutagenesis (ISM)	HPLC	Stereoselectivities of *p*-substituted styrene derivatives were increased to 98.5–99.5% ee	[33]
17	$P450_{SP\alpha}$ from *Sphingomonas paucimobilis*	Site-directed mutagenesis	N/A[a]	Mutant F288G showed reverses stereoselectivity	[111]
18	Propene monooxygenases from *Mycobacterium* sp. strain M156	Site-directed mutagenesis	N/A[a]	Site 188 and site 94 affect stereoselectivity	[142]
Stability					
19	$P450_{BM3}$ from *B. megaterium*	Random mutagenesis Saturation mutagenesis Site-directed mutagenesis	UV–vis[b]	Mutants showed significantly resistance toward acetone, acetonitrile, dimethylformamide, and ethanol	[143]
20	CPO from *C. fumago*	Random mutagenesis	UV–vis	The mutant B117 was completely resistant to the suicide inactivation reaction	[144]

[a] Not applicable.
[b] Not engineered for epoxidation.

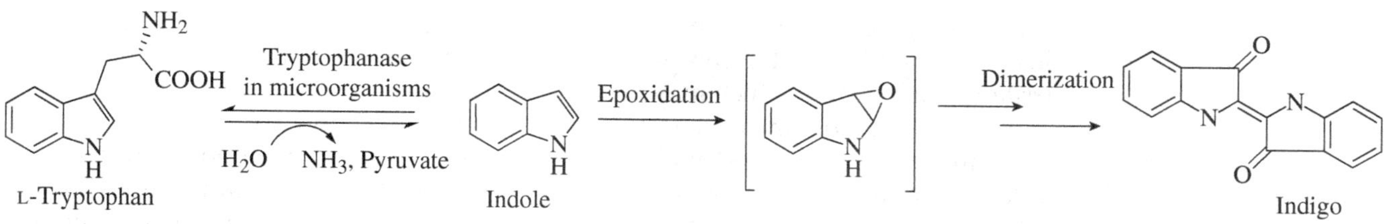

SCHEME 13.15

Postulated mechanism of enzyme-catalyzed indigo formation.

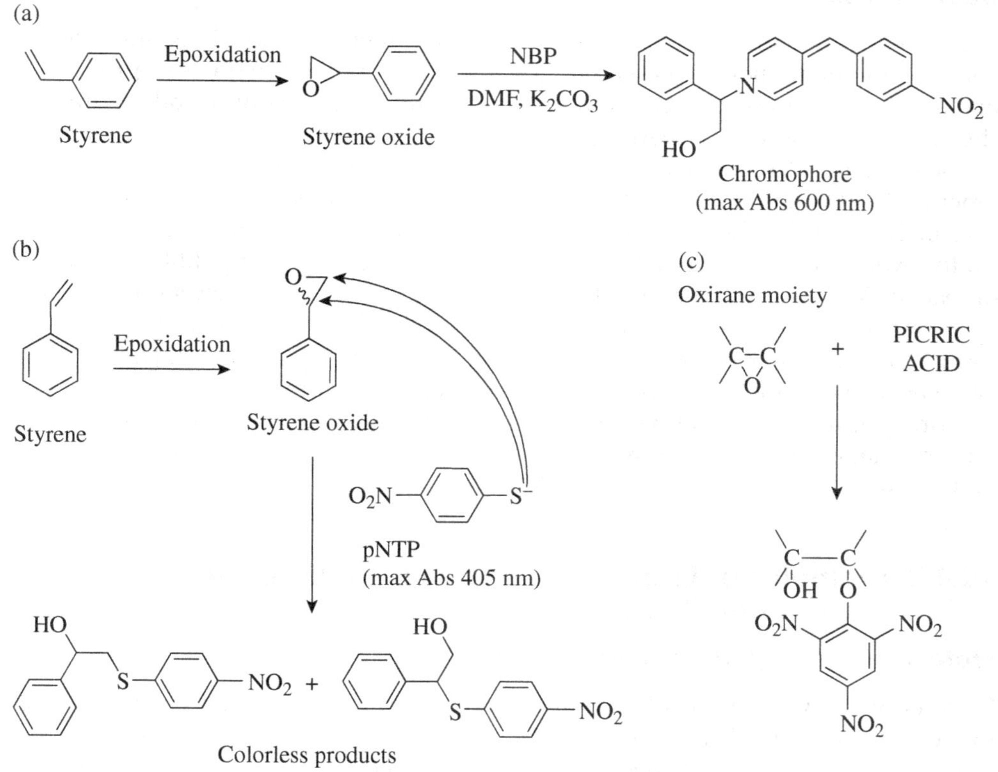

SCHEME 13.16

Mechanisms of NBP assay (a), pNTP assay (b), and picric acid assay (c).

that epoxidation activity can be indirectly quantified (Scheme 13.16a). In the case of styrene, the sensitivity of NBP assay reaches 50–100 μM of epoxidation product, a favorable detection sensitivity to a screening assay of directed evolution of epoxygenases [147]. However, the testing samples need a series of strict treatments before spectrophotometric analysis. This method has been applied to screen engineered cytochrome P450 BM-3 variants with improved catalytic turnovers and epoxidation selectivities [90].

Another chromophore for the analysis of epoxidation activity is *p*-nitrothiophenolate (pNTP), a yellow colorimetric reagent exhibiting absorbance maxima at 405 nm. The reaction between pNTP and epoxides results in a colorless product and a related linear decrease in absorbance at 405 nm (Scheme 13.16b) [150]. Compared with NBP assay, pNTP assay displays a clear advantage in technical simplicity, as some heating and cooling procedures are not required. The sensitivity limit of pNTP assay is approximately 400 μM for styrene oxide, which can be enhanced to 140 μM by adding methylated β-cyclodextrins to the reaction system as inclusion host for styrene to reduce the evaporation [151].

Recently, a modified picric acid assay has been employed in the screening of cytochrome P450 BM-3 mutants with epoxidation reactivity toward progesterone in a medium-throughput screening form [138]. Picric acid is a chromophore reagent that can add exclusively to the oxirane moiety. The reaction between this reagent and epoxides results in an orange-red-colored picryl ether derivative, which, under basic conditions gives an absorption maximum at 490 nm (Scheme 13.16c) [152]. However, like other highly nitrated compounds, picric acid is an explosive, as well as a strong irritant and allergen, which may prevent its general use.

In general, those microtiter plate-based assays are relatively time consuming and low throughput compared with solid-phase assays, but they provide quantitative data and represent the most effective means for evolving enzymes of epoxidation activity. The practical limit on the number of colonies that can be screened in microtiter plates is usually tens of thousands of colonies.

GC/HPLC Assay

For screening of stereoselectivity or substrate specificity, GC or HPLC analysis is a common method. They are based on the direct detection of substrates or products in epoxidation reactions. Those methods demonstrate high sensitivity and accuracy. In the case of propene as screening substrate for the alkene monooxygenase from *R. rhodochrous* B-276, the detection limit of the chiral GC method for a single enantiomer is about 0.01 mM in the ether extract to detect differences in ee of as low as 5%. The analysis rate reaches ten samples per hour, which is considered suitable for scale up to several thousand mutants [153]. GC/HPLC method is compatible to a diverse of reaction systems, and the sample treatment procedure is relatively simple, usually by extraction with organic solvents. Moreover, GC/HPLC assay based on epoxides formation is a reliable and convenient method for secondary screening. However, the GC/HPLC-based methods also demand strict equipment requirements. To increase the throughput, a multichannel sprayer / injection system and sophisticated program for rapid data analysis may be required besides standard GC/HPLC chromatography systems with appropriate columns, detectors, and procedures.

13.4.2 Examples of Engineered Enzymes for Biocatalytic Epoxidation Reactions

Protein Engineering of Cytochrome P450s

Great efforts have been devoted to the laboratory evolution of P450s, which has led to the identification of numerous variants with increased epoxidation reactivity [90, 129, 132, 133, 154], altered specificity [91], and enhanced stability [143] (Table 13.1, entries 1–5, 9, 10, 12–17, 19).

A representative is *B. megaterium* $P450_{BM3}$, from which many improved mutants with increasing catalytic performance have been identified and characterized. Variants $P450_{BM3}$ 139-3 [129, 130] and 9-10A [131], although not engineered for improving epoxidation reactivity, demonstrated surprisingly increased epoxidation activity compared to the wild-type enzyme and became convenient platforms for future engineering works to optimizing $P450_{BM3}$ [90, 91, 138, 140, 141, 143]. In addition, $P450_{CAM}$ from *P. putida* [132, 133] and $P450_{pyr}$ from *Sphingomonas* sp. HXN-200 [33, 134] have been improved for epoxidation by site-directed mutagenesis (Table 13.1, entries 3–5).

Both directed evolution and rational design are useful strategies for altering substrate specificity of P450s. By creating a model structure of the lowest energy transition state complex for amorphadiene in the $P450_{BM3}$ active site, Keasling and coworkers found two key mutations (F87A and A328L) that could enable the binding of the substrate amorphadiene and introduced them into the wild-type enzyme. Two additional mutations (R47L and Y51F) were also introduced into $P450_{BM3}$ to eliminate the competitive inhibition of other intracellular metabolites. Consequently, a $P450_{BM3}$ mutant G4 with all the four mutations mentioned previously has been constructed and used for the development of the semibiosynthetic route for artemisinin [91]. On the other hand, Wang and coworkers have constructed a $P450_{BM3}$ 139-3 random mutant library by error-prone PCR and screened the mutant for the epoxidation of progesterone using a modified picric acid assay [138].

Manipulating stereoselectivity is also an important field in the engineering of P450s. Generally, rational or semirational analysis of key positions (such as positions 82 [90, 141], 87 [90, 113], 184 [140], 328 [90], and 438 [141] in $P450_{BM3}$) based on protein structure has been used for the mutant design. The engineered P450 mutants achieve high [33, 112, 113, 141] or inverse [90, 111, 140] stereoselectivity. The practical applicability of $P450_{BM3}$ has been further increased due to the isolation of variants with significantly increased resistance to organic solvents. Starting with a activity-improved $P450_{BM3}$ mutant F87A, Schwaneberg and coworkers performed two rounds of random mutagenesis and a final bank mutation of position 87 to obtain those resistant variants (Table 13.1, entry 19) [143].

Protein Engineering of SMOs

With the aid of indigo-formation-based screening method, the SMO from *P. putida* CA-3 has been engineered to be more active for styrene and indene epoxidation after three rounds of random mutagenesis [135]. The third-generation mutant R3-10 exhibits eight-fold and 12-fold improvement in styrene and indene oxidation rates compared to the wild-type enzyme [135]. The rates of styrene oxide and indene oxide formation by recombinant *E. coli* cells expressing mutant R3-10 reach 266 and 433 $\mu mol\, min^{-1}\, g^{-1}$ CDW, respectively.

On the other hand, rational protein design has been employed to the engineering of SMO from *Pseudomonas* sp. LQ26. Based on the X-ray crystal structure and molecular docking, three mutants (Y73V, Y73F, and S96A) have been found to exhibit higher enzymatic activities than the wild type in the epoxidation of styrene, while retaining excellent stereoselectivity. The specific epoxidation activity of the most active mutant S96A toward styrene and trans-beta-methylstyrene are 2.6 and 2.3-fold of the wild type, respectively. In addition, the Y73V mutant showed an unexpected reversal of enantiomeric preference toward 1-phenylcyclohexene [136]. A similar approach has led to the identification of mutant L45A which exhibits an altered substrate preference toward the bulkier substrate alpha-ethylstyrene rather than styrene [139].

Protein Engineering of CPO

CPO has been optimized by directed evolution to overcome the restrictions of suicide inactivation and relatively low epoxidation rate. Through four generations of PCR-based random mutagenesis and screening for resistance to inactivation by allylbenzene, Hager and coworkers have successfully engineered CPO from *C. fumago* to completely resistant to the mechanism-based suicide inactivation caused by hydrogen peroxide and primary olefins [144]. The mutant showed evidently enhanced styrene epoxidation activity at longer time periods, even though the initial rate is only slightly higher than the wild-type CPO [144]. Subsequently, they have isolated other CPO mutants exhibiting approximately eight-fold increased epoxidation activity and 20-fold increased stability toward the toluene–isopropanol–water ternary solvent system after three rounds of random mutagenesis [137].

13.5 CONCLUSIONS AND OUTLOOK

Bioepoxidation reactions possess great potential for green synthesis. They can be routinely carried out at ambient temperature under atmospheric pressure and utilize molecular oxygen as a clean reactant in the majority of cases. Furthermore, they often exhibit exceptionally high chemo-, regio-, and stereoselectivity and can provide substrate spectrum complementary to chemocatalysis for certain types of molecules. However, despite of many successful examples of chiral epoxides production using enzymes or microbes, the wider implementation of bioepoxidation in chemical synthesis falls far behind classic chemocatalysis, and the industrial application has remained extremely limited, mainly due to insufficient enzyme availability, a narrow substrate spectrum, poor stability and low catalytic efficiency, as well as complicated electron relay system involving multiple coenzymes and cofactors. In addition, the high reactivity of the oxirane moiety can easily lead to product inhibition or denaturization of the biocatalysts, imposing extra challenges on the process optimization.

Recent developments in this field demonstrates that the toolbox of enzymes catalyzing epoxidation is expanding gradually. Firstly, whole genome sequencing and genome database mining are expected to facilitate the exploiting of new "epoxygenase." However, a more efficient heterologous expression system will be needed for the recombinant production of those multicomponent enzymes. Secondly, protein engineering is developing rapidly, which has opened up unlimited access to a variety of enzymes with improved characteristics, although a specific laboratory evolution platform should be set up to rapidly identify beneficial mutants. Thirdly, synthetic biology enables the combination of the downstream enzymatic reactions to epoxidation, which uses the epoxides *in situ*, resulting in higher yields and more versatile and diversified products. Additional approaches to biocatalysts stabilization, such as immobilization and recycling, should be investigated, which would further provide process flexibility.

ACKNOWLEDGMENTS

We thank the National Natural Science Foundation of China (20802073, 21072183, 21572220, and 21372216), the 100 Talents Program of the Chinese Academy of Sciences, and the Open Fund of Key Laboratory of Environmental and Applied Microbiology (KLCAS-2013-06) for financial support.

REFERENCES

[1] E. J. de Vries and D. B. Janssen, *Curr. Opin. Biotechnol.* **14** (2003) 414–420.
[2] J. G. Smith, *Synthesis* **1984** (1984) 629–656.
[3] M. Breuer, K. Ditrich, T. Habicher, B. Hauer, M. Kesseler, R. Sturmer, and T. Zelinski, *Angew. Chem. Int. Ed.* **43** (2004) 788–824.
[4] V. Farina, J. T. Reeves, C. H. Senanayake, and J. H. J. Song, *Chem. Rev.* **106** (2006) 2734–2793.
[5] G. Sello, T. Fumagalli, and F. Orsini, *Curr. Org. Synth.* **3** (2006) 457–476.
[6] S. Hwang, C. Y. Choi, and E. Y. Lee, *J. Ind. Eng. Chem.* **16** (2010) 1–6.
[7] H. Lin, J. Y. Liu, H. B. Wang, A. A. Q. Ahmed, and Z. L. Wu, *J. Mol. Catal. B Enzym.* **72** (2011) 77–89.
[8] D. L. Chang, J. Zhang, B. Witholt, and Z. Li, *Biocatal. Biotransform.* **22** (2004) 113–130.
[9] K. B. Sharpless, *Angew. Chem. Int. Ed.* **41** (2002) 2024–2032.
[10] F. Hollmann, I. Arends, K. Buehler, A. Schallmey, and B. Buhler, *Green Chem.* **13** (2011) 226–265.
[11] M. Tokunaga, J. F. Larrow, F. Kakiuchi, and E. N. Jacobsen, *Science* **277** (1997) 936–938.
[12] P. Kumar, V. Naidu, and P. Gupta, *Tetrahedron* **63** (2007) 2745–2785.
[13] W. J. Choi, *Appl. Microbiol. Biotechnol.* **84** (2009) 239–247.
[14] G. W. Zheng and J. H. Xu, *Curr. Opin. Biotechnol.* **22** (2011) 784–792.
[15] Z. Li, J. B. van Beilen, W. A. Duetz, A. Schmid, A. de Raadt, H. Griengl, and B. Witholt, *Curr. Opin. Chem. Biol.* **6** (2002) 136–144.
[16] M. Schrewe, M. K. Julsing, B. Buhler, and A. Schmid, *Chem. Soc. Rev.* **42** (2013) 6346–6377.
[17] Z.-X. Wang, Y. Tu, M. Frohn, J.-R. Zhang, and Y. Shi, *J. Am. Chem. Soc.* **119** (1997) 11224–11235.
[18] W. Zhang, J. L. Loebach, S. R. Wilson, and E. N. Jacobsen, *J. Am. Chem. Soc.* **112** (1990) 2801–2803.
[19] T. Katsuki and K. B. Sharpless, *J. Am. Chem. Soc.* **102** (1980) 5974–5976.
[20] R. Noyori, M. Aoki, and K. Sato, *Chem. Commun.* (2003) Issue 10: 1977–1986.

[21] E. N. Jacobsen, W. Zhang, A. R. Muci, J. R. Ecker, and L. Deng, *J. Am. Chem. Soc.* **113** (1991) 7063–7064.
[22] Y. Tu, Z.-X. Wang, and Y. Shi, *J. Am. Chem. Soc.* **118** (1996) 9806–9807.
[23] S. W. May, and R. D. Schwartz, *J. Am. Chem. Soc.* **96** (1974) 4031–4032.
[24] J. Crosby, *Tetrahedron* **47** (1991) 4789–4846.
[25] W.-R. Abraham, Microbial epoxidation: Application in biotechnology. In: R. N. Patel (Ed.), *Stereoselective Biocatalysis*, CRC Press, Boca Raton, 2000, pp. 181–204.
[26] D. E. T. Pazmino, M. Winkler, A. Glieder, and M. W. Fraaije, *J. Biotechnol.* **146** (2010) 9–24.
[27] M. Sono, M. P. Roach, E. D. Coulter, and J. H. Dawson, *Chem. Rev.* **96** (1996) 2841–2888.
[28] E. M. Isin and F. P. Guengerich, *Biochim. Biophys. Acta Gen. Subj.* **1770** (2007) 314–329.
[29] F. P. Guengerich, *Arch. Biochem. Biophys.* **409** (2003) 59–71.
[30] P. R. Ortiz de Montellano, J. A. Fruetel, J. R. Collins, D. L. Camper, and G. H. Loew, *J. Am. Chem. Soc.* **113** (1991) 3195–3196.
[31] C. F. Harford-Cross, A. B. Carmichael, F. K. Allan, P. A. England, D. A. Rouch, and L.-L. Wong, *Protein Eng.* **13** (2000) 121–128.
[32] A. Archelas and R. Furstoss, *Top. Curr. Chem.* **200** (1999) 159–191.
[33] A. T. Li, J. Liu, S. Q. Pham, and Z. Li, *Chem. Commun.* **49** (2013) 11572–11574.
[34] J. D. Pikus, J. M. Studts, C. Achim, K. E. Kauffmann, E. Munck, R. J. Steffan, K. McClay, and B. G. Fox, *Biochemistry* **35** (1996) 9106–9119.
[35] A. G. Katopodis, K. Wimalasena, J. Lee, and S. W. May, *J. Am. Chem. Soc.* **106** (1984) 7928–7935.
[36] K. Furuhashi, A. Taoka, S. Uchida, I. Karube, and S. Suzuki, *Eur. J. Appl. Microbiol. Biotechnol.* **12** (1981) 39–45.
[37] K. Furuhashi, M. Shintani, and M. Takagi, *Appl. Microbiol. Biotechnol.* **23** (1986) 218–223.
[38] H. Saeki and K. Furuhashi, *J. Ferment. Bioeng.* **78** (1994) 399–406.
[39] A. Miura and H. Dalton, *Biosci. Biotechnol. Biochem.* **59** (1995) 853–859.
[40] S. C. Gallagher, R. Cammack, and H. Dalton, *Eur. J. Biochem.* **247** (1997) 635–641.
[41] J. G. Leahy, P. J. Batchelor, and S. M. Morcomb, *FEMS Microbiol. Rev.* **27** (2003) 449–479.
[42] F. J. Small and S. A. Ensign, *J. Biol. Chem.* **272** (1997) 24913–24920.
[43] N. Y. Zhou, A. Jenkins, C. K. N. C. K. Chion, and D. J. Leak, *Appl. Environ. Microbiol.* **65** (1999) 1589–1595.
[44] V. Champreda, Y. J. Choi, N. Y. Zhou, and D. J. Leak, *Appl. Microbiol. Biotechnol.* **71** (2006) 840–847.
[45] K. McClay, B. G. Fox, and R. J. Steffan, *Appl. Environ. Microbiol.* **66** (2000) 1877–1882.
[46] M. Kok, R. Oldenhuis, M. P. van der Linden, P. Raatjes, J. Kingma, P. H. van Lelyveld, and B. Witholt, *J. Biol. Chem.* **264** (1989) 5435–5441.
[47] B. Buhler, A. Schmid, B. Hauer, and B. Witholt, *J. Biol. Chem.* **275** (2000) 10085–10092.
[48] J. Shanklin, E. Whittle, and B. G. Fox, *Biochemistry* **33** (1994) 12787–12794.
[49] M. G. Wubbolts, P. Reuvekamp, and B. Witholt, *Enzyme Microb. Technol.* **16** (1994) 608–615.
[50] A. Kantz, F. Chin, N. Nallamothu, T. Nguyen, and G. T. Gassner, *Arch. Biochem. Biophys.* **442** (2005) 102–116.
[51] K. Otto, K. Hofstetter, M. Rothlisberger, B. Witholt, and A. Schmid, *J. Bacteriol.* **186** (2004) 5292–5302.
[52] G. Bestetti, P. Di Gennaro, A. Colmegna, I. Ronco, E. Galli, and G. Sello, *Int. Biodeterior. Biodegrad.* **54** (2004) 183–187.
[53] D. Kuhn, F. S. O. Fritzsch, X. M. Zhang, V. F. Wendisch, L. M. Blank, B. Buhler, and A. Schmid, *J. Biotechnol.* **163** (2013) 194–203.
[54] A. Schmid, K. Hofstetter, H. J. Feiten, F. Hollmann, and B. Witholt, *Adv. Synth. Catal.* **343** (2001) 732–737.
[55] S. Bernasconi, F. Orsini, G. Sello, A. Colmegna, E. Galli, and G. Bestetti, *Tetrahedron Lett.* **41** (2000) 9157–9161.
[56] J. H. Han, M. S. Park, J. W. Bae, E. Y. Lee, Y. J. Yoon, S. G. Lee, and S. Park, *Enzyme Microb. Technol.* **39** (2006) 1264–1269.
[57] H. Lin, J. Qiao, Y. Liu, and Z. L. Wu, *J. Mol. Catal. B Enzym.* **67** (2010) 236–241.
[58] H. Toda, R. Imae, T. Komio, and N. Itoh, *Appl. Microbiol. Biotechnol.* **96** (2012) 407–418.
[59] D. Tischler, D. Eulberg, S. Lakner, S. R. Kaschabek, W. J. H. van Berkel, and M. Schlomann, *J. Bacteriol.* **191** (2009) 4996–5009.
[60] D. Tischler, R. Kermer, J. A. D. Groning, S. R. Kaschabek, W. J. H. van Berkel, and M. Schlomann, *J. Bacteriol.* **192** (2010) 5220–5227.

[61] D. Tischler, J. A. D. Groning, S. R. Kaschabek, and M. Schlomann, *Appl. Biochem. Biotechnol.* **167** (2012) 931–944.

[62] P. Di Gennaro, A. Colmegna, E. Galli, G. Sello, F. Pelizzoni, and G. Bestetti, *Appl. Environ. Microbiol.* **65** (1999) 2794–2797.

[63] U. E. Ukaegbu, A. Kantz, M. Beaton, G. T. Gassner, and A. C. Rosenzweig, *Biochemistry* **49** (2010) 1678–1688.

[64] A. Kantz and G. T. Gassner, *Biochemistry* **50** (2011) 523–532.

[65] M. J. Nuell, G. H. Fang, M. J. Axley, P. Kenigsberg, and L. P. Hager, *J. Bacteriol.* **170** (1988) 1007–1011.

[66] J. R. Beckwith and L. P. Hager, *J. Biol. Chem.* **238** (1963) 3091.

[67] J. Geigert, T. D. Lee, D. J. Dalietos, D. S. Hirano, and S. L. Neidleman, *Biochem. Biophys. Res. Commun.* **136** (1986) 778–782.

[68] L. P. Hager, F. J. Lakner, and A. Basavapathruni, *J. Mol. Catal. B Enzym.* **5** (1998) 95–101.

[69] V. M. Dembitsky, *Tetrahedron* **59** (2003) 4701–4720.

[70] M. Sundaramoorthy, J. Terner, and T. L. Poulos, *Structure (Lond.)* **3** (1995) 1367–1377.

[71] X. W. Yi, A. Conesa, P. J. Punt, and L. P. Hager, *J. Biol. Chem.* **278** (2003) 13855–13859.

[72] H. A. Wagenknecht and W. D. Woggon, *Chem. Biol.* **4** (1997) 367–372.

[73] K. Kuhnel, W. Blankenfeldt, J. Terner, and I. Schlichting, *J. Biol. Chem.* **281** (2006) 23990–23998.

[74] H. Tian, X. She, L. Shu, H. Yu, and Y. Shi, *J. Am. Chem. Soc.* **122** (2000) 11551–11552.

[75] M. Mahmoudian and A. Michael, *Appl. Microbiol. Biotechnol.* **37** (1992) 23–27.

[76] C. A. G. M. Weijers, C. G. Vanginkel, and J. A. M. Debont, *Enzyme Microb. Technol.* **10** (1988) 214–218.

[77] A. Q. H. Habetscrutzen, S. J. N. Carlier, J. A. M. Debont, D. Wistuba, V. Schurig, S. Hartmans, and J. Tramper, *Enzyme Microb. Technol.* **7** (1985) 17–21.

[78] A. Liese, K. Seelbach, A. Buchholz, and J. Haberland, Processes. In: A. Liese, K. Seelbach, and C. Wandrey (Eds.), *Industrial Biotransformations*, Wiley-VCH, Weinheim, 2006, p. 147.

[79] S. W. May and B. J. Abbott, *Biochem. Biophys. Res. Commun.* **48** (1972) 1230–1234.

[80] S. W. May, M. S. Steltenkamp, R. D. Schwartz, and C. J. McCoy, *J. Am. Chem. Soc.* **98** (1976) 7856–7858.

[81] M.-J. de Smet, J. Kingma, H. Wynberg, and B. Witholt, *Enzyme Microb. Technol.* **5** (1983) 352–360.

[82] E. J. Allain, L. P. Hager, L. Deng, and E. N. Jacobsen, *J. Am. Chem. Soc.* **115** (1993) 4415–4416.

[83] A. Zaks and D. R. Dodds, *J. Am. Chem. Soc.* **117** (1995) 10419–10424.

[84] S. H. Hu and L. P. Hager, *Tetrahedron Lett.* **40** (1999) 1641–1644.

[85] A. F. Dexter, F. J. Lakner, R. A. Campbell, and L. P. Hager, *J. Am. Chem. Soc.* **117** (1995) 6412–6413.

[86] F. J. Lakner, K. P. Cain, and L. P. Hager, *J. Am. Chem. Soc.* **119** (1997) 443–444.

[87] F. J. Lakner and L. P. Hager, *J. Org. Chem.* **61** (1996) 3923–3925.

[88] S. H. Hu, P. Gupta, A. K. Prasad, R. A. Gross, and V. S. Parmar, *Tetrahedron Lett.* **43** (2002) 6763–6766.

[89] D. J. Bougioukou and I. Smonou, *Tetrahedron Lett.* **43** (2002) 339–342.

[90] T. Kubo, M. W. Peters, P. Meinhold, and F. H. Arnold, *Chem. Eur. J.* **12** (2006) 1216–1220.

[91] J. A. Dietrich, Y. Yoshikuni, K. J. Fisher, F. X. Woolard, D. Ockey, D. J. McPhee, N. S. Renninger, M. C. Y. Chang, D. Baker, and J. D. Keasling, *ACS Chem. Biol.* **4** (2009) 261–267.

[92] H. Toda, R. Imae, and N. Itoh, *Tetrahedron Asymmetry* **23** (2012) 1542–1549.

[93] S. Peter, M. Kinne, R. Ullrich, G. Kayser, and M. Hofrichter, *Enzyme Microb. Technol.* **52** (2013) 370–376.

[94] A. Archelas and R. Furstoss, *Tetrahedron Lett.* **33** (1992) 5241–5242.

[95] B. D. Brandes and E. N. Jacobsen, *Tetrahedron Asymmetry* **8** (1997) 3927–3933.

[96] M. Palucki, G. J. McCormick, and E. N. Jacobsen, *Tetrahedron Lett.* **36** (1995) 5457–5460.

[97] K. Hofstetter, J. Lutz, I. Lang, B. Witholt, and A. Schmid, *Angew. Chem. Int. Ed.* **43** (2004) 2163–2166.

[98] F. Hollmann, K. Hofstetter, T. Habicher, B. Hauer, and A. Schmid, *J. Am. Chem. Soc.* **127** (2005) 6540–6541.

[99] R. Ruinatscha, C. Dusny, K. Buehler, and A. Schmid, *Adv. Synth. Catal.* **351** (2009) 2505–2515.

[100] D. Lim, Y. H. Kim, J. C. Joo, and Y. J. Yoo, *Enzyme Microb. Technol.* **47** (2010) 313–321.

[101] R. Ruinatscha, K. Buehler, and A. Schmid, *J. Mol. Catal. B Enzym.* **103** (2014) 100–105.

[102] H. Lin, Y. Liu, and Z. L. Wu, *Tetrahedron Asymmetry* **22** (2011) 134–137.
[103] S. Panke, M. Held, M. G. Wubbolts, B. Witholt, and A. Schmid, *Biotechnol. Bioeng.* **80** (2002) 33–41.
[104] B. Buhler, J. B. Park, L. M. Blank, and A. Schmid, *Appl. Environ. Microbiol.* **74** (2008) 1436–1446.
[105] D. Kuhn, M. A. Kholiq, E. Heinzle, B. Buhler, and A. Schmid, *Green Chem.* **12** (2010) 815–827.
[106] Y. Xu, X. Jia, S. Panke, and Z. Li, *Chem. Commun.* (2009) 1481–1483.
[107] S. K. Wu, Y. Z. Chen, Y. Xu, A. T. Li, Q. S. Xu, A. Glieder, and Z. Li, *ACS Catal.* **4** (2014) 409–420.
[108] B. Buhler, B. Witholt, B. Hauer, and A. Schmid, *Appl. Environ. Microbiol.* **68** (2002) 560–568.
[109] S. Colonna, N. Gaggero, L. Casella, G. Carrea, and P. Pasta, *Tetrahedron Asymmetry* **4** (1993) 1325–1330.
[110] S. Cheung, V. McCarl, A. J. Holmes, N. V. Coleman, and P. J. Rutledge, *Appl. Microbiol. Biotechnol.* **97** (2013) 1131–1140.
[111] T. Fujishiro, O. Shoji, N. Kawakami, T. Watanabe, H. Sugimoto, Y. Shiro, and Y. Watanabe, *Chem. Asian. J.* **7** (2012) 2286–2293.
[112] S. Eiben, L. Kaysser, S. Maurer, K. Kuhnel, V. B. Urlacher, and R. D. Schmid, *J. Biotechnol.* **124** (2006) 662–669.
[113] Q. S. Li, J. Ogawa, R. D. Schmid, and S. Shimizu, *FEBS Lett.* **508** (2001) 249–252.
[114] A. Shehzad, S. Panneerselvam, M. Linow, M. Bocola, D. Roccatano, J. Mueller-Dieckmann, M. Wilmanns, and U. Schwaneberg, *Chem. Commun.* **49** (2013) 4694–4696.
[115] A. Li, S. Wu, J. P. Adams, R. Snajdrova, and Z. Li, *Chem. Commun.* **50** (2014) 8771–8774.
[116] V. S. Martin, S. S. Woodard, T. Katsuki, Y. Yamada, M. Ikeda, and K. B. Sharpless, *J. Am. Chem. Soc.* **103** (1981) 6237–6240.
[117] M. E. Jung and Y. H. Jung, *Tetrahedron Lett.* **30** (1989) 6637–6640.
[118] H. Lin, Y. Liu, and Z. L. Wu, *Chem. Commun.* **47** (2011) 2610–2612.
[119] W. Zhang, A. Basak, Y. Kosugi, Y. Hoshino, and H. Yamamoto, *Angew. Chem. Int. Ed.* **44** (2005) 4389–4391.
[120] H. Fu, M. Newcomb, and C. H. Wong, *J. Am. Chem. Soc.* **113** (1991) 5878–5880.
[121] S. L. Johnstone, G. T. Phillips, B. W. Robertson, P. D. Watts, M. A. Bertola, H. S. Koger, and A. F. Marx, Stereoselective synthesis of S-(−)-beta-blocks via microbially produced epoxide intermediates. In: C. Laane, J. Tramper, and M. D. Lilly (Eds.), *Biocatalysis in Organic Media*, Elsevier Science Publishers, Amsterdam, 1987, pp. 387–392.
[122] M. Woodland, C. Matthews, and D. Leak, *Arch. Microbiol.* **163** (1995) 231–234.
[123] S. Prichanont, D. J. Leak, and D. C. Stuckey, *Enzyme Microb. Technol.* **22** (1998) 471–479.
[124] Y. Li and P. C. Cirino, *Biotechnol. Bioeng.* **111** (2014) 1273–1287.
[125] M. A. Fisher and D. Tullman-Ercek, *Curr. Opin. Biotechnol.* **24** (2013) 1010–1016.
[126] M. T. Reetz, *Angew. Chem. Int. Ed.* **50** (2011) 138–174.
[127] E. M. Brustad and F. H. Arnold, *Curr. Opin. Chem. Biol.* **15** (2011) 201–210.
[128] P. A. Romero and F. H. Arnold, *Nat. Rev. Mol. Cell Biol.* **10** (2009) 866–876.
[129] E. T. Farinas, M. Alcalde, and F. Arnold, *Tetrahedron* **60** (2004) 525–528.
[130] A. Glieder, E. T. Farinas, and F. H. Arnold, *Nat. Biotechnol.* **20** (2002) 1135–1139.
[131] M. W. Peters, P. Meinhold, A. Glieder, and F. H. Arnold, *J. Am. Chem. Soc.* **125** (2003) 13442–13450.
[132] D. P. Nickerson, C. F. HarfordCross, S. R. Fulcher, and L. L. Wong, *FEBS Lett.* **405** (1997) 153–156.
[133] M. P. Mayhew, V. Reipa, M. J. Holden, and V. L. Vilker, *Biotechnol. Prog.* **16** (2000) 610–616.
[134] S. Q. Pham, G. Pompidor, J. Liu, X.-D. Li, and Z. Li, *Chem. Commun.* **48** (2012) 4618–4620.
[135] L. J. Gursky, J. Nikodinovic-Runic, K. A. Feenstra, and K. E. O'Connor, *Appl. Microbiol. Biotechnol.* **85** (2010) 995–1004.
[136] H. Lin, D.-F. Tang, A. A. Q. Ahmed, Y. Liu, and Z.-L. Wu, *J. Biotechnol.* **161** (2012) 235–241.
[137] G. P. Rai, S. Sakai, A. M. Florez, L. Mogollon, and L. P. Hager, *Adv. Synth. Catal.* **343** (2001) 638–645.
[138] D. Jiang, R. Tu, P. Bai, and Q. Wang, *Biotechnol. Lett.* **35** (2013) 1663–1668.
[139] A. A. Qaed, H. Lin, D.-F. Tang, and Z.-L. Wu, *Biotechnol. Lett.* **33** (2011) 611–616.
[140] K. L. Tee, and U. Schwaneberg, *Angew. Chem. Int. Ed.* **45** (2006) 5380–5383.

[141] W. C. Huang, P. M. Cullis, E. L. Raven, and G. C. K. Roberts, *Metallomics* **3** (2011) 410–416.
[142] C. Chion, S. E. Askew, and D. J. Leak, *Appl. Environ. Microbiol.* **71** (2005) 1909–1914.
[143] T. S. Wong, F. H. Arnold, and U. Schwaneberg, *Biotechnol. Bioeng.* **85** (2004) 351–358.
[144] G. P. Rai, Q. Zong, and L. P. Hager, *Isr. J. Chem.* **40** (2000) 63–70.
[145] B. D. Ensley, B. J. Ratzkin, T. D. Osslund, M. J. Simon, L. P. Wackett, and D. T. Gibson, *Science* **222** (1983) 167–169.
[146] E. W. van Hellemond, D. B. Janssen, and M. W. Fraaije, *Appl. Environ. Microbiol.* **73** (2007) 5832–5839.
[147] M. Alcalde, E. T. Farinas, and F. H. Arnold, *J. Biomol. Screen.* **9** (2004) 141–146.
[148] K. Hemminki, T. Heinonen, and H. Vainio, *Arch. Toxicol.* **49** (1981) 35–41.
[149] J. H. Kim and J. J. Thomas, *Bull. Environ. Contam. Toxicol.* **49** (1992) 879–885.
[150] S. C. Agarwal, B. L. Van Duuren, J. J. Solomon, and S. A. Kline, *Environ. Sci. Technol.* **14** (1980) 1249–1253.
[151] K. L. Tee, O. Dmytrenko, K. Otto, A. Schmid, and U. Schwaneberg, *J. Mol. Catal. B Enzym.* **50** (2008) 121–127.
[152] J. Fioriti, A. Bentz, and R. Sims, *J. Am. Oil Chem. Soc.* **43** (1966) 37–41.
[153] A. Perry and T. J. Smith, *J. Biomol. Screen.* **11** (2006) 553–556.
[154] S. T. Jung, R. Lauchli, and F. H. Arnold, *Curr. Opin. Biotechnol.* **22** (2011) 809–817.

CHAPTER 14

Dynamic Kinetic Resolution via Hydrolase–Metal Combo Catalysis

Pilar Hoyos[1], Vittorio Pace[2], María J. Hernáiz[1], and Andrés R. Alcántara[1]

[1] Organic and Pharmaceutical Chemistry Department, Faculty of Pharmacy, Complutense University of Madrid, Madrid, Spain

[2] Department of Pharmaceutical Chemistry, Faculty of Life Sciences, University of Vienna, Vienna, Austria

14.1 INTRODUCTION

Due to the importance of chirality for the activity and properties of many compounds, synthesis of optically pure compounds is increasingly in demand in the pharmaceutical, fine chemicals, and agroalimentary industries. The exquisite regio-, chemo-, and enantioselectivity commonly displayed by enzymes have allowed the employment of these biocatalysts in the production of biologically active compounds, offering the development of more sustainable and economically attractive strategies for the synthesis of optically pure molecules [1–6]. Biocatalysis satisfies many of the 12 principles of green chemistry [7], such as the employment of gentle reaction conditions (physiological pH, temperature and water as the usual reaction medium, although many green solvents can also be used) or the evasion of protection and deprotection steps, producing less waste and shortening the production processes. Subsequently, the production of fine chemicals and drugs by using biotechnological methodologies is becoming widely implemented on an industrial scale [8–11].

Enzyme-catalyzed kinetic resolution (KR) of racemates have been widely employed for the preparation of enantiomerically pure compounds. In particular, the use of serine hydrolases as catalysts of the enantioselective hydrolysis of esters and amides or the stereoselective alcoholysis or aminolysis of esters is a common strategy to afford chiral compounds through KR processes [12–14]. However, a maximum 50% yield of chiral product can be afforded through this methodology, which restricts the real applicability of the process. Among the different approaches that have been proposed to circumvent this limitation, dynamic kinetic resolution (DKR) appears as a promising strategy to achieve maximum theoretical yield of 100% of optically pure compounds. The appropriate *in situ* combination of an enzyme-catalyzed KR and the racemization of the remnant substrate allow the complete transformation of a racemate into an enantiomerically pure product, as shown in Figure 14.1.

Green Biocatalysis, First Edition. Edited by Ramesh N. Patel.

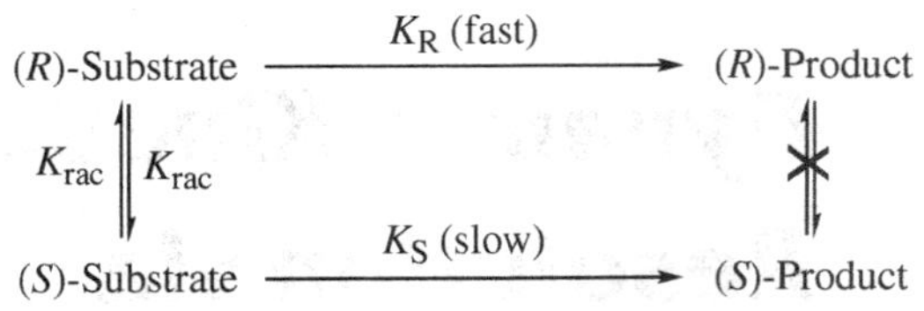

FIGURE 14.1
Dynamic kinetic resolution (DKR) process.

In this way, different enzyme-catalyzed KR processes have been successfully coupled with the racemization of the nonreactant substrate mediated by the action of a transition metal catalyst for synthesizing many chiral secondary alcohols and amines [15–19]. The most commonly employed strategy to develop a DKR involves the use of a hydrolase, commonly a lipase, as a biocatalyst for the enantioselective transesterification process, in combination with a transition metal-catalyzed racemization of the remnant substrate. As it is well known, lipases display high enantioselectivity toward a wide variety of substrates in transesterification reactions and high activity in organic solvents, which allows the synthesis of nonwater-soluble chiral molecules [20]. In this context, several requirements have to be fulfilled to perform an efficient DKR [21]: (i) the KR process has to be very stereoselective, (ii) the racemization must be faster than the transformation of the slow-reacting enantiomer ($K_{rac} \gg K_S$ slow; Figure 14.1), (iii) the racemization catalyst must not interact with the reaction product in any way, and (iv) the KR and the racemization process must be compatible under the same reaction conditions, including the reaction temperature and the acyl donor employed in the KR, as it could interfere with the racemization catalyst.

This chapter attempts to provide an update on the usefulness of hydrolases in combination with metal-based catalysts in DKR processes to afford relevant chiral molecules, precursors of biologically active compounds.

14.2 DKR OF SECONDARY ALCOHOLS

Due to the importance of functionalized *sec*-alcohols as intermediates in the preparation of biologically active compounds, there is a great interest in their preparation in a chiral manner. In this section we will discuss the implementation of a DKR of a *sec*-alcohol as the key step in the synthetic route of chiral compounds with interesting biological properties.

14.2.1 Racemization Catalysts for DKR of *sec*-Alcohols

Racemization of the remnant substrate in a DKR process can be performed either spontaneously or by the employment of a chemo- or biocatalyst, which must be compatible with the reaction conditions used for the KR reaction. In the case of *sec*-alcohols, most of successful DKRs have been carried out by the use of ruthenium complex catalysts, soluble in the organic reaction media, which promote racemization through redox processes. The first examples describe the resolution of 1-phenylethanol (***rac*-1**) by the combination of a rhodium catalyst ($Rh_2(OAc)_4$) with *Pseudomonas fluorescens* lipase [22], although more effective results were afforded by Bäckvall and coworkers [23], who developed the DKR of the same substrate and derivatives catalyzed by *Candida antarctica* lipase (CALB) and a ruthenium complex (Shvo's catalyst, **2** (Figure 14.2)), affording excellent conversions and enantiomeric excess (ee) values [24].

The later metal complex needs to be activated by heat (Figure 14.3), thus requiring the use of thermostable lipases, active at high temperatures.

It should be highlighted that this catalyst has not only been employed in the racemization of *sec*-alcohols but also primary alcohols with an unfunctionalized stereogenic center in the β-position [25]. In this case the metal complex causes a previous

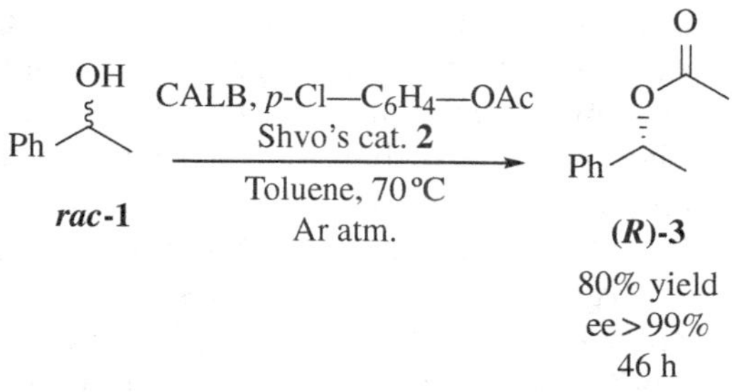

FIGURE 14.2 DKR of ***rac*-2** catalyzed by CALB and ruthenium catalyst **1**.

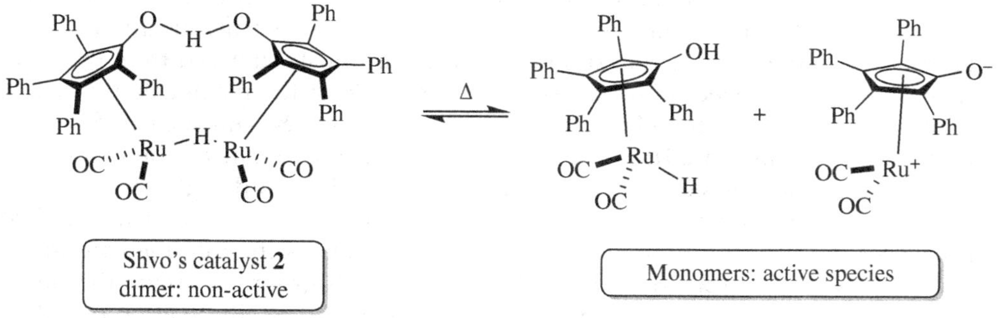

FIGURE 14.3 Thermal activation of Shvo's catalyst **2**.

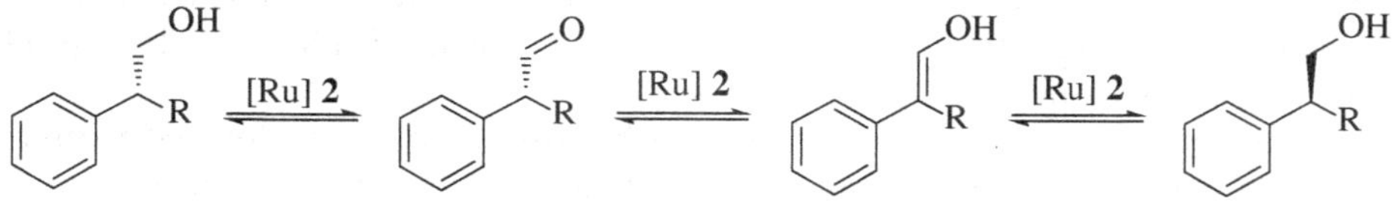

FIGURE 14.4 Ruthenium-catalyzed racemization of primary alcohols.

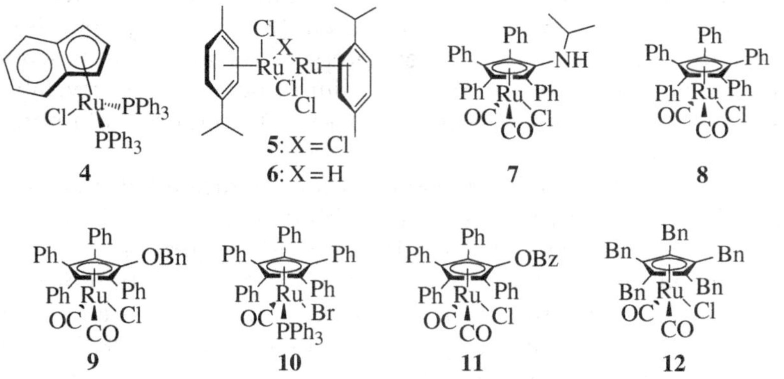

FIGURE 14.5 Ruthenium complexes employed in the DKR of *sec*-alcohols.

dehydrogenation of the alcohol to furnish an intermediate aldehyde, followed by an enolization, and finally the ruthenium catalyzes the readdition of hydrogen to the aldehyde (Figure 14.4). This process has been successfully coupled with the enzyme-catalyzed resolution in the DKR of primary alcohols [25].

However, not only CALB but also other lipases have been fruitfully coupled in the deracemization of a wide variety of substrates, such as diols [26]; α- [27], β- [28], and δ-hydroxyesters [29]; benzoins [30]; hydroxynitriles [31]; haloalcohols [32]; hydroxyalkanephosphonates [33]; γ-hydroxyamides [34]; hydroxyacids and hydroxyaldehydes protected with bulky groups [35]; or cyclic allylic alcohols [36]. From those pioneer examples, many efforts have been attempted in order to design other ruthenium catalysts, which could decrease the reaction time and temperature, and improve the reaction conditions, to extend the applicability of this strategy to the resolution of other substrates. Some of those ruthenium catalysts that have led to relevant results are shown in Figure 14.5.

The first optimizations were achieved by Park and coworkers, who employed the indenyl ruthenium complex **4**, activated in the presence of molecular O_2, combined with immobilized lipase from *Pseudomonas cepacia* in the DKR of ***rac*-1** [37]. The same group successfully applied cymene ruthenium complexes **5** and **6** in the DKR of allylic alcohols, furnishing the corresponding chiral acetates in high yields (over 80%) at room temperature [38]; later on, it was shown that these complexes could be used as catalysts in the DKR of *sec*-alcohols in ionic liquids, allowing the reuse of the catalysts [39], and very recently Agrawal *et al.* have employed catalyst **5** in combination with the bidentate ligand 1,4-bis(diphenylphosphino)butane as the racemization catalyst of the DKR of benzoins at room temperature [40].

The development of the cyclopentadienyl ruthenium catalysts **7** and **8** entailed the possibility of carrying out DKR processes at room temperature, decreasing reaction times, and the formation of secondary products, mostly derived from the accumulation of the oxidized substrate as an intermediate of the metal catalyst action. Catalyst **7** was first described by Kim, Park, and coworkers, which carried out the racemization of **(*S*)-1** in 30 min after activation with potassium *tert*-butoxide [41], although the time needed to reach maximum conversions and high enantiopurity values in a lipase-catalyzed DKR was still long. However, it has been further applied in the DKR of different aromatic and aliphatic alcohols as the intermediate in the synthesis of macrolide stagonolide-E [42, 43]. Bäckvall's group described the ruthenium complex **8**, which catalyzed the racemization of **(*S*)-1** in 10 min after activation with potassium *tert*-butoxide and was completely compatible with CALB activity [44, 45], leading to fast DKRs of a broad range of secondary alcohols as allylic alcohols (section "DKR of Allylic Alcohols") [44, 46, 47] or the DYKAT of diols (section "DYKAT of Diols") [48–52].

During the following years, different ruthenium complexes derived from catalysts **7** and **8** were designed and employed in DKR processes, optimizing their activity and stability. Ruthenium catalysts **9** [53], **10** [54], and **11** [55] are air-stable complexes, which can be used in DKR processes under ambient conditions. Kanerva and coworkers have recently reported the ruthenium complex **12**, which presents a more simple and cost-efficient preparation, although its scope is not as broad as the scope of catalyst **8**. With the aim of obtaining a recyclable catalyst, Park described a polymer-supported derivative **13** (Figure 14.6) obtained from ruthenium catalyst **9** [53].

Some other metals have been also used for the racemization of *sec*-alcohols, showing high activity and compatibility with biocatalysts. It is worthy of mention that the vanadium oxo-compounds **14** and **15** (Figure 14.6) are able to catalyze the racemization of allylic alcohols with the transposition of the hydroxyl group through the formation of allyl vanadate intermediates [56, 57]. These racemization reactions were combined with lipases in the DKR of allylic alcohols, achieving very high conversions and ee values [56–58]. Very recently, Akai and coworkers have prepared a novel mesoporous silica-immobilized oxovanadium catalyst (**V-MPS**) by combining $O{=}V(OSiPh_3)_3$ **14** and mesoporous silica (MPS). Thus, the oxovanadium species are confined inside a solid carrier, minimizing their interactions with the biocatalyst. The combined lipase–V-MPS catalyst is reusable and allows the achievement of DKR processes for a wide range of racemic alcohols [59].

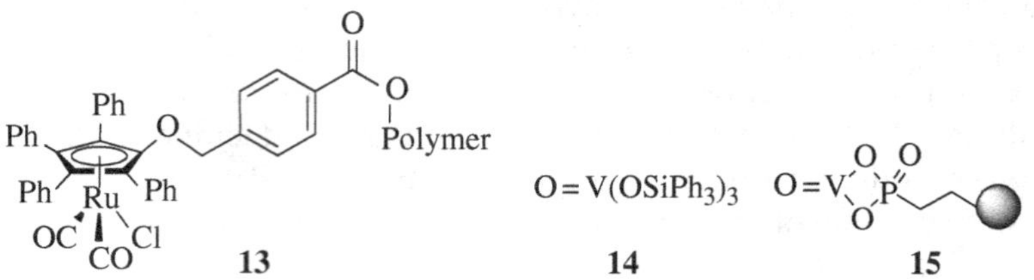

13 14 15

FIGURE 14.6

Immobilized and reusable ruthenium catalyst **13**; oxovanadium compounds **14** and **15** as catalysts for the racemization of allylic alcohols.

14.2.2 Synthetic Applications of the DKR of *sec*-Alcohols

As mentioned in the Section 14.1, chiral *sec*-alcohols are valuable building blocks for the production of many pharmaceuticals and fine chemicals. The DKR technique has emerged as an attractive alternative to develop sustainable and economically attractive routes for obtaining optically pure compounds. The enzymatic selectivity and the compatibility between the action of the metal catalyst and the biocatalyst are probably the limiting factors in the process. Actually the optimization of these processes is still an important research area, in order to increase the spectrum of substrates that can be resolved. However, today a broad scope of alcohols has been described as DKR substrates, including this strategy in the preparation of biologically active compounds.

DKR of Benzylic Alcohols

As commonly employed lipases present high enantioselectivity toward the resolution of (***rac*-1**), this simple substrate has been generally used as standard to test the viability of the lipase–metal combo catalysis, so that the scope of the reaction could be extended to other *sec*-alcohols afterwards. Enantiopure acetate **(*R*)-3** was first obtained in high yields and ee through a DKR process catalyzed by CALB and the ruthenium Shvo's complex **1** (Figure 14.2) [23], and this methodology has been employed on an industrial scale by DSM for the production of **(*R*)-2** [60]. Since then, different approaches have been described for optimizing the reaction, by decreasing the reaction times and/or the reaction temperatures, and for expanding the transformation to other benzylic alcohols [44, 53, 54, 61].

In this context, Bäckvall and coworkers employed complex **8** in combination with CALB in the DKR of 1-(6-chloropyridin-3-yl)ethanol **16**. This DKR process is presented as the key step of an innovative and scalable route to prepare chiral Me-imidacloprid **(S)-17**, a chloronicotinyl insecticide (Figure 14.7) [62]. DKR parameters were optimized, reaching maximum conversion and ee at 50 °C in toluene after 36 h. Although the ruthenium catalyst works at room temperature, increasing the temperature enhances the reaction rate and the enantioselectivity.

The group of Kim and Park described a five-step strategy for the preparation of enantiomerically pure rivastigmine **21**, an acetylcholinesterase inhibitor of the carbamate type, including the chemoenzymatic DKR of the benzylic alcohol

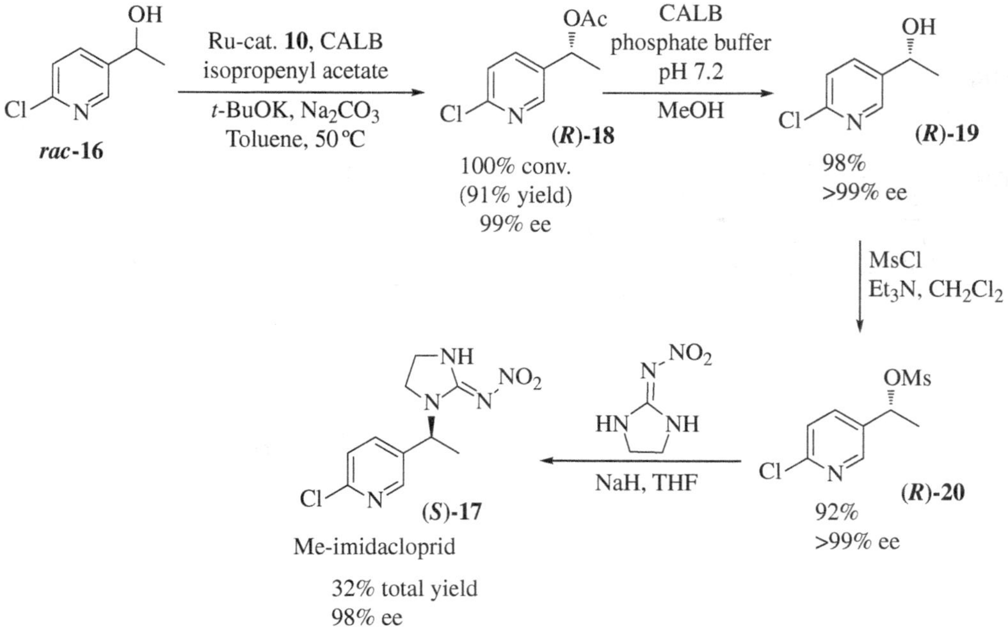

FIGURE 14.7

DKR of 1-(6-chloropyridin-3-yl) ethanol as the key step in the synthesis of chiral Me-imidacloprid.

intermediate **22** as the key step (Figure 14.8) [63]. Immobilized CALB was coupled with the polymer-bound racemization catalyst **13** at room temperature to achieve the chiral acetate in very high yield and ee in 24 h.

Other functionalized benzylic alcohols, such as β-haloalcohols, β-hydroxynitriles, or azido alcohols, useful in the preparation of very interesting and versatile building blocks, have been also obtained through DKR processes. Thus, chiral β-haloalcohols constitute precursors of chiral epoxides and β- and γ-amino alcohols, widely used as adrenergic receptor blockers [64]. The first examples of DKR processes applied in the preparation of enantiomerically pure β-haloalcohols were described by Pàmies and Bäckvall [32] through the coupling of Shvo's catalyst **2** and *P. cepacia* lipase. Excellent results were achieved in the preparation of optically active β-chloro alcohols, but not for the corresponding bromo derivatives. In a further step, the hydrolysis of the obtained chiral esters gave the corresponding enantiopure epoxide, which presents a wide range of synthetic applications [32]. The preparation of these chlorohydrins was further optimized by Träff *et al.* by the development of a DKR mediated by the combination of *P. cepacia* lipase and the ruthenium catalyst **8** at room temperature [65]. Later on, Johnston *et al.* reported the DKR of a β-chloro alcohol as the key step in the preparation of enantiomerically pure (*R*)-bufuralol ((***R***)**-26**; Figure 14.9), a nonselective β-adrenoceptor blocking agent of comparable

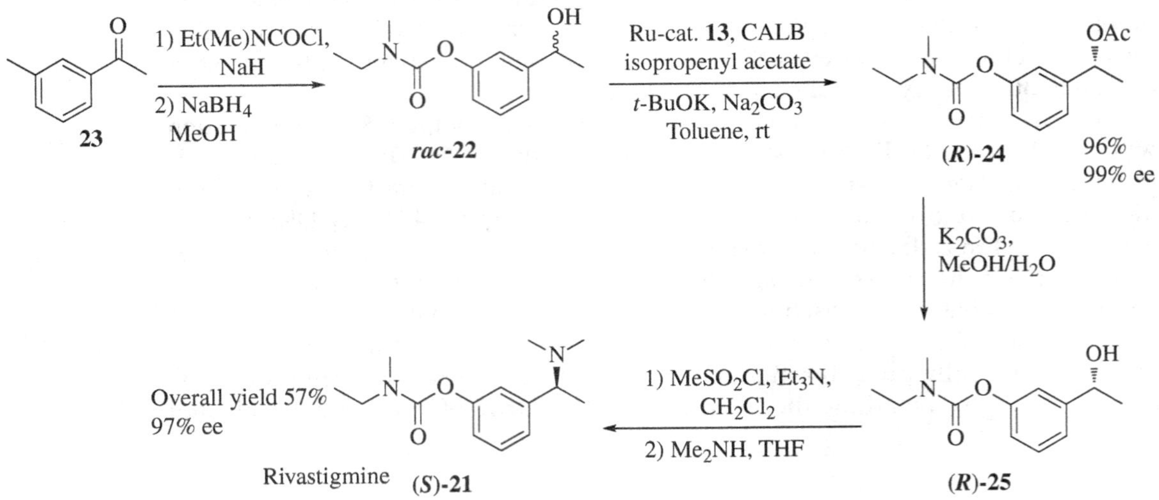

FIGURE 14.8

Chemoenzymatic synthesis of rivastigmine.

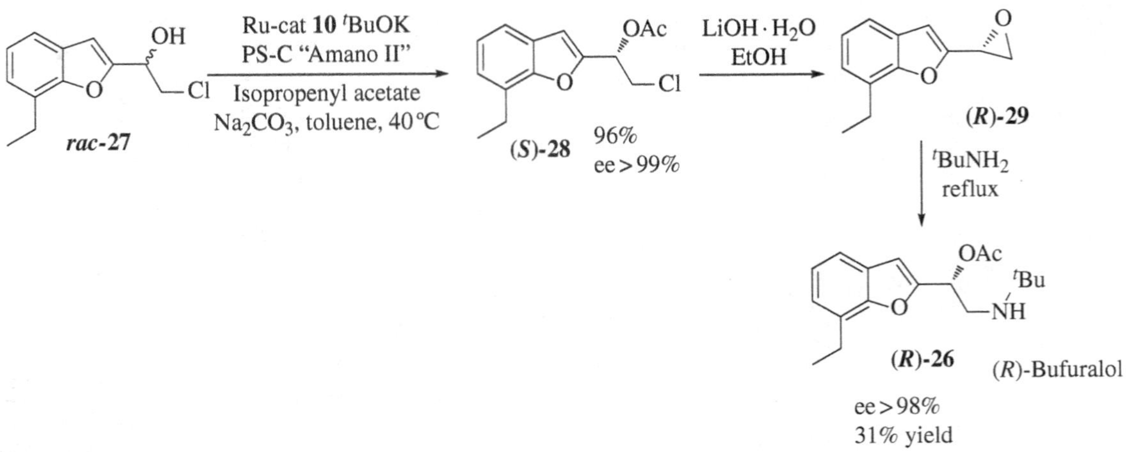

FIGURE 14.9

Application of the DKR of chlorohydrins to the synthesis of (*R*)-bufuralol.

potency as propranolol, very effective for the treatment of hypertension [66]. DKR conditions, described in Figure 14.9, provided full conversion of chiral acetate **(S)-28** after 24 h, with an excellent enantiopurity. Finally isolated **(S)-28** was successfully converted into the desired molecule ***(R)*-26**.

Optically pure azido alcohols are very important intermediates in the preparation of chiral aziridines, included in the structure of many DNA-alkylating antitumoral agents [67, 68], or β-amino alcohols, important structural units of biologically active compounds such as ephedrine and α- or β-adrenergic blockers [64]. In fact, both building blocks are also included in many organic synthesis reactions [69–71]. Pámies and Bäckvall reported the first DKR of aromatic β-azido alcohols by using Shvo's catalyst and immobilized CALB, affording the corresponding chiral acetates with high yields and ee's [72]. To demonstrate the importance of this synthetic route, they applied this methodology to the synthesis of the enantiopure β-adrenergic blocker (*R*)-propranolol (***(R)*-30**; Figure 14.10).

β-Hydroxynitriles are crucial building blocks in organic and medicinal chemistry, due to the high versatility of the cyano group, which can be easily transformed into other functional groups, therefore leading to β-hydroxyamides, β-hydroxyacids, β-hydroxyesters, diols, or amino alcohols. Pámies and Bäckvall described the preparation of enantiopure aromatic β-hydroxynitriles in a DKR process catalyzed by immobilized CALB and the Shvo's catalyst **2** [31]. Kamal *et al.* reported the synthesis of the enantiopure *R*-distomer of the antidepressant duloxetine **33** through a lipase-catalyzed KR of a racemic β-hydroxynitrile and the subsequent stereoinversion to render the eutomer (*S*)-duloxetine via a Mitsunobu reaction [73]. More recently, Träff *et al.* have described a similar process, but the yields were drastically enhanced due to the application of a DKR process. In fact, as shown in Figure 14.11,

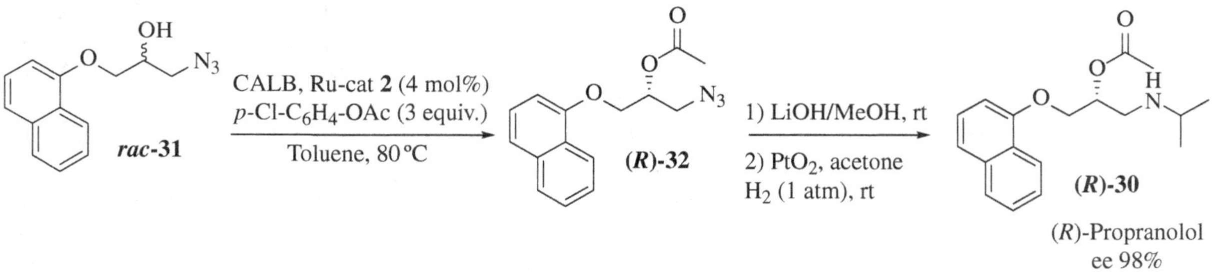

FIGURE 14.10

DKR of β-azido alcohols as the key step in the synthesis of (*R*)-propranolol.

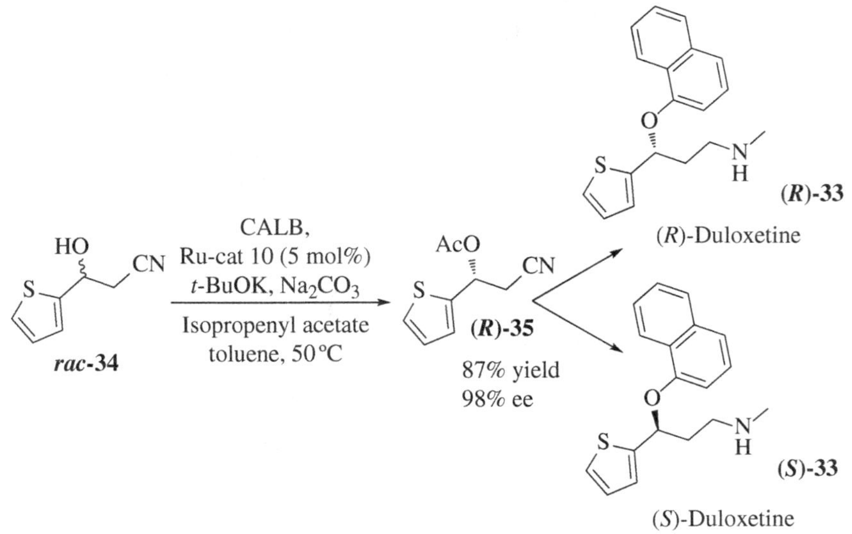

FIGURE 14.11

DKR of β-hydroxynitriles as the key step of the synthesis of duloxetine.

the DKR of the starting β-hydroxynitrile (***rac*-34**) using CALB and ruthenium catalyst **8** afforded the corresponding β-cyano acetate (**(*R*)-35**) in 87% yield and 98% ee. Subsequent synthetic steps allowed the preparation of both (*R*)- and (*S*)-duloxetine (Figure 14.11) [74].

DYKAT of Diols

The enzyme-catalyzed kinetic asymmetric transformation (KAT) of a diastereomeric 1:1 *syn*:*anti* mixture is limited to a maximum theoretical yield of 25% of one enantiomer. This important drawback has been overcome by the combination of the actions of a ruthenium complex and a lipase in a dynamic kinetic asymmetric transformation (DYKAT), the desymmetrization of racemic or diastereomeric mixtures involving interconverting diastereomeric intermediates, implying different equilibration rates of the stereoisomers. Thus, this strategy allows the preparation of optically active diols, widely employed in organic and medicinal chemistry, as they are an important source of chiral auxiliaries and ligands and they can be easily employed as precursors of much other functionality.

In this sense, Persson *et al.* carried out the first DYKAT process of symmetrical diols via coupled Shvo's catalyst **2** and CALB catalysis. Thus, 1:1 mixtures of *meso*- and *R,S*-diols were transformed to enantiomerically pure diacetates in moderate yields and high ee values (Figure 14.12a), although undesired *meso*-diacetates were formed by an intramolecular acyl-transfer pathway [26].

Few years later, very important improvements were reported in the DYKAT of symmetrical *sec*-diols by the implementation of ruthenium catalysts working at lower temperatures [44, 61]. Hence, Martín-Matute *et al.* reported a highly efficient DYKAT of diols performed by the combination of ruthenium catalyst **8** and lipase action. In the DYKAT of 2,5-hexanediol (**36**; Figure 14.12b), the faster epimerization and lower reaction temperature reduced the concentration of unwanted intermediates, such as alcohols bearing a carbonyl group or an (*R*)-acetoxy group at the δ position affording the *meso*-diacetates [75]. The fast epimerization catalyzed by **8** in the DYKAT of 2,4-pentanediol **38** avoided unwanted acyl migrations (Figure 14.12b) [75]. Thus, starting from *R,S-meso* mixtures (1:1) of the diols, the corresponding enantiopure diacetylated products were obtained in high yields and ee of >99%.

The applicability of this methodology was further demonstrated by Borén *et al.*, who described the DYKAT of a series of aliphatic 1,4-diols, catalyzed by CALB and ruthenium catalyst **8**. Through these reaction conditions, the corresponding optically

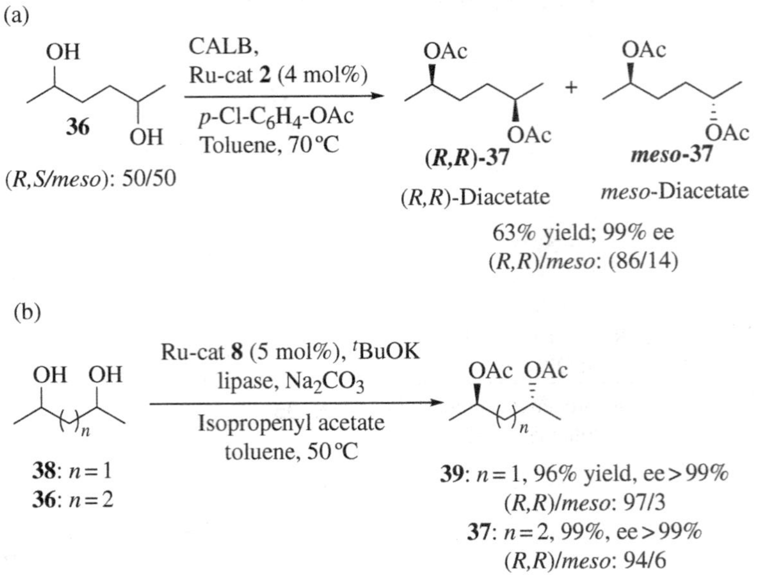

FIGURE 14.12

(a) Synthesis of chiral diacetylated diols by ruthenium catalyst **2**-mediated DYKAT process. (b) Synthesis of chiral diacetylated diols by ruthenium catalyst **8**-mediated DYKAT process.

active diacetates were afforded and subsequently employed in the synthesis of enantiopure 2,5-disubstituted pyrrolidines (Figure 14.13) [50].

Ruthenium catalyst **8** has been also coupled with CALB in the DYKAT of 1,5-diols **46** and **47** (Figure 14.14a), affording the enantiomerically pure diacetates in high yields, precursors of chiral 2,6-disubstituted piperidine and 3,5-disubstituted morpholine, respectively [51]. A similar procedure has been reported for the preparation of enantiopure antiangiogenic (+)-solenopsin A (Figure 14.14b) [48].

Krumlinde *et al.* have applied the DYKAT methodology for preparing bicyclic diol derivatives [49]. The corresponding diacetates were achieved in high yields and enantiopurity through the DYKAT process, and optically pure diols were further obtained through simple hydrolysis. Those diols were selectively oxidized to the corresponding keto alcohols in a ruthenium-catalyzed Oppenauer reaction, employing

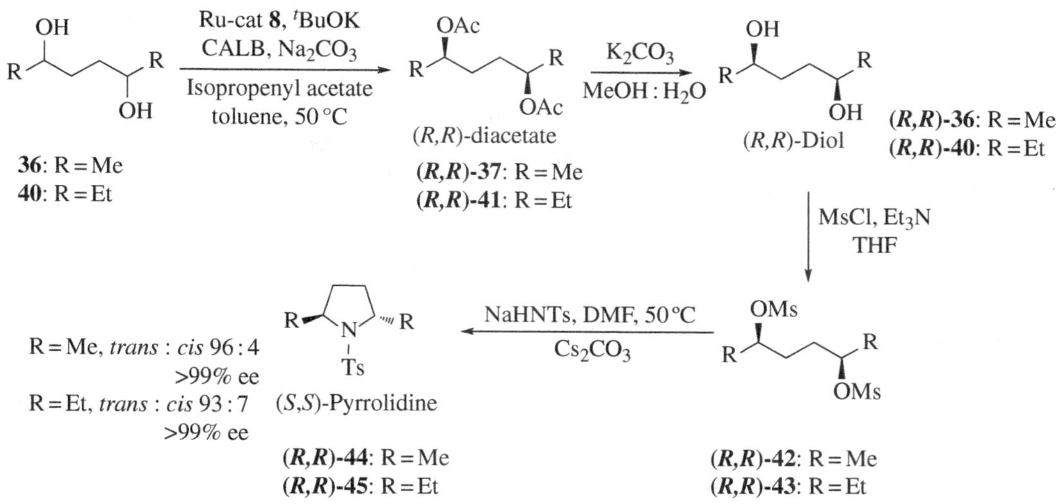

FIGURE 14.13

DYKAT of 1,4-diols as the key step in the preparation of 2,5-disubstituted pyrrolidines.

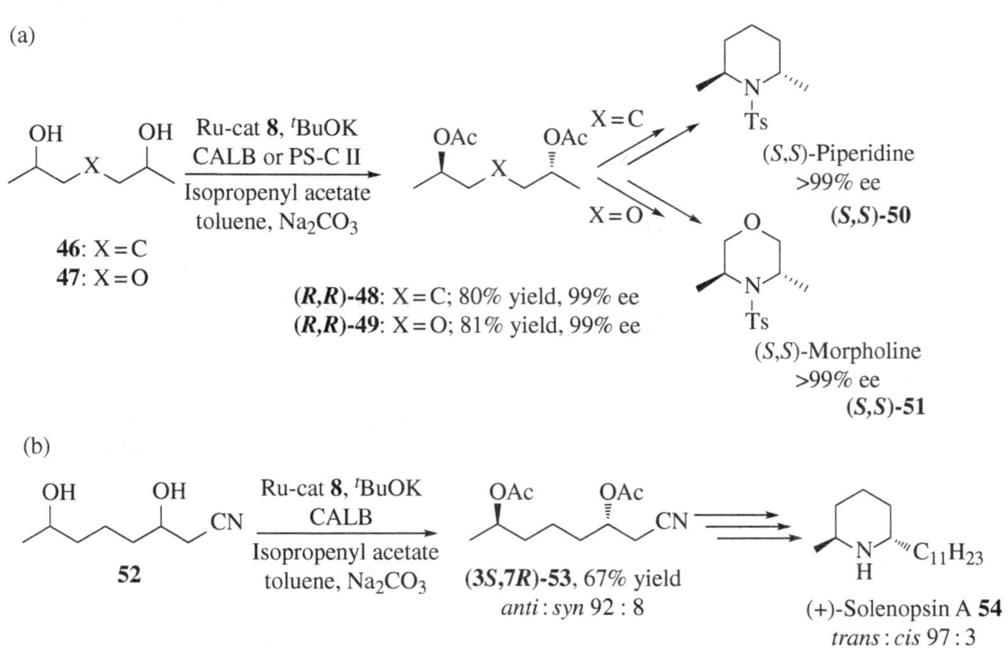

FIGURE 14.14

Application of the DYKAT of 1,5-diols. (a) Synthesis of 2,6-disubstituted and 3,5-disubstituted six-membered chiral heterocycles. (b) Preparation of enantiopure (+)-solenopsin A.

the product **(*R*)-61** as the starting material for the synthesis of the antidepressant sertraline, as shown in Figure 14.15.

Bäckvall and coworkers reported an efficient and innovative methodology to prepare *syn*- and *anti*-1,3-aminoalcohols based on a two-step procedure combining organocatalysis and organometallic and enzymatic catalysis [76]. This type of compound is very attractive both in organic and medicinal chemistry as a great variety of biologically active molecules contain these motifs in their structure [77–79] or because they are valuable building blocks in asymmetric synthesis [80, 81]. Hence, *N*-Boc-protected β-aminoketones, obtained through an enantioselective proline-catalyzed Mannich reaction in a first step, were subjected to reduction and subsequent DYKAT to furnish enantio- and diastereomerically pure *N*-Boc-protected 1,3-aminoacetates (Figure 14.16). CALB was combined with ruthenium complex **2**, working in toluene at 90 °C. This complex **2** could act as a dual catalyst for both reduction of the ketone and epimerization of the alcohol. Hydrolysis of the acetate and the *tert*-butyl carbamate yielding the free aminoalcohols was subsequently carried out without any loss of enantio- or diastereoselectivity.

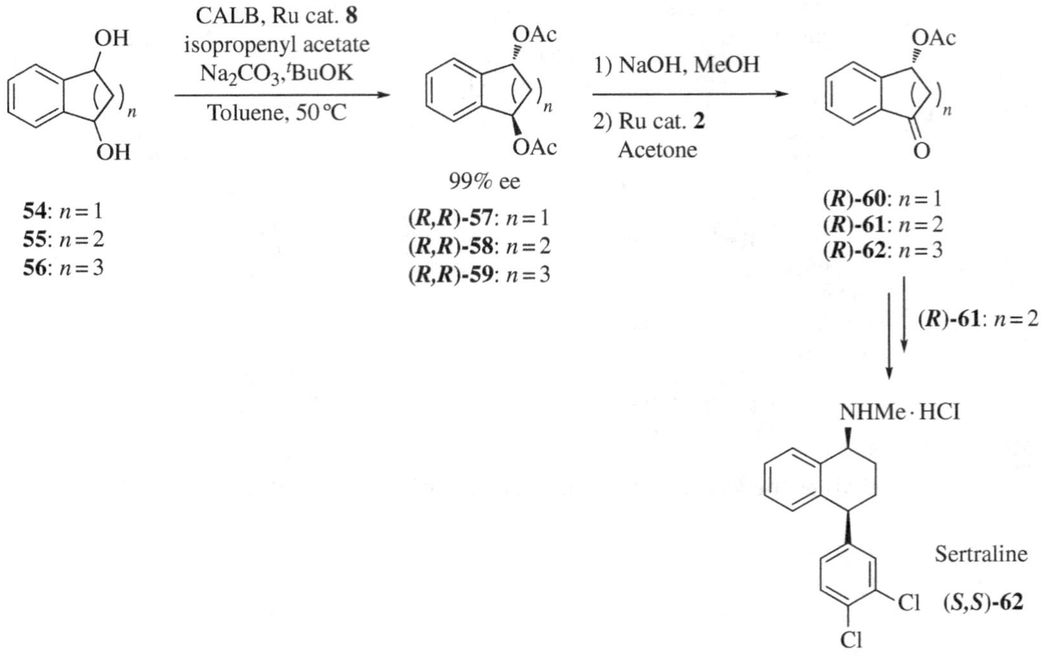

FIGURE 14.15
DYKAT of bicyclic diols for the preparation of sertraline.

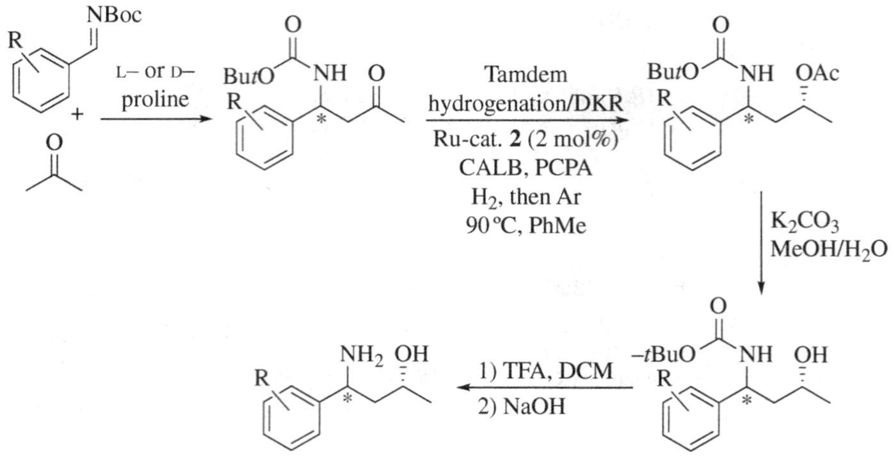

FIGURE 14.16
Enantioselective synthesis of *syn*- and *anti*-1,3-aminoalcohols starting from aminoketones and subsequent reduction/DYKAT.

DKR of Allylic Alcohols

The chiral synthesis of allylic alcohols has been the focus of many research works due to the high versatility of these molecules in the preparation of many active compounds [58, 82]. Allen and Williams reported the first example of DKR of allylic alcohols via lipase–palladium catalyst coupling deracemization of cyclic allylic acetates [83]. However, the accumulation of secondary products, as well as the long reaction times required, limited the use of this strategy.

In this respect, ruthenium complexes have also been coupled with hydrolases in the efficient DKR of allylic alcohols. Lee *et al.* reported complexes **5** and **6** as the racemization catalysts of the DKR of allylic alcohols in combination with a CALB-mediated transesterification to give the corresponding allylic acetates in yields higher than 80% [38]. Ruthenium complex **7** was successfully employed in the DKR of a variety of allylic alcohols, although very long reaction times were required [61]. Better results were obtained by Bögar *et al.* by the combination of the actions of the ruthenium complex **8** and CALB. As an example of the synthetic usefulness of these compounds, these authors employed the chiral allylic acetates in a further step, as substrates of an oxidative cleavage, affording enantiomerically pure acetylated acyloins (Figure 14.17a) [47].

Bäckvall and coworkers also employed the DKR of allylic alcohols as the key step in the preparation of chiral α-methyl-substituted carboxylic acids, a very important building block of 2-arylpropionic acids belonging to the nonsteroidal anti-inflammatory drug (NSAID) family [84]. α-Methyl-substituted allylic **63** alcohols were employed as substrate of a DKR developed by the combination of the metal complex **8** and CALB or subtilisin Carlsberg, providing then the *R*- or the *S*-enantiomer of the ester, respectively. These products were transformed through a copper-catalyzed allylic substitution reaction to give the corresponding olefin, which underwent a ruthenium-catalyzed oxidative cleavage to afford the target acids (Figure 14.17b). This methodology was further optimized by the same group, starting from readily available an inexpensive *trans*-cinnamaldehyde [46].

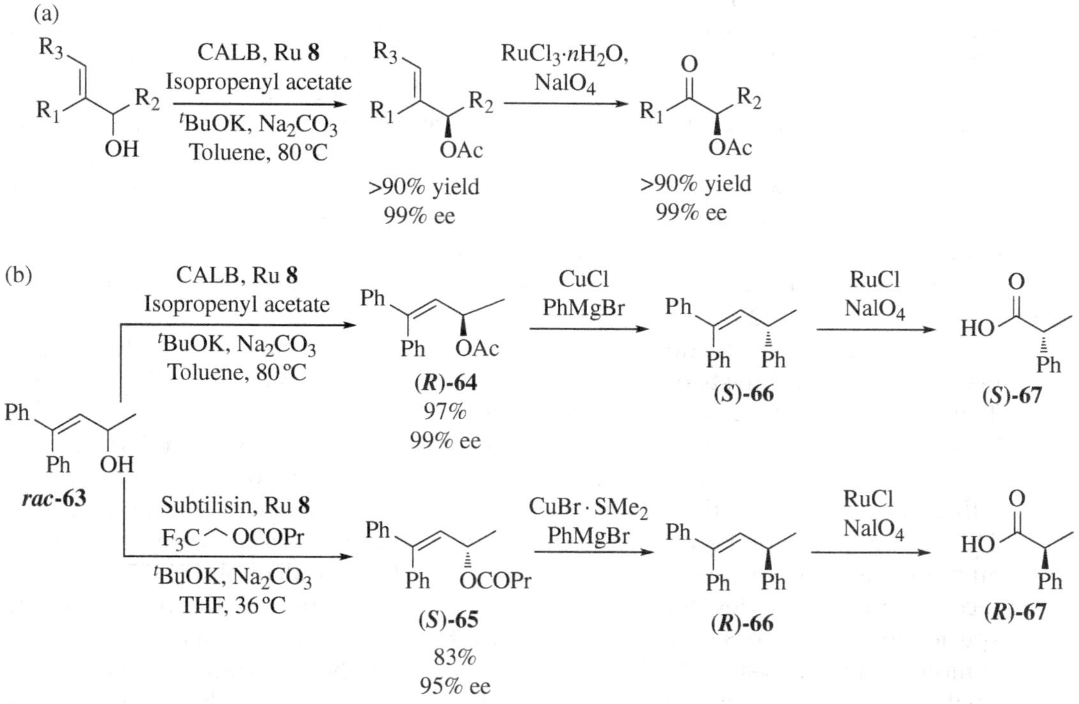

FIGURE 14.17

(a) DKR of allylic alcohols as the key step of optically active acyloin synthesis. (b) DKR of allylic alcohols as the key step of optically active α-methyl carboxylic acid synthesis.

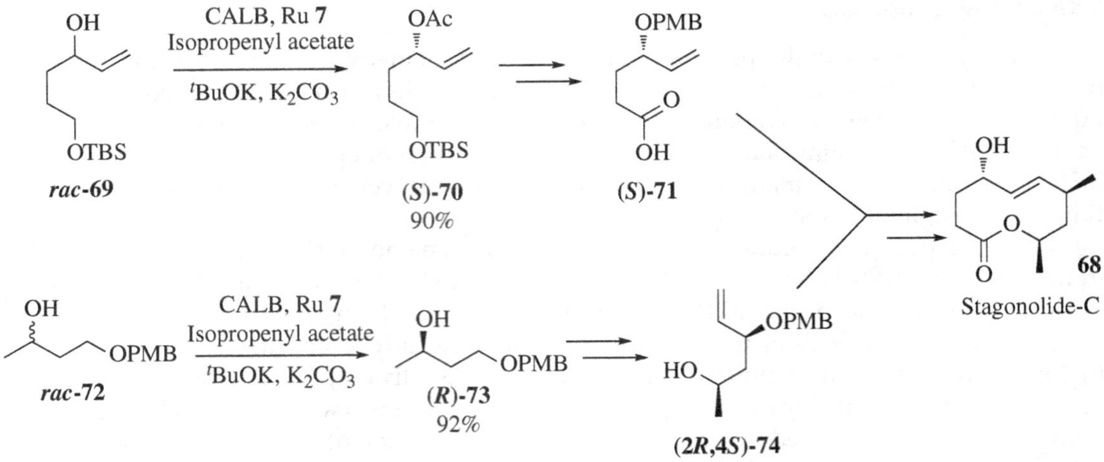

FIGURE 14.18
DKR applied in the synthesis of stagonolide-C.

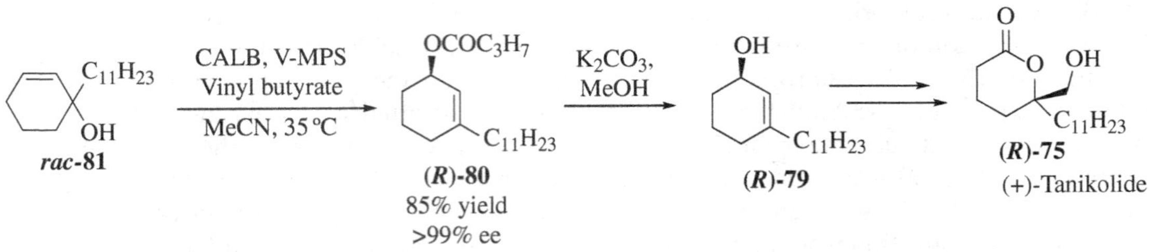

FIGURE 14.19
V-MPS–CALB-catalyzed DKR of allylic alcohols as the key step of the synthesis of (+)-tanikolide.

The DKR of *sec*-alcohols through hydrolase–metal combo catalysis, particularly the DKR of allylic alcohols, has been thoroughly investigated by this group. They have recently described the DKR of very useful functionalized allylic alcohols, obtaining the desired esters in high yields and enantiopurity [36], as well as a very interesting route for the synthesis of chiral α-substituted ketones and lactones, based on the DKR of exocyclic allylic alcohols [85]. Other research groups have also taken advantage of the DKR technique for the synthesis of other molecules of interest. Thus, Nanda and coworkers have employed the ruthenium complex **7**-based DKR of a *sec*-alcohol and an allylic alcohol to obtain the chiral intermediates needed for the synthesis of the phytotoxic stagonolide-C (**68**; Figure 14.18) [42].

As discussed earlier, not only ruthenium but also other metal complexes have been applied as catalysts of the racemization of allylic alcohols in DKR processes. Thus, Akai *et al.* first reported a vanadium-catalyzed DKR combining complex $[VO(OSiPh_3)_3]$ (**14**) and immobilized CALB or immobilized lipase from *Burkholderia cepacia* (PS-D). Very high conversions and ee values were obtained, employing either vinyl acetate or ethoxyvinyl acetate as acyl donors at room temperature [57]. This group has carried out a detailed investigation in this research area, leading to an optimization of the oxovanadium-based DKR of allylic alcohols [56, 57, 59]. Very recently, they have developed the covalent immobilization of the oxovanadium species inside an MPS support (**V-MPS**), resulting in a very efficient catalyst to combine with lipases in the DKR of a variety of allylic substrates, which can be easily recovered and reused without losing activity. In fact, they have applied this strategy in the preparation of a key synthetic intermediate of the antifungal (+)-tanikolide ((***R***)**-75**; Figure 14.19) and in the total synthesis of (*R*)-imperanene ((***R***)**-76**; Figure 14.20).

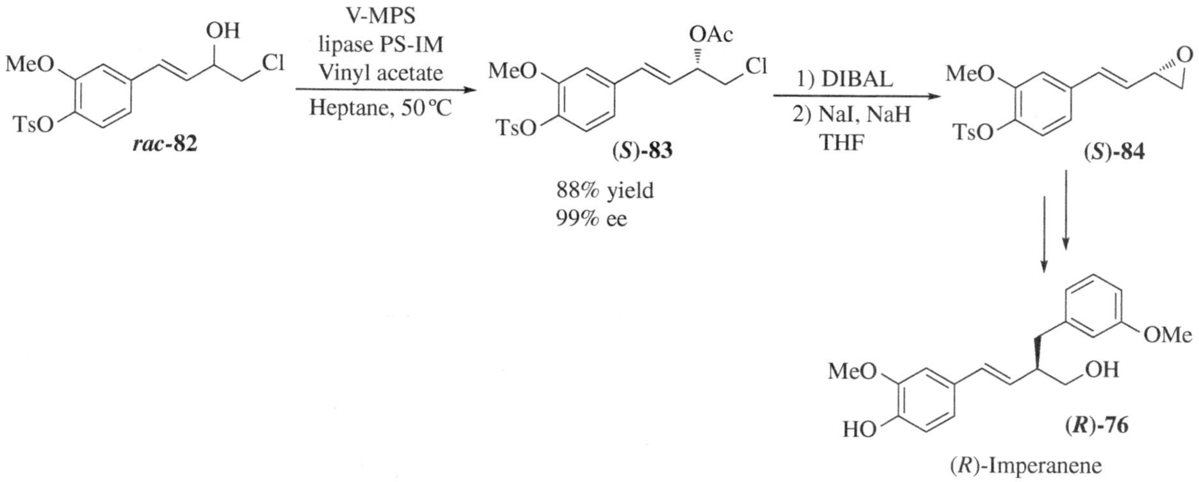

FIGURE 14.20

V-MPS–CALB-catalyzed DKR of allylic alcohols as the key step of the synthesis of (*R*)-imperanene.

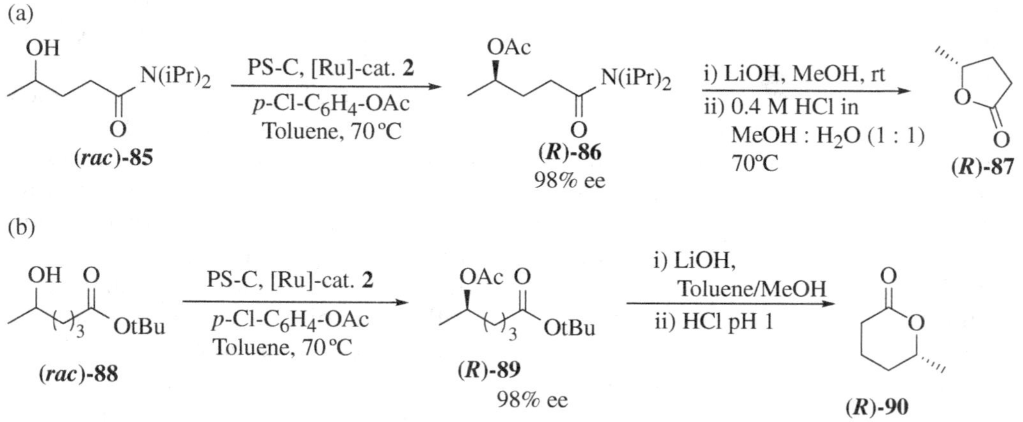

FIGURE 14.21

(a) DKR of γ-hydroxyacid derivatives. (b) DKR of δ-hydroxyesters.

DKR of Hydroxycarbonyl Compounds

Hydroxycarbonyl compounds are widely employed building blocks in organic and medicinal chemistry due to the high versatility of these functional groups [15]. There is a great interest in their chiral preparation, as they can be successfully employed as enantiopure starting material in the synthesis of chiral compounds. In this context, the DKR of a great variety of hydroxyketones has been described. One of the first works was carried out by Huerta *et al.*, who combined Shvo's catalyst **2** with lipase from *P. cepacia* in cyclohexane at 60 °C and used *p*-chlorophenyl acetate to deracemize α-hydroxyesters, achieving the corresponding optically active acetylated products in moderate to high yields and excellent ee's [28]. A similar procedure was further reported by Runmo *et al.* for obtaining chiral acetylated γ-hydroxy derivatives, applied in the synthesis of chiral lactones (Figure 14.21a) [86], and by Pamies *et al.* in the synthesis of chiral δ-lactones starting from δ-hydroxyesters (Figure 14.21b) [29].

α-Hydroxyketones (acyloins) are a group of hydroxycarbonylic compounds of special interest as the acyloin core can be converted in many other functionalities, with a wide number of applications for the fine chemistry sector, as well as pharmaceuticals [87]. However, the applicability of the lipase–ruthenium complex-catalyzed DKR strategy is not so obvious for these compounds, due to the formation of an

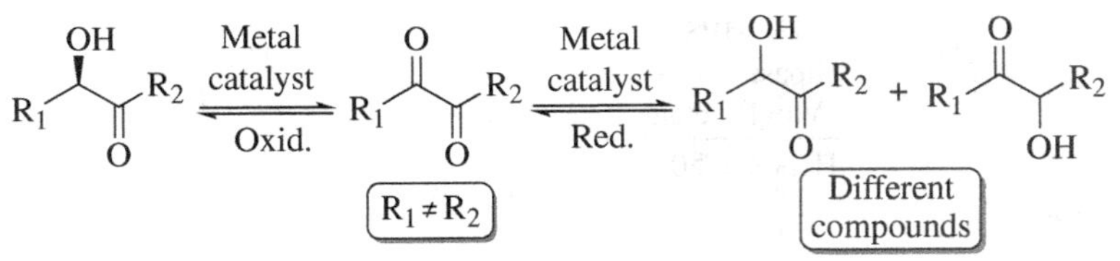

FIGURE 14.22
Possible mixture of products afforded in the racemization of unsymmetrical α-hydroxyketones.

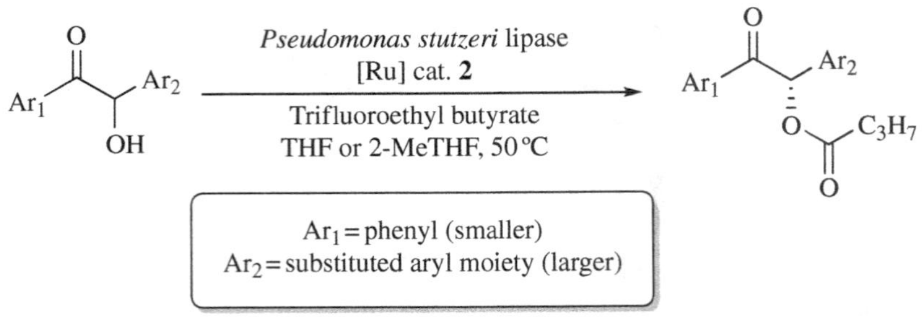

FIGURE 14.23
DKR of benzoin derivatives.

intermediate diketone by the action of the metal complex and the subsequent rearrangement of the starting molecule, forming two different compounds (Figure 14.22) in the DKR of unsymmetrical molecules. Thus, different methodologies have been elaborated in order to circumvent this limitation.

In fact, Ödman *et al.* reported a two-compartment DKR of acyloins on which CALB-catalyzed enantioselective transesterification was confined in a first compartment and the simultaneous racemization of the remnant alcohol took place in a second compartment, mediated by the acidic resin Amberlyst 15 [88]. Through this strategy, 91% yield and high ee value were described. As it was commented earlier, Bogár *et al.* designed a methodology to obtain optically pure acyloins through the DKR of allylic alcohols, followed by the oxidative cleavage of the C—C double bond (Figure 14.17a) [47]. Hoyos *et al.* reported the synthesis of chiral benzoins (1,2-diaryl-2-hydroxyethanones) by a lipase–ruthenium combo catalysis process, combining Shvo's catalyst **2** and *Pseudomonas stutzeri* lipase [30]. The procedure was optimized by the lipase immobilization [89], as well as the conduction of first steps for a possible scale-up of the reaction in a more environmentally friendly solvent, 2-methyltetrahydrofuran [90]. The special catalytic behavior of the lipase was studied in depth [91], allowing the design of a DKR procedure to obtain chiral unsymmetrical benzoins, containing different aryl moieties on both sides of the acyloin core [92]. In fact, lipase selectivity was enhanced by increasing the size of the phenyl ring that is supposed to be placed in the large pocket; the lipase should show higher preference toward the stereoselective esterification of those benzoins containing no substituents at the aromatic ring of the benzoyl moiety (Figure 14.23). In this manner, a broad range of enantiomerically pure benzoin esters have been prepared. Very recently, Martín-Matute and coworkers have applied the action of the ruthenium complex **5** (Figure 14.5) and 1,4-bis(diphenylphosphino)butane in this process, allowing the reduction of the temperature to ambient temperature [40].

14.3 DKR OF AMINES

Enantiomerically pure amines constitute very useful building blocks in the preparation of biologically active compounds. The amino function can be found in the structure of a broad range of natural products, playing a crucial role in their biological activity. Thus, the synthesis of chiral amines is still an open challenge in the pharmaceutical industry, and the chemoenzymatic DKR technique appears as a promising alternative for their preparation [17, 93, 94]. However, the *one-pot* coupling of a hydrolase-catalyzed KR with a metal-mediated racemization in the DKR of an amino compound is more complex than the DKR of *sec*-alcohols, because amines can act as strong ligands for active metal intermediates and furthermore amine C–N bonds can also be cleaved by transition metal catalysts [95].

14.3.1 Racemization Catalyst for the DKR of Amines

The first example of racemization catalyst applied to the DKR of amines was described by Reetz and Schimossek, contemporary to the report of first cases of *sec*-alcohol DKRs [96]. Nevertheless, reported racemization catalysts for amines are not so numerous, due to the severe conditions required for the racemization, which frequently are not compatible with the activity of the enzyme. That pioneering work reported the use of 5% Pd/C as heterogeneous racemization catalyst and combined its action with CALB activity in the DKR of 1-phenylethylamine (***rac*-91**). Although just moderate yields could be achieved after 8 days, palladium was, since then, considered a very attractive catalyst in the racemization of primary amines.

Park and coworkers have developed a palladium nanocatalyst, applicable to the DKR of primary amines as well as amino acid amides [97]. This nanocatalyst, Pd/AlO(OH), prepared as palladium nanoparticles entrapped in aluminum hydroxide required 70 °C for the racemization of benzylic amines and higher temperature (100 °C) in the case of aliphatic amines. For that reason, DKR processes were developed in combination with thermostable immobilized CALB. In fact, these authors have described the coupling of this catalyst with the activity of *P. stutzeri* lipase in the DKR of β-amino acid amides being necessary to perform the enzymatic resolution and the racemization at different temperatures and to add fresh lipase after the racemization due to the thermal enzyme deactivation [98].

On the other hand, Parvulescu *et al.* have reported a deep study about the use of other different supports, the best results involving employing this metal on nonacidic supports such as alkaline earth metal salts, $BaSO_4$, $CaCO_3$, etc. [99–101]. The proposed reaction mechanism for the Pd-catalyzed racemization of **91**, via reversible dehydrogenation–hydrogenation, is shown in Figure 14.24, indicating the formation of by-products **93** and **94** [100, 102]. Thus, Pd on basic supports was coupled with immobilized CALB in the DKR of primary benzylic amines [99, 100]. This same group has also reported the activity of other metal catalysts, such as Raney nickel and cobalt catalysts for the DKR of primary amines, in combination with CALB activity [103].

More recently, Bäckvall and coworkers have described a very effective catalyst for the racemization of amines: highly dispersed Pd nanoparticles supported in the large pores of silica-based mesocellular foam (MCF) [104]. This catalyst proved to be a more robust catalyst in amine racemization than any other catalysts reported before and is completely compatible with CALB activity.

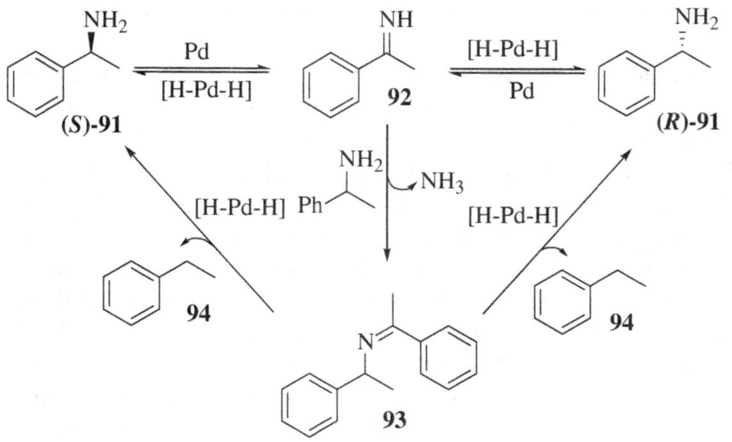

FIGURE 14.24
Palladium-catalyzed racemization of primary amines.

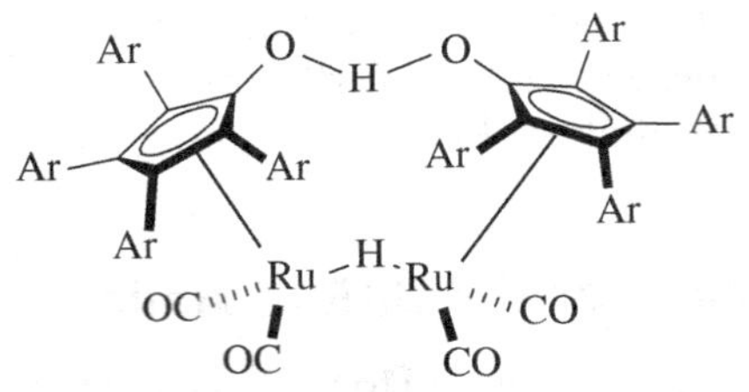

FIGURE 14.25
Ruthenium-based catalyst employed in the racemization of amines.

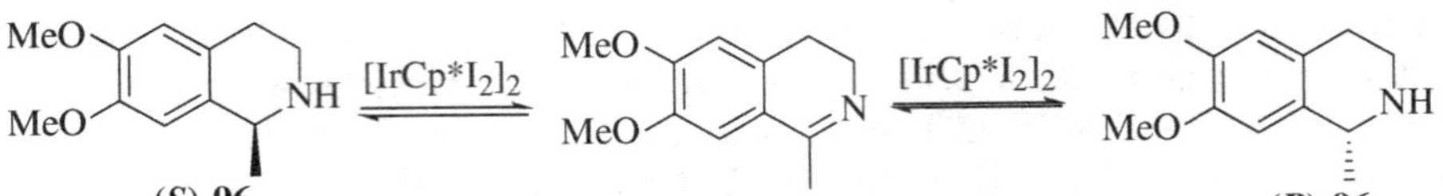

FIGURE 14.26
Iridium-based amine racemization catalyst.

As in the case of *sec*-alcohols, ruthenium complex has also been investigated as a catalyst in the racemization of primary amines. In fact, Shvo's complex **2** (Figure 14.3) was employed by the Bäckvall's group as the catalyst of the racemization of amines under transfer hydrogenation conditions [105]. However, temperatures up to 110 °C were required for amine racemization, incompatible with the lipase resolution, and furthermore, side products were formed in the medium and a hydrogen source was needed. To avoid these drawbacks, the racemization at high temperature was carried out after a first lipase-catalyzed KR, followed by a second KR process, and a hydrogen source such as 2,4-dimethylpentan-3-ol was employed.

A few years later, Paetzold and Bäckvall reported and optimized the DKR of primary amines based on the substrate racemization mediated by an improved Shvo's catalyst variant **95** (Figure 14.25) [106]. This complex was efficiently employed in the DKR of different benzylic and aliphatic primary amines [106–108].

While very fruitfully results can be accomplished in the resolution of primary amines, not many examples have been reported for the resolution of secondary amines. It is worthy of mention that the racemization of secondary amines was successfully achieved by the employment of a novel iridium-based racemization catalyst. In this sense, Page and coworkers reported the pentamethylcyclopentadienyl iridium(III) iodide dimer $[IrCp^*I_2]_2$, which is very active for the racemization of secondary amines under mild conditions (Figure 14.26) and completely compatible with enzyme-catalyzed KR in a DKR process [109, 110].

14.3.2 Synthetic Applications of the DKR of Amines

As it was mentioned earlier, the first example of Pd lipase-catalyzed DKR of primary amines was reported by Reetz and Schimossek in 1996, affording moderate yields after 8 days [96]. These pioneer results were the basis of the latter successful DKR procedures.

The group of Kim proposed another strategy to afford chiral amides based on the development of a DKR process of ketoximes (Figure 14.27a), and Pd/C was chosen as the reduction/racemization catalyst in combination with CALB and ethyl acetate as the acyl donor. Different acyclic and cyclic ketoximes were tested as substrates, achieving the corresponding enantiopure amides in moderate to high yields and high ee values, although long reaction times were required [111]. Park and coworkers described a practical procedure for the DKR of amines using a palladium nanocatalyst (Pd/AlO(OH)) for amine racemization, coupled with immobilized CALB, and employing methoxyethyl acetate or ethyl acetate as acyl donors at 70 °C in 3 days [97]. Both strategies were combined some years later in the synthesis of the potent calcimimetic (+)-NPS R-568 (**98**; Figure 14.27b). The improved palladium nanocatalyst Pd/AlO(OH) was coupled with CALB resolution to afford the enantiomerically pure

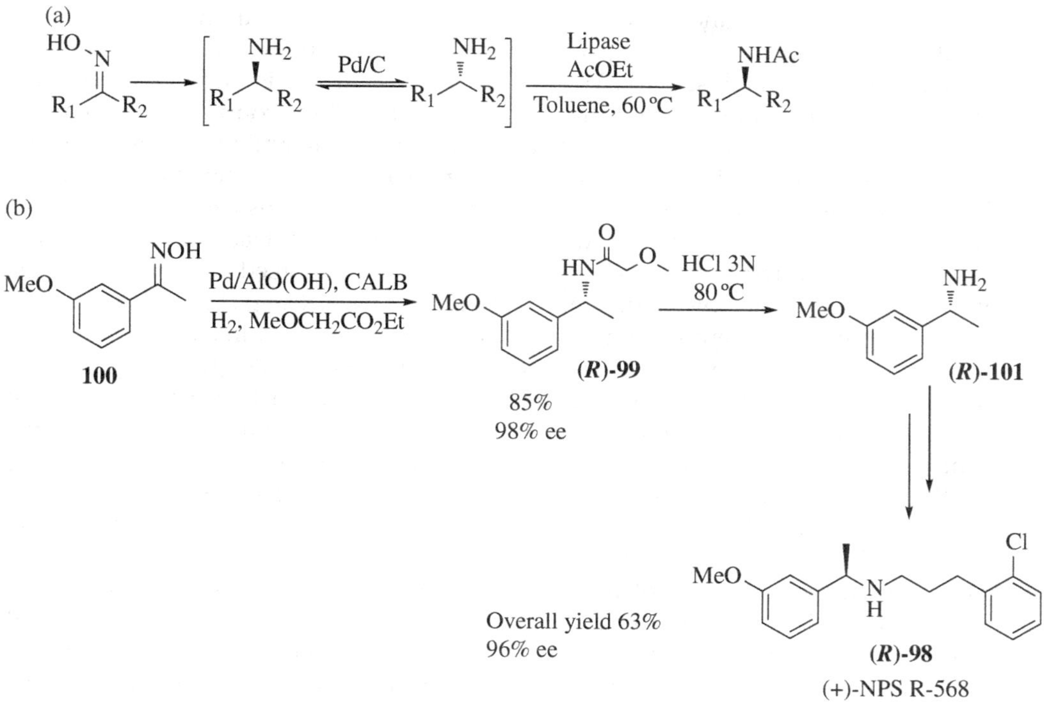

FIGURE 14.27

(a) Lipase/palladium-catalyzed transformation of ketoximes to optically active amines. (b) DKR as the key step of the synthesis of the calcimimetic compound (+)-NPS R-568.

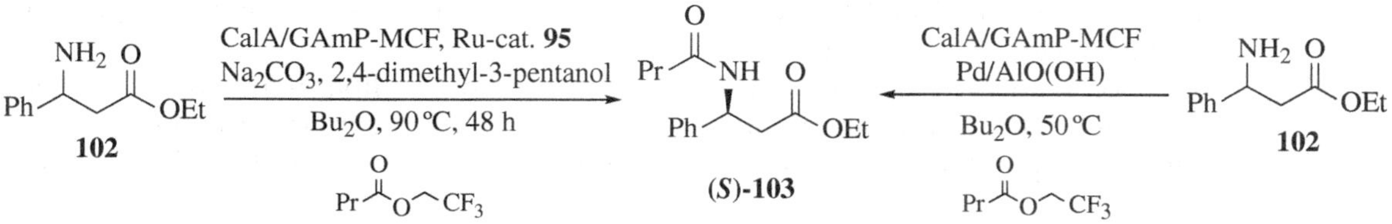

FIGURE 14.28

DKR of β-amino esters.

amide **(*R*)-99** through a three-step procedure: Pd-catalyzed reduction of ketoximes to amines, Pd-catalyzed racemization of amines, and lipase-catalyzed enantioselective acylation of amines to amides [112].

The palladium nanocatalyst has also been coupled successfully with lipases in the synthesis of chiral amino acid derivatives. Bäckvall's group reported the DKR of β-amino esters, important building blocks in the pharmaceutical industry, as well as β-amino acids, for the synthesis of anticancer drugs, antibiotics, and HIV inhibitors [113]. Pd/AlO(OH) was combined with CalA immobilized in functionalized mesocellular foam (CalA/GAmP-MCF) (Figure 14.28) [113]. The latest immobilized lipase was previously described by this group and employed in combination with the ruthenium complex **95** in the deracemization of β-amino esters (Figure 14.28) [114], but high temperatures were required. This new catalytic system allowed the performance of the DKR of β-amino esters at 50 °C, yielding the desired amides in high yields and enantiopurity.

Bäckvall's group has taken advantage of the catalyst immobilization in mesocellular silica foams (MCFs) to coimmobilize palladium nanoparticles and CALB in MCF, so that each cavity of the support will contain both the lipase and the nanopalladium, affording a single and recyclable hybrid catalyst of the DKR for primary amines [115]. The immobilization technique and the DKR of **91** were optimized, obtaining the chiral amide target in maximum yields and ee.

The group of Kim and Park has also employed the palladium nanocatalyst Pd/AlO(OH) in the DKR of phenylglycine amide and derivatives, employing immobilized CALB as biocatalyst [116]. Working at 60 °C high yields and ee values were obtained, but very long reaction times were needed. This group also applied this methodology in the synthesis of di- or tripeptides by using amino acid esters as the acyl donors for the DKR (Figure 14.29). As commented before, they have extended the DKR to the resolution of β-amino acid amides [98]. In this case, lipase from *P. stutzeri* was employed as the resolution catalyst, which was deactivated at the high temperature required for the palladium nanocatalyst-catalyzed racemization. This problem was overcome by sequential KR–racemization–KR steps [116].

Alternatively, ruthenium complex has also been employed in combination with lipases in the DKR of amines. As commented earlier, Bäckvall's group was the first describing the coupling of the ruthenium complex **2** and a lipase as catalysts of DKRs of amines [105]. A broad range of primary and secondary amines were efficiently used as substrates of the racemization in the presence of a hydrogen source, but the temperature required was too high to perform a *one-pot* procedure. Some years later, Paetzold and Bäckvall employed the Shvo's catalyst derivative **95** (Figure 14.25) working at 90 °C coupled with immobilized CALB, affording the transformation of a variety of unfunctionalized primary amines into one enantiomer in high yield and high enantioselectivity [106]. Later on, a similar protocol was employed by Thalen *et al.* in the DKR of readily available 1,2,3,4-tetrahydro-1-naphthylamine, included in a novel synthetic route of norsertraline ((**1*R*,4*S*)-106**; Figure 14.30) [108].

The same group has described an optimized protocol of the DKR of ***rac*-91** for use on multigram scale [117]. CALB and the ruthenium complex **95** were employed, and it was found that the catalyst loading could be decreased down to 1.25 mol% if alkyl methoxyacetates were used as acyl donors. (*R*)-2-Methoxy-*N*-(1-phenylethyl)acetamide was afforded in 83% yield and 98% ee, in a 45 mmol scale, employing methyl methoxyacetate as the acyl donor.

Gotor and coworkers have efficiently applied the lipase–ruthenium combo-catalyzed DKR to prepare a variety of chiral amines, precursors of molecules with interesting pharmacological activity. As described in Figure 14.31, CALB and Shvo's

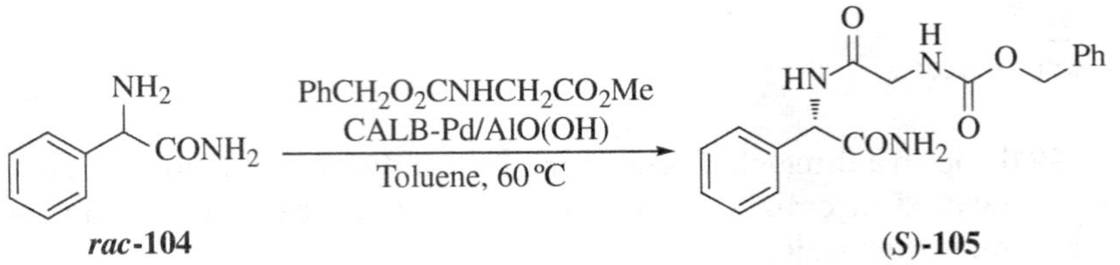

FIGURE 14.29
Synthesis of a dipeptide based on a chemoenzymatic DKR.

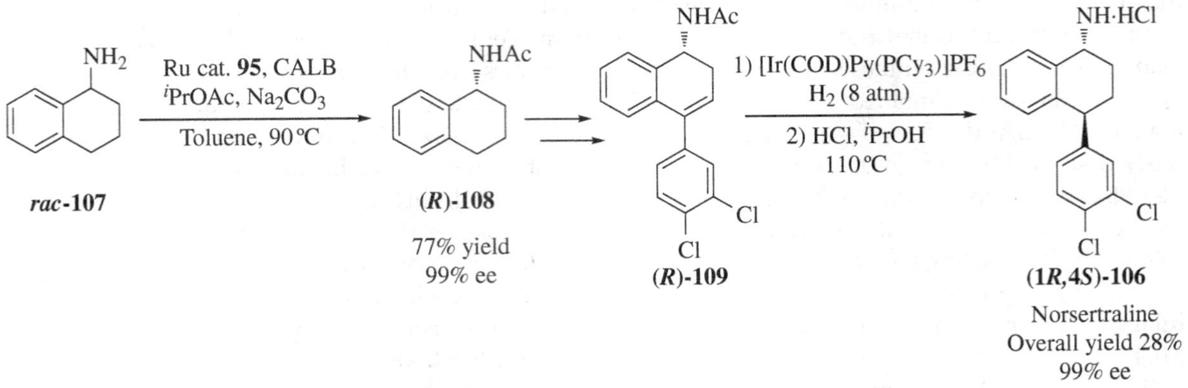

FIGURE 14.30
Synthesis of norsertraline including a DKR process as the key step.

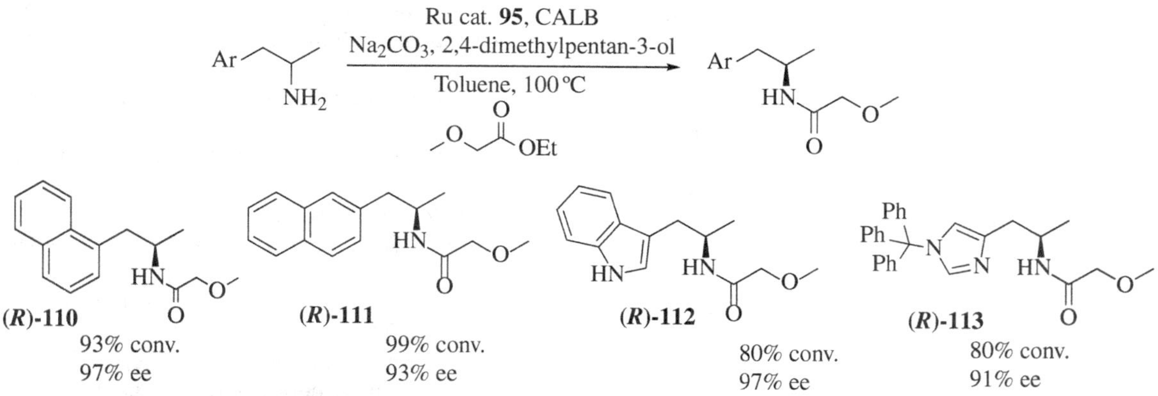

FIGURE 14.31

Shvo's catalyst/lipase-catalyzed DKR of 1-aryl- and heteroarylpropan-2-amines.

complex **2** were coupled employing ethyl methoxyacetate as the acyl donor and 2,4-dimethylpentan-3-ol as the hydrogen donor [118]. Thus, optically pure 1-aryl- and 1-heteroarylpropan-2-amines were prepared: the naphthyl derivatives **110** and **111** are efficient monoamine oxidase (MAO) inhibitors; the indolyl derivative **112** is a precursor of potent β-adrenergic receptor agonists, and compound **113** is the reference histamine H3-receptor agonist most extensively used, whose (*R*)-enantiomer is significantly more potent than its (*S*)-counterpart.

14.4 CONCLUSION

Chemoenzymatic DKR using hydrolase–metal combo catalysis can be catalogued, without any doubt, as one of the most attractive and challenging areas of research for synthesizing enantiomerically pure compounds. In this sense, different metal complexes have been shown to be active in the racemization of chiral alcohols and amines and also compatible with enzymatic resolution processes, leading to the efficient preparation of chiral building blocks, which are very useful for the pharmaceutical industry. New optimized processes are increasingly being explored in order to improve the sustainability and economy of the procedure, as well as to extend the applicability of this strategy to novel substrates.

REFERENCES

[1] Nestl, B. M., Hammer, S. C., Nebel, B. A., and Hauer, B. (2014). New generation of biocatalysts for organic synthesis. *Angew. Chem. Int. Ed.*, **53**, 3070–3095.

[2] Patel, R. N., *Biocatalytic Routes to Chiral Intermediates for Development of Drugs, Biocatalysis for Green Chemistry and Chemical Process Development*, John Wiley & Sons, Inc., Hoboken, NJ, 2011, pp. 89–149.

[3] Patel, R. N. (2011). Biocatalysis: Synthesis of key intermediates for development of pharmaceuticals. *ACS Catal.*, **1**, 1056–1074.

[4] Hoyos, P., Pace, V., and Alcantara, A. R. (2013). Biocatalyzed on water synthesis of chiral building blocks for the preparation of anti-cancer drugs: A greener approach. *Curr. Org. Chem.*, **17**, 1132–1157.

[5] Bornscheuer, U. T., Huisman, G. W., Kazlauskas, R. J., Lutz, S., *et al.* (2012). Engineering the third wave of biocatalysis. *Nature*, **485**, 185–194.

[6] Rajagopalan, A. and Kroutil, W. (2011). Biocatalytic reactions: Selected highlights. *Mater. Today*, **14**, 144–152.

[7] Anastas, P. and Eghbali, N. (2010). Green chemistry: Principles and practice. *Chem. Soc. Rev.*, **39**, 301–312.

[8] Muñoz Solano, D., Hoyos, P., Hernáiz, M. J., Alcántara, A. R., and Sánchez-Montero, J. M. (2012). Industrial biotransformations in the synthesis of building blocks leading to enantiopure drugs. *Bioresour. Technol.*, **115**, 196–207.

[9] Sanchez, S. and Demain, A. L. (2011). Enzymes and bioconversions of industrial, pharmaceutical, and biotechnological significance. *Org. Process Res. Dev.*, **15**, 224–230.

[10] Nestl, B. M., Nebel, B. A., and Hauer, B. (2011). Recent progress in industrial biocatalysis. *Curr. Opin. Chem. Biol.*, **15**, 187–193.

[11] Wagner, J., Grill, H., Choleretic medicine. US Patent 3888994, 1975.

[12] Liese, A., Seebach, K., and Wandrey, C., *Industrial Biotransformations*, Wiley-VCH Verlag GmbH & Co. KGaA, Weinheim, 2006.

[13] Buchholz, K., Kasche, V., and Bornscheuer, U. T., *Biocatalysts and Enzyme Technologies*, Wiley-VCH Verlag GmbH & Co. KGaA, Weinheim, 2005.

[14] Faber, K., *Biotransformations in Organic Chemistry*, Springer, Berlin Heidelberg, 2004.

[15] Hoyos, P., Pace, V., and Alcantara, A. R. (2012). Dynamic kinetic resolution via hydrolase-metal combo catalysis in stereoselective synthesis of bioactive compounds. *Adv. Synth. Catal.*, **354**, 2585–2611.

[16] Rebolledo, F., González-Sabín, J., and Gotor, V., *Enzymatic Dynamic Kinetic Resolution in Stereoselective Synthesis, Stereoselective Synthesis of Drugs and Natural Products*, John Wiley & Sons, Inc., Hoboken, NJ, 2013, pp. 1683–1711.

[17] Kim, Y., Park, J., and Kim, M. J. (2011). Dynamic kinetic resolution of amines and amino acids by enzyme-metal cocatalysis. *ChemCatChem*, **3**, 271–277.

[18] Pellissier, H. (2011). Recent developments in dynamic kinetic resolution. *Tetrahedron*, **67**, 3769–3802.

[19] Martín-Matute, B. and Bäckvall, J.-E., Dynamic Kinetic Resolutions, in: Gotor, V., Alfonso, I., and García-Urdiales, E. (Eds.), *Asymmetric Organic Synthesis with Enzymes*, Wiley-VCH Verlag GmbH & Co. KGaA, Weinheim, 2008, pp. 89–113.

[20] Gotor-Fernandez, V., Brieva, R., and Gotor, V. (2006). Lipases: Useful biocatalysts for the preparation of pharmaceuticals. *J. Mol. Catal. B: Enzym.*, **40**, 111–120.

[21] Martín-Matute, B. and Bäckvall, J.-E. (2007). Dynamic kinetic resolution catalyzed by enzymes and metals. *Curr. Opin. Chem. Biol.*, **11**, 226–232.

[22] Dinh, P. M., Howarth, J. A., Hudnott, A. R., Williams, J. M. J., and Harris, W. (1996). Catalytic racemisation of alcohols: Applications to enzymatic resolution reactions. *Tetrahedron Lett.*, **37**, 7623–7626.

[23] Larsson, A. L. E., Persson, B. A., and Bäckvall, J.-E. (1997). Enzymatic resolution of alcohols coupled with ruthenium-catalyzed racemization of the substrate alcohol. *Angew. Chem. Int. Ed. Engl.*, **36**, 1211–1212.

[24] Persson, B. A., Larsson, A. L. E., Le Ray, M., and Bäckvall, J.-E. (1999). Ruthenium- and enzyme-catalyzed dynamic kinetic resolution of secondary alcohols. *J. Am. Chem. Soc.*, **121**, 1645–1650.

[25] Strübing, D., Krumlinde, P., Piera, J., and Bäckvall, J.-E. (2007). Dynamic kinetic resolution of primary alcohols with an unfunctionalized stereogenic center in the β-position. *Adv. Synth. Catal.*, **349**, 1577–1581.

[26] Persson, B. A., Huerta, F. F., and Bäckvall, J.-E. (1999). Dynamic kinetic resolution of secondary diols via coupled ruthenium and enzyme catalysis. *J. Org. Chem.*, **64**, 5237–5240.

[27] Huerta, F. F., Laxmi, Y. R. S., and Bäckvall, J.-E. (2000). Dynamic kinetic resolution of alpha-hydroxy acid esters. *Org. Lett.*, **2**, 1037–1040.

[28] Huerta, F. F. and Bäckvall, J.-E. (2001). Enantioselective synthesis of beta-hydroxy acid derivatives via a one-pot aldol reaction-dynamic kinetic resolution. *Org. Lett.*, **3**, 1209–1212.

[29] Pàmies, O. and Bäckvall, J.-E. (2002). Enzymatic kinetic resolution and chemoenzymatic dynamic kinetic resolution of delta-hydroxy esters. An efficient route to chiral delta-lactones. *J. Org. Chem.*, **67**, 1261–1265.

[30] Hoyos, P., Fernández, M., Sinisterra, J. V., and Alcántara, A. R. (2006). Dynamic kinetic resolution of benzoins by lipase-metal combo catalysis. *J. Org. Chem.*, **71**, 7632–7637.

[31] Pàmies, O. and Bäckvall, J.-E. (2001). Efficient lipase-catalyzed kinetic resolution and dynamic kinetic resolution of beta-hydroxy nitriles. A route to useful precursors for gamma-amino alcohols. *Adv. Synth. Catal.*, **343**, 726–731.

[32] Pàmies, O. and Bäckvall, J.-E. (2002). Chemoenzymatic dynamic kinetic resolution of beta-halo alcohols. An efficient route to chiral epoxides. *J. Org. Chem.*, **67**, 9006–9010.

[33] Pàmies, O. and Bäckvall, J.-E. (2003). An efficient route to chiral α- and β-hydroxyalkanephosphonates. *J. Org. Chem.*, **68**, 4815–4818.

[34] Fransson, A.-B. L., Borén, L., Pàmies, O., and Bäckvall, J.-E. (2005). Kinetic resolution and chemoenzymatic dynamic kinetic resolution of functionalized γ-hydroxy amides. *J. Org. Chem.*, **70**, 2582–2587.
[35] Kim, M. J., Choi, Y. K., Choi, M. Y., Kim, M. J., and Park, J. (2001). Lipase/ruthenium-catalyzed dynamic kinetic resolution of hydroxy acids, diols, and hydroxy aldehydes protected with a bulky group. *J. Org. Chem.*, **66**, 4736–4738.
[36] Lihammar, R., Millet, R., and Bäckvall, J.-E. (2013). Enzyme- and ruthenium-catalyzed dynamic kinetic resolution of functionalized cyclic allylic alcohols. *J. Org. Chem.*, **78**, 12114–12120.
[37] Koh, J. H., Jung, H. M., Kim, M.-J., and Park, J. (1999). Enzymatic resolution of secondary alcohols coupled with ruthenium-catalyzed racemization without hydrogen mediator. *Tetrahedron Lett.*, **40**, 6281–6284.
[38] Lee, D., Huh, E. A., Kim, M.-J., Jung, H. M., *et al.* (2000). Dynamic kinetic resolution of allylic alcohols mediated by ruthenium- and lipase-based catalysts. *Org. Lett.*, **2**, 2377–2379.
[39] Kim, M. J., Kim, H. M., Kim, D., Ahn, Y., and Park, J. (2004). Dynamic kinetic resolution of secondary alcohols by enzyme-metal combinations in ionic liquid. *Green Chem.*, **6**, 471–474.
[40] Agrawal, S., Martinez-Castro, E., Marcos, R., and Martin-Matute, B. (2014). Readily available ruthenium complex for efficient dynamic kinetic resolution of aromatic alpha-hydroxy ketones. *Org. Lett.*, **16**, 2256–2259.
[41] Choi, J. H., Kim, Y. H., Nam, S. H., Shin, S. T., *et al.* (2002). Aminocyclopentadienyl ruthenium chloride: Catalytic racemization and dynamic kinetic resolution of alcohols at ambient temperature. *Angew. Chem. Int. Ed.*, **41**, 2373–2376.
[42] Jana, N., Mahapatra, T., and Nanda, S. (2009). Chemo-enzymatic asymmetric total synthesis of stagonolide-C. *Tetrahedron: Asymmetry*, **20**, 2622–2628.
[43] Das, T. and Nanda, S. (2012). Chemoenzymatic total synthesis of stagonolide-E. *Tetrahedron Lett.*, **53**, 256–258.
[44] Martín-Matute, B., Edin, M., Bogár, K., and Bäckvall, J.-E. (2004). Highly compatible metal and enzyme catalysts for efficient dynamic kinetic resolution of alcohols at ambient temperature. *Angew. Chem. Int. Ed.*, **43**, 6535–6539.
[45] Martín-Matute, B. and Bäckvall, J.-E. (2004). Ruthenium- and enzyme-catalyzed dynamic kinetic asymmetric transformation of 1,4-diols: Synthesis of gamma-hydroxy ketones. *J. Org. Chem.*, **69**, 9191–9195.
[46] Thalén, L. K., Sumic, A., Bogár, K., Norinder, J., *et al.* (2010). Enantioselective synthesis of alpha-methyl carboxylic acids from readily available starting materials via chemoenzymatic dynamic kinetic resolution. *J. Org. Chem.*, **75**, 6842–6847.
[47] Bogar, K., Vidal, P. H., Alcántara León, A. R., and Bäckvall, J.-E. (2007). Chemoenzymatic dynamic kinetic resolution of allylic alcohols: A highly enantioselective route to acyloin acetates. *Org. Lett.*, **9**, 3401–3404.
[48] Leijondahl, K., Boren, L., Braun, R., and Bäckvall, J.-E. (2009). Enzyme- and ruthenium-catalyzed dynamic kinetic asymmetric transformation of 1,5-diols. Application to the synthesis of (+)-Solenopsin A. *J. Org. Chem.*, **74**, 1988–1993.
[49] Krumlinde, P., Bogar, K., and Backvall, J. E. (2010). Asymmetric synthesis of bicyclic diol derivatives through metal and enzyme catalysis: Application to the formal synthesis of sertraline. *Chem. Eur. J.*, **16**, 4031–4036.
[50] Borén, L., Leijondahl, K., and Bäckvall, J.-E. (2009). Dynamic kinetic asymmetric transformation of 1,4-diols and the preparation of trans-2,5-disubstituted pyrrolidines. *Tetrahedron Lett.*, **50**, 3237–3240.
[51] Leijondahl, K., Boren, L., Braun, R., and Bäckvall, J.-E. (2008). Enantiopure 1,5-diols from dynamic kinetic asymmetric transformation. Useful synthetic intermediates for the preparation of chiral heterocycles. *Org. Lett.*, **10**, 2027–2030.
[52] Edin, M., Martín-Matute, B., and Bäckvall, J.-E. (2006). Ruthenium- and lipase-catalyzed DYKAT of 1,2-diols: An enantioselective synthesis of syn-1,2-diacetates. *Tetrahedron: Asymmetry*, **17**, 708–715.
[53] Kim, N., Ko, S. B., Kwon, M. S., Kim, M. J., and Park, J. (2005). Air-stable racemization catalyst for dynamic kinetic resolution of secondary alcohols at room temperature. *Org. Lett.*, **7**, 4523–4526.
[54] Ko, S.-B., Baburaj, B., Kim, M.-J., and Park, J. (2007). Air-stable racemization catalysts for the dynamic kinetic resolution of secondary alcohols. *J. Org. Chem.*, **72**, 6860–6864.

[55] Kim, M. J., Choi, Y. K., Kim, S., Kim, D., *et al.* (2008). Highly enantioselective dynamic kinetic resolution of 1,2-diarylethanols by a lipase-ruthenium couple. *Org. Lett.*, **10**, 1295–1298.

[56] Akai, S., Hanada, R., Fujiwara, N., Kita, Y., and Egi, M. (2010). One-pot synthesis of optically active allyl esters via lipase-vanadium combo catalysis. *Org. Lett.*, **12**, 4900–4903.

[57] Akai, S., Tanimoto, K., Kanao, Y., Egi, M., *et al.* (2006). A dynamic kinetic resolution of allyl alcohols by the combined use of lipases and [VO(OSiPh3)(3)]. *Angew. Chem. Int. Ed.*, **45**, 2592–2595.

[58] Akai, S. (2014). Dynamic kinetic resolution of racemic allylic alcohols via hydrolase-metal combo catalysis: An effective method for the synthesis of optically active compounds. *Chem. Lett.*, **43**, 746–754.

[59] Egi, M., Sugiyama, K., Saneto, M., Hanada, R., *et al.* (2013). A mesoporous-silica-immobilized oxovanadium cocatalyst for the lipase-catalyzed dynamic kinetic resolution of racemic alcohols. *Angew. Chem. Int. Ed.*, **52**, 3654–3658.

[60] Verzijl, G. K. M., de Vries, J. G., and Broxterman, Q. B., Process for the preparation of enantiomerically enriched esters and alcohols. WO Patent 0190396, 2003.

[61] Choi, J. H., Choi, Y. K., Kim, Y. H., Park, E. S., *et al.* (2004). Aminocyclopentadienyl ruthenium complexes as racemization catalysts for dynamic kinetic resolution of secondary alcohols at ambient temperature. *J. Org. Chem.*, **69**, 1972–1977.

[62] Krumlinde, P., Bogár, K. N., and Bäckvall, J.-E. (2009). Synthesis of a neonicotinoide pesticide derivative via chemoenzymatic dynamic kinetic resolution. *J. Org. Chem.*, **74**, 7407–7410.

[63] Han, K., Kim, C., Park, J., and Kim, M.-J. (2010). Chemoenzymatic synthesis of rivastigmine via dynamic kinetic resolution as a key step. *J. Org. Chem.*, **75**, 3105–3108.

[64] Agustian, J., Kamaruddin, A. H., and Bhatia, S. (2010). Single enantiomeric beta-blockers: The existing technologies. *Process Biochem.*, **45**, 1587–1604.

[65] Träff, A., Bogár, K., Warner, M., and Bäckvall, J.-E. (2008). Highly efficient route for enantioselective preparation of chlorohydrins via dynamic kinetic resolution. *Org. Lett.*, **10**, 4807–4810.

[66] Johnston, E. V., Bogár, K., and Bäckvall, J.-E. (2010). Enantioselective synthesis of (*R*)-bufuralol via dynamic kinetic resolution in the key step. *J. Org. Chem.*, **75**, 4596–4599.

[67] Hargreaves, R. H., Hartley, J. A., and Butler, J. (2000). Mechanisms of action of quinone-containing alkylating agents: DNA alkylation by aziridinylquinones. *Front. Biosci.*, **5**, 172–180.

[68] Müller, P. and Fruit, C. (2003). Enantioselective catalytic aziridinations and asymmetric nitrene insertions into CH bonds. *Chem. Rev.*, **103**, 2905–2920.

[69] Cimarelli, C., Fratoni, D., and Palmieri, G. (2009). A convenient synthesis of new diamine, amino alcohol and aminophosphines chiral auxiliaries based on limonene oxide. *Tetrahedron: Asymmetry*, **20**, 2234–2239.

[70] Padwa, A., Aziridines and Azirines: Monocyclic, in: Katritzky, A. R., Ramsden, C. A., Scriven, E. F. V., and Taylor, R. J. K. (Eds.), *Comprehensive Heterocyclic Chemistry III*, Elsevier, Oxford, 2008, pp. 1–104.

[71] Fujiwara, Y., Katagiri, T., and Uneyama, K. (2003). Trifluoromethylated amino alcohols as chiral ligands for highly enantioselective Reformatsky reaction. *Tetrahedron Lett.*, **44**, 6161–6163.

[72] Pàmies, O. and Bäckvall, J.-E. (2001). Dynamic kinetic resolution of beta-azido alcohols. An efficient route to chiral aziridines and beta-amino alcohols. *J. Org. Chem.*, **66**, 4022–4025.

[73] Kamal, A., Khanna, G. B. R., Ramu, R., and Krishnaji, T. (2003). Chemoenzymatic synthesis of duloxetine and its enantiomer: Lipase-catalyzed resolution of 3-hydroxy-3-(2-thienyl) propanenitrile. *Tetrahedron Lett.*, **44**, 4783–4787.

[74] Träff, A., Lihammar, R., and Bäckvall, J.-E. (2011). A chemoenzymatic dynamic kinetic resolution approach to enantiomerically pure (*R*)- and (*S*)-duloxetine. *J. Org. Chem.*, **76**, 3917–3921.

[75] Martín-Matute, B., Edin, M., and Bäckvall, J.-E. (2006). Highly efficient synthesis of enantiopure diacetylated C-2-symmetric diols by ruthenium- and enzyme-catalyzed dynamic kinetic asymmetric transformation (DYKAT). *Chem. Eur. J.*, **12**, 6053–6061.

[76] Millet, R., Traff, A. M., Petrus, M. L., and Backvall, J. E. (2010). Enantioselective synthesis of syn- and anti-1,3-aminoalcohols via beta-aminoketones and subsequent reduction/dynamic kinetic asymmetric transformation. *J. Am. Chem. Soc.*, **132**, 15182–15184.

[77] Gomes, C. R. B., Moreth, M., Cardinot, D., Kopke, V., *et al.* (2011). Synthesis and antimycobacterial activity of novel amino alcohols containing central core of the anti-HIV drugs lopinavir and ritonavir. *Chem. Biol. Drug Des.*, **78**, 1031–1034.

[78] Sham, H. L., Zhao, C., Li, L., Betebenner, D. A., *et al.* (2002). Novel lopinavir analogues incorporating non-Aromatic P-1 side chains—Synthesis and structure–Activity relationships. *Bioorg. Med. Chem. Lett.*, **12**, 3101–3103.

[79] Carlier, P. R., Lo, M. M. C., Lo, P. C. K., Richelson, E., *et al.* (1998). Synthesis of a potent wide-spectrum serotonin-, norepinephrine-, dopamine-reuptake inhibitor (SNDRI) and a species-selective dopamine-reuptake inhibitor based on the gamma-amino alcohol functional group. *Bioorg. Med. Chem. Lett.*, **8**, 487–492.

[80] Lait, S. M., Rankic, D. A., and Keay, B. A. (2007). 1,3-Aminoalcohols and their derivatives in asymmetric organic synthesis. *Chem. Rev.*, **107**, 767–796.

[81] Cossy, J., Pardo, D. G., Dumas, C., Mirguet, O., *et al.* (2009). Rearrangement of β-amino alcohols and application to the synthesis of biologically active compounds. *Chirality*, **21**, 850–856.

[82] Hussain, M. M. and Walsh, P. J. (2008). Tandem reactions for streamlining synthesis: Enantio- and diastereoselective one-pot generation of functionalized epoxy alcohols. *Acc. Chem. Res.*, **41**, 883–893.

[83] Allen, J. V. and Williams, J. M. J. (1996). Dynamic kinetic resolution with enzyme and palladium combinations. *Tetrahedron Lett.*, **37**, 1859–1862.

[84] Norinder, J., Bogár, K., Kanupp, L., and Bäckvall, J.-E. (2007). An enantioselective route to α-methyl carboxylic acids via metal and enzyme catalysis. *Org. Lett.*, **9**, 5095–5098.

[85] Warner, M. C., Nagendiran, A., Bogár, K., and Bäckvall, J.-E. (2012). Enantioselective route to ketones and lactones from exocyclic allylic alcohols via metal and enzyme catalysis. *Org. Lett.*, **14**, 5094–5097.

[86] Runmo, A.-B. L., Pàmies, O., Faber, K., and Bäckvall, J.-E. (2002). Dynamic kinetic resolution of gamma-hydroxy acid derivatives. *Tetrahedron Lett.*, **43**, 2983–2986.

[87] Hoyos, P., Sinisterra, J. V., Molinari, F., Alcántara, A. R., and De Maria, P. D. (2010). Biocatalytic strategies for the asymmetric synthesis of alpha-hydroxy ketones. *Accounts Chem. Res.*, **43**, 288–299.

[88] Ödman, P., Wessjohann, L. A., and Bornscheuer, U. T. (2005). Chemoenzymatic dynamic kinetic resolution of acyloins. *J. Org. Chem.*, **70**, 9551–9555.

[89] Hoyos, P., Buthe, A., Ansorge-Schurmacher, M. B., Sinisterra, J. V., and Alcántara, A. R. (2008). Highly efficient one pot dynamic kinetic resolution of benzoins with entrapped *Pseudomonas stutzeri* lipase. *J. Mol. Catal. B: Enzym.*, **52**(3), 133–139.

[90] Hoyos, P., Quezada, M. A., Sinisterra, J. V., and Alcántara, A. R. (2011). Optimised dynamic kinetic resolution of benzoin by a chemoenzymatic approach in 2-MeTHF. *J. Mol. Catal. B: Enzym.*, **72**, 20–24.

[91] Maraite, A., Hoyos, P., Daniel Carballeira, J., Cabrera, A. C., *et al.* (2013). Lipase from *Pseudomonas stutzeri*: Purification, homology modelling and rational explanation of the substrate binding mode. *J. Mol. Catal. B: Enzym.*, **87**, 88–98.

[92] Hoyos, P., Pace, V., Sinisterra, J. V., and Alcántara, A. R. (2011). Chemoenzymatic synthesis of chiral unsymmetrical benzoins. *Tetrahedron*, **67**, 7321–7329.

[93] Ghislieri, D. and Turner, N. J. (2014). Biocatalytic approaches to the synthesis of enantiomerically pure chiral amines. *Top. Catal.*, **57**, 284–300.

[94] Höhne, M. and Bornscheuer, U. T. (2009). Biocatalytic routes to optically active amines. *ChemCatChem*, **1**, 42–51.

[95] Lee, J. H., Han, K., Kim, M. J., and Park, J. (2010). Chemoenzymatic dynamic kinetic resolution of alcohols and amines. *Eur. J. Org. Chem.*, **6**, 999–1015.

[96] Reetz, M. T. and Schimossek, K. (1996). Lipase-catalyzed dynamic kinetic resolution of chiral amines: Use of palladium as the racemisation catalyst. *Chimia*, **50**, 668–669.

[97] Kim, M. J., Kim, W. H., Han, K., Choi, Y. K., and Park, J. (2007). Dynamic kinetic resolution of primary amines with a recyclable Pd nanocatalyst for racemization. *Org. Lett.*, **9**, 1157–1159.

[98] Choi, E., Kim, Y., Ahn, Y., Park, J., and Kim, M.-J. (2013). Highly enantioselective enzymatic resolution of aromatic β-amino acid amides with Pd-catalyzed racemization. *Tetrahedron: Asymmetry*, **24**, 1449–1452.

[99] Parvulescu, A., Janssens, J., Vanderleyden, J., and De Vos, D. (2010). Heterogeneous catalysts for racemization and dynamic kinetic resolution of amines and secondary alcohols. *Top. Catal.*, **53**, 931–941.

[100] Parvulescu, A. N., Jacobs, P. A., and De Vos, D. E. (2007). Palladium catalysts on alkaline-earth supports for racemization and dynamic kinetic resolution of benzylic amines. *Chem. Eur. J.*, **13**, 2034–2043.
[101] Parvulescu, A. N., Jacobs, P. A., and De Vos, D. E. (2009). Support influences in the Pd-catalyzed racemization and dynamic kinetic resolution of chiral benzylic amines. *Appl. Catal. Gen.*, **368**, 9–16.
[102] Parvulescu, A., Vos, D. D., and Jacobs, P. (2005). Efficient dynamic kinetic resolution of secondary amines with Pd on alkaline earth salts and a lipase. *Chem. Commun.*, Issue 42, 5307–5309.
[103] Parvulescu, A. N., Jacobs, P. A., and De Vos, D. E. (2008). Heterogeneous Raney nickel and cobalt catalysts for racemization and dynamic kinetic resolution of amines. *Adv. Synth. Catal.*, **350**, 113–121.
[104] Shakeri, M., Tai, C.-W., Göthelid, E., Oscarsson, S., and Bäckvall, J.-E. (2011). Small Pd nanoparticles supported in large pores of mesocellular foam: An excellent catalyst for racemization of amines. *Chem. Eur. J.*, **17**, 13269–13273.
[105] Pàmies, O., Éll, A. H., Samec, J. S. M., Hermanns, N., and Bäckvall, J.-E. (2002). An efficient and mild ruthenium-catalyzed racemisation of amines: Application to the synthesis of optically pure amines. *Tetrahedron Lett.*, **3**, 4699–4702.
[106] Paetzold, J. and Bäckvall, J.-E. (2005). Chemoenzymatic dynamic kinetic resolution of primary amines. *J. Am. Chem. Soc.*, **127**, 17620–17621.
[107] Hoben, C. E., Kanupp, L., and Bäckvall, J.-E. (2008). Practical chemoenzymatic dynamic kinetic resolution of primary amines via transfer of a readily removable benzyloxycarbonyl group. *Tetrahedron Lett.*, **49**, 977–979.
[108] Thalen, L. K., Zhao, D. B., Sortais, J. B., Paetzold, J., *et al.* (2009). A chemoenzymatic approach to enantiomerically pure amines using dynamic kinetic resolution: Application to the synthesis of norsertraline. *Chem. Eur. J.*, **15**, 3403–3410.
[109] Blacker, A. J., Stirling, M. J., and Page, M. I. (2007). Catalytic racemisation of chiral amines and application in dynamic kinetic resolution. *Org. Process Res. Dev.*, **11**, 642–648.
[110] Stirling, M., Blacker, J., and Page, M. I. (2007). Chemoenzymatic dynamic kinetic resolution of secondary amines. *Tetrahedron Lett.*, **48**, 1247–1250.
[111] Choi, Y. K., Kim, M. J., Ahn, Y., and Kim, M.-J. (2001). Lipase/palladium-catalyzed asymmetric transformation of ketoximes to optically active amines. *Org. Lett.*, **3**, 4099–4101.
[112] Han, K., Kim, Y., Park, J., and Kim, M.-J. (2010). Chemoenzymatic synthesis of the calcimimetics (+)-NPS R-568 via asymmetric reductive acylation of ketoxime intermediate. *Tetrahedron Lett.*, **51**, 3536–3537.
[113] Engström, K., Shakeri, M., and Bäckvall, J.-E. (2011). Dynamic kinetic resolution of β-amino esters by a heterogeneous system of a palladium nanocatalyst and *Candida antarctica* lipase A. *Eur. J. Org. Chem.*, **2011**, 1827–1830.
[114] Shakeri, M., Engstrom, K., Sandstrom, A. G., and Bäckvall, J.-E. (2010). Highly enantioselective resolution of beta-amino esters by *Candida antarctica* lipase A immobilized in mesocellular foam: Application to dynamic kinetic resolution. *ChemCatChem*, **2**, 534–538.
[115] Engstrom, K., Johnston, E. V., Verho, O., Gustafson, K. P. J., *et al.* (2013). Co-immobilization of an enzyme and a metal into the compartments of mesoporous silica for cooperative tandem catalysis: An artificial metalloenzyme. *Angew. Chem. Int. Ed.*, **52**, 14006–14010.
[116] Choi, Y. K., Kim, Y., Han, K., Park, J., and Kim, M. J. (2009). Synthesis of optically active amino acid derivatives via dynamic kinetic resolution. *J. Org. Chem.*, **74**, 9543–9545.
[117] Thalén, L. K. and Bäckvall, J.-E. (2010). Development of dynamic kinetic resolution on large scale for (±)-1-phenylethylamine. *Beilstein J. Org. Chem.*, **6**, 823–829.
[118] Rodriguez-Mata, M., Gotor-Fernández, V., González-Sabin, J., Rebolledo, F., and Gotor, V. (2011). Straightforward preparation of biologically active 1-aryl- and 1-heteroarylpropan-2-amines in enantioenriched form. *Org. Biomol. Chem.*, **9**, 2274–2278.

CHAPTER 15

Discovery and Engineering of Enzymes for Peptide Synthesis and Activation

Ana Toplak, Muhammad I. Arif, Bian Wu, and Dick B. Janssen

Biochemical Laboratory, Groningen Biomolecular Sciences and Biotechnology Institute, University of Groningen, Groningen, the Netherlands

15.1 INTRODUCTION

In chemoenzymatic peptide synthesis, chemically modified amino acids or small peptides are coupled enzymatically to obtain dipeptides or larger peptides. The enzymatic coupling comprises an acyl donor, which is either an amino acid or peptide that becomes the N-terminal part of the ligation product, and an amine donor (or acyl acceptor), which becomes the C-terminal part. Coupling is catalyzed by peptidases, which are used in the reverse direction (Figure 15.1) [1]. The synthesis can be driven thermodynamically if the equilibrium of the coupling reaction is on the side of the synthetic product. The equilibrium can often be shifted in the synthetic direction, for example by using conditions that cause product precipitation. If that is not the case, a synthetic reaction can still occur if the conversion is performed with a high-energy activated precursor and formation of a coupling product by the enzyme is kinetically preferred over hydrolysis. The high-energy precursor is the acyl donor, activated at its carboxylate group, for example in the form of an ester, and often also N-protected to prevent self-coupling. Synthetic yields can be improved by increasing the concentration of the amine donor that acts as a nucleophile, by modifying reaction conditions, and by optimizing the enzyme [2].

Besides the peptide coupling reactions themselves, also some of the required modification reactions, such as carboxylate group activation and N-terminal amine protection, can also be catalyzed by enzymes. The use of enzymes in peptide synthesis and modification is attractive since it makes side-chain protection superfluous, allows mild reaction conditions that prevent racemization, and can introduce a desired regio-, stereo-, and substrate selectivity.

The success of chemoenzymatic peptide synthesis is critically dependent on the availability of effective peptide coupling and modification enzymes, both in the case of kinetically controlled synthesis with an activated acyl donor and in the case of thermodynamically controlled synthesis starting with free amino acids or peptides. If the coupling reaction case is kinetically controlled, critical properties are a high synthesis/hydrolysis

Green Biocatalysis, First Edition. Edited by Ramesh N. Patel.

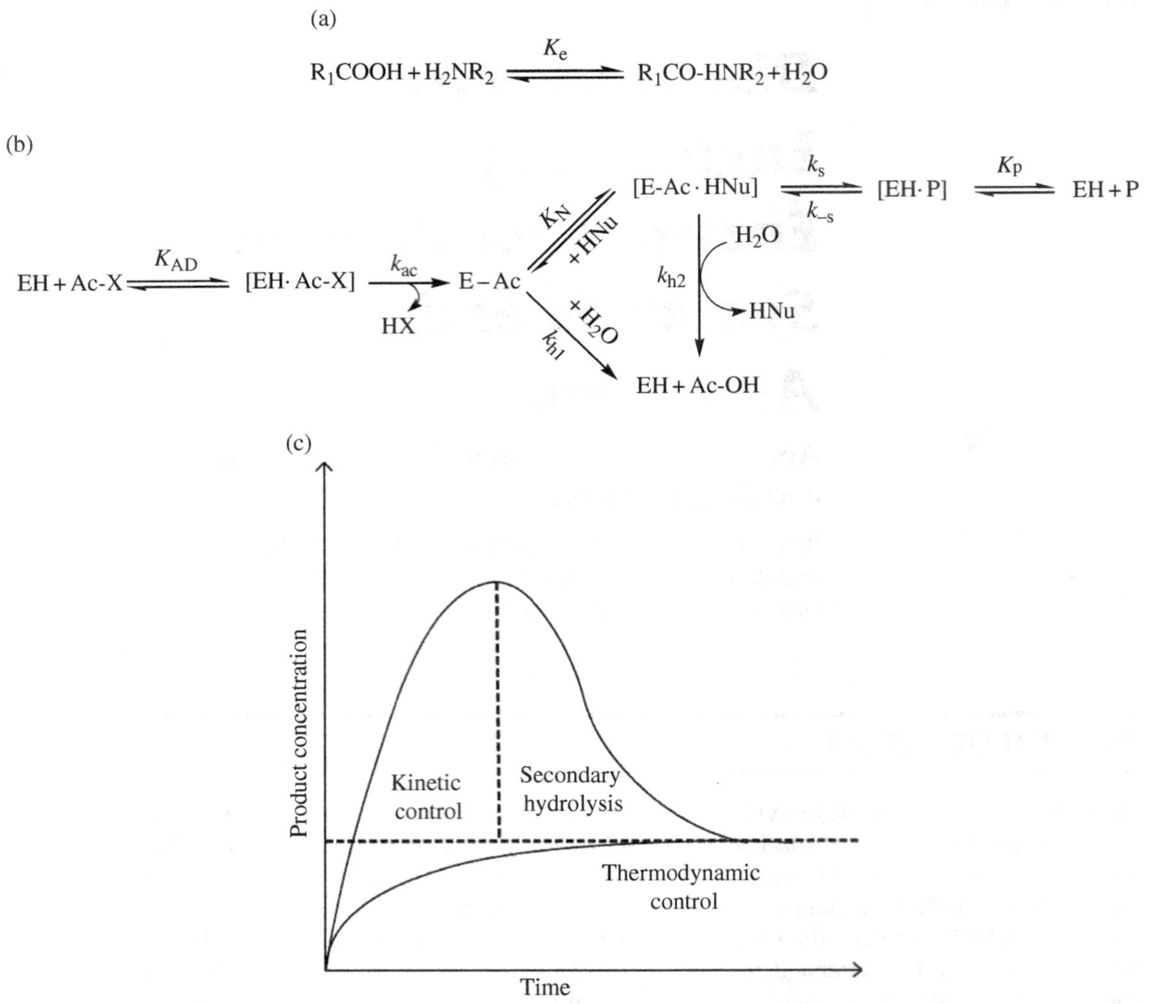

FIGURE 15.1

Principle of enzymatic peptide synthesis. (a) Synthesis under thermodynamic control is based on microscopic reversibility. The overall equilibrium constant K_e will be strongly dependent on ionization constants of the different species and the reaction pH. (b) Synthesis under kinetic control. The activated acyl donor (Ac-X) forms a Michaelis complex [EH·Ac-X] in the first step that collapses into the covalent acyl-enzyme intermediate E-Ac. The deacylation of this intermediate can be performed either by water yielding hydrolytic product Ac-OH and free enzyme or by an added nucleophile (HNu) that leads to the synthetic product P and free enzyme via a second Michaelis complex [E-Ac·HNu]. This complex may still hydrolyze if water competes with bound nucleophile. Synthetic yields are determined by rate constants k_{ac}, k_{h1}, k_{h2}, k_s, and k_{-s} and binding constants K_{AD}, K_N, and K_P. (c) Time course of the formation of synthetic products.

ratio with the activated acyl donor and a broad substrate range. Since the reactions proceed via acyl-enzyme intermediates, the amine donor has to outcompete water if the conversion is carried out in an aqueous environment. Furthermore, the coupling enzyme should not cleave internal peptide bonds in precursor peptides or in the coupling product. Alternatively, reactions are performed in the presence of organic solvent, which may be added as cosolvent, as a second phase in a two-phase system, or be used as a neat organic phase. The kinetic and selectivity criteria for enzymes to be used in an aqueous medium and the enzyme robustness needed for reactions in organic solvents make the availability of suitable enzymes a critical success factor.

This chapter highlights classical and recent work on enzymes used in chemoenzymatic peptide synthesis, with an emphasis on peptidases with beneficial characteristics such as broad applicability and high stability that are important for application in peptide synthesis.

15.2 CLASSIFICATION OF ENZYMES FOR PEPTIDE COUPLING

Peptidases suitable for peptide synthesis are found in a wide variety of organisms (Table 15.1). Many reports describe the use of commercially available enzymes isolated from biological sources, such as trypsin and chymotrypsin from mammalian pancreas, pepsin and chymosin from the stomach, papain from plants, and subtilisin and thermolysin secreted by bacteria. Genome analysis shows that peptidases are present in all kingdoms of life: from archaea and viruses to the higher eukaryotes. Moreover, peptidase-encoding genes comprise about 2% of the human genome, most of them having very specific functions, for example in hormone processing, but with unknown utility in peptide synthesis [45]. The omnipresence and tremendous diversity of peptidases and the myriad of biological functions all contribute to the continued scientific interest in this enzyme family.

According to the enzyme classification system, peptidases belong to subgroup 4 of the hydrolases (EC 3.4.X.X). Based on the work of Rawlings and Barrett [46] the current MEROPS peptidase database counts more than 2400 peptidases and provides a description of their specificity, structural information, and literature references. Peptidases are frequently classified according to their reaction mechanism using the key groups involved in catalysis as an identifier. This gives aspartic (A), cysteine (C), glutamic (G), metallo (M), asparagine (N), serine (S), threonine (T), and unknown (U) proteases, with the letter indicated followed by an assigned number. Peptidases can also be classified in terms of families and clans, which are based on structure or sequence comparison. This is a phylogenetic classification. The name of a clan is formed in most cases by a letter of the catalytic type, followed by a serial capital letter. For example SB represents the subtilisin clan of serine peptidases and contains two families: S8 (subtilisin family) and S53 (sedolisin family). Even though members of the same clan have a common ancestor, a sequence may have diverged so much that relatedness can only be detected by structural comparison. Not surprisingly, some phylogenetic groups comprise multiple catalytic types, for which the abbreviation P was proposed. More information on peptidase classification is given by Rawlings and Salvesen [47].

One can distinguish two general types of peptidase catalytic mechanisms: catalysis via formation of a covalent intermediate and catalysis by activation of a water molecule that directly attacks the peptide bond. In the case of covalent catalysis, the nucleophile attacking the peptide bond is a side-chain hydroxyl or sulfhydryl, as in the serine, threonine, and cysteine peptidases. It is a part of a classical nucleophile–base–acid catalytic triad, as in the serine hydrolases that provide the textbook example of an enzyme catalytic mechanism, or it is in a functionally equivalent but structurally different variant thereof. For example, in papain the nucleophile is a cysteine occurring as a thiolate–imidazolium ion pair formed by a histidine acting as the base and a water molecule replacing the acidic group present in most serine protease catalytic triads. In enzymes of the Ntn-hydrolase superfamily (Ntn, N-terminal nucleophile), such as penicillin acylases and the catalytic subunits of the proteasome, the role of the histidine base is adopted by the free NH_2-group of the nucleophilic residue that is at the N-terminus of the β-chain in these dimeric enzymes [48]. The formation of acyl-enzyme intermediates in serine hydrolases and related enzymes makes them suitable for application in kinetically controlled synthesis reactions (Figure 15.1).

Metalloproteases and glutamic and aspartic proteases, on the other hand, do not form covalent intermediates, but activate a water molecule that directly attacks the carbonyl carbon of the peptide bond and displaces the amide nitrogen [49]. These enzymes are often the preferred catalysts for thermodynamically controlled coupling (Figure 15.1). An example of a metalloprotease applied for peptide synthesis is thermolysin, which contains a HExxH+E sequence motif that coordinates a zinc ion and a water molecule. Zinc polarizes the carbonyl group and facilitates deprotonation of the

TABLE 15.1 Peptidases used for Chemoenzymatic Enzymatic Peptide Synthesis

Enzyme	Source	Specificity	Application
Serine peptidases			
α-Chymotrypsin, clan PA, EC 3.4.21.1	Pancreas, recombinant yeast system available for rat enzyme	P1 = Tyr, Trp, Phe, Leu P1′ nonspecific	Used in the synthesis of bioactive peptides in water–organic solvent mixtures and ionic liquids; substrate-mimetic-mediated coupling of long fragments [3–7]; amino acid polymers for biomaterials [8]
Trypsin clan PA, EC 3.4.21.4	Bovine, porcine, murine pancreas, recombinant production in *Escherichia coli*, *Pichia pastoris*, *Saccharomyces cerevisiae*	P1 = Arg, Lys P1′ nonspecific	Synthesis of bioactive peptides in water and water–cosolvent mixtures [6], insulin transpeptidation [9], substrate-mimetic approach [10]
Subtilisin clan SB, EC 3.4.21.62	Various *Bacillus* sp., recombinant production in *E. coli*	P1 = large, nonbranched hydrophobic residues preferred P1′ = Gly > Ser, Lys, Ala > Arg > Gln, Thr, Val, Met, Asn > His > Ile, Trp > Glu > Asp, Leu, Pro	Bioactive peptide synthesis in neat organic solvent and water–organic solvent mixtures; amino acid oligomerization [11–13]; engineered thiol variants for enhanced synthesis [14, 15]
Proteinase K clan SB, EC 3.4.21.64	Secreted by the *Tritirachium album*	P1 = aromatic and hydrophobic residues P1′ = similar to subtilisins	[Leu] enkephalin peptide synthesis in organic solvent in a multistage reactor [16]; synthesis of branched peptides [17]
Protease V8 clan PA, EC 3.4.21.19	Extracellular peptidase from *Staphylococcus aureus*	P1 = Asp, Glu	Substrate-mimetic-mediated coupling of dipeptides and peptide fragments [18, 19]
Carboxypeptidase Y clan SC, EC 3.4.16.5	produced by *S. cerevisiae*	P1′ = hydrophobic residues	C-terminal extension in water–organic solvent medium [20]
Prolyl aminopeptidase (PAP) clan SC, EC 3.4.11.5	*Streptomyces* sp.	P1 = Pro, P1′ = broad	Synthesis of peptides and short polymers containing Pro [21]
Cysteine peptidases			
Papain clan CA, EC 3.4.22.2	Papaya plant latex *Carica papaya*, recombinant production in *S. cerevisiae*, *P. pastoris* and *E. coli*	P1 = Arg, Lys; P2 = aromatic hydrophobic P1′ = large hydrophobic	Kinetically controlled peptide synthesis; synthesis of bioactive peptides in organic solvent (dermorphin, enkephalin); amino acid oligomerization [22]; di- and tripeptide synthesis with peptidomimetic leaving group [23–27]
Bromelain clan CA, EC 3.4.22.33	Pineapple fruit (*Ananas comosus*)	P1 and P1′ = polar amino acids	[Met] and [Leu] enkephalin fragment condensation (2 + 3), [28]
Ficain clan CA, EC 3.4.22.3	Extraction from *Ficus glabrata* latex	P1 = Gly, Ser, Glu, Tyr, Phe; P2 = hydrophobic side chains	Lys-Met ethyl ester oligomer synthesis; short peptide synthesis in frozen aqueous medium [29]

Clostripain clan CD, EC 3.4.22.8	Clostridium histolyticum, recombinant production in *E. coli* and *Bacillus subtilis*	P1 = Arg, P1′ = D-amino acids, Pro	Di-and tripeptide synthesis in frozen aqueous medium; synthesis of peptide isosters [30]
Cathepsin L clan CA, EC 3.4.22.15	*Fasciola hepatica*, recombinant in yeast	P2 = large hydrophobic residues, P1′ = Ser > Ala > Lys > Asn > Gln	Cbz-Phe-Arg-Ser-NH_2 synthesis under kinetic control [31]
Sortase A clan CL, EC 3.4.22.70	*S. aureus*, *E. coli* expression system	LPXT-G, X = any amino acid	Peptide and protein labeling, cyclization and immobilization, protein–protein fusion, cell surface labeling [32, 33]
Metallopeptidases			
Thermolysin clan MA zinc endopeptidase, EC 3.4.24.27	Secreted by *Bacillus thermoproteolyticus*, recombinant in *E. coli*	P1′ = large hydrophobic, also polar and charged residues	Aspartame precursor synthesis, dipeptide synthesis [34–36]
		P1 = hydrophobic residues, P2, P2′ = Leu > Ala > Phe > Gly, P3′ = basic residues	
Stearolysin clan MA zinc endopeptidase	Secreted by *Geobacillus stearothermophilus*, recombinant in *E. coli*	Similar to thermolysin, P1′ = Phe	Thermostable variant boilysin in dipeptide Cbz-Asp-X-OMe (X = Phe, D-Phe, Ala, Ile, Leu, Met, Tyr, or Val) synthesis [36]
Pseudolysin (elastase) clan MA, EC 3.4.24.26	Secreted by *Pseudomonas aeruginosa*, recombinant in *E. coli* and *Pichia*	P1′ = hydrophobic or aromatic residues; Phe > Leu > Tyr > Val, Ile > Ala	Dipeptide synthesis, including substrates as norvaline and norleucine, Cbz-Ala-Phe-NH_2 synthesis [37]
		P1,P2′ = Ala	
Vimelysin clan MA	Secreted by *Vibrio* sp. T1800; recombinant *E. coli*	P1′ = Phe	Aspartame precursor synthesis in the presence of 30% DMSO and ethanol [38]
Carboxypeptidase A clan MC, EC 3.4.17.1	*Bos taurus*; recombinant yeast	P1′ = large, hydrophobic residues	Dipeptide synthesis in low water systems, 27% highest yield obtained [39]
D,D-Dipeptidase	*Brevibacillus borstelensis* BCS-1; recombinant *E. coli*	P1′ = D-amino acid	Cbz-L-Asp–D-AlaOBzl alitame sweetener precursor synthesis [40]
Aminopeptidase SSAP	*Streptomyces septatus*, recombinant *E. coli*	Broad specificity, inactive toward N-protected peptides	Various dipeptide methyl esters starting from free acyl donor and amino acid methyl ester as nucleophile in methanol [21]
Aspartic proteases			
Pepsin clan AA, EC 3.4.23.1	Porcine, recombinant *E. coli*	P1 and P1′ = hydrophobic residues, P2 = hydrophilic residues	Various di- and tripeptides in two-phase systems, hexapeptide synthesis at pH 4.6 in 20% (v/v) DMF [41–43]
Chymosin clan AA, EC 3.4.23.4	Calf stomach (bovine, camel), recombinant *E. coli*, yeast, and filamentous fungi	P1 and P1′ = hydrophobic residues	Tetra- and hexapeptide synthesis at pH 5.3 in 20% v/v DMF [44]

water nucleophile that attacks the carbonyl group of the substrate. Catalysis is facilitated by the glutamic acid, which accepts a proton from the zinc-bound water and transfers a proton to the leaving group. In the majority of the aspartic peptidases, a pair of aspartic residues acts together to activate a water molecule. In glutamic peptidases, a pair of glutamic acid residues plays a similar role. The water directly attacks the peptide bond, and the reaction in peptide synthesis follows the reverse mechanism.

The serine, cysteine, and aspartic proteases and the metalloproteases have all been applied in peptide synthesis. Nevertheless, shortcomings still exist and at an industrial scale chemoenzymatic peptide synthesis is certainly not always the preferred method. Since the properties of the catalyst to a large extent determine the feasibility of industrial application, the discovery and engineering of better variants are an intensive field of research.

15.3 SERINE AND CYSTEINE PROTEASES FOR PEPTIDE SYNTHESIS

15.3.1 Chymotrypsin, Trypsin, and Related Enzymes

Classical enzymes employed for peptide coupling of the serine hydrolase family are chymotrypsin, trypsin, and subtilisin. Chymotrypsin and trypsin are secreted in the mammalian gut as inactive precursors, which are activated by autoproteolysis and structural reorganization. The use of chymotrypsin for peptide synthesis has been reported since the 1930s [50]. Most early examples concern peptide synthesis using amides or (m)ethyl esters as acyl donors and free amino acids and their amides, short peptides, or short peptide amides as nucleophilic acyl acceptors [3]. These studies revealed that a high pH, high nucleophile concentration, and low product solubility stimulate the formation of synthetic product. Ethyl esters appeared suitable acyl donors in kinetically controlled conversions, and amino acid amides act better as nucleophilic acyl acceptors than the free amino acids [4]. Furthermore, tripeptides often performed better than dipeptides or amino acids as acyl acceptors.

Trypsin and chymotrypsin are often used in the presence of cosolvents, which are added to increase peptide substrate solubility, both in the case of thermodynamic coupling and for kinetically controlled synthesis, especially for longer peptides. For example the condensation of tri- and heptapeptide fragments, forming a Trp^3–Ser^4 linkage, in a kinetically controlled manner using α-chymotrypsin in 48% (v/v) DMF allowed over 97% conversion to a D-Phe-containing analog of gonadotropin-releasing hormone ([D-Phe^6]GnRH) on multigram scale (Figure 15.2) [5]. Reactions with trypsins in neat organic solvent have also been explored. The thermodynamically controlled fragment condensation of octa- and pentapeptide fragments yielded an α-melanocyte-stimulating hormone precursor with an Arg^8–Trp^9 ligation site. This conversion was catalyzed by trypsin in an organic solvent mixture (Figure 15.3) [6].

Chymotrypsin and trypsin have restricted substrate specificity. This can be an advantage, since it reduces the number of possibilities for hydrolysis of a synthetic product, but also a disadvantage, since it restricts the diversity of ligation sites that can be generated. Overall, the leaving group specificity (P1′, P2′) of trypsin and chymotrypsin in hydrolytic reactions matches that of the nucleophile selectivity in synthetic conversions [3, 7]. Chymotrypsin-catalyzed synthetic reactions are less sensitive to the nature of the acyl donor than to the nature of the nucleophile. Trypsin accepts positively charged amino acids at the P1 site (Arg, Lys), whereas chymotrypsin prefers bulky hydrophobic groups at the P1 and P1′ positions. At the P1′ position Val, Leu, Ile, and Phe are well accepted by chymotrypsin, whereas Glu and Pro are poorly accepted at the P1′ position, and Pro also at the P2′ position [3]. D-Amino acids also react poorly as nucleophiles. Addition of salts may improve nucleophilicity of peptides with charges at P2′ or P3′.

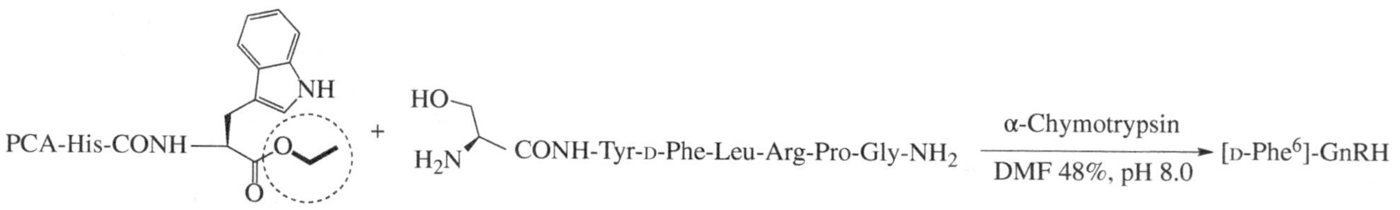

FIGURE 15.2

Example of bioactive peptide synthesis under kinetic control using α-chymotrypsin in water–cosolvent medium [5]. The gonadotropin-releasing hormone (PCA-His-Trp-Ser-Tyr-Gly-Leu-Arg-Pro-Gly-NH_2) analogue [D-Phe[6]]GnRH was obtained in 97.5% yield. Abbreviations: PCA, pyroglutamic acid.

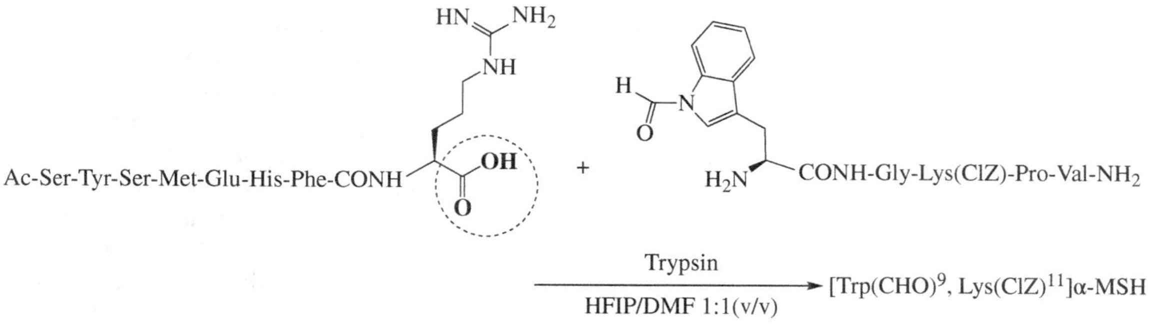

FIGURE 15.3

Example of bioactive peptide synthesis under thermodynamic control using trypsin in water–cosolvent medium. The α-melanocyte-stimulating hormone precursor was synthesized in 95% yield using trypsin in an almost dry mixture of DMF and hexafluoroisopropanol [6]. Protective groups: CHO, formyl group; ClZ, 2-chlorobenzyloxycarbonyl.

Recognizing the importance of binding of the acyl donor for synthetic performance in kinetically controlled coupling, modifications of the leaving group so that it matches enzyme selectivity have been proposed [10, 51] (Table 15.2). In the so-called inverse substrate or substrate-mimetic approach, a reactive group such as a *p*-guanidinophenyl (OGp) ester forces the formation of an acyl-enzyme intermediate, even in the absence of an amino acid at the P1 position that matches the S1 pocket specificity. Instead, the S1 pocket is used to bind the leaving group, which is designed to mimic the side chain that fits in the S1 subsite, and a reactive acyl-enzyme is formed. This approach enabled the acceptance of amino acids and peptide sequences

TABLE 15.2 Strongly Activating Groups for Kinetic Peptide Coupling

Type	Activating Group	Abbreviation	Structure	Related Enzymes
Activating ester	Carbamoylmethyl ester	Cam ester	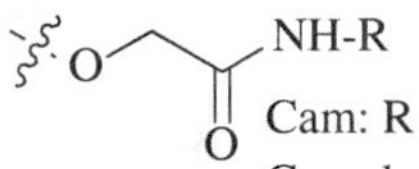 Cam: R = H Cam derivative: R = amino acid	Serine peptidases, papain, subtiligase
	2,2,2-Trifluoroethyl ester	Tfe ester	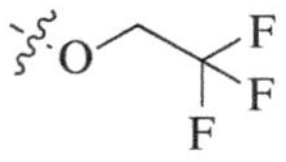	Serine peptidases
Substrate mimetic	*p*-Guanidinophenyl ester	OGp ester	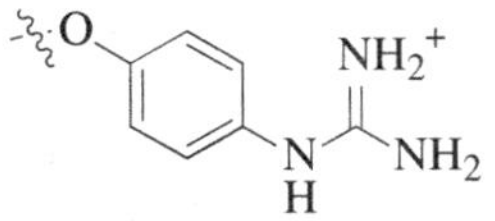	Arginine-specific peptidases (trypsin, chymotrypsin, and papain)
	Carboxymethyl thioesters	SCm ester	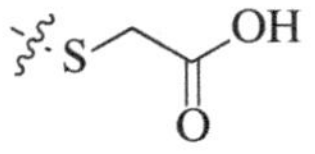	Nonarginine-specific peptidases

that are normally not recognized by chymotrypsin or trypsin [52, 53]. Work with the Glu-specific endopeptidases (V8 from *Staphylococcus aureus* and BL-GSE from *Bacillus licheniformis*) showed the use of the carboxymethyl thioester group (2-mercaptoacetate thioester) as substrate mimetic for a protease that is not active with arginine at the P1 position [18, 19].

Proteases of the trypsin family have recently been applied in the synthesis of peptides that self-assemble to nanofibrous structures and for preparing biomaterials. An example is the chymotrypsin-catalyzed polymerization of the amphiphilic peptide Lys-Leu-OEt to an alternating oligomer that forms a hydrogel [8]. Furthermore, chymotrypsin was used for the synthesis of poly-L-cysteine, a material for the selective chelation of SH-reactive metal ions. A high yield (80%) was obtained under mild reaction conditions [54].

A trypsin-related bacterial enzyme that was tested in coupling reactions is the lysine-specific serine protease I from *Achromobacter lyticus*. The enzyme is secreted and widely used in sequence analysis of proteins [55]. It was also used in a chemo-enzymatic route for the production of human insulin from porcine insulin. Since this endopeptidase cleaves only after Lys, it could be applied for replacing the C-terminus of the insulin B-chain from -Lys-Ala (porcine C-terminal sequence) to -Lys-Thr (human C-terminus) in a two-step or a single-step reaction. Using the B-chain as the acyl donor and Thr-OBu^t in DMF–ethanol mixtures as the nucleophile, a high conversion (85–90%) was obtained [9, 56]. Trypsin could also be applied in this biotransformation but required higher enzyme loading.

15.3.2 Subtilisin-Like Enzymes

Subtilisins have a similar catalytic mechanism as chymotrypsin, but the catalytic triad is ordered in a different way along the sequence, and the enzymes are not phylogenetically related. Subtilisins are secreted by various soil-dwelling *Bacillus* strains. They comprise the largest group of commercial proteases and account for more than half of the world's total sales of enzymes, mainly for use in detergents. Well known commercial variants include subtilisin Carlsberg (also known as alcalase), subtilisin BPN′ (used in many protein engineering studies), and subtilisin E. Siezen and Leunissen [28] introduced the term "subtilases" for subtilisin-like peptidases. Genome analysis revealed that subtilases occur in all kingdoms of life. Subtilases from the phylogenetic group of subtilisins have been widely explored for the synthesis of short peptides. This is much less the case for subtilases belonging to the thermitase and proteinase K groups. However, subtilisins were reported to be less suitable for peptide synthesis than chymotrypsin and trypsin due to their high hydrolytic activity. This triggered research on subtilisin thiol variants, which form a thioester intermediate with reduced sensitivity to hydrolysis as compared to the aminolysis reaction required for peptide coupling. Furthermore, subtilisin has a much broader substrate range, implying that it detects a larger number of proteolytic target sites in substrates and products than chymotrypsin and trypsin.

Due to the high stability of these enzymes, peptide synthesis using subtilisin BPN′ and subtilisin Carlsberg can be carried out in neat organic solvent instead of water and water–cosolvent mixtures, effectively avoiding hydrolysis of the activated precursor and of the product [57]. The use of anhydrous organic solvents also suppresses unwanted side reactions and can influence enzyme stereoselectivity, regiospecificity, and chemoselectivity, which may allow unexpected conversion [57, 58]. For example, a dipeptide containing D–L, L–D, or D–D peptide bonds could be synthesized using subtilisin in anhydrous solvent, whereas these bonds are usually not sensitive to peptidase-mediated hydrolysis [57]. The tolerance to hydrophilic solvents that dissolve long protected peptides, such as THF and acetonitrile, allowed the coupling of activated 9-aa and 10-aa peptides carrying protecting groups from chemical peptide synthesis, resulting in the formation of a protected 19-amino acid peptide [11, 13].

Like chymotrypsin, subtilases can catalyze reactions leading to special peptides. Proteinase K catalyzed the polymerization of L-Phe in the presence of tris(2-aminoethyl) amine to form a branched oligo(L-phenylalanine), which showed self-assembly to form fluorescent fibers [17].

15.3.3 Other Serine Hydrolases

Of the serine exopeptidases, carboxypeptidase Y from *Saccharomyces cerevisiae*, in particular, has been used in synthetic applications. This carboxypeptidase is an α/β-hydrolase fold enzyme with a Ser–Asp–His catalytic triad of a topology that is different from that of trypsin and subtilisin family enzymes. Carboxypeptidase Y catalyzes dipeptide coupling in an aqueous medium [59]. As with chymotrypsin, such synthetic reactions are best performed at basic pH. Protein engineering studies have revealed that the enzyme has a rather low selectivity for the P1′ site, preferring hydrophobic groups, but Lys and Arg are also accepted. Selectivity can be influenced by mutations. Reducing the number of enzyme–substrate hydrogen bonds by mutating residue Glu65 to Ala was beneficial for the aminolysis reaction [20].

15.3.4 Aminopeptidases

Prolyl aminopeptidases (PAP) are exopeptidases that hydrolytically cleave off an N-terminal Pro from peptides. The enzymes belong to the α/β hydrolase fold proteins. A PAP from *Streptomyces thermoluteus* carrying the active site nucleophile mutation S144C was used as a catalyst for the synthesis of proline-containing peptides. Dipeptide synthesis was obtained with an amino acid methyl or benzyl ester as the acyl donor and prolyl-OBz as the nucleophile [21]. Under alkaline conditions, cyclization and polymerization of prolyl-OBz was observed.

15.3.5 Peptidases Accepting β-Amino Acids

In view of the growing importance of β-peptides, Kohler and coworkers explored the synthetic applicability of β-peptide-specific aminopeptidases produced by strains of *Sphingosinicella* and *Ochrobactrum anthropi*. These enzymes (DmpA) belong to the P1 serine peptidase family, are distantly related to Ntn-hydrolase family, and cleave N-terminal β-amino acids from β- and α/β-peptides. In synthetic applications, oligomerization of β-amino acids occurred to produce up to octameric peptides. The enzymes could also form α- and β-amino acid containing dipeptides with an N-terminal β-amino acid. Furthermore, the BapA enzyme from *Sphingosinicella xenopeptidilytica* 3-2 W4 could couple various β-amino acids to the N-terminus of short peptide [60]. A similar incorporation of N-terminal D-amino acids during small peptide synthesis was reported with a *Streptomyces* aminopeptidase [61].

15.3.6 D-Amino Acid-Specific Peptidases

Few microbial proteases acting on D-peptides are known. The alkaline D-peptidase (ADP) from *Bacillus cereus* is related to DD-carboxypeptidase and β-lactamases. These enzymes have an accessible groove in which the nucleophilic serine and other catalytic amino acids are located. This D-peptidase could be applied for the synthesis of the 92-amino acid peptidyl prolyl *cis–trans* isomerase from *Escherichia coli* by condensation of two peptide fragments, of which the 35-amino acid acyl donor was activated as the OGp ester [62]. Thus the D-amino acid-selective enzyme was used for preparing a protein composed of L-amino acids and making the product insensitive to hydrolysis by the coupling enzyme.

15.3.7 Sulfhydryl Peptidases

The application of papain in peptide synthesis is well established [23–25]. Papain can be used for the preparation of di- and tripeptides in an aqueous medium with cosolvent addition (up to 40%) and at high pH to promote synthetic activity. The enzyme is a sulfhydryl protease with no homology to the trypsin or subtilase families of hydrolases. Since the catalytic nucleophile is a cysteine and because thioesters are relatively more prone to aminolysis than oxo-esters, the enzyme could be very attractive for synthesis. However, unlike the case with the thiol variants of some serine hydrolases, the proteolytic activity is still high, and the broad substrate range of proteolysis makes peptide substrate and product hydrolysis more problematic than trypsin or chymotrypsin. Extensive enzyme engineering studies on papain are lacking, probably due to the laborious procedure for isolation of active papain from inclusion bodies formed in *E. coli*.

The enzyme has a rather broad substrate range with a slight preference for Lys and Arg in the S1 pocket and for hydrophobic aromatic residues in the S2 pocket. The S1′ pocket is not selective, which allowed the synthesis of dipeptides with nonproteinogenic amino acids [24]. To overcome the S1 preference, the substrate-mimetic approach described under Section 15.3.1 was explored with the 4-guanidinophenyl (OGp) leaving group in dipeptide synthesis [25].

Fragment condensations with papain have also been reported, for example enkephalin fragment coupling (2+3) was achieved in 50% yield using equimolar amounts of acyl donor (PhAc-Tyr-Gly-OMe) and nucleophile (H-Gly-Phe-Leu-O^{t}Bu) at pH 9.0 in buffer with 20% methanol [26]. Similar yields were obtained in low water systems such as buffer containing acetonitrile 4% (v/v).

The synthetic potential of papain in low water systems has been explored with acyl donors bearing the carbamoylmethyl (Cam) leaving group. Dipeptide synthesis proceeded in >80% yield, and yields up to 60% were obtained for the synthesis of bioactive peptides like dermorphin-(1-4) (Boc-Tyr-D-Ala-Phe-Gly-NH_2). Papain is also used as a versatile protease in polymer chemistry for the synthesis of amino acid oligomers and cooligomers of α-hydroxy acids and amino acids [22].

Homologues of the classical papaya papain that have been tested for peptide synthesis are plant-derived enzymes such as bromelain and ficain. Bromelain extracted from pineapple could be applied in Met- and Leu-enkephalin synthesis in a low water system, reaching 97% yield [27]. Ficain isolated from *Ficus glabrata* latex was tested in di- and tripeptide synthesis in frozen water medium [23] and was used in the synthesis of Lys and Met oligomers starting from the respective ethyl esters [29].

Another papain-like enzyme that has been tested in peptide synthesis is clostripain from *Clostridium histolyticum*. It has a remarkable specificity for arginine in the S1 pocket and for proline in the S1′ pocket and can be produced using a recombinant system in *E. coli* [23]. Synthesis of dipeptides such as Cbz-Arg-Pro-NH_2 and Cbz-Arg-D-Leu-NH_2 has been reported. Moreover, the substrate-mimetic approach using the OGp activating group broadened acyl donor specificity to nonproteinogenic acyl donors such as β-Ala and 4-phenylbutyric acid ester, with yields of over 86% when coupled with different nucleophiles in an aqueous medium [30]. The broad specificity of the S1′ pocket allowed the synthesis of various peptide isosteres using Bz-Arg-OH and a range of acyl acceptor amines [63].

Animal tissue is a source of papain-related lysosomal cysteine peptidases that have been used in di- and tripeptide synthesis. Due to the insufficient homogenic preparations of cathepsin B, earlier reports of its synthetic activity have to be taken with caution [16]. The availability of a recombinant production system for human cathepsin B will allow experiments with pure enzyme. The endopeptidase cathepsin L from parasite *Fasciola hepatica* was expressed in yeast and used in the kinetically controlled synthesis of Cbz-Phe-Arg-Ser-NH_2 starting from Cbz-Phe-Arg-OMe and H-Ser-NH_2 in an aqueous medium [31].

15.3.8 Sortase

For coupling of larger peptide substrates, sortases, which are responsible for covalent anchoring of surface proteins to the peptidoglycan of Gram-positive bacteria, can be employed. Sortase A (SrtA) (MEROPS peptidase family C60) is a cysteine peptidase isolated from *S. aureus* that has emerged as a powerful tool for protein bioconjugation and transpeptidation reactions [32]. During catalysis, recognition of the R-LPXTG motif (R = protein, X = any amino acid) is followed by thiolate nucleophile attack and cleavage of the Thr-Gly amide bond. The acyl-enzyme complex (R-LPXT-SrtA) is cleaved by a peptide with an N-terminal glycine of the pentapeptide groups in peptidoglycan, yielding a covalent protein–peptidoglycan bond. This sortagging reaction can serve in applications such as protein labeling, protein–protein fusion, and protein cyclization and immobilization. Recently, the use of sortase for the conversion of a QALPETGEE peptide to its hydrazide derivative QALPET-$NHNH_2$ was reported. Such a peptide hydrazide can act as an acyl donor in peptide coupling. Via the formation of a hydrazide, this sortase reaction allowed fusion of a deoxy-D-ribose 5-phosphate aldolase fragment (amino acids 1–239) with the 17-amino acid C-terminal fragment, producing catalytically active enzyme connected by an LPAA linker. Similar reactions were possible with substituted hydrazines, allowing the labeling of proteins containing the sortase motif with fluorescent probes [64].

The efficiency of the sortase-mediated reactions depends on the flexibility and accessibility of the region comprising the recognition motif. Sortase engineering and mutant selection for better motif recognition have been reported [65]. Details regarding sortase mechanism and applications of the enzyme can be found in an excellent review by Ritzefeld [32].

15.3.9 Metalloproteases in Peptide Synthesis

The metalloprotease thermolysin has been widely used for peptide synthesis [66, 67]. Thermolysin is produced and secreted by the thermophilic bacterium *Bacillus thermoproteolyticus*. Catalysis involves water activation by a zinc ion that is coordinated by side chains of two histidines and a glutamate. Thermolysin specificity is defined by its S1′ pocket accepting large hydrophobic, polar, and charged residues. In the S1 pocket hydrophobic residues are preferred, and Leu > Ala > Phe > Gly is the preference order for the S2 and S2′ pockets. This hydrolytic specificity of thermolysin is reflected in its synthetic specificity.

Many related so-called thermolysin-like proteinases (TLPs) from various Gram-positive strains have been described [47], including neutral proteases from *Bacillus subtilis*, and some of these variants are applied in peptide synthesis. Several metalloenzymes acting as carboxy- or aminopeptidase have also been characterized, but these variants have not been extensively used in peptide synthesis. A bovine carboxypeptidase A [39] and orange carboxypeptidase C [68] have been applied for dipeptide synthesis in water–organic solvent mixtures, both under thermodynamic and under kinetic control.

With thermolysin and related metalloproteases, peptide synthesis is carried out under thermodynamic control where the reversal of hydrolysis is achieved by modifying reaction conditions in such a way that the equilibrium shifts to synthesis. Thermolysin has been employed in numerous di- and tripeptide coupling reactions. For example, thermolysin has been applied for the synthesis of artificial sweetener aspartame (L-α-Asp-L-Phe-OMe) on multiton level. Unlike the chemical route, which requires optically pure starting materials and gives substantial formation of by-products, the enzymatic synthesis is an elegant and environmentally friendly method. The reaction is performed under mild conditions and is both regioselective (only the α-carboxyl group reacts) and stereoselective (for L-Phe), thus allowing the use of racemic acyl acceptor (D,L-Phe-OMe). The D-isomer of Phe forms a precipitate with Cbz-L-α-Asp-L-Phe-OMe, thereby changing the equilibrium in favor of synthesis. After

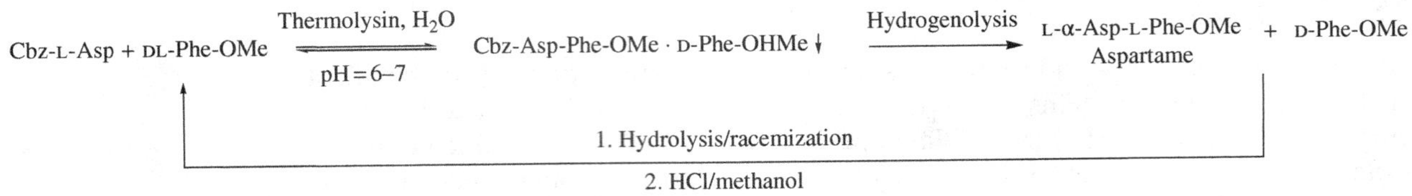

FIGURE 15.4

Chemoenzymatic peptide synthesis of aspartame using thermolysin under thermodynamic control. The coupling reaction is stereo- and regioselective. The unreacted isomer D-Phe-OMe forms a precipitate with the product, shifting the equilibrium toward the synthesis. After precursor isolation and hydrogenolysis, the D-isomer is chemically racemized and can be reused [69, 70].

deprotection, D-Phe can be racemized and recycled (Figure 15.4). This is an example of substrate-induced precipitation for thermodynamic control of peptide synthesis.

Another example of thermolysin-catalyzed peptide synthesis is the production of precursors of enkephalins [57], in particular, coupling of nonnatural amino acids, such as halophenylalanines, is also possible [34]. Similar to α-chymotrypsin and papain, thermolysin has been used for peptide polymerization [71]. Using the reversibility of thermolysin-catalyzed peptide bond formation, dynamic combinatorial libraries of peptides could be established and screened for the formation of nanostructures with special properties [72].

15.3.10 Aspartic Proteases in Peptide Synthesis

Aspartic proteases form a group of proteolytic enzymes that catalyze peptide bond cleavage by acid–base catalysis and activation of a water molecule for nucleophilic attack on the amide carbon. Crystal structures of mammalian and fungal enzymes are known. In pepsin, the best studied aspartic protease, catalysis proceeds by water activation and leaving group protonation. Both involve an aspartate, which explains the low pH optimum of around 4. The two aspartic acid residues are situated around a hydrophobic cleft that can accommodate seven amino acids (the S4–S3′ subsites). The active site is covered by a flexible flap, which contributes to S1 subsite specificity.

Since the catalytic cycle does not include an acyl-enzyme intermediate, attempts to use pepsin for synthesis have been focused on thermodynamically controlled conversions [41]. Conversions can be done with an addition of organic solvent to increase medium hydrophobicity and shift the equilibrium to synthesis. Synthesis of several tri- and hexapeptides by pepsin in buffer containing 18% DMF reached yields of 30–78%. Pepsin can also be used in two-phase systems. The synthetic yields are determined by enzyme specificity, by equilibrium of the coupling reactions, as well as by partitioning of substrates and products over the two phases. Pepsin has a substrate preference for hydrophobic peptides, but in coupling reactions of Cbz-Xaa-Phe-OH and Phe-OMe, a better yield was observed with hydrophilic groups in the P2 position [42, 43]. Hydrophobic peptides and some nonproteinogenic substrates are well accepted in the S1′ pocket, which render pepsin a useful catalyst for the synthesis of nonnatural peptides [41]. If immobilized enzyme is used, it is preferable to keep the substrate and products in solution to avoid problems with separation of products from the catalyst. For this reason, DMF has been added as a solubilizing agent, benefiting from the fact that pepsin is highly resistant to denaturing conditions.

A prominent aspartic protease related to pepsin is chymosin, produced in the calf stomach for casein hydrolysis. Its use as a catalyst in peptide synthesis has been described, again in thermodynamically driven coupling reactions, where linkages are introduced in hydrophilic regions of the target peptides. Carboxybenzoyl (Cbz) and *p*-nitroanilides were used as amine and acyl protecting groups, respectively. Using DMF to keep the substrates in solution and acetate buffer at low pH (5.3) to stimulate synthesis, coupling products were formed and precipitated [43]. This way, several tetra- and hexapeptides were produced. The recombinant overexpression of chymosin in *E. coli*, yeast, and, for industrial production, filamentous fungi is well established. Consequently, chymosin variants could be obtained by protein engineering, and mutants with higher pH optimum and higher thermostability have been described [44]. The use of these enzymes in peptide synthesis remains underexplored.

15.4 PROTEASE DISCOVERY

The restricted applicability of commercially available proteases for peptide coupling reactions warrants the continuous importance of mining natural resources for enzymes with novel properties, such as a broad substrate range, acceptance of nonproteinogenic amino acids, high coupling efficiency, and robustness under reaction conditions.

15.4.1 Metagenomics

Since, with current protocols, <1% of the existing microbial diversity can be isolated in the laboratory, metagenomic approaches comprising cloning and sequence- or function-based screening can contribute to rapid biocatalyst discovery. Genes can be discovered that are very different from those with an established function. For example, using metagenomics a metalloprotease was discovered that has <30% similarity to any known protease [73]. The potential of metagenomics was recently also demonstrated for variants of subtilisin. In total 51 different subtilisins were discovered, each carrying two to eight substitutions that are distant from the active site but affect temperature stability and substrate specificity [74]. It seems likely that these approaches can discover variants with altered characteristics in peptide synthesis.

15.4.2 Proteases from Thermophiles

Whereas subtilisin Carlsberg, subtilisin BNP′, and subtilisin E display moderate temperature stabilities, several thermostable proteases of the subtilase superfamily have been obtained from extremophilic organisms. Examples are aqualysin I from *Thermus aquaticus* and thermitase from *Thermoactinomyces vulgaris*. Although over 30 of such characterized thermophilic peptidases are known, only few have been applied in peptide synthesis. The serine peptidase from *T. aquaticus* Rt41A (PreTaq) was applied in kinetically controlled synthesis of Bz-Ala-Tyr-NH_2 with modest yield (26%) using immobilized enzyme in DMF (90% v/v), Bz-Ala-OMe as the acyl donor, and a large excess of Tyr-NH_2 as the nucleophile at 40 °C and pH 10 [74]. Two serine peptidases of the subtilase family from the hyperthermophilic organisms *T. aquaticus* and *Deinococcus geothermalis* were also applied in peptide synthesis under harsh conditions such as high temperature (60 and 80 °C) in anhydrous solvent with the addition of DMF as cosolvent (40% v/v) [76].

15.4.3 Solvent-Tolerant Proteases

The hypothesis that extracellular enzymes of organic solvent-tolerant microorganisms should exert resistance to such organic solvents led to the discovery of a solvent-tolerant metalloprotease called pseudolysin produced by the bacterium *Pseudomonas aeruginosa* [77]. Organic solvent tolerance of this metalloprotease was attributed to a disulfide bond (Cys30—Cys58) and to the presence of charged residues at the enzyme surface (Y45, N201). In the thermodynamic approach using medium with 50% DMF, the enzyme could effectively synthesize the dipeptide Cbz-Arg-Leu-NH_2 (78% yield) and the aspartame precursor Cbz-Asp-Phe-OMe (89% yield) in the presence of 50% DMSO [77]. Similarly, the opioid peptides endomorphin-1 (Tyr-Pro-Trp-Phe-NH_2) [78] and endomorphin-2 (Tyr-Pro-Phe-Phe-NH_2) [79] were synthesized by the organic solvent-tolerant metalloproteases from *P. aeruginosa* and *B. cereus*, respectively. Both peptidases tolerated up to 50% (v/v) of DMSO and DMF, and conversions gave endomorphin-2 with 80% overall yield. These enzymes are all structurally related to thermolysin.

15.4.4 Proteases from Salt-Resistant Organisms

Halophilic microorganisms that evolved to survive low water activity conditions are promising for the discovery of peptidases that survive under harsh conditions. A halophilic serine peptidase from *Halobacterium salinarum* was reported to catalyze the synthesis of di- and tripeptides, for example producing Ac-Phe-Gly-Leu-NH_2 from Ac-Phe-OEt as acyl donor in 76% yield in the presence of 33% (v/v) DMF, 20 equivalents of the nucleophile dipeptide amide, and 2.8 M NaCl [80]. The acyl donor specificity of the enzyme is quite broad, but the nucleophile preferred at S1′ is Gly, Ala, or another small hydrophobic group. Although promising, the synthetic potential of halophilic peptidases is still poorly exploited, and biochemical information on such enzymes is scarce.

15.5 PROTEASES ENGINEERED FOR IMPROVED SYNTHESIS

Protein engineering of enzymes for peptide synthesis has mainly focused on improving the synthesis/hydrolysis ratio of serine hydrolases used in kinetically controlled peptide synthesis and on studies aimed at engineering enzyme specificity. Other targets are improving the resistance to organic solvents and tailoring the hydrolytic activity, but only a few of these engineered enzymes were tested later as catalysts in peptide coupling.

15.5.1 Solvent-Resistant and Thermostable Subtilase Mutants

Extensive work on stabilizing subtilisins by mutagenesis has been carried out. Thermostability and organic cosolvent tolerance are often positively correlated. Stabilizing mutations include introduction of groups that form salt bridges or internal hydrogen bonds, improvement of hydrophobic interactions, and formation of internal cross-links by disulfide bonds [81–83]. An excellent overview on subtilisin BPN′ engineering is given by P. Bryan [84]. The synthetic potential of a DMF-tolerant subtilisin BPN′ mutant was tested [85]. This enzyme, subtilisin 8350, carries mutations N218S (improved hydrogen bonding), G169A (improved hydrophobic interaction and conformational restriction), M50F (improved hydrophobic interaction), Q206C (oxidized to Cys-SH during posttranslational modification, better van der Waals interactions), N76D (improved Ca^{2+} binding and hydrogen bonding), and Y217K (better H bonding). The enzyme catalyzed the synthesis of several bioactive peptides in 50% DMF under kinetic control, including enkephalin-releasing factor (Tyr-D-Arg, 70% yield), chemotactic peptide (Met-Leu-Phe-$NHCH_2C_6H_5$, 95% yield), and Leu-enkephalinamide and dermorphin in yields above 70%. The enzyme also catalyzed oligomerization of Met-OMe (formation of a 50-mer) where the degree of polymerization was influenced by the amount of DMF added [85].

An even more stable variant of subtilisin BPN′, subtilisin 8397 (M50F/N76D/G169A/Q206C/N218S), was designed and used in glycopeptide synthesis [14] and in the synthesis of a peptide heptamer in 50% DMF [86]. Further mutagenesis to increase the stability in the presence of DMF revealed K43N and K256Y as important mutations. The mutations M222A and Y217W broadened the P1′ specificity of this subtilisin. The authors also report regioselective transesterification, transacylation, and amidation of peptide esters [87].

15.5.2 Thermostable Thermolysin Variants

Thermolysin has been subjected to extensive protein engineering studies aimed at enhancing thermostability [88]. Such studies yielded mutants (D150A, D150E, D150W, I168A, and N227H) that had an improved activity in aspartame synthesis as compared to the wild type [35]. Engineering the thermostability of stearolysin, a protease from *Geobacillus stearothermophilus* CU21 that is 86% sequence identical to thermolysin, resulted in an extremely stable eightfold variant that was active at 100 °C and in the presence of denaturants. This enzyme, termed boilysin, had a specificity comparable to as thermolysin in the synthesis of dipeptides (Cbz-Asp-X-OMe (X = Phe, D-Phe, Ala, Ile, Leu, Met, Tyr, or Val)) and an increased activity at elevated temperatures and low calcium concentrations [36].

15.5.3 Increasing Aminolysis to Hydrolysis Ratio by Protein Engineering

Protein engineering methods aimed at increasing the aminolysis to hydrolysis ratio of peptidases have focused on replacing the active site serine by a cysteine. Early studies on thiol-subtilisin [89] and seleno-subtilisin [90] demonstrated increased aminolysis to hydrolysis ratio in peptide synthesis in aqueous medium or in DMF-containing buffers. Thiol-subtilisin is a weak peptidase and can only rapidly cleave esters. Since thiol- and selenol-acyl-enzyme intermediates undergo aminolysis faster than oxo-esters, these enzymes are particularly useful for the synthesis of amide bonds. However, the esterolytic activity is lower than that of the wild-type enzyme, and the modified enzymes are oxidant sensitive. Mutagenesis of subtilisin BPN′ with concomitant introduction of the mutations S221C and P225A yielded a variant more suitable for peptide synthesis [15]. This variant of subtilisin mutant was termed subtiligase. It forms acyl-enzyme intermediates only with acyl donors activated as esters, and the glycolate phenylalanyl amide functionality was selected as an activating group. The specificity of subtiligase has been studied, but the ligation efficiency is still not well predictable, as it does not directly reflect subsite specificities [91]. By screening tripeptides as acyl acceptor, it was found that especially the P1′ specificity influences ligation efficiency. The potential of subtiligase is clear from the chemo-enzymatic synthesis of intact and catalytically active ribonuclease A (124 amino acids) by sequential ligation of six fragments in the C → N direction with 10% overall yield on milligram scale [92].

Further mutagenesis of subtiligase included the introduction of stabilizing mutations (M50F, N76D, N109S, K213R, N218), yielding a denaturant-resistant variant [93]. In addition, random mutagenesis at 25 positions around the active site of subtiligase and screening the coupling activity of a phage-displayed enzyme library yielded variants with modified selectivity. By screening for the attachment of a labeled peptide to its N-terminus, two subtiligase mutants were identified (M124L/L126V and M124L/S125A) with an improved ligase activity [94]. The applicability of subtiligase is broad, and it has been used for selective labeling of proteins by coupling a peptide to the N-terminus of a target protein [95]. Subtiligase has also been used for kinetically controlled peptide backbone cyclization. The cyclization of peptides of length 12–31 amino acids activated as the C-terminal glycolate phenylalanylamide ester was performed in an aqueous solution [96].

15.5.4 Protein Engineering of Trypsin-Like Proteases

Variants of trypsin with increased aminolysis rate in substrate-mimetic-mediated peptide synthesis have been obtained by mutating active site residues and residues in the S1 subsite. A variant called trypsiligase was described by Bordusa and coworkers [97]. It carries the mutations D189K (reduces Arg and tolerates Tyr specificity in the

S1 subsite), K60E (introduces Arg selectivity in the S1′ subsite), and N143H/E151H (introduces Zn-dependent His specificity in the S2′ subsite), and thus recognizes the Y↓RH sequence motif. The enzyme could be applied to process proteins carrying a YRH linker at the N-terminus, and the resulting RH- N-terminus could be labeled by acting as a nucleophile towards the acyl-enzyme formed by reacting trypsiligase with an OGp-activated acyl donor that carries a fluorescent group [15]. Trypsiligase adopts a partially disordered inactive zymogen-like conformation and is activated by addition of the peptide substrate and zinc ions. Trypsiligase has also been applied in C-terminal protein modification, including PEG or dye labeling of a protein C-terminus containing the YRH trypsiligase recognition motif [98]. A Fab fragment derived from a therapeutic antibody used in breast cancer treatment was extended at the C-terminus of its heavy chain with a tag containing the YRH recognition sequence. This was followed by functionalization using PEG- or dye-labeled RHAK peptide as a nucleophile in a transpeptidation reaction. Trypsiligase-mediated protein modification offers an alternative to other enzyme-mediated peptide modification reactions, such as those catalyzed by subtiligase, SrtA, and transglutaminase.

Mutagenesis studies on the chymotrypsin-like protease B from *Streptomyces griseus* (SGBP) revealed ligation activity when the active site serine was replaced by Ala or Gly (S195A/G) [99]. The ligation was proposed to proceed through a histidine-involved acyl-enzyme intermediate. The stability of the enzyme was improved by introducing mutations with stabilizing effects (S195G and T213L), yielding streptoligase [100].

15.5.5 Computational Design

Alongside structure-based engineering and directed evolution, computational protein design is becoming a powerful tool for tailoring enzymes to specific biotechnological applications. Recently, a *de novo* designed peptide ligase based on a 33 α-helical peptide that forms a coiled coil has been reported [101]. Furthermore, a peptidase with designed specificity toward α-gliadin has been developed using computational methods by D. Baker and coworkers [102]. The rapid developments in this field suggest that computational methods will soon contribute to the redesign of peptidases for synthetic applications.

15.6 ENZYMES FOR PEPTIDE TERMINAL MODIFICATION

Enzymatic modification of peptides to prepare them for coupling reactions, such as C-terminal activation, provides apparent advantages over chemical methods. This includes higher specificity and efficiency, reduced requirement for protection of side chains, and absence of racemization. The terminal functionalization of peptides can also have significant effects on their biological properties and afford modified peptides for a broad array of biophysical studies and applications. Modification of the N-terminus can be attractive because it reduces degradation by aminopeptidases. Moreover, modified peptides can be used for diagnostic purposes or as labels [103]. Here we focus on enzyme-catalyzed reactions for (de)protection/activation of peptide termini that are relevant for chemoenzymatic peptide synthesis.

15.6.1 Subtilisins for C-Terminal Peptide Modification

Subtilisins can catalyze a variety of C-terminal modification reactions [11–13] (Figure 15.5). Quaedflieg and coworkers investigated the use of subtilisin Carlsberg in anhydrous solvents for converting Cbz-protected amino acids and Cbz dipeptides to methyl or ethyl esters. They also studied the conversion of peptide *t*-butyl esters, which are readily obtained by solid-phase peptide synthesis, to methyl, ethyl, or benzyl esters [11].

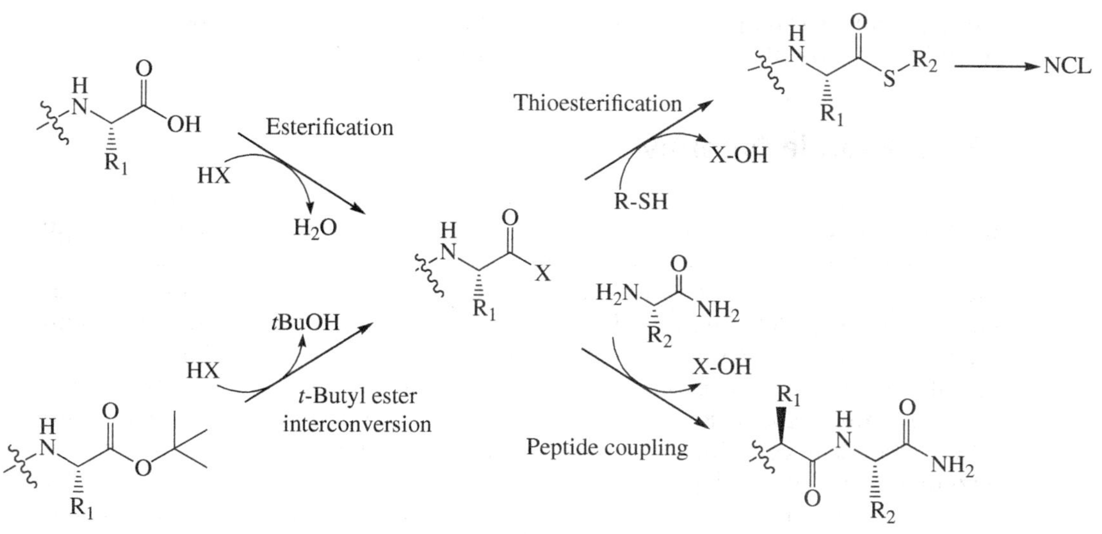

FIGURE 15.5
Applications of subtilisin in peptide modification.

The products were obtained in high yields and without racemization. Thus, subtilisin catalyzed methyl ester synthesis and transesterification reactions that yielded activated acyl donors that could be applied in further coupling reactions, including production of anti-inflammatory peptides (For-Met-Leu-Phe-OMe and Boc-Phe-Leu-Phe-Leu-Phe-OH). In a multistep synthesis scheme, subtilisin catalyzed the formation of activated methyl esters, the initial coupling reactions, transesterification of a *t*Bu ester to methyl esters, a second coupling, and the final hydrolysis to the desired product. A similar strategy resulted in the synthesis of a thermolysin tripeptide assay substrate, Cbz-Phe-Leu-OH, where subtilisin catalyzed initial coupling and later hydrolyzed the peptide to obtain the product [11]. Using neat organic solvent, cross-linked enzyme aggregates of subtilisin Carlsberg could be used for the preparation of peptides modified at their C-terminus with chromogenic and fluorogenic moieties. Such compounds are applied as substrates in enzymatic assays and need to be optically pure. The synthesis of various dipeptides with C-terminal arylamide groups was demonstrated, giving products in high yields and high enantiomeric purity, in particular for widely used *p*-nitroanilides and 7-amino-4-methylcoumarins [104].

15.6.2 C-Terminal Activation by Lipase

Nuijens *et al.* explored the use of subtilisin cross-linked enzyme aggregates and lipase for the activation of amino acids and peptides to obtain acyl donors for coupling reactions. This included amidation [104], and conversion to esters by attachment of the carboxamidomethyl (Cam) or trifluoroethyl (Tfe) group [11–13]. Interestingly, the esters could be formed and coupled to peptides by the same alcalase. Ester synthesis could also be catalyzed by lipase (CalB), and esterification and peptide bond formation could be catalyzed in the same reaction mixture. In these coupled reactions, the best results were obtained with a mixture of CalB and subtilisin.

Peptide thioesters are key intermediates in protein and bioconjugate synthesis, for example in native chemical ligation for peptide synthesis and in ligation and macrolactamization reactions. Facile enzymatic thioesterification of synthetically produced peptide esters that proceeds under mild conditions and without racemization would enhance the applicability of the native chemical ligation in peptide synthesis. Quaedflieg and Merkx applied subtilisin Carlsberg in neat organic solvent for the transesterification reaction of N-terminally protected peptide esters by a thiol to obtain the corresponding peptide thioester [105]. Thioesterification also proceeded under aqueous conditions using subtiligase [106], which also catalyzed the synthesis of aliphatic thioesters and peptide thioacids, albeit with some hydrolytic side activity [107].

15.6.3 Peptide Deformylase

The formyl group is a cost-efficient protective group for amine functionalities in peptide synthesis. However, chemical removal of formyl protective group requires acidic conditions that can lead to hydrolysis of peptide bonds. Enzymatic removal of the formyl group using peptide deformylases (PDFs) under mild conditions is an alternative [108]. PDFs are found predominantly in prokaryotes, where they remove the formyl group from the first methionine of the nascent peptide chain during ribosomal translation. They belong to the zinc metalloproteases, and more than 50 crystal structures are available. Recently, an improved aspartame process has been designed, where chemical coupling of the For-L-Asp and L-Phe-OMe is followed by the enzymatic resolution/deprotection step of the desired For-α-Asp-Phe-OMe (80%) isomer and its byproduct β-isomer (20%) using a PDF (Figure 15.6). With this enzymatic approach, the cheaper N-terminal protection strategy and coupling are used, and higher yields of aspartame are obtained [69].

15.6.4 Peptide Amidases for C-Terminal Modification

Peptide amidases specifically hydrolyze the C-terminal amide bond of peptide amides, a reaction of interest for the deprotection of the C-terminus of such peptide amides. Recently, the application of the peptide amidase from *Stenotrophomonas maltophilia* in peptide amidation was explored [109]. The enzyme can be recombinantly produced in *E. coli* and is rather stable in organic solvents. Peptide amidase from the flavedo of oranges catalyzes C-terminal amidation of di- , tri-, and tetrapeptides [110] and also the interconversion of peptide amides into methyl esters [111]. This enzyme has not been well characterized, but the *Stenotrophomonas* peptide amidase belongs to the amidase signature (AS) family superfamily, characterized by a large β-sheet core domain containing the active site residues and covered by α-helices [112]. The *Stenotrophomonas* enzyme was used for conversion of di- and tripeptide amides to methyl ester in neat organic solvent. The methyl ester (Cbz-Gly-Tyr-OMe) was subsequently used as an acyl donor and converted by a peptidase to Cbz-Gly-Tyr-Phe-NH_2 [113]. The complete absence of endopeptidase activity of peptide amidases might make the enzymes more attractive for application in peptide modification than subtilisins.

Peptide amidation using carboxypeptidases has also been successful, although the carboxypeptidases may lack the broad specificity observed for peptide amidases. To overcome this, a strategy combining enzymatic addition of a modified amino acid and a photochemical reaction was designed [114]. The authors used the transacylation ability of carboxypeptidase Y to prepare peptide conjugates with *N*-(2-nitrobenzoyl)

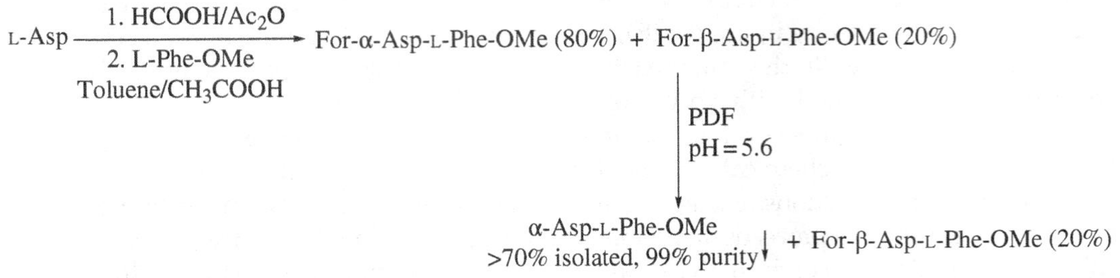

FIGURE 15.6
Use of peptide deformylase in aspartame synthesis. The cheap formyl group is used for protection and coupling proceeds chemically. After coupling, the formyl group of the desired product is selectively removed by peptide deformylase.

amido groups. Subjecting the peptides to ultraviolet light resulted in the removal of 2-nitrobenzyl protective ligand to give peptide amides in high yield.

15.6.5 Enzymes for N-Terminal Modification

Chymotrypsin and V8 peptidase, members of the trypsin family of serine proteases, as well as an engineered trypsin variant (D189K/K60E), were employed for the N-terminal introduction of 2-aminobenzoyl- and 3-(4-hydroxy-phenyl)-propionyl groups, and biotin in the form of substrate mimetics into peptides [115].

15.6.6 Enzymes for Peptide Cyclization

Naturally occurring bioactive cyclic peptides include bacterial microcins, fungal cyclosporins, θ-defensins, and retrocyclins of animal and plant cyclotides. The absence of a free C- and N-termini makes these peptides resistant to exopeptidase degradation. Furthermore, they are rigid, thermostable, and resistant to denaturing agents and may act as sturdy templates for epitope display or receptor interactions. Biological cyclic peptides are either genetically encoded with post-translational cyclization or formed through nonribosomal peptide synthesis and cyclization. It is no surprise that these molecules and their structural scaffolds are of particular interest for the pharmaceutical applications.

Enzymatic peptide cyclization can be catalyzed by thioesterase domains of nonribosomal peptide synthetase (NRPS). However, a thioester substrate anchored on an NRPS module is usually required for this reaction. As an exception, an isolated tyrocidine thioesterase domain (TycTE) expressed in *E. coli* was used for cyclization and dimerization of long peptides, with a P1 and P1′ specificity for D-Phe and Leu [116]. Other examples of enzyme-catalyzed cyclization of peptides *in vitro* have been reported [117]. The majority of enzymes considered need a recognition motif and are N- or C-terminus specific. For instance, staphylococcal sortase A, discussed earlier, needs a large recognition motif (LPXTG, X = any amino acid). After cleavage between Tyr and Gly, the enzyme catalyzes the transfer of the R-LPXT group to a nucleophilic acceptor peptide, which should have an N-terminal glycine. This way, macrocyclic peptides and glycopeptides could be formed, with cyclization efficiency dependent on chain length [33].

Another recently discovered enzyme, butelase-1 from *Clitoria ternatea* (butterfly pea), was used for the cyclization of linear peptides, including the human salivary antimicrobial peptide histatin. The enzyme required a C-terminal Asx-His-Val recognition motif, thereby releasing His-Val dipeptide and acting as an Asx ligase [118]. This C-terminal specificity and the lack of a recombinant expression system limit the applicability of this enzyme. Immobilized trypsin was used in an effective combination of macrolactamization with on-bead affinity purification in the synthesis of its cyclic inhibitor cyclotide MCoTI-I [119]. The reaction was performed by first binding the trypsin inhibitor, after which a shift to pH 3 gave elution of the folded cyclized protein with correctly positioned disulfide bonds. Cyclization occurs without C-terminal activation and accompanies the thio-zip reaction that drives correct folding of these natural cyclic peptides.

15.7 CONCLUSIONS

Success in chemoenzymatic peptide synthesis and peptide modification is critically dependent on the discovery of suitable enzyme–substrate combinations. Many classical proteases have used successfully used in peptide synthesis, and key factors are medium engineering and substrate engineering. The latter includes the selection of coupling sites for enzyme-mediated fragment condensation and the selection of leaving groups in kinetically controlled synthesis. Strongly activating leaving groups that allow the formation of high-energy intermediates that yield stable products in subsequent aminolysis reactions expand the application scope. Many reactions can be done in the presence of cosolvents to increase peptide solubility or in neat organic solvent to suppress peptide hydrolysis. This requires highly stable enzymes, which have been obtained by selection from extremophiles and by protein engineering. Neat organic solvents may change enzyme selectivity and broaden the range of reactions that can be performed. When working in aqueous conditions, the use of thiol-subtilisins is advantageous and allows efficient fragment coupling. An important trend is the development of highly specific proteases that, in combination with genetically or chemically introduced sequence motifs or tags in target proteins, can be applied for selective modification reactions.

REFERENCES

[1] Jakubke, H.-D., Kuhl, P., Konnecke, A. (1985). Basis of protease-catalyzed peptide bond formation. *Angew. Chem. Int. Ed. Engl.*, **24**, 85–93.
[2] Schellenberger, V., Jakubke, H.-D. (1991). Protease-catalyzed kinetically controlled peptide synthesis. *Angew. Chem. Int. Ed. Engl.*, **30**, 1437–1449.
[3] Schellenberger, V., Schellenberger, U., Mitin, Y. V., Jakubke, H. D. (1990). Characterization of the S′-subsite specificity of bovine pancreatic alpha-chymotrypsin via acyl transfer to added nucleophiles. *Eur. J. Biochem.*, **187**, 163–167.
[4] Morihara, K., Oka, T. (1977). Alpha-chymotrypsin as the catalyst for peptide synthesis. *Biochem. J.*, **163**, 531–542.
[5] Schellenberger, V., Schellenberger, U., Jakubke, H.-D., Hansicke, A., Bienert, M., Krause, E. (1990). Chymotrypsin-catalyzed fragment coupling synthesis of D-Phe(6)-GnRH. *Tetrahedron Lett.*, **31**, 7305–7306.
[6] Nishino, N., Xu, M., Mihara, H., Fujimoto, T. (1992). Use of hexafluoroisopropyl alcohol in tryptic condensation for partially protected precursor of α-melanocyte stimulating hormone. *Tetrahedron Lett.*, **33**, 3137–3140.
[7] Schellenberger, V., Braune, K., Hofmann, H. J., Jakubke, H.-D. (1991). The specificity of chymotrypsin: a statistical analysis of hydrolysis data. *Eur. J. Biochem.*, **199**, 623–636.
[8] Qin, X., Xie, W., Tian, S., Cai, J., Yuan, H., Yu, Z., Butterfoss, G. L., Khuong, A. C., Gross, R. A. (2013). Enzyme-triggered hydrogelation via self-assembly of alternating peptides. *Chem. Commun.*, **49**, 4839–4841.
[9] Morihara, K. (1990). Enzymatic semisynthesis of human insulin: an update. *J. Mol. Recognit.*, **3**, 181–186.
[10] Bordusa, F., Ullmann, D., Elsner, C., Jakubke, H.-D. (1997). Substrate mimetic mediated peptide synthesis: an irreversible ligation strategy that is independent of substrate specificity. *Angew. Chem. Int. Ed. Engl.*, **36**, 2473–2475.
[11] Nuijens, T., Cusan, C., van Dooren, T. J. G. M., Moody, H. M., Rijkers, D. T., Liskamp, R. M., Quaedflieg, P. J. L. M. (2010). Fully enzymatic peptide synthesis using C-terminal *tert*-butyl ester interconversion. *Adv. Synth. Catal.*, **352**, 2399–2404.
[12] Nuijens, T., Schepers, A. H. M., Cusan, C., Kruijtzer, J. A. W., Rijkers, D. T., Liskamp, R. M., Quaedflieg, P. J. L. M. (2013). Enzymatic fragment condensation of side chain-protected peptides using subtilisin A in anhydrous organic solvents: a general strategy for industrial peptide synthesis. *Adv. Synth. Catal.*, **355**, 287–293.
[13] Nuijens, T., Piva, E., Kruijtzer, J. A. W., Rijkers D. T. S., Liskamp, R. M., Quaedflieg, P. J. L. M. (2011). Fully enzymatic N→C-directed peptide synthesis using C-terminal peptide α-carboxamide to ester interconversion. *Adv. Synth. Catal.*, **353**, 1039–1044.
[14] Wong, C., Schuster, M., Wang, P., Searst, P. (1993). Enzymatic synthesis of N- and 0-linked glycopeptides. *J. Am. Chem. Soc.*, **115**, 5893–5901.
[15] Abrahmsén, L., Tom, J., Burnier, J., Butcher, K. A., Kossiakoff, A., Wells, J. A. (1991). Engineering subtilisin and its substrates for efficient ligation of peptide bonds in aqueous solution. *Biochemistry*, **30**, 4151–4159.
[16] Fruton, J. S. (1982). Proteinase-catalyzed synthesis of peptide bonds. *Adv. Enzymol. Relat. Areas Mol. Biol.*, **53**, 239–306.
[17] Ageitos, J. M., Baker, P. J., Sugahara, M., Numata, K. (2013). Proteinase K-catalyzed synthesis of linear and star oligo(L-phenylalanine) conjugates. *Biomacromolecules*, **14**, 3635–3642.

[18] Wehofsky, N., Wissmann, J., Alisch, M., Bordusa, F. (2000). Engineering of substrate mimetics as novel-type substrates for glutamic acid-specific endopeptidases: design, synthesis, and application. *Biochim. Biophys. Acta*, **1479**, 114–122.
[19] Wehofsky, N., Bordusa, F. (1999). Programming of enzyme specificity by substrate mimetics: investigations on the Glu-specific V8 protease reveals a novel general principle of biocatalysis. *FEBS Lett.*, **443**, 220–224.
[20] Mortensen, U. H., Stennicke, H. R., Raaschou-Nielsen, M., Breddam, K. (1994). Mechanistic study on carboxypeptidase Y-catalyzed transacylation reactions. Mutationally altered enzymes for peptide synthesis. *J. Am. Chem. Soc.*, **116**, 34–41.
[21] Yamamoto, Y., Usuki, H., Iwabuchi, M., Hatanaka, T. (2010). Prolyl aminopeptidase from *Streptomyces thermoluteus* subsp. *fuscus* strain NBRC14270 and synthesis of proline-containing peptides by its S144C variant. *Appl. Environ. Microbiol.*, **76**, 6180–6185.
[22] Schwab, L. W., Kloosterman, W. M. J., Konieczny, J., Loos, K. (2012). Papain catalyzed (co) oligomerization of α-amino acids. *Polymers*, **4**, 710–740.
[23] Hänsler, M., Ullmann, G., Jakubke, H. D. (1995). The application of papain, ficin and clostripain in kinetically controlled peptide synthesis in frozen aqueous solutions. *J. Pept. Sci.*, **1**, 283–287.
[24] Barbas, C. F., Wong, C.-H. (1987). Papain catalysed peptide synthesis: control of amidase activity and the introduction of unusual amino acids. *Chem. Commun.*, Issue 8, 533–534.
[25] de Beer, R. J., Zarzycka, B., Mariman, M., Amatdjais-Groenen, H. I., Mulders, M. J., Quaedflieg, P. J. L. M., van Delft, F. L., Nabuurs, S. B., Rutjes, F. P. (2012). Papain-specific activating esters in aqueous dipeptide synthesis. *ChemBioChem*, **13**, 1319–1326.
[26] Didziapetris Didziapetris, R., Drabnig, B., Schellenberger, V., Jakubke, H. D., Svedas, V. (1991). Penicillin acylase-catalyzed protection and deprotection of amino groups as a promising approach in enzymatic peptide synthesis. *FEBS Lett.*, **287**, 31–33.
[27] Clapés, P., Torres, J., Adlercreutz, P. (1995). Enzymatic peptide synthesis in low water content systems: preparative enzymatic synthesis of [Leu]- and [Met]-enkephalin derivatives. *Bioorg. Med. Chem.*, **3**, 245–255.
[28] Siezen, R. J., Leunissen, J. A. M. (1997). Subtilases: the superfamily of subtilisin-like serine proteases. *Protein Sci.*, **6**, 501–523.
[29] Andre, M., Kühl, B., Brenner-Weiss, G., Syldatk, C., Rudat, J. (2014). Cationic heterooligopeptides by ficain-catalyzed co-oligomerization of lysine and methionine ethylesters. *J. Pept. Sci.*, **20**, 625–629.
[30] Günther, R., Stein, A., Bordusa, F. (2000). Investigations on the enzyme specificity of clostripain: a new efficient biocatalyst for the synthesis of peptide isosteres. *J. Org. Chem.*, **65**, 1672–1679.
[31] Ruth, D. M., McMahon, G., O'Fágáin, C. (2006). Peptide synthesis by recombinant *Fasciola hepatica* cathepsin L1. *Biochimie*, **88**, 117–120.
[32] Ritzefeld, M. (2014). Sortagging: a robust and efficient chemoenzymatic ligation strategy. *Chemistry*, **20**, 8516–8529.
[33] Wu, Z., Guo, X., Guo, Z. (2011). Sortase A-catalyzed peptide cyclization for the synthesis of macrocyclic peptides and glycopeptides. *Chem. Commun.*, **47**, 16–19.
[34] Imaoka, Y., Kawamoto, T., Ueda, M., Tanaka, A. (1994). Peptide synthesis with halophenylalanines by thermolysin. *Appl. Microbiol. Biotechnol.*, **40**, 653–656.
[35] Kusano, M., Yasukawa, K., Inouye, K. (2010). Synthesis of *N*-carbobenzoxy-l-aspartyl-l-phenylalanine methyl ester catalyzed by thermolysin variants with improved activity. *Enzyme Microb. Technol.*, **46**, 320–325.
[36] Kühn, D., Dürrschmidt, P., Mansfeld, J., Ulbrich-Hofmann, R. (2002). Boilysin and thermolysin in dipeptide synthesis: a comparative study. *Biotechnol. Appl. Biochem.*, **36**, 71–76.
[37] Rival, S., Saulnier, J., Wallach, J. (2000). On the mechanism of action of pseudolysin: kinetic study of the enzymatic condensation of Z-Ala with Phe-NH_2. *Biocatal. Biotransform.*, **17**, 417–429.
[38] Kunugi, S., Koyasu, A., Takahashi, S., Oda, K. (1997). Peptide condensation activity of a neutral protease from *Vibrio* sp. T1800 (Vimelysin). *Biotechnol. Bioeng.*, **53**, 387–390.
[39] Vértesi, A., Simon, L. M. (1998). Carboxypeptidase A-catalyzed dipeptide synthesis in organic media. *J. Biotechnol.*, **66**, 75–82.
[40] Baek, D. H., Song, J. J., Kwon, S. J., Park, C., Jung, C. M., Sung, M. H. (2004). Characteristics of a new enantioselective thermostable dipeptidase from *Brevibacillus borstelensis* BCS-1 and its application to synthesis of a D-amino-acid-containing dipeptide. *Appl. Environ. Microbiol.*, **70**, 1570–1575.
[41] Bemquerer, M. P., Adlercreutz, P., Tominaga, M. (1994). Pepsin-catalyzed peptide synthesis in organic media: studies with free and immobilized enzyme. *Int. J. Pept. Protein Res.*, **44**, 448–456.

[42] Malak, C. A., Filippova, I. Y., Lysogorskaya, E. N., Anisimova, V. V., Lavrenova, G. I., Stepanov, V. M. (1992). Pepsin as a catalyst of peptide synthesis. Enzyme co-precipitation with emerging peptide products. *Int. J. Pept. Protein Res.*, **39**, 443–449.
[43] Malak, C. A. (1999). Pepsin as a catalyst for peptide synthesis: formation of peptide bonds not typical for pepsin substrate specificity. *J. Pept. Res.*, **53**, 606–610.
[44] Malak, C. A. (1992). Calf chymosin as catalyst of peptide synthesis. *Biochem. J.*, **288**, 941–943.
[45] Marnett, A. B., Craik, C. S. (2005). Papa's got a brand new tag: advances in identification of proteases and their substrates. *Trends Biotechnol.*, **23**, 59–64.
[46] Rawlings, N. D., Barrett, A. J., Bateman, A. (2010). MEROPS: the peptidase database. *Nucleic Acids Res.*, **38**, D227–D233.
[47] Rawlings, N. D., Salvesen, G. (2013). *Handbook of Proteolytic Enzymes* (Academic Press: London/Boston).
[48] Suresh, C. G., Pundle, A. V., SivaRaman, H., Rao, K. N., Brannigan, J. A., McVey, C. E., Verma, C. S., Dauter, Z., Dodson, E. J., Dodson, G. G. (1999). Penicillin V acylase crystal structure reveals new Ntn-hydrolase family members. *Nat. Struct. Biol.*, **6**, 5–7.
[49] Lipscomb, W. N., Sträter, N. (1996). Recent advances in zinc enzymology. *Chem. Rev.*, **96**, 2375–2433.
[50] Bergmann, M., Fraenkel-Conrat, H. (1937). The role of specificity in the enzymatic synthesis of proteins: syntheses with intracellular enzymes. *J. Biol. Chem.*, **119**, 707–720.
[51] Cerovský, V., Bordusa, F. (2000). Protease-catalyzed fragment condensation via substrate mimetic strategy: a useful combination of solid-phase peptide synthesis with enzymatic methods. *J. Pept. Res.*, **55**, 325–329.
[52] Günther, R., Thust, S., Hofmann, H. J., Bordusa, F. (2000). Trypsin-specific acyl-4-guanidinophenyl esters for α-chymotrypsin-catalysed reactions. *Eur. J. Biochem.*, **267**, 3496–3501.
[53] Grünberg, R., Domgall, I., Günther, R., Rall, K., Hofmann, H. J., Bordusa, F. (2000). Peptide bond formation mediated by substrate mimetics. *Eur. J. Biochem.*, **267**, 7024–7030.
[54] Narai-Kanayama, A., Hanaishi, T., Aso, K. (2012). α-Chymotrypsin-catalyzed synthesis of poly-l-cysteine in a frozen aqueous solution. *J. Biotechnol.*, **157**, 428–436.
[55] Kishimoto, T., Kondo, J., Takai-Igarashi, T., Tanaka, H. (2011). Accurate mass comparison coupled with two endopeptidases enables identification of protein termini. *Proteomics*, **11**, 485–489.
[56] Morihara, K., Oka, T., Tsuzuki, H. (1980). *Achromobacter* protease I-catalyzed conversion of porcine insulin into human insulin. *Biochem. Biophys. Res. Commun.*, **92**, 396–402.
[57] Koskinen, A. M. P., Klibanov, A. M. (1996). *Enzymatic Reactions in Organic Media* (Springer: Dordrecht, the Netherlands).
[58] Margolin, A. L., Tai, D., Klibanov, A. M. (1987). Incorporation of D-amino acids into peptides via enzymatic condensation in organic solvents. *J. Am. Chem. Soc.*, **109**, 7885–7887.
[59] Christensen, U. (1994). Kinetic characterization of carboxypeptidase-Y-catalyzed peptide semisynthesis prediction of yields. *Amino Acids*, **6**, 177–187.
[60] Heck, T., Limbach, M., Geueke, B., Zacharias, M., Gardiner, J., Kohler, H. P., Seebach, D. (2007). Enzyme-catalyzed formation of beta-peptides: beta-peptidyl aminopeptidases BapA and DmpA acting as beta-peptide-synthesizing enzymes. *Chem. Biodivers.*, **4**, 2016–2030.
[61] Arima, J., Morimoto, M., Usuki, H., Mori, N., Hatanaka, T. (2010). Beta-alanyl peptide synthesis by *Streptomyces* S9 aminopeptidase. *J. Biotechnol.*, **147**, 52–58.
[62] Wehofsky, N., Pech, A., Liebscher, S., Schmidt, S., Komeda, H., Asano, Y., Bordusa, F. (2008). D-amino acid specific proteases and native all-L-proteins: a convenient combination for semisynthesis. *Angew. Chem. Int. Ed. Engl.*, **47**, 5456–5460.
[63] Günther, R., Bordusa, F. (2000). Protease catalysis mediated by a substrate mimetic: a novel enzymatic approach to the synthesis of carboxylic acid amides. *Chemistry*, **6**, 463–467.
[64] Li, Y. M., Li, Y. T., Pan, M., Kong, X. Q., Huang, Y. C., Hong, Z. Y., Liu, L. (2014). Irreversible site-specific hydrazinolysis of proteins by use of sortase. *Angew. Chem. Int. Ed. Engl.*, **17**, 2198–2202.
[65] Chen, I., Dorr, B. M., Liu, D. R. (2011). A general strategy for the evolution of bond-forming enzymes using yeast display. *Proc. Natl. Acad. Sci. U. S. A.*, **108**, 11399–11404.
[66] Oka, T., Morihara, K. (1980). Peptide bond synthesis catalyzed by thermolysin. *J. Biochem.*, **88**, 807–813.
[67] Wayne, S. I., Fruton, J. S. (1983). Thermolysin-catalyzed peptide bond synthesis. *Proc. Natl. Acad. Sci. U. S. A.*, **80**, 3241–3244.
[68] Steinke, D., Schwarz, A., Wandrey, C., Kula, M. R. (1991). Formation of peptide bonds by carboxypeptidase *c* from orange leaves. *Enzyme Microb. Technol.*, **13**, 262–266.
[69] Quaedflieg, P. J. L. M., Sonke, T., Wagner, A. F. V. (2003). Synthesis and recovery of aspartame involving enzymatic deformylation step. US Patent US006617127B2.

[70] Bommarius, A. S., Riebel, B. R. (2004). *Biocatalysis: Fundamentals and Applications* (Wiley-VCH: Weinheim, Germany).

[71] Williams, R. J., Smith, A. M., Collins, R., Hodson, N., Das, A. K., Ulijn, R. V. (2009). Enzyme-assisted self-assembly under thermodynamic control. *Nat. Nanotechnol.*, **4**, 19–24.

[72] Nalluri, S. K. M., Ulijn, R. V. (2013). Discovery of energy transfer nanostructures using gelation-driven dynamic combinatorial libraries. *Chem. Sci.*, **4**, 3699–3705.

[73] Gupta, R., Beg, Q. K., Lorenz, P. (2002). Bacterial alkaline proteases: molecular approaches and industrial applications. *Appl. Microbiol. Biotechnol.*, **59**, 15–32.

[74] Gabor, E., Niehaus, F., Aehle, W., Eck, J. (2012). Zooming in on metagenomics: molecular microdiversity of Subtilisin Carlsberg in soil. *J. Mol. Biol.*, **418**, 16–20.

[75] Wilson, S.-A., Daniel, R. M., Peek, K. (1994). Peptide synthesis with a proteinase from the extremely thermophilic organism *Thermus* Rt41A. *Biotechnol. Bioeng.*, **44**, 337–346.

[76] Toplak, A., Nuijens, T., Quaedflieg, P. J. L. M., Wu, B., Janssen, D. B. (2015). Peptide synthesis in neat organic solvents with novel thermostable proteases. *Enzyme Microb. Technol.*, **73–74**, 20–28.

[77] Tsuchiyama, S., Doukyu, N., Yasuda, M., Ishimi, K., Ogino, H. (2007). Peptide synthesis of aspartame precursor using organic-solvent-stable PST-01 protease in monophasic aqueous-organic solvent systems. *Biotechnol. Progr.*, **23**, 820–823.

[78] Sun, H., He, B., Xu, J., Wu, B., Ouyang, P. (2011). Efficient chemo-enzymatic synthesis of endomorphin-1 using organic solvent stable proteases to green the synthesis of the peptide. *Green Chem.*, **13**, 1680–1685.

[79] Xu, J., Sun, H., He, X., Bai, Z., He, B. (2013). Highly efficient synthesis of endomorphin-2 under thermodynamic control catalyzed by organic solvent stable proteases with in situ product removal. *Bioresour. Technol.*, **129**, 663–666.

[80] Ryu, K., Kim, J., Dordick, J. S. (1994). Catalytic properties and potential of an extracellular protease from an extreme halophile. *Enzyme Microb. Technol.*, **16**, 266–275.

[81] Almog, O., Gallagher, D. T., Ladner, J. E., Strausberg, S., Alexander, P., Bryan, P., Gilliland, G. L. (2002). Structural basis of thermostability: analysis of stabilizing mutations in subtilisin BPN′. *J. Biol. Chem.*, **277**, 27553–27558.

[82] Zhao, H., Arnold, F. H. (1999). Directed evolution converts subtilisin E into a functional equivalent of thermitase. *Protein Eng.*, **12**, 47–53.

[83] Chen, K. Q., Robinson, A. C., Van Dam, M. E., Martinez, P., Economou, C., Arnold, F. H. (1991). Enzyme engineering for nonaqueous solvents. II. Additive effects of mutations on the stability and activity of subtilisin E in polar organic media. *Biotechnol. Prog.*, **7**, 125–129.

[84] Bryan, P. N. (2000). Protein engineering of subtilisin. *Biochim. Biophys. Acta*, **1543**, 203–222.

[85] Wong, C. H., Chen, S. T., Hennen, W. J., Bibbs, J. A., Wang, Y. F., Liu, J. L. C., Pantoliano, M. W., Whitlow, M., Bryan, P. N. (1990). Enzymes in organic synthesis: use of subtilisin and a highly stable mutant derived from multiple site-specific mutations. *J. Am. Chem. Soc.*, **112**, 945–953.

[86] Zhong, Z., Liu, L. J. C., Dinterman, L. M., Finkelman, M. A. J., Mueller, W. T., Rollence, M. L., Whitlow, M., Wong, C. H. (1991). Engineering subtilisin for reaction in dimethylformamide. *J. Am. Chem. Soc.*, **113**, 683–684.

[87] Sears, P., Schuster, M., Wang, P., Witte, K., Wong, C.-H. (1994). Engineering subtilisin for peptide coupling: studies on the effects of counterions and site-specific modifications on the stability and specificity of the enzyme. *J. Am. Chem. Soc.*, **116**, 6521–6530.

[88] Van den Burg, B., Vriend, G., Veltman, O. R., Venema, G., Eijsink, V. G. (1998). Engineering an enzyme to resist boiling. *Proc. Natl. Acad. Sci. U. S. A.*, **95**, 2056–2060.

[89] Nakatsuka, T., Sasaki, T., Kaiser, E. (1987). Peptide segment coupling catalyzed by the semisynthetic enzyme thiolsubtilisin. *J. Am. Chem. Soc.*, **109**, 3808–3810.

[90] Wu, Z.-P., Hilvert, D. (1989). Conversion of a protease into an acyl transferase: selenolsubtilisin. *J. Am. Chem. Soc.*, **111**, 4513–4514.

[91] Braisted, A. C., Judice, J. K., Wells, J. A. (1997). Synthesis of proteins by subtiligase. *Methods Enzymol.*, **289**, 298–313.

[92] Jackson, D. Y., Burnier, J., Quan, C., Stanley, M., Tom, J., Wells, J. A. (1994). A designed peptide ligase for total synthesis of ribonuclease A with unnatural catalytic residues. *Science*, **266**, 243–247.

[93] Chang, T. K., Jackson, D. Y., Burniert, J. P., Wells, J. A. (1994). Subtiligase: a tool for semisynthesis of proteins. *Proc. Natl. Acad. Sci. U. S. A.*, **91**, 12544–12548.

[94] Atwell, S., Wells, J. A. (1999). Selection for improved subtiligases by phage display. *Proc. Natl. Acad. Sci. U. S. A.*, **96**, 9497–9502.

[95] Wells, J. A., Mahrus, S. (2008). Specific N-terminal labelling of peptides and proteins in complex mixtures. Canadian Patent Application WO2008/092030A2.

[96] Jackson, D. Y., Burnier, J. P., Wells, J. A. (1995). Enzymatic cyclization of linear peptide esters using subtiligase. *J. Am. Chem. Soc.*, **117**, 819–820.

[97] Liebscher, S., Schöpfel, M., Aumüller, T., Sharkhuukhen, A., Pech, A., Höss, E., Parthier, C., Jahreis, G., Stubbs, M. T., Bordusa, F. (2014). N-terminal protein modification by substrate-activated reverse proteolysis. *Angew. Chem. Int. Ed. Engl.*, **53**, 3024–3028.

[98] Liebscher, S., Kornberger, P., Fink, G., Trost-Gross, E. M., Höss, E., Skerra, A., Bordusa, F. (2014). Derivatization of antibody Fab fragments: a designer enzyme for native protein modification. *ChemBioChem*, **15**, 1096–1100.

[99] Elliott, R. J., Bennet, A. J., Braun, C. A., MacLeod, A. M., Borgford, T. J. (2000). Active-site variants of *Streptomyces griseus* protease B with peptide-ligation activity. *Chem. Biol.*, **7**, 163–171.

[100] Joe, K., Borgford, T. J., Bennet, A. J. (2004). Generation of a thermostable and denaturant-resistant peptide ligase. *Biochemistry*, **43**, 7672–7677.

[101] Kennan, A. J., Haridas, V., Severin, K., Lee, D. H., Ghadiri, M. R. (2001). A *de novo* designed peptide ligase: a mechanistic investigation. *J. Am. Chem. Soc.*, **123**, 1797–1803.

[102] Gordon, S. R., Stanley, E. J., Wolf, S., Toland, A., Wu, S. J., Hadidi, D., Mills, J. H., Baker, D., Pultz, I. S., Siegel, J. B. (2012). Computational design of an α-gliadin peptidase. *J. Am. Chem. Soc.*, **134**, 20513–20520.

[103] Sunbul, M., Yin, J. (2009). Site specific protein labeling by enzymatic posttranslational modification. *Org. Biomol. Chem.*, **7**, 3361–3371.

[104] Nuijens, T., Piva, E., Kruijtzer, J. A. W., Rijkers D. T. S., Liskamp, R. M. J., Quaedflieg, P. J. L. M. (2012). Enzymatic C-terminal amidation of amino acids and peptides. *Tetrahedron Lett.*, **53**, 3777–3779.

[105] Quaedflieg, P. J. L. M., Merkx, N. S. M. (2009). Chemo-enzymatic synthesis of a C-terminal thioester of an amino acid or a peptide. International Patent Application WO2009/047354A1.

[106] Tan, X.-H., Yang, R., Wirjo, A., Liu, C.-F. (2008). Subtiligase as a hydrothiolase for the synthesis of peptide thioacids. *Tetrahedron Lett.*, **49**, 2891–2894.

[107] Tan, X.-H., Wirjo, A., Liu, C.-F. (2007). An enzymatic approach to the synthesis of peptide thioesters: mechanism and scope. *ChemBioChem*, **8**, 1512–1515.

[108] Di Toma, C., Sonke, T., Quaedflieg, P. J. L. M., Volker Wagner, A. F., Janssen, D. B. (2013). Purification and use of *E. coli* peptide deformylase for peptide deprotection in chemoenzymatic peptide synthesis. *Protein Expr. Purif.*, **89**, 73–79.

[109] Cerovsky, V., Kula, M.-R. (2001). Studies on peptide amidase-catalysed C-terminal peptide amidation in organic media with respect to its substrate specificity. *Biotechnol. Appl. Biochem.*, **33**, 183–187.

[110] Cerovsky, V., Kula, M.-R. (1998). C-Terminal peptide amidation catalyzed by orange flavedo peptide amidase. *Angew. Chem. Int. Ed. Engl.*, **37**, 1885–1887.

[111] Quaedflieg, P. J. L. M., Sonke, T., Verzijl, G., Wiertz, R. (2009). Enzymatic conversion of oligopeptide amides to oligopeptide alkylesters. US Patent Application Publication US20090298118A1.

[112] Labahn, J., Neumann, S., Büldt, G., Kula, M. R., Granzin, J. (2002). An alternative mechanism for amidase signature enzymes. *J. Mol. Biol.*, **322**, 1053–1064.

[113] Arif, M. I., Toplak, A., Szymanski, W., Feringa, B. L., Nuijens, T., Quaedflieg, P. J. L. M., Wu, B., Janssen, D. B. (2014). One-step C -terminal deprotection and activation of peptides with peptide amidase from *Stenotrophomonas maltophilia* in neat organic solvent. *Adv. Synth. Catal.*, **356**, 2197–2202.

[114] Henriksen, D., Breddam, K., Moller, J., Buchardt, O. (1992). Peptide amidation by chemical protein engineering. A combination of enzymatic and photochemical synthesis. *J. Am. Chem. Soc.*, **114**, 1876–1877.

[115] Bordusa, F., Jakubke, H.-D. (2004). Method for the selective modification of peptides and proteins. US Patent Application Publication US20040077037A1.

[116] Lin, H., Thayer, D. A., Wong, C.-H., Walsh, C. T. (2004). Macrolactamization of glycosylated peptide thioesters by the thioesterase domain of tyrocidine synthetase. *Chem. Biol.*, **11**, 1635–1642.

[117] Bordusa, F. (2001). Enzymes for peptide cyclization. *ChemBioChem*, **2**, 405–409.

[118] Nguyen, G. K., Wang, S., Qiu, Y., Hemu, X., Lian, Y., Tam, J. P. (2014). Butelase 1 is an Asx-specific ligase enabling peptide macrocyclization and synthesis. *Nat. Chem. Biol.*, **10**, 732–738.

[119] Thongyoo, P., Roqué-Rosell, N., Leatherbarrow, R. J., Tate, E. W. (2008). Chemical and biomimetic total syntheses of natural and engineered MCoTI cyclotides. *Org. Biomol. Chem.*, **6**, 1462–1470.

CHAPTER 16

Biocatalysis for Drug Discovery and Development

Youyun Liang[1], Mingzi M. Zhang[1], Ee Lui Ang[1], and Huimin Zhao[1,2]

[1] Metabolic Engineering Research Laboratory, Science and Engineering Institutes, Agency for Science, Technology and Research, Singapore, Singapore

[2] Department of Chemical and Biomolecular Engineering, University of Illinois at Urbana-Champaign, Urbana, IL, USA

16.1 INTRODUCTION

The process of drug discovery and development is a lengthy one (Figure 16.1). The prediscovery process involves an in-depth study of the disease to acquire an understanding at genetic, protein, and cell levels [1]. Next, therapeutic targets at different levels, for example, a single protein or a single gene, are identified in the target identification phase. This then leads to the target validation phase where further studies are conducted to confirm involvement of the identified targets in the disease. The next phase of the process, the drug discovery phase, entails the discovery and evaluation of candidate molecules or leads that interact with the identified targets. This is often an iterative process that requires several rounds of lead compound optimization before subsequent preclinical and clinical trials. Following the trials, the safety and efficacy of the drug are then analyzed to determine if the new drug should be approved and mass-produced. Each phase in the entire process is faced with its unique challenges and obstacles. In the drug discovery and optimization phase, it is well known that the optical purity of the candidate compounds has major effects on the activity, efficacy, and safety of many compounds. However, synthesizing optically pure compounds remains a major challenge in organic synthesis. Following approval, the mass production of approved drugs is also nontrivial. The ability to synthesize optically pure compounds, efficiency of the synthetic routes, and environmental friendliness of synthetic processes are all major considerations in the production process.

With these considerations, it is becoming increasingly apparent that biocatalysis may hold the key to many of the existing challenges faced in the synthesis of pharmaceutical compounds. Biocatalysts, which encompasses enzymes and whole cells, are inherently highly selective and often offer better regio-, chemo- and stereo-control than chemical catalysts [2–4]. They are also non-toxic and work under mild aqueous conditions without the need for high temperatures and pressure [2–4]. Many biocatalysts are able to catalyze complex reactions in a single step, whereas many more steps involving repeated protection and deprotection may be required in an equivalent synthetic process. As biocatalysts generally have similar optimal reaction

Green Biocatalysis, First Edition. Edited by Ramesh N. Patel.

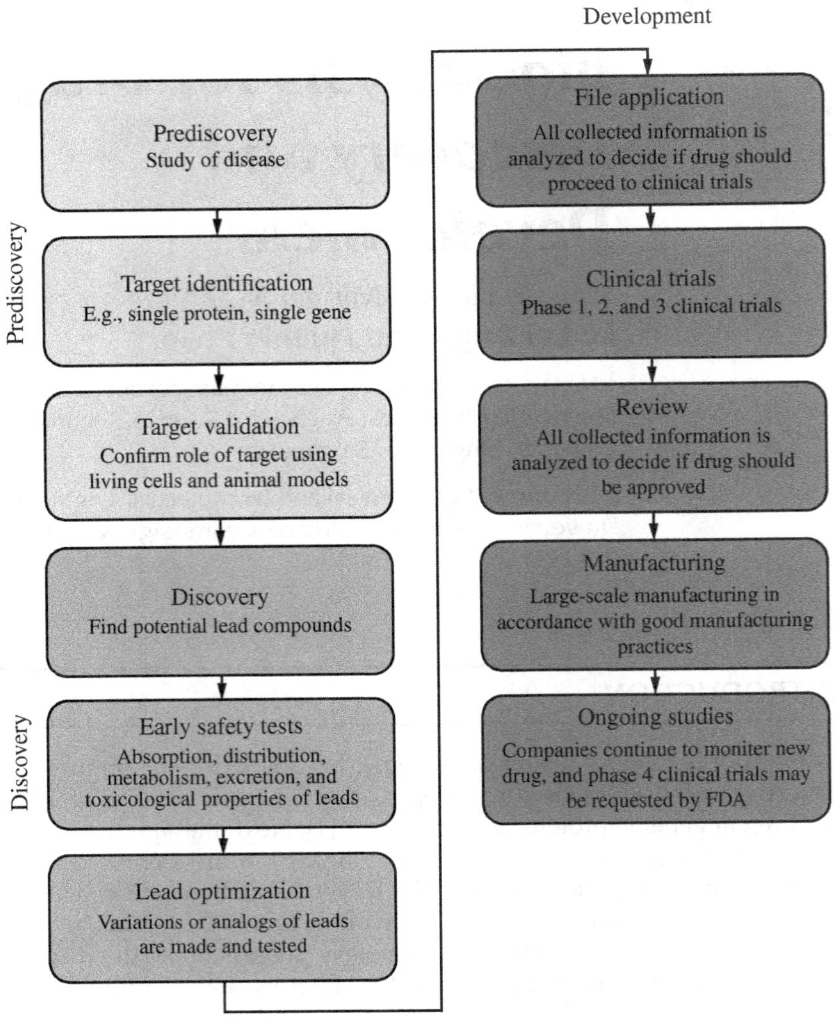

FIGURE 16.1

Phases in the drug discovery and development process. The entire process typically takes 10 to 15 years and costs approximately US$800 million to $1 billion. Biocatalysis can be employed in the discovery, lead optimization, and manufacturing phases of the workflow.

conditions, they offer the possibility of one-pot or tandem reactions, thus minimizing the cost of processing and purification.

This being said, biocatalysts have their drawbacks. Many enzymes, especially naturally occurring ones, exhibit limited robustness in typical production settings [2–4]. Hence, they are limited to a tight operating range in terms of temperature, pH, and solvent composition. Furthermore, the activities of enzymes in natural hosts are often tightly regulated through substrate or product inhibition [3]. Such inhibitions may negatively affect activities in production settings. Due to the high substrate specificities, there is also a limit on the range of reactions that can be catalyzed by existing enzymes. Furthermore, the scale-up of a biocatalytic process is non-trivial as other considerations such as choice of reactors and risk of contamination come into play [4]. Fortunately, biocatalysts exhibiting greater stabilities, activities, and substrate scopes have been developed through enzyme engineering techniques. Every so often, enzymes that catalyze new reactions are also being discovered. The increasing adoption of biocatalysis and the application of biocatalysts to the synthesis of blockbuster drugs such as sitagliptin and atorvastatin at production scales speak for the vast potential and commercial viability of biocatalytic processes.

Before switching to a biocatalytic process, several decisions have to be made. First, the reactions and their corresponding biocatalysts have to be identified. For some reactions, there might be several biocatalysts that could do the job. The relative substrate scopes, activities, and selectivities of the different biocatalysts have to be considered for the choice of biocatalyst. The context of the reaction is also a major consideration. There is the option of carrying out single enzymatic reactions, tandem reactions with several enzymes in one-pot, or whole-cell reactions. For reactions involving both synthetic and biocatalytic steps, there is also a choice of whether to keep these reactions separate or in one pot. With separate reactions, there is less concern about compatibility of reaction conditions and solvents. However, the additional processing and separation required would increase the cost of the process.

This chapter aims to provide chemists and biotechnologists with an overview of the state of biocatalysis and its application to different reactions in the drug discovery and development process. In Section **16.2**, biocatalysts that catalyze different chemical reactions will be introduced. In Section **16.3**, selected examples of multistep enzymatic reactions will be highlighted. Finally the chapter will be wrapped up with a discussion on current trends and future perspectives.

16.2 SINGLE ENZYMATIC REACTIONS

In this section, different wild-type and engineered enzymes capable of catalyzing different chemical reactions are described, with special attention placed on pharmaceutical applications since 2009. These examples are summarized in Table 16.1.

16.2.1 Hydrolytic Reaction

Hydrolases catalyze the hydrolytic cleavage of single bonds and are classified according to the functional groups that they act on. Some examples of hydrolases include lipases, esterases, phosphatases, sulfatases, and epoxide hydrolases. Hydrolases, in particular lipases, are widely employed in biotransformations for the chiral resolution of pharmaceutical products or their intermediates. The popularity of lipases stems from the robustness of these enzymes and their activities in both aqueous and organic solvents [42]. One of the more commonly adopted lipases is a commercially available preparation—lipase B from *Candida antarctica* (CALB).

One recent application of CALB was in the synthesis of (*S*,*S*)-reboxetine succinate, a candidate drug for the treatment of fibromyalgia [5]. (*S*,*S*)-Reboxetine succinate is the enantiomer of reboxetine mesylate, an antidepressant under the trade name Edronax™. A number of biological assays have demonstrated the superior activity of (*S*,*S*)-reboxetine succinate over the racemic reboxetine mesylate. A crucial step in the synthesis of (*S*,*S*)-reboxetine succinate was the regioselective protection of a single primary alcohol in the enantioenriched diol intermediate. However, initial attempts to carry out the regioselective protection with both trimethylsilyl and acetyl protecting groups resulted in suboptimal regioselectivities and yields. To overcome the poor regiocontrol, the CALB enzyme was evaluated for the selective monoacetylation of the enantioenriched diol intermediate (Scheme 1). With optimization of the reaction conditions and solvents, >99% regioselectivity and *in situ* yield were achieved with the CALB-catalyzed protection step. The process was subsequently scaled up to a 1 kg input with an overall yield of 70%. This represented a 58% improvement in overall yield with significant savings in production cost and reduction in waste. To further enhance the stability of CALB, the lipase was also immobilized on solid substrates [6]. Merck scientists immobilized CALB on five resins with different compositions and functional groups through both covalent and hydrophobic immobilization. The best preparation was then chosen by comparing the relative enzyme activities and stabilities on the different resins. Using the immobilized CALB, a continuous process for the dynamic kinetic resolution (DKR) of an azlactone using a ring-opening ethanolysis to produce (*S*)-γ-fluoroleucine ethyl ether was developed (Scheme 2). (*S*)-γ-Fluoroleucine ethyl ether is a chiral building block used in the manufacture of odanacatib, a selective cathepsin K inhibitor currently investigated in clinical trials for the treatment of osteoporosis. In this study, the best immobilized

TABLE 16.1 Summary of Enzyme-Catalyzed Reactions

Reaction	Biocatalyst	Target	Reaction Scheme	Production Parameters
Scheme 1: Hydrolytic reaction	CALB	(*S*,*S*)-Reboxetine succinate (for treatment of fibromyalgia) [5]	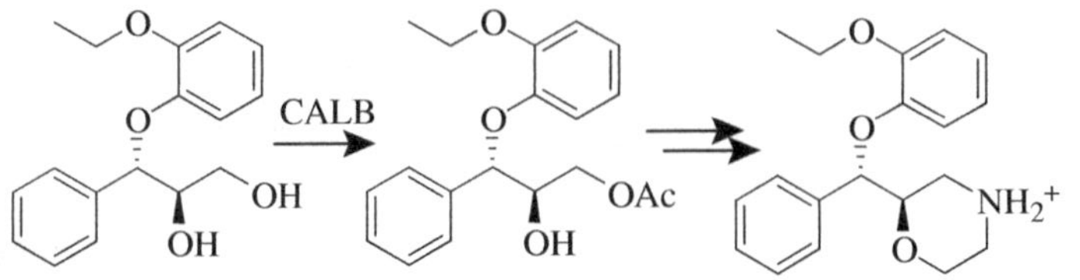 (*S*,*S*)-Reboxetine succinate	>99% yield >98% regioselectivity 1 kg scale
Scheme 2: Hydrolytic reaction	Immobilized CALB	(*S*)-γ-Fluoroleucine ethyl ether chiral starting material for odanacatib (for treatment of osteoporosis) [6]	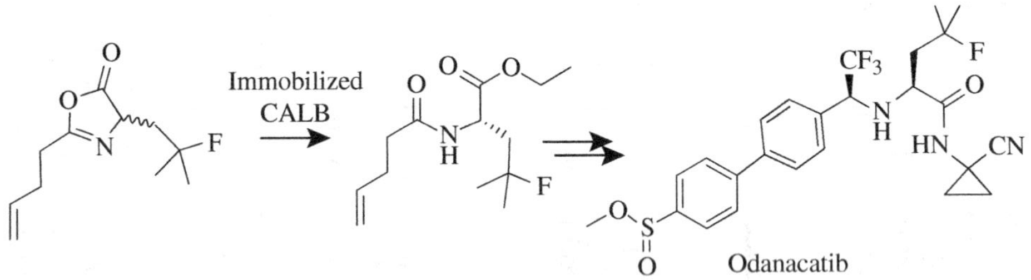 Odanacatib	Run in continuous mode with 100-fold higher substrate-to-catalyst ratio compared to batch process
Scheme 3: Alkene reduction	Panel of enoate reductases	(*R*)-3-Hydroxy-2-methylpropanoate (Roche ester) and derivatives [7]	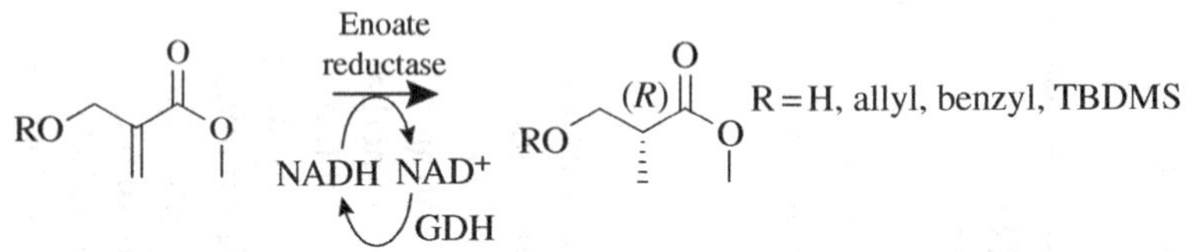 R = H, allyl, benzyl, TBDMS	>99% *ee*
Scheme 4: Alkene reduction	OYE2/3 (enoate reductase)	(*S*)-2-Ethoxy-3-(*p*-methoxyphenyl) propanoate, precursor of tesaglitazar (for treatment of type 2 diabetes) [8]	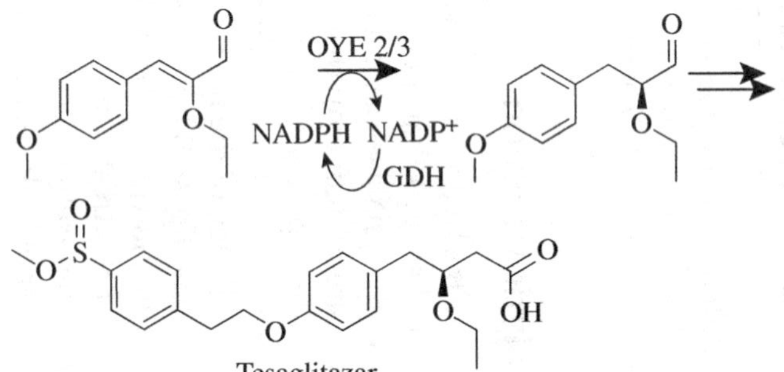 Tesaglitazar	94% yield >98% *ee* Gram scale

Scheme 5: Aldehyde and ketone reduction	Evolved *L. kefir* KRED	Montelukast (Singulair™, for treatment of asthma and allergies) [9]	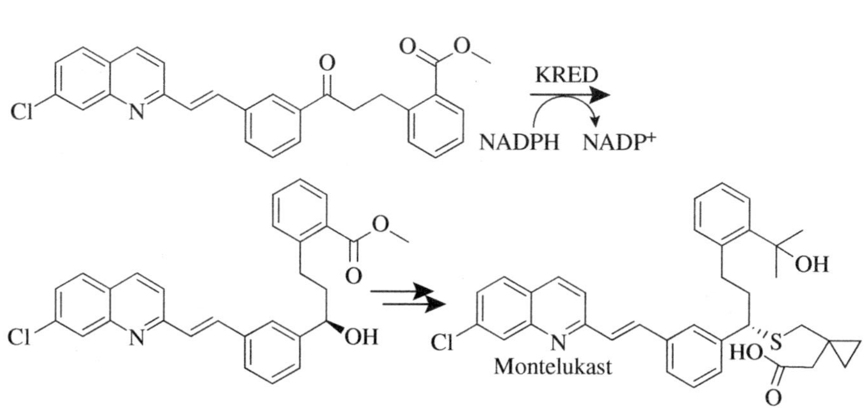	>99.9% *ee* 70% organic solvent >200 kg scale
Scheme 6: Aldehyde and ketone reduction	Evolved *L. kefir* KRED	Ezetimibe, the active ingredient used in Zetia™ (for lowering of low-density lipoprotein cholesterol) [10]	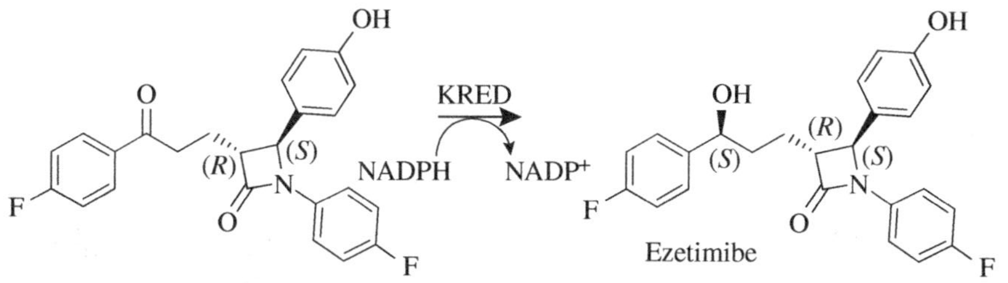	99% *ee* 90% conversion in 24 h
Scheme 7: C—H activation	Engineered $P450_{BM3}$	Artemisinic-11(*S*), 12-epoxide, intermediate in artemisinin synthesis (for treatment of malaria) [11]	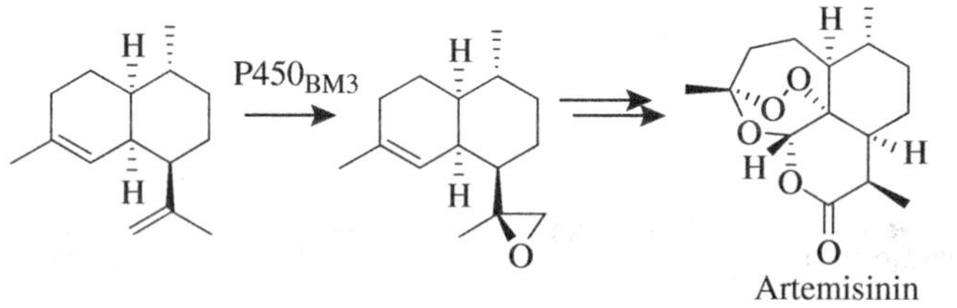	Titers for epoxide intermediate >250 mg/L
Scheme 8: C—H activation	Engineered $P450_{BM3}$	Activate organic molecules for deoxofluorination [12]	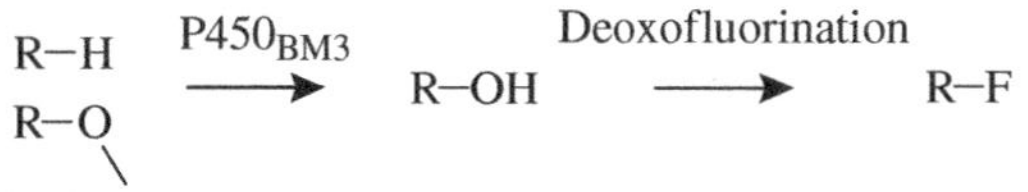	55–100% regioselectivity 100–300 mg preparative scales
Scheme 9: C—H activation	Variants of *Aspergillus niger* MAO-N	Telaprevir (for treatment of hepatitis C) [13, 14]	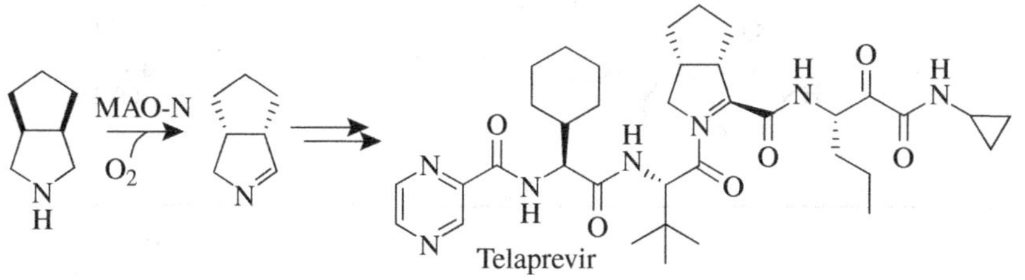	94% *ee* for MAO-N-catalyzed step 45% overall yield

(*Continued*)

TABLE 16.1 (*Continued*)

Reaction	Biocatalyst	Target	Reaction Scheme	Production Parameters
Scheme 10: C—H activation	Variants of *A. niger* MAO-N	Boceprevir (for treatment of hepatitis C) [15]	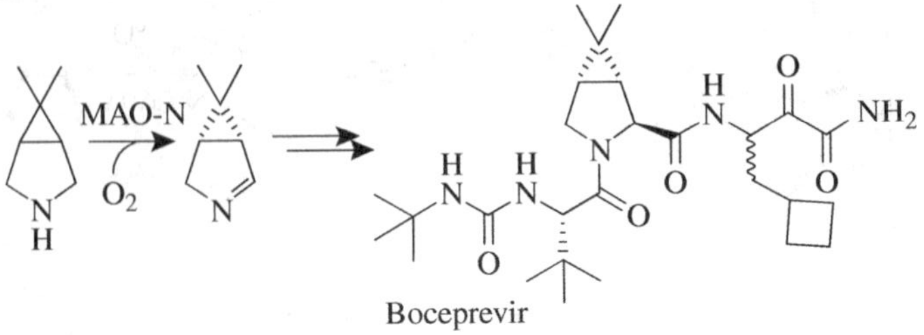	>99% *ee* 90% yield for MAO-N-catalyzed step
Scheme 11: Oxidative insertion across C—C bonds	Panel of BVMOs	β-Aminoalkyl acetates and β-amino alcohols [16, 17]	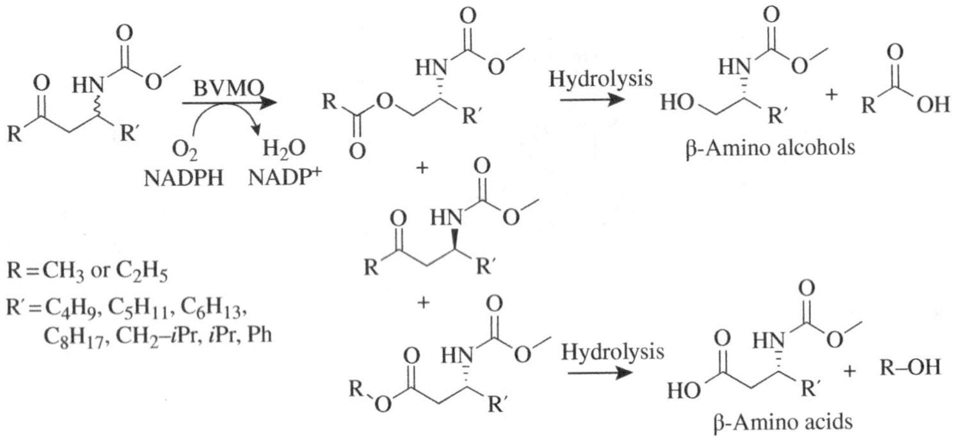	*E* > 200 for some reactions
Scheme 12: Oxidation of heteroatoms	Thermostable PAMO	Oxidative kinetic resolution of a variety of 2-aryl- and 2-alkylcyclohexanones [18]	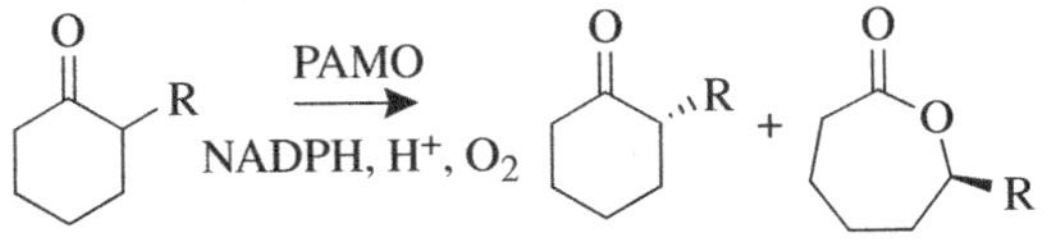	>86% *ee* >82% conversion
Scheme 13: Oxidation of heteroatoms	CHMO variants	Esomeprazole (for treatment of gastroesophageal reflux disease) [19]	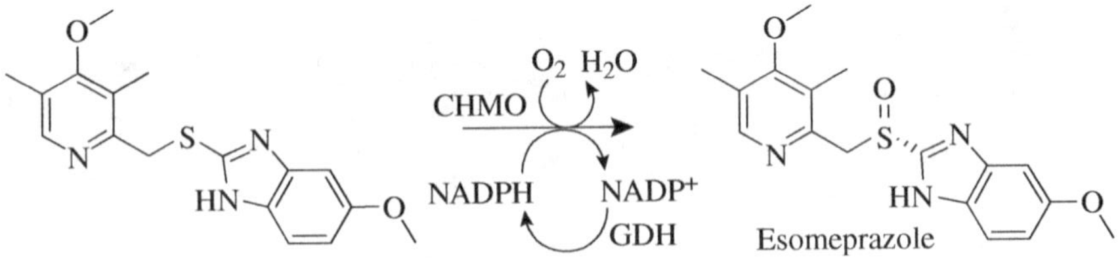	>90% *ee* 87% yield
Scheme 14: Oxidation of heteroatoms	CHMO variants	Armodafinil (for treatment of narcolepsy) [20]	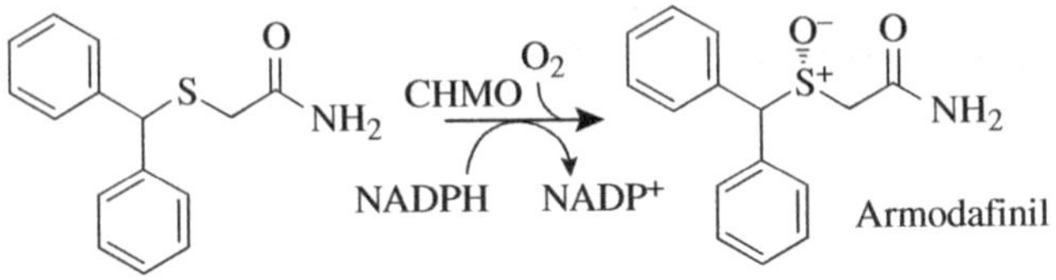	>90% *ee* 92.4% yield

Scheme 15: Dioxygen insertion across C=C bonds	Toluene dioxygenase	Oseltamivir (Tamiflu™, antiviral targeting influenza A and B viruses) [21]	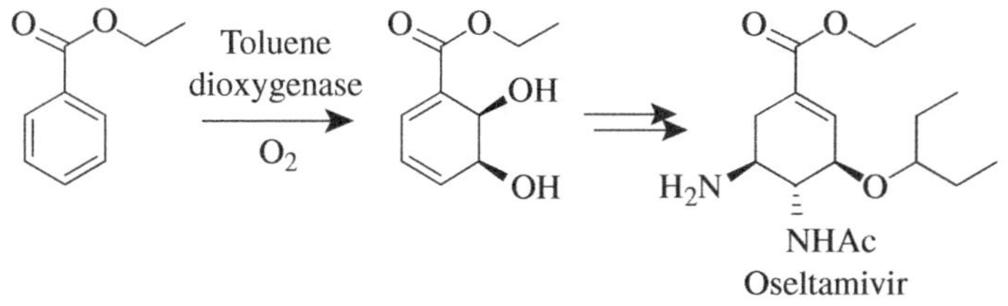	1–2 g/L for toluene dioxygenase-catalyzed step
Scheme 16: Aldol addition	*Escherichia coli* FSA	Imino sugars and other polyhydroxylated compounds [22]	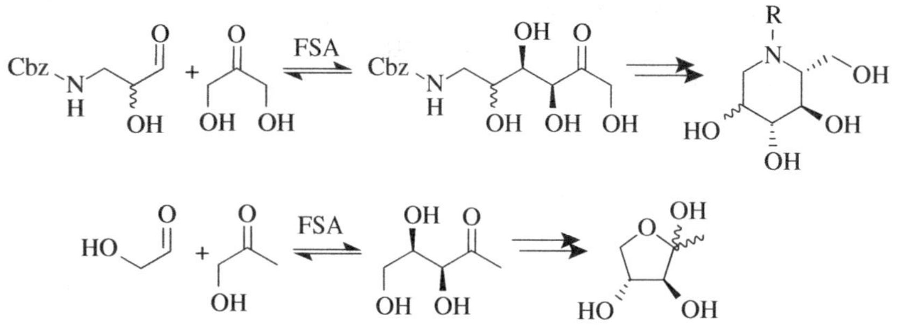	Majority of reactions had >70% conversion
Scheme 17: Aldol addition	*E. coli* FSA	A variety of polyol frameworks [23]	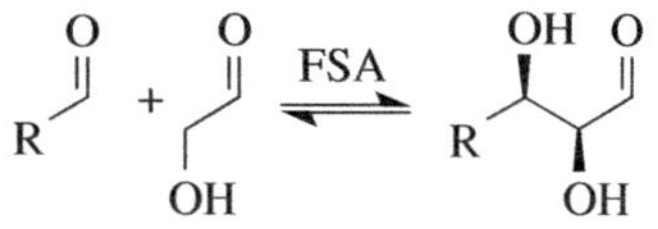 R = H, CH_2OH, C_2H_4NHCbz, CH_2Ph, CH_2OBn, $CHOHCH_2OBn$, $CHOHCH_2NHCbz$	28–68% yields
Scheme 18: Nitroaldol addition	*S. griseoverticillatum* MTG	Henry reaction of nitroalkanes with aliphatic, aromatic, and heteroaromatic aldehydes [24]	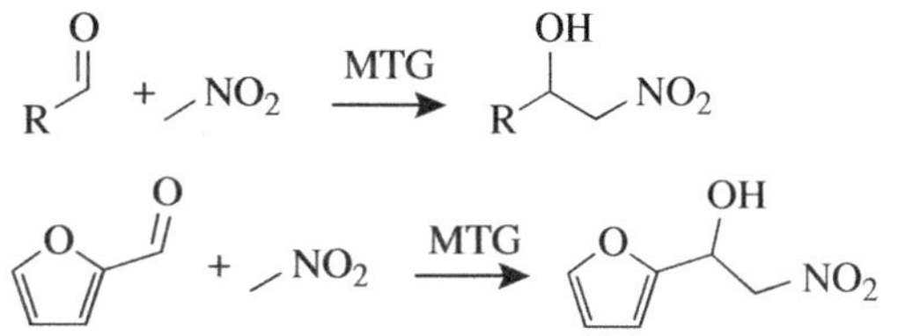	>62% yields for aromatic aldehydes bearing electron-withdrawing substituent
Scheme 19: Michael-type reaction	CALA, CALB, *M. miehei* lipase, and *T. lanuginosus* lipase	Michael addition with acidic esters, α,β-unsaturated ketones, and nitroalkenes [25–27]	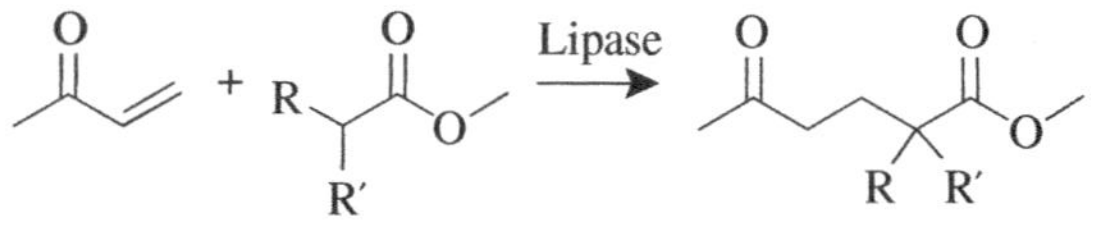	Modest enantioselectivity
Scheme 20: Michael-type reaction	4-OT	Michael addition of acetaldehyde to different nitrostyrenes [28]	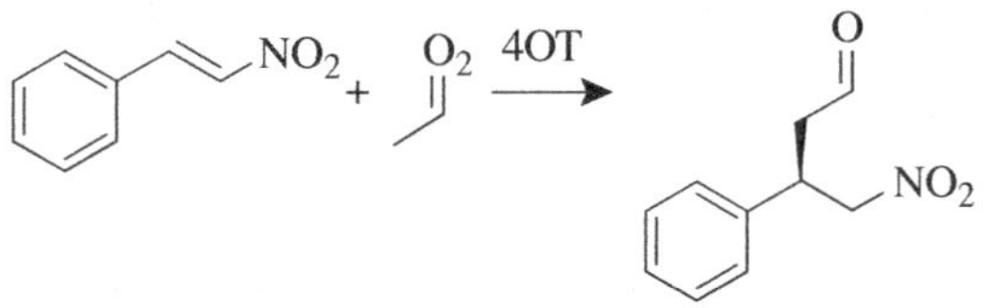	89% *ee*

(*Continued*)

TABLE 16.1 (*Continued*)

Reaction	Biocatalyst	Target	Reaction Scheme	Production Parameters
Scheme 21: Diels–Alder reaction	*De novo* designed Diels–Alderase	Diels–Alder reaction [29]	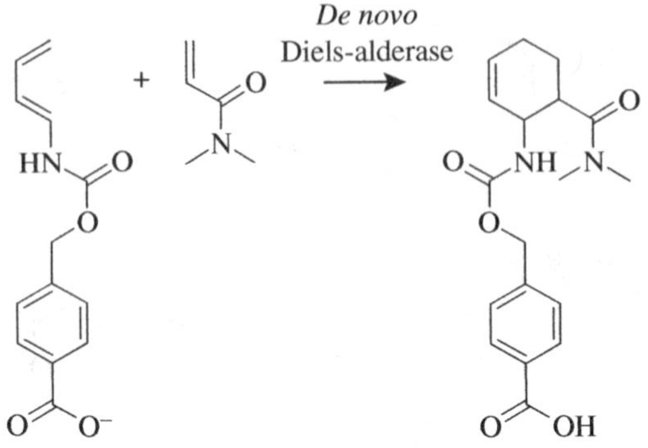	Up to 94% *ee*
Scheme 22: Diels–Alder reaction	*S. spinosa* SpnF	Diels–Alder reaction [30]	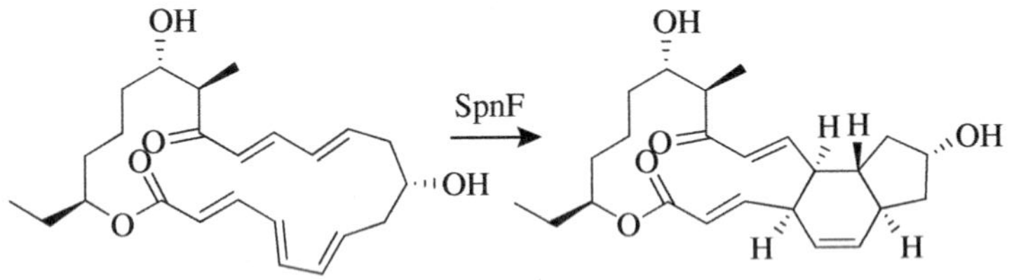	500-fold rate enhancement over uncatalyzed process
Scheme 23: Pictet–Spengler reaction	Norcolaurine synthase	(*S*)-Norcoclaurine (antiaallergic, β-adrenergic compound) and other tetrahydroisoquinolines [31]	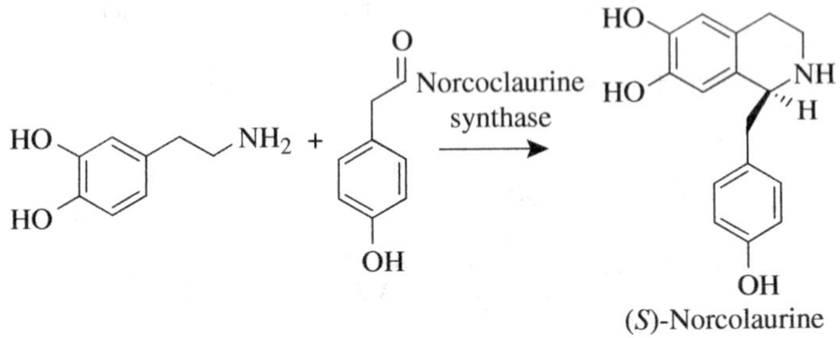	Conversions as high as 71% after 3 h
Scheme 24: Transamination	Prositagliptin ketone transaminase evolved from methyl-ketone transaminase	Sitagliptin (for treatment of diabetes) [32]	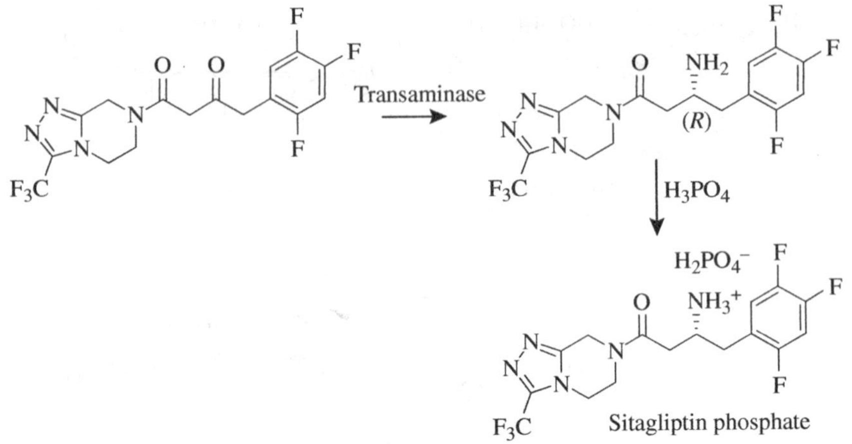	99.95% *ee* 92% yield

Scheme 25: Transamination	Library of (*R*)-selective transaminases	MK-7246 (selective CRTH2 antagonist for treatment of respiratory disease) [33]	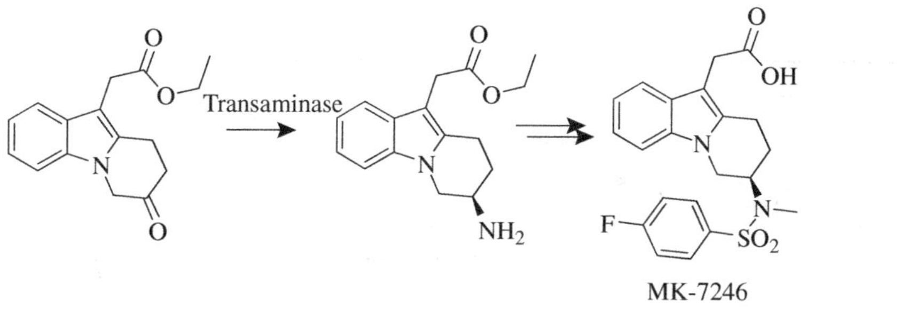	98% *ee* 80% yield
Scheme 26: Transamination	Prositagliptin ketone transaminase	Suvorexant (brain penetrant dual orexin receptor antagonist for treatment of insomnia) [34]	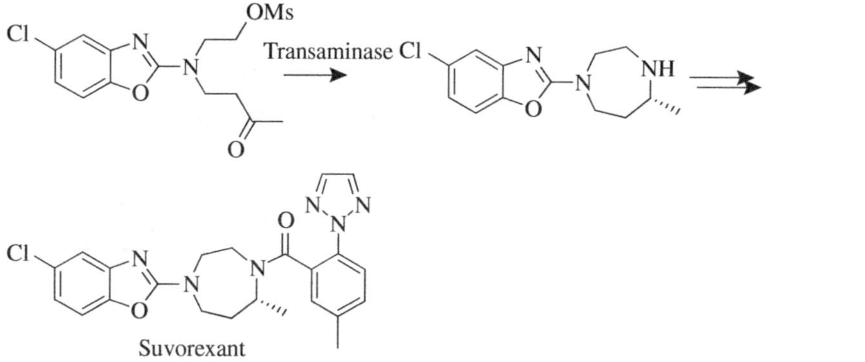	99% *ee* 71% yield
Scheme 27: Transamination	*V. fluvialis* ω-transaminase	IPI-926 (for treatment of cancer) [35]	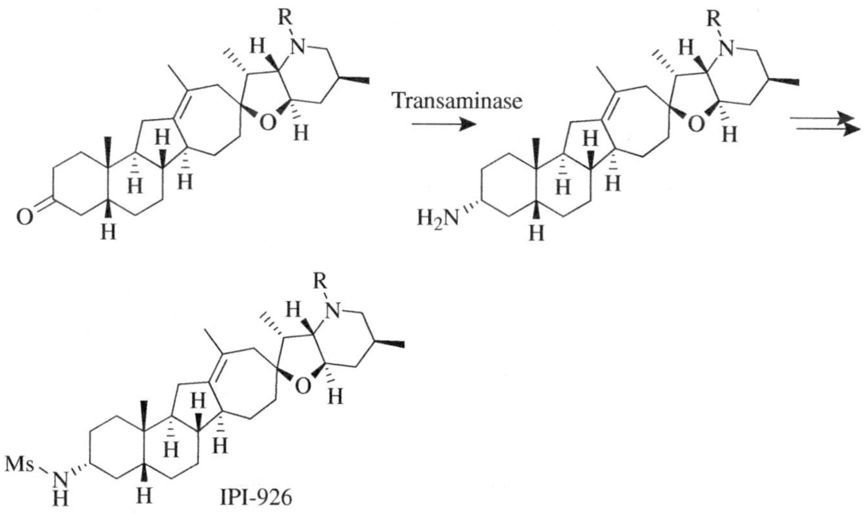	>70% *ee* 48.7% conversion

(*Continued*)

TABLE 16.1 (*Continued*)

Reaction	Biocatalyst	Target	Reaction Scheme	Production Parameters
Scheme 28: Transamidation	MTG	PEGylation of proteins [36]	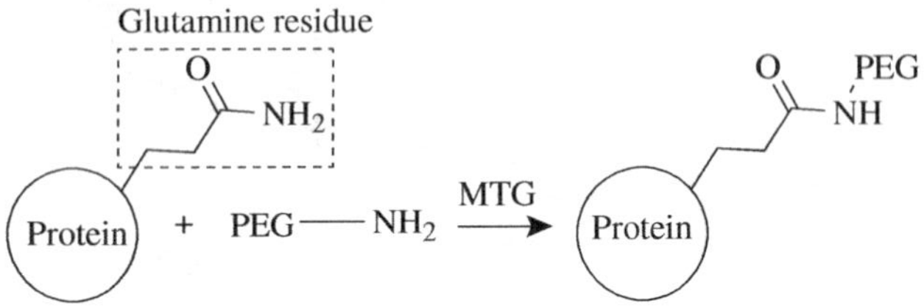	Transamidation of a selected subset of glutamine residues
Scheme 29: Fluorination	Polyketide synthase	Fluorination of natural products [37]	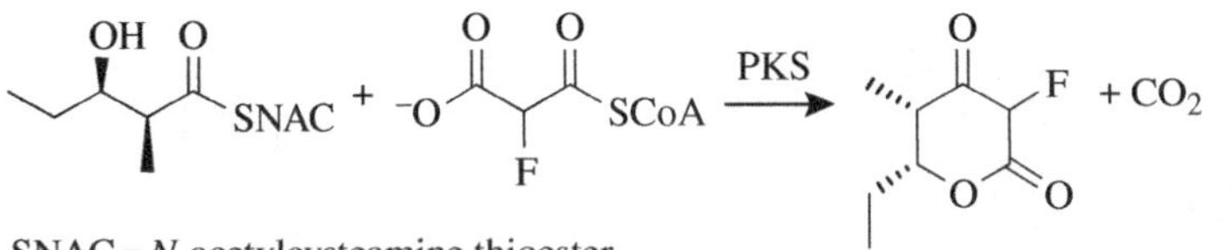 SNAC = *N*-acetylcysteamine thioester	Comparable yields for fluorinated and nonfluorinated products
Scheme 30: Halohydrin ring closure, epoxide ring opening	HHDH	Atorvastatin chloride (active ingredient in cholesterol-lowering drug Lipitor™) [38]	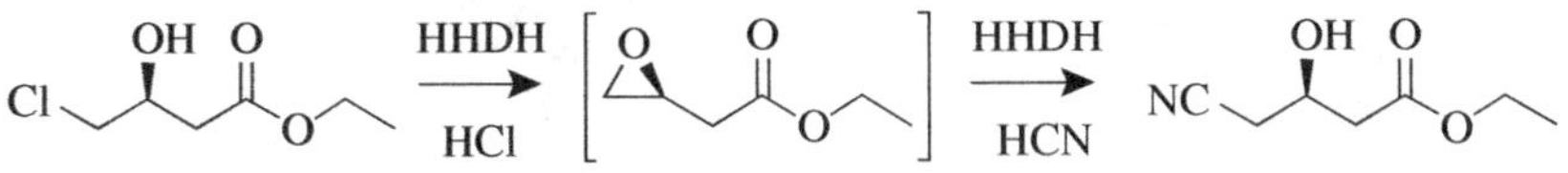	92% yield Turnover number $>5 \times 10^4$
Scheme 31: Spiroepoxide azidolysis	HheA, HheC	β-Substituted *tert*-azido alcohols from spiroepoxide [39]	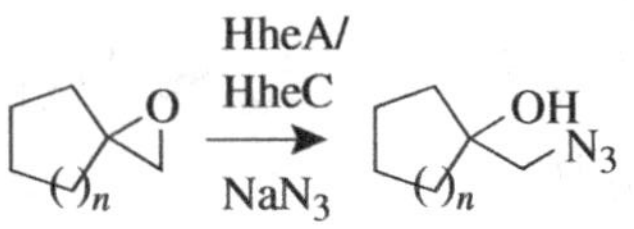	>94% *ee* 75–96% yields
Scheme 32: Hydration, oxidation	*A. denitrificans* Michael hydratase	3-Hydroxy carbonyl and 1,3-dicarbonyl compounds [40]	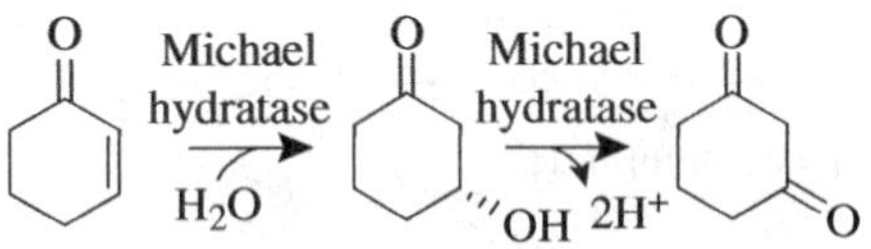	Oxidation step is (*R*)-selective
Scheme 33: Asymmetric C—H activation, C—C formation reactions	Artificial metalloenzyme	Dihydroisoquinolones [41]	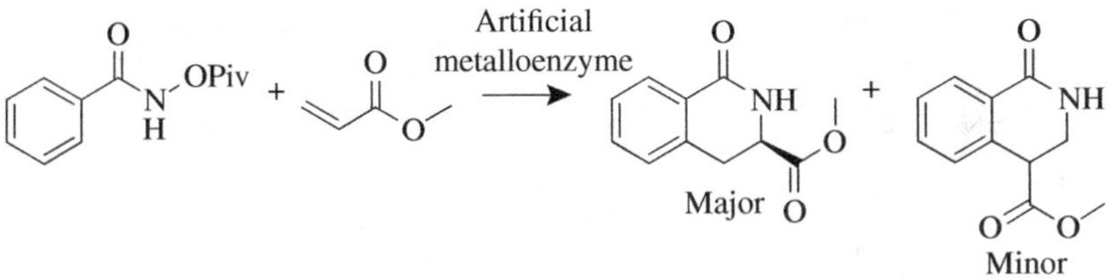	95% yield 82% *ee*

enzyme was found to be more active and 15 times more stable than previous preparations, thus enabling a continuous DKR process that is significantly more cost effective than the original batch process, and with a threefold reduction in the process *E*-factor (total waste (kg)/total product (kg)).

16.2.2 Reduction

A large percentage of industrial processes carried out by biocatalysts are reduction and oxidation reactions. This stems from the fact that biocatalysts provide an attractive and green alternative to their transition metal counterparts for redox reactions. One major consideration when carrying out reduction reactions is the regeneration of redox species. This can be achieved by coupling reduction and oxidation reactions *in vitro* or by using whole-cell biocatalysis. In this section, emphasis will be placed on single reduction reactions. More elaboration will be made on tandem reactions and whole-cell biocatalysis in Section **16.3**.

Alkene Reduction

In alkene reduction, asymmetric reduction of carbon–carbon double bonds results in the generation of two chiral centers. Examples of enzymes that are able to catalyze this reaction include flavin- and NADPH-dependent enoate reductases [3]. These enzymes typically have wide substrate scopes and are able to reduce α,β-unsaturated carbonyls, nitroalkenes, and carboxylic acid derivatives [43]. In a demonstration of their industrial importance, enoate reductases were used for the synthesis of (*R*)-3-hydroxy-2-methylpropanoate, commonly known as "Roche ester," an important chiral building block used in the synthesis of antibiotics, vitamins, and natural products (Scheme 3). In the study, a panel of 11 enoate reductases was employed to reduce methyl 2-hydroxymethylacrylate derivatives bearing different protecting groups on the hydroxyl moieties [7]. In this work, enoate reductases catalyzed the asymmetric reduction of methyl 2-hydroxymethylacrylate and its *O*-allyl, *O*-benzyl, and *O*-tert-butyldimethylsilyl derivatives to generate the corresponding (*R*)-methyl-3-hydroxy-2-methylpropionate products with up to 99% enantiomeric excess (*ee*). In another example, enoate reductases from the old yellow enzyme (OYE) family were used for the gram-scale conversion of 2-alkoxyenal to ethyl (*S*)-2-ethoxy-3-(*p*-methoxyphenyl)propanoate (EEHP), an important precursor of tesaglitazar, a peroxisome proliferator-activated receptor antagonist examined for the treatment of type 2 diabetes (Scheme 4) [8]. The use of the enoate reductases in this transformation, together with a cofactor regeneration step, gave a final yield of 94% with >98% *ee*. The productivity of the *in vitro* reaction was two orders of magnitude higher than the original whole-cell-catalyzed process.

Aldehyde and Ketone Reduction

Ketoreductases (KREDs) and other carbonyl reductases are important biocatalysts for the regio- and stereoselective reduction of carbonyl compounds. These enzymes provide green alternatives to the traditional boron- or phosphine-containing ligand-based chemical routes [3]. In an impressive demonstration of its potential, wild-type *Lactobacillus kefir* KRED was evolved by Codexis through a combination of molecular biology, bioinformatics, process chemistry, and high-throughput screening strategies to generate a KRED variant for the production of montelukast (Scheme 5) [9]. Montelukast, which goes by the trade name Singulair™, is a leukotriene receptor antagonist used in the treatment of asthma and allergies. The evolved KRED was used to reduce a bulky and hydrophobic ketone precursor to the desired (*S*)-alcohol with >99.9% *ee* in 70% organic solvent. This process currently runs at scales of >200 kg and is not possible with any available wild-type KRED. Following the success of this project, Codexis further evolved *L. kefir* KRED for the production of ezetimibe, the active ingredient used in Zetir™ (Scheme 6) [10]. Zetir is used for the lowering of low-density lipoprotein cholesterol and is currently being produced by chemical synthesis. With the evolved KRED, a greener and more efficient process with 99% diastereomeric excess was achieved with 90% conversion in 24 h. One major consideration when using KRED is the supply of NADPH as a cofactor. As mentioned earlier, the requisite NADPH can be supplemented by coupling the reactions with other dehydrogenases or by carrying out whole-cell biocatalysis. These strategies will be addressed in Section **16.3**.

16.2.3 Oxidation

Oxidation reactions are particularly important to organic chemists. Through oxidation reactions, it is possible to introduce functionality into relatively cheap starting materials such as alkanes, alkenes, or aromatic compounds. Using biocatalysts for oxidation reactions reduces the need for harsh oxidative agents such as chromium and nickel while providing better chemo- and stereoselectivity. In this section, the biocatalytic oxidation of different functionalities is described.

C–H Activation

Cytochrome P450 enzymes are a widely studied class of enzymes. In the pharmaceutical industry, P450 enzymes are routinely used to generate human drug metabolites for evaluation of drug toxicities, pharmacokinetics, and other pharmacological properties. This class of enzyme has also been useful for the activation of unreactive carbon centers and is used in the synthesis of an array of pharmaceutical products and intermediates [3]. One of the more prominent examples in recent years is the use of engineered P450 enzymes in artemisinin biosynthesis [11]. Artemisinin is one of the most effective drugs for treating malaria. However, naturally derived and chemically synthesized artemisinins are prohibitively expensive in many areas. One of the most challenging steps in artemisinin synthesis is the selective oxidation and reduction of amorphadiene to dihydroartemisinic acid. To overcome this challenge, wild-type *Bacillus megaterium* P450 enzyme ($P450_{BM3}$) was engineered with the help of computational models to enable the selective oxidation of amorphadiene (Scheme 7). This reaction was not possible with any wild-type P450 enzymes. Other strategies, such as whole-cell biocatalysis, have also been employed for the production of artemisinin (Section **16.3.2**). The utility of P450 enzymes has also been demonstrated in the selective fluorination of pharmaceutical targets. In order to achieve selective fluorination at unreactive carbon centers, a simple yet elegant two-step approach was adopted (Scheme 8) [12]. First, the substrates of interest were activated through P450-catalyzed oxygenation. This was followed by deoxofluorination of the activated sites to generate the target fluorinated compounds. To extend the reaction to a variety of organic scaffolds, a panel of 96 $P450_{BM3}$ variants was assembled to catalyze reactions with assorted substrates. Apart from C—H activation, other strategies have been employed for the fluorination of pharmaceutical targets. These strategies will be discussed in Section **16.2.10**.

Another class of enzyme capable of catalyzing direct activation of C—H bonds is monoamine oxidase (MAO). Flavin-dependent MAO activates sp^3 C—H bonds that are α to a nitrogen atom to generate imines. These enzymes can be used to desymmetrize prochiral meso-amines and have been explored in feasible manufacturing routes for different pharmaceutical compounds such as telaprevir (Scheme 9) [13, 14] and boceprevir (Scheme 10) [15]. Both compounds are used for the treatment of hepatitis C and require lengthy synthesis using standard peptide chemistry. To enable more efficient synthesis of these drugs, improved variants of *Aspergillus niger* MAO-N obtained through directed evolution have been combined with different chemical synthesis methods to greatly shorten the synthetic routes. In the case of telaprevir, a combination of enzymatic oxidation by MAO and an Ugi-type three-component reaction greatly shortened the number of synthetic steps from 24 to 11 [14]. In the synthesis of boceprevir, the addition of the biocatalytic step to the synthetic process also led to 150% improvement in product yields and 60% reduction in raw material use [15].

Oxidative Insertion Across C—C Bonds

Baeyer–Villiger reactions, which involve cleavage of C—C bonds and insertion of oxygen, are traditionally catalyzed by toxic and explosive compounds such as *m*-chloroperbenzoic acid and trifluoroacetic acid [3]. These reactions are also plagued by limited specificities. Since their discovery, flavin-dependent Baeyer–Villiger monooxygenases (BVMOs) have provided safer and more specific alternative routes for oxidative insertions across C—C bonds. The most widely used BVMO is the *Acinetobacter calcoaceticus* NCIMB9871 cyclohexanone monooxygenase [44]. More than 100 substrates have been reported for this BVMO, thus illustrating its promiscuity and utility in a myriad of chemical transformations. In a recent study, an in-house panel of 16 BVMOs was screened to identify suitable candidates for the kinetic resolution of different racemic substituted aliphatic β-amino ketones to synthesize β-aminoalkyl acetates and β-amino alcohols (Scheme 11) [16, 17]. These products were not typically formed through Baeyer–Villiger oxidation reactions. Within the assembled panel, enzymes of excellent enantioselectivities ($E > 200$) were identified. Interestingly, some enzymes also showed opposite enantioselectivity toward the same substrates. The wide substrate scope, excellent enantioselectivities, and ability to generate different enantiomers further demonstrate the usefulness of BVMOs in generating chiral building blocks for pharmaceuticals.

Oxidation of Heteroatoms

Another important application of BVMOs is the oxidation of sulfides to chiral sulfoxides. The chemical oxidation of sulfides is prone to overoxidations, which often result in formation of sulfone by-products with low chiral purities [2]. Using an optimized BVMO, the enantioselectivity and chemospecificity of oxidation reactions can be controlled to generate pure products with high *ee*. The utility of BVMOs for sulfide oxidation was recently exemplified by engineered thermostable phenylacetone monooxygenases (PAMOs) [18]. The wild-type enzymes had very limited substrate scopes, accepting only phenylacetone and similar linear phenyl-substituted analogs. Through laboratory evolution, mutant enzymes that were able to oxidize 2-aryl- and 2-alkylcyclohexanones with high activity and enantioselectivity were generated (Scheme 12). Another two BVMOs, derived from *Acinetobacter* cyclohexanone monooxygenases (CHMOs), were recently employed in the synthesis of two pharmaceutical compounds: the first being esomeprazole (Scheme 13) [19], a proton-pump inhibitor for treatment of gastroesophageal reflux disease, and the second being armodafinil (Scheme 14) [20], an analeptic drug for the treatment of narcolepsy. Through directed evolution of the wild-type enzymes, variants that are capable of generating chiral sulfoxides of high enantiopurity with fewer processing steps were generated.

Dioxygen Insertion Across C=C Bonds

Regiospecific hydroxylation of aromatic compounds is difficult to achieve chemically. However, dioxygenases readily catalyze such reactions with great regio-, stereo-, and enantiospecificity. One notable application of dioxygenases is the use of a dioxygenase in the synthesis of oseltamivir (Scheme 15) [21]. This drug, marketed under the trade name Tamiflu™, is an antiviral targeting influenza A and B viruses. In the novel approach, a toluene dioxygenase was overexpressed in *Escherichia coli* for the whole-cell fermentation of ethyl benzoate to its diol which was then extracted from the fermentation broth and subjected to further chemical transformation. This newly developed process eliminated the need for azides and allowed the process to start from ethyl benzoate as a raw material instead of the more expensive homochiral *cis*-dihydrodiol.

16.2.4 C—C Bond-Forming Reaction

The controlled formation of carbon–carbon bonds remains a challenge in organic synthesis. In this section, a diverse range of enzymes capable of catalyzing a variety of C—C bond-forming reactions will be discussed.

Aldol Addition

The catalysis of aldol reactions via metallo- and organocatalyst is often complicated by the need for stoichiometric quantities of reagents. In addition, aldol reactions are extremely sensitive to reaction conditions and often require the extensive protection of multifunctional polar groups to reduce undesired cross-reactivity. In view of these limitations, aldolases form an attractive alternative for the catalysis of aldol reactions. Aldolases catalyze the reversible addition reaction between a nucleophilic donor and an electrophilic acceptor, leading to the formation of new stereogenic centers. A good example would be the D-fructose-6-phosphate aldolase (FSA) from *E. coli*. Aldol reactions catalyzed by this enzyme offer an attractive route for the synthesis of a variety of polyhydroxylated products. In one study, FSA was employed to catalyze the stereoselective aldolase addition of hydroxyacetone and dihydroxyacetone to a range of aldehyde acceptors in good yields and excellent enantioselectivities (Scheme 16) [22]. In a follow-up study, it was further discovered that FSA had the ability to catalyze the stereoselective self- and cross-aldol-type reactions with glycolaldehyde to generate a variety of polyol frameworks (Scheme 17) [23]. Aside from their importance as artificial sweeteners, polyols also play an important role in the pharmaceutical industry as osmotic diuretics. They are used in the treatment of constipation, brain swelling, and acute kidney failure.

Nitroaldol Addition

Nitroaldol addition, also known as the Henry reaction, represents another class of important reactions in organic synthesis. Nitroaldol reaction involves the formation of new C—C bonds through the coupling of a nitroalkane with a carbonyl compound. The reaction is typically catalyzed by strong bases such as lithium diisopropylamide (LDA), butyllithium, and $Ba(OH)_2$. As a milder and greener alternative, biocatalysis has recently been explored for the catalysis of such reactions. In one particular study, the reaction of 4-nitrobenzaldehyde and nitromethane was catalyzed by microbial transglutaminases (MTGs) from *Streptoverticillium griseoverticillatum* (Scheme 18) [24]. The reaction, carried out in a cyclohexane-water solvent system at room temperature, yielded the corresponding β-nitro alcohol without any additional side products. Small-scale screening also revealed that MTG accepted a broad range of aliphatic, aromatic, and heteroaromatic aldehyde substrates.

16.2.5 Michael-Type Reaction

The Michael-type reaction, which involves the addition of a nucleophile to an α,β-unsaturated carbonyl compound, is an important conjugate addition reaction used for the formation of new C—C and carbon–heteroatom bonds. The reaction traditionally requires either strongly basic or acidic conditions, leading to the generation of hazardous waste products and unwanted side reactions [3]. To discover a greener and more controlled alternative to the traditional Michael-type addition reaction, a range of lipases have been screened for their ability to catalyze such reactions [25–27]. By assaying for reactivity with acidic esters, α,β-unsaturated ketones, and nitroalkenes, four lipases—namely, *C. antarctica* lipase A, CALB, *Mucor miehei* lipase, and *Thermomyces lanuginosus* lipase—were found to be suitable biocatalysts for Michael-type addition reactions (Scheme 19). However, further evolution and engineering of these enzymes are required as most of the enzymes screened had modest or no enantioselectivity. Another enzyme with demonstrated activity for Michael-type reaction is 4-oxalocrotonate tautomerase (4-OT) (Scheme 20). In a recent study, the enzyme was used to catalyze the Michael-type addition of acetaldehyde to different nitrostyrenes with respectable moderate yields and respectable *ee* [28]. It was postulated that the reaction likely occurred via enamine formation of the amino-terminal proline residue of 4-OT with acetaldehyde. Although further improvements have to be made before the practical application of the enzyme, the study represents another step toward biocatalyzed Michael-type addition reactions.

16.2.6 Diels–Alder Reaction

Another important conjugate addition reaction in organic synthesis is the Diels–Alder reaction. This involves the reaction of a conjugated diene with an alkene to form four new stereogenic centers. In a groundbreaking work, *de novo* computational design was used to create enzymes capable of catalyzing a biomolecular Diels–Alder reaction (Scheme 21) [29]. Using the Rosetta methodology, 84 designs were generated for experimental verification, which involved expression of these enzymes in *E. coli*, enzyme purification, and use of a liquid chromatography–tandem mass spectrometry assay to detect any Diels–Alder activity. Out of the 50 soluble enzymes, two were found to exhibit Diels–Alder activity. Further mutations enabled the activity, stereoselectivity, and substrate specificity of the engineered enzymes to be improved, and it is likely that further enhancements can be made through directed evolution. In another major breakthrough, a naturally occurring enzyme capable of catalyzing a Diels–Alder reaction was recently identified [30]. This enzyme, SpnF from *Saccharopolyspora spinosa*, is involved in the formation of spinosyn A, a tetracyclic polyketide-derived insecticide (Scheme 22). Kinetic studies confirmed that SpnF's only known function was to catalyze the specific acceleration of a [4 + 2] cycloaddition reaction, bringing about a 500-fold rate enhancement.

16.2.7 Pictet–Spengler Reaction

The Pictet–Spengler reaction involves ring closure of β-arylethylamine after condensation with an aldehyde or ketone [31]. The reaction is typically catalyzed by strong acids such as hydrochloric and trifluoroacetic acid. In a recent study, norcolaurine synthase, which is involved in alkaloid synthesis in plants, was used to catalyze the Pictet–Spengler reaction using mild reaction conditions at pH 7 (Scheme 23) [31]. The enzyme facilitated cyclization of dopamine with various acetaldehydes, yielding a diverse range of substituted tetrahydroisoquinolines. Examples of pharmaceutically relevant tetrahydroisoquinolines include noscapine, which acts an antitussive, michellamine B, which possesses antihuman immunodeficiency virus (HIV) activity, and solifenacin, which has a urinary antispasmodic effect.

16.2.8 Terpene Cyclization

The terpene cyclase or synthase is another class of enzyme that is capable of synthesizing new C–C bonds. Terpenoids is one of the most diverse classes of natural products. Their syntheses involve the cyclization of linear polyisoprene precursors by terpene cyclases. Terpene cyclases have been used for the production of the anticancer compound taxol (paclitaxel) [45] and the antimalarial compound artemisinin [46]. The production of these two compounds will be discussed in Section **16.3.2**.

16.2.9 Transfer Reaction

The most notable examples of transfer reactions in organic synthesis are transamination and transamidation reactions. These reactions will be discussed in this section.

Transamination

Transamination reactions have great utility in the synthesis of chiral amines. Transaminases, or aminotransferases, have been used for the synthesis of pharmaceutical targets such as sitagliptin, a compound used in the treatment of diabetes [32]. In the study, a methyl-ketone transaminase that lacked any activity toward its intended substrate, prositagliptin ketone, was altered through a combination of strategies including substrate walking, modeling, mutation, and directed evolution to create a prositagliptin ketone transaminase that could be applied in manufacturing setting. The final enzyme variant was able to catalyze the transamination reaction with excellent *ee* of 99.95% and demonstrated great potential for the replacement of the existing high-pressure rhodamine-catalyzed process (Scheme 24). Owing to its wide substrate scope, the same transaminase was also evolved for the synthesis of MK-7246 (Scheme 25) [33] and suvorexant (Scheme 26) [34]. MK-7246 is a CRTH2 inhibitor explored for the treatment of asthma, allergic rhinitis, and atopic dermatitis, while suvorexant is a dual orexin inhibitor explored for the treatment of insomnia. An ω-transaminase from *Vibrio fluvialis* has also been explored in the synthesis of IPI-926, a drug that underwent several clinical trials for cancer treatment (Scheme 27) [35]. The enzyme could catalyze transamination with different amine donors to yield the desired (*R*)-amine intermediate, thus potentially shortening the existing synthetic process.

Transamidation

Transamidation reactions involve the transfer of an amine from a carboxamide to a suitable acceptor. Unlike its chemical counterpart, enzymatic catalysis of transamidation does not require additional protection and deprotection steps. An enzyme capable of catalyzing a transamidation reaction is MTG. MTG catalyzes the reaction between glutamine amide and the amine of lysine and has broad utility in the pharmaceutical industry (Scheme 28) [36]. The enzyme is able to accept a diverse range of amine substrates and was used for the addition of polyethylene glycol to proteins in a bid to reduce immunogenicity and renal clearance. In structures containing multiple glutamines, selective transamidation with MTG could also be achieved by adding different cosolvents to the mixture.

16.2.10 Fluorination

Fluorination reactions have great importance in the pharmaceutical industry. Through selective fluorination of active pharmaceutical compounds, more potent drugs with greater stability and reduced toxicity can be generated. To date, there are only a few examples of enzyme-catalyzed fluorination reactions. These include use of a fluorinase from *Streptomyces cattleya* [47] and P450 enzymes to activate unreactive carbons for deoxofluorination (section "C—H Activation") [12]. To further expand the scope of fluorine chemistry, a recent study has demonstrated the possibility of incorporating fluorine into a polyketide backbone both *in vitro* and *in vivo* by using fluoromalonyl-CoA as extenders for polyketide synthases (Scheme 29) [37]. In this work, the terminal thioesterase module of the 6-deoxyerythronolide B synthase (DEBS) was shown to accept fluoromalonyl-CoA in chain-extension catalysis to generate 2-fluoro-2-desmethyltriketide lactone. This reaction can be applied for the fluorination of important natural products such as the antibiotic, erythromycin, and the immunosuppressant, rapamycin.

16.2.11 Other Reactions

Another important enzyme that has been applied in organic synthesis is halohydrin dehalogenase (HHDH). This enzyme is able to catalyze ring closure of vicinal halohydrins to generate chiral epoxides, as well as the reverse reaction involving ring opening of epoxides by halides. HHDHs have been applied in the synthesis of atorvastatin calcium, the active ingredient in the cholesterol-lowering drug, Lipitor™ (Scheme 30) [38]. Following the initial reduction reaction to generate (*S*)-ethyl-4-chloro-3-hydroxybutyrate, an HHDH variant was used to catalyze the replacement of the chloro-substituent with a cyano group by reacting with hydrogen cyanide at ambient temperature and pressure. Through directed evolution, the volumetric productivity of the HHDH-catalyzed reaction was improved 2500-fold, with significant reductions in production cost and waste. More recently, the novel activity of HHDHs toward spiroepoxides has also been characterized (Scheme 31) [39]. In the study, HHDH HheA from *Arthrobacter* species AD2 and HheC from *Agrobacterium radiobacter* were used to catalyze regioselective azidolysis of spiroepoxides containing five-, six-, and seven-membered cycloalkane rings, scaffolds found in many natural and synthetic bioactive compounds. Apart from exhibiting high regioselectivities, many of the reactions proceeded with high yields and enantioselectivity, thus demonstrating an attractive and simple route for the preparation of β-substituted *tert*-azido alcohols.

16.2.12 Bifunctional Enzymes

In addition to the aforementioned enzymes, bifunctional enzymes, which are capable of catalyzing two different chemical reactions, have also been reported. A putative Michael hydratase from *Alicycliphilus denitrificans* that first hydrates α,β-unsaturated carbonyl substrates and then oxidizes the intermediate to the dicarbonyl product was recently postulated (Scheme 32) [40]. In the absence of oxidation reagents, the enzyme was observed to only add water to C=C yielding the 3-hydroxy carbonyl compound. However, under an oxidative environment, the dicarbonyl product was obtained through a further oxidation step. The broad substrate scope of this enzyme would render it useful for the manufacture of 3-hydroxy carbonyl and 1,3-dicarbonyl compounds. A second example of a bifunctional enzyme is an artificial metalloenzyme, which combined a streptavidin component containing a glutamic acid or aspartic acid residue with a docked biotinylated rhodium(III) complex to enable catalytic asymmetric C—H activation and C—C formation reactions (Scheme 33) [41]. With the artificial metalloenzyme, the coupling of benzamides and alkenes to yield dihydroisoquinolones was accelerated 100-fold when compared with the isolated rhodium complex with up to 86% *ee*. This artificial bifunctional enzyme demonstrates the potential of *de novo* biocatalyst design and hints at a next generation of build-to-order biocatalysts.

16.3 MULTIENZYME BIOCATALYTIC REACTIONS

With a rapidly expanding biocatalytic toolkit that is capable of carrying out key chemical transformations in the discovery and development of pharmaceuticals (Table 16.1), combining multiple biocatalysts to carry out a series of transformations without isolation of intermediates is the logical extension. Multienzyme reactions can be catalyzed by different combinations of soluble enzymes, whole cells, permeabilized whole cells, and immobilized enzymes.

16.3.1 One-Pot Cascade Reactions

In addition to their exquisite chemo-, regio-, and enantioselectivity, biocatalysts are particularly suitable for one-pot cascade reactions given their mutual compatibility under the range of relatively mild and "green" operating conditions. This is not surprising given that enzymes have evolved to function cooperatively and orthogonally within the same physical environment of living cells. By reducing the number of purification steps, "one-pot" cascade reactions are generally more cost and time efficient. A major advantage of multistep one-pot cascades is the concurrent *in situ* conversion of intermediates or removal of inhibitory by-products. This increases conversion rates and yields by shifting the equilibria of reversible reactions toward product formation and, in the case of unstable or poorly soluble intermediates, by circumventing the need for costly intermediate purification steps. Overall, the ease of operation and improvements in space–time yields conferred by one-pot cascade reactions translate into less waste and lower raw material and operational costs (e.g., product separation, equipment operation). With volume–time–output identified as the most critical factor defining a good chemical manufacturing process [48], it is likely that one-pot multienzyme biocatalytic procedures will be more relevant for industrial production and preferred over nonintegrated ones.

There is an increasing interest in using enzymes to improve the yield of enantiomerically pure compounds, including alcohols and amines, in the pharmaceutical industry. While the technology for preparation of optically active alcohols using lipases or alcohol dehydrogenases is well established [49, 50], a variety of one-pot enzymatic strategies for production of enantiopure amines and amino acids by deracemization of racemic mixtures has only recently been developed. This section highlights multienzymatic deracemization of amino acids and amines by DKR and stereoinversion. Cyclic deracemization strategies are chemoenzymatic and will not be covered here. Further examples of enzymatic, whole-cell-mediated as well as chemoenzymatic deracemization of racemic compounds are reviewed [51–54].

DKR, which involves the conversion of one enantiomer into the desired product in conjunction with *in situ* racemization of the other enantiomer, is one of the most widely used approaches to produce compounds of a single chirality. Compared to conventional kinetic resolutions with a maximum theoretical yield of 50%, DKR potentially allows 100% yield of the optically active product. The Asano laboratory has developed a one-pot three-step enzymatic DKR of α-aminonitriles to form (*S*)- and (*R*)-α-amino acids of a single chirality (>99% *ee*) at >99% yields [55]. The first and most critical step is the hydrolysis of racemic α-aminonitrile by a non-stereoselective nitrile hydratase, which is followed by the stereoselective hydrolysis of the α-amino acid amide with concurrent racemization of the α-amino acid amide (Figure 16.2a). This approach is dependent upon the availability of a nitrile hydratase that is able to act on both α-aminonitrile enantiomers. In a different three-step biocatalytic cascade, DKR of racemic mandelic acid yielded enantiopure L-phenylglycine [56]. D-Mandelic acid is oxidized to form the corresponding keto acid before being reduced to L-phenylglycine by stereoselective D-mandelate dehydrogenase and L-amino acid dehydrogenase, respectively (Figure 16.2b). The NADH/NAD^+ cofactor is recycled by the mandelate dehydrogenase, and a lower substrate concentration is needed to achieve high conversion (>97%). While this redox-neutral cascade justifies the one-pot strategy by circumventing the need for other cofactor regeneration systems, it also revealed a downside to having multiple potentially interfering reactions in the same vessel.

Another deracemization strategy is stereoinversion, which involves one-pot two-step generation of stable nonchiral intermediates followed by asymmetric synthesis.

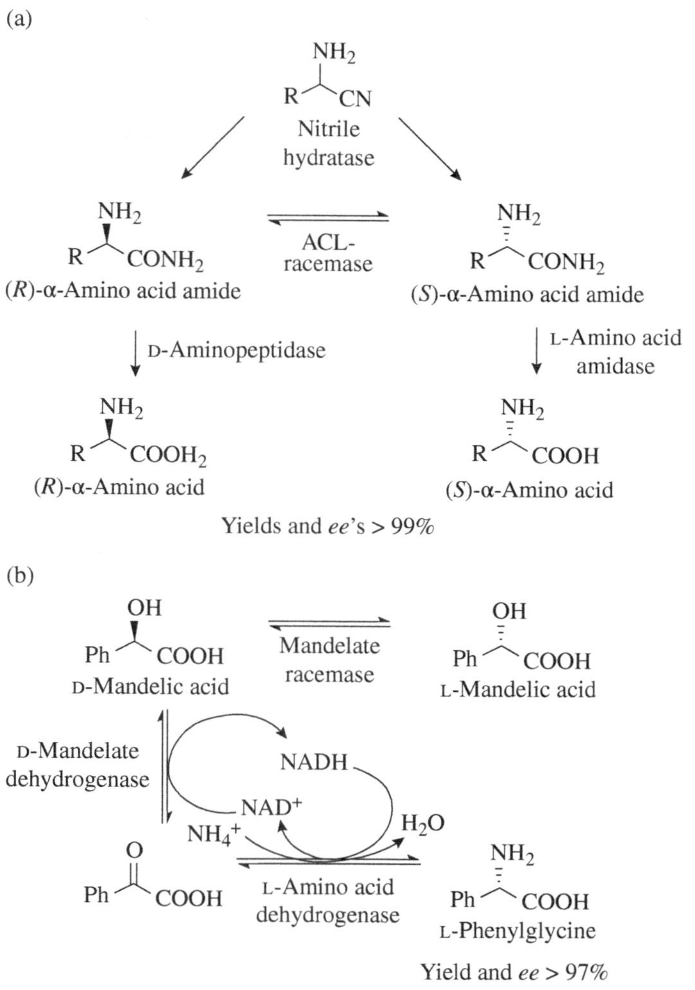

FIGURE 16.2

Enzymatic dynamic kinetic resolution (DKR) for deracemization of amino acids. (a) DKR of α-aminonitriles to α-amino acids. (b) DKR of mandelic acid to L-phenylglycine.

For preparation of enantiopure amino acids, irreversible oxidation by amino acid oxidases generates keto acids that are then substrates for enantioselective transaminases [57–60] (Figure 16.3a). As described in the section on transamination, ω-transaminases have emerged as powerful industrial biocatalysts because of their ability to introduce amino groups into ketones and keto acids with high stereoselectivity [61]. Since the thermodynamics of reversible transaminase reactions generally strongly favors ketone/amino acid over amine/keto acid, cascade designs involving ω-transaminases for asymmetric syntheses are critical to shift the reaction equilibrium toward product formation. This is generally achieved by removal of pyruvate, the by-product of the transamination reaction, using a variety of systems involving pyruvate decarboxylases, lactate, glucose, and amino acid dehydrogenases [61] (Figure 16.4). Stereoselective reduction of keto acids to the corresponding amino acids by amino acid dehydrogenases has also been demonstrated [57]. Deracemization of primary amines has been demonstrated using two ω-transaminases with opposing stereoselectivity in a one-pot two-step reaction [62]. The first transaminase step involves the kinetic resolution of the racemic mixture by enantioselective deamination. The second transaminase step completes the stereoinversion by stereoselective reductive amination of the ketone. The configuration of the enantiomer produced is determined by the order of transaminase application (Figure 16.3b).

The variations and combinations of cascade design involving these core biocatalysts are vast. In addition, biocatalysts with complementary activities continue to be uncovered. For example, ammonia lyases and aminomutases have been demonstrated to be viable biocatalysts for the synthesis of enantiopure α- and β- amino acids

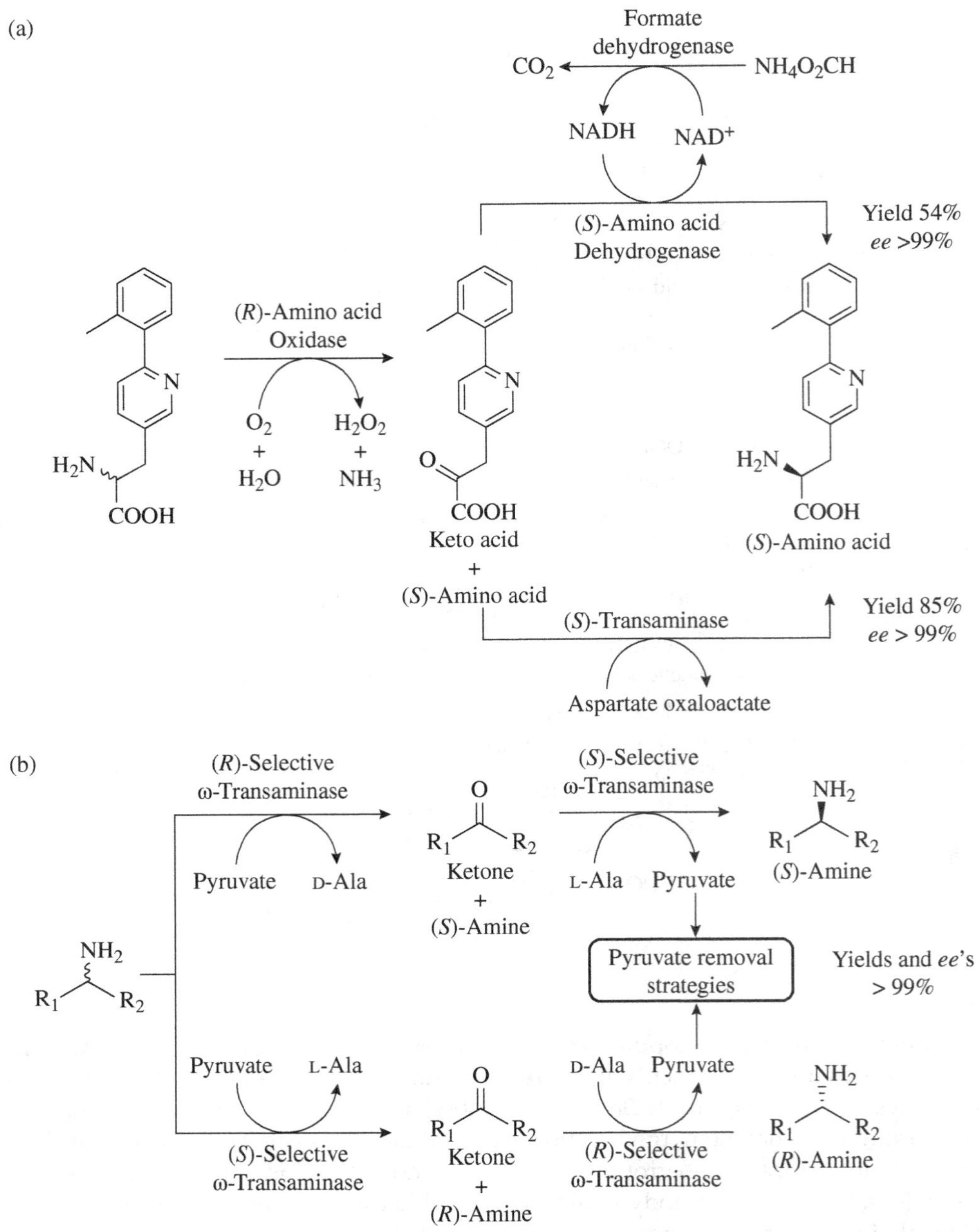

FIGURE 16.3

Enzymatic stereoinversion for deracemization of (a) amino acids and (b) amines.

[63–65]. Overall, compared to chemical methods, these one-pot biocatalytic cascades enable "green" racemization of amines and amino acids under mild reaction conditions.

16.3.2 Whole-Cell Biocatalysts

Whole-cell biocatalysts are basically one-pot cascade reactions, with various enzymatic reactions being carried out concurrently within individual cells. Compared to purified enzymes, whole-cell biocatalysts are inexpensive and easily scalable and can be stably stored indefinitely. Certain enzymes are unstable and lose activity when purified from cells. Living cells also contain and regenerate otherwise expensive redox cofactors and, with metabolic engineering, can produce desired chemicals from inexpensive carbon and nitrogen sources [66, 67]. On the other hand, a major downside of using whole cells for biotransformations is the increased cost of product extraction and purification from fermentation broths. One also has to consider the

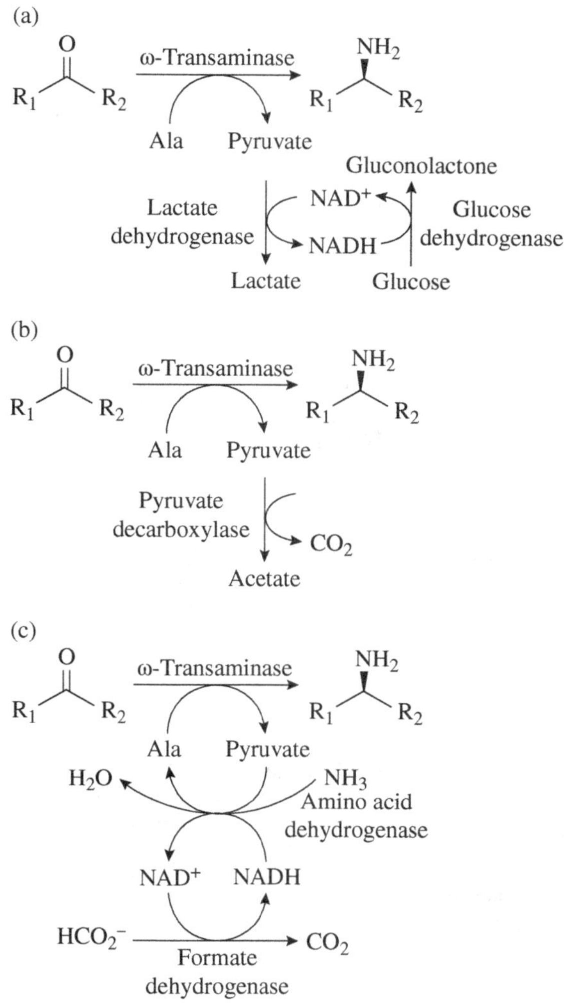

FIGURE 16.4

Cascade strategies for ω-transaminases catalyzed asymmetric synthesis of chiral amines with alanine as the amine donor. (a) Pyruvate removal using lactate dehydrogenase. This is coupled to a glucose/glucose dehydrogenase cofactor regeneration reaction. (b) Pyruvate removal using pyruvate decarboxylase. (c) Regeneration of alanine from pyruvate using an amino acid dehydrogenase. This is coupled to a formate/formate dehydrogenase cofactor regeneration system.

existence of enzymes with undesirable activities in the cell. With glucose and redox cofactors being one of the cheapest and most expensive starting raw materials respectively, bioconversions using whole cells or permeabilized cells can be extremely economically competitive from an industrial perspective.

Optically pure (*S*)- and (*R*)-mandelic acids are important for the synthesis of antibiotics and other pharmaceuticals [68–71]. Generally synthesized chemically [72, 73], mandelic acid stereoisomers have also been prepared from racemates biocatalytically with stereoselective nitrilases or microbes capable of enantioselective degradation [74–77]. Nonetheless, these strategies rely on nonrenewable petroleum-based feedstocks such as benzaldehyde and mandelonitrile. Synthesis of mandelic acid from glucose using the shikimate aromatic amino acid biosynthesis pathway offers a greener and potentially more cost-effective alternative [78] (Figure 16.5). By deleting competing pathways to redirect flux to phenylpyruvate and introducing a hydroxymandelate synthase, up to 0.74 g/L of (*S*)-mandelic acid can be produced from 20 g/L glucose, representing an eightfold improvement in yield compared to the unoptimized strain. Two other enzymes, hydroxymandelate oxidase and D-mandelate dehydrogenase, convert (*S*)-mandelic acid to the (*R*)-enantiomer. While promising, the industrial relevance of this fermentative route will require additional strain and process optimization to increase the yield and productivity of this process. Likewise hijacking the shikimate pathway, bioproduction of *trans-p*-hydroxycinnamic acid (pHCA) from glucose has been demonstrated in *Saccharomyces cerevisiae*, *E. coli*, *Pseudomonas*, and *Streptomyces* [79–81]. By introducing a bifunctional phenylalanine ammonia lyase/tyrosine ammonia lyase (PAL/TAL) enzyme [82],

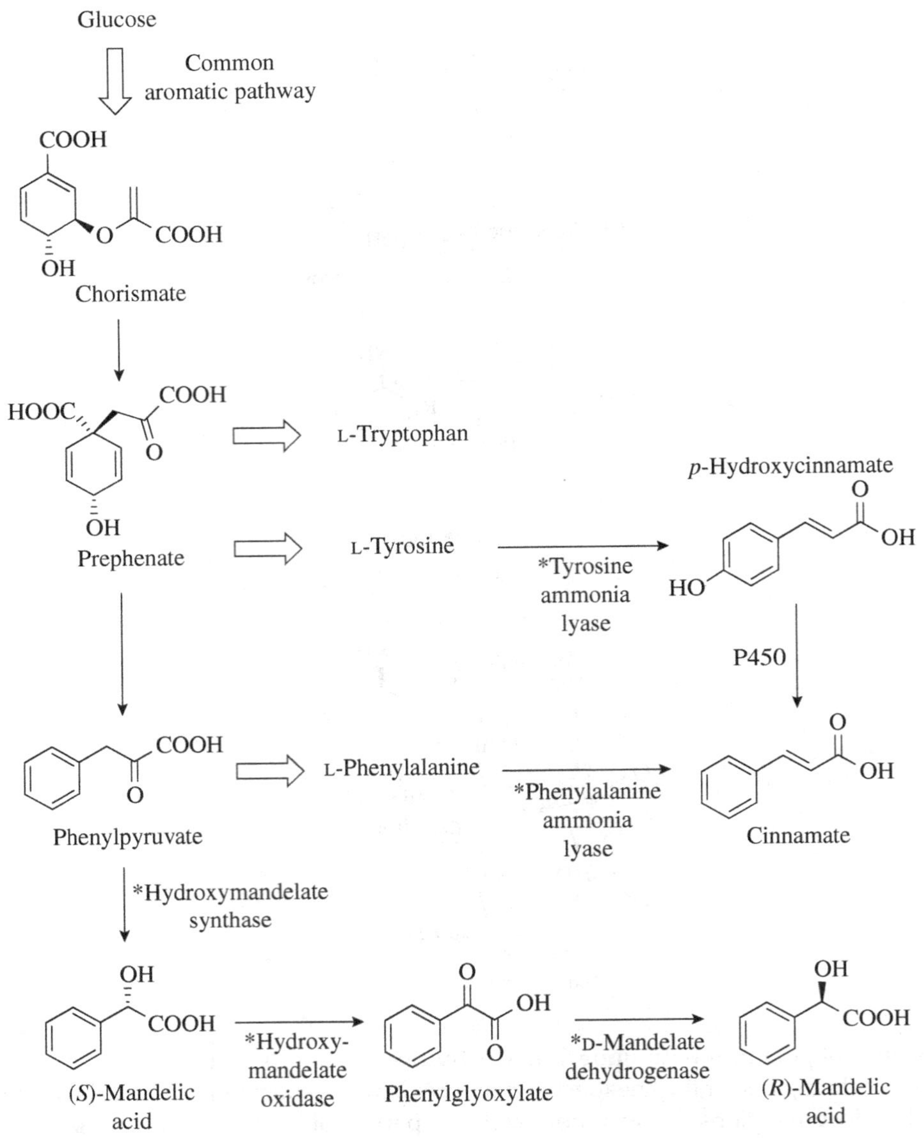

FIGURE 16.5

Production of aromatic *p*-hydroxycinnamic acid and mandelic acid from glucose by whole-cell biocatalysts. Enzymes marked (*) are heterologously expressed. Depending on the desired product chemical, enzymes mediating competing pathways will be disrupted.

phenylalanine and tyrosine in cells will be deaminated to form cinnamic acid and pHCA, respectively (Figure 16.5). Cinnamic acid can be further hydroxylated to pHCA by a cytochrome P450 monooxygenase. Ongoing efforts include enzyme engineering to increase the TAL to PAL activity so that pHCA synthesis goes through the more direct tyrosine-based route [80]. Other than glucose, microorganisms can be engineered to produce these chemicals from different sources of renewable feedstocks such as cellulose [79].

Whole-cell biocatalysts are particularly advantageous in the biosynthesis of natural products, which are often challenging to synthesize chemically [83]. This is epitomized by the microbial production of artemisinic acid, a key precursor of antimalarial drug artemisinin (Figure 16.6). As mentioned earlier, artemisinin supply based on extraction from *Artemisia annua* plant is costly, inconsistent, and inadequate to address the global demand. To tackle this problem, Keasling and coworkers transplanted the biosynthetic genes from *A. annua* into yeast and rewired its metabolic pathways to overproduce artemisinic acid [46]. Further pathway and process engineering efforts by private and public institutions led to a milestone in 2013, when Sanofi and PATH announced the large-scale production of artemisinin based on the engineered yeast that ferments glucose into artemisinic acid, which provides a more stable artemisinin supply chain to patients worldwide.

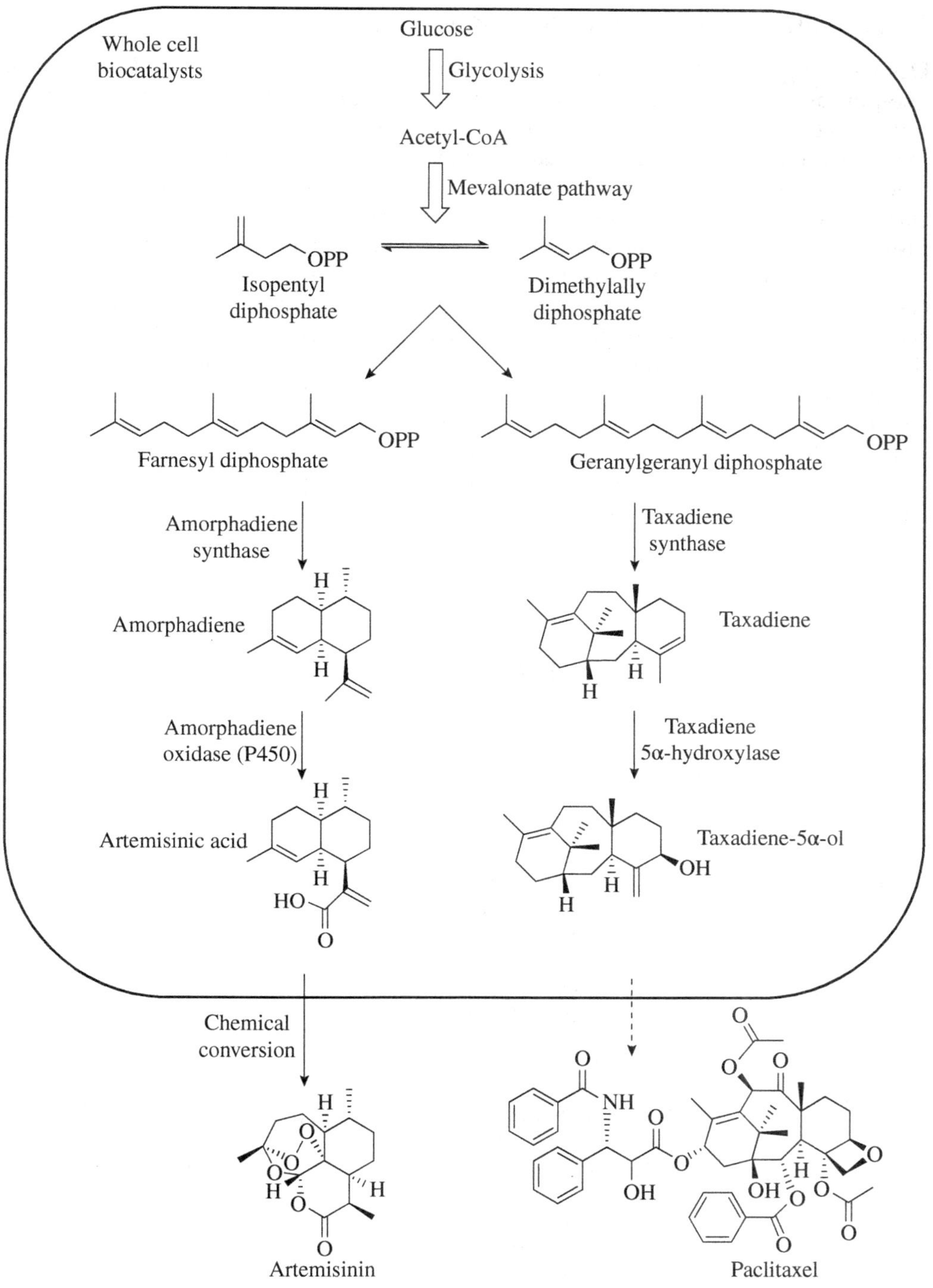

FIGURE 16.6

Production of artemisinin and paclitaxel precursors by engineered whole-cell biocatalysts from glucose. Introduction of biosynthetic genes from *Artemisia annua* encoding the amorphadiene synthase and amorphadiene oxidase yielded microbial strains that produce artemisinic acid. Artemisinic acid can be chemically converted into artemisinin. Introduction of the *Taxus* genes encoding taxadiene synthase and taxadiene 5α-hydroxylase resulted in *E. coli* strains that produce key paclitaxel intermediates. The biosynthetic pathway for paclitaxel has not been fully elucidated.

A similar approach was employed to produce an intermediate of anticancer agent paclitaxel (Taxol™) [45]. Coupling a heterologous terpenoid biosynthetic pathway and an engineered *Taxus* cytochrome P450 to the upstream isoprenoid pathway in *E. coli* cells resulted in regioselective production of taxadien-5α-ol from simple sugars by fermentation (Figure 16.6). This is the first of many steps toward the biosynthesis of baccatin III, an intermediate from which paclitaxel can be synthesized chemically. If achieved, it is likely that the resulting bioconversion using engineered microorganisms will be an economically competitive route to current paclitaxel production methods by botanic extraction and plant cell fermentations [84].

16.3.3 Multistep Biocatalytic Conversions

Despite the general compatibility between biocatalysts and apparent advantages of one-pot cascades, there are downsides to performing multiple transformations in the same reaction vessel. With different reactions running simultaneously, it is impossible to obtain an operating condition that is optimal for all reactions. In other words, most reactions in the cascade will be compromised in terms of reaction rates or enzyme stability. Moreover, side products from some reactions may be inhibitory to others as previously mentioned in the DKR of mandelic acid to form L-phenylglycine.

This is demonstrated by the various synthetic routes to 12-ketoursodeoxycholic acid (12-keto-UDCA), the optically active component and immediate precursor of secondary bile acid, ursodeoxycholic acid (UDCA), which is an important therapeutic for treatment of gallstones and liver diseases as well as a possible preventive drug against colon cancer [85–87]. Chemical synthesis of UDCA via reduction of cholic acid to chenodeoxycholic acid and its subsequent epimerization is low yielding due to poor selectivity of the redox steps and the need for protecting groups. The overall yield of the laborious seven-step procedure is <30% [88–90]. Alternatively, several chemoenzymatic routes from cholic acid to 12-keto-UDCA involving hydroxysteroid

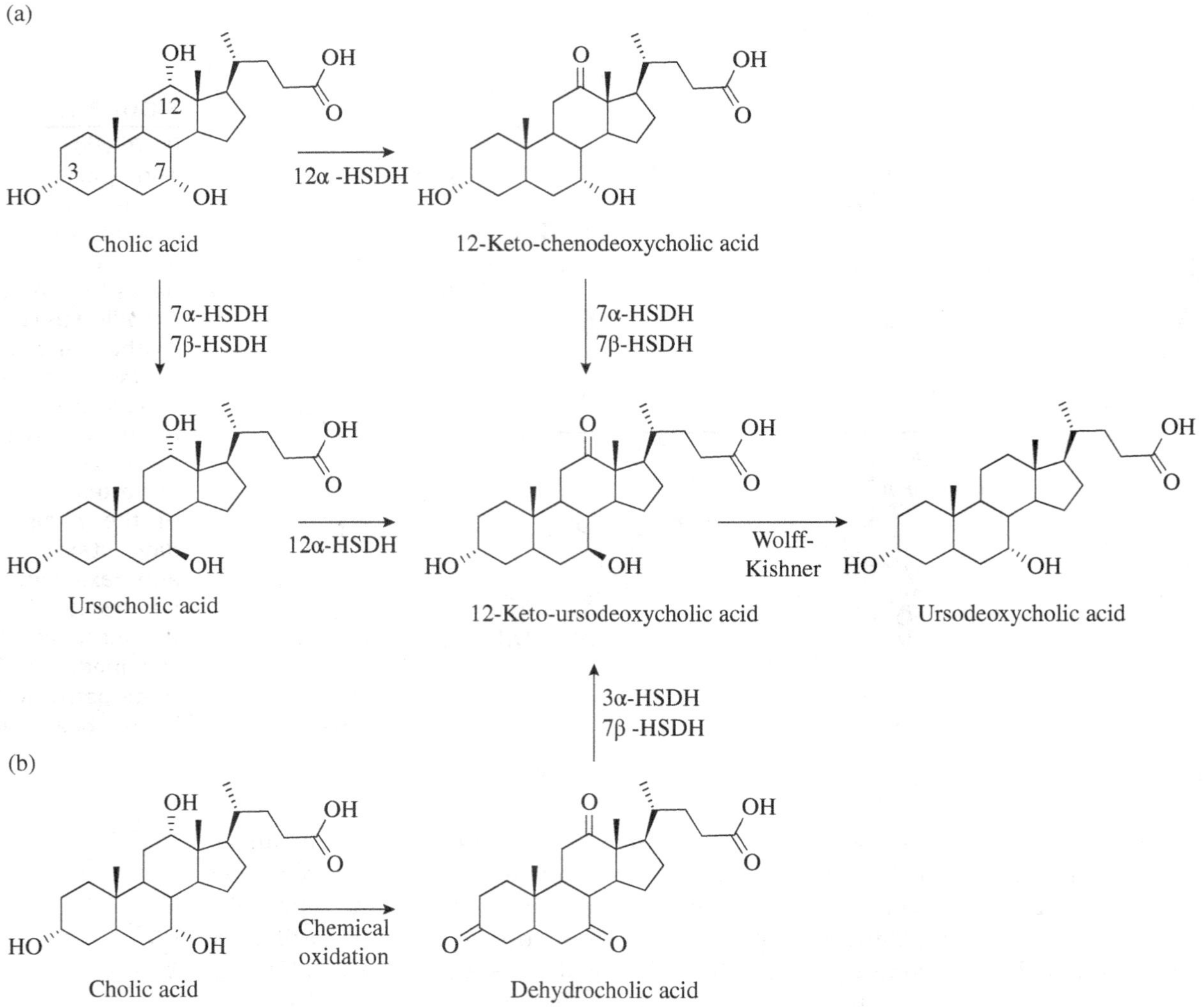

FIGURE 16.7
Chemoenzymatic routes for preparation of ursodeoxycholic acid. (a) Enzymatic conversion of cholic acid to 12-ketodeoxycholic acid. (b) Enzymatic conversion of dehydrocholic acid to 12-ketodeoxycholic acid. HSDH, hydroxysteroid dehydrogenases.

dehydrogenases (HSDHs) [91–93] and cofactor regeneration systems from various bacterial sources have been demonstrated. These routes can be broadly grouped into two main strategies (Figure 16.7). The first strategy involves the epimerization of the hydroxyl group at the C7 position and the oxidation of the C12 hydroxyl to a ketone, yielding 12-keto-UDCA (Figure 16.7a). Epimerization at C7 was achieved using alternating oxidative and reductive steps catalyzed by a NADH-dependent 7α-HSDH and a NADPH-dependent 7β-HSDH, respectively, while a NADH-dependent 12α-HSDH catalyzed the oxidation of the C12 hydroxyl. Coupled cofactor regeneration systems provided the driving force for the oxidative and reductive steps. The five-enzyme one-pot synthesis of 12-keto-UDCA is possible, but physical compartmentalization of the oxidative and reductive enzymes is needed to avoid undesired reduction of 12-keto-UDCA and lowering of the overall yield [94]. The second strategy involves an initial step consisting of chemical oxidation of cholic acid to dehydrocholic acid (DHCA) [95], followed by regio- and stereoselective reduction of the carbonyls at positions C3 and C7 by an NADH-dependent 3α-HSDH and a NADPH-dependent 7β-HSDH, respectively, to give 12-keto-UDCA (Figure 16.7b). Combining these two enzymes with a cofactor regeneration enzyme that accepts both NAD and NADPH in one-pot converted DHCA into 12-keto-UDCA with 99% yield and purity at a preparative scale without additional purification steps [96]. More than 90% conversion was also achieved at the 300 mg preparative scales using engineered whole-cell biocatalysts expressing these three enzymes [97, 98]. Altogether, the best overall yields for the bioconversion of cholic acid to 12-keto-UDCA range from 70 to 80% for both strategies depending on the source of enzymes, cofactor regeneration systems, and cascade designs [94–99]. Taking into account the 70–90% yield of the Wolff–Kishner reduction of 12-keto-UDCA [100], it is evident that the both biocatalytic strategies have significantly improved the production of UDCA.

16.4 FUTURE PERSPECTIVE: BIOCATALYSTS FOR THE PHARMACEUTICAL INDUSTRY

Biocatalysis is rapidly becoming a valuable technology in the pharmaceutical industry for drug discovery and development, where there is a critical need for pure enantiomers of complex molecules. Amidst the ongoing green chemistry revolution to develop sustainable chemicals manufacturing with little or no waste, a major synthetic challenge is C—H functionalization, which historically requires toxic reagents and is low yielding. The ability to selectively break unactivated C—H bonds and form C—C bonds, known as dehydrogenative cross-coupling, will significantly shorten synthetic routes and streamline chemical manufacturing processes [101]. Another chemical reaction of particular interest to the pharmaceutical industry is targeted late-stage fluorination of complex molecules. Selective addition of the highly electronegative fluorine to a drug can considerably improve its pharmaco-properties including pharmacokinetics, bioavailability, and delivery or binding to a drug target. Fluorinated drugs include vancomycin, chloramphenicol, Lipitor, and Prozac™. The fact that the number of fluorinated drugs has grown from 2% of the market in the 1970s to 25% today is a testament to the positive impact and opportunities provided by fluorine in the pharmaceutical industry [102]. The development of selective and efficient approaches to fluorinate complex molecules will aid development of drug candidates as well as industrial production of marketed drugs. Nonetheless, fluorination reactions require harsh, costly reagents and are often difficult to scale up. Notably, many improvements in C—H functionalization and fluorination require toxic precious metals like palladium [103–105]. While alternative metal catalysts that are abundant and less toxic are being explored, biocatalysts such as the P450 enzymes and fluorinase described in Section **16.2.10** may provide green, practical solutions to major synthetic challenges in the pharmaceutical industry.

16.4.1 Biocatalyst Discovery: New Enzymes, New Chemistries

Nature remains an important source of enzymes with new physicochemical and biological properties. Advances in genomics technologies have played a key role in allowing access to the diverse catalytic capabilities that Earth's biodiversity has to offer [106–108]. Next-generation sequencing technologies [109], and more recently, third-generation single-molecule sequencing methods [110], dramatically lowered the cost–time barriers for genomic and metagenomic studies, leading to exponentially increasing genetic information that can be mined for new natural products, biosynthetic pathways, and enzymes. Homology-based and functional screens of metagenomic libraries have identified nitrilases, esterases, oxidoreductases, and lipases with different substrate specificity and enantio- and stereoselectivity [111–114]. Transcriptomics, which connects genome to gene function, can be a powerful tool in the discovery of new enzymes and biocatalysts. RNA-seq using next-generation sequencing allows characterization of dynamic transcriptomes at single nucleotide resolution without the need for genome information [115]. Focusing on coregulated genes in the iridoid biosynthetic pathway, Geu-Flores and coworkers identified an iridoid synthase that forms the iridoid core using a mechanism that is distinct from known monoterpene synthases [116]. The discovery of this novel enzyme reveals an alternative biocatalytic route to generate cyclic scaffolds on structurally diverse bioactive products [116]. Likewise, functional proteomics can be useful in the discovery of enzymes for biotechnological applications. Activity-based protein profiling, or activity-based proteomics, uses chemical probes to isolate and identify enzymes of a desired mechanistic class and substrate specificity [117, 118]. A major upside of this approach is that it relies on activity instead of abundance or sequence homology, which can be critical when sifting through complex proteomes or metaproteomes for enzymes with high levels of the desired activity. With the advent of bioorthogonal chemistries and expanding repertoire of activity-based probes, activity-based proteomics has been used to discover and functionally annotate enzyme activities including serine and cysteine proteases, lipases, kinases, and glycosidases [117, 118].

Besides mining for naturally occurring enzymes, it is now possible to design and create functional enzymes *de novo*. Building on improved structural and mechanistic understanding of existing enzymes, bioinformatics tools allow the design of novel enzymes from known protein scaffolds based on the ideal active site geometry for a reaction of interest. This approach was successfully employed by David Baker's laboratory to design and construct enzymes capable of catalyzing Kemp elimination [119], retro-aldol [29], and Diels–Alder [29] reactions. Notably, these are reactions for which no naturally occurring enzymes exist. The first naturally occurring Diels–Alderase as described in Section **16.2.6** was later discovered [30, 120]. For organometallic catalysis, strategic placement of a carboxylate side chain to the metallocenter at the active site transformed a noncatalytic streptavidin protein into an artificial rhodium(III) metalloenzyme that is capable of catalyzing asymmetric C—H functionalization with comparable enantioselectivity to chiral cyclopentadienyl ligands [41, 121].

Technologies that expand the chemical space of enzymes beyond that of the side chains of the 20 canonical amino acids will be useful toward generating biocatalysts with novel catalytic capacities. In nature, enzymes acquire additional chemical functionalities by means of post-translational modifications or cofactors [122–124]. By reallocating nonsense and frameshift quadruplet codons, scientists have expanded the genetic code to site-specifically incorporate unnatural amino acids with diverse side chain modifications and reactivities into enzymes [125, 126]. This technology is likely to benefit from a breakthrough in expanding nature's genetic code, where scientists have demonstrated the incorporation and stable propagation of unnatural nucleotides in *E. coli* cells [127]. Additionally, unnatural chemical moieties can be selectively introduced posttranslationally using bioorthogonal labeling strategies [118, 128, 129] as well as intein-mediated protein splicing [130–132]. Collectively, these emerging technologies promise to expand the protein function space beyond that of naturally occurring enzymes [133–135].

16.4.2 Biocatalyst Development: Improvement of Desired Properties

Naturally occurring and *de novo* designed enzymes are impractical as industrial biocatalysts and need to be engineered for high performance under harsh conditions of the manufacturing process. While enzyme stability can be enhanced by immobilization, biocatalysts will also have to be optimized for activity and selectivity toward nonnatural substrates as well as compatibility with the entire chemical process. Large changes in enzyme properties require correspondingly significant alterations in the protein structure and amino acid sequence [32, 136]. Supported by advanced DNA sequencing and synthesis technologies as well as bioinformatics tools, protein engineering strategies such as rational design, directed evolution, and semirational design have been successfully applied to tailor enzymes to fit process specifications [137]. In contrast with rational design, which requires structural understanding of the enzyme, directed evolution involves the iterative process of creating enzyme variants and isolating desired variants. While the latter strategy explores a bigger functional space, it is more labor intensive given the significantly larger number of mutants to be tested and is highly dependent on the selection or screens available [138]. Semirational design combines rational design with directed evolution by focusing efforts on generating targeted subsets of functional enzyme mutants, thus maintaining diversity while reducing library sizes. Guided by bioinformatics, and sometimes deep sequencing, to map sequence–function landscapes and enrich for functional variants in enzyme libraries [136, 139–142], the semirational approach is to date the most widely used and most successful strategy for developing industrial biocatalysts [137]. Section **16.2** covers the development of individual biocatalysts for different chemical transformations.

Whole-cell biocatalysts will also have to be optimized for industrial processes. Protein engineering strategies may be used to tailor the stability, activity, selectivity, and substrate preference of individual enzymes in the heterologous hosts. In addition to transplanting biosynthetic pathways into industrial platform microorganisms, metabolic pathways will need to be rewired and rebalanced to minimize by-products and maximize flux toward the desired products [143, 144]. These modifications can be guided by the increasing number of genome-scale metabolic models, which can now be reconstructed semiautomatically [145, 146]. Alternatively, directed evolution can also be applied to cell-based biocatalysts. Technologies for combinatorial diversification and assembly of pathways generate variants that can then be screened or selected for high yields and productivity [147–153]. Breakthroughs in the development and applications of complementary next-generation genome editing technologies such as zinc finger nuclease, transcription activator-like effector nuclease (TALEN), and/or clustered regularly interspaced short palindromic repeats (CRISPR)-based genome editing systems [154–156] will greatly enhance the scope and utility of less conventional platform organisms that are currently limited by the lack of tools for efficient genetic manipulation.

Underpinning enzyme- and cell-based biocatalyst development are the rapid advances in DNA synthesis [157–161]. Entire genes, pathways, and genomes can now be synthesized *de novo* faster, cheaper, and more accurately than ever before. Notable milestones include the synthesis and assembly of the first bacterial genome as well as the first artificial eukaryotic chromosome, both of which are functional in cellular hosts [157, 162]. The development and use of biocatalysts can be expensive, and biocatalytic processes are currently the exception rather than the rule for pharmaceutical synthesis. As DNA synthesis becomes more scalable and cost effective, we envision reaching a tipping point at which customized gene or pathway libraries are routinely generated for screening or directed evolution and "designer" enzymes and cells are synthesized for desired chemical biotransformations.

16.4.3 Integration of Biocatalytic Processes

Despite the indubitable advantages of biocatalysts, it is unlikely that chemical synthesis will be put out of business. Discovering and/or developing enzymes with activities toward desired substrates remains a significant bottleneck in biocatalysis. For many chemical transformations, industrially applicable biocatalysts are not available. As such, integrating enzyme-based or whole-cell biocatalysts with the more conventional chemical catalysts opens up more possibilities. Training chemists in biocatalysis will help them to develop better routes and processes to the desired compounds. A key challenge, however, would be to find compatible conditions for the different catalytic reactions and the most efficient process for the desired chemical transformations.

To harness fully the potential of biocatalysts for complex syntheses of molecules, the integration of enzymatic with chemical catalytic steps is imperative. Improving the tolerance of enzymes to chemical reaction conditions bridges the transition between biological and chemical steps. Besides obtaining enzymes that are tolerant to harsh conditions via mutagenesis or directed evolution, enzymes can be further stabilized by encapsulation, physical immobilization, cross-linking, or chemical modifications such as PEGylation [163–165]. By allowing repeated and sustained use, immobilized enzymes are frequently employed in continuous biocatalytic processes, which circumvent production challenges associated with product inhibition and unstable intermediates, significantly increasing catalytic efficiency and overall productivity [166–168]. Membrane reactors are also often applied to separate enzymes from substrates and products and may be operated in batch or continuous processes [167]. Solvent engineering involving the introduction of water-miscible and immiscible solvents has been used to improve the efficiencies of biocatalytic processes. Aqueous–organic biphasic systems help regulate concentrations of inhibitory molecules of the enzymatic reactions [169–172]. Since the mid-2000s, ionic liquids have emerged as attractive solvents or cosolvents for biocatalytic processes as they allow organic substrates to dissolve without denaturing and deactivating enzymes [173, 174]. An added benefit of ionic solvents is the tunability of their physiochemical properties by changing the cation–anion composition, thereby allowing control of enzyme stability, activity, and the overall reaction.

One-pot reactions involving chemical and biological catalysts are particularly challenging due to the drastically different and incompatible conditions required for each to operate as well as mutual inactivation of both catalysts. Recently, considerable progress has been made in the development of concurrent one-pot chemoenzymatic processes, but the scope and number of enzymes and chemical reactions that can be carried out remain limited [175, 176]. Nonetheless, emerging strategies such as artificial metalloenzymes [41, 177], biometallic whole cells [178], and supramolecular assemblies [179] offer opportunities to bring together metal catalysis and organo- and biocatalysis to access more complex transformations.

16.5 CONCLUSION

Biocatalysis has been established as an environmentally friendly and sustainable alternative to conventional chemical processes in the pharmaceutical industry in the recent decade. Advances in recombinant DNA and genomics technologies, protein and metabolic engineering, as well as bioinformatics are critical in the discovery and development of biocatalysts for industrial processes. Significant progress has also been made in our ability to design biocatalytic cascades and to combine them with traditional chemocatalytic routes. The opportunities and combinations are endless.

Continued technological and conceptual advances in these areas will help address current synthetic challenges and realize the full potential of biocatalysts for drug discovery and development in the pharmaceutical industry.

REFERENCES

[1] PHRMA, Biopharmaceutical research and development: The process behind new medicines. Pharmaceutical Research and Manufacturers of America 2015; Available from: http://www.phrma.org/sites/default/files/pdf/rd_brochure_022307.pdf (accessed January 6, 2015).

[2] Huisman, G.W. and S.J. Collier, On the development of new biocatalytic processes for practical pharmaceutical synthesis. *Curr. Opin. Chem. Biol.*, 2013. **17**(2): 284–292.

[3] Clouthier, C.M. and J.N. Pelletier, Expanding the organic toolbox: A guide to integrating biocatalysis in synthesis. *Chem. Soc. Rev.*, 2012. **41**(4): 1585–1605.

[4] Muñoz Solano, D., et al., Industrial biotransformations in the synthesis of building blocks leading to enantiopure drugs. *Bioresour. Technol.*, 2012. **115**: 196–207.

[5] Hayes, S.T., et al., Commercial synthesis of (*S*,*S*)-reboxetine succinate: A journey to find the cheapest commercial chemistry for manufacture. *Org. Process Res. Dev.*, 2011. **15**(6): 1305–1314.

[6] Truppo, M.D. and G. Hughes, Development of an improved immobilized CAL-B for the enzymatic resolution of a key intermediate to odanacatib. *Org. Process Res. Dev.*, 2011. **15**(5): 1033–1035.

[7] Stueckler, C., et al., Asymmetric synthesis of (*R*)-3-hydroxy-2-methylpropanoate ('roche ester') and derivatives via biocatalytic C═C-bond reduction. *Adv. Synth. Catal.*, 2010. **352**(14–15): 2663–2666.

[8] Bechtold, M., et al., Biotechnological development of a practical synthesis of ethyl (*S*)-2-ethoxy-3-(p-methoxyphenyl)propanoate (EEHP): Over 100-fold productivity increase from yeast whole cells to recombinant isolated enzymes. *Org. Process Res. Dev.*, 2011. **16**(2): 269–276.

[9] Liang, J., et al., Development of a biocatalytic process as an alternative to the (−)-DIP-Cl-mediated asymmetric reduction of a key intermediate of montelukast. *Org. Process Res. Dev.*, 2009. **14**(1): 193–198.

[10] Crowe, M.A., et al., February 2013. Biocatalysts for ezetimibe synthesis. US patent 20130052699.

[11] Dietrich, J.A., et al., A novel semi-biosynthetic route for artemisinin production using engineered substrate-promiscuous $P450_{BM3}$. *ACS Chem. Biol.*, 2009. **4**(4): 261–267.

[12] Rentmeister, A., F.H. Arnold, and R. Fasan, Chemo-enzymatic fluorination of unactivated organic compounds. *Nat. Chem. Biol.*, 2009. **5**(1): 26–28.

[13] Köhler, V., et al., Enantioselective biocatalytic oxidative desymmetrization of substituted pyrrolidines. *Angew. Chem. Int. Ed.*, 2010. **49**(12): 2182–2184.

[14] Znabet, A., et al., A highly efficient synthesis of telaprevir by strategic use of biocatalysis and multicomponent reactions. *Chem. Commun.*, 2010. **46**(42): 7918–7920.

[15] Li, T., et al., Efficient, chemoenzymatic process for manufacture of the boceprevir bicyclic [3.1.0]proline intermediate based on amine oxidase-catalyzed desymmetrization. *J. Am. Chem. Soc.*, 2012. **134**(14): 6467–6472.

[16] Rehdorf, J., et al., Enzymatic synthesis of enantiomerically pure β-amino ketones, β-amino esters, and β-amino alcohols with Baeyer–Villiger monooxygenases. *Chem. Eur. J.*, 2010. **16**(31): 9525–9535.

[17] Rehdorf, J., M.D. Mihovilovic, and U.T. Bornscheuer, Exploiting the regioselectivity of Baeyer–Villiger monooxygenases for the formation of β-amino acids and β-amino alcohols. *Angew. Chem. Int. Ed.*, 2010. **49**(26): 4506–4508.

[18] Reetz, M.T. and S. Wu, Laboratory evolution of robust and enantioselective Baeyer–Villiger monooxygenases for asymmetric catalysis. *J. Am. Chem. Soc.*, 2009. **131**(42): 15424–15432.

[19] Bong, Y.K., et al., January 2013. Synthesis of prazole compounds. US patent 20130017580.

[20] Ang, E.L., et al., October 2013. Biocatalysts and methods for the synthesis of armodafinil. US patent 20130260426.

[21] Werner, L., A. Machara, and T. Hudlicky, Short chemoenzymatic azide-free synthesis of oseltamivir (Tamiflu): Approaching the potential for process efficiency. *Adv. Synth. Catal.*, 2010. **352**(1): 195–200.

[22] Concia, A.L., et al., D-Fructose-6-phosphate aldolase in organic synthesis: Cascade chemical-enzymatic preparation of sugar-related polyhydroxylated compounds. *Chem. Eur. J.*, 2009. **15**(15): 3808–3816.
[23] Garrabou, X., et al., Asymmetric self- and cross-aldol reactions of glycolaldehyde catalyzed by D-fructose-6-phosphate aldolase. *Angew. Chem. Int. Ed.*, 2009. **48**(30): 5521–5525.
[24] Tang, R.-C., et al., Enzyme-catalyzed Henry (nitroaldol) reaction. *J. Mol. Catal. B: Enzym.*, 2010. **63**(1–2): 62–67.
[25] Cai, J.-F., Z. Guan, and Y.-H. He, The lipase-catalyzed asymmetric C—C Michael addition. *J. Mol. Catal. B: Enzym.*, 2011. **68**(3–4): 240–244.
[26] de Souza, R.O.M.A., et al., Michael additions of primary and secondary amines to acrylonitrile catalyzed by lipases. *Tetrahedron Lett.*, 2009. **50**(17): 2017–2018.
[27] Strohmeier, G.A., et al., Investigation of lipase-catalyzed Michael-type carbon–carbon bond formations. *Tetrahedron*, 2009. **65**(29–30): 5663–5668.
[28] Zandvoort, E., et al., Bridging between organocatalysis and biocatalysis: Asymmetric addition of acetaldehyde to β-nitrostyrenes catalyzed by a promiscuous proline-based tautomerase. *Angew. Chem. Int. Ed.*, 2012. **51**(5): 1240–1243.
[29] Siegel, J.B., et al., Computational design of an enzyme catalyst for a stereoselective bimolecular Diels–Alder reaction. *Science*, 2010. **329**(5989): 309–313.
[30] Kim, H.J., et al., Enzyme-catalysed [4 + 2] cycloaddition is a key step in the biosynthesis of spinosyn A. *Nature*, 2011. **473**(7345): 109–112.
[31] Ruff, B.M., S. Bräse, and S.E. O'Connor, Biocatalytic production of tetrahydroisoquinolines. *Tetrahedron Lett.*, 2012. **53**(9): 1071–1074.
[32] Savile, C.K., et al., Biocatalytic asymmetric synthesis of chiral amines from ketones applied to sitagliptin manufacture. *Science*, 2010. **329**(5989): 305–309.
[33] Molinaro, C., et al., CRTH2 antagonist MK-7246: A synthetic evolution from discovery through development. *J. Org. Chem.*, 2012. **77**(5): 2299–2309.
[34] Mangion, I.K., et al., Enantioselective synthesis of a dual orexin receptor antagonist. *Org. Lett.*, 2012. **14**(13): 3458–3461.
[35] Austad, B., et al., April 2014. Enzymatic transamination of cyclopamine analogs. US patent 8703448.
[36] Mero, A., et al., A new method to increase selectivity of transglutaminase mediated PEGylation of salmon calcitonin and human growth hormone. *J. Control. Release*, 2011. **154**(1): 27–34.
[37] Walker, M.C., et al., Expanding the fluorine chemistry of living systems using engineered polyketide synthase pathways. *Science*, 2013. **341**(6150): 1089–1094.
[38] Ma, S.K., et al., A green-by-design biocatalytic process for atorvastatin intermediate. *Green Chem.*, 2010. **12**(1): 81–86.
[39] Majerić Elenkov, M., et al., Catalytic activity of halohydrin dehalogenases towards spiroepoxides. *Org. Biomol. Chem.*, 2012. **10**(26): 5063–5072.
[40] Jin, J., et al., MhyADH catalysed Michael addition of water and in situ oxidation. *Chem. Commun.*, 2010. **46**(45): 8588–8590.
[41] Hyster, T.K., et al., Biotinylated Rh(III) complexes in engineered streptavidin for accelerated asymmetric C—H activation. *Science*, 2012. **338**(6106): 500–503.
[42] Secundo, F. and G. Carrea, Lipase activity and conformation in neat organic solvents. *J. Mol. Catal. B: Enzym.*, 2002. **19–20**: 93–102.
[43] Stuermer, R., et al., Asymmetric bioreduction of activated C=C bonds using enoate reductases from the old yellow enzyme family. *Curr. Opin. Chem. Biol.*, 2007. **11**(2): 203–213.
[44] Mihovilovic, M.D., Enzyme mediated Baeyer-Villiger oxidations. *Curr. Org. Chem.*, 2006. **10**(11): 1265–1287.
[45] Ajikumar, P.K., et al., Isoprenoid pathway optimization for Taxol precursor overproduction in *Escherichia coli*. *Science*, 2010. **330**(6000): 70–74.
[46] Ro, D.-K., et al., Production of the antimalarial drug precursor artemisinic acid in engineered yeast. *Nature*, 2006. **440**(7086): 940–943.
[47] O'Hagan, D., Recent developments on the fluorinase from *Streptomyces cattleya*. *J. Fluor. Chem.*, 2006. **127**(11): 1479–1483.
[48] Dach, R., et al., The right criteria defining a good chemical manufacturing process. *Org. Process Res. Dev.*, 2012. **16**(11): 1697–1706.
[49] Ni, Y. and J.H. Xu, Biocatalytic ketone reduction: A green and efficient access to enantiopure alcohols. *Biotechnol. Adv.*, 2012. **30**(6): 1279–1288.
[50] Goldberg, K., et al., Biocatalytic ketone reduction—A powerful tool for the production of chiral alcohols—Part I: Processes with isolated enzymes. *Appl. Microbiol. Biotechnol.*, 2007. **76**(2): 237–248.

[51] Rachwalski, M., N. Vermue, and F.P. Rutjes, Recent advances in enzymatic and chemical deracemisation of racemic compounds. *Chem. Soc. Rev.*, 2013. **42**(24): 9268–9282.
[52] Ricca, E., B. Brucher, and J.H. Schrittwieser, Multi-enzymatic cascade reactions: Overview and perspectives. *Adv. Synth. Catal.*, 2011. **353**: 2239–2262.
[53] Schrittwieser, J.H., et al., Recent biocatalytic oxidation-reduction cascades. *Curr. Opin. Chem. Biol.*, 2011. **15**(2): 249–256.
[54] Zheng, G.W. and J.H. Xu, New opportunities for biocatalysis: Driving the synthesis of chiral chemicals. *Curr. Opin. Biotechnol.*, 2011. **22**(6): 784–792.
[55] Yasukawa, K., R. Hasemi, and Y. Asano, Dynamic kinetic resolution of α-aminonitriles to form chiral α-amino acids. *Adv. Synth. Catal.*, 2011. **353**: 2328–2332.
[56] Resch, V., W.M.F. Fabian, and W. Kroutil, Deracemisation of mandelic acid to optically pure non-natural L-phenylglycine via a redox-neutral biocatalytic cascade. *Adv. Synth. Catal.*, 2010. **352**(6): 993–997.
[57] Chen, Y., et al., Enzymatic preparation of an (*S*)-amino acid from a racemic amino acid. *Org. Process Res. Dev.*, 2011. **15**: 241–248.
[58] Koszelewski, D., et al., Deracemisation of α-chiral primary amines by a one-pot, two-step cascade reaction catalysed by ω-transaminases. *Eur. J. Org. Chem.*, 2009. **2009**(14): 2289–2292.
[59] Truppo, M.D., N.J. Turner, and J.D. Rozzell, Efficient kinetic resolution of racemic amines using a transaminase in combination with an amino acid oxidase. *Chem. Commun.*, 2009(**16**): 2127–2129.
[60] Koszelewski, D., et al., Enzymatic racemization of amines catalyzed by enantiocomplementary ω-transaminases. *Chem. Eur. J.*, 2011. **17**(1): 378–383.
[61] Simon, R.C., et al., Recent developments of cascade reactions involving ω-transaminases. *ACS Catal.*, 2014. **4**(1): 129–143.
[62] Koszelewski, D., et al., Deracemization of mexiletine biocatalyzed by omega-transaminases. *Org. Lett.*, 2009. **11**(21): 4810–4812.
[63] Cox, B.M., J.B. Bilsborrow, and K.D. Walker, Enhanced conversion of racemic alpha-arylalanines to (*R*)-beta-arylalanines by coupled racemase/aminomutase catalysis. *J. Org. Chem.*, 2009. **74**(18): 6953–6959.
[64] Turner, N.J., Ammonia lyases and aminomutases as biocatalysts for the synthesis of α-amino and β-amino acids. *Curr. Opin. Chem. Biol.*, 2011. **15**(2): 234–240.
[65] Wu, B., et al., Efficient tandem biocatalytic process for the kinetic resolution of aromatic β-amino acids. *Adv. Synth. Catal.*, 2010. **352**(9): 1409–1412.
[66] Carballeira, J.D., et al., Microbial cells as catalysts for stereoselective red-ox reactions. *Biotechnol. Adv.*, 2009. **27**(6): 686–714.
[67] Lee, W.H., et al., Engineering of NADPH regenerators in *Escherichia coli* for enhanced biotransformation. *Appl. Microbiol. Biotechnol.*, 2013. **97**(7): 2761–2772.
[68] Ward, M., et al., Anti-HIV-1 activity of poly(mandelic acid) derivatives. *Biomacromolecules*, 2007. **8**(11): 3308–3316.
[69] Whitesell, J.K. and D. Reynolds, Resolution of chiral alcohols with mandelic acid. *J. Org. Chem.*, 1983. **48**: 3548–3551.
[70] Bhushan, R. and C. Agarwal, Direct enantiomeric TLC resolution of DL-penicillamine using (*R*)-mandelic acid and L-tartaric acid as chiral impregnating reagents and as chiral mobile phase additive. *Biomed. Chromatogr.*, 2008. **22**(11): 1237–1242.
[71] Inghardt, T., January 2011. Mandelic acid derivatives and their use as thrombin inhibitors. Canada patent 2456426.
[72] Brittain, H.G., *Profiles of Drug Substances, Excipients and Related Methodology*. Vol. **34**. 2011, Cambridge, MA: Elsevier/Academic Press.
[73] Blixt, H.J., et al., June 2011. Process for the preparation and resolution of mandelic acid derivatives. US patent 7960582.
[74] Rustler, S., et al., Simultaneous expression of an arylacetonitrilase from *Pseudomonas fluorescens* and a (*S*)-oxynitrilase from *Manihot esculenta* in *Pichia pastoris* for the synthesis of (*S*)-mandelic acid. *Appl. Microbiol. Biotechnol.*, 2008. **80**(1): 87–97.
[75] Banerjee, A., P. Kaul, and U.C. Banerjee, Enhancing the catalytic potential of nitrilase from *Pseudomonas putida* for stereoselective nitrile hydrolysis. *Appl. Microbiol. Biotechnol.*, 2006. **72**(1): 77–87.
[76] He, Y.C., et al., Preparation of (*R*)-(−)-mandelic acid and its derivatives from racemates by enantioselective degradation with a newly isolated bacterial strain *Alcaligenes* sp. ECU0401. *Bioprocess Biosyst. Eng.*, 2008. **31**(5): 445–451.

[77] Huang, H.-R. and J.-H. Xu, Preparation of (S)-mandelic acid from racemate using growing cells of *Pseudomonas putida* ECU1009 with (*R*)-mandelate degradation activity. *Biochem. Eng. J.*, 2006. **30**(1): 11–15.
[78] Sun, Z., et al., Metabolic engineering of the L-phenylalanine pathway in *Escherichia coli* for the production of *S*- or *R*-mandelic acid. *Microb. Cell Fact.*, 2011. **10**: 71.
[79] Kawai, Y., et al., p-Hydroxycinnamic acid production directly from cellulose using endoglucanase- and tyrosine ammonia lyase-expressing *Streptomyces lividans*. *Microb. Cell Fact.*, 2013. **12**(1): 45.
[80] Vannelli, T., et al., Production of p-hydroxycinnamic acid from glucose in *Saccharomyces cerevisiae* and *Escherichia coli* by expression of heterologous genes from plants and fungi. *Metab. Eng.*, 2007. **9**(2): 142–151.
[81] Nijkamp, K., et al., Optimization of the solvent-tolerant *Pseudomonas putida* S12 as host for the production of p-coumarate from glucose. *Appl. Microbiol. Biotechnol.*, 2007. **74**(3): 617–624.
[82] Xue, Z., et al., Identification, characterization and functional expression of a tyrosine ammonia-lyase and its mutants from the photosynthetic bacterium *Rhodobacter sphaeroides*. *J. Ind. Microbiol. Biotechnol.*, 2007. **34**(9): 599–604.
[83] Chang, M.C. and J.D. Keasling, Production of isoprenoid pharmaceuticals by engineered microbes. *Nat. Chem. Biol.*, 2006. **2**(12): 674–681.
[84] Bringi, V., et al., September 2007. Enhanced production of taxol and taxanes by cell cultures of Taxus species. US patent 7264951.
[85] Combes, B., et al., A randomized, double-blind, placebo-controlled trial of ursodeoxycholic acid in primary biliary cirrhosis. *Hepatology*, 1995. **22**(3): 759–766.
[86] Khare, S., et al., Ursodeoxycholic acid inhibits Ras mutations, wild-type Ras activation, and cyclooxygenase-2 expression in colon cancer. *Cancer Res.*, 2003. **63**(13): 3517–3523.
[87] Gleeson, D., et al., Final outcome of ursodeoxycholic acid treatment in 126 patients with radiolucent gallstones. *QJM: Int. J. Med.*, 1990. **76**(279): 711–729.
[88] Hofmann, A.F., The preparation of chenodeoxycholic acid and its glycine and taurine conjugates. *Acta Chem. Scand.*, 1963. **17**: 173–186.
[89] Fieser, L.F. and S. Rajagopalan, Oxidation of steroids. III. Selective oxidations and acylations of steroids in the bile acid series. *J. Am. Chem. Soc.*, 1950. **72**: 5530–5536.
[90] Fieser, L.F., et al., Cathylation (carboethoxylation) of steroid alcohols. *J. Am. Chem. Soc.*, 1952. **74**: 3309–3313.
[91] Weuster-Botz, D., et al., March 2014. Novel 7 beta-hydroxysteroid dehydrogenase mutants and process for the preparation of ursodeoxycholic acid. US patent 20140087421.
[92] Schmid, R., et al., August 2013. Novel 7 alpha-hydroxysteroid dehydrogenase knockout mutants and use thereof. US patent 20130224792.
[93] Aigner, A., et al., April 2011. Novel 12 alpha-hydroxysteroid dehydrogenases, production and use thereof. US patent 20110091921.
[94] Monti, D., et al., One-pot multienzymatic synthesis of 12-ketoursodeoxycholic acid: Subtle cofactor specificities rule the reaction equilibria of five biocatalysts working in a row. *Adv. Synth. Catal.*, 2009. **351**(9): 1303–1311.
[95] Bert, S. and D.F. Hinkley, December 1960. Oxidation of cholic acid with chlorine. US patent 2966499.
[96] Liu, L., et al., One-step synthesis of 12-ketoursodeoxycholic acid from dehydrocholic acid using a multienzymatic system. *Appl. Microbiol. Biotechnol.*, 2013. **97**(2): 633–639.
[97] Braun, M., et al., Novel whole-cell biocatalysts with recombinant hydroxysteroid dehydrogenases for the asymmetric reduction of dehydrocholic acid. *Appl. Microbiol. Biotechnol.*, 2012. **95**(6): 1457–1468.
[98] Sun, B., et al., Multi-enzymatic one-pot reduction of dehydrocholic acid to 12-keto-ursodeoxycholic acid with whole-cell biocatalysts. *Biotechnol. Bioeng.*, 2013. **110**(1): 68–77.
[99] Fossati, E., et al., Exploitation of the alcohol dehydrogenase-acetone NADP-regeneration system for the enzymatic preparative-scale production of 12-ketochenodeoxycholic acid. *Biotechnol. Bioeng.*, 2006. **93**(6): 1216–1220.
[100] Sutherland, J.D., I.A. Macdonald, and T.P. Forrest, The enzymic and chemical synthesis of ursodeoxycholic and chenodeoxycholic acid from cholic acid. *Prep. Biochem.*, 1982. **12**(4): 307–321.
[101] Wang, D.H., et al., Ligand-enabled reactivity and selectivity in a synthetically versatile aryl C—H olefination. *Science*, 2010. **327**(5963): 315–319.
[102] Wang, J., et al., Fluorine in pharmaceutical industry: Fluorine-containing drugs introduced to the market in the last decade (2001–2011). *Chem. Rev.*, 2014. **114**(4): 2432–2506.

[103] Ackermann, L., Carboxylate-assisted transition-metal-catalyzed C—H bond functionalizations: Mechanism and scope. *Chem. Rev.*, 2011. **111**(3): 1315–1345.
[104] Yeung, C.S. and V.M. Dong, Catalytic dehydrogenative cross-coupling: Forming carbon-carbon bonds by oxidizing two carbon-hydrogen bonds. *Chem. Rev.*, 2011. **111**(3): 1215–1292.
[105] Lee, H.G., P.J. Milner, and S.L. Buchwald, Pd-catalyzed nucleophilic fluorination of aryl bromides. *J. Am. Chem. Soc.*, 2014. **136**(10): 3792–3795.
[106] Challis, G.L., Mining microbial genomes for new natural products and biosynthetic pathways. *Microbiology*, 2008. **154**(Pt 6): 1555–1569.
[107] Nett, M., H. Ikeda, and B.S. Moore, Genomic basis for natural product biosynthetic diversity in the actinomycetes. *Nat. Prod. Rep.*, 2009. **26**(11): 1362–1384.
[110] Schloss, P.D. and J. Handelsman, Metagenomics for studying unculturable microorganisms: Cutting the Gordian knot. *Genome Biol.*, 2005. **6**(8): 229.
[111] Metzker, M.L., Sequencing technologies—The next generation. *Nat. Rev. Genet.*, 2010. **11**(1): 31–46.
[112] Ozsolak, F., Third-generation sequencing techniques and applications to drug discovery. *Expert Opin. Drug Discovery*, 2012. **7**(3): 231–243.
[113] Fernandez-Arrojo, L., et al., Metagenomic era for biocatalyst identification. *Curr. Opin. Biotechnol.*, 2010. **21**(6): 725–733.
[114] Steele, H.L., et al., Advances in recovery of novel biocatalysts from metagenomes. *J. Mol. Microbiol. Biotechnol.*, 2009. **16**(1–2): 25–37.
[115] Iqbal, H.A., Z. Feng, and S.F. Brady, Biocatalysts and small molecule products from metagenomic studies. *Curr. Opin. Chem. Biol.*, 2012. **16**(1–2): 109–116.
[116] Lorenz, P. and J. Eck, Metagenomics and industrial applications. *Nat. Rev. Microbiol.*, 2005. **3**(6): 510–516.
[117] Martin, J.A. and Z. Wang, Next-generation transcriptome assembly. *Nat. Rev. Genet.*, 2011. **12**(10): 671–682.
[118] Geu-Flores, F., et al., An alternative route to cyclic terpenes by reductive cyclization in iridoid biosynthesis. *Nature*, 2012. **492**(7427): 138–142.
[119] Heal, W.P., T.H. Dang, and E.W. Tate, Activity-based probes: Discovering new biology and new drug targets. *Chem. Soc. Rev.*, 2011. **40**(1): 246–257.
[120] Willems, L.I., et al., Bioorthogonal chemistry: Applications in activity-based protein profiling. *Acc. Chem. Res.*, 2011. **44**(9): 718–729.
[121] Rothlisberger, D., et al., Kemp elimination catalysts by computational enzyme design. *Nature*, 2008. **453**(7192): 190–195.
[122] Townsend, C.A., A "diels-alderase" at last. *ChemBioChem*, 2011. **12**(15): 2267–2269.
[123] Ye, B. and N. Cramer, Chiral cyclopentadienyl ligands as stereocontrolling element in asymmetric C—H functionalization. *Science*, 2012. **338**(6106): 504–506.
[124] Silverman, R.B., *Organic Chemistry of Enzyme-Catalyzed Reactions*. 2nd ed. 2002, Cambridge, MA: Elsevier/Academic Press.
[125] Walsh, C.T., Posttranslational modification of proteins: Expanding nature's inventory. *Angew. Chem. Int. Ed.*, 2005. **45**(7): 1020.
[126] Yukl, E.T. and C.M. Wilmot, Cofactor biosynthesis through protein post-translational modification. *Curr. Opin. Chem. Biol.*, 2012. **16**(1–2): 54–59.
[127] Liu, C.C. and P.G. Schultz, Adding new chemistries to the genetic code. *Annu. Rev. Biochem.*, 2010. **79**: 413–444.
[128] Wang, L., J. Xie, and P.G. Schultz, Expanding the genetic code. *Annu. Rev. Biophys. Biomol. Struct.*, 2006. **35**: 225–249.
[129] Malyshev, D.A., et al., A semi-synthetic organism with an expanded genetic alphabet. *Nature*, 2014. **509**(7500): 385–388.
[130] Bertozzi, C.R., A decade of bioorthogonal chemistry. *Acc. Chem. Res.*, 2011. **44**(9): 651–653.
[131] Jing, C. and V.W. Cornish, Chemical tags for labeling proteins inside living cells. *Acc. Chem. Res.*, 2011. **44**(9): 784–792.
[130] Lockless, S.W. and T.W. Muir, Traceless protein splicing utilizing evolved split inteins. *Proc. Natl. Acad. Sci. U. S. A.*, 2009. **106**(27): 10999–11004.
[131] Muralidharan, V. and T.W. Muir, Protein ligation: An enabling technology for the biophysical analysis of proteins. *Nat. Methods*, 2006. **3**(6): 429–438.
[132] Vila-Perello, M. and T.W. Muir, Biological applications of protein splicing. *Cell*, 2010. **143**(2): 191–200.
[133] Brustad, E.M. and F.H. Arnold, Optimizing non-natural protein function with directed evolution. *Curr. Opin. Chem. Biol.*, 2011. **15**(2): 201–210.

[134] Liu, C.C., et al., Mutagenesis and evolution of sulfated antibodies using an expanded genetic code. *Biochemistry*, 2009. **48**(37): 8891–8898.
[135] Liu, C.C., et al., Protein evolution with an expanded genetic code. *Proc. Natl. Acad. Sci. U. S. A.*, 2008. **105**(46): 17688–17693.
[136] Fox, R.J., et al., Improving catalytic function by ProSAR-driven enzyme evolution. *Nat. Biotechnol.*, 2007. **25**(3): 338–344.
[137] Bornscheuer, U.T., et al., Engineering the third wave of biocatalysis. *Nature*, 2012. **485**(7397): 185–194.
[138] Dietrich, J.A., A.E. McKee, and J.D. Keasling, High-throughput metabolic engineering: Advances in small-molecule screening and selection. *Annu. Rev. Biochem.*, 2010. **79**: 563–590.
[139] Meyer, M.M., L. Hochrein, and F.H. Arnold, Structure-guided SCHEMA recombination of distantly related beta-lactamases. *Protein Eng. Des. Sel.*, 2006. **19**(12): 563–570.
[140] Whitehead, T.A., et al., Optimization of affinity, specificity and function of designed influenza inhibitors using deep sequencing. *Nat. Biotechnol.*, 2012. **30**(6): 543–548.
[141] Fowler, D.M., et al., High-resolution mapping of protein sequence-function relationships. *Nat. Methods*, 2010. **7**(9): 741–746.
[142] Weinreich, D.M., et al., Darwinian evolution can follow only very few mutational paths to fitter proteins. *Science*, 2006. **312**(5770): 111–114.
[143] Keasling, J.D., Manufacturing molecules through metabolic engineering. *Science*, 2010. **330**(6009): 1355–1358.
[144] Pirie, C.M., et al., Integrating the protein and metabolic engineering toolkits for next-generation chemical biosynthesis. *ACS Chem. Biol.*, 2013. **8**(4): 662–672.
[145] Henry, C.S., et al., High-throughput generation, optimization and analysis of genome-scale metabolic models. *Nat. Biotechnol.*, 2010. **28**(9): 977–982.
[146] Thiele, I. and B.O. Palsson, A protocol for generating a high-quality genome-scale metabolic reconstruction. *Nat. Protoc.*, 2010. **5**(1): 93–121.
[147] Li, Y., et al., Multiplex iterative plasmid engineering for combinatorial optimization of metabolic pathways and diversification of protein coding sequences. *ACS Synth. Biol.*, 2013. **2**(11): 651–661.
[148] Du, J., et al., Customized optimization of metabolic pathways by combinatorial transcriptional engineering. *Nucleic Acids Res.*, 2012. **40**(18): e142.
[149] Colloms, S.D., et al., Rapid metabolic pathway assembly and modification using serine integrase site-specific recombination. *Nucleic Acids Res.*, 2014. **42**(4): e23.
[150] Coussement, P., et al., One step DNA assembly for combinatorial metabolic engineering. *Metab. Eng.*, 2014. **23**: 70–77.
[151] Gibson, D.G., et al., Enzymatic assembly of DNA molecules up to several hundred kilobases. *Nat. Methods*, 2009. **6**(5): 343–345.
[152] Quan, J. and J. Tian, Circular polymerase extension cloning for high-throughput cloning of complex and combinatorial DNA libraries. *Nat. Protoc.*, 2011. **6**(2): 242–251.
[153] Santos, C.N. and G. Stephanopoulos, Combinatorial engineering of microbes for optimizing cellular phenotype. *Curr. Opin. Chem. Biol.*, 2008. **12**(2): 168–176.
[154] Qi, L.S., et al., Repurposing CRISPR as an RNA-guided platform for sequence-specific control of gene expression. *Cell*, 2013. **152**(5): 1173–1183.
[155] Sun, N. and H. Zhao, Transcription activator-like effector nucleases (TALENs): A highly efficient and versatile tool for genome editing. *Biotechnol. Bioeng.*, 2013. **110**(7): 1811–1821.
[156] Gersbach, C.A., T. Gaj, and C.F. Barbas, 3rd, Synthetic zinc finger proteins: The advent of targeted gene regulation and genome modification technologies. *Acc. Chem. Res.*, 2014. **47**(8): 2309–2318.
[157] Gibson, D.G., et al., Creation of a bacterial cell controlled by a chemically synthesized genome. *Science*, 2010. **329**(5987): 52–56.
[158] Quan, J., et al., Parallel on-chip gene synthesis and application to optimization of protein expression. *Nat. Biotechnol.*, 2011. **29**(5): 449–452.
[159] Matzas, M., et al., High-fidelity gene synthesis by retrieval of sequence-verified DNA identified using high-throughput pyrosequencing. *Nat. Biotechnol.*, 2010. **28**(12): 1291–1294.
[160] LeProust, E.M., et al., Synthesis of high-quality libraries of long (150mer) oligonucleotides by a novel depurination controlled process. *Nucleic Acids Res.*, 2010. **38**(8): 2522–2540.

[161] Richmond, K.E., et al., Amplification and assembly of chip-eluted DNA (AACED): A method for high-throughput gene synthesis. *Nucleic Acids Res.*, 2004. **32**(17): 5011–5018.
[162] Annaluru, N., et al., Total synthesis of a functional designer eukaryotic chromosome. *Science*, 2014. **344**(6179): 55–58.
[163] Bornscheuer, U.T., Immobilizing enzymes: How to create more suitable biocatalysts. *Angew. Chem. Int. Ed.*, 2003. **42**(29): 3336–3337.
[164] Gupta, M.N. and S. Raghava, Enzyme stabilization via cross-linked enzyme aggregates. *Methods Mol. Biol.*, 2011. **679**: 133–145.
[165] Laurent, N., R. Haddoub, and S.L. Flitsch, Enzyme catalysis on solid surfaces. *Trends Biotechnol.*, 2008. **26**(6): 328–337.
[166] Nadeau, L.J., H.R. Luckarift, and J.C. Spain, December 2011. Biocatalytic process for the production of ortho-aminophenols from chloramphenicol and analogs. US patent 8071340.
[167] Rao, N.N., Continuous biocatalytic processes. *Org. Process Res. Dev.*, 2009. **13**: 1202–1205.
[168] Jones, E., et al., Biocatalytic oxidase: Batch to continuous. *Chem. Eng. Res. Des.*, 2012. **90**(6): 726–731.
[169] Ni, Y., et al., Highly stereoselective reduction of prochiral ketones by a bacterial reductase coupled with cofactor regeneration. *Org. Biomol. Chem.*, 2011. **9**(15): 5463–5468.
[170] Wu, X., et al., Preparation of ethyl 3R,5S-6-(benzyloxy)-3,5-dihydroxy-hexanoate by recombinant diketoreductase in a biphasic system. *Bioresour. Technol.*, 2011. **102**(3): 3649–3652.
[171] Wang, L.J., et al., Highly efficient synthesis of chiral alcohols with a novel NADH-dependent reductase from *Streptomyces coelicolor*. *Bioresour. Technol.*, 2011. **102**(14): 7023–7028.
[172] de Gonzalo, G., et al., Enzymatic reduction of ketones in "micro-aqueous" media catalyzed by ADH-A from *Rhodococcus ruber*. *Org. Lett.*, 2007. **9**(11): 2163–2166.
[173] Gorke, J., F. Srienc, and R.J. Kazlauskas, Towards advanced ionic liquids. Polar, enzyme-friendly solvents for biocatalysis. *Biotechnol. Bioprocess Eng.*, 2010. **15**: 40–53.
[174] Domínguez de Maria, P. and Z. Maugeri, Ionic liquids in biotransformations: From proof-of-concept to emerging deep-eutectic-solvents. *Curr. Opin. Chem. Biol.*, 2011. **15**(2): 220–225.
[175] Denard, C.A., J.F. Hartwig, and H. Zhao, Multistep one-pot reactions combining biocatalysts and chemical catalysts for asymmetric synthesis. *ACS Catal.*, 2013. **3**: 2856–2864.
[176] Denard, C.A., et al., Cooperative tandem catalysis by an organometallic complex and a metalloenzyme. *Angew. Chem. Int. Ed.*, 2014. **53**(2): 465–469.
[177] Köhler, V., et al., Synthetic cascades are enabled by combining biocatalysts with artificial metalloenzymes. *Nat. Chem.*, 2013. **5**(2): 93–99.
[178] Foulkes, J.M., et al., Engineering a biometallic whole cell catalyst for enantioselective deracemization reactions. *ACS Catal.*, 2011. **1**(11): 1589–1594.
[179] Wang, Z.J., et al., A supramolecular approach to combining enzymatic and transition metal catalysis. *Nat. Chem.*, 2013. **5**(2): 100–103.

CHAPTER 17

Application of Aromatic Hydrocarbon Dioxygenases

Watumesa A. Tan[1,2] and Rebecca E. Parales[1]

[1]Department of Microbiology and Molecular Genetics, College of Biological Sciences, University of California, Davis, CA, USA

[2]Faculty of Biotechnology, Atma Jaya Catholic University of Indonesia, Jenderal Sudirman, Jakarta, Indonesia

17.1 INTRODUCTION

The first reported aromatic hydrocarbon dioxygenase, toluene dioxygenase, was identified to catalyze *cis*-dihydroxylation of benzene and toluene in *Pseudomonas putida* F1 [1, 2]. Since then, many more dioxygenases have been studied, and the crystal structures of several have been determined [3]. This group of enzymes plays an important role in the initial step in the biodegradation of both naturally occurring and man-made aromatic compounds, including aromatic acids, polycyclic aromatic hydrocarbons, polychlorinated biphenyls, and nitroaromatic, aminoaromatic and halogenated aromatic compounds [4, 5].

Dioxygenases are nonheme Rieske-type NAD(P)H-dependent enzymes that introduce both atoms of molecular oxygen into their substrates. The multicomponent enzyme systems are composed of a catalytic oxygenase component and one- or two-electron transfer proteins, including a flavoprotein reductase, and in some cases a ferredoxin that mediates electron transfer from the reductase to the oxygenase component ([3]; Figure 17.1).

In general, Rieske dioxygenases are capable of oxidizing a broad range of substrates, well beyond the range of compounds that serves as growth substrates for the host bacterium [4, 5, 7]. For example, toluene dioxygenase from *P. putida* F1 facilitates the oxidation of over 200 different substrates [7], and naphthalene dioxygenase from *Pseudomonas* sp. NCIB 9816-4 has been documented to oxidize over 60 different substrates [8]. In addition, these enzymes catalyze diverse types of reactions, including *cis*-dihydroxylations, O- and N-dealkylations, desaturations, and the formation of chiral sulfoxides from sulfides ([7–9]; Figure 17.2). In many cases the products are chiral compounds that are high in enantiomeric purity [8, 10, 11]. For the aforementioned reasons, dioxygenases are attractive biocatalysts for bioremediation and industrial applications.

17.2 CHALLENGES IN AROMATIC HYDROCARBON DIOXYGENASE APPLICATIONS

The practical use of dioxygenases in industry is generally limited to the use of whole-cell biocatalysts due to several challenges. The catalytic activity of dioxygenases is dependent on reducing equivalents provided by NADH or NADPH, which require regeneration by electron transfer proteins [12, 13]. The regeneration process is typically a limiting step

Green Biocatalysis, First Edition. Edited by Ramesh N. Patel.

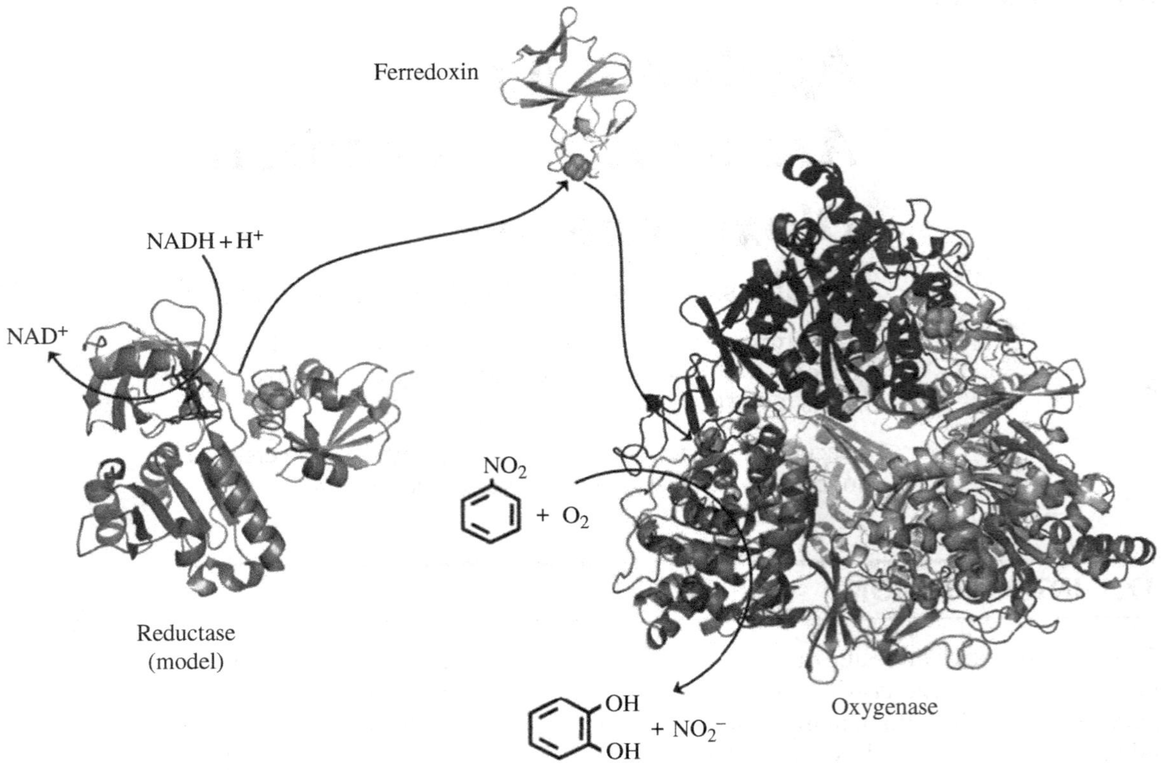

FIGURE 17.1
Components of the nitrobenzene dioxygenase of *Comamonas* sp. strain JS765. Structure of nitrobenzene dioxygenase from Ref. [6].

in the industrial applications as it reduces cost effectiveness [14]. Dioxygenases are also composed of multiple components that are not covalently linked, thus negatively impacting the enzymatic activity *in vitro* [15]. In addition, bioprocess environments may require high temperatures, extreme pH, high substrate and product concentrations, oxidants, and organic cosolvents. During a continuous manufacturing process, an enzyme may need to tolerate such conditions for months [12, 13].

In one traditional approach, immobilization of enzymes or whole cells has been used to improve enzyme stability, catalytic activity, and optimum temperature and to prevent inhibition by high substrate content or metal ions. For example, catechol 1,2-dioxygenase from *Stenotrophomonas maltophilia* KB2 entrapped on calcium alginate gel and catechol 2,3-dioxygenase from the thermophilic *Bacillus stearothermophilus* immobilized on glyoxal agarose gel retained their structural rigidity and showed resistance toward inactivation by high substrate concentrations [16, 17]. Different agents of immobilization may result in various types of enzymatic improvement, as shown in a comparison study of protocatechuate 3,4-dioxygenase immobilization [18]. Aside from the increased enzyme stability, immobilization of the dioxygenase in both types of media resulted in the decrease of optimum temperature by 5–10 °C [18]. The ability to maintain lower temperature during catalysis is a desirable feature for industrial applications, as it lowers the overall production costs [13].

Despite the advances of enzyme immobilization to improve the stability and enzymatic properties of enzymes such as the single-component dioxygenases that catalyze the cleavage of dihydroxylated aromatic compounds described earlier, this approach has yet to be successfully applied for 2- or 3-component enzymes such as Rieske-type aromatic ring hydroxylating dioxygenases. However, whole-cell immobilization remains a promising option to improve the aromatic compound degradation process. In one study, toluene dioxygenase-producing *P. putida* F1 cells were

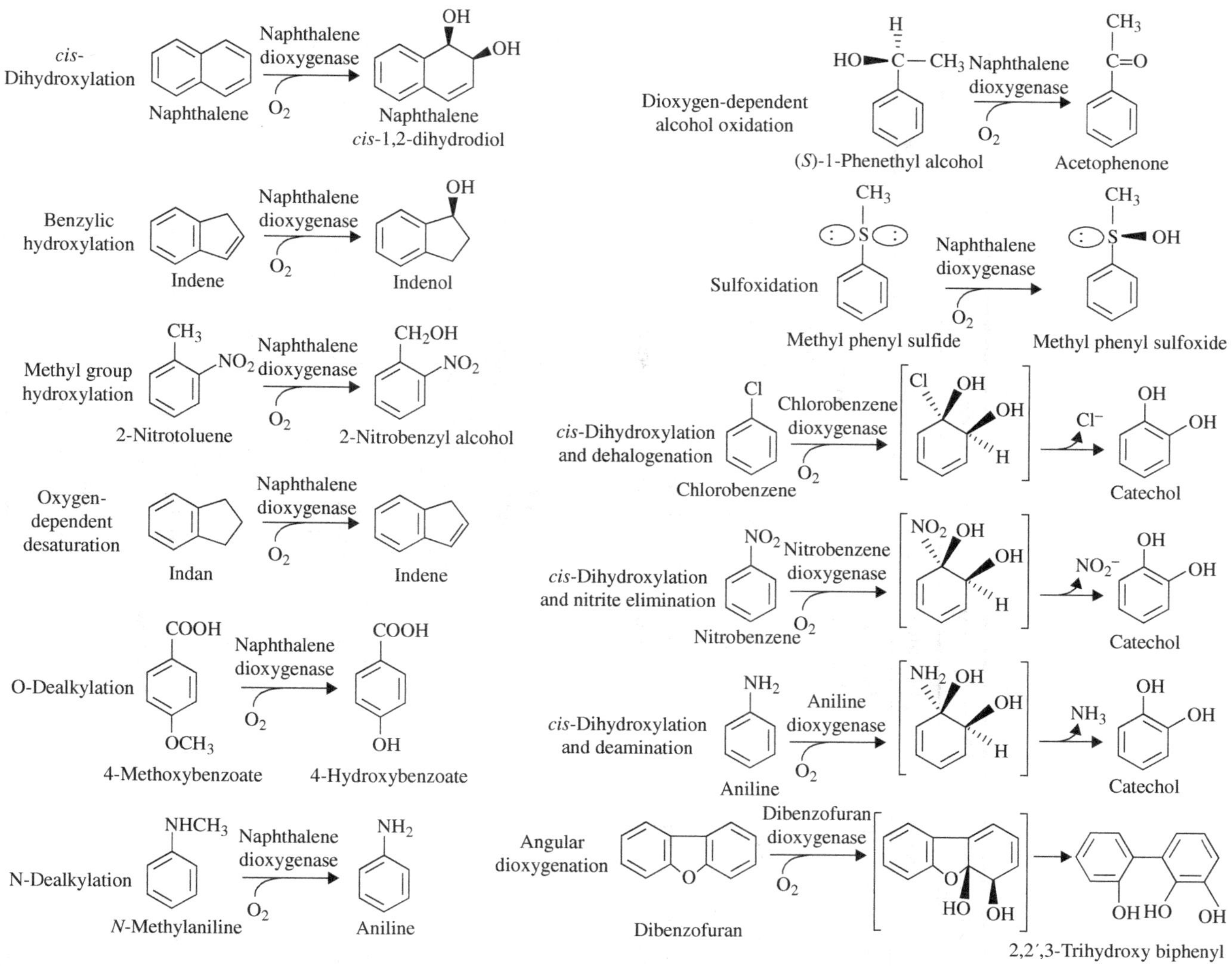

FIGURE 17.2

Reactions catalyzed by aromatic hydrocarbon dioxygenases.

entrapped in a mixture of recycled agave-fiber and polymer foam composites, and a lower rate of feedback inhibition by intermediary catechol produced in benzene degradation was observed [19].

17.3 PROTEIN ENGINEERING TO IMPROVE ENZYMATIC ACTIVITY AND ALTER SUBSTRATE SPECIFICITY

Various protein engineering strategies have been used to improve the enzymatic rate, regioselectivity, and substrate specificity of dioxygenases for a variety of applications (Table 17.1). Combined with current knowledge of the structure and function of the enzymes, engineering of variant dioxygenases is a promising strategy to optimize/expand the biocatalytic capability of microorganisms [23, 30, 32]. Some specific examples are described below.

A variant enzyme with a V205A substitution in the AtdA3 subunit of the aniline dioxygenase produced by *Acinetobacter* sp. strain YAA was found to oxidize 2-isopropylaniline, a substrate that is not recognized by the wild-type enzyme [33]. However, a decrease in enzymatic activity was observed as a trade-off for the increased substrate range [33]. Saturation and random mutagenesis yielded a variant

TABLE 17.1 Examples of Protein Engineering Approaches Used to Improve or Modify Aromatic Hydrocarbon Dioxygenases

Enzyme	Organism	Method	Amino Acid Substitution(s)	Feature Changes	Reference
Aniline dioxygenase	*Acinetobacter* sp. strain YAA	Saturation mutagenesis and error-prone PCR	V205A/I248L/S404C	Increased activity Widened substrate specificity to include 2,4-dimethylaniline and 2-isopropylaniline	[20]
Biphenyl dioxygenase	*Burkholderia xenovorans* LB400	Random mutagenesis of region III	T335A/F336M/I	Altered regiospecificity toward 2,2′-dichlorobiphenyl producing dead end product	[21]
Biphenyl dioxygenase	*B. xenovorans* LB400	Family shuffling of *bphA* from polychlorinated biphenyl-contaminated soil DNA	Multiple mutations (variant S100)	Attack on carbon 5 and 6 of 2,2′-dichlorobiphenyl; new ability to oxidize benzene, ethylbenzene, and toluene	[22]
Biphenyl dioxygenase	*Pseudomonas pseudoalcaligenes* KF707	Site-directed mutagenesis based on molecular structure	I335F	Acquired the ability to degrade 2,5,2′,5′-tetrachlorobiphenyl by 3,4-dioxygenation and showed bifunctional 2,3-dioxygenase and 3,4-dioxygenase activities for 2,5,2′-trichlorobiphenyl and 2,5,4′-trichlorobiphenyl	[23]
Biphenyl dioxygenase	*P. pseudoalcaligenes* KF707	Site-directed mutagenesis	T376A/V/F/L T376F	Novel ability to oxidize dibenzofuran Enhanced (KF707 enzyme) activities toward various polychlorinated biphenyls	[24]
2,4-Dinitrotoluene dioxygenase	*Burkholderia cepacia* R34	Saturation mutagenesis	V350F	Increased activity for *o*-nitrophenol, *m*-nitrophenol, *o*-methoxyphenol, and naphthalene; new activity against *m*-methoxyphenol, *o*-cresol, and *m*-cresol	[25]
			V350M	Increased activity for *o*-nitrophenol, *o*-methoxyphenol, and naphthalene	
2,4-Dinitrotoluene dioxygenase	*Burkholderia* sp. DNT	Saturation mutagenesis	I204Y/L	Increased activities against 2-nitrotoluene and 2,4- and 2,6-dinitrotoluene; new activity against 2,3- and 2,5-dinitrotoluene	[26]
			I204Y	Increased oxidation rate of aminonitrotoluenes	[27]

Naphthalene dioxygenase	*Ralstonia* sp. U2	Saturation mutagenesis	F350T	Increased activity against 2,6-dinitrotoluene; new ability to produce 3-amino-4-methyl-5-nitrocatechol and 2-amino-4,6-dinitrobenzyl alcohol from 2-amino-4,6-dinitrotoluene	[28]
		DNA shuffling and saturation mutagenesis	G407S/F350T	Increased oxidation of 2,6-/2,3-/2-amino-4, 6-dinitrotoluene	
		DNA shuffling	G50S/L225R/A269T	Increased oxidation of 4-amino-2-nitrotoluene into 4-amino-2-nitrocresol and 4-amino-2-nitrobenzyl alcohol as new products; increased oxidation of 2,3-dinitrotoluene into 2,3-dinitrobenzyl alcohol	
			L225R	Faster generation of 4-amino-2-nitrocresol and 4-amino-2-nitrobenzyl alcohol from 4-amino-2-nitrotoluene and of 2,3-nitrobenzyl alcohol from 2,3-dinitrotoluene	
Nitrobenzene dioxygenase	*Comamonas* sp. JS765	Site-directed mutagenesis	N258/I350F	Increased production of the dead-end metabolite nitrobenzyl alcohol	[29]
			F293H/Q/I	Increased production of 4-nitrobenzyl alcohol from 4-nitrotoluene	
			F293Q	Increased 2,6-dinitrotoluene oxidation rate	
			I293F	Decreased 2-nitrotoluene oxidation rate	
Tetrachlorobenzene dioxygenase	*Ralstonia* sp. PC12	Site-directed mutagenesis	F366L	Changes in regioselectivity; reduced activity with chlorotoluenes	[30]
			L272F/W	Increased product formation rate with dichlorotoluenes; changed oxidation of trichlorotoluenes	
			F366L/L272F/W	New products formed from 2,4,5-trichlorotoluene: 3,4-Dichloro-6-methylcatechol 3,6-Dichloro-4-methylcatechol 4,6-Dichloro-3-methylcatechol 1,2-Dihydro-1,2-dihydroxy-1-methyl-3,4,6-trichlorohexa-3,5-diene	
o-Xylene dioxygenase	*Rhodococcus* sp. DK17	Site-directed mutagenesis based on crystal structures. Expressed in *Escherichia coli* BL21(DE3)	L266F	Increased hydroxylation of biphenyl into 2- or 3-hydroxybiphenyl Improved hydroxylation of *o*-xylene to 3, 4-dimethylphenol	[31]

All amino acid substitutions listed are in the alpha subunit of the dioxygenase component.

enzyme with elevated activity and the ability to hydroxylate a wider range of aromatic amines, including 2-isopropylaniline and the carcinogenic aromatic amine pollutant 2,4-dimethylaniline [20].

Biphenyl and polychlorinated biphenyls are oxidized to dihydrodiol compounds by biphenyl dioxygenase during the initial step in the degradation pathway of biphenyl-degrading bacteria [34]. To reduce the size of library that needed to be screened, a mutation strategy was designed to target only the specific region of the dioxygenase gene known to encode the catalytic site [35]. A stretch of seven amino acids in region III of the biphenyl dioxygenase oxygenase component from *Burkholderia xenovorans* LB400 was shown to influence substrate specificity and determine regiospecificity toward chlorobiphenyls [35]. The T335A/F336M/I mutation altered regiospecificity toward 2,2′-dichlorobiphenyl, producing the undesired dead end product 3,4-dihydro-3,4-dihydroxy-2,2′-dichlorobiphenyl, thus impairing the mutant's ability to utilize the compound for its growth [21]. In the biphenyl dioxygenase of *Pseudomonas pseudoalcaligenes* KF707, a substitution from isoleucine to phenylalanine at the corresponding amino acid residue 335 also affected enzyme regiospecificity toward 2,2′-dichlorobiphenyl [23]. The mutations allowed KF707 to degrade 2,5,2′,5′-tetrachlorobiphenyl via 3,4-dioxygenation and resulted in bifunctional 2,3-dioxygenase and 3,4-dioxygenase activities for 2,5,2′-trichlorobiphenyl and 2,5,4′-trichlorobiphenyl [23]. In another experiment, shuffled fragments of *bphA* (encoding the oxygenase alpha subunit) amplified from polychlorinated biphenyl-contaminated soil were used to the replace corresponding portion of *bphA* in *B. xenovorans* LB400 [22]. The variant enzyme S100, which carried multiple amino acid substitutions, including those at residues 335 and 336, attacked carbons 5 and 6 of 2,2′-dichlorobiphenyl to produce exclusively *cis*-5,6-dihydro-5,6-dihydroxy-2,2′-dichlorobiphenyl, a reaction that wild-type LB400 biphenyl dioxygenase was unable to perform [22, 35]. In addition, variant S100 exhibited new abilities to oxygenate benzene, toluene, and ethylbenzene [22].

The T376 residue near the catalytic center in the oxygenase component of the *P. pseudoalcaligenes* KF707 biphenyl dioxygenase has been demonstrated to control substrate specificity and regiospecificity with various polychlorinated biphenyls [23, 24]. Enzymes with T376A/V/F/L substitutions were able to oxidize dibenzofuran, a substrate not recognized by the wild-type KF707 enzyme [24]. In addition, the T376F variant showed enhanced and expanded activities with various polychlorinated biphenyls [24].

In some related dioxygenases, substitutions at amino acid residue 204 located at the enzyme active site have been demonstrated to increase the oxidation rate or allow the recognition of new substrates. The I204Y and I204L mutations in 2,4-dinitrotoluene dioxygenase from *Burkholderia* sp. strain DNT expanded the substrate specificity of the enzyme, allowing the oxidation of the 2,3- and 2,5- isomers of dinitrotoluene. Increased activities against 2-nitrotoluene and 2,4- and 2,6-dinitrotoluene were also observed for both variant enzymes [26]. In addition, the I204L variant demonstrated higher oxidation rate with 4-nitrotoluene compared to the wild-type enzyme [26]. A subsequent study indicated that the I204Y substitution also resulted in an increased oxidation rate with aminonitrotoluenes [27]. Substitutions at the corresponding amino acid residue were observed in the 2-nitrotoluene dioxygenase of *Acidovorax* sp. strain JS42 variants that were forced to grow on 3-nitrotoluene [36]. Such mutations increased the affinity of the enzyme for 3-nitrotoluene and led to a change in enzyme regiospecificity that favored the production of 4-methylcatechol rather than 3-methylcatechol from 3-nitrotoluene [36].

The crystal structure of oxygenase component of naphthalene dioxygenase from *Pseudomonas* sp. strain NCIB 9816-4 [37] and amino acid alignments of this enzyme with those of nitrobenzene dioxygenase from *Comamonas* sp. strain JS765 [38], 2-nitrotoluene dioxygenase from *Acidovorax* sp. strain JS42 [39], 2,4-dinitrotoluene dioxygenases from *Burkholderia* sp. strain DNT [40] and *Burkholderia cepacia* R34 [41], from naphthalene dioxygenase and *Ralstonia* sp. strain U2 [42] were utilized to pinpoint three amino acid residues—N258, F293, and F350—that were predicted to affect nitrobenzene dioxygenase regiospecificity [29]. When tested with nitrotoluene isomers as substrates, the N258V and I350F substitutions resulted in the production of increased amounts of nitrobenzyl alcohols, which are dead-end metabolites, rather than methylcatechols, which serve as growth substrates [29]. The amino group of residue N258 was subsequently shown to form hydrogen bonds with the nitro group of nitroarenes based on the crystal structure of nitrobenzene dioxygenase [6], highlighting the importance of this residue in positioning substrates in the active site for ring oxidation [29].

Another study demonstrated that a F350T substitution in the naphthalene dioxygenase from *Ralstonia* sp. strain U2 enabled the oxidation of carcinogenic dinitrotoluenes [28]. Substitutions of valine to methionine or phenylalanine at the corresponding position in the 2,4-dinitrotoluene dioxygenase from *B. cepacia* R34 resulted in increased activity with nitrophenols and naphthalene, with the latter introducing new activity against *m*-methoxyphenol and *o*- and *m*-cresol [25]. The corresponding F366L substitution in the *Ralstonia* sp. PC12 tetrachlorobenzene dioxygenase (TecA) altered enzyme regioselectivity, although there was a trade-off with decreased activity [30]. Meanwhile, a significant increase in dichlorotoluene oxidation was observed with TecA variants harboring L272F/W substitutions, although changes at this position did not affect enzyme regioselectivity [30]. Dual changes at both positions in TecA resulted in a switch of the site of oxidation of 2,4,5-trichlorotoluene, resulting in 3,4,6-trichloro-1-methyl-1,2-dihydroxy-1,2-dihydrocyclohexan-3,5-diene, 4,6-dichloro-3-methylcatechol, 3,6-dichloro-4-methylcatechol, and 3,4-dichloro-6-methylcatechol as new products [30].

Residue 293 is located at a hydrophobic cleft on the surface of nitrobenzene dioxygenase that mediates aromatic substrate binding. The specificity of enzymes with substitutions at position 293 varied depending on the residue present as well as the substrate [29]. For example, F293H/Q/I substitutions in the alpha subunit of nitrobenzene dioxygenase did not cause any significant changes in the ratios of products formed from 3-nitrotoluene (3-methylcatechol, 4-methylcatechol, and 3-nitrobenzyl alcohol), but they increased the amount of the dead-end metabolite 4-nitrobenzyl alcohol relative to 4-methylcatechol from 4-nitrotoluene [29]. The F293Q substitution appeared to be most beneficial for enzyme activity toward 2,6-dinitrotoluene [29]. In addition, the I293F substitution decreased the rates of 2-nitrotoluene and nitrobenzene oxidation, but did not affect oxidation of 2,4- or 2,6-dinitrotoluene [29, 43]. Overall, these findings indicate that changes in the volume of the active site pocket due to the reduced size of the hydrophobic residue at position 293 are unfavorable for positioning of small substrates but not for larger substrates [29]. In addition, molecular docking and active site volume calculations indicated that the hydrophobic interactions and steric considerations coordinated the location of prochiral sulfide substrate in the enzyme active site, affecting the activity and enantioselectivity of nitrobenzene dioxygenase [44].

DNA shuffling led to the identification of two amino acid residues at positions 225 and 407 of naphthalene dioxygenase from *Ralstonia* sp. strain U2 that influenced enzyme regiospecificity and increased activity against dinitrotoluenes [28]. A subsequent study proposed the construction of hybrid enzymes expressed heterologously in *Escherichia coli* to expand the range of aromatic compound oxidation [45]. A hybrid between naphthalene dioxygenase and dinitrotoluene dioxygenase allowed simultaneous oxidation of naphthalene and 2,4-dinitrotoluene [45]. Meanwhile, the dinitrotoluene dioxygenase/nitrobenzene dioxygenase system allowed dual catalysis of the explosive 2,4,6-trinitrotoluene reduction products, 2-amino-4,6-dinitrotoluene and 4-amino-2,6-dinitrotoluene [45].

The *o*-xylene dioxygenase from *Rhodococcus* sp. DK17 is able to hydroxylate biphenyl exclusively to *cis*-2,3-biphenyl dihydrodiol despite the strain's inability to use biphenyl as a sole source of carbon and energy [46]. Based on the crystal structure of *o*-xylene dioxygenase, enzymes with substitutions at four different amino acid positions that were predicted to affect substrate positioning (A218L, D262L, L266F, and V297L) were generated [47, 48]. A subsequent study investigated the hydroxylation of biphenyl by these variant enzymes expressed in *E. coli* [31]. GC/MS analysis of biotransformation products indicated that only the L226F mutant enzyme had significantly increased production of 2-hydroxybiphenyl and 3-hydroxybiphenyl from biphenyl by 24-fold and 66-fold, respectively [31]. In addition, the L226F mutant enzyme hydroxylated *o*-xylene at much higher rates than the wild-type enzyme [31].

17.4 PROTEIN ENGINEERING FOR THE PRODUCTION OF SPECIFIC CHEMICALS

The ability of aromatic hydrocarbon dioxygenases to catalyze reactions with a wide range of substrates has made them particularly useful for the production of complex chiral chemicals that would otherwise be difficult, expensive, and environmentally unfriendly to produce.

Chiral compounds play an important role as intermediates in the production of chemicals and pharmaceuticals. For years, aromatic dioxygenases have been used to produce chiral compounds from an array of substrates through the generation of *cis*-dihydrodiol intermediates [9, 49]; some specific examples are shown in Table 17.2. *cis*-Dihydrodiol intermediates are generally stable, yet they are rapidly removed in wild-type strains due to further metabolism [49, 55]. Strategies such as the generation of dehydrogenase mutants to block the next step of catabolism or the utilization of recombinant *E. coli* harboring genes encoding the aromatic dioxygenase have been developed to allow accumulation of *cis*-dihydrodiols [49, 54].

Examples of biologically active compounds produced through whole-cell oxidative reactions at an industrial scale were reviewed by Muñoz-Solano *et al.* [56] and Matsui *et al.* [13]. A few are described in detail here and additional examples are shown in Table 17.2.

Recombinant *E. coli* expressing the nitrobenzene dioxygenase from *Comamonas* sp. JS765 with amino acid substitutions at position 258 or 293 was used to increase the production of chiral sulfoxides from thioanisole, *p*-tolyl, Cl-thioanisole, and Br-thioanisole [44]. Chiral sulfoxides are mostly known as important precursors in the pharmaceutical industry [13]. For example, modafinil, marketed under the name Provigil© by Cephalon and prescribed for the treatment of uncontrollable sleepiness caused by narcolepsy or sleep apnea, has a chiral sulfoxide center [57, 58]. Esomeprazole, another chiral sulfoxide, is marketed by AstraZeneca under the name Nexium© as a proton pump inhibitor that reduces gastric acid secretion for treatment of gastroesophageal reflux disease [59].

Hydroxytyrosol is a potent antioxidant naturally present in virgin olive oil that has been reported to have anticancer, antiatherogenic, antimicrobial, and anti-inflammatory activities [60, 61]. A biocatalytic strategy for its production is desirable as it prevents fluctuating yields often encountered in hydroxytyrosol recovery strategies from olives and olive-processing wastewater [62]. In addition, hydroxytyrosol can be produced from multiple starting materials and biocatalysts, which should allow the development of an economically feasible process, preferably using inexpensive substrates [43, 63]. Family shuffling generated a new variant of nitrobenzene dioxygenase with three substitutions—F222C, F251L, and G253S—which allowed the oxidation of 3-nitrophenethyl alcohol, a lower-cost choice compared to the mostly used substrate tyrosol [63], to the antioxidant hydroxytyrosol 375-fold better than wild type and with very high regioselectivity, reducing the production of unwanted side product [43]. Homology modeling results suggested

TABLE 17.2 Products of Dioxygenase Catalysis Used as Pharmaceutical and Industrial Precursors

Enzyme	Organism	Substrate	*cis*-Dihydrodiol Intermediate	Potential End Product	Reference
Toluene dioxygenase	*Pseudomonas putida* UV4	Toluene	Toluene 2,3-*cis*-dihydrodiol	6C-Methyl-D-mannose	[50]
		Chlorobenzene	Chlorobenzene 2,3-*cis*-dihydrodiol	(−)-Cladospolide A	[50]
		Bromobenzene	Bromobenzene 2,3-*cis*-dihydrodiol	(−)-*ent*-Bengamide E	[50]
		o-Methoxyphenol	Cyclohexenone *cis*-diol	(+)-Clivonine (natural prod) (+)-Trianthine (+)-Palintantin (antibiotic)	[51]
	P. putida F1	Indene	*cis*-(1*S*,2*R*)-Dihydroxyindan	Indinavir sulfate (Crixivan)	[52]
Toluene dioxygenase	*P. putida* F1	*p*-Xylene	*cis*-1,2-Dihydroxy-3,6-dimethyl-3,5-cyclohexadiene	Strawberry furanone	[53]
Tetrachlorobenzene dioxygenase	*Burkholderia* sp. PS12				
Naphthalene dioxygenase	*P. putida* 9816/11	Naphthalene	Naphthalene 1,2-*cis*-dihydrodiol	(+)-Goniodiol	[49]
	Pseudomonas sp. NCIB9816-4	2-Chlorophenol	Chlorohydroquinone	Pharmaceuticals	[54]
2,4-Dinitrotoluene dioxygenase	*Burkholderia cepacia* R34	*o*-Nitrophenol	Nitrohydroquinone	Dephostatin	[25]
		o-Cresol	Methylhydroquinone	Helibisabonol A Puraquinoic acid	
		3-Methoxyphenol	Methoxyhydroquinone	Triptycene quinones	
Nitrobenzene dioxygenase	*Comamonas* sp. JS765	Thioanisole and aromatic sulfides	Chiral sulfoxides	Esomeprazole Armodafinil	[44]
		3-Nitrophenethyl alcohol	Hydroxytyrosol	Hydroxytyrosol	[43]

that these amino acid substitutions increase activity by increasing the size of the active site to allow access of the larger substrate and also by hydrogen bonding to the substrate [43].

Indigo is a textile dye naturally produced by *Indigofera* plants. Currently, the indigo industry relies on chemical synthesis, yet there is a potential in utilizing microbial process to produce indigo. The first evidence of microbial oxidation of indole to indigo was demonstrated in recombinant *E. coli* strains expressing naphthalene dioxygenase from *Pseudomonas* sp. NCIB 9816-4 [64]. Since then, many more dioxygenase-expressing bacteria have been identified to possess similar ability to produce indigo from indole [65–69]. Indigo can be biocatalytically produced from tryptophan, which is converted into indole by tryptophanase, a naturally occurring enzyme in *E. coli* ([64]; Figure 17.3). Dioxygenase activity on indole generates 3-hydroxyindole, 2-oxindole, and isatin [64]. Indigo is produced through spontaneous dimerization of two 3-hydroxyindole molecules, while the reaction between 3-hydroxyindole and 2-oxindole or isatin creates indirubin ([64, 66]; Figure 17.3).

Multiple strategies have been utilized to engineer the production of indigo from glucose in *E. coli* [71]. Genes encoding naphthalene dioxygenase from *P. putida* were introduced into a recombinant *E. coli* strain with a tryptophan pathway that had been modified to generate a high level of indole [71]. This strategy was combined with expression of a gene encoding isatin hydrolase, which catalyzes the conversion of isatin to isatic acid, thereby blocking the formation of indirubin and increasing the indigo production to more than 18 g/l [71]. In another study, chromosomal integration of tetralin dioxygenase-encoding *thnA1A2A3A4* genes from *Sphingomonas macrogolitabida* strain TFA in *E. coli* resulted in a strain that maintained long-term indigo production and solved the instability issue often encountered with plasmid-based expression systems [72].

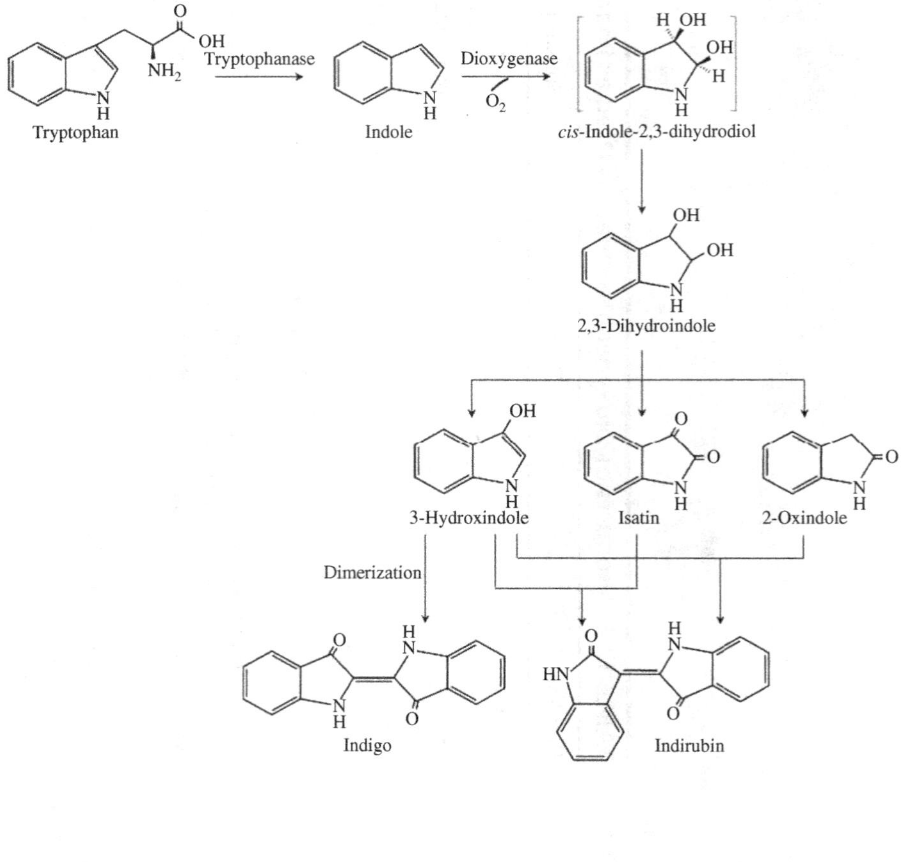

FIGURE 17.3

Microbial synthesis pathway for indigo based on Ref. [64].

17.5 STRAIN MODIFICATION FOR THE DEVELOPMENT OF NEW BIODEGRADATION PATHWAYS

Rieske dioxygenases are key enzymes in many bacterial aromatic compound biodegradation pathways, and researchers have taken advantage of their broad specificities and malleability by modifying enzymes and using them to create new biodegradation pathways.

As discussed previously, amino acid residues A258, F293, and F350 had been identified to control substrate specificity of nitrobenzene dioxygenase from *Comamonas* sp. JS765 [29]. This information was further used to engineer a strain with the ability to degrade chloronitrobenzenes [70]. As observed for the activity with nitrobenzene and nitrotoluenes, substitutions at positions 258, 293, and 350 altered the activity and regiospecificity of nitrobenzene dioxygenase toward chloronitrobenzene isomers [70]. Substitutions at positions 258 and 350 increased oxidation at the chloro-substituted carbon, while changes at position 293 led to increased specificity with 2- and 4-chloronitrobenzene, in which nitrobenzyl alcohol production was eliminated and only chlorocatechols were generated [70].

The wild-type nitrobenzene dioxygenase and the best variant, F293Q, were expressed in *Ralstonia* sp. JS705, a natural degrader of chlorobenzene that was isolated from contaminated groundwater [73, 74]. In this strain, chlorobenzene is converted into 3-chlorocatechol, which is processed through the modified *ortho*-cleavage pathway to produce TCA cycle intermediates [73, 74]. Introduction of wild-type nitrobenzene dioxygenase and its F293Q variant led to the generation of variants of *Ralstonia* sp. JS705 that were able to grow on and degrade all three chloronitrobenzene isomers [70]. As opposed to known chloronitrobenzene degradation pathways that required multiple steps for aromatic ring cleavage [75–78], the constructed pathway involved a single enzymatic step to convert chloronitrobenzene to chlorocatechols ([70]; Figure 17.4). A similar naturally occurring pathway was subsequently identified in a *Pseudomonas stutzeri* isolate capable of growing on 2-chloronitrobenzene [79].

Short- and long-term laboratory evolution experiments have been utilized to select for variants of *Acidovorax* sp. strain JS42 that are able to grow on and degrade alternative nitroarene substrates [80, 81]. Wild-type strain JS42 is capable of using 2-nitrotoluene as a sole carbon, nitrogen, and energy source [82]. The initial oxidation of 2-nitrotoluene is catalyzed by 2-nitrotoluene-2,3-dioxygenase, which produces 3-methylcatechol and nitrite [82]. 3-Methylcatechol then undergoes *meta*-cleavage

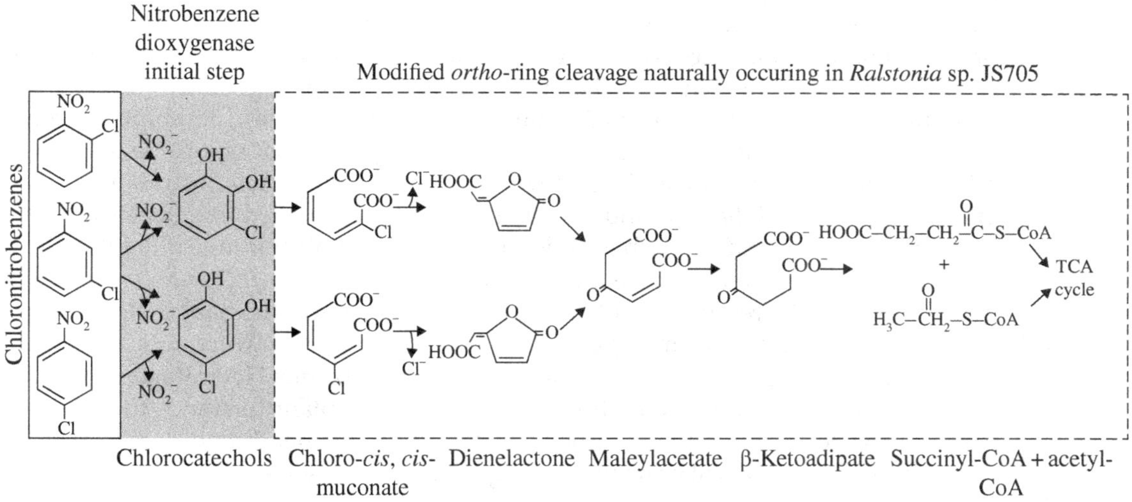

FIGURE 17.4
Chloronitrobenzene degradation pathway in the engineered *Ralstonia* sp. JS705 strain [70].

and its products enter central metabolism [82]. Despite the ability of 2-nitrotoluene dioxygenase to recognize and oxidize 3-nitrotoluene and 4-nitrotoluene, *Acidovorax* sp. strain JS42 is unable to grow on these substrates [80].

In two different studies, Ju and Parales [80] and Mahan *et al.* [81] evolved variants of *Acidovorax* sp. strain JS42 that are able to degrade various mononitrotoluene isomers by forcing the wild-type strain to grow on 3-nitrotoluene or 4-nitrotoluene. Both the 3-nitrotoluene$^+$ and 4-nitrotoluene$^+$ variants carried mutations in the *ntdAc* gene encoding the alpha subunit of the oxygenase of the multicomponent dioxygenase, resulting in amino acid residue changes at various positions. Interestingly, the 4-nitrotoluene$^+$ mutants retained their ability to grow on 2-nitrotoluene [80], while the 2-nitrotoluene-degrading ability of most of the 3-nitrotoluene$^+$ variants was reduced. In contrast to the 4-nitrotoluene$^+$ dioxygenases, which were sufficient to confer the ability to grow on 4-nitrotoluene to a JS42 mutant lacking a functional 2-nitrotoluene dioxygenase, introduction of the mutant dioxygenase genes from the evolved 3-nitrotoluene$^+$ strains did not confer the ability to grow on all substrates [81]. Together, these results indicate that additional beneficial mutation(s) elsewhere in the genome must have been acquired during selection on 3-nitrotoluene [81]. These types of laboratory selections have allowed the isolation of bacterial variants that are capable of degrading multiple pollutants and provide insight into the selection processes that occur in nature.

17.6 PHYTOREMEDIATION: THE EXPRESSION OF BACTERIAL DIOXYGENASES IN PLANT SYSTEMS FOR BIOREMEDIATION PURPOSES

The widespread presence of polycyclic aromatic hydrocarbons and their potential harm to various organisms have generated interest in efficiently eliminating these compounds from the environment. The combination of plant phytoremediation and bacterial aromatic dioxygenases is one of the options that has been explored to clean up polycyclic aromatic hydrocarbon-contaminated sites. Previously, the construction of transgenic plants for aromatic compound degradation was generally limited to the introduction of bacterial genes involved in the downstream pathway [83, 84], as the proper positioning of the Rieske center to obtain a functional enzyme complex created an obstacle to the incorporation of multicomponent dioxygenases into plants [85]. Mohammadi *et al.* [86] attempted to individually clone *bphAE*, *bphF*, and *bphG*, encoding the oxygenase, ferredoxin, and reductase components of biphenyl dioxygenase, respectively, into tobacco plants. These genes belong to the *bph* operon of *B. xenovorans* LB400, which are required for initial dioxygenation of biphenyl. A combination of the purified enzymes from the plants was able to catalyze the oxidation of 4-chlorobiphenyl to 2,3-dihydro-2,3-dihydroxy-4′-chlorobiphenyl [86]. Unfortunately, the authors were unable to generate plants that were able to express all four genes simultaneously [86]. However, a recent study demonstrated the generation of *Arabidopsis thaliana* and rice plant variants that carried four bacterial genes including *nidA* and *nidB*, which encode the large and small subunits of naphthalene dioxygenase from *Mycobacterium vanbaalenii* PYR-1, as well as *nahAa* and *nahAb*, the genes encoding the reductase and ferredoxin components of the naphthalene dioxygenase enzyme system from *P. putida* G7 [87]. All four genes were simultaneously expressed in transgenic plants, resulting in an active enzyme. Heterologous production of the hybrid naphthalene dioxygenase enhanced plant tolerance toward 2–4-ring polycyclic aromatic hydrocarbons and allowed phenanthrene degradation *in vivo* [87]. Further work will be necessary to optimize such engineered plants for efficient phytoremediation. Another approach that bypasses plant genetic engineering hurdles and possible issues with the use of genetically modified plants involves phytoremediation combined with soil bioaugmentation using natural bacterial isolates that degrade the contaminants of interest. Such a study demonstrated that *B. xenovorans* LB400 could enhance phytoremediation of polychlorinated biphenyls by switchgrass [88].

17.7 CONCLUDING REMARKS

Despite their limitations, aromatic dioxygenases are attractive for both environmental and industrial purposes. Detailed understanding of dioxygenase structure and mechanism will continue to direct protein engineering strategies to improve enzymatic activities and substrate ranges and create new products, many of which are likely to be chiral. The use of aromatic dioxygenase-producing bacteria in bioremediation should be coordinated with other aspects of aromatic compound degradation, such as bacterial detection of aromatic compounds and chemotaxis toward contaminated sites [89].

ACKNOWLEDGMENTS

We thank Kristina M. Mahan for preparing Figure 17.1. Dioxygenase research in the Parales laboratory has been supported by the National Science Foundation (MCB 02627248 and MCB 1022362).

REFERENCES

[1] Gibson, D. T.; Hensley, M.; Yoshioka, H.; Mabry, T. *J. Biochem.* 1970, **9**, 1626.
[2] Gibson, D. T.; Koch, J. R.; Kallio, R. E. *Biochemistry* 1968, **7**, 2653.
[3] Ferraro, D. J.; Gakhar, L.; Ramaswamy, S. *Biochem. Biophys. Res. Commun.* 2005, **338**, 175.
[4] Gibson, D. T.; Parales, R. E. *Curr. Opin. Biotechnol.* 2000, **11**, 236.
[5] Parales, R. E.; Resnick, S. M. In *Biodegradation and Bioremediation*; Singh, A., Ward, O. P., Eds.; Springer-Verlag: Berlin, 2004, p. 175.
[6] Friemann, R.; Ivkovic-Jensen, M. M.; Lessner, D. J.; Yu, C.-L.; Gibson, D. T.; Parales, R. E.; Eklund, H.; Ramaswamy, S. *J. Mol. Biol.* 2005, **348**, 1139.
[7] Boyd, D. R.; Sharma, N. D.; Allen, C. C. R. *Curr. Opin. Biotechnol.* 2001, **12**, 564.
[8] Resnick, S. M.; Lee, K.; Gibson, D. T. *J. Ind. Microbiol.* 1996, **17**, 438.
[9] Hudlicky, T.; Gonzalez, D.; Gibson, D. T. *Aldrichimica Acta* 1999, **32**, 35.
[10] Boyd, D. R.; Sharma, N. D.; Malone, J. F.; Allen, C. C. R. *Chem. Commun.* 2009, **28**, 3633.
[11] Boyd, D. R.; Sharma, N. D.; Stevenson, P. J.; Blain, M.; McRoberts, C.; Hamilton, J. T.; Argudo, J. M.; Mundi, H.; Kulakov, L. A.; Allen, C. C. *Org. Biomol. Chem.* 2011, **9**, 1479.
[12] Dror, A.; Fishman, A. *Comput. Struct. Biotechnol. J.* 2012, **2**, e201209011.
[13] Matsui, T.; Dekishima, Y.; Ueda, M. *Appl. Microbiol. Biotechnol.* 2014, **98**, 7699.
[14] Torres Pazmiño, D. E.; Winkler, M.; Glieder, A.; Fraaije, M. W. *J. Biotechnol.* 2010 **146**, 9.
[15] Allen, C. C. R. *J. Biocatal. Biotrans.* 2012, **1**, 1.
[16] Fernandez-Lafuente, R.; Guisan, J. M.; Ali, S.; Cowan, D. *Enzyme Microb. Technol.* 2000, **26**, 568.
[17] Guzik, U.; Hupert-Kocurek, K.; Marchlewicz, A.; Wojcieszyńska, D. *Electron. J. Biotechnol.* 2014, **17**, 83.
[18] Guzik, U.; Hupert-Kocurek, K.; Krysiak, M.; Wojcieszyńska, D. *Biomed. Res. Int.* 2014, **2014**, 138768.
[19] Robledo-Ortíz, J. R.; Ramírez-Arreola, D. E.; Pérez-Fonseca, A. A.; Gómez, C.; González-Reynoso, O.; Ramos-Quirarte, J.; González-Núñez, R. *Int. Biodeterior. Biodegrad.* 2011, **65**, 549.
[20] Ang, E. L.; Obbard, J. P.; Zhao, H. *Appl. Microbiol. Biotechnol.* 2009, **81**, 1063.
[21] Barriault, D.; Sylvestre, M. *J. Biol. Chem.* 2004, **279**, 47480.
[22] Vézina, J.; Barriault, D.; Sylvestre, M. *J. Bacteriol.* 2007, **189**, 779.
[23] Suenaga, H.; Watanabe, T.; Sato, M.; Ngadiman; Furukawa, K. *J. Bacteriol.* 2002, **184**, 3682.
[24] Suenaga, H.; Sato, M.; Goto, M.; Takeshita, M.; Furukawa, K. *Biosci. Biotechnol. Biochem.* 2006, **70**, 1021.

[25] Keenan, B. G.; Leungsakul, T.; Smets, B. F.; Wood, T. K. *Appl. Environ. Microbiol.* 2004, **70**, 3222.
[26] Leungsakul, T.; Keenan, B. G.; Yin, H.; Smets, B. F.; Wood, T. K. *Biotechnol. Bioeng.* 2005, **92**, 416.
[27] Leungsakul, T.; Keenan, B. G.; Mori, M. A.; Morton, M. D.; Stuart, J. D.; Smets, B. F.; Wood, T. K. *Biotechnol. Bioeng.* 2006, **93**, 231.
[28] Keenan, B. G.; Leungsakul, T.; Smets, B. F.; Mori, M. A.; Henderson, D. E.; Wood, T. K. *J. Bacteriol.* 2005, **187**, 3302.
[29] Ju, K.-S.; Parales, R. E. *Appl. Environ. Microbiol.* 2006, **72**, 1817.
[30] Pollmann, K.; Wray, V.; Hecht, H.-J.; Pieper, D. H. *Microbiology* 2003, **149**, 903.
[31] Yoo, M.; Kim, D.; Zylstra, G. J.; Kang, B. S.; Kim, E. *Res. Microbiol.* 2011, **162**, 724.
[32] Kim, D.; Yoo, M.; Choi, K. Y.; Kang, B. S.; Kim, E. *Biores. Technol.* 2013, **145**, 123.
[33] Ang, E. L.; Obbard, J. P.; Zhao, H. M. *FEBS J.* 2007, **274**, 928.
[34] Furukawa, K.; Suenaga, H.; Goto, M. *J. Bacteriol.* 2004, **186**, 5189.
[35] Barriault, D.; Lépine, F.; Mohammadi, M.; Milot, S.; Leberre, N.; Sylvestre, M. *J. Biol. Chem.* 2004, **279**, 47489.
[36] Mahan, K. M.; Penrod, J. T.; Ju, K.-S; Al Kass, N.; Tan, W. A.; Truong, R.; Parales, J. V.; Parales, R. E. *Appl. Environ. Microbiol.* 2015, **81**, 309.
[37] Kauppi, B.; Lee, K.; Carredano, E.; Parales, R. E.; Gibson, D. T.; Eklund, H.; Ramaswamy, S. *Structure* 1998, **6**, 571.
[38] Lessner, D. J.; Johnson, G. R.; Parales, R. E.; Spain, J. C.; Gibson, D. T. *Appl. Environ. Microbiol.* 2002, **68**, 634.
[39] Parales, J. V.; Kumar, A.; Parales, R. E.; Gibson, D. T. *Gene* 1996, **181**, 57.
[40] Suen, W.-C.; Haigler, B. E.; Spain, J. C. *J. Bacteriol.* 1996, **178**, 4926.
[41] Johnson, G. R.; Jain, R. K.; Spain, J. C. *J. Bacteriol.* 2002, **184**, 4219.
[42] Fuenmayor, S. L.; Wild, M.; Boyles, A. L.; Williams, P. A. *J. Bacteriol.* 1998, **180**, 2522.
[43] Bernath-Levin, K.; Shainsky, J.; Sigawi, L.; Fishman, A. *Appl. Microbiol. Biotechnol.* 2014, **98**, 4975.
[44] Shainsky, J.; Bernath-Levin, K.; Isaschar-Ovdat, S.; Glaser, F.; Fishman, A. *Protein Eng. Des. Sel.* 2013, **26**, 335.
[45] Keenan, B. G.; Wood, T. K. *Appl. Microbiol. Biotechnol.* 2006, **73**, 827.
[46] Kim, D.; Lee, J. S.; Choi, K. Y.; Kim, Y.-S.; Choi, J. N.; Kim, S.-K.; Chae, J.-C.; Zylstra, G. J.; Lee, C. H.; Kim, E. *Enzym. Microb. Technol.* 2007, **41**, 221.
[47] Kim, D.; Choi, K. Y.; Yoo, M.; Choi, J. N.; Lee, C. H.; Zylstra, G. J.; Kang, B. S.; Kim, E. *Appl. Microbiol. Biotechnol.* 2010, **86**, 1841.
[48] Kim, D.; Chae, J.-C.; Zylstra, G. J.; Kim, Y.-S.; Kim, S.-K.; Nam, M. H.; Kim, Y. M.; Kim, E. *Appl. Environ. Microbiol.* 2004, **70**, 7086.
[49] Boyd, D. R.; Bugg, T. D. H. *Org. Biomol. Chem.* 2006, **4**, 181.
[50] Nolan, L. C.; O'Connor, K. E. *Biotechnol. Lett.* 2008, **30**, 1879.
[51] Boyd, D. R.; Sharma, N. D.; Malone, J. F.; McIntyre, P. B.; McRoberts, C.; Floyd, S.; Allen, C. C.; Gohil, A.; Coles, S. J.; Horton, P. N.; Stevenson, P. J. *J. Org. Chem.* 2015, **80**, 3429.
[52] Zhang, N.; Stewart, B. G.; Moore, J. C.; Greasham, R. L.; Robinson, D. K.; Buckland, B. C.; Lee, C. *Metab. Eng.* 2000, **2**, 339.
[53] Newman, L. M.; Garcia, H.; Hudlicky, T.; Selifonov, S. A. *Tetrahedron* 2004, **60**, 729.
[54] Lee, K. *FEMS Microbiol. Lett.* 2006, **255**, 316.
[55] Boyd, D. R.; Sharma, N. D.; Coen, G. P.; Gray, P. J.; Malone, J. F.; Gawronski, J. *Chemistry* 2007, **13**, 5804.
[56] Muñoz Solano, D.; Hoyos, P.; Hernáiz, M. J.; Alcántara, A. R.; Sánchez-Montero, J. M. *Bioresour. Technol.* 2012, **115**, 196.
[57] O'Mahony, G. E.; Kelly, P.; Lawrence, S. E.; Maguire, A. R. *ARKIVOC* 2011, **2011**, 1.
[58] Kumar, R. *Drugs* 2012, **68**, 1803.
[59] Bentley, R. *Chem. Soc. Rev.* 2005, **34**, 609.
[60] Cicerale, S.; Lucas, L. J.; Keast, R. S. *Curr. Opin. Biotechnol.* 2012, **23**, 129.
[61] Rezaei-Sadabady, R.; Akbarzadeh, A. *Toxin Rev.* 2015, **34**, 7.
[62] Allouche, N.; Fki, I.; Sayadi, S. *J. Agric. Food Chem.* 2004, **52**, 267.
[63] Achmon, Y.; Fishman, A. *Appl. Microbiol. Biotechnol.* 2015, **99**, 1119.
[64] Ensley, B. D.; Ratzkin, B. J.; Osslund, T. D.; Simon, M. J.; Wackett, L. P.; Gibson, D. T. *Science* 1983, **222**, 167.
[65] Murdock, D.; Ensley, B. D.; Serdar, C.; Thalen, M. *Bio/Technology* 1993, **11**, 381.
[66] Eaton, R. W.; Chapman, P. J. *J. Bacteriol.* 1995, **177**, 6983.
[67] Kim, J. Y.; Lee, K.; Kim, Y.; Kim, C.-K.; Lee, K. *Lett. Appl. Microbiol.* 2003, **36**, 343.

[68] Pathak, H.; Madamwar, D. *Appl. Biochem. Biotechnol.* 2010, **160**, 1616.
[69] Qu, Y.; Xu, B.; Zhang, X.; Ma, Q.; Zhou, H.; Kong, C.; Zhang, Z.; Zhou, J. *Biochem. Eng. J.* 2013, **72**, 54.
[70] Ju, K.-S.; Parales, R. E. *Microb. Biotechnol.* 2009, **2**, 241.
[71] Berry, A.; Dodge, T. C.; Pepsin, M.; Weyler, W. *J. Ind. Microbiol. Biotechnol.* 2002, **28**, 127.
[72] Royo, J. L.; Moreno-Ruiz, E.; Cebolla, A.; Santero, E. *J. Biotechnol.* 2005, **116**, 113.
[73] Muller, T. A.; Werlen, C.; Spain, J. C.; Van Der Meer, J. R. *Environ. Microbiol.* 2003, **5**, 163.
[74] van der Meer, J. R.; Werlen, C.; Nishino, S. F.; Spain, J. C. *Appl. Environ. Microbiol.* 1998, **64**, 4185.
[75] Hinteregger, C.; Loidl, M.; Streichsbier, F. *J. Basic Microbiol.* 1994, **34**, 77.
[76] Kuhlmann, A.; Hegemann, W. *Acta Hydrochim. Hydrobiol.* 1997, **25**, 298.
[77] Park, H.-S.; Lim, S.-J.; Chang, Y. K.; Livingston, A. G.; Kim, H.-S. *Appl. Environ. Microbiol.* 1999, **65**, 1083.
[78] Xiao, Y.; Wu, J. F.; Liu, H.; Wang, S. J.; Liu, S. J.; Zhou, N. Y. *Appl. Microbiol. Biotechnol.* 2006, **73**, 166.
[79] Liu, H.; Wang, S.-J.; Zhang, J.-J.; Dai, H.; Tang, H.; Zhou, N.-Y. *Appl. Environ. Microbiol.* 2011, **77**, 4547.
[80] Ju, K.-S.; Parales, R. E. *Mol. Microbiol.* 2011, **82**, 355.
[81] Mahan, K. M.; Penrod, J. T.; Ju, K.-S.; Al Kass, N.; Tan, W. A.; Truong, R.; Parales, J. V.; Parales, R. E. *Appl. Environ. Microbiol.* 2015, **81**, 301.
[82] Haigler, B. E.; Wallace, W. H.; Spain, J. C. *Appl. Environ. Microbiol.* 1994, **60**, 3466.
[83] Uchida, E.; Ouchi, T.; Suzuki, Y.; Yoshida, T.; Habe, H.; Yamaguchi, I.; Omori, T.; Nojiri, H. *Environ. Sci. Technol.* 2005, **39**, 7671.
[84] Novakova, M.; Mackova, M.; Chrastilova, Z.; Viktorova, J.; Szekeres, M.; Demnerova, K. *Biotechnol. Bioeng.* 2009, **102**, 29.
[85] Chebrou, H.; Hurtubise, Y.; Barriault, D.; Sylvestre, M. *J. Bacteriol.* 1999, **181**, 4805.
[86] Mohammadi, M.; Chalavi, V.; Novakova-Sura, M.; Laliberté, J. F.; Sylvestre, M. *Biotechnol. Bioeng.* 2007, **97**, 496.
[87] Peng, R. H.; Fu, X. Y.; Zhao, W.; Tian, Y. S.; Zhu, B.; Han, H. J.; Xu, J.; Yao, Q. H. *Environ. Sci. Technol.* 2014, **48**, 12824.
[88] Liang, Y.; Meggo, R.; Hu, D.; Schnoor, J. L.; Mattes, T. E. *Ecol. Eng.* 2014, **71**, 215.
[89] Parales, R. E.; Luu, R. A.; Hughes, J. G.; Ditty, J. L. *Curr. Opin. Biotechnol.* 2015, **33**, 318.

CHAPTER 18

Ene-reductases and their Applications

Tanja Knaus, Helen S. Toogood, and Nigel S. Scrutton

Faculty of Life Sciences, Manchester Institute of Biotechnology, University of Manchester, Manchester, UK

18.1 INTRODUCTION

Ene-reductases (ERs) constitute a class of enzymes capable of catalyzing the asymmetric reduction of electronically activated alkenes at the expense of nicotinamide (NAD(P)H) coenzymes. They have become highly important in biocatalysis for the production of enantiopure pharmaceutical, fine chemical and agrochemical products, as up to two stereogenic centers can be created in one single step. At least four different classes of ERs have been explored for their biocatalytic applicability, which differ in the reaction mechanism, substrate scope, stereo-, and/or enantioselectivity: (i) ERs from the old yellow enzyme (OYE) family, (ii) enoate reductases, (iii) a family of medium-chain dehydrogenase/reductase (MDR) oxidoreductases, and (iv) short-chain dehydrogenase/reductases (SDRs) from plants [1].

Enoate reductases (EC 1.3.1.31) are flavin-dependent and iron–sulfur-containing proteins found among others in *Clostridium* species. Members of this NADH:flavin oxidoreductase/NADH oxidase family are distinguished from other ERs due to their high stereospecificity and strict regioselectivity for the reduction of double bonds of monoacids and monoesters, as well as reducing classical substrates such as α,β-unsaturated aldehydes, cyclic ketones, and methyl ketones. However, these enzymes are extremely oxygen sensitive, so they have not been employed in biocatalysis so far [2, 3].

A recently investigated class of ERs is zinc-independent members of the MDR superfamily, more specifically the leukotriene B4 dehydrogenase (LTD, MDR002) subfamily [4, 5]. For example, a flavin-free double bond reductase (DBR) from *Nicotiana tabacum* (NtDBR) was characterized and was capable of reducing a wide variety of α,β-unsaturated aldehydes, ketones, and nitroalkenes [6]. Very few SDR ERs have been investigated, such as isopiperitenone reductase (IPR) from *Mentha piperita* [7]. This enzyme catalyzes the reduction of the monoterpenoid isopiperitenone to isopulegone during the biosynthesis of menthols.

However, the most extensively investigated class of ERs is members of the OYE family of flavin oxidoreductases (EC 1.6.99.1). There is detailed information known about OYEs, such as their structure, reaction mechanism, substrate scope, kinetic properties, and biocatalytic approaches. Therefore, this chapter will focus on this latter class of enzymes. They have been intensively studied over the past decade in view of their applicability in preparative-scale biotransformations [1, 8–12]. These FMN-containing enzymes catalyze the asymmetric reduction of α,β-unsaturated

Green Biocatalysis, First Edition. Edited by Ramesh N. Patel.

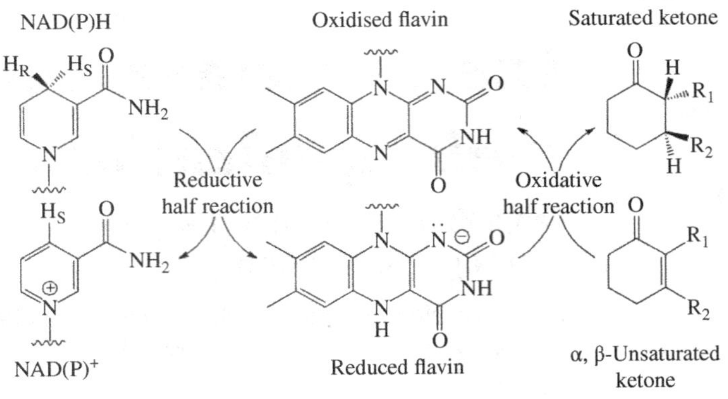

SCHEME 18.1 Asymmetric bioreduction of activated ketones catalyzed by members of the old yellow enzyme family showing the reductive (nicotinamide-dependent) and oxidative (alkene-dependent) half reactions [13].

activated alkenes in a ping-pong bi-bi mechanism (Scheme 18.1): In the reductive half reaction, a hydride is transferred from the nicotinamide moiety of NAD(P)H to the enzyme-bound flavin. After release of the oxidized $NAD(P)^+$, the hydride is further transferred from N5 of the reduced flavin to the activated alkene (oxidative half reaction) [13, 14]. The alkene substrate is bound in the active site of the enzyme through a hydrogen-bonding network via a His/His or His/Asn pair [15], which enhances the C=C bond polarization and supports the hydride transfer from $FMNH_2$ onto Cβ of the substrate [14]. Simultaneously, a proton transfer, usually delivered from a conserved Tyr residue [13] (Cys in the case of morphinone reductase from *Pseudomonas putida* [16]) from the opposite side or water, completes the reduction process. The reaction resembles an asymmetric conjugate Michael-type addition of a hydride onto activated alkenes and results in exclusively relative *trans*-stereospecificity [15].

18.2 SUBSTRATE CLASSES AND INDUSTRIAL APPLICATIONS

Biocatalytic hydrogenations performed by ERs require an electron-withdrawing substituent conjugated to the C=C double bond that activates the substrate for the hydride transfer. Nonactivated (isolated) alkenes, so-called borderline substrates, bearing a single conjugated ester moiety or a free carboxylic group, are only very poorly reduced or not converted at all using ERs from the OYE family [17–19]. Here, a few examples of OYE homologue catalyzed reactions employed for industrial applications are shown, categorized by the substrate-activating group:

1. α,β-Unsaturated aldehydes (enals) are in general good substrates yielding almost quantitative conversion if purified ERs are used. However, whole cell biotransformations often result in the formation of significant amounts of saturated alcohols due to the presence of *prim*-alcohol dehydrogenases (ADHs) in the cells [20, 21]. Moreover, α-substituted aldehydes show relatively low chemical stability, resulting in spontaneous chemical racemization and therefore low optical purity [22].

   A convenient strategy for the production of α-methyldihydrocinnamaldehyde derivatives, the fragrances in the perfumes Lilial™ (**1b**) and Helional™ (**2b**), was developed via enzymatic bioreduction of the corresponding cinnamaldehyde precursors (**1a**, **2a**). The (*S*)-antipodes were produced with OYE1 (*Saccharomyces pastorianus*), OYE2 (*Saccharomyces cerevisiae*), and OYE3 (*S. cerevisiae*) in an aqueous–organic biphasic system (containing 20% *t*-BuOMe) at >95% *ee* and quantitative yields (Figure 18.1a) [23]. Several studies have been performed for the production of (*R*)- and (*S*)-citronellal (**3b**), which are key intermediates in the synthesis of menthol, starting from

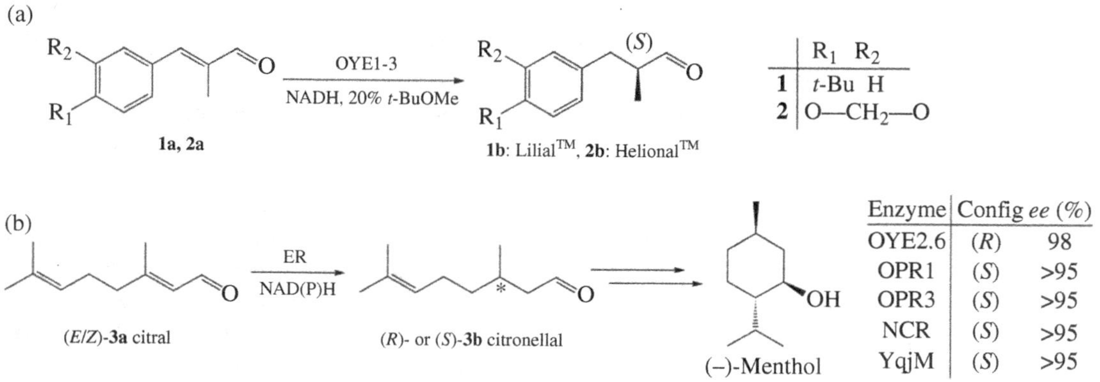

FIGURE 18.1

Ene-reductase-catalyzed reaction for the production of (a) fragrances Lilial™ (**1b**) and Helional™ (**2b**) and (b) citronellal (**3b**).

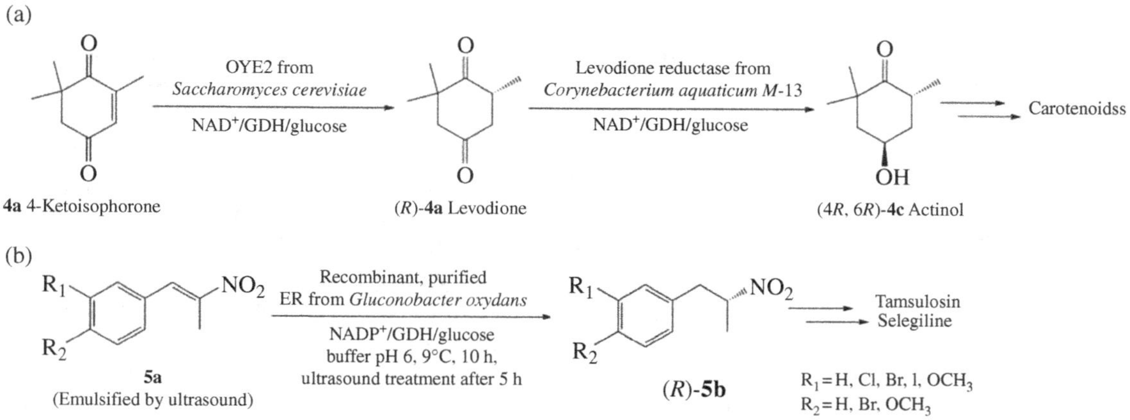

FIGURE 18.2

(a) One-pot two-step enzymatic cascade reaction for the conversion of 4-ketoisophorone (**4a**) to (*R*)-levodione (**4b**) and (4*R*,6*R*) actinol (**4c**) catalyzed by OYE2 and levodione reductase [29]. (b) Enzymatic reduction of nitro precursors for drug synthesis.

(*E*/*Z*)-citral (**3a**). Using isolated OYEs, the stereoselectivities obtained were excellent for both enantiomers (>95%). Whereas (*S*)-citronellal could be achieved in quantitative amounts, the conversion for (*R*)-citronellal was only 69% (Figure 18.1b) [24–27].

2. A second broad class of substrates accepted by ERs is α,β-unsaturated ketones (enones). The most studied candidate of this class is the cyclic ketone 4-ketoisophorone (**4a**), which is converted by a large number of OYE homologues to the commercially useful (*R*)-levodione [11]. Therefore, the reduction of ketoisophorone is commonly used as a standard test substrate for the screening of new ERs. Up to now, all characterized OYE family members afford the (*R*)-enantiomer with stereoselectivities up to 99% [11]. The reduction of this ketone leads to a very important key intermediate for the synthesis of various carotenoids [28] and was first applied in 2003 by Wada and coworkers [29] for the production of actinol (**4c**) in a two-enzyme cascade reaction (Figure 18.2a). In the first step, the unsaturated ketone (**4a**) was hydrogenated by OYE2 (recombinant *E. coli* cell extract) followed by keto group reduction by levodione reductase from *Corynebacterium aquaticum* M-13. Recycling of the NADH coenzyme was

achieved by employing the glucose dehydrogenase/glucose (GDH/glucose) recycling system, thereby yielding quantitative (4*R*,6*R*)-actinol with an *ee* value of 94% [29].

3. Enzymatic asymmetric reduction of α- or β-substituted nitroalkenes provides an attractive method to prepare highly enantiopure nitroalkenes, as they can easily be converted into the corresponding amines, aldehydes, carboxylic acids, oximes, hydroxylamines, and denitrated compounds [30]. Therefore the hydrogenation of alkyl- and aryl-substituted nitroolefins has been extensively studied by various groups, including the determination of the effects of reaction conditions (e.g., solvent conditions and source of coenzymes) on product yield and enantiopurity [26, 27, 31–34]. In particular, the OYE homologue PETNR (*Enterobacter cloacae*) was subjected to intensive mutagenesis studies to avoid the formation of the oxime side product and for switching enantiopreference for the target nitro components [25, 35, 36]. Chiral amines are important for the synthesis of commercial drugs like tamsulosin and selegiline [37]. Only recently a strategy for the highly enantioselective asymmetric reduction of α-substituted nitroalkanes by an ER from *Gluconobacter oxydans* [38] was developed with conversions up to >99% and *ee* values of up to 95% (Figure 18.2b) [39].
4. Carboxylic acids, and derivatives such as esters, are known as weak electron-withdrawing groups yielding little to no reduction by OYE family members [17–19, 40]. However, the addition of a second activating group can increase the C=C bond polarization and facilitate the hydrogenation. Several studies have been performed on α,β-unsaturated dicarboxylic acids and esters [41, 42] and α-halo-substituted unsaturated acids and esters [19, 40, 43–45] for determining the reaction rate and/or stereochemical outcome of the asymmetric hydrogenation using ERs. For example, an additional halogenated substituent in the α-position was shown to be beneficial for ER activity, whereas β-alkyl or β-aryl substituents were detrimental for the reactivity of nonhalogenated substrates [40, 43]. Two or more chlorine substituents caused a secondary spontaneous β-elimination reaction of hydrohalous acids after reduction of the alkene function, giving rise to sequential biodegradation hence generating the saturated carboxylic esters. This allowed for the production of both enantiomers of methyl 2-chloro-iodopropionate, 2-bromo-iodopropionate, and 2-iodopropionate in good to excellent enantiopurity via enzyme-based stereocontrol using OYEs (Figure 18.3a) [45]. Enantiopure α-haloesters are, in particular, valuable synthons for the synthesis of chiral active pharmaceutical ingredients used for the treatment of noninsulin-dependent type-2 diabetes mellitus (T2DM) [22]. Additionally, the use of a carboxylic function in the starting material can offer advantages, in contrast to aldehydes, such as increased stability of the stereogenic center toward spontaneous chemical racemization or the inertness of the acid group toward bioreduction. A practical synthesis of an enantioenriched stereogenic unit, needed for the inclusion in the structural skeleton of relevant pharmaceutical ingredients for the treatment of T2DM, was developed in Brenna's group [22]. Excellent stereoselectivities were obtained with (*S*)-configured α-halo-β-substituted acids and methyl esters. Methyl (*Z*)-2-bromocrotonate (**15a**) afforded, respectively, (*S*)-2-bromobutanoic acid (**15b**, *ee* = 97%) and (*S*)-methyl-2-bromobutanoate (**15c**, *ee* = 97%) by baker's yeast fermentation and OYE1-3 (Figure 18.3b).
5. Roche ester, (*R*)-3-hydroxy-2-methylpropionic acid methyl ester (**16b**), is a chiral building block for the synthesis of vitamins (α-tocopherol), natural products (spiculoic acid A), antibiotics (calcimycin), and fragrance components (muscone) and therefore of high interest for industry. Faber and coworkers employed ERs for the asymmetric synthesis of methyl 2-hydroxymethylacrylate derivatives

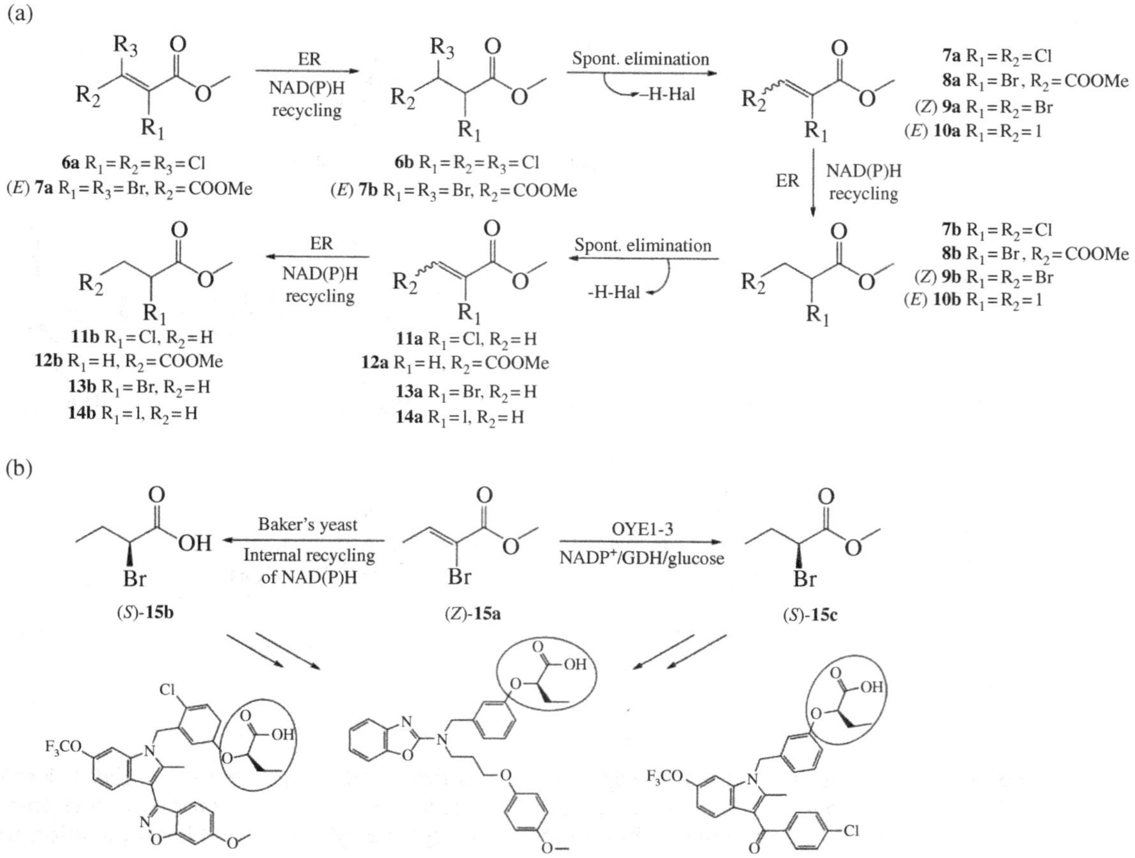

FIGURE 18.3

ER-catalyzed (a) reductive dehydrohalogenation of α,β-halogenated carboxylic ester derivatives and (b) production of (*S*)-configured α-halo-β-substituted acids and methyl ester and structural skeletons of relevant pharmaceutical ingredients (the building block is circled).

with enantioselectivities >99%. All the ERs tested strictly furnished the (*R*)-enantiomer. Hydroxyl group protection (allyl, benzyl, or TBDMS ethers) did not show any influence in stereochemical outcome; however, the reaction rate was affected. Direct access to the *O*-protected Roche esters could be achieved, generating convenient intermediates for further synthesis, with conversions up to >99% (Figure 18.4a) [46].

6. Enantiopure nitriles are important building blocks as they can easily be converted to amines, aldehydes, and carboxylic acids [1]. A series of α,β-unsaturated nitriles were converted by commercially available ERs (ERED 101–114; Merck and Co) to their optically active nitrile products in high yields (Figure 18.4b). All enzymes tested showed (*R*)-selectivity and up to 99% enantioselectivity (the absolute configuration of the pharmaceutical building block, compound **18b**, was not specified) [47]. Recently published studies describe the hydrogenation of a library of β-cyano-α,β-unsaturated acids and esters using ERs from the OYE family [48, 49]. For example, a biocatalytic route for GABA analogues, such as pregabalin could be provided (Figure 18.4c). Both enantiomers were obtained with *ee* values up to >99% by using either pure (*Z*)- or (*E*)-isomers and by varying the size of the ester moiety [49]. Moreover, deuterium-labeled studies showed that the nitrile group was the activating group for the ER-catalyzed reaction over the carboxylic acid and/or the ester moiety [50].

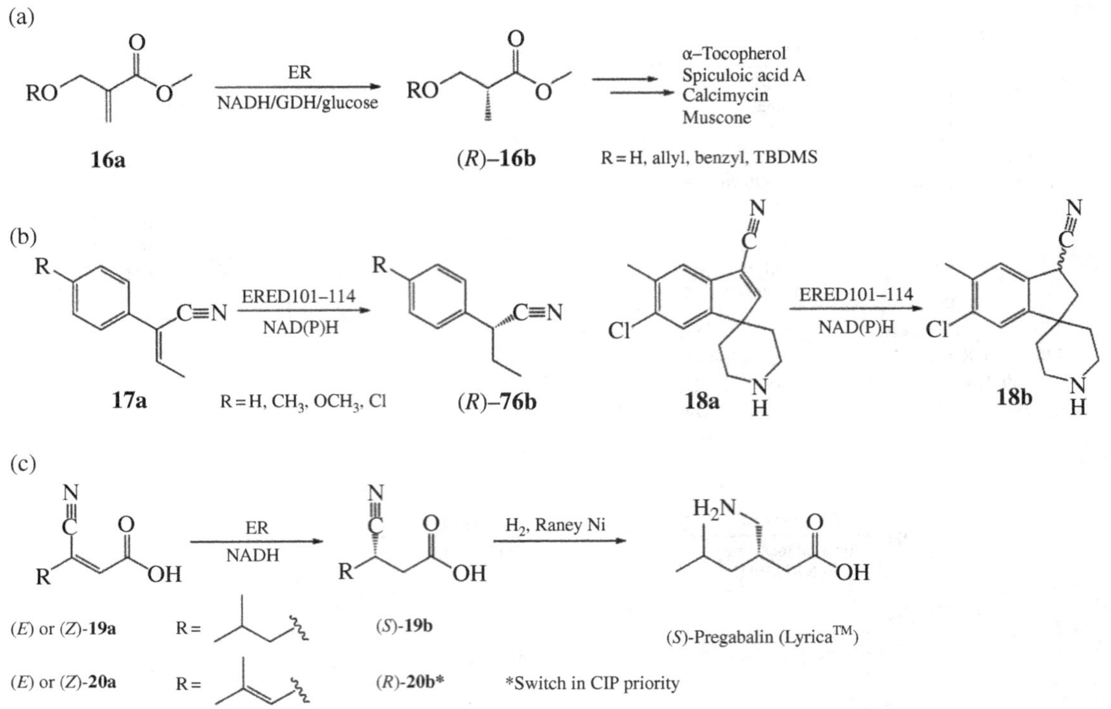

FIGURE 18.4

(a) Roche ester production using ene-reductases from OYE1-3, YqjM (*Bacillus subtilis*), NCR (*Zymomonas mobilis*), NerA (*Agrobacterium radiobacter*), OPR1 (*Arabidopsis thaliana*), and XenA (*Pseudomonas putida II-B*). (b) Asymmetric bioreduction of unsaturated nitriles by commercially available ERs. (c) Asymmetric bioreduction of β-cyanoacrylic acids and hydrogenation to pregabalin.

18.3 MULTIENZYME REACTIONS

An increasing trend is to perform the synthesis of fine chemicals and/or pharmaceuticals either by using multiple enzymatic reactions or through chemoenzymatic steps. Enantio- and diastereomerically pure γ-butyrolactones, building blocks in the synthesis of natural products such as milbemycin β2, jasplakinolide, and amphidinolides, were synthesized using a one-pot, two-enzyme cascade reaction combining an OYE with an ADH. In the first step, the double bond of ethyl 2-methyl-4-oxopent-2-enoate (**21a**) was stereoselectively reduced using OYE1. Several ADHs were trialed for the subsequent carbonyl reduction to the γ-hydroxy ester, which then underwent spontaneous lactonization [51]. Asymmetric bioreduction of the two (*E*/*Z*)-isomers of the starting material was stereoconvergent and yielded the (*R*)-enantiomer. Carbonyl reduction proceeded with enzyme-based stereocontrol, where—depending on the choice of ADH used—both stereoisomers (2*R*,4*S*)-**21c** (Prelog-type ADH-T from *Thermoanaerobacter* species) and (2*R*,4*R*)-**21c** (anti-Prelog-type ADH-LK from *Lactobacillus kefir*) were produced in yields up to 80% with stereoselectivities up to >99% (Figure 18.5a) [51].

Recently, the stereoselective synthesis of bicyclic primary or secondary alcohols was reported by Brenna *et al.* [52]. Again, the combination of an ER with an ADH was used to catalyze reactions on substrates based on tetralin or chroman backbones, which led to precursors of important active pharmaceutical ingredients (rotigotine, robalzotan, and ebalzotan) with optical purities up to >99% (Figure 18.5b) [52]. In this case, the addition of an ADH allowed for a minimization of the spontaneous chemical racemization of the unstable saturated aldehyde intermediate.

A one-pot three-enzyme system for the production of a valuable chiral building block has been developed to a 70 g scale leading to 89% isolated yield [(*R*)-enantiomer].

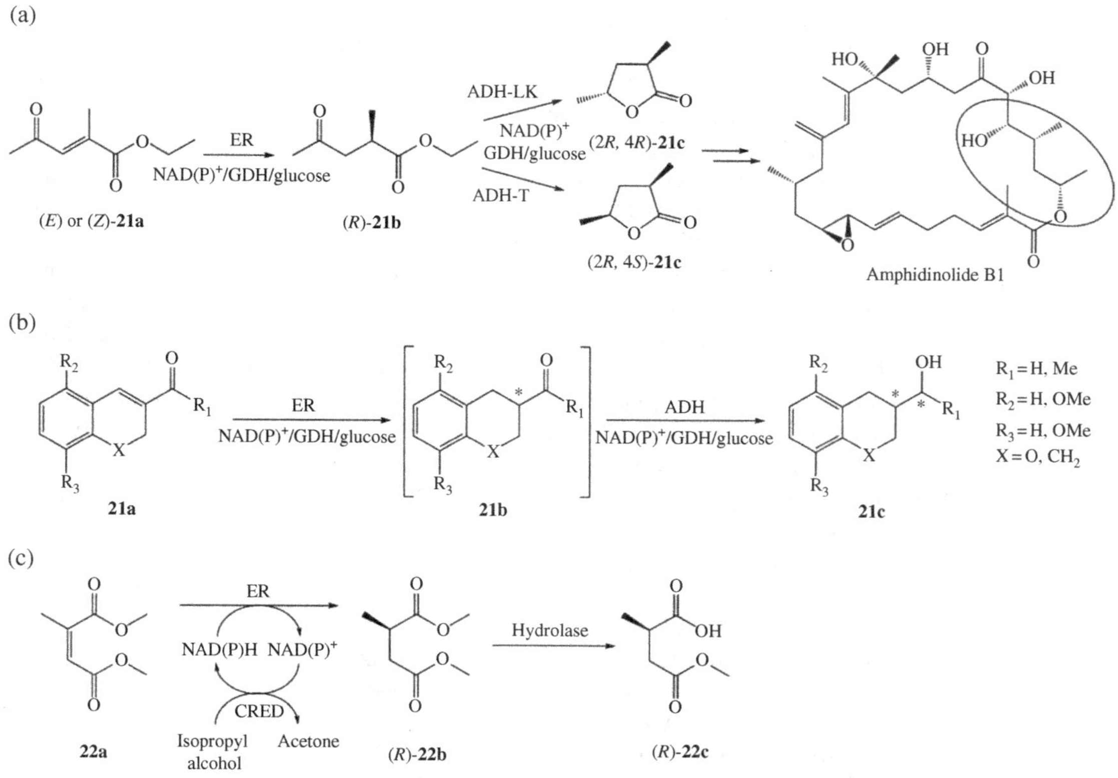

FIGURE 18.5

Multienzyme cascade synthesis combining ene-reductases with alcohol dehydrogenases. (a) Synthesis of optically pure γ-butyrolactones (the building block is circled) and (b) optically pure primary or secondary tetralin or chroman derivatives. (c) Three-enzyme system for the generation of a valuable chiral building block.

This multienzyme process includes the ER-catalyzed reduction of an α,β-unsaturated diester (**22a**) and subsequent hydrolysis of **22b** by a hydrolase to the monoester (**22c**). Cofactor recycling was performed with an $NAD(P)^+$/carbonyl reductase (CRED) including simultaneous coproduct (acetone) removal (Figure 18.5c) [53].

18.4 ALTERNATIVE HYDRIDE SOURCES

A disadvantage of nicotinamide-dependent enzymes is the high cost of the coenzyme (if used in stoichiometric amounts) making their use cost prohibitive for many synthetic industrial purposes. Therefore, alternative strategies have been adopted to supply the necessary hydride equivalents:

1. One method is the coupling of an *in situ* recycling system, where $NAD(P)^+$ is continuously regenerated to NAD(P)H. Although only catalytic levels of $NAD(P)^+$ are needed, stoichiometric amounts of a cosubstrate are required to drive the recycling system to overcome thermodynamic limitations. Various systems have been adopted for ER-catalyzed biocatalytic reactions such as the (i) GDH/glucose, (ii) glucose-6-phosphate dehydrogenase/glucose-6-phosphate (G6PDH/G6P), (iii) formate dehydrogenase/formate (FDH/formate), and (iv) phosphite dehydrogenase/phosphite (PTDH/phosphite) system [54]. Unfortunately, the nature of the substrate has a high influence on the efficiency of the recycling system. For example, the metal-dependent enzymes GDH, G6PDH, and FDH were inactive in the presence of *cis*-configurated α,β-unsaturated dicarboxylic acids as these substrates act as strong chelating

agents. However, the deactivation of the recycling enzymes could be overcome by the addition of metal ions (e.g., Ca^{2+}, Mg^{2+}, or Zn^{2+}) to the reaction medium [41]. Moreover product yield and/or enantioselectivity can vary dramatically, depending on the recycling system which has been employed [26, 31, 55]. Recent examples of cofactor recycling systems include the solvent-stable ADH from *Rhodococcus ruber* (ADH-A) [56] which uses only two equivalents of 2-propanol as H-donor or the selectAZyme CRED [53] that needs isopropanol as cosubstrate, both producing acetone. However, in the case of ADH-A, α,β-unsaturated aldehydes are not suitable substrates as the carbonyl function gets "overreduced" to the alcohol moiety [56]. The recycling enzymes can either form highly volatile products (CO_2 in the case of FDH, ADH and CRED form acetone) or unstable compounds that hydrolyze immediately (gluconolactone/gluconic acid or 6-phosphogluconate for GDH and G6PDH, respectively). Therefore, all four systems work practically irreversibly and shift the equilibrium toward reduction [10].

2. Photoenzymatic reduction of the C=C double bond using YqjM—a OYE homologue from *Bacillus subtilis*—was investigated by Reetz and coworkers in 2008 [57]. Illumination of YqjM with visible light in the presence of free flavin and EDTA as a sacrificial electron donor quantitatively converted 4-ketoisophorone (**4a**) to (*R*)-levodione (**4b**, Figure 18.6a) [57]. Similar studies were performed by Hollmann *et al.* [58] for the photoenzymatic flavin reduction using YqjM and NEMR (*N*-ethylmaleimide reductase from *E. coli*) [58]. The alternative hydride sources used were EDTA, formate, or phosphite, and the reduction was performed under strict anaerobic conditions to avoid rapid reoxidation of the highly reactive reduced flavin by molecular oxygen. For the conversion of **4a** to **4b** up to 65% of the specific activities in comparison to the NADH-driven reaction were obtained when flavin reduction occurred with light while preserving the enantioselectivity. Additionally, undesired side reactions, such as reduction of the carbonyl groups by contaminating ADHs, could be circumvented [58].
3. Electrochemical regeneration of nicotinamide-dependent coenzymes has been investigated, such as a direct cathodic reduction of $NAD(P)^+$. In this process, the hydride donors (electrons) are delivered in ideal atom efficiency as they are basically mass free. Moreover no cosubstrate is needed, and electrochemical power is among the cheapest redox equivalent available. However, despite these advantages, this method is hampered by a low selectivity. Moreover, after transfer of the first electron, a highly reactive NAD·radical is generated, forming enzymatically inactive NAD dimers [59].
4. Recently both electrochemistry and mediator redox chemistry have been combined in a biphasic miniaturized bioreactor for the NAD(P)H-free biocatalytic reduction catalyzed by PETNR and TOYE (*Thermoanaerober pseudethanolicus*) [60]. In this case, the electrons are transferred through an artificial mediator *N*,*N*,dimethyl-4-4-bipyridinium (dimethyl viologen (MV)), which involves the reduction of MV^{2+} by a reversible one-electron reaction to its blue radical monocation MV^+ (Figure 18.6b). The schematic of the bioreactor is shown in Figure 18.6c. A single peristaltic pump draws the solution from the reservoir through the bioreactor. The aqueous phase, containing the mediator and enzyme, passes through the electrochemical cell wherein reduction of the mediator takes place. The aqueous phase gets mixed with the substrate-containing organic phase and passes through 50 parallel reaction channels. The reactor was kept strictly anaerobic to avoid MV oxidation by molecular oxygen. Successful substrate conversions and easy product recovery, although with lower efficiency than reactions catalyzed with NADH as coenzyme, have been demonstrated with five different α,β-unsaturated ketones, aldehydes, and maleimides. For example, a rate of 1.25 and 0.7 $mM\,h^{-1}$ was obtained for the bioreduction of 2-cyclohexenone (**23a**) with TOYE and

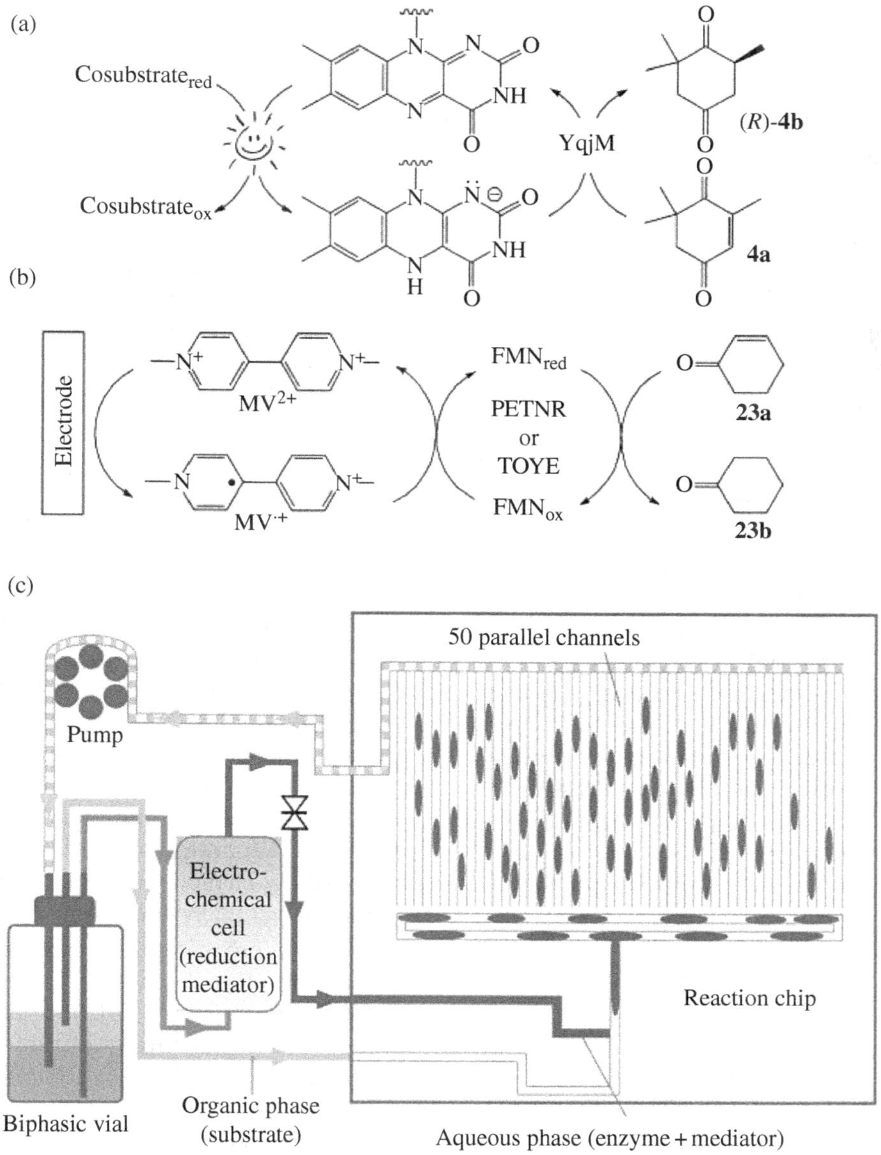

FIGURE 18.6

(a) Photoenzymatic flavin reduction using YqjM. (b) Electroenzymatic conversion of 2-cyclohexenone (**23a**) to 2-cyclohexanone (**23b**) catalyzed by PETNR and TOYE using methyl viologen as electron-shuttling system. (c) Schematic representation of the anaerobic biphasic bioreactor.

PETNR, respectively, in comparison to 3.1 mM h^{-1} in the presence of a two-fold access of NADPH. Additionally, enzyme and mediator stability and reusability were maintained over 12 h.

5. Asymmetric bioreduction of activated C=C bonds can be achieved via direct electron transfer by a single flavoprotein between two enone substrates, yielding an oxidized and reduced product in equimolar amounts. This disproportionation of enones is therefore nicotinamide independent, although a maximum of 50% alkane yields can be obtained with one enone substrate. The reductive half reaction proceeds via the desaturation of 2-cyclohexenone (**23a**) forming $FMNH_2$ and cyclohexa-1,4-dienone, which irreversibly tautomerizes to phenol (**23c**). The tautomerization step provides a strong driving force for the overall process. In the oxidative half reaction, the flavin-bound electrons are transferred to a second equivalent of the enone, generating the equivalent alkane [61]. However, electron-rich phenols are often inhibitors of OYEs as they can form stable charge-transfer complexes with the electron-deficient flavin in the active site [14]. Recently, Faber and coworkers have overcome this inhibition by optimizing the reaction parameter (pH and temperature) and employing phenolic coproduct removal using a solid-phase

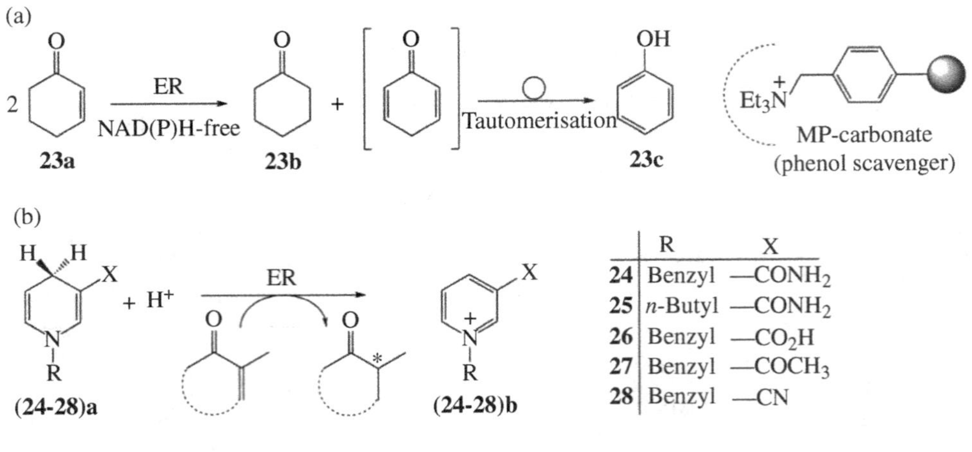

FIGURE 18.7

(a) ER-catalyzed disproportionation of 2-cyclohexenone using MP-carbonate as a phenol scavenger. (b) Asymmetric bioreduction of activated C—C double bonds using synthetic nicotinamide mimics.

organic resin (Figure 18.7a) [62]. After optimizing the reaction conditions and adsorbing the phenolic product using macroporous triethylammonium methylpolystyrene carbonate (MP-carbonate), the conversion for 2-cyclohexenone (**23a**) could be increased from 55 to 97% using the enzyme chromate reductase (CrS) from *Thermus scotoductus* SA-01. The moderate conversion (48%) of dimethyl citraconate to dimethyl (*R*)-2-methylsuccinate using the cosubstrate 1,4-cyclohexanedione was optimized to 92% with an enantioselectivity of >99%. However, this technique is not applicable to all substrates as, in some cases, substrate and product were sequestered by MP-carbonate, and there were examples of undesired racemization of chiral sensitive α-substituted ketones [62]. Further investigations identified cheap and commercially available cosubstrates that resulted in conversions and enantioselectivities comparable to those obtained in the presence of an excess of a nicotinamide coenzyme or in combination with traditional NAD(P)H recycling systems [63].

6. Synthetic nicotinamide coenzymes (mNADs) were recently described by Hollmann and coworkers as substitutes for the more expensive traditional nicotinamide coenzymes NADH and NADPH (Figure 18.7b) [64]. Near identical conversions and product enantiopurities in comparison to NAD(P)H could be obtained in biotransformations with three different OYE homologues (YqjM, TsER from *T. scotoductus*, and RmER from *Ralstonia metallidurans*). A very interesting fact is that other nicotinamide-dependent enzyme classes, such as ADHs or monooxygenases, showed very poor activity toward mNADs. This allows the use of poorly purified ERs, as undesired side reactions such as reduction of the keto- or aldehyde moiety are eliminated [64]. Unfortunately, there is currently no *in situ* regeneration method for reduced mNADs available, which makes their use less cost-effective; however, they offer great promise for future developments.

18.5 IMPROVEMENTS OF PRODUCTIVITY, STEREOSELECTIVITY, AND/OR CONVERSION

ER-catalyzed reactions are often limited by low productivity and/or low enantiopurity. Moreover, enzymes are often not optimally suited for industrial applications due to their insufficient stability (pH, temperature, organic solvents) under process conditions. Another weakness is that, in nature, the counterpart of a particular enzyme is often missing. Several studies have been performed for overcoming those drawbacks by optimizing reaction parameters or engineering enzymes. A few selected examples are summarized as follows:

1. Typically, ER-catalyzed reactions are performed at 5 mM substrate concentration since these enzymes are plagued by substrate and/or product inhibition [24, 65, 66]. The implementation of an *in situ* substrate feeding product removal (SFPR) technology was successfully applied to the OYE-catalyzed reaction of highly enantiopure ethyl (*S*)-2-ethoxy-3-(*p*-methoxyphenyl)propanoate (EEHP; **29c**), an important precursor of several PPAR-α/γ agonists, like tesaglitazar [65, 67]. SFPR works on the adsorption of substrate and products on a hydrophobic resin, maintaining soluble concentrations at levels that are not toxic for the microorganism or enzyme. By combining isolated OYE3 instead of baker's yeast whole cells with SFPR technology, the oxidation of the reduced aldehyde **29b** to the carboxylic acid **29c** on a 1 g scale (substrate loading 30 g L^{-1}) yielded an excellent productivity of 55.6 g l^{-1} day^{-1}, with 94% conversion and 98% *ee* (over 100-fold increased productivity from whole cells to isolated enzymes; Figure 18.8a) [67].
2. Site saturation mutagenesis of Trp116 was performed for OYE1 from *S. pastorianus* by the group of Stewart [68]. The original hypothesis was that smaller amino acid residues in position 116 would allow for the binding of more bulky substituents and hence increase the activity or broaden the substrate specificity for this enzyme. Two variants Trp116Phe and Trp116Ile have been identified, but both had lower specific activities for the asymmetric hydrogenation of 3-methylcyclohex-2-enone and 3-methylcyclohex-2-enone in comparison to the wild-type enzyme, although the stereoselectivities remained the same (>98% for the (*S*)-enantiomer). However, the reduction of (*R*)- and (*S*)-carvone (**30a** and **31a**, respectively) showed interesting results with these variants (Figure 18.8b). (*R*)-Carvone (**30a**) was reduced by all three enzymes to the same *trans*-product (**30b**) with high diastereoselectivity. The reduction of (*S*)-carvone yielded *cis*-**31b** when the reduction was performed with the wild-type enzyme and the Trp116Phe variant, whereas Trp116Ile afforded the *trans*-product **31c**, which is the enantiomer of the product that this protein made from **30a**. Studies using deuterated (4*R*)-NADPH indicated that the Trp116Ile variant carries out net *trans*-addition of H_2 across the

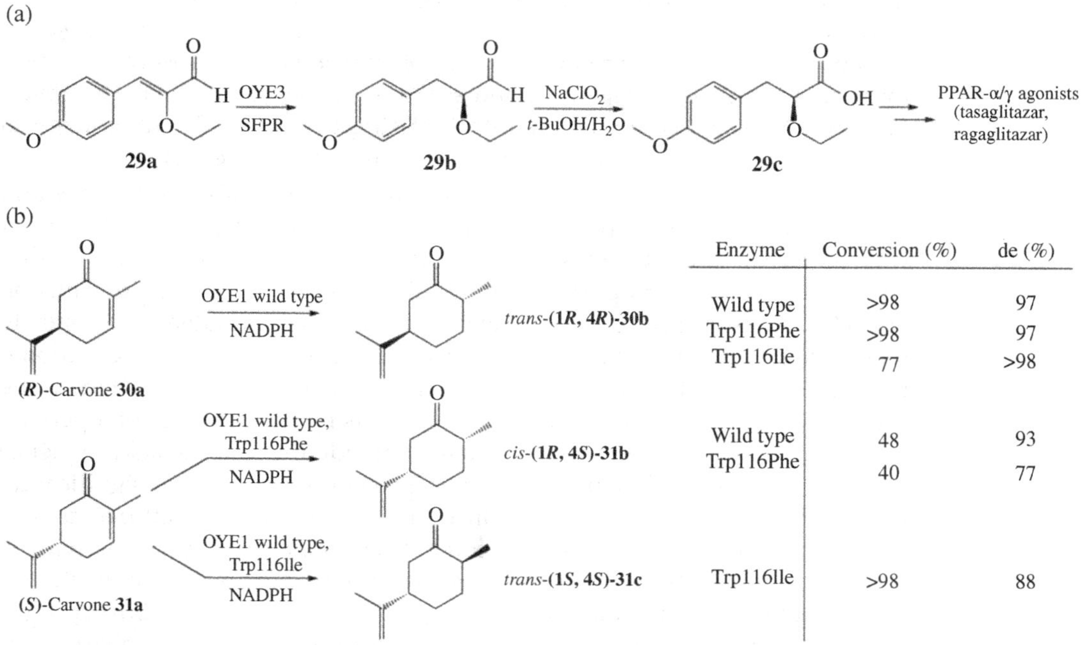

Enzyme	Conversion (%)	de (%)
Wild type	>98	97
Trp116Phe	>98	97
Trp116Ile	77	>98
Wild type	48	93
Trp116Phe	40	77
Trp116Ile	>98	88

FIGURE 18.8

(a) OYE3-catalyzed reduction employing SFPR technology for the synthesis of EEHP precursor [67]. (b) Asymmetric reduction of carvone enantiomers (**30a** and **31a**) using OYE1 and variants [68].

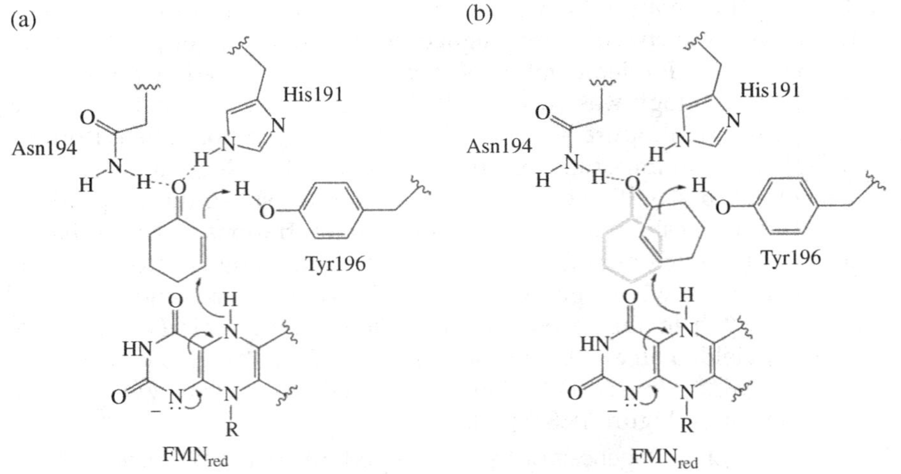

FIGURE 18.9
Catalytic mechanism of *Saccharomyces pastorianus* OYE1. (a) "Normal" binding of the substrate in the active site of the enzyme. Net *trans*-addition of the hydride occurs by a hydride attack on the β-carbon with concomitant α-protonation by Tyr196. (b) "Flipped" orientation on the Trp116Ile variant. The gray outline represents the "normal" binding.

double bond of (*S*)-**31a**. The substrate, however, binds in a "flipped" orientation (Figure 18.9b) compared to that of (*R*)-**30a**. The 2-cyclohexenone backbone can be flipped while maintaining a similar relative position of the β-carbon. Also the orientation of the carbonyl oxygen is similar compared to the conventional substrate binding (Figure 18.9a) [68].

3. YqjM was subjected to iterative saturation mutagenesis studies (ISM; CASTing) to generate variants with improved catalytic ability [69]. Initially, 20 residues were targeted for saturation mutagenesis on selected residues near the binding pocket based on the X-ray structure of the inhibitor-bound enzyme, with the asymmetric reduction of 3-methylcyclohexenone to 3-methylcyclohexanone chosen as the model reaction. A few positive hits were used as templates for selective residue randomization in a second round of mutagenesis. The goal was to test iterative CASTing as a suitable method to increase the activity of an enzyme while maintaining control over its enantioselectivity. The study successfully generated variants that produced either the (*S*)- or the (*R*)-product from 3-methylcyclohexenone, with *ee* values up to 97% and good activities. Moreover, some of the best variants also proved to be excellent catalysts with other cyclopentenone and cyclohexenone derivatives without performing additional mutagenesis and screening experiments [69]. This study proved ISM in the combination with CASTing to be an efficient strategy in directed evolution of ERs while minimizing the screening effort.
4. Circular permutation (CIP) was used to improve the catalytic efficiency and expand the substrate scope of OYE1 by Lutz and coworkers [70]. This study was based on the hypothesis that noncatalytic structural regions within enzymes may play an important and rate-limiting role in the catalytic function of OYEs. To test this, a variety of flexible regions were mutated to investigate their potential functional role. Covalently linking the amino and the carbonyl termini with a peptide and the introduction of new termini elsewhere in the protein structure generated a library of variants by the breaking of a peptide bond. This generated a library of proteins with identical amino acid sequence but shifted termini. They possess altered protein dynamics that significantly impacted on the catalytic performance of the enzyme. The cell-free cpOYE library variants were prepared by combining whole-gene synthesis with *in vitro* transcription/translation and were tested with three reference substrates, 4-ketoisophorone (**4a**), cinnamaldehyde, and (*S*)-carvone (**31a**). Surprisingly, over 70 functional OYE1 variants were identified, with several biocatalysts exhibiting over an order of magnitude improved catalytic activity toward at least one of the tested substrates. Interestingly, for all three substrates, the locations of the new protein termini in the active variants fall into the same

four structural loop/lid regions near the active site of the enzyme. This research demonstrates the importance of these structural elements in enzyme function and supports the hypothesis that conformational flexibility may be rate limiting in OYE1 [70].

5. Rational enzyme loop modification was performed for the mesophilic ER NCR from *Zymomonas mobilis* to create variants with higher temperature and solvent stability [71]. Sequence and structural alignments were performed between the two classes of ERs [11], namely, the classical (e.g., OYE1, MR from *P. putida*, and NCR) and thermophilic-like OYEs (e.g., YqjM, GkOYE from *Geobacillus kaustophilus*, and TOYE). Two loop variants were created, which were shortened by several amino acid residues, as thermophilic-like ER typically contains shorter, more rigid surface loop regions. The modulation of one specific loop segment near the active site of NCR showed an increased tolerance to organic solvents along with a significant enhancement in its thermostability [71].
6. Single generation site-saturated mutagenesis studies have been performed for the OYE homologue PETNR to increase the substrate scope, improve the enantioselectivity, and reduce the oxime formation for the asymmetric reduction of α,β-unsaturated nitroolefins [35, 36, 72]. Residues known to be involved in the catalytic activity and residues lining the active site cavity were selected for mutagenesis studies. Moreover, all possible variants from two active-site residue libraries His181X and His184X were identified and analyzed, and crystal structures of ligand-bound wild-type enzyme and few of its variants were performed to get a better understanding of the consequences of each mutation [36]. Variants could be identified, showing a switch in enantioselectivity for the reduction of (*E*)-1-nitro-2-phenylpropene (**32a**) (Figure 18.10) from the (*S*)- to the (*R*)-enantiomer (Tyr26Ser). Additional variants showed an improved *ee* value for (*S*)-**32b** (His181Asn and His184Asn/Ala) or a significant reduction of byproduct formation **32c** (His184Arg); however, major enhancements in the activity of PETNR toward nonnatural substrates were not detected [36, 71].
7. Molecular dynamics studies in combination with sequence analysis of twelve OYE homologues were carried out to predict the stereopreference for the asymmetric reduction of 1-nitro-2-phenylpropene **32a** based on the amino acid sequence and structure [73]. A proline residue in the third position of a sequence pattern motif in loop β2 should thereby lead to high *S*-selectivity. In contrast, moderately selective enzymes exhibit a nonpolar short-chain amino acid, and all *R*-selective OYE homologues were predicted to contain long polar and charged amino acids in this position (Table 18.1) [73].

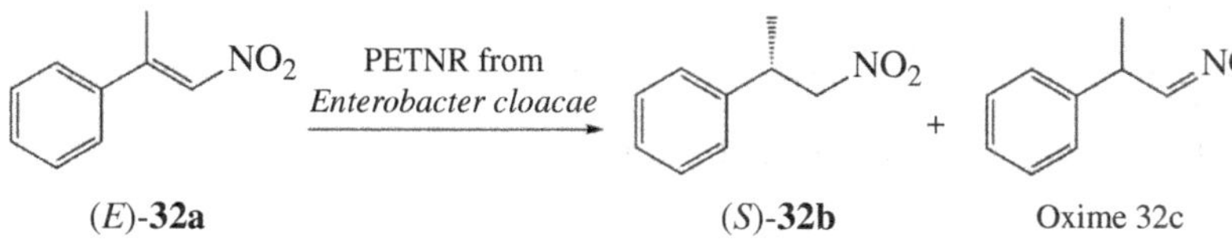

FIGURE 18.10 Asymmetric reduction of 1-nitro-2-phenylpropene by PETNR.

TABLE 18.1 Sequence Motifs Identified in Loop β2 for Prediction of the Stereoselective Outcome on the Reduction of 1-Nitro-2-phenylpropene in OYE Homologues

Stereoselectivity	Sequence Motif in Loop2
High *S*	G-[FYW]-**P**-X(2)-P-G-[ILV]-[FHYW]
Moderate *R*/*S*	G-[FYW]-[**AV**]-X(2)-P-G-[ILV]-[FHYW]
High *R*-	G-[FYW]-[**ENK**]-X(2)-P-G-[ILV]-[FHYW]

Data from Ref. [73].

This hypothesis was tested by another group for a previously uncharacterized ER from *Photorhabdus luminescens* subsp *laumondii* NCR_{PL} identified by BLAST analysis, which should be highly *R*-selective according to the sequence pattern found (GYENTPGIW) [74]. Additionally several variants of the *S*-selective ER from *Z. mobilis* NCR_{ZM}—which has a proline residue in position three in the β2 loop—were performed in order to invert its stereoselectivity. Unfortunately, neither the newly cloned ER nor one of the NCR_{ZM} variants (Pro67GluHisLysAsnGln) produced the (*S*)-enantiomer in the hydrogenation of **32a**. Not even a drop in stereoselectivity could be seen for the NCR_{ZM} variants in comparison to the wild-type enzyme [73]. This study shows the difficulties in predicting global properties on ER based on a primary sequence.

REFERENCES

[1] Toogood, H.; Mansell, D.; Gardiner, J. M.; Scrutton, N. S. Enantioselective Bioreduction of Carbon-Carbon Double Bonds. In *Comprehensive Chirality*, Turner, N. J., Ed. Elsevier: Amsterdam, 2012; Vol. **1**; pp. 216–255.

[2] Tischer, W.; Bader, J.; Simon, H. *Eur. J. Biochem.* 1979, **97**, 103–112.

[3] Rohdich, F.; Wiese, A.; Feicht, R.; Simon, H.; Bacher, A. *J. Biol. Chem.* 2001, **276**, 5779–5787.

[4] Nordling, E.; Jörnvall, H.; Persson, B. *Eur. J. Biochem.* 2002, **269**, 4267–4276.

[5] Hirata, T.; Matsushima, A.; Sato, Y.; Iwasaki, T.; Nomura, H.; Watanabe, T.; Toyoda, S.; Izumi, S. *J. Mol. Catal. B: Enzym.* 2009, **59**, 158–162.

[6] Mansell, D. J.; Toogood, H. S.; Waller, J.; Hughes, J. M. X.; Levy, C. W.; Gardiner, J. M.; Scrutton, N. S. *Adv. Synth. Catal.* 2013, **3**, 370–379.

[7] Ringer, K. L.; McConkey, M. E.; Davis, E. M.; Rushing, G. W.; Croteau, R. *Arch. Biochem. Biophys.* 2003, **418**, 80–92.

[8] Toogood, H. S.; Scrutton, N. S. *Curr. Opin. Chem. Biol.* 2014, **19**, 107–115.

[9] Toogood, H. S.; Scrutton, N. S. *Catal. Sci. Technol.* 2013, **3**, 2182–2194.

[10] Winkler, C. K.; Tasnadi, G.; Clay, D.; Hall, M.; Faber, K. *J. Biotechnol.* 2012, **162**, 381–389.

[11] Toogood, H. S.; Gardiner, J. M.; Scrutton, N. S. *ChemCatChem* 2010, **2**, 892–914.

[12] Stuermer, R.; Hauer, B.; Hall, M.; Faber, K. *Curr. Opin. Chem. Biol.* 2007, **11**, 203–213.

[13] Kohli, R. M.; Massey, V. *J. Biol. Chem.* 1998, **273**, 32763–32770.

[14] Brown, B. J.; Deng, Z.; Karplus, P. A.; Massey, V. *J. Biol. Chem.* 1998, **273**, 32753–32762.

[15] Williams, R. E.; Bruce, N. C. *Microbiology* 2002, **148**, 1607–1614.

[16] Barna, T.; Messiha, H. L.; Petosa, C.; Bruce, N. C.; Scrutton, N. S.; Moody, P. C. E. *J. Biol. Chem.* 2002, **277**, 30976–30983.

[17] Ferraboschi, P.; Grisenti, P.; Casati, R.; Fiecchi, A.; Santaniello, E. *J. Chem. Soc. Perkin Trans.* 1987, **1**, 1743–1748.

[18] Koul, S.; Crout, D. H. G.; Errington, W.; Tax, J. *J. Chem. Soc. Perkin Trans.* 1995, **1**, 2969–2988.

[19] Utaka, M.; Konishi, S.; Mizuoka, A.; Ohkubo, T.; Sakai, T.; Tsuboi, S.; Takeda, A. *J. Org. Chem.* 1989, **54**, 4989–4992.

[20] Hall, M.; Hauer, B.; Stuermer, R.; Kroutil, W.; Faber, K. *Tetrahedron Asymmetry* 2006, **17**, 3058–3062.

[21] Müller, A.; Hauer, B.; Rosche, B. *J. Mol. Catal. B: Enzym.* 2006, **38**, 126–130.

[22] Brenna, E.; Gatti, F. G.; Manfredi, A.; Monti, D.; Parmeggiani, F. *Org. Process Res. Dev.* 2012, **16**, 262–268.

[23] Stueckler, C.; Mueller, N. J.; Winkler, C. K.; Glueck, S. M.; Gruber, K.; Steinkellner, G.; Faber, K. *Dalton Trans.* 2010, **39**, 8472–8476.

[24] Bougioukou, D. J.; Walton, A. Z.; Stewart, J. D. *Chem. Commun.* 2010, **46**, 8558–8560.

[25] Fryszkowska, A.; Toogood, H.; Sakuma, M.; Gardiner, J. M.; Stephens, G. M.; Scrutton, N. S. *Adv. Synth. Catal.* 2009, **351**, 2976–2990.

[26] Hall, M.; Stueckler, C.; Ehammer, H.; Pointner, E.; Oberdorfer, G.; Gruber, K.; Hauer, B.; Stuermer, R.; Kroutil, W.; Macheroux, P.; Faber, K. *Adv. Synth. Catal.* 2008, **350**, 411–418.

[27] Mueller, N. J.; Stueckler, C.; Hauer, B.; Baudendistel, N.; Housden, H.; Bruce, N. C.; Faber, K. *Adv. Synth. Catal.* 2010, **352**, 387–394.
[28] Demole, E.; Enggist, P. *Helv. Chim. Acta* 1974, **57**, 2087–2091.
[29] Wada, M.; Yoshizumi, A.; Noda, Y.; Kataoka, M.; Shimizu, S.; Takagi, H.; Nakamori, S. *Appl. Environ. Microbiol.* 2003, **69**, 933–937.
[30] Ono, N. *The Nitro Group in Organic Synthesis*. Wiley-VCH: Weinheim, 2002; p. 392.
[31] Hall, M.; Stueckler, C.; Hauer, B.; Stuermer, R.; Friedrich, T.; Breuer, M.; Kroutil, W.; Faber, K. *Eur. J. Org. Chem.* 2008, **2008**, 1511–1516.
[32] Adalbjornsson, B. V.; Toogood, H. S.; Fryszkowska, A.; Pudney, C. R.; Jowitt, T. A.; Leys, D.; Scrutton, N. S. *ChemBioChem* 2010, **11**, 197–207.
[33] Yanto, Y.; Winkler, C. K.; Lohr, S.; Hall, M. l.; Faber, K.; Bommarius, A. S. *Org. Lett.* 2011, **13**, 2540–2543.
[34] Brenna, E.; Fronza, G.; Fuganti, C.; Gatti, F. G. *Eur. J. Org. Chem.* 2010, 5077–5084.
[35] Toogood, H. S.; Fryszkowska, A.; Hulley, M.; Sakuma, M.; Mansell, D.; Stephens, G. M.; Gardiner, J. M.; Scrutton, N. S. *ChemBioChem* 2011, **12**, 738–749.
[36] Hulley, M. E.; Toogood, H. S.; Fryszkowska, A.; Mansell, D.; Stephens, G. M.; Gardiner, J. M.; Scrutton, N. S. *ChemBioChem* 2010, **11**, 2433–2447.
[37] Bohlmann, R. *Angew. Chem. Int. Ed.* 2001, **40**, 4101.
[38] Richter, N.; Groger, H.; Hummel, W. *Appl. Microbiol. Biotechnol.* 2011, **89**, 79–89.
[39] Burda, E.; Ress, T.; Winkler, T.; Giese, C.; Kostrov, X.; Huber, T.; Hummel, W.; Groger, H. *Angew. Chem. Int. Ed. Engl.* 2013, **52**, 9323–9326.
[40] Tasnadi, G.; Winkler, C. K.; Clay, D.; Sultana, N.; Fabian, W. M. F.; Hall, M.; Ditrich, K.; Faber, K. *Chem. Eur. J.* 2012, **18**, 10362–10367.
[41] Stueckler, C.; Hall, M.; Ehammer, H.; Pointner, E.; Kroutil, W.; Macheroux, P.; Faber, K. *Org. Lett.* 2007, **9**, 5409–5411.
[42] Brenna, E.; Gatti, F. G.; Manfredi, A.; Monti, D.; Parmeggiani, F. *Adv. Synth. Catal.* 2012, **354**, 2859–2864.
[43] Brenna, E.; Fronza, G.; Fuganti, C.; Monti, D.; Parmeggiani, F. *J. Mol. Catal. B: Enzym.* 2011, **73**, 17–21.
[44] Brenna, E.; Gatti, F. G.; Manfredi, A.; Monti, D.; Parmeggiani, F. *Eur. J. Org. Chem.* 2011, Issue 20–21 4015–4022.
[45] Tasnádi, G.; Winkler, C. K.; Clay, D.; Hall, M.; Faber, K. *Catal. Sci. Technol.* 2012, **2**, 1548–1552.
[46] Stueckler, C.; Winkler, C. K.; Bonnekessel, M.; Faber, K. *Adv. Synth. Catal.* 2010, **352**, 2663–2666.
[47] Kosjek, B.; Fleitz, F. J.; Dormer, P. G.; Kuethe, J. T.; Devine, P. N. *Tetrahedron Asymmetry* 2008, **19**, 1403–1406.
[48] Brenna, E.; Gatti, F. G.; Manfredi, A.; Monti, D.; Parmeggiani, F. *Catal. Sci. Technol.* 2013, **3**, 1136–1146.
[49] Winkler, C. K.; Clay, D.; Davies, S.; O'Neill, P.; McDaid, P.; Debarge, S.; Steflik, J.; Karmilowicz, M.; Wong, J. W.; Faber, K. *J. Org. Chem.* 2013, **78**, 1525–1533.
[50] Winkler, C. K.; Clay, D.; Turrini, N. G.; Lechner, H.; Kroutil, W.; Davies, S.; Debarge, S.; O'Neill, P.; Steflik, J.; Karmilowicz, M.; Wong, J. W.; Faber, K. *Adv. Synth. Catal.* 2014, **356**, 1878–1882.
[51] Korpak, M.; Pietruszka, J. *Adv. Synth. Catal.* 2011, **353**, 1420–1424.
[52] Brenna, E.; Gatti, F. G.; Malpezzi, L.; Monti, D.; Parmeggiani, F.; Sacchetti, A. *J. Org. Chem.* 2013, **78**, 4811–4822.
[53] Mangan, D.; Miskelly, I.; Moody, T. S. *Adv. Synth. Catal.* 2012, **354**, 2185–2190.
[54] Toogood, H. S.; Knaus, T.; Scrutton, N. S. *ChemCatChem* 2013, **6**, Issue 4 951–954.
[55] Hall, M.; Stueckler, C.; Kroutil, W.; Macheroux, P.; Faber, K. *Angew. Chem. Int. Ed. Engl.* 2007, **46**, 3934–3937.
[56] Tauber, K.; Hall, M.; Kroutil, W.; Fabian, W. M.; Faber, K.; Glueck, S. M. *Biotechnol. Bioeng.* 2011, **108**, 1462–1467.
[57] Taglieber, A.; Schulz, F.; Hollmann, F.; Rusek, M.; Reetz, M. T. *ChemBioChem* 2008, **9**, 565–572.
[58] Grau, M. M.; van der Toorn, J. C.; Otten, L. G.; Macheroux, P.; Taglieber, A.; Zilly, F. E.; Arends, I. W. C. E.; Hollmann, F. *Adv. Synth. Catal.* 2009, **351**, 3279–3286.
[59] Hollmann, F.; Arends, I. W. C. E.; Buehler, K. *ChemCatChem* 2010, **2**, 762–782.
[60] Fisher, K.; Mohr, S.; Mansell, D.; Goddard, N. J.; Fielden, P. R.; Scrutton, N. S. *Catal. Sci. Technol.* 2013, **3**, 1505–1511.
[61] Stueckler, C.; Reiter, T. C.; Baudendistel, N.; Faber, K. *Tetrahedron* 2010, **66**, 663–667.

[62] Winkler, C. K.; Clay, D.; van Heerden, E.; Faber, K. *Biotechnol. Bioeng.* 2013, **110**, 3085–3092.
[63] Winkler, C. K.; Clay, D.; Entner, M.; Plank, M.; Faber, K. *Chem. Eur. J.* 2014, **20**, 1403–1409.
[64] Paul, C. E.; Gargiulo, S.; Opperman, D. J.; Lavandera, I.; Gotor-Fernandez, V.; Gotor, V.; Taglieber, A.; Arends, I. W. C. E.; Hollmann, F. *Org. Lett.* 2013, **15**, 180–183.
[65] Brenna, E.; Gatti, F. G.; Monti, D.; Parmeggiani, F.; Sacchetti, A. *Chem. Commun.* 2012, **48**, 79–81.
[66] Brenna, E.; Gatti, F. G.; Monti, D.; Parmeggiani, F.; Sacchetti, A. *ChemCatChem* 2012, **4**, 653–659.
[67] Bechtold, M.; Brenna, E.; Femmer, C.; Gatti, F. G.; Panke, S.; Parmeggiani, F.; Sacchetti, A. *Org. Process Res. Dev.* 2012, **16**, 269–276.
[68] Padhi, S. K.; Bougioukou, D. J.; Stewart, J. D. *J. Am. Chem. Soc.* 2009, **131**, 3271–3280.
[69] Bougioukou, D. J.; Kille, S.; Taglieber, A.; Reetz, M. T. *Adv. Synth. Catal.* 2009, **351**, 3287–3305.
[70] Daugherty, A. B.; Govindarajan, S.; Lutz, S. *J. Am. Chem. Soc.* 2013, **135**, 14425–14432.
[71] Reich, S.; Kress, N.; Nestl, B. M.; Hauer, B. *J. Struct. Biol.* 2014, **185**, 228–233.
[72] Fryszkowska, A.; Toogood, H.; Sakuma, M.; Stephens, G. M.; Gardiner, J. M.; Scrutton, N. S. *Catal. Sci. Technol.* 2011, **1**, 948–957.
[73] Oberdorfer, G.; Steinkellner, G.; Stueckler, C.; Faber, K.; Gruber, K. *ChemCatChem* 2011, **3**, 1562–1566.
[74] Classen, T.; Pietruszka, J.; Schuback, S. M. *ChemCatChem* 2013, **5**, 711–713.

CHAPTER 19

Recent Developments in Aminopeptidases, Racemases, and Oxidases

Yasuhisa Asano[1,2], Seiji Okazaki[1,2], and Kazuyuki Yasukawa[1,2]

[1] Asano Active Enzyme Molecule Project, ERATO, JST, Imizu, Japan

[2] Department of Biotechnology and Biotechnology Research Center, Toyama Prefectural University, Imizu, Japan

19.1 AMINOPEPTIDASE

19.1.1 Discovery of D-Stereospecific Aminopeptidase and its Utilization for Dynamic Kinetic Resolution

In order to develop an enzymatic method to synthesize D-alanine *N*-alkyl amide, which is found in the structure of an artificial sweetener, alitame (L-α-aspartyl-*N*-(2,2,4,4-tetramethyl-3-thietanyl)-D-alanine amide), we screened microorganisms for an enzyme that catalyzes D-stereospecific amino acid amide hydrolysis [1].

Hydrolase acting on D-amino acid amide and ester has been suggested to catalyze the synthesis of D-amino acid *N*-alkyl amide from D-amino acid amide or esters by deacylating the possible intermediate amino acyl-enzyme complex formed. This enzyme may also be applicable to the kinetic resolution of racemic amino acid amides to form D-amino acids, which are building blocks for the synthesis of several pharmaceuticals and pesticides. Four groups of hydrolytic enzymes have been suggested to act on amino acid derivatives based on the functional groups and the stereochemistry of the amino acids. L-Aminoacylase, which acts on *N*-acyl L-amino acids, is one of the first enzymes utilized in enzyme engineering to produce L-amino acids industrially through the kinetic resolution of racemic *N*-acyl L-amino acids [2]. D-Aminoacylase from *Alcaligenes faecalis* DA1 has also been identified [3].

19.1.2 Discovery of D-Aminopeptidase, D-Amino Acid Amidase, and Alkaline D-Peptidase

Since D-stereospecific amino acid amides and peptide hydrolases were previously unknown and were not targets of enzymology, we started to screen for these enzymes and subsequently discovered three kinds that exhibited D-stereoselectivities for D-amino acid derivatives: D-aminopeptidase [4], D-amino acid amidase [5], and alkaline D-peptidase [6] (Table 19.1). *Ochrobactrum anthropi* C1-38 was isolated through an enrichment culture technique as a utilizer of D-alanine amide (D-Ala-NH_2) as the sole

Green Biocatalysis, First Edition. Edited by Ramesh N. Patel.

TABLE 19.1 D-Stereospecific Amino Acid Amides and Peptides Hydrolases

Enzyme Name	Origin	Active Site	Application
D-Aminopeptidase (DAP)	*Ochrobactrum anthropi* C1-38	Ser	Dynamic kinetic resolution, peptide bond formation
D-Amino acid amidase (DaaA)	*O. anthropi* SV3	Ser	Dynamic kinetic resolution
Alkaline D-peptidase (ADP)	*Bacillus cereus* DF4-B	Ser	Peptide bond formation

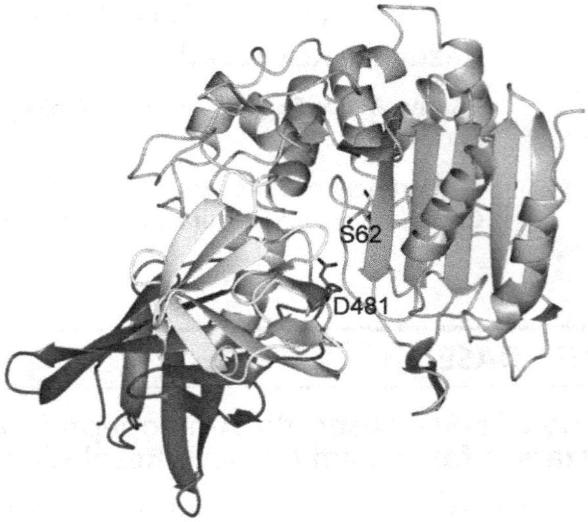

FIGURE 19.1
Overall structure of DAP. Ser62 and Asp481 are represented as cylinders.

nitrogen source. The identification of this enzyme was carefully performed using intensive purification to remove coexisting L-amino acid hydrolases hindering the detection of D-stereoselective hydrolase in cell-free extracts of bacteria. This enzyme was named D-aminopeptidase (DAP, EC 3.4.11.19) because it catalyzes the hydrolysis of D-amino acid-containing peptides, such as $(\text{D-Ala})_4$, as the most suitable substrates with superior kinetic parameters, compared with those of D-alanine amide [4].

Ochrobactrum anthropi SV3 was isolated as a D-valine amide degrader after a 4-month acclimation of bacterial culture. The strain produced an enzyme that was characterized as D-amino acid amidase (DaaA) [5]. Furthermore, *Bacillus cereus* DF4-B excreted alkaline D-peptidase (ADP), which acted on the synthetic oligopeptide D-phenylalanine tetramer $(\text{D-Phe})_4$. The enzyme was characterized as the first endopeptidase acting on D-amino acid-containing peptides by recognizing the second amino acid from its *N*-terminus under alkaline conditions [6].

19.1.3 Structure of D-Aminopeptidase (DAP)

The crystal structure of DAP was solved at a resolution of 1.9 Å [7]. The enzyme was found to fold into three domains: A (residues 3–331), B (residues 341–418), and C (residues 422–520) (Figure 19.1). Domain A, which contains conserved catalytic residues, was shown to fold in a similar manner to the class C penicillin-binding proteins (PBPs) of type AmpH [8]. Domains B and C were both antiparallel eight-stranded β barrels. Asp481 on the γ-loop (residues 476–486) of domain C protruded into the substrate binding site of DAP, and a previous study suggested that the free amino group of the N-terminal amino acid of substrate peptide was recognized by the side chain of Asp481 [7].

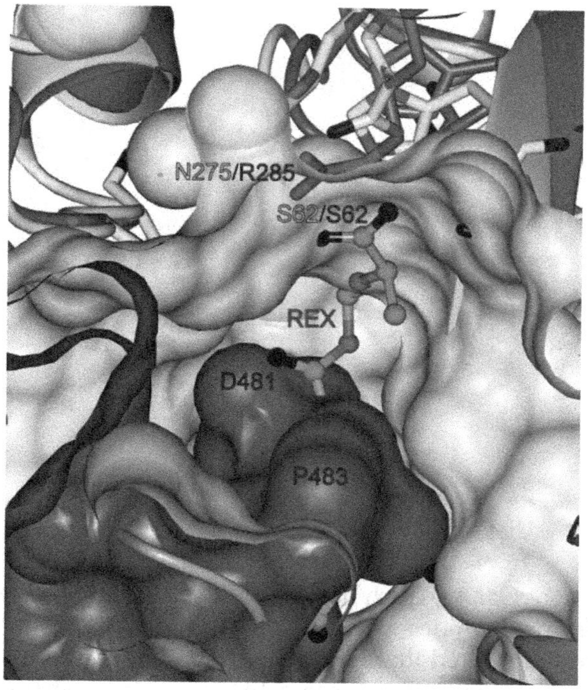

FIGURE 19.2
Surface representation of the substrate binding site of DAP. The substrate binding site of DD-peptidase (PDB ID 1IKG) complexed with REX (ball and stick representation) is also superposed to that of DAP.

Although DAP exhibits aminopeptidase activity, which hydrolyzes the N-terminal residue of peptides, it does not have carboxypeptidase activity, which hydrolyzes the C-terminal residue of peptides. The presence or absence of these peptidase activities has been correlated with the shape of the active site cavity of DAP. It has a cavity for aminopeptidase activity similar to DD-peptidase [9] (PDB ID 1IKG) complexed with glycyl L-α-amino-ε-pimelyl-D-Ala-D-Ala (REX), in which the cavity allows for the existence of the C-terminal D-Ala of REX (Figure 19.2). Due to the presence of this cavity, the N-terminal residue of the substrate peptide may approach the nucleophile Ser62 O^{γ}. Although DD-peptidase has a cavity for carboxypeptidase activity, in which the cavity allows for the existence of the N-terminal L-α-amino-ε-pimelyl group of REX, the cavity for carboxypeptidase activity in DAP was found to be blocked by the steric hindrance caused by the existence of residues 480–483 on the γ-loop (Figure 19.2). This is further supported by the increase in carboxypeptidase activity by the polyglycine mutant, 475G487, in which there is less steric hindrance by the γ-loop [10].

19.1.4 Structure of D-Amino Acid Amidase (DaaA)

The crystal structures of the apo form and D-Phe complexed form of DaaA were solved at resolutions of 2.1 Å and 2.4 Å, respectively [11]. DaaA has also been shown to fold in a similar manner to the class C PBPs of type AmpH [8] such as DAP (Figure 19.3). DaaA is composed of an α/β domain (residues 1–60, 243–363; an antiparallel β-sheet) and helical domain (residues 61–242; a Ω-loop (residues 207–223) and L3-loop (residues 107–126) and remaining helices). In the complex structure under an artificially packed environment around the Ω-loop, DaaA forms a pseudo acyl-enzyme intermediate between Ser60 O^{γ} and the carbonyl moiety of D-Phe. The complex structure indicates the residues critical for D-Phe recognition. The carboxyl group of Glu114, which corresponds to Asp481 of DAP, formed a salt bridge with the free amino group of the substrate D-Phe (Figure 19.3). The phenyl ring of D-Phe was located in the hydrophobic pocket constructed by Ala59, Phe113, Glu114, Trp215, Phe234, Ala239, Ala242, Gly243, and Ile311. These environments correspond to the

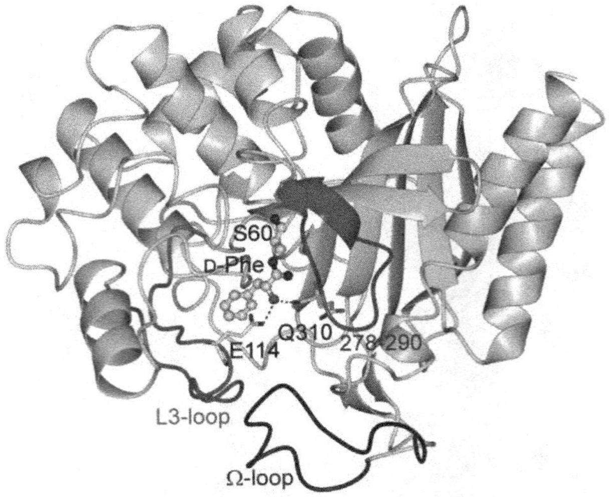

FIGURE 19.3
Overall structure of DaaA complexed with D-Phe (ball and stick representation). Ser60 is represented as ball and stick, and E114 and Q310 are represented as cylinders. Salt bridge and hydrogen bond are represented as broken lines. Ω- and L3-loops and residues 278–290 are represented as black.

substrate specificity of DaaA, which has been shown to catalyze the D-stereospecific hydrolysis of amino acid amides with bulky hydrophobic side chains [5, 12].

A previous study reported that DaaA lacked aminopeptidase and carboxypeptidase activities [5]. The absence of these activities may be attributed to the steric hindrance of the substrate-binding pocket by residues 278–290 and the Ω-loop.

19.2 RACEMASE

19.2.1 Synthesis of D-Amino Acids by Optical Resolution and Dynamic Kinetic Resolution

Racemases are enzymes that catalyze the inversion of the chiral center by deprotonation of the C^{α}, followed by reprotonation on the opposite face of the planar carbanionic transition-state species [13, 14]. In order to overcome the high energetic barrier of racemization, for example, on α-amino acids, some racemases employ pyridoxal phosphate (PLP) as a cofactor to use the resonance-stabilized amino acid complex as an electron sink because the estimated pK_a values for the C^{α} of amino acids are high, in the range 21–32 [14, 15]. The formation of an imine PLP–substrate covalent bond makes the pK_a value of α-hydrogen of amino acids low. The second class of enzymes includes proline, aspartate, and glutamate racemases and diaminopimelate epimerase, with a cofactor-independent two-base mechanism [14].

In the kinetic resolution of amino acid amides with the use of amidases, such as DAP and DaaA, it is possible to synthesize D-amino acids by kinetic resolution, selectively from racemic acid amides [16]. An *Escherichia coli* transformant highly expressing DAP catalyzed the synthesis of 2.5 M (about 220 g/l) D-alanine from 5 M racemic alanine amide in a 4.5-h reaction. D-2-Amino butyric acid, D-methionine, D-norvaline, and D-norleucine were synthesized in a similar manner. We have been successful in the evolution of DAP [17] and DaaA by mutations [18].

Even in the optical resolution of amino acid amides with these new enzymes, the theoretical yield of the synthesis of one of the enantiomers of amino acids cannot exceed 50%. If amino acid amide-racemizing enzymes (racemase) existed, it may be possible to synthesize chiral amino acids through the racemization of amino acid amides (Figure 19.4).

α-Amino ε-caprolactam (ACL) is a heterocyclic compound that has been used as a starting material for the enzymatic production of L-lysine by Toray Company [19]. Soda *et al.* previously reported that ACL racemase acted on ACL, α-amino δ-valerolactam, and α-amino-β-thio-ε-caprolactam [20, 21].

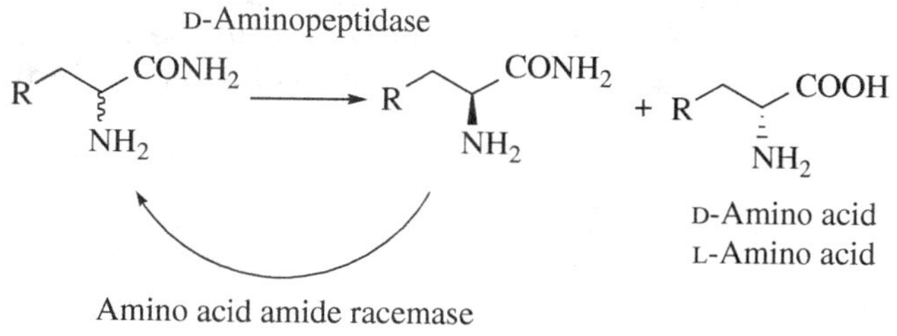

FIGURE 19.4

Dynamic kinetic resolution of amino acid amide by DAP in the presence of ACL racemase.

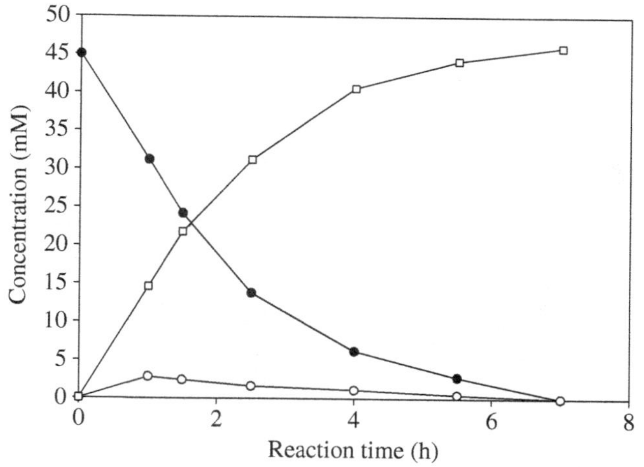

FIGURE 19.5

Production of D-alanine from L-alanine amide by DAP in the presence of ACL racemase. (○), D-alanine amide; (●), L-alanine amide; (□), D-alanine.

Based on the structural similarities of amino acid amides with ACL, we predicted that ACL racemase catalyzed the racemization of not only the cyclic amino lactam compounds but also amino acid amides. We soon discovered that ACL racemase from *Achromobacter obae* catalyzed the racemization of amino acid amides, such as 2-aminobutyramide and alanine amide [22]. A previous study demonstrated that D-alanine was produced efficiently from L-alanine amide by a dynamic kinetic resolution reaction when the enzyme was used together with DAP from *O. anthropi* [23]. It had previously been unclear whether ACL racemase catalyzed the racemization of simple amino acid amides.

Specific activity for L-2-amino butyric acid amide was 9.5 U/mg, which was 2.7% that for L-ACL (350 U/mg with a substrate concentration of 100 mM). Activities toward alanine amide, threonine amide, norvaline amide, and norleucine amide were all <2.1% that for ACL. The enzyme did not act on α-amino acid, peptides consisting of alanine, or alanine methyl esters. The K_{eq} values for 2-aminobutyramide and alanine amide were both calculated to be 1.0, which identified them as typical racemase-catalyzed reactions [23]:

$$K_{eq} = [K_{cat}/K_m]_{\text{D-isomer}} / [K_{cat}/K_m]_{\text{L-isomer}} = 1.0$$

By combining ACL racemase from *A. obae*, expressed in *E. coli*, with DAP or DaaA from *O. anthropi*, optically active D-alanine and other D-amino acids were synthesized stoichiometrically from amino acid amides by dynamic kinetic resolution [22, 23]. Figure 19.5 shows that D-alanine was synthesized from L-alanine amide. L-Alanine amide was racemized to DL-alanine amide by ACL racemase, and only D-alanine amide was hydrolyzed to optically active D-alanine by the actions of DAPS. D-2-aminobutyric acid, D-serine, and D-methionine were synthesized in a similar procedure. By using L-stereoselective hydrolase together with ACL racemase, L-amino acids have also been synthesized stoichiometrically. For example, L-alanine, L-leucine, and L-methionine were synthesized by the dynamic kinetic resolution reaction when L-amino acid amidase from *Pseudomonas azotoformans* IAM 1603 was used together with ACL racemase (Table 19.2) [24]. Thus, it is possible to synthesize the enantiomer of amino acids

TABLE 19.2 Synthesis of Various L-Amino Acids from the Corresponding D-Amino Acid Amides by L-Amino Acid Amidase in the Presence of ACL Racemase

Substrate	Product	Yield (%)	ee (%)
D-Alanine amide	L-Alanine	100	>99
D-Leucine amide	L-Leucine	100	>99
D-Methionine amide	L-Methionine	100	>99

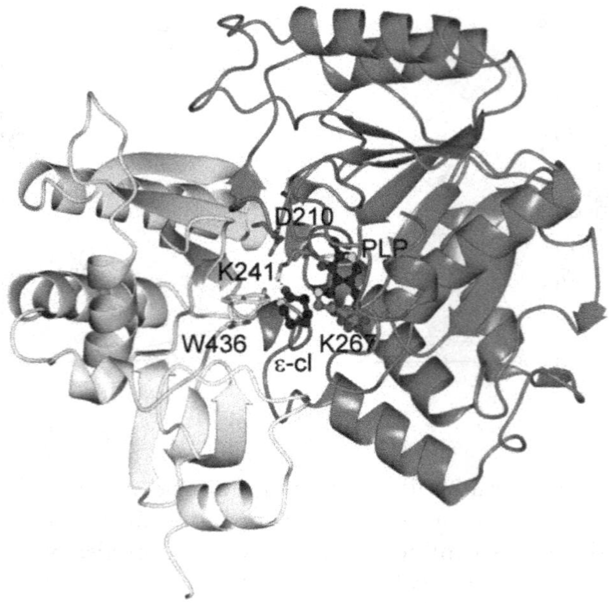

FIGURE 19.6 Overall structure of ACL racemase complexed with ε-caprolactam (ε-cl, ball and stick representation). PLP and Lys267 are represented as ball and stick. Hydrogen bond is represented as broken line.

stoichiometrically from amino acid amides by this procedure. We have been successful in expanding the substrate specificity of ACL racemase to accommodate amino acids with bulky substituents. D-Phenylalanine was synthesized from racemic phenylalanine amide with the use of DaaA and its mutant [25]. When the reaction was carried out with nonstereoselective nitrile hydratase, it was possible to synthesize chiral amino acids from racemic α-phenylalaninonitrile.

19.2.2 Structure of ACL Racemase

The crystal structures of the apo form and ε-caprolactam complexed form of ACL racemase were solved at resolutions 2.21 Å and 2.40 Å, respectively [26]. The structure of ACL racemase was found to be composed of an N-terminal domain (residues 3–43), C-terminal domain (residues 321–436), and large PLP-binding domain (residues 48–319) (Figure 19.6). PLP bound in a cleft between these domains, and the C4′ atom of PLP connected covalently to the ε-amino group of Lys267 to form the internal aldimine (Schiff base) linkage.

In the crystalline state, two subunit molecules in the asymmetric unit formed a tightly packed homodimer (a contact area of ~4000 Å) stabilized by 48 interchain hydrogen bonds. This long list underscores the strength of the intersubunit interactions in this dimer and indicates that ACL racemase forms a dimer in solution similar to other fold-type I enzymes [13].

ε-Caprolactam was found to be located in the active site of the complexed structure (Figure 19.6). The structures of ε-caprolactam and ACL are similar, and ε-caprolactam has been identified as a competitive inhibitor of ACL racemase [20]. Therefore, the binding modes of substrates such as ACL may be predicted from that of ε-caprolactam in the complex structure. The carbonyl O of ε-caprolactam is known to form a hydrogen bond with Lys241 N^{ζ}. The ring of ε-caprolactam is sandwiched between the aromatic side chains of Trp49 and Tyr137. The location of the side chain of Asp210 is suitable for approaching substrate by interactions with the nitrogen of amides or lactams. A recently conducted mutational analysis on Asp210 identified Asp210 as an acid/base catalytic residue on the *re* face of the PLP ring [27].

19.2.3 *In Silico* Identification of ACL Racemases

Two groups recently achieved *in silico* identification for ACL racemases by identifying Lys241 as a key amino acid residue [27, 28]. ACL and amino acid amide-racemizing activities were detected among tens of candidates [27]. These newly discovered ACL racemases may be useful for chiral amino acid synthesis by dynamic kinetic resolution and for determining the unknown physiological functions of ACL racemases.

19.3 AMINO ACID OXIDASE

19.3.1 Development of Novel *R*-Stereoselective Amine Oxidase

Chiral amines have been attracting attention as an important composition, particularly for pharmaceutical products. The organic synthetic methods of optically active amine compounds have been developed through the traditional resolution of racemic amines with the formation of diastereomer salts using an optically active mandelic acid or tartaric acid. Enzymatic synthesis has mainly used lipase and *S*- or *R*-stereoselective amine transaminase (AT) [29–31] (Figure 19.7). Turner *et al.* successfully synthesized chiral (*R*)- and (*S*)-amines by kinetic resolution using a combination of stereoselective AT and D- or L-amino acid oxidase (AAOx) [32] (Figure 19.7). However, the theoretical yield of the products has been limited to 50% in the kinetic resolution.

Dynamic kinetic resolution is regarded as a useful synthetic method because the theoretical yield of products reaches 100%. Kim *et al.* previously reported that the dynamic kinetic resolution of amines formed *N*-protecting amines through the coupling of lipase-catalyzed resolution and racemization with Pd nanocatalysis in toluene [33] (Figure 19.7). These methods must separate the remaining substrate and product, and difficulties have been associated with the purification process due to its complexity. In contrast the deracemization method has provided optical active products in a theoretical yield of 100%. In addition, the complicated separation of the remaining substrate and product is not needed because the molecular compositions of the starting substrate and product are the same. Deracemization methods utilizing AT [34] or amine oxidase (AOx) [35] have been reported for the synthesis of chiral amines. A transaminase-catalyzed deracemization method consists of a complicated two-sequence reaction: (i) production of ketones from (*S*)-mexiletine by *S*-stereoselective AT together with an amino acceptor regeneration system and (ii) *R*-stereoselective transamination of ketones by the combination of (*R*)-AT and a byproduct-removing system (Figure 19.8). On the other hand, the latter method is a simple procedure because stereoselective AOx catalyzes one side of racemic amines to produce imine intermediates and a chemical reductant then reduces imines to racemic amines. However, AOx preferentially accepts simple straight chain primary amines, such as butylamine, phenylethylamine, and dopamine, as substrates rather than chiral amines at the alpha position such as α-methylbenzylamine (MBA). 2,4,5-Trihydroxyphenylalanine quinone (TPQ)-dependent AOx from *E. coli* and *Klebsiella oxytoca* has been shown to catalyze the oxidization of the *R*-enantiomer of amphetamine with moderate enantiomeric ratios (*E*-value = ~15) [36]; however, they are not considered suitable for the deracemization reaction because cofactor TPQ and intermediate imines form covalent bonds.

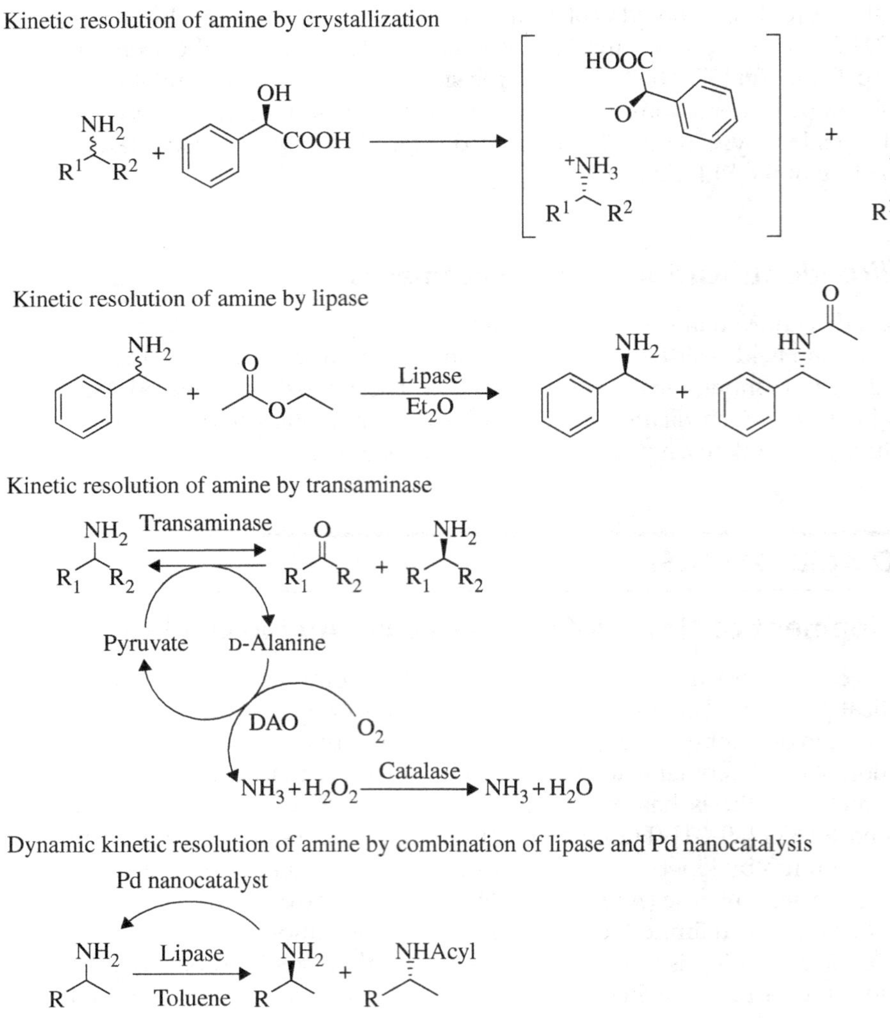

FIGURE 19.7
Kinetic resolution of amine.

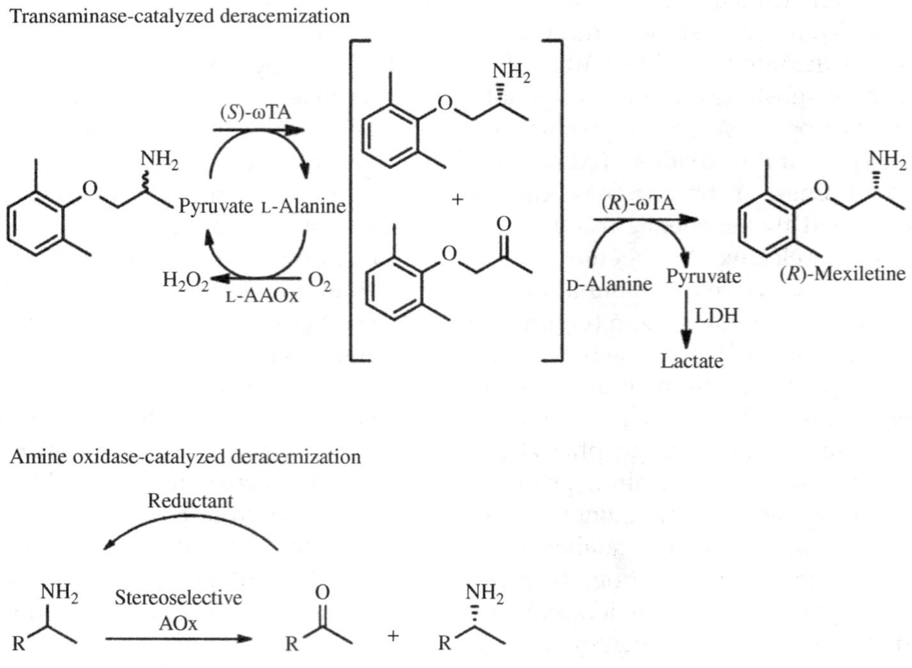

FIGURE 19.8
Deracemization of amine.

Turner *et al.* first reported engineered *S*-stereoselective flavin-dependent monoamine oxidase variants from *Aspergillus niger* for the deracemization of racemic amines producing the *S*-enantiomers of primary, secondary, and tertiary amines (Figure 19.8) [35]. On the other hand, Leisch *et al.* showed that wild-type cyclohexylamine oxidase from *Brevibacterium oxydans* IH-35A, which had been isolated and characterized by Hasegawa *et al.*, oxidized the *S*-enantiomer of amines and successfully synthesized (*R*)-amine through a deracemization reaction [37].

However, *R*-stereoselective AOx applicable to the deracemization method have not yet been identified.

We described the development of a valuable *R*-stereoselective amine oxidase for the synthesis of (*S*)-amine through a deracemization reaction [38].

19.3.2 Design of *R*-Stereoselective Amine Oxidase

Porcine kidney D-amino acid oxidase (pkDAAOx) was identified as the first mammalian flavoprotein to catalyze the oxidative deamination of α-amino acids with strict *R*-stereoselectivity. pkDAAOx is widely used as a general reagent in research and bioassays of D-amino acids. Although many studies have been conducted on its purification, characterization, gene cloning, expression, and X-ray crystal structure, pkDAAOx was not shown to catalyze the oxidation of amines or even LAAOx, which belongs to the AOx protein family.

A previous study revealed that the structure of flavin-dependent pkDAAOx complexed with an inhibitor benzoate (PDB: 1VE9) [39]. The overall structure of pkDAAOx was markedly different from that of the AOx family of proteins (Figure 19.9) [40]. On the other hand, the substrate binding sites of pkDAAOx and LAAOx were found to have several similarities such as the mode of acceptance of the substrate in the active site (Figure 19.9). The carboxylate group of benzoate bound in the carboxylate binding site by the Arg283 and Tyr228 formed in pkDAAOx. This

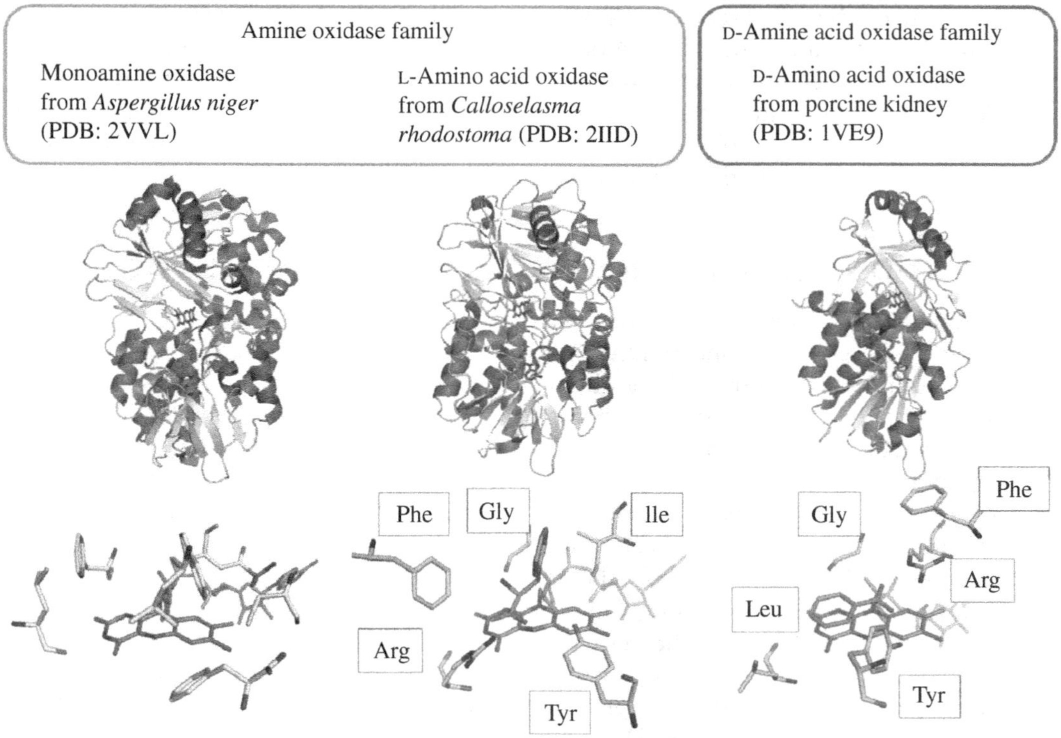

FIGURE 19.9

Comparison of structure of AOx, LAAOx, and DAAOx.

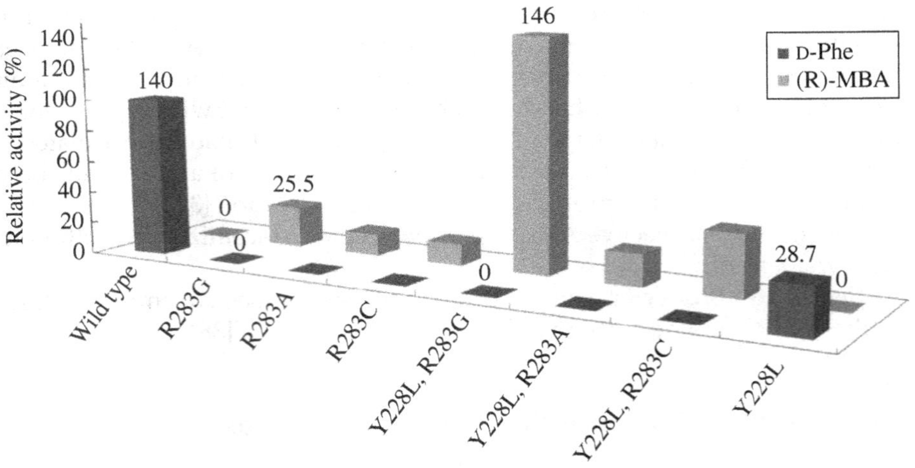

FIGURE 19.10
Screening result of (*R*)-amine oxidase. The activity of (*R*)-phenylalanine (D-Phe), corresponding to 0.11 U/mg, was taken as 100%. The reaction mixture (total volume 1.0 ml) was composed of 100 mM KPB (pH 8.0), 10 mM substrate, and an appropriate amount of the cell-free extract.

arginine residue was found to be conserved in several DAAOx or LAAOx from different sources. Therefore, the residues Tyr228 and Arg283 in the catalytic site were selected as the targets of mutation to improve substrate specificity. The single saturation mutagenesis of residues Tyr228 and Arg283 was performed and the resulting mutant libraries were screened by a colorimetric assay to measure AOx activity toward MBA. Positive clones were determined by the mutants R283G, R283A, and R283C, which catalyzed the oxidation of the *R*-enantiomer of MBA. These mutants were used as parents for the second round of saturation mutagenesis of Tyr228 and screening. In the final variant, the Y228L/R283G mutant was shown to improve MBA oxidative activity markedly (Figure 19.10). Furthermore, the Y228L mutant and other variants did not act on MBA with decreased oxidative activity toward (*R*)-phenylalanine.

We purified and characterized the Y228L/R283G mutant from recombinant *E. coli*. The specific activity of this mutant was 21.5 U/mg toward (*R*)-MBA and lost its ability to oxidize the *R*-enantiomer of amino acids such as phenylalanine and methionine. This mutant oxidized MBA derivatives and cyclic amines, but could not accept simple straight chain amines, which natural AOx oxidized preferentially (Figure 19.11). AAOx and AOx exhibited completely different substrate specificity without crossing each other but were able to convert enzyme activity by mutations of the one point amino acid residue in the active site of pkDAAOx.

19.3.3 Deracemization Reaction with *R*-Stereoselective AOx

Hafner *et al.* first reported an oxidase-catalyzed deracemization method using amino acids as the substrate and pkDAAOx or LAAOx from *Crotalus adamanteus* together with sodium borohydride as the chemical reductant in 1971 [42]. A procedure for the successful deracemization of amino acids was previously reported by Soda *et al.* [43]. They focused on proline and pipecolic acid as substrates for the production of L-enantiomer by deracemization because these substrates formed stable imines rather than unfavorable keto acids in water by DAAOx. However, the enzyme was denatured by the chemical reaction with sodium borohydride. Turner *et al.* developed an effective production method for (*R*)- or (*S*)-amino acids and (*S*)-amines by a deracemization method using milder chemical reducing reagents such as sodium cyanoborohydride and artificial transfer hydrogenase [44, 45].

We optimized the reaction for the deracemization of MBA such as pH, temperature, and concentrations of chemical reductants with mutant pkDAAOx (Y228L, R283G). When using $NaCNBH_3$ and NH_3—BH_3 as reducing agents, the undesired product acetophenone mainly appeared. Therefore, the purified mutant (300 U) may be applied for the deracemization of racemic MBA (5 mM, 0.24 g) with sodium

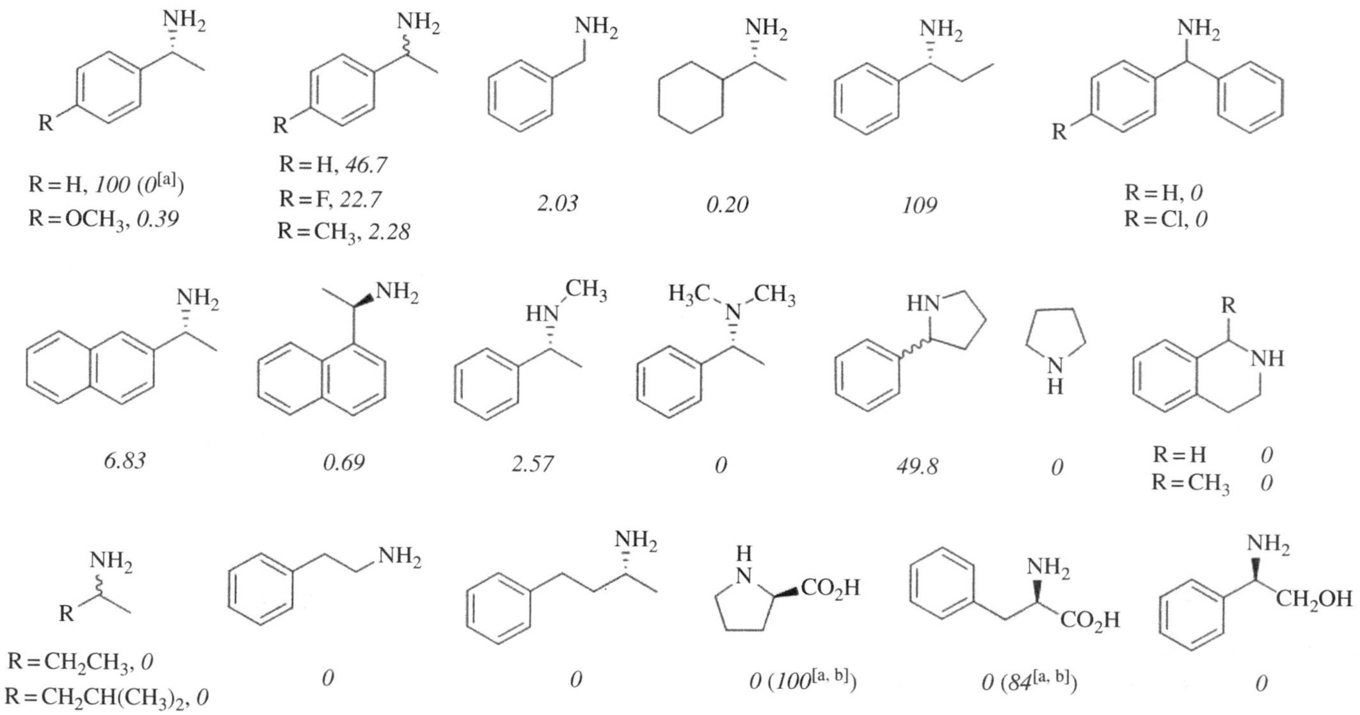

FIGURE 19.11

Substrate specificity of mutant pkDAAOx (Y228L, R283G). The activity for (*R*)-MBA corresponding to 21.5 U/mg was taken as 100%. The italic numbers below the structures indicate relative activity. [a]Substrate specificity of wild-type pkDAAOx; [b]Reference [41].

borohydride (100 mM) in order to produce (*S*)-MBA with an isolation yield of 65% and enantiomeric purity of 99% ee [38].

19.3.4 Structure of the Mutant Porcine Kidney D-Amino Acid Oxidase (Y228L, R283G)

We determined the crystal structure of the (*R*)-MBA binding pkDAAOx mutant (Y228L, R283G) (PDB: 3WGT) [38]. The electron density of (*R*)-MBA was confirmed on the *re* side of FAD (Figure 19.12). In the wild type, the carboxylic acid of benzoate bound to the active site by forming a hydrogen bonding interaction between the guanidinium group of Arg283 and the phenyl ring was located on the uracil ring. However, the phenyl ring was located on the xylene ring in the mutant due to the formation of a hydrophobic cavity by the mutation. The phenyl group of benzoate was located at the opposite side of (*R*)-MBA in the mutant. Concerning other notable structural differences between the mutant and wild-type enzymes, the structure of Phe242 was different in the following residues: the side chain of Phe242 rotated ~60° between the wild type and mutant and formed part of the active site wall, while the phenyl ring of (*R*)-MBA formed a hydrophobic interaction with this rotation and bound to the active site.

We described newly evolved (*R*)-AOx from pkDAAOx and its applicability to the deracemization reaction for (*S*)-amine synthesis.

The oxidase-catalyzed deracemization reaction has merits in chiral synthesis, because no protection of the amino group is necessary, and the purification of the product is simple. Therefore, the creation of novel biocatalysts for a new, efficient, and convenient reaction processes is challenging and necessary in order to develop a sustainable society in the future.

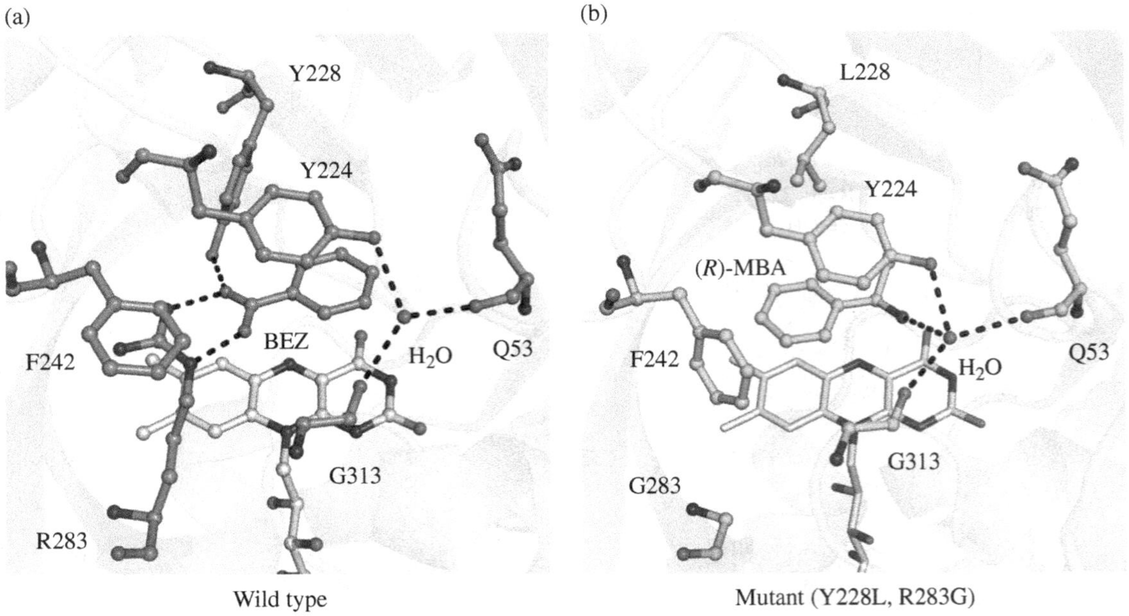

FIGURE 19.12
Active site of wild-type (a) and mutant pkDAAOx (b).

REFERENCES

[1] Glowaky, R. C., Hendrick, M. E., Smiles, R. E., and Torres, A., *ACS Symposium Series Sweeteners 450*, American Chemical Society, New York, 1991, pp. 57–67.

[2] Tosa, T., Mori, T., Fuse, N., and Chibata, I., "Studies on continuous enzyme reaction. IV. Preparation of a DEAE-sephadex-aminoacylase column and continuous optical resolution of acyl-DL-amino acids." *Biotechnol. Bioeng.*, **9**, 603–615 (1967).

[3] Yang, Y. B., Lin, C. S., Tseng, C. P., Wang, Y. J., and Tsai, Y. C., "Purification and characterization of D-aminoacylase from *Alcaligenes faecalis* DA1." *Appl. Environ. Microbiol.*, **57**, 1259–1260 (1991).

[4] Asano, Y., Nakazawa, A., Kato, Y., and Kondo, K., "Properties of a novel D-stereospecific aminopeptidase from *Ochrobactrum anthropi*." *J. Biol. Chem.*, **264**, 14233–14239 (1989).

[5] Komeda, H. and Asano, Y., "Gene cloning, nucleotide sequencing, and purification and characterization of the D-stereospecific amino-acid amidase from *Ochrobactrum anthropi* SV3." *Eur. J. Biochem.*, **267**, 2028–2036 (2000).

[6] Asano, Y., Kato, Y., Nakazawa, A., and Kondo, K., "Structural similarity of D-aminopeptidase to carboxypeptidase DD and β-lactamases." *Biochemistry*, **31**, 2316–2328 (1992).

[7] Gilles, C. B., Remaut, H., Villeret, V., Prangé, T., Fanuel, L., Delmarcelle, M., Joris, B., Frère, J. M., and Beeuman, J. V., "Crystal structure of a D-aminopeptidase from *Ochrobactrum anthropi*, a new member of the 'penicillin-recognizing enzyme' family." *Structure*, **8**, 971–980 (2000).

[8] Sauvage, E., Kerff, F., Terrak, M., Ayala, J. A., and Charlier, P., "The penicillin-binding proteins: structure and role in peptidoglycan biosynthesis." *FEMS Microbiol. Rev.*, **32**, 234–258 (2008).

[9] McDonough, M. A., Anderson, J. W., Silvaggi, N. R., Pratt, R. F., Knox, J. R., and Kelly, J. A., "Structures of two kinetic intermediates reveal species specificity of penicillin-binding proteins." *J. Mol. Biol.*, **322**, 111–122 (2002).

[10] Delmarcelle, M., Boursoit, M. C., Filée, P., Baurin, S. L., Frère, J. M., and Joris, B., "Specificity inversion of *Ochrobactrum anthropi* D-aminopeptidase to a D,D-carboxypeptidase with new penicillin binding activity by directed mutagenesis." *Protein Sci.*, **14**, 2296–2303 (2005).

[11] Okazaki, S., Suzuki, A., Komeda, H., Yamaguchi, S., Asano, Y., and Yamane, T., "Crystal structure and functional characterization of a D-stereospecific amino acid amidase from *Ochrobactrum anthropi* SV3, a new member of the penicillin-recognizing proteins." *J. Mol. Biol.*, **368**, 79–91 (2007).

[12] Asano, Y., Mori, T., Hanamoto, S., Kato, Y., and Nakazawa, A., "A new D-stereospecific amino acid amidase from *Ochrobactrum anthropi*." *Biochem. Biophys. Res. Commun.*, **162**, 470–474 (1989).

[13] Eliot, A. C. and Kirsch, J. F., "Pyridoxal phosphate enzymes: mechanistic, structural, and evolutionary considerations." *Annu. Rev. Biochem.*, **73**, 383–415 (2004).

[14] Asano, Y. and Hölsch, K., "Isomerizations." In *Enzyme Catalysis in Organic Synthesis*, Ed. Drauz, K., Gröger, H., and May, O., Wiley-VCH, Weinheim, Vol. **3**, 2012, pp. 1607–1684.

[15] Tanner, M. E, "Understanding nature's strategies for enzyme-catalyzed racemization and epimerization." *Acc. Chem. Res.*, **35**, 237–246 (2002).

[16] Asano, Y., Kishino, K., Yamada, A., Hanamoto, S., and Kondo, K., "Plasmid-based, D-aminopeptidase-catalysed synthesis of (*R*)-amino acids." *Recl. Trav. Chim. Pays-Bas*, **110**, 206–208 (1991).

[17] Asano, Y. and Yamaguchi, K., "Mutants of D-aminopeptidase with increased thermal stability." *J. Ferment. Bioeng.*, **79**, 614–616 (1995).

[18] Komeda, H., Ishikawa, N., and Asano, Y., "Enhancement of the thermostability and catalytic activity of D-stereospecific amino-acid amidase from *Ochrobactrum anthropi* SV3 by directed evolution." *J. Mol. Catal. B Enzym.*, **21**, 283–290 (2003).

[19] Fukumura, T., "Conversion of D- and DL-α-amino-ε-caprolactam into L-lysine using both yeast cells and bacterial cells." *Agric. Biol. Chem.*, **41**, 1327–1330 (1977).

[20] Ahmed, S. A., Esaki, N., Tanaka, H., and Soda, K., "Properties of α-amino-ε-caprolactam racemase from *Achromobacter obae*." *Agric. Biol. Chem.*, **47**, 1887–1893 (1983).

[21] Ahmed, S. A., Esaki, N., Tanaka, H., and Soda, K., "L-α-amino-β-thio-ε-caprolactam, a new sulfur containing substrate for α-amino-ε-caprolactam racemase." *FEBS Lett.*, **174**, 76–79 (1984).

[22] Asano, Y. and Yamagichi, S., "Dynamic kinetic resolution of amino acid amide catalyzed by D-aminopeptidase and α-amino-ε-caprolactam racemase." *J. Am. Chem. Soc.*, **127**, 7696–7697 (2005).

[23] Asano, Y. and Yamaguchi, S., "Discovery of amino acid amides as new substrates for α-amino-ε-caprolactam racemase from *Achromobacter obae*." *J. Mol. Catal. B Enzym.*, **36**, 22–29 (2005).

[24] Yamaguchi, S., Komeda, H., and Asano, Y., "New enzymatic method of chiral amino acid synthesis by dynamic kinetic resolution of amino acid amides: use of stereoselective amino acid amidases in the presence of α-amino-ε-caprolactam racemase." *Appl. Environ. Microbiol.*, **73**, 5370–5373 (2007).

[25] Yasukawa, K. and Asano, Y., "Enzymatic synthesis of chiral phenylalanine derivatives by a dynamic kinetic resolution of corresponding amide and nitrile substrates with a multi-enzyme system." *Adv. Synth. Catal.*, **354**, 3327–3332 (2012).

[26] Okazaki, S., Suzuki, A., Mizushima, T., Kawano, T., Komeda, H., Asano, Y., and Yamane, T., "The novel structure of a pyridoxal 5′-phosphate-dependent fold-type I racemase, α-amino-ε-caprolactam racemase from *Achromobacter obae*." *Biochemistry*, **48**, 941–950 (2009).

[27] Payoungkiattikun, W., Okazaki, S., Nakano, S., Ina, A., H-Kittikun, A., and Asano, Y., "*In silico* identification for α-amino-ε-caprolactam racemases by using information of the structure and function relationship." *Appl. Biochem. Biotechnol.*, **176**, 1303–1314 (2015).

[28] Steffen-Munsberg, F., Vickers, C., Kohls, H., Land, H., Mallin, H., Nobili, A., Skalden, L., van den Bergh, T., Joosten, H. J.,Berglund, P., Höhne, M., and Bornscheuer, U. T., "Bioinformatic analysis of a PLP-dependent enzyme superfamily suitable for biocatalytic applications." *Biotechnol. Adv.*, **33**, 506–604 (2015).

[29] Reetz, M. T., & Dreisbach, C., "Highly efficient lipase-catalyzed kinetic resolution of chiral amines." *Chimia*, **48**, 570 (1994).

[30] Choi, Y. K., Kim, M. J., Ahn, Y., and Kim, M. J., "Lipase/palladium-catalyzed asymmetric transformations of ketoximes to optically active amines." *Org. Lett.*, **25**, 4099–4101 (2001).

[31] Koszelewski, D., Tauber, K., Faber, K., and Kroutil, W., "ω-Transaminases for the synthesis of non-racemic α-chiral primary amines." *Trends Biotechnol.*, **28**, 324–332 (2010).

[32] Truppo, M. D., Turner, N. J., and Rozzell, J. D., "Efficient kinetic resolution of racemic amines using a transaminase in combination with an amino acid oxidase." *Chem. Commun.*, **16**, 2127–2129 (2009).

[33] Kim, M. J., Kim, W. H., Han, K., Choi, Y. K., and Park, J., "Dynamic kinetic resolution of primary amine with a recyclable Pd nanocatalysis for racemization." *Org. Lett.*, **9**, 1157–1159 (2007).

[34] Koszelewski, D., Pressnitz, D., Clay, D., and Kroutil, W., "Deracemization of mexiletine biocatalyzed by ω-transaminases." *Org. Lett.*, **11**, 4810–4812 (2009).

[35] Alexeeva, M., Enright, A., Dawson, M. J., Mahmoudian, M., and Turner, N. J., "Deracemization of α-methylbenzylamine using an enzyme obtained by in vitro evolution." *Angew. Chem. Int. Ed.*, **41**, 3177–3180 (2002).
[36] Hacisalihoglu, A., Jongejan, A., Jongejan, J. A., and Duine, J. A., "Enantioselective oxidation of amphetamine by copper-containing quinoprotein amine oxidases from *Escherichia coli* and *Klebsiella oxytoca*." *J. Mol. Catal. B: Enzym.*, **11**, 81–88 (2000).
[37] Leisch, H., Grosse, S., Iwaki, H., Hasegawa, Y., and Lau, P. C. K., "Cyclohexylamine oxidase as a useful biocatalyst for the kinetic resolution and dereacemization of amines." *Can. J. Chem.*, **90**, 39–45 (2012).
[38] Yasukawa, K., Nakano, S., and Asano, Y., "Tailoring D-Amino acid oxidase from the pig kidney to *R*-stereoselective amine oxidase and its use in the deracemization of α-methylbenzylamine." *Angew. Chem. Int. Ed.*, **53**, 4428–4431 (2014).
[39] Setoyama, C., Miura, R., Nishina, Y., Shiga, K., Mizutani, H., Miyahara, I., and Hirotsu, K., "Crystallization of expressed porcine kidney D-amino acid oxidase and preliminary X-ray crystallographic characterization." *J. Biochem.*, **119**, 1114–1117 (1996).
[40] Pawelek, P. D., Cheah, J., Coulombe, R., Macheroux, P., Ghisla, S., and Vrielink, A., "The structure of L-amino acid oxidase reveals the substrate trajectory into an enantiomerically conserved active site." *EMBO J.*, **19**, 4204–4215 (2000).
[41] Tishkov, V. I. and Khoronenkova, S. V., "D-Amino acid oxidase: structure, catalytic mechanism, and practical application." *Translated from Biokhimiya*, **70**, 51–67 (2005).
[42] Hafner, E. W. and Wellner, D., "Demonstration of imino acids as products of the reactions catalyzed by D- and L-amino acid oxidases." *Proc. Natl. Acad. Sci. U. S. A.*, **68**, 987–991 (1971).
[43] Huh, J. W., Yokoigawa, K., Esaki, N., and Soda, K., "Synthesis of L-proline from the racemate by coupling of enzymatic enantiospecific oxidation and chemical non-enantiospecific reduction." *J. Biosci. Bioeng.*, **74**, 189–190 (1992).
[44] Alexandre, F. R., Pantaleone, D. P., Taylor, P. P., Fotheringham, I. G., Ager, D. J., and Turner, N. J., "Amine-borane: effective reducing agents for the deracemization of DL-amino acids using L-amino acid oxidase from *Proteus myxofaciens*." *Tetrahedron Lett.*, **43**, 707–710 (2002).
[45] Kohler, V., Wilson, Y. M., Durrenberger, M., Ghislieri, D., Churakova, E., Quinto, T., Knorr, L., Haussinger, D., Hollmann, F., Turner, N. J., and Ward, T. R., "Synthetic cascades are enabled by combining biocatalysts with artificial metalloenzymes." *Nat. Chem.*, **5**, 93–99 (2013).

CHAPTER 20

Biocatalytic Cascades for API Synthesis

John M. Woodley

Department of Chemical and Biochemical Engineering, Technical University of Denmark (DTU), Kongens Lyngby, Denmark

20.1 INTRODUCTION

Today, several hundred industrial processes for chemical manufacture, mostly in the pharmaceutical and fine chemical sectors, employ enzymes as biocatalysts to accelerate chemical reactions under mild conditions. The precise rationale for implementation depends upon the particular industrial sector, but in all cases the exquisite selectivity of enzymes is a primary driver. In the pharmaceutical sector where application has been the most widespread, such selectivity has been used in particular to assist in the synthesis of optically pure chiral molecules (either by resolution or preferentially by direct asymmetric synthesis). Likewise, biocatalysis has proved particularly useful for the synthesis of target molecules containing more than one chiral center, which is a growing trend. In all cases, use of mild reaction conditions, combined with high selectivity, means that protection of sensitive functional groups (and the subsequent deprotection steps) can frequently be eliminated. This results in synthetic schemes with fewer reaction steps, higher yields, and better use of reagents [1]. Hence there are multiple arguments for the implementation of biocatalytic steps in the synthesis of active pharmaceutical ingredients (APIs) [2–6], and indeed the earliest examples can be traced back to the 1960s. More significantly, since the 1990s, technology has also been developed to allow the screening of a range of altered enzymes (swapping amino acids by protein engineering) for their ability to accept a wider range of substrates (justifying generic application) and have a better kinetic profile (reducing substrate and product inhibition) or increased stability under reaction conditions (enabling them to be used for longer) [7]. This special ability to alter the properties of the enzyme, combined with an ever-increasing range of commercially available enzymes, makes biocatalysis one of the most important technologies for medicinal and process chemists alike in the pharmaceutical industry both today and in the future. The many examples in this text support this tremendous development over recent decades.

From a synthetic perspective, the future of biocatalysis will develop in two main directions. Firstly, the application of existing biocatalysts will be widened to include many novel compounds and reactions. To some extent this is already possible, but we are only at the start of exploring many new synthetic possibilities. Secondly, entirely

Green Biocatalysis, First Edition. Edited by Ramesh N. Patel.

new biocatalysts will be created for reactions without parallel in nature. Alongside these synthetic developments the ability to operate biocatalysts in a range of different conditions (either by engineering the biocatalyst or the process) will also enable new opportunities to be forthcoming [8]. However, potentially even more exciting would be the possibility of integrating biocatalysis into new synthetic schemes by creating cascades of either chemoenzymatic or multienzymatic reactions [9]. This will ultimately enable entirely new retrosynthetic possibilities to create novel target molecules [10], also inspired by nature [11]. In this chapter the focus will be on multistep enzymatic processes for API synthesis and their future development. This will require a completely new set of considerations to develop, implement, and evaluate suitable processes. In this chapter these concepts will be discussed and some examples given of these next-generation processes.

20.2 MULTIENZYMATIC BIOCATALYSIS

20.2.1 Rationale

One feature of the synthesis of (small-molecule) pharmaceuticals is that the molecular complexity, so important for precise pharmacological activity, needs to be built in stages. This means that syntheses involve multiple reaction steps, from starting structure to the active compound. Simpler molecules are combined to make larger structures, which are subsequently decorated with functional groups. Characteristic of such synthetic sequences is that there are losses of material through the process and, perhaps even more importantly, byproducts that are formed along the way may need to be separated (and intermediates purified) between reaction steps. Finally, the optimal conditions (e.g., pH, temperature, pressure) for each reaction may also vary significantly. In the early work applying biocatalysis to such schemes, it was important to use enzymes only for select reactions where chirality was introduced. For example, in a ten step synthesis, perhaps just one step would be biocatalytic to introduce chirality [3] and the remainder, chemical steps to add functionality. In particular for the synthesis of pharmaceuticals the use of a single biocatalytic reaction step surrounded by neighboring chemical reaction steps [12] has become established, with the consequence that changes in reaction conditions are frequently required in order to accommodate the biocatalytic step. Today the far wider availability of enzymes means that the situation has changed and, combined with genetic engineering to allow effective expression, the cost of enzymes has been reduced to such an extent that we can easily imagine a multistep API synthesis using not just one enzymatic step, but several. If the enzymatic steps are neighboring each other, there are further advantages, as the conditions under which all biocatalytic reactions work are not so dissimilar. This means that changes of pH, temperature, and pressure through the process will be minimized. Of course, even with neighboring biocatalytic steps, some optimization is still required, but the process can be very much simplified. Indeed, nature uses this strategy to great effect inside microbial cells, where entire pathways are built of multiple enzymatic reactions, without any by-product separation or intermediate purification or isolation [13–16]. Interestingly in organocatalysis such approaches are also being explored to create multicomponent, domino, and tandem reactions [17]. This logic can be extended to biocatalytic synthesis so that multiple biocatalytic reactions can be integrated in a single reactor (so-called one-pot operations). These types of catalytic reaction schemes are frequently referred to as cascade catalysis (see Figure 20.1). There is a significant advantage in the production of APIs using this approach leading to potential process intensification and avoiding the need for intermediate separation and purification. Furthermore, the need for different solvents and auxiliary chemicals can be significantly reduced, improving greatly the environmental profile of such processes [18].

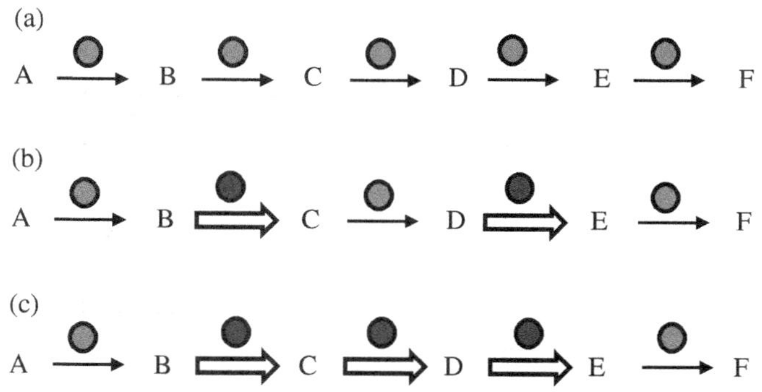

FIGURE 20.1

Basic concept of cascade biocatalysis. (a) Conventional multistep chemical synthesis of compound A–F, via intermediates B–E. (b) Replacement of steps B–C and D–E by biocatalysis, maintaining interstep recovery. (c) Replacement of adjacent steps B–C, C–D, D–E by biocatalysis, with the potential for eliminating interstep recovery.

20.2.2 Biocatalytic Cascade Concepts

From a synthetic perspective two basic biocatalytic cascade schemes can be envisioned. First are those cascades that add value to an existing biocatalytic step, by enabling synthesis from a cheaper starting point or producing a higher value product. Either can be achieved through the addition of one of more enzymes. To be called a cascade, the catalysts should in principle be together in a single reactor from the start, although this does not necessarily imply that all catalytic reactions run simultaneously. Examples of such one-pot reaction schemes have been reported for multienzymatic cascades, but this may not always prove desirable. The optimal operating conditions for each enzyme will sometimes dictate separation of the respective biocatalysts. Separation can be in time or space, leading to different reactor options, as explained later.

The second approach is to assist an existing biocatalytic reaction step by adding one or more ancillary enzymes with the objective of making the first enzyme reaction more effective. In nature we also find enzymes not only as primary catalysts for synthesis but also ancillary helpers (e.g., for cofactor recycling and to shift equilibrium). These ancillary enzymes are sometimes termed secondary catalysts [19–22]. Importantly, such enzymes must have appropriate activity relative to the primary enzyme and preferably use cheaper substrates than the main reaction. Historically such systems have been used for assay and analysis [23], but their role in chemical synthesis is potentially even more powerful.

Using these two groups of multienzymatic systems as a basis, there have been several attempts to classify such cascade reaction schemes [19, 24]. Three basic cascade schemes can be distinguished, namely, (1) linear, (2) parallel and cyclic, and (3) orthogonal. The motivation for implementation in each case is a little different:

1. **Linear cascade**. Those cascades which add value to the reaction by virtue of extending the reaction sequence (either combining the previous step or the subsequent step) are referred to as linear cascades (Figure 20.2a).
2. **Parallel/cyclic cascade**. The second scheme is where one or more additional enzymes are used to assist a primary enzyme. This is done in order to achieve a degree of self-sufficiency in cofactors or reagents. There are two types of scheme, referred to either as parallel cascades (Figure 20.2b) or cyclic cascades (Figure 20.2c). Such schemes are also be found in nature to balance redox in microbial metabolic pathways.
3. **Orthogonal cascade**. The final scheme is where substrates are supplied, or products removed, *in situ* via an additional enzyme. Such schemes are referred to as orthogonal cascades (Figure 20.2d). Such cascades are also found in nature, for example, to enable thermodynamically unfavorable reactions to take place.

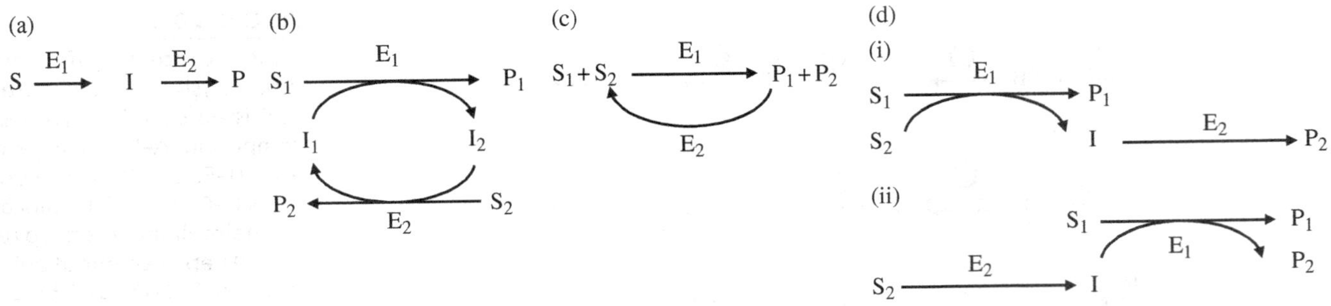

FIGURE 20.2
Examples of cascade schemes for two-enzyme combinations. (a) Linear. (b) Parallel. (c) Cyclic. (d) Orthogonal. E_1, E_2 represent enzymes; I, I_1,I_2 represent intermediates; P, P_1,P_2 represent products; and S, S_1,S_2 represent substrates.

TABLE 20.1 Rationale for Implementation of Different Cascade Types

Cascade	Rationale	Process Benefit
Linear	Addition of preceding step	Added value of reaction scheme
	Addition of following step	Added value of reaction scheme
Parallel	Cofactor self-sufficiency	Reduce cofactor costs
Cyclic	Cofactor self-sufficiency	Reduce cofactor costs
	Reagent self-sufficiency	Reduce reagent costs
Orthogonal	*In situ* reactant supply	Use of inhibitory substrate
	In situ product removal	Shift equilibrium
	In situ coproduct removal	Shift equilibrium

Table 20.1 lists the motivation for the different cascade types.

It is clear from the preceding discussion that there are many benefits for reaction chemistry, biocatalyst, and process using multienzymatic reactions (see also the excellent review by Ricca and coworkers [25]). Most importantly, the option of using multiple enzymes enables entirely new pathways or synthetic routes to become possible. Such a concept builds on what Nature already does today [13]. However, while this can provide inspiration, *de novo* pathways will require attention to be paid not only to biocatalyst compatibility and kinetics but also to reaction thermodynamics. Nature normally already allows for this with large negative Gibbs free energy values for reactions at the start and end of a pathway.

Likewise, several important technical questions are now raised by these developments, such as what format these multiple enzyme systems should take, what types of reactor should be used, and whether reagents should be added or intermediates removed. These questions will be discussed in the following sections.

20.3 PROCESS ASPECTS FOR MULTISTEP BIOCATALYSIS

It will be essential to integrate correctly the development of new biocatalytic routes into chemical process development using retrosynthetic approaches [10, 26], but it will also be important to consider process issues from the start. These include balancing of reaction schemes, selection of reactors, and process intensification and integration.

20.3.1 Balancing Reaction Schemes

Despite the fact that when converting natural compounds many enzymes are very effective catalysts, the use of nonnatural substrates in an industrial synthetic situation means that the rate of an enzyme reaction when first tested in the laboratory is frequently not high enough. In the context of multienzyme reactions this may lead to a buildup or shortage of some intermediates, dependent on the relative rate of adjacent enzymes. Low concentrations of intermediates may limit the reaction rate of the following reaction, and high concentrations of reagents, intermediates, and products can also limit the reaction rate (inhibition) and reduce enzyme stability (denaturation). While examples exist of many operationally stable enzymes, in general, high concentrations of reagents can reduce enzyme stability. The stability of enzymes varies with type, storage, and operating conditions, and therefore, rather than the stability alone, the cost contribution from the enzyme catalyst should be considered [27]. The balancing of enzyme activities was also highlighted in a recent review where the application of process modeling was also proposed as a means to evaluate such schemes [28], as will be discussed later. The important point is to balance activities and stabilities at suitable concentrations in such a way as to enable effective product recovery downstream, from reasonable-sized reactors.

20.3.2 Biocatalytic Reactor Options

The basic reactor choice is between a packed bed, stirred tank, or combinations thereof (with or without possibilities for enzyme retention by membranes or immobilization) and compartmentalization. The packed bed reactor can only handle immobilized enzyme(s) and operates in plug-flow mode (mixing only in the radial, rather than axial direction). For many enzymes their kinetics are such that operation in plug-flow hydrodynamic mode is favored. Hence, in cases where mixing is required (e.g., for pH control via addition of a neutralizing acid or alkali or addition of an inhibitory substrate), then the use of multiple CSTRs can be used (Figure 20.3).

For multiple enzymes, combinations of reactor configuration and/or operation are also possible, dependent upon the kinetic characteristics of each enzyme and the cost contribution of each enzyme, relative to the other components [20]. Such

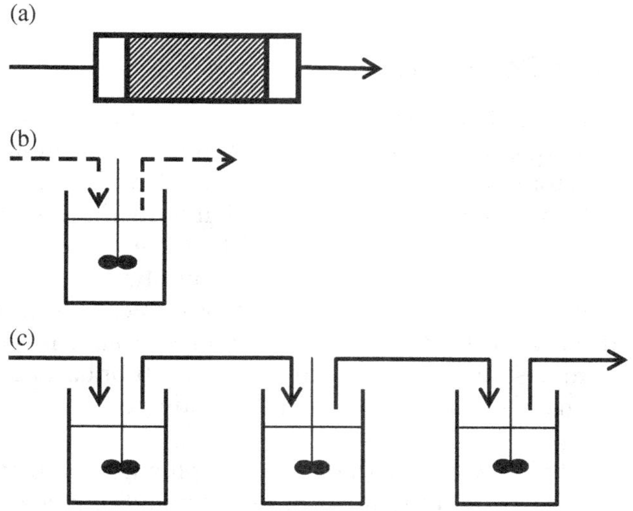

FIGURE 20.3
Basic reactor options for biocatalysis. (a) Plug-flow packed bed reactor. (b) Well-mixed stirred tank. (c) Three continuous well-mixed stirred tanks in series (simulating plug flow).

combinations of reactor configuration and operation can also bring extra flexibility to cope with the different characteristics of each enzyme. For example, each reactor can operate with different enzymes, different reactor hydrodynamics (e.g., plug flow, well mixed), and/or different conditions (e.g., pH, T). An alternative approach is to use a single reactor with variations in conditions dependent upon position (spatial variation) or time (temporal variation) in the reactor. In both strategies the need for flexibility needs to be addressed in process development.

20.3.3 Process Intensification

Many implemented examples of single-step biocatalytic processes have clearly shown the need for innovative process engineering solutions, as well as highly sophisticated molecular biology to engineer the biocatalyst. It is particularly interesting is that no single objective function (e.g., lowest production cost or lowest development cost) or development route (e.g., protein engineering or process engineering) or solution (e.g., operating with enzyme immobilization for 100 recycles by microfiltration or operating with soluble enzyme for five recycles by ultrafiltration) exists, even in a given case. A clear need, from the perspective of process engineers, is to develop a means to navigate the solution space in an effective way. In the pharmaceutical sector the time limitations (as a result of a limited patent lifetime) mean that the emphasis is on speed of development. It is clear that automated, systematic methods of data collection, linked with design of experiments and process models, will have huge benefits in much the same way they have already in other sectors of the chemical industry. Process engineering strategies, such as feeding of substrates and removal of products during the reaction, *in situ* product removal (ISPR) [29–32], will also be required in multienzyme processes. In such cases an added degree of freedom can also come from spatial (for continuous) or temporal (for batch) changes in the reactor system via compartmentalization, adding significance to the complexity of the problem.

Compartmentalization is a concept that can be used widely in multienzymatic processes whereby different parts of the reactor operate under different conditions (e.g., two-liquid phase biocatalysis) or catalysts are separated (e.g., by immobilization). The two compartments may be separated by a phase boundary (most likely solid–liquid or liquid–liquid). The compartments will selectively contain enzymes and reaction components such that not all enzymes and components are present in all parts of the reactor at the same concentration at a given time. This has benefits not only for the reaction itself (e.g., reducing product inhibition) but also downstream processing (e.g., separation of enzymes).

20.3.4 Continuous Processes

Batch operation implies that all the reagents are added with the catalyst to the reactor together and subsequently, following reaction, removed. A more sophisticated version of such a reactor operates with feeding (fed batch) or sequential supply of reagents, change of conditions, and removal of intermediate and by-products. In this way a simple reactor setup can be used to achieve great flexibility. Most examples at laboratory scale reported in the scientific literature follow such schemes. Such a system can also be made continuous (which is of great interest currently for the synthesis of APIs [33, 34]). Now the time variable is replaced by residence time in the process, in a series of plug flow and tubular operations. In this way a flow system can be built for a biocatalytic cascade (see the example presented in Figure 20.4).

The advantage of the flow system (if operated in plug flow) is that excellent control of conditions can be assured (nothing mixes with anything before or following since the fluid flows as a plug through the process).

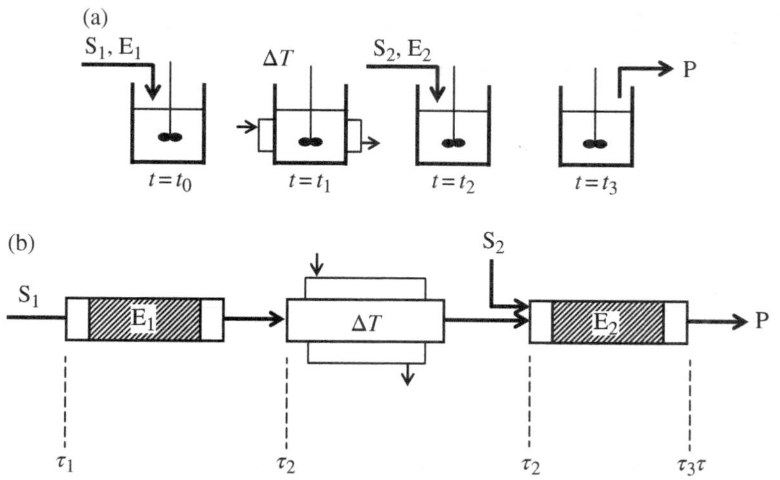

FIGURE 20.4

Conversion of (a) batch operating scheme into (b) flow operating scheme. E_1, E_2 represent enzymes; P represents product; S_1,S_2 represent substrates; t, t_0, t_1, t_2, t_3 represents process time; and τ, τ_0, τ_1, τ_2, τ_3, represents residence time. At process time/residence time 0, S_1 is added; at 1, the temperature is changed (ΔT); at 2, S_2 and E_2 are added; and at 3, reactors are emptied.

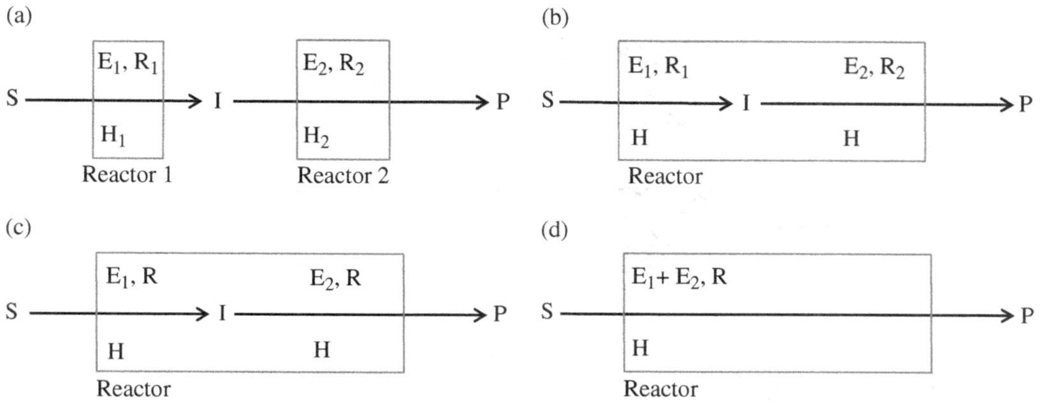

FIGURE 20.5

Integration options for linear cascade. (a) Process-level integration. (b) Reactor-level integration. (c) Reaction-level integration. (d) Catalyst-level integration. E_1, E_2 represent enzymes; H, H_1,H_2 represent reactor hydrodynamics; P represents product; R, R_1,R_2 represent reaction conditions; and S represents substrate.

20.3.5 Process Integration

A primary aim of integration is to ensure that the reaction yield is as high as possible. This is of utmost importance to reduce the process mass intensity and capitalize upon the reactant supplied. One way to achieve this is by linking reactions together without intermediate separation. In the case of a two-enzyme linear cascade, the two reactions can each be carried out in two separate reactors, each with separate conditions and separated catalysts. This makes the system very flexible while ensuring a minimization of waste due to excessive separating reagents as a result of unwanted recovery operations [18]. Such integration could be referred to as process-level integration (Figure 20.5a).

A second level of integration (reactor integration) is to run the reactions in a single reactor. This implies that the hydrodynamics (plug flow or well mixed) are the same, but not necessarily the reaction conditions. These could still be separated in time (in a batch reactor) or space (in a continuous reactor). Compartmentalization is also possible, meaning the two catalysts could be separated and replaced at different times as required (Figure 20.5b).

TABLE 20.2 Conditions Required for Different Degrees of Integration for Multienzymatic Processes

Integration Level	No Interstep Separation	Same Hydrodynamics	Same Conditions	Same Stability
Process	X			
Reactor	X	X		
Reaction	X	X	X	
Catalyst	X	X	X	X

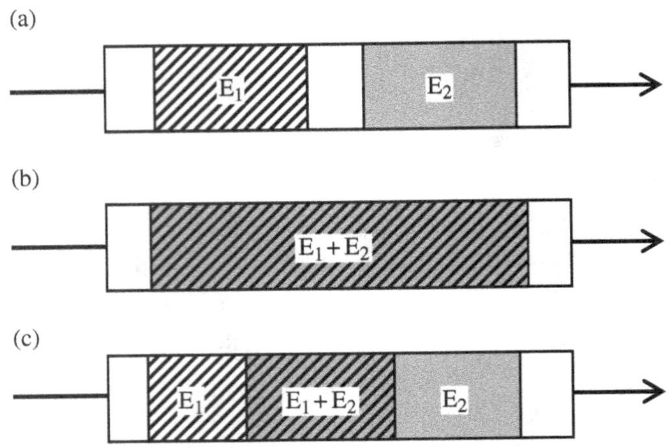

FIGURE 20.6
Catalyst integration in flow systems. (a) Separation of enzymes. (b) Complete integration of enzymes. (c) Partial integration of enzymes.

The third level of integration (reaction integration) is to combine reactions together within the single reactor, meaning also that the conditions are the same. The biocatalysts can still be separated to afford replacement dependent on format (Figure 20.5c).

The final level of integration is at the level of the catalyst. Here liquid enzyme can be mixed into a homogeneous solution of multiple enzymes or immobilized enzymes combined together on a single support. Likewise they can be coexpressed in a common microbial host for whole-cell conversions. In all cases similar stability of the catalysts are required as well as matching rates and conditions. There are some interesting examples reported in the scientific literature presenting such concepts, but for practical purposes much of the flexibility is lost (see Figure 20.5d). See also Table 20.2, which summarizes the conditions required for each level of integration.

One of the advantages of continuous operation is that in flow mode catalyst integration can be achieved to differing extents. For example, for linear cascades different concepts can be considered (Figure 20.6).

For parallel, cyclic, and orthogonal cascades more complex systems can be devised using a modular approach. The modules could also contain recovery options (for products or by-products), using liquid–liquid extraction or resin adsorption operations as required [35]. Such a modular approach can not only provide optimal conditions for each part of the process but can also enable operation in a flexible way to allow for the different time constants of operations and phenomena. A requirement for implementation of such technology is to find suitable methods of immobilization and retention of the catalysts in the reactor modules [35–37].

20.4 PROCESS DEVELOPMENT

Perhaps the first decision to be made in process development is at the most constrained level of integration, meaning the catalyst. Such issues do not arise with conventional single biocatalytic steps but are highly important in a biocatalytic cascade. One of the key criteria here is whether the enzymes can be operated together without compromise to any of the individual enzyme's activity or stability. An interaction matrix (see process modeling) can be used to assist such decision making. In cases where the cost of one or more of the enzyme(s) is not critical, it will be possible to combine in a one-pot operation. In other cases, where the cost of an individual enzyme becomes critical, then it may be necessary to separate the catalysts, such that each can operate under optimal conditions. Likewise, selection of the basic reactor type (packed bed, stirred tank, or combinations thereof) and biocatalyst recovery (mesh, microfiltration, ultrafiltration, or combinations thereof) will determine the structure of the process flowsheet and therefore is an early consideration in the development of any bioprocess. The criterion for selection of the final type of biocatalyst and reactor combination is primarily economic and may best be evaluated by the four metrics in common use to assess the economic feasibility of biocatalytic processes [27]:

1. Reaction yield (g product/g substrate)
2. Biocatalyst yield (g product/g biocatalyst)
3. Product concentration (g product/l)
4. Space–time yield (g product/l/h)

The balance between the four metrics is dependent upon the relative costs in a process. For example, a process with a high cost of biocatalyst will require a high biocatalyst yield, while those with a high cost of process plant will require a high space–time yield, and those with a high downstream processing cost will require a high product concentration to leave the reactor.

Process modeling is increasingly be implemented as a means of mathematically describing enzymatic reactions [38, 39]. Two types of models need to be developed—those that describe the reaction phase and those that describe the associated unit operations and process via mass balances. Much progress has already been made, but more sophisticated models are still required to enable a more predictive approach for scale-up and design. This will also be an important need in the future as we move from empirical to more mechanistically based models. Alongside this it will be necessary to build property databases of suitable reagents and chemicals. In many cases predictive tools for the properties of many of these molecules would perhaps be even more useful to save valuable experimental time, especially since many of the most interesting molecules are not available in databases. The chemistry, in particular of many processes where biological catalysts can best be exploited, is complex, and the building of a suitable database and predictive tools will be important in the future. Kinetic, thermodynamic, and process models for multienzyme processes are particularly valuable [39] because of the complexity of the choices and decisions, which characterize process development of such systems. Using the models it becomes possible to focus experimental efforts and develop the process faster than would otherwise have been the case.

For a multienzyme process, this evaluation is critically important to achieve a better theoretical understanding of the process and to achieve useful modeling and process design. The reaction considerations describe the key characteristics necessary if one is to understand how the interaction between enzymes and components can be interpreted for modeling. Furthermore, such information forms the basis for the formulation of reaction rates for the different enzymes that are involved in a biocatalytic cascade. The key information required is the following:

Reagent data: Physical and chemical properties of the compounds involved in the reaction must be known (e.g., density, water solubility, viscosity, boiling point, Henry's constant, pK_a, pH stability).

Cascade architecture: A graphical representation of all reactions in the multienzyme process is the basis for describing the final model structure. It includes the primary reactions, secondary reactions, and competing reactions. For a single enzyme, reaction mechanisms are well developed, and they are then included into the full model to describe the multienzyme process by combining the effect of the individual enzymes. In this way, the different possible reaction schemes are generated to give the cascade structure.

Interaction matrix: This matrix is used to identify the different interactions that can exist between compounds and enzymes in the process. Here, the structure defined in the previous step is required. In order to build the matrix, the compounds involved in the process (i.e., substrates, intermediates, by-products, products, etc.) are arranged in rows, and the enzymes are arranged in columns. In this way, the matrix is filled defining the relationship between each compound and enzyme in turn, that is, substrate, product, inhibitor, or no interaction between a given compound and enzyme. This compiled information is extremely powerful to help make decisions about the relevant terms or kinetic parameters that must be added or removed from the reaction rate expressions and process model. The position of the new term/parameter in the final expression is defined by the enzyme kinetic mechanism, which shows how the compound inhibits the enzyme, for example, competitive, uncompetitive, noncompetitive, or mixed inhibition. Similarly the matrix helps define the process configuration options.

Process configuration: Data from the interaction matrix is also used to help formulate the process configurations that are possible. Using process models here can also enable alternatives to be tested ahead of experimental validation.

Process control: In biocatalytic cascades, in batch mode variables such as pH [13] and temperature are often controlled during the process. In flow mode the control may be easier but still requires sufficient understanding. Process control can be divided into two basic control layers [40]. The first is a regulatory layer, which controls the process variables. The second is a supervisory layer, which manages variables with greater impact on the entire process. In this case a more detailed controller design is required. For biocatalytic cascades, this issue is highly important especially to achieve a sufficient degree of process intensification.

20.5 BIOCATALYTIC CASCADE EXAMPLES

Much progress has been made using the concepts described in the previous section to implement new biocatalytic cascades. The vast majority of these have been examined at laboratory scale, but the following list gives a representative picture of the current state of the art. The cascades are divided into classes, according to the definitions described earlier.

20.5.1 Linear Cascades

The simplest but most valuable cascades are the linear cascades. For example, Herter and coworkers [41] report an excellent example of a linear cascade using galactose oxidase mutants to oxidize amino alcohols and amino diols to aldehydes, which are subsequently used in one-pot cascades with aldolase or xanthine dehydrogenase to yield amino sugars or alternatively using the cyclized imine of the aldehyde with aldehyde oxidase to yield lactams. A second example of a linear cascade concerns the development of a bienzymatic system using (*S*)-hydroxynitrile lyase and a nitrile hydratase or nitrilase to covert aldehydes, via hydrocyanation and subsequent hydration or hydrolysis, into the corresponding (*S*)-2-hydroxycarboxylic amides or acids [42]. In this linear system the biocatalysts were integrated via a form of coimmobilization using a so-called combi-CLEA [43] and compared with coexpressed enzymes in a common *Escherichia coli* host. Further examples of linear cascades were recently reported for the synthesis of chiral amines (specifically 1,2-amino alcohols) using ω-transaminase now preceded by either a lipase, a carboligase, or alternatively a transketolase [44].

20.5.2 Parallel Cascades

For practical implementation the use of parallel cascades is of great value for cofactor recycle. For example, van Hecke and coworkers [45] report the synthesis of lactobionic acid from lactose using cellobiose dehydrogenase. The enzyme requires an electron acceptor, and using ABTS and laccase in a parallel cascade, such a system was effectively operated and modeled.

20.5.3 Cyclic Cascades

Cyclic cascades are used for reagent recycle to afford a degree of self-sufficiency. An excellent example of the use of a cyclic cascade concerns the use of catalase to remove the by-product hydrogen peroxide from oxidase-catalyzed conversions, which is otherwise toxic at very low concentrations to enzymes. Catalase converts hydrogen peroxide to water and oxygen, which are not only harmless reagents but in the case of oxidation-based primary enzymes can serve as a cosubstrate, thereby reducing the overall demand for oxygen. Such a scheme has been reported to enhance primary oxidative enzymes such as amino acid oxidase [46] and also recently aldehyde-deformylating oxygenase (formerly known as aldehyde decarbonylase) [47]. In the latter case the catalase was fused with the alkane-producing enzyme to enhance enzymatic activity fivefold. Another elegant example of a cyclic cascade has been reported by Babich and coworkers [48]. Here three reaction steps are carried out with two enzymes. Starting from dihydroxyacetone and various aldehydes, the application of either rabbit muscle aldolase or rhamnulose-1-phosphate aldolase enables the synthesis of various complex chiral carbohydrate analogues. Phosphorylated dihydroxyacetone can be used by the aldolases, together with the recycle of the phosphate group via the supply of (cheap) pyrophosphate in combination with a further enzyme, acid phosphatase. The system was run with immobilized enzymes in continuous flow mode.

20.5.4 Orthogonal Cascades

Orthogonal cascades have been used to date for the removal of by-products in order to shift equilibrium. An excellent example of an orthogonal cascade is based on using alanine as an amino donor for the ω-transaminase-catalyzed synthesis of chiral amines [49].

A second enzyme, pyruvate decarboxylase, converts the by-product of the first reaction (pyruvate) and creates a new by-product (CO_2), making the thermodynamics of the second reaction very favorable and driving the first reaction (which is otherwise thermodynamically unfavorable). Alternative systems could use lactate dehydrogenase, but this would then require an additional parallel system to recycle NAD^+.

20.5.5 Linear–Parallel

More complex cascades involve more than one concept. For example, Shanmuganathan and coworkers [50] report a two-enzyme linear cascade using benzaldehyde lyase and glucose dehydrogenase to synthesize optically active α-aryl vicinal diols from aromatic aldehydes. However, the glucose dehydrogenase requires recycling the NADH, which was achieved by use of formate dehydrogenase, affording CO_2 as a by-product and thus driving the reaction. In this way the parallel cascade enabled the linear cascade to take place effectively.

20.5.6 Linear–Cyclic

Another cascade where two concepts are used was reported some years ago but still serves as an excellent illustration. This application demonstrates a four-step, one-pot cascade for the synthesis of 5-deoxy-5-ethyl-D-xylulose from glycerol using a phytase, glycerol phosphate oxidase, and fructose-1,6-bisphosphate aldolase and reapplication of the phytase [51]. Here too the trick of phosphorylation using pyrophosphate was applied. The linear reaction requires assistance though because the glycerol phosphate oxidase produces hydrogen peroxide, which is detrimental to enzyme activity at low concentrations. This was removed by addition of catalase, coproducing water and half a mole of oxygen, which was then reused by the oxidase.

20.5.7 Complex Cascades

Further complexity can be built. For example, Tauber and coworkers [52] report multiple examples of enzymatic cascades for the asymmetric amination of secondary alcohols building on the application of ω-transaminases for the synthesis of chiral amines. They also give cases where multiple strategies are employed. As an example, to add value to the ω-transaminase reaction, the starting ketone can be synthesized from an alcohol using an alcohol dehydrogenase. However such a linear cascade has two basic problems. The first is that alcohol dehydrogenase is $NAD(P)^+$ dependent, and second is that, dependent upon the amine donor, the transaminase reaction may be thermodynamically limited. The authors give examples of cofactor recycling and equilibrium shifting by the application of parallel and orthogonal cascades, respectively, using, for example, NADPH oxidase and alanine dehydrogenase (which itself requires NADH recycle with formate dehydrogenase in a further parallel cascade). Further more elaborate examples are given, creating a redox neutral cascade by using an NAD^+-dependent ADH, thus giving a combined cyclic and parallel system.

Divergent reaction conditions are well exemplified in the case of reduction and oxidation steps. Consequently it is easier to run such systems sequentially [53]. This leads to the possibility of a linear–cyclic–parallel system for oxidation and reduction, proposed by Oberleitner and coworkers [54]. Here oxidative and reductive catalysts were combined to balance redox, followed by a further oxidation reaction. For example, alcohol dehydrogenase and enoate reductase, followed by Baeyer–Villiger monooxygenase, have been demonstrated in such a system. The last oxidation needs to be run with a parallel system or alternatively in a whole cell for cofactor recycle.

20.5.8 Convergent Parallel Cascade

Recently a very interesting concept was introduced for a convergent parallel reaction, meaning that both the substrate of the main reaction and the parallel reaction lead to the same product (Figure 20.7).

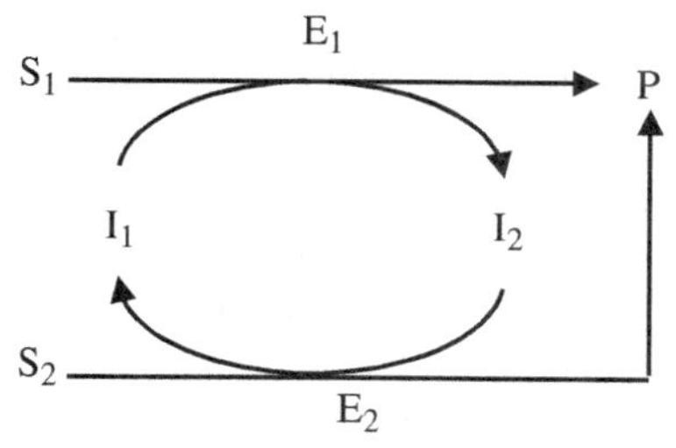

FIGURE 20.7

Concept of convergent parallel reaction. E_1, E_2 represent enzymes; I_1, I_2 represent intermediates; P represents product; and S_1, S_2 represent substrates.

In this case the concept was exemplified with a Baeyer–Villiger monooxygenase and an alcohol dehydrogenase for the synthesis of ε-caprolactone from cyclohexanone and the so-called "double-smart cosubstrate" 1,6-hexanediol [55].

20.6 FUTURE OUTLOOK

20.6.1 Protein Engineering

The development of recombinant DNA (rDNA) technology enables several possibilities for the exploitation of biocatalysis. For example, it has provided a cheap way to produce a given biocatalyst. The desired enzyme (or enzymes) can now be overexpressed, meaning that they represent a far larger fraction of the available protein in the cell. This not only reduces the required scale of the fermentation but for isolated enzyme applications the downstream recovery. Additionally a synthetically interesting enzyme found in nature may be expressed in a poor host for production (e.g., the host may be pathogenic). Such a situation can be also overcome by cloning into a suitable "production" host such as *E. coli*. Of equal importance is that rDNA technology also enables an alteration of the properties of the biocatalyst and in recent years there has been enormous progress in the development of new protein engineering tools in particular for activity enhancement (also at high concentration of substrates) [56–58]. These tools will also have an impact on biocatalytic cascades. For example, in a two-enzyme scheme, the enzymes may be engineered to work under compromised conditions for both or at the optimum for one or alternatively the other enzyme, dependent upon the relative costs. The ability to swap the amino acids either in the active site or even at remote positions of the protein has been found capable of altering and controlling substrate repertoire, stability, activity (reaction rate), and selectivity. Today, synthetic chemists routinely use so-called directed evolution combined with rational strategies based on structure–function relationships to engineer proteins [59].

An important strategy available to process chemists and engineers seeking to effectively integrate biocatalysis into a synthetic pathway is to modify the biocatalyst to improve the matching of conditions between adjacent (enzymatic) reactions. There are two requirements. First, there is a need to match the rates of the reactions (without the need for an excess of one or other biocatalyst), and second, there is a need to improve the tolerance of each enzyme to the reaction conditions of the other enzymes.

To date three types of improvement with respect to integration have been attempted. Firstly, engineering enzymes against product inhibition has been achieved using directed evolution for a tyrosine aminotransferase [60], superoxide dismutase, and ω-transaminase. Secondly, improving the tolerance of enzymes to organic solvent is frequently required. Given the low water solubility of many compounds of industrial interest, it is often the case that enzymes need to be used in the presence of organic solvents. This can often be achieved by solubilization of the enzyme through modification of the enzyme with polyethylene glycol (PEG), fatty acids, or surfactants. However, most enzymes are rendered inactive by high concentrations of organic solvent [61]. In some previous examples, directed evolution has enabled an improvement in the tolerance of enzyme activity to the presence of cosolvent [62].

Additionally, in some cases, tolerance to organic solvent may be achieved indirectly through an increase in enzyme thermostability, although this correlation is not always predictable. Thirdly, in some cases the enzyme is required to use reactive substrates (e.g., aldehydes) or generate reactive side products (e.g., H_2O_2), which react with and slowly deactivate the enzyme. Tolerance to oxidation by H_2O_2 exposure has been previously addressed by directed evolution for manganese peroxidase in which an enzyme variant achieved a ninefold greater stability to peroxide than the wild-type enzyme [63]. More recently, the irreversible deactivation of 2-deoxy-D-ribose 5-phosphate aldolase at high concentrations of the aldehyde reactant was minimized by identifying mutants from a library generated by error-prone polymerase chain reaction (PCR) [64]. It is often assumed that aldehydes would react mainly with the lysine primary amines to form imines, which then react further to create an irreversible modification. Indeed some of the mutants that had increased tolerance to inactivation by chloroacetaldehyde were found to be mutated at one of the lysine residues. However, targeted random mutagenesis of each of the lysine residues individually using saturation mutagenesis did not identify any further mutants with increased tolerance to chloroacetaldehyde.

The application of an integrated multienzymatic process at the catalytic level creates a much greater burden on the compatibility requirements of the enzymes. Future applications of directed evolution are expected to address these issues, although this will also require suitable screening assays that enable enzyme activity to be assessed over time and in the presence of the targeted chemical reagents [65, 66]. Computer-based methods are increasingly being used [67] and will doubtless expand in the future, but it remains the case that modifications inside the active site can be better targeted than those in the rest of the protein [7, 68].

Most complex is that improvements in both the biocatalyst(s) and the process need to go hand in hand. Consequently process engineers have an important role here in integrating the targets required for a cost-effective process together with the possibilities provided by the "biocatalyst engineers" [69].

20.6.2 Flow Chemistry and Process Intensification

The development of multienzyme processes and the associated technology represent an exciting scientific and engineering endeavor for the future. New challenges are raised in terms of process selection (process synthesis) alongside protein engineering. Likewise opportunities for new products and also new routes to existing products will be forthcoming. Process design will need to incorporate the extra elements of biocatalyst integration, alongside combinations of reactor configuration and operation to afford maximum flexibility and optimization. There is also a need for further definition of more standardized processes. In an analogous way, the trade-off between one-pot and complete intermediate purification will need to be evaluated. Not only will research in areas such as protein engineering and high-intensity biocatalysis need to be examined, but also mathematical modeling tools need to be developed to examine the trade-offs and alternative strategies. Already new approaches using multienzyme–nanoparticle assemblies [70] and also synthetic cascades with artificial metalloenzymes [71] are being developed in the laboratory. For the future, it will be important to consider all approaches to biocatalytic cascades, inspired by nature [11, 72].

REFERENCES

[1] Hoyos, P.; Pace, V.; Hernaiz, M.J.; Alcantara, A.R. *Curr. Green Chem.* 2014, **1**, 155–181.
[2] Breuer, M.; Ditrich, K.; Habicher, T.; Hauer, B.; Keßeler, M.; Stürmer, R.; Zelinski, T. *Angew. Chem. Int. Ed.* 2004, **43**, 788–824.
[3] Pollard, D.J.; Woodley, J.M. *Trends Biotechnol.* 2007, **25**, 66–73.

[4] Wohlgemuth, R. *Curr. Opin. Biotechnol.* 2010, **21**, 713–724.
[5] Wells, A.S.; Finch, G.L.; Michels, P.C.; Wong, J.W. *Org. Process Res. Dev.* 2012, **16**, 1986–1993.
[6] Nestl, B.M.; Hammer, S.C.; Nebel, B.A.; Hauer, B. *Angew. Chem. Int. Ed.* 2014, **53**, 3070–3095.
[7] Bornscheuer, U.T.; Huisman, G.W.; Kazlauskas, R.J.; Lutz, S.; Moore, J.C.; Robins, K. *Nature* 2012, **485**, 185–194.
[8] Yang, G.; Ding, Y. *Bioorg. Med. Chem.* 2014, **22**, 5604–5612.
[9] Nestl, B.M.; Nebel, B.A.; Hauer, B. *Curr. Opin. Chem. Biol.* 2011, **15**, 187–193.
[10] Turner, N.J.; O'Reilly, E. *Nat. Chem. Biol.* 2013, **9**, 285–288.
[11] Sigrist, R.; da Costa, B.Z.; Marsaioli, A.J.; de Oliveira, L.G. *Biotechnol. Adv.* 2015, **33**, 394–411.
[12] Schmid, A.; Dordick, J.S.; Hauer, B.; Kiener, A.; Wubbolts, M.; Witholt, B. *Nature* 2001, **409**, 258–268.
[13] Bruggink, A.; Schoevaart, R.; Kieboom, T. *Org. Process Res. Dev.* 2003, **7**, 622–640.
[14] Pellissier, H. *Tetrahedron* 2006, **62**, 2143–2173.
[15] Findrik, Z.; Vasic-Racki, D. *Chem. Biochem. Eng. Q.* 2009, **23**, 545–553.
[16] Lowry, B.; Walsh, C.T.; Khosla, C. *Synlett* 2015, **26**, 1008–1025.
[17] Wende, R.C.; Schreiner, P.R. *Green Chem.* 2012, **14**, 1821–1849.
[18] Ni, Y.; Holtmann, D.; Hollmann, F. *ChemCatChem* 2014, **6**, 930–943.
[19] Santacoloma, P.A.; Sin, G.; Gernaey, K.V.; Woodley, J.M. *Org. Process Res. Dev.* 2011, **15**, 203–212.
[20] Xue, R.; Woodley, J.M. *Bioresour. Technol.* 2012, **115**, 183–195.
[21] Gardner, T.S. *Trends Biotechnol.* 2013, **31**, 123–125.
[22] Sagt, C.M.J. *Appl. Microbiol. Biotechnol.* 2013, **97**, 2319–2326.
[23] Monošík, R.; Ukropcová, D.; Stredanský, M.; Šturdík, E. *Anal. Biochem.* 2012, **421**, 256–261.
[24] Simon, R.C.; Richter, N.; Busto, E.; Kroutil, W. *ACS Catal.* 2013, **4**, 129–143.
[25] Ricca, E.; Brucher, B.; Schrittwieser, J.H. *Adv. Synth. Catal.* 2011, **353**, 2239–2262.
[26] Tao, J.; Zhao, L.; Ran, N. *Org. Process Res. Dev.* 2007, **11**, 259–267.
[27] Tufvesson, P.; Lima-Ramos, J.; Nordblad, M.; Woodley, J.M. *Org. Process Res. Dev.* 2011, **15**, 266–274.
[28] Muschiol, J.; Peters, C.; Oberleitner, N.; Mihovilovic, M.D.; Bornscheuer, U.T.; Rudroff, F. *Chem. Commun.* 2015, **51**, 5798–5811.
[29] Stark, D.; von Stockar, U. *Adv. Biochem. Eng. Biotechnol.* 2003, **80**, 149–175.
[30] Woodley, J.M.; Bisschops, M.; Straathof, A.J.J.; Ottens, M. *J. Chem. Technol. Biotechnol.* 2008, **83**, 121–123.
[31] Pabby, A.K.; Sastre, A.M. *J. Membr. Sci.* 2013, **430**, 263–303.
[32] van Hecke, W.; Kaur, G.; de Wever, H. *Biotechnol. Adv.* 2014, **32**, 1245–1255.
[33] Wegner, J.; Ceylan, S.; Kirschning, A. *Adv. Synth. Catal.* 2012, **354**, 17–57.
[34] Hessel, V.; Kralisch, D.; Kockmann, N.; Noël, T.; Wang, Q. *ChemSusChem* 2013, **6**, 746–789.
[35] Andrade, L.H.; Kroutil, W.; Jamison, T.F. *Org. Lett.* 2014, **16**, 6092–6095.
[36] Schoffelen, S.; van Hest, J.C.M. *Curr. Opin. Struct. Biol.* 2013, **23**, 613–621.
[37] Jia, F.; Harasimhan, B.; Mallapragada, S. *Biotechnol. Bioeng.* 2013, **111**, 209–222.
[38] Vasić-Rački, D.; Findrik, Z.; Vrsalović Presečki, A. *Appl. Microbiol. Biotechnol.* 2011, **91**, 845–856.
[39] Ardao, I.; Zeng, A.P. *Chem. Eng. Sci.* 2013, **87**, 183–193.
[40] Skogestad, S. *Comput. Chem. Eng.* 2004, **28**, 219–234.
[41] Herter, S.; McKenna, S.M.; Frazer, A.R.; Leimkühler, S.; Carnell, A.J.; Turner, N.J. *ChemCatChem* 2015, **7**, 2313–2317.
[42] van Rantwijk, F.; Stolz, A. *J. Mol. Catal. B Enzym.* 2015, **114**, 25–30.
[43] Chmura, A.; Rustler, S.; Paravidino, M.; van Rantwijk, F.; Stolz, A.; Sheldon, R.A. *Tetrahedron Asymmetry* 2013, **24**, 1225–1232.
[44] Sehl, T.; Maugeri, Z.; Rother, D. *J. Mol. Catal. B Enzym.* 2015, **114**, 65–71.
[45] van Hecke, W.; Bhagwat, A.; Ludwig, R.; Dewulf, J.; Haltrich, D.; Van Langenhove, H. *Biotechnol. Bioeng.* 2008, **102**, 1475–1482.
[46] Tindal, S.R.; Carr, R.; Archer, I.V.J.; Woodley, J.M. *Chem. Today* 2011, **29**, 50–51.
[47] Andre, C.; Kim, S.W.; Yu, X.-H.; Shanklin, J. *Proc. Natl. Acad. Sci.* 2013, **110**, 3191–3196.
[48] Babich, L.; Hartog, A.F.; van Hemert, L.J.C.; Rutjes, F.P.J.T.; Wever, R. *ChemSusChem* 2012, **5**, 2348–2353.

[49] Höhne, M.; Kühl, S.; Robins, K.; Bornscheuer, U.T. *ChemBioChem* 2008, **9**, 363–365.
[50] Shanmuganathan, S.; Natalia, D.; Greiner, L.; Dominguez de Maria, P. *Green Chem.* 2012, **14**, 94–97.
[51] Schoevaart, R.; van Rantwijk, F.; Sheldon, R.A. *J. Org. Chem.* 2000, **65**, 6940–6943.
[52] Tauber, K.; Fuchs, M.; Sattler, J.H.; Pitzer, J.; Pressnitz, D.; Koszelewski, D.; Faber, K.; Pfeffer, J.; Haas, T.; Kroutil, W. *Chem. Eur. J.* 2013, **19**, 4030–4035.
[53] Schrittwieser, J.H.; Sattler, J.; Resch, V.; Mutti, F.; Kroutil, W. *Curr. Opin. Chem. Biol.* 2011, **15**, 249–256.
[54] Oberleitner, N.; Peters, C.; Muschiol, J.; Kadow, M.; Saβ, S.; Bayer, T.; Schaaf, P.; Iqbal, N.; Rudroff, F.; Mihovilovic, M.D.; Bornscheuer, U.T. *ChemCatChem* 2013, **5**, 3524–3528.
[55] Bornadel, A.; Hatti-Kaul, R.; Hollmann, F.; Kara, S. *ChemCatChem* 2015, **7**, 2442–2445.
[56] Bommarius, A.S.; Blum, J.K.; Abrahamson, M.J. *Curr. Opin. Chem. Biol.* 2011, **15**, 194–200.
[57] Quin, M.B.; Schmidt-Dannert, C. *ACS Catal.* 2011, **1**, 1017–1021.
[58] Davids, T.; Schmidt, M.; Böttcher, D.; Bornscheuer, U.T. *Curr. Opin. Chem. Biol.* 2013, **17**, 215–220.
[59] Strohmeier, G.A.; Pichler, H.; May, O.; Gruber-Khadjawi, M. *Chem. Rev.* 2011, **111**, 4141–4164.
[60] Rothman, S.C.; Voorhies, M.; Kirsch, J.F. *Protein Sci.* 2004, **13**, 763–772.
[61] Ogino, H.; Ishikawa, H. *J. Biosci. Bioeng.* 2001, **91**, 109–116.
[62] Wong, T.S.; Arnold, F.H.; Schwaneberg, U. *Biotechnol. Bioeng.* 2004, **85**, 351–358.
[63] Miyazaki-Imamura, C.; Oohira, K.; Kitagawa, R.; Nakano, H.; Yamane, T.; Takahashi, H. *Protein Eng.* 2003, **16**, 423–428.
[64] Jennewein, S.; Schürmann, M.; Wolberg, M.; Hilker, I.; Luiten, R.; Wubbolts, M.; Mink, D. *Biotechnol. J.* 2006, **1**, 537–548.
[65] Schoffelen, S.; van Hest, J.C.M. *Soft Matter* 2012, **8**, 1736–1746.
[66] Lin, J.-Y.; Palomec, L.; Wheeldon, I. *ACS Catal.* 2014, **4**, 505–511.
[67] Damborsky, J.; Brezovsky, J. *Curr. Opin. Chem. Biol.* 2014, **19**, 8–16.
[68] Lutz, S. *Curr. Opin. Biotechnol.* 2010, **21**, 734–743.
[69] Woodley, J.M. *Curr. Opin. Chem. Biol.* 2013, **17**, 310–316.
[70] Kang, W.; Liu, J.; Wang, J.; Nie, Y.; Guo, Z.; Xia, J. *Bioconjug. Chem.* 2014, **25**, 1387–1394.
[71] Köhler, V.; Wilson, Y.M.; Dürrenberger, M.; Ghislieri, D.; Churakova, E.; Quinto, T.; Knörr, L.; Häussinger, D.; Hollmann, F.; Turner, N.J.; Ward, T.R. *Nat. Chem.* 2013, **5**, 93–99.
[72] Lopez-Gallego, F.; Schmidt-Dannert, C. *Curr. Opin. Chem. Biol.* 2010, **14**, 174–183.

CHAPTER 21

Yeast-Mediated Stereoselective Synthesis

René Csuk

Organische Chemie, Martin-Luther Universität Halle-Wittenberg, Halle, Germany

21.1 INTRODUCTION

Mixed cultures of a broad variety of microorganisms, especially of bacteria and yeasts in particular have been used for millenia to produce dairy products, bread, and alcoholic beverages. The foundations for a scientific approach, however, were laid by L. Pasteur [1] in 1862 by using a pure culture of *Bacterium xylinum* to transform [2] ethanol into acetic acid. In 1874, Dumas [3] reported the reduction of sulfur to hydrogen sulfide by fermenting yeast, *Saccharomyces cerevisiae*. The first reduction ("phytochemical reduction") of an organic molecule under anaerobic conditions was performed by Windisch [4] in 1898, and furfuryl alcohol **2** was obtained from furfural **1** (Figure 21.1).

During these early days of chemistry, developments in organic chemistry were often closely associated with those in biology, biochemistry, and a discipline that is called nowadays biotechnology. Hence, the use and application of microorganisms for synthesis is definitely not an invention of the twentieth century but a rather well known way of performing reactions. The foundation of important technologies such as the hydrolysis of starch was followed by the fermentation of the resulting monosaccharides. As early as 1816, Kirchhoff [5] described the saccharification of starch, and Döbereiner [6] used yeast for the production of "un liqueur vineuse."

Also, these early applications of yeast (as well as of other microorganism) were not "small scale" performed to satisfy the burning curiosity of a minority of scientists. The brewing industry was already a large industry. As early as in 1796 ~30.000 m^3 of beer had been brewed by Samuel Whitbread. As S. M. Roberts [7] pointed out, biotechnology may have an ancient history, but the combination of fermentation and the manufacture of ethanol (by using multiplate countercurrent condenser as developed by Coffey in Dublin, 1830) can be regarded as the first biotechnological process of modern chemistry.

"Biochemical transformations" were used in these early days quite regularly—albeit they were not understood. Technological improvements followed whenever a more detailed insight into these transformations occurred, and as early as 1838 this "knowledge" was found in textbooks such as Berzelius' *Textbook of Chemistry* (4th German edition).

Numerous microbial transformations (including their technological applications) followed, and during 1970–1990 these yeast (or microorganism or enzymatic)-mediated bioconversions, biotransformations, or biodegradations were hailed as an ultimate tool for synthetic transformations. They seemed to be suited [8–16] to displace traditional organic chemistry. This opinion was revised, and an initial enthusiasm was

Green Biocatalysis, First Edition. Edited by Ramesh N. Patel.

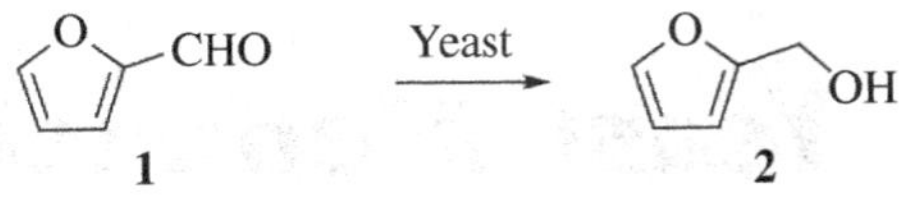

FIGURE 21.1
Reduction of furfural by baker's yeast—performed by W. Windisch as early as 1898.

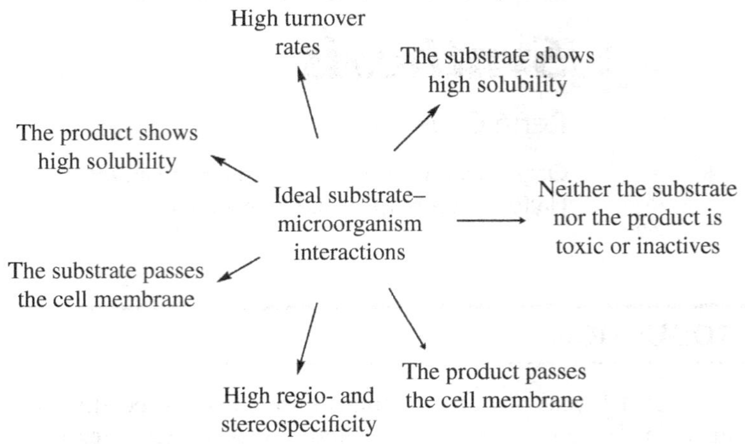

FIGURE 21.2
Summary of ideal substrate–microorganism interactions.

replaced by defining yeast-mediated reactions as to supporting rather than supplanting organic synthesis, and these reactions should be used (or considered to be used) whenever a given reaction is not easily accomplished by classical chemical methods [14, 17–23]. Thus, yeast-mediated transformations have become a method of choice for producing fine chemicals, pharmaceuticals [24–29] as well as cosmetic ingredients (including flavor and fragrances) [30–37], and bioactive compounds [38–41] under mild conditions in high yields and excellent selectivity [42].

Still these whole-cell biotransformations show some advantages as compared to an enzymatic approach. For the use of yeast as well as for the use of isolated enzymes, the problems of biocompatibility, stability, availability of the biocatalyst and its price, and special design criteria have to be considered very early, since these properties are crucial for choosing the best reaction setup. An ideal substrate, however, will be found very rarely. Some of the properties for an ideal substrate are depicted in Figure 21.2.

A situation represented by an ideal interaction between the yeast and an (ideal) substrate is almost never found in daily life, and several problems arise every time when performing "nonideal" reactions; some conceivable solutions to these basic problems are shown in Figure 21.3.

It is undisputed and evident that the use of isolated enzymes has a number of advantages. They are often more specific, and their handling is easier than dealing with living cells needing attendance, nutrition [43], care, and larger-sized equipment. Work-up of an enzyme-catalyzed reaction is easier than that of a whole-cell-mediated transformation, but enzymes are often more expensive, and the addition or recycling of a cofactor might be necessary.

From our experience, baker's yeast (*S. cerevisiae*)-mediated reactions are easily performed on lab scale, and there is plenty of knowledge for their scaling up for industrial use already available. The use of certain other microorganisms might be a problem. Initial experiments with baker's yeast can be performed very easily; the yeast can be obtained from the next brewery (we have spoken in the past to several master brewers very friendly, and help has never been denied) as a slurry—hence avoiding sterile growing of the cells from a small culture. Major drawbacks are the rather messy and time-consuming work-up procedures (the product has to be separated from the biomass) and the isolation and purification of the desired product. In specific and isolated cases, reaction times may be long; this problem might be solved by using genetically modified yeasts or recombinant yeast enzymes [44–52]. For special systems even the use of a continuous solid–gas bioreactor has been suggested [53].

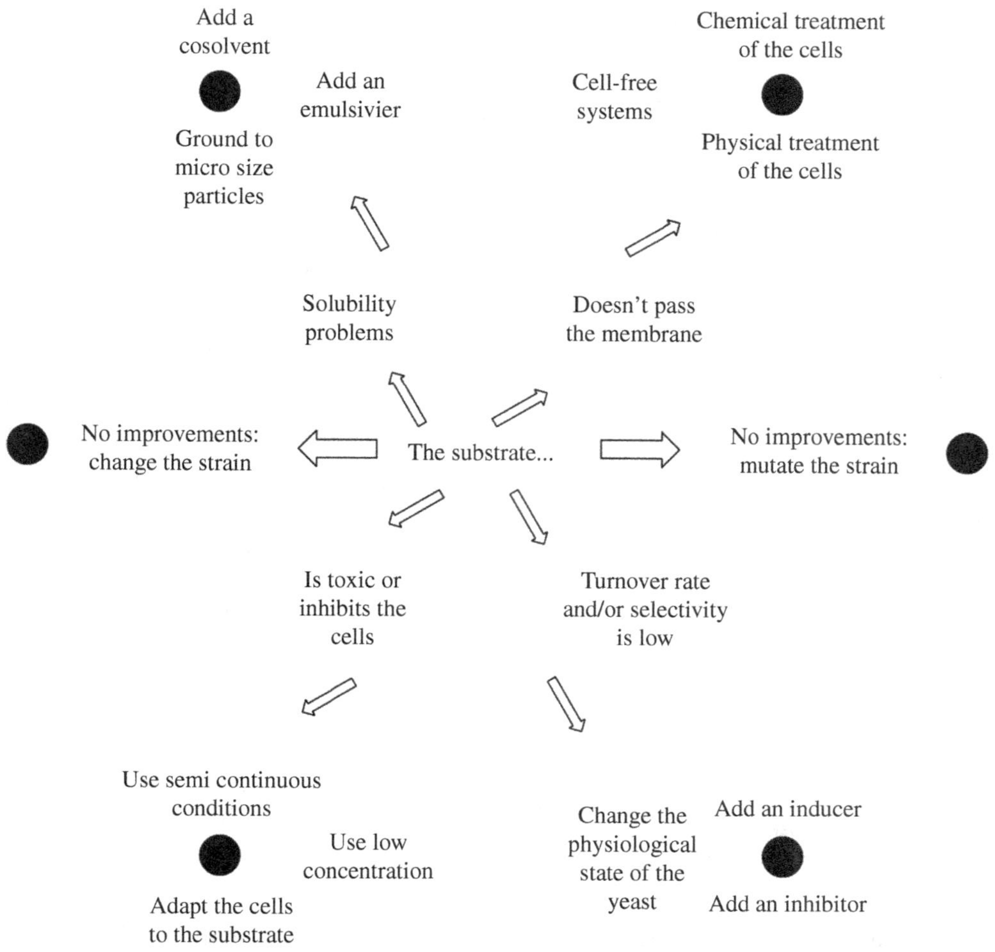

FIGURE 21.3
Problems to be dealt with and how to solve them.

21.2 REDUCTIONS OF ALDEHYDES AND KETONES

Unsaturated compounds can be reduced by baker's yeast, and a new stereogenic center is created [54]. These microbial desymmetrizations of a prochiral substrate have proven to be a powerful methodology [55, 56]. The reduction of a carbonyl group by baker's yeast is a very old reaction; it has originally been described in 1898 for the synthesis of furfuryl alcohol from furfural (cf. Figure 21.1) [4, 57].

Since then, many different ketones **3** were reduced by baker's yeast, and the secondary alcohols **4** were obtained (Figure 21.4) [14, 22, 58–86]. Reduction of an aldehyde furnished the corresponding primary alcohol even with some complex structures [87–91].

The high stereoselectivity of these reductions has often been explained by applying Prelog's rule [92] involving a hydrogen transfer [93] onto the *re* face of a prochiral ketone, but this calls for the utmost caution since more than one enzyme might be involved in an intact organism. As shown by a comparative study, even the reduction of well-known compounds might be tuned to afford the formation of products of an anti-Prelog enantiopreference [94].

The reduction of aldehydes and ketones has been discussed in several excellent reviews [17, 95–97], including the electromicrobial reduction [98] of carbonyl compounds. In general, larger amounts of yeast are required for the reduction of ketones [35, 99] than for the reduction of aldehydes [96, 100]. The reduction of *para*-substituted acetophenones **5** (Figure 21.5) gave (*S*)-1-arylethanols **6** showing an enantiomeric excess (ee) between 82 and 96% [54, 101–107]. Scaling up of these reductions is possible without loss of yields or selectivity [65].

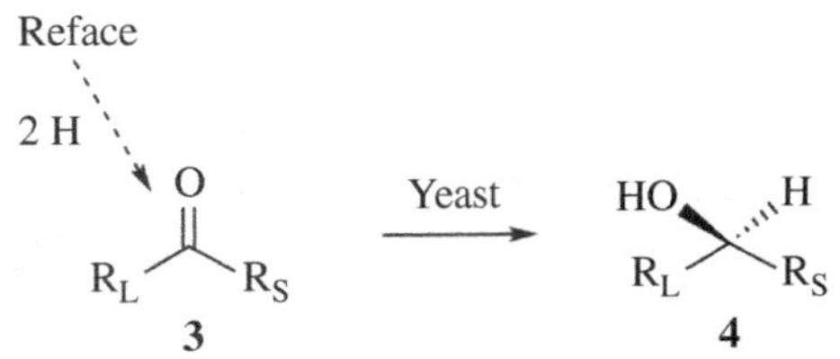

FIGURE 21.4
Reduction of ketones (R_L, large group; R_S, small group) to yield secondary alcohols following *Prelog's* rule.

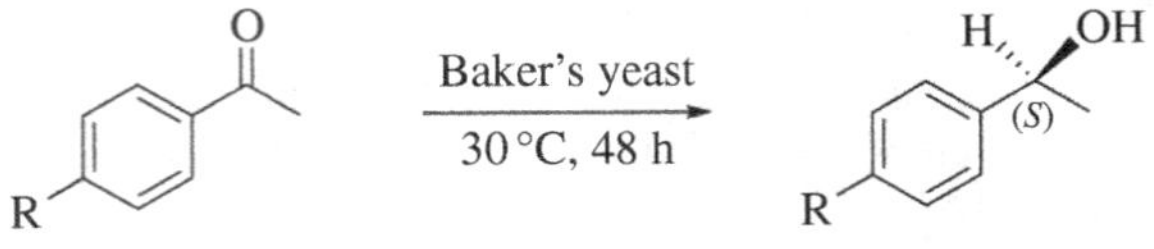

FIGURE 21.5
Reduction of substituted acetophenones by baker's yeast.

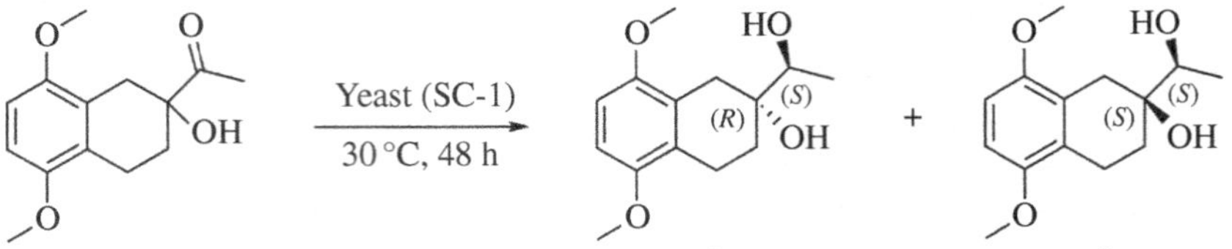

FIGURE 21.6
Yeast-mediated key step in the total synthesis of anthracyclinones.

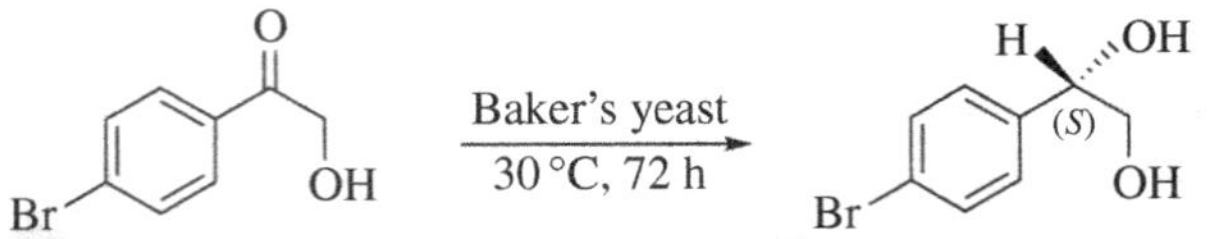

FIGURE 21.7
Synthesis of (*S*)-1-(4-bromophenyl)-ethane-1,4-diol (**11**).

The rate of these reductions was decreased by electron-donating substituents, but there was no influence of the substituents onto the steric course of the reaction [102, 103]. *Erythro*-configurated products were obtained when additional substituents being capable of coordination (either by lone-pair interaction or by hydrogen bonding) were present in the molecule. The reduction of dichlorofluoroacetophenone, however, proceeded with low selectivity and low yields [108]. Very interesting intermediates **8** and **9** for the total synthesis of anthracyclinones [109] have been obtained by reducing racemic **7** [110] using *S. cerevisiae* Sigma-YSC-1 (Figure 21.6).

Reduction of **10** for 3 days under anaerobic conditions gave (*S*)-**11** (Figure 21.7) showing an ee of excellent 99.4% [111].

As an alternative to baker's yeast, the use of plant tissues (e.g., from apple, carrot, cucumber, onion, or potato) has been suggested [112, 113]. It remains unclear, however, whether endophytic microorganisms are involved in these reductions [114]. Heterocyclic ketones [111, 115–117] as well as many α-hydroxy ketones [96, 118–121] have been the subject of yeast-mediated reductions. For the latter compounds very high ee values have been obtained, albeit the yields were moderate. Monobenylated (*S*)-1,*n*-diols have been prepared using highly diluted yeast suspensions, and ee values up to 95% could be achieved [122]. Yeast-mediated reductions of acyclic ketones were key steps in the total synthesis of the pheromone sulcatol [123], brefeldin A [124], and *endo*-brevicomin [125].

A nice application in pharmaceutical synthesis has been reported. Thus, from the reduction of pentoxifylline a new methylxanthine could be obtained, and a high ee of 98% was observed for this transformation [126, 127]. Since the baker's yeast reduction of 3-oxo-3-phenylpropanenitrile has been difficult to achieve, a library of baker's yeast reductases was screened. As a result, this approach allowed the synthesis of both antipodes of antidepressants fluoxetine, atomoxetine, and nisoxetine [128].

Albeit there are many reports on the reduction of steroidal cycloalkanones [15, 129–132] and their precursors [133] by yeast (or rather often by bacterial contaminants of the yeast), less examples are known for the reduction of simple cycloalkanones [94, 134–136]. Addition of extra yeast nutrient and riboflavin to the reaction mixture allowed the smooth reduction of racemic **12** (Figure 21.8). Commercially available baker's yeast gave low enantioselectivity in these reactions; this could be improved by selecting special strains. The ratio between products **13**:**14** changed upon prolonged incubation of **12** but was kept constant by controlling the concentration of glucose [17]. Reduction of racemic **15** gave 25% of **18** (ee = 92%), as a side product **19** was obtained (ee = 40%). Finally, from dichlorinated racemic **16** compounds **20** (ee = 88%) and **21** (ee > 99%) were obtained, while the reduction of monochloro racemic **17** gave a mixture of **22** and **23**. Thus, the reduction of these bicyclo[4.2.0]octenones is completely diastereoselective from the ketone's *exo* face [137, 138].

Similarly, reductions of different racemic 2-oxabicyclo[3.2.0]heptan-6-ones [139], 2-oxabicyclo[2.2.1]heptane-7-carboxylates [140], norbornenone [102], and bicyclo[2.2.2]octan-2-one [141] have been performed; the ee values, however, were moderate. For the reduction [141] of a 4-twistanone (tricyclo[4.4.0.0^{3,8}]decan-4-one), the alternative use of *Rhodotorula rubra* has been suggested. Kinetic resolution of racemic 5,6-epoxy-bicyclo[2.2.1]heptane-2-one using whole cells of genetically engineered *S. cerevisiae* allowed the synthesis of (+)-5,6-epoxy-bicyclo[2.2.1] heptane-2-ol [142].

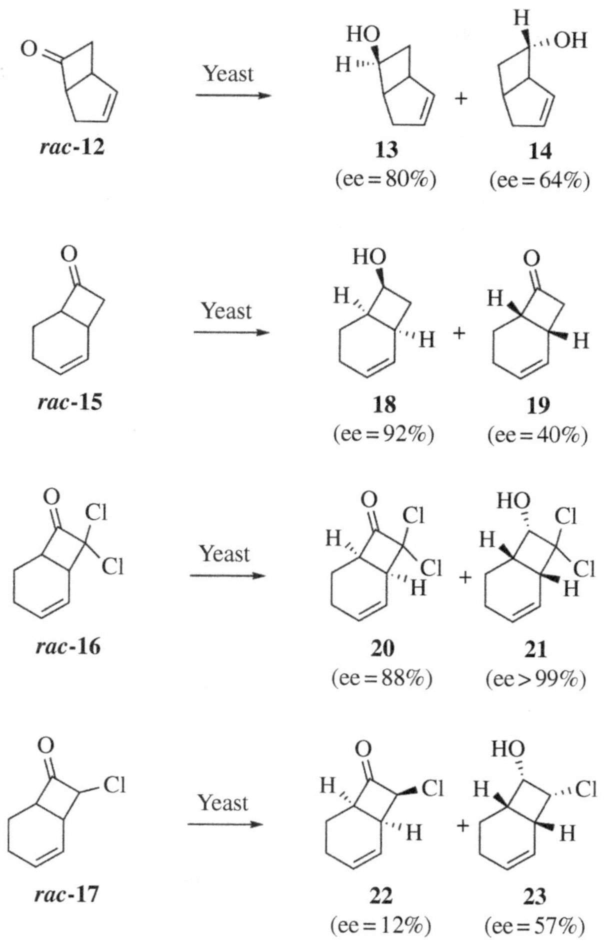

FIGURE 21.8

Stereoselective reductions of bicyclic ketones by yeast.

21.3 REDUCTION OF THIOCARBONYLS OR SULFUR-CONTAINING COMPOUNDS

Microbial reduction of thioaldehydes resulted in the formation of thiols; thus thioacetaldehyde gave ethanethiol [143, 144], and from thiobutyraldehydes the corresponding thiols were obtained [145]. While thioethers were cleaved [146], thioacetals were stable. Thus, compound **24** (Figure 21.9) was reduced with an ee of 99% to (*S*)-1-(1,3-dithian-2-yl)-2-propanol **25**, a central intermediate for the synthesis of (*S*,*S*)-grahamimycin A1 [147].

Similarly, 1,1-bis(*p*-tolylthio)ketones [148, 149] as well as β-keto dithioesters have been reduced by baker's yeast; the yields were low, but high ee values (>95%) could be achieved. The reduction of β-keto sulfides gave the corresponding alcohols with ee values ranging between 70 and 94% [150, 151]. In general, the reduction of these compounds with yeast seems difficult and proceeds best only at low concentrations of the substrate [150]. The reduction of carbonyl groups holding adjacent sulfur substituents seems critically dependent upon the substituents attached adjacent to the carbonyl group and to the sulfur-containing group. Thus, the ease of reduction increases from β-keto sulfides to β-keto sulfoxides to β-keto sulfones [150, 152–154]. In addition, a number of 2-keto sulfones **26** have been treated with yeast, and the (*S*)-2-hydroxy sulfones **27** (Figure 21.10) have been obtained in 10–90% yield and ee values ranging between 10 and 100% [17, 150, 155–158]. Reduction of a cyclohexanone **28** by baker's yeast followed by acid-catalyzed cyclization furnished the sulfide lactone **29** with excellent enantiopurity [159].

21.4 REDUCTION OF FUNCTIONALIZED CARBONYL AND DICARBONYL COMPOUNDS

When α-heterocyclic substituted ketones have been subjected to yeast-mediated reductions, yields and enantiospecificity were low for 2-acylthiazoles [149], ketoisoxazoles [160], and 5-acetyl-2-isoxazolines [161].

The reduction of aromatic nitro compounds has been performed as early as 1914 by A. Neuberg [162, 163] and is well documented [17, 96, 164], but there seems to be no example for the reduction of an aliphatic nitro group; this functional group was stable during the reduction of carbonyl groups, with an exception of those nitro

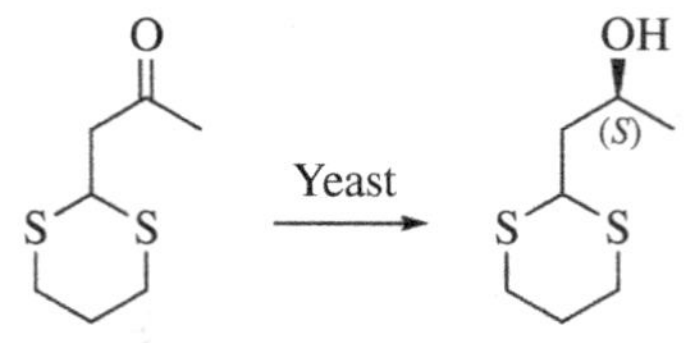

24 25

FIGURE 21.9
Reduction of a thioacetal-protected dicarbonyl compound.

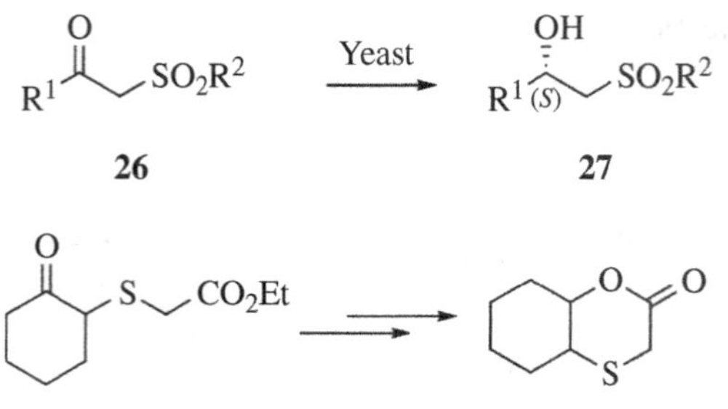

26 27

28 29

FIGURE 21.10
Synthesis of (*S*)-2-hydroxy sulfones from 2-keto sulfones and sulfide lactone.

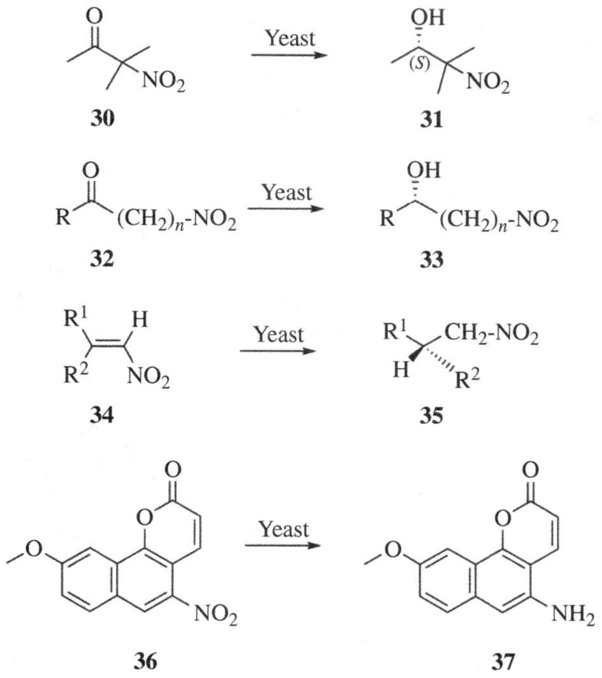

FIGURE 21.11 Synthesis of nitro-substituted alkanes, alkenes, and aromatic compounds.

alcohols where a retro-nitroaldol reaction took place [165, 166]. Usually the reduction of aromatic nitro compounds resulted in the formation of amines; the reduction of aromatic nitro compounds bearing electron-withdrawing groups, however, gave hydroxylamines [167, 168]. A hydroxylamine has also been obtained by the reduction of an aldoxime [169]. Reduction of substituted nitroalkenes gave the corresponding nitroalkanes [170], and the reduction of nitrostyrenes in an organic solvent by dried baker's yeast furnished the corresponding nitroalkanes [171].

Compound **30** (Figure 21.11) gave a 57% yield of (*S*)-**31** [165], and from ketones **32** alcohols **33** were obtained [172]. Nitroalkenes **34** gave the nitroalkanes **35** with ee values ranging between 66 and 98% and yields between 20 and 60% [173, 174]. 9-Methoxytariacuripyrone **36** was reduced by baker's yeast [175] to yield 5-amino-9-methoxy-3,4-dihydro-2*H*-benzo[h]chromen-2-one **37** being highly active as an antituberculosis agent.

The reduction of (*Z*)-2-chloromethyl-3-arylacrylaldehydes with baker's yeast gave (*R*)-2-methyl-3-arylpropanols showing high ee; the mechanism of this reaction was explored by labeling experiments [176]. The enantioselectivity of this reaction could be switched by substrate engineering. Similarly, enantiomerically enriched α-halo-β-arylpropionic acids were obtained by an enantioselective reduction of the C=C bond of some (*Z*)-methyl α-keto-β-arylacrylates [177]. Biocatalytic reduction of 2-fluoro-2-alken-1-ols by baker's yeast gave (*S*)-2-fluorinated alkanols with high enantioselectivities. The progress of these reductions was shown to depend on the configuration of the double bond and the substituent in β-position [178].

The reduction of cyclohexane-1,2-dione **38** (Figure 21.12) gave racemic *trans*-cyclohexane-1,2-diol **39** [179, 180]. From the reduction of camphorquinone **40**, a 63% yield [96] of *exo*-2-hydroxy camphor **41** and 3-hydroxy camphor **42** was obtained. In addition, the reduction of substituted acenaphthenequinones afforded the corresponding 2-hydroxyacenaphthenones in 24–48% yields with 10–93% ee [181]. The Wieland–Miescher ketone has also been reduced with high enantioselectivity [182].

A series of 2,2-disubstituted cyclopentane-1,3-diketones **43** (Figure 21.13) has been reduced, and by monoreduction the products **44** and **45** were obtained in 9–75% yield with high diastereoselectivity [183–188].

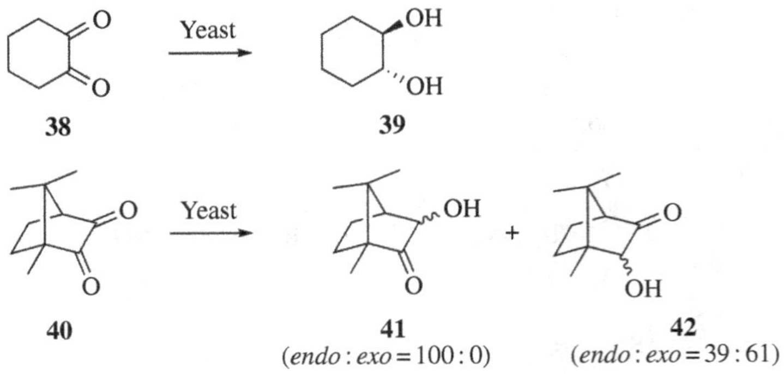

38 **39** **40** **41** **42**

FIGURE 21.12
Reduction of cyclic 1,2-dicarbonyl compounds.

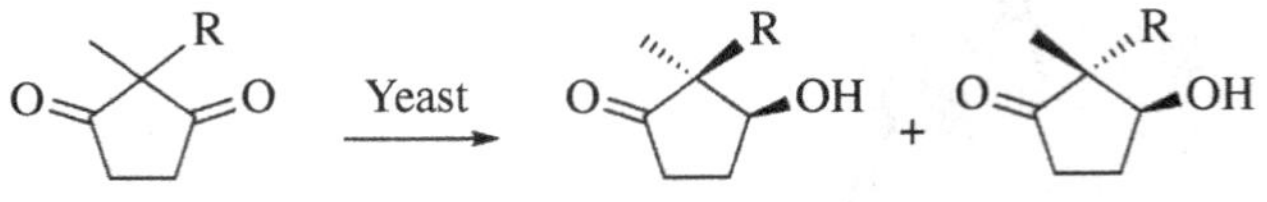

43 **44** **45**

FIGURE 21.13
Reduction of cyclic 1,3-dicarbonyl compounds.

46 **47**

FIGURE 21.14
Reduction of an acyclic 1,4-dicarbonyl compound.

Several of these products [189–191] have been prepared and used for the total synthesis of natural products [185–187]. The formation of by-products was observed quite often during these reductions. This might be due to the slow rates of the reaction. Hence, the addition of activators has been suggested [191]. These "activators" (most often α,β-unsaturated alcohols, ketones, or nitriles) seem to act as suicide substrates for several oxidoreductases [192, 193], whereas, for the reduction of cyclopentanoid 1,3-diketones, stereoselectivity was high, and most often [194] low selectivity was observed for cyclohexanoid 1,3-diketones [184, 194–197].

High recovery yield was observed, however, for the reduction of cyclic 1,3-diketones being part of a medium-sized ring [198]. Low yields and fair to moderate ee and de values have been obtained for the reduction of cyclic 1,4-diketones or quinones [199–203].

Better results were obtained for the reduction of acyclic diketones; acyclic 1,2-diketones are excellent substrates for baker's yeast. The selectivity, however, is often rather low [121, 179, 204–207]. An effective way to improve stereoselectivity of the reduction process was the introduction of a sterically demanding sulfur-containing group [208, 209].

Acyclic 1,3- and 1,4-diketones gave only poor results; their reduction proceeded slowly but is incomplete, and often the de and ee values were only moderate [210–213]. An excellent result, however, was achieved for the reduction of 2,5-hexanedione **YY255** (Figure 21.14); upon treatment of **46** with baker's yeast (Budweiser) for 144 h a 55% yield of **47** was obtained [214]. The main product showed an (*S*,*S*)-configuration (2*S*,5*S*) : (2*R*,5*S*) : (2*R*,5*R*) = 49.8 : 1.04 : 1 (ee 96%, 2% *meso*); recrystallization gave the (2*S*,5*S*)-configurated material displaying an ee > 98% containing <1% of the *meso* compound.

Similarly, the reduction of bicyclo[2.2.2]octane-2,6-dione to (1*R*,4*S*,6*S*)-6-hydroxy-bicyclo[2.2.2]octane-2-one has been performed using an engineered recombinant strain of baker's yeast with excellent results (complete conversion, 97% de, >99% ee, isolated yield 84%) [215].

21.5 REDUCTION OF KETO ESTERS

The reduction of α-keto esters has also been performed using baker's yeast [216]. Thus, different 2-oxo-2-arylacetic acid derivatives gave optically pure α-hydroxy acid derivatives [217, 218]. Ethyl pyruvate gave (*R*)-ethyl lactate efficiently [217]. Of particular interest seems the synthesis of enantiomerically pure (*R*)-pantoyllactone **49** (Figure 21.15) via enantiospecific reduction of ketopantolactone **48**.

3-Carboxyalkyl-γ-butyrolactone has also been synthesized by combining pure enzymes and whole microbial cells in the same process. Thus, the corresponding 2-hydroxy esters and lactones were obtained from 2-oxoglutaric acid [219]. Low ee values, however, were obtained for the reduction of α,γ-diketo esters and keto α,γ-diesters [220–224].

It seems that the enantioselectivity of the reduction of keto esters with the keto group being part of a five-membered ring proceeds better than that of open-chain keto esters substituted at carbon C-2 [225, 226]. The reduction of **50** (Figure 21.16) gave 80% of (1*R*,2*S*)-**51** showing both an ee and a de of 100%. In another experiment, however, **51** was obtained with a de of only 60% [217, 227]. Several mold strains have been shown to perform the same reduction with high enantio- and diastereoselection, whereas from the baker's yeast-mediated reactions quite often mixtures of the stereoisomeric products were obtained [227]. More complex substrates have also been studied [228–232], but yields and/or stereoselectivity were often only moderate.

Better results were obtained for six-membered rings, and the reduction of these six-membered cyclic keto esters is more widespread. Thus, from the reduction of racemic **52** using baker's yeast a 65–68% yield of ethyl (1*R*,2*S*)-2-hydroxy-cyclohexanone carboxylate **53** was obtained; depending on conditions and strains ee values between 86 and 99% and de values between 76 and 99% were obtained [217, 227, 233]. Sterically demanding substituents lowered the yields and selectivity of these reactions [234–238]. Better results were achieved for the reduction of benzo-anellated oxo-esters [233] using "starving conditions" and for the reduction of cyclic keto esters possessing a heteroatom [239] instead of a carbon in the five- or six-membered ring [205, 225, 226, 233]. Two representative examples have been depicted in Figure 21.17. By this reduction **55** was obtained from **54** with an ee of 85%, and **57** (from **56**) showed an ee > 99% [225, 233, 240].

Several α-hydroxy amides have been prepared from α-oxo-esters using a double sequence reaction: In a first step a highly enantioselective bioreduction with

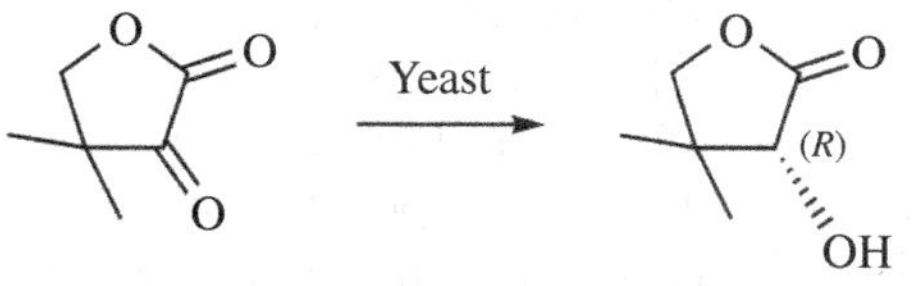

48 **49**

FIGURE 21.15
Synthesis of (*R*)-pantoyllactone by yeast-mediated reduction.

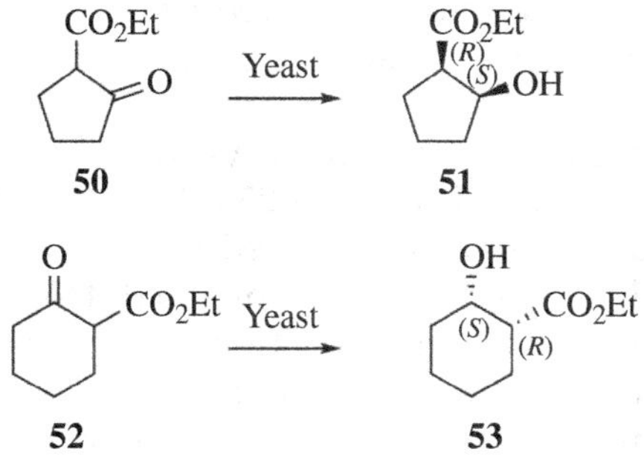

50 **51**

52 **53**

FIGURE 21.16
Reduction of cyclic keto esters.

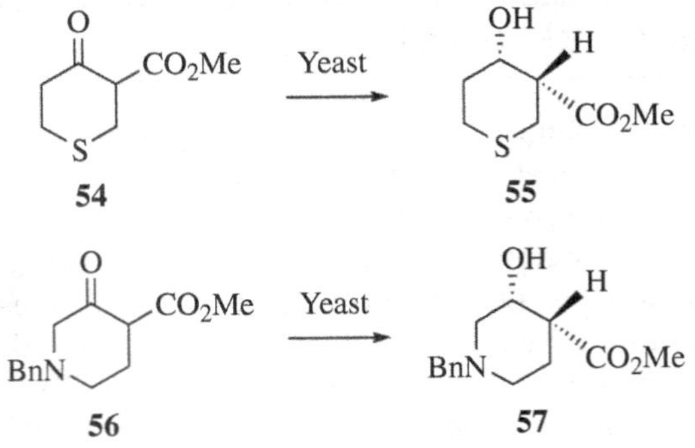

FIGURE 21.17
Reduction of heterocyclic keto esters.

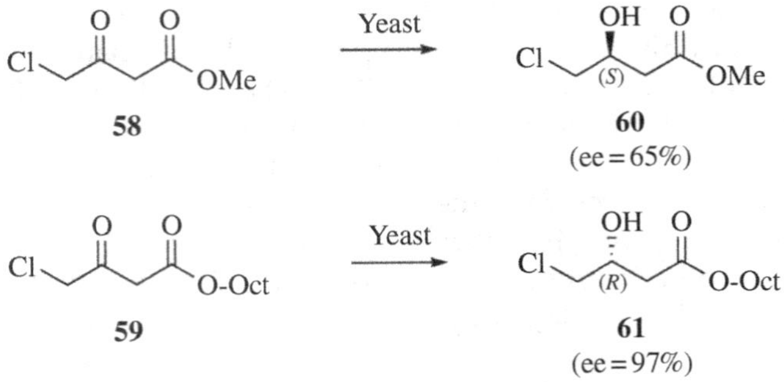

FIGURE 21.18
Reduction of γ-chloroacetoacetates.

baker's yeast gave the corresponding α-hydroxy esters. The esters were transformed by a nonenantiospecific lipase-catalyzed aminolysis to yield butylamides [241].

Numerous examples have been reported for the reduction of acyclic β-keto esters, and their results were explained by applying Prelog's rule [95, 97, 242–246]. When exceptions to this rule were observed, it was assumed that either a different oxidoreductase other than baker's yeast genuine alcohol dehydrogenase gave this reaction. Products of low enantioselectivity were explained as a result of more than one enzyme being involved in the biohydrogenation [122, 242, 247, 248]. In general, the optical purity of the product as well as its absolute configuration depends strongly on the size and nature of the substituents adjacent to the carbonyl group and of the ester moiety. The concentration of the glucose (as a nutrient), the substrate, the pH value, the presence of additives [67, 154, 249], and the cultivation of the yeast also influence these reactions [247, 250–254].

The reduction of different γ-chloroacetoacetates **58** and **59** (Figure 21.18) has been studied quite often and in more detail, and its stereochemical course (to yield products **60** and **61**, respectively) can be altered by changing the size of the ester moiety [17, 248, 255, 256]. The addition of a nonpolar resin to the reaction mixture also increased yields and selectivity [257]. Recently, enantioselective keto ester reductions in water have been accomplished either by baker's yeast or by ruthenium-catalyzed reactions; interestingly enough, the highest ee values have been obtained using *S. cerevisiae* [258]. In addition, a metabolic *in vivo* study has been performed. These data revealed that under aerobic conditions the reduction occurs preferentially in the mitochondrial matrix, while under anaerobic conditions the bioreduction occurs in the cytosol [259].

The reduction of ethyl acetoacetate **62** to yield ethyl (*S*)-hydroxybutanoate **63** (Figure 21.19) has also been studied [119, 193, 250, 260–266]. The yields for this reduction were between 30 and 80%, and the ee ranged between 77 and >98%. The different results of this reduction have been explained by the different conditions: the carbon source was different, but "starving conditions" were applied, and sometimes additives were added for improving the reaction, pretreatment of the cells with ethanol was performed, the reaction was quenched before reaching complete reduction

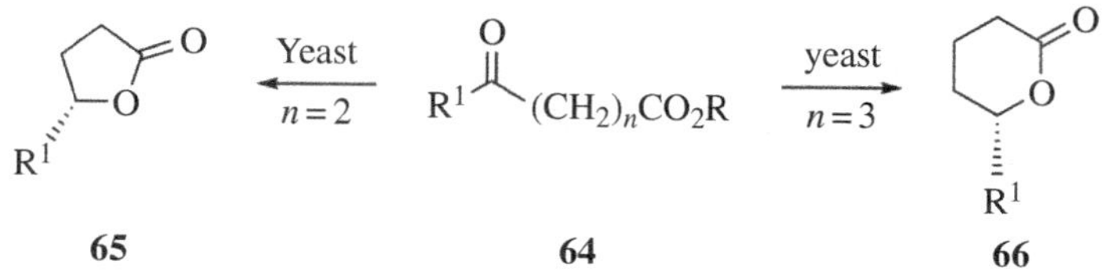

FIGURE 21.19

Reduction of ethyl acetoacetate.

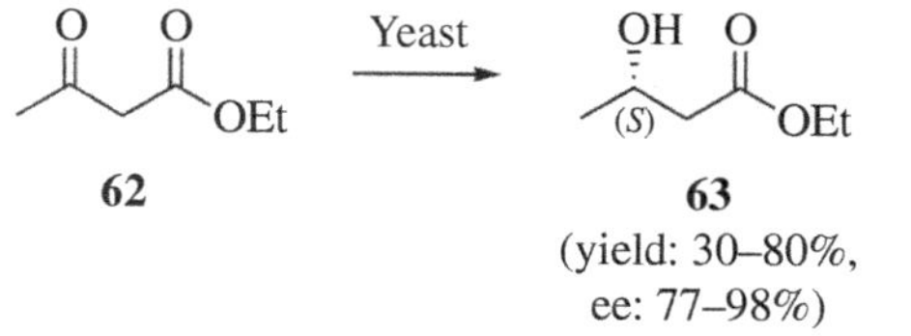

FIGURE 21.20

Reduction of γ- and δ-keto esters.

[233, 264, 267, 268], and many other microorganisms have been used instead of baker's yeast [107, 117].

The best results, however, were obtained for the reduction of **62** to (*R*)-3-hydroxy-acetoacetate using *Thermoanaerobium brockii* [269]. While acetoacetates differing in their ester moiety were reduced by baker's yeast predominantly to (*S*)-3-hydroxybutanoates, all other 3-keto alkanoates with $R>CH_3$ gave products of (*R*)- or (*S*)-configuration [217, 253, 262, 263, 265, 270–275].

Many β-keto esters possessing one or more heteroatoms containing moieties attached to C-4 have been reduced by baker's yeast, and the yeast-to-substrate ratio was found to be the most important parameter for obtaining good results [193, 250, 251, 256, 265, 276–282]. Yields and stereoselectivity were dropped for β-keto esters carrying an additional stereogenic center [283, 284], and the reduction of α-thio-substituted β-keto esters has been suggested as an alternative to the reduction of unsubstituted β-keto esters to improve yields and selectivity. The yields, however, remained low [285] as for many other β-keto esters possessing a substituent in the α-position [155, 264, 286–292]. In addition, several racemic β-keto amides have been reduced [293], and the substrate acceptance and the enantioselectivity of the yeast reductases have been assessed [294]. Similarly, the bioreduction of α-keto β-lactams allowed the synthesis of precursors of novel paclitaxel analogues [295–297].

From the reduction of γ- and δ-keto esters the corresponding γ- or δ-lactones (Figure 21.20; transformation of starting material **64** to yield the corresponding lactones **65** and **66**, respectively) were obtained [298–305]. A two-step mechanism has been suggested for this transformation, and it was assumed that first the ester was hydrolyzed by a nonspecific esterase followed by a reductive step of either the acid or of a corresponding CoA-thioester intermediate. It seems that the acids are more rapidly reduced than the esters [298].

A different approach (immobilization of the substrate instead of the catalyst) has been chosen for the synthesis of ethyl (*S*)-3-hydroxybutanoate by the bioreduction of an polyethylene glycol (PEG) acetoacetate with dry baker's yeast in toluene with a small amount of water. After washings and derivatization, the product was obtained with an excellent ee of 97% and an overall yield of 70% [306].

21.6 HYDROLYSIS OF ESTERS

The hydrolysis of an ester moiety is regarded by many scientists as a most simple reaction, even sometimes annoying. This was also true for baker's yeast-mediated hydrolyses [307, 308]. The first report of an ester hydrolysis has been published in the steroid field, and it was reported as an undesired side reaction [309]. Many different enzymes of baker's yeast may be responsible for the hydrolysis of ester moieties: not

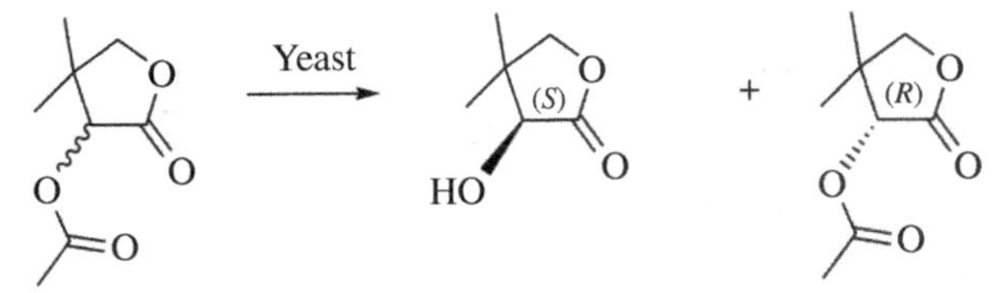

rac-67 68 69

FIGURE 21.21
Synthesis of (S)-pantoyllactone by yeast-mediated hydrolysis.

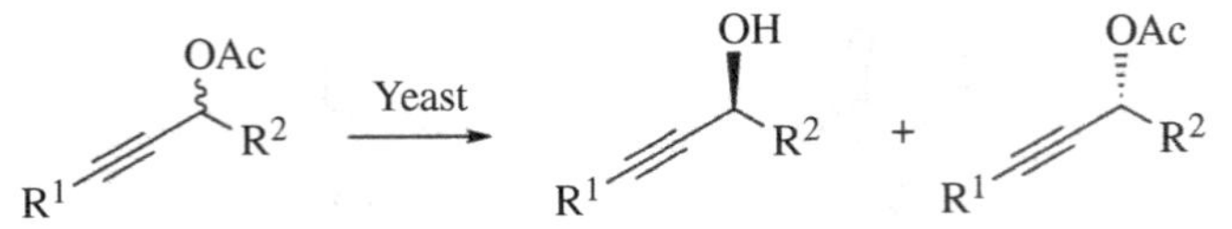

rac-71 70 71

FIGURE 21.22
Synthesis of optically active α-alkyne-3-ols by yeast-mediated hydrolysis.

only proteinases [310] and esterases [311–315] but also lipases [316–318], phospholipases [319–322], and tributyrinases [323]. Nowadays these hydrolyses are not any longer regarded as "undesired" or even "unwanted" but very useful, especially in the synthesis of many natural products—for example, prostaglandins [324–327], amino acids [68, 328, 329], other α-substituted carboxylic esters like acyloxyesters and lactones [328, 330, 331], or precursors for the synthesis of β-lactam antibiotics [332].

Thus, the selective hydrolysis of 2-O-acetyl-pantoyllactone (**67**; Figure 21.21) with fermenting baker's yeast gave under anaerobic conditions 28% of (*S*)-**68** (86% ee), while the use of the lipase from *Aspergillus* sp. allowed the isolation of (*R*)-**69** showing an ee of 96%; many other isolated enzymes failed in this reaction [331].

Also, optically active α-alkyne-3-ols **70** (Figure 21.22) were obtained from the hydrolysis of the corresponding racemic acetates **71**. This reaction could be extended to other substrates, but an unsubstituted CH_2 unit adjacent to the stereogenic center is mandatory for achieving excellent enantioselection [333].

21.7 IMMOBILIZED BAKER'S YEAST

As an alternative to the use of actively fermenting cells, the use of immobilized baker's yeast has been suggested. Immobilized baker's yeast, however, exhibits a significant lower activity as compared to the same amount of living cells, but the isolation of the products is facilitated, and the operational stability of the biocatalyst is higher. Usually, the product formation rates are high but somewhat lower than in suspension [334]. Although the process of immobilization imposes stress on the metabolic behavior [335] and the metabolic properties [336, 337] are changed, the tolerance to additives and solvents [338] was better, but diffusional limitations, for example, in oxygen transfer rates, have been observed.

The use of immobilized yeast allows the use of solvents different from water [81, 339]—for example, hexane [340, 341], while the use of a water-miscible solvent (THF, DMSO, or 1,4-dioxane) or of a nonpolar solvent saturated with water sometimes deactivated the cells [321, 322, 342, 343]; as an exception to this rule, fluorenones have been bioreduced in DMSO/water mixtures to fluorenols with good to excellent enantioselectivity [344].

Calcium alginate [345–347] as well as κ-carrageenan [348–350] has been used for immobilizing baker's yeast, and a sodium alginate/calcium chloride-immobilized baker's yeast has been used during the synthesis of D- and L-armentomycin [351]. This system has also been used for the reduction of several β-keto esters [352], and the antiselectivity as well as the yields was improved compared to reductions using classical conditions [217, 353–355]. Calcium alginate-immobilized cells of *S. cerevisiae* were pretreated at 50 °C for 30 min, and then the cells showed an increased performance (80%) and an excellent selectivity (ee 99%) in the reduction of 3-chloropropiophenone to yield (*S*)-3-chloro-1-phenyl-propanol [356]. Similarly, propranolol and analogs have been accessed [357, 358]. Baker's yeast has also been immobilized in meso/nano-porous silicates; this system has been used for the regio- and chemoselective reduction of aromatic nitro and carbonyl compounds [359].

21.8 WHOLE-CELL BIOCATALYSIS IN IONIC LIQUIDS AND DEEP EUTECTIC SOLVENTS

Despite many efforts the use of classical organic solvents for biocatalytic transformations using whole cells remains limited. Water as a natural reaction medium limits the number of applications, since only a small number of substrates are sufficiently water soluble. As an alternative the use of an ionic liquid may be considered. A very early example for such a reaction has been reported by Cull in 2000 [360].

The ionic liquid, however, has to be biocompatible; it must not be toxic to the yeast, and its biocompatibility strongly depends on the nature of the composing cation and anion. For cations possessing alkyl substituents, the length of these alkyl chains seems crucial: the longer the alkyl chains, the lower is the tolerance of the microorganism to the ionic liquid [361–364]. This might be due to an insertion of the cation into the yeast's membrane followed by its disintegration or by an accumulation of toxic molecules inside the cells [360, 363].

The use of certain anions, for example $[BF_4]$, decreases the biocompatibility (obviously by forming hydrofluoric acid due to the presence of water) [365, 366]. Effects of residual water content [367–370], pH value [371], miscibility with water [372], as well as the kind of microorganism must not be neglected.

Ionic liquids are often hygroscopic [371], and considerable amounts of water might "contaminate" the ionic liquid affecting its properties, especially the viscosity of the system [373–375]. Impurities in the ionic liquid (from its synthesis) might be a reason for its toxicity to the cells [376]. Biocompatibility tests using 1-alkyl-3-methylpyridinium- or 1-alkyl-3-methylimidazolium-based ionic liquids and baker's yeast have been performed [361], and their toxic effect for the cells was attributed to the amphiphilic nature of the cation [362, 364, 367, 371, 377–381]. *Saccharomyces cerevisiae* tolerates $[PF_6]$ as a counterion better than $[BF_4]$, which was more toxic to the cells [382]. Ionic liquids have also been used with immobilized baker's yeast [364], and the toxicity of 1-alkyl-3-methylimidazolium-based ionic liquids increased with the exposure time.

Despite these limitations, yeast-mediated reactions have been performed in ionic liquids quite successfully. Several (a)-cyclic ketones, 1,3-diketones, and (a)-cyclic β-keto esters were reduced to yield the corresponding alcohols by alginate-immobilized baker's yeast in ionic liquid/water mixtures [360]. The yields were moderate, and the enantioselectivity was similar to that from "classical" reactions or lower [383, 384]. However, the process of optimizing the conditions for these reactions might be lengthy but rewarding [385]. The alternative use of *Geotrichum candidum* has been suggested, and for several substrates excellent enantioselectivity was reported [386].

The reduction of acetyltrimethylsilane [387, 388] by immobilized baker's yeast using two different ionic liquid buffer systems was studied in more detail, and the effects of temperature and pH were investigated. Finally high yields (>99%) and superior enantioselectivity (ee > 99.9%) could be obtained [382].

As an alternative to ionic liquids, "deep eutectic solvents" (DES) have been investigated. DES are mixtures of metal or ammonium salts with hydrogen-bond donors; they show low melting points and are cheap, more biocompatible, and regarded as safe [389, 390]. Also, liquid polymers have been used in combination with supercritical carbon dioxide [391].

21.9 C—C BOND-FORMING AND BREAKING REACTIONS

Acyloin-type condensation [392] reactions [393, 394] have been published quite long ago by Dirscherl [395] and Neuberg [396, 397]. In the mid-1970s these reactions were investigated in more detail by Fuganti and coworkers [398]. Thus, substituted benzaldehydes **72** (Figure 21.23) finally afforded upon treatment with baker's yeast products **73** in circa 20–30% yield but with high enantioselectivity (ee > 97%) and a good *anti*/*syn* ratio with *anti*-1-aryl-propane-1,2-diols being the main product [399].

By this technique D-(−)-ephedrine was prepared in an industrial process combining chemical and microbiological synthesis in a very elegant way [400]. Many other aldehydes were subjected to this reaction [17]; the reaction failed due to the toxicity of several aldehydes, especially salicylaldehyde, and in the aliphatic series, heptanal showed the highest cell toxicity [401].

This reaction has a broad tolerance for the added aldehyde, but only (internal) acetaldehyde is accepted as a substrate; other aldehydes were not incorporated at all. Deuterium labeling allowed insights into the course of the reaction [402, 403]. This acyloin-type condensation reaction has been used quite successfully in the total synthesis of vitamin E [404, 405], rare sugars [406], and several chiral synthons [251, 398, 402, 404–424].

Aldol reactions have been catalyzed by different strains of yeast; depending on the reaction conditions, products of either (*R*)- or (*S*)-configuration were obtained [425, 426]. These reactions can also be performed in an organic medium as shown for the reaction between 4-nitrobenzaldehyde and acetone [426].

Treatment of (*Z*)-3-methyl-2,4-pentadienal **74** (Figure 21.24) with baker's yeast for 10 days gave 65% of enantiomerically pure (*S*)-3-methyl-4-penten-1-ol **75** [427]; several more examples for biohydrogenations have been reported [428–433], but the yields were moderate to low. These low yields are typical for the baker's yeast-mediated reductions of prochiral α,β-unsaturated aldehydes [434]. An improvement could be achieved by applying an *in situ* substrate-feeding removal technique [435, 436].

The reduction of α-methylene ketones gave α-methyl ketones, albeit the enantioselectivity was moderate [437]. C_{10} synthons (as present not only in insect pheromones and sesquiterpenoids but also in tocopherol) have been accessed from geranial and analogues by baker's yeast-mediated reductions [265, 431, 432, 438]. By this approach (*R*)-citronellol could be obtained in enantiomerically pure form [432, 439]. The reaction was optimized, and a two liquid-phase system has been developed [440].

Baker's yeast-mediated decarboxylations [441, 442] of substituted cinnamic acids **76** (Figure 21.25) gave products **77** paralleling earlier reports [443] for the same reaction using *Bacillus pumilus*.

An (*E*)-configuration of the alkenic bond seems mandatory; the corresponding (*Z*)-stereoisomers were nonsubstrates in these reactions [444]. (*E*)-Cinnamaldehyde was decarbonylated and styrene was obtained as a product [445]. Decarboxylative

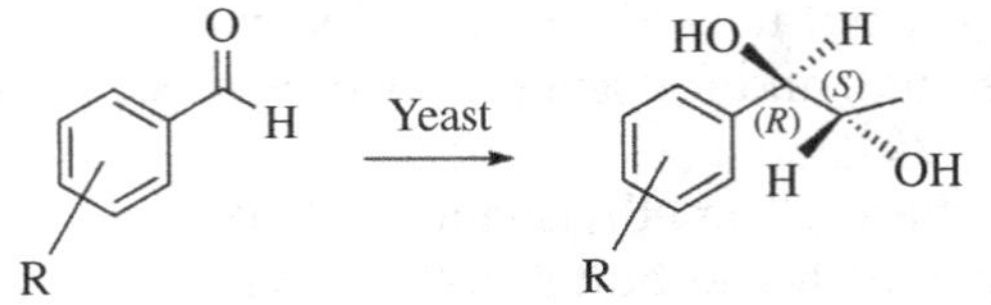

72 73

FIGURE 21.23
Acyloin-type condensation mediated by baker's yeast.

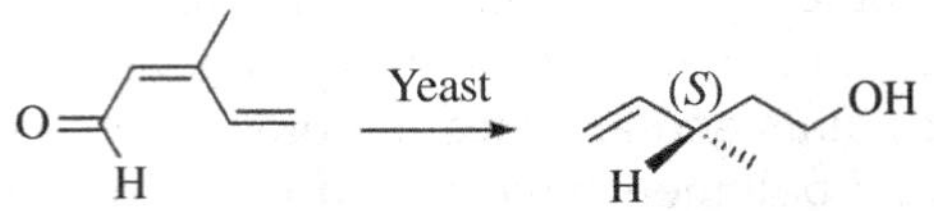

74 75

FIGURE 21.24
Reduction of (*Z*)-3-methyl-2,4-pentadienal by baker's yeast.

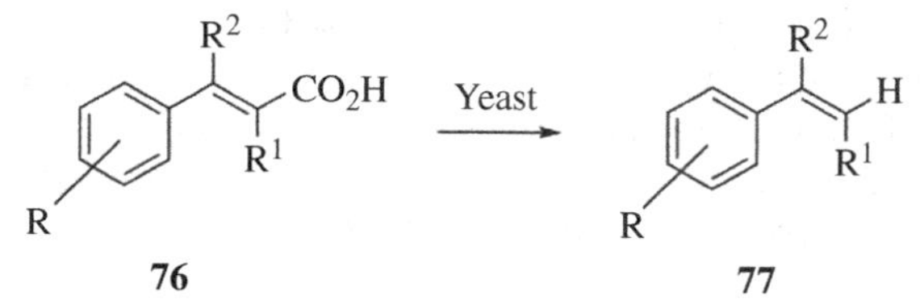

FIGURE 21.25
Baker's yeast-mediated decarboxylation of substituted cinnamic acids.

incorporation of linear C3–5 α-oxo-carboxylic acids led to the formation of (*R*)-α-hydroxy ketones [446].

Baker's yeast-mediated C—C bond-forming reactions could be improved by sonification. Thus, use of ultrasonically activated baker's yeast resulted in the formation of lanosterol from (*S*)-2,3-oxidosqualene in 83% yield [447, 448]. Rearrangement reactions have been observed as side reactions for analogues [449, 450]. A yeast-mediated oxidative intramolecular cyclization leading to a xanthone skeleton has been used for the synthesis of oxy-guttiferone A [451]. Pyrans have been prepared in an environmentally friendly one-pot synthesis in water [452]. Polyfunctionalized 4*H*-pyrans were obtained in a baker's yeast-catalyzed one-pot three-component cyclocondensation of aryl aldehydes, malononitrile, and β-dicarbonyl compounds in DMF as a solvent [453]. Similarly, from the cyclocondensation of aryl aldehydes, amines, and thioglycolic acid 2,3-diaryl-4-thiazolidinones [454] and from the cyclocondensation of 2-aminothiophenol and aldehydes, 2-aryl-/hetarylbenzothiazoles were obtained [455]. A baker's yeast-catalyzed three-component Biginelli reaction was used to synthesize 3,4-dihydropyrimidin-2(1*H*)-ones in good yields [456], and from a Hantzsch four-component reaction, polyhydroquinolines and 1,4-dihydropyridines were obtained [457, 458].

21.10 MISCELLANEOUS REACTIONS

Baker's yeast seems to be compatible with several metal-containing or organometallic species, as shown for the reduction of ferrocenyl derivatives [459], arylketone-$Cr(CO)_3$ complexes [460], indanone-$Cr(CO)_3$ complexes [460], and a planar chiral metallocene aldehyde [461]. This approach allowed the synthesis of all four enantiopure stereoisomers of 1-ferrocenyl-1,3-butanediol [462] and of 1,1-disubstituted ferrocenyl amino alcohols [463]. The reduction of porphyrins and hemoglobins by baker's yeast is long known; there are only a few reports for the reduction of inorganic materials [17].

Even though fluorine-containing molecules are important building blocks for the synthesis of drugs, the number of baker's yeast-mediated transformations (mostly reductions) remains small [104, 278, 464–468] although trifluoromethyl ketones [469–471] are reduced faster than methyl ketones but slower than their bromomethyl analogues [104]. (*E*)-2-Alkyl-4,4,4-trichloro-2-butenals were regiospecifically converted to the corresponding saturated alcohols with high enantioselectivity, whereas the (*Z*)-isomers were reduced to afford allyl alcohols [472]. Polyfluorinated β-keto esters were efficiently reduced, and the products were usually obtained in good yields [276, 278]. The reduction of 2-fluorocinnamyl alcohols by baker's yeast furnished (*S*)-2-fluoro-3-aryl-propanols in high yields with up to 92% ee [473]. A comparative study on the influence of substituents onto the sterical course of these reactions has been performed [474].

There are only a few examples for the transformations of phosphorus-containing compounds with baker's yeast [475, 476]. Thus baker's yeast has been used for the synthesis of α-aminophosphonates in a one-pot reaction from aldehyde, diethyl phosphite, and an amine [477]. The reduction of (3-chloro-2-oxo-propyl)-phosphonic acid diethyl ester by baker's yeast gave (2*R*)-3-chloro-2-hydroxypropyl-phosphoric acid diethyl ester, which could be used for the synthesis of (*R*)-carnitine [478–480]. The reduction of diketo-n-butyl-phosphonates furnished optically active hydroxyketo-*n*-butyl-phosphonates (17–58% yield, ee 55–90%) [481]. Lower ee values were achieved for the reduction of dialkyl 4-(dialkyloxyphosphoryl)-3-oxo-butanoates by baker's yeast [482], whereas the reduction of cyclic dialkyl (3-oxo-1-cycloalkyl)-phosphonates with baker's yeast proceeded quite smoothly [41].

21.11 CONCLUSIONS

Many different reactions have been performed using baker's yeast in recent decades. While in the beginning yeast-mediated reactions (especially reductions) were seen as a method of choice and a method able to substitute classical organic synthesis, this enthusiasm made place for a more pragmatic view. Nowadays, whole-cell-mediated reactions are seen as one of many tools in the toolbox of a chemist. Close inspection of the results from recent literature retrieval showed that the numbers of publications using baker's yeast for synthesis declined since the mid-2000s. But the number of publications increased where whole cells were used in a nonaqueous environment, for example, cells were used in multiphase systems, in supercritical fluids, in ionic liquids; and in combinations thereof. Thus, the yeast story that began in 1862 with L. Pasteur has not come to an end; we have just turned some pages and are starting to open a new chapter for yeast-mediated stereoselective transformations in synthesis.

REFERENCES

[1] L. Pasteur, *C. R. Hebd. Seances Acad. Sci.* 1862, **55**, 28.
[2] A. J. Brown, *J. Chem. Soc.* 1886, **49**, 172–187.
[3] J. B. Dumas, *Ann. Chim. Phys.* 1874, **Ser. 5**, **3**, 57.
[4] W. Windisch, *Wochenschr. Brau.* 1898, **15**, 189–191.
[5] G. S. Kirchhoff, *J. Pharm. Sci. Access.* 1816, **2**, 250.
[6] J. W. Döbereiner, *J. Pharm. Sci. Access.* 1815, **1**, 342–345.
[7] S. M. Roberts, N. J. Turner, A. J. Willetts, M. K. Turner, *Introduction to Biocatalysis Using Enzymes and Microorganism*, Cambridge University Press, Cambridge, 1995, pp. 1–42.
[8] R. S. Chaleff, *Pure Appl. Chem.* 1988, **60**, 821–824.
[9] K. Mori, T. Sugai, *J. Synth. Org. Chem. Jpn.* 1983, **41**, 1044–1053.
[10] T. H. Maugh, *Science* 1983, **221**, 351–354.
[11] D. E. Eveleigh, *Sci. Am.* 1981, **245**, 155–178.
[12] D. Perlman, *Dev. Ind. Microbiol.* 1980, **21**, 15–23.
[13] R. Azerad, *Adv. Org. Synth.* 2005, **1**, 455–518.
[14] E. Santaniello, P. Ferraboschi, A. Manzocchi, *NATO Sci. Ser., 1* 2000, **33**, 95–115.
[15] M. Ferrero, V. Gotor, Biocatalytic synthesis of steroids, in R. N. Patel (Ed.) *Stereoselective Biocatalysis*, Marcel Dekker, Inc., New York, 2000, pp. 579–631.
[16] M. Ferrero, V. Gotor, *Monatsh. Chem.* 2000, **131**, 585–616.
[17] R. Csuk, B. I. Glänzer, *Chem. Rev.* 1991, **91**, 49–97.
[18] H. G. W. Leuenberger, in: K. Kieslich (Ed.) *Biotransformations*, Vol. **8**, Verlag Chemie, Weinheim, 1984, pp. 5–29.
[19] R. Csuk, B. I. Glänzer, Yeast-mediated stereoselective biocatalysis, in R. N. Patel (Ed.) *Stereoselective Biocatalysis*, Marcel Dekker, Inc., New York, 2000, pp. 527–578.
[20] S. M. Roberts, *J. Chem. Educ.* 2000, **77**, 344–348.
[21] T. Sugai, *Curr. Org. Chem.* 1999, **3**, 373–406.
[22] P. Besse, J. Bolte, C. Demuynck, L. Hecquet, H. Veschambre, *Recent Res. Dev. Org. Chem.* 1997, **1**, 191–228.
[23] E. Santaniello, P. Ferraboschi, *Adv. Asymmetric Synth.* 1997, **2**, 237–283.
[24] M. Chartrain, J. Armstrong, L. Katz, S. King, J. Reddy, Y.-J. Shi, D. Tschaen, R. Greasham, *Ann. N. Y. Acad. Sci.* 1996, **799**, 612–619.
[25] M. Chartrain, R. Greasham, J. Moore, P. Reider, D. Robinson, B. Buckland, *J. Mol. Catal. B: Enzym.* 2001, **11**, 503–512.

[26] Z. Wei, Z. Li, G. Lin, M. Tang, L. Xia, *Sci. China, Ser. B: Chem.* 2000, **43**, 331–336.
[27] S. P. R. De, *Crit. Rev. Biotechnol.* 1998, **18**, 25–83.
[28] A. Zaks, D. R. Dodds, *Drug Discov. Today* 1997, **2**, 513–531.
[29] R. N. Patel, A. Banerjee, C. G. McNamee, D. B. Brzozowski, L. J. Szarka, *Tetrahedron: Asymmetry* 1997, **8**, 2547–2552.
[30] S. Serra, Opportunities for biocatalysis in the flavor, fragrance, and cosmetic industry, in J. Tao, R. Kazlauskas (Eds.) *Biocatalysis for Green Chemistry and Chemical Process Development*, John Wiley & Sons, Inc., Hoboken, 2011, pp. 223–254.
[31] E. Brenna, *Curr. Org. Chem.* 2003, **7**, 1347–1367.
[32] A. Abate, E. Brenna, N. C. Dei, C. Fuganti, S. Serra, *Tetrahedron: Asymmetry* 2002, **13**, 899–904.
[33] T. Raab, T. Hauck, A. Knecht, U. Schmitt, U. Holzgrabe, W. Schwab, *Chirality* 2003, **15**, 573–578.
[34] E. Mariani, M. Bruno, C. Villa, C. Neuhoff, *Int. J. Cosmet. Sci.* 2000, **22**, 227–234.
[35] B. Bollbuck, W. Tochtermann, *Tetrahedron* 1999, **55**, 7191–7208.
[36] S. Tsuboi, J.-I. Sakamoto, H. Yamashita, T. Sakai, M. Utaka, *J. Org. Chem.* 1998, **63**, 1102–1108.
[37] M. Uchiyama, N. Katoh, R. Mimura, N. Yokota, Y. Shimogaichi, M. Shimazaki, A. Ohta, *Tetrahedron: Asymmetry* 1997, **8**, 3467–3474.
[38] R. Azerad, *Adv. Biochem. Eng./Biotechnol.* 1999, **63**, 169–218.
[39] S. Geresh, T. J. Valiyaveettil, Y. Lavie, A. Shani, *Tetrahedron: Asymmetry* 1998, **9**, 89–96.
[40] D. H. Dao, Y. Kawai, K. Hida, S. Hornes, K. Nakamura, A. Ohno, M. Okamura, T. Akasaka, *Bull. Chem. Soc. Jpn.* 1998, **71**, 425–432.
[41] M. Attolini, F. Bouguir, G. Iacazio, G. Peiffer, M. Maffei, *Tetrahedron* 2001, **57**, 537–543.
[42] M. Breuer, K. Ditrich, T. Habicher, B. Hauer, M. Kesseler, R. Stuermer, T. Zelinski, *Angew. Chem. Int. Ed.* 2004, **43**, 788–824.
[43] R. de Souza Pereira, *FEBS Lett.* 2000, **475**, 43–46.
[44] M. Bechtold, E. Brenna, C. Femmer, F. G. Gatti, S. Panke, F. Parmeggiani, A. Sacchetti, *Org. Process Res. Dev.* 2012, **16**, 269–276.
[45] M. Katz, T. Frejd, B. Hahn-haegerdal, M. F. Gorwa-Grauslund, *Biotechnol. Bioeng.* 2003, **84**, 573–582.
[46] E. Bialecka-Florjanczyk, A. U. Kapturowska, Genetically modified baker's yeast *Saccharomyces cerevisiae* in chemical synthesis and biotransformations, in D. Ekinci (Ed.) *Chemical Biology*, InTech, Rijeka, Croatia, 2012, pp. 211–234.
[47] B. Pscheidt, A. Glieder, *Microb. Cell Fact.* 2008, **7**, 25; doi:10.1186/475-2859-7-25.
[48] S. Rodriguez, M. M. Kayser, J. D. Stewart, *J. Am. Chem. Soc.* 2001, **123**, 1547–1555.
[49] M. M. Kayser, *J. Heterocycl. Chem.* 1999, **36**, 1533–1537.
[50] M. M. Kayser, *Tetrahedron* 2009, **65**, 947–974.
[51] M. Kayser, G. Chen, J. Stewart, *Synlett* 1999, 153–158.
[52] J. D. Stewart, *Biotechnol. Genet. Eng. Rev.* 1997, **14**, 67–143.
[53] I. Goubet, T. Maugard, S. Lamare, M. D. Legoy, *Enzyme Microb. Technol.* 2002, **31**, 425–430.
[54] R. Macleod, H. Prosser, L. Fikentscher, H. S. Mosher, J. Lanyi, *Biochemistry* 1964, **3**, 838–846.
[55] E. Garcia-Urdiales, I. Alfonso, V. Gotor, *Chem. Rev.* 2011, **111**, PR110–PR180.
[56] E. Garcia-Urdiales, I. Alfonso, V. Gotor, *Chem. Rev.* 2005, **105**, 313–354.
[57] C. J. Lintner, H. J. von Liebig, *Z. Physiol.* 1911, **72**, 449–454.
[58] H. Gröger, W. Hummel, R. Metzner, Reduction: asymmetric biocatalytic reduction of ketones, in E. M. Carreira, H. Yamamoto (Eds.) *Synthetic Methods/Comprehensive Chirality*, Elsevier B.V., Amsterdam, 2012, pp. 181–215.
[59] H. Gröger, W. Hummel, S. Borchert, M. Krausser, Reduction of ketones and aldehydes to alcohols, in K. Drauz, H. Gröger, O. May (Eds.) *Enzyme Catalysis in Organic Synthesis*, Vol. **2**, 3rd edition, Wiley-VCH/Verlag GmbH & Co. KGaA, Weinheim, 2012, pp. 1037–1110.
[60] H. Gröger, S. Borchert, M. Krausser, W. Hummel, Enzyme-catalyzed asymmetric reduction of ketones, in M. C. Flickinger (Ed.) *Encyclopedia of Industrial Biotechnology, Bioprocess, Bioseparation, and Cell Technology*, John Wiley & Sons, Inc., Hoboken, 2010, pp. 2094–2110.
[61] S. Buchholz, H. Gröger, Enantioselective biocatalytic reduction of ketones for the synthesis of optically active alcohols, in R. N. Patel (Ed.) *Biocatalysis in the Pharmaceutical and Biotechnology Industries*, CRC Press, New York, 2007, pp. 757–790.
[62] S. De Wildeman, N. Sereinig, Enzymatic reduction of carbonyl groups, in E. M. Carreira, C. P. Decicco, P. A. Evans, A. Fürstner, G. A. Molander, P. J. Reider, E. Schaumann, M. Shibasaki, E. J. Thomas, B. M. Trost, J. G. de Vries (Eds.) *Science of Synthesis: Stereoselective Synthesis*, Vol. **2**, Georg Thieme Verlag, Stuttgart, 2011, pp. 133–208.

[63] C. Cheng, H.-R. Tsai, *J. Chem. Technol. Biotechnol.* 2008, **83**, 1479–1485.
[64] K. Goldberg, K. Schroer, S. Luetz, A. Liese, *Appl. Microbiol. Biotechnol.* 2007, **76**, 249–255.
[65] R. N. Patel, A. Goswami, L. Chu, M. J. Donovan, V. Nanduri, S. Goldberg, R. Johnston, P. J. Siva, B. Nielsen, J. Fan, W. He, Z. Shi, K. Y. Wang, R. Eiring, D. Cazzulino, A. Singh, R. Mueller, *Tetrahedron: Asymmetry* 2004, **15**, 1247–1258.
[66] B. K. Pchelka, A. Loupy, A. Petit, *Tetrahedron: Asymmetry* 2006, **17**, 2516–2530.
[67] N. W. Fadnavis, S. K. Vadivel, U. T. Bhalerao, *Tetrahedron: Asymmetry* 1997, **8**, 2355–2359.
[68] N. W. Fadnavis, N. P. Reddy, U. T. Bhalerao, *J. Org. Chem.* 1989, **54**, 3218–3221.
[69] N. W. Fadnavis, K. R. Radhika, *Tetrahedron: Asymmetry* 2004, **15**, 3443–3447.
[70] X. Liu, T.-S. Zhu, P.-D. Sun, J.-H. Xu, *Synth. Commun.* 2001, **31**, 1521–1526.
[71] X. L. Gao, G. Zhou, Y. K. Guan, W. D. Li, Y. L. Li, *Chin. Chem. Lett.* 2001, **12**, 291–292.
[72] F. Molinari, C. Bertolini, F. Aragozzini, *Biocatal. Biotransform.* 1998, **16**, 87–94.
[73] F. Molinari, R. Gandolfi, R. Villa, E. G. Occhiato, *Tetrahedron: Asymmetry* 1999, **10**, 3515–3520.
[74] I. Chin-Joe, P. M. Nelisse, A. J. J. Straathof, J. A. Jongejan, J. T. Pronk, J. J. Heijnen, *Biotechnol. Bioeng.* 2000, **69**, 370–376.
[75] O. Bortolini, G. Fantin, M. Fogagnolo, A. Medici, P. Pedrini, *Recent Res. Dev. Pure Appl. Chem.* 1999, **3**, 137–152.
[76] P. Besse, G. Baziard-Mouysset, K. Boubekeur, P. Palvadeau, H. Veschambre, M. Payard, G. Mousset, *Tetrahedron: Asymmetry* 1999, **10**, 4745–4754.
[77] B.-L. Zhang, S. Pionnier, *Chem.-Eur. J.* 2003, **9**, 3604–3610.
[78] M. Katz, B. Hahn-Hagerdal, M. F. Gorwa-Grauslund, *Enzyme Microb. Technol.* 2003, **33**, 163–172.
[79] W. M. Clark, A. J. Kassick, M. A. Plotkin, A. M. Eldridge, I. Lantos, *Org. Lett.* 1999, **1**, 1839–1842.
[80] V. N. Odinokov, L. P. Botsman, L. V. Saranova, E. V. Syutkina, G. A. Emel'yanova, *Chem. Nat. Compd.* 1998, **33**, 687–688.
[81] A. J. Smallridge, M. A. Trewhella, *Methods Biotechnol.* 2001, **15**, 417–422.
[82] A. J. Smallridge, A. Ten, M. A. Trewhella, *Tetrahedron Lett.* 1998, **39**, 5121–5124.
[83] W. F. H. Sybesma, A. J. J. Straathof, J. A. Jongejan, J. T. Pronk, J. J. Heijnen, *Biocatal. Biotransform.* 1998, **16**, 95–134.
[84] R. Tanikaga, Y. Obata, K.-I. Kawamoto, *Tetrahedron: Asymmetry* 1997, **8**, 3101–3106.
[85] C. Forzato, P. Nitti, G. Pitacco, E. Valentin, *Tetrahedron: Asymmetry* 1997, **8**, 1811–1820.
[86] M. Chartrain, D. Mathre, R. A. Reamer, S. Patel, I. Shinkai, R. Greasham, *J. Ferment. Bioeng.* 1997, **83**, 395–396.
[87] F.-D. Irimie, C. Paizs, C. Majdik, M. Tosa, R. Misca, R. Silaghi-Dumitrescu, *Heterocycl. Commun.* 2002, **8**, 489–492.
[88] F. D. Irimie, C. Paizs, M. Tosa, C. Majdik, P. Moldovan, R. Misca, M. Caprioara, M. Neacsu, *Stud. Univ. Babes-Bolyai, Chem.* 2000, **45**, 23–28.
[89] F. D. Irimie, C. Paizs, F. Joo, R. Silaghi-Dumitrescu, M. Tosa, C. Majdik, *Roum. Biotechnol. Lett.* 1999, **4**, 71–74.
[90] F. D. Irimie, C. Paisz, F. Joo, R. Silaghi-Dumitrescu, M. Tosa, C. Majdik, *Prog. Catal.* 1999, **8**, 70–73.
[91] S. Roy, V. Alexandre, M. Neuwels, T. L. Le, *Adv. Synth. Catal.* 2001, **343**, 738–743.
[92] V. Prelog, *Pure Appl. Chem.* 1964, **9**, 119–130.
[93] H. K. Kölbl, H. Hildebrand, N. Piel, T. Schröder, W. Zitzmann, *Pure Appl. Chem.* 1988, **60**, 825–831.
[94] J. Silva, J. Alarcon, S. A. Aguila, J. B. Alderete, *Z. Naturforsch., C: J. Biosci.* 2010, **65**, 1–9.
[95] H. Simon, J. Bader, H. Gunther, S. Neumann, J. Thanos, *Angew. Chem. Int. Ed.* 1985, **24**, 539–553.
[96] C. Neuberg, *Adv. Carbohydr. Chem.* 1949, **4**, 75–117.
[97] T. Fujisawa, T. Sato, T. Itoh, J. Synth. *Org. Chem., Jpn.* 1986, **44**, 519–531.
[98] H. Günther, C. Frank, H. J. Schuetz, J. Bader, H. Simon, *Angew. Chem. Int. Ed.* 1983, **22**, 322–323.
[99] Y. Zhao, G. B. DeLancey, *Biotechnol. Bioeng.* 1999, **64**, 442–451.
[100] C. Neuberg, H. Lustig, R. Dresel, *Arch. Biochem.* 1948, **19**, 163–169.
[101] O. Cervinka, L. Hub, *Collect. Czech. Chem. Commun.* 1966, **31**, 2615–2618.
[102] G. Eichberger, K. Faber, H. Griengl, *Monatsh. Chem.* 1985, **116**, 1233–1236.
[103] D. R. Deardorff, D. C. Myles, K. D. Macferrin, *Tetrahedron Lett.* 1985, **26**, 5615–5618.
[104] M. Bucciarelli, A. Forni, I. Moretti, G. Torre, *Synthesis* 1983, 897–898.

[105] K. Nakamura, K. Ushio, S. Oka, A. Ohno, S. Yasui, *Tetrahedron Lett.* 1984, **25**, 3979–3982.
[106] C. O. Meese, *Justus Liebigs Ann. Chem.* **1986**, 2004–2007.
[107] F. Aragozzini, E. Maconi, R. Craveri, *Appl. Microbiol. Biotechnol.* 1986, **24**, 175–177.
[108] B. Barkakaty, Y. Takaguchi, S. Tsuboi, *Tetrahedron* 2006, **63**, 970–976.
[109] O. Achmatowicz, B. Szechner, *Top. Curr. Chem.* 2008, **282**, 143–186.
[110] S. Terashima, K. Tamoto, *Tetrahedron Lett.* 1982, **23**, 3715–3718.
[111] B. Koppenhoefer, W. Winter, E. Bayer, *Justus Liebigs Ann. Chem.* 1983, 1986–1995.
[112] Z.-H. Yang, R. Zeng, G. Yang, Y. Wang, L.-Z. Li, Z.-S. Lv, M. Yao, B. Lai, *J. Ind. Microbiol. Biotechnol.* 2008, **35**, 1047–1051.
[113] J. S. Yadav, S. Nanda, P. T. Reddy, A. B. Rao, *J. Org. Chem.* 2002, **67**, 3900–3903.
[114] P. Rodriguez, M. Barton, V. Aldabalde, S. Onetto, P. Panizza, P. Menendez, D. Gonzalez, S. Rodriguez, *J. Mol. Catal. B: Enzym.* 2007, **49**, 8–11.
[115] P. J. Davis, S. K. Yang, R. V. Smith, *Appl. Environ. Microbiol.* 1984, **48**, 327–331.
[116] T. Sato, T. Mizutani, Y. Okumura, T. Fujisawa, *Tetrahedron Lett.* 1989, **30**, 3701–3702.
[117] F. Aragozzini, E. Maconi, D. Potenza, C. Scolastico, *Synthesis* 1989, 225–227.
[118] J. Barry, H. B. Kagan, *Synthesis* 1981, 453–455.
[119] D. D. Ridley, M. Stralow, *Chem. Commun.* 1975, 400.
[120] J. P. Guette, N. Spassky, D. Boucherot, *Bull. Soc. Chim. Fr.* 1972, 4217–4224.
[121] K. Kieslich, *Microbial Transformations of Non-steroid Cyclic Compounds*, Thieme Verlag, Stuttgart, 1976.
[122] A. Manzocchi, A. Fiecchi, E. Santaniello, *Synthesis* 1987, 1007–1009.
[123] A. Belan, J. Bolte, A. Fauve, J. G. Gourcy, H. Veschambre, *J. Org. Chem.* 1987, **52**, 256–260.
[124] C. Ledrian, A. E. Greene, *J. Am. Chem. Soc.* 1982, **104**, 5473–5483.
[125] S. Ramaswamy, A. C. Oehlschlager, *J. Org. Chem.* 1989, **54**, 255–257.
[126] E. Pekala, T. Wojcik, *Acta Pol. Pharm.* 2007, **64**, 109–113.
[127] H. Akgun, A. Balkan, B. Berk, H. Urgan, E. Karaaslan, *FABAD J. Pharm. Sci.* 2001, **26**, 67–71.
[128] R. J. Hammond, B. W. Poston, I. Ghiviriga, B. D. Feske, *Tetrahedron Lett.* 2007, **48**, 1217–1219.
[129] L. Eignerova, Z. Prochazka, *Collect. Czech. Chem. Commun.* 1974, **39**, 2828–2834.
[130] D. H. R. Barton, R. B. Boar, D. A. Widdowson, *J. Chem. Soc. C* 1970, 1208–1213.
[131] C. Cheng, H.-R. Tsai, *Anal. Chim. Acta* 2008, **623**, 168–177.
[132] V. Crocq, C. Masson, J. Winter, C. Richard, G. Lemaitre, J. Lenay, M. Vivat, J. Buendia, D. Prat, *Org. Process Res. Dev.* 1997, **1**, 2–13.
[133] S. Fujieda, M. Tomita, K.-I. Fuhshuku, S. Ohba, S. Nishiyama, T. Sugai, *Adv. Synth. Catal.* 2005, **347**, 1099–1109.
[134] R. L. Crumbie, D. D. Ridley, G. W. Simpson, *Chem. Commun.* 1977, 315–316.
[135] Z. Wimmer, M. Budesinsky, T. Macek, A. Svatos, D. Saman, S. Vasickova, M. Romanuk, *Collect. Czech. Chem. Commun.* 1987, **52**, 2326–2337.
[136] T. Oritani, S. Kudo, K. Yamashita, *Agric. Biol. Chem.* 1982, **46**, 757–760.
[137] D. J. Kertesz, A. F. Kluge, *J. Org. Chem.* 1988, **53**, 4962–4968.
[138] H. G. Davies, T. C. C. Gartenmann, J. Leaver, S. M. Roberts, M. K. Turner, *Tetrahedron Lett.* 1986, **27**, 1093–1094.
[139] G. Lowe, S. Swain, *J. Chem. Soc., Perkin Trans. 1* 1985, 391–398.
[140] I. Stibor, I. Vesely, J. Palecek, J. Mostecky, *Synthesis* 1986, 640–642.
[141] M. Nakazaki, H. Chikamatsu, K. Naemura, M. Asao, *J. Org. Chem.* 1980, **45**, 4432–4440.
[142] M. Carlquist, C. Olsson, B. Bergdahl, E. W. J. van Niel, M. F. Gorwa-Grauslund, T. Frejd, *J. Mol. Catal. B: Enzym.* 2009, **58**, 98–102.
[143] C. Neuberg, F. F. Nord, *Ber. Dtsch. Chem. Ges.* 1914, **47**, 2264–2271.
[144] C. Neuberg, F. F. Nord, *Biochem. Z.* 1914, **67**, 46–50.
[145] F. F. Nord, *Ber. Dtsch. Chem. Ges.* 1919, **52**, 1207–1211.
[146] C. Neuberg, E. Schwenk, *Biochem. Z.* 1915, **71**, 118–121.
[147] D. Ghiringhelli, *Tetrahedron Lett.* 1983, **24**, 287–290.
[148] G. Guanti, L. Banfi, A. Guaragna, E. Narisano, *Chem. Commun.* 1986, 138–140.
[149] G. Guanti, L. Banfi, E. Narisano, *Tetrahedron Lett.* 1986, **27**, 3547–3550.
[150] R. L. Crumbie, B. S. Deol, J. E. Nemorin, D. D. Ridley, *Aust. J. Chem.* 1978, **31**, 1965–1980.
[151] T. Yamazaki, M. Asai, T. Ohnogi, J. T. Lin, T. Kitazume, *J. Fluorine Chem.* 1987, **35**, 537–555.
[152] A. R. Maguire, N. O'Riordan, *Tetrahedron Lett.* 1999, **40**, 9285–9288.
[153] N. L. Di, C. Franchini, A. Nacci, A. Scilimati, M. S. Sinicropi, *Tetrahedron: Asymmetry* 1999, **10**, 1913–1926.

[154] R. Hayakawa, K. Nozawa, K. Kimura, M. Shimizu, *Tetrahedron* 1999, **55**, 7519–7528.
[155] T. Itoh, Y. Yonekawa, T. Sato, T. Fujisawa, *Tetrahedron Lett.* 1986, **27**, 5405–5408.
[156] A. P. Kozikowski, B. B. Mugrage, C. S. Li, L. Felder, *Tetrahedron Lett.* 1986, **27**, 4817–4820.
[157] R. Tanikaga, K. Hosoya, A. Kaji, *Chem. Lett.* 1987, 829–832.
[158] T. Sato, Y. Okumura, J. Itai, T. Fujisawa, *Chem. Lett.* 1988, 1537–1540.
[159] A. R. Maguire, S. G. Collins, A. Ford, *ARKIVOC (Gainesville, FL, U. S.)* 2003, 96–109.
[160] G. Bianchi, G. Comi, I. Venturini, *Gazz. Chim. Ital.* 1984, **114**, 285–286.
[161] C. Ticozzi, A. Zanarotti, *Tetrahedron Lett.* 1988, **29**, 6167–6170.
[162] C. Neuberg, E. Welde, *Biochem. Z.* 1914, **60**, 472–479.
[163] C. Neuberg, E. Welde, *Biochem. Z.* 1914, **62**, 470–471.
[164] F.-D. Irimie, C. Paizs, M.-I. Tosa, C. Alforoaei, V. Miclaus, *Heterocycl. Commun.* 1997, **3**, 549–553.
[165] T. Fujisawa, H. Hayashi, Y. Kishioka, *Chem. Lett.* 1987, 129–132.
[166] T. Hafner, H. U. Reissig, *Justus Liebigs Ann. Chem.* 1989, 937–938.
[167] F. Li, J. Cui, X. Qian, R. Zhang, *Chem. Commun.* 2004, 2338–2339.
[168] A. Navarro-Ocana, L. F. Olguin, H. Luna, M. Jimenez-Estrada, E. Barzana, *J. Chem. Soc., Perkin Trans. 1* 2001, 2754–2756.
[169] S. S. Chimni, R. J. Singh, *World J. Microbiol. Biotechnol.* 1998, **14**, 247–250.
[170] Y. Kawai, Y. Inaba, N. Tokitoh, *Tetrahedron: Asymmetry* 2001, **12**, 309–318.
[171] R. R. Bak, A. F. McAnda, A. J. Smallridge, M. A. Trewhella, *Aust. J. Chem.* 1996, **49**, 1257–1260.
[172] K. Nakamura, Y. Inoue, J. Shibahara, S. Oka, A. Ohno, *Tetrahedron Lett.* 1988, **29**, 4769–4770.
[173] H. Ohta, N. Kobayashi, K. Ozaki, *J. Org. Chem.* 1989, **54**, 1802–1804.
[174] H. Ohta, K. Ozaki, G. Tsuchihashi, *Chem. Lett.* 1987, 191–192.
[175] P. Alvarez-Fitz, L. Alvarez, S. Marquina, J. Luna-Herrera, V. M. Navarro-Garcia, *Molecules* 2012, **17**, 8464–8470.
[176] E. Brenna, G. Fronza, C. Fuganti, F. G. Gatti, A. Manfredi, F. Parmeggiani, P. Ronchi, *J. Mol. Catal. B: Enzym.* 2012, **84**, 94–101.
[177] E. Brenna, F. G. Gatti, A. Manfredi, D. Monti, F. Parmeggiani, *Eur. J. Org. Chem.* 2011, **2011**, 4015–4022.
[178] F. Luo, D. Lu, Y. Gong, *J. Mol. Catal. B: Enzym.* 2011, **70**, 101–107.
[179] R. Chenevert, S. Thiboutot, *Chem. Lett.* 1988, 1191–1192.
[180] R. Chenevert, S. Thiboutot, *Can. J. Chem.* 1986, **64**, 1599–1601.
[181] X. Y. Wang, J. N. Cui, W. M. Ren, F. Li, C. L. Lu, X. H. Qian, *Chin. Chem. Lett.* 2007, **18**, 681–684.
[182] K.-I. Fuhshuku, M. Tomita, T. Sugai, *Adv. Synth. Catal.* 2003, **345**, 766–774.
[183] H. Kosmol, K. Kieslich, R. Vossing, H. J. Koch, K. Petzoldt, H. Gibian, *Justus Liebigs Ann. Chem.* 1967, **701**, 198–205.
[184] L. M. Kogan, V. E. Gulaya, I. V. Torgov, *Tetrahedron Lett.* 1967, **8**, 4673–4676.
[185] D. W. Brooks, P. G. Grothaus, H. Mazdiyasni, *J. Am. Chem. Soc.* 1983, **105**, 4472–4473.
[186] D. W. Brooks, P. G. Grothaus, W. L. Irwin, *J. Org. Chem.* 1982, **47**, 2820–2821.
[187] D. W. Brooks, P. G. Grothaus, J. T. Palmer, *Tetrahedron Lett.* 1982, **23**, 4187–4190.
[188] Z. L. Wei, Z. Y. Li, G. Q. Lin, *Tetrahedron: Asymmetry* 2001, **12**, 229–233.
[189] Z. P. Zhuang, W. S. Zhou, *Tetrahedron* 1985, **41**, 3633–3641.
[190] W. M. Dai, W. S. Zhou, *Tetrahedron* 1985, **41**, 4475–4482.
[191] R. P. Lanzilotta, D. G. Bradley, C. C. Beard, *Appl. Micrbiol.* 1975, **29**, 427–429.
[192] R. R. Rando, *Biochem. Pharmacol.* 1974, **23**, 2328–2331.
[193] K. Nakamura, K. Inoue, K. Ushio, S. Oka, A. Ohno, *Chem. Lett.* 1987, 679–682.
[194] D. W. Brooks, H. Mazdiyasni, P. G. Grothaus, *J. Org. Chem.* 1987, **52**, 3223–3232.
[195] K. Mori, H. Mori, M. Yanai, *Tetrahedron* 1986, **42**, 291–294.
[196] M. Yanai, T. Sugai, K. Mori, *Agric. Biol. Chem.* 1985, **49**, 2373–2377.
[197] A. Murai, N. Tanimoto, N. Sakamoto, T. Masamune, *J. Am. Chem. Soc.* 1988, **110**, 1985–1986.
[198] D. W. Brooks, H. Mazdiyasni, P. Sallay, *J. Org. Chem.* 1985, **50**, 3411–3414.
[199] J. Dangelo, G. Revial, R. Azerad, D. Buisson, *J. Org. Chem.* 1986, **51**, 40–45.
[200] H. G. W. Leuenberger, W. Boguth, E. Widmer, R. Zell, *Helv. Chim. Acta* 1976, **59**, 1832–1849.
[201] R. Zell, E. Widmer, T. Lukac, H. G. W. Leuenberger, P. Schonholzer, E. A. Broger, *Helv. Chim. Acta* 1981, **64**, 2447–2462.

[202] G. Hoffmann, R. Wiartalla, *Tetrahedron Lett.* 1982, **23**, 3887–3888.
[203] Y.-Y. Zhu, D. J. Burnell, *Tetrahedron: Asymmetry* 1996, **7**, 3295–3304.
[204] R. Belrhlid, A. Fauve, H. Veschambre, *J. Org. Chem.* 1989, **54**, 3221–3223.
[205] D. Buisson, S. Elbaba, R. Azerad, *Tetrahedron Lett.* 1986, **27**, 4453–4454.
[206] M. Imuta, H. Ziffer, *J. Org. Chem.* 1978, **43**, 3319–3323.
[207] I. Deschamps, W. J. King, F. F. Nord, *J. Org. Chem.* 1949, **14**, 184–187.
[208] T. Fujisawa, E. Kojima, T. Sato, *Chem. Lett.* 1987, 2227–2228.
[209] T. Fujisawa, E. Kojima, T. Itoh, T. Sato, *Tetrahedron Lett.* 1985, **26**, 6089–6092.
[210] J. Bolte, J. G. Gourcy, H. Veschambre, *Tetrahedron Lett.* 1986, **27**, 565–568.
[211] H. Ohta, K. Ozaki, G. Tsuchihashi, *Chem. Lett.* 1987, 2225–2226.
[212] H. Ohta, K. Ozaki, J. Konishi, G. Tsuchihashi, *Agric. Biol. Chem.* 1986, **50**, 1261–1266.
[213] A. Fauve, H. Veschambre, *J. Org. Chem.* 1988, **53**, 5215–5219.
[214] J. K. Lieser, *Synth. Commun.* 1983, **13**, 765–767.
[215] T. Johanson, M. Carlquist, C. Olsson, A. Rudolf, T. Frejd, M. F. Gorwa-Grauslund, *Appl. Microbiol. Biotechnol.* 2008, **77**, 1111–1118.
[216] Y. Kawai, K. Hida, M. Tsujimoto, S.-I. Kondo, K. Kitano, K. Nakamura, A. Ohno, *Bull. Chem. Soc. Jpn.* 1999, **72**, 99–102.
[217] B. S. Deol, D. D. Ridley, G. W. Simpson, *Aust. J. Chem.* 1976, **29**, 2459–2467.
[218] H. Suemune, Y. Mizuhara, H. Akita, T. Oishi, K. Sakai, *Chem. Pharm. Bull.* 1987, **35**, 3112–3118.
[219] E. M. Rustoy, E. N. Pereyra, S. Moreno, A. Baldessari, *Tetrahedron: Asymmetry* 2004, **15**, 3763–3768.
[220] H. Akita, H. Matsukura, T. Oishi, *Chem. Pharm. Bull.* 1986, **34**, 2656–2659.
[221] H. Akita, A. Furuichi, H. Koshiji, K. Horikoshi, T. Oishi, *Chem. Pharm. Bull.* 1983, **31**, 4384–4390.
[222] L. Poppe, L. Novak, P. Kolonits, A. Bata, C. Szantay, *Tetrahedron* 1988, **44**, 1477–1487.
[223] S. Tsuboi, E. Nishiyama, H. Furutani, M. Utaka, A. Takeda, *J. Org. Chem.* 1987, **52**, 1359–1362.
[224] S. Tsuboi, E. Nishiyama, M. Utaka, A. Takeda, *Tetrahedron Lett.* 1986, **27**, 1915–1916.
[225] R. W. Hoffmann, W. Ladner, W. Helbig, *Justus Liebigs Ann. Chem.* 1984, 1170–1179.
[226] R. W. Hoffmann, W. Helbig, W. Ladner, *Tetrahedron Lett.* 1982, **23**, 3479–3482.
[227] D. Buisson, R. Azerad, *Tetrahedron Lett.* 1986, **27**, 2631–2634.
[228] D. W. Brooks, M. Wilson, M. Webb, *J. Org. Chem.* 1987, **52**, 2244–2248.
[229] K. Mori, M. Tsuji, *Tetrahedron* 1988, **44**, 2835–2842.
[230] K. Mori, M. Tsuji, *Tetrahedron* 1986, **42**, 435–444.
[231] Y. Tirilly, J. Kloosterman, G. Sipma, J. J. Kettenesvandenbosch, *Phytochemistry* 1983, **22**, 2082–2083.
[232] Z. F. Xie, K. Funakoshi, H. Suemune, T. Oishi, H. Akita, K. Sakai, *Chem. Pharm. Bull.* 1986, **34**, 3058–3060.
[233] D. Seebach, S. Roggo, T. Maetzke, H. Braunschweiger, J. Cercus, M. Krieger, *Helv. Chim. Acta* 1987, **70**, 1605–1615.
[234] T. Kitahara, H. Kurata, K. Mori, *Tetrahedron* 1988, **44**, 4339–4349.
[235] T. Kitahara, K. Mori, *Tetrahedron Lett.* 1985, **26**, 451–452.
[236] S. Tanaka, K. Wada, S. Marumo, H. Hattori, *Tetrahedron Lett.* 1984, **25**, 5907–5910.
[237] K. Mori, H. Tamura, *Justus Liebigs Ann. Chem.* 1988, 97–105.
[238] G. Frater, *Helv. Chim. Acta* 1980, **63**, 1383.
[239] D. W. Knight, N. Lewis, A. C. Share, D. Haigh, *J. Chem. Soc., Perkin Trans. 1* 1998, 3673–3684.
[240] R. Hayakawa, M. Shimizu, *Synlett* 1999, 1298–1300.
[241] Y. Salinas, R. M. Oliart, M. Ramirez-Lepe, A. Navarro-Ocana, G. Valerio-Alfaro, *Appl. Microbiol. Biotechnol.* 2007, **75**, 297–302.
[242] C. J. Sih, C. S. Chen, *Angew. Chem. Int. Ed.* 1984, **23**, 570–578.
[243] H. Ohta, *J. Syn. Org. Chem., Jpn.* 1983, **41**, 1018–1030.
[244] A. Comini, C. Forzato, P. Nitti, G. Pitacco, E. Valentin, *Tetrahedron: Asymmetry* 2004, **15**, 617–625.
[245] J. Nowacki, J. Szychowski, K. Wojtasiewicz, I. Matuszewska, M. Jaremko, Z. Czarnocki, *Heterocycles* 2002, **57**, 705–714.
[246] R. Hayakawa, M. Shimizu, *Synlett* 1999, 1328–1330.
[247] W. R. Shieh, A. S. Gopalan, C. J. Sih, *J. Am. Chem. Soc.* 1985, **107**, 2993–2994.
[248] B. N. Zhou, A. S. Gopalan, F. Vanmiddlesworth, W. R. Shieh, C. J. Sih, *J. Am. Chem. Soc.* 1983, **105**, 5925–5926.

[249] R. Hayakawa, K. Nozawa, M. Shimizu, T. Fujisawa, *Tetrahedron Lett.* 1998, **39**, 67–70.
[250] K. Nakamura, M. Higaki, K. Ushio, S. Oka, A. Ohno, *Tetrahedron Lett.* 1985, **26**, 4213–4216.
[251] C. Fuganti, P. Grasselli, P. Casati, M. Carmeno, *Tetrahedron Lett.* 1985, **26**, 101–104.
[252] K. Nakamura, Y. Kawai, S. Oka, A. Ohno, *Bull. Chem. Soc. Jpn.* 1989, **62**, 875–879.
[253] K. Ushio, K. Inouye, K. Nakamura, S. Oka, A. Ohno, *Tetrahedron Lett.* 1986, **27**, 2657–2660.
[254] J. Heidlas, K. H. Engel, R. Tressl, *Eur. J. Biochem.* 1988, **172**, 633–639.
[255] C. J. Sih, B. N. Zhou, A. S. Gopalan, W. R. Shieh, C. S. Chen, G. Girdaukas, F. Vanmiddlesworth, *Ann. N.Y. Acad. Sci.* 1984, **434**, 186–193.
[256] E. Santaniello, R. Casati, F. Milani, J. Chem. *Res. Synop.* 1984, 132–133.
[257] M. Wolberg, I. A. Kaluzna, M. Mueller, J. D. Stewart, *Tetrahedron: Asymmetry* 2004, **15**, 2825–2828.
[258] S. Zeror, J. Collin, J.-C. Fiaud, L. A. Zouioueche, *Tetrahedron: Asymmetry* 2010, **21**, 1211–1215.
[259] C. E. Perles, P. J. S. Moran, P. L. O. Volpe, *J. Mol. Catal. B: Enzym.* 2008, **52–53**, 82–87.
[260] K. Mori, K. Tanida, *Tetrahedron* 1981, **37**, 3221–3225.
[261] E. Hungerbuhler, D. Seebach, D. Wasmuth, *Helv. Chim. Acta* 1981, **64**, 1467–1487.
[262] R. U. Lemieux, J. Giguere, *Can. J. Chem.* 1951, **29**, 678–690.
[263] G. Frater, *Helv. Chim. Acta* 1979, **62**, 2829–2832.
[264] J. Ehrler, F. Giovannini, B. Lamatsch, D. Seebach, *Chimia* 1986, **40**, 172–173.
[265] M. Hirama, M. Shimizu, M. Iwashita, *Chem. Commun.* 1983, 599–600.
[266] B. Seuring, D. Seebach, *Helv. Chim. Acta* 1977, **60**, 1175–1181.
[267] K. Mori, *Tetrahedron* 1981, **37**, 1341–1342.
[268] B. Wipf, E. Kupfer, R. Bertazzi, H. G. W. Leuenberger, *Helv. Chim. Acta* 1983, **66**, 485–488.
[269] D. Seebach, M. F. Züger, F. Giovannini, B. Sonnleitner, A. Fiechter, *Angew. Chem. Int. Ed.* 1984, **23**, 151–152.
[270] R. Muller, F. Lingens, *Angew. Chem. Int. Ed.* 1986, **25**, 779–789.
[271] D. Seebach, M. F. Züger, *Helv. Chim. Acta* 1982, **65**, 495–503.
[272] P. Deshong, M. T. Lin, J. J. Perez, *Tetrahedron Lett.* 1986, **27**, 2091–2094.
[273] T. Sato, *Can. J. Chem.* 1987, **65**, 2732–2733.
[274] M. Hirama, M. Uei, *J. Am. Chem. Soc.* 1982, **104**, 4251–4253.
[275] M. Utaka, H. Higashi, A. Takeda, *Chem. Commun.* 1987, 1368–1369.
[276] D. Seebach, P. Renaud, W. B. Schweizer, M. F. Züger, M. J. Brienne, *Helv. Chim. Acta* 1984, **67**, 1843–1853.
[277] D. Seebach, M. Eberle, *Synthesis* 1986, 37–40.
[278] T. Kitazume, N. Ishikawa, *Chem. Lett.* 1983, 237–238.
[279] D. W. Brooks, R. P. Kellogg, C. S. Cooper, *J. Org. Chem.* 1987, **52**, 192–196.
[280] M. Hirama, T. Nakamine, S. Ito, *Chem. Lett.* 1986, 1381–1384.
[281] A. Furuichi, H. Akita, H. Matsukura, T. Oishi, K. Horikoshi, *Agric. Biol. Chem.* 1987, **51**, 293–299.
[282] K. Mori, H. Mori, T. Sugai, *Tetrahedron* 1985, **41**, 919–925.
[283] P. Raddatz, H. E. Radunz, G. Schneider, H. Schwartz, *Angew. Chem. Int. Ed.* 1988, **27**, 426–427.
[284] M. Hirama, T. Nakamine, S. Ito, *Tetrahedron Lett.* 1986, **27**, 5281–5284.
[285] T. Fujisawa, T. Itoh, T. Sato, *Tetrahedron Lett.* 1984, **25**, 5083–5086.
[286] M. F. Züger, F. Giovannini, D. Seebach, *Angew. Chem. Int. Ed.* 1983, **22**, 1012.
[287] K. Nakamura, T. Miyai, K. Nozaki, K. Ushio, S. Oka, A. Ohno, *Tetrahedron Lett.* 1986, **27**, 3155–3156.
[288] H. Akita, A. Furuichi, H. Koshiji, K. Horikoshi, T. Oishi, *Chem. Pharm. Bull.* 1983, **31**, 4376–4383.
[289] G. Frater, U. Muller, W. Gunther, *Tetrahedron* 1984, **40**, 1269–1277.
[290] D. Buisson, S. Henrot, M. Larcheveque, R. Azerad, *Tetrahedron Lett.* 1987, **28**, 5033–5036.
[291] H. Akita, H. Koshiji, A. Furuichi, K. Horikoshi, T. Oishi, *Tetrahedron Lett.* 1983, **24**, 2009–2010.
[292] H. Akita, H. Matsukura, T. Oishi, *Tetrahedron Lett.* 1986, **27**, 5397–5400.
[293] R. E. Saxon, H. Leisch, T. Hudlicky, *Tetrahedron: Asymmetry* 2008, **19**, 672–681.
[294] Y. Yang, M. M. Kayser, F. D. Rochon, S. Rodriguez, J. D. Stewart, *J. Mol. Catal. B: Enzym.* 2005, **32**, 167–174.
[295] M. D. Mihovilovic, M. Spina, P. Stanetty, *ARKIVOC (Gainesville, FL, U. S.)* 2005, 33–44.
[296] M. D. Mihovilovic, M. M. Kayser, S. Rodriguez, J. D. Stewart, A novel approach to the synthesis of the taxol side chain: adapting *S. cerevisiae* as enantioselective reagent for the reduction of α-keto-β-lactams, ECHET98: Electronic Conference on Heterocyclic Chemistry, June 29–July 24, 1998, Imperial College Press, 1998, pp. 162–170.

[297] M. M. Kayser, M. D. Mihovilovic, J. Kearns, A. Feicht, J. D. Stewart, *J. Org. Chem.* 1999, **64**, 6603–6608.
[298] M. Utaka, H. Watabu, A. Takeda, *J. Org. Chem.* 1987, **52**, 4363–4368.
[299] J. P. Vigneron, R. Meric, M. Dhaenens, *Tetrahedron Lett.* 1980, **21**, 2057–2060.
[300] M. Gessner, C. Gunther, A. Mosandl, *Z. Naturforsch., Teil C* 1987, **42**, 1159–1164.
[301] M. Utaka, H. Watabu, A. Takeda, *Chem. Lett.* 1985, 1475–1476.
[302] Y. Naoshima, H. Ozawa, H. Kondo, S. Hayashi, *Agric. Biol. Chem.* 1983, **47**, 1431–1434.
[303] K. Mori, T. Otsuka, *Tetrahedron* 1985, **41**, 547–551.
[304] C. Forzato, R. Gandolfi, F. Molinari, P. Nitti, G. Pitacco, E. Valentin, *Tetrahedron: Asymmetry* 2001, **12**, 1039–1046.
[305] N. O. Mahmoodi, R. Yousefi-Malekroudi, *Russ. J. Org. Chem.* 2006, **42**, 365–368.
[306] G. M. Bonora, S. Drioli, C. Forzato, P. Nitti, G. Pitacco, *Lett. Org. Chem.* 2005, **2**, 89–91.
[307] E. Bialecka-Florjanczyk, J. Krzyczkowska, I. Stolarzewicz, *Biocatal. Biotransform.* 2010, **28**, 288–291.
[308] E. Bialecka-Florjanczyk, E. Majewska, *Synth. Commun.* 2010, **40**, 1264–1269.
[309] L. Mamoli, *Ber. Dtsch. Chem. Ges.* 1938, **71**, 2696–2698.
[310] T. Achstetter, O. Emter, C. Ehmann, D. H. Wolf, *J. Biol. Chem.* 1984, **259**, 3334–3343.
[311] F. H. Schermers, J. H. Duffus, A. M. Macleod, *J. Inst. Brew., London* 1976, **82**, 170–174.
[312] E. Parkkinen, H. Suomalainen, *J. Inst. Brew., London* 1982, **88**, 34–38.
[313] E. Parkkinen, H. Suomalainen, *J. Inst. Brew., London* 1982, **88**, 98–101.
[314] E. Parkkinen, *Cell. Mol. Biol.* 1980, **26**, 147–154.
[315] E. Parkkinen, E. Oura, H. Suomalainen, *J. Inst. Brew., London* 1978, **84**, 5–8.
[316] E. J. F. Demant, *FEBS Lett.* 1978, **85**, 109–113.
[317] G. Schaffner, P. Matile, *Biochem. Physiol. Pflanz.* 1979, **174**, 811–821.
[318] S. Miura, S. Kurozumi, T. Toru, T. Tanaka, M. Kobayashi, S. Matsubara, S. Ishimoto, *Tetrahedron* 1976, **32**, 1893–1898.
[319] T. Nurminen, H. Suomalainen, *Biochem. J.* 1970, **118**, 759–763.
[320] T. Nurminen, E. Oura, H. Suomalainen, *Biochem. J.* 1970, **116**, 61–69.
[321] W. Witt, M. E. Schweingruber, A. Mertsching, *Biochem. Biophys. Acta* 1984, **795**, 108–116.
[322] W. Witt, A. Mertsching, E. Konig, *Biochem. Biophys. Acta* 1984, **795**, 117–124.
[323] J. Bours, D. A. A. Mossel, *Arch. Lebensmittelhyg.* 1973, **24**, 197–203.
[324] C. J. Sih, R. Sood, R. G. Salomon, P. Price, M. Casey, G. Peruzzotti, *J. Am. Chem. Soc.* 1972, **94**, 3643–3644.
[325] C. J. Sih, R. G. Salomon, P. Price, R. Sood, G. Peruzzotti, *J. Am. Chem. Soc.* 1975, **97**, 857–864.
[326] C. J. Sih, R. G. Salomon, R. Sood, P. Price, G. Peruzzotti, *Chem. Commun.* 1972, 240–241.
[327] C. S. Chen, Y. Fujimoto, G. Girdaukas, C. J. Sih, *J. Am. Chem. Soc.* 1982, **104**, 7294–7299.
[328] B. I. Glänzer, K. Faber, H. Griengl, *Tetrahedron* 1987, **43**, 771–778.
[329] R. Csuk, B. I. Glänzer, *J. Fluorine Chem.* 1988, **39**, 99–106.
[330] B. I. Glänzer, K. Faber, H. Griengl, M. Rohr, W. Wöhrer, *Enzyme Microb. Tech.* 1988, **10**, 744–749.
[331] B. I. Glänzer, K. Faber, H. Griengl, *Enzyme Microb. Tech.* 1988, **10**, 689–690.
[332] S. S. Bari, S. Madan, M. K. Sethi, *Indian J. Chem., Sect. B* 1999, **38B**, 10–17.
[333] B. I. Glänzer, K. Faber, H. Griengl, *Tetrahedron* 1987, **43**, 5791–5796.
[334] K. Burg, O. Mauz, S. Noetzel, K. Sauber, *Angew. Makromol. Chem.* 1988, **157**, 105–121.
[335] B. Mattiasson, V. Hahn-Haegerdahl, *J. Appl. Microbiol. Biotechnol.* 1982, **16**, 52–55.
[336] P. M. Doran, J. E. Bailey, *Biotechnol. Bioeng.* 1986, **28**, 73–87.
[337] P. M. Doran, J. E. Bailey, *Biotechnol. Bioeng.* 1987, **29**, 892–897.
[338] T. K. Dutta, T. B. Samanta, *Bioorg. Med. Chem. Lett.* 1997, **7**, 629–632.
[339] Z.-H. Yang, R. Zeng, Y. Wang, W. Li, Z.-S. Lv, *Asia-Pac. J. Chem. Eng.* 2008, **3**, 217–222.
[340] K. Nakamura, K. Inoue, K. Ushio, S. Oka, A. Ohno, *J. Org. Chem.* 1988, **53**, 2589–2593.
[341] D. R. Griffin, J. L. Gainer, G. Carta, *Biotechnol. Prog.* 2001, **17**, 304–310.
[342] K. Horikoshi, A. Furuichi, H. Koshiji, H. Akita, T. Oishi, *Agric. Biol. Chem.* 1983, **47**, 435–436.
[343] H. Akita, A. Furuichi, H. Koshiji, K. Horikoshi, T. Oishi, *Tetrahedron Lett.* 1982, **23**, 4051–4054.
[344] F. Li, J. Cui, X. Qian, W. Ren, X. Wang, *Chem. Commun.* 2006, 865–867.
[345] M. Kierstan, C. Bucke, *Biotechnol. Bioeng.* 1977, **19**, 387–397.
[346] M. Utaka, S. Onoue, A. Takeda, *Chem. Lett.* 1987, 971–972.
[347] T. Murakata, H. Honma, S. Nakazato, C. Kuroda, S. Sato, *J. Chem. Eng. Jpn.* 2001, **34**, 299–305.

[348] T. Tosa, T. Sato, T. Mori, K. Yamamoto, I. Takata, Y. Nishida, I. Chibata, *Biotechnol. Bioeng.* 1979, **21**, 1697–1709.
[349] Y. Naoshima, H. Hasegawa, T. Saeki, *Agric. Biol. Chem.* 1987, **51**, 3417–3419.
[350] Y. Naoshima, H. Hasegawa, *Chem. Lett.* 1987, 2379–2382.
[351] M. Utaka, S. Konishi, T. Okubo, S. Tsuboi, A. Takeda, *Tetrahedron Lett.* 1987, **28**, 1447–1450.
[352] A. R. Kumbhar, V. V. Kulkarni, M. B. Pande, K. Karande, S. S. Laddha, *Proceedings of ECSOC-11, The 11th International Electronic Conference on Synthetic Organic Chemistry*, November 1–30, 2007, J. A. Seijas, M. P. Vázquez Tato (Eds.) Santiago de Compostella, Spain, pp. c002/001–c002/007.
[353] T. Sakai, T. Nakamura, K. Fukuda, E. Amano, M. Utaka, A. Takeda, *Bull. Chem. Soc. Jpn.* 1986, **59**, 3185–3188.
[354] M. Kawai, K. Tajima, S. Mizuno, K. Niimi, H. Sugioka, Y. Butsugan, A. Kozawa, T. Asano, Y. Imai, *Bull. Chem. Soc. Jpn.* 1988, **61**, 3014–3016.
[355] K. Nakamura, Y. Kawai, S. Oka, A. Ohno, *Tetrahedron Lett.* 1989, **30**, 2245–2246.
[356] G. Yang, Z. Ou, S. Yao, J. Xu, *J. Mol. Catal. B: Enzym.* 2009, **57**, 83–88.
[357] F. Martinez-Lagos, J. V. Sinisterra, *J. Mol. Catal. B* 2005, **36**, 1–7.
[358] L. F. Martinez, J. D. Carballeira, J. L. Bermudez, E. Alvarez, J. V. Sinisterra, *Tetrahedron: Asymmetry* 2004, **15**, 763–770.
[359] S. K. Mohapatra, P. Selvam, *Lett. Org. Chem.* 2006, **3**, 901–904.
[360] S. G. Cull, J. D. Holbrey, V. Vargas-Mora, K. R. Seddon, G. J. Lye, *Biotechnol. Bioeng.* 2000, **69**, 227–233.
[361] K. M. Docherty, C. F. Kulpa, *Green Chem.* 2005, **7**, 185–189.
[362] M. Matsumoto, K. Mochiduki, K. Kondo, *J. Biosci. Bioeng.* 2004, **98**, 344–347.
[363] K. O. Evans, *Colloid Surf. A* 2006, **274**, 11–17.
[364] Z. H. Yang, R. Zeng, Y. Wang, X. K. Li, Z. S. Lv, B. Lai, S. Q. Yang, J. G. Liao, *Food Technol. Biotechol.* 2009, **47**, 62–66.
[365] R. P. Swatloski, J. D. Holbrey, R. D. Rogers, *Green Chem.* 2003, **5**, 361–363.
[366] F. van Rantwijk, R. A. Sheldon, *Chem. Rev.* 2007, **107**, 2757–2785.
[367] W. Wang, M. H. Zong, W. Y. Lou, *J. Mol. Catal. B* 2009, **56**, 70–76.
[368] Y. N. Li, X. A. Shi, M. H. Zong, C. Meng, Y. Q. Dong, Y. H. Guo, *Enzyme Microb. Technol.* 2007, **40**, 1305–1311.
[369] F. Zhang, Y. Ni, Z. H. Sun, P. Zheng, W. Q. Lin, P. Zhu, N. F. Ju, *Chin. J. Catal.* 2008, **29**, 577–582.
[370] J. Y. He, L. M. Zhou, P. Wang, L. Zu, *Process Biochem.* 2009, **44**, 316–321.
[371] W. Y. Lou, W. Wang, R. F. Li, M. H. Zong, *J. Biotechnol.* 2009, **143**, 190–197.
[372] S. M. Lee, W. J. Chang, A. R. Choi, Y. M. Koo, *Korean J. Chem. Eng.* 2005, **22**, 687–690.
[373] K. R. Seddon, A. Stark, M. J. Torres, *Pure Appl. Chem.* 2000, **72**, 2275–2287.
[374] J. G. Huddleston, A. E. Visser, W. M. Reichert, H. D. Willauer, G. A. Broker, R. D. Rogers, *Green Chem.* 2001, **3**, 156–164.
[375] K. N. Marsh, J. A. Boxall, R. Lichtenthaler, *Fluid Phase Equilib.* 2004, **219**, 93–98.
[376] P. J. Scammells, J. L. Scott, R. D. Singer, *Aust. J. Chem.* 2005, **58**, 155–169.
[377] J. Pernak, I. Goc, I. Mirska, *Green Chem.* 2004, **6**, 323–329.
[378] J. Pernak, J. Rogoza, I. Mirska, *Eur. J. Med. Chem.* 2001, **36**, 313–320.
[379] J. Pernak, M. Smiglak, S. T. Griffin, W. L. Hough, T. B. Wilson, A. Pernak, J. Zabielska-Matejuk, A. Fojutowski, K. Kita, R. D. Rogers, *Green Chem.* 2006, **8**, 798–806.
[380] J. Pernak, K. Sobaszkiewicz, I. Mirska, *Green Chem.* 2003, **5**, 52–56.
[381] W. Y. Lou, W. Wang, T. J. Smith, M. H. Zong, *Green Chem.* 2009, **11**, 1377–1384.
[382] W. Y. Lou, M. H. Zong, T. J. Smith, *Green Chem.* 2006, **8**, 147–155.
[383] H. Pfruender, R. Jones, D. Weuster-Botz, *J. Biotechnol.* 2006, **124**, 182–190.
[384] D. Weuster-Botz, S. Stevens, A. Hawrylenko, *Biochem. Eng. J.* 2002, **11**, 69–72.
[385] Y. G. Shi, Y. Fang, Y. P. Ren, H. P. Wu, H. L. Guan, *J. Ind. Microbiol. Biotechnol.* 2008, **35**, 1419–1424.
[386] T. Matsuda, Y. Yamagishi, S. Koguchi, N. Iwai, T. Kitazume, *Tetrahedron Lett.* 2006, **47**, 4619–4622.
[387] A. F. Patrocinio, P. J. S. Moran, *J. Braz. Chem. Soc.* 2001, **12**, 7–31.
[388] P. Zani, *J. Mol. Catal. B* 2001, **11**, 279–285.
[389] J. T. Gorke, F. Srienc, R. J. Kazlauskas, *Chem. Commun.* 2008, 1235–1237.
[390] M. C. Gutierrez, M. L. Ferrer, L. Yuste, F. Rojo, F. del Monte, *Angew. Chem. Int. Ed.* 2010, **49**, 2158–2162.
[391] D. J. Heldebrant, H. N. Witt, S. M. Walsh, T. Ellis, J. Rauscher, P. G. Jessop, *Green Chem.* 2006, **8**, 807–815.
[392] O. Ward, A. Singh, Recent developments in enzymatic acyloin condensation, in R. N. Patel (Ed.) *Biocatalysis in the Pharmaceutical and Biotechnology Industries*, CRC Press LLC, Boca Raton, 2007, pp. 605–621.

[393] A. D. A. M. Curtis, Carbon–carbon bond formation using enzymes, in H.-J. Rehm, G. Reed (Eds.) *Biotechnology: A Multi-Volume Comprehensive Treatise*, Vol. **10**, Wiley-VCH, Weinheim, 2001, pp. 5–40.
[394] G. Seoane, *Curr. Org. Chem.* 2000, **4**, 283–304.
[395] W. Dirscherl, *Z. Physiol. Chem.* 1931, **201**, 47–77.
[396] C. Neuberg, J. Hirsch, *Biochem. Z.* 1921, **115**, 282–310.
[397] C. Neuberg, E. Reinfurth, *Biochem. Z.* 1923, **143**, 553–565.
[398] C. Fuganti, P. Grasselli, F. Spreafico, C. Zirotti, P. Casati, *J. Org. Chem.* 1984, **49**, 543–546.
[399] C. Neuberg, L. Liebermann, *Biochem. Z.* 1921, **120**, 311–325.
[400] A. H. Rose, *Industrial Microbiology*, Butterworth, Washington, DC, 1961.
[401] H.-P. Schmauder, D. Gröger, *Pharmazie* 1968, **23**, 320–331.
[402] C. Fuganti, D. Ghiringhelli, P. Grasselli, *Chem. Commun.* 1975, 846–847.
[403] G. Fronza, C. Fuganti, P. Grasselli, G. Poli, S. Servi, *J. Org. Chem.* 1988, **53**, 6153–6154.
[404] C. Fuganti, P. Grasselli, *Chem. Commun.* 1979, 995–997.
[405] C. Fuganti, P. Grasselli, *Chem. Commun.* 1982, 205–206.
[406] C. Fuganti, P. Grasselli, G. Marinoni, *Tetrahedron Lett.* 1979, **20**, 1161–1164.
[407] C. Fuganti, P. Grasselli, *Chem. Ind.* 1977, **24**, 983.
[408] C. Fuganti, P. Grasselli, *Chem. Commun.* 1978, 299–300.
[409] R. Bernardi, C. Fuganti, P. Grasselli, G. Marinoni, *Synthesis* 1980, 50–52.
[410] G. Bertolli, G. Fronza, C. Fuganti, P. Grasselli, L. Majori, F. Spreafico, *Tetrahedron Lett.* 1981, **22**, 965–968.
[411] G. Fronza, C. Fuganti, P. Grasselli, L. Majori, G. Pedrocchifantoni, F. Spreafico, *J. Org. Chem.* 1982, **47**, 3289–3296.
[412] C. Fuganti, P. Grasselli, S. Servi, C. Zirotti, *Tetrahedron Lett.* 1982, **23**, 4269–4272.
[413] G. Fronza, C. Fuganti, P. Grasselli, G. Pedrocchifantoni, *J. Carbohydr. Chem.* 1983, **2**, 225–248.
[414] C. Fuganti, P. Grasselli, S. Servi, *J. Chem. Soc., Perkin Trans. 1* 1983, 241–244.
[415] C. Fuganti, P. Grasselli, G. Pedrocchifantoni, S. Servi, C. Zirotti, *Tetrahedron Lett.* 1983, **24**, 3753–3756.
[416] C. Fuganti, P. Grasselli, S. Servi, F. Spreafico, C. Zirotti, P. Casati, *J. Chem. Res. Synop.* 1984, 112–113.
[417] C. Fuganti, P. Grasselli, S. Servi, F. Spreafico, C. Zirotti, P. Casati, *J. Org. Chem.* 1984, **49**, 4087–4089.
[418] G. Fronza, C. Fuganti, P. Grasselli, G. Pedrocchifantoni, *Carbohydr. Res.* 1985, **136**, 115–124.
[419] G. Fronza, C. Fuganti, P. Grasselli, S. Servi, *Tetrahedron Lett.* 1985, **26**, 4961–4964.
[420] C. Fuganti, P. Grasselli, F. Spreafico, C. Zirotti, P. Casati, *J. Chem. Res. Synop.* 1985, 22–23.
[421] G. Fronza, C. Fuganti, P. Grasselli, S. Servi, *Tetrahedron Lett.* 1986, **27**, 4363–4366.
[422] C. Fuganti, P. Grasselli, P. F. Seneci, P. Casati, *Tetrahedron Lett.* 1986, **27**, 5275–5276.
[423] G. Fronza, C. Fuganti, P. Grasselli, S. Servi, *J. Org. Chem.* 1987, **52**, 2086–2089.
[424] C. Fuganti, P. Grasselli, G. Poli, S. Servi, A. Zorzella, *Chem. Commun.* 1988, 1619–1621.
[425] C. Andreu, M. del Olmo, *J. Mol. Catal. B: Enzym.* 2013, **92**, 57-61.
[426] M. del Olmo, C. Andreu, G. Asensio, *J. Mol. Catal. B: Enzym.* 2011, **72**, 90–94.
[427] P. Gramatica, P. Manitto, D. Monti, G. Speranza, *Tetrahedron* 1988, **44**, 1299–1304.
[428] P. Ferraboschi, P. Grisenti, R. Casati, A. Fiecchi, E. Santaniello, *J. Chem. Soc., Perkin Trans. 1* 1987, 1743-1748.
[429] P. Ferraboschi, A. Fiecchi, P. Grisenti, E. Santaniello, *J. Chem. Soc., Perkin Trans. 1* 1987, 1749–1752.
[430] H. G. W. Leuenberger, W. Boguth, R. Barner, M. Schmid, R. Zell, *Helv. Chim. Acta* 1979, **62**, 455–463.
[431] P. Gramatica, P. Manitto, L. Poli, *J. Org. Chem.* 1985, **50**, 4625–4628.
[432] P. Gramatica, P. Manitto, D. Monti, G. Speranza, *Tetrahedron* 1987, **43**, 4481–4486.
[433] Y. Chu, B. L. Zhang, V. Silvestre, J. P. Cheng, *Bioorg. Chem.* 2006, **34**, 158–166.
[434] P. G. Dumanski, P. Florey, M. Knettig, A. J. Smallridge, M. A. Trewhella, *J. Mol. Catal. B* 2001, **11**, 905–908.
[435] F. Parmeggiani, *Chim. Oggi* 2011, **29**, 13–15.
[436] E. Brenna, F. G. Gatti, D. Monti, F. Parmeggiani, A. Sacchetti, *Chem. Commun.* 2012, **48**, 79–81.
[437] F. E. P. Siqueira, J. A. R. Rodrigues, P. J. S. Moran, *Tetrahedron: Asymmetry* 2001, **12**, 847–852.
[438] M. Hirama, T. Noda, S. Ito, *J. Org. Chem.* 1985, **50**, 127–129.
[439] P. Gramatica, P. Manitto, B. M. Ranzi, A. Delbianco, M. Francavilla, *Experientia* 1982, **38**, 775–776.

[440] S. D. Doig, A. T. Boam, D. J. Leak, A. G. Livingston, D. C. Stuckey, *Biocatal. Biotransform.* 1998, **16**, 27–44.

[441] H. Ohta, T. Sugai, Enzyme-mediated decarboxylation reactions in organic synthesis, in R. N. Patel (Ed.) *Stereoselective Biocatalysis*, Marcel Dekker, Inc., New York, 2000, pp. 467–526.

[442] H. Iding, P. Siegert, K. Mesch, M. Pohl, *Biochim. Biophys. Acta. Protein Struct. Mol. Enzymol.* 1998, **1385**, 307–322.

[443] S. R. Indahl, R. R. Scheline, *Appl. Microbiol.* 1968, **16**, 667.

[444] P. Gramatica, B. M. Ranzi, P. Manitto, *Bioorg. Chem.* 1981, **10**, 14–21.

[445] S. L. Chen, H. J. Peppler, *J. Biol. Chem.* 1956, **221**, 101–106.

[446] T. Kitazume, N. Ishikawa, *Chem. Lett.* 1984, 1815–1818.

[447] J. Bujons, R. Guajardo, K. S. Kyler, *J. Am. Chem. Soc.* 1988, **110**, 604–606.

[448] O. Rotthaus, M. Demuth, *Tetrahedron* 2002, **58**, 7291–7293.

[449] J. C. Medina, R. Guajardo, K. S. Kyler, *J. Am. Chem. Soc.* 1989, **111**, 2310–2311.

[450] A. Kamal, P. B. Sattur, *Chem. Commun.* 1989, 835–836.

[451] Y. Fromentin, P. Grellier, J. D. Wansi, M.-C. Lallemand, D. Buisson, *Org. Lett.* 2012, **14**, 5054–5057.

[452] M. Saha, A. K. Pal, *Synth. Commun.* 2013, **43**, 1708–1713.

[453] U. R. Pratap, D. V. Jawale, P. D. Netankar, R. A. Mane, *Tetrahedron Lett.* 2011, **52**, 5817–5819.

[454] U. R. Pratap, D. V. Jawale, M. R. Bhosle, R. A. Mane, *Tetrahedron Lett.* 2011, **52**, 1689–1691.

[455] U. R. Pratap, J. R. Mali, D. V. Jawale, R. A. Mane, *Tetrahedron Lett.* 2009, **50**, 1352–1354.

[456] A. Kumar, R. A. Maurya, *Tetrahedron Lett.* 2007, **48**, 4569–4571.

[457] A. Kumar, R. A. Maurya, *Tetrahedron Lett.* 2007, **48**, 3887–3890.

[458] J. H. Lee, *Tetrahedron Lett.* 2005, **46**, 7329–7330.

[459] V. I. Sokolov, L. L. Troitskaia, O. A. Reutov, *Dokl. Akad. Nauk SSSR* 1977, **237**, 1376–1379.

[460] J. Gillois, D. Buisson, R. Azerad, G. Jaouen, *Chem. Commun.* 1988, 1224–1225.

[461] S. Top, G. Jaouen, J. Gillois, C. Baldoli, S. Maiorana, *Chem. Commun.* 1988, 1284–1285.

[462] A. Patti, S. Pedotti, *Tetrahedron: Asymmetry* 2006, **17**, 778–785.

[463] L. Veum, H. Brouard, P. Meffre, M. Larcheveque, D. Buisson, E. Demousseau, R. Azerad, *Tetrahedron: Asymmetry* 2000, **11**, 4055–4059.

[464] M. Bucciarelli, A. Forni, I. Moretti, G. Torre, *Chem. Commun.* 1978, 456–457.

[465] T. Kitazume, T. Yamazaki, N. Ishikawa, *Nippon Kagaku Kaishi* 1983, 1363–1368.

[466] T. Kitazume, N. Ishikawa, *Chem. Lett.* 1984, 587–590.

[467] T. Fujisawa, M. Shimizu, Biocatalytic approaches to enantiomerically enriched fluoro-organic compounds, in V. A. Soloshonok (Ed.) *Enantiocontrolled Synthesis of Fluoro-Organic Compounds: Stereochemical Challenges and Biomedical Targets*, John Wiley & Sons, Inc., Hoboken, 1999, pp. 307–348.

[468] P. V. Ramachandran, H. C. Brown, Asymmetric reduction of fluorine-containing carbonyl compounds, in V. A. Soloshonok (Ed.) *Enantiocontrolled Synthesis of Fluoro-Organic Compounds: Stereochemical Challenges and Biomedical Targets*, John Wiley & Sons, Inc., Hoboken, 1999, pp. 179–228.

[469] T. Fujisawa, S. Tanaka, Y. Onogawa, M. Shimizu, *Tetrahedron Lett.* 1999, **40**, 1953–1956.

[470] K. Nakamura, T. Matsuda, M. Shimizu, T. Fujisawa, *Tetrahedron* 1998, **54**, 8393–8402.

[471] T. Fujisawa, Y. Onogawa, A. Sato, T. Mitsuya, M. Shimizu, *Tetrahedron* 1998, **54**, 4267–4276.

[472] N. Li, F. Zhang, Y. Gong, *Tetrahedron Lett.* 2007, **48**, 1895–1898.

[473] F. Luo, P. Wang, Y. Gong, *Tetrahedron Lett.* 2010, **51**, 1693–1695.

[474] G. Fronza, C. Fuganti, S. Serra, *Eur. J. Org. Chem.* 2009, 6160–6171.

[475] M. Ordonez, V. Labastida-Galvan, S. Lagunas-Rivera, *Tetrahedron: Asymmetry* 2010, **21**, 129–147.

[476] E. Zymanczyk-Duda, P. Kafarski, B. Lejczak, *Enzyme Microb. Technol.* 2000, **26**, 265–270.

[477] A. K. Bhattacharya, M. Mujahid, *Synth. Commun.* 2013, **43**, 2583–2589.

[478] C.-Y. Yuan, K. Wang, J. Li, Z.-Y. Li, *Phosphorus, Sulfur Silicon Relat. Elem.* 2002, **177**, 2391–2397.

[479] K. Wang, Z. Li, C. Yuan, *Phosphorus, Sulfur Silicon Relat. Elem.* 2002, **177**, 1797–1800.

[480] C.-Y. Yuan, K. Wang, Z.-Y. Li, *Heteroat. Chem.* 2001, **12**, 551–556.

[481] K. Wang, J.-F. Li, C.-Y. Yuan, Z.-Y. Li, *Chin. J. Chem.* 2002, **20**, 1379–1387.

[482] C. Yuan, K. Wang, J. Li, Z. Li, *Heteroat. Chem.* 2002, **13**, 153–156.

CHAPTER 22

Biocatalytic Introduction of Chiral Hydroxy Groups using Oxygenases and Hydratases

Jun Ogawa[1], Makoto Hibi[2], and Shigenobu Kishino[1]

[1] Division of Applied Life Science, Graduate School of Agriculture, Kyoto University, Kyoto, Japan

[2] Laboratory of Industrial Microbiology, Graduate School of Agriculture, Kyoto University, Kyoto, Japan

22.1 INTRODUCTION

To widen the practical use of enzymes for the chemical industry, it is important to diversify available reaction types by developing multicomponent enzymatic processes, which are functional combinations of enzymatic reactions and complex biological systems providing reducing power or cosubstrates for the reaction. Recombinant microorganisms expressing the genes coding oxygenases or hydratases were constructed, and their potentials for chiral alcohol synthesis were evaluated in combination with their accessory components.

Cytochrome P450 monooxygenases act as hydroxylases on the inactive carbon–hydrogen bonds of alkanes, fatty acids, terpenes, and steroids, and are expected to be potential catalysts for fine chemical synthesis. However, the need for the electron-donating cofactor NADPH is one of the main barriers prohibiting the practical use of cytochrome P450 monooxygenases. In this chapter, the example of cytochrome P450 BM-3 from *Bacillus megaterium* for the asymmetric hydroxylation of propylbenzene and 3-chlorostyrene supported by coupling with NADPH regeneration system is described.

Fe(II)/α-ketoglutarate-dependent dioxygenases (Fe/αKG-DOs) also act as hydroxylases on the inactive carbon–hydrogen bonds. Fe/αKG-DOs are novel hydroxylases acting on hydrophilic substrates in contrast with cytochrome P450 monooxygenases that prefer hydrophobic substrates. However, it needs α-ketoglutarate (αKG) as a cosubstrate. We established a hydroxylation system using recombinant *Escherichia coli* expressing Fe/αKG-DOs and providing αKG via a modified central metabolic pathway. Many Fe/αKG-DOs that are applicable for the hydroxylation system are introduced in this chapter.

We recently found a novel fatty acid hydratase with regioselectivity to carbon–carbon double bonds and stereoselectivity to hydroxy group introduction. The hydratase needs FAD and NADH as a cofactor and an activator, respectively.

Green Biocatalysis, First Edition. Edited by Ramesh N. Patel.

The use of this enzyme for chiral hydroxylated fatty acid synthesis coupled with NADH-providing system from glucose via glycolytic pathway of *E. coli* is also described in this chapter.

22.2 REGIO- AND STEREOSELECTIVE HYDROXYLATION OF PROPYLBENZENE AND 3-CHLOROSTYRENE BY CYTOCHROME P450 BM-3 AND ITS MUTANT

Some cytochrome P450s exhibit high regio- and stereoselective monooxygenation activity. Therefore, cytochrome P450s are expected to be potential catalysts for fine chemical synthesis. Cytochrome P450 BM-3 wild type (WT) and a mutant F87G showed high activity toward propylbenzene and 3-chlorostyrene with considerable stereoselectivity [1]. Furthermore, reaction analysis using cytochrome P450 BM-3 mutants revealed that the stereoselectivity is under the control of the residue size at position 87 of this enzyme [1].

We reported the oxidation of propylbenzene and 3-chlorostyrene by WT with high turnover (479 nmol 1-phenyl-1-propanol/min/nmol P450 and 300 nmol 3-chlorostyrene oxide/min/nmol P450). Furthermore, the residue size at position 87 of P450 BM-3 was found to play a critical role in determining stereoselectivity in the oxidation of propylbenzene and 3-chlorostyrene. Replacement of Phe87 with Val, Ala, and Gly resulted in decreases in the optical purity of the (*R*)-(+)-1-phenyl-1-propanol produced from 90.0 to 37.4, 26.0 and −15.6% ee, respectively, and in increases in those of the (*R*)-(+)-3-chlorostyrene oxide produced from −61.0 to −38.0, 67.0, and 94.6% ee, respectively [1]. A WT and a mutant with replacement of Phe87 with Gly (F87G) were applied to asymmetric hydroxylation coupled with NADPH regeneration system using glucose dehydrogenase (Figure 22.1). These results indicated the high potential of cytochrome P450 BM-3 and its mutants for the preparation of useful chiral compounds.

22.3 REGIO- AND STEREOSELECTIVE HYDROXYLATION OF ALIPHATIC AMINO ACIDS BY FE(II)/α-KETOGLUTARATE-DEPENDENT DIOXYGENASES

Various types of monooxygenation reactions are carried out by the enzymes belonging to the Fe/αKG-DO superfamily [2]. Fe/αKG-DOs catalyzes the monooxygenation of a wide range of "primary" substrates accompanied by the oxidative decarboxylation of αKG, a common "secondary" substrate, into succinate (Figure 22.2). With respect to their use as biocatalysts, Fe/αKG-DOs are characterized by the ability to react with various hydrophilic compounds. Amino acids are typical substrates of Fe/αKG-DOs and are usually converted into hydroxy amino acids. Among the Fe/αKG-DOs acting on free amino acids, L-proline hydroxylases have been well studied for industrial applications, because the resulting hydroxyprolines are important intermediates used to prepare various kinds of chiral pharmaceuticals and chiral antibiotics [3, 4]. To date, three different types of the L-proline hydroxylases have been identified from microorganisms: L-proline *cis*-4-hydroxylase (*cis*-P4H) [5], L-proline *trans*-4-hydroxylase (*trans*-P4H) [6], and L-proline *cis*-3-hydroxylase (*cis*-P3H) [7]. Of these, *trans*-P4H and *cis*-P3H were used in the large-scale productions of *trans*-4-hydroxy-L-proline and *cis*-3-hydroxy-L-proline, respectively [8, 9].

Recently, we identified several novel types of Fe/αKG-DOs that catalyze the hydroxylation of free aliphatic amino acids. Some hydroxy aliphatic amino acids have several chiral carbons and may be important as precursors and chiral auxiliaries in the chemical synthesis of other compounds [10, 11]. In this section, we describe these novel Fe/αKG-DOs and reviewed the homology of their amino acid sequences, catalytic characteristics, and process development toward their industrial applications.

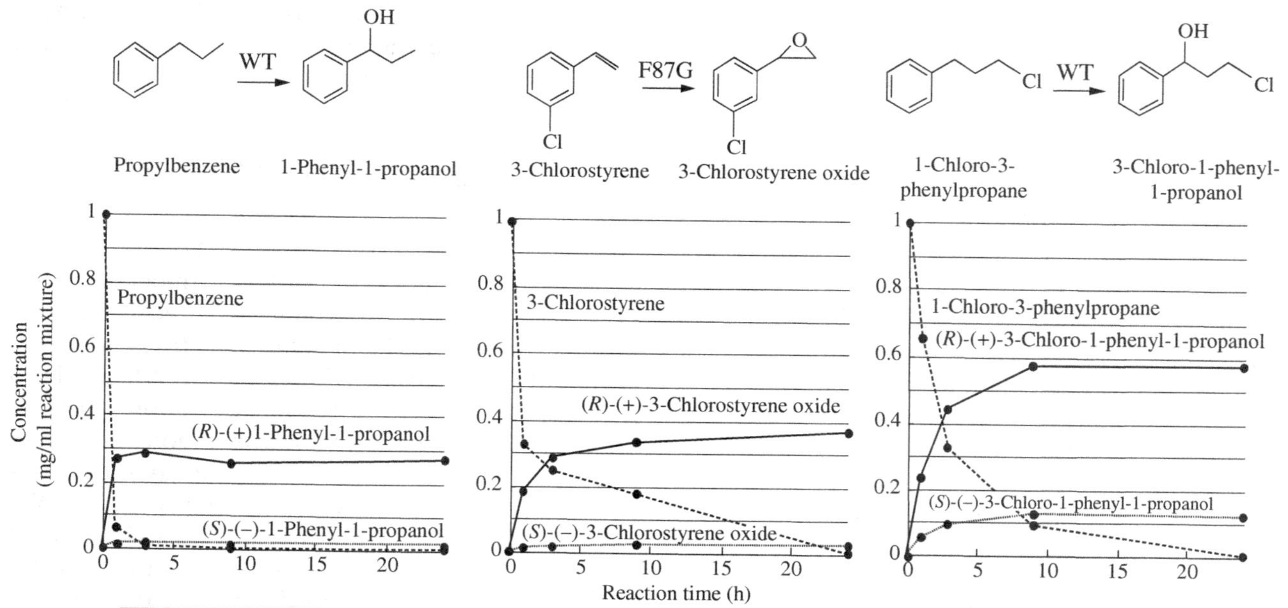

Yield (%)	28.6
Optical purity of (*R*)-product (% ee)	90.0
Activity (nmol/min/nmol P450)	0.429

Yield (%)	36.5
Optical purity of (*R*)-product (% ee)	90.6
Activity (nmol/min/nmol P450)	3.10

Yield (%)	68.9
Optical purity of (*R*)-product (% ee)	65.3
Activity (nmol/min/nmol P450)	1.65

FIGURE 22.1

Regio- and stereoselective transformation of propylbenzene and 3-chlorostyrene by cytochrome P450 BM-3 WT and a mutant F87G.

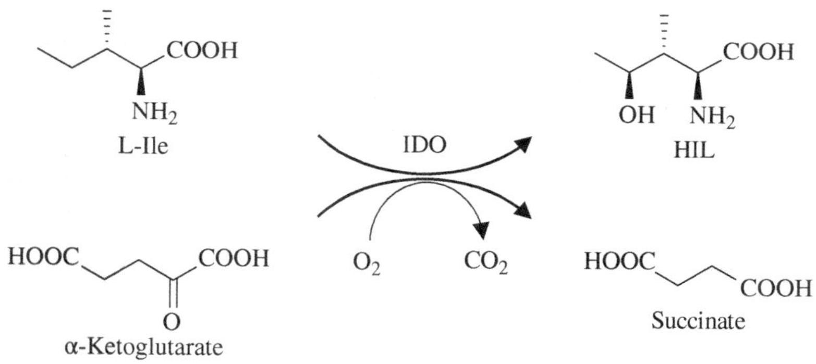

FIGURE 22.2

Representative reaction for asymmetric monooxygenation of L-amino acids with Fe/αKG-DOs: L-Ile hydroxylation by IDO.

22.3.1 L-Isoleucine 4-Hydroxylase

Recently, we reported a novel metabolic pathway for L-isoleucine via (2*S*,3*R*,4*S*)-4-hydroxyisoleucine (HIL) and (2*S*,3*R*)-2-amino-3-methyl-4-ketopentanoic acid (AMKP) in *Bacillus thuringiensis* 2e2 [12, 13]. HIL synthesis was found to be catalyzed stereoselectively by a novel Fe/αKG-DO, L-isoleucine 4-hydroxylase (IDO) (Figure 22.2). This metabolic pathway functions as an effective bypass pathway that compensates for the incomplete TCA cycle in *Bacillus* species [14].

Moreover, IDO stereoselectively hydroxylates several aliphatic L-amino acids as well as L-isoleucine [15]. It converts L-leucine and L-norvaline into (2*S*)-4-hydroxyleucine and (2*S*,4*S*)-4-hydroxynorvaline, respectively. When L-norleucine is used as the substrate, single diastereomers of (2*S*)-4-hydroxynorleucine and

(2*S*)-5-hydroxynorleucine are produced together at a ratio of 10:1. Interestingly, when IDO reacts with L-*allo*-isoleucine, it does not produce a C4-hydroxylated product but rather a C3-hydroxylated form of L-*allo*-isoleucine.

22.3.2 Fe/αKG-DOs Closely Homologous with L-Isoleucine 4-Hydroxylase

IDO is a member of the Pfam family PF10014 (the former DUF 2257) of uncharacterized conserved bacterial proteins. By *in silico* screening of PF10014 member enzymes, five function-unknown Fe/αKG-DOs were found to have an amino acid sequence homology with IDO: AVI of *Agrobacterium vitis* S4, BPE of *Bordetella petrii* DSM 12804, GOX of *Gluconobacter oxydans* 621H, MFL of *Methylobacillus flagellatus* KT, and HilB of *Pantoea ananatis* AJ13355 (Table 22.1). According to the results of substrate specificity analysis, all of these Fe/αKG-DOs have the sulfoxidation activity of L-methionine in common. AVI, BPE, GOX, and MFL convert L-leucine into (2*S*)-4-hydroxyleucine as well as IDO, but they do not react with L-isoleucine. As for AVI and BPE, they additionally have unique terminal C4-hydroxylation activity toward L-threonine and produce the corresponding diol. Although HilB was shown to catalyze C4-hydroxylation of L-isoleucine and L-leucine, the primary role of HilB was recently elucidated in *P. ananatis* AJ13355. There is another gene-encoded Fe/αKG-DO, HilA, next to HilB gene in the genome of the microorganism. HilA has no amino acid sequence homology with IDO and has the specific ability to terminally hydroxylate the methyl group of L-isoleucine and L-valine to produce (2*S*,3*S*)-4′-hydroxyisoleucine and (2*S*)-4-hydroxyvaline. Because HilB also has C4-hydroxylation activity of (2*S*,3*S*)-4′-hydroxyisoleucine, a novel metabolic pathway converting L-isoleucine into (2*S*,3*S*,4*S*)-4,4′-hydroxyisoleucine is carried out by HilA and HilB in *P. ananatis* AJ13355 (Figure 22.3). Thus, some IDO-like Fe/αKG-DOs categorized as PF10014 are highly regioselective hydroxylases acting on the C4-position of aliphatic L-amino acid substrates (Figure 22.4).

TABLE 22.1 Fe(II)/αKG-Dependent Dioxygenases (Fe/αKG-Dos) Acting as Stereoselective Aliphatic Amino Acid Hydroxylases

Fe/αKG-DOs	Gene Bank Accession No.	Origin	AA Homology with IDO (%)	
			Identity	Similarity
IDO	ADJ94127.1	*Bacillus thuringiensis* 2e2	100	100
IDO-like Fe/αKG-DO				
HilB	BAK13117.1	*Pantoea ananatis* AJ13355	37	54
GOX	YP_192070.1	*Gluconobacter oxydans* 621H	26	47
BPE	YP_001629976.1	*Bordetella petrii* DSM 12804	26	44
MFL	YP_546733.1	*Methylobacillus flagellatus* KT	24	46
AVI	ABG82019.1	*Agrobacterium vitis* S4	23	45
Other Fe/αKG-DO				
HilA	BAK13116.1	*P. ananatis* AJ13355	n.d.	n.d.
LdoA	ACC80786.1	*Nostoc punctiforme* PCC 73102	n.d.	n.d.
SadA	YP_777923.1	*Burkholderia ambifaria* AMMD	n.d.	n.d.

n.d., not detected.

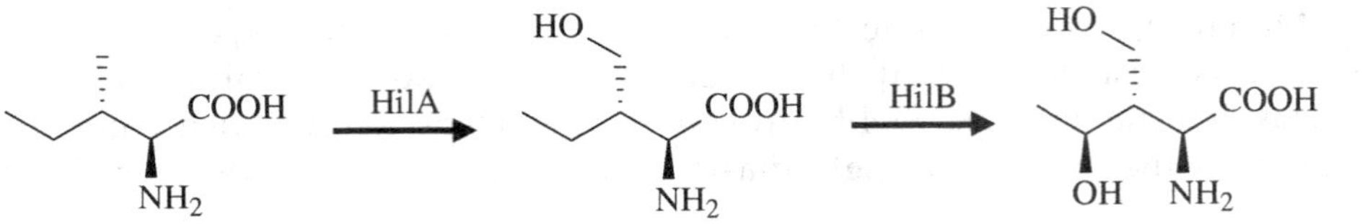

FIGURE 22.3 Cascade hydroxylation reactions of L-Ile catalyzed by HilA and HilB.

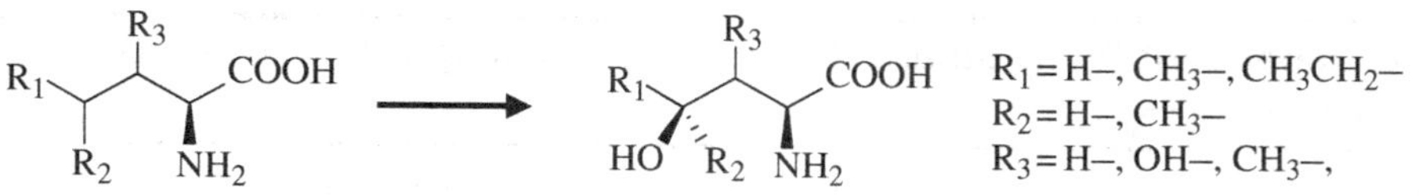

FIGURE 22.4 Regio- and stereoselective hydroxylation of aliphatic L-amino acids catalyzed by IDO and its homologs.

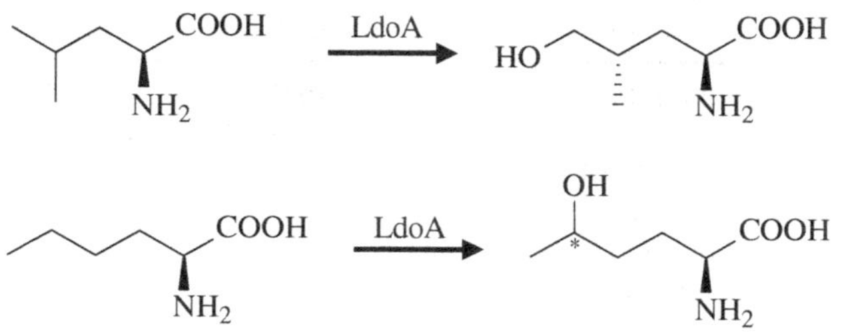

FIGURE 22.5 Regio- and stereoselective hydroxylation of L-Leu catalyzed by LdoA.

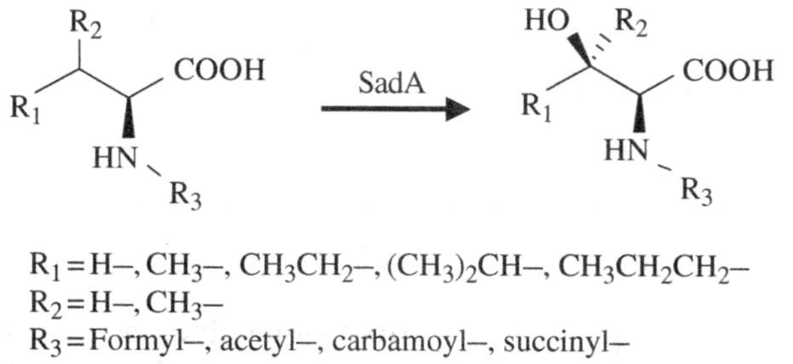

FIGURE 22.6 Regio- and stereoselective hydroxylation of *N*-substituted aliphatic L-amino acids catalyzed by SadA.

22.3.3 L-Leucine 5-Hydroxylase

L-Leucine 5-hydroxylase (LdoA) found in *Nostoc punctiforme* PCC 73102 is a unique type of Fe/αKG-DO [16]. LdoA has no amino acid sequence homology with, and shows distinct regioselectivity in hydroxylation when compared to, the IDO-like Fe/αKG-DOs and HilA (Table 22.1). LdoA catalyzes regio- and stereoselective C5-hydroxylation of L-leucine and L-norleucine into (2*S*,4*S*)-5-hydroxyleucine and (2*S*)-5-hydroxynorleucine, respectively (Figure 22.5). (2*S*,4*S*)-5-Hydroxyleucine is an intermediate in the conversion pathway of L-leucine into (2*S*,4*S*)-4-methylproline, which is known to be a common component of nostopeptolides and nostocyclopeptides produced by several *Nostoc* species [17–19]. LdoA is the enzyme responsible for this conversion pathway in *N. punctiforme* PCC 73102 [20]. Nostopeptolide A1 and A3 produced by *Nostoc* sp. Lukešová 30/93 were found to be weak inhibitors of the expression and production of proinflammatory mediators in response to TNF treatment in human lung microvascular cells [21]. Thus, (2*S*,4*S*)-4-methylproline biosynthetic enzymes like LdoA could be integral for the industrial production of these bioactive peptides.

22.3.4 *N*-Succinyl L-Leucine 3-Hydroxylase

As another unique type of Fe/αKG-DO, *N*-succinyl L-leucine 3-hydroxylase (SadA) was found in *Burkholderia ambifaria* AMMD [22]. SadA has no amino acid sequence homology with, and shows distinct regioselectivity in hydroxylation, when compared to the IDO-like Fe/αKG-DOs, HilA, and LdoA (Table 22.1). SadA has catalytic activity toward several *N*-substituted L-amino acids, such as *N*-succinyl, *N*-acetyl, *N*-formyl, and *N*-carbamoyl ones (Figure 22.6). Among them, *N*-succinyl L-leucine is the best substrate for SadA, and it is converted into *N*-succinyl L-*threo*-β-hydroxyleucine with over 99% diastereoselectivity. Similarly, SadA catalyzes stereoselective C3-hydroxylation of several *N*-succinyl L-amino acids, such as *N*-succinyl L-valine, *N*-succinyl L-2-aminobutyrate, *N*-succinyl L-isoleucine, and *N*-succinyl L-leucine into corresponding *N*-succinyl β-hydroxy L-amino acids, such as *N*-succinyl L-β-hydroxyvaline, *N*-succinyl

TABLE 22.2 Properties of Fe(II)/αKG-Dependent Dioxygenases (Fe/αKG-Dos) Acting as Stereoselective Aliphatic Amino Acid Hydroxylases

	Hydroxylation		Sulfoxidation	
Fe/αKG-DOs	Substrate (Regio-/Stereoselectivity)	k_{cat}/K_m (min^{-1} mM)	Substrate (Stereoselectivity)	k_{cat}/K_m (min^{-1} mM)
IDO	L-Ile, L-norvaline (4/*S*); L-norleucine (4,5/n.d.); L-Leu (4/-); L-*allo*-Ile (3/*R*)	120 (L-Ile)	L-Met, L-ethionine, *S*-methyl-L-Cys, *S*-ethyl-L-Cys, *S*-allyl-L-Cys (*S*)	8.4 (L-Met)
AVI	L-Leu (4/*S*); L-Thr (4/-)	6.4 (L-Leu)	L-Met (n.d.)	1.4 (L-Met)
BPE	L-Leu (4/*S*); L-Thr (4/-)	10 (L-Leu)	L-Met (n.d.)	1.6 (L-Met)
GOX	L-Leu (4/*S*)	66 (L-Leu)	L-Met (n.d.)	8.5 (L-Met)
MFL	L-Leu (4/*S*)	>39 (L-Leu)	L-Met (n.d.)	1.9 (L-Met)
HilA	L-Ile (4′/-); L-Val (4/-)	n.d.	—	—
HilB	L-Ile, L-Leu, 4′-OH-L-Ile (4/*S*)	4.7 (L-Leu)	L-Met (n.d.)	2.3 (L-Met)
LdoA	L-Leu (5/4*S*), L-norleucine (5/n.d.)	9.4 (L-Leu)	L-Met, L-ethionine (n.d.)	5.8 (L-Met)
SadA	*Ns*-L-Leu, *Ns*-L-Ile, *Ns*-L-aminobutyrate, *Ns*-L-Phe (3/*R*); *Ns*-L-*allo*-Ile (3/*S*); *Ns*-L-norleucine, *Ns*-L-norvaline (n.d./n.d.)	14 (*Ns*-L-Leu)	*Ns*-L-Met (n.d.)	n.d.

-, non chiral; n.d., not determined; *Ns*, *N*-Succinyl.

L-threonine, (2*S*,3*R*)-*N*-succinyl L-β-hydroxyisoleucine, and *N*-succinyl L-*threo*-β-hydroxyleucine. L-*threo*-β-Hydroxyleucine is a promising target material for the preparation of cyclic depsipeptides that possess useful physiological activities [23, 24].

Using a combination of *N*-succinyl L-leucine 3-hydroxylase (SadA) and LasA, *N*-succinyl L-amino acid desuccinylase (LasA) found in *B. ambifaria* strain AMMD, *N*-succinyl L-leucine was successfully converted into L-*threo*-β-hydroxyleucine with 93% molar yield and over 99% diastereomeric excess [25]. Consequently, this new biocatalytic production system has advantages in optical purity and reaction efficiency for application in the industrial mass production of several useful β-hydroxy α-amino acids.

22.3.5 Catalytic Properties of the Aliphatic Amino Acid Hydroxylases

Substrate specificities, regio-/stereoselectivities, and specificity constants (k_{cat}/K_m) for hydroxylation and sulfoxidation reactions using IDO-like Fe/αKG-DOs, HilA, LdoA, and SadA are summarized in Table 22.2. Most of these enzymes, except for HilA, are confirmed to have sulfoxidation activity as well as hydroxylation activity, and in these reactions, the side chains of L-methionine and L-leucine, respectively, are common substrates. Based on the comparison of their specificity constants, the hydroxylation reaction is more effective than the sulfoxidation reaction in all cases, and the hydroxylation of L-isoleucine with IDO is the most efficient among the reactions of these Fe/αKG-DOs. Additionally, preference for weak acidic and low-temperature conditions is common to the hydroxylation reactions of these Fe/αKG-DOs. For example, the highest activities were observed at pH 6.0 and 30 °C for IDO [12], at pH 5.0 and 20 °C for LdoA [16], and at pH 6.5 and 30 °C for SadA [22].

22.3.6 Practical Use of Fe(II)/α-Ketoglutarate-Dependent Dioxygenases Coupled with Cosubstrate Generation System

HIL, which was originally found in the seeds of fenugreek, the annual herb *Trigonella foenum-graecum* [26], is a promising compound for drug and functional food development because of its antidiabetes and antiobesity activity. We have recently developed an efficient production system for HIL using a genetically manipulated *E. coli* 2Δ strain

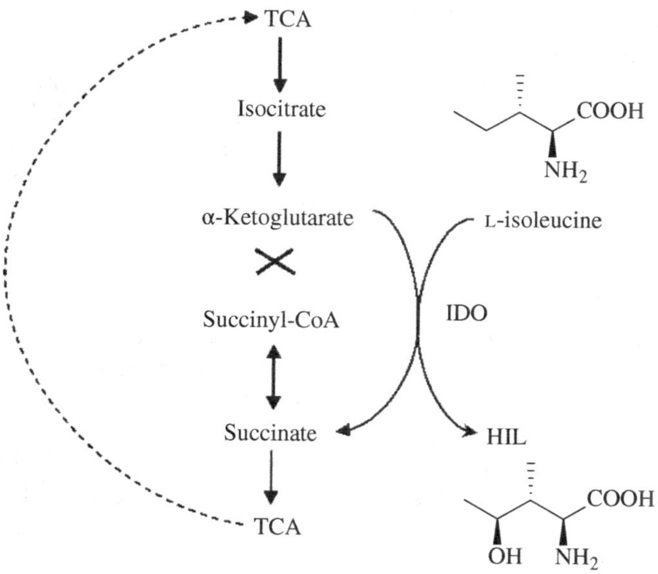

FIGURE 22.7
Modified TCA cycle in the *Escherichia coli* 2Δ strain resulting from the simultaneous oxidation of L-isoleucine and αKG.

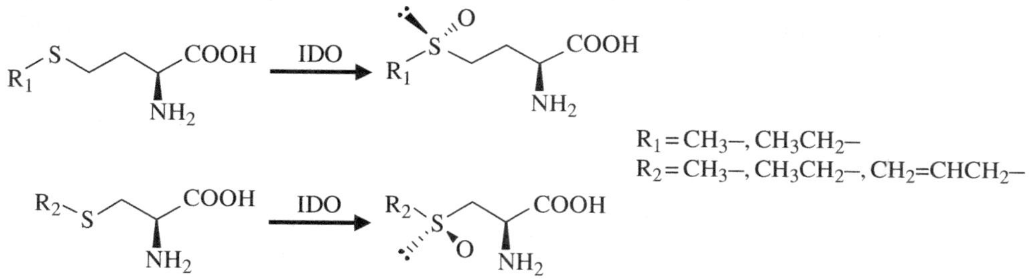

FIGURE 22.8
Asymmetric sulfoxidation of sulfur-containing L-amino acids catalyzed by IDO.

heterologously expressing IDO [27]. Because the *E. coli* 2Δ strain lacks the activities of αKG dehydrogenase, isocitrate lyase, and isocitrate dehydrogenase kinase/phosphatase, it could not grow in a minimal medium due to the blockage of TCA during succinate synthesis. The IDO activity in the 2Δ strain was able to "shunt" destroyed TCA, thereby coupling L-isoleucine hydroxylation and cell growth (Figure 22.7). Using this strain, we performed the direct biotransformation of L-isoleucine into HIL with 82% yield and over 99% optical purity. This system could easily be applied to the production of other IDO-producible hydroxy amino acids described earlier.

Furthermore, IDO was found to catalyze asymmetric sulfoxidation of L-methionine, L-ethionine, *S*-methyl-L-cysteine, *S*-ethyl-L-cysteine, and *S*-allyl-L-cysteine into corresponding (*S*)-configured sulfoxides such as L-methionine (*S*)-sulfoxide, L-ethionine (*S*)-sulfoxide, (+)-methiin, (+)-ethiin, and (+)-alliin (Figure 22.8) [28]. (+)-Methiin and (+)-alliin are promising materials for use in functional foods or drugs since they have antibiotic, antioxidant, anti-inflammatory, antidiabetic, anti-Alzheimer's, and anticholesterolemic effects. By using the dried cell powder of IDO-expressing *E. coli* as a biocatalyst, preparative scale productions of these chiral sulfoxides with high enantiomeric purity could easily be accomplished via bioconversion, allowing further development of their industrial applications.

22.4 REGIO- AND STEREOSELECTIVE HYDRATION OF UNSATURATED FATTY ACIDS BY A NOVEL FATTY ACID HYDRATASE

Fatty acid saturation metabolism, so called biohydrogenation, is considered to be a detoxifying metabolism of gut bacteria to transform toxic-free polyunsaturated fatty acids to less toxic-free saturated fatty acids. We have revealed the complex metabolic pathway of the biohydrogenation in lactic acid bacteria in detail [29]. The enzyme system of biohydrogenation was found to consist of four enzymes. The required action of these enzymes, that is, hydratase, dehydrogenase, isomerase, and enone reductase, accomplished the saturation of a carbon–carbon double bond and generated unique PUFA molecular species such as hydroxy fatty acids, oxo fatty acids, and conjugated fatty acids as intermediates. Chiral hydroxy fatty acids were efficiently produced by the hydratase under optimized conditions.

22.4.1 Linoleic Acid Δ9 Hydratase

Linoleic acid Δ9 hydratase, which is involved in the linoleic acid saturation metabolism of *Lactobacillus plantarum* AKU 1009a, was cloned as his-tagged recombinant enzyme, purified with affinity column, and characterized [30]. The enzyme required FAD as a cofactor for its activity, and the activity was enhanced by NADH. The maximum activities for hydration of linoleic acid and for dehydration of 10-hydroxy-*cis*-12-octadecenoic acid (HYA) were observed at 37 °C, pH 5.5, with 0.5 M NaCl. C16 and C18 free fatty acids with *cis*-9 double bond served as substrates for hydration with C10 regiospecificity and (*S*) stereospecificity (Figure 22.9). 10-Hydroxy fatty acids served as substrates for dehydration reactions. The apparent K_m value for linoleic acid was estimated to be 92 μM with its k_{cat} values at $2.6 \times 10^{-2}\,s^{-1}$, and Hill factor was 3.3. The apparent K_m value for HYA was estimated to be 98 μM with its k_{cat} values at $1.2 \times 10^{-3}\,s^{-1}$.

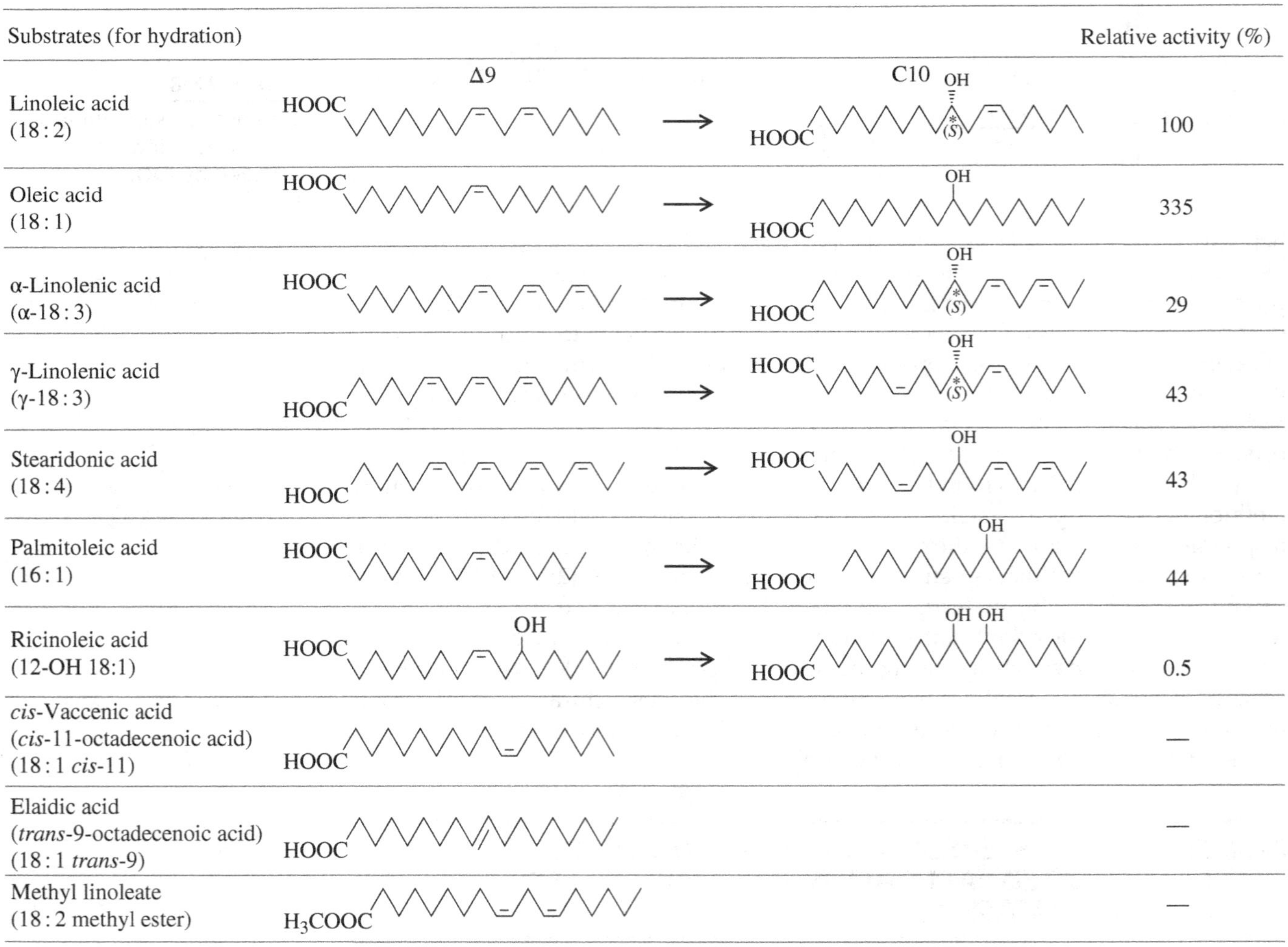

FIGURE 22.9

Substrate specificity of hydration reaction catalyzed by linoleic acid Δ9 hydratase.

22.4.2 Efficient Enzymatic Production of Hydroxy Fatty Acids by Linoleic Acid Δ9 Hydratase

Escherichia coli overexpressing linoleic acid Δ9 hydratase from *L. plantarum* AKU 1009a was applied to produce hydroxy fatty acids with industrial potential [30]. 280 g/l of linoleic acid (1 M) was converted into HYA with a high conversion rate, 98% (mol/mol), by the cells of the recombinant *E. coli* in the presence of FAD (0.1 mM) and NADH (5 mM). Lowering the reaction temperature reduced the solubility of the products and resulted in a high yield by controlling the reaction equilibrium. Oleic acid, α-linolenic acid, γ-linolenic acid, stearidonic acid, *cis*-9-hexadecenoic acid, and 12-hydroxy-*cis*-9-octadecenoic acid were also converted into corresponding 10-hydroxy fatty acids with conversion rates of 98, 96, 95, 93, 98, and 99% (mol/mol) to 280 g/l of each substrate, respectively. Although the hydratase requires NADH for its maximal activity, the reaction catalyzed by the recombinant *E. coli* cells proceeded well with glucose instead of NADH, suggesting that the *E. coli* cells supplied NADH via glucose metabolism. HYA was produced with high accumulation (289 g/l) and high yield (97 mol%) in the reaction mixture containing glucose instead of NADH.

22.5 CONCLUSION

Bioprocesses, which involve hydroxylation biocatalysts for the production of useful chiral alcohols, are expected to become a leading player in green chemistry. The first step in bioprocess development is the creation of useful catalysts. This chapter introduces some examples of enzyme library expansion stemming from screening of novel enzymes. Recently, rational methods for creating new biocatalysts have been rapidly developed. Modern gene technology, crystal structure analysis, and bioinformatics make possible the modulation of enzyme functions through site-directed mutagenesis, DNA shuffling, and so on. In this chapter, we presented one example of modification of the substrate specificity of cytochrome P450 BM-3. These results proved the potentials of the artificially modified enzymes for fine chemical synthesis.

ACKNOWLEDGMENT

This work was supported in part by the Japanese Ministry of Education, Science, Sports and Culture Grants-in-Aid for Scientific Research (KAKENHI) 23248014 and 22658027 (to Jun Ogawa) and 30432347 and 21780070 (to Makato Hibi), the Bio-oriented Technology Research Advancement Institution of Japan (to Jun Ogawa), and NEDO Innovation Commercialization Venture Support Project (for collaboration of NITTO PHARMA and Jun Ogawa).

REFERENCES

[1] Li, Q.-S., Ogawa, J., Schmid, R. D., Shimizu, S. (2001). Residue size at position 87 of cytochrome P450 BM-3 determines its stereoselectivity in propylbenzene and 3-chlorostyrene oxidation. *FEBS Lett.*, **508**, 249–252.

[2] Hausinger, R. P. (2004). Fe(II)/α-ketoglutarate-dependent hydroxylases and related enzymes. *Crit. Rev. Biochem. Mol. Biol.*, **39**, 21–68.

[3] Berkes, D., Kolarovic, A., Povazanec, F. (2000). Stereoselective sodium borohydride reduction, catalyzed by manganese(II) chloride, of γ-oxo-α-amino acids. A practical approach to *syn*-γ-hydroxy-α-amino acids. *Tetrahedron Lett.*, **41**, 5257–5260.

[4] Remuzon, P. (1996). *Trans*-4-hydroxy-L-proline, a useful and versatile chiral starting block. *Tetrahedron*, **52**, 13803–13835.

[5] Hara, R., Kino, K. (2009). Characterization of novel 2-oxoglutarate dependent dioxygenases converting L-proline to *cis*-4-hydroxy-L-proline. *Biochem. Biophys. Res. Commun.*, **379**, 882–886.

[6] Shibasaki, T., Mori, H., Chiba, S., Ozaki, A. (1999). Microbial proline 4-hydroxylase screening and gene cloning. *Appl. Environ. Microbiol.*, **65**, 4028–4031.

[7] Mori, H., Shibasaki, T., Yano, K., Ozaki, A. (1997). Purification and cloning of a proline 3-hydroxylase, a novel enzyme which hydroxylates free L-proline to *cis*-3-hydroxy-L-proline. *J. Bacteriol.*, **179**, 5677–5683.

[8] Johnston, R. M., Chu, L. N., Liu, M., Goldberg, S. L., Goswami, A., Patel, R. N. (2009). Hydroxylation of L-proline to *cis*-3-hydroxy-L-proline by recombinant *Escherichia coli* expressing a synthetic L-proline-3-hydroxylase gene. *Enzyme Microb. Technol.*, **45**, 484–490.

[9] Shibasaki, T., Mori, H., Ozaki, A. (2000). Enzymatic production of *trans*-4-hydroxy-L-proline by regio- and stereospecific hydroxylation of L-proline. *Biosci. Biotechnol. Biochem.*, **64**, 746–750.

[10] Blaskovich, M. A., Evindar, G., Rose, N. G. W., Wilkinson, S., Luo, Y., Lajoie, G. A. (1998). Stereoselective synthesis of *threo* and *erythro* β-hydroxy and β-disubstituted-β-hydroxy α-amino acids. *J. Org. Chem.*, **63**, 3631–3646.

[11] Palomo, C., Arrieta, A., Cossio, F. P., Aizpurua, J. M., Mielgo, A., Aurrekoetxea, N. (1990). Highly stereoselective synthesis of α-hydroxy β-amino acids through β-lactams: Application to the synthesis of the taxol and bestatin side chains and related systems. *Tetrahedron Lett.*, **31**, 6429–6432.

[12] Kodera, T., Smirnov, S. V., Samsonova, N. N., Kozlov, Y. I., Koyama, R., Hibi, M., Ogawa, J., Yokozeki, K., Shimizu, S. (2009). A novel L-isoleucine hydroxylating enzyme, L-isoleucine dioxygenase from *Bacillus thuringiensis*, produces (2*S*,3*R*,4*S*)-4-hydroxyisoleucine. *Biochem. Biophys. Res. Commun.*, **390**, 506–510.

[13] Ogawa, J., Kodera, T., Smirnov, S. V., Hibi, M., Samsonova, N. N., Koyama, R., Yamanaka, H., Mano, J., Kawashima, T., Yokozeki, K., Shimizu, S. (2011). A novel L-isoleucine metabolism in *Bacillus thuringiensis* generating (2*S*,3*R*,4*S*)-4-hydroxyisoleucine, a potential insulinotropic and anti-obesity amino acid. *Appl. Microbiol. Biotechnol.*, **89**, 1929–1938.

[14] Perlman, D., Perlman, K. L., Bodanszky, M. (1977). Microbial production of vitamin B_{12} antimetabolites. II. 2-amino-4-keto-3-methylpentanoic acids from *Bacillus cereus* 439. *Bioorg. Chem.*, **6**, 263–271.

[15] Hibi, M., Kawashima, T., Kodera, T., Smirnov, S. V., Sokolov, P. M., Sugiyama, M., Shimizu, S., Yokozeki, K., Ogawa, J. (2011). Characterization of *Bacillus thuringiensis* L-isoleucine dioxygenase for production of useful amino acids. *Appl. Environ. Microbiol.*, **77**, 6926–6930.

[16] Hibi, M., Kawashima, T., Sokolov, P. M., Smirnov, S. V., Kodera, T., Sugiyama, M., Shimizu, S., Yokozeki, K., Ogawa, J. (2012). L-Leucine 5-hydroxylase of *Nostoc punctiforme* is a novel type of Fe(II)/α-ketoglutarate-dependent dioxygenase that is useful as a biocatalyst. *Appl. Microbiol. Biotechnol.*, **97**, 2467–2472.

[17] Becker, J. E., Moore, R. E., Moore, B. S. (2004). Cloning, sequencing, and biochemical characterization of the nostocyclopeptide biosynthetic gene cluster: Molecular basis for imine macrocyclization. *Gene*, **325**, 35–42.

[18] Hoffmann, D., Hevel, J. M., Moore, R. E., Moore, B. S. (2003). Sequence analysis and biochemical characterization of the nostopeptolide A biosynthetic gene cluster from *Nostoc* sp. GSV224. *Gene*, **311**, 171–180.

[19] Jokela, J., Herfindal, L., Wahlsten, M., Permi, P., Selheim, F., Vasconcelos, V., Doskeland, S. O., Sivonen, K. (2010). A novel cyanobacterial nostocyclopeptide is a potent antitoxin against microcystins. *ChemBioChem*, **11**, 1594–1599.

[20] Luesch, H., Hoffmann, D., Hevel, J. M., Becker, J. E., Golakoti, T., Moore, R. E. (2003). Biosynthesis of 4-methylproline in cyanobacteria: Cloning of *nosE* and *nosF* genes and biochemical characterization of the encoded dehydrogenase and reductase activities. *J. Org. Chem.*, **68**, 83–91.

[21] Pflüger, M., Kapuscik, A., Lucas, R., Koppensteiner, A., Katzlinger, M., Jokela, J., Eger, A., Jacobi, N., Wiesner, C., Hofmann, E., Önder, K., Kopecky, J., Schütt, W., Hundsberger, H. (2013). A combined impedance and alfaLISA-based approach to identify anti-inflammatory and barrier-protective compounds in human endothelium. *J. Biomol. Screen.*, **18**, 67–74.

[22] Hibi, M., Kawashima, T., Kasahara, T., Sokolov, P. M., Smirnov, S. V., Kodera, T., Sugiyama, M., Shimizu, S., Yokozeki, K., Ogawa, J. (2012). A novel Fe(II)/α-ketoglutarate-dependent dioxygenase from *Burkholderia ambifaria* has β-hydroxylating activity of *N*-succinyl L-leucine. *Lett. Appl. Microbiol.*, **55**, 414–441.

[23] Taniguchi, M., Suzumura, K. I., Nagai, K., Kawasaki, T., Saito, T., Takasaki, J., Suzuki, K. I., Fujita, S., Tsukamoto, S. I. (2003). Structure of YM-254890, a novel $G_{q/11}$ inhibitor from *Chromobacterium* sp. QS3666. *Tetrahedron*, **59**, 4533–4538.

[24] Tymiak, A. A., McCormick, T. J., Unger, S. E. (1989). Structure determination of lysobactin, a macrocyclic peptide lactone antibiotic. *J. Org. Chem.*, **54**, 1149–1157.

[25] Hibi, M., Kasahara, T., Kawashima, T., Yajima, H., Kozono, S., Smirnov, S. V., Kodera, T., Sugiyama, M., Shimizu, S., Yokozeki, K., Ogawa, J. (2015). Multi-enzymatic synthesis of optically pure β-hydroxy α-amino acids. *Adv. Synth. Catal.*, **357**, 767–774.

[26] Fowden, L., Pratt, H. M., Smith, A. (1973). 4-Hydroxyisoleucine from seed of *Trigonella foenum-graecum*. *Phytochemistry*, **12**, 1707–1711.

[27] Smirnov, S. V., Kodera, T., Samsonova, N. N., Kotlyarova, V. A., Rushkevich, N. Y., Kivero, A. D., Sokolov, P. M., Hibi, M., Ogawa, J., Shimizu, S. (2010). Metabolic engineering of *Escherichia coli* to produce (2*S*,3*R*,4*S*)-4-hydroxyisoleucine. *Appl. Microbiol. Biotechnol.*, **88**, 719–726.

[28] Hibi, M., Kawashima, T., Yajima, H., Smirnov, S. V., Kodera, T., Sugiyama, M., Shimizu, S., Yokozeki, K., Ogawa, J. (2013). Enzymatic synthesis of chiral amino acid sulfoxides by Fe(II)/α-ketoglutarate-dependent dioxygenase. *Tetrahedron: Asymmetry*, **24**, 990–994.

[29] Kishino, S., Takeuchi, M., Park, S. B., Hirata, A., Kitamura, N., Kunisawa, J., Kiyono, H., Iwamoto, R., Isobe, Y., Arita, M., Arai, H., Ueda, K., Shima, J., Takahashi, S., Yokozeki, K., Shimizu, S., Ogawa, J. (2013). Polyunsaturated fatty acid saturation by gut lactic acid bacteria affecting host lipid composition. *Proc. Natl. Acad. Sci. U. S. A.*, **110**, 17808–17813.

[30] Takeuchi, M., Kishino, S., Hirata, A., Park, S. B., Kitamura, N., Ogawa, J. (2015). Characterization of the linoleic acid Δ9 hydratase catalyzing the first step of polyunsaturated fatty acid saturation metabolism in *Lactobacillus plantarum* AKU 1009a. *J. Biosci. Bioeng.*, **119**, 636–641.

CHAPTER 23

Asymmetric Synthesis with Recombinant Whole-Cell Catalysts

Harald Gröger, Werner Hummel, and Severin Wedde

Faculty of Chemistry, Bielefeld University, Bielefeld, Germany

23.1 INTRODUCTION

Asymmetric catalytic transformations play a central role in the field of organic chemistry and are of great interest for applications in the chemical and pharmaceutical industry. Among those, hydrolytic and reverse reactions to form carboxylate derivatives as well as C—C bond-forming and redox reactions are of particular importance. For a long time this field was dominated by the use of chemocatalysts [1]. For example, numerous chiral heavy-metal complexes have been found to be highly efficient catalysts, on a technical scale, in particular, for the asymmetric reduction of imines, ketones, and enamides. A representative success story is imine reduction for the manufacture of metolachlor as one of the largest industrially applied asymmetric chemocatalytic processes to date, and the widely technically applied hydrogenation technology for the reduction of ketones and enamides, for which Noyori and Knowles were awarded the Nobel Prize [2].

An alternative for the production of chiral building blocks, with a continuously increasing number of examples, also on an industrial scale, is biocatalysis. Known for a long time, the use of biocatalytic alternatives has also been limited for a long time, which has been due to—among others—the following two reasons: (i) wild-type whole cells such as baker's yeast require a large amount of biomass (due to the formation of the desired enzymes in the cell only in low amounts) and often give unsatisfying enantioselectivities (due to the action of more than one enzyme), and (ii) the use of isolated enzymes requires additional costs for cell disruption and enzyme isolation as well as the addition of an external amount of cofactor. Due to impressive progress in the field of molecular biology, these limitations have often been overcome. High overexpression of the desired enzyme in recombinant host organism and high cell-density fermentation technology provide efficient and economically attractive access to these microorganisms (so-called designer cells), which contain exclusively the desired biocatalyst (enzyme) in large amounts. Such designer cells can be used directly in organic synthetic reactions without the need to isolate the enzyme in additional downstream operation steps. The presence of the cofactor in the cells requires no addition of an external amount of cofactors or supply thereof in very low amount only. Due to these advantages, recombinant whole cells overexpressing cofactor-dependent enzymes have become very attractive catalysts for asymmetric synthesis. This is underlined by an increasing number of recent organic biotransformations based on the use of such "designer cells."

Green Biocatalysis, First Edition. Edited by Ramesh N. Patel.

In this chapter, a summary is given about biotransformations using recombinant whole-cell organisms. This chapter is subdivided into a section about the design and construction of whole-cell catalysts, followed by an overview about examples of whole-cell-catalyzed biotransformations in organic synthesis. Although the focus of this chapter will be on the design, construction, and application of recombinant whole-cell catalysts, some examples of transformations based on the use of wild-type whole cells will be also given (in particular for transformations that are also applied on an industrial scale).

23.2 THE DESIGN/CONSTRUCTION OF WHOLE-CELL CATALYSTS

Biotransformations with whole cells as the catalyst represent an attractive production process as an alternative to the use of isolated enzymes. In the broader sense, naturally occurring native strains and recombinant strains can be regarded and utilized as whole-cell catalysts. Obviously, it is to be expected and is preferable that highly efficient processes can be reached using fast-growing recombinant strains such as *Escherichia coli* with the ability to employ them for a high-level expression of recombinant wild-type or mutated proteins, as compared to using wild-type strains. Thus, only the former kind of whole cells will be considered in this subchapter. In particular, the application of low-cost and high-efficiency *E. coli* strains as host organisms shortens the time to develop a novel useful strain.

Eukaryotic recombinant whole-cell factories were developed, too, mainly based on *Saccharomyces cerevisiae* or other yeast strains. These works were supported by the publication of the complete genome sequence of *S. cerevisiae* in 1996 [3]. For reduction and oxidation reactions, the overexpression of various reductases was typically combined with the deletion of competing enzymes, since *S. cerevisiae* contains a large number of reductase enzymes with overlapping substrate specificities but differing stereoselectivities. For example, the Stewart's group developed multiple modified ("second-generation") strains of *S. cerevisiae* combining the knockout of disturbing genes and overexpressing desired reductases for the enantioselective reduction of β-ketoesters [4].

In order to construct a whole-cell catalyst, the genes coding for the desired proteins can be generated by amplifying the full-length gene from genomic DNA by means of a polymerase chain reaction (PCR), using oligonucleotide primers with appropriate restriction sites. After purification the PCR product is cloned into a suitable vector followed by the transformation of this plasmid into a host strain. Usually, commercially available plasmids and host strains are used (Table 23.1). The individual work steps are supported by a lot of valuable kits useful for molecular biology operations. Moreover, companies provide the possibility to synthesize complete genes with the additional advantage of optimizing the codons according to the codon usage of the host.

By using commercially available expression vectors it is possible to develop strains that overexpress one or more proteins. Expression of just one protein occurs, for example, in hydrolases or in enzymes forming or splitting C—C bonds such as aldolases. However, a large number of publications utilizing whole-cell catalysts deal with reduction or, in some cases, oxidation reactions. Typically, these reactions require the coupling of a second enzyme to regenerate the consumed redox cofactor. This means that two proteins need to be expressed, at best even in a specific ratio of activities. For coexpression of two or more proteins, different strategies must be evaluated to develop an optimized process. One important parameter is the activity ratio of the enzymes involved.

To coexpress two proteins, for example, a reductase and a coenzyme-regenerating enzyme, frequently used vectors are pET vectors or the so-called "Duet" vectors (pETDuet, pACYCDuet, and others). One important parameter

TABLE 23.1 Examples of Whole-Cell Biocatalysts Based on *E. coli* Expressing One or Two Enzymes

Host	Gene (Insert)	Origin of Gene	Bioconversion Product	Ligated Into R = Resistance I = Inductor	Reference
Whole-cell biocatalyst with one gene					
E. coli BL21(DE3)	L-Threonine aldolase	*Streptomyces avelmitilis*	L-*threo*-3,4-Dihydroxyphenylserine	pET21(a) (Novagen; USA)	[5]
E. coli (surface display)	Alditol oxidase	*Streptomyces coelicolor*	C1 oxidation of polyols (e.g., xylitol and sorbitol)	pBAD (Invitrogen) R: ampicillin I: L-arabinose	[6]
E. coli SG13009 (Qiagen)	Benzaldehyde lyase	*Pseudomonas fluorescens*	2-Hydroxy ketones	pKK233-2 R: ampicillin I: IPTG	[7]
E. coli SG13009 (Qiagen)	Benzoylformate decarboxylase	*Pseudomonas putida*	2-Hydroxy ketones	pKK233-2 R: ampicillin I: IPTG	[7]
E. coli BL21(DE3)	Lipase	Bacterial isolate strain K107	Biodiesel	pET28a R: kanamycin I: ITPG	[8]
E. coli BL21(DE3)	Simvastatin synthase LovD (acyltransferase)	*Aspergillus terreus*	Simvastatin	pET28a R: kanamycin I: IPTG	[9]
Two or more genes					
E. coli JM109	Oxynitrilase and nitrilase	*Havea esculenta* (oxynitrilase) and *P. fluorescens* (nitrilase)	Mandelic acid and mandeloamide	pJOE5361.1 R: ampicillin I: rhamnose	[10]
E. coli JM109	Aldehyde reductase (AR) and glucose dehydrogenase (GDH)	*Sporobolomyces salmonicolor* (AR) *Bacillus megaterium* (GDH)	Ethyl (*R*)-4-chloro-3-hydroxybutanoate ((*R*)-CHBE)	pKK223-3 R: ampicillin I: IPTG	[11]
E. coli BL21(DE3)	Alcohol dehydrogenase and glucose dehydrogenase	*Lactobacillus kefir* (ADH) *Bacillus subtilis* (GDH)	(*R*)-Phenylethanol	pET21a(+) R: ampicillin I: IPTG	[12]

concerns the order of the inserted genes. Weckbecker and Hummel [12] compared the biotransformation efficiencies of two whole-cell catalysts with two different plasmids based on the commercially available vector pET21a. Each plasmid contained both genes (coding for an alcohol dehydrogenase (ADH) and for glucose dehydrogenase (GDH)) in reverse order. Cells containing the plasmid with the ADH gene as first gene showed about a ten times higher conversion rate than the plasmid with the genes in the reverse order. For further optimization either a single plasmid with two promoters could be employed so that each gene has its own promoter or two different but compatible plasmids could be applied to adjust the amount of enzyme activities.

In pioneering publications, the group of Shimizu and Kataoka presented a strategy for the construction of a universally applicable whole-cell reduction biocatalyst [11, 13, 14]. They developed an *E. coli* transformant expressing the GDH gene from *Bacillus megaterium* as a cofactor regenerator and an aldehyde reductase from *Sporobolomyces salmonicolor*. This strain was applied for the asymmetric reduction of ethyl 4-chloro-3-oxobutanoate (COBE) to ethyl (*R*)-4-chloro-3-hydroxybutanoate ((*R*)-CHBE), but it is applicable to the production of many other useful chiral alcohols by replacing the carbonyl reductase gene by other appropriate gene coding for the

NAD(P)H-dependent catalytic reduction reactions such as carbonyl or enoate reductases and hydroxy acid or amino acid dehydrogenases [14].

Recently, Schwaneberg *et al.* published results to optimize the whole-cell double oxidation of aliphatic alkanes catalyzed by a BM3 monooxygenase and an ADH [15]. Compared to the activity of the ADH used in this process ((*S*)-specific one from *Rhodococcus erythropolis* with about 47 U/mg), the activities of P450 monooxygenases are significantly lower (0.01–10 U/mg). In order to ensure a high concentration of P450 monooxygenase in the whole-cell catalyst to avoid a rate limitation in the initial oxidation step, three different coexpression strategies were evaluated: a two-plasmid system, a commercially available pACYCDuet-1, and pALXtreme-1a with a novel artificial operon system. The latter plasmid contains the gene coding for P450 BM3 monooxygenase and the gene coding for the ADH. The two-plasmid system contains the plasmid with the novel operon system and the gene coding for the monooxygenase as well as a pACYC-based plasmid with the ADH gene. In shake-flask experiments, the two-plasmid system, the pACYCDuet-1 system, and the artificial operon system yielded 14, 1, and 8 mg product per gram cell dry weight (cdw), respectively.

Another kind of whole-cell biocatalyst is represented by bacterial cell surface display systems [16, 17]. They were first introduced by Henning *et al.* in 1986, expressing the protein OmpA on the surface of *E. coli* [18]. Meanwhile, broad spectrums of enzymes such as hydrolases, oxidoreductases, or electron transfer proteins have been displayed at the cell surface. Unlike enzymes expressed intracellularly, the proteins displayed at the cell surface are freely accessible to the substrate. The cell envelope that acts as a matrix and the peptide-based immobilization method are suited to support stabilization of the biocatalyst. At the cell surface correct protein folding must occur in the absence of folding catalysts such as molecular chaperones while expression within the cell is accompanied by cell's own or recombinant chaperones. Like whole-cell catalysts with intracellular enzymes, this kind of catalyst can be simply removed by centrifugation. Sorbitol dehydrogenase (SDH) from *Rhodobacter sphaeroides* was the first example of a dehydrogenase to be expressed on the cell surface using autodisplay [19]. SDH is a dimer with a subunit molecular mass of 29 kDa. Apart from sorbitol this enzyme can utilize other substrates including galactitol or L-arabinol, too [20]. Applying a specific labeling method, the number of SDH molecules per cell was determined to be 150 000. The whole-cell biocatalyst obtained by autodisplay of SDH was used for the efficient synthesis of sorbitol, fructose, D-tagatose, and L-ribulose.

Biodiesel can be prepared by an environmentally friendly method, applying lipase-catalyzed transesterification. Kondo *et al.* [21] developed a cell-surface display system based on *S. cerevisiae* for a lipase from *Rhizopus oryzae*. This was achieved by fusing the N-terminus of the lipase (with a prosequence) to the flocculation functional domain of the FLO1 gene. This gene encodes a lectin-like cell-wall protein and is composed of several domains, including a secretion signal, a flocculation functional domain, a GPI anchor attachment signal, and a membrane-anchoring domain. Recently, a similar whole-cell catalyst was developed by Lin *et al.* [22] using *Pichia pastoris* as the host strain.

Recently, Fraaije *et al.* [6] published the development of an *E. coli*-based recombinant strain expressing the gene coding for alditol oxidase (AldO). In order to improve the substrate availability, AldO was directed either to the periplasm or the cell surface of an *E. coli*. For export into the periplasm, AldO was fused N-terminally to *E. coli* signal sequences known to direct proteins into the SecB or twin-arginine translocation (Tat) pathway. The MalE [23] or the DsbA [24] signal sequences are used for the periplasmic transport of recombinant proteins. For surface display, AldO was fused to an ice nucleation protein (INP)-based anchoring motif. Both the Tat-exported enzyme and the INP-surface-displayed AldO are active leading to a whole-cell AldO biocatalyst system, thereby eliminating the necessity for enzyme purification and enabling the full biotechnological accessibility of AldO.

The key information to characterize the efficiency of a whole-cell catalyst concerns the productivity of the cells. This can be expressed as product formed per hour or day related to gram wet (or dry) weight or related to cell density expressed as turbidity (OD). This time-related productivity is also known as space–time yield (STY) (mM (or g) of formed product per l (or ml) and h (or day)). In order to correlate these data with the activity of the catalyst, cells are disrupted and centrifuged to obtain the cell-free crude extract followed by measurement of the enzyme activity in the crude extract. Additionally, SDS polyacrylamide gel electrophoresis was performed to visualize the amounts of soluble and insoluble enzyme protein. In particular, if processes are catalyzed by two or more enzymes, these values are important parameters when compared to the productivity of the cells. On the occurrence of insoluble recombinant proteins [25] it is possible to express additional molecular chaperones. It is not clear which single chaperone or which combination of chaperones is the most suitable to improve the protein folding process optimally, thus a set of genes coding for different chaperones needs to be tested. A more targeted strategy was described by Tang *et al.* [9]. Applying a combination of homology structural prediction and site-directed mutagenesis, they identified two cysteine residues in the simvastatin synthase LovD that are responsible for nonspecific intermolecular crosslinking, leading to the formation of oligomers and aggregation. Replacement of Cys40 and Cys60 by alanine residues resulted in a significant enhancement of protein solubility and whole-cell biocatalytic activities.

23.3 BIOTRANSFORMATIONS WITH WHOLE-CELL CATALYSTS

In the following section, a summary about selected examples of applications of whole-cell-catalyzed biotransformations in organic synthesis will be given. In particular, synthetic applications in the presence of recombinant whole-cell catalysts will be described. Examples of transformations utilizing wild-type whole cells will be provided, in particular when those biotransformations have resulted in technical applications.

23.3.1 Hydrolysis Reactions

(Dynamic) Kinetic Resolution of Hydantoins Using Hydantoinases

An industrially important conversion is the transformation of racemic hydantoins into enantiomerically pure α-amino acids. For a long time the D-α-amino acids have been more easily accessible, and an industrial process for the production of D-*p*-hydroxyphenylglycine, D-**3**, which serves as a side chain in the semisynthetic β-lactam antibiotic amoxicillin, was established decades ago (Scheme 23.1) [26–28]. In these (early) processes wild-type microorganisms were utilized. In this case enzymatic racemization might not be a specific need for obtaining a sufficient dynamic kinetic resolution due to the easy racemization of this aryl-substituted hydantoins, which proceeds under rather basic conditions. Thus, the use of wild-type whole cells of *Bacillus brevis* in immobilized form enabled an efficient process for the synthesis of *N*-carbamoyl D-*p*-hydroxyphenylglycine, D-**2**, as demonstrated by Kaneka Corporation (Scheme 23.1). A subsequent cleavage of the carbamoyl moiety was then done chemically. At a later stage, the final cleavage of the *N*-carbamoyl moiety was done enzymatically, leading to a production process on an annual 300–700 tons scale utilizing the enzyme components in immobilized form.

In contrast to D-α-amino acids, the analog access for L-α-amino acids remained a challenge for a long time. Recently, however, such an efficient process technology was established at Degussa AG (now: Evonik Industries AG),

SCHEME 23.1

Synthesis of *N*-carbamoyl D-*p*-hydroxyphenylglycine with an immobilized wild-type whole-cell catalyst from *Bacillus brevis* as key step in a process for D-*p*-hydroxyphenylglycine.

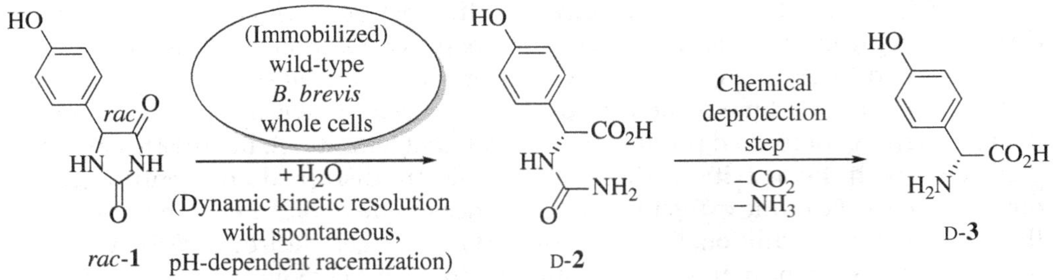

SCHEME 23.2

Synthesis of L-α-amino acids via dynamic kinetic resolution of hydantoins with a recombinant whole-cell catalyst containing a racemase, L-hydantoinase, and L-carbamolyase.

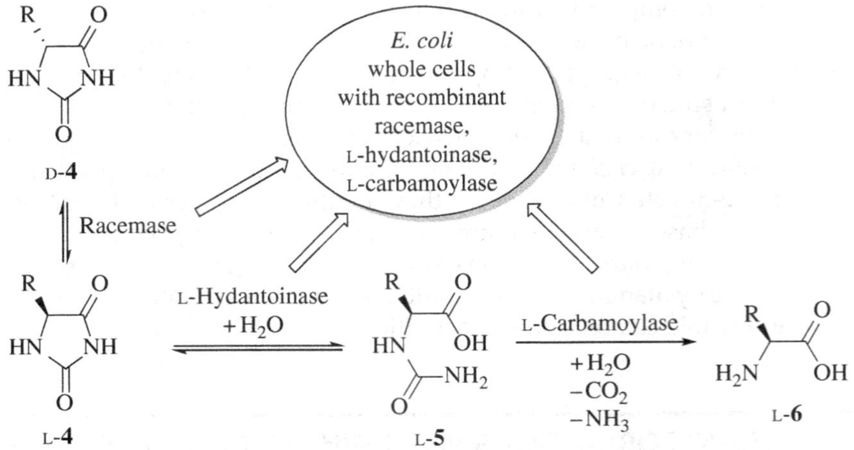

and again whole-cell catalysis played a crucial role. For this methodology, however, the developed solution was based on the use of recombinant whole-cell catalysts. In this type of biocatalysts, for which also the term "designer cells" was coined [29], all of the three required enzymes are present in an overexpressed form (Scheme 23.2) [30, 31].

The dynamic kinetic resolution process catalyzed by this whole-cell catalyst is based on the following reaction steps (Scheme 23.2) [30, 31]. First, a racemase ensures racemization of the hydantoins *rac*-**4** used as starting materials whereas the hydantoinase enantioselectively cleaves the hydantoin under formation of the *N*-carbamoyl L-α-amino acid, L-**5**. The irreversible step, thus shifting the equilibrium toward the direction of the final α-amino acid product L-**6**, is the subsequent cleavage of the *N*-carbamoyl moiety under the formation of the desired L-α-amino acid L-**6** as well as carbon dioxide and ammonia as by products (Scheme 23.2). It is noteworthy that these three required enzymes, namely, a racemase, hydantoinase, and carbamoylase, can originate from completely different strains, thus enabling the preparation of a whole-cell catalyst containing the most suitable enzyme for each reaction independent of its origin. A representative application of this type of whole-cell catalyst with three overexpressed enzymes in it is the dynamic kinetic resolution of (3′-methylmercaptoethyl) hydantoin under formation of the enantiomerically pure α-amino acid L-methionine [30].

Resolution of Mandelonitrile with a Nitrilase

A further hydrolytic process in which whole-cell catalysis turned out to be very suitable is the transformation of a racemic nitrile into the corresponding acid exemplified for the dynamic kinetic resolution of mandelonitrile into (*R*)-mandelic acid, (*R*)-**10**. This reaction is catalyzed by means of a nitrilase, which is known as highly enantioselective enzyme. As early as 1991, researchers from Asahi Chemical Industry Ltd. reported such a reaction utilizing wild-type whole cells from *Alcaligenes faecalis* bearing a suitable nitrilase [32]. When starting from racemic mandelonitrile, *rac*-**7**, the desired (*R*)-mandelic acid, (*R*)-**10**, was formed in 91% yield and with an excellent enantioselectivity of >99% ee (Scheme 23.3). The mandelonitrile is permanently racemized by spontaneous cleavage, and formation of this cyanohydrin under the reaction conditions applied, thus enabling—in combination with the nitrilase-catalyzed hydrolysis of one cyanohydrin enantiomer—a dynamic kinetic resolution process. Accordingly, the process also runs efficiently when starting from benzaldehyde and hydrogen cyanide instead of mandelonitrile as a substrate component. The whole-cell-catalyzed mandelonitrile hydrolysis has been optimized by applying a two-phase solvent system. When using a water–toluene biphasic reaction media in combination with recombinant *E. coli* cells containing a nitrilase from *Alcaligenes* sp., a high substrate concentration of 500 mM was tolerated, leading to an impressive STY of 352.6 g/(l·day) and 98% ee [33].

It should be added that the biotransformation of mandelonitrile into (*R*)-mandelic acid has also gained industrial interest and commercial production of at least several tons per year at BASF and Mitsubishi Rayon was reported [34].

23.3.2 Hydration and Dehydration Reactions

Hydration of Nitriles Forming Amides Using a Nitrile Hydratase

Whereas biocatalysis emerged as a leading technology in the field of (chiral) fine chemicals and pharmaceuticals, examples in the field of bulk chemicals are rare up to now. A major challenge in biocatalysis is still the development of industrial processes in the field of bulk chemicals with an annual production volume being in the range of multi-10 000 or multi-100 000 tons per year. One of the exceptions is the production of the bulk chemical acrylamide, **12**, which is done by means of a nitrile hydratase starting from acrylonitrile, **11** (Scheme 23.4) [35, 36]. Although this reaction does not belong to the reaction types of asymmetric transformations as the focus of this chapter, major achievements associated with it will be summarized in the following as it demonstrates the power of whole-cell catalysis to challenge industrial reactions. Interestingly, even as a technical production strain

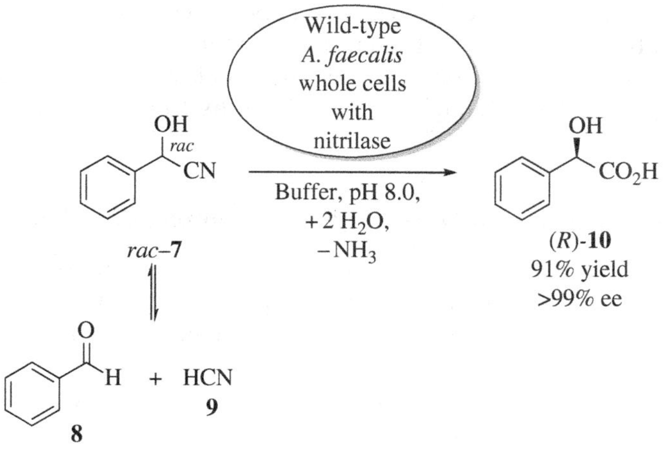

SCHEME 23.3

Synthesis of (*R*)-mandelic acid via dynamic kinetic resolution of mandelonitrile with a wild-type whole-cell catalyst from *Alcaligenes faecalis* containing a nitrilase.

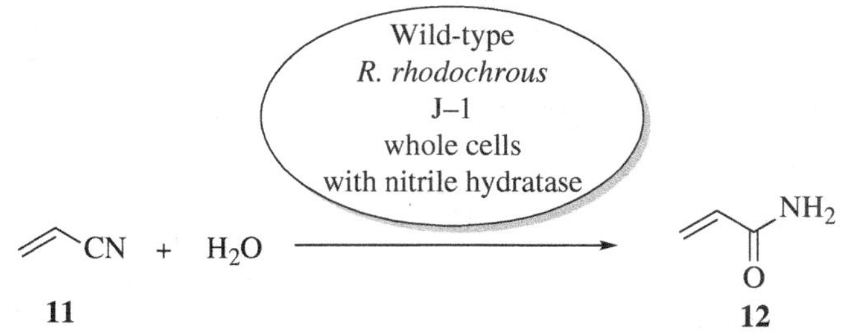

SCHEME 23.4

Synthesis of acrylamide via hydration of acrylonitrile with a wild-type whole-cell catalyst from *Rhodococcus rhodochrous* J-1 containing a nitrile hydratase.

a wild-type-derived whole-cell catalyst is used, which is in contrast to most of today's whole-cell-catalyzed processes, which are typically based on recombinant strains. The first industrial production based on a nitrile hydratase was established in 1985 by Nitto Chemical (later: Mitsubishi-Rayon Co. Ltd). At a later stage, the strains *Pseudomonas chlororaphis* B23 and *Rhodococcus rhodochrous* J-1 became production strains. The high efficiency of this production process utilizing wild-type cells with a nitrile hydratase is underlined by an impressive (estimated) annual production volume of about 400 000 tons for acrylamide, **12**.

Dehydration of Aldoximes Forming (Chiral) Nitriles Using an Aldoxime Dehydratase

An enzyme being complementary to a nitrile hydratase, which catalyzes water addition to a nitrile, is an aldoxime dehydratase as a biocatalyst being capable to eliminate water from an oxime moiety, thus leading to the formation of a nitrile. Such aldoxime dehydratases are also synthetically useful as demonstrated by the Asano group who applied an aldoxime dehydratase from a *Bacillus* sp., for example, for the dehydration of Z-phenylacetaldoxime to produce phenylacetonitrile [37–41]. For this substrate, the overexpressed aldoxime dehydratase showed a high activity of 14 000 U/l of fermentation culture of a developed recombinant *E. coli* strain [37]. In addition, numerous other transformations of aromatic and aliphatic aldehydes have been reported [37–41].

The aldoxime dehydratase also turned out to be useful for the enantioselective synthesis of chiral nitriles as demonstrated recently jointly by the Asano and Gröger groups [42]. An *E*- and *Z*-mixture of racemic aldoximes serves as starting materials, and the aldoxime dehydratase is preferably used directly in the form of its recombinant *E. coli* strain with high overexpression of this enzyme. In the presence of this recombinant whole-cell biocatalyst, the kinetic resolution of racemic (*E*-enriched) aldoxime *E*-*rac*-**13** with a high *E*/*Z* ratio of 99 : 1 through enantioselective dehydration proceeds under the formation of nitrile (*S*)-**14** with high conversion of 50% and an excellent enantioselectivity of 98% ee (Scheme 23.5). Thus, the aldoxime dehydratase is highly enantioselective for the *E*-racemate, whereas in the case of the *Z*-racemate both enantiomers have been shown to be converted. This process concept is of great preparative interest since chiral nitriles are formed directly, without the need to use highly toxic cyanide, starting from aldehydes via aldoximes as easily prepared intermediates (by condensation of aldehydes with hydroxylamine). As a byproduct only water is formed.

Hydration of Nonactivated Alkenes (Styrenes) Forming Chiral Alcohols

A so-called dream reaction of high synthetic value is the asymmetric addition of water to a nonactivated C—C double bond, for example, in a styrene substrate molecule. Whereas an efficient chemocatalyst for this reaction is still unknown [43], a combination

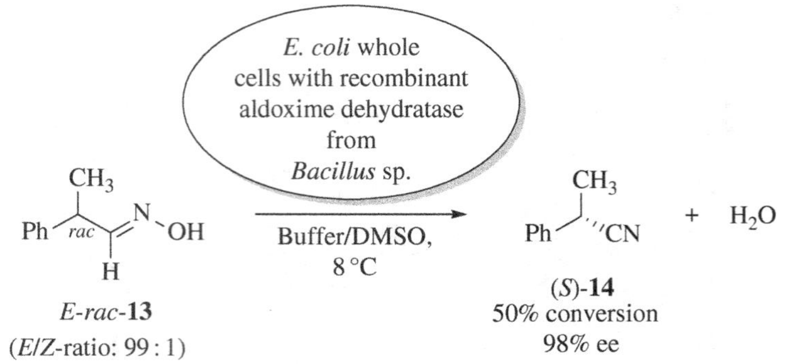

SCHEME 23.5

Synthesis of a chiral nitrile via kinetic resolution of a racemic *E*-aldoxime with a recombinant whole cell-catalyst containing an aldoxime dehydratase.

of a chemo- and biocatalyst enables a formal hydration by means of a two-step process based on initial Wacker oxidation and enzymatic reduction of the ketone intermediate [44, 45]. A first direct asymmetric biocatalytic process for the hydration of styrenes has been reported by Faber and Glueck and coworkers [46], who found that several phenolic acid decarboxylases are suitable for this unusual reaction. It is noteworthy that in the presence of a whole-cell catalyst bearing such a suitable decarboxylase, a range of *p*-hydroxy-substituted styrenes, for example **15**, were converted successfully into the corresponding *p*-hydroxy-substituted phenylethan-1-ols with up to 77% conversion and enantioselectivities of up to 71%ee. A representative example is shown in Scheme 23.6 with the enantioselective hydration of **15** to (*S*)-**16**. The enzymes applied were utilized in the form of recombinant whole-cell catalysts with an *E. coli* as a host organism.

23.3.3 C—C Bond-Forming Reactions

"Umpolung" Reactions: Benzoin Condensation and Related Syntheses

Whole-cell-catalyzed transformations have also been reported in the field of carbon–carbon bond-forming reactions, and the efficiency of such transformations is underlined by several impressive processes (see also subsequent sections of this chapter). The first example of an industrial enzymatic carbon–carbon bond-forming reaction, which has been conducted by means of a whole-cell catalyst, goes back to the 1930s when Knoll AG (later a subsidiary of BASF) applied a whole-cell-catalyzed synthesis of (*R*)-phenylacetylcarbinol required as an intermediate in the industrial production of L-ephedrine [47–49]. The industrial process itself is based again on early developments in this field already reported by Neuberg and Hirsch in the 1920s [47]. As a biocatalyst a yeast is used, and molasses serves as an economically attractive starting material. The yeasts *S. cerevisiae* and *Candida utilis* turned out to be suitable. From molasses pyruvate, **18** is formed, which is the substrate for a pyruvate decarboxylase being present in the yeast whole cells. Pyruvate can be considered as an "activated acetaldehyde" and undergoes a pyruvate decarboxylase-catalyzed decarboxylation and subsequent "umpolung" reaction under C—C bond formation with benzaldehyde, **17**, which is added as a second substrate. As a cofactor thiamine diphosphate, being present in the yeast whole cells, is required for both steps (decarboxylation, carboligation). In more detail, the industrial process starts with an initial fed-batch production phase of biomass and pyruvate from molasses, followed by the biotransformation phase involving the addition of sugars and the feeding of benzaldehyde to produce the carboligation product (*R*)-**19** by means of the pyruvate decarboxylase being present in the yeast (Scheme 23.7). The resulting (*R*)-phenylacetylcarbinol, (*R*)-**19**, is then further converted in a chemical process to the final product L-ephedrine. It is noteworthy (thus underlining the high efficiency of this process) that this early microbial process based on the use of wild-type yeast cells is still in operational use by several companies [49]. In more recent work with a wild-type strain of

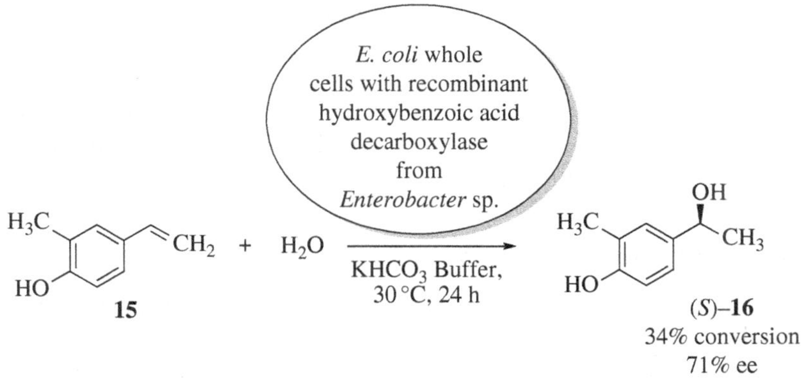

SCHEME 23.6
Synthesis of a substituted *p*-hydroxyphenylethan-1-ol via asymmetric hydration of its corresponding styrene with a recombinant whole-cell catalyst containing a decarboxylase.

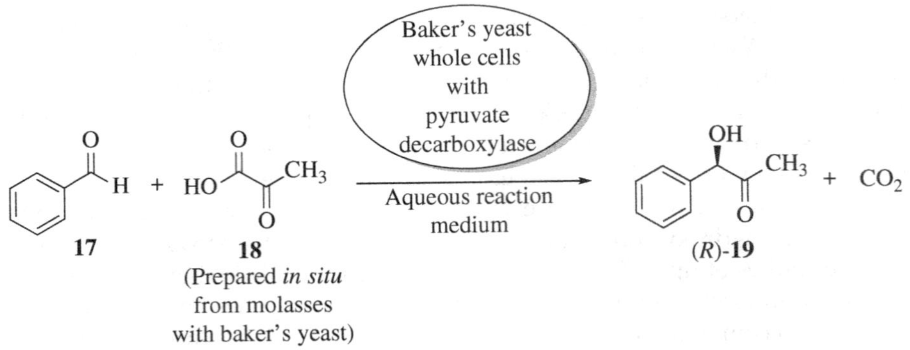

SCHEME 23.7
Synthesis of (*R*)-phenylacetcarbinol via asymmetric carboligation with a wild-type yeast-cell catalyst containing a pyruvate decarboxylase.

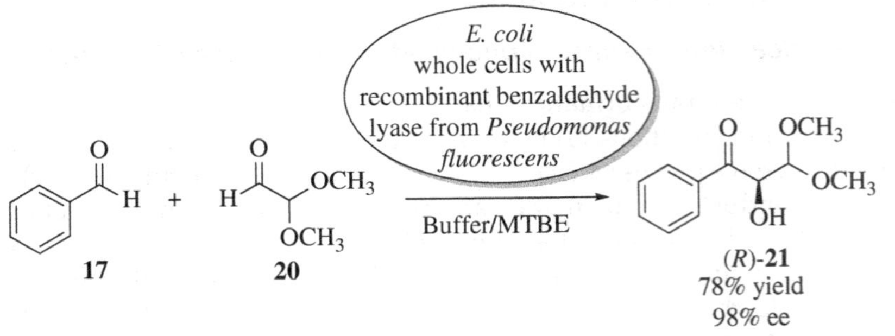

SCHEME 23.8
Synthesis of a 2-hydroxy ketone via asymmetric carboligation with a recombinant whole-cell catalyst containing a benzaldehyde lyase.

C. utilis in biphasic reaction media, such yeast-catalyzed carboligations have been reported to proceed under formation of the carboligated product in concentrations of up to 58 g/l when starting from benzaldehyde and pyruvate [50, 51].

In recent work, a major focus has been on the development of recombinant lyases for such benzoin-type condensations and carboligations via "umpolung" [52]. Typically, these biocatalysts have been used in an isolated form with addition of external amount of the required cofactor thiamine diphosphate. The first whole-cell-based carboligations of this type of umpolung reactions, which run in the presence of recombinant strains, were recently reported by Degussa researchers and academic partners, in particular the Liese and Pohl groups (Scheme 23.8) [7]. As a biocatalyst, recombinant *E. coli* whole cells overexpressing either a benzaldehyde lyase from *Pseudomonas fluorescens* or a benzoylformate decarboxylase from *Pseudomonas putida* were utilized. The resulting biocatalytic syntheses of enantiomerically pure 2-hydroxy ketones led to the formation of a range of 2-hydroxy ketones in high yields of up to 90% and with excellent enantioselectivities of up to >99% ee. For example, the whole-cell-catalyzed conversion of benzaldehyde, **17**, and 2,2-dimethoxyacetaldehyde, **20**, gave the desired ligation product (*R*)-**21** in 78% yield and with 98% ee (Scheme 23.8). A solvent mixture consisting of methyl *tert*-butyl ether (MTBE) and an aqueous buffer turned out to represent a suitable (biphasic) reaction medium for these whole-cell-catalyzed processes and enabled a high substrate loading with concentrations being in the range of 0.3–1 M as well as an efficient extractive downstream processing. Furthermore, the cell-internal pool of the required cofactor thiamine diphosphate is sufficient for the biotransformation. Thus, an external addition of this expensive cofactor is not required when using such *E. coli* whole-cell catalysts, thus overcoming the limitation of analogous reactions in the presence of isolated enzymes.

Oxynitrilase-Catalyzed Hydrocyanation of Aldehydes

When Rosenthaler reported the oxynitrilase-catalyzed hydrocyanation by using almond meal in an aqueous medium in 1908 [53], this contribution not only represented one of the oldest biotransformations in organic chemistry but also the starting point for the (long-term) development of a leading industrial technology for the enantioselective production of chiral cyanohydrins, which are produced on a technical scale for various applications in the field of fine chemicals, agrochemicals, and pharmaceuticals, respectively [54–56]. A limitation of the early stage reactions, however, has been related to enantioselectivity, which was typically only in a modest range. A breakthrough has been achieved by the Effenberger group when conducting the oxynitrilase-catalyzed hydrocyanation in organic medium, thus suppressing the undesired side reaction of a noncatalytic hydrocyanation leading to a racemic cyanohydrin [57]. In the presence of an immobilized oxynitrilase a highly enantioselective process has been achieved. Another strategy to suppress the undesired reaction is conducting the enzymatic hydrocyanation at low pH values as demonstrated by the Kula group [58]. In recent years a range of recombinant oxynitrilases have been developed and utilized typically in an isolated form. A further milestone was achieved by the Griengl group by developing an enzymatic hydrocyanation in a two-phase reaction medium utilizing recombinant oxynitrilases [54–56]. This process technology also found technical applications at DSM, and, for example, high STY of 2.1 mol/(l·h) together with excellent enantioselectivity of 99% ee was reported for the synthesis of the pyrethroid intermediate (*S*)-3-phenoxybenzaldehyde cyanohydrin [56].

It is noteworthy, however, that not only recombinant strains but also wild-type biomass still play a role as interesting biocatalysts. This—at first glance—surprising fact is due to readily available and cheap biomass containing sufficient amount of oxynitrilase. Typical sources are plant meals, in particular almond meal that is utilized after treatment with an organic solvent (for defatting) and a drying step. Thus, in spite of the availability of recombinant enzymes and related processes with isolated enzymes, a range of efficient processes with defatted almond meal powder as biocatalyst have even more recently been reported. An advantage of this type of biocatalyst is its simple accessibility and low cost. These applications with plant-material derived biomass containing oxynitrilase also comprises the use of defatted almond meal in microaqueous media, which simplifies workup by enabling direct filtration of the reaction mixture from the biomass without the need for an extraction step. By means of this methodology the cyanohydrin formed, (*R*)-**22**, was obtained in yields of up to 100% and with excellent enantioselectivities of up to 99% ee when operating at a reaction temperature of 4 °C (Scheme 23.9), whereas at an elevated reaction temperature of 30 °C a yield of 97 and 98% ee was found [59].

It should be added that these biocatalytic transformations with dried plant meal powders represents one of the rare examples in modern biocatalysis in which plant biomass is directly utilized as a catalyst component and also leads to attractive synthetic processes.

Threonine Aldolase-Catalyzed Aldol Reactions Using Glycine and an Aldehyde

Enzymatic carbon–carbon bond formation based on the use of recombinant whole cells was reported by Baik *et al.* for the synthesis of L-*threo*-3,4-dihydroxyphenylserine, L-*threo*-**25** (Scheme 23.10) [5]. This type of L-amino acid is of interest due to its use for the treatment of Parkinson's disease. The reaction is straightforward starting from 3,4-dihydroxybenzaldehyde and glycine, which are utilized in nonprotected form. In

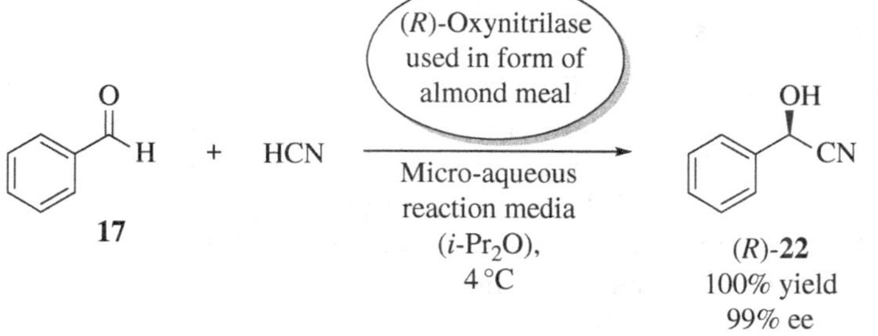

SCHEME 23.9
Synthesis of (*R*)-mandelonitrile via asymmetric hydrocyanation of benzaldehyde with almond meal containing an (*R*)-oxynitrilase.

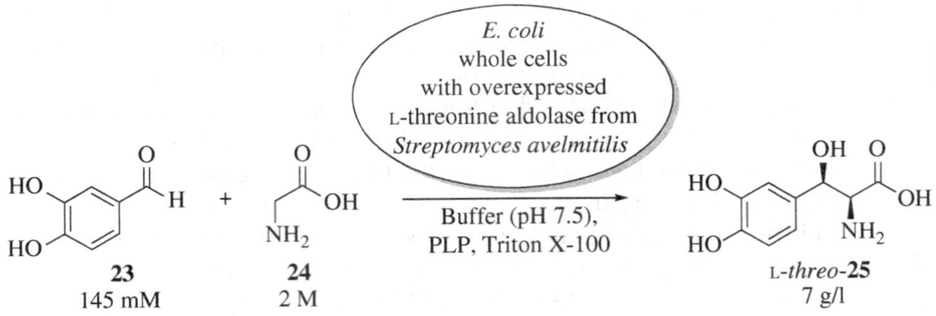

SCHEME 23.10
Synthesis of L-*threo*-3,4-dihydroxyphenylserine via asymmetric aldol reaction of 3,4-dihydroxybenzaldehyde and glycine with a recombinant whole-cell catalyst containing a threonine aldolase.

the presence of a recombinant *E. coli* whole-cell catalyst bearing an L-threonine aldolase from *Streptomyces avelmitilis* in overexpressed form, the desired product L-*threo*-**25** is formed directly in one reaction step with 7 g/l after 114 h (at a whole-cell concentration of 5 g/l) when running the biotransformation in a continuously operating reactor (data for enantioselectivity and diastereoselectivity are not given). A hollow-fiber membrane filter was used to retain and recycle the enzyme, and the addition of Triton-X as an additive as well as a low reaction temperature turned out to be beneficial for the productivity. However, the process that operates at a substrate concentration of 145 mM of aldehyde **23** requires a large excess of glycine (**24**, 2 M, corresponding to an about 14-fold excess) under optimized conditions.

23.3.4 Reduction Reactions

Reduction of Ketones Forming Chiral Alcohols

The asymmetric reduction of ketones for the synthesis of chiral alcohols has emerged as one of the leading technologies in the field of whole-cell applications at both lab and industrial scale. This fact is noteworthy from two perspectives: first, it shows the power of whole-cell biocatalysis for the production of chiral alcohols required in the field of fine chemicals and pharmaceuticals; second, since in the field of chiral alcohols "traditionally" Noyori hydrogenation has been regarded for a long time as the leading key technology, it also underlines the high competitiveness of biocatalysis with even excellent alternative "classic chemical" or chemocatalytic methods. Today asymmetric ketone reduction using biocatalysts is highly competitive with the chemical Nobel Prize technology asymmetric hydrogenation and numerous cases, in particular recently developed industrial processes, are based on biocatalytic whole-cell transformations with *in situ* cofactor regeneration as the method of choice. Besides different concepts of cofactor regeneration, there are also different forms of whole-cell biocatalyst formulations like growing cells, resting cells (nongrowing but metabolically active cells), lyophilized cells, immobilized cells, and permeabilized cells. Since the field of asymmetric biocatalytic whole-cell reductions of ketones has developed toward a broadly investigated research area, in the following only selected examples can be given. For a comprehensive overview of this field, extensive recently published reviews are available [60].

Wild-Type Whole Cells

Besides recombinant whole cells, microbial reduction using wild-type cells interestingly still plays an important role, which is underlined by recently reported technical processes. Cofactor regeneration is conducted by means of D-glucose, which is metabolized by wild-type microorganisms, thus providing the required reduced cofactor forms NADH or NADPH. One of the most "classic" microbial reductions of ketones is based on the use of baker's yeast, and numerous examples in this field covering a broad substrate range are known [61–69]. In addition, wild-type microorganisms such as *Geotrichum candidum* [70], *Lactobacillus kefir* [71], *Rhizopus arrhizus* [72], *Phaseolus aureus* [73], and *Pisum sativa* [74] turned out to be suitable biocatalysts for asymmetric ketone reduction. In the following section, technical applications of microbial reduction of ketones with wild-type whole cells are given. Researchers of Rohner Ltd. reported a technical process for the (*S*)-enantioselective reduction of β-ketoester **26** utilizing baker's yeast as a catalyst, thus leading to the alcohol (*S*)-**27** in 60–75% yield and with >98% ee (Scheme 23.11) [75, 76]. A further analogous microbial process by Rohner Ltd. was reported for the production of ethyl (1*R*,2*S*)-*cis*-2-hydroxycyclohexane carboxylic acid ethyl ester (70% yield, >97% de, >93% ee).

Further microbial reductions with wild-type organisms comprise the synthesis of a chiral C-13 side-chain intermediate of paclitaxel by BMS researchers utilizing *Hansenula fabianii* cells, which were produced in a 15l-scale fermentation [77], as well as the synthesis of an *N*-acylated aminoalcohol with excellent conversion (>99% ee) and enantioselectivity (>98% ee) on a 280 l scale by researchers from Merck & Co. based on the use of *Candida sorbophila* whole cells [78, 79]. A combination of ketone reduction with wild-type organisms and the use of XAD-7 resin to decrease the solubility of a substrate being toxic for the microorganism was reported by Eli Lilly researchers, thus obtaining the desired alcohol product in high yield (96%) and with >99.9% ee, even when operating at a high substrate loading of 80 g/l [80–82]. Scalability of this process was demonstrated with a batch process operating on a 300 l (reactor-volume) scale.

A *Rhodococcus ruber* strain reported by Faber and Kroutil *et al.* represents a further interesting wild-type organism since it shows a broad substrate spectrum as well as a (synthetically interesting) high isopropanol tolerance (up to 50% (v/v)), which contributes to a better solubility of hydrophobic ketones and serves at the same time as a cosubstrate for *in situ* cofactor recycling [83–85].

Substrate-Coupled Cofactor Regeneration with Recombinant Whole Cells

Besides wild-type strains, more recently the use of recombinant whole cells has gained increasing popularity for application in asymmetric ketone reduction. When overexpressing the ADH only, *in situ* cofactor recycling based on a "substrate-coupled approach" represents a favorite approach as demonstrated in an early contribution by the Itoh group [86] utilizing a recombinant ADH from a *Corynebacterium* overexpressed in *E. coli*. This concept has been also applied by Daicel researchers in the presence of an *E. coli* catalyst with recombinant ADH from *Candida parapsilosis*. This biocatalyst catalyzes the reduction of β-ketoester **28** at a 36.6 g/l substrate loading and furnished the alcohol (*R*)-**29** in 95.2% yield and with 99% ee (Scheme 23.12) [87].

As selected examples of more recently developed ketone reductions with recombinant whole cells following the concept of substrate-coupled *in situ* cofactor recycling, the work by the Schmid and Buehler group with a recombinant ADH from *Thermus* sp. [88] and by the Kroutil group with a DMSO-tolerant recombinant ADH from *Paracoccus pantotrophus* [89] shall be mentioned here. This recombinant whole-cell

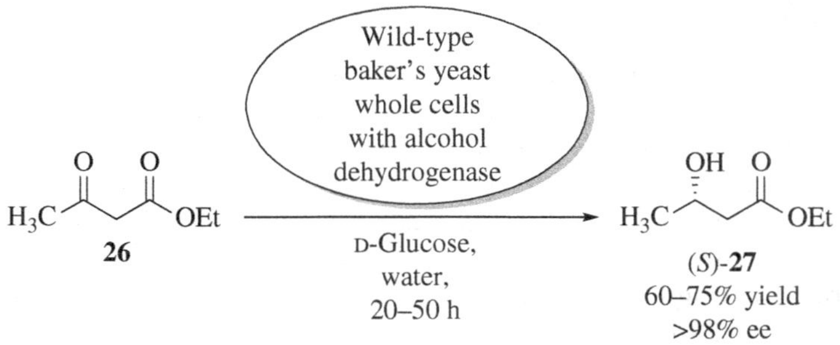

SCHEME 23.11
Synthesis of an (*S*)-3-hydroxybutyrate via asymmetric reduction of its corresponding β-ketoester with a baker's yeast-cell catalyst containing alcohol dehydrogenases.

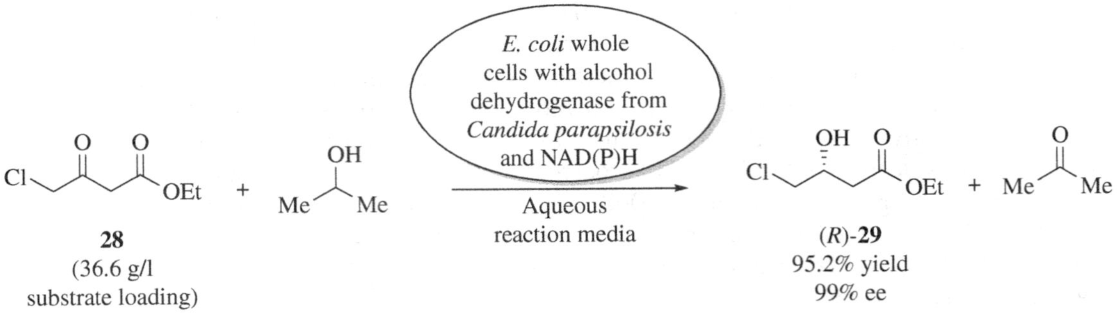

SCHEME 23.12

Synthesis of an (*S*)-4-chloro-3-hydroxybutyrate via asymmetric reduction of its corresponding β-ketoester with a recombinant whole-cell catalyst containing an alcohol dehydrogenase.

concept further turned out to be suitable for biotransformations running in a continuous mode as demonstrated by Liese and Lütz *et al.* for the reduction of methyl acetoacetate [90]. It is in particular noteworthy that by means of *E. coli* whole cells with an overexpressed ADH from *Lactobacillus brevis*, this process runs at a high substrate concentration of 2.5 mol/l, leading to the product in an STY of 700 g/(l·day) and with >99% ee.

An example of the efficient application of wild-type microorganisms in this field is the asymmetric reduction of different ketones like 2-decanones to the corresponding (*S*)-alcohols by the usage of lyophilized cells of *R. ruber* DSM 44541 encoding a *sec*-ADH and 22% (v/v) 2-propanol as cosubstrate for cofactor regeneration [84]. In this case it has been further demonstrated that the recombinant expression of the same ADH can increase the reaction rates [91]. As shown in this example by the Kroutil and Liese groups, the higher expression of the catalytic enzyme and thereby higher activity of the whole-cell catalyst are advantageous so that the application of recombinant whole-cell catalyst appears to be preferable. As both reduction of ketone and oxidation of the cosubstrate 2-propanol for recycling of nicotinamide coenzyme are catalyzed by the same enzyme, the maximum achievable conversion is determined by the thermodynamic equilibrium of all four components. For achieving conversions of 99%, an *in situ* product removal strategy was developed [91].

As enzymes naturally work in aqueous medium typically also this type of whole-cell biocatalysis utilizing isopropanol for *in situ* cofactor regeneration in asymmetric ketone reduction is performed with wet cells. However, recently it was shown jointly by the Schwaneberg, Ansorge-Schumacher, and Dominguez de Maria groups that the reduction of ketones by *E. coli* cells with an overexpressed carbonyl reductase can be done in neat substrates with isopropanol-coupled cofactor regeneration [92]. A selected example is given in Scheme 23.13. It is noteworthy that high productivity of up to 500 g/l was achieved in combination with >98% conversion and >99% ee for the reduction of acetophenone, **30**. As a biocatalyst, wet cells were active but rapidly deactivated within a few hours whereas lyophilized cell remain stable for a couple of days.

FDH-Based Cofactor Regeneration

Recombinant whole cells bearing an ADH and a formate dehydrogenase (FDH) offer a further alternative for cofactor recycling, in which formate is converted into carbon dioxide while reducing NAD^+ to NADH. Researchers from Daicel Chemical Industries Ltd. realized such an approach utilizing a recombinant *E. coli* W3110 strain, which coexpresses an ADH from *Pichia finlandica* and an FDH from *Mycobacterium*, and led to the enantioselective reduction ethyl 4-chloro-3-oxobutanoate, **28**, at a 32 g/l substrate loading [87]. The desired alcohol (*S*)-**29** was obtained in 98.5% yield and 99% ee (Scheme 23.14).

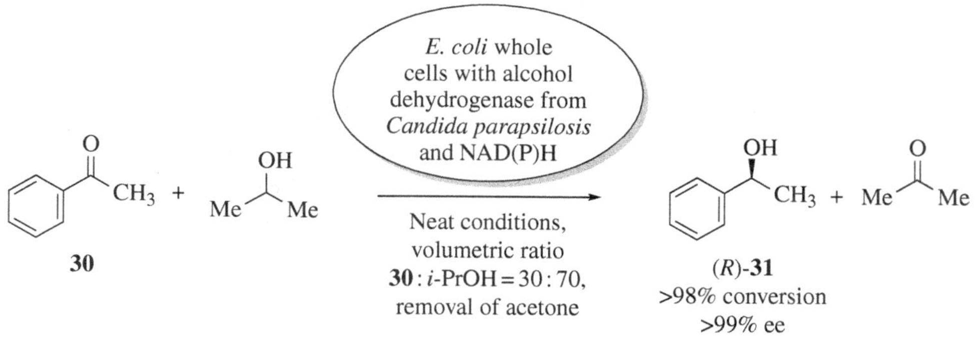

SCHEME 23.13

Synthesis of (S)-phenylethan-1-ol via asymmetric reduction of acetophenone with a recombinant whole-cell catalyst containing an alcohol dehydrogenase under neat conditions.

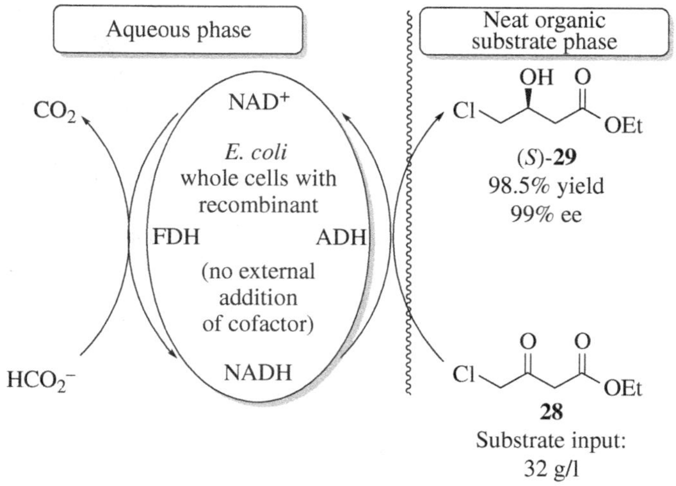

SCHEME 23.14

Synthesis of an (S)-4-chloro-3-hydroxybutyrate via asymmetric reduction of its corresponding β-ketoester with a recombinant whole-cell catalyst containing an alcohol dehydrogenase and a formate dehydrogenase.

GDH-Based Cofactor Regeneration

Another interesting option of recombinant whole-cell biocatalysis is the use of *E. coli* cells, which overexpress a glucose dehydrogenase (GDH) besides the ADH. The highly specific activities of GDHs as well as the robustness toward organic (hydrophobic) molecules are what make this approach particularly attractive. Industrial applications of this reduction approach have been presented, among others, by Kaneka Corporation and Degussa AG (now: Evonik Industries AG).

For example, such types of recombinant *E. coli* whole cells based on the overexpression of a GDH from *B. megaterium* were developed by the Shimizu group and applied for the reduction of particularly β-ketoester **28** in an *n*-butyl acetate/water two-phase solvent system [11, 13, 14, 93, 94]. In addition to glucose as a cosubstrate, a low amount of $NADP^+$ is added. For example, in the presence of an *E. coli* whole-cell catalyst, bearing a recombinant ADH from *S. salmonicolor* and GDH from *B. megaterium*, a substrate loading of 300 g/l of organic phase gave an (*R*)-enantioselective reduction of the β-ketoester **28** with 94% conversion and 92% ee (Scheme 23.15) [11]. In addition, a whole-cell catalyst based on the same ADH and GDH concept was also developed for the (*S*)-enantioselective reduction of β-ketoester **28** by the same group [95]. Reduction of other substrates was also reported by the Shimizu group jointly with researchers from Kaneka [14]. Furthermore, commercial-scale production of this enzymatic reduction concept based on the use of recombinant whole cells at Kaneka was also reported, utilizing this method since 2000 for the industrial production of ethyl (*S*)-4-chloro-3-hydroxybutanoate [14].

An efficient and technically feasible reduction technology based on the use of recombinant whole cells was developed jointly by Degussa researchers and the Hummel group. In the presence of recombinant whole cells overexpressing an ADH and a GDH, efficient enantioselective ketone reductions were achieved in a pure aqueous reaction medium (Scheme 23.16) [96, 97]. The substrate loading exceeds 150 g/l

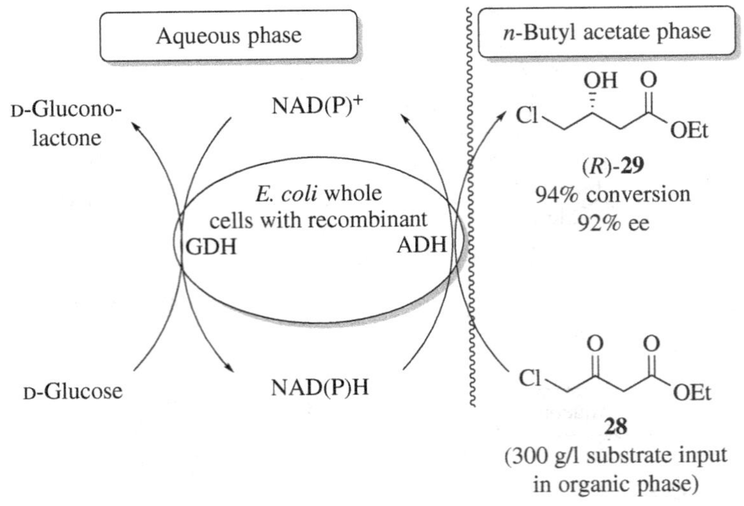

SCHEME 23.15

Synthesis of an (*S*)-4-chloro-3-hydroxybutyrate via asymmetric reduction of its corresponding β-ketoester with a recombinant whole-cell catalyst containing an alcohol dehydrogenase and a glucose dehydrogenase.

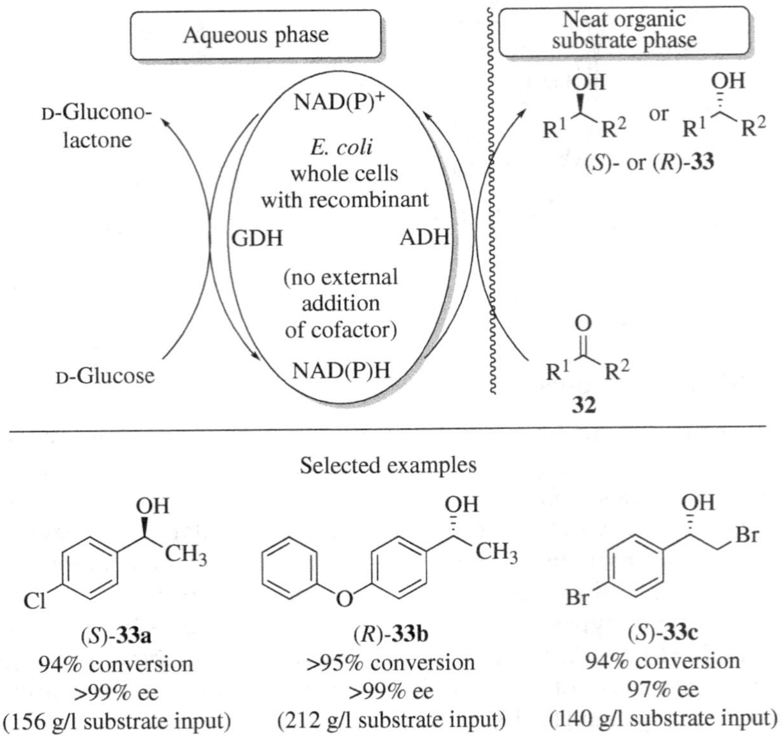

SCHEME 23.16

Synthesis of chiral alcohols via asymmetric reduction of their corresponding ketones with recombinant whole-cell catalysts containing an alcohol dehydrogenase and a glucose dehydrogenase.

for a broad range of substrates **32**. In general, there is no need for external cofactor supply, and this methodology provides synthetic access to (*R*)- as well as (*S*)-enantiomers of the resulting alcohols **33**. It is noteworthy that these processes proceed with excellent conversions (up to >95%) and enantioselectivities (up to >99% ee). After further optimization, this process technology has also been applied on an industrial scale at Degussa AG. Furthermore, a detailed description on the development of a recombinant *E. coli* whole-cell catalyst overexpressing an (*R*)-selective ADH from *L. kefir* and a GDH from *Bacillus subtilis* has been reported by the Hummel group [12].

The use of a recombinant whole-cell catalyst based on an ADH and a GDH also turned out to be a valuable approach for the synthesis of the α-hydroxy ester (*R*)-**35**, which serves as a key intermediate for clopidogrel (Scheme 23.17) [98]. To this end, Ema and Sakai *et al.* designed *E. coli* cells overexpressing an ADH from *S. cerevisiae* and a GDH for *in situ* cofactor recycling. This reduction runs at a high substrate loading of 198 g/l and leads to (*R*)-**35** with 86% conversion, in 82% yield and with

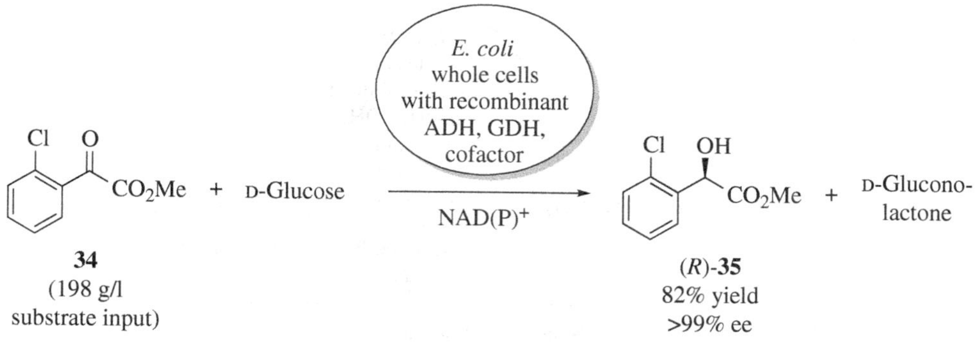

SCHEME 23.17

Synthesis of an (*R*)-*ortho*-chloromandelate via asymmetric reduction of its corresponding α-ketoester with a recombinant whole-cell catalyst containing an alcohol dehydrogenase and a glucose dehydrogenase.

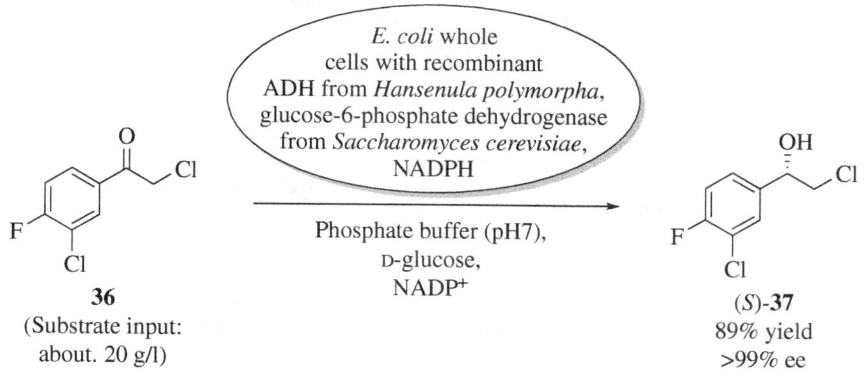

SCHEME 23.18

Synthesis of a substituted (*S*)-α-chloro-phenylethan-1-ol via asymmetric reduction of its corresponding ketone with a recombinant whole-cell catalyst containing an alcohol dehydrogenase and a glucose-6-phosphate dehydrogenase.

>99% ee. Besides this clopidogrel intermediate (*R*)-**35**, Ema and Sakai *et al.* also applied this whole-cell reduction technology for the enantioselective synthesis of a broad range of other chiral alcohols [99].

In all the examples described earlier, ADH and GDH have been overexpressed in one *E. coli* cell. The alternative, using two different types of microbial cells, which overexpress either the ADH or the GDH, has been studied by Li and coworkers [100, 101]. A prerequisite is that both cells are permeabilized, thus enabling the transport of the cofactor between the cells. This concept has been applied successfully in the enantioselective reduction of ethyl 3-oxo-4,4,4-trifluorobutyrate, which proceeds with 91% ee.

G6P-DH-Based Cofactor Regeneration

A further option for the construction of whole-cell catalysts for enantioselective ketone reduction is based on the combined use of an ADH and a glucose-6-phosphate dehydrogenase (G-6-PDH). The construction of such cells, and their application in the reduction of 2,3′-dichloro-4′-fluoroacetophenone, **36**, was reported by BMS researchers [102]. The desired product (*S*)-**37** was obtained in 89% yield and with >99% ee (Scheme 23.18). It is noteworthy that cheap glucose (instead of expensive glucose-6-phosphate) can be fed as a co-substrate due to the cell-internal phosphorylation of glucose toward glucose-6-phosphate. On the other hand, this also required (nonpermeabilized) intact cells as in a fermentation process. This might be a reason why the substrate loading was somewhat lower compared to the processes based on permeabilized whole cells (see previous sections). In detail, the reduction of **36** was carried out at a substrate loading of about 20 g/l.

Reductive Amination of α-Keto Acids Forming α-Amino Acids

A valuable synthetic approach toward enantiomerically pure nonnatural α-amino acids is based on enzymatic reductive amination of α-keto acids [103]. The most prominent industrial example in this field is the synthesis of L-*tert*-leucine, which has

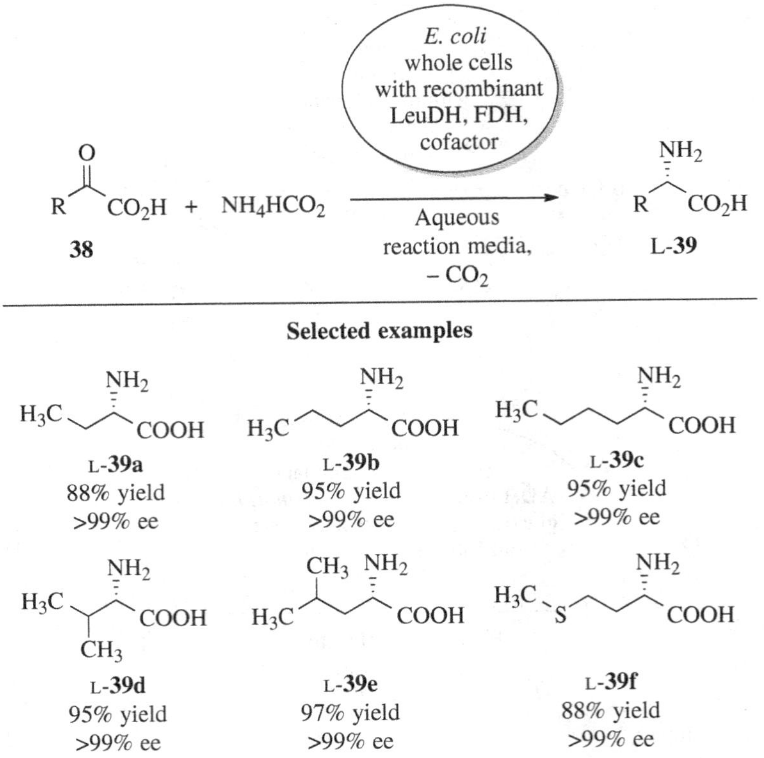

SCHEME 23.19

Synthesis of L-α-amino acids via asymmetric reductive amination of their corresponding α-keto acids with a recombinant whole-cell catalyst containing an amino acid dehydrogenase and a formate dehydrogenase.

been implemented for regular production scale at the end of the 1990s at Degussa AG [104–106]. This process is based on a reductive amination process in the presence of a leucine dehydrogenase (LeuDH) and an FDH for *in situ* cofactor recycling. Whereas the LeuDH requires the reduced form of the cofactor, NADH, and oxidizes it to NAD^+, the FDH regenerates the reduced form, NADH, by oxidizing formate to carbon dioxide simultaneously. Since the latter step proceeds irreversibly, the reductive amination can proceed under full conversion. It is noteworthy that this process represents the first process with *in situ* cofactor recycling with isolated enzymes. However, the direct use of whole cells offers an attractive alternative process option, and such a concept has been studied first by the Soda group who demonstrated that whole-cell reductive amination works with a range of substrates **38** efficiently [107]. Selected examples are shown in Scheme 23.19. The reductive amination was done in the presence of several types of *E. coli*-type whole cells, bearing an amino acid dehydrogenase in combination with an FDH. As an amino acid dehydrogenase component, a LeuDH, alanine dehydrogenase, and phenylalanine dehydrogenase, respectively, were utilized, thus enabling the reductive amination of a broad substrate range due to the complementary substrate spectra of such amino acid dehydrogenases. The desired L-amino acids were formed in high yields exceeding 85% in most cases and with excellent enantioselectivity of >99% ee in the presence of a LeuDH and phenylalanine dehydrogenase and somewhat lower 80% ee in the presence of an alanine dehydrogenase. It is noteworthy that the reductive aminations were conducted utilizing only the whole-cell intracellular "cofactor pool." The substrate concentrations varied from 0.05 to 0.6 M of α-keto acid. In addition, an approach toward D-amino acids has been realized by constructing a recombinant whole-cell catalyst based on a plasmid encoding for a D-selective amino acid transaminase (utilizing D-alanine as amine donor) and an alanine racemase, L-alanine dehydrogenase, and FDH. The latter three enzymes enable the *in situ*–regeneration of the amine donor D-alanine by consuming formate under the formation of carbon dioxide.

The development of an efficient whole-cell-catalyzed approach toward the production of the industrially required L-*tert*-leucine, (*S*)-**39g** (L-**39g**), was reported by Degussa

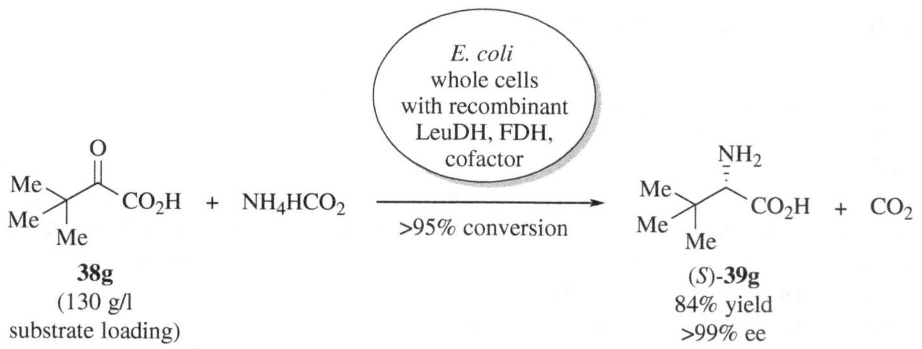

SCHEME 23.20
Synthesis of L-*tert*-leucine via asymmetric reductive amination of trimethylpyruvic acid with a recombinant whole-cell catalyst containing a leucine dehydrogenase and a formate dehydrogenase.

researchers jointly with the Altenbuchner group (Scheme 23.20) [29, 108]. The recombinant *E. coli*-type whole-cell catalyst consists of two plasmids. The high-copy plasmid has been utilized for the overexpression of a mutant of the FDH from *Candida boidinii* due to its lower specific activity of 6 U/mg, whereas the gene encoding for the highly active LeuDH from *Bacillus cereus* (specific activity: about 400 U/mg) was inserted into a medium-copy plasmid. Thus, the remaining enzyme activities could be adjusted closer to the desired ratio, and in the presence of this whole-cell catalyst reductive amination of trimethyl pyruvate, **38g**, proceeds at a high overall substrate loading of 1.0 M (with a continuous addition of the substrate), leading to the desired L-*tert*-leucine, L-**39g**, with >95% conversion, in 84% isolated yield and with >99% ee. It should be added that this process does not require an additional amount of "external" cofactor and fully proceeds with the "internal" cofactor being present in the whole cells.

A further application of the whole-cell-catalyzed reductive amination reported by Degussa researchers is a process for the preparation of L-neopentylglycine, a further nonnatural L-amino acid of pharmaceutical interest. The whole-cell-catalyzed process operates at an overall substrate loading of 88 g/l (corresponding to a substrate concentration of 0.53 M), and the desired L-neopentylglycine was formed with >95% conversion and excellent enantioselectivity of >99% ee [109]. As a biocatalyst, in this case again recombinant *E. coli* whole cells overexpressing an LeuDH from *B. cereus* and a mutant of an FDH from *C. boidinii* were utilized. After the biotransformation, the further unit operations of this process include an ion exchange chromatography for the purification of the formed L-neopentylglycine, which then was obtained in 83% yield.

Reductive Amination of Ketones with Transaminases Forming Chiral Amines

The asymmetric transamination of ketones, which represents an elegant route toward chiral amines as demonstrated in the synthesis of sitagliptin in the presence of isolated enzyme [110], has also been conducted by means of whole cells as catalysts. Such a whole-cell application in combination with a stabilization of whole-cell catalysts by immobilization of cells was reported by the Woodley group [111]. In the case of whole *E. coli* cells overexpressing ω-transaminase in a commercially available matrix called LentiKats®, biocatalysts can be reused for five reaction cycles in the synthesis of the aromatic chiral 3-amino-1-phenylbutane from 4-phenyl-2-butanone and isopropyl amine, retaining about 80% of original activity without diffusional limitations or loss of activity [111]. Depending on the respective substrate and product, the cell membrane represents a barrier, so that biotransformation may be limited by diffusion. That this parameter can influence the productivity of the whole-cell-mediated process to a large extent was shown in the synthesis of 3-amino-1-phenylbutane: permeabilization of the cells with 0.1% cetrimonium bromide resulted in an increase in the reaction rate of 40%. Immobilization of these permeabilized cells in LentiKats also stabilized these cells [111].

Reduction of Activated Alkenes

A further redox transformation for which enzymes turned out to be suitable is the reduction of an activated C=C bond. Thus, besides reductive transformations of carbonyl moieties, the enzymatic reduction of activated alkenes has also been recently studied very extensively [112, 113]. From a retro-synthetic perspective such transformations are of great interest as well, since a broad range of electron-withdrawing groups are conceivable as substituents at the C=C double bond, thus leading to the required activated alkene. In recent years this reaction has attracted a lot of interest due to numerous synthetic applications.

Besides such a synthetic diversity, to which many groups contributed recently, impressive process development in terms of volumetric productivity was also conducted. The Shimizu group reported a highly efficient process for the reduction of the C=C double bond in ketoisophorone, **40**, under the formation of chiral diketone (*R*)-**41**, which is carried out in the presence of a recombinant whole-cell biocatalyst [114, 115]. The recombinant whole cells are based on two plasmids containing the genes encoding for an ene reductase from *Candida macedoniensis* and a GDH for *in situ* cofactor recycling, respectively. High overexpression was obtained for both enzymes after introducing the two plasmids into *E. coli* BL21 (DE3) cells. Subsequent application of these whole cells led to a C=C reduction process, which runs at a substrate loading of 98 g/l and led to the formation of the desired product (*R*)-**41** with 97% conversion and an excellent enantioselectivity of >99% ee (Scheme 23.21). This process represents one of the most efficient transformations in the field of enzymatic alkene reductions so far.

Furthermore, asymmetric reduction of activated C=C double bonds of enones and nitro-olefins by recombinant *S. cerevisiae* whole cells has been carried out by the Raimondi group [116]. It is noteworthy that the recombinant *S. cerevisiae* strains utilized as biocatalysts lack endogenous old yellow enzyme (OYE) but express separate OYE genes from (nine) nonconventional yeasts.

23.3.5 Oxidation Reactions

Epoxidation of Styrenes Forming Epoxides

Asymmetric epoxidation represents a further oxidation reaction of broad interest due to the versatile utilization of chiral epoxides in the preparation of a range of difunctionalized organic molecules via ring opening with heteroatom nucleophiles. An efficient (*S*)-enantioselective epoxidation of the styrene **42**, by means of a recombinant *E. coli* whole-cell catalyst overexpressing a styrene monooxygenase from *Pseudomonas* sp., has been developed by the Schmid group (Scheme 23.22) [117]. The epoxidation runs at a high substrate concentration of 695 mM, leading to a product concentration of the desired (*S*)-styrene oxide, (*S*)-**43** of 604 mM, and a volumetric productivity of

SCHEME 23.21

Synthesis of (6*R*)-2,2,6-trimethylcyclohexane-1,4-dione via asymmetric reduction of ketoisophorone with a recombinant whole-cell catalyst containing an ene reductase and a glucose dehydrogenase.

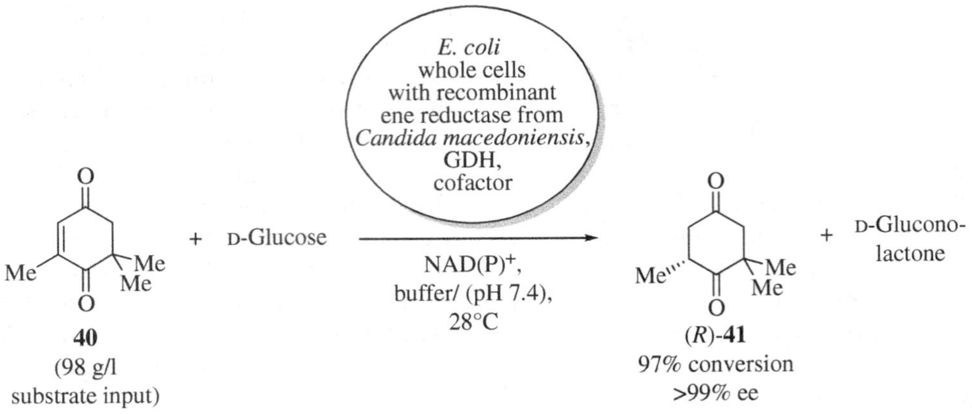

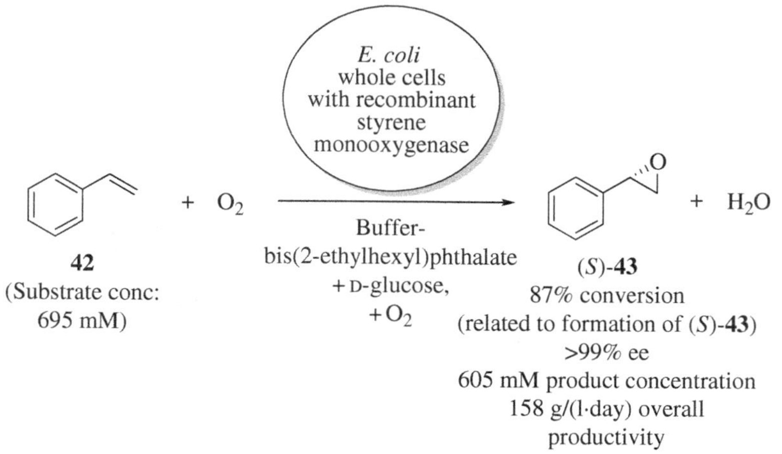

SCHEME 23.22

Synthesis of (*S*)-styrene oxide via asymmetric epoxidation of styrene with a recombinant whole-cell catalyst containing a styrene monooxygenase.

6.57 g/(l·h). The cofactor regeneration is carried out by means of the metabolization of D-glucose within the fermentation process. As a reaction medium for this process, a two-phase system consisting of an aqueous buffer phase and bis-(2-ethylhexyl) phthalate as the organic phase was used [117]. In this case of the epoxidation of styrene to (*S*)-styrene oxide, (*S*)-**43**, by recombinant *E. coli* expressing styrene monooxygenase StyAB cell growth and the biotransformation can compete with each other [118]. This situation can occur, in particular, if glucose is used for cofactor regeneration. Thus, resting cells are preferred as far as the stability of the catalyst is sufficient [118]. In addition, excellent enantioselectivity of >99% ee for the transformation of styrene into (*S*)-styrene oxide in the presence of a recombinant *E. coli* expressing styrene monooxygenase StyAB was obtained in an earlier study [117, 119, 120].

Baeyer–Villiger Oxidations Forming Lactones

The oxidation of cyclic ketones to lactones was initially described in 1899 by Baeyer and Villiger and became one of the major reactions in organic chemistry [121]. With respect to asymmetric synthesis, transition metal-based methods [122–125] and metal-free organocatalytic methods [126] have been successfully developed, but often also show limitations in the production of optically pure esters and lactones. In addition, the use of peroxyacids or organic peroxides, respectively, raises concerns in terms of safety issues, in particular when using them on an elevated scale. The application of Baeyer–Villiger monooxygenases (BVMOs) therefore represents an attractive strategy, since molecular oxygen is utilized for the oxygenation of ketones and high regio- as well as enantioselectivities can be obtained with a wide range of natural and synthetic products [121, 127]. Although BVMOs are produced by numerous bacteria and fungi, asymmetric Baeyer-Villiger oxidation has been often done with cyclohexanone monooxygenase (CHMO) from *Acinetobacter calcoaceticus* NCIMB 9871 (CHMO), which is an NADPH- and oxygen-dependent enzyme [121, 127].

However, the application of purified enzymes and the addition of costly cofactors are economically not favorable options. Thus, there has been an industrial interest in developing suitable recombinant whole-cell catalysts. Since *A. calcoaceticus* is a pathogen belonging to class 2 microorganisms [128], the use of the wild-type strain is not an option on an industrial scale. Another drawback of the wild-type organism is the constitutive expression of lactone hydrolase, leading to product degradation, which can be circumvented by the use of recombinant whole cells [121]. However, in the case of CHMO from *A. calcoaceticus* it was shown not to be trivially achieved [121]. The biotransformation of bicyclo[3.2.0]hept-2-en-6-one, **44**, to its corresponding regioisomeric lactones (1*S*,5*R*)-**45** and (1*R*,5*S*)-**46** (Scheme 23.23), which are precursors of the Corey lactone as an important key intermediate in pharmaceutically relevant prostaglandin synthesis [129], shows not only the great potential

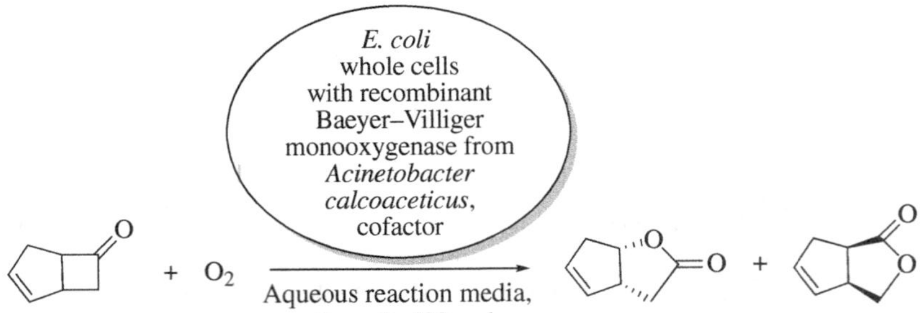

rac-cis-**44** (Substrate input: 20 g/l)

70% overall yield for (1*S*,5*R*)-**45** and (1*R*,5*S*)-**46** (after purification)

(1*S*,5*R*)-**45** >98% ee

(1*R*,5*S*)-**46** >98% ee

SCHEME 23.23

Synthesis of lactones via regio- and enantioselective Baeyer–Villiger oxidation of a racemic cyclic ketone with a recombinant whole-cell catalyst containing a Baeyer–Villiger monooxygenase.

of whole-cell biocatalyst but also the challenges due to substrate and product inhibitions [121, 127]. Substrate inhibitions can be easily circumvented by substrate feeding. In the case of the bioconversion of bicyclo[3.2.0]hept-2-en-6-one, it was also shown that the substrate can be supplied adsorbed on a solid resin as a substrate reservoir in a two-phase system. Simultaneously the product could be removed from the reaction mixture via adsorption onto these resins, thereby preventing product inhibition [127]. By means of this method, the combined overall yield for lactones (1*S*,5*R*)-**45** and (1*R*,5*S*)-**46** was 70% after purification, and both lactones were formed with excellent enantiomeric excesses of >98% ee.

In some cases the lactones formed can be degraded by hydrolases or the efficiency of biotransformation can be decreased by the reduction of substrates by dehydrogenases present in whole cells [121]. By recombinant strain development the side reaction can be prevented by gene knockout of the corresponding genes if cell growth and overexpression of the protein of interest are not affected. Although the cell membrane of the biocatalysts protects the enzymes against organic solvents and prevents cofactor leaching, it also represents a barrier for the mass transfer of substrates and products. This was shown by a nearly tenfold lower activity of whole cells (50–60 U/g) compared with sonicated cells for conversion of bicyclo[3.2.0]hept-2-en-6-one (500 U/g) [121]. Since molecular oxygen is used as sole oxidant, a sufficient mixing and oxygen transfer to the reaction mixture has to be ensured. The Furstoss group took this parameter into account by a specially designed vessel with high aeration rates. As a result of improved and upscaled biocatalyst production combined with the strategies shown for process development, the production of regioisomeric lactones (1*S*,5*R*)-2 oxabicyclo[3.3.0]oct-6-en-3-one and (1*R*,5*S*)-3-oxabicyclo[3.3.0]oct-6-en-2-one could be improved from 1 g/l (9.3 mM) with *A. calcoaceticus* whole-cell catalyst to 20 g/l (185 mM) with recombinant *E. coli* catalysts with an overall yield of 83% with each lactone in enantiomeric excess of >98% ee [127], thus underlining the tremendous potential of BVMO-based whole-cell catalysis in asymmetric synthesis.

Hydroxylation of Alkanes Forming Alcohols and Further Oxidized Products

A broad range of applications of whole cells as biocatalysts have been reported in the field of hydroxylation reactions. The direct use of wild-type cells is still widely applied and popular, in particular when getting a recombinant strain is difficult due to limitations in expressing the desired enzyme. This is true, for example, in the case of membrane-bound monooxygenases. In cases when recombinant monooxygenases were accessible, such enzymes have not only been used for a single transformation but were also combined with other enzymes in recombinant whole-cell catalysts for reaction cascades. In the following section, selected organic synthetic processes for these catalytic concepts will be given.

To start with, the hydroxylation based on the use of wild-type cells, the Adam group reported the suitability of a *B. megaterium* wild-type strain for the conversion of alkanes as well as cycloalkanes into the corresponding alcohols and ketones [130]. The hydroxylations proceed with, in part, excellent enantioselectivities of up to >99% ee. However, when using *n*-alkanes as substrates, reactions proceed with modest regioselectivity leading to mixtures of ω-1-, ω-2-, and ω-3-alcohols. As a selected example, the hydroxylation of *n*-octane, **47**, is shown in Scheme 23.24.

Besides the functionalization of alkanes and cycloalkanes, from an early stage on enzymatic steroid hydroxylation with wild-type whole cells was a broadly studied research field, leading to a range of industrial processes [131]. An early example is the production of hydrocortisone by microbial 11α-hydroxylation starting from bile acid. Further examples are microbial hydroxylations running at a scale of up to 200 m^3 at the company Schering AG.

Besides applications of whole-cell-catalyzed oxidation reactions in the field of fine chemicals, a recent impressive development addressed the production of specialty and bulk chemicals on a multithousand-ton scale. Although not a chiral molecule, the synthesis of methyl ω-aminolaurate by means of whole-cell catalysis will be summarized in the following as an elegant approach and a breakthrough in producing bulk chemicals via biocatalysis. The Schmid and Bühler group jointly with researchers from Evonik Industries AG designed a recombinant whole-cell catalyst for the multi step transformation of methyl dodecanoate into methyl ω-aminododecanoate (Scheme 23.25) [132]. The synthetic concept is based on efficient initial selective ω-hydroxylation by means of *E. coli* whole cells, which contain a monooxygenase and an

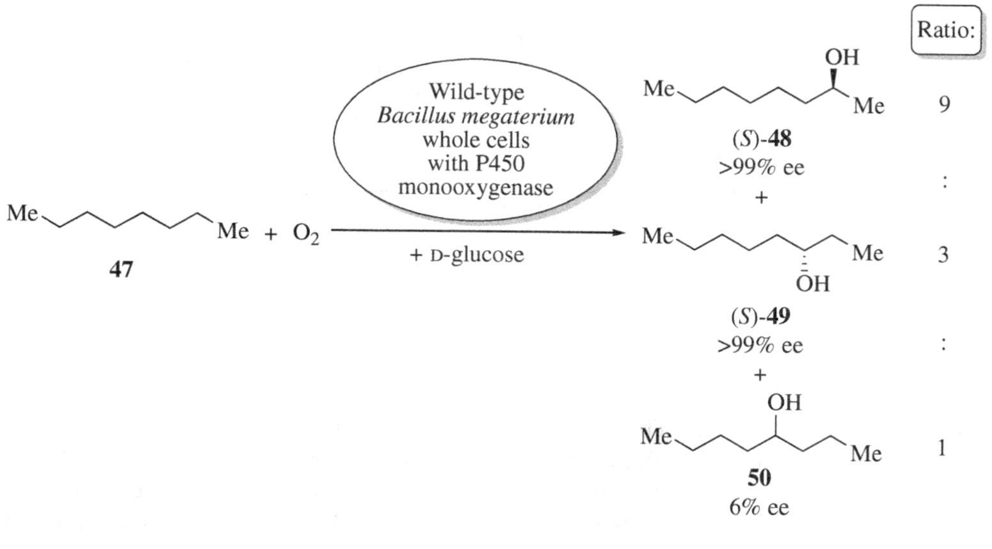

SCHEME 23.24

Synthesis of octanol regioisomers via asymmetric hydroxylation of *n*-octane with a wild-type whole-cell catalyst containing a P450-monooxygenase.

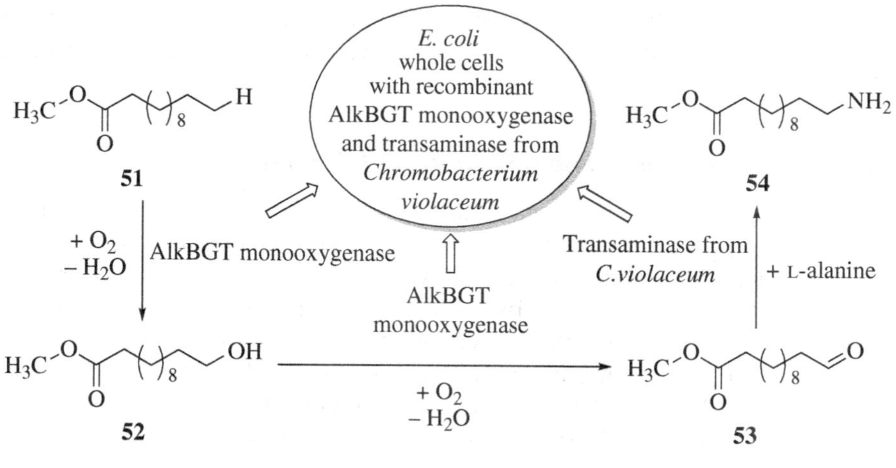

SCHEME 23.25

Synthesis of methyl ω-aminododecanoate via a three-step transformation with a recombinant whole-cell catalyst containing an AlkBGT-monooxygenase, an AlkL-membrane protein, and a transaminase.

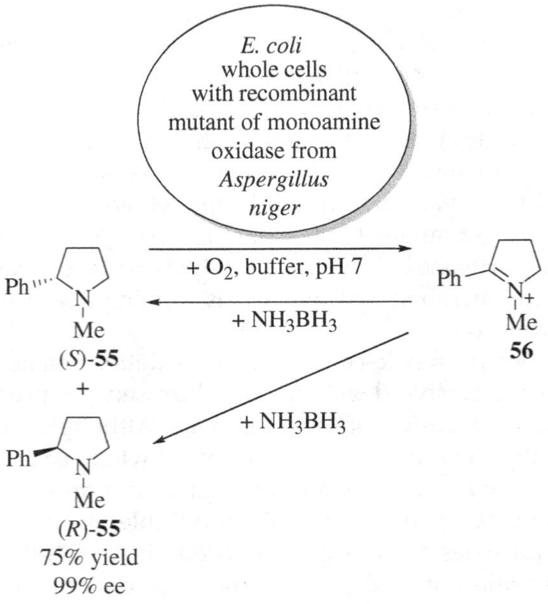

SCHEME 23.26

Synthesis of (*R*)-*N*-methyl-2-phenylpyrrolidine via chemoenzymatic deracemization of its corresponding racemate with a recombinant whole-cell catalyst containing a monoamine oxidase.

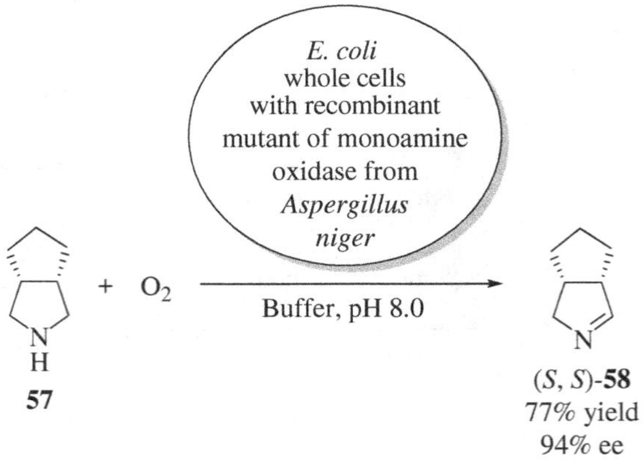

SCHEME 23.27

Synthesis of a chiral cyclic imine via desymmetrization of its corresponding prochiral cyclic secondary *meso*-type amine with a recombinant whole-cell catalyst containing a monoamine oxidase.

AlkL-membrane protein. The latter contributes to the transport of the ester into the cell through the hydrophobic membrane. The hydroxylation proceeds very efficiently leading to the hydroxylated product **52** as well as the aldehyde product **53** through further oxidation. When integrating a recombinant transaminase into the whole-cell catalyst, conversion of the aldehyde **53** then leads to the formation of the target product **54**. As an amine donor, L-alanine was used. It is noteworthy that this process technology has been applied on pilot plant scale by Evonik Industries AG in connection with the production of ω-amino-lauric acid for polyamide-12 from palm kernel oil [133].

Oxidative Deracemizations and Resolutions of Amines

Asymmetric synthesis of amines is a synthetically important research area due to the broad range of applications of chiral amines in the field of pharmaceuticals [134]. Amine oxidases are a versatile class of catalysts, which turned out to be very suitable for the preparation of chiral amines by means of desymmetrization or deracemization reactions. With respect to the latter, a racemic amine is enantioselectively oxidized by the amine oxidase, and simultaneously the *in situ*-formed imines or iminium ions, respectively, are reduced in a nonenantioselective chemical reduction process (with typically a borohydride) back to the racemic amine [135]. The key catalyst for such a transformation is an enantioselective amine oxidase, and it is noteworthy that this enzyme has been used in the form of a whole-cell catalyst. For example, in the presence of an *E. coli* whole-cell catalyst with an overexpressed optimized mutant of the monoamine oxidase from *Aspergillus niger*, tertiary amines have been deracemized furnishing the desired product in both high conversion and excellent enantiomeric excess. The deracemization of the cyclic tertiary amine *N*-methyl-2-phenylpyrrolidine, *rac*-**55**, through enzymatic (*S*)-enantioselective oxidation and chemical reduction of the *in situ*-formed iminium species **56** proceeds efficiently, leading to 99% ee of the (*R*)-enantiomer of this compound, (*R*)-**55**, after completion of the process (Scheme 23.26).

Besides deracemization, recombinant *E. coli* whole cells with an overexpressed amine oxidase also have been utilized by Turner *et al.* in the enantioselective desymmetrization of pyrrolidines [136]. For example, when starting from the prochiral pyrrolidine **57**, the desired imine (*S*,*S*)-**58** was synthesized with >98% conversion and obtained in 77% yield and with 94% ee. Other prochiral pyrrolidines were enantioselectively converted into the corresponding imines as well. A selected example is given in Scheme 23.27. It is noteworthy that this whole-cell-catalyzed amine oxidase-based desymmetrization reaction of **57** has been successfully integrated into a chemoenzymatic multistep synthesis of the corresponding 3,4-substituted L-proline derivative used for preparation of the drug molecule telaprevir.

23.4 CONCLUSION

The direct use of whole cells as catalysts turned out as a versatile, broadly applicable, and highly efficient strategy for a range of different types of (enantioselective) biotransformations. Recombinant microorganisms are a particularly valuable type of whole cells, which (i) contain the desired enzyme or enzymes in an overexpressed form, and (ii) can be produced in an economically attractive fashion by means of high-cell-density fermentations. Having such economically attractive biocatalysts in hand, the next step consists in their use for organic synthesis. It is noteworthy that up to now many such syntheses have been realized comprising hydrolytic and hydration as well as dehydration reactions, C—C bond-forming reactions, and redox reactions in both the reductive and oxidative mode. Often the desired products are obtained in high yields and with excellent enantiomeric excesses of up to >99% ee. Furthermore, numerous of such processes run at high substrate input, thus fulfilling another desired key feature for a technically attractive process. Thus, today, a range of such whole-cell-catalyzed transformations are also conducted on an industrial scale.

REFERENCES

[1] I. Ojima (ed.), *Catalytic Asymmetric Synthesis*, vol. 3, John Wiley & Sons, Inc., Hoboken, 2010.

[2] E. Schmidt, H. U. Blaser (eds.), *Large-Scale Asymmetric Catalysis*, Wiley-VCH, Weinheim, 2004.

[3] A. Goffeau, B. G. Barrell, H. Bussey, R. W. Davis, B. Dujon, H. Feldmann, F. Galibert, J. D. Hoheisel, C. Jacq, M. Johnston, E. J. Louis, H. W. Mewes, Y. Murakami, P. Philippsen, H. Tettelin, S. G. Oliver, *Science* 1996, **274**, 546–567.

[4] S. Rodriguez, M. M. Kayser, J. D. Stewart, *J. Am. Chem. Soc.* 2001, **123**, 1547–1555.

[5] S.-H. Baik, H. Yoshioka, *Biotechnol. Lett.* 2009, **31**, 443–448.

[6] E. van Bloois, R. T. Winter, D. B. Janssen, M. W. Fraaije, *Appl. Microbiol. Biotechnol.* 2009, **83**, 679–687.

[7] P. Dominguez de Maria, T. Stillger, M. Pohl, M. Kiesel, A. Liese, H. Gröger, H. Trauthwein, *Adv. Synth. Catal.* 2008, **350**, 165–173.

[8] B. Gao, E. Su, J. Lin, Z. Jiang, Y. Ma, D. Wei, *J. Biotechnol.* 2009, **139**, 169–175.

[9] X. Xie, I. Pashkov, X. Gao, J. L. Guerrero, T. O. Yeates, Y. Tang, *Biotechnol. Bioeng.* 2009, **102**, 20–28.
[10] S. Baum, F. van Rantwijk, A. Stolz, *Adv. Synth. Catal.* 2012, **354**, 113–122.
[11] M. Kataoka, K. Yamamoto, H. Kawabata, M. Wada, K. Kita, H. Yanase, S. Shimizu, *Appl. Microbiol. Biotechnol.* 1999, **51**, 486–490.
[12] A. Weckbecker, W. Hummel, in: *Methods in Biotechnology, Microbial Enzymes and Biotransformations* (ed.: J. L. Barredo), vol. 17, Humana Press, Totowa, 2005, pp. 225–238.
[13] M. Kataoka, L. P. S. Rohani, M. Wada, K. Kita, H. Yanase, I. Urabe, S. Shimizu, *Biosci. Biotechnol. Biochem.* 1998, **62**, 167–169.
[14] M. Kataoka, K. Kita, M. Wada, Y. Yasohara, J. Hasegawa, S. Shimizu, *Appl. Microbiol. Biotechnol.* 2003, **62**, 437–445.
[15] C. A. Müller, A. Dennig, T. Welters, T. Winkler, A. J. Ruff, W. Hummel, H. Gröger, U. Schwaneberg, *J. Biotechnol.* 2014, **191**, 196–204.
[16] J. Jose, T. F. Meyer, *Microbiol. Mol. Biol. Rev.* 2007, **71**, 600–619.
[17] P. Samuelson, E. Gunneriusson, P. A. Nygren, S. Stahl, *J. Biotechnol.* 2002, **96**, 129–154.
[18] R. Freudl, S. MacIntyre, M. Degen, U. Henning, *J. Mol. Biol.* 1986, **188**, 491–494.
[19] J. Jose, S. von Schwichow, *ChemBioChem* 2004, **5**, 100–108.
[20] S. Schauder, K. H. Schneider, F. Giffhorn, *Microbiology* 1995, **141**, 1857–1863.
[21] T. Matsumoto, H. Fukuda, M. Ueda, A. Tanaka, A. Kondo, *Appl. Environ. Microbiol.* 2002, **68**, 4517–4522.
[22] Z. Jin, S. Y. Han, L. Zhang, S. P. Zheng, Y. Wang, Y. Lin, *Bioresour. Technol.* 2013, **130**, 102–109.
[23] F. J. Mergulhao, D. K. Summers, G. A. Monteiro, *Biotechnol. Adv.* 2005, **23**, 177–202.
[24] M. Marrichi, L. Camacho, D. G. Russell, M. P. DeLisa, *J. Biol. Chem.* 2008, **283**, 35223–35235.
[25] F. Baneyx, M. Mujacic, *Nat. Biotechnol.* 2004, **22**, 1399–1408.
[26] M. Ikemi, *Bioprocess Technol.* 1994, **19**, 797–813.
[27] J. Ogawa, S. Shimizu, *J. Mol. Catal. B: Enzym.* 1997, **2**, 163–176.
[28] A. Liese, K. Seelbach, C. Wandrey (eds.), *Industrial Biotransformations*, 2nd ed., Wiley-VCH, Weinheim, 2006, p. 411f.
[29] O. May, H. Gröger, *Spec. Chem. Mag.* 2004, (5), 24–25.
[30] O. May, S. Verseck, A. Bommarius, K. Drauz, *Org. Process Res. Dev.* 2002, **6**, 452–457.
[31] O. May, *Elements–Degussa Sci. Newsl.* 2003, (2), 24–27.
[32] K. Yamamoto, K. Oishi, I. Fujimatsu, K.-I. Komatsu, *Appl. Environ. Microbiol.* 1991, **57**, 3028–3032.
[33] Z.-J. Zhang, J. Pan, J.-F. Liu, J.-H. Xu, Y.-C. He, Y.-Y. Liu, *J. Biotechnol.* 2011, **152**, 24–29.
[34] J.-S. Gong, Z.-M. Lu, H. Li, J.-S. Shi, Z.-M. Zhou, Z.-H. Xu, *Microb. Cell Fact.* 2012, **11**, 142.
[35] H. Yamada, M. Kobayashi, *Biosci. Biotechnol. Biochem.* 1996, **60**, 1391–1400.
[36] Y. Asano, P. Kaul, in: *Comprehensive Chirality* (eds.: E. M. Carreira, H. Yamamoto), vol. 7, Elsevier, Amsterdam, 2012, pp. 122–142.
[37] Y. Asano, Y. Kato, *FEMS Microbiol. Lett.* 1998, **158**, 185–190.
[38] Y. Kato, K. Nakamura, H. Sakiyama, S. G. Mayhew, Y. Asano, *Biochemistry* 2000, **39**, 800–809.
[39] Y. Asano, *J. Biotechnol.* 2002, **94**, 65.
[40] S.-X. Xie, Y. Kato, H. Komeda, S. Yoshida, Y. Asano, *Biochemistry* 2003, **42**, 12056–12066.
[41] Y. Kato, S. Yoshida, S.-X. Xie, Y. Asano, *J. Biosci. Bioeng.* 2004, **97**, 250–259.
[42] R. Metzner, S. Okazaki, Y. Asano, H. Gröger, *ChemCatChem* 2014, **6**, 3105–3109.
[43] L. Hintermann, *Top. Organomet. Chem.* 2010, **31**, 123–155.
[44] I. Schnapperelle, W. Hummel, H. Gröger, *Chem. Eur. J.* 2012, **18**, 1073–1076.
[45] H. Sato, W. Hummel, H. Gröger, *Angew. Chem.* 2015, **127**, 4570–4574; *Angew. Chem. Int. Ed.* 2015, **54**, 4488–4492.
[46] C. Wuensch, J. Gross, G. Steinkellner, K. Gruber, S. M. Glueck, K. Faber, *Angew. Chem.* 2013, **125**, 2349–2353; *Angew. Chem. Int. Ed.* 2013, **52**, 2293–2297.
[47] C. Neuberg, J. Hirsch, *Biochem. Z.* 1921, **115**, 282–310.
[48] A. Liese, K. Seelbach, C. Wandrey (eds.), *Industrial Biotransformations*, 2nd ed., Wiley-VCH, Weinheim, 2006, p. 447f.
[49] B. Schäfer, *Naturstoffe der chemischen Industrie*, Elsevier, München, 2007, p. 389.
[50] C. Gunawan, G. Satianegara, A. K. Chen, M. Breuer, B. Hauer, P. L. Rogers, B. Rosche, *FEMS Yeast Res.* 2007, **7**, 33–39.
[51] B. Rosche, M. Breuer, B. Hauer, P. L. Rogers, *Biotechnol. Lett.* 2005, **27**, 575–581.

[52] M. Pohl, C. Dresen, M. Beigi, M. Müller, in: *Enzyme Catalysis in Organic Synthesis* (eds.: K. Drauz, H. Gröger, O. May), 3rd ed., vol. 2, Wiley-VCH, Weinheim, 2012, pp. 919–945.
[53] I. Rosenthaler, *Biochem. Z.* 1908, **14**, 238–253.
[54] M. Khadjawi, M. H. Fechter, H. Griengl, in: *Enzyme Catalysis in Organic Synthesis* (eds.: K. Drauz, H. Gröger, O. May), 3rd ed., vol. 2, Wiley-VCH, Weinheim, 2012, pp. 947–990.
[55] T. Purkarthofer, W. Skranc, C. Schuster, H. Griengl, *Appl. Microbiol. Biotechnol.* 2007, **76**, 309–320.
[56] A. Liese, K. Seelbach, C. Wandrey (eds.), *Industrial Biotransformations*, 2nd ed., Wiley-VCH, Weinheim, 2006, p. 311.
[57] F. Effenberger, T. Ziegler, S. Förster, *Angew. Chem.* 1987, **99**, 491–492; *Angew. Chem. Int. Ed. Engl.* 1987, **26**, 458–460.
[58] U. Niedermeyer, M.-R. Kula, *Angew. Chem.* 1990, **102**, 423–425; *Angew. Chem. Int. Ed. Engl.* 1990, **29**, 386–387.
[59] G. Lin, S. Han, Z. Li, *Tetrahedron* 1999, **55**, 3531–3540.
[60] H. Gröger, W. Hummel, S. Borchert, M. Kraußer, in: *Enzyme Catalysis in Organic Synthesis* (eds.: K. Drauz, H. Gröger, O. May), 3rd ed., vol. 2, Wiley-VCH, Weinheim, 2012, pp. 1037–1110.
[61] (a) S. Servi, *Synthesis* 1990, 1–25; (b) R. Csuk, B. I. Glaenzer, *Chem. Rev.* 1991, **91**, 49–97.
[62] D. Acetti, E. Brenna, C. Fuganti, *Tetrahedron: Asymmetry* 2007, **18**, 488–492.
[63] K. Ahmad, S. Kould, S. C. Taneja, A. P. Singh, M. Kapoor, R.-U. Hassan, V. Verma, G. N. Qazi, *Tetrahedron: Asymmetry* 2004, **15**, 1685–1692.
[64] T. Katoh, S. Mizumoto, M. Fudesaka, Y. Nakashima, T. Kajimoto, M. Node, *Synlett* 2006, **14**, 2176–2182.
[65] T. Johanson, M. Carlquist, C. Olsson, A. Rudolf, T. Frejd, M. F. Gorwa-Grauslund, *Appl. Microbiol. Biotechnol.* 2008, **77**, 1111–1118.
[66] C. J. Wallentin, E. Orentas, E. Butkus, K. Wärnmark, *Synthesis* 2009, 864–867.
[67] M. Katzberg, K. Wechler, M. Müller, P. Dünkelmann, J. Stohrer, W. Hummel, M. Bertau, *Org. Biomol. Chem.* 2009, **7**, 304–314.
[68] M. O. Tosa, P. V. Podea, C. Paizs, F. D. Irimie, *Tetrahedron: Asymmetry* 2008, **19**, 2068–2071.
[69] Y. Chen, H. Lin, X. Xu, S. Xia, L. Wang, *Adv. Synth. Catal.* 2008, **350**, 426–430.
[70] K. Nakamura, T. Matsuda, *Curr. Org. Chem.* 2006, **10**, 1217–1246.
[71] M. Amidjojo, D. Weuster-Botz, *Tetrahedron: Asymmetry* 2005, **16**, 899–901.
[72] N. A. Salvi, S. Chattopadhyay, *Tetrahedron* 2001, **57**, 2833–2839.
[73] G. Kumaraswamy, S. Ramesh, *Green Chem.* 2003, **5**, 306–308.
[74] J. S. Yadav, B. V. S. Reddy, C. Sreelakshmi, A. B. Rao, *Synthesis* 2009, **9**, 1881–1885.
[75] M. Bertau, M. Bürli, *Chimia* 2000, **54**, 503–507.
[76] M. Bertau, M. Bürli, E. Hungerbühler, *Chim. Oggi* 1998, **16**, 58–61.
[77] R. Patel, A. Banerjee, J. Howell, C. McNamee, D. Brozozowski, D. Mirfakhrae, V. Nanduri, J. Thottathil, L. Szarka, *Tetrahedron: Asymmetry* 1993, **4**, 2069–2084.
[78] M. Charttrain, C. Roberge, J. Chung, J. McNamara, D. Zhao, R. Olewinski, G. Hunt, P. Salmon, D. Roush, S. Yamazaki, T. Wang, E. Grabowski, B. Buckland, R. Greasham, *Enzyme Microb. Technol.* 1999, **25**, 489–496.
[79] A. Liese, K. Seelbach, C. Wandrey (eds.), *Industrial Biotransformations*, 2nd ed., Wiley-VCH, Weinheim, 2006, pp. 191–194.
[80] B. A. Anderson, M. M. Hansen, A. R. Harkness, C. L. Henry, J. T. Vicenzi, M. J. Zmijewski, *J. Am. Chem. Soc.* 1995, **117**, 12358–12359.
[81] J. T. Vicenzi, M. J. Zmijewski, M. R. Reinhard, B. E. Landen, W. L. Muth, P. G. Marler, *Enzyme Microb. Technol.* 1997, **20**, 494–499.
[82] A. Liese, K. Seelbach, C. Wandrey (eds.), *Industrial Biotransformations*, 2nd ed., Wiley-VCH, Weinheim, 2006, pp. 164–166.
[83] W. Stampfer, B. Kosjek, C. Moitzi, W. Kroutil, K. Faber, *Angew. Chem.* 2002, **114**, 1056–1059; *Angew. Chem. Int. Ed.* **2002**, *41*, 1014–1017.
[84] W. Stampfer, B. Kosjek, K. Faber, W. Kroutil, *J. Org. Chem.* 2003, **68**, 402–406.
[85] W. Stampfer, K. Edegger, B. Kosjek, K. Faber, W. Kroutil, *Adv. Synth. Catal.* 2004, **346**, 57–62.
[86] N. Itoh, M. Matsuda, M. Mabuchi, T. Dairi, J. Wang, *Eur. J. Biochem.* 2002, **269**, 2394–2402.
[87] A. Matsuyama, H. Yamamoto, Y. Kobayashi, *Org. Process Res. Dev.* 2002, **6**, 558–561.
[88] V. Höllrigl, F. Hollmann, A. C. Kleeb, K. Buehler, A. Schmid, *Appl. Microb. Biotechnol.* 2008, **81**, 263–273.
[89] I. Lavandera, A. Kern, M. Schaffenberger, J. Gross, A. Glieder, S. De Wildeman, W. Kroutil, *ChemSusChem* 2008, **1**, 431–436.

[90] K. Schroer, U. Mackfeld, I. A. W. Tan, C. Wandrey, F. Heuser, S. Bringer-Meyer, A. Weckbecker, W. Hummel, T. Daußmann, R. Pfaller, A. Liese, S. Lütz, *J. Biotechnol.* 2007, **132**, 438–444.

[91] K. Goldberg, K. Edegger, W. Kroutil, A. Liese *Biotechnol. Bioeng.* 2006, **95**, 192–198.

[92] A. Jakoblinnert, R. Mladenov, A. Paul, F. Sibilla, U. Schwaneberg, M. B. Ansorge Schumacher, P. Dominguez de María *Chem. Commun.* 2011, **47**, 12230–12232.

[93] M. Kataoka, L. P. S. Rohani, K. Yamamoto, M. Wada, H. Kawabata, K. Kita, H. Yanase, S. Shimizu, *Appl. Microbiol. Biotechnol.* 1997, **48**, 699–703.

[94] S. Shimizu, M. Kataoka, M. Katoh, T. Morikawa, T. Miyoshi, H. Yamada, *Appl. Environ. Microbiol.* 1990, **56**, 2374–2377.

[95] N. Kizaki, Y. Yasohara, J. Hasegawa, M. Wada, M. Kataoka, S. Shimizu, *Appl. Microbiol. Biotechnol.* 2001, **55**, 590–595.

[96] H. Gröger, C. Rollmann, F. Chamouleau, I. Sebastien, O. May, W. Wienand, K. Drauz, *Adv. Synth. Catal.* 2007, **349**, 709–712.

[97] A. Berkessel, C. Rollmann, F. Chamouleau, S. Labs, O. May, H. Gröger, *Adv. Synth. Catal.* 2007, **349**, 2697–2704.

[98] T. Ema, N. Okita, S. Ide, T. Sakai, *Org. Biomol. Chem.* 2007, **5**, 1175–1176.

[99] T. Ema, H. Yagasaki, N. Okita, K. Nishikawa, T. Korenga, T. Sakai, *Tetrahedron: Asymmetry* 2005, **16**, 1075–1078.

[100] J. Zhang, B. Witholt, Z. Li, *Chem. Commun.* 2006, 398–400.

[101] W. Zhang, K. O´Connor, D. I. C. Wang, Z. Li, *Appl. Environm. Microbiol.* 2009, **75**, 687–694.

[102] R. L. Hanson, S. Goldberg, A. Goswami, T. P. Tully, R. N. Patel, *Adv. Synth. Catal.* 2005, **347**, 1073–1080.

[103] W. Hummel, H. Gröger, in: *Enzyme Catalysis in Organic Synthesis* (eds.: K. Drauz, H. Gröger, O. May), 3rd ed., vol. **2**, Wiley-VCH, Weinheim, 2012, pp. 1165–1203.

[104] G. Krix, A. S. Bommarius, K. Drauz, M. Kottenhahn, M. Schwarm, M. R. Kula, *J. Biotechnol.* 1997, **53**, 29–39.

[105] A. S. Bommarius, K. Drauz, W. Hummel, M.-R. Kula, C. Wandrey, *Biocatalysis* 1994, **10**, 37–47.

[106] A. Bommarius, M. Schwarm, K. Drauz, *J. Mol. Catal. B: Enzym.* 1998, **5**, 1–11.

[107] A. Galkin, L. Kulakova, T. Yoshimura, K. Soda, N. Esaki, *Appl. Environ. Microbiol.* 1997, **63**, 4651–4656.

[108] A. Menzel, H. Werner, J. Altenbuchner, H. Gröger, *Eng. Life Sci.* 2004, **4**, 573–576.

[109] H. Gröger, O. May, H. Werner, A. Menzel, J. Altenbuchner, *Org. Process Res. Dev.* 2006, **10**, 666–669.

[110] C. K. Savile, J. M. Janey, E. C. Mundorff, J. C. Moore, S. Tam, W. R. Jarvis, J. C. Colbeck, A. Krebber, F. J. Fleitz, J. Brands, P. N. Devine, G. W. Huisman, G. J. Hughes, *Science* 2010, **329**, 305–309.

[111] M. Cardenas-Fernandez, W. Neto, C. López, G. Alvaro, P. Tufvesson, J. M. Woodley, *Biotechnol. Prog.* 2012, **28**, 693–398.

[112] R. Stuermer, B. Hauer, M. Hall, K. Faber, *Curr. Opin. Chem. Biol.* 2007, **11**, 203–213.

[113] D. J. Bougioukou, J. D. Stewart, in: *Enzyme Catalysis in Organic Synthesis* (eds.: K. Drauz, H. Gröger, O. May), vol. **2**, 3rd ed., Wiley-VCH, Weinheim, 2012, pp. 1111–1163.

[114] M. Kataoka, A. Kotaka, R. Thiwthong, M. Wada, S. Nakamori, S. Shimizu, *J. Biotechnol.* 2004, **114**, 1–9.

[115] M. Kataoka, A. Kotaka, A. Hasegawa, M. Wada, A. Yoshizumi, S. Nakamori, S. Shimizu, *Biosci. Biotechnol. Biochem.* 2002, **66**, 2651–2657.

[116] D. Romano, M. L. Contente, F. Molinari, I. Eberini, E. Ruvutuso, C. Sensi, A. Amaretti, M. Rossi, S. Raimondi, *Microb. Cell Fact.* 2014, **13**, 60.

[117] D. Kuhn, M. A. Kholiq, E. Heinzle, B. Bühler, A. Schmid, *Green Chem.* 2010, **12**, 815–827.

[118] M. Schrewe, M. K. Julsing, B. Bühler, A. Schmid *Chem. Soc. Rev.* 2013, **42**, 6346–6377.

[119] S. Panke, B. Witholt, A. Schmid, M. G. Wubbolts, *Appl. Environ. Microbiol.* 1998, **64**, 2032–2043.

[120] A. Schmid, K. Hofstetter, H. J. Feiten, F. Hollmann, B. Witholt, *Adv. Synth. Catal.* 2001, **343**, 732–737.

[121] V. Alphand, G. Carrea, R. Wohlgemuth, R. Furstoss, J. M. Woodley, *Trends Biotechnol.* 2003, **21**, 318–323.

[122] G. Strukul, *Angew. Chem. Int. Ed.* 1998, **37**, 1198–1209.

[123] (a) C. Bolm, O. Beckmann, A. Cosp, C. Palazzi, *Synlett* 2001, **9**, 1461–1463; (b) A. Watanabe, T. Uchida, R. Irie, T. Katsuki, *Tetrahedron Lett.* 2002, **43**, 4481–4485.

[124] Y. Peng, X. Feng, K. Yu, Z. Li, Y. Jiang, C. H. Yeung, *J. Organomet. Chem.* 2001, **619**, 204–208.
[125] A. Watanabe, T. Uchida, R. Irie, T. Katsuki, *Tetrahedron Lett.* 2002, **43**, 4481–4485.
[126] S.-I. Murahashi, S. Ono, Y. Imada, *Angew. Chem.* 2002, **114**, 2472–2474.
[127] H. D. Simpson, V. Alphand, R. Furstoss, *J. Mol. Catal. B: Enzym.* 2001, **16**, 101–108.
[128] S. M. Roberts, R. W. H. Wan, *J. Mol. Catal. B: Enzym.* 1998, **4**, 111–136.
[129] K. Miyaji, Y. Ohara, Y. Takahashi, T. Tsuruda, K. Arai, *Tetrahedron Lett.* 1991, **32**, 4557–4560.
[130] W. Adam, Z. Lukacs, C. R. Saha-Möller, B. Weckerle, P. Schreier, *Eur. J. Org. Chem.* 2000, 2923–2926.
[131] S. Kardinahl, D. Rabelt, M. Reschke, *Chem. Ing. Tech.* 2006, **78**, 209–217.
[132] M. Schrewe, N. Ladkau, B. Bühler, A. Schmid, *Adv. Synth. Catal.* 2013, **355**, 1693–1697.
[133] Evonik Industries AG, press release "An alternative raw material for polyamide 12: Evonik is operating a pilot plant for bio-based ω-amino lauric acid", July 30, 2013.
[134] A. Kleemann, J. Engels, B. Kutscher, D. Reichert, *Pharmaceutical Substances: Syntheses, Patents, Applications*, 4th ed., Thieme, Stuttgart, 2001.
[135] C. J. Dunsmore, R. Carr, T. Fleming, N. J. Turner, *J. Am. Chem. Soc.* 2006, **128**, 2224–2225.
[136] V. Köhler, K. R. Bailey, A. Znabet, J. Raftery, M. Helliwell, N. J. Turner, *Angew. Chem.* 2010, **122**, 2228–2230.

CHAPTER 24

Lipases and Esterases as User-Friendly Biocatalysts in Natural Product Synthesis

Kenji Mori

Department of Applied Biological Chemistry, Faculty of Agriculture, The University of Tokyo, Tokyo, Japan

24.1 INTRODUCTION

In 1978, when I first employed microorganisms for my enantioselective synthesis of natural products, there was still a clear borderline between organic synthesis and enzyme-mediated synthesis. It was a time long before the advent of directed evolution and synthetic biology [1]. What I dreamed of in 1978 was to use bottled enzymes just like common reagents in organic laboratories. Some of my colleagues in the fields of microbiology and enzymology regarded my idea to be premature at that time.

My first paper in chemoenzymatic synthesis appeared in the fall of 1978 under the title "Synthesis of optically active alkynyl alcohols by microbial asymmetric hydrolysis of the corresponding acetates" [2, 3]. As shown in Figure 24.1, asymmetric hydrolysis of (±)-**1** with *Bacillus subtilis* var. *niger* was only moderately successful, yielding the recovered (*R*)-**1** and the hydrolyzed (*S*)-**2** of 7–96% ee [2, 3]. After 35 years of cooperative endeavor by biochemists as the suppliers of enzymes and synthetic chemists as the users, it is now possible to employ various lipases and esterases as commercially available and bottled reagents for enantioselective synthesis of natural products, pharmaceuticals, and agrochemicals.

This chapter illustrates the application of lipases and esterases as user-friendly biocatalysts in (i) desymmetrization of prochiral or *meso*-diols and diacetates, (ii) kinetic resolution of racemic alcohols, and (iii) preparation of enantiopure intermediate(s) from a mixture of stereoisomers by enzymatic differentiation. All the examples were taken from our own works in natural products synthesis.

24.2 DESYMMETRIZATION OF PROCHIRAL OR *meso*-DIOLS AND DIACETATES

Desymmetrization of a prochiral or *meso*-compound is in principle more efficient than kinetic resolution of a racemate, because the former converts all of the starting material to the desired product, while kinetic resolution of a racemate gives the desired product in 50% yield at most. It is essential to design a good substrate

Green Biocatalysis, First Edition. Edited by Ramesh N. Patel.

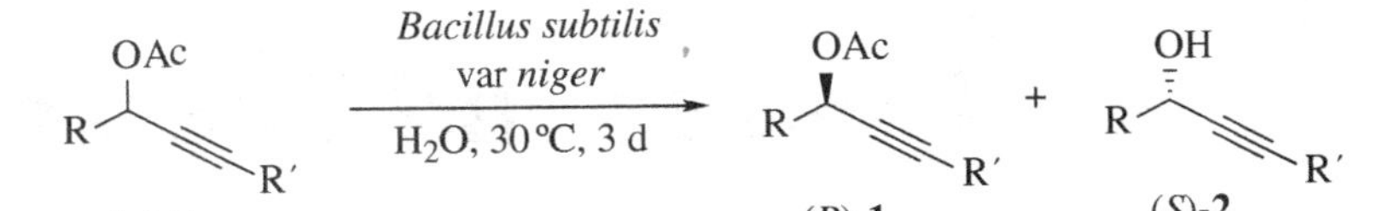

(±)-**1** (*R*)-**1** (*S*)-**2**

	R	R′	Yield (%)	ee	Yield (%)	ee
a:	*n*-Pr	H	22	96	36	68
b:	*n*-Oct	H	53	7	25	16
c:	Me_2CHCH_2	Me	25	74	36	54

FIGURE 24.1

Asymmetric hydrolysis of (±)-alkynyl acetate.

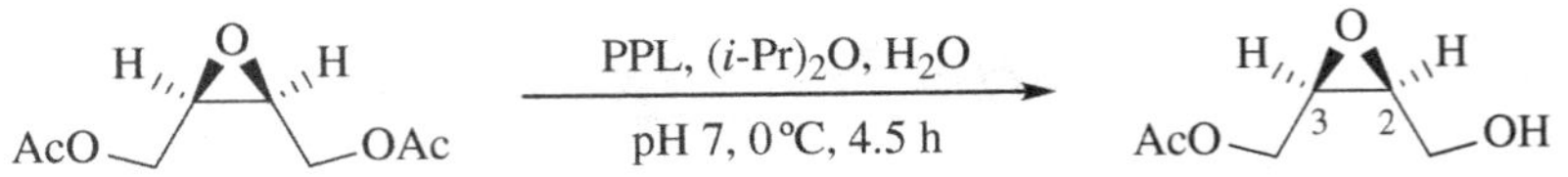

3

(2*S*,3*R*)-**4**
71% yield
(90% ee)

FIGURE 24.2

Preparation of epoxy half-acetate (2*S*,3*R*)-**4**.

for lipases and esterases to secure a highly enantiomerically pure product. Derivatization of the product to a crystalline compound is often necessary to enhance the enantiomeric purity by its recrystallization. It is also important to design a product of broad utility for the synthesis of natural products with a common structural motif.

24.2.1 Desymmetrization of *meso*-Compounds with 1,2-Stereogenic Centers

(2S,3R)-4-Acetoxy-2,3-Epoxy-1-Butanol (4)

A number of insect pheromones are epoxides. Asymmetric hydrolysis of *meso*-diacetate **3** (Figure 24.2) with pig pancreatic lipase (PPL) gives (2*S*,3*R*)-4-acetoxy-2,3-epoxy-1-butanol (**4**, 90% ee) in 71% yield [4]. Various epoxide pheromones were synthesized from (2*S*,3*R*)-**4**, which could be purified to give enantiopure material by recrystallizing the corresponding 3,5-dinitrobenzoate.

In 2003, (2*R*,3*S*)-4-*t*-butyldiphenylsilyloxy-2,3-epoxy-1-butanol (**5**) with 99.5% ee was prepared and found to be a better building block than (2*S*,3*R*)-**4** (see section "(2*R*,3*S*)-4-*t*-Butyldiphenylsilyloxy-2,3-Epoxy-1-Butanol (**5**)") [5]. These two epoxide building blocks **4** and **5** were utilized for the synthesis of not only epoxide pheromones but also pheromone alcohols and ketones after alkylative cleavage of the epoxy ring (Figure 24.3).

(1S,2R)-1-Acetoxymethyl-2-Hydroxymethylcyclopropane (7)

There are many cyclopropane-containing compounds among natural products. For example, plakoside A (Figure 24.4), an immunosuppressive marine galactosphingolipid, contains two cyclopropane rings. Synthesis of plakoside A was achieved by employing (1*S*,2*R*)-**7** as the key building block. As shown in Figure 24.4, acetylation of diol **6** with vinyl acetate and lipase AK (Amano) gives (1*S*,2*R*)-**7** as an almost pure enantiomer [6, 7]. This is a simple and efficient approach to obtain chiral and nonracemic 1,2-disubstituted cyclopropanes.

(2S,3R)-4-Acetoxy-2,3-Isopropylidenedioxy-1-Butanol (9)

Erythritol-derived diol **8** was acetylated with vinyl acetate and lipase AK (Amano) to give (2*S*,3*R*)-**9** (Figure 24.5) [8]. This half-acetate **9** was converted to (+)-*endo*-brevicomin, a male-produced component of the southern pine beetle pheromone.

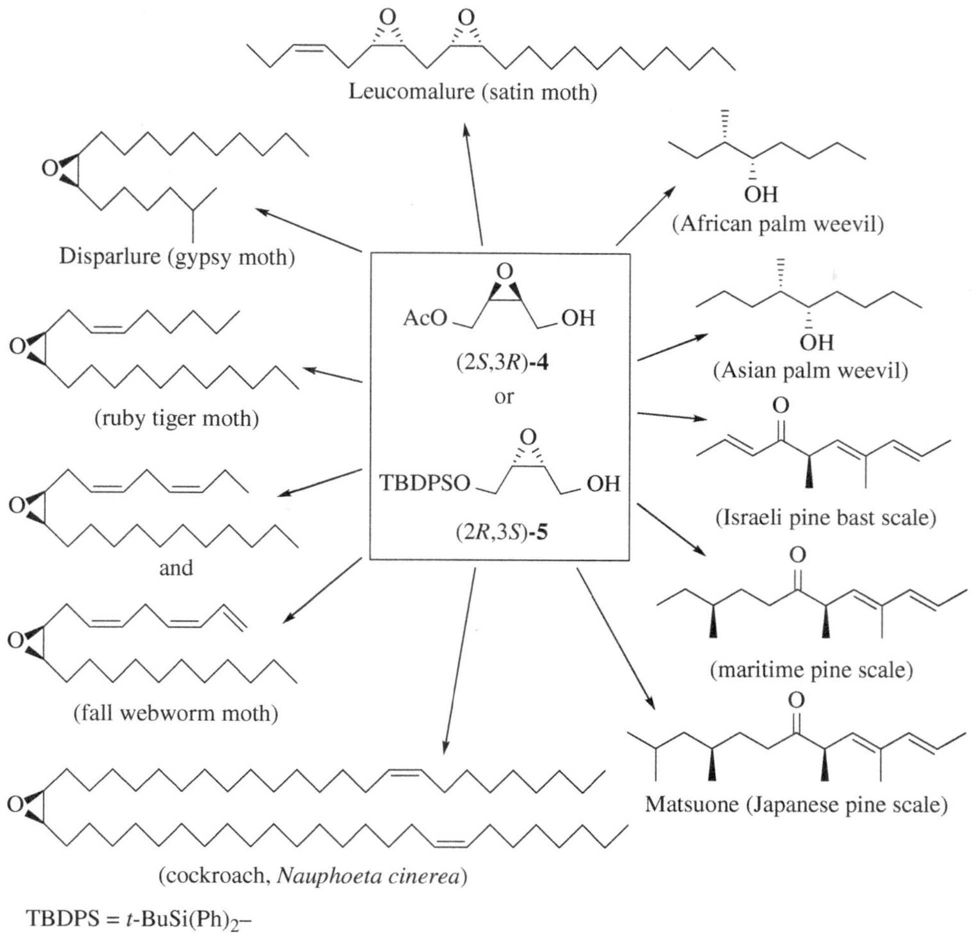

FIGURE 24.3

Pheromones synthesized from (2*S*,3*R*)-**4** and (2*R*,3*S*)-**5**.

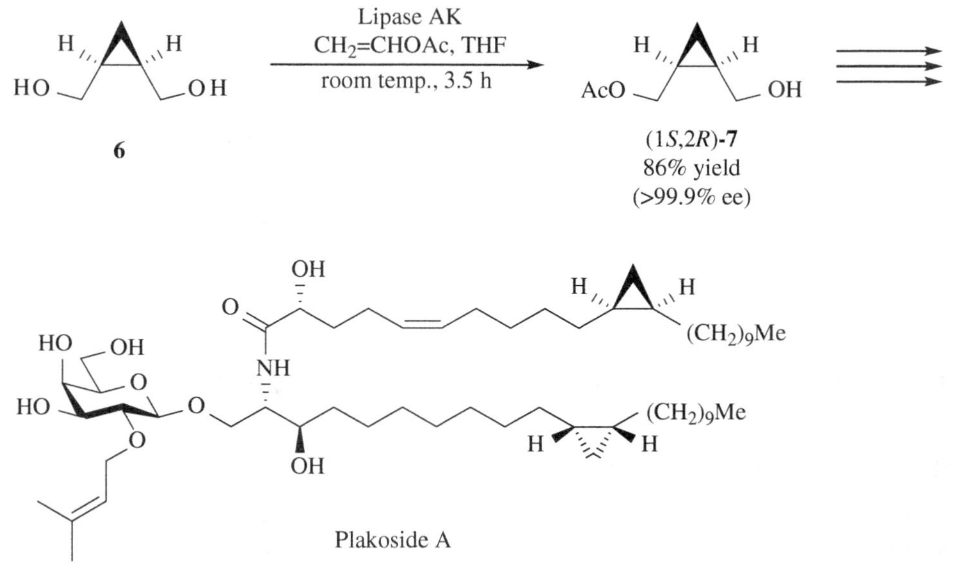

FIGURE 24.4

Preparation of half-acetate (1*S*, 2*R*)-**7**, which was converted to plakoside A.

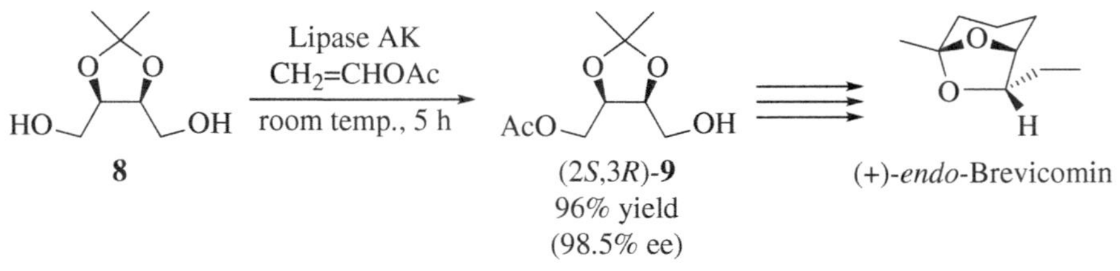

FIGURE 24.5

Preparation of half-acetate (2*S*, 3*R*)-**9**, which was converted to (+)-*endo*-brevicomin.

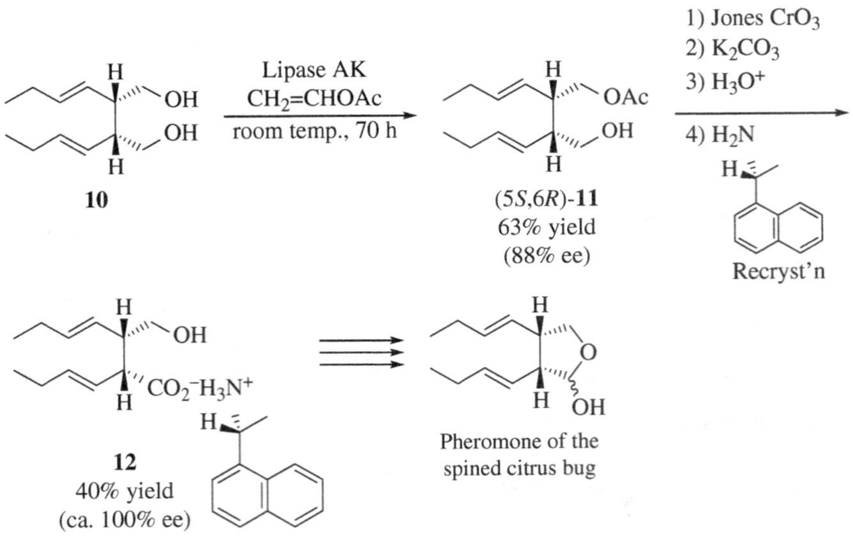

FIGURE 24.6

Preparation of half-acetate (5*S*, 6*R*)-**11**, which was converted to the pheromone of the spined citrus bug.

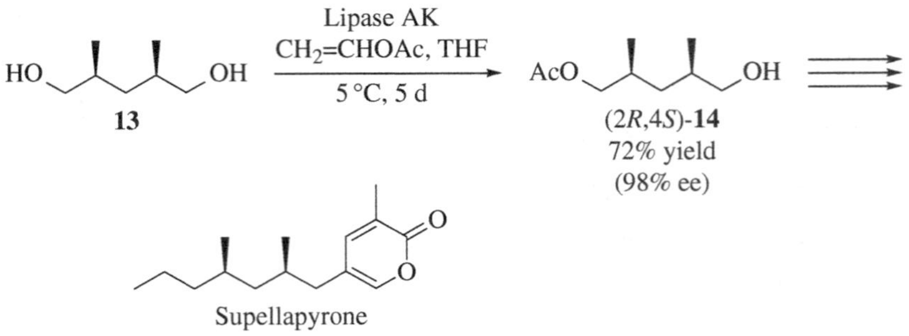

FIGURE 24.7

Preparation of half-acetate (2*R*, 4*S*)-**14**, which was converted to supellapyrone (the pheromone of the brown-banded cockroach).

***(3E,5S,6R,7E)-5-Acetoxymethyl-6-Hydroxymethyl-3,7-Decadiene* (11)**

Asymmetric acetylation of *meso*-diol **10** with vinyl acetate and lipase AK gave (5*S*,6*R*)-**11** of 88% ee. This was purified by derivatization to **12**, which could be recrystallized. The resulting pure **12** was converted to the pheromone of the spined citrus bug (Figure 24.6) [9].

24.2.2 Desymmetrization of *meso*-Compounds with 1,3- and 1,5-Stereogenic Centers

***(2R,4S)-5-Acetoxy-2,4-Dimethyl-1-Pentanol* (14)**

Asymmetric acetylation of *meso*-diol **13** with vinyl acetate and lipase AK gave (2*R*,4*S*)-**14**, which was converted to supellapyrone, the pheromone of the brown-banded cockroach (Figure 24.7) [10].

***(2R,6S)-7-Acetoxy-2,6-Dimethyl-1-Heptanol* (16)**

Asymmetric acetylation of *meso*-diol **15** with isopropenyl acetate and lipase PS (Amano) gave (2*R*,6*S*)-**16**, which served as the key building block for the synthesis of four insect pheromones (Figure 24.8) [11].

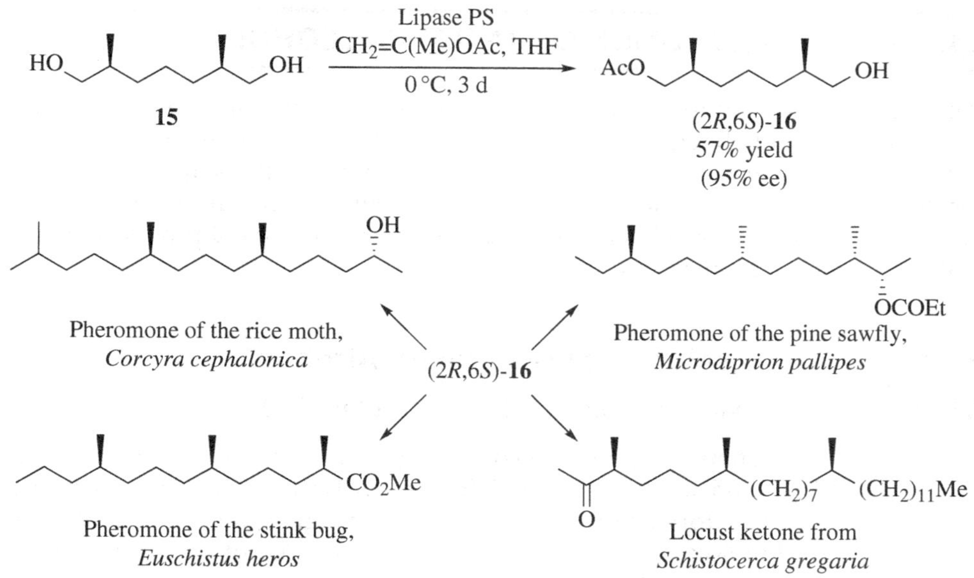

FIGURE 24.8
Preparation of half-acetate (2*R*, 6*S*)-**16**, which was converted to four insect pheromones.

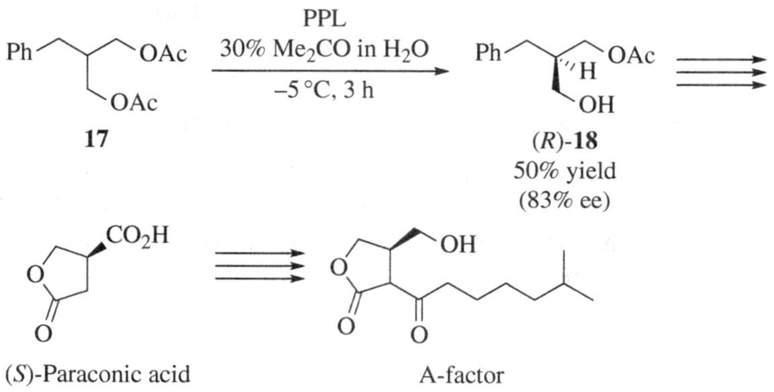

FIGURE 24.9
Preparation of half-acetate (*R*)-**18**, which was converted to A-factor (a microbial hormone).

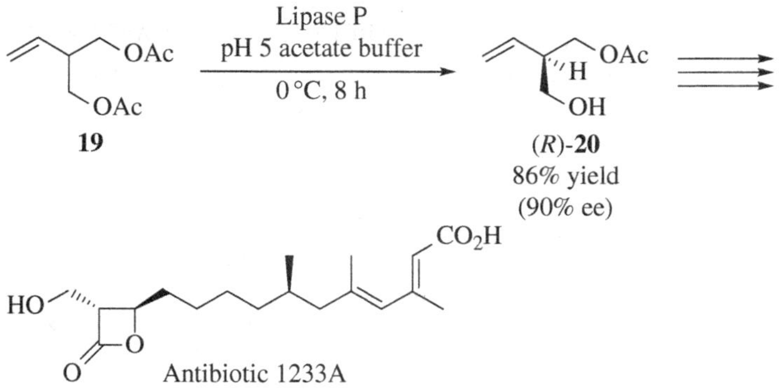

FIGURE 24.10
Preparation of half-acetate (*R*)-**20**, which was converted to an antibiotic 1233A.

24.2.3 Desymmetrization of Prochiral Compounds with a Single Stereogenic Center

(R)-2-Hydroxymethyl-3-Phenylpropyl Acetate (18)

Asymmetric hydrolysis of prochiral diacetate **17** with pig pancreatic lipase (PPL) yielded (*R*)-**18** (Figure 24.9) [12]. This was converted to (*S*)-paraconic acid, purified as its amine salt, and eventually furnished A-factor, a microbial hormone.

(R)-2-Hydroxymethyl-3-Butenyl Acetate (20)

Asymmetric hydrolysis of prochiral diacetate **19** with lipase P (Amano) gave (*R*)-**20** (Figure 24.10) [13]. Antibiotic 1233A was synthesized from (*R*)-**20**.

24.3 KINETIC RESOLUTION OF RACEMIC ALCOHOLS

Although kinetic resolution provides the desired enantiomer in 50% yield at most, this is a method of choice due to the simplicity of the experimental procedure. If both the enantiomers are required for structure–activity relationship studies, kinetic resolution of racemic alcohols is a convenient method to secure the enantiomers. A drawback of this method is time-consuming separation of the products of enzymatic reaction by chromatography.

24.3.1 Kinetic Resolution of (±)-Primary Alcohols

(2R,3S)-4-t-Butyldiphenylsilyloxy-2,3-Epoxy-1-Butanol (5)

Asymmetric acetylation of (±)-epoxy alcohol **5** with vinyl acetate and lipase PS gave acetate (2*S*,3*R*)-**21** and alcohol (2*R*,3*S*)-**5** (Figure 24.11) [5]. These products were obtained with satisfactory enantiomeric purities through a single enzymatic acetylation. When *t*-butyldimethylsilyl protective group was employed instead of the *t*-butyldiphenylsilyl group, the enantiomeric purity of the products was less satisfactory. The recovered alcohol (2*R*,3*S*)-**5** was converted to leucomalure, the satin moth pheromone.

(2S,3R)-3,7-Dimethyl-2,3-Epoxy-6-Octenyl Acetate (23)

Asymmetric acetylation of (±)-2,3-epoxynerol (**22**) with vinyl acetate and lipase PS was repeated three times to give (2*S*,3*R*)-**23** and (2*R*,3*S*)-**22** (Figure 24.12) [14]. Both were obtained with greater than 98% enantiomeric purity. The acetate (2*S*,3*R*)-**23** was converted to the pheromone of the Colorado potato beetle.

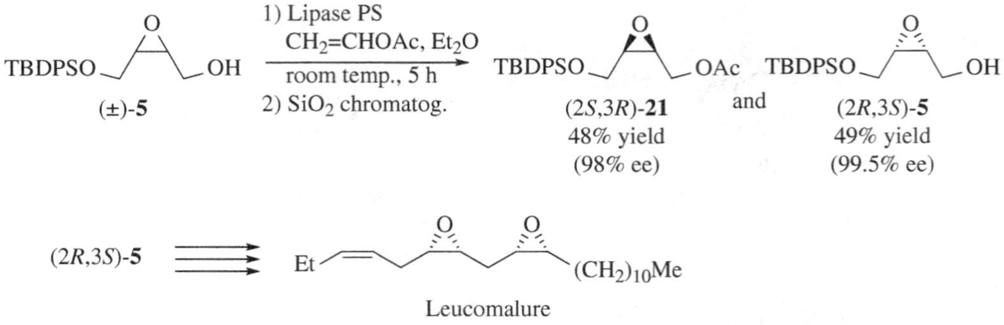

FIGURE 24.11

Preparation of epoxide (2*R*,3*S*)-**5**, which was converted to leucomalure (the pheromone of the satin moth).

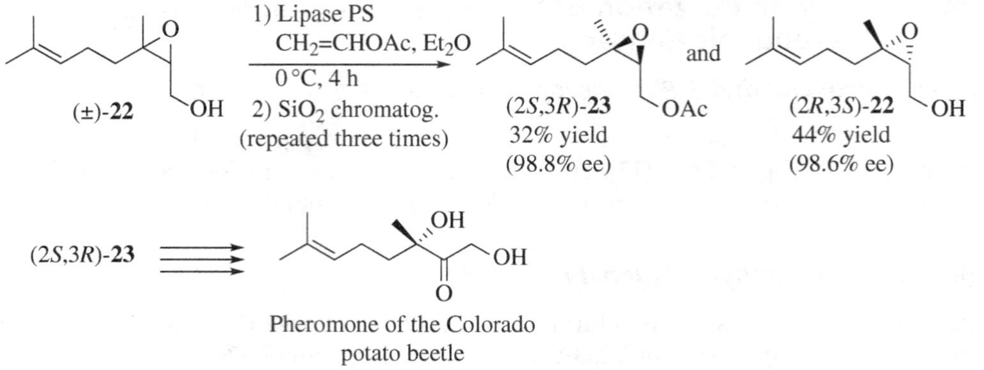

FIGURE 24.12

Preparation of epoxide (2*S*,3*R*)-**23**, which was converted to the pheromone of the Colorado potato beetle.

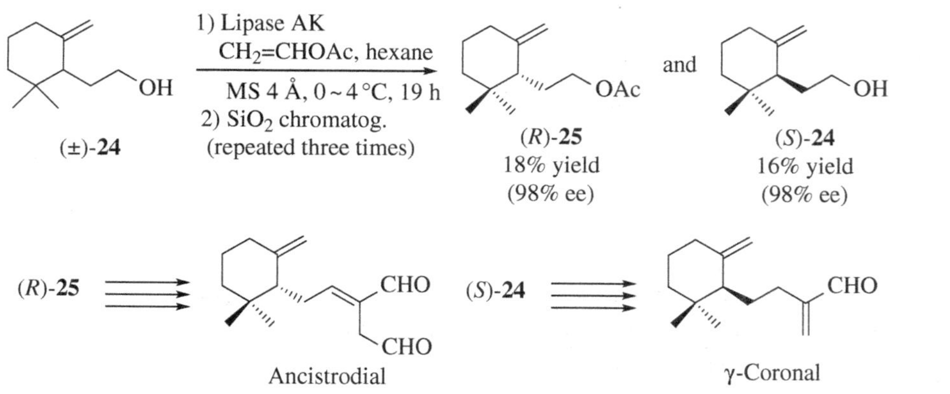

FIGURE 24.13

Preparation of the enantiomers of γ-cyclohomogeraniol (**24**), which was converted to two terpenoids.

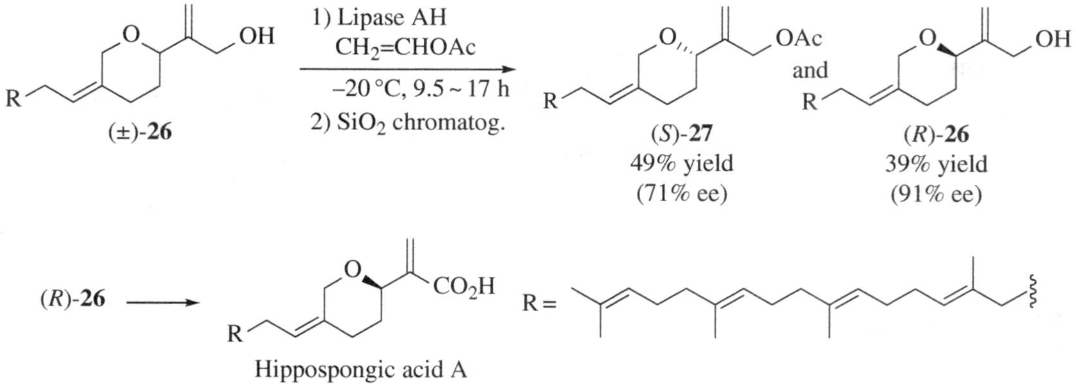

FIGURE 24.14

Preparation of (*R*)-**26**, which was converted to (+)-hippospongic acid (a marine triterpenoid).

(R)-γ-Cyclohomogeranyl Acetate (25)

Asymmetric acetylation of (±)-γ-cyclohomogeraniol (**24**) with vinyl acetate and lipase AK gave (*R*)-γ-cyclohomogeranyl acetate (**25**) and (*S*)-γ-cyclohomogeraniol (**24**) (Figure 24.13) [15, 16]. These were converted to ancistrodial (a defense substance of a termite) and γ-coronal (an ambergris odorant), respectively.

(R)-2-[5′-(4,″9,″13,″17″-Tetramethyl-4,″8,″12,″16″-Octadecatetraenylidene) Tetrahydropyran-2′-yl]-2-Propen-1-ol (26)

Asymmetric acetylation of tetrahydropyranyl alcohol (±)-**26** with vinyl acetate and lipase AH gave (*S*)-**27** and (*R*)-**26**. The latter was converted to (+)-hippospongic acid A, a metabolite of a marine sponge (Figure 24.14) [17].

24.3.2 Kinetic Resolution of Acyclic (±)-Secondary Alcohols

(2S,6Z)-6,8-Nonadien-2-ol (29)

Asymmetric hydrolysis of (±)-**28** with lipase AK in acetone/water at pH7.5 yielded (*S*)-**28** (87% ee) and the hydrolyzed alcohol (*R*)-**29** (80% ee). The (*S*)-**28** was further treated with lipase AK to improve the ee to 94%, which afforded (*S*)-**29**, the pheromone of the leaf miner, *Nepticula melella* (Figure 24.15) [18].

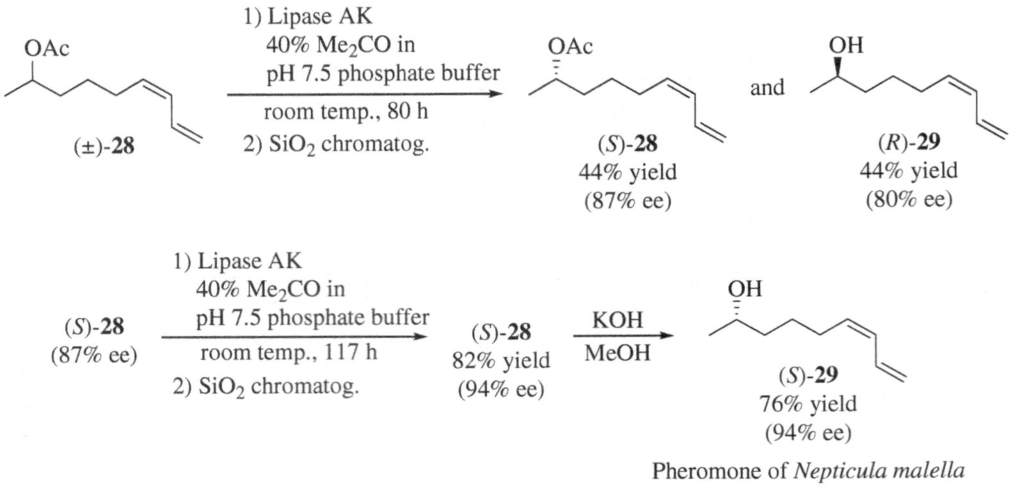

FIGURE 24.15

Preparation of (2*S*,6*Z*)-6,8-nonadien-2-ol (**29**), the pheromone of a leaf miner *Nepticula melella*.

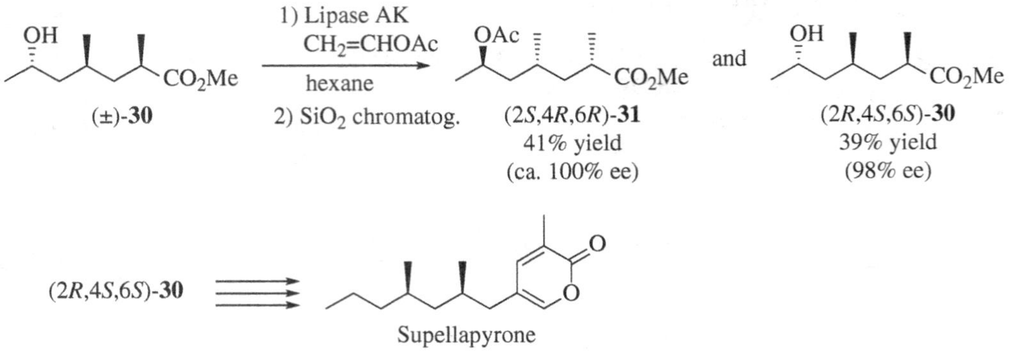

FIGURE 24.16

Preparation of hydroxyl ester (2*R*,4*S*,6*S*)-**30**, which was converted to supellapyrone.

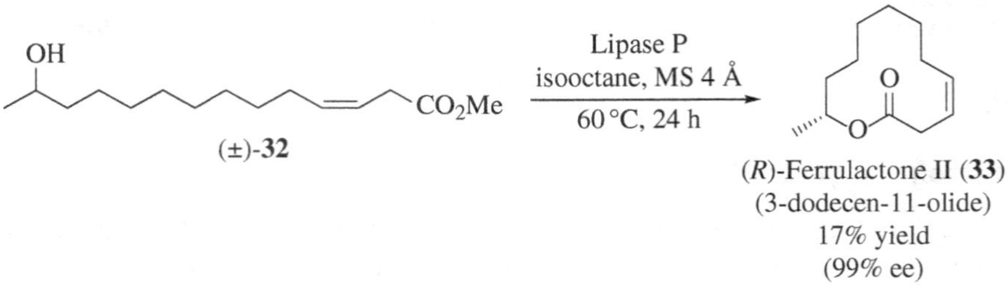

FIGURE 24.17

Preparation of (3*Z*,11*R*)-3-dodecen-11-olide (**33**), the pheromone of the rusty grain beetle.

Methyl (2R,4S,6S)-6-Hydroxy-2,4-Dimethylheptanoate (30)

Asymmetric acetylation of (±)-**30** with vinyl acetate and lipase AK left (2*R*,4*S*,6*S*)-**30** intact, while its enantiomer was acetylated to give (2*S*,4*R*,6*R*)-**31**. The hydroxy ester (2*R*,4*S*,6*S*)-**30** was converted to the cockroach pheromone, supellapyrone (Figure 24.16) [19].

(3Z,11R)-3-Dodecen-11-Olide (33)

Treatment of (±)-hydroxy ester **32** with lipase P (Amano) affected the enantioselective macrolactonization to give the pheromone of the rusty grain beetle (*R*)-**33**, although the yield was only 17% (Figure 24.17) [20].

(S)-6-Methyl-2-Heptyn-4-ol (35)

Lipase A (Amano, from *Aspergillus* sp.) was employed for the asymmetric hydrolysis of (±)-**34** to give (*S*)-**35**, which was converted to the pheromone of the American palm weevil, *Rhynchophorus palmarum* (Figure 24.18) [21].

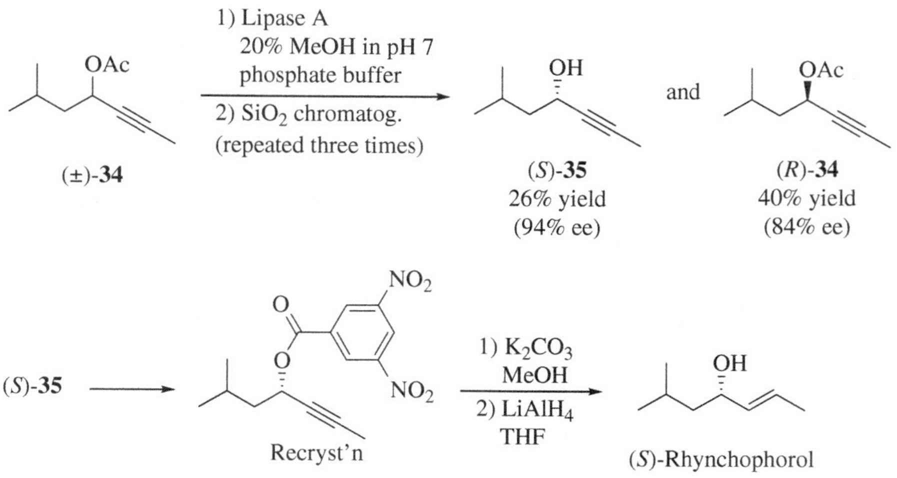

FIGURE 24.18

Preparation of (*S*)-6-methyl-2-heptyn-4-ol (**35**), which was converted to rhynchophorol (the pheromone of the American palm weevil).

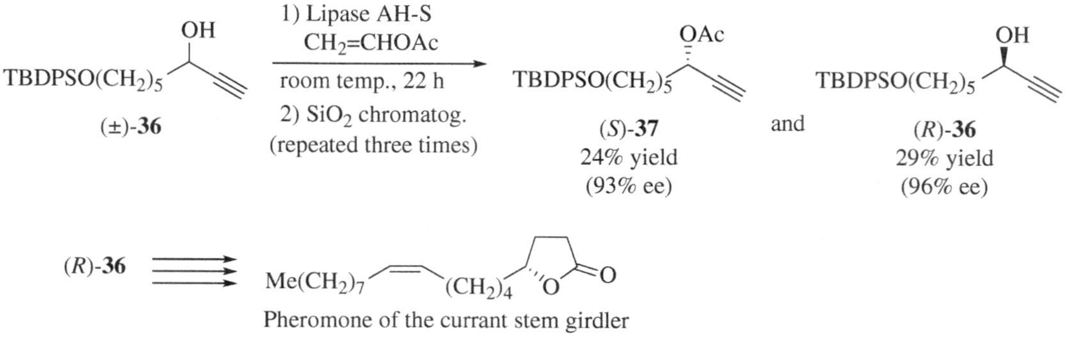

FIGURE 24.19

Preparation of (*R*)-8-*t*-butyldiphenylsilyloxy-1-octyn-3-ol (**36**), which was converted to (4*R*,9*Z*)-9-octadecen-4-olide (the pheromone of the currant stem girdler).

(R)-8-t-Butyldiphenylsilyloxy-1-Octyn-3-ol (36)

Asymmetric acetylation of (±)-**36** with a terminal ethynyl group was examined with several lipases. After a single enzymatic acetylation, the ee of the recovered (*R*)-**36** was 82% with lipase AH-S, 71% with lipase PS-C, and 69% with lipase AK-20. The acetylation was less sluggish with lipase AH-S (22h for 50% acetylation), while 101 and 76h were necessary in the cases of lipases PS-C and AK-20, respectively [22]. (*R*)-Alcohol **36** was obtained by repeating the acetylation three times with lipase AH-S (Figure 24.19) [22]. This was converted to the pheromone of the currant stem girdler.

(R)-3-Acetoxy-1-Trimethylsilyl-1-Pentadecyne (39)

Kinetic resolution of 1-alkyn-3-ols employing lipase AH-S was rather tedious to repeat the acetylation and chromatographic separation three times. Fortunately, in 1997, Anastasia and coworkers found that (±)-1-trimethylsilyl-1-alkyn-3-ols can be resolved into (*R*)-acetates and (*S*)-alcohols in a single step without repeating the enzymatic kinetic resolution [23].

Indeed, asymmetric acetylation of (±)-**38** with vinyl acetate and lipase PS-C gave (*R*)-acetate **39** and the recovered (*S*)-alcohol **38** of satisfactory enantiomeric purities after a single enzymatic reaction. It must be noted that the trimethylsilylethynyl group of **38** was recognized by lipase PS-C as a group bulkier than the *n*-decyl group.

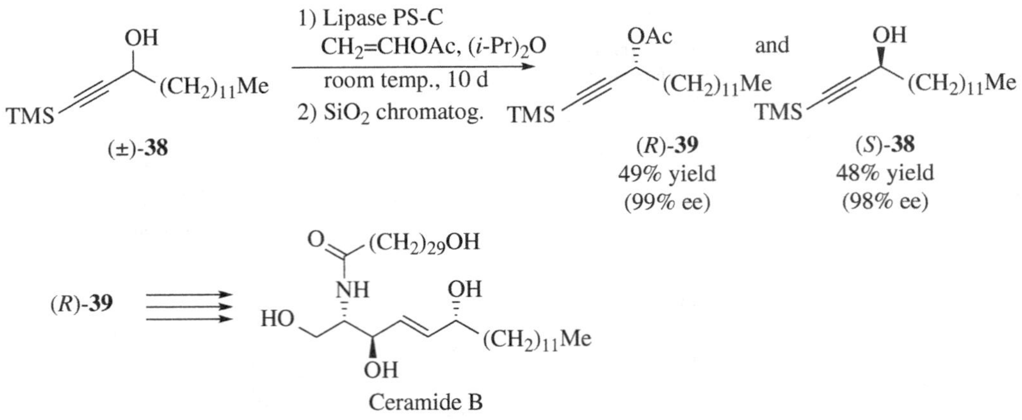

FIGURE 24.20
Preparation of (*R*)-3-acetoxy-1-trimethylsilyl-1-pentadecyne (**39**), which was converted to ceramide B.

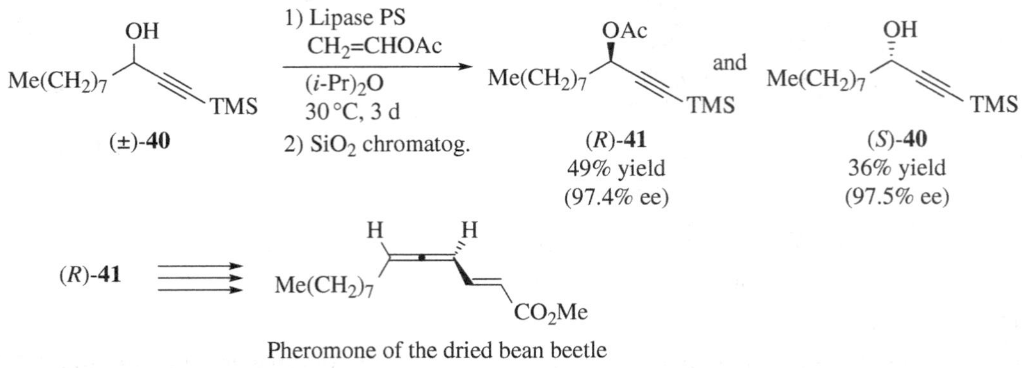

FIGURE 24.21
Preparation of (*R*)-3-acetoxy-1-trimethylsilyl-1-undecyne (**41**), which was converted to an allene pheromone.

Comparison of Figure 24.19 with Figure 24.20 clearly shows the reversed stereochemical outcome of the two enzymatic reactions. The acetate (*R*)-**39** was converted to ceramide B, a protein-bound ceramide in human skin (Figure 24.20) [24].

(R)-3-Acetoxy-1-Trimethylsilyl-1-Undecyne (41)

Asymmetric acetylation of (±)-**40** with vinyl acetate and lipase PS furnished (*R*)-acetate **41** and (*S*)-alcohol **40**. The former was converted to the axially chiral pheromone of the dried bean beetle (Figure 24.21) [25].

24.3.3 Kinetic Resolution of Cyclic (±)-Secondary Alcohols

(R)-2,4,4-Trimethyl-2-Cyclohexen-1-ol (43)

Asymmetric hydrolysis of (±)-2,4,4-trimethyl-2-cyclohexenyl acetate (**42**) with pig liver esterase (PLE) in methanol/phosphate buffer (pH 7.5) yielded enantiopure (*R*)-alcohol **43** in 26% yield together with 67% of the recovered (*S*)-**42** (Figure 24.22) [26]. Two degraded carotenoids, (−)-α-damascone and (−)-loliolide, were synthesized from (*R*)-**43**.

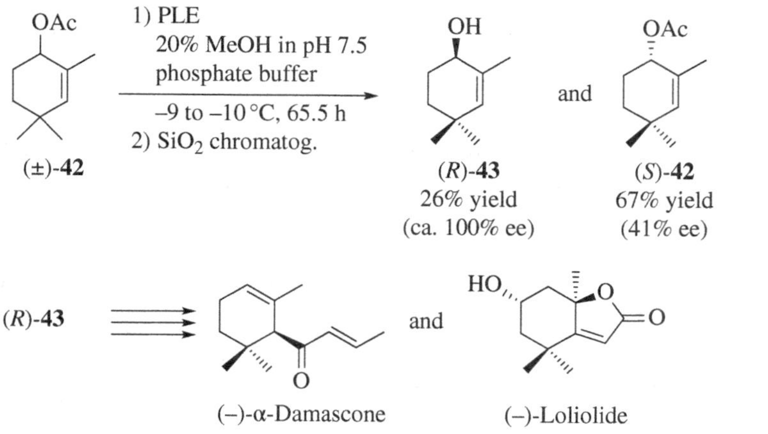

FIGURE 24.22

Preparation of (*R*)-2,4,4-trimethyl-2-cyclohexen-1-ol (**43**), which was converted to degraded carotenoids.

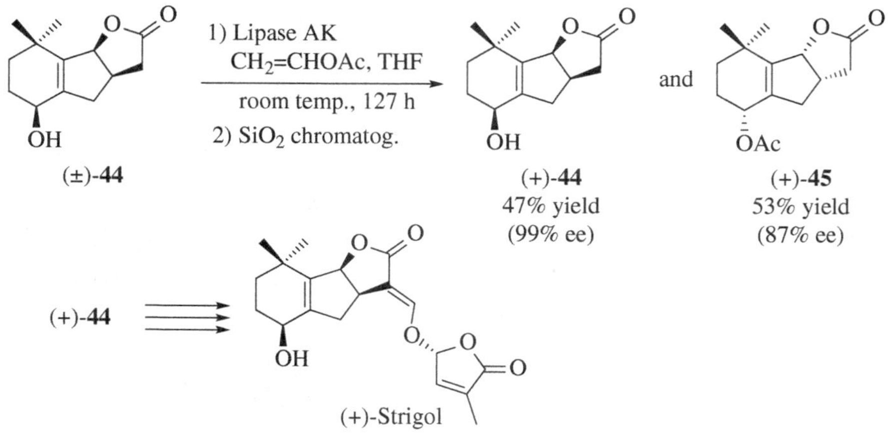

FIGURE 24.23

Preparation of hydroxyl lactone (+)-**44**, which was converted to strigol.

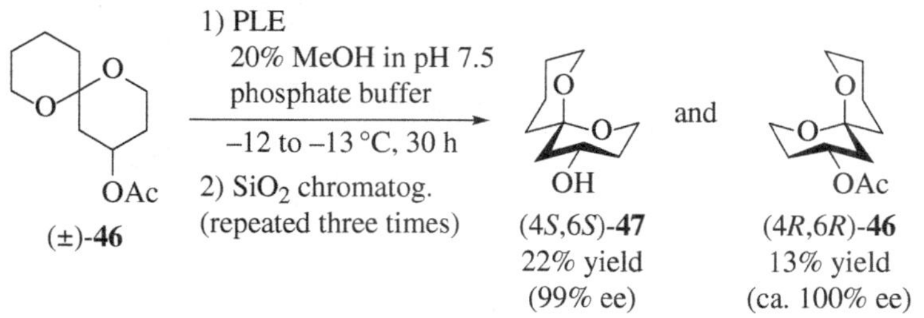

FIGURE 24.24

Preparation of (4*S*,6*S*)-4-hydroxy-4,9-dioxaspiro[5.5]undecane (**47**), a pheromone component of the olive fruit fly.

(3aR,5S,8bS)-5-Hydroxy-8,8-Dimethyl-3,3a,4,5,6,7,8,8b-Octahydroindeno[1.2-b]furan-2-One (44)

Asymmetric acetylation of (±)-**44** with lipase AK afforded (+)-**44** (99% ee, 47% yield) and (+)-**45** (87% ee) (Figure 24.23) [27]. The alcohol (+)-**44** was converted to (+)-strigol, a potent stimulant for the germination of parasitic weeds.

(4S,6S)-4-Hydroxy-4,9-Dioxaspiro[5.5]undecane (47)

Asymmetric hydrolysis of acetate (±)-**46** with a spiroacetal ring system was achieved with pig liver esterase (PLE) to give alcohol (4*S*,6*S*)-**47** and the unreacted acetate (4*R*,6*R*)-**46** (Figure 24.24) [28]. The alcohol (4*S*,6*S*)-**47** is a component of the sex pheromone of the olive fruit fly.

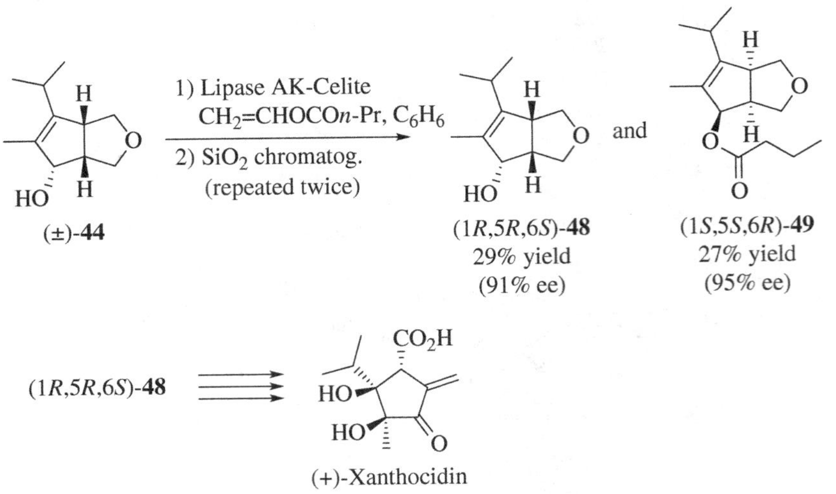

FIGURE 24.25

Preparation of bicyclic alcohol (1*R*,5*R*,6*S*)-**48**, which was converted to (+)-xanthocidin.

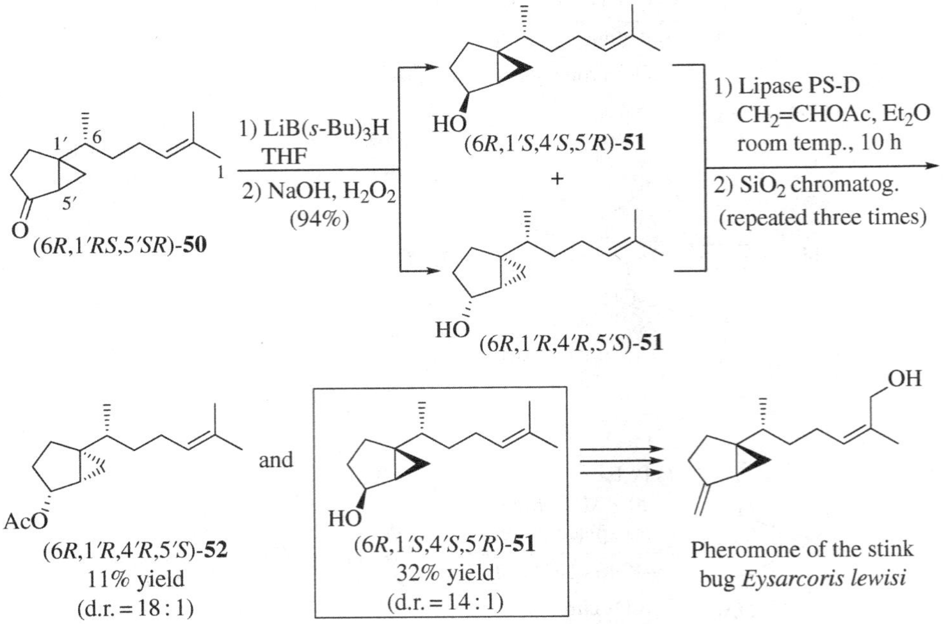

FIGURE 24.26

Preparation of bicyclic alcohol (6*R*,1′*S*,4′*S*,5′*R*)-**51**, which was converted to the pheromone of *Eysarcoris lewisi*.

(1R,5R,6S)-8-Isopropyl-7-Methyl-3-Oxabicyclo[3.3.1]oct-7-en-6-ol (48)

Asymmetric acylation of (±)-**44** with vinyl butanoate and lipase AK gave butanoate (+)-**49** and the recovered (1*R*,5*R*,6*S*)-**48** (Figure 24.25) [29]. The latter was converted to an antibiotic (+)-xanthocidin.

(6R,1′S,4′S,5′R)-2-Methyl-6-(4′-Hydroxybicyclo[3.3.0]hexyl)hept-2-ene (51)

Reduction of a 1:1 diastereomeric mixture of (6*R*,1′*RS*,5′*SR*)-**50** with L-Selectride® gave a mixture of two alcohols (6*R*,1′*S*,4′*S*,5′*R*)- and (6*R*,1′*R*,4′*R*,5′*S*)-**51**, which served as the substrate for asymmetric acetylation. Treatment of the mixture with vinyl acetate and lipase PS-D in Et_2O was followed by SiO_2 chromatography to give acetate (6*R*,1′*R*,4′*R*,5′*S*)-**52** and recovered (6*R*,1′*S*,4′*S*,5′*R*)-**51**. The latter was converted to the pheromone of the stink bug, *Eysarcoris lewisi* (Figure 24.26) [30].

24.4 PREPARATION OF ENANTIOPURE INTERMEDIATE(S) FROM A MIXTURE OF STEREOISOMERS

This section describes three examples enantiopure intermediates are secured by enzymatic kinetic resolution of a mixture of four stereoisomers. These successful cases are utilized to achieve syntheses of natural products with plural stereogenic centers.

24.4.1 (1*S*,4*R*)-4-*t*-Butyldimethylsilyloxy-3-Chloro-2-Cyclopenten-1-ol (54)

Asymmetric hydrolysis of a stereoisomeric mixture of (±)-**53** with pig pancreatic lipase (PPL) afforded (1*S*,4*R*)-**54** in 25% yield, which was indeed the theoretical yield. Other stereoisomers remained as the recovered **53** (Figure 24.27) [31]. Punaglandin 4, a marine prostanoid, was synthesized from (1*S*,4*R*)-**54**.

24.4.2 (4*R*,5*S*)-5-Hydroxy-4-Methyl-3-Hexanone (55)

Asymmetric acetylation of hydroxyl ketone (±)-**55** containing a small amount of (±)-**56** with vinyl acetate and novozyme 435 gave the acetylation products (4*S*,5*R*)-**57** and (4*R*,5*S*)-**55** contaminated with a small amount of (4*S*,5*S*)-**56** (Figure 24.28) [32]. (−)-Stegobinone, the pheromone of the drugstore beetle, was synthesized from (4*R*,5*S*)-**55** [32].

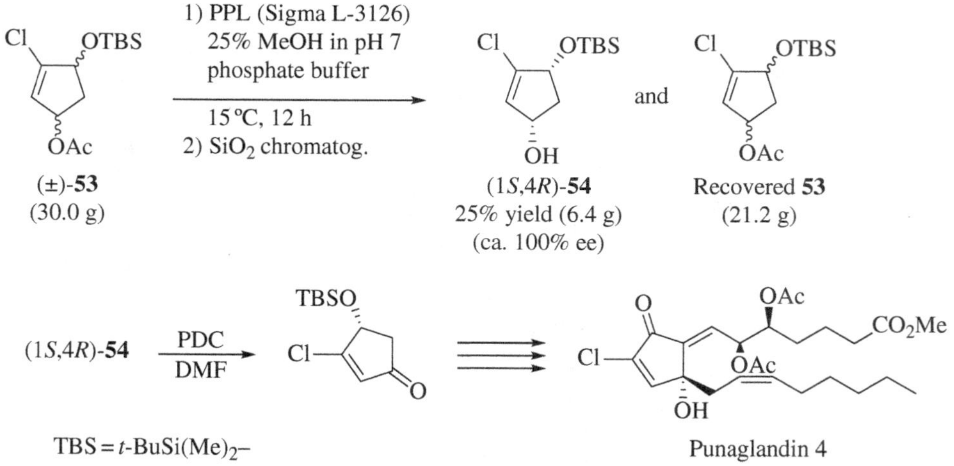

FIGURE 24.27 Preparation of (1*S*,4*R*)-4-*t*-butyldimethylsilyloxy-3-chloro-2-cyclopenten-1-ol (**55**), which was converted to punaglandin 4.

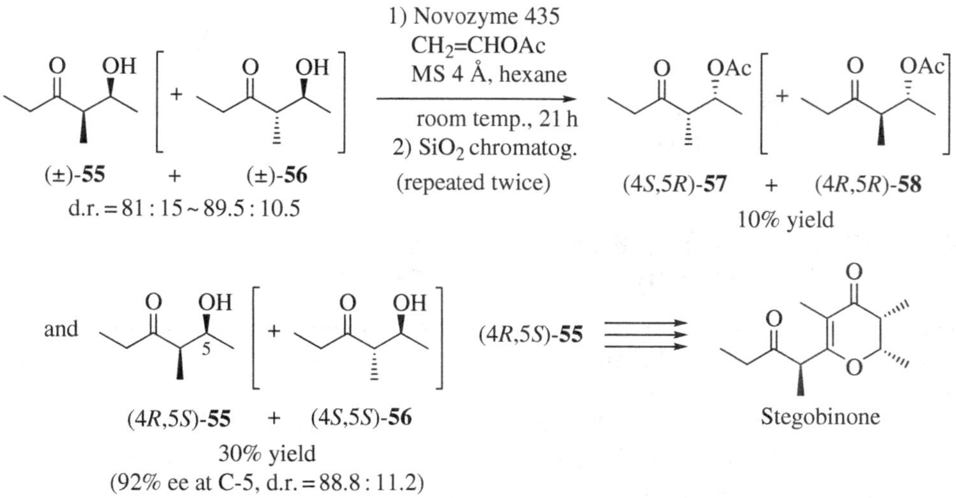

FIGURE 24.28 Preparation of hydroxyl ketone (4*R*,5*S*)-**55**, which was converted to stegobinone (the pheromone of the drugstore beetle).

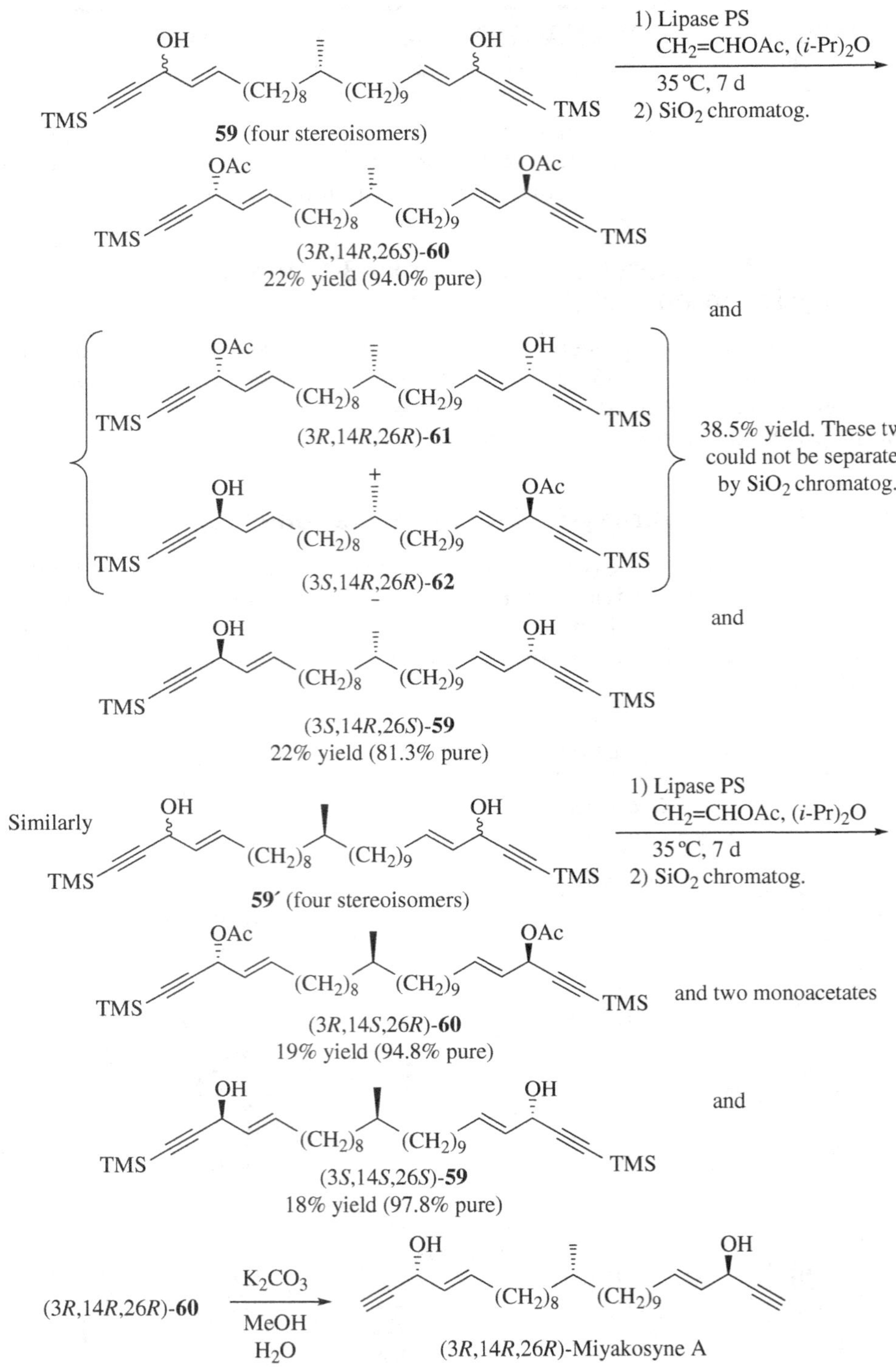

FIGURE 24.29
Preparation of diacetate (3*R*,14*R*,26*R*)-**60**, which was converted to miyakosyne A (a cytotoxic metabolite of a marine sponge).

24.4.3 (3*R*,14*R*,26*R*)-3,26-Diacetoxy-14-Methyl-1,2-bis(trimethylsilyl)octacosa-4,24-Diene-1,27-Diyne (60)

Asymmetric acetylation of a mixture of four stereoisomeric diols (**59**) with vinyl acetate and lipase PS afforded a mixture of four compounds. Chromatographic separation of the mixture afforded diacetate (3*R*,14*R*,26*R*)-**60**, a mixture of monoacetates (3*R*,14*R*,26*S*)-**61** and (3*S*,14*R*,26*R*)-**62**, and diol (3*S*,14*R*,26*S*)-**59**. The diacetate (3*R*,14*R*,26*R*)-**60** yielded (3*R*,14*R*,26*R*)-miyakosyne A, a cytotoxic metabolite of a marine sponge *Petrosia* sp. (Figure 24.29) [33].

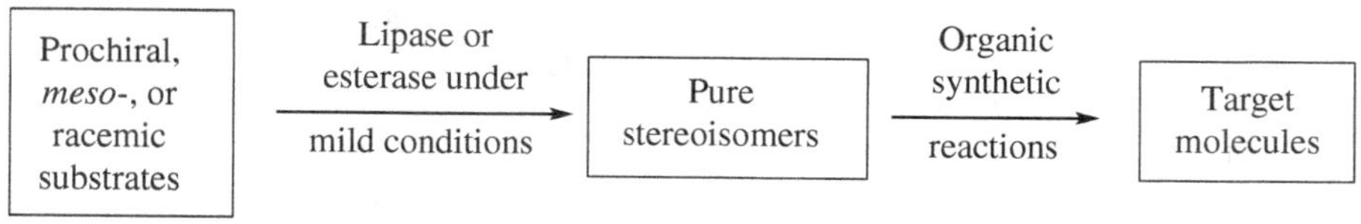

FIGURE 24.30
Summary of Mori's chemoenzymatic syntheses.

24.5 CONCLUSION

Nearly 30 examples of natural products syntheses were discussed in this chapter, all of which were based on lipase or esterase-assisted preparation of enantiopure alcohols as the key steps. The concept common to all of the discussed syntheses is shown in Figure 24.30.

The key enzymatic step converts prochiral, *meso*-, or racemic substrates into pure stereoisomers by means of lipase or esterase-assisted asymmetric processes under mild and environmentally benign conditions. The resulting pure enantiomers can be converted by synthetic reactions to the target molecules such as pheromones, hormones, and other bioregulators.

The examples in this chapter will hopefully help readers to adopt an enzymatic reaction when they execute their own retrosynthetic analysis of their target molecules. Biocatalytic retrosynthesis will become an important tool in preparative organic chemistry [34], especially in the synthesis of natural products, medicinals, and agrochemicals [35].

ACKNOWLEDGMENTS

I thank my past coworkers, whose names are given in the references, for their enthusiasm for chemoenzymatic syntheses. My thanks are due to Dr. Takuya Tashiro (RIKEN) and my wife Keiko Mori for their help in preparing the figures and typing the text, respectively.

REFERENCES

[1] M.T. Reetz, *J. Am. Chem. Soc.* **135**, 12480–12496 (2013).
[2] K. Mori, H. Akao, *Tetrahedron Lett.* 4127–4130 (1978).
[3] K. Mori, H. Akao, *Tetrahedron* **36**, 91–96 (1980).
[4] J.-L. Brevet, K. Mori, *Synthesis* 1007–1012 (1992).
[5] S. Muto, K. Mori, *Eur. J. Org. Chem.* 1300–1307 (2003).
[6] M. Seki, K. Mori, *Eur. J. Org. Chem.* 3797–3809 (2001).
[7] T. Tashiro, K. Akasaka, H. Ohrui, E. Fattorusso, K. Mori, *Eur. J. Org. Chem.* 3659–3665 (2002).
[8] K. Mori, H. Kiyota, *Liebigs Ann. Chem.* 989–992 (1992).
[9] K. Mori, M. Amaike, H. Watanabe, *Liebigs Ann. Chem.* 1287–1294 (1993).
[10] K. Fujita, K. Mori, *Eur. J. Org. Chem.* 493–502 (2001).
[11] Y. Nakamura, K. Mori, *Eur. J. Org. Chem.* 2175–2182 (1999).
[12] K. Mori, N. Chiba, *Liebigs Ann. Chem.* 957–962 (1989).
[13] K. Mori, Y. Takahashi, *Liebigs Ann. Chem.* 1057–1065 (1991).
[14] T. Tashiro, K. Mori, *Tetrahedron: Asymmetry* **16**, 1801–1806 (2005).
[15] S. Horiuchi, H. Takikawa, K. Mori, *Eur. J. Org. Chem.* 2851–2854 (1998).
[16] S. Horiuchi, H. Takikawa, K. Mori, *Bioorg. Med. Chem.* **7**, 723–726 (1999).
[17] M. Ichihashi, H. Takikawa, K. Mori, *Biosci. Biotechnol. Biochem.* **65**, 2569–2572 (2001).
[18] K. Mori, H. Ogita, *Liebigs Ann. Chem.* 1065–1068 (1994).
[19] K. Mori, Y. Takeuchi, *Proc. Jpn. Acad. Ser. B* **70**, 143–145 (1994).
[20] K. Mori, H. Tomioka, *Liebigs Ann. Chem.* 1011–1018 (1992).
[21] K. Mori, H. Ishigami, *Liebigs Ann. Chem.* 1195–1198 (1992).
[22] C. Shibata, K. Mori, *Eur. J. Org. Chem.* 1083–1088 (2004).

[23] P. Allevi, P. Ciuffreda, M. Anastasia, *Tetrahedron: Asymmetry* **8**, 93–99 (1997).
[24] Y. Masuda, K. Mori, *Eur. J. Org. Chem.* 4789–4800 (2005).
[25] K. Mori, *Tetrahedron* **68**, 1936–1946 (2012).
[26] K. Mori, P. Puapoomchareon, *Liebigs Ann. Chem.* 1053–1056 (1991).
[27] K. Hirayama, K. Mori, *Eur. J. Org. Chem.* 2467–2477 (1999).
[28] Y. Yokoyama, H. Takikawa, K. Mori, *Bioorg. Med. Chem.* **4**, 409–412 (1996).
[29] K. Mori, A. Horinaka, M. Kido, *Liebigs Ann. Chem.* 817–825 (1994).
[30] T. Tashiro, K. Mori, *Tetrahedron: Asymmetry* **19**, 1215–1223 (2008).
[31] K. Mori, T. Takeuchi, *Tetrahedron* **44**, 333–342 (1988).
[32] K. Mori, S. Sano, Y. Yokoyama, M. Bando, M. Kido, *Eur. J. Org. Chem.* 1135–1142 (1998).
[33] K. Mori, K. Akasaka, S. Matsunaga, *Tetrahedron* **70**, 392–401 (2014).
[34] N.J. Turner, E. O'Reilly, *Nat. Chem. Biol.* **9**, 285–288 (2013).
[35] K. Mori, *Chemical Synthesis of Hormones, Pheromones and Other Bioregulators*, John Wiley & Sons, Ltd, Chichester, 2010, pp. 1–299.

CHAPTER 25

Hydroxynitrile Lyases for Biocatalytic Synthesis of Chiral Cyanohydrins

Romana Wiedner[1], Helmut Schwab[1,2], and Kerstin Steiner[1]

[1] acib GmbH, Graz, Austria

[2] Institute of Molecular Biotechnology, Graz, Austria

25.1 INTRODUCTION

Biocatalytic processes have become very important in the chemical industry [1–4]. Of particular importance is one property of enzymes—their stereoselectivity—which enables either of the two enantiomers to be reacted or formed preferentially in chemical reactions with chiral or prochiral compounds. Thus, resolution of racemic mixtures and more importantly the direct synthesis of enantiomerically pure products can be achieved without the need to protect group chemistry. Biocatalytic reactions are usually carried out under mild conditions, thus avoiding unwanted side reactions [5, 6]. Particularly in combination with enzyme immobilization, which enables easy work-up of the product and reusability of the catalyst, biocatalysis is a promising approach in green chemistry. Moreover, the availability of protein engineering techniques also makes evolved or tailor-made biocatalysts more suitable for meeting the requirements of industrial applications.

One important class of enzymes, which are nowadays commercially available and used at industrial scale, comprises hydroxynitrile lyases (HNLs) (EC 4.1.2.10, EC 4.1.2.11, EC 4.1.2.46, and EC 4.1.2.47). In nature, hydroxynitrile lyases catalyze the enantioselective cleavage of α-cyanohydrins (hydroxynitriles; Scheme 25.1).

Most importantly, they are valuable biocatalysts due to their ability to synthesize chiral α-cyanohydrins by a C—C bond-forming condensation reaction. A new chiral center is formed, the carbon chain is prolonged by one carbon atom, and an additional versatile functional group—the nitrile—is introduced into the molecule. Enantiopure cyanohydrins are versatile building blocks and intermediates that serve as starting material for many enzymatic and chemical follow-up reactions, which find application in the pharmaceutical, agrochemical, and cosmetic industries [7–10].

In a recent review [2] different waves of biocatalysis were defined. Hydroxynitrile lyases are part of all three waves. Almost two centuries ago, in 1837, Liebig and Wöhler observed the formation of hydrocyanic acid (HCN) during the cleavage of amygdalin using "emulsin," a protein extract derived from bitter almonds (*Prunus amygdalus*) [11] without knowing that an enzyme is the reason for the reaction. The first wave of biocatalysis is defined by the recognition that components of living cells could be applied to chemical transformations, which is what Rosenthaler did

Green Biocatalysis, First Edition. Edited by Ramesh N. Patel.

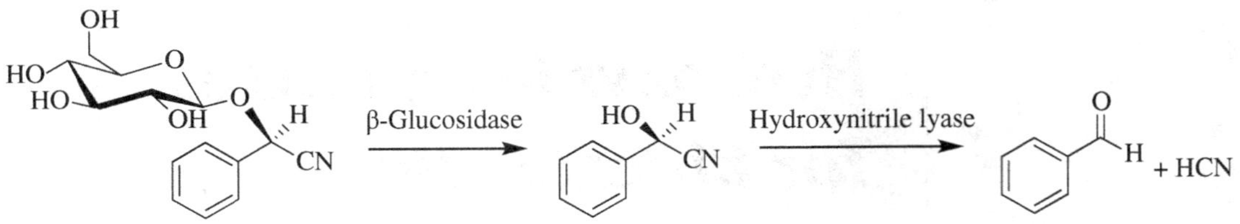

SCHEME 25.1
Cyanogenesis of prunasin.

when he used emulsin for the first synthesis of (*R*)-mandelonitrile [12]. Essential requirements for utilizing enzymes in industrial applications are their availability in sufficient amounts with constant quality and batch-to-batch reproducibility at low cost, a broad substrate scope, high reactivity, selectivity, and high stability under the realistic conditions of the industrial process. As unfortunately this is not always the case in naturally occurring enzymes, the second and the third waves of biocatalysis are characterized by protein engineering technologies using random and structure-based methodologies [2], both of which have been applied for HNLs.

In our review we will present recent examples for the discovery and improvement of HNLs and their application in the synthesis of various industrially relevant substrates.

25.2 DISCOVERY OF HYDROXYNITRILE LYASES: BIOPROSPECTING

Cyanogenesis protects many plants from small herbivores and against microbial attack [13]. Cyanogenic glycosides like amygdalin, linamarin, or prunasin are secondary plant metabolites and comprise an α-hydroxynitrile-type aglycone and a sugar moiety (mainly D-glucose) [14, 15]. They are stored in vacuoles, and on maceration of the plant cells they are hydrolyzed by β-glucosidases, which are located in a different compartment of the cell. The subsequent release of HCN is either spontaneous or catalyzed by HNLs (Scheme 25.1) [16]. Cyanogenesis has not only been reported in more than 3000 plant species like Rosaceae, Linaceae, Clusiaceae, Olacaceae, Euphorbiaceae, Gramineae, and Polypodiaceae but also in other organisms like bacteria, fungi, lichen, arthropods, or insects [17–20]. Interestingly, HNLs were also discovered in noncyanogenic plants such as *Arabidopsis thaliana* [21]. On the other hand, as far as is known, cyanogenesis in bacteria usually follows a different biochemical route, in which an HCN synthase oxidizes glycine to produce HCN and CO_2 and does not involve an HNL [22].

There are different methods for the discovery of novel enzymes, either activity or sequence based [23, 24]. For a long time the main method for the discovery of new HNLs was activity based, usually screening of different plants and plant parts for their cyanogenesis potential. However, many cyanogenic plants do not contain an HNL, and the cyanogenesis can be the result of chemical degradation of the cyanogenic compound, and on the other hand, HNLs have also been discovered in noncyanogenic plants [7]. Thus, the presence of an HNL needs to be subsequently confirmed preferably in the enantioselective synthesis of cyanohydrins using gas chromatography (GC) and high-performance liquid chromatography (HPLC). In contrast, sequence-based methods can help to identify HNLs in unexpected organisms, but only if the sequence is similar to a known HNL sequence. As HNLs represent a very diverse family with no sequence similarities between some members, some HNLs would have been missed by this method.

25.2.1 Screening Plants Based on Detection of Activity

Already more than a century ago Rosenthaler extensively investigated the distribution of emulsin-like enzymes in plants [25]. Subsequently, many plants have been successfully screened for HNL activity (for detailed tables please refer to [25–32]). The fastest assays detect HCN, which is released during cyanogenesis, based on a method by Feigl and Anger [33–35], but often plant extracts are also screened for their potential in the synthesis of certain interesting products. Many HNLs were subsequently isolated and/or characterized in more details. For some of them, the gene sequences are known and therefore they can be heterologously expressed (Table 25.1).

25.2.2 Isolation of HNL Proteins and Identification of the Encoding Genes

Large-scale applications of HNLs were limited for a long time as only crude preparations of plant parts or purified enzymes from plants were available. In addition to their overall limited availability, the amount of HNLs varies strongly depending on the part of the plant, on the season, on the location of growth, etc. The identification of genes encoding HNLs now enables the heterologous production of HNLs in industrially relevant expression systems such as *Escherichia coli* and *Pichia pastoris* (see Table 25.1). Thus, sufficient quantities of proteins can be produced with constant quality and batch-to-batch reproducibility at low cost.

Pfeil and his group were the first to successfully isolate the R-HNL protein from remains of bitter almond oil production by extraction and precipitation [36]. Soon afterward, they found out by disk electrophoresis that the enzyme preparation was actually a heterogeneous mixture of isoenzymes [76]. The same group subsequently isolated HNLs from several other Rosaceae (see, e.g., Refs. [27, 49]). In most Rosaceae the HNLs are present as isoenzymes [29]. Screening of cDNA and genomic libraries of *Prunus serotina* yielded five ORFs encoding different *Ps*HNL isoenzymes [40, 41]. Based on the sequences of the *mdl5* gene from *P. serotina* [41] and the *mdl1* gene of *P. amygdalus* [77], primers were designed, and the *hnl5* gene from *P. amygdalus* was cloned from genomic DNA [37, 78]. Similarly, the isoenzyme PaHNL4 was derived using primers based on the sequence homology of the gene *mdl4* from *P. serotina* [79].

Nowadays, HNLs from various plants have been isolated by chromatographic methods. In many cases the N-terminal sequence was established. Based on the N-terminal sequence and a highly conserved segment within the sequences of various *Prunus* species, the gene encoding *Pm*HNL2 was amplified from genomic and cDNA from *Prunus mume* by different PCR methods [38].

Another option for the identification of the gene is the immunoscreening of cDNA libraries with antibodies produced specifically using the respective purified proteins (e.g., *Hb*HNL from *Hevea brasiliensis* [80], *Sb*HNL from *Sorghum bicolor* [61]).

25.2.3 Database Mining

With an increasing number of genome sequences available not just from bacteria but also from plants and with the availability of several known HNL sequences, database searches can identify interesting, sometimes unexpected target sequences. Several sequences similar to the (*S*)-selective HNLs *Me*HNL and *Hb*HNL were discovered in the genome of the noncyanogenic model plant *A. thaliana*, which were subsequently cloned from cDNA and expressed in *E. coli* [21]. Interestingly, one of them showed high (*R*)-selective HNL activity and was further characterized.

Proteomics data derived from one- and two-dimensional gel electrophoresis and mass spectrometry [81] combined with the genome sequence of the plant pathogen *Xylella fastidiosa* [82] revealed a putative protein, which showed some

TABLE 25.1 Sources and Discovery of Hydroxynitrile Lyases

Enzyme Source	Discovery	Substrate Range[a]	Size Monomer (kDa)	Spec	Expression Host	References[b]
Prunus amygdalus (almond), *Pa*HNL	First HNL identified in emulsin, protein isolated with various methods, five isoenzymes, *mdl1* genes isolated from cDNA library, *hnl4* and *5* derived with primers designed based on the homology to *Prunus serotina* and *mdl1*	Aliphatic, aromatic, and heteroaromatic α,β-unsaturated aldehydes and methyl ketones	61	(*R*)	*Pichia pastoris, Schizosaccharomyces pombe*	[25, 36, 37]
Prunus mume (Japanese apricot), *Pm*HNL	Screening for HNL activity in cyanogenic plants [26], protein purified by chromatography, N-terminal sequencing, primers designed based on homology to *Pa*HNL and *Ps*HNL, inverted PCR, gene identified	Aliphatic, aromatic, and heteroaromatic aldehydes and methyl ketones	58	(*R*)	*P. pastoris*	[38]
Prunus lyonii (Catalina cherry), *Ply*HNL	Extraction from seeds, purified by chromatography, one isoenzyme	Only cyanogenesis of mandelonitrile tested	59	(*R*)	—	[39]
P. serotina (black cherry), *Ps*HNL	Screening for HNL activity in Rosaceae [27], protein isolated by chromatography and chromatofocusing, five isoenzymes, *mdl1*, *mdl2*, *mdl4*, and *mdl5* genes isolated from cDNA expression library; screening of genomic library resulted *mdl3*	Aliphatic and aromatic aldehydes	57–59	(*R*)	Tobacco	[27, 40–42]
Prunus armeniaca (apricot), *Par(s)*HNL	Screening for HNL activity in Rosaceae [25], protein isolated by chromatography, four isoenzymes	Sterically demanding aromatic aldehydes, γ,δ-unsaturated aldehydes	58	(*R*)	—	[27, 43, 44]
Prunus pseudoarmeniaca (wild apricot)	Partially purified	Only benzaldehyde tested		(*R*)	—	[45]
Prunus communis L. var. *dulcis Borkh* (almond), *Pc*HNL	Screening for activity on 4-methylsulfonylbenzaldehyde, not purified	4-Methylsulfonylbenzaldehyde		(*R*)	—	[46]
P. amygdalus turcomanica Lincz. (a wild almond), *Pat*HNL	Purified by $(NH_4)_2SO_4$ saturation and chromatography	Aromatic aldehydes	25	(*R*)	—	[47]
Prunus laurocerasus	Screening for HNL activity in Rosaceae [25], protein isolated, three isoenzymes, N-terminal sequencing	Aromatic aldehydes	60	(*R*)	—	[48]
Prunus persica (peach)	Screening for HNL activity in Rosaceae [25], protein isolated, three isoenzymes	Aliphatic and aromatic aldehydes	60	(*R*)	—	[49]
Prunus avium (wild cherry)	Screening for HNL activity in Rosaceae [25], protein isolated, four isoenzymes	Aliphatic and aromatic aldehydes	58	(*R*)	—	[49]

Prunus domestica (plum)	Screening for HNL activity in Rosaceae [25], protein isolated, six isoenzymes	Aliphatic and aromatic aldehydes	58	(*R*)	—	[49]
Prunus spinosa (blackthorn)	Screening for HNL activity in Rosaceae [25], protein isolated, three isoenzymes	Only cyanogenesis of mandelonitrile tested	58	(*R*)	—	[49]
Malus communis (apple)	Screening for HNL activity in Rosaceae [25], protein isolated, two isoenzymes	Only aromatic aldehydes tested	75	(*R*)	—	[49]
Eriobotrya japonica (loquat), *Ej*HNL	Screening for HNL activity in cyanogenic plants [25, 26], purified by $(NH_4)_2SO_4$ fractionation and chromatography, N-terminal sequencing, primers designed based on homology to *Pa*HNL and *Ps*HNL, gene identified	Aliphatic, aromatic, and heteroaromatic aldehydes	57	(*R*)	*P. pastoris*	[50, 51]
Sorbus aucuparia (rowan)	Screening for HNL activity in cyanogenic plants [26], protein isolated by chromatography, N-terminal sequencing			(*R*)		[52]
Chaenomeles speciosa (quince), *Cs*HNL	Screening for HNL activity in cyanogenic plants [26], not purified	Aliphatic and aromatic ketones		(*S*)	—	[53]
Ximenia americana (yellow plum), *Xa*HNL	Purified by chromatography and chromatofocusing, two isoenzymes	Aromatic aldehydes	37	(*S*)	—	[54]
Hevea brasiliensis (rubber tree), *Hb*HNL	Screening for HNL activity in *Hevea* plants, purified, gene identified by immunoscreening of cDNA library	Aliphatic, aromatic, and heteroaromatic α,β-unsaturated aldehydes and methyl ketones	29.2	(*S*)	*Escherichia coli, P. pastoris, Saccharomyces cerevisiae*	[55–58]
Manihot esculenta (cassava), *Me*HNL	Protein purified by chromatography from the cyanogenic plant, N-terminal sequencing, peptide sequencing, degenerate primer design based on these sequences, cDNA library, gene identified	Aliphatic, aromatic, and heteroaromatic α,β-unsaturated aldehydes and methyl ketones and aromatic ketones	29.5	(*S*)	*E. coli, P. pastoris, S. cerevisiae, Leishmania tarentolae*	[59, 60]
Sorghum bicolor (millet), *Sb*HNL	Discovered during a study of the biosynthesis of dhurrin and purified by different methods (e.g., chromatography), screening of cDNA library using monoclonal antibodies, a single gene encoded both subunits	Aromatic and heteroaromatic aldehydes and methyl ketones	Heterotetramer, 18 and 33	(*S*)	*E. coli* (inactive)	[61–63]
Linum usitatissimum (flax), *Lu*HNL	Purified by $(NH_4)_2SO_4$ fractionation chromatography and preparative slab gel electrophoresis from the cyanogenic plant, N-terminal sequencing, RACE-PCR, gene identified	R: aliphatic aldehydes and methyl ketones S: phenyl ketones (with at least one CH_2 separating aromatic ring and keto group)	42	(*R, S*)	*E. coli, P. pastoris*	[64–66]

(*Continued*)

TABLE 25.1 (*Continued*)

Enzyme Source	Discovery	Substrate Range[a]	Size Monomer (kDa)	Spec	Expression Host	References[b]
Arabidopsis thaliana (cress), *At*HNL	Gene sequence identified in the genome sequence based on similarity to *Hb*HNL and *Me*HNL	Aliphatic, aromatic, and heteroaromatic α,β-unsaturated aldehydes and methyl ketones	29	(*R*)	*E. coli*	[21]
Baliospermum montanum (red physic nut), *Bm*HNL	Screening for HNL activity in cyanogenic plants [26], purified by $(NH_4)_2SO_4$ fractionation and chromatography, N-terminal sequencing, RACE-PCR, gene identified	Aliphatic and aromatic aldehydes and ketones	29.5	(*S*)	*E. coli*	[67]
Passiflora edulis (passion fruit), *Pe*HNL	Screening for HNL activity in cyanogenic plants [26], purified by $(NH_4)_2SO_4$ fractionation and chromatography, N-terminal sequencing	Aromatic and heterocyclic aldehydes	15	(*R*)	—	[68]
Adenia racemosa, *Ar*HNL	$(NH_4)_2SO_4$ fractionation and chromatography, tryptic digest, sequencing, back-translation, codon usage of yeast		12.4	(*S*)	*E. coli*, *P. pastoris*	[69]
Phlebodium aureum (fern), *Pha*HNL	Extraction and chromatography and chromatofocusing from the cyanogenic plant, at least three isoenzymes	Aromatic and heterocyclic aldehydes	20	(*R*)	—	[70]
Pouteria sapota (mamey)	Screening for HNL activity in cyanogenic plants, not purified	Aromatic, heteroaromatic, and aliphatic α,β-unsaturated aldehydes				[71]
Abrus precatorius (rosary pea)	Screening for HNL activity in cyanogenic plants, purified by chromatography	Only benzaldehyde tested	Heterotetramer, 42.0 and 36.5	?	—	[31]
Annona muricata (guanabana), *Am*HNL	Screening for HNL activity in cyanogenic plants, not purified	Aromatic, heteroaromatic, and α,β-unsaturated aldehydes		(*S*)	—	[72]
Vicia sativa (common vetch)	Not purified	Aromatic, heteroaromatic, and fluoro-substituted aromatic aldehydes		(*R*)	—	[73]
Burkholderia phytofirmans PsJN, *Bb*HNL	Screening for HNL activity of bacterial (endophytic) gene libraries, database search	Aromatic aldehydes	14	(*R*)	*E. coli*	[74]
Granulicella tundricola, *Gt*HNL	Database search based	Aromatic aldehydes	14	(*R*)	*E. coli*	[17]
Acidobacterium capsulatum, ATCC 51196 *Ac*HNL	Database search based	Aromatic aldehydes	14	(*R*)	*E. coli*	[75]

[a] Substrate range, not exclusive, depends on the broadness of the applied screening.
[b] Example references are given including the first publication on the discovery of the HNL, on the sequence and heterologous expression.

similarity to HNLs with an alpha/beta-hydrolase fold. The respective gene was cloned, and the resulting protein, which was expressed in *E. coli*, displayed HNL activity in the cyanogenesis of racemic mandelonitrile [19]. However, enantioselectivity and the reverse reaction, the proof that the enzyme actually can synthesize cyanohydrins, were not reported. Considering that sequence comparison with other HNLs with alpha/beta-hydrolase folds showed that the enzyme is lacking several of the important residues for HNL activity (i.e., Thr11 and Lys236 in *S*-selective HNLs and Asn12 in *R*-selective HNLs; for details see Section 25.4 below) besides the catalytic triad, which is also present in esterases and lipases, these data would be crucial to confirm that the enzyme is indeed an HNL and that the observed activity is not the result of some unspecific reaction as it has been reported for EstB from *Pseudomonas marginata* [29].

25.2.4 Heterologous Expression

E. coli is an expression host frequently used for the expression of prokaryotic proteins. It is safe and easy to handle, has a fast generation time, and is cost efficient [83], but in many cases it is not suitable for the expression of eukaryotic proteins due to insufficient folding of complex proteins from higher organisms and a lack of certain post-translational modifications such as glycosylation, which results in insoluble and inactive proteins. *P. pastoris* is an ideal alternative expression host for eukaryotic proteins as it can grow on inexpensive media at a high growth rate, reaching high cell densities [84]. The overexpressed proteins can also be secreted to the medium, and thus the purification effort is reduced [85]. For example, the glycosylated and disulfide bridge containing *Pa*HNL was so far only efficiently expressed as a soluble and active protein in *P. pastoris* [37]. Although the HNLs that contain an α/β-hydrolase fold can be expressed intracellularly and soluble in *E. coli* (Table 25.1), the expression levels are sometimes not high enough for industrial application, and the *P. pastoris* expression system is used instead [56]. Alternatively, the HNL from *Manihot esculenta* (*Me*HNL), which was originally expressed mainly in inclusion bodies in *E. coli*, is one example for which the expression level was significantly improved simply by codon optimization and optimizing the expression conditions [86]. In another approach, by adding the inducer from the beginning and cultivation at low temperature, 850 times higher yield and activity of *Me*HNL were achieved at 17 °C compared to 37 °C [60].

25.3 APPLICATIONS OF HYDROXYNITRILE LYASES

25.3.1 Cyanohydrins

Cyanohydrins can be directly synthesized using hydroxynitrile lyases, which offer high enantioselectivity, broad substrate variability, and high yield in the synthesis of cyanohydrins or their derivatives. The starting materials, aldehydes and prochiral ketones, are readily available and cheap. Cyanohydrin formation is a base-catalyzed equilibrium reaction. While aldehydes are good substrates for HNLs, reaching in many cases high yields of enantiopure cyanohydrins, ketones are less reactive, and in equilibrium the cyanogenesis reaction can be favored over the carboligation reaction, depending on the chemical structure of the substrate, thereby reducing the yield and enantiopurity of the product. In such unfavorable cases, the use of a large excess of HCN and the constant removal of the cyanohydrin product by coupling the reaction to a second irreversible step can in some cases solve the problem [87]. Many examples of aldehydes and ketones already used in the synthesis of cyanohydrins can be found in the tables of a recent review by Griengl and coworkers [8]. Apart from free hydrogen cyanide, many different cyanide sources such as alkali cyanides, acetone cyanohydrin, trialkylsilyl cyanides, and alkyl cyanoformates can be used (e.g., Refs. [88–92]).

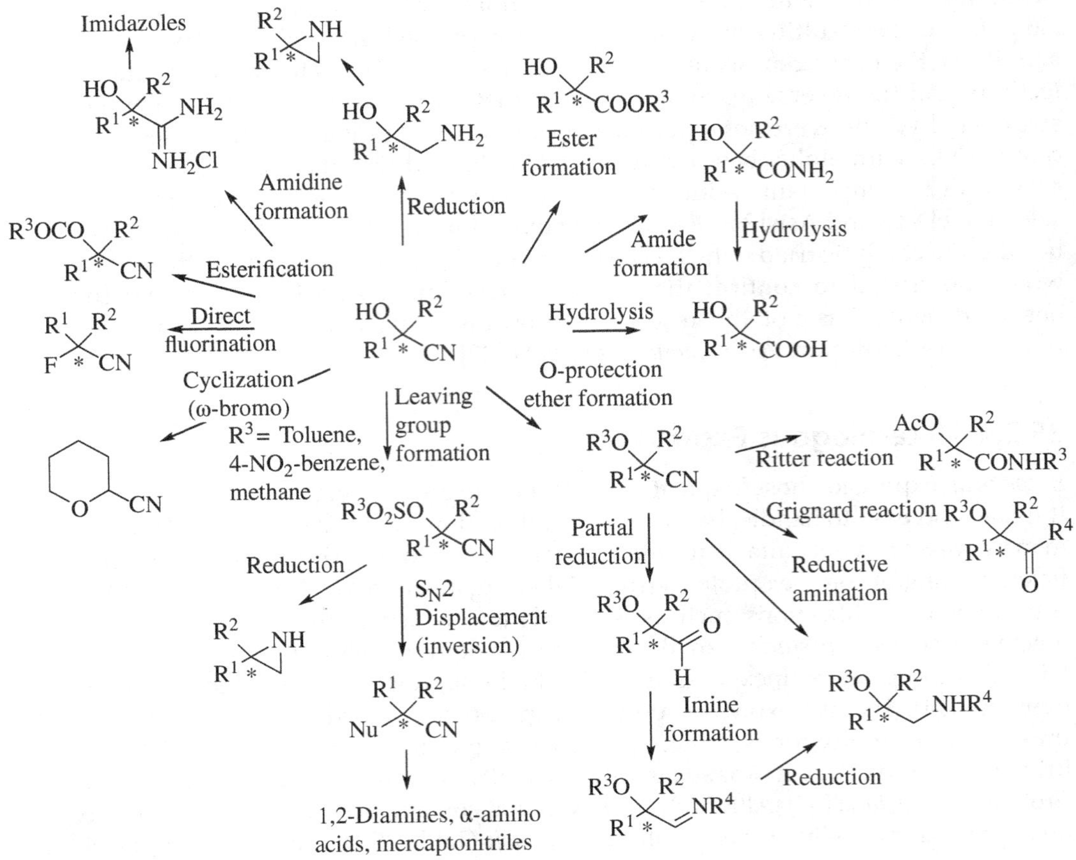

SCHEME 25.2
Follow-up reactions of chiral cyanohydrins.

Cyanohydrins are bifunctional molecules and therefore constitute a particularly useful class of compounds for synthetic purposes. A hydroxyl and a nitrile functional group are available for chemical and enzymatic follow-up reactions (Scheme 25.2 [7, 93–95]), resulting in hydroxy carboxylic acids [96–99], carbamates [100], hydroxyamides [101], primary and secondary hydroxyamines [46, 102–104], aziridines [105, 106], aminonitriles [107, 108], diamines [108], azidonitriles [108], α-fluoronitriles [109], hydroxy ketones [110], and many more. The (*S*)-selective *Hb*HNL and the (*R*)-selective *Pa*HNL, in particular are used on large scale either for the synthesis of (*S*)-3-phenoxybenzaldehyde cyanohydrin, a precursor for pyrethroids, a class of insecticides; (*R*)- and (*S*)-mandelonitrile, which can be further converted to mandelic acids; (*R*)-chloromandelonitrile, a precursor for an anticoagulant; (*R*)-2-hydroxy-4-phenylbutyronitrile, which serves as intermediate for the production of angiotensin-converting enzyme inhibitors (ACEi); or (*R*)-2-amino-1-(2-furyl)ethanol [94]. Several HNLs are commercially available as free or immobilized enzymes.

25.3.2 β-Nitro Alcohols

In the mid-2000s, Griengl and coworkers reasoned that a small molecule with a similar pK_a as HCN, for example, nitromethane, could act as nucleophile for addition to carbonyl compounds (nitroaldol or Henry reaction; Scheme 25.3) [111]. The Henry reaction is a classical name reaction in organic chemistry for the formation of C–C bonds. The resulting β-nitro alcohols can be transformed to nitroalkenes, 2-nitroketones, α-hydroxycarboxylic acids, and 1,2-amino alcohols. Although several other enzymes and proteins such as hydrolases and lipases [112], transglutaminase

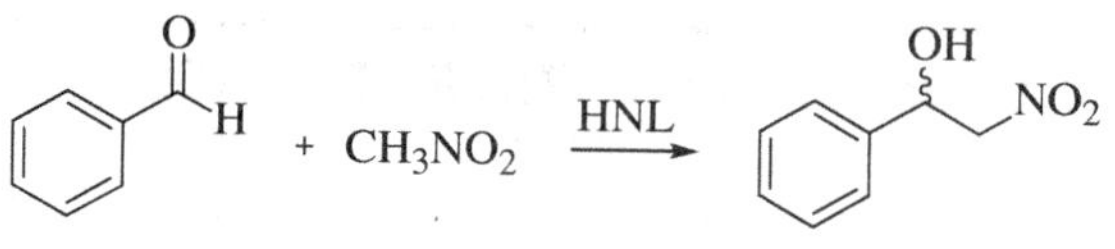

SCHEME 25.3
Biocatalytic Henry reaction.

[113], and a glucoamylase [114] and even BSA [115] can catalyze the nitroaldol reaction, hydroxynitrile lyases are the only enzymes that are enantioselective. The first HNL that was discovered to enantioselectively catalyze the reaction was the (*S*)-selective *Hb*HNL [111], accepting a broad range of aliphatic, aromatic, and heteroaromatic aldehydes and even 2-heptanone [116]. Not only nitromethane but also nitroethane and 2-nitropropane were accepted by *Hb*HNL, which led to two newly generated chiral centers in the product.

*Me*HNL is also capable of catalyzing the nitroaldol reaction, albeit with lower activity and selectivity, whereas *Pa*HNL showed no activity at all [116]. The only (*R*)-selective enzyme with a α/β-hydrolase fold (like *Hb*HNL and *Me*HNL), *At*HNL, catalyzed the Henry reaction of aromatic aldehydes and $MeNO_2$ [117].

25.4 STRUCTURAL AND MECHANISTIC ASPECTS

Hydroxynitrile lyases are considered as an example of convergent evolution [118] as they belong to at least four different unrelated structural folds—α/β-hydrolase fold proteins, oxidoreductases, cupins, and alcohol dehydrogenases—and thus also their reaction mechanisms varies. HNLs from Rosaceae are a conserved group of flavin adenine dinucleotide (FAD)-containing (*R*)-selective glycoproteins. All other known HNLs are FAD independent. They are more versatile and vary in their amino acid sequence, molecular weight, structural fold, oligomerization state, glycosylation pattern, enantioselectivity, and substrate range. Thus, although HNLs all catalyze the same reaction, their detailed reaction mechanisms differ significantly due to their different structures [119]. Five of the HNLs with known structure, the HNLs from *H. brasiliensis* [120–125], *M. esculenta* [126–129], *B. montanum* [130], *S. bicolor* [131, 132], and *A. thaliana* [133] display an α/β-hydrolase fold. Interestingly, although for a long time it was thought that the α/β-hydrolase fold contains only (*S*)-selective HNLs, the discovery of the (*R*)-selective HNL from *A. thaliana* added another surprising aspect to the diversity of HNLs. As expected, the reaction mechanism of the (*R*)- and the (*S*)-selective α/β-hydrolase fold enzymes vary, although both contain a Ser–His–Asp catalytic triad. While in *Hb*HNL the His residue acts as general base deprotonating the serine residues, which serves as mediator and simultaneously deprotonates the cyanohydrin, the His residue of *At*HNL directly abstracts a proton from the substrate. Another difference is the stabilization of the released CN^-, which is proposed to interact with the positively charged Lys236 in *Hb*HNL, which is missing in *At*HNL, and is subsequently protonated by the His residue [123, 134]. Instead in *At*HNL, the negative charge of the CN^- is stabilized by an oxyanion hole and finally protonated indirectly by the Ser, which serves as mediator to the His residue [133, 135].

The structures of the HNLs from *P. amygdalus* and *P. mume* resemble a glucose–methanol–choline (GMC) oxidoreductase [136–138]. Other HNLs with known sequence from Rosaceae, such as from *P. serotina* (*Ps*HNL) and *E. japonica* (*Ej*HNL), also belong to this family. Again, a His residue might act as a general base and abstract a proton from mandelonitrile. The cyanide could be stabilized by an overall positive charge in the active site caused by several positively charged amino acids and might be protonated by a second histidine residue, but the exact mechanism is still under discussion.

*Lu*HNL from *L. usitatissimum* shows high sequence similarity to the Zn^{2+}-alcohol dehydrogenase superfamily [66, 139]. However, no structure has been published so far. HNL activity is also found in the cupin fold superfamily, which was discovered recently, and a novel group of metal-dependent bacterial HNLs has been reported [17, 74, 75].

25.5 ENGINEERING OF HYDROXYNITRILE LYASES

Different strategies for enzyme engineering are routinely applied in laboratories to create biocatalysts featuring the desired properties required for a certain biotransformation (e.g., Refs. [140–143]). Directed enzyme evolution is based on the generation of a large pool of muteins and subsequent screening and selection of the best variants exhibiting the required characteristics. In contrast, rational design presupposes a more detailed knowledge about an enzyme like amino acid sequence and structural data (3D structure or homology model). However, if applicable for a special enzyme and a certain goal, the combination of both approaches might be the most promising strategy (semirational approach, designed evolution) (e.g., Refs. [144–148]). Targeted properties are, for example, improved activity, substrate scope, stability (pH, thermal, solvent), or selectivity. Many successful examples of molecular engineering of hydroxynitrile lyases are described in literature (Table 25.2). Often their application as industrial biocatalysts is restricted due to limited acceptance of unnatural substrates or low stability under the required reaction conditions. In particular, pH stability is one of the most important features of HNLs and a requirement for their technical applicability, as the unselective chemical background reaction, which competes with the enantioselective enzyme-catalyzed reaction, increases at pH values greater than 4.5, and destroys the enantiopurity of the product. An important prerequisite for protein engineering is the availability of high-throughput assays, which allow the analysis of a large number of variants. While the use of *E. coli* in high-throughput screenings is an established routine, *P. pastoris* growth conditions had to be adapted for uniform growth on a small scale to enable reliable applicability of high-throughput screenings [165].

Different methods are available for the screening of HNL activity. Most of them follow the cyanohydrin cleavage reaction. Two colony-based assays have been developed by our group [34, 166]. A direct substrate-independent assay developed by Krammer *et al.* is based on the colorimetric detection of liberated HCN on an HCN-sensitive detection paper. The indirect screening method established by Reisinger and coworkers is based on the coupling of the benzaldehyde released from mandelonitrile to an NAD^+-dependent benzaldehyde dehydrogenase and tracking the increase of NADH fluorescence of this second enzyme. Further HNL screening procedures are performed in microtiter plates and are either based on the direct spectrophotometrical detection of aromatic aldehydes and ketones released from cyanohydrins [167] or the indirect spectrophotometrical detection of HCN [168], which can be applied to a broad substrate range. The latter assay involves the oxidation of the CN^- by N-chlorosuccinimide (stabilized by succinimide) to CN^+, which in the following reacts with isonicotinic acid. The reaction product forms a dye with barbituric acid that is measured at 600 nm. A high-throughput assay that allows the screening in the cyanohydrin synthesis reaction was developed by Pscheidt *et al.* [32]. The method is based on the detection of the remaining substrate aldehyde after performing the bioconversion in microtiter plates. The aldehyde reacts with 4-hydrazino-7-nitrobenzofurazan (NBDH), and the decrease in the reaction product, the strongly fluorescent hydrazone, is monitored [169].

25.5.1 Substrate Scope, Activity, and Enantioselectivity

HNLs often exhibit only low activity toward industrially relevant compounds. Thus, a lot of effort was (and still is) made to engineer HNLs for the acceptance of unfavorable substrates by rational design and directed evolution.

TABLE 25.2 Summary of Engineered HNL Variants Exhibiting Improved Properties

Enzyme	Mutation(s)	Property	References
*Pa*HNL5	L1Q	Improved expression in *P. pastoris*	[37]
*Pa*HNL5	L1Q/A111G	Improved yield and *ee* in the synthesis of (*R*)-2-chloromandelonitrile	[37]
*Pa*HNL5	L1Q/A111G/4E10[a] L1Q/N3I/A111G L1Q/I108M/A111G and combinatorial mutant L1Q/N3I/I108M/A111G/4E10[a]	Improved yield and *ee* in the synthesis of (*R*)-2-chloromandelonitrile	[149]
*Pa*HNL5	L1Q/V360I	Improved yield and *ee* in the synthesis (*R*)-2-hydroxy-4-phenylbutyronitrile and (*R*)-2-hydroxy-4-phenyl-3-butene nitrile	[150]
*Pa*HNL5	V317A	Improved yield and *ee* in the synthesis of (*R*)-hydroxypivaldehyde cyanohydrin and (*R*)-pivaldehyde cyanohydrin	[151]
*Gt*HNL	A40H/V42T/Q110H	Improved activity, enantioselectivity, and substrate scope	[152]
*Hb*HNL	W128A	Increased acceptance of bulky substrates	[34, 153, 154]
*Hb*HNL	W128A/H103L/P187L and W128A/Q215H	Further increased activity and selectivity compared to the starting variant W128A	[153]
*Hb*HNL	L121Y	Enhanced reaction rate toward mandelonitrile	[155]
*Me*HNL	W128A, W128C, W128Y, and W128L	Increased acceptance of bulky substrates	[129, 156]
*Me*HNL	G113S	Improved solubility during expression in *Escherichia coli*, improved thermal stability	[157]
*Me*HNL	G165D/E, V173L, T163D/E/S, K21E/D/N	Improved thermal stability and organic solvent stability	[158]
*Me*HNL	K176P, K199P, K224P, and combinations thereof	Improved solubility during expression in *E. coli*	[159]
*Me*HNL	H103L, H103M, H103V, H103I, H103C	Improved solubility during expression in *E. coli*	[159, 160]
*Me*HNL	V2K, V2N, V2I, V2R, V2Q	Improved solubility during expression in *E. coli*	[161]
*Me*HNL	V2I/H103L	Improved solubility during expression in *E. coli*	[161]
*At*HNL	P48Q/Q50E/A51Q/E53N/K60E/E64T/K67E/I93R/E138T/R140I/N141T	Improved stability at acidic pH	[162]
SABP2	G12T/M239K	Conversion of an esterase to an HNL, low enantioselectivity	[163]
*Bg*EstC	S276K	Conversion of an esterase to an HNL, low enantioselectivity	[164]

[a] 4E10: the mutant harbors two silent mutations at positions 85 and 432.

Structure-guided design was applied by Glieder *et al.* to increase the activity of *Pa*HNL5-L1Q toward the commercially interesting substrate (*R*)-2-chlorobenzaldehyde (Scheme 25.4) [37]. The corresponding cyanohydrin (*R*)-2-chloromandelonitrile is hydrolyzed under acidic conditions to (*R*)-2-chloromandelic acid, which is a building block for the platelet-aggregation inhibitor clopidogrel [96]. Molecular modeling of *Pa*HNL5 and substrate docking studies revealed unfavorable steric interactions of (*R*)-2-chloromandelonitrile with the side chains of Val317 and especially Ala111. Site-directed mutagenesis of Ala111 to glycine resulted in a variant that, on the one hand, showed drastically decreased activity in the cleavage of mandelonitrile but, on the other hand, exhibited strongly enhanced activity in the synthesis of (*R*)-2-chloromandelonitrile. The purified mutant *Pa*HNL5-L1Q/A111G (0.5 mg)

SCHEME 25.4

*Pa*HNL5-catalyzed synthesis of (*R*)-2-chloromandelonitrile.

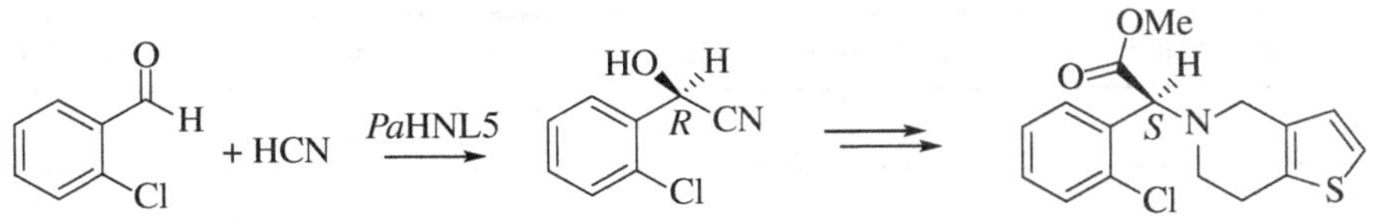

TABLE 25.3 Improved Conversions and Enantiomeric Excesses (*R*) Obtained in the Hydrocyanation Reaction of Various Aldehydes Catalyzed by *Pa*HNL5 Variants

Entry	*Pa*HNL5 Variant	Substrate	Conv. (%)	*ee* (%)	Reaction Conditions	References
1	L1Q/A111G	2-Chlorobenzaldehyde	69.5 95.9[a]	96.5	33.3 μg enzyme/mmol substrate, biphasic system, pH 3.4, 2 h, 7 h[a]	[37]
2	L1Q/A111G/4E10[b]	2-Chlorobenzaldehyde	72.9	98.8	33.3 μg enzyme/mmol substrate, biphasic system, pH 3.4, 2 h	[149]
3	L1Q/N3I/A111G		72.8	98.8		
4	L1Q/I108M/A111G		88.3	98.9		
5	L1Q/N3I/I108M/A111G/4E10[b]	2-Chlorobenzaldehyde	94.6 92.7[a]	98.6	33.3 μg enzyme/mmol substrate, biphasic system, pH 3.4, 2 h, 4 h[a]	[149]
6	V360I	3-Phenylpropanal	98	96.7	67 μg enzyme/mmol substrate, biphasic system, pH 3.4, 4 h	[150]
7	V360I	3-Phenylpropenal	97	97.6	27 μg enzyme/mmol substrate, biphasic system, pH 3.4, 3 h	[150]
8	V317A	Hydroxypivaldehyde	88[a]	96	218 U/mmol substrate, aqueous system, pH 2.4, 20 h	[151]
9	V317A	Pivaldehyde	100	97.2	1.1 mg enzyme/mmol substrate, aqueous system, pH 3.0, 6 h 10 min	[151]

[a] Recovery yield, preparative scale.
[b] 4E10 contains two silent mutations at positions 85 and 432.

catalyzed the bioconversion of 15 mmol 2-chlorobenzaldehyde in an emulsion system with a specific activity of 409 U/mg, which corresponds to a more than six-fold increased specific activity in comparison to the starting variant *Pa*HNL5-L1Q. Performing the synthesis reaction on a preparative scale, *Pa*HNL5-L1Q/A111G (~5 mg) converted 150 mmol of 2-chlorobenzaldehyde at pH 3.4 into (*R*)-2-chloromandelonitrile with a recovery yield of 95.9% and an enantiomeric excess (ee) of 96.5% after 7 h (Table 25.3, entry 1).

*Pa*HNL5-L1Q/A111G, which was the first recombinant *Pa*HNL that was industrially used for multiton-scale production of (*R*)-2-chloromandelonitrile, was subjected to directed evolution in order to further improve its catalytic activity [149]. About 10 000 transformants were screened spectrophotometrically for their activity in the cleavage of (*R*)-2-chloromandelonitrile at 300 nm. The mutations found in the best variants from the first random library were recombined, resulting in a second small library of approximately 500 transformants. Interestingly, one mutant (*Pa*HNL5-L1Q/A111G/4E10) that contained no additional amino acid mutation but two silent mutations exhibited an improved *ee* of 98.8% and a slightly increased conversion of 72.9% at pH 3.4 after 2 h in comparison to 69.5% conversion and 96.1% *ee* of the starting variant. The variant *Pa*HNL5-L1Q/N3I/A111G exhibited a similar conversion and *ee* value as *Pa*HNL5-L1Q/A111G/4E10. The mutant

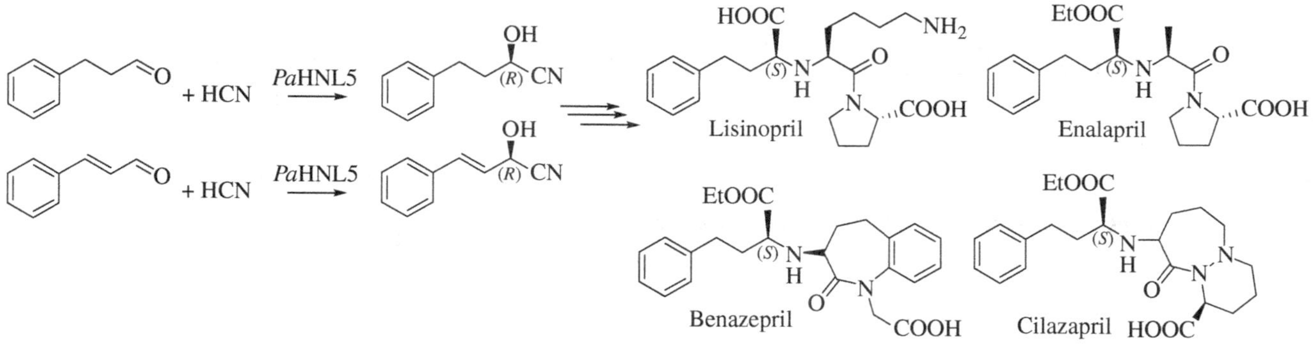

SCHEME 25.5

*Pa*HNL5-catalyzed synthesis of (*R*)-2-hydroxy-4-phenylbutyronitrile and (*R*)-2-hydroxy-4-phenyl-3-butene nitrile from 3-phenylpropanal and 3-phenylpropenal.

*Pa*HNL5-L1Q/I108M/A111G converted even 88.3% of the substrate with an *ee* (*R*) of 98.9% under the same conditions after 2h. The best variant, combinatorial mutant *Pa*HNL5-L1Q/N3I/I108M/A111G/4E10, catalyzed the synthesis of (*R*)-2-chloromandelonitrile on a preparative scale in a biphasic system using 150 mmol substrate and 5 mg enzyme, with a recovery yield of 92.7% and an *ee* of 98.6% at pH 3.4 after 4h (Table 25.3, entries 2–5).

Two other substrates of industrial interest are 3-phenylpropanal and 3-phenylpropenal. Their corresponding cyanohydrins (*R*)-2-hydroxy-4-phenylbutyronitrile and (*R*)-2-hydroxy-4-phenyl-3-butene nitrile are building blocks for the production of so-called prils, ACEi. The Glieder group investigated the conversion of 3-phenylpropanal with *Pa*HNL5 (Scheme 25.5) [150]. The enzyme catalyzed the reaction with an ee (*R*) of 89.4% (93% conversion) after 4h, leaving room for improvement especially in terms of enantioselectivity. Hence, a structure-guided approach was chosen to increase the enzyme's enantioselectivity. They modeled the homology models of *Pa*HNL5 in complex with both enantiomers of 2-hydroxy-4-phenylbutyronitrile and found that the alkyl chain of the substrate either interacted with Ala111 (in the case of the (*S*)-enantiomer) or with Val360 (in the case of the (*R*)-enantiomer). Based on these findings, the authors redesigned the substrate binding site of *Pa*HNL5 preserving its hydrophobicity. Out of 24 preliminary designed *in silico* muteins, 12 variants were prepared and investigated for their performance in the biosynthesis of (*R*)-2-hydroxy-4-phenylbutyronitrile. The best variant, V360I, converted 98% of the substrate with an *ee* of approximately 97% in a biphasic system at pH 3.4 after 4h reaction time using 67 µg enzyme/mmol substrate (Table 25.3, entry 6). Turnover frequencies (TOF) were six times higher in comparison to the wild type. Consequently, the required amount of enzyme for the production of (*R*)-2-hydroxy-4-phenylbutyric acid and derivatives was reduced 10- to 30-fold, respectively. In addition, the variant V360I was superior in the hydrocyanation reaction of 3-phenylpropenal. Fifteen millimol of substrate was converted by 0.4 mg of the biocatalyst at pH 3.4 with an *ee* of approximately 98 and 97% conversion after 3h (Table 25.3, entry 7).

Another successful example for the improvement of the substrate specificity of *Pa*HNL5 was described by the Glieder group, employing a semirational strategy to improve the activity toward the sterically demanding hydroxypivaldehyde [151]. The respective (*R*)-cyanohydrin is of major industrial importance, since it is a precursor of (*R*)-pantolactone, a crucial building block for the synthesis of (*R*)-pantothenic acid (vitamin B_5) (Scheme 25.6). Site-saturation libraries of all seven hydrophobic amino acids in the substrate binding pocket of *Pa*HNL5 (F72, V113, V317, V329, L331, L343, and V360) were generated. From each library more than 200 transformants were tested for improved conversion of hydroxypivaldehyde

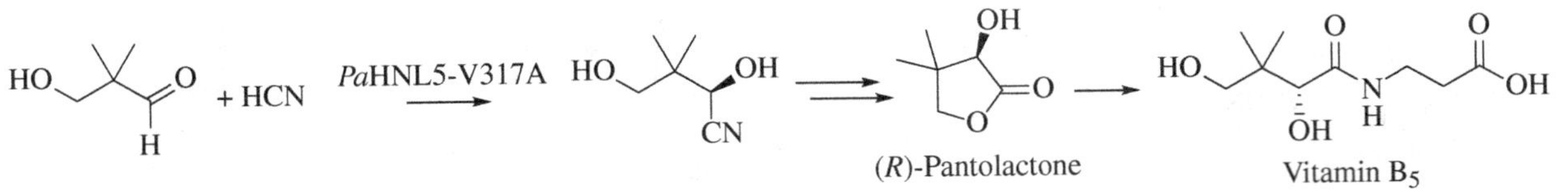

SCHEME 25.6
Enantioselective synthesis of vitamin B5 from hydroxypivaldehyde.

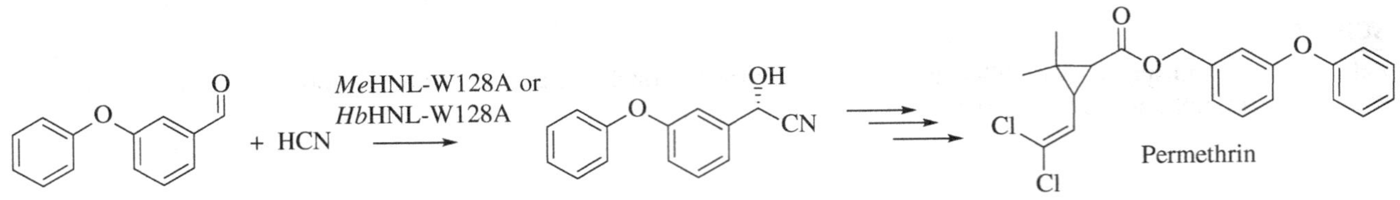

SCHEME 25.7
Synthesis of 3-phenoxybenzaldehyde cyanohydrin.

at pH 2.4. The best variant *Pa*HNL5-V317A resulted in approximately 82% *ee* (*R*) and 86.5% conversion after 20.5 h using 0.85 mg enzyme/mmol substrate, which corresponds to an approximately fourfold improvement of enantioselectivity in comparison to the wild type. The authors attributed the improved enantioselectivity to a reduction of stabilizing interactions with the "wrong" (*S*)-enantiomer. This hypothesis was confirmed by comparison of potential energy differences between wild type and variant. Determination of TOF values revealed an approximately 13-fold higher reaction rate of *Pa*HNL5-V317A. Performing the synthesis in an aqueous reaction system on preparative scale with 218 U of purified *Pa*HNL5-V317A/mmol substrate yielded 88% (*R*)-hydroxypivaldehyde cyanohydrin with 96% *ee* after 20 h (Table 25.3, entry 8). *Pa*HNL5-V317A was also found to be beneficial for the conversion of pivaldehyde (Table 25.3, entry 9).

*Pa*HNL5 is not the only (*R*)-selective HNL whose substrate scope, activity, and enantioselectivity were improved by enzyme engineering. Random and site-saturation mutagenesis of the bacterial *Gt*HNL from *Granulicella tundricola* originated a remarkable number of variants with enhanced HNL activity in the cyanogenesis of (*R*)-mandelonitrile and, more importantly, in the synthesis of cyanohydrins [152]. *Gt*HNL-A40H/V42T/Q110H, a combinatorial variant of three beneficial amino acid exchanges in the active site of *Gt*HNL, was superior to all the other variants with 490-fold increased specific activity in comparison to the wild type at the same reaction conditions. *Gt*HNL-A40H/V42T/Q110H is highly active and enantioselective in the synthesis of chiral cyanohydrins, such as 2-chlorobenzaldehyde cyanohydrin, (*R*)-2-hydroxy-4-phenylbutyronitrile, and (*R*)-2-hydroxy-4-phenyl-3-butene nitrile reaching full conversion and excellent *ee*. Moreover, the general importance of these amino acids in the active site of HNLs with cupin fold was confirmed by the successful transfer of these amino acid exchanges to another cupin HNL, *Ac*HNL from *Acidobacterium capsulatum*.

Several attempts were also made to increase the substrate scope of (*S*)-selective HNLs. The active sites of *Hb*HNL from *H. brasiliensis* and *Me*HNL from *M. esculenta* are accessible via a narrow hydrophobic tunnel harboring a tryptophan residue (W128), which hampers the passage of large substrates. Replacement of W128 by smaller residues led to improved conversions of sterically hindered substrates [154] like the industrially interesting 3-phenoxybenzaldehyde, whose (*S*)-cyanohydrin is a pyrethroid precursor (Scheme 25.7).

TABLE 25.4 Improved Conversion and Enantiomeric Excesses Obtained in the Hydrocyanation Reaction of Various Substrates Catalyzed by *Hb*HNL and *Me*HNL Variants

Entry	Enzyme Variant(s)	Substrate	Conv. (%)	*ee* (%)	Reaction Conditions	References
	*Hb*HNL					
1	W128A	4-Methoxycyclohex-3-ene carbaldehyde	92.3	1:1.7[a]	100 U enzyme/mmol substrate, biphasic system, pH 5.5, 6 h	[153]
2	W128A/H103L/P187L	4-Methoxycyclohex-3-ene carbaldehyde	95[b]	95:5[a]	100 U enzyme/mmol substrate, biphasic system, pH 5.5, 1 h	[153]
3	W128A/Q215H	4-Trimethylsilyloxycyclohex-3-ene carbaldehyde	92[b]	93:7[a]	200 U enzyme/mmol substrate, biphasic system, pH 5.5, 2 h	[153]
	*Me*HNL					
4	W128A	3-Phenoxybenzaldehyde	98	90	DIPE with 4.6 mg immobilized enzyme/1 mmol substrate	[156]
5	W128A	4-Methylbenzaldehyde	93	97	DIPE with 4.6 mg immobilized enzyme/1 mmol substrate	[156]
6	W128L	3-Heptanone	87	96	DIPE with 4.6 mg immobilized enzyme/1 mmol substrate	[156]

[a] Ratio (2*S*,1′*RS*)/(2*R*, 1′*RS*).
[b] Recovery yield of crude cyanohydrin, preparative scale.

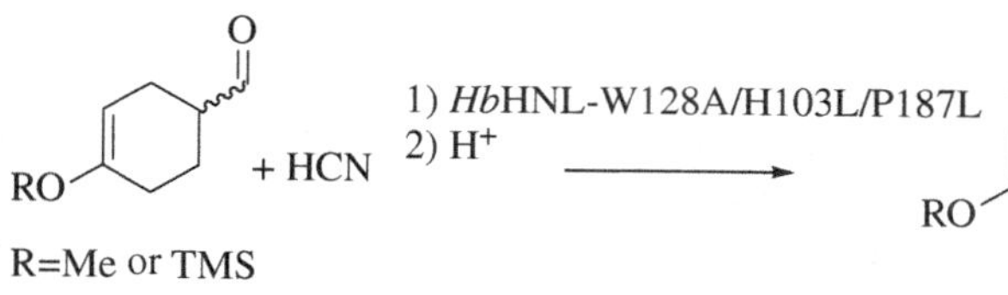

SCHEME 25.8 Conversion of 4-methoxycyclohex-3-ene carbaldehyde.

Interestingly, the W128A mutation was also discovered during the screening of a random mutant library of *Hb*HNL toward (*S*)-3-phenoxybenzaldehyde cyanohydrin [34]. The screening afforded an improved variant, *Hb*HNL-E11K/W128A/T132A, with 1.8-fold higher cleavage activity compared to the wild type. The W128A mutation was assumed to be responsible for the improved activity on (*S*)-3-phenoxybenzaldehyde cyanohydrin. Moreover, *Hb*HNL-W128A was shown to exhibit higher yields and selectivities in the synthesis of 2-hydroxy-(4′-methoxycyclohex-3′-enyl)acetonitrile and 2-hydroxy-(4′-trimethylsilyloxycyclohex-3′-enyl)acetonitrile in comparison to *Hb*HNL wild type (Table 25.4, entry 1, [153]). Directed evolution by epPCR based on *Hb*HNL-W128A resulted in two further improved variants. The mutations H103L and Q215H were identified as responsible for enhanced yields and selectivities with both substrates. In addition, mutant P187L exhibited improved conversion of 4-methoxycyclohex-3-ene carbaldehyde. The conversion of 4-methoxycyclohex-3-ene carbaldehyde was catalyzed by *Hb*HNL-W128A/H103L/P187L (a combinatorial mutant) with 95% yield and a ratio (2*S*,1′*RS*)/(2*R*,1′*RS*) of 95:5 (Scheme 25.8). The variant *Hb*HNL-W128A/Q215H converted 4-trimethylsilyloxycyclohex-3-ene carbaldehyde with 92% yield and a ratio (2*S*,1′*RS*)/(2*R*,1′*RS*) of 93:7 (Table 25.4, entries 2 and 3).

Effenberger and coworkers performed a detailed structural study of the rationally designed mutant *Me*HNL-W128A [129]. The X-ray structure of *Me*HNL-W128A with and without substrate (4-hydroxybenzaldehyde) confirmed the enlargement of the active site entrance by the replacement of the bulky tryptophan residue for the smaller

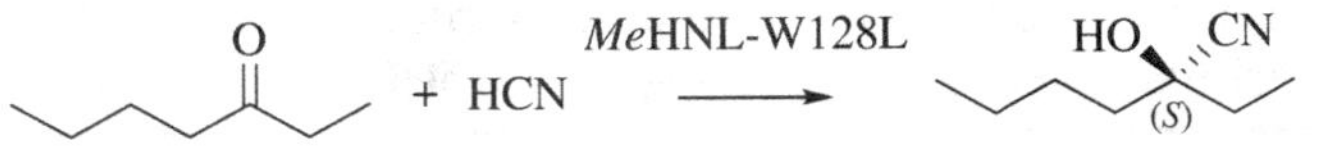

SCHEME 25.9

*Me*HNL-W128L-catalyzed addition of HCN to 3-heptanone.

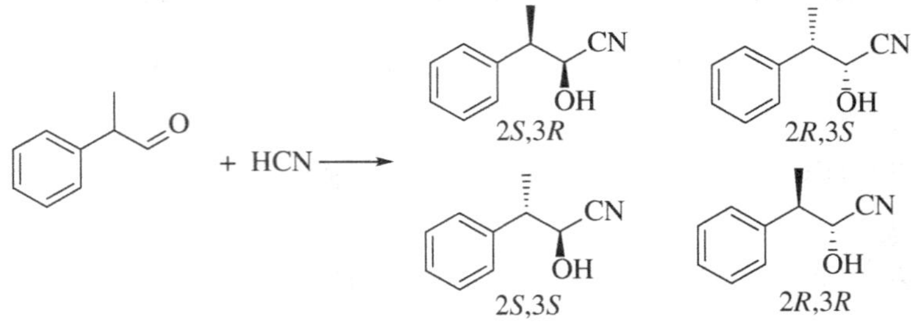

SCHEME 25.10

*Me*HNL-catalyzed addition of HCN to 2-phenylpropanal.

alanine. In a subsequent study, W128 was substituted by amino acids with site chains of decreasing sizes: tyrosine, leucine, cysteine, and alanine, respectively [156]. All variants exhibited comparable specific activities toward the natural substrate acetone cyanohydrin. In contrast, specific activities toward the bulky 4-hydroxymandelonitrile were improved 85-fold (W128Y), 180-fold (W128L), 445-fold (W128A), and 935-fold (W128C). The muteins were investigated for their activity on a number of bulky substituted benzaldehyde derivatives, aliphatic saturated and unsaturated aldehydes, as well as methyl and ethyl ketones. All variants were superior to *Me*HNL wild type in the hydrocyanation of 4-methylbenzaldehyde and 3-phenoxybenzaldehyde, but did not beat *Me*HNL-W128A, which produced 4-methylbenzaldehyde cyanohydrin with an increased conversion of 93% (*Me*HNL wild type: 50%) and an *ee* of 97% (wild type: 99%). For 3-phenoxybenzaldehyde cyanohydrin the conversion was improved from 47 to 98% using mutant *Me*HNL-W128A (Scheme 25.7, Table 25.4, entries 4 and 5).

For sterically demanding unsaturated and saturated aldehydes, no dependence on the size of the amino acid at position 128 in the channel entrance was found. The already good conversion (80%) of the unsaturated 3-phenylpropenal and *ee* (95%) achieved with *Me*HNL wild type were slightly improved with the variants (except W128C). *Me*HNL-W128A as catalyst resulted in 87% conversion and 97% *ee*. The same mutant exhibited higher conversion rates (97%) and an increased *ee* (82%) with 3-phenylpropanal in comparison to *Me*HNL wild type (90% conversion, 67% *ee*). Interestingly, the channel mutant *Me*HNL-W128L also showed strongly improved conversions and *ee* values toward ethyl ketones (Table 25.4, entry 6, Scheme 25.9).

However, using the channel variants as catalysts for the conversion of α- and β-substituted aldehydes containing two stereocenters led to drastic changes or even inversion of the stereoselectivity in comparison to wild-type *Me*HNL [170]. Using 2-phenylpropanal as substrate for the HNL-catalyzed HCN addition in DIPE, the preference for the formation of (2*R*,3*S*)- and (2*S*,3*R*)-configured products (Scheme 25.10) increased with decreasing size of the side chain of the amino acid at position W128. While the wild type exhibited (*S*)-selectivity, *Me*HNL-W128A and *Me*HNL-W128V were (*R*)-selective. Interestingly, the conversion of substrates with the bulky substituent at a larger distance from the carbonyl group (3-phenyl-substituted aldehydes) delivered (*S*)-configured products with all variants and the wild type.

Recently the reaction rate of *Hb*HNL toward mandelonitrile was improved, applying an active site redesign approach [155] based on the comparison of the active sites of *Hb*HNL and the esterase salicylic acid-binding protein 2 (SABP2) (from *Nicotiana tabacum*), both belonging to the α/β-hydrolase superfamily and sharing 45% sequence identity. Since the natural substrate of SAPB2 is the aromatic methyl salicylate, it was suspected that the activity of *Hb*HNL toward the unnatural substrate mandelonitrile could be improved by providing its active site with amino acids essential for substrate binding in SAPB2. Structural comparison of the two enzymes identified 11 amino acids supposedly relevant for substrate binding, which were subsequently introduced into *Hb*HNL by site-directed mutagenesis. The best mutant, *Hb*HNL-L121Y, exhibited a 4.5 times higher specific activity (149 U/mg) in the cyanogenesis of mandelonitrile in comparison to the wild type. Moreover, the authors showed by saturation mutagenesis at position L121 that the activity of the mutated *Hb*HNL increased with increasing size of a hydrophobic amino acid site chain at the respective position (L121Y>L121F>L121M>L121I; but L121W was too large).

25.5.2 Stability

The substitution of Gly113 to serine, which is located on the surface of *Me*HNL, was found to have great impact on protein folding [157]. Circular dichroism analysis revealed major differences in secondary structure contents of wild-type *Me*HNL and mutant *Me*HNL-G113S. In consequence, the authors found two to three times higher expression level of the mutated enzyme in *E. coli* in the soluble fraction. Moreover, *Me*HNL-G113S showed increased thermal stability. While the purified wild type only exhibited 50% of its initial specific activity after heating at 60 °C for 50 min, the purified mutant *Me*HNL-G113S lost only 10% of its specific activity under the same conditions. This result was attributed to the mentioned change in secondary structure, that is, an increase of β-strands. Furthermore, *Me*HNL-G113S exhibited slightly improved pH stability. Several site-directed variants of *Me*HNL were reported to possess improved thermal stability and higher organic solvent tolerance in comparison to the wild type [158]. *Me*HNL-G165D/E and *Me*HNL-V173L are characterized by an increased heat tolerance of approximately 5 °C and an increased stability in ethanol and ethyl acetate. A further improvement in thermal and solvent stability was achieved through combination of the mutations. *Me*HNL-G165E/V173L/M174L retained 90% of its initial activity after heat treatment at 70 °C for 30 min and moreover only lost 20% activity after incubation in ethyl acetate for 24 h.

*Me*HNL wild type is already superior to other HNLs with α/β-hydrolase fold with respect to pH and temperature stability. The highly similar (*R*)-HNL from *A. thaliana* (*At*HNL) is less stable at acidic pH values (pH < 5.4), which is a clear disadvantage for industrial application [171]. In order to improve the pH stability of *At*HNL, the Pohl group chose a rational approach and substituted 11 amino acids on the protein surface for the corresponding residues in *Me*HNL [162] with the goal to favor the formation of a tetramer as in *Me*HNL instead of a dimer as in *At*HNL. The resulting surface-modified variant exhibited 14-fold higher stability at pH 5.0 in comparison to the wild type. In the cyanogenesis of mandelonitrile at pH 4.5, the mutant showed a specific activity of 23.6 U/mg, whereas *At*HNL wild type was inactive under the same conditions. Using the variant as catalyst in the HCN addition to benzaldehyde at pH 4.5 (0 °C) in a two-phase system, (*R*)-mandelonitrile was obtained with a yield of 79% and an *ee* of 99.3%.

In a different approach the pH stability of *At*HNL was improved by the introduction of a C-terminal bacterial flavin-based fluorescent protein (FbFP) fusion tag [172]. The increased half-life of cFbFP-*At*HNL at acidic pH values (about a 43-fold increase at pH 4.75) was attributed to a change in the quaternary structure in comparison to the wild-type enzyme. The fusion protein was investigated in the synthesis of mandelonitrile and 2-chloromandelonitrile in a two-phase system with MTBE. At pH 4.75, (*R*)-mandelonitrile was produced with 75% yield and an *ee* > 96% after 60 min. Maximum conversion of 2-chlorobenzaldehyde (78%) was achieved after 5 min with an *ee* of 99%.

25.5.3 Expression

The expression level of *Pa*HNL5 in *P. pastoris* was significantly improved by substitution of the first amino acid leucine for glutamine and replacement of the native secretion signal by the signal sequence of the alpha-mating factor from

S. cerevisiae [37]. Moreover the same group showed that the N-glycosylation is required for the correct folding of the nascent polypeptide chain of *Pa*HNL5 and subsequent secretion of the protein [173, 174]. Particularly for *Me*HNL, there are several further examples of improved expression and solubility as a result of enzyme engineering (Table 25.2). Asano's group employed rational design and directed evolution to yield *Me*HNL muteins, which were mainly expressed in the soluble fraction of *E. coli*, while the main part of the wild-type enzyme was located in the insoluble pellet [160, 161]. Especially the exchange of the hydrophilic H103 to the hydrophobic leucine or methionine resulted in highly soluble and active enzyme variants.

25.5.4 New Catalytic Activities

Besides the "typical" properties targeted by enzyme engineering, the catalytic mechanism can also be addressed by engineering strategies. An example was reported for the SABP2 from *N. tabacum*, an esterase that was converted to an HNL by substitution of only two amino acids [163]. Like some HNLs, SABP2 belongs to the α/β-hydrolase superfamily and contains the same catalytic triad (Ser, His, and Asp) in the active site, although it catalyzes a completely different reaction, that is, ester hydrolysis. Kazlauskas and coworkers compared the structurally similar enzymes and identified amino acid residues that could switch the esterase into an HNL. One of the resulting variants, SAPB2-G12T/M239K, which was created by site-directed mutagenesis catalyzed the cyanogenesis of racemic mandelonitrile with a specific activity of approximately 20 mU/mg, while its original esterase activity was practically lost. The mutated enzyme also catalyzed the synthesis of mandelonitrile from benzaldehyde and HCN in a biphasic system with DIPE, with approximately 10% conversion and an *ee* (*S*) of 20% after 3 h. Vice versa, the same group succeeded in converting *Hb*HNL into an esterase by substitution of three amino acids [175]. Structural comparison of the active sites of *Hb*HNL and SAPB2 and site-directed mutagenesis resulted in a triple mutant, *Hb*HNL-T11G/K236G/E79H, that exhibited a more than 200-fold decreased HNL activity toward mandelonitrile and a 92-fold increase of specific esterase activity toward *p*-nitrophenyl acetate (35 mU/mg in comparison to 0.88 mU/mg of the wild type).

25.6 REACTION ENGINEERING AND REACTION SYSTEMS

As discussed in the previous section, enzyme engineering has been used extensively for HNLs to optimize their applicability as industrial biocatalysts for certain envisaged processes. Besides, rather than adapting the enzyme, an alternative starting point is the modification and optimization of the reaction conditions by reaction engineering. Different parameters, such as temperature, pH, buffer compounds, and reaction systems that influence the course of a reaction, might be addressed [23]. For industrial application the process needs to be cost and time efficient as well as eco-friendly.

25.6.1 Reaction Systems

One of the major issues that need to be controlled is the nonspecific chemical addition of HCN to the substrate, which negatively impacts the enantiopurity of the product, as enantiopurity is the result of the ratio of enzymatic to nonenzymatic cyanohydrin formation. This nonenzymatic background reaction is pH and temperature dependent: performing the conversions at low pH values (<4–5) and low temperature (5–8 °C) allows the suppression of the undesired racemic product formation, thereby enabling the synthesis of cyanohydrins with higher enantiopurity. However, as an aqueous reaction system at a close to neutral pH and temperatures between 20 and 40 °C would represent the optimal environment for most enzymes, the enzyme activity and especially also the stability and thus conversion usually decrease by lowering the pH and/or temperature. The enzyme stability at low pH can be increased by additives. *Hb*HNL can be stabilized by addition of polyols [167], and *At*HNL was stabilized by saccharose and sorbitol to increase the half-life time 36-fold at pH 5.4 [171], whereas *Me*HNL could not be stabilized with these reagents. Several HNLs, for example, *Me*HNL and *At*HNL, were more stable in cell-free crude extract than as purified enzyme preparation. At pH 4.0, the *Me*HNL activity was found in the aggregated protein fraction, indicating that *Me*HNL was stabilized by entrapment in aggregated *E. coli* proteins [171]. The choice of the buffer system also influenced the enzyme stability and activity. One prominent example was the behavior of different HNLs in acetate buffer. While the enzyme activities of *Me*HNL and *Hb*HNL were inhibited, *At*HNL activity was not affected [171]. The enzyme inhibition was reversible when the buffer was exchanged, as the enzymes' stability was not affected. *Me*HNL showed significantly higher K_m values in acetate than in other buffers [129, 157], and together with the finding that an acetate ion was bound in the active site of the *Me*HNL structure (PDB-code 1DWP [127]), it was suggested that acetate acted as a competitive inhibitor, which was in agreement with the fact that acetone cyanohydrin is the natural substrate for *Me*HNL and *Hb*HNL, whereas *At*HNL showed poor conversion.

The low solubility of many substrates in water, which leads to low space–time yields, is another drawback of aqueous reaction systems. The most widely used reaction system for HNL-catalyzed transformations is a biphasic system consisting of an aqueous phase and a water-immiscible organic phase [23, 176]. While the enzyme is present in the aqueous phase, substrates and products can be dissolved in high concentrations in the separate organic phase. Thus, due to the low concentrations of substrate and product in the aqueous phase, substrate or product inhibitions are avoided; the unselective chemical background reaction is reduced, and moreover, product separation is simplified, and reuse of the enzyme in the aqueous phase is easily accomplished [10]. On the downside, for substrates with too high partition coefficients, substrate concentrations will be very low in the aqueous phase, resulting in low reaction rates of the enzyme. This is relevant in particular for substrates with high K_m values [23]. Thus, biphasic systems are usually stirred intensively to increase the surface area between the two layers, which in turn can decrease the enzyme stability. Parameters such as pH, temperature, and buffer composition play an important role in this system as well, and in addition, the choice of solvent is crucial [23, 176, 177].

Alternatively, an organic solvent-free reaction system can also be used, in which only the substrate represents the organic phase, to which an aqueous phase containing the enzyme in buffer is added. Thus, the substrate concentration is only limited by the solubility of the substrate in the aqueous phase. This approach was successfully applied for the conversion of ketones which are notoriously difficult substrates due to thermodynamic limitations [178]. Conversions of acetophenone up to 78% with greater than 99% *ee* were obtained using *Me*HNL (Scheme 25.11).

In general, reactions can be performed not only as a batch process but also in a continuous flow reactor, which enables a generally enhanced control and an improved mass and heat transfer processes [179]. A monophasic continuous flow process was reported by Chen *et al.* [180], who used a column filled with crude almond meal, through which the liquid phase flowed, providing the substrates for the synthesis of chiral cyanohydrins. While this is a convenient approach for production processes,

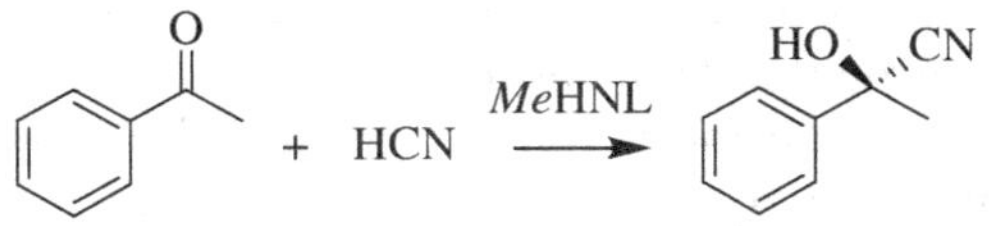

SCHEME 25.11
*Me*HNL-catalyzed synthesis of (*S*)-acetophenone cyanohydrin.

microreactors have been designed for development and optimization purposes. Recently, the applicability of microreactors for crude cell lysates of *Hb*HNL and *Pa*HNL has been studied in an aqueous–organic biphasic system [179, 181].

25.6.2 Immobilization of HNLs

The necessity for the usage of immobilized HNLs arises from different facts: an important point that was repeatedly discussed earlier is the unselective chemical background reaction which occurs at pH values above approximately 4.5 in aqueous– and biphasic aqueous–organic solvent reaction systems. Moreover, aqueous buffer and biphasic systems are critical from an economical point of view and with regard to multistep synthesis. The reagents needed for derivatization of cyanohydrins decompose in water; the extraction of the products with organic solvents is therefore necessary. This implies an enormous waste production because large amounts of organic solvents are needed and the aqueous phase has to be disposed. The use of immobilized HNLs enables the bioconversion in pure organic solvents; thus, the unwanted chemical HCN addition can be suppressed, and moreover, waste production is diminished. Furthermore, stabilization of the enzyme by immobilization and recyclability of the biocatalyst are relevant factors [182, 183]. A detailed review about HNL immobilization was recently published by Hanefeld [184], who classified the strategies applied for immobilization as follows: either (i) noncovalent or (ii) covalent immobilization by (iii) encapsulation techniques. Noncovalent strategies involve immobilization via ionic interactions or via hydrogen bonds. Due to the different properties of HNLs and requirements of the different applications, no general immobilization protocol for HNLs exists, and comparison between results is difficult. Thus, Hanefeld's group set out to systematically investigate and compare the activity and recyclability of differently immobilized *Me*HNL [185]. The best recyclability was obtained for *Me*HNL immobilized on Celite. For the pH-sensitive *At*HNL, which was deactivated below pH 5.0 [171], absorption on Celite (Celite-R633) was found to be the best strategy for immobilization [186]. Performing the conversion of benzaldehyde in buffer-saturated MTBE, an excellent *ee* of 99.8 and 96.8% conversion were obtained after 45 min, in comparison to 97.9% conversion and 99.7% *ee* afforded with precipitated *At*HNL. As a big advantage, Celite-immobilized *At*HNL could be reused up to five times—applied in a solvent-resistant nylon mesh "tea bag"—with barely reduced activity and constant high *ee* values. This "tea-bag" approach was subsequently also applied for *E. coli* whole cells expressing *At*HNL in a monophasic microaqueous reaction system with MTBE enabling the recycling of the cells [187].

Covalent immobilization of HNLs was realized on different materials like silica beads or Eupergit [176]. The reusability of Eupergit-immobilized HNL from *Prunus pseudoarmeniaca* was impressively demonstrated by Tükel and coworkers [45]. Immobilized *Pp*HNL (on Eupergit C and Eupergit C 250 L) was employed in the cyanogenesis of mandelonitrile in a batch-type column reactor and retained about 97% of its initial activity after 20 reaction cycles. Performing the synthesis reaction of (*R*)-mandelonitrile, the immobilized enzyme could be reused up to 10 times, with 90–96% residual activity and with a constantly high *ee* of 99%.

Carrier-free covalent immobilization techniques comprise cross-linked enzyme crystals (CLECS) and cross-linked enzyme aggregates (CLEAS), and have the advantage that no additional carrier material is needed. Nevertheless, the necessity for enzyme crystallization to obtain CLECS is too laborious for technical applications, and moreover this approach was, for example, unsuccessfully investigated with *Me*HNL, which was finally deactivated during the cross-linking [188]. In contrast, preparing enzyme aggregates from aqueous solutions and subsequently cross-linking them to CLEAs are a more convenient strategy [189, 190]. An interesting attempt is the combi-CLEA approach [191]. The use of a bienzymatic CLEA of

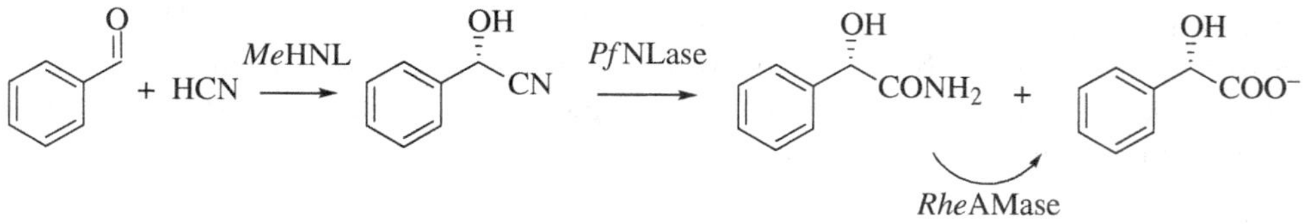

SCHEME 25.12
Trienzymatic synthesis of (*S*)-2-mandelic acid.

*Me*HNL and a nitrilase from *Pseudomonas fluorescens* EBC 191 (*Pf*NLase) enabled the synthesis of (*S*)-mandelic acid from benzaldehyde via *Me*HNL-catalyzed HCN addition and dynamic kinetic resolution of the hydroxynitrile by the unselective and less pH-sensitive (to pH 5.5) nitrilase [192]. (*S*)-Mandelic acid was produced with high enantiomeric purity (98%); however, large amounts of (*S*)-mandelic amide were obtained as a byproduct. To overcome this drawback, the combi-CLEA strategy was extended, and a third enzyme, an amidase from *Rhodococcus erythropolis* (*Rhe*AMase), was included, forming a triple CLEA [193]. Thus, applying the *Me*HNL–*Pf*NLase–*Rhe*AMase-triple CLEA in DIPE buffer (90 : 10, pH 5.5, RT), the synthesis of (*S*)-mandelic acid was afforded with 90% yield and an *ee* of greater than 99% (Scheme 25.12).

A further immobilization strategy, as mentioned earlier, is encapsulation. An interesting approach was shown with *Me*HNL, which was immobilized as biocatalytic active static emulsions (BASE) [194]. The enzyme was present in small aqueous pools (droplets) containing buffer and salts and thus was located in an optimal environment. The droplets were distributed in spherical matrices made of polydimethylsiloxane (PDMS). A major benefit of this strategy is the possibility of using the enzyme under optimum conditions (pH), without loss of enantiopurity of the product. *Me*HNL immobilized in BASE spheres (citrate phosphate buffer, pH 6.0, 10 °C; solvent: DIPE) was investigated in the conversion of diverse benzaldehyde derivatives as well as aliphatic ketones. Especially for substituted benzaldehydes, very good to excellent conversions and excellent *ee*'s were obtained.

25.7 CONCLUSION

Hydroxynitrile lyases were discovered more than hundred years ago. While in the beginning only crude plant extracts containing HNLs were available and the focus of the research was on the discovery of new HNL sources as well as on the investigation of reaction conditions and their improvement in order to obtain optimal product formation (high productivity and enantioselectivity), nowadays gene and protein sequences of several HNLs are available, and enzyme engineering by either random or rational methods is applied to improve biocatalytic transformations. High-throughput activity and selectivity screening assays allow enzyme improvements in reasonable time frames.

A significant number of (*R*)- and (*S*)-selective enzymes are known for the synthesis of cyanohydrins, which have been characterized to varying extents concerning their substrate scope, specificity, structure, and reaction mechanisms. Several of them are available in sufficient amounts either as plant extract or as recombinant protein for commercial use. HNL catalysis offers exciting new opportunities for broad applications in industrial production.

ACKNOWLEDGMENT

This work has been supported by the Federal Ministry of Science, Research and Economy (BMWFW), the Federal Ministry of Traffic, Innovation and Technology (bmvit), the Styrian Business Promotion Agency (SFG), the Standortagentur Tirol, and ZIT—The Technology Agency of the City of Vienna through the COMET Funding

Program managed by the Austrian Research Promotion Agency (FFG). The research leading to these results has received funding from the European Union's Seventh Framework Programme for research, technological development, and demonstration under grant agreement no. 289646 (Kyrobio).

REFERENCES

[1] K. Drauz, H. Gröger, O. May (Eds.), *Enzyme Catalysis in Organic Synthesis*, Wiley-VCH, Weinheim, 2012.

[2] U. T. Bornscheuer, G. W. Huisman, R. J. Kazlauskas, S. Lutz, J. C. Moore, K. Robins, *Nature* 2012, **485**, 185–194.

[3] A. Schmid, J. S. Dordick, B. Hauer, A. Kiener, M. Wubbolts, B. Witholt, *Nature* 2001, **409**, 258–268.

[4] M. T. Reetz, *J. Am. Chem. Soc.* 2013, **135**, 12480–12496.

[5] H. Gröger, W. Hummel, *Curr. Opin. Chem. Biol.* 2014, **19**, 171–179.

[6] Y. Ni, D. Holtmann, F. Hollmann, *ChemCatChem* 2014, **6**, 930–943.

[7] M. Dadashipour, Y. Asano, *ACS Catal.* 2011, **1**, 1121–1149.

[8] M. Gruber-Khadjawi, M. H. Fechter, H. Griengl, in *Enzyme Catalysis in Organic Synthesis* (Eds.: K. Drauz, H. Gröger, O. May), Wiley-VCH, Weinheim, 2012, pp. 947–990.

[9] E. Lanfranchi, K. Steiner, A. Glieder, I. Hajnal, R. A. Sheldon, S. van Pelt, M. Winkler, *Recent Pat. Biotechnol.* 2013, **7**, 197–206.

[10] M. Winkler, A. Glieder, K. Steiner, in *Comprehensive Chirality, Synthetic Methods VI—Enzymatic and Semi-Enzymatic*, Vol. **7** (Eds.: E. M. Carreira, H. Yamamoto), Elsevier, Amsterdam, 2012, pp. 350–371.

[11] F. Wöhler, J. Liebig, *Annalen der Pharmacie* 1837, **22**, 1–24.

[12] L. Rosenthaler, *Biochem. Z.* 1908, **14**, 238–253.

[13] D. S. Seigler, in *Herbivores: Their Interactions with Secondary Plant Metabolites, the Chemical Participants*, Vol. **1** (Eds.: G. A. Rosenthal, M. R. Berenbaum), Academic Press, San Diego, CA, 1991, pp. 35–77.

[14] N. Bjarnholt, F. Rook, M. S. Motawia, C. Cornett, C. Jorgensen, C. E. Olsen, J. W. Jaroszewski, S. Bak, B. L. Moller, *Phytochemistry* 2008, **69**, 1507–1516.

[15] B. L. Moller, *Curr. Opin. Plant Biol.* 2010, **13**, 338–347.

[16] J. Vetter, *Toxicon* 2000, **38**, 11–36.

[17] I. Hajnal, A. Lyskowski, U. Hanefeld, K. Gruber, H. Schwab, K. Steiner, *FEBS J.* 2013, **280**, 5815–5828.

[18] J. Poulton, *Plant Physiol.* 1990, **94**, 401.

[19] C. Sulzbacher Caruso, R. de Fatima Travensolo, R. Campus Bicudo, E. G. Macedo Lemos, A. P. Ulian de Araujo, E. Carrilho, *Microb. Pathog.* 2009, **47**, 118–127.

[20] M. Zagrobelny, S. Bak, B. Moeller, *Phytochemistry* 2008, **69**, 1457–1468.

[21] J. Andexer, J. von Langermann, A. Mell, M. Bocola, U. Kragl, T. Eggert, M. Pohl, *Angew. Chem. Int. Ed.* 2007, **46**, 8679–8681.

[22] C. Blumer, D. Haas, *Arch. Microbiol.* 2000, **173**, 170–177.

[23] J. N. Andexer, J. V. Langermann, U. Kragl, M. Pohl, *Trends Biotechnol.* 2009, **27**, 599–607.

[24] G. A. Behrens, A. Hummel, S. K. Padhi, S. Schätzle, U. T. Bornscheuer, *Adv. Synth. Catal.* 2011, **353**, 2191–2215.

[25] L. Rosenthaler, *Arch. Pharm.* 1913, **251**, 56–84.

[26] Y. Asano, K. Tamura, N. Doi, T. Ueatrongchit, A. Kittikun, T. Ohmiya, *Biosci. Biotechnol. Biochem.* 2005, **69**, 2349–2357.

[27] E. Gerstner, V. Matzke, E. Pfeil, *Naturwissenschaften* 1968, **55**, 561–563.

[28] L. Hernandez, H. Luna, F. Ruiz-Teran, A. Vazquez, *J. Mol. Catal. B: Enzym.* 2004, **30**, 105–108.

[29] A. Hickel, G. Heinrich, H. Schwab, H. Griengl, *Biotechnol. Tech.* 1997, **11**, 55–58.

[30] M. A. Kassim, S. A. Sooklal, R. Archer, K. Rumbold, *S. Afr. J. Bot.* 2014, **93**, 9–13.

[31] A. Pratush, M. Sharma, A. Seth, T. C. Bhalla, *J. Biochem. Tech.* 2011, **3**, 274–279.

[32] B. Pscheidt, M. Avi, R. Gaisberger, F. S. Hartner, W. Skranc, A. Glieder, *J. Mol. Catal. B: Enzym.* 2008, **52**, 183–188.

[33] F. Feigl, V. Anger, *Analyst* 1966, **91**, 282–284.

[34] B. Krammer, K. Rumbold, M. Tschemmernegg, P. Pöchlauer, H. Schwab, *J. Biotechnol.* 2007, **129**, 151–161.

[35] A. Takos, D. Lai, L. Mikkelsen, M. Abou Hachem, D. Shelton, M. S. Motawia, C. E. Olsen, T. L. Wang, C. Martin, F. Rook, *Plant Cell* 2010, **22**, 1605–1619.
[36] W. Becker, E. Pfeil, *Biochem. Z.* 1965, **346**, 301–321.
[37] A. Glieder, R. Weis, W. Skranc, P. Pöchlauer, I. Dreveny, S. Majer, M. Wubbolts, H. Schwab, K. Gruber, *Angew. Chem. Int. Ed.* 2003, **42**, 4815–4818.
[38] Y. Fukuta, S. Nanda, Y. Kato, H. Yurimoto, Y. Sakai, H. Komeda, Y. Asano, *Biosci. Biotechnol. Biochem.* 2011, **75**, 214–220.
[39] L. L. Xu, B. K. Singh, E. E. Conn, *Arch. Biochem. Biophys.* 1986, **250**, 322–328.
[40] I. P. Cheng, J. E. Poulton, *Plant Cell Physiol.* 1993, **34**, 1139–1143.
[41] Z. Hu, J. E. Poulton, *Plant Physiol.* 1999, **119**, 1535–1546.
[42] R. S. Yemm, J. E. Poulton, *Arch. Biochem. Biophys.* 1986, **247**, 440–445.
[43] R. Bhunya, N. Jana, T. Das, S. Nanda, *Synlett* 2009, **8**, 1237–1240.
[44] R. Bhunya, T. Mahapatra, S. Nanda, *Tetrahedron: Asymmetry* 2009, **20**, 1526–1530.
[45] S. S. Tükel, D. Yildirim, D. Alagöz, Ö. Alptekin, G. Yücebilgic, R. Bilgin, *J. Mol. Catal. B: Enzym.* 2010, **66**, 161–165.
[46] W. Y. Lu, P. Chen, G. Q. Lin, *Tetrahedron* 2008, **64**, 7822–7827.
[47] D. Alagöz, S. S. Tukel, D. Yildirim, *J. Mol. Catal. B: Enzym.* 2014, **101**, 40–46.
[48] E. Gerstner, U. Kiel, *Hoppe Seylers Z. Physiol. Chem.* 1975, **356**, 1853–1857.
[49] E. Gerstner, E. Pfeil, *Hoppe Seylers Z. Physiol. Chem.* 1972, **353**, 271–286.
[50] T. Ueatrongchit, A. Kayo, H. Komeda, Y. Asano, A. Kittikun, *Biosci. Biotechnol. Biochem.* 2008, **72**, 1513–1522.
[51] G. J. Zhao, Z. Q. Yang, Y. H. Guo, *J. Biosci. Bioeng.* 2011, **112**, 321–325.
[52] T. Friedrich, M. Kesseler, B. Hauer, WO03093467.
[53] F. Fang, A. Ji, Z. Meng, *React. Kinet. Catal. Lett.* 2008, **93**, 233–239.
[54] G. W. Kuroki, E. E. Conn, *Proc. Natl. Acad. Sci. U. S. A.* 1989, **86**, 6978–6981.
[55] M. Hasslacher, C. Kratky, H. Griengl, H. Schwab, S. Kohlwein, *Proteins* 1997, **27**, 438–449.
[56] M. Hasslacher, M. Schall, M. Hayn, R. Bona, K. Rumbold, J. Lückl, H. Griengl, S. Kohlwein, H. Schwab, *Protein Expr. Purif.* 1997, **11**, 61–71.
[57] N. Klempier, H. Griengl, M. Hayn, *Tetrahedron Lett.* 1993, **34**, 4769–4772.
[58] D. Selmar, R. Lieberei, B. Biehl, E. E. Conn, *Physiol. Plant.* 1989, **75**, 97–101.
[59] J. Hughes, J. P. C. Decarvalho, M. A. Hughes, *Arch. Biochem. Biophys.* 1994, **311**, 496–502.
[60] H. Semba, E. Ichige, T. Imanaka, H. Atomi, H. Aoyagi, *Methods Mol. Biol.* 2008, **643**, 133–144.
[61] H. Wajant, K. W. Mundry, K. Pfizenmaier, *Plant Mol. Biol.* 1994, **26**, 735–746.
[62] C. Bove, E. E. Conn, *J. Biol. Chem.* 1961, **236**, 207–210.
[63] H. Wajant, K. W. Mundry, *Plant Sci.* 1993, **89**, 127–133.
[64] H. Breithaupt, M. Pohl, W. Bonigk, P. Heim, K. Schimz, M. Kula, *J. Mol. Catal. B: Enzym.* 1999, **6**, 315–332.
[65] K. Trummler, H. Wajant, *J. Biol. Chem.* 1997, **272**, 4770–4774.
[66] K. Trummler, J. Roos, U. Schwaneberg, F. Effenberger, S. Förster, K. Pfizenmaier, H. Wajant, *Plant Sci.* 1998, **139**, 19–27.
[67] M. Dadashipour, M. Yamazaki, K. Momonoi, K. Tamura, K. I. Fuhshuku, Y. Kanase, E. Uchimura, G. Kaiyun, Y. Asano, *J. Biotechnol.* 2011, **153**, 100–110.
[68] T. Ueatrongchit, K. Tamura, T. Ohmiya, A. Kittikun, Y. Asano, *Enzyme Microb. Technol.* 2010, **46**, 456–465.
[69] T. Friedrich, B. Hauer, M. Kesseler, WO2004090124.
[70] H. Wajant, S. Förster, D. Selmar, F. Effenberger, K. Pfizenmaier, *Plant Physiol.* 1995, **109**, 1231–1238.
[71] A. Solis, H. Luna, N. Manjarrez, H. Perez, *Tetrahedron* 2004, **60**, 10427–10431.
[72] A. Solis, C. Luna, H. I. Perez, N. Manjarrez, *Tetrahedron: Asymmetry* 2003, **14**, 2351–2353.
[73] S. Q. Han, P. K. Ouyang, P. Wei, Y. Hu, *Biotechnol. Lett.* 2006, **28**, 1909–1912.
[74] Z. Hussain, R. Wiedner, K. Steiner, T. Hajek, M. Avi, B. Hecher, A. Sessitsch, H. Schwab, *Appl. Environ. Microbiol.* 2012, **78**, 2053–2055.
[75] R. Wiedner, M. Gruber-Khadjawi, H. Schwab, K. Steiner, *Comput. Struct. Biotechnol. J.* 2014, **10**, 58–62.
[76] H. J. Aschhoff, E. Pfeil, *Hoppe Seylers Z. Physiol. Chem.* 1970, **351**, 818–826.
[77] M. Suelves, P. Puigdomenech, *Planta* 1998, **206**, 388–393.
[78] R. Bona, I. Dreveny, A. Glieder, C. Kratky, H. Mayrhofer, R. Neuhofer, P. Pöchlauer, H. Schwab, W. Skranc, I. Wirth, EP1223220.
[79] A. Glieder, K. Gruber, R. Luiten, W. Skranc, R. Weis, WO2004083424.

[80] M. Hasslacher, M. Schall, M. Hayn, H. Griengl, S. D. Kohlwein, H. Schwab, *J. Biol. Chem.* 1996, **271**, 5884–5891.

[81] M. B. Smolka, D. Martins, F. V. Winck, C. E. Santoro, R. R. Castellari, F. Ferrari, I. J. Brum, E. Galembeck, H. D. Coletta, M. A. Machado, S. Marangoni, J. C. Novello, *Proteomics* 2003, **3**, 224–237.

[82] A. J. G. Simpson, F. C. Reinach, P. Arruda, *et al.*, *Nature* 2000, **406**, 151–157.

[83] S. C. Makrides, *Microbiol. Rev.* 1996, **60**, 512–538.

[84] J. M. Cregg, I. Tolstorukov, A. Kusari, J. Sunga, K. Madden, T. Chappell, *Methods Enzymol.* 2009, **463**, 169–189.

[85] L. Näätsaari, B. Mistlberger, C. Ruth, T. Hajek, F. S. Hartner, A. Glieder, *PLoS One* 2012, **7**, e39720.

[86] H. Semba, EP1016712.

[87] F. Effenberger, S. Heid, *Tetrahedron: Asymmetry* 1995, **6**, 2945–2952.

[88] N. Li, M. H. Zong, H. S. Peng, H. C. Wu, C. Liu, *J. Mol. Catal. B: Enzym.* 2003, **22**, 7–12.

[89] M. I. Monterde, S. Nazabadioko, F. Rebolledo, R. Brieva, V. Gotor, *Tetrahedron: Asymmetry* 1999, **10**, 3449–3455.

[90] S. Nanda, Y. Kato, Y. Asano, *Tetrahedron* 2005, **61**, 10908–10916.

[91] V. I. Ognyanov, V. K. Datcheva, K. S. Kyler, *J. Am. Chem. Soc.* 1991, **113**, 6992–6996.

[92] T. Purkarthofer, W. Skranc, H. Weber, H. Griengl, M. Wubbolts, G. Scholz, P. Pöchlauer, *Tetrahedron* 2004, **60**, 735–739.

[93] M. North, *Tetrahedron: Asymmetry* 2003, **14**, 147–176.

[94] T. Purkarthofer, W. Skranc, C. Schuster, H. Griengl, *Appl. Microbiol. Biotechnol.* 2007, **76**, 309–320.

[95] M. Sharma, N. N. Sharma, T. C. Bhalla, *Enzyme Microb. Technol.* 2005, **37**, 279–294.

[96] M. Breuer, K. Ditrich, T. Habicher, B. Hauer, M. Kesseler, R. Stürmer, T. Zelinski, *Angew. Chem. Int. Ed.* 2004, **43**, 788–824.

[97] H. Gröger, *Adv. Synth. Catal.* 2001, **343**, 547–558.

[98] O. Sosedov, K. Matzer, S. Bürger, C. Kiziak, S. Baum, J. Altenbuchner, A. Chmura, F. Van Rantwijk, A. Stolz, *Adv. Synth. Catal.* 2009, **351**, 1531–1538.

[99] P. Zandbergen, J. Brussee, A. Vandergen, C. G. Kruse, *Tetrahedron: Asymmetry* 1992, **3**, 769–774.

[100] F. L. Cabirol, A. E. C. Lim, U. Hanefeld, R. A. Sheldon, *Org. Process Res. Dev.* 2010, **14**, 114–118.

[101] S. van Pelt, F. Van Rantwijk, R. A. Sheldon, *Adv. Synth. Catal.* 2009, **351**, 397–404.

[102] F. Effenberger, B. Gutterer, J. Jäger, *Tetrahedron: Asymmetry* 1997, **8**, 459–467.

[103] T. Purkarthofer, T. Pabst, C. van den Broek, H. Griengl, O. Maurer, W. Skranc, *Org. Process Res. Dev.* 2006, **10**, 618–621.

[104] L. Veum, S. R. M. Pereira, J. C. van der Waal, U. Hanefeld, *Eur. J. Org. Chem.* 2006, 1664–1671.

[105] F. Effenberger, U. Stelzer, *Tetrahedron: Asymmetry* 1995, **6**, 283–286.

[106] B. Ritzen, M. C. M. van Oers, F. L. van Delft, F. P. J. T. Rutjes, *J. Org. Chem.* 2009, **74**, 7548–7551.

[107] F. Effenberger, U. Stelzer, *Angew. Chem. Int. Ed.* 1991, **30**, 873–874.

[108] F. Effenberger, A. Kremser, U. Stelzer, *Tetrahedron: Asymmetry* 1996, **7**, 607–618.

[109] U. Stelzer, F. Effenberger, *Tetrahedron: Asymmetry* 1993, **4**, 161–164.

[110] J. Brussee, E. C. Roos, A. Vandergen, *Tetrahedron Lett.* 1988, **29**, 4485–4488.

[111] T. Purkarthofer, K. Gruber, M. Gruber-Khadjawi, K. Waich, W. Skranc, D. Mink, H. Griengl, *Angew. Chem. Int. Ed.* 2006, **45**, 3454–3456.

[112] S. E. Milner, T. S. Moody, A. R. Maguire, *Eur. J. Org. Chem.* 2012, 3059–3067.

[113] R. C. Tang, Z. Guan, Y. H. He, W. Zhu, *J. Mol. Catal. B: Enzym.* 2010, **63**, 62–67.

[114] N. Gao, Y. L. Chen, Y. H. He, Z. Guan, *RSC Adv.* 2013, **3**, 16850–16856.

[115] E. Busto, V. Gotor-Fernandez, V. Gotor, *Org. Process Res. Dev.* 2011, **15**, 236–240.

[116] M. Gruber-Khadjawi, T. Purkarthofer, W. Skranc, H. Griengl, *Adv. Synth. Catal.* 2007, **349**, 1445–1450.

[117] K. I. Fuhshuku, Y. Asano, *J. Biotechnol.* 2011, **153**, 153–159.

[118] E. Pichersky, E. Lewinsohn, *Annu. Rev. Plant Biol.* 2011, **62**, 549–566.

[119] K. Gruber, C. Kratky, *J. Polym. Sci. Polym. Chem.* 2004, **42**, 479–486.

[120] G. Gartler, C. Kratky, K. Gruber, *J. Biotechnol.* 2007, **129**, 87–97.

[121] K. Gruber, M. Gugganig, U. G. Wagner, C. Kratky, *Biol. Chem.* 1999, **380**, 993–1000.

[122] K. Gruber, G. Gartler, B. Krammer, H. Schwab, C. Kratky, *J. Biol. Chem.* 2004, **279**, 20501–20510.

[123] A. Schmidt, K. Gruber, C. Kratky, V. S. Lamzin, *J. Biol. Chem.* 2008, **283**, 21827–21836.
[124] U. Wagner, M. Hasslacher, H. Griengl, H. Schwab, C. Kratky, *Structure* 1996, **4**, 811.
[125] J. Zuegg, K. Gruber, M. Gugganig, U. Wagner, C. Kratky, *Protein Sci.* 1999, **8**, 1990–2000.
[126] H. Lauble, K. Decanniere, H. Wajant, S. Förster, F. Effenberger, *Acta Crystallogr. D* 1999, **55**, 904–906.
[127] H. Lauble, S. Förster, B. Miehlich, H. Wajant, F. Effenberger, *Acta Crystallogr. D* 2001, **57**, 194–200.
[128] H. Lauble, B. Miehlich, S. Förster, H. Wajant, F. Effenberger, *Protein Sci.* 2001, **10**, 1015–1022.
[129] H. Lauble, B. Miehlich, S. Förster, C. Kobler, H. Wajant, F. Effenberger, *Protein Sci.* 2002, **11**, 65–71.
[130] S. Nakano, M. Dadashipour, Y. Asano, *Biochim. Biophys. Acta* 2014, **1844**, 2059–2067.
[131] H. Lauble, S. Knodler, H. Schindelin, S. Forster, H. Wajant, F. Effenberger, *Acta Crystallogr. D* 1996, **52**, 887–889.
[132] H. Lauble, B. Miehlich, S. Förster, H. Wajant, F. Effenberger, *Biochemistry* 2002, **41**, 12043–12050.
[133] J. N. Andexer, N. Staunig, T. Eggert, C. Kratky, M. Pohl, K. Gruber, *ChemBioChem* 2012, **13**, 1932–1939.
[134] F. C. Cui, X. L. Pan, J. Y. Liu, *J. Phys. Chem. B* 2010, **114**, 9622–9628.
[135] W. Zhu, Y. Liu, R. Zhang, *Proteins* 2014, DOI:10.1002/prot.24648.
[136] I. Dreveny, C. Kratky, K. Gruber, *Protein Sci.* 2002, **11**, 292–300.
[137] I. Dreveny, A. S. Andryushkova, A. Glieder, K. Gruber, C. Kratky, *Biochemistry* 2009, **48**, 3370–3377.
[138] H. Lauble, K. Müller, H. Schindelin, S. Förster, F. Effenberger, *Proteins* 1994, **19**, 343–347.
[139] C. Roberge, F. Fleitz, D. Pollard, P. Devine, *Tetrahedron: Asymmetry* 2007, **18**, 208–214.
[140] U. T. Bornscheuer, *Synlett* 2013, **24**, 150–156.
[141] K. Steiner, H. Schwab, *Comput. Struct. Biotechnol. J.* 2012, **2**, e201209010.
[142] G. Strohmeier, H. Pichler, O. May, M. Gruber-Khadjawi, *Chem. Rev.* 2011, **111**, 4141–4164.
[143] H. J. Wijma, D. B. Janssen, *FEBS J.* 2013, **280**, 2948–2960.
[144] A. S. Bommarius, J. K. Blum, M. J. Abrahamson, *Curr. Opin. Chem. Biol.* 2011, **15**, 194–200.
[145] E. M. Brustad, F. H. Arnold, *Curr. Opin. Chem. Biol.* 2011, **15**, 201–210.
[146] S. Lutz, *Curr. Opin. Biotechnol.* 2010, **21**, 734–743.
[147] M. T. Reetz, D. Kahakeaw, R. Lohmer, *ChemBioChem* 2008, **9**, 1797–1804.
[148] M. T. Reetz, *Angew. Chem. Int. Ed.* 2013, **52**, 2658–2666.
[149] Z. Liu, B. Pscheidt, M. Avi, R. Gaisberger, F. S. Hartner, C. Schuster, W. Skranc, K. Gruber, A. Glieder, *ChemBioChem* 2008, **9**, 58–61.
[150] R. Weis, R. Gaisberger, W. Skranc, K. Gruber, A. Glieder, *Angew. Chem. Int. Ed.* 2005, **44**, 4700–4704.
[151] B. Pscheidt, Z. B. Liu, R. Gaisberger, M. Avi, W. Skranc, K. Gruber, H. Griengl, A. Glieder, *Adv. Synth. Catal.* 2008, **350**, 1943–1948.
[152] R. Wiedner, B. Kothbauer, T. Pavkov-Keller, M. Gruber-Khadjawi, K. Gruber, H. Schwab, K. Steiner, *ChemCatChem* 2014, DOI:10.1002/cctc.201402742.
[153] M. Avi, R. M. Wiedner, H. Griengl, H. Schwab, *Chem. Eur. J.* 2008, **14**, 11415–11422.
[154] H. Bühler, F. Effenberger, S. Förster, C. Kratky, P. Lauble, H. Schwab, E. Steiner, U. Wagner, H. Wajant, EP0969095.
[155] J. von Langermann, D. M. Nedrud, R. J. Kazlauskas, *ChemBioChem* 2014, **15**, 1931–1938.
[156] H. Bühler, F. Effenberger, S. Förster, J. Roos, H. Wajant, *ChemBioChem* 2003, **4**, 211–216.
[157] G. Yan, S. Cheng, G. Zhao, S. Wu, Y. Liu, W. Sun, *Biotechnol. Lett.* 2003, **25**, 1041–1047.
[158] E. Ichige, H. Semba, T. Shijuku, S. Harayama, US20080124784.
[159] Y. Asano, M. Dadashipour, M. Yamazaki, N. Doi, H. Komeda, *Protein Eng. Des. Sel.* 2011, **24**, 607–616.
[160] M. Dadashipour, Y. Fukuta, Y. Asano, *Protein Expr. Purif.* 2011, **77**, 92–97.
[161] Y. Asano, T. Akiyama, F. Yu, E. Sato, US8030053.
[162] D. Okrob, J. Metzner, W. Wiechert, K. Gruber, M. Pohl, *ChemBioChem* 2012, **13**, 797–802.
[163] S. K. Padhi, R. Fujii, G. A. Legatt, S. L. Fossum, R. Berchtold, R. J. Kazlauskas, *Chem. Biol.* 2010, **17**, 863–871.
[164] S. Feichtenhofer, W. W. Höffken, H. Schwab, *Poster at Enzyme Engineering XX*, University of Groningen, Groningen, the Netherlands, 2009.

[165] R. Weis, R. Luiten, W. Skranc, H. Schwab, M. Wubbolts, A. Glieder, *FEMS Yeast Res.* 2004, **5**, 179–189.
[166] C. Reisinger, F. van Assema, M. Schürmann, Z. Hussain, P. Remler, H. Schwab, *J. Mol. Catal. B: Enzym.* 2006, **39**, 149–155.
[167] A. Hickel, M. Graupner, D. Lehner, A. Hermetter, O. Glatter, H. Griengl, *Enzyme Microb. Technol.* 1997, **21**, 361–366.
[168] J. Andexer, J. K. Guterl, M. Pohl, T. Eggert, *Chem. Commun.* 2006, 4201–4203.
[169] M. Konarzycka-Bessler, U. T. Bornscheuer, *Angew. Chem. Int. Ed.* 2003, **42**, 1418–1420.
[170] H. Bühler, B. Miehlich, F. Effenberger, *ChemBioChem* 2005, **6**, 711–717.
[171] J. K. Guterl, J. N. Andexer, T. Sehl, J. von Langermann, I. Frindi-Wosch, T. Rosenkranz, J. Fitter, K. Gruber, U. Kragl, T. Eggert, M. Pohl, *J. Biotechnol.* 2009, **141**, 166–173.
[172] K. E. Scholz, B. Kopka, A. Wirtz, M. Pohl, K. E. Jaeger, U. Krauss, *Appl. Environ. Microbiol.* 2013, **79**, 4727–4733.
[173] R. Gaisberger, R. Weis, R. Luiten, W. Skranc, M. Wubbolts, H. Griengl, A. Glieder, *J. Biotechnol.* 2007, **129**, 30–38.
[174] R. Weis, R. Gaisberger, K. Gruber, A. Glieder, *J. Biotechnol.* 2007, **129**, 50–61.
[175] D. M. Nedrud, H. Lin, G. Lopez, S. K. Padhi, G. A. Legatt, R. J. Kazlauskas, *Chem. Sci.* 2014, **5**, 4265–4277.
[176] M. Avi, H. Griengl, in *Organic Synthesis with Enzymes in Non-Aqueous Media* (Eds.: G. Carrea, S. Riva), Wiley-VCH, Weinheim, 2008, pp. 211–226.
[177] J. von Langermann, J. Guterl, M. Pohl, H. Wajant, U. Kragl, *Bioprocess Biosyst. Eng.* 2008, **31**, 155–161.
[178] J. von Langermann, A. Mell, E. Paetzold, T. Daußmann, U. Kragla, *Adv. Synth. Catal.* 2007, **349**, 1418–1424.
[179] K. Koch, R. van den Berg, P. Nieuwland, R. Wijtmans, H. Schoemaker, J. van Hest, F. Rutjes, *Biotechnol. Bioeng.* 2008, **99**, 1028–1033.
[180] P. R. Chen, S. Q. Han, G. Q. Lin, Z. Y. Li, *J. Org. Chem.* 2002, **67**, 8251–8253.
[181] K. Koch, R. van den Berg, P. Nieuwland, R. Wijtmans, M. Wubbolts, H. Schoemaker, F. Rutjes, J. van Hest, *Chem. Eng. J.* 2008, **135**, S89–S92.
[182] R. A. Sheldon, S. van Pelt, *Chem. Soc. Rev.* 2013, **42**, 6223–6235.
[183] U. Hanefeld, L. Q. Cao, E. Magner, *Chem. Soc. Rev.* 2013, **42**, 6211–6212.
[184] U. Hanefeld, *Chem. Soc. Rev.* 2013, **42**, 6308–6321.
[185] G. Torrelo, N. van Midden, R. Stloukal, U. Hanefeld, *ChemCatChem* 2014, **6**, 1096–1102.
[186] D. Okrob, M. Paravidino, R. V. A. Orru, W. Wiechert, U. Hanefeld, M. Pohl, *Adv. Synth. Catal.* 2011, **353**, 2399–2408.
[187] K. E. Scholz, D. Okrob, B. Kopka, A. Grünberger, M. Pohl, K. E. Jaeger, U. Krauss, *Appl. Environ. Microbiol.* 2012, **78**, 5025–5027.
[188] D. Costes, E. Wehtje, P. Adlercreutz, *J. Mol. Catal. B: Enzym.* 2001, **11**, 607–612.
[189] R. A. Sheldon, *Org. Process Res. Dev.* 2011, **15**, 213–223.
[190] R. A. Sheldon, in *Comprehensive Chirality, Synthetic Methods VI—Enzymatic and Semi-Enzymatic*, Vol. **9** (Eds.: E. M. Carreira, H. Yamamoto), Elsevier, Amsterdam, 2012, pp. 353–366.
[191] F. Van Rantwijk, A. Stolz, *J. Mol. Catal. B: Enzym.* 2014, DOI:10.1016/j.molcatb.2014.08.012.
[192] C. Mateo, A. Chmura, S. Rustler, F. van Rantwijk, A. Stolz, R. A. Sheldon, *Tetrahedron: Asymmetry* 2006, **17**, 320–323.
[193] A. Chmura, S. Rustler, M. Paravidino, F. Van Rantwijk, A. Stolz, R. A. Sheldon, *Tetrahedron: Asymmetry* 2013, **24**, 1225–1232.
[194] J. von Langermann, S. Wapenhensch, *Adv. Synth. Catal.* 2014, **356**, 2989–2997.

CHAPTER 26

Biocatalysis: Nitrilases in Organic Synthesis

Jin-Song Gong, Jin-Song Shi, and Zheng-Hong Xu

School of Pharmaceutical Science, Jiangnan University, Wuxi, P.R. China

26.1 INTRODUCTION

Nitriles are important chemicals in organic synthesis because of their universal existence and versatile transformations. They can be chemically synthesized using several approaches. Aromatic nitriles can be prepared by the Sandmeyer reaction and gas-phase ammoxidation, whereas nucleophilic substitution of alkyl halides by sodium cyanide is the most common method for preparing aliphatic nitriles [1]. Unsaturated nitriles can also be generated by Wittig-type reactions for aldehydes or by the Knoevenagel reaction [2]. These compounds are widely applied in the production of pesticides, pharmaceuticals, and fine chemicals; they can also be used as the intermediates in organic synthesis for producing amides, amidines, carboxylic acids, aldehydes, ketones, etc. [2, 3]. Among them, carboxylic acid synthesis from nitriles has received significant attention in industry over the past several years.

Nitrile compounds are generally synthetically more accessible than their corresponding carboxylic acids [4]. However, chemical synthesis of carboxylic acids from nitriles usually requires harsh reaction conditions and produces large amounts of byproducts. Therefore, the biocatalytic approach using nitrilases as the catalyst for organic synthesis is of great interest due to its "green" characteristics; it can directly convert nitriles into acids under mild conditions with low energy consumption, low pollution, and high production yield [5, 6]. Nevertheless, the application potential is still limited by several factors, such as unsatisfactory specific activity and poor thermostability. This chapter mainly focuses on the nitrilase improvement strategies including fermentation methods for increasing nitrilase production and molecular modification methods for improving catalytic properties. Moreover, nitrilases are widespread in the natural environment, and to date the enzymes have been found in bacteria, filamentous fungi, yeasts, and plants. Further discovery and exploitation of potential nitrilases is still necessary for future applications. On the other hand, nitrilases applied in organic synthesis for production of commodity chemicals and pharmaceutical intermediates have been demonstrated in detail.

Green Biocatalysis, First Edition. Edited by Ramesh N. Patel.

26.2 NITRILASE DISCOVERY

26.2.1 Conventional Screening

Nitrilase-producing microorganisms have been found in diverse habitats such as agricultural wastewater, industrial wastes, and marine sponges. Strains from genera of *Rhodococcus*, *Pseudomonas*, and *Alcaligenes* are one of the major resources for nitrilase production in the literature. A simple and reliable method for hunting nitrilase activity in microorganisms is frequently described, which employs nitriles as the sole nitrogen or/and carbon source. Xue *et al*. screened a *Alcaligenes faecalis* using phenylacetonitrile as the sole source of carbon and nitrogen, which bears enantioselective nitrilase activity and can hydrolyze (*R*,*S*)-mandelonitrile into (*R*)-(−)-mandelic acid [7]. Recently, a cyanopyridine-degrading strain of *Pseudomonas putida* CGMCC3830 was obtained utilizing 3-cyanopyridine as sole nitrogen source, and its hydrolysis reaction was used for efficient synthesis of nicotinic/isonicotinic acids from cyanopyridine [6]. In addition, screening of nitrilase-producing strains is usually done by using a pH indicator on agar plates to improve screening efficiency of positive clones. An acrylonitrile-converting *Rhodococcus ruber* was isolated in petroleum-contaminated sludge samples from several oil companies using 200 mM acrylonitrile as the substrate and bromothymol blue as an indicator [8], which can mark the nitrilase-producing colonies with yellow color from the initial state of dark blue. Fungal strains of *Gibberella*, *Penicillium*, *Aspergillus*, and *Fusarium* are also common producers of nitrilase. The Brazilian marine fungal strain *Aspergillus sydowii* CBMAI 934 was isolated from the marine sponge *Chelonaplysilla erecta*; the results showed that it can degrade the methylphenylacetonitriles to synthesize the corresponding acids [9].

26.2.2 Metagenomic Mining

The natural environment is a treasure of stored huge amounts of microbial resources and strains possessing novel enzymes; most of which, however, demonstrate difficulty in isolation and cultivation through traditional methods [10]. Metagenomic mining presents an opportunity to access these new enzymes using the procedures of genome extraction, library construction, and targeted enzyme screening. This method can avoid conventional cultivation of individual microorganisms but directly obtain their encoding genes and screen the potential recombinant strains. Bayer *et al*. successfully mined an aliphatic nitrilase Nit1 from metagenomic libraries including millions of individual clones using cinnamonitrile and a mixture of six different nitriles as the substrates [11]. BLAST analysis results indicated its low identities (<66%) with reported sequences in GenBank and literatures. The Nit1 showed dominant activity toward dinitriles. Nevertheless, it almost could not hydrolyze the aliphatic mononitriles such as acetonitrile and propionitrile.

26.2.3 Genome Mining

Bioinformatics and advanced DNA sequencing technology (with massive scale and reduced cost) as well as molecular biology techniques made it possible to explore potential genes in GenBank database [12]. Over the past decades, the increasing number of genomes with different origins has been available online, which is regarded as potential resource for searching novel nitrilases. On this basis, genome mining, which can be performed by total gene synthesis (usually combined with codon optimization) or amplification from the DNA of corresponding microorganisms, is used for exploring these potential genes. In the recent years, this approach has attracted increasing attention in the area of nitrilase-mediated biocatalysis. Zhu *et al*. found two nitrilase genes (bll6402 [13] and blr3397 [14]) from *Bradyrhizobium japonicum*

USDA 110 and overexpressed them in *Escherichia coli*. The two genes showed very low sequence identity of 33%; in addition, corresponding recombinant enzymes displayed significantly different characteristics and structures. bll6402 has a molecular weight of 37 kDa on SDS-PAGE and 455 kDa by light scattering analysis, while blr3397 is proven to be 34.5 and 340 kD. Substrate affinity analysis demonstrated that the former exhibited high activity toward mandelonitrile and the latter preferred to hydrolyze aliphatic nitriles but almost inactivated with mandelonitrile. Seffernick *et al.* [15] predicted nitrilase genes with high specificity to mandelonitrile through sequence data and distant superfamily structure analysis. A nitrilase from *Burkholderia xenovorans* LB400 (gi91784632) was obtained and proved to be capable of utilizing the glutaronitrile or adiponitrile as the sole nitrogen source. It showed high nitrilase activity toward mandelonitrile with a catalytic efficiency (k_{cat}/K_m) of 3.6×10^4 per M/s. This method exhibits certain potential in exploring novel enzymes with required specificities from unannotated sequences.

26.3 NITRILASE IMPROVEMENT

Nitrilase production as well as its catalytic properties can be improved through culture optimization at the fermentation level and gene modification at the molecular level.

26.3.1 Culture Optimization

Effect of Carbon Sources

The carbon source, nitrogen source, and inducer are considered as the most crucial medium components for nitrilase production. Glucose or glycerol is commonly used as the preferred carbon sources for nitrilase-producing strains. The highest nitrilase activity of *A. faecalis* MTCC 10757 was supported by 10 g/l glucose, whereas fructose as its carbon source resulted in good biomass production but negligible nitrilase formation [16]. However, another *Alcaligenes* strain was found to utilize 10 g/l sodium acetate as the optimum carbon source; and it could only obtain low activity with glycerol and glucose as the carbon sources [17]. The optimal carbon source of *P. putida* CGMCC3830 for producing nitrilase proved to be 13.5 g/l glycerol [18]. Fungal nitrilases from *Fusarium proliferatum* AUF-2 [19] and *Fusarium oxysporum* H3 [20] were determined to exhibit optimal activity with 50 g/l and 10 g/l glucose for cultivation, respectively. Liu *et al.* [21] used glycerol feeding strategies to improve the nitrilase production in the fermentation of *E. coli* JM109/pNLE. They observed moderate nitrilase activity at a glycerol feeding rate of 9 g/h; however, a higher feeding rate led to faster growth but lower activity.

Effect of Nitrogen Sources

The nitrogen sources have significant effects on cell growth and nitrilase activity in the literature. Various nitrogen sources, both organic and inorganic, such as peptone, beef extract, yeast extract, urea, nitrites, and ammonium salts, were often used as the nitrogen sources for cultivating nitrilase-producing strains. A one-factor-at-a-time approach was used in nitrilase production optimization for *Arthrobacter nitroguajacolicus* ZJUTB06-99, and 5 g/l yeast extract was selected as the nitrogen source [22]. A combination of nitrogen sources are usually used as the medium components in nitrilase formation. The composite of peptone (10 g/l) plus yeast extract (5 g/l) was found to be an optimum nitrogen source of *Alcaligenes* sp. ECU0401 nitrilase [17]. However, the combination nitrogen sources (peptone plus beef extract or peptone plus yeast extract) only favored a heavy biomass of *F. oxysporum* H3 but inhibited its nitrilase production [20]. Through a culture optimization strategy using response surface methodology, 2.31 g/l sodium nitrate was used as the nitrogen source of *F. proliferatum* nitrilase [19].

Effect of Inducers

Nitrilases are generally inducible in nature, and there are few reports on the constitutive nitrilases reported in the literature. Nitrile substrates and products and ε-caprolactam are the most common inducers for nitrilase synthesis. For instance, acetonitrile is proven to be the most suitable inducer for nitrilase production in *P. putida* MTCC 5110 [23]. Maximum activity was observed with addition of acetonitrile at the beginning. The author also attempted acetonitrile feeding at every 4 h to ensure continuous induction; however, feeding of acetonitrile had no positive impact on the enzyme activity. In the cultivating process of *F. oxysporum* H3, caprolactam was the most efficient inducer and showed a significant advantage over other inducers, such as 2-cyanopyridine, acetonitrile, and valeronitrile [20]. Interestingly, *Bacillus subtilis* ZJB-063 still displayed nitrilase activity without the addition of any inducers in the fermentation medium; however, the addition of ε-caprolactam could lead to the formation of nitrile hydratase and amidase activity [24]. 2-Cyanopyridine turned out to be a powerful nitrilase inducer for several strains of *Aspergillus*, *Fusarium*, and *Penicillium*, which supported high specific activity of nitrilase but only generated a low concentration of target protein [25]; the reverse was the case with the addition of its analogs, 3- and 4-cyanopyridine, which favored extremely high protein concentration but observed very low nitrilase activity. Certain nitrilase activity has been still detected in *P. putida* CGMCC3830 without addition of inducers, while addition of 1 g/l urea could increase the nitrilase activity by 70% [18].

Effect of Culture Conditions

Besides the medium components, culture conditions including temperature and pH are also important factors for enzyme formation in the culture broth. The Maximum enzyme activity of *F. oxysporum* was observed at 30 °C and initial pH of 7.2 [20]. *Alcaligenes* sp. could provide moderate biomass and enzyme activity within the pH range of 6.0–9.0, and pH 6.0 supported the highest growth and enzyme production [17]. The optimum temperature and pH for the production of a thermophilic nitrilase from *Streptomyces* sp. were determined to be 45 °C and pH 7.0, respectively [26]. *Pseudomonas putida* gave the highest biomass and enzyme production when the initial pH of the medium was adjusted to 6.0, and the pH of the medium gradually increased during the fermentation course. Thus the effect of controlled pH on nitrilase formation was investigated, and results showed that there was no substantial improvement through pH control [23]. Furthermore, agitation may also influence on the cell morphology, product formation efficiency, substrate utilization, and growth rate. The effect of agitation on *P. putida* nitrilase was further evaluated, and similar activities were found with agitator speeds of 100 and 150 rpm, while the nitrilase production at 150 rpm showed a slightly higher efficiency than that of 100 rpm [23].

26.3.2 Nitrilase Reengineering

Semirational Design

Although culture optimization can obviously improve the nitrilase production or increase the protein concentration from a macroscopic view, enzyme reengineering at the gene level may substantially amend the unsatisfactory enzymatic properties and synthesis requirements of nitrilase, from a microscopic view. In recent years, enzyme reengineering approaches, mainly including semirational and rational design, have emerged as useful tools for improving the application performance of natural biocatalysts such as catalytic activity, stability, and by product formation so that they can be more suitable for industrial biocatalysis [27]. The former method is usually performed based on sequence information including sequence alignment and analysis as well as published data about nitrilase modification, whereas the latter method combines protein structure information with sequences. This part mainly focuses on the major achievements made in the last few years on protein design of nitrilase, but research advances in directed evolution and other novel strategies are also included.

Heterologous expression of genes in host cells such as *E. coli* was essential for further study but has become a routine technique at present. Kiziak *et al.* [28] amplified a 1052 bp DNA sequence (consisted of 350 amino acid residues) from *Pseudomonas fluorescens* EBC191 and expressed this gene in *E. coli* JM109. Substrate specificity analysis showed that the enzyme displayed high nitrilase activities toward (substituted) phenylacetonitriles, such as mandelonitrile, 2-phenylvaleronitrile, phenylacetonitrile, and 2-phenylpropionitrile. Nitrile conversion results found that amide compounds were simultaneously formed with the release of carboxylic acids. For instance, mandelonitrile was transformed into mandelic acid and mandeloamide with a ratio of 80:20, while conversion of 2-phenylpropionitrile, phenylacetonitrile, and 2-phenylglycinonitrile generated considerably lower amounts of amides. In the catalytic mechanism of nitrilase, it is widely proposed that the active site of nitrilase is composed of a cysteine, a glutamate, and a lysine residue, and the nitrile carbon is probably nucleophilic, attacked by the sulfhydryl group on catalytic cysteine residues of nitrilase [29]. On this basis, the effects of several C-terminal mutations on enzymatic properties of *P. fluorescens* nitrilase were subsequently investigated, and it was demonstrated that 47–67 amino acid deletion from C-terminal decreased nitrilase activities and increased the amide formation, with 2-phenylpropionitrile and mandelonitrile as the substrates [30]. In order to analyze the effects of mutations in close proximity to the catalytic active cysteine residue on nitrilase, amino acid exchanges close to the catalytic center were carried out through site-directed mutagenesis. The authors analyzed the amino acid residues in direct proximity to the catalytic center by sequence alignments; they found that Ala165 directly adjacent to cysteine distinguished *P. fluorescens* nitrilase from all other nitrilases. Mutation results indicated that variants carrying large amino acid residues at this position increased the enantioselectivity for the formation of (*R*)-acids from racemic mandelonitrile and 2-phenylpropionitrile, whereas variants carrying an aromatic substituent in this position formed reduced amounts of amide from mandelonitrile [31]. In another study, Petříčková *et al.* [32] performed sequence alignments of *Aspergillus niger* CBS 513.88 and *Neurospora crassa* OR74A nitrilases with several bacterial, plant, and fungal enzymes. Their results showed that several residues located next to the catalytically active cysteine residue (C162 and C167) differed from the residues in the corresponding position of some other nitrilases. Thus various variants were constructed by site-directed mutagenesis, and among them the W168A variant of *N. crassa* nitrilase was found to produce larger amounts of mandelamide and 2-phenylpropionamide from nitriles than wild type.

Rational Design

Enzyme functions are closely related to the three-dimensional structure of the protein. The increasing number of protein crystal structures in the PDB database and advances in homology modeling afforded unprecedented opportunities for scholars to locate and modify more effectively the crucial residues adjacent to the active center and at domain interfaces or hinge regions so that the catalysts can obtain required properties for commercial use. Raczynska *et al.* [33] analyzed the first protein structure of nitrilase derived from a hyperthermophilic archaeon *Pyrococcus abyssi*. Its crystal structure was found without ligands, with an acetate ion bound and a bromide ion in the active site. Docking calculations were performed using fumaro- and malononitriles as the substrates, and for the catalytic mechanism it was proposed that Cys146 acts as the nucleophile, Glu42 as the general base, and Lys113/Glu42 as the general acid. However, no further study was performed on this nitrilase. Recently, nine nitrilases with different origins were mined from GenBank database (blasted with *A. faecalis* nitrilase as the template) and used for conversion of iminodiacetonitrile [34]. Among them, nitrilases from *Acidovorax facilis*, *A. faecalis*, and *Rhodococcus rhodochrous* showed suitable activity toward iminodiacetonitrile, while *A. facilis* nitrilase was proven to exhibit the highest activity, and thus it was selected for subsequent characterization. Based on homology modeling (with the structures of *P. abyssi* nitrilase and several other enzymes as the template), docking studies (elaborating the in silico interactions between the enzyme and substrate), and sequence alignment (the nine sequences as aforementioned), *A. facilis* nitrilase variants including single and multiple mutations were constructed to further improve the catalytic activity. Only three variants were inactivated toward iminodiacetonitrile and all other variants showed increased k_{cat} value and decreased K_m value to varying degrees. A multiple mutant F168V/L201N/S192F displayed the highest conversion rate of 96%, which was 1.48-fold higher than that of wild type.

Irrational Design

Directed evolution, also known as *in vitro* molecular evolution, belongs to the methodology of irrational protein design. It can rapidly modify the biocatalysts in laboratory through an *in vitro* version of Darwinian evolution [12]. So far, there have been limited crystal structures of nitrilase available in the PDB database, which may hamper the research on rational design and thus promote the development of directed evolution. *Alcaligenes faecalis* nitrilase was treated by error-prone PCR to construct the random mutation libraries for screening of nitrilase variants with improved activity and stability [35]. Enzyme screening was conducted on a 96 well flat-bottom plate for cultivating nitrilase-producing cells and measuring the absorbance of reaction mixture by Berthelot assay after cell lysis. Several rounds of error-prone PCR as well as further combination mutations by site-directed mutagenesis were performed, and variants that showed increased nitrile conversion rate and stability under low pH conditions were obtained. Bioconversion results showed that (*R*)-2-chloro-mandelonitrile could be fully hydrolyzed into (*R*)-2-chloro-mandelic acid with ee > 99% by a nitrilase variant at pH 4.5. Very recently, *P. fluorescens* EBC191 nitrilase, as described earlier, has been further modified by random mutations via error-prone PCR method to improve its mandeloamide formation [36]. The enzyme screening system was constructed based on detecting colored iron(III)-hydroxamate complexes from mandeloamide through the addition of hydroxylamine and ferric iron ions utilizing simultaneously expressed amidase in the host cells. Using this method, Sosedov and coworkers screened 30 nitrilase variants showing increased amide formation ability, which carried various deletions at the carboxy-terminus or amino acid exchanges in certain positions. A nitrilase variant from combination of positive mutations displayed significant nitrile hydratase activity, and the formation of mandeloamide from mandelonitrile reaches up to 19-fold higher than that of mandelic acid. On the other hand, *Arabidopsis thaliana* nitrilase 2 was engineered by directed evolution to enhance its activity toward various substrates [37]. Three rounds of error-prone PCR as well as combination mutations were performed, and thousands of variants were constructed and screened by Berthelot assay. A final variant carrying six amino acid mutations was selected, which showed a fourfold increase of specific activity compared with wild type. Substrate spectrum analysis demonstrated that the final variant displayed higher nitrilase activity toward various substrates; in particular, it could even hydrolyze nitriles that were hardly accepted by the wild-type enzyme. However, on the whole, there have been relatively fewer reports of the directed evolution approach on nitrilase modification to date. This might be due to fewer simple and highly sensitive studies of high-throughput screening methods in the literature.

Other Strategies of Nitrilase Reengineering

A newly isolated nitrilase-producing strain *A. faecalis* WBX11 was modified by low-energy ion beam implantation techniques to improve the nitrilase activity [38]. This method is a novel microbial breeding approach contributing to multiple areas including gene transfer and cell modification, and it exhibits higher mutation frequency and wider mutation spectrum than conventional mutagenic sources such as X-ray, UV ray, and microwave. The screening of *A. faecalis* nitrilase mutants was carried out using a selective minimal salt medium supplemented with a pH indicator, and colonies with obvious color change were selected for further study. The nitrilase activity and biomass of a positive nitrilase mutant were found to be improved by 45% and 75%, respectively, while no obvious changes occurred for its optimal reaction conditions.

Li *et al.* [39] recently reported fusing the *A. faecalis* nitrilase expressed in *E. coli* with an amphipathic peptides. Their studies found that ~90% of nitrilase exists in the form of active inclusion bodies (aggregates) in the host cells. The thermal stability was significantly improved by 6.8- and 4.3-fold at 45 °C and 50 °C, respectively. The nitrilase aggregates were purified and immobilized with sodium alginate. The stability was further increased by 1.5-fold. Immobilized enzymes demonstrated an improvement of substrate tolerance from 30 mM to 200 mM with mandelonitrile as the substrate.

26.4 APPLICATIONS IN ORGANIC SYNTHESIS

26.4.1 Production of Glycolic Acid

Glycolic acid is a principal intermediate for preparing household and industrial cleaners, personal care products, automotive oil additives, and polyglycolic acid (applied for dissolvable sutures, drug delivery materials, and gas barrier packaging materials) [17, 40]. Enzymatic synthesis of glycolic acid from glycolonitrile (Scheme 26.1) under mild conditions has increasingly become an alternative process to the corresponding chemical method, which needs high pressure and high temperature under acidic conditions. Panova *et al.* obtained a recombinant *A. facilis* 72 W nitrilase in *E. coli* with high activity toward glycolonitrile, and its specific activity was improved by 33-fold via protein engineering and protein expression optimization [40]. Glycolonitrile solution of >99% yield and purity was firstly generated through the reaction of formaldehyde and hydrogen cyanide, and this substrate, without further purification, was directly used in glycolic acid production with glutaraldehyde/polyethylenimine cross-linked carrageenan-immobilized cells as the catalyst. The biocatalyst productivity of >1000 g acid per gram cell was observed, and 3.2 M ammonium glycolate could be prepared in consecutive batch reactions or in a continuous stirred-tank reactor.

26.4.2 Production of Iminodiacetic Acid

Iminodiacetic acid has important applications in the chemical industry for the manufacture of chelating agents, herbicide, and surfactants, especially for the production of herbicide glyphosate, for which there is a huge demand in the field of agriculture. Bioconversion of iminodiacetonitrile using nitrilase is an attractive method for

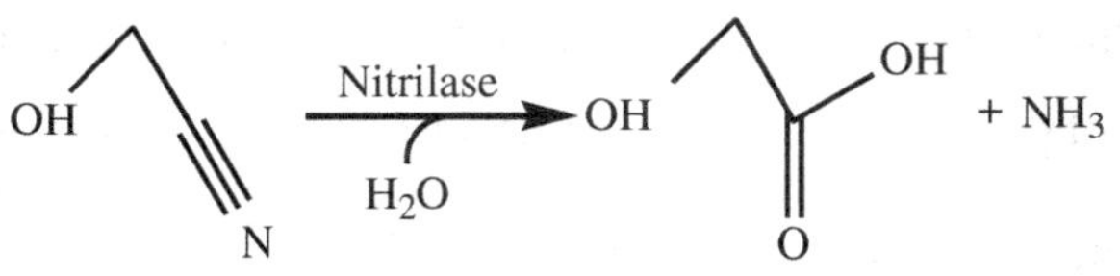

SCHEME 26.1
Enzymatic synthesis of glycolic acid from glycolonitrile.

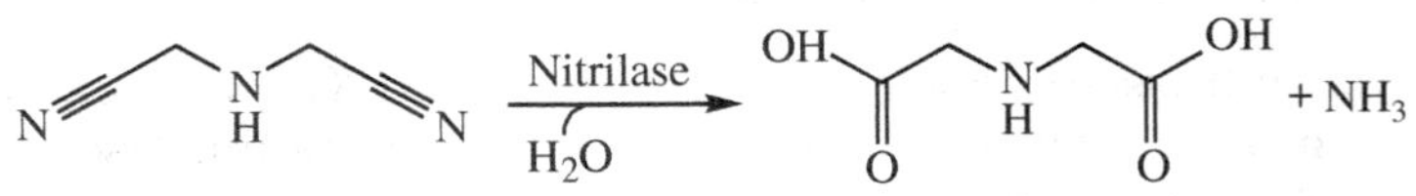

SCHEME 26.2
Enzymatic synthesis of iminodiacetic acid from iminodiacetonitrile.

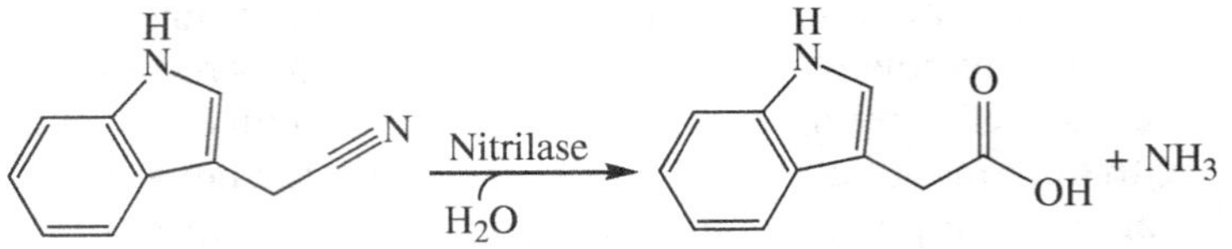

SCHEME 26.3
Enzymatic synthesis of indole-3-acetic acid from indole-3-acetonitrile.

iminodiacetic acid production (Scheme 26.2). Zhang *et al.* [41] used a sensitive and specific high-throughput screening model with iminodiacetonitrile as the sole nitrogen source and bromocresol purple as the indicator to isolate iminodiacetonitrile-degrading strains. A strain of *A. faecalis* ZJUTBX11 was proven potential in nitrile hydrolyzing after culture optimization. Iminodiacetonitrile (158 mM) could be converted to iminodiacetic acid (125 mM) with more than 95% of conversion yield through 500 min. Subsequently, this strain was subjected to low-energy ion beam implantation, as demonstrated previously, to improve its catalytic properties [38]. Finally, substrate inhibition of the modified nitrilase was found to be substantially alleviated, and no obvious inhibition can be observed even with 250 mM substrate for conversion. Moreover, the *A. faecalis* cells were immobilized using the encapsulation method in alginate–chitosan–alginate membrane liquid-core capsules to reduce the mass transfer resistance and increase its substrate tolerance [42]. The immobilized cells could be recycled for ten times and still retain 90% of the nitrilase activity, whereas the corresponding free cells only retained 35% of the activity. Immobilized recombinant *A. facilis* nitrilase entrapped in polyvinyl alcohol and sodium alginate copolymer was also used by Liu *et al.* for iminodiacetic acid production [43]. About 35% of activity remained after ten reuse cycles of iminodiacetonitrile biotransformation, while free cells completely inactivated in the ninth reuse cycle.

26.4.3 Production of Indole-3-Acetic Acid

Indole-3-acetic acid is regarded as the most abundant natural auxin and was conventionally isolated from plants. It can be used for regulating the plant growth and development processes as it promotes cell enlargement and phototropism and induces flowering [44]. In *Cruciferae*, indole-3-acetic acid is synthesized by nitrilase from indole-3-acetonitrile, which is formed from tryptophan through indole-3-acetaldoxime or indole glycosinolates as the intermediate [45]. The nitrilases from *Zea mays*, *A. thaliana*, and *A. faecalis* JM3 were proven efficient in indole-3-acetonitrile hydrolysis for biosynthesis of indole-3-acetic acid (Scheme 26.3), and their encoding genes were successfully cloned and expressed [45–47]. Agarwal *et al.* [44] used a thermostable nitrilase-producing strain *Streptomyces* sp. MTCC 7546 as the catalyst for producing indole-3-acetic acid from indole-3-acetonitrile. Its optimal conditions for the maximum conversion were 60 °C and pH 7.5, and the kinetic parameters K_m and V_{max} were calculated to be 1.3 mM and 28 mM/min, respectively. The determination results of reaction mixture by HPLC illustrated about 40% conversion of indole-3-acetonitrile into corresponding acid.

26.4.4 Conversion of Phenylacetonitrile and its Derivates

Biotransformation of phenylacetonitrile and its derivates using nitrilases has become an attractive approach due to its selectivity and mild reaction conditions in recent years. Oliveira *et al.* obtained several strains belonging to *Aspergillus*, *Penicillium*, *Cladosporium*,

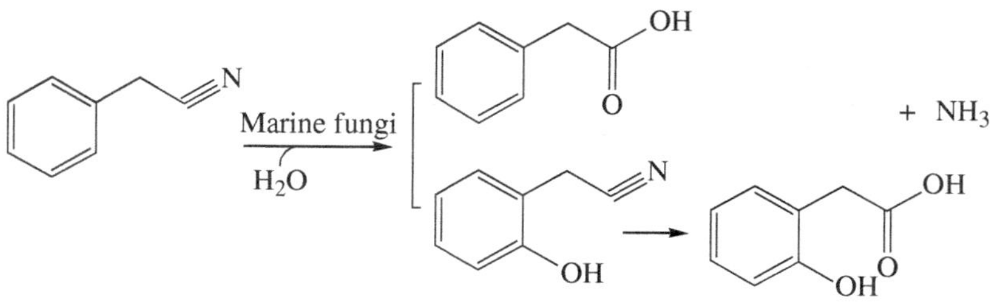

SCHEME 26.4
Enzymatic conversion of phenylacetonitrile into hydroxyphenylacetic acid.

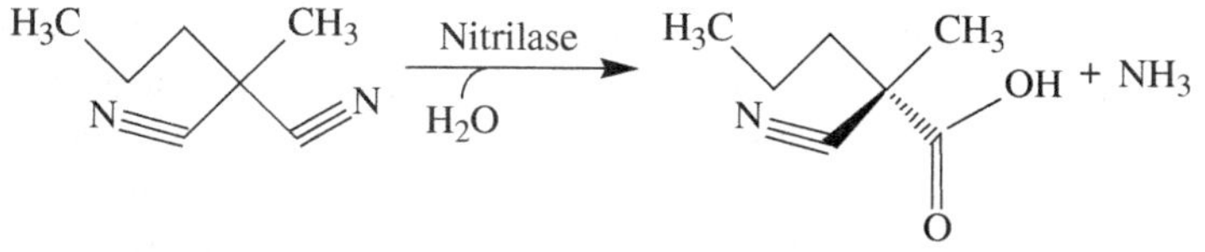

SCHEME 26.5
Enzymatic hydrolysis of 2-methyl-2-propylmalononitrile into CMPA.

and *Bionectria* genera from marine environment with the addition of phenylacetonitrile in mineral medium, which can biotransform the phenylacetonitrile into hydroxyphenylacetic acid with 51% yield (Scheme 26.4) [48]. It is speculated that the phenylacetonitrile was firstly hydrolyzed to carboxylic acid, and subsequently the ortho position of aromatic ring was hydroxylated. Furthermore, in another study, the 4-fluorophenylacetonitrile was only transformed into 4-fluorophenylacetic acid by a strain of *A. sydowii* Ce19 with yield of 51%, while no hydroxylated product was detected in the reaction mixture. The selective hydroxylation is frequently observed in the literature. Rustler *et al.* successfully isolated a black yeast of *Exophiala oligosperma* from nature with phenylacetonitrile as sole nitrogen source under acidic conditions of pH4.0 [49]. Conversion of phenylacetonitrile by resting cells showed that both phenylacetic acid and 2-hydroxyphenylacetic acid were generated as the product. The marine filamentous fungus *A. sydowii* CBMAI 934 was adopted for biotransformation of methylphenylacetonitriles [9]. Experimental results showed that the phenylacetonitrile, 2-methylphenylacetonitrile, 3-methylphenylacetonitrile, and 4-methylphenylacetonitrile can be directly converted into 2-hydroxyphenylacetic acid, 2-methylphenylacetic acid, 3-methylphenylacetic acid, and 4-methylphenylacetic acid by using *A. sydowii* cells as the biocatalyst.

26.4.5 Regioselective Hydrolysis of Dinitriles

Regioselectivity is a particularly important characteristic of nitrilase. To date, there have been numerous studies focused on the bioconversion of dinitriles by using whole cells or purified enzymes of nitrilase as the catalyst. Yoshida *et al.* [5] expressed *R. rhodochrous* J1 nitrilase gene in *E. coli*, and the recombinant nitrilase was used for enantioselectively converting 2-methyl-2-propylmalononitrile into (*S*)-2-cyano-2-methylpentanoic acid (CMPA) (Scheme 26.5). Finally, 563mM CMPA (80g/l) was generated in the reaction system with a molar conversion yield of 97% and ee value of 97% within 24h under the conditions of pH8.0 and 30°C. A recombinant nitrilase obtained from metagenomic libraries was utilized in regioselective conversion of the dinitrile 2-methylglutaronitrile [11]. The enzyme displayed high activities toward aliphatic dinitriles including succinonitrile, adiponitrile, glutaronitrile, and 2-methylglutaronitrile. In particular, the regioselectivity of nitrilase toward 2-methylglutaronitrile was determined by HPLC, methylation, and GC–MS analysis. Both the cyanocarboxylic acid and dicarboxylic acid were detected in the final products. In addition, the recombinant *P. abyssi* nitrilase was proven highly specific toward aliphatic dinitriles [50]. Fumaronitrile and malononitrile were the best substrates with K_m of 9.48mM and 3.47mM, respectively; both were regioselectively hydrolyzed into monoacid mononitrile by *P. abyssi* nitrilase.

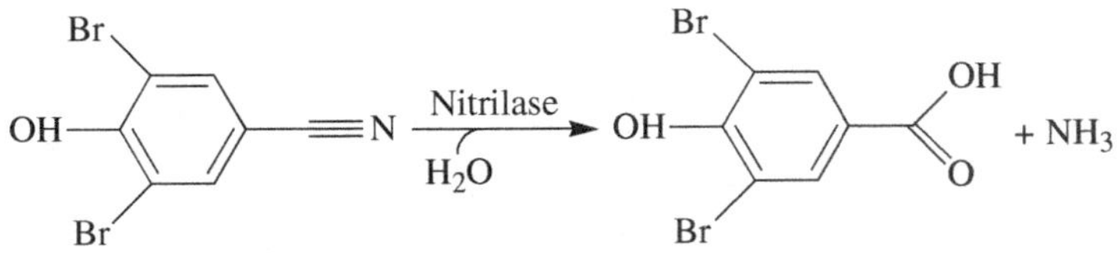

SCHEME 26.6
Enzymatic conversion of bromoxynil into 3,5-dibromo-4-hydroxybenzoic acid.

26.4.6 Degradation of Benzonitrile Herbicides

Dihalogenated benzonitrile analogues are usually used as the herbicides like bromoxynil, ioxynil, and dichlobenil for weed control [51]. Thousands of tons of these herbicides were applied for cereal crops per year. However, these compounds are toxic and harmful to soils and groundwater due to the presence of a cyano substituent (nitrile group) on the benzene ring [52]. The utilization of nitrilases for degrading the toxic benzonitrile herbicides into less toxic carboxylic acids is regarded as a sustainable approach and has attracted increasing attention in the literature. A *Klebsiella ozaenae* nitrilase, which can efficiently convert bromoxynil into 3,5-dibromo-4-hydroxybenzoic acid (Scheme 26.6), was cloned by Stalker *et al.* and heterologously expressed in *E. coli* [53]. The recombinant nitrilase was specific toward bromoxynil, and reaction kinetic parameters K_m and V_{max} were determined to be 0.31 mM and of 15 μmol of NH_3 released/minute/milligram protein, respectively. Detzel *et al.* constructed a recombinant *E. coli* strain harboring *Klebsiella pneumoniae* subsp. *ozaenae* nitrilase on the cell surface using the autodisplay system [54], which is a surface display system that allows the display of heterologous enzymes on the cell surface of *E. coli* with about 10^5 molecules per cell and can be used for the cost-efficient construction of whole-cell biocatalysts applied in the field of bioremediation. Localization of the nitrilase in the *E. coli* cell envelope was monitored by SDS-PAGE. The optimum reaction conditions turned out to have pH 8.0 and temperature of 45 °C. Conversion results indicated that the whole-cell biocatalyst was specific toward the three benzonitrile herbicides and exhibited catalysis preference in the order of chloroxynil > bromoxynil > ioxynil with product generation of 1.67, 0.89, and 0.13 mM within 72 h, respectively.

26.5 CONCLUSIONS AND FUTURE PROSPECTS

Over the past years, the possibility of applying nitrilases in organic synthesis has been increasingly recognized. Nitrilases have been found to be widely present in the natural environment. In particular, uncharacterized genetic data in the GenBank and unculturable microorganisms in the natural environment were precious resources owning various nitrilases, and some novel techniques including metagenomic technology and genome mining provided possibilities for acquiring these enzymes. From the limited literature, the two techniques have been proven to exhibit certain potential in exploration of novel and powerful nitrilases [27]. The academic progress in increasing nitrilase production by culture optimization and improving catalytic properties by enzyme reengineering has stimulated the utilization of nitrilases in the chemical and agricultural industry. Several nitrilases have been applied effectively in organic synthesis for the production of commodity chemicals and pharmaceutical intermediates. With the development of the third wave of biocatalysis, molecular biology methods as well as protein engineering provided numerous opportunities for nitrilase improvement [12]. The requirement for these improved catalysts in industry promoted nitrilases to be developed as an increasingly important tool in organic synthesis.

ACKNOWLEDGMENTS

This work was supported by the National High Technology Research and Development Program of China (No. 2012AA022204C), the Natural Science Foundation of Jiangsu Province (No. BK20140133 and BK2012127), and the National Natural Science Foundation of China (No. 21206055).

REFERENCES

[1] Oishi T, Yamaguchi K, Mizuno N. Catalytic oxidative synthesis of nitriles directly from primary alcohols and ammonia [J]. *Angew Chem Int Ed* 2009, **121** (34): 6404–6406.

[2] Debabov V, Yanenko A. Biocatalytic hydrolysis of nitriles [J]. *Rev J Chem* 2011, **1** (4): 385–402.

[3] Velankar H, Clarke KG, du Preez D, Cowan DA, Burton SG. Developments in nitrile and amide biotransformation processes [J]. *Trends Biotechnol* 2010, **28** (11): 561–569.

[4] Osswald S, Wajant H, Effenberger F. Characterization and synthetic applications of recombinant AtNIT1 from *Arabidopsis thaliana* [J]. *Eur J Biochem* 2002, **269** (2): 680–687.

[5] Yoshida T, Mitsukura K, Mizutani T, Nakashima R, Shimizu Y, Kawabata H, Nagasawa T. Enantioselective synthesis of (*S*)-2-cyano-2-methylpentanoic acid by nitrilase [J]. *Biotechnol Lett* 2013, **35** (5): 685–688.

[6] Zhu X-Y, Gong J-S, Li H, Lu Z-M, Shi J-S, Xu Z-H. Bench-scale biosynthesis of isonicotinic acid from 4-cyanopyridine by *Pseudomonas putida* [J]. *Chem Pap* 2014, **68** (6): 739–744.

[7] Xue Y-P, Xu S-Z, Liu Z-Q, Zheng Y-G, Shen Y-C. Enantioselective biocatalytic hydrolysis of (*R,S*)-mandelonitrile for production of (*R*)-(−)-mandelic acid by a newly isolated mutant strain [J]. *J Ind Microbiol Biotechnol* 2011, **38** (2): 337–345.

[8] Kamal A, Kumar MS, Kumar CG, Shaik TB. Bioconversion of acrylonitrile to acrylic acid by *Rhodococcus ruber* strain AKSH-84 [J]. *J Microbiol Biotechnol* 2011, **21** (1): 37–42.

[9] Oliveira J, Seleghim M, Porto A. Biotransformation of methylphenylacetonitriles by Brazilian marine fungal strain *Aspergillus sydowii* CBMAI 934: eco-friendly reactions [J]. *Mar Biotechnol* 2014, **16** (2): 156–160.

[10] Lorenz P, Eck J. Metagenomics and industrial applications [J]. *Nat Rev Microbiol* 2005, **3** (6): 510–516.

[11] Bayer S, Birkemeyer C, Ballschmiter M. A nitrilase from a metagenomic library acts regioselectively on aliphatic dinitriles [J]. *Appl Microbiol Biotechnol* 2011, **89** (1): 91–98.

[12] Bornscheuer UT, Huisman GW, Kazlauskas RJ, Lutz S, Moore JC, Robins K. Engineering the third wave of biocatalysis [J]. *Nature* 2012, **485** (7397): 185–194.

[13] Zhu D, Mukherjee C, Biehl ER, Hua L. Discovery of a mandelonitrile hydrolase from *Bradyrhizobium japonicum* USDA110 by rational genome mining [J]. *J Biotechnol* 2007, **129**: 645–650.

[14] Zhu D, Mukherjee C, Yang Y, Rios BE, Gallagher DT, Smith NN, Biehl ER, Hua L. A new nitrilase from *Bradyrhizobium japonicum* USDA 110: gene cloning, biochemical characterization and substrate specificity [J]. *J Biotechnol* 2008, **133** (3): 327–333.

[15] Seffernick JL, Samanta SK, Louie TM, Wackett LP, Subramanian M. Investigative mining of sequence data for novel enzymes: a case study with nitrilases [J]. *J Biotechnol* 2009, **143** (1): 17–26.

[16] Nageshwar YVD, Sheelu G, Shambhu RR, Muluka H, Mehdi N, Malik MS, Kamal A. Optimization of nitrilase production from *Alcaligenes faecalis* MTCC 10757 (IICT-A3): effect of inducers on substrate specificity [J]. *Bioprocess Biosyst Eng* 2011, **34** (5): 515–523.

[17] He Y-C, Xu J-H, Su J-H, Zhou L. Bioproduction of glycolic acid from glycolonitrile with a new bacterial isolate of *Alcaligenes* sp. ECU0401 [J]. *Appl Biochem Biotechnol* 2009, **160** (5): 1428–1440.

[18] Zhu X-Y, Gong J-S, Li H, Lu Z-M, Zhou Z-M, Shi J-S, Xu Z-H. Screening, identification and culture optimization of a newly isolated aromatic nitrilase-producing bacterium—*Pseudomonas putida* CGMCC3830 [J]. *Chin J Biotechnol* 2014, **30** (3): 412–424.

[19] Yusuf F, Chaubey A, Raina A, Jamwal U, Parshad R. Enhancing nitrilase production from *Fusarium proliferatum* using response surface methodology [J]. *Springerplus* 2013, **2** (1): 1–7.

[20] Gong J-S, Lu Z-M, Shi J-S, Dou W-F, Xu H-Y, Zhou Z-M, Xu Z-H. Isolation, identification, and culture optimization of a novel glycinonitrile—hydrolyzing fungus—*Fusarium oxysporum* H3 [J]. *Appl Biochem Biotechnol* 2011, **165** (3–4): 963–977.

[21] Liu J-F, Zhang Z-J, Li A-T, Pan J, Xu J-H. Significantly enhanced production of recombinant nitrilase by optimization of culture conditions and glycerol feeding [J]. *Appl Microbiol Biotechnol* 2011, **89** (3): 665–672.

[22] Shen M, Liu Z-Q, Zheng Y-G, Shen Y-C. Enhancing endo-nitrilase production by a newly isolated *Arthrobacter nitroguajacolicus* ZJUTB06-99 through optimization of culture medium [J]. *Biotechnol Bioprocess Eng* 2009, **14** (6): 795–802.

[23] Naik SC, Kaul P, Barse B, Banerjee A, Banerjee UC. Studies on the production of enantioselective nitrilase in a stirred tank bioreactor by *Pseudomonas putida* MTCC 5110 [J]. *Bioresour Technol* 2008, **99** (1): 26–31.

[24] Zheng YG, Chen J, Liu ZQ, Wu MH, Xing LY, Shen YC. Isolation, identification and characterization of *Bacillus subtilis* ZJB-063, a versatile nitrile-converting bacterium [J]. *Appl Microbiol Biotechnol* 2008, **77**: 985–993.

[25] Kaplan O, Vejvoda V, Charvátová-Pišvejcová A, Martínková L. Hyperinduction of nitrilases in filamentous fungi [J]. *J Ind Microbiol Biotechnol* 2006, **33** (11): 891–896.

[26] Khandelwal AK, Nigam VK, Choudhury B, Mohan MK, Ghosh P. Optimization of nitrilase production from a new thermophilic isolate [J]. *J Chem Technol Biotechnol* 2007, **82** (7): 646–651.

[27] Gong J-S, Lu Z-M, Li H, Zhou Z-M, Shi J-S, Xu Z-H. Metagenomic technology and genome mining: emerging areas for exploring novel nitrilases [J]. *Appl Microbiol Biotechnol* 2013, **97** (15): 6603–6611.

[28] Kiziak C, Conradt D, Stolz A, Mattes R, Klein J. Nitrilase from *Pseudomonas fluorescens* EBC191: cloning and heterologous expression of the gene and biochemical characterization of the recombinant enzyme [J]. *Microbiology* 2005, **151**: 3639–3648.

[29] Kobayashi M, Komeda H, Yanaka N, Nagasawa T, Yamada H. Nitrilase from *Rhodococcus rhodochrous* J1 [J]. *J Biol Chem* 1992, **267** (29): 20746–20751.

[30] Kiziak C, Klein J, Stolz A. Influence of different carboxy-terminal mutations on the substrate-, reaction- and enantiospecificity of the arylacetonitrilase from *Pseudomonas fluorescens* EBC191 [J]. *Protein Eng Des Sel* 2007, **20** (8): 385–396.

[31] Kiziak C, Stolz A. Identification of amino acid residues responsible for the enantioselectivity and amide formation capacity of the arylacetonitrilase from *Pseudomonas fluorescens* EBC191 [J]. *Appl Environ Microbiol* 2009, **75** (17): 5592–5599.

[32] Petříčková A, Sosedov O, Baum S, Stolz A, Martínková L. Influence of point mutations near the active site on the catalytic properties of fungal arylacetonitrilases from *Aspergillus niger* and *Neurospora crassa* [J]. *J Mol Catal B: Enzym* 2012, **77**: 74–80.

[33] Raczynska JE, Vorgias CE, Antranikian G, Rypniewski W. Crystallographic analysis of a thermoactive nitrilase [J]. *J Struct Biol* 2011, **173** (2): 294–302.

[34] Liu Z-Q, Baker PJ, Cheng F, Xue Y-P, Zheng Y-G, Shen Y-C. Screening and improving the recombinant nitrilases and application in biotransformation of iminodiacetonitrile to iminodiacetic acid [J]. *PLoS One* 2013, **8** (6): e67197.

[35] Schreiner U, Hecher B, Obrowsky S, Waich K, Klempier N, Steinkellner G, Gruber K, Rozzell JD, Glieder A, Winkler M. Directed evolution of *Alcaligenes faecalis* nitrilase [J]. *Enzyme Microb Technol* 2010, **47** (4): 140–146.

[36] Sosedov O, Stolz A. Random mutagenesis of the arylacetonitrilase from *Pseudomonas fluorescens* EBC191 and identification of variants, which form increased amounts of mandeloamide from mandelonitrile [J]. *Appl Microbiol Biotechnol* 2014, **98** (4): 1595–1607.

[37] Schreiner U, Steinkellner G, Rozzell JD, Glieder A, Winkler M. Improved fitness of *Arabidopsis thaliana* nitrilase 2 [J]. *ChemCatChem* 2010, **2** (3): 263–267.

[38] Zhang J-F, Liu Z-Q, Zheng Y-G. Improvement of nitrilase production from a newly isolated *Alcaligenes faecalis* mutant for biotransformation of iminodiacetonitrile to iminodiacetic acid [J]. *J Taiwan Inst Chem Eng* 2012, **44** (2): 169–176.

[39] Li S, Yang X. Catalytic properties improvements of *Alcaligenes faecalis* nitrilase by self-assembly induced aggregation [J]. 2014, 31 (Supplement): S44.

[40] Panova A, Mersinger LJ, Liu Q, Foo T, Roe DC, Spillan WL, Sigmund AE, Ben-Bassat A, Wagner LW, O'Keefe DP, Wu S, Petrillo KL, Payne MS, Breske ST, Gallagher FG, DiCosimo R. Chemoenzymatic synthesis of glycolic acid [J]. *Adv Synth Catal* 2007, **349** (8–9): 1462–1474.

[41] Zhang J-F, Liu Z-Q, Zheng Y-G, Shen Y-C. Screening and characterization of microorganisms capable of converting iminodiacetonitrile to iminodiacetic acid [J]. *Eng Life Sci* 2012, **12** (1): 69–78.

[42] Zhang J-F, Liu Z-Q, Zhang X-H, Zheng Y-G. Biotransformation of iminodiacetonitrile to iminodiacetic acid by *Alcaligenes faecalis* cells immobilized in ACA-membrane liquid-core capsules [J]. *Chem Pap* 2014, **68** (1): 53–64.

[43] Liu ZQ, Zhou M, Zhang XH, Xu JM, Xue YP, Zheng YG. Biosynthesis of iminodiacetic acid from iminodiacetonitrile by immobilized recombinant *Escherichia coli* harboring nitrilase [J]. *J Mol Microbiol Biotechnol* 2012, **22** (1): 35–47.

[44] Agarwal A, Nigam VK. Nitrilase mediated conversion of indole-3-acetonitrile to indole-3-acetic acid [J]. *Biocatal Agric Biotechnol* 2014, **3** (4): 351–357.

[45] Kobayashi M, Izui H, Nagasawa T, Yamada H. Nitrilase in biosynthesis of the plant hormone indole-3-acetic acid from indole-3-acetonitrile: cloning of the *Alcaligenes* gene and site-directed mutagenesis of cysteine residues [J]. *Proc Natl Acad Sci U S A* 1993, **90** (1): 247–251.

[46] Park WJ, Kriechbaumer V, Müller A, Piotrowski M, Meeley RB, Gierl A, Glawischnig E. The nitrilase ZmNIT2 converts indole-3-acetonitrile to indole-3-acetic acid [J]. *Plant Physiol* 2003, **133** (2): 794–802.

[47] Bartling D, Seedorf M, Mithöfer A, Weiler EW. Cloning and expression of an *Arabidopsis* nitrilase which can convert indole-3-acetonitrile to the plant hormone, indole-3-acetic acid [J]. *Eur J Biochem* 1992, **205** (1): 417–424.

[48] Oliveira J, Mizuno C, Seleghim M, Javaroti D, Rezende M, Landgraf M, Sette L, Porto A. Biotransformation of phenylacetonitrile to 2-hydroxyphenylacetic acid by marine fungi [J]. *Mar Biotechnol* 2013, **15** (1): 97–103.

[49] Rustler S, Stolz A. Isolation and characterization of a nitrile hydrolysing acidotolerant black yeast—*Exophiala oligosperma* R1 [J]. *Appl Microbiol Biotechnol* 2007, **75**: 899–908.

[50] Mueller P, Egorova K, Vorgias CE, Boutou E, Trauthwein H, Verseck S, Antranikian G. Cloning, overexpression, and characterization of a thermoactive nitrilase from the hyperthermophilic archaeon *Pyrococcus abyssi* [J]. *Protein Expr Purif* 2006, **47** (2): 672–681.

[51] Veselá A, Franc M, Pelantová H, Kubáč D, Vejvoda V, Šulc M, Bhalla T, Macková M, Lovecká P, Janů P, Demnerová K, Martínková L. Hydrolysis of benzonitrile herbicides by soil actinobacteria and metabolite toxicity [J]. *Biodegradation* 2010, **21** (5): 761–770.

[52] Holtze MS, Sørensen SR, Sørensen J, Aamand J. Microbial degradation of the benzonitrile herbicides dichlobenil, bromoxynil and ioxynil in soil and subsurface environments—insights into degradation pathways, persistent metabolites and involved degrader organisms [J]. *Environ Pollut* 2008, **154** (2): 155–168.

[53] Stalker DM, Malyj LD, McBride KE. Purification and properties of a nitrilase specific for the herbicide bromoxynil and corresponding nucleotide sequence analysis of the bxn gene [J]. *J Biol Chem* 1988, **263** (13): 6310–6314.

[54] Detzel C, Maas R, Tubeleviciute A, Jose J. Autodisplay of nitrilase from *Klebsiella pneumoniae* and whole-cell degradation of oxynil herbicides and related compounds [J]. *Appl Microbiol Biotechnol* 2013, **97** (11): 4887–4896.

CHAPTER 27

Biotechnology for the Production of Chemicals, Intermediates, and Pharmaceutical Ingredients

Hans-Peter Meyer

Institute of Life Technologies, University of Applied Sciences and Arts Western Switzerland, Sion, Switzerland

27.1 INTRODUCTION

What a challenge: to "feed, heal and fuel the world" with a biobased economy! This vision, to replace our oil-based economy, converting biomass into fuels, energy, and chemicals [1] requires nothing less than reinventing global value chains. The scale and challenge of this endeavor dwarfs the project of landing the first human being on the moon launched by President John F. Kennedy in 1961. The "space race" was characterized by national pride and secrecy and fired by the Cold War mentality. After Cold War we face global warming, but this time we need to work together on a global scale for true and effective results. Biotechnology has always been propagated as one solution for several of our sustainability problems. As one consequence, large programs of first-, second-, and third-generation biofuels for mass transportation have been established in the United States, Europe, and Asia, and some of these programs have been running for well over 50 years. Huge amounts of tax money were used to develop and then to subsidize these products to keep them on the market. Considering the three axes of sustainability (society, ecology, and economics), biofuels are probably not a good model or example with which to demonstrate and prove the sustainability of biotechnological processes and products. This chapter suggests other areas, in which biotechnology should provide new manufacturing approaches. One such area is the application of biotechnology for the manufacture of chemicals, intermediates, and pharmaceutical ingredients. However, these are, technically speaking, high-hanging fruits in some ways. If we consider that the foundation for biotransformations of fine and specialty chemicals was laid some 80 years ago with the development of vitamin C or steroids, for example, the currently available commercial enzyme toolbox is respectable but still a far cry from where it should and could be. Times to market are still long, and biotechnology typically delivers second-generation processes, if any. In 1991, Richard L. Hinman of Pfizer wrote that biotechnology will remain a niche technology [2]. "Why, after so much promise, and in the face of the prodigious flow of new products for medicinal and agricultural markets, has the

Green Biocatalysis, First Edition. Edited by Ramesh N. Patel.

TABLE 27.1 Selected Historic Event with Respect to Biotechnology Used in the Production of Chemicals

1893 Commercial lactic acid fermentation by Boehringer Ingelheim in Europe
1916 Acetone, butanol, and ethanol large-scale fermentation in England
1923 Pfizer opens a fermentation production plant for citric acid used mainly for soft drinks
1934 Large-scale vitamin C production via biooxidation (Reichstein synthesis) by Hoffmann-La Roche
1940 Discovery of bacterial dextran and start of production development, the first of many polysaccharides
1943 Large-scale penicillin fermentation developed in the United States
1950s Bioconversion of steroid hormones at Upjohn and Schering
1956 Fermentation process for L-glutamic acid developed
1967 Industrial use of glucose isomerase for high fructose corn syrup production
1970 Hydrolysis of penicillin to 6-aminopenicillanic acid
1980s Start of large-scale biopolymer (e.g., polyhydroxyalkanoates) fermentation mainly in Europe, the United States, and China
1983 Recombinant production of indigo by Genencor
1985 Production of the red color Shikonin from plant cell suspension culture for lipstick in Japan
1985 Enzymatic process for the production of acrylamide
1989 Large-scale biotransformation to L-carnitine by Lonza
1990 Semisynthesis of Taxol from baccatin
1991 Phytase (a phosphohydrolase) produced with recombinant yeasts launched in markets as a feed additive
1993 Paclitaxel (cytotoxic) by plant cell culture in 75 m^3 bioreactors by Phyton in Germany
1995 Industrial biotransformation process to nicotinamide by Lonza
2001 Enzymatic wax removal from cotton by Novozymes
2002 Lactic acid for polymers by Nature Works
2003 1,3-Propanediol for polymers by fermentation by DuPont
2005 Low trans fats for nutrition by ADM and Novozymes
2006 Presidential Green Chemistry award for the biocatalysis of atorvastatin of Pfizer
2008 Cooperation (DSM, Roquette) on a biosuccinic demonstration fermentation plant, which opened in 2010
2010 Biocatalysis to sitagliptin (diabetes) by Merck and Codexis
2011 1,4-Butanediol by fermentation process by Genomatica
2012 Commercial artemisinic acid (precursor of artemisinin) using a microbial strain after massive pathway engineering

harvest (of biotechnology) in the chemical area been so thin?" "The overall impact on the chemical process industry will be slight. Organic chemistry will continue as its mainstream." Should we care at all for biotechnology as a tool for chemicals, intermediates, and pharmaceutical ingredients? Table 27.1 lists a few selected highlights, historic events, and achievements in biotechnology that were particularly important for the organic chemical industry.

In hindsight, one can understand Richard L. Hinman's statement. However, considering the frequency of observed implementations of new biotechnological manufacturing steps in organic chemical synthesis and the confirmed commitment of large global chemical players to actually using biotechnology as an established manufacturing tool today, it is evident that his fundamental conclusions were not correct. Notwithstanding, it cannot be denied that progress should have been more rapid. This chapter will address some of the bottlenecks and give a thumbnail sketch of where and how problems should be tackled.

The author would start with as bold statement: by merging the best of the two technologies, organic chemical synthesis and biotechnology, chemical manufacturing will and must be revolutionized in the next 10–20 years. The main challenge for mankind today stems from the fact that a rapidly growing world population aspires to live a long, healthy, and comfortable life. However, shrinking resources, growing waste problems, and climate extremes are factors that limit these aspirations and create a lingering danger of social unrest. The world population rose 22-fold over the last millennium, per capita income 13-fold, and world gross domestic product (GDP) nearly 300-fold [3]. For the moment, politicians continue to rely on old-fashioned growth models of consumption to keep GDP growing and unemployment low. However, in order to avoid a collapse, we need different value chains, new technologies, and new products. These new technologies must encompass the whole value chain, with radically new manufacturing methods for affordable and sustainable products to assure economic growth and personal prosperity for 8–9 billion people. But one step at the time and let's start from the beginning of the value chain.

27.2 VALUE CHAINS AND MARKETS

Chemistry has always used two sources for its raw materials: petrochemical and plant-based raw and starting materials. Especially for life science products, chemistry uses refined plant-based raw materials to produce mainly fine chemicals but also commodities (steroids, antibiotics, vitamins, solvents) produced by biocatalysis or fermentation. Furthermore, 70% of current small molecule therapeutics derives their structure from plant origins, but the bulk of raw materials are assured by petrochemicals. Coal, which was at the start of the manufacturing value chain for organic chemical synthesis until 1950, has been replaced by oil and gas for the production of basic chemicals such as propylene, ethylene, carbon monoxide, and hydrogen. By the mid-2030s, about one-third of oil and gas should now be replaced by biomaterials to be transformed in biorefineries. As described in Section 27.2.2, biobased products include a vast variety of products for many different applications and markets including pharmaceutical, personal care, and energy. A fully biobased economy, with a changing paradigm from oil to biomass, first needs a strong agricultural basis and a different refining approach. Biomass is overfunctionalized and overoxidized when compared to the low functionality of hydrocarbons and aromatics from crude oil, and both require different conversion methods. Functionality is introduced in oil-based chemistry but removed or used selectively when working with renewable biomass. Cofeeding crude oil and biomass-derived molecules into existing petroleum refineries would be the most cost-efficient and reasonable approach in the near future but difficult unless the oxygen content of the biomass is reduced in advance.

Another problem is that the supply of agricultural commodities is not predictable. Between 2005 and the summer of 2008, the price of wheat and corn tripled, and the price of rice climbed fivefold. Tensions because of increasing food prices have also triggered social unrest in Egypt, Bangladesh, and Haiti. The FAO World Food Index [4] calculated that the cost of important agro and food raw materials reached all-time records in 2008, 2009, and 2010, especially in developing countries. Due to increased global demand and low food stocks, weather-induced reductions in harvest immediately impacted prices leading to the highest spikes ever in these three recent years. It is evident that biomass production for biorefining purposes will also have an impact on food production unless massive wood clearing takes place. Considering the amounts of biomass needed and the sensitivity of markets to price and supply fluctuations, the goal of feeding, healing, and fueling the world is not realistic considering actual technologies. Even if we were to use a destructive "Soviet"-style, monoculture agriculture and regulate every hectare under cultivation, it would be impossible for the actual growing global population, due to lack of land, water, and sooner or later phosphorus resources. What we will see is more likely a hybrid global economy: renewable resources, precision farming, and genetically modified plants (GMO) must assure a steady quantitative and qualitative supply of agro-based raw materials complementing fossil resources. As in the petrochemical industry, the biomass industry must learn to squeeze out every drop from its raw biomass materials by using all parts of the plant. Whatever we produce, agriculture, forestry, and aquaculture will

TABLE 27.2 Comparison of the Approximate Share of Costs of the Steps during the Production of Higher-Value Chemicals and Commodities by Fermentation

	Higher-Value Chemical (%)	Commodity (%)
Raw materials	10	70
Other variable costs	10	15
Personnel, maintenance, and depreciation	80	15
Proportion of costs USP (vs. DSP)	<25	>75

Also indicated are the average proportions of costs of upstream processing (USP), which includes fermentation and cell separation, and downstream processing (DSP) for the isolation and purification of the product. Pharmaceutical applications are driven foremost by the cost of development, regulatory approval, and compliance. In contrast, the primary market driver for commodities is raw material costs and productivity.

be the basis of a bioeconomy value chain and the first contributor. For biotechnological applications, these raw materials have to be refined to a certain degree. Glucose, glucose syrup, starch, corn steep liquor, cotton seed, linseed, soy bean meal, and lately lignocellulose are examples of partially refined agricultural commodities used in fermentation. Refined agroproducts are also used for the production of the enzymes that convert a petrochemical intermediate by biocatalysis or biotransformation. Companies like Cargill [5], with 75 businesses structured in four segments (agro, food, financial, industrial) and three segments involved in biotechnology, have quite some bargaining power, as they control large parts of value chains since moving also into biosynthesis and fermentation (e.g., citric, itaconic, lactic acids, ethanol). It was mentioned above that a steady supply of raw materials with low price fluctuations and steady quality is a prerequisite. But different markets are affected differently by price or quality variations. Commodity products are very sensitive to price fluctuations; higher value products are more sensitive to quality fluctuations. Table 27.2 gives a comparative breakdown of fermentation costs for the production of high value versus commodities.

The number of steps needed to transform an agriculturally refined commodity into a chemical is much smaller for a biofuel than for pharmaceuticals. Consequently, a large part of the cost, especially for a second-generation or lignocellulosic biofuel or commodity chemical, is generated by the pretreatment necessary before enzymatic hydrolysis, the cost for the enzyme, and fermentation. For a pharmaceutical product, on the other hand, the majority of the costs are generated by the isolation and purification of the final product. Thus, price and quality sensitivity strongly depend on markets, products, and applications. A frequently used way to classify these markets, products, and applications is the color code [6] shown in Table 27.3.

While the red biotechnology market is a relatively homogeneous market with well-defined products (a majority of injectable protein products and monoclonal antibodies), white biotechnology is characterized by a multitude of markets served with very different products produced with many different eukaryotic and prokaryotic microorganisms, which have to be compliant with different regulations. This heterogeneity is one of the problems in attracting investors, as venture capital is deterred by confusing messages. The importance of organic chemistry is ubiquitous and affects all aspects of our society. There are hardly any activities or products that are not dependent on organic chemistry. But organic chemistry and biotechnology are two strongly related disciplines linked by biochemistry and (bio)chemical engineering. How important this relationship already is is shown by the selection of markets products presented in the next chapter, which is far from being exhaustive but shows the broad application potential of both chemistry and biotechnology.

TABLE 27.3 Classification of the Applications of Biotechnology with Respect to Markets and Estimates of Current Sales Volumes, Compound Annual Growth Rate, and the Number of Companies Globally Active in the Field

Market	Color Code	Type of Products	Market Size	Companies
Pharma biotechnology	**Red biotechnology**	Monoclonal antibodies, other therapeutic proteins, vaccines, insulins, pDNA	>170 billion US$ CAGR 12%	>6000
Industrial biotechnology	**White biotechnology**	Small molecule pharma and fine chemicals, flavor and fragrance, enzymes, bulk chemicals a.o.	>60 billion US$ (without biofuels) CAGR 6%	>4000
Agro biotechnology	**Green biotechnology**	Transgenic or genetically modified (GM) seeds and plants	15 billion US$ CAGR 11%	100
Environmental biotechnology	**Gray biotechnology**	Environmental biotechnology, services, and solution for bioremediation and waste treatment	<1 billion US$ CAGR 5–10%	<50
Marine biotechnology	**Blue biotechnology**	Products and lead substances from the marine environment	2 billion US$ CAGR 4%	>50
Total Biotechnology Market 2012	**n.a.**	All compounds produced by means of biotechnologies	~250 billion US$ CAGR 11%	10600

27.2.1 Pharmaceuticals

The pharmaceutical market is the most important driver for innovation for large as well as for small molecule process technology, and it is appropriate to understand the burning issues of this market. Although monoclonal antibodies and other injectable therapeutic proteins (and soon biosimilars) are the shining biotechnology stars and the biggest contributors within the red biotechnology sector, one should not forget that well over 90% of all sales in pharmaceuticals are still achieved with small molecule drugs [7]. In 2012, the global market for pharmaceuticals was 990 billion US$. 170 billion US$ (17%) was so-called biopharmaceuticals including vaccines and insulins. Over 210 billion US$ worth of products are produced using biotechnology. Small molecule drugs produced or coproduced using biotechnology are assigned to white biotechnology. Simply considering the numbers in Table 27.3 and the previous statement reveals the extent to which the potential remains untapped. The American Chemical Society's Green Chemistry Institute (GCI) Pharmaceutical Roundtable was established to encourage green chemistry and green engineering into the small molecule pharmaceutical industry [8]. The roundtable informs and influences the research agenda by outlining key green chemistry research areas from a pharmaceutical perspective. The work of this group can be an inspiration for biotechnologists for areas where chemistry needs support! About 70% of the current small molecule therapeutics derives their structure from plants used in traditional medicine, and ~40% of clinically used pharmaceuticals are natural products or their derivatives [9]. Oncology is the fastest-growing pharma market with a considerable number of cytotoxic natural products and their derivates on the market. Large and small molecule products merge with the clinical use of antibody small molecule drug conjugates. From microbial oxidation using *Gluconobacter oxydans* for the biooxidation steps of glucamine derivative drugs for Gaucher's disease, from old products like steroids or antibiotics to newer ones, as listed in Table 27.4, they all benefit from biotechnology [10, 11].

One remarkable achievement of red biotechnology was that, although costs in drug development exploded, the costs for process development of manufacturing processes for proteins were reduced over time. This should also be the goal for small molecule pharmaceuticals, which represent the majority of active pharmaceutical ingredients (APIs). The sustainable production of affordable pharmaceuticals and health-care products should be a priority for biotechnology. These objectives are

TABLE 27.4 Examples of Small Molecule Pharmaceuticals that are Accessible to Biocatalysis or for which a Chemoenzymatic Processes has been Developed Replacing Chemo- and Metal Catalysis

Drug (Trade Name)	Indication	Company	Enzymes Used
Abacavir (Ziagen)	HIV/AIDS	GlaxoSmithKline	Lactamases
Aliskiren (Tekturna)	Hypertension	Novartis	Lipase
Atazanavir (Reyataz)	HIV/AIDS	Bristol-Myers Squibb/Codexis	LDH
Atorvastatin (Lipitor)	Cholesterol	Pfizer/Codexis	KRED, HHDH, nitrilase, aldolase
Boceprevir (Victrelis)	HIV/AIDS	Merck	MAO and TA
Captopril (Capoten)	Hypertension	Bristol-Myers Squibb	Lipase
Duloxetine (Cymbalta)	Depression	Merck	KRED
Imagabalin	Anxiolytic	Codexis	TA
Montelukast (Singulair)	Asthma	Codexis	KRED
Esomeprazole (Nexium)	Dyspepsia	Nexium/Codexis	BVMO
Perindopril (Coversyl)	Hypertension	Servier	PAL
Pregabalin (Lyrica)	Neuropathic pain	Pfizer	Lipase
Pseudoephedrine	Stimulant	Daichi	KRED, aldolase
Simvastatin (Zocor)	Cholesterol	Merck/Codexis	Acylase
Sitagliptin (Januvia)	Diabetes	Merck/Codexis	T A
Telaprevir (Incivek)	Hepatitis	J & J/Vertex Pharma/Codexis	MAO
Crizotinib (Xalkori)	Anticancer	Codexis	Lipase
Ezetimibe (Zetia, Vytorin)	Cholesterol	Codexis	KRED
Zanamivir (Relenza)	Antiflu	GlaxoSmithKline	Aldolase
Sertraline (Zoloft)	Anxiolytic	Pfizer	ADH

ω-TA, ω-transaminases; BVMO, Baeyer–Villiger monooxygenase; HHDH, halohydrin dehalogenase; KRED, ketoreductase, leucine dehydrogenase; MAO, monoamine oxidase; PAL, phenylalanine ammonia-lyase.

technically difficult, but the required breakthroughs will benefit other areas as well. Nowadays, synthetic chemists are able to synthesize even the most complex molecule. However, this often comes at a prohibitive economic and ecological price. This is demonstrated particularly well by the "oligo" pharmaceuticals such as oligopeptides, oligonucleotides, and oligosaccharides.

Since the synthesis of oxytocin in 1953, the first therapeutic peptide to be chemically synthesized, chemical liquid and solid phase peptide synthesis (LPPS, SPPS) have made tremendous progress. About 50 peptide drugs are on the market and 400 are in different phases of clinical development. Peptides that are forecasted to reach blockbuster status with annual sales over 1 billion US$ are leuprorelin and goserelin acetate, both indicated against hormone-responsive cancers or exenatide and liraglutide indicated against type-2 diabetes [12]. Cell-penetrating peptides, peptide signal transmitters, antimicrobial peptides, and peptide drugs and painkillers from venoms are under investigation. A number of drugs on the market were also derived from peptide venoms as shown in Table 27.5, and this is probably only the tip of the iceberg [13].

It is now possible to synthesize polypeptides with more than 200 amino acids. As for nucleotides and oligosaccharides, peptide synthesis is almost exclusively chemical. All three types of molecules can be produced by chemical synthesis methods but at exorbitant cost and with enormous waste generation. Chemical synthesis beyond certain limits of marketed volumes and molecular complexity is not possible (Figure 27.1). Peptides are being increasingly investigated for uses not only in the pharmaceutical sector but also for animal health, cosmetics, and other applications. However, for such products, as higher market volumes (>1 to) and larger peptides (>40 amino acids) are required, they are not within reach of chemical synthesis. The larger the

TABLE 27.5 List of Venom-Derived Drugs which are Currently on the Market

Drug (Trade Name)	Company	Indication	Source
Exenatide (Byetta)	Amylin Pharmaceuticals, Lilly	Diabetes	Lizard
Captopril (Capoten)	Bristol-Myers Squibb	Hypertension	Snake
Eptifibatide (Integrilin)	Millennium Pharmaceuticals Cor Therapeutics	Anticoagulant	Snake
Ziconotide (Prialt)	Elan Corporation	Chronic pain	Cone snail
Tirofiban (Aggrastat)	Merck & Medicure Pharma	Angina pectoris	Snake

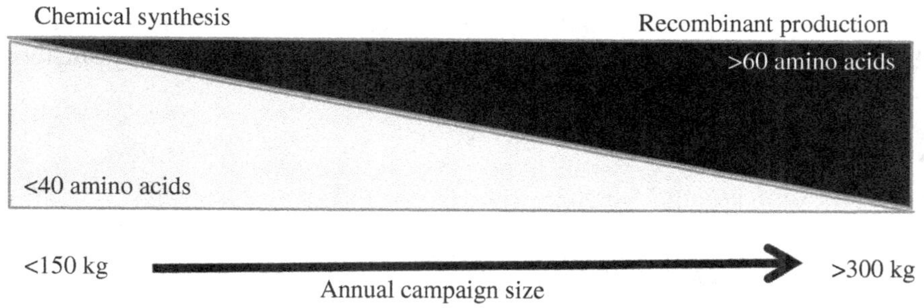

FIGURE 27.1 Production of peptides by chemical synthesis or recombinant technology? The larger the peptide and the larger annual production volume, the better the chance to succeed using recombinant technologies.

peptide and the production volume, the lower the chance of succeeding with chemical synthesis.

Recombinant peptide manufacturing offers many advantages over chemical synthesis and will be instrumental in the near future for economic and sustainable production of peptides [14]. Oligonucleotides and oligosaccharides are other compounds where biotechnological production methods are required. Sugar-based APIs such as acarbose (diabetes), Arixtra (thrombosis), topiramate (anticonvulsant), zanamivir (antiviral), and oseltamivir (antiviral) all need tedious chemical synthetic steps. Arixtra (fondaparinux), a heavily O- and N-glycosylated pentasaccharide, is produced by a 55-step process (or 25 chemical steps for its sibling drug idraparinux) for which biotechnology was not able to design a better process, although it is a natural molecule [15]. Sulfated oligosaccharides have diverse biological functions and present a significant potential. Jennewein Biotechnologie GmbH in Germany manufactures the human milk sugar oligosaccharide fucosyllactose using microbes genetically converted into cell factories and has entered into collaboration with Pfizer [16].

The pharmaceutical market is prone to major changes. Medical practice today is reactive but will become predictive, preventive, personalized, and participative in the future—the so-called 4P medicine. The time of "one-size-fits-all" blockbuster drugs is coming to an end, in favor of "nichebusters" fueled by personalized medicine with the working principle that every human being is genetically unique as is the required treatment. Tissue, bone, and organ repair are increasingly applied in clinics. As a result of the innovation gap in large pharma companies with small molecule drugs, small companies take advantage of a "repurposing" of drug candidates that failed for different ailments, which is cheaper than testing new ones. Novel drug screening and testing using differentiated human tissues allow a much better identification of new and old small molecule compounds. Such methods considerably increase the possibility of existing small molecule libraries, such as the Natural Products Discovery Institute (NIPD), home of the former Merck and Schering-Plough natural products libraries. Pharma alien companies enter the pharmaceutical market and are refreshing an otherwise risk-averted industry. Such a new player is the company Calico, a California life science company established in 2013 by Google Inc., targeting aging and associated diseases using big data technology to drive research [17]. All these events will affect the type of molecules and synthetic processes needed in the future. What we will most likely also experience is a merger of the qualities between small and large molecules. Large molecules such as therapeutic proteins and monoclonal antibodies will become smaller and "neater" for formulation, storage, and therapeutic application, while small molecules will increase in their complexity, meaning functionalities, and chirality. These drug molecules will be of great variety, for which we need the cost-efficient and sustainable production methods.

27.2.2 Medical Technology (MedTech)

Tissue, organ repair, and gene therapy are future and to some extent already existing options in the clinic. The autologous (with the patient's own cells) or allogeneic reconstruction of soft tissues, bones, or organs with differentiated or stem cells requires appropriate, biodegradable polymer scaffolds for 2D or 3D tissue reconstructions. The chemical nature of these scaffolds depends upon the application, but the following organic materials are most frequently used:

- Collagen
- Chondroitin sulfate
- Silk
- Polyhydroxyalkanoates
- Polylactic acid
- Polycaprolactone

To avoid problems with infections, nonanimal-derived materials are preferred for scaffold production, although the chemical isolation process from animal slaughterhouse residues is so harsh, that very few viruses or other potential biohazards could survive the procedure. Thus, biogenic polymers are increasingly produced by fermentation. These matrices are mostly functionalized with small and large molecules for better adherence and growth of cells. Some of these materials have already been used on patients. Chondroitin sulfate is a glycosamine mucopolysacharide used in ophthalmology or cartilage repair, where the supply from sharks or other animal sources is replaced by recombinant production. Polyhydroxy 4 butyrate (P4HBA) is used for sutures (see also Section 27.2.6). Another application under investigation is bioabsorbable polyhydroxyalkanoate drug-eluting medical devices. In early 2011 Evonik acquired the reabsorbable polylactide polymer Resomer®, a portfolio of polymers for medical device and controlled release applications, from Boehringer Ingelheim. Fibrin glue as a conduit for peripheral nerves worked in rats for the initial phase of nerve regeneration [18]. It can be expected that other biodegradable biopolymers have been tested including specific functionalization. Differentiated human tissues on 2D and 3D scaffolds can also be used for drug screening and drug testing [19]. Their realization requires a joint approach using chemistry and biotechnology to produce such functionalized matrices. Nature-inspired materials will also need nature-inspired solutions for their production.

27.2.3 Food and Feed

The food and feed industry has been using enzymes in manufacturing processes for a long time, and ~30% of the food we consume daily is fermented in one way or another. Biosynthetic or chemoenzymatic processes for amino acid and vitamin production are well established and implemented. The volume of industrial vitamin production in tons is about 10 times smaller than that of amino acids (>4 000 000 tons). Of the 13 vitamins, representing a market volume of about 2.5 billion US$, seven are produced using chemical synthesis only. Of the remaining six, three are produced by hybrid processes (B3, B5, C) and three are produced using a purely biosynthetic or fermentative approach (B2, B12, K3). The impressive advantages that can be achieved with

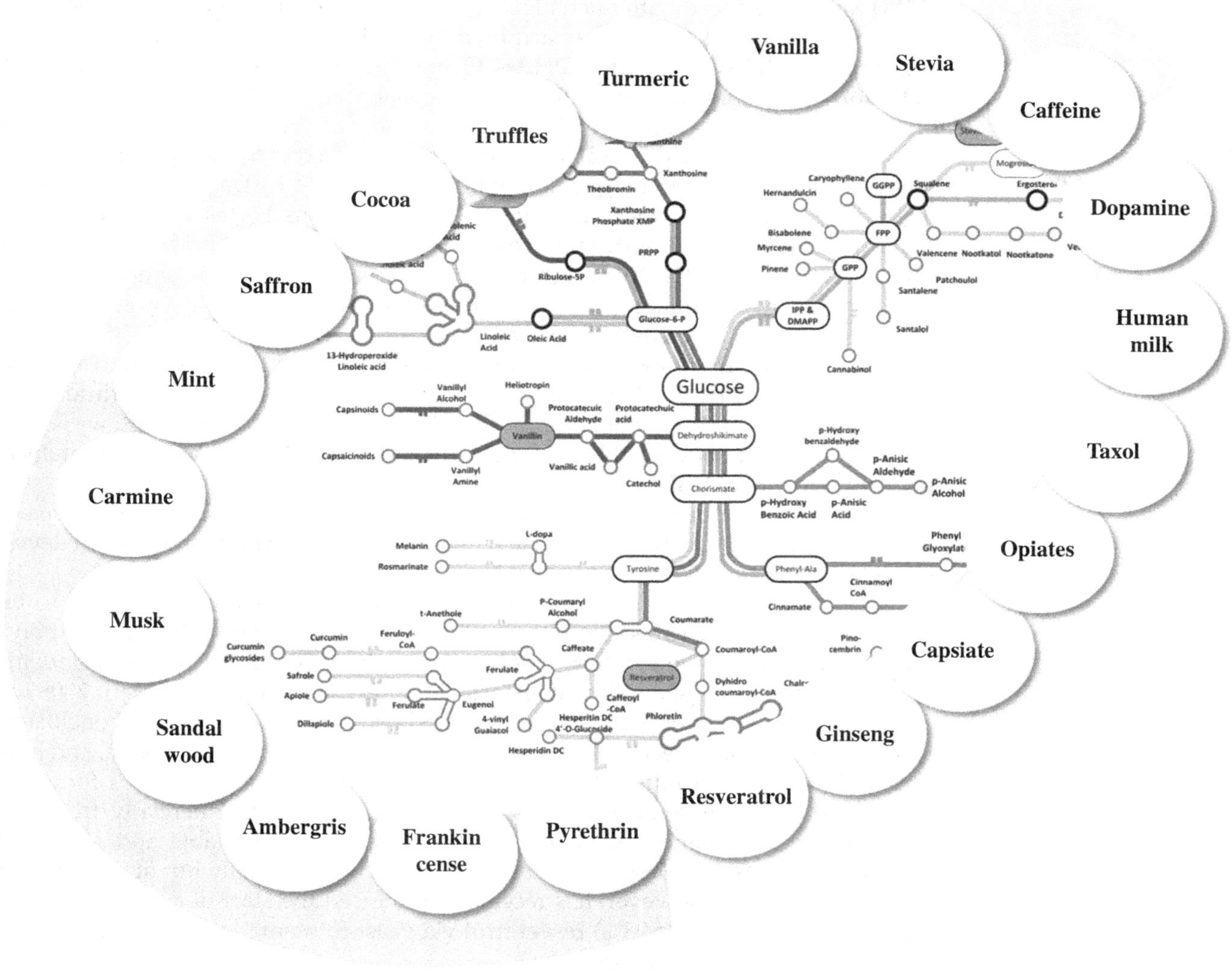

FIGURE 27.2

The use of yeast as production platform for the production of functional food ingredients and nutraceuticals, by combining different genes from different organisms into *Saccharomyces cerevisiae*. This type of metabolic engineering and recoding of microbial pathways to design the "perfect" organism for the industrial manufacturing of organic chemical products will rapidly gain importance. Courtesy of Evolva.

biotechnological production were demonstrated with the vitamin B2 fermentation of DSM, resulting in considerable waste and cost reduction when compared to chemical production [20]. New enzyme applications are being investigated such as using asparaginases from c-LEcta to reduce traces of the potentially carcinogenic chemical acrylamide during food pretreatment [21]. Evolva develops sustainable, fermentation-based ingredients for health, wellness, and nutrition products and focuses on ingredients that have sourcing and production "issues" (see also Sections 27.2.3 and 27.2.4). These are ingredients whose current supply has taste, environmental impact, or sustainability issues or are overharvested, endangered, or too expensive when farmed. The companies' alternative production process is via biosynthesis and yeast fermentation with an initial focus on resveratrol, vanillin, stevia, saffron, and agarwood oil (Figure 27.2).

The trend continues with the biotechnological production of nutraceuticals and functional food ingredients such as beta-glucans, carotenoids, collagen hydrolysates, colorants, flavonoids, polyunsaturated fatty acids (PUFAs), sweeteners, and many more. The PUFAs such as docosahexaenoic acid (DHA) or eicosapentaenoic acid (EPA) can be produced with microalgae and fungal fermentation instead of fish liver oil extraction. Martek Biosciences, acquired by DSM in 2010, has been one of the first companies with PUFA products DHASCO and ARASCO favorably reviewed by the FDA for infant formulas in 2001. DuPont collaborated with Amarin on a yeast (*Yarrowia lipolytica*)-based ω-3 PUFA process capable of producing EPA alone, using metabolic engineering. Through the acquisition of Danisco, DuPont acquired considerable additional biotechnological know-how. Souvenaid® by Danone is a nutritional approach to support memory in aging adults; it contains Fortasyn Connect™, the nutrient combination of Souvenaid containing also DHA, EPA, uridine monophosphate, and choline [22]. In November 2004, the FDA approved Omacor™ from GlaxoSmithKline, containing an omega-3-acid ethyl ester as a lipid-lowering agent. Lonza is producing ω-3 DHA by fermentation.

The sales volume of food coloring of 1.6 billion US$ in 2011 is expected to increase to 2.1 billion US$ by 2018, growing at 4.3% per annum. In 2011 the natural color market overtook the synthetic color market. The importance of this switch was demonstrated by the removal of blue smarties by Nestle in 2006, because they contained a synthetic colorant [23].

Truvia™ is a natural sweetener from the stevia plant composed of different sweet and bitter glycosides, extracted from leaves but with fermentation processes in development (see Section 27.3.2).

However, health claims are sometimes difficult to prove and turn out to be unsupported. A drug called SRT501 was built based on resveratrol, a natural polyphenol found in red wine, which was believed to have beneficial effects for inflammation, carcinogenesis, and longevity. GlaxoSmithKline acquired Sirtris in 2008 for 270 million US$ only to stop the SRT501 trial shortly thereafter [24]. The conclusion is that the beneficial effect of the consumption of red wine, dark chocolate, and berries on inflammation must come from other polyphenols or substances found in those foodstuffs. Resveratrol is currently largely derived commercially from the invasive species Japanese knotweed, which suffers from variable and generally insufficient purity. To a lesser extent it is synthesized chemically but at a very high price point. Evolva, however, has recently announced the start of commercial production of high-purity (>98%) resveratrol via yeast fermentation, the only resveratrol produced using that process.

27.2.4 Flavor and Fragrance

The flavor and fragrance industry usually sources its raw materials by extraction of secondary metabolites from plants and animals. However, this industry sector is looking increasingly toward alternatives as natural sources are at risk from climate change, mismanagement through societal pressure, and seasonal variation leading to price pressure and potential shortages. In some cases, organic synthesis has been developed for a number of products, although some synthetic products such as flavor notes do not have the desired performance characteristics. As a consequence, the flavor and fragrance industry is looking toward biocatalysis and biotechnology for sustainable production as structurally complex molecules are not efficiently accessible by chemistry. The manipulation of enzymes such as cyclases and synthases, used in processing complex terpenes, is starting to provide alternative, more sustainable routes. Synthetic biology is increasingly being applied to incorporate pathways from fragile natural organisms into industrial producer strains such as *Escherichia coli* and yeasts (see also Section 27.3.2). These newly engineered biosynthetic pathways will lead to lower costs, higher purity, and more sustainable and reliable supply chains. Terpenes are one example where a regular supply can be difficult and which is difficult to access by chemical synthesis. Santalol, distilled from sandalwood tree, *Santalum* sp., is an example of a commercial raw material used for perfumes, where the demand greatly exceeds the very limited supply from trees. The situation strongly resembles the situation of Taxol, a cytotoxic plant compound that stabilizes microtubules and has been approved since the early 1990s as an anticancer drug. But the commercial Taxol requirements by far exceeded the possible supply by extraction from the pacific yew tree, and chemoenzymatic and biosynthetic alternatives had to be developed.

Givaudan, IFF, and Firmenich are the three biggest players, covering about 50% of the global market. All of them are investing in sustainable, cost-effective, and reliable sourcing of key fragrance ingredients, using one or the other biotechnological options. Firmenich, together with academic and private partners, is working on massive pathway grafting to produce terpenes for the flavor and fragrance industry. The biosynthesis of several sesquiterpenes has been elucidated and will be cloned into an appropriate host [25, 26]. Ambrox diterpene from the sperm whale can be produced using recombinant *E. coli* high-density fermentation producing the intermediate sclareol using pathway constituents of the clary sage plant. The sclareol is then chemically converted into Ambrox [27].

Allylix is another company developing cost-effective terpene fermentation and purification for sustainable supply of the isoprenoids such as nootkatone and valencene (grapefruit and orange smell) yeast as a host organism [28]. Givaudan created a novel class of octahydroazulene derivatives, which possess a very strong and pleasant floral and woody odor using a laccase-mediator system on an olefinic Patchouli distillate [29, 30]. This perfume material, produced using an innovative enzyme process, was recently launched on the market as Akigalawood™ [31]. In June 2014, the luxury brand Urbano launched a men's perfume with this biotechnology fragrance product [32]. Givaudan also very recently acquired the French company Soliance, a small enterprise producing high-value active ingredients by extraction from plants, microorganisms, and algae. It has been said that plants are the world's best chemists. Also Amyris manufactures and sells the fragrance patchouli oil by recombinant yeast fermentation, besides the renewable building block hydrocarbon farnesene [33]. Thus, it seems natural to determine the metabolic pathways of industrially relevant plants, which produce high-value-added metabolites, to reproduce these pathways in organisms which can be easily grown in fermenters. For this purpose, PhytoMetaSyn [34], a consortium in Canada, is identifying the biosynthetic genes responsible for the immense chemical diversity of plant metabolism in microbial systems and in engineered yeasts. Prototype natural products are nootkatone, abietic acid, morphine, strictosidine, betulinic acid, and xanthohumol.

27.2.5 Cosmetics and Personal Care

The most significant applications of natural ingredients are in food and beverage applications, followed by the cosmetics and the toiletry sector. To remain in business, producers must include natural products in their formulations and the cosmetic industry must be attentive to green sourcing and green chemistry. Consumers may be lenient with respect to "dirty" manufacturing of pharmaceuticals, but not with cosmetics. Extraction from plants remains an important source of raw materials, but similar to the fragrance and flavor industry, there are efforts to transfer sourcing to more reliable fermentation methods. The industry has recognized the requirement and opportunities and is actively collaborating to produce existing ingredients in more sustainable ways. Optibiocat [35], for example, is the acronym for "Optimized esterase biocatalysts for cost-effective industrial production," an EU project with 16 participants, the largest being the Greek producer of natural cosmetics Korres [36]. The goal of the EU project is to reduce the number of chemical steps and required temperatures from up to 160 °C to 50 °C or 60 °C using feruloyl and glucuronyl esterases.

There are several cosmetic formulations that already contain new small and large molecules issued from fermentation. Ectoin®, hydroxyectoine, and Glycoin® are small molecules harvested from the fermentation of extremophilic microorganisms [37] marketed by Merck for Bitop AG. These molecules help microorganisms to survive extreme environmental conditions, and this protection principle is also used in cosmetic formulations. Ceramides, CoQ10, hyaluronic acid, or small antiacne peptides are other recent examples of fermentation products used in cosmetic and skin care products. Cysteine has become popular in the cosmetics sector, as a free radical scavenger in antiaging preparations. Cysteine used to be obtained by extraction from human hair, but an *E. coli* fermentation process has been developed by Wacker Chemie AG. There is a large market for fatty acids, alcohols, aldehydes, wax, and esters, which are used predominantly in soaps, detergents, cosmetic additives, and flavoring compounds. Fatty acids are produced by hydrogenation of plant oil-derived fatty acid methyl esters or by synthesis from petrochemical precursors. The microbial fatty acid metabolism is an alternative for the overproduction of fatty acids, esters, alcohols, and other products. Carbohydrate complexes can be esterified with fatty acids such as linoleic acid using enzymes to produce moisturizers. Alpha hydroxyl acids like glyoxylic acid and polyhydroxy acids like gluconolactone, glucoheptonolactone, maltobionic acid, or lactobionic acid are other compounds used in cosmetic formulations. Libragen Cosmetics [38] markets natural plant polyphenols, which are grafted with sugars by means of an enzymatic reaction to make them stable, bioavailable, and highly water soluble. An industrial consortium of 15 companies and organizations including cosmetics companies (e.g., L'Oreal, Michelin, Roquette, Sofiproteol, Syral, Total, and Veolia) is building a 20 million euro preindustrial demonstration plant named Toulouse White Biotechnology (TWB) for biopolymers, biofuels chemical intermediates [39]. There was a myth that protein–protein interactions could not be inhibited by small molecule drugs because their surfaces were too large. Hybrigenics teamed up with L'Oreal to discover new interactions among skin proteins and small molecule inhibitors were identified for innovative cosmetics. The enzyme preservation system called Biovert [40] employs two enzymes, glucose oxidase and lactoperoxidase, in combination with glucose in skin and baby care products Biovert mimics a naturally occurring antimicrobial–antioxidant protection system generating a steady supply of H_2O_2 from O_2 and H_2O, driven by the enzyme glucose oxidase and its substrate glucose. This is the same basic principle as the *in situ* generation of peracetic with a perhydrolase, developed by DuPont (Section 27.2.10), except that this Arch Chemicals (now Lonza) product is applicable for biological material due to its milder character. Luteolin from the snow white (*Leontopodium alpinum*) or active ingredients extracted from basilikum have UV protection capabilities. However, snow white in particular is a difficult production plant for commercial amounts. Therefore, Rootec [41] has proposed an interesting alternative, which can be applied for higher value and complex chemicals: manufacturing in hairy roots (Figure 27.3).

Nestle paid 3.6 billion US$ in 2014 to take full control of Galderma, the dermatology business that revives skin [42]. Galderma will become part of a new division known as Nestle Skin Health SA. This is the most recent push by Nestle into health care that promises faster growth and wider profit margins. This is another example where the boundaries with pharma are getting smaller and consequently, because of the validation of health claims, requires the financial staying power of large companies.

27.2.6 Polymers

Biopolymer is not synonymous with biodegradable [43]. The first biobased polymers were cellulose-based bioplastics such as cellulose acetate (CA), cellulose acetate butyrate (CAB), and cellulose acetate propionate (CAP) produced by the esterification of cellulose and in use for the last 100 years. Biopolymers are used for a plethora of applications ranging from salad dressing to medical devices. The following are a number of newer biopolymeric compounds that have commercial potential:

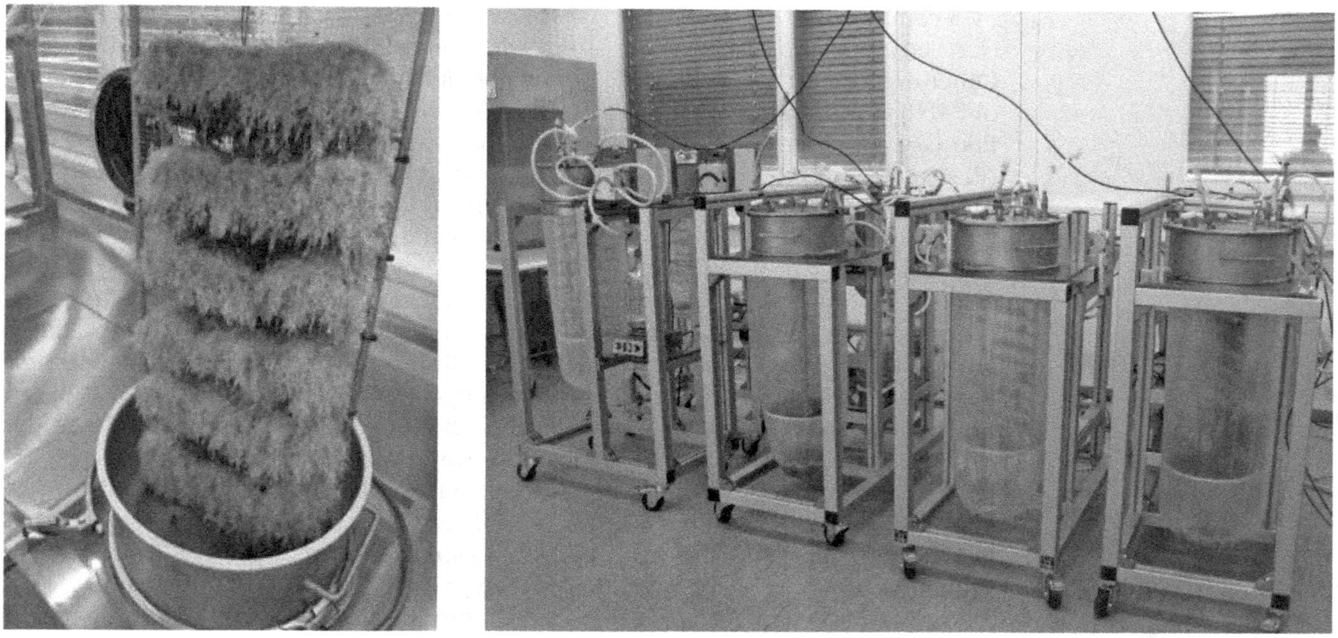

FIGURE 27.3
Production of complex and high-value small molecules using hairy root tissue culture technology. Root biomass of clonal origin (left) is cultivated in a sterile and controlled environment (right). But instead of growing biomass suspended in a liquid culture broth in a fermenter, the hairy root tissue mass is growing on a support (left) and sprayed with sterile growth medium, resulting in maximal cell density and potentially very high product concentrations in a sterile, confined, and scalable environment. Courtesy of ROOTec.

Proteins: Some examples of elastomeric proteins in biological materials with interesting properties are abductin, elastin, fibroin, resilin, silk fibroin, titin, and byssal thread proteins. The tensile strength of spider silk is higher than that of steel. Amsilk [44] has developed a unique process for producing bioengineered spider silk analogs with a recombinant process. The protein can be shipped and stored as a white powder in barrels and spun into a strong fiber called "Biosteel." The applications of spider silk are mainly foreseen for medical devices, as it will not replace other high-strength polymers such as Kevlar in the near future. Shigeyoshi Osaki of Japan's Nara Medical University has used spider dragline silk to make violin strings [45]. "Shrilk" is a mixture of two natural polymers, silk and chitin, from discarded shrimp cells [46]. Proteins from marine organisms are also a source of interesting bioinspired materials. Marine mussels have very robust proteins that help them to affix themselves to rocks [47], which have potential industrial applications ranging from pharma to tire manufacturing. Using proteins, the possibility of an allergenic immunological response must be considered.

Polyhydroxyalkanoates (PHA): PHA exhibit thermoplastic and elastomeric properties and are divided into three subgroups [48]:

1. Short chain <5 monomers
2. Medium chain 6–14 monomers
3. Long chain >14 monomers

The biodegradable PHA biopolymers can be produced with over 50 monomers, but production is not well established yet. The situation is different with poly(4-hydroxybutyrate) (P4HB), which is the first and only PHA-based product approved by the FDA for clinical use. The most remarkable property of P4BH is its high elasticity, comparable to ultrahigh molecular weight polyethylene. It is used for sutures in surgery with one product on the market called TephaFLEX®, produced by Tepha Medical Devices [49]. This P4HB homo- and the heterobiopolymers are produced by recombinant *E. coli* fermentation and a solvent-based purification process. Le Meur *et al.* tested a number of *E. coli* strains for their ability to accumulate P4HB [50]. Other organisms do not allow the production of a homopolymer. The biodegradation of P4HB in the human body takes place via surface erosion instead of burst degradation (see also Section 27.2.2). This avoids local inflammation as with polylactic acid (PLA) or polyglycolic acid when used as scaffolds or in medical devices. The use of PHA for mass polymer application is more difficult from a cost perspective. Metabolix purchased the PHA technology from Monsanto in 2001, and started collaboration with Archer Daniels Midland (ADM). However, this joint venture to produce the bioplastic Mirel™, a PHA-based biopolymer, in an Iowa plant with a capacity of 50 000 tons per year, ended because of lack of results within a reasonable timeframe. The cost of PHA are an issue, and recombinant production is with *E. coli* is preferred as the natural producer *Ralstonia eutropha* can hardly reach cost targets. The Spanish company Antibioticos SA has replaced ADM in the endeavor, after the cooperation between ADM and the MIT spin-off Metabolix accumulated losses of over 200 million US$ in 6 years. The capacity of the Antibioticos SA plant in Leon is about five times lower. The Spanish Leon plant can only produce 10 000 tons per year. Nodax™ is a branched PHA thermoplastic produced by fermentation with *Pseudomonas* sp. and a cost target of 1–2 US$ per kg. DSM invested in China-based PHA developer Tianjin GreenBio Materials Co., Ltd, which established one of the world's largest production plants for PHA in the Binhai District, with DSM Venture Capital participating in the investment in China [51].

Costs of biobased biopolymers are high; waste materials are investigated as carbon and energy sources for production. Biomass, methane from flares, CO or CO_2 are in discussion as raw materials for the production of biopolymers and other products by fermentation. Fundamental strategies for reducing PHA production costs include: (i) polyhydroxybutyrate (PHB) production from CO_2 by Knallgas (hydrogen-oxidizing) bacteria using H_2 as electron donor. The problem of hydrogen-oxidizing bacteria fermentation is the explosive manufacturing environment. (ii) PHB production from waste methane, which is flared or released to the atmosphere in oil fields. A large EU consortium [52] with six companies and eight academic partners approaches this approach by pyrolysis of wastes followed by microbial fermentation of the resulting syngas for the biosynthesis of the PHA biopolymers.

Polyethylene: Polyethylene is the most common plastic and can be produced with ethylene from dehydrated bioethanol, an option considered by Dow Chemical and Braskem with a joint venture for the production of polyethylene from sugarcane in a 350 000 tons per year facility in Brazil [53]. With the goal to "decarbonize" their supply chain, several other companies are also reconsidering their source of polymers. Tetra Pack has an agreement with Braskem for Brazilian sugarcane-based polyethylene to be used in all Tetra Pack packages produced in Brazil. Coca-Cola, Ford, Heinz, Nike and Procter & Gamble have formed the Plant PET Technology Collaborative, a working group focused on the development and use of 100% plant-based PET materials and fibers for bottling, packaging or other purposes. Classically polyethylene terephthalate (PET) is made by reacting ethylene glycol (EG) with purified terephthalic acid (PTA). The PET in the "PlantBottle," however, is biobased with sugarcane as raw material.

Polylactic acid (PLA): PLA biodegradable thermoplastic polyester produced from lactic acid is the commercially most advanced bioplastic. Biophan produces foils made of PLA produced by Nature Works in Blair, Nebraska, for vegetable and fruit packaging. Sulzer [54] has developed a new process to convert lactates to PLA with an entirely skid-mounted unit. A 1000 ton continuous demonstration pilot plant for PLA was built in Winterthur and Pfäffikon in Switzerland in cooperation with Purac (a lactic acid producer) and Synbra [55], which started commercial PLA polymerization in 2011. Teijin in Matsuyama, Japan, developed a PLA-based heat-resistant biopolymer called Biofront [56]. But it remains an economically challenging business. Cereplast (OK), a maker of various bioplastics and production capacities of over 80 000 tons per years with plants in Assisi (Italy) and Seymour (United States), had to move its US and European headquarters to save costs. But this did not help as they had to file for Chapter 11 protection of the US bankruptcy code [57].

Furanics: Avantium Technologies in Amsterdam, a spin-off of Royal Dutch Shell, developed furanics for bioplastics and biofuels from biomass. This technology uses proprietary catalytic systems for the conversion of biomass into carbohydrates that are then converted to furanics.

Isoprenoids: The largest isoprenoid market is polyisoprene rubber, which exceeds 1 000 000 metric tons per year at a price of ±1 € per kg. CEA, Michelin, Protéus (PCAS Group), and SDTech joined forces to develop a novel recycling procedure for used tires [58]. This Tire Recycling Project (TREC) driven by Michelin is regenerating rubber compounds to make new tires. Used tires undergo gasification and the syngas is used in a fermentation process to produce alcohol and butadiene. There is a research agreement between Genencor (a DuPont company) and The Goodyear Tire & Rubber Company to develop the BioIsoprene™ monomer as an alternative to petroleum-derived isoprene [59]. 60 g of isoprene per liter of sugar solution is produced using microbes expressing isoprene synthase. Lanxess invested 10 million US$ in cooperation with Gevo for the production of butyl rubber from sugar by blocking competing pathways for ethanol and acetic acid followed by a dehydration process to convert isobutyl alcohol to isobutene, which can then be polymerized to butyl rubber [60]. Lanxess is the biggest rubber producer with 15 000 employees. Soviet researchers identified a dandelion species *Taraxacum kok-saghyz* from Kazakhstan in the 1930s as a possible source of a rubber polymer. The plant was cultivated during the war by German troops. The EU has now started an 11 million euro project to investigate and improve the commercial potential together with Russian scientists. A pilot plant at the University of Münster will demonstrate proof of concept to use the plant biomass for tire manufacturing. A similar project exists between Bridgestone and the Ohio State University [61].

27.2.7 Surfactants and Lubricants

Biosurfactants: Environmental concerns about the use of surfactants triggered an interest in biobased and natural products. The applications of these biosurfactants in different markets range from wound cleaning to drag reduction in pipelines, when used in combination with (bio)polymers. Sugars (in glycolipids) and amino acids (in lipopeptides) are the two most important examples of surfactant polar headgroups of natural origin, and they are accessible by fermentation or biotransformation, such as enzymatic glycosylation to replace chemical glycosylation by Fischer esterification. Sophorolipid, a disaccharide linked to long-chain hydroxy fatty acid, has been known for almost 50 years. Kawaken Fine Chemicals offers the silk-derived surfactants "Kwasilk" [62]. Surfactins are cyclic lipopeptide biosurfactants that are produced by various strains of *Bacillus* sp. Pepfactants™ are rationally designed biodegradable peptide surfactants allowing the creation of a stable emulsion for specialized applications [63]. The proposed production is based on a recombinant *E. coli* fusion protein approach [64]. Mannosetriol lipids (MEL) are biosurfactants isolated from yeasts. Rhamnolipid, another glycolipid biosurfactant produced by a *Pseudomonas* sp. strain, has been well known for over 40 years and has been proposed and tested for a range of applications from contact lens cleaning to bioremediation of water-insoluble contaminants. Modular Genetics, Inc. developed a fermentation process for the production of acyl glutamate by fermentation for the high-end personal care "green" market [65]. Butyl glycoside is used in dishwashers and for cosmetics and can be produced using an Aspergillus transglucosidase. France seems to be successful in the valorization of its agricultural and marine sector by using these feedstocks not only to produce commodities, such as succinic acid (e.g., Roquette), but also cosmetics, pharma products, or biobased surfactants. Wheatoleo is a French company that develops biobased surfactants [66].

Biolubricants: Wind turbines use around 1000 l of lubricating oils and 200 kg of grease per year. It is inevitable that some of that oil and grease is released into the environment during operation and maintenance, and there is an interest in using less damaging biogenic oils and greases. Sofiprotéol [67], formed on the initiative of the French federation of oilseed and protein crop producers, develops biolubricants made from vegetable oils. In 2013, Amyris and Cosan expanded their joint venture, Novvi, to include finished lubricants for industrial, commercial, and automotive uses in addition to the joint venture's original scope of renewable, base oils for such markets and picked Albemarle to fabricate a lubricant base out of farnesene [68].

27.2.8 Commodity Chemicals

The production of acetone, *n*-butanol, and ethanol by fermentation (ABE fermentation) was developed in the United Kingdom during World War I to assure solvent supply for the production of gunpowder. Isopropanol, *n*-butanol, and ethanol (IBE) can also be produced by an anaerobic fermentation process. As for biofuels and biopolymers with bulk character, the agricultural raw material supply is the key ingredient in the value chain. For this reason, companies whose core business is in trading and processing plant-based raw materials such as corn, wheat, and potatoes are partnering with companies having their core competency in microbial fermentation. A selection of chemical commodities that can be produced by biotechnology is presented in the following.

Acrylic acid: OPX Biotechnologies [69], a biobased company in Boulder, CO, United States, brings a commercial biobased acrylic acid plant online in partnership with Dow Chemical. The plant will have an annual capacity of 100–300 million tons of acrylic acid, first using corn, and later lignocellulosics as carbon end energy source. Cargill and Novozymes have been jointly developing a process to produce acrylic acid from glucose or other carbohydrate sources.

Adipic acid: Adipic acid is used to make nylon and several companies are working on biobased nylon manufacturing. Verdezyne is one of them, producing this nylon key component and thermoplastic polyurethanes by fermentation with *Candida tropicalis* with renewable plant oils as carbon and energy source.

Butanediol: BioAmber has licensed DuPont's hydrogenation catalyst technology to convert biobased 1,4-butanediol into other chemicals. Genomatica from San Diego produces biobased 1,4-butanediol by fermentation for applications in thermoplastic copolymers [70]. Genomatica and Novamont announced a joint venture by converting a Novamont facility in Italy for the production of about 18 000 tons of butanediol per annum. Genomatica also licensed its biobased 1,4-butanediol processing technology to BASF. Earlier this year, DuPont and BP announced a partnership targeting multiple butanol molecules. Another industrial process developed by Genomatica is for methyl ethyl ketone (MEK), a solvent for paints.

Dodecanoic acid: Aliphatic α,ω-dicarboxylic acids (DCA) are used in a wide variety of plastics, high moisture resistant nylon, and other chemical applications. Diacids can be enzymatically produced from long-chain fatty acids. *Yarrowia* sp. biocatalyst is able to convert long-chain fatty acids directly to long-chain diacids. DCA12 is produced in the largest quantity (>40 MM pound per year).

Glucaric acid: DTI formerly known as DanChem Technologies has started their biobased glucaric acid in Danville, VA, United States. Rivertop from Montana, United States, uses oxidation technology for glucaric acid to replace phosphates in laundry detergents.

Isobutanol: GEVO wants to produce isobutanol at 50% of the cost of petrochemical processes, with animal feed as a byproduct. The isobutanol can be further used directly or dehydrated to isobutylene [71]. The recombinant *E. coli* process produces higher chain alcohols including isobutanol, 2-methyl-1-butanol, and 2-phenylethanol from glucose. Isobutanol is less volatile and less corrosive when blended into gasoline.

Lactic acid: Lactic acid is an attractive target for biobased production besides ethanol, acrylamide, and 1,3-propanediol. Global production of lactic acid is between 300 000 and 400 000 tons a year. Lactic acid can be produced from sugars from starchy or cellulosic waste, the second, however, only by simultaneous hydrolysis and fermentation. Two major producers are CCA Biochem BV in the Netherlands using fermentation and Sterling Chemicals in Texas using chemistry [72]. Other producers are Purac, Galactic, ADM, and Cargill with Dow Chemicals, several of them in cooperation with Nature Works for the production of the polylactic acid biopolymer.

1,3-Propanediol (PDO): Sorona® is a DuPont polymer that contains 37% annually renewable plant-based ingredients. DuPont is building the largest renewable materials facility in the world. UK sugar producer Tate & Lyle operates another propanediol plant for Bio-PDO™ using corn sugar as raw material instead of ethylene oxide. The plant in Loudon, Tennessee, United States, will be operated by the DuPont Tate & Lyle Bio Products, a joint venture founded in 2004. PDO will primarily be used to produce Sorona. Novamont is a European polymer manufacturer marketing a family of bioplastics called Mater-Bi® obtained from renewable materials. Sorana is a biopolymer of DuPont consisting of 1,3-propanediol and terephthalic acid.

Polyethylene: In 2010, Brazilian Braskem was the first company to put a biobased polyethylene plant into operation in Triunfo with a capacity of 200 000 tons per year [73]. The same company was first considering partnering with Dow Chemical for polyethylene resins from sugarcane-based ethanol.

Succinic acid: Succinic acid is widely used as a chemical building block in polymers, food, and pharmaceuticals, but the biobased succinic acid will nevertheless remain small, as only 25 000–35 000 tons is produced derived from petroleum by oxidation of butane. But the potential of biosuccinic acid is much bigger, since it can be easily converted chemically into 1,4-butanediol for which the market is much larger (1.3 million tons annually). Several companies and consortia are developing biobased manufacturing processes. Starting from starch can lead to a 40% reduction in energy requirements plus a reduction in CO_2 emissions. DSM agreed to develop a fermentative process for the production of biorenewable succinic acid (Biosuccinium™) with Roquette, a French starch company, in a 50 : 50 joint venture called Reverdia [74]. The production organism is *Saccharomyces cerevisiae* producing 100 g/l from starch at a pH of 3, to avoid the salt form of the acid. Reverdia runs a demonstration plant in Lestrem (France) and 10 000 tons a year 100 m^3 plants in Cassano Spinola, Italy. BASF and Dutch Corbion Purac [75] cooperate with production of ~10 000 tons biosuccinic acid per year in Spain. BioAmber [76] received several awards for developing a biosynthetic route to commodities including succinic acid at lower cost than existing petroleum-based routes and reached a licensing agreement on derivatives of biobased succinic acid with DuPont.

27.2.9 Energy

The mobility acquired with horses had its price in the nineteenth century. The surface of arable land required to feed the horses reached about the same surface required to feed the human population. The disposal of manure was a considerable logistical problem in large cities like London, despite the fact that only a small fraction of the European population could afford to own a horse for mobility. All this changed with the mass manufacture of affordable cars operating with oil. Global hydrocarbon consumption now stands at about 218 million barrels of oil equivalent energy a day, according to the BP Statistical Review of World Energy, which includes 83 million barrels of oil as well as 75 million barrels of oil equivalent from coal and about 60 million barrels of oil equivalent from natural gas. Oil has provided unprecedented wealth to many people over the last 200 years. Yet the impact of the imbalanced wealth created during the oil era becomes increasingly apparent. The era of cheap oil is over, and instead of having to dispose of horse droppings on streets, we have a problem of carbon dioxide accumulation in the air. But how can we replace all that oil? Bioenergy also comes in three forms: as a solid in its original form such as wood, as a liquid in biofuel (ethanol and biodiesel), or as a biogas. Liquid biofuels have become a <80 billion US$ business, albeit heavily dependent on subsidies, tax exemptions, and protective tax barriers. The most important producer of biofuels is Brazil, with a total bioethanol production for 2012/2013 of 23.64 billion liters in 405 sucrose processing plants. From sugar cane, 153 kg of fermentable sugar can be extracted, which is then converted into 90 l of ethanol [77]. Practically all of this biofuel is of the first-generation type (from starch). Several second-generation (lignocellulose) biofuel plants are under construction; the largest will be operated by Iogen Corporation and DuPont in North America. Third-generation (algae) biofuels are still far from commercialization. The first of the two main problems with biofuels concerns agriculture. Biofuels (as well as other bulk biocommodities) need a stable and massive biomass supply chain without short-term price fluctuations, and this in turn requires a strictly regulated or "Soviet-style" agriculture. Economically viable production also requires the biorefining of the entire crop. The second problem is that the biological conversion of recalcitrant lignocellulosic biomaterial is not competitive, making the nonbiotechnological alternative biomass-to-liquids (BTL) approach more attractive. Biomass is converted and "concentrated" into a liquid "BioSynCrude" by pyrolysis at 500 °C and in a second catalytic step converted into the "Bioliq." Economic metrics of the gasification step of the biomass are not widely available, but it seems that the Carbo-V®-Process of Choren, recently purchased by Linde, is one of the most promising [78]. The coexpression of hydrolytic enzymes in GMO biomass to improve the biodegradability may be a long-term option to solve this need for costly pretreatment, but it requires public acceptance of large-scale use of GMO. Furthermore, the phototrophic mass cultivation of algae, which is a very attractive biofuels possibility, especially with marine microalgae, is far from being close to commercial realization anywhere. One big hurdle is the inevitable limitation of light when cell densities are high. Most of the biofuels today are produced in South and North America. ADM, British Petroleum, Conoco Phillips, Abengoa [79], Shell, Cargill, and DuPont are global companies which work with innovators such as Verenium, Gevo, Mendel Biotechnologies, Dyadic, and others. George Soros and Richard Branson are two investors who have put their money in biofuels. BASF very recently acquired San Diego-based Verenium to strengthen BASF's footprint in the strategic enzyme market. There are very few European industrial biofuel projects and their commercial future is rather difficult. The Glencore daughter Biopetrol, for example, with two biodiesel manufacturing sites in Germany and the Netherlands was delisted from the stock market after losses exceeded 100 million Swiss Francs per year. There is also a difference in the motivation for biofuel between the United States and Europe. The United States considered biofuels as one strategic option to reduce the nation's dependency on foreign oil. The main motivation in Europe, however, is to use biofuels as a strategic option for CO_2 reduction. Today, the United States is on its way to become an energy exporter, thanks to fracking, and the Europeans have realized that some biofuels can be very unsustainable indeed. It took thousands and thousands of years during prehistoric times to convert sunlight into oil, gas and coal, which mankind is now consuming in a very short time span. This indicates that annual biomass available for energy can only cover a very small fraction of our energy needs.

27.2.10 Other Markets and Products

Hydrophobins are low molecular weight hydrophobic proteins secreted by fungi that are able to assemble spontaneously into amphipathic monolayers [80]. This protein protects fungal biomass (raincoat of fungi) from external influences and makes water roll off very efficiently. BASF produces recombinant hydrophobic proteins (trade name H Star proteins) in industrial amounts for various applications. Enzymes can be used to detoxify nerve agents. Paraoxonases are enzymes capable of hydrolyzing organophosphate esters of phosphoric acid which are used in insecticides, herbicides, or nerve agents. PON1 is a gene expressing an aromatic esterase. The enzyme is naturally produced by the liver and is able to break down poisonous G-type nerve agents such as sarin, tabun, soman, and cyclosarin but is largely ineffective. Directed evolution is used by introducing artificial mutations to bring the activity of these enzymes to activity levels needed for a practical application in decontamination kits. CSIRO received the DuPont award for their phosphatase enzyme bioremediation formulation, which is already on the market. DuPont developed an enzyme kit based on *Thermotoga maritima* wild-type esterases for the *in situ* generation of peracetic acid by the perhydrolysis of acyl esters. The product is applied mainly on, for example, poultry farms.

27.3 THE TOOLBOX

27.3.1 The Current Toolbox

There are three approaches to perform large-scale production of chemicals, intermediates, and pharmaceutical ingredients: use a (i) purely chemical strategy, use a (ii) mixed chemoenzymatic route, or use a (iii) purely biotechnological route with a de novo biosynthesis by fermentation or multistep biotransformation [81]. Table 27.6 summarizes the options of the biotechnological toolbox and briefly explains the terms. The biotechnological methods are unmatched in their chemo-, regio-, and stereoselectivity.

The potential of biotechnology for sustainable manufacturing of organic chemicals may still be far from where it actually could be, but there is steady progress [10, 82]. This can be observed when analyzing the research and development presented at the "International Symposium on Biocatalysis and Biotransformation" (the Biotrans conference series), an event with a long tradition, assembling the important academic and industrial players. The results are summarized in Table 27.7.

There has been a change over the last few years: hydrolytic enzymes are no longer the most frequently used enzyme group in R&D. Among the lectures and posters presented in 2013, oxidoreductases became the current stars, followed by transferases.

TABLE 27.6 Overview of Biotechnological Methods that can be Applied for Chemistry

Biosynthesis. De novo production of an entire organic chemical molecule by a living microorganism cultivated in special vessels (fermenters or bioreactors), which allows the "monoseptic" propagation of the desired organism. Molecules of the primary (e.g., amino acids) or secondary metabolism (e.g., cytotoxics) can be the product, and the microorganism itself can be the desired product, for example, for a biotransformation or an enzyme isolated from this organism and used for biocatalysis
Biocatalysis. Chemical conversion of a substance with the aid of an isolated or immobilized enzyme. One or several enzymes (enzyme cascade) can be used in a one-pot reaction. Some enzymes allow the reaction to be run in organic solvents
Biotransformation. Chemical conversion of a substance into a desired chemical product with the aid of a (usually) living cell. The biotransformation can take place during the fermentation process in parallel with biomass formation or after fermentation, as the biotransformation itself does not necessarily need to be operated under sterile conditions
Cell-free methods. Prokaryotic as well as eukaryotic cell content machineries or hydrogels used for protein production can be used for toxic or nonnatural products and proteins. The productivity of such systems is, however, very limited when compared to fermentation

TABLE 27.7 Overview of the Enzyme Classes Presented as Oral or Poster Presentations at the Last Six "Biotrans Symposia"

Enzyme Class	2013	2011	2009	2007	2005	2003
Oxidoreductase	43	39	32	34	24	28
Transferases	13	13	10	8	6	3
Hydrolases	25	35	46	41	55	58
Lyases	9	13	9	12	12	10
Isomerases	10	3	3	2	2	1
Ligases	1	1	0	1	1	0

In order to identify the research focus in biocatalysis, all abstracts submitted to the "International Symposium on Biocatalysis and Biotransformations" for the years 2003–2013 have been analyzed and the relative numbers indicated in percentages.

As an example, 34% of all presentations in 2007 dealt with oxidoreductases. The sum must not add up to 100% since some of the presented papers did not deal with biotransformations.

Alcohol dehydrogenases (ADHs) and ketoreductases (KREDs) have become quite established in organic chemical synthesis, with test kits and a few enzymes available today in amounts required for industrial biotransformations. Transaminases are probably the next enzyme class (besides hydrolases and oxidoreductases) establishing themselves in organic chemical synthesis. Two thirds of all the transferases studied were transaminases, the remaining catalyzed transglycosylation reactions (14%) or a whole range of different aryl- and alkyl-group transfer reactions (24% of all transferases). Forty-six contributions were on lipases and three on esterase. Together they represent 12% of all contributions and 47% of the hydrolytic enzymes in 2013. In earlier events, in 2007, for example, two thirds of all contributions on hydrolytic enzymes were lipases/esterases. Hydrolytic enzymes are losing overall importance but lipases/esterases still play a role, which is not surprising considering their commercial availability and the practical experiences accumulated with them over many years. *In situ* product recovery (ISPR), ultrasound-assisted biocatalysis, tube-in-tube membrane reactors for biphasic processes, cofactor recycling using electrodes, decision guidance, immobilization techniques, or computational design methods were engineering and process design topics presented during the last Biotrans 2013. However, only one topic in engineering and process design topic was a little more prominent: multistep biocatalysis. However, the (only) 14 papers dealing with multienzyme cascades show that one pot, one enzyme is still the classic approach in R&D. One weakness of conferences is often the low degree of active participation of industry. This is not different for the "Biotrans" series, which had only 34 contributions (i.e., <10%) from industry. The vast majority of papers were from academic institutions. Concrete examples of industrial reactions are given in Table 27.8.

27.3.2 The Future Toolbox

Biotechnology creates value upstream in a production process for chemicals, intermediates, and pharmaceutical ingredients. This created value must be conserved in the following isolation and purification or downstream process. Innovation is needed both in the upstream process (USP) and in the downstream process (DSP) to minimize losses, remove side products, and still be environmentally tenable. Bioreactions are mostly carried out in aqueous solutions and product isolation can be problematic because of the similarity of solubility and physicochemical properties of the product, substrate, and side product. However, effective methods exist to address most isolation and purification problems. Alternative DSP practices are increasingly used and

TABLE 27.8 Examples of Reactions Developed and Carried Out by Lonza, One of the Pioneering Companies Using Biotechnology for Organic Chemical Synthesis of Fine Chemicals, Intermediates, and Pharmaceuticals [83]

	Reaction Class	Substrate Class	A Lonza Example
Oxidation	Methyl group oxidation	Heteroaromatic compounds	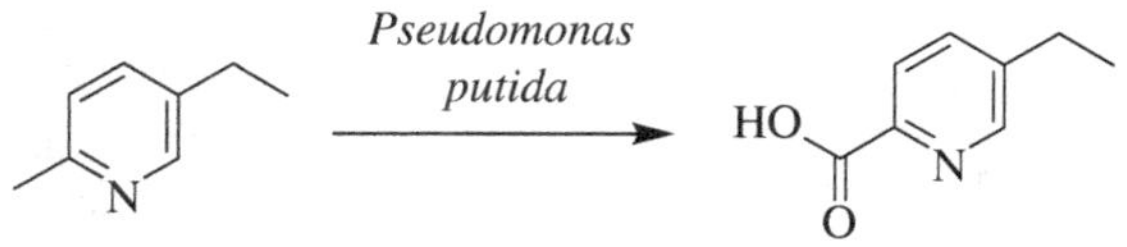
	Ethyl group oxidation	Heteroaromatic compounds	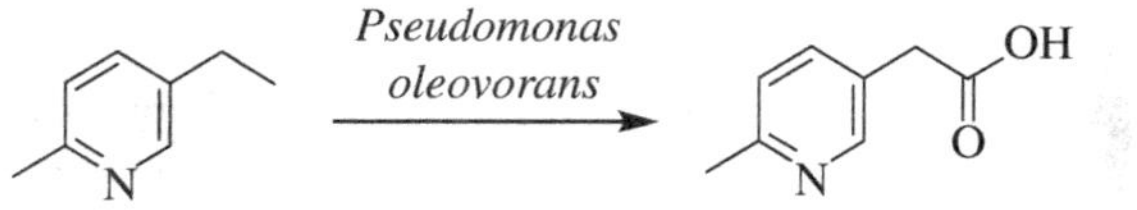
	Selective oxidation	Polyols	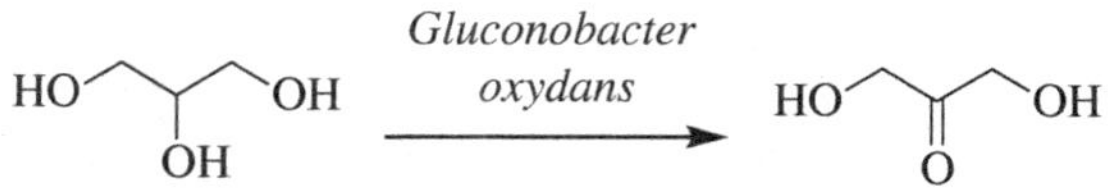
	Hydroxylation	Heteroaromatic compounds	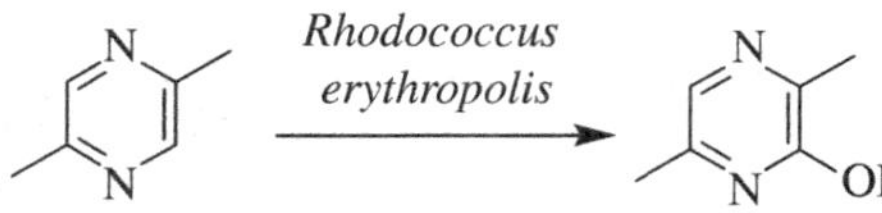
Reduction	Carbonyl reduction	β-Oxo esters	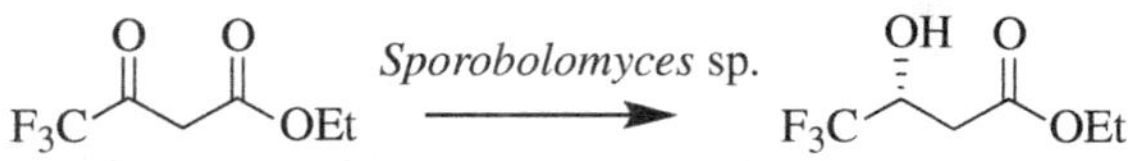
	Double-bond reduction	Activated enols	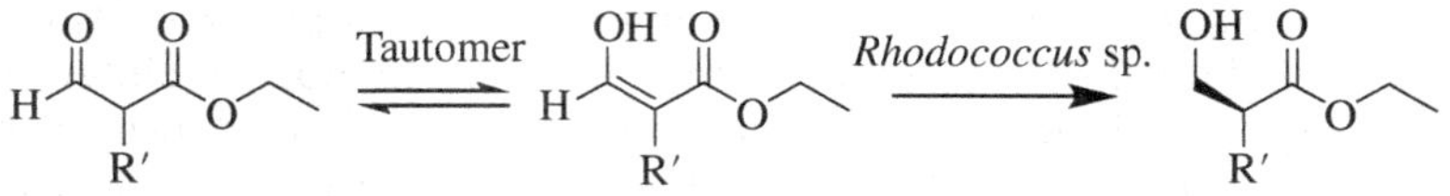
Hydrolysis	Acetyl hydrolysis	α-Amino acids penicillin	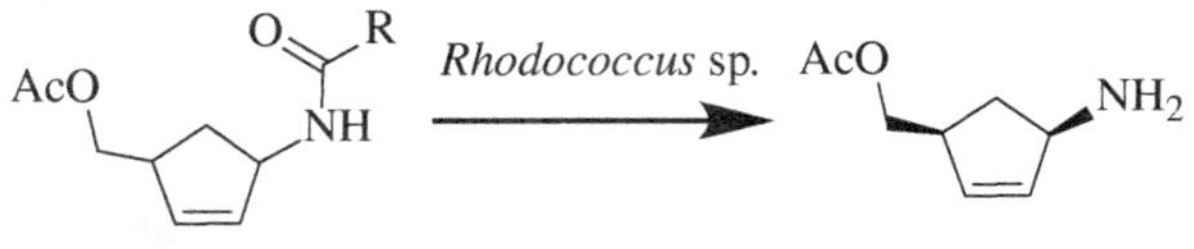
	Ester hydrolysis	α-Hydroxy carboxylic acid esters	
	Nitrile partial hydrolysis	*N*-containing heterocyclic cyanides	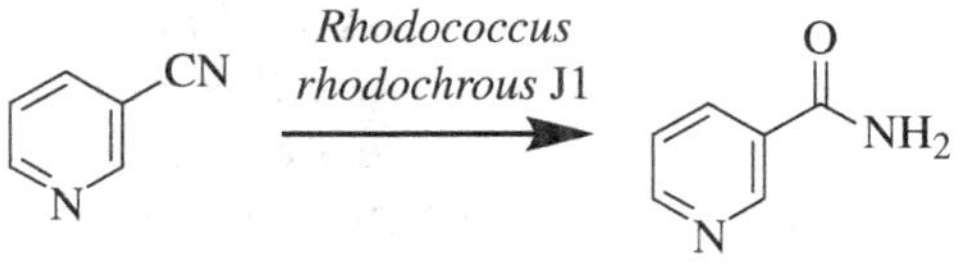
	Amide hydrolysis	2-Carboxamides	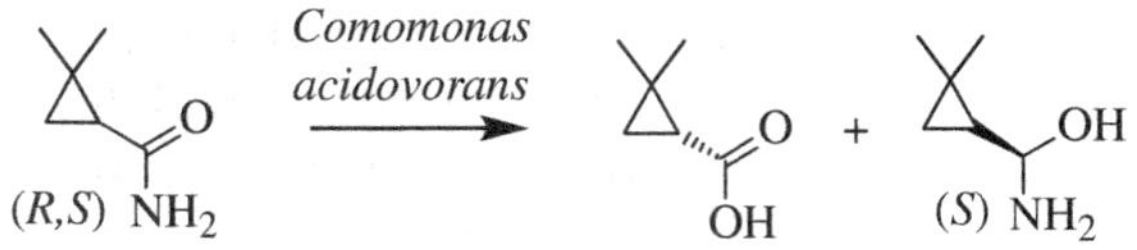
Amination	Transamination (*R*- and *S*-specific)	*N*-containing-3-amino heterocycles (asymmetric synthesis or racemic resolution)	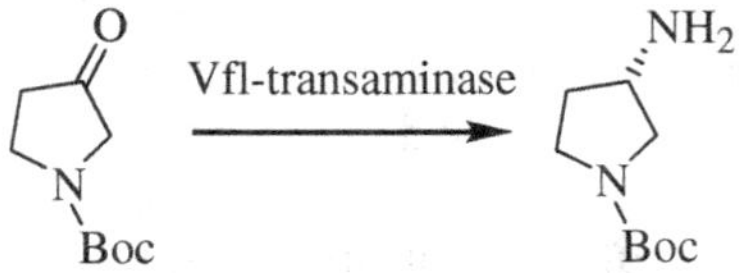

Examples of market participants are AB Enzymes, Iogen, Verenium Corporation, Dyadic, Specialty Enzymes, Biochemical, Codexis, c-LEcta, Evocatal, DSM, Lonza, Promega, Solvay Pharmaceuticals, and others.

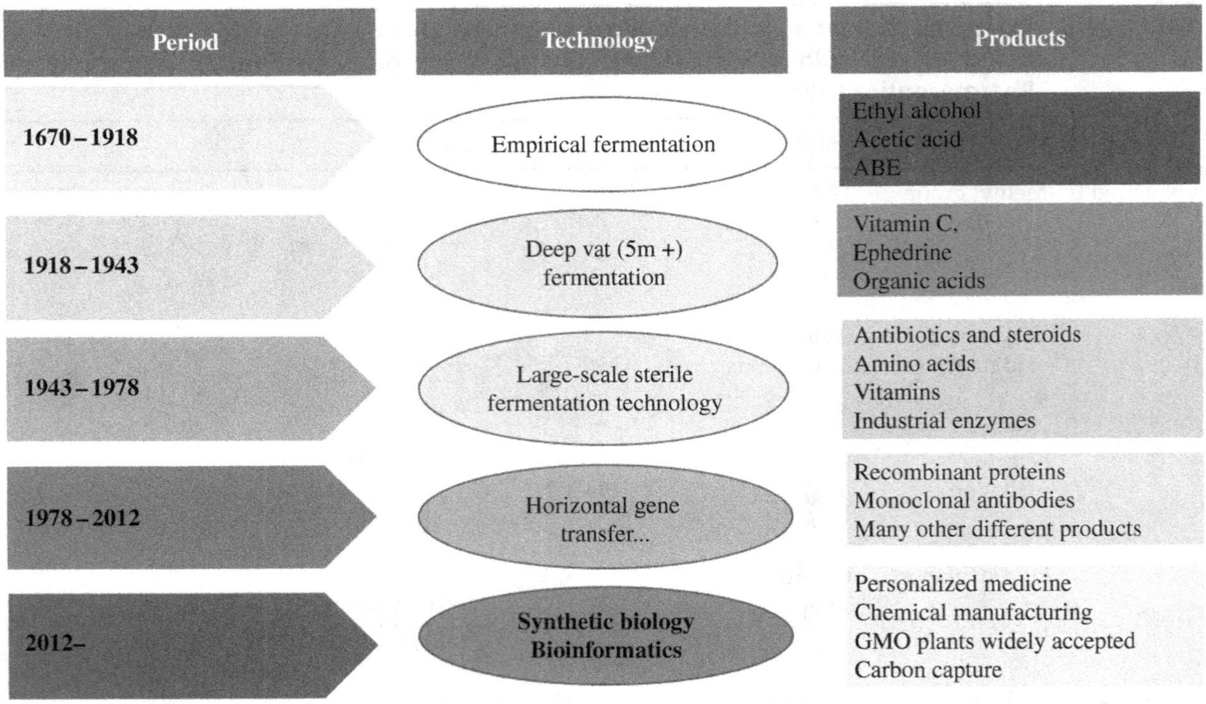

FIGURE 27.4

Overview of the historic key events and technical breakthroughs leading to modern biotechnology applied for the production of chemicals, intermediates, and pharmaceutical ingredients. Horizontal gene transfer or genetic engineering was the groundbreaking innovation that opened the door to modern biotechnology. Synthetic biology in combination with bioinformatics will fuel the next breakthrough with similar importance, changing the way organic chemicals will be produced in the future. However, the full realization of true synthetic biology is a slow uphill battle, but its impact will be similar to molecular biology in the 1970s and the 1980s. However, we can already harvest the first achievements today.

cutting-edge technologies, which might be applied in the future, are in development. For example, the thinnest ever membrane with defined pores perforated with focused ion beam technology was produced, with only two layers of graphene and precise pore diameters, and these new membranes will allow for very fast and precise filtration purposes for gases and liquids on an industrial scale if commercialized [84]. Enzymatic processes are finely tuned and product inhibition is a common phenomenon. ISPR addresses this product inhibition by removing the inhibiting product during the bioprocess as it is formed. Unfortunately, no generic method is available yet. The opposite problem, educt, or substrate inhibition is effectively solved using fed-batch technology.

However, the need for innovation when producing chemicals, intermediates, and pharmaceuticals with biotechnological methods is mainly in the upstream processing part. Diversity is not only needed in the biosphere but also in our enzyme toolbox, and this must be achieved through better use of the natural biodiversity. We do not need more of the same but more of the different. Enzymes for reductive amination, fluorination of nonactivated C, ω-transaminases, monooxygenases, or in general C–N and C–C bond enzyme classes are frequently mentioned reactions and enzymes on wish lists. Once an enzyme with the desired activity is available, any negative attributes with respect to industrial synthetic organic processing can be rapidly engineered out.

Synthetic biology and its derivatives (synthetic biotechnology, systems biocatalysis, cell-free synthetic biology) are a much (ab) used term. However, these developments, in essence, will fundamentally change the way we produce chemicals in the future (Figure 27.4). Synthetic biology, in the strict sense, is about redesigning, recoding, reprogramming, and reconstructing genes as well as alternative biological systems. This bottom-up approach, assembling cellular components to design and construct the perfect organism for the production of a chemical, intermediate, or pharmaceutical is in its infancy. Traditional genetic and metabolic engineering is often simply rebranded as synthetic biology. However, its impact will be similar to that of molecular biology in the 1970s and the 1980s, and the first products are already being harvested today. Artemisinin (Amyris, Sanofi) and 1,3-propanediol (DuPont) are two early examples where synthetic biology principles were successfully applied for commercial purposes. Evolva, founded in 2004, is successfully applying synthetic biology principles in *S. cerevisiae*. Through massive pathway transplantation and genetic pathway engineering, they have four products on the market or which are soon to be launched, namely, resveratrol, stevia, vanilla, and saffron, all produced in recombinant yeast. Bioinformatics is at the center of this, enabling the handling of the "big data deluge" from the "omics" world. Massive process data handling requires computer power on a much bigger scale than today. Similarly, systems biocatalysis is a new approach consisting of organizing enzymes in vitro to generate an artificial metabolism for synthetic purposes. The strategy of this key new platform deals with the analysis of enzymatic systems in vivo, the development and discovery of new biocatalysts, and their assembly in vitro into novel synthetic metabolic pathways. This network aims at the controllable construction of metabolic pathways for the efficient synthesis of valuable chemical products.

27.4 SUSTAINABILITY, GREEN PREMIUM PRICING, AND SUBSIDIES

To address a growing public concern about the burden of chemical manufacturing, the chemical industry founded "Responsible Care" in 1985 as a voluntary initiative [85]. The 12 principles of green chemistry, formulated by Anastas and Warner in 1998 to prevent pollution [86] and reduce resource consumption, have been followed consciously or subconsciously by synthetic and process engineers for one simple reason: these principles make a lot of sense from an economic point of view. However, biotechnology is needed to bring green chemical synthesis to the next level. There are various definitions and a vision of a bioeconomy and its relations to sustainability varies [87], but a positive impact of bioeconomy and biotechnology is not self-evident. Sustainability is more than just cleaner processes, less waste produced, less energy consumed, or decreased greenhouse gas emissions. Sustainability has three axes:

1. Ecology
2. Economy
3. Society

Ecology: Ecology is a major driver for the use of biotechnology as one tool among others for a "greener chemistry." The E-factor, introduced by Roger Sheldon over 20 years ago, defines the amount of consumables used per kilogram of product produced. It may not be a true measure of the environmental impact, as the E-factor considers only the amount but not the nature of the product used or waste produced. But it gives a good quantitative idea and shows that small molecule pharmaceutical drugs have one of the highest E-factors in the organic chemical industry, with 25–100 kg of material consumed per kilogram of product. The petrochemical industry, on the other hand, is one of the "cleanest" industries. Of course their low E-factors are offset by the partly gigantic amounts produced in the petrochemical industry. But from an ecological point of view, especially the pharmaceutical industry needs to reduce its manufacturing footprint. Numerous other green metrics (e.g., atom economy, stoichiometric factor, reaction mass intensity) were developed since the introduction of the E-factor, which remains one of the most frequently quoted. Eco-footprint, for example, is a new visualization tool for green metrics which gathers ten indicators to assess the environmental impact of a process. The proof of concept of the tool was carried out with two cosmetic products that L'Oreal produced chemically [88].

Using greenhouse gases (CO_2 from combustion processes, methane from flares) as feedstock, two birds can be killed with one stone. Carbon dioxide is captured and a product produced. With this form of biological carbon capture, carbon dioxide trade certificates can be sold. This approach theoretically allows the highest margins and difference of price between the starting material and the final product, if manufacturing costs can be kept low. Chemical and biological solutions are possible. Novomer chose contract manufacturer Albemarle to produce polyols and other polymers that use CO_2 as raw material. Novomer develops catalysts for the production of low-cost chemicals and polymers by converting abundant feedstocks like carbon dioxide and carbon monoxide plus an epoxide (e.g., ethylene oxide) as starting materials to acrylic acid, 1,4-butanediol, and ethylene carbonate. Microbial production with C1 sources such as CO_2 and methane can be used for the production of PHA [89], as proposed by the Spanish SME Biopolis leading a European consortium developing the conversion of CO_2 and methane to PHB. Methane is used to reduce CO_2 to PHA. There are other biotechnological and purely chemical opportunities for direct CO_2 capture that are not discussed in detail. Codexis works on a low-cost biomimetic CO_2 emission capture from power generation or aluminum processing plants using enzyme catalysis with a thermophilic carbonic anhydrase, which allows capture of CO_2. The Gram-positive bacterial genus *Clostridium* shows a great metabolic diversity and is also able to grow on a spectrum of waste gases from industry or on syngases CO and H_2. LanzaTech, together with Siemens and Oakblo, is working on the microbial conversion of CO from the steel industry [90]. Joule has created photosynthetic microorganisms that use sunlight to convert carbon dioxide efficiently into ethanol or diesel photobioreactors.

Economy: A bioprocess substituting a chemical process must be economically viable, as long-term subsidies or tax breaks are economically not attractive. Jobs must be created and the activities must contribute to the gross national product and blend into a national and local economy. In many Western countries, the tertiary or service sector has been enlarged (often with an excessive proportion in the financial service sector) at the cost of the secondary or manufacturing sector. After the bank bailout by taxpayers and the transfer of jobs to low-wage countries leading to a creeping deindustrialization, people have started to realize that the secondary sector and manufacturing really matters! Biotechnology is one possibility to bring back attractive jobs with a long-term prospect. They may not have the excessive gains per full-time employee of the financial sector, but they are based on real and not virtual products. This, however, requires that banks play their inherent role and invest money into the manufacturing sector. Moreover, a bioeconomy is an opportunity for the primary or agricultural sector.

Green or premium price is an additional price paid by the consumer or the government for an additional technical, emotional, and/or strategic performance for a biotechnology product. For example, 72% of consumers would be prepared to pay a 10–20% premium price increase for bioplastics. But premium pricing will become increasingly difficult and a premium pricing against a gradient cannot be maintained for too long time if tax money is needed. Irresponsible behavior by the financial sector and excessive spending by public households lead to a lack of funds. The decision on where to spend money should be driven by standardized objectives and not lobbying. The past 30 years of globalization have led to an easy-money environment but we are coming to an end of that era and we need to get prepared for a bumpy road.

Society: The question is "what can biotechnology do for society?" and not the other way round, meaning the funding for our next research project. Gains for society are, for example, reducing inequalities, technologies contributing solutions for a nation and future generations. The UN climate change report of its international panel Intergovernmental Panel on Climate Change (IPCC), presented in Yokohama on March 28, 2014, did not look good [91]. However, the problem is that cranking up domestic consumption is what are politicians generally proposed in order to reduce unemployment and to sustain economic growth. Good news usually reads, "GDP is on the rise as consumers start to spend more again." Unless we come up with new manufacturing methods and consider life cycles of products, we are spiraling to certain exhaustion and forced negative economic growth. Biotechnology must provide economists and politicians with sustainable products and manufacturing processes.

27.5 REGULATORY ASPECTS AND PUBLIC PERCEPTION

Regulatory: For biofuels or commodities, volumetric productivity and raw material costs are the important drivers. Bioproduction of fine chemicals, intermediates, or pharmaceuticals, on the other hand, are driven by the cost of development, regulatory approval, and compliance. Time constraints, due to dates stipulated by clinical trials, are an additional difficulty for pharmaceuticals. As the products, manufacturing methods, markets, and applications are very different, so are practices regulating manufacturing processes to assure safe products for consumers. During the German biotechnology days in May 2012, a pharma executive said during his presentation, "anybody can do research, but only big pharmas can develop." What he meant was getting a drug candidate through clinical trials onto the market. Although the chemoenzymatic synthesis of a cytotoxic parenteral product or a biofuel, for example, is an entirely different issue, all necessary rules and regulations for the application of enzymes and microbial cells for manufacturing are in place!

Table 27.9 summarizes the biotechnological methods used in the production of chemicals, intermediates, and pharmaceuticals. Although rules are clear, problems and product recalls have been reported. In 2013 product issues arose with imported enzyme preparations contaminated with chloramphenicol used in feed additives. One important aspect to be compliant is to use trustworthy suppliers of raw materials, enzymes, and cells. It is important to ensure that a supplier operates with validated processes, even if an enzyme is used as a consumable several steps upstream from the finished product. If an enzyme is used in a critical step and close to the end product, one should consider personally auditing the supplier. Enzymes are ideally produced with generally recognized as safe (GRAS) organism or accepted recombinant strains and vectors, also used for the injectable therapeutic proteins. If a Gram-negative organism is used, enzyme preparations need to be endotoxin-free, except if used in processes not intended for human consumption. It is essential to use bovine spongiform encephalopathy/transmissible spongiform encephalopathy (BSE/TSE) free raw materials also in post-fermentation operation such as DNase steps and assure no negative impact from residuals from the biocatalysts or leachates, for example, with immobilized enzymes. Certified suppliers and custom manufacturers with flawless and long-term track record should also be used. To cut a long story short, there are no uncertainties about rules and regulations on how to use biotechnological processes for the production of chemicals, intermediates, and pharmaceutical ingredients, but the rules and methods need to be known and interpreted correctly. The only areas where regulatory aspects might not yet be completely clear are novel applications and in combination with novel products, such as functionalized biopolymers in medical devices.

Public perception: The rate of change is starting to outpace the capacity of man, society, and politics to follow these developments. The contrast between science push and consumer aversion is partly debilitating the transfer from science to the market. It took 2000 years to develop microbial antibiotics—as molds were used as anti-infectives by ancient cultures. The helicopter took 400 years from the first concept by

TABLE 27.9 The Use of Biotechnology for the Production of Chemicals, Intermediates, and Pharmaceuticals Is Clearly Regulated

BSL 1–4	Biosafety levels 1–4 for a containment used to grow biomass. Harmless organisms are in BLS 1, whereas BLS 4 is required for work with them most dangerous agents. Most applications discussed in this chapter require BLS 1, eventually BLS 2. Learn about BLS on www.cdc.gov/training/quicklearns/biosafety/ of the Centers for Disease Control and Prevention
CDER	The FDA's Center for Drug Evaluation and Research, reviewing new molecular entities (NMEs) and new chemical entities (NCEs). CBER is FDA's Center for Biologics Evaluation and Research (CBER) for the regulation of complex biological products
CPPS	Critical Process Parameters and Critical Quality Attributes (CQAs)
cGMP	Current Good Manufacturing Practice standards required by the FDA for the production of products and consumables intended for human consumption
EFSA	European Food and Safety Authority (www.efsa.europa.eu). Safety assessment of microorganisms used in food and feed can be downloaded
EMA	European Medicines Agency (www.ema.europa.eu) publishes harmonized guidelines between Europe, Japan, and the United States
FAMI-QS	European Feed Additives and PreMixtures Quality System Feed
FDA	The US Food and Drug Administration (www.fda.gov) and regulatory authority for drugs, vaccines, veterinary products, food and dietary supplements, cosmetics, medical devices, and other products. Other countries have their own regulatory agencies such as The State Food and Drug Administration in China, the Central Drugs Standard Control Organization (CDSCO) in India, or the Pharmaceutical and Food Safety Bureau of the Ministry of Health, Labor and Welfare (MHLW) in Japan. Because of the importance of the US market, its regulatory authority has corresponding influence
FSMA	FDA's Food Safety Modernization Act of 2011, shifting the focus from responding to contamination to prevention
GRAS	Generally recognized as safe. The GRAS list includes close to 50 microorganisms
HACCP	Hazard Analysis and Critical Control Points (HACCP) for food safety from raw material production to the finished product
HFFIA	Halal Feed and Food Inspection Authority according to Islamic law. The term covers and designates not only food and drink but also all matters of daily life
ICH	International Conference on Harmonisation (www.ich.org) harmonizing guidelines on quality, safety, efficacy, and multidisciplinary guidelines since the 1960s
ICH Q10	Guidance for the implementation of a pharmaceutical quality system
ISO	International Organization for Standardization (www.iso.org). Quality systems for the manufacturing and service industry with ISO 9000 standards series since the late 1980s
Kosher	Companies producing enzyme and/or food products by biotechnology can become certified through an inspection by a local rabbi or an inspecting organization such as OK (www.ok.org) or others
PAT	Process analytical technology to design, analyze, and control processes through measurement of CPPS
QbD	Quality by design. Systematic approach to achieve product quality by thorough process understanding, monitoring, and control
SOP	Standard operating procedure

The table lists some of the relevant abbreviations, quality standards, and inspecting authorities in alphabetical order.

Leonardo da Vinci to the first commercial Sikorsky helicopter. The telephone took about 40 years in the nineteenth century. It took exactly 20 years to bring DNA technology into the manufacturing plant and the Internet needed about 20 years to change the world. Today some future inventions are announced before anybody has set a foot into a laboratory to start the corresponding experiments. Society needs time to adapt to new technologies and it seems as if the rate of change is exceeding its capacity to follow these developments, and biotechnology is no exception. Societal perception of biotechnology is ambivalent through sensation mongering, and as a result of media reporting, this technology is often portrayed as either an enemy or an ally. Even today the public does not fully understand what biotechnology is and is often needlessly alarmed. The fact that we address different markets, different products, different processes, different technologies, and different regulatory bodies complicates communication. We ourselves make things worse by giving mixed signals nurtured by many different interests. Funding and venture capital is hard to come by, as products and technologies are frequently presented in overoptimistic manner. We need to better manage the delicate balance between getting people excited about the possibilities of biotechnology and managing expectations. Biofuels have dominated the white biotechnology picture in society. We need to show and prove that biotechnology can provide sustainable processes in other areas as well. Industry now realizes that regulatory processes are not the sole hurdle prior to commercialization but that national and international public opinion and gender is an important determinant of people's attitudes toward science and technology [92].

27.6 INNOVATION (NOT ONLY IN THE LABORATORY!)

Executives frequently underline the importance of R&D and innovation but are reluctant to commit financially and are hesitant to collaborate with academic institutions, because the latter are less interested in incremental innovation and short-term gains but are looking for long-term commitments needed to solve difficult technical problems and to access high-hanging fruits. The industry is extremely risk averse. "Not only large global players, also new companies become eventually conservative and risk-averse, which makes it difficult to develop something orthogonal to technology they have become familiar with. To do that one sometimes has to start the next company," said Andreas Plückthun, serial entrepreneur and professor at the University of Zürich. Obviously, only a wet baby appreciates a change. However, investors also become involved at ever later stages. More than a few suggest that the government should take the initial risk and companies take over when the risk is more easily calculable. This is one reason why universities and public research institutions are becoming more and more involved in the commercialization of their own research results, leading to an increasing number of university patents [93] and spin-offs generated by them. This is not a bad development per se, but the danger is that there is ever less fundamental research leading to unexpected true breakthrough, as in the past when the molecular tools for recombination we use today were identified. Cooperation is not yet really rewarded in the industrial environment. However, some positive changes can be observed, using, for example, open-access approaches and electronic knowledge-sharing tools to foster collaboration. This is also a necessity as the boundaries between the traditional distinctive worlds of science and technology and information technology are becoming more and more porous. The Swiss company Lonza, for example, established the Lonza Innovation for Future Technologies (LIFT) to assure the long-term perspective for the company, with a remarkable budget and external cooperation as the key element of success, in the search for disruptive technologies. But the fate of this or other programs with radical approaches is that they are typically among the first victims of CEO changes. Fortunately, several industrial precompetitive collaborations to develop technology platforms and derive mutual advantages from them have been established. The US National Intelligence Council (USNIC) and the EU Institute for Security Studies (EUISS) identified biotechnology as one of three over-the-horizon issues that is likely to rise in importance and will demand a higher level of global cooperation. Sage Bionetworks [94] was founded in 2009 as a nonprofit research organization and open biology information platform to link genetics and drug discovery through pattern recognition [95]. Chem21 and Swiss Industrial Biocatalysis Consortium (SIBC) are two examples [96, 97] of cooperation platforms aiming at crossing classic barriers. The first one, Chem21, is a European Federation of Pharmaceutical Industries Association—Innovative Medicines Initiative consortium. It is Europe's largest public–private partnership dedicated to the development of sustainable manufacturing. As mentioned, the number of reaction types that can rely on a broad panel of commercial enzymes, remains limited. Thus, one major goal of the second cooperation platform, the SIBC, was explicitly to exchange not only information but also living strains and enzymes among its eight industrial partners in order to improve the availability of characterized enzymes and strains. Some of the Swiss partners are competitors on the global market, but they realized that this cooperation helps them survive better in very competitive environments. Two conditions have to be met to make such a consortium work. (1) An appropriate legal framework needs to be established, which by the way took SIBC 2 years to realize. This is an example where one has to be innovative not only in the laboratory but also in legal frameworks, for example, for unusual cooperation models. (2) Only a limited number of active industrial partners are accepted and the trust level must be high to make the consortium work. Cloud manufacturing linking chemical manufacturing to computational intelligence to facilitate sustainable biobased chemical processing is another interdisciplinary approach of which we will see more in the future. It is often stated that chemists should become more familiar with the use and application of enzymes in organic chemical synthesis. It is a valid request but the opposite is also true. The author's personal experience is that chemists, especially in the academic world, become acquainted with molecular biology and genetic engineering and, even use it, much more easily than biologists become familiar with organic chemistry.

27.7 CONCLUSIONS

The application of biotechnology for chemistry, intermediates, and pharmaceuticals can look into a promising future if we do things right. The opportunities in numbers: the sales of global chemical markets are expected to grow over 5 trillion in 2020. Only about 3–6% of all chemical sales have been generated with some help from biotechnology in the past, but this figure is anticipated to grow to ~20% in 2020, which translates into something like 1 trillion US$! That means an order of magnitude increase from today's figures. The share of biotechnologically produced fine chemicals is expected to grow even faster from 8 to 60% between 2001 and 2010. Estimates and definitions may vary but the message is clear: the proportion of products manufactured using biotechnology is expected to increase over proportionally. Section 27.2, although neither exhaustive nor complete, gives an idea on how multifaceted the application of biotechnology already is today. Actually, by the time this book will be printed and published, many more applications, collaborations, and new options will be available. However, this multiplication of the possibilities for the manufacturing of chemicals using biotechnology has led to strange priorities and to a dilution of available resources. This dilution of funds in combination with insufficient focus in times of abundant opportunities can be a toxic cocktail. Not only are new process technologies and products needed but also new models of focused collaboration and networking to facilitate progress and to drive the implementation of biotechnologies. The fine chemical intermediate and pharmaceutical sectors are also leading the innovation in "white" or "industrial" biotechnology, solving scientifically and technically demanding problems, mostly for nonsubsidized product sold on the free market. We need to pay more attention to the processing of these high-value-added products in the biorefinery concept using prokaryotes and eukaryotes including microalgae.

The ranking of the industries using biotechnology in terms of value added actually goes more or less in the same order: (i) pharma, (ii) food industry, (iii) agrochemicals, (iv) cosmetics, (v) polymers, and (vi) others. The ranking of enzymes used has shifted to enzymes important for higher-value compounds: oxidoreductases EC 1 (50%), hydrolases EC 3 (35%), transferases EC 2 (>5%), lyases EC 4 (>5%), and isomerases EC 5 (>1%). The diversity of the industrially available enzyme toolbox, however, remains a problem that must be solved; especially as there is no doubt today, that biotechnology and biocatalysis are key technologies for a sustainable (green) chemistry. In 2013, for example, GlaxoSmithKline [98] CEO Witty communicated, among other things, that a shift from synthetic chemical reactions to enzymatic reactions would take place with the corresponding investments. It almost seems that we are privileged in having too many choices, like a child in the candy store who cannot decide what to choose. But keeping the stretched financial situations and required corrective measures of governments to reduce financial liabilities in mind, there is no time for scatter shooting and nonselective funding. The focus must be where organic chemical synthesis has the greatest difficulty in reducing its manufacturing footprint: high-value complex and chiral molecules.

ACKNOWLEDGMENTS

I would like to thank Miriam Guider (HES SO Sion) and Gareth Griffiths (Luna) for their redaction work and Oleg Werbitzky (JTI), Diego Schmidhalter (Lonza), and Manfred Zinn (HES SO Sion) for their valuable comments.

REFERENCES

[1] King, D. (2010) The future of industrial biorefineries. World Economic Forum, Geneva, Switzerland. REF 210610. www.weforum.org (accessed December 24, 2015).
[2] Hinman, L.R. (1991) A role for biotech in producing chemicals? *Nat. Biotechnol.*, **9**, 533–534.
[3] The Contours of World Development OECD. http://www.theworldeconomy.org/ (accessed June 10, 2014).
[4] FAO's Food Price Index Revisited. http://www.fao.org/fileadmin/templates/worldfood/Reports_and_docs/FO-Expanded-SF.pdf (accessed May 29, 2014).
[5] http://www.cargill.com (accessed December 24, 2015).
[6] Meyer, H.-P., Schmidhalter, D. (2014) The History and Economic Relevance of Industrial Scale Suspension Culture of Living Cells, in *Industrial Scale Suspension Culture of Living Cells* (Eds. H.-P. Meyer and D. Schmidhalter), Wiley Blackwell, Weinheim, Germany, pp. 1–37.
[7] Bayer Pharma AG (2014) Small and large molecules: drugs on a chemical and biological basis. http://www.bayerpharma.com/en/research-and-development/technologies/small-and-large-molecules/index.php (accessed June 26, 2014).
[8] Andrews, I., Dunn, P., Hayler, J., Hinkley, B., Hughes, D., Kaptein, B., Lorenz, K., Mathew, S., Rammeloo, T., Wang, L., Well, A., White, T.D. (2011) Green chemistry articles of interest to the pharmaceutical industry. *Org. Process Res. Dev.*, **15**, 748–756.
[9] Bade, R., Chan, H.F., Reynisson, J. (2010) Characteristics of known drug space. Natural products, their derivatives and synthetic drugs. *Eur. J. Med. Chem.*, **45**(12), 5646–5652.
[10] Patel, R.N. (Ed.) (2010) *Biocatalysis in the Pharmaceutical and Biotechnological Industries*, CRC Press, Boca Raton, FL.
[11] Bornscheuer, U.T., Huisman, G.W., Kazlauskas, R.J., Lutz, S., Moore, J.C., Robins, K. (2012) Engineering the third wave of biocatalysis. *Nature*, **485**, 185–194.
[12] Meyer, H.-P. Werbitzky, O., Signorell, G.A. (2014) Bridging the gap between chemistry and biotechnology—large molecules with potential, how could biotechnology complement chemistry. *Curr. Org. Chem.*, **18**, 944–959.
[13] Werbitzky, O., Giraud, M. (2014) Manufacturing of Venom-Derived Therapeutic Peptides, in *Venoms to Drugs* (Ed. G.F. King), RSC Publishing, Cambridge, UK, pp. 290–305.

[14] Wegmüller, S., Schmid, S. (2014) Recombinant peptide production in microbial cells. *Curr. Org. Chem.*, **18**, 1005–1019.
[15] Petitou, M., van Boeckel, C.A.A. (2004) A synthetic antithrombin III binding pentasaccharide is now a drug! What comes next? *Angew. Chem. Int. Ed.*, **43**(3), 3118–3133.
[16] Biotechnologie.de (2014) Milk sugar for healthy baby foods. https://www.biotechnologie.de/BIO/Navigation/EN/Funding/foerderbeispiele,did=171048.html (accessed June 28, 2014).
[17] Larry Page, Google. http://plus.google.com/+LarryPage/posts/Lh8SKC6sED1 (accessed June 22, 2014).
[18] Kalbermatten, D.F., Pettersson, J., Kingham, P.J., Pierer, G., Wiberg, M., Terenghi, G. (2008) New fibrin conduit for peripheral nerve repair. *J. Reconstr. Microsurg.*, **25**(1), 27–33.
[19] Competence Centre TEDD. http://www.project.zhaw.ch/en/science/tissue-engineering-for-drug-development.html (accessed December 10, 2013).
[20] EuropaBio (2008) How industrial biotechnology can tackle climate change. http://www.europabio.org/how-industrial-biotechnology-can-tackle-climate-change-0 (accessed June 3, 2013).
[21] c-LEcta (2014) c-LEcta enzymes and trains—products—asparaginase. http://www.c-lecta.com/?lang=en&category=products&page=asparaginase (accessed June 21, 2014).
[22] Schelten, P., Twisk, J.W.R., Blesa, R., Scarpini, E., von Arnim, C.A.F., Bongers, A., Harrison, J., Swinkels, S.H.N., Stam, C.J., de Waal, H., Wurtman, R.J., Wieggers, R.L. Vellas, B., Kamphuis, J.G.H. (2012) Efficacy of Souvenaid in mild Alzheimer's disease: results from a randomized, controlled trial. *J. Alzheimers Dis.*, **31**, 225–236.
[23] Crowley, L. (February 11, 2008) Blue Smarties are back thanks to Spirulina. http://www.foodnavigator.com/Market-Trends/Blue-Smarties-are-back-thanks-to-Spirulina (accessed June 22, 2014).
[24] Timmerman, L. (March 12, 2013) GlaxoSmithKline shuts down sirtris. http://www.xconomy.com/boston/2013/03/12/glaxosmithkline-shuts-down-sirtris-five-years-after-720m-buyout/ (accessed June 15, 2014).
[25] Deguerry, F., Pastore, L., Wu, S., Clark, A., Chappell, J., Schalk, M. (2006) The diverse sesquiterpene profile of patchouli, *Pogostemon cablin*, is correlated with a limited number of sesquiterpene synthases. *Arch. Biochem. Biophys.*, **454**(2), 123–136.
[26] Daviet, L., Schalk, M. (2010) Biotechnology in plant essential oil production: progress and perspective in metabolic engineering of the terpene pathway. *Flavour Fragr. J.*, **25**(3), 123–127.
[27] Schalk, M., Pastore, L., Mirata, M.A., Khim, S., Schouwey, M., Deguerry, F., Pineda, V., Rocci, L., Daviet, L. (2012) Toward a biosynthetic route to sclareol and amber odorants. *J. Am. Chem. Soc.*, **134**(46), 18900–18903.
[28] http://www.allylix.com (accessed June 3, 2014).
[29] Granier, T., Locher, E., Schilling, B. (2012) 1-Hydroxy-octahydroazulenes as fragrances. WO 2012001018 A1 Patent application number PCT/EP2011/060862 (January 5, 2012).
[30] Meyer, H.-P., Eichhorn, E., Hanlon, S., Lütz, S., Schürmann, M., Wohlgemuth, R., Coppolecchia, R. (2013) The use of enzymes in organic synthesis and the life sciences: perspectives form the Swiss Industrial Biocatalysis Consortium (SIBC). *Catal. Sci. Technol.*, **3**, 29–40.
[31] Givaudan, S.A. (2013) Givaudan Annual Report 2012. http://asp-gb.secure-zone.net/v2/index.jsp?id=1134/2952/6465&startPage=21 (accessed June 22, 2014).
[32] Vam aí o #urbano. http://blogconsultoria.natura.net/vem-ai-o-urbano/ (accessed June 22, 2014).
[33] Chatsko, M. (June 18, 2014) What's the value of Amyris' molecule development strategy? http://www.fool.com/investing/general/2014/06/18/whats-the-value-of-amyris-molecule-development-str.aspx (accessed June 29, 2014).
[34] http://www.phytometasyn.ca (accessed December 24, 2015).
[35] http://www.optibiocat.eu (accessed December 24, 2015).
[36] http://www.korres.com (accessed December 24, 2015).
[37] http://www.bitop.de; http://www.induchem.com (accessed December 24, 2015).
[38] http://www.induchem.com (accessed December 24, 2015).
[39] http://www.toulouse-white-biotechnology.com (accessed December 24, 2015).
[40] Arch Personal Care Products: Biovert Enzyme and Substrate. http://www.archchemicals.com/Fed/PC/Products/FunctionalIngredients/BiovertEnzymeSubstrate.htm (accessed June 18, 2014).
[41] http://www.rootec.com (accessed December 24, 2015).

[42] Bloomberg Business (2014) Nestle's galderma skin-care buy shows health care challenge. http://www.bloomberg.com/news/2014-02-12/nestle-s-galderma-skin-care-buy-shows-health-challenge.html (accessed June 22, 2014).
[43] Ravenstijn, J. (2010) Bio-based polymers: a revolutionary change. *JEC Compos. Mag.*, **59**, 17–20.
[44] http://www.amsilk.com (accessed December 24, 2015).
[45] Spider Silk Makes Plucky Violin Strings. http://news.discovery.com/tech/biotechnology/spider-silk-makes-violin-strings-120306.htm (accessed June 20, 2014).
[46] Harvards Gazette (2014) As strong as an insect's shell. http://news.harvard.edu/gazette/story/2014/05/promising-solution-to-plastic-pollution/ (accessed June 20, 2014).
[47] Wilker, J.J. (2010) The iron-fortified adhesive system of marine mussels. *Angew. Chem. Int. Ed.*, **49**, 8076–8078.
[48] Zinn, M., Witholt, B., Egli, T. (2001) Occurrence, synthesis and medical application of bacterial polyhydroxyalkanoate. *Adv. Drug Deliv. Rev.*, **53**, 5–21.
[49] http://www.tepha.com (accessed December 24, 2015).
[50] Le Meur, S., Zinn, M., Egli, T., Thöni-Meyer, L., Ren, Q. (2013) Poly(4-hydroxybutyrate) (P4HB) production in recombinant *Escherichia coli*: P4HB synthesis is uncoupled with cell growth. *Microb. Cell Fact.*, **12**, 1–11.
[51] http://www.tjgreenbio.com (accessed December 24, 2015).
[52] http://www.synpol.org (accessed December 24, 2015).
[53] http://greenchemicalsblog.com/2013/05/29/braskem-launches-green-ldpe/ (accessed June 22, 2014).
[54] http://www.sulzer.com/en/Products-and-Services/Process-Technology/Polymer-Production-Technology/Poly-Lactic-Acid-PLA (accessed June 20, 2014).
[55] http://www.synbratechnology.nl (accessed December 24, 2015).
[56] http://www.teijin.com/rd/technology/bioplastic/ (accessed June 20, 2014).
[57] http://www.prweb.com/releases/2014/02/prweb11582642.htm (accessed June 19, 2014).
[58] MICHELIN (2014) The CEA, Michelin, Protéus (PCAS Group) and SDTech join forces to develop two innovative uses for used tires. http://www.michelin.com/eng/media-room/press-and-news/press-releases/Group/The-CEA-Michelin-Proteus-PCAS-Group-and-SDTech-join-forces-to-develop-two-innovative-uses-for-used-tires (accessed June 23, 2014).
[59] http://biosciences.dupont.com/about-us/collaborations/goodyear/ (accessed June 20, 2014).
[60] http://www.biofuelsdigest.com/bdigest/2010/05/27/lanxess-invests-10-million-in-gevo-signs-pact-to-produce-isobutene/ (accessed June 28, 2014).
[61] http://www.polymersolutions.com/blog/tire-manufacturer-testing-rubber-from-dandelion/ (accessed June 22, 2014).
[62] http://www.kawakenfc.co.jp/e/household.html (accessed December 24, 2015).
[63] http://www.aibn.uq.edu.au/pepfactants (accessed June 20, 2014).
[64] Huang, H. (2008) Life cycle assessment comparison between Pepfactant and chemical surfactant production. Master Thesis, Beijing University of Chemical Technology, Beijing, China.
[65] http://www.businesswire.com/news/home/20130715006213/en/Modular-Genetics-Demonstrates-Production-Bio-Derived-Glycinate-Ultra-Mild (accessed June 22, 2014).
[66] http://www.wheatoleo.com (accessed June 6, 2014).
[67] http://www.sofiproteol.com (accessed June 6, 2014).
[68] http://investors.amyris.com/releasedetail.cfm?releaseid=752562 (accessed June 16, 2014).
[69] http://www.opxbio.com (accessed December 24, 2015).
[70] Yim, H., Haselbeck, R., Niu, W., Pujol-Baxley, C., Burgard, A., Boldt, J., Khandurina, J., Trawick, J.D., Osterhout, R.E. Stephen, R., Estadilly, J., Teisan, S., Schreyer, H.B. Andrea, S., Yang, T.H., Lee, S.Y.L., Burk, M.J., van Dien, S. (2011) Metabolic engineering of *Escherichia coli* for direct production of 1,4-butanediol. *Nat. Chem. Biol.*, **7**, 445–452.
[71] Kolodziej, R., Scheib, J. (2012) Bio-isobutanol: the next-generation biofuel. *Hydrocarb. Process.*, **9**, 79–86.
[72] UKEssays.com (2014) An overview of lactic acid environmental sciences essay. http://www.ukessays.com/essays/environmental-sciences/an-overview-of-lactic-acid-environmental-sciences-essay.php#ixzz36hgUqqyt (accessed June 15, 2014).
[73] European Bioplastics (2014) Braskem's Triunfo plant for biobased PE in Brazil. http://en.european-bioplastics.org/blog/2012/07/13/dropping-in-bioplastics-same-performance-but-renewable/ (accessed June 6, 2014).

[74] Reverdia (2014) Commercial Plants for the Production of Biosuccinium™. http://www.reverdia.com/technology/commercial-plants/ (accessed June 24, 2014).

[75] Williams, D. (2014) BASF-Purac JV starts bio-succinic acid plant in Spain. http://www.chemanager-online.com/en/news-opinions/headlines/basf-purac-jv-starts-bio-succinic-acid-plant-spain (accessed June 21, 2014).

[76] http://www.bio-amber.com (accessed December 24, 2015).

[77] Da Silva, S.S., Antunes, F.A.F., Marcelino, P.F., Freitas, W.L.C., Brumano, L.P., Gonçalves, B.M.C., Philipini, R.R., Soler, M., Martiniano, S.E., Chandel, A.K. (2014) Applied biotechnology for the production of biofuels and bioproducts from sugarcane bagasse. Biotech 2014 and 6th Czech-Swiss Symposium, June 11–12, Prague, Czech Republic.

[78] Linde Buys Choren's Carbo-V Technology (February 15, 2012). http://www.renewablesinternational.net/linde-buys-chorens-carbo-v-technology/150/515/33097/ (accessed June 2014).

[79] http://www.abengoabioenergy.com (accessed December 24, 2015).

[80] Bayry, J., Aimanianda, V., Guijarro, J.I., Sunde, M., Latge, J.-P. (2012) Hydrophobins—unique fungal proteins. *PLoS Pathog.*, **8**(5), e1002700.

[81] Ghisalba, O., Wohlgemuth, R., Meyer, H.-P. (2010) Industrial Biotransformation, in *Encyclopedia of Industrial Biotechnology: Bioprocess, Bioseparation, and Cell Technology* (Ed. M.C. Flickinger), John Wiley & Sons, Inc., Hoboken, NJ, pp. 1–18.

[82] Whittall, J., Sutton, P. (Eds.) (2012) *Practical Methods for Biocatalysis and Biotransformations 2*, 1st ed., John Wiley & Sons, Ltd., Chichester, UK.

[83] Mirata, M.A. (2015) *Lonza AG, Specialty Ingredients R&T Fermentation and Biocatalysis*, Lonza Ltd, Basel, Switzerland.

[84] Celebi, K., Buchheim, J., Wyss, R.M., Droudian, A., Gasser, P., Shorubalko, I., Kye, J.I., Lee, C., Park, H.G. (2014) Ultimate permeation across atomically thin porous graphene. *Science*, **344**, 289–344.

[85] ICCA Responsible Care Initiative. http://www.icca-chem.org/en/Home/Responsible-care/ (accessed March 14, 2014).

[86] Anastas, P.T., Warner, J.C. (1998) *Green Chemistry: Theory and Practice*, Oxford University Press, New York.

[87] Pfau, S.F., Hagens, J.E. Dankbaar, B., Smiths, A.J.M. (2014) Visions of sustainability in bioeconomy research. *Sustainability*, **6**(3), 1222–1249.

[88] Leseurre, L., Merea, C., Duprat de Paule, S., Pinchart, A. (2014) Eco-footprint: a new tool for the "Made in Chimex"considered approach. *Green Chem.*, **16**, 1139–1148.

[89] Khosravi-Darani, K., Zahra-Beigom, M., Tomohito, A., Kenji, T. (2013) Microbial production of poly(hydroxybutyrate) from C1 carbon sources. *Appl. Microbiol. Biotechnol.*, **97**, 1407–1424.

[90] Daniell, J., Köpke, M., Simpson, S.D. (2012) Commercial biomass syngas fermentation. *Energies*, **5**, 5372–5417.

[91] Sessions—IPCC—Intergovernmental Panel on Climate. http://www.ipcc.ch/meeting_documentation/meeting_documentation_ipcc_sessions_and_ipcc_wgs_sessions.shtml (accessed February 24, 2014).

[92] Simon, R.M. (2012) Technological Salience and the Use of Knowledge to Generate Attitudes, in *Innovations in Biotechnology* (Ed. C. Agbo), InTech Open Science, Rijeka, Croatia. www.intechopen.com (accessed December 24, 2015).

[93] Simeth, M., Raffo, J. (2013) What makes companies pursue an open science strategy? Working Paper No. 6, August 2013, WIPO (World Intellectual Property Organization) Economic and Statistic Series.

[94] http://www.sagebase.org (accessed December 24, 2015).

[95] Hoffman, W. (2014) The shifting currents of bioscience innovation. *Glob. Policy*, **5**(1), 76–84.

[96] Wells, A., Meyer, H.-P. (2014) Biocatalysis as a strategic green technology for the chemical industry. *ChemCatChem*, **6**, 918–920.

[97] Wirz, B., Kittelmann, M., Meyer, H.-P., Wohlgemuth, R. (2010) Swiss Industrial Biocatalysis Consortium. *Chimia*, **64**, 780–781.

[98] GlaxoSmithKline (2014) Full Year Results 2012. http://www.gsk.com/content/dam/gsk/globals/documents/pdf/Investors/quarterly-results/2012/Q4-2012-analyst-presentation-transcript.pdf (accessed January 14, 2014).

CHAPTER 28

Microbial Transformations of Pentacyclic Triterpenes

Robert Azerad

Laboratoire de Chimie et Biochimie Pharmacologiques et Toxicologiques, Université Paris Descartes, Paris, France

BDM Department, Bertin Pharma, Montigny-le-Bretonneux, France

28.1 INTRODUCTION

Pentacyclic triterpenes (PTs) are widely distributed in higher plant kingdom. Most of them are known as constituents of crude tree gum materials known since ancient times for their aromatic properties, such as incense and myrrh (from Asia and Africa), or their resinous properties such as mastic or dammars (from Asia and South America) used for the calking of hulls [1]. Their structure, originally formed from the cyclization and rearrangement of oxidosqualene, essentially derives from a small number of basic skeletons such as lupeol and α- or β-amyrin (Figure 28.1), with additional modifications, such as regio- and stereoselective hydroxylations and further oxidation on the ring carbons or on methyl groups, including double bond migrations and rearrangements, thus defining a considerable number of compounds families resulting from the combination of such modifications. Moreover, as usual in plant products, single or multiple glycosylation of hydroxyl or carboxylic groups of the aglycones (also called sapogenins) may add to the structural complexity and diversification of these compounds (saponins or saponosides). Most of them are bioactive compounds, sometimes already employed in folk medicine, and exhibiting a wide range of biological benefits, including antibiotic, anti-HIV, antioxidant, anti-inflammatory, and antitumoral properties.

Previous general reviews on structure and biological activities of PTs [2–5] are mainly centered on their anti-inflammatory properties [6, 7], apoptosis induction [8, 9], metabolic and vascular diseases [10], and anticancer activity [11, 12], and more recently antidiabetes activity [13, 14], covering the literature up to 2010–2012. The revival of interest demonstrated by the increasing number of similar products obtained by extraction or semisynthesis and the description of their biological activities *in vivo* and *in vitro* have prompted a large number of new publications in the last years. A detailed quantified triterpene distribution within various plant materials (39 plants) using GC/MS techniques can be found in a paper by Jäger *et al.* [15] (Figure 28.2).

However poor bioavailability and some adverse effects may have limited their use in medicinal applications. Biotransformation by microbial methods has been used for producing new derivatives, expanding their molecular diversity and

Green Biocatalysis, First Edition. Edited by Ramesh N. Patel.

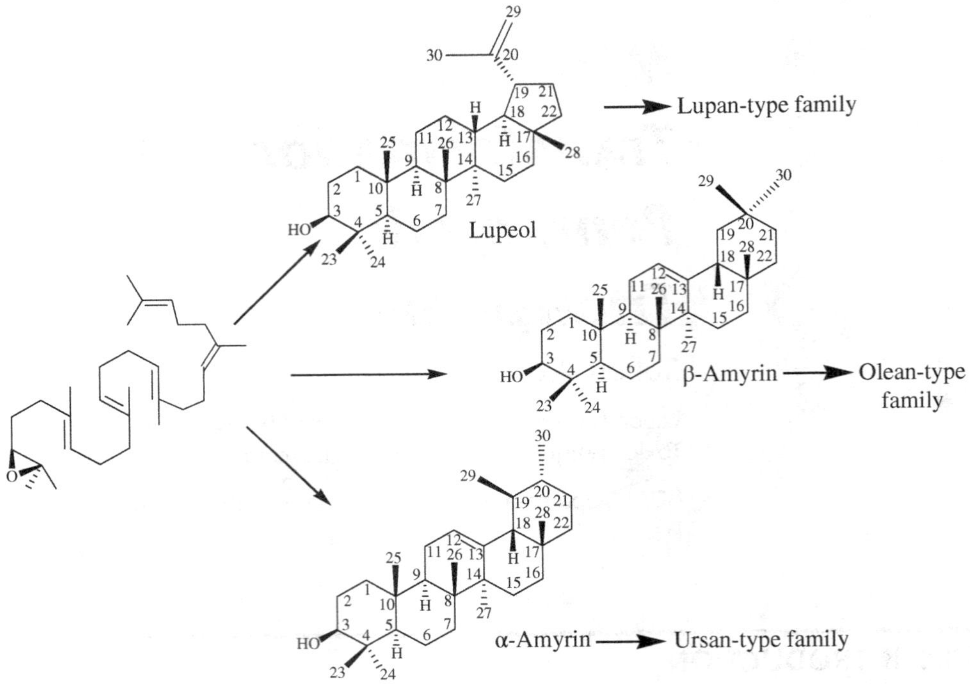

FIGURE 28.1
Biosynthetic pathways leading from oxidosqualene to the main families of PTs skeletons.

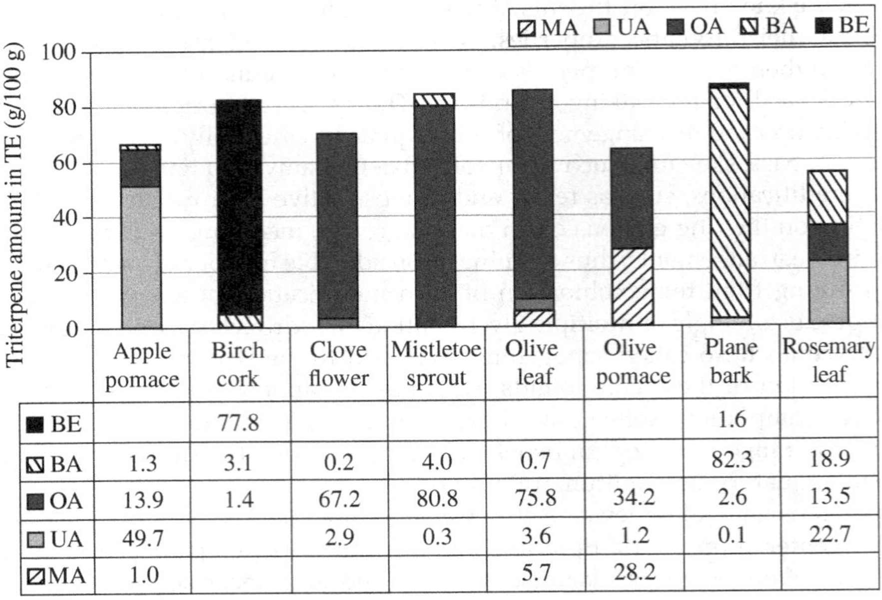

	Apple pomace	Birch cork	Clove flower	Mistletoe sprout	Olive leaf	Olive pomace	Plane bark	Rosemary leaf
■ BE		77.8					1.6	
▧ BA	1.3	3.1	0.2	4.0	0.7		82.3	18.9
■ OA	13.9	1.4	67.2	80.8	75.8	34.2	2.6	13.5
□ UA	49.7		2.9	0.3	3.6	1.2	0.1	22.7
▨ MA	1.0				5.7	28.2		

FIGURE 28.2
Triterpene amount within triterpene dry extracts (TEs) from various plant materials. BA, betulinic acid (**6**); BE, betulin; MA, maslinic acid (**82**); OA, oleanolic acid (**30**); and UA, ursolic acid (**87**). From Jäger *et al.* [15], © 2009 by the authors; with permission of Molecular Diversity Preservation International, Basel, Switzerland.

eventually promoting better physicochemical properties and tolerability. Microbial transformation is a useful tool for organic chemists looking for new compounds, as a consequence of the variety of reactions observed with natural products. Obviously, the major advantage of biological catalysts (whole cells or enzymatic extracts) versus chemical methods to modify such complex structures is their capacity to promote novel reactions and directly introduce functional groups with high regio- and stereoselectivity. A considerable number of such biotransformations of PTs has been carried out and partially reviewed in the since the mid-2000s [16–20]. This number will probably extend in the next years from this rich source of secondary plant metabolites (more than 1250 compounds described up to 2007 [21] from about 4000 triterpene compounds).

This review summarizes, up to mid 2014, the most representative microbial transformations and the derivatives of special interest obtained from the lupan-type (lupeol), olean-type (β-amyrin), and ursan-type (α-amyrin) pentacyclic triterpenoids (Figure 28.1) in the last 20 years.

28.2 TYPICAL BIOTRANSFORMATIONS IN THE LUPANE FAMILY

Lupeol (**1**) is found in vegetables such as white cabbage, pepper, cucumber, tomato; in fruits such as olive, fig, mango, strawberry, red grapes; and in a number of medicinal plants all around the world. The last 15 years have seen tremendous efforts by researchers worldwide to develop this wonderful molecule for its clinical use in the treatment of a variety of disorders. These studies also provide insight into the mechanism of action of lupeol and suggest that it is a multitarget agent with a large anti-inflammatory potential, targeting key molecular pathways in a variety of cells [22].

The biotransformation of lupeol (Figure 28.3) by *Aspergillus ochraceus* and *Mucor rouxii* afforded each two compounds (**2–3** and **4–5**) characterized by mass spectrometry fragmentations in GC-MS but not by NMR [23]. The authors deduced that *A. ochraceus* is a good biocatalyst to introduce double bonds into the lupeol structure, whereas *M. rouxii* exhibits ability to biocatalyze oxygen insertions in that PT.

Betulin (3β,28-dihydroxy-20(29)-lupene) and betulinic acid (3β-hydroxy-lup-20(29)-en-28-oic acid) (**6**) are widely distributed in the plant kingdom, mainly in the bark of some trees such as white birch (*Betula pubescens*), from which it gets its name, but also the ber tree (*Ziziphus mauritiana*). Betulinic acid has been shown a long time ago to exhibit a variety of biological activities, including inhibition of human immunodeficiency virus (HIV), replication in H9 lymphocyte cells, blockage of HIV type 1 entry into cells, and inhibition of DNA polymerase b [24, 25]. Synthetic derivatives of betulinic acid have also been investigated as specific inhibitors of HIV type 1 [26] and selective apoptosis inducers [27]. In addition, betulinic acid has been reported to be a melanoma-specific cytotoxic agent in both *in vitro* cell culture and *in vivo* studies [28]. Due to its high level of antitumor activity and lack of toxicity, betulinic acid is an attractive and promising new lead compound for use against human melanoma and is currently undergoing preclinical development for treatment or

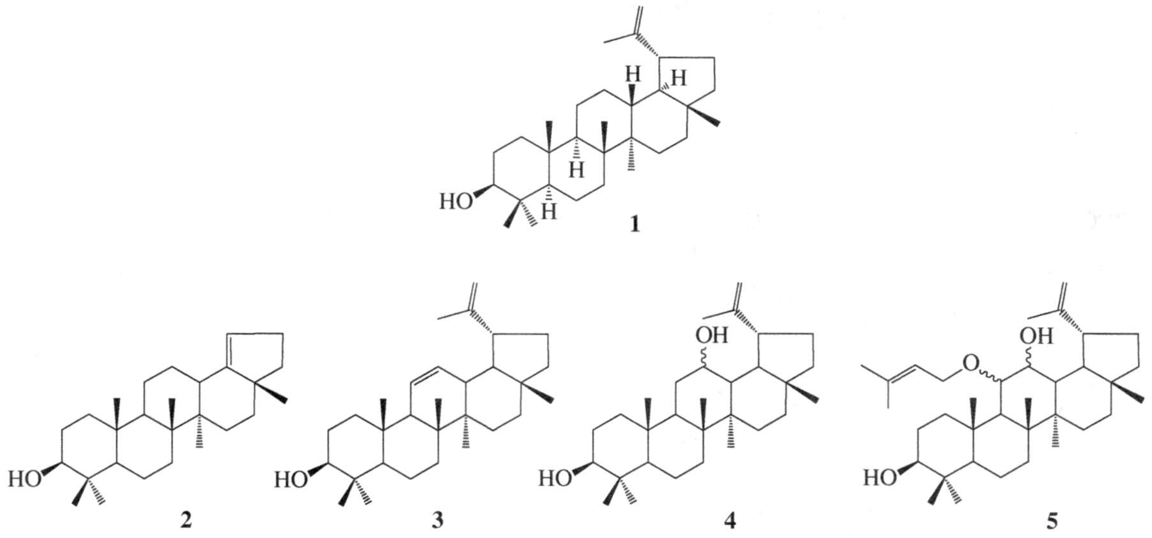

FIGURE 28.3
Proposed structures for the biotransformation products of lupeol **1**, produced by *Aspergillus ochraceus* (**2** and **3**) and *Mucor rouxii* (**4** and **5**) [23].

prevention of malignant melanoma. The closely related betulonic acid (**7**), its 3-keto derivative, has similar properties.

Several studies have been devoted to the one-step oxidative conversion of betulin to betulinic acid with selected strains of fungal microorganisms [29–31], as well to the microbial conversion of betulin or betulinic acid to their 3-oxo counterparts (betulone or betulonic acid, respectively) [27, 32–34].

Initially the microbial transformation of the antimelanoma agent betulinic acid was studied with the main objective of utilizing microorganisms as *in vitro* models to predict and prepare potential mammalian metabolites of this compound [35–37]. Microbial transformation studies showed in screening experiments a number of microorganisms capable of biotransforming betulinic acid [38, 39]. Several of these cultures, *Bacillus megaterium* ATCC 14581 or ATCC 13368, *Cunninghamella elegans* ATCC 9244, and *Mucor mucedo* UI-4605, were selected for preparative-scale transformations (Figure 28.4).

Bioconversion of **6** with resting-cell suspensions of phenobarbital-induced *B. megaterium* ATCC 14581 or ATCC 13368 resulted in the production of the known betulonic acid (**7**) and six new metabolites: 3β,7β-dihydroxy-lup-20(29)-en-28-oic acid (**8**), 3-oxo-1β-hydroxy-lup-20(29)-en-28-oic acid (**9**), 3-oxo-11α-hydroxy-lup-20(29)-en-28-oic acid (**10**), 1β,3β,7β-trihydroxylup-20(29)-en-28-oic acid (**11**), 3β,6α,7β-trihydroxylup-20(29)-en-28-oic acid (**12**), and 3β,7β,15α-trihydroxylup-20(29)-en-28-oic acid (**13**). Biotransformation of **6** with growing cultures of *C. elegans* ATCC 9244 produced one new metabolite, characterized as 1β,3β,7β-trihydroxy-lup-20(29)-en-28-oic acid (**11**). Incubation of **6** with growing cultures of *M. mucedo* UI-4605 afforded metabolite **10**. Structure elucidation of all metabolites was based on NMR and HRMS analyses. In addition, the antimelanoma activity of metabolites **8–11** was evaluated against two human melanoma cell lines, Mel-1 (lymph node) and Mel-2 (pleural fluid). Some metabolites were more active than the parent compound (betulinic acid) against Mel-2. The cytotoxicity results indicated that the *in vitro* antimelanoma activity of betulinic acid is significantly affected by oxidation at different sites of the molecule. These results also demonstrate that the structural requirements for cytotoxicity in betulinic acid and its metabolites differed depending on the type of melanoma cell line used [39].

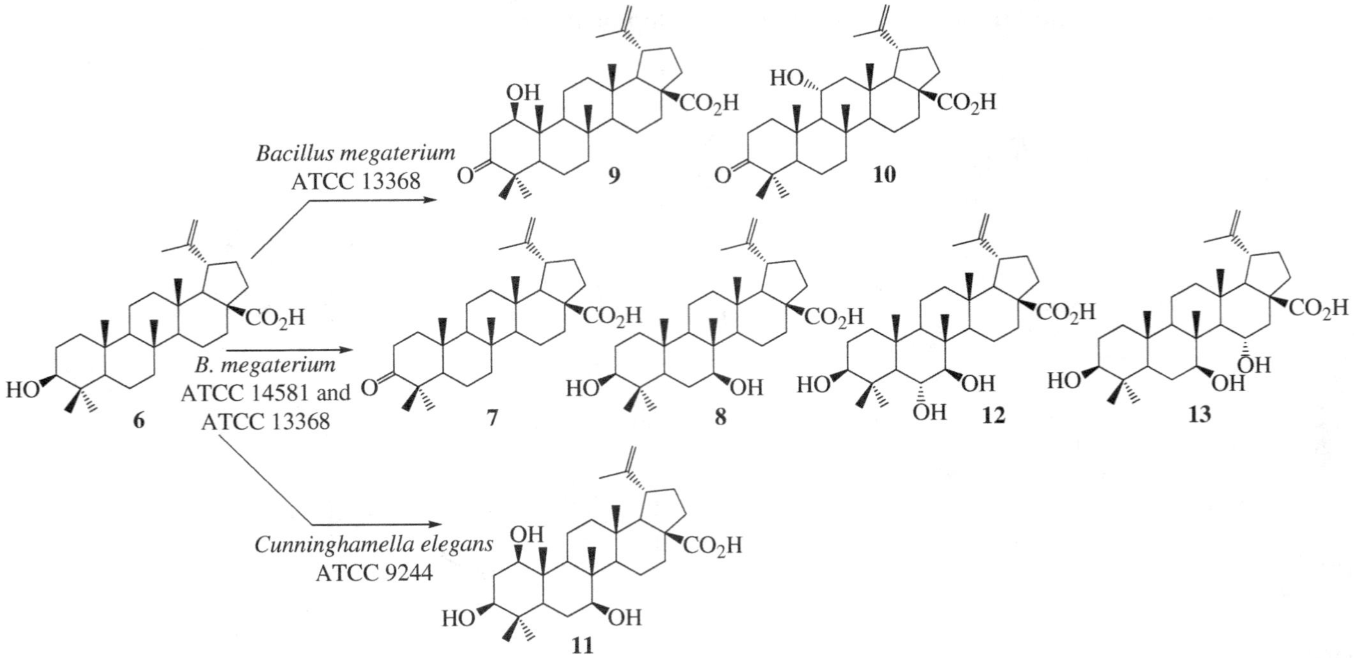

FIGURE 28.4

Some typical microbial transformations of betulinic acid (**6**).

Betulinic acid (**6**) and the closely related betulonic acid (**7**), which has similar properties, were also submitted to incubations with selected fungi [40] such as *Arthrobotrys* DPB134, *Chaetophoma* DPB125 and *Dematium*, fungal endophytes species isolated from the bark of *Platanus orientalis* (a producer tree for betulinic acid), as well as *Colletotrichum* sp. DPB136 isolated from corn leaves (Figure 28.5). Biotransformations with *Arthrobotrys* converted betulonic acid (**7**) into 3-oxo-7β-hydroxylup-20(29)-en-28-oic acid (**14**), 3-oxo-7β,15α-dihydroxylup-20(29)-en-28-oic acid (**15**), and 3-oxo-7β,30-dihydroxylup-20(29)-en-28-oic acid (**16**). *Colletotrichum* converted betulinic acid (**6**) into 3-oxo-15α-hydroxylup-20(29)-en-28-oic (**17**) acid whereas betulonic acid (**7**) was converted into the same product and 3-oxo-7β,15α-dihydroxylup-20(29)-en-28-oic acid (**15**). *Chaetophoma* converted betulonic acid (**7**) into 3-oxo-25-hydroxy-lup-20(29)-en-28-oic acid (**18**), and both *Chaetophoma* and *Dematium* converted betulinic acid (**6**) into betulonic acid (**7**). Those fungi, therefore, are useful for mild, selective oxidations of lupane substrates at positions C-3β, C-7β, C-15α, C-25, and C-30 (Figure 28.5). Similar results had been reported earlier with betulonic acid (**7**) using another fungal microorganism *Chaetomium longirostre* IFO 9873, which produced metabolites hydroxylated at C-7 and C-15, together with a *seco*-derivative whose ring A was opened at C-3, C-4 [41]. Several of these compounds were shown to be potential cancer chemopreventive agents.

Incubation of betulinic acid (**1**) and 23-hydroxybetulinic acid (**19**) (Figure 28.6) with *Nocardia* sp. NRRL 5646 afforded their corresponding methyl esters, betulinic acid methyl ester (**20**), and 23-hydroxybetulinic acid methyl ester (**21**), respectively [42]. However, microbial transformation of betulonic acid (**7**) by the same *Nocardia* sp. NRRL 5646 at a preparative scale [43] resulted in addition in the isolation of an unexpected α-hydroxylation product, methyl 2α-acetoxy-3-oxo-lup-20(29)-en-28-oate (**22**) (6.5% yield), beside the expected compound methyl 3-oxo-lup-20(29)-en-28-oate (**23**) (50% yield). The structures of metabolites were elucidated unambiguously by ESI-MS and 2D NMR spectroscopy.

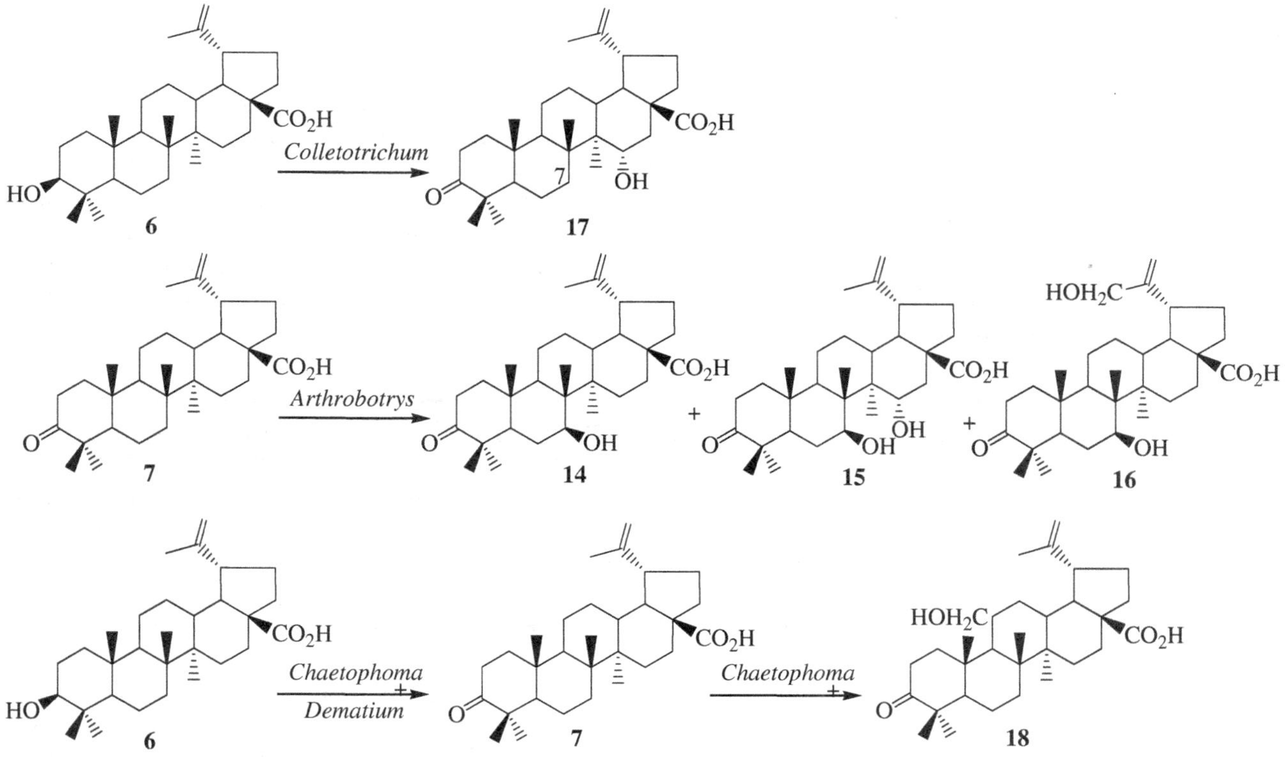

FIGURE 28.5
Microbial transformations of betulinic acid (**6**) and betulonic acid (**7**) by endophytic fungi [39, 40].

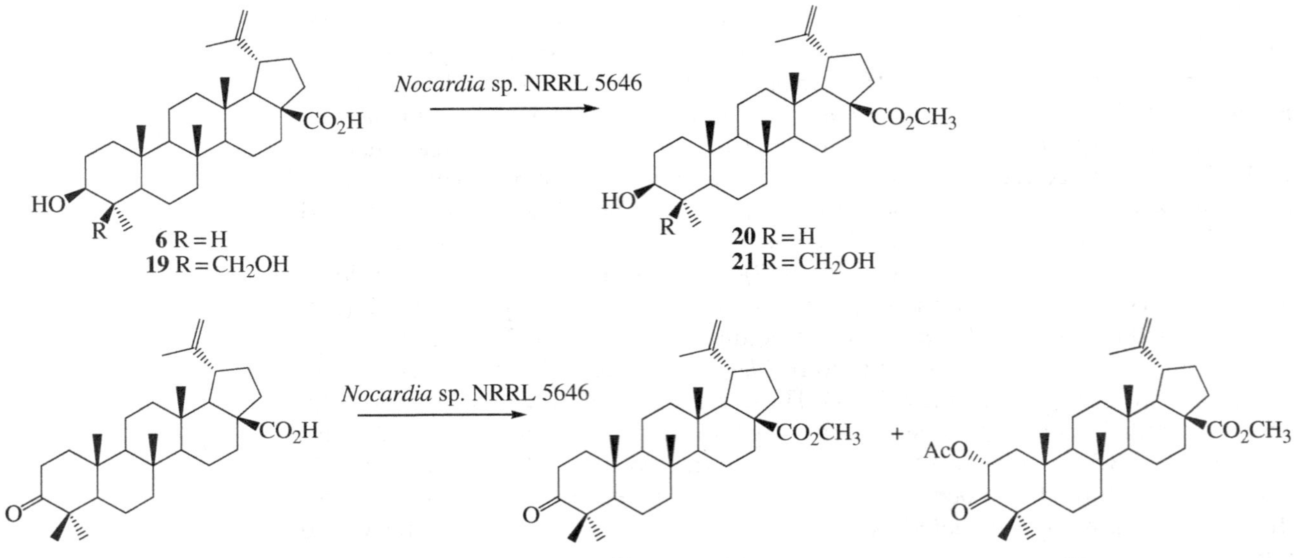

FIGURE 28.6
Biotransformations of betulinic acid (**6**) and betulonic acid (**7**) by *Nocardia* sp. NRRL 5646 [42, 43].

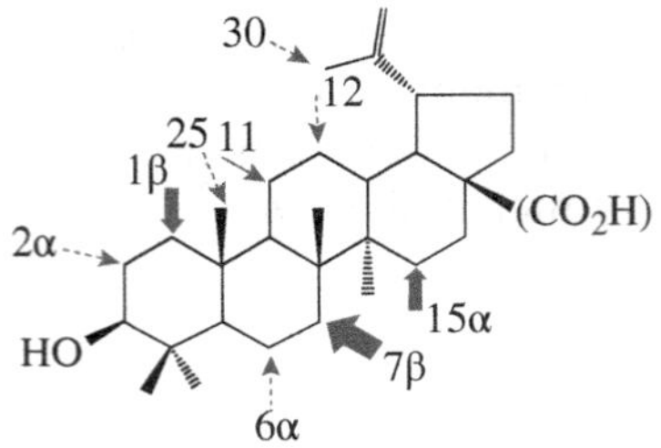

FIGURE 28.7
Main hydroxylation sites observed in the microbial transformations of lupan derivatives.

In summary, microbial hydroxylation of betulin, lupeol, and betulinic–betulonic acids was thus useful to produce hydroxylated lupane derivatives at C-7β, C-15α, C-1β, but also frequently in other positions: C-11, C-12, C-6, C-2α, C-30, C-25 (in order of priority, see Figure 28.7), with sometimes several combinations of hydroxylated sites. A summary of these results with yields is presented in Table 28.1. Besides the existing chemical methods, a variety of metabolites with potential new or increased biological activities may be obtained. Some of them have been more recently isolated in plants as natural products.

28.3 TYPICAL BIOTRANSFORMATIONS IN THE OLEANE FAMILY

β-Amyrin (**24**) is found in various plant resins in mixture with the α-isomer and especially in bee pollen oil of lotus *Nelumbo nucifera* Gaertn (Indian Lotus), 3 g/kg [44], or the bark of *Amphipterygium adstringens*, 2.4 g/kg [45]. β-Amyrin acetate (**25**) was deacetylated by *Rhodobacter sphaeroides*, a photosynthetic bacterium, and converted to olean-12-ene-3β,24-diol (**26**), and erythrodiol (olean-12-ene-3β,28-diol, **27**) [46] (Figure 28.8).

Microbial transformation of β-amyrin (**24**) by growing cells of the fungus *Lecanicillium muscarinium* (formerly *Cephalosporium aphidicola*) was successfully accomplished after 15 days of incubation [47]. Two products (Figure 28.9) purified by column chromatography were identified by 1H and 13C mono- and bidimensional nuclear magnetic resonance as 3β-hydroxy-olean-12-en-11-one (**28**) and

TABLE 28.1 Microbial Transformation Products and Yields Obtained from Lupan-Family PTs

Substrate PTs	Microorganisms	Products	Yields (%)
Lupeol (1)	*Aspergillus ochraceus* [23]	3β-Hydroxy-18(19)-deisopropenyl-lupene (**2**)	19
		3β-Hydroxy-11(12)-dehydro-20(29)-lupene (**3**)	11
	Mucor rouxii [23]	3β,12-Dihydroxy-20(29)-lupene (**4**)	26.5
		3β,12-Dihydroxy-11-dimethylallyloxy-20(29)-lupene (**5**)	16
Betulin	*Armillaria luteo-virens* Sacc QH [26]	Betulinic acid (**6**)	6–9.3
	Cunninghamella blakesleeana AS 3.910 [27]	Betulinic acid (**6**)	2.5
	Chaetomium longirostre IFO 9873 [37]	3,4-*seco*-4,28-Dihydroxy-lup-20(29)-en-3-oic acid	3
	Microsporum canis [28]	Betulinic acid (**6**)	—
	Trichophyton tonsurans [28]	Betulinic acid (**6**)	—
Betulinic acid (6)	*Bacillus megaterium* ATCC 14581 [37, 38]	3-Oxo-20(29)-lupen-28-oic acid (**7**)	0.45
		3β,7β-Dihydroxy-20(29)-lupen-28-oic acid (**8**)	0.1
		3β,6α,7β-Trihydroxy-20(29)-lupen-28-oic acid (**12**)	0.45
		1β,3β,7β-Trihydroxy-20(29)-lupen-28-oic acid (**11**)	—
	B. megaterium ATCC 13368 [37, 38]	3-Oxo-20(29)-lupen-28-oic acid (**7**)	4.1
		3-Oxo-1β-hydroxy-20(29)-lupen-28-oic acid (**9**)	0.15
		3-Oxo-11α-hydroxy-20(29)-lupen-28-oic acid (**10**)	0.2
	Cunninghamella elegans ATCC9244 [37, 38]	3β,7β,15α-Trihydroxy-20(29)-lupen-28-oic acid (**13**)	0.5
		1β,3β,7β-Trihydroxy-20(29)-lupen-28-oic acid (**11**)	0.4
	Mucor mucedo UI-4605 [37, 38]	3-Oxo-11α-hydroxy-20(29)-lupen-28-oic acid (**10**)	0.2
	Colletotrichum sp. [39]	3-Oxo-15α-hydroxy-20(29)-lupen-28-oic acid (**17**)	2.3
	Chaetophoma sp. [39]	3-Oxo-20(29)-lupen-28-oic acid (**7**)	0.95
	Dematium sp. [39]	3-Oxo-20(29)-lupen-28-oic acid (**7**)	0.6
	Nocardia sp. NRRL5646 [41]	3β-Hydroxy-20(29)-lupen-28-oic acid methyl ester (**20**)	—
Betulonic acid (7)	*C. longirostre* IFO 9873 [40]	3,4-*seco*-4-Hydroxy-lup-20(29)-ene-3,28-dioic acid	4
		3-Oxo-7β,15α-Dihydroxy-lup-20(29)-en-28-oic acid (**15**)	6
		3,4-*seco*-4,7β,17-Trihydroxy-28-norlup-20(29)-en-3-oic acid	12
	Nocardia sp. NRRL5646 [42]	3-Oxo-20(29)-lupen-28-oic acid methyl ester (**23**)	50
		2α-Acetoxy-3-oxo-20(29)-lupen-28-oic acid methyl ester (**22**)	6.5
		3-Oxo-7β-hydroxy-20(29)-lupen-28-oic acid (**14**)	1.6
	Arthrobotrys sp. [39]	3-Oxo-7β,15α-dihydroxy-20(29)-lupen-28-oic acid (**15**)	0.6
		3-Oxo-7β,30-dihydroxy-20(29)-lupen-28-oic acid (**16**)	1.3
		3-Oxo-25-hydroxy-20(29)-lupen-28-oic acid (**18**)	4.2
	Chaetophoma sp. [39]	3-Oxo-15α-hydroxy-20(29)-lupen-28-oic acid (**17**)	3
	Colletotrichum sp. [39]	3-Oxo-7β,30-dihydroxy-20(29)-lupen-28-oic acid (**16**)	1.7
23-Hydroxy betulinic acid (19)	*Nocardia* sp. NRRL5646 [41]	3β,23-Dihydroxy-20(29)-lupen-28-oic acid methyl ester (**20**)	—

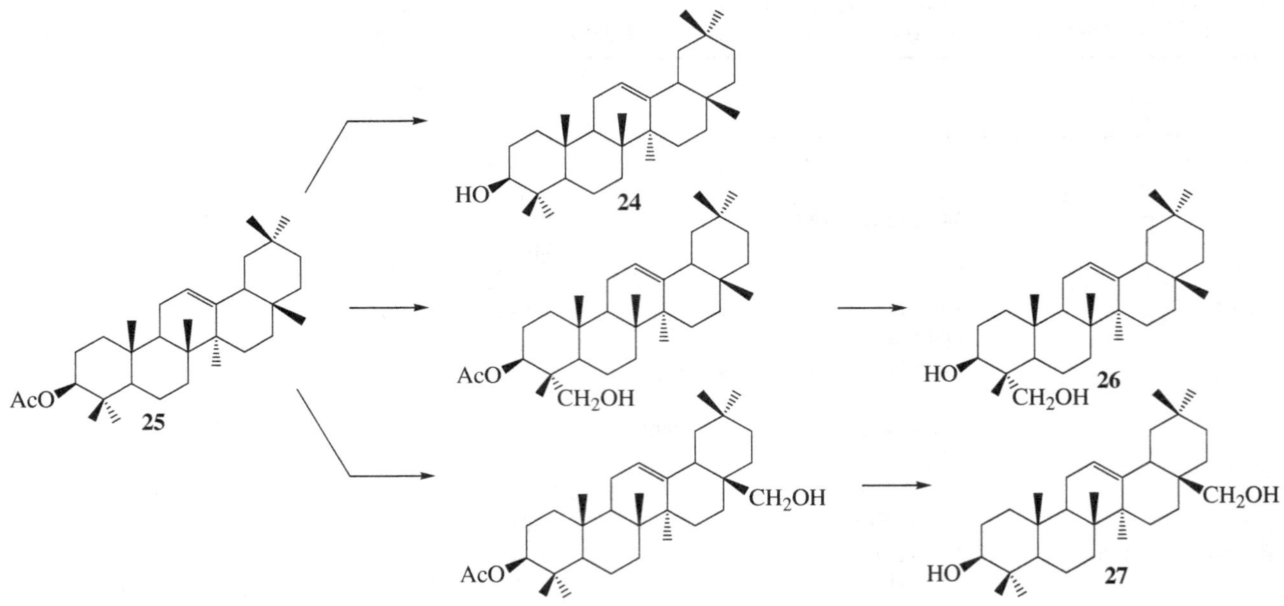

FIGURE 28.8
Bioconversion of β-amyrin acetate (**25**) to metabolites **26** and **27** by *Rhodobacter sphaeroides* [46].

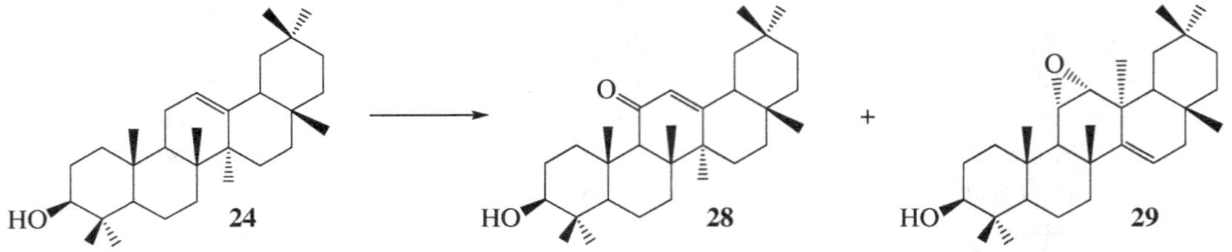

FIGURE 28.9
Microbial transformation of β-amyrin (**24**) by *Lecanicillium muscarinium* [47].

11α,12α-oxidotaraxerol (**29**); this latter is a new compound formed by an interesting oxidative rearrangement of the starting material, probably via an initial hydroxylation at C-11.

Oleanolic acid (**30**) is a widespread pentacyclic sapogenin. It is reported to possess interesting bioactivities including anti-inflammatory, hepatoprotective, antitumor, antiviral, and antimicrobial activities. In the years 1969–1972, Hikino *et al.* [48–51] incubated oleanolic acid (**30**) with several fungal strains such as *Fusarium lini*, *Colletotrichum phomoides*, or *Cunninghamella blakesleeana* and obtained in low yields four products, 3β-hydroxy-11-oxo-olean-12-ene-28-oic acid (**31**), 3β,15α-dihydroxy-11-oxo-olean-12-ene-28-oic acid (**32**), 3β,12-dihydroxy-11-oxo-olean-28(13)-olide (oleandroside, **33**), and 3β-hydroxy-olean-11,13-dien-28-oic acid (**34**), or three different products, 3β,6β-dihydroxy-olean-12-ene-28-oic acid (sumaresinolic acid, **35**), 3β,6β,21β-trihydroxy-olean-12-ene-28-oic acid (**36**), and 3β-hydroxy-21-oxo-olean-12-ene-28-oic acid (**37**), or six new products, 3β-hydroxy-olean-11,13(18)-dien-28-oic acid (**38**), 3β,7β-dihydroxy-olean-12-ene-28-oic acid (**39**), 3β-hydroxy-11-oxo-olean-12-ene-28-oic acid (**40**), 3β,7β-dihydroxy-olean-11,13(18)-dien-28-oic acid (**41**), 3β,7β,13-trihydroxy-olean-11-ene-28(13)-olide (**42**), 1β,3β-dihydroxy-olean-12-ene-28-oic acid (**43**), respectively. *Penicillium chrysogenum* afforded three products: **36**, **37**, and 3-oxo-21β-hydroxy-olean-12-ene-28-oic acid (**44**) (Figure 28.10). Starting from 3β,7β,13-trihydroxy-olean-12-ene-28-oic acid (erythrodiol, **45**), *C. blakesleeana* afforded 3β,7β,13-trihydroxy-olean-11-ene-28(13)-olide (**46**), 3β,7β,13,15α-tetrahydroxy-olean-11-ene-28(13)-olide (**47**), 3β,28-dihydroxy-olean-12-ene (**48**), and 3β,28-dihydroxy-11-oxo-olean-12-ene (**49**) (Figure 28.11).

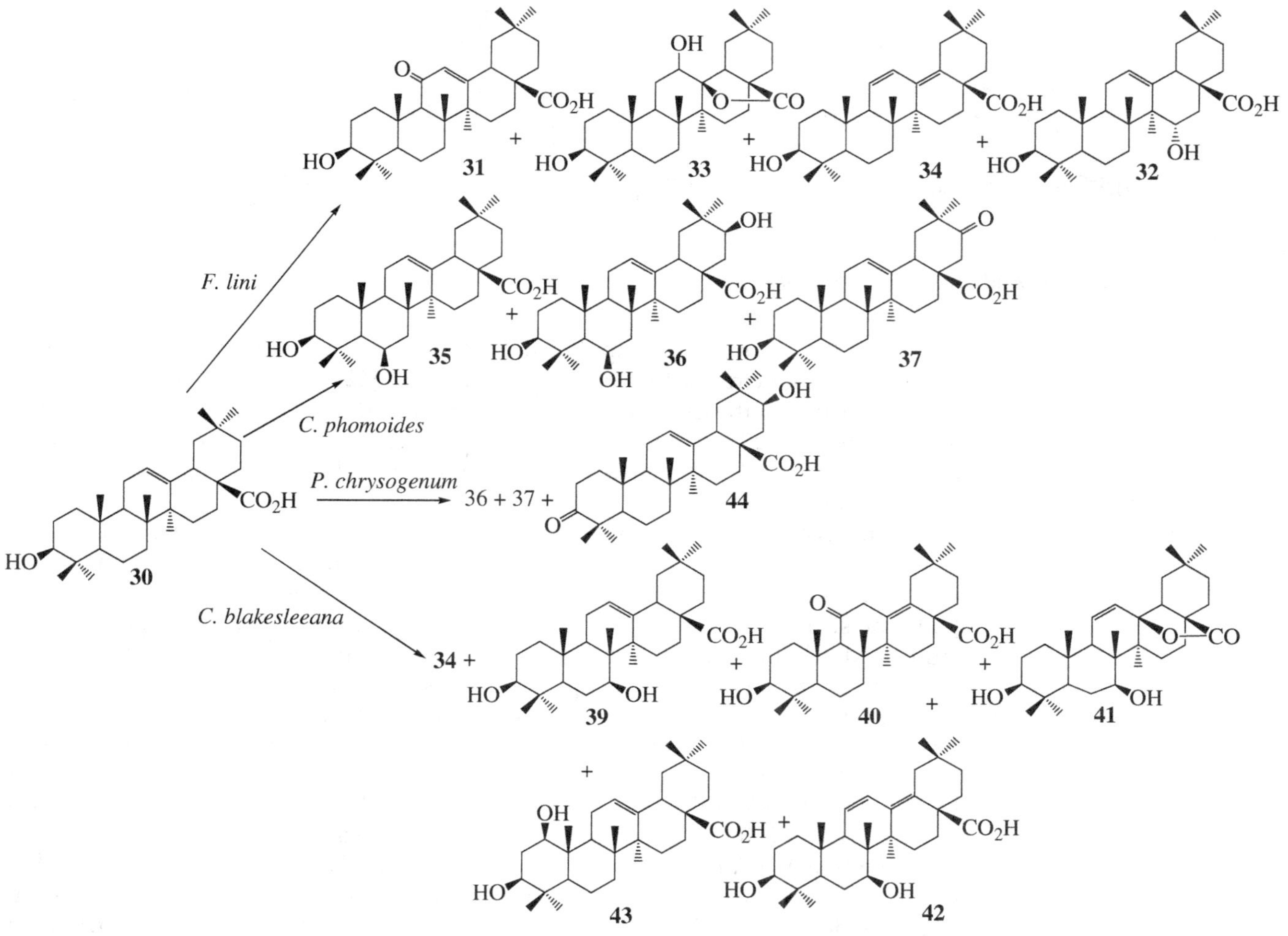

FIGURE 28.10

Microbial transformations of oleanolic acid (**30**) by *Fusarium lini*, *Colletotrichum phomoides*, *Penicillium chrysogenum*, or *Cunninghamella blakesleeana* [48–51].

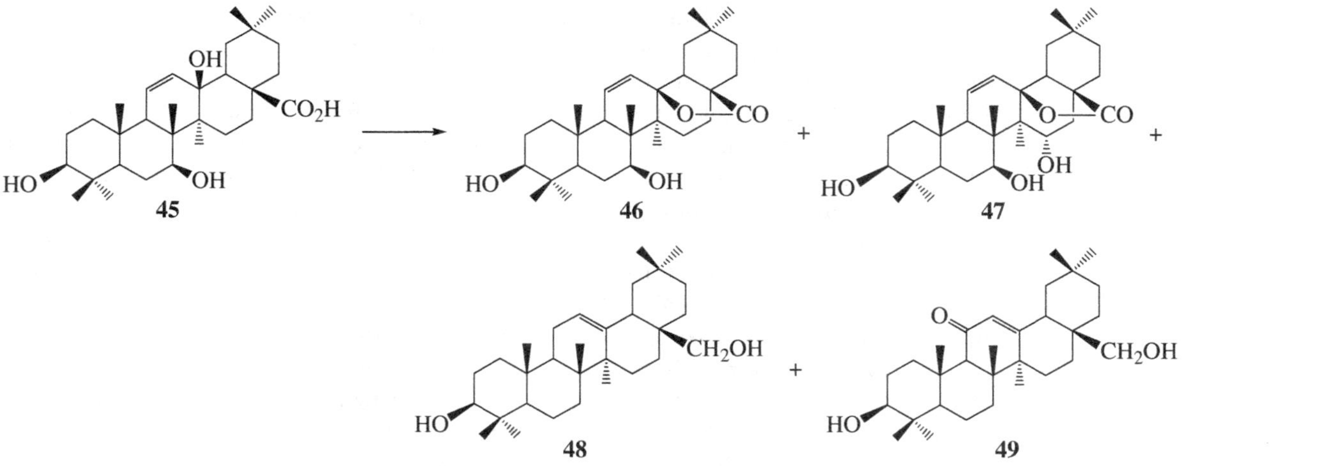

FIGURE 28.11

Microbial transformations of erythrodiol (**45**) by *Cunninghamella blakesleeana* [52].

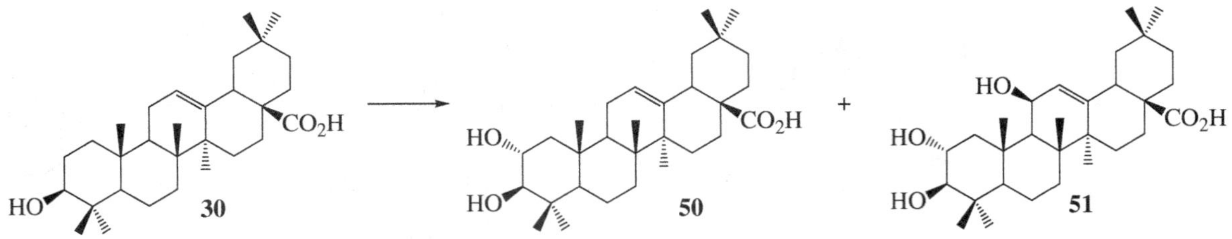

FIGURE 28.12
Transformation of oleanolic acid (**30**) by *Fusarium lini* [53].

Using the same *Nocardia* sp. NRRL 5646 already utilized for the biotransformation of betulinic acid (see Section 28.1). Zhang *et al.* [42] showed that the same methylation reaction of the C-28 carboxylic group occurred with oleanolic acid and ursolic acid (see Section 28.3). Interestingly, the ursolic acid methyl ester was also irreversibly rearranged to oleanolic methyl ester by migration of the C-19 methyl group to the C-20 position.

More recently, it was shown that the biotransformation of oleanolic acid (**30**), with *F. lini* afforded, in low amounts, two oxidative metabolites, 2α,3β-dihydroxy-olean-12-en-28-oic acid (**50**) and 2α,3β,11β-trihydroxy-olean-12-en-28-oic acid (**51**) [52]. The structures (Figure 28.12) were characterized on the basis of spectroscopic studies. Metabolites exhibited a potent inhibition of α-glucosidase from *Saccharomyces* sp. using *p*-nitrophenyl α-D-glucopyranoside as a substrate.

Zhu *et al.* [53] have investigated the microbial transformation of three olean-type PTs, oleanolic acid (**30**), 3-oxo oleanolic acid (**52**), and esculentoside A, a 2β,3β,24α-trihydroxy-29-carboxymethyl oleanolic acid 3-*O*-glycoside (**53**). A screening of 12 microbial strains led to identified two strains for preparative biotransformations, *Streptomyces griseus* ATCC 13273 and *A. ochraceus* CICC 40330, which resulted in the isolation of a total of ten metabolites including various glycosyl derivatives (Figure 28.13). Highly efficient deglycosylation (see Section 28.5) and regioselective 23- and 30-methyl oxidation was demonstrated, providing an alternative method to expand the structural diversity of oleanolic PTs. All the structures of the metabolites were elucidated unambiguously by ESI-MS, 1H NMR, 13C NMR, and 2D NMR spectroscopy.

Microbial transformations of oleanolic acid (**30**) by *Alternaria longipes* and *Penicillium adametzi* were carried out by Liu *et al.* [54]. Six transformed products in the form of glycosidic derivatives and three other transformed products, including one acyl glucosylated product, were isolated with new hydroxylation positions in 2α, 7β, 19α, and 21β (**60–63**). Their structures (Figure 28.14) were elucidated as 21β-hydroxy-oleanolic acid (**61**), and 7β,21β-dihydroxyl oleanolic acid (**62**), plus 2α-, 3β-, and 19α-hydroxylated oleanolic acids, specifically HeLa cell lines β-D-glucosylated at the C-3 or the C-28 position (**59a–c, 60, 63**) based on extensive NMR studies. Among them, **62** was a new compound and had stronger cytotoxic activities against than the substrate. At the same time, it was reported for the first time in this paper that the skeleton of compound **60** was changed from oleane to ursane as a result of a methyl migration (see Section 28.3), and two 2α,19α-hydroxy-ursene derivatives were formed in small amounts as glucosylation products (see Section 28.5).

The fungus *Absidia glauca* [55], used as suspended cells in buffer, transformed 3-oxo-olean-12-en-28-oic acid (**52**) into three metabolites: 1β,11α-dihydroxy-3-oxo-olean-12-en-28-oic acid (**64**), 1β,11α,21β-trihydroxy-3-oxo-olean-12-en-28-oic acid (**66**), and 1β-hydroxy-3-oxo-olean-11-en-28(13)-olide (**65**) (Figure 28.15).

The transformation of 3-oxo-olean-12-en-28-oic acid (**52**) by *C. longirostre* [56] followed a different course, as the first reaction was probably a Baeyer–Villiger-like reaction on cycle A, followed by opening of the 7-membered lactone and further hydroxylation in the 22β-position (Figure 28.15).

In a biotransformation process using *M. rouxii* incubated with oleanolic acid (**30**) [57], three new derivatives, 7β-hydroxy-3-oxo-olean-12-en-28-oic acid

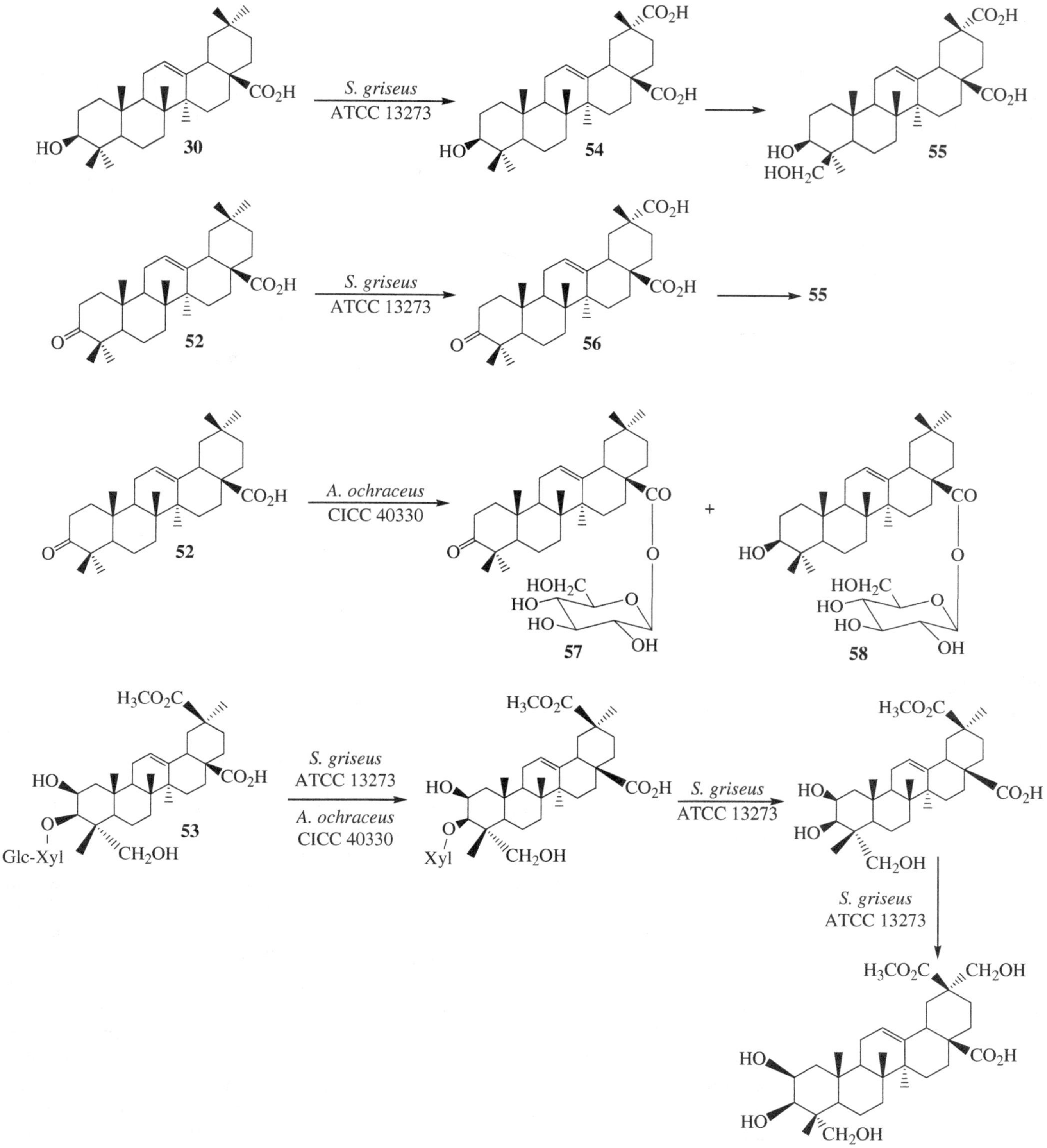

FIGURE 28.13

Microbial transformations of oleanolic acid (**30**), 3-oxo oleanolic acid (**52**), and esculentoside A (**53**) by *Streptomyces griseus* or *Aspergillus ochraceus* [54].

(**67**), 7β,21β-dihydroxy-3-oxo-olean-12-en-28-oic acid (**68**), and 3β,7β,21β-trihydroxy-olean-12-en-28-oic acid (**69**), and one known compound, 21β-hydroxy-3-oxo-olean-12-en-28-oic acid (**70**), were isolated, and the structures (Figure 28.16) were elucidated on the basis of spectroscopic analyses.

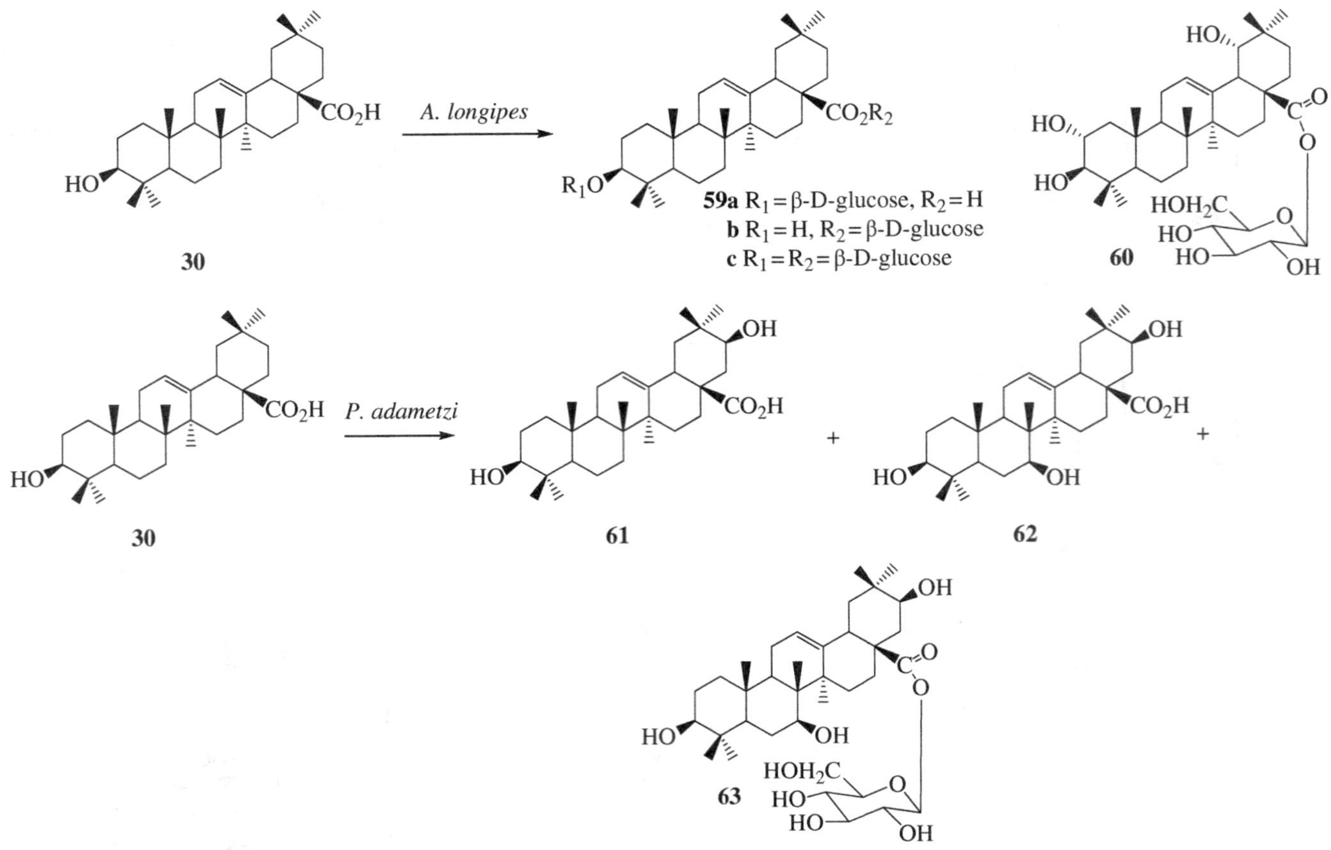

FIGURE 28.14
Some typical biotransformation products of oleanolic acid (**30**) by *Alternaria longipes* or *Penicillium adametzi* [55].

Microbial transformation of oleanolic acid (**30**) using a strain of *Trichothecium roseum* [58] resulted in the isolation of two new hydroxylated metabolites, characterized as 15α-hydroxy-3-oxo-olean-12-en-28-oic acid (**71**) and 7β,15α-dihydroxy-3-oxo-olean-12-en-28-oic acid (**72**) (Figure 28.17).

Microbial transformation of oleanolic acid (**30**) using *Rhizomucor miehei* resulted in the isolation of three metabolites [59] (Figure 28.18): 3β,29-dihydroxy-olean-12-en-28-oic acid (queretaroic acid, **73** a natural product already isolated from Mexican cacti), 3β,7β,29-trihydroxy-olean-12-en-28-oic acid (**74**), and 1β,3β,29-trihydroxy-olean-12-en-28-oic acid (**75**).

Echinocystic acid (3β,16α-dihydroxy-olean-12-en-28-oic acid, **76**) was first isolated from *Echinocystis fabaceae* a long time ago. Since its discovery, in recent years, echinocystic acid and its glucoside have been found in many plants and shown to have many bioactivities such as cytotoxic effects *in vitro* against different cell lines. Echinocystic acid and its saponosides have other bioactivities, including anti-inflammatory, antifungal and anti-HIV activities, inhibition of pancreatic lipase, an immunostimulatory effect, and interleukin inhibitory activities. Microbial transformation of echinocystic acid by *Nocardia corallina* CGMCC 4.1037 [60] afforded the oxidized metabolite 3-oxo-16α-hydroxy-olean-12-en-28-oic acid (**77**), the corresponding C-28 acyl glucosylated metabolites 3β,16α-dihydroxy-olean-12-en-28-oic acid 28-*O*-β-D-glucopyranoside (**78**) and 3-oxo-16α-hydroxy-olean-12-en-28-oic acid 28-*O*-β-D-glucopyranoside (**79**) (Figure 28.19).

The oleane-type PT aglycone, phytolaccagenin (2β,3β,23-trihydroxy-olean-12-ene-28,29-dioic acid 29-methyl ester) (**80**), is the major saponin aglycone extracted from the roots of the plant *Phytolacca esculenta*, widely distributed in East Asia and used for its high activity in the treatment of acute inflammation. Phytolaccagenin was used as substrate in a preparative-scale biotransformation by *S. griseus* ATCC 13273 for developing new bioactive derivatives. An oxidized metabolite, with a regiospecific hydroxylation

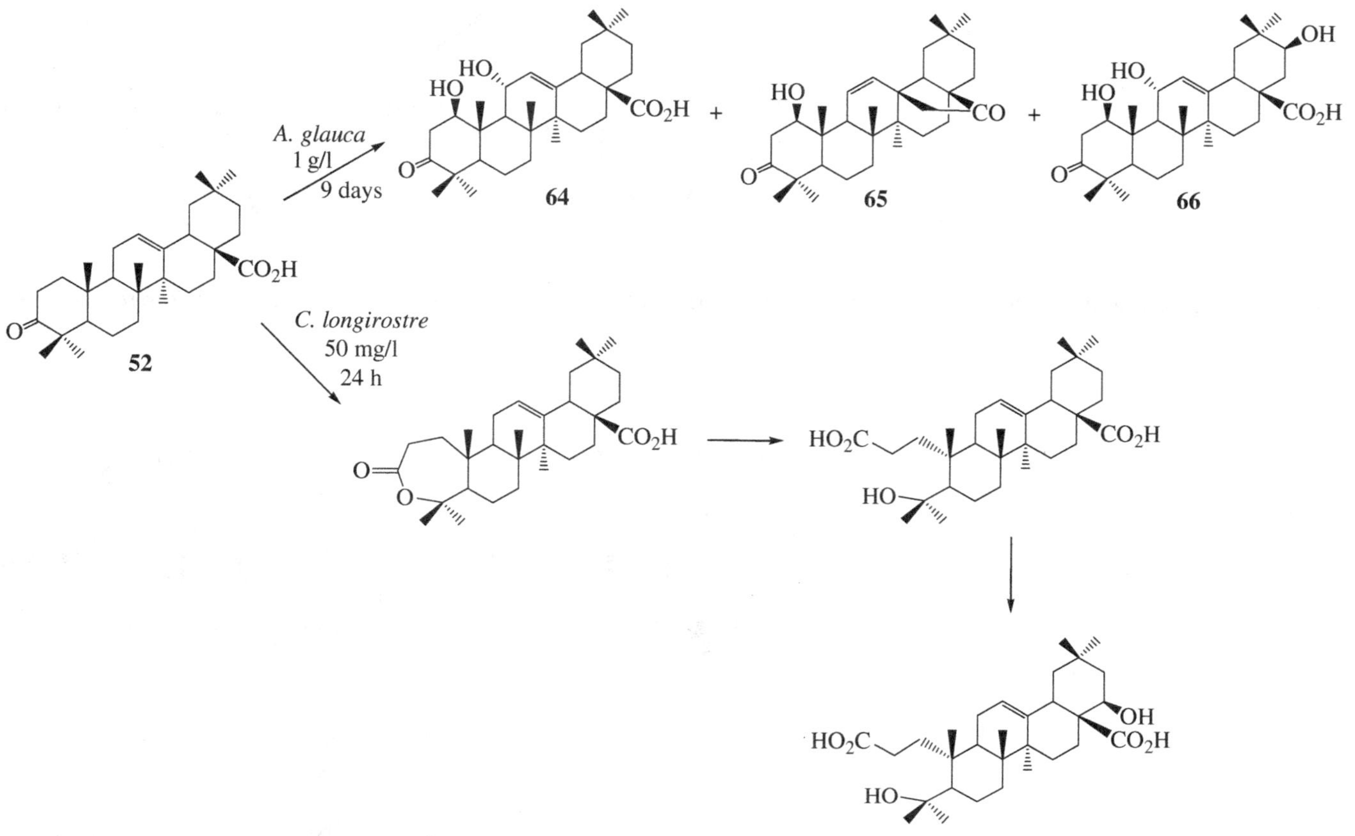

FIGURE 28.15

Microbial transformation of 3-oxo-olean-12-en-28-oic (**52**) acid by *Absidia glauca* [56] and *Chaetomium longirostre* [57].

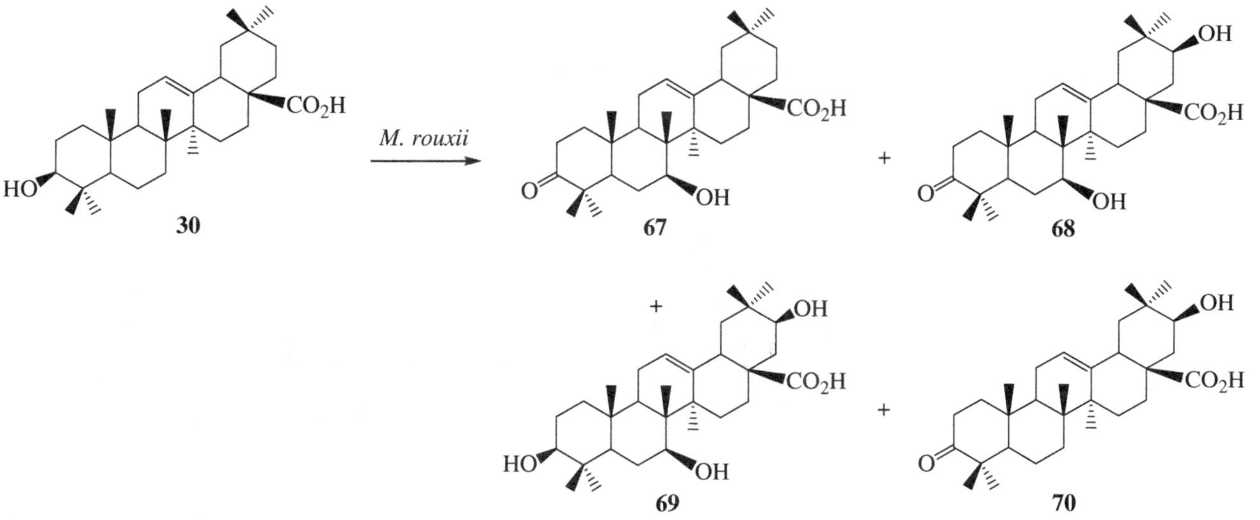

FIGURE 28.16

Microbial transformation of oleanolic acid (**30**) by *Mucor rouxii* [58].

on the C-29 methyl group, was obtained from a preparative-scale biotransformation. The metabolite was identified as 2β,3β,24,30-tetrahydroxy-olean-12-ene-28,29-dioic acid 29-methyl ester (**81**) by mass and 2D NMR spectra [61] (Figure 28.20).

An industrial procedure has been established for the isolation of 2α,3β-dihydroxy-olean-12-en-28-oic acid (maslinic acid, **82**) on a large scale, starting from olive fruits (*Olea europaea*) [62]. Maslinic acid was bioconverted by *Cunninghamella blakesleana*

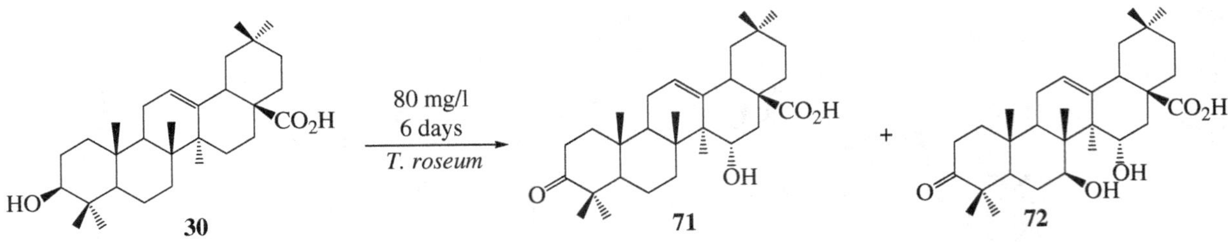

FIGURE 28.17

Structures of two new hydroxylated metabolites of oleanolic acid (**30**) obtained by incubation with *Trichothecium roseum* [59].

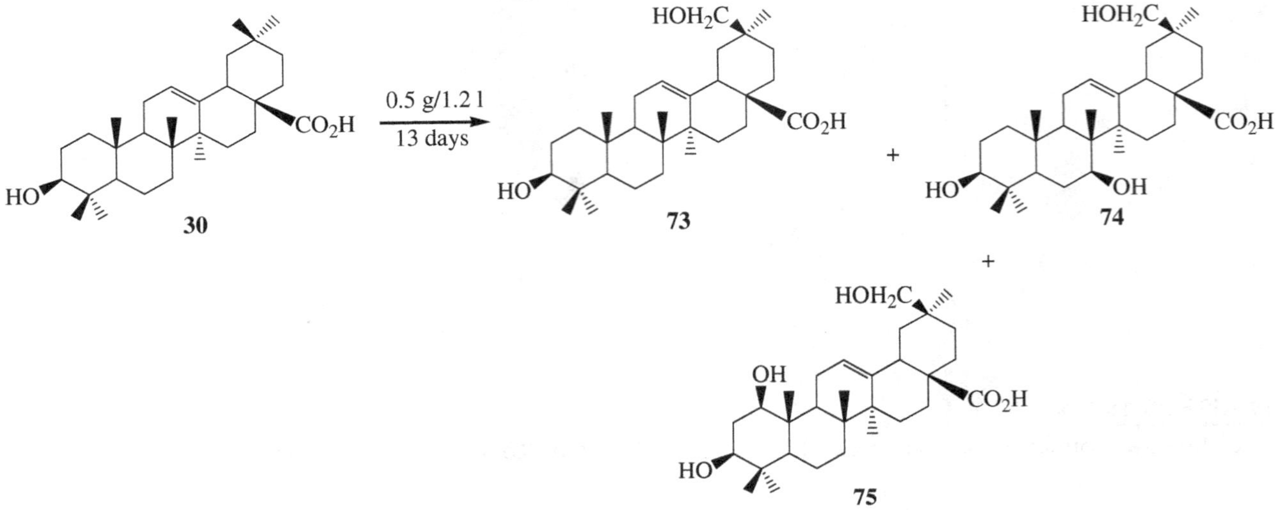

FIGURE 28.18

Microbial transformation of oleanolic acid (**30**) by *Rhizomucor miehei* [60].

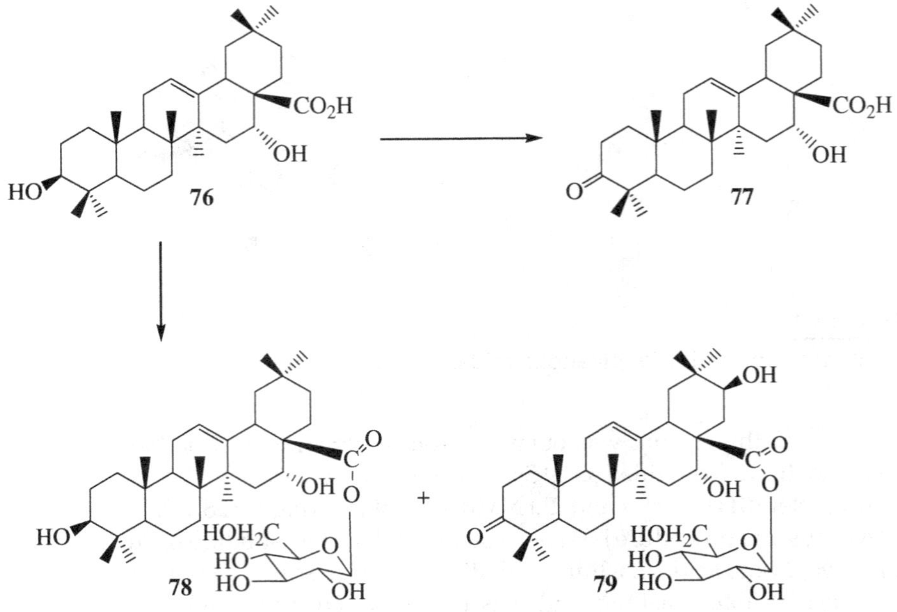

FIGURE 28.19

Microbial transformation of echinocystic acid (**76**) by *Nocardia corallina* CGMCC 4.1037 [61].

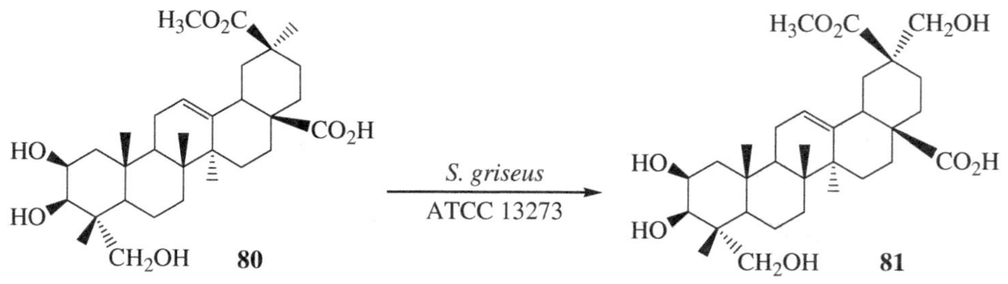

FIGURE 28.20
Microbial hydroxylation of phytolaccagenin (**80**) by *Streptomyces griseus* ATCC 13273 [62].

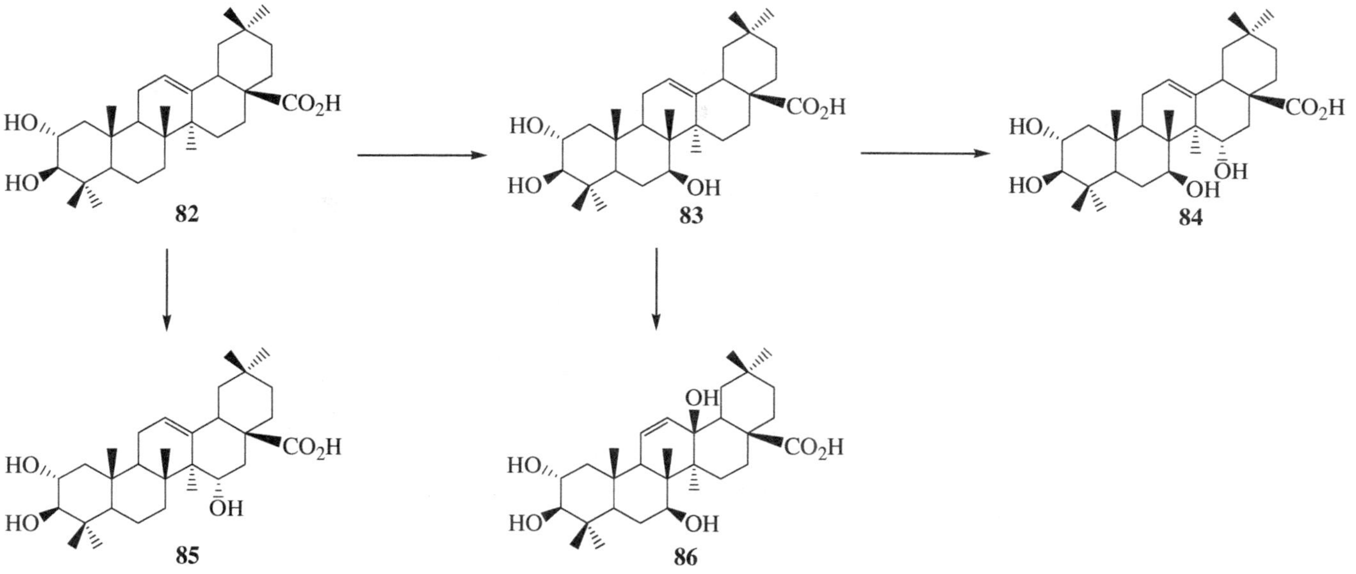

FIGURE 28.21
Microbial transformation of maslinic acid (**82**) by *Cunninghamella blakesleeana* CGMCC 3.910 [64].

CGMCC 3.910 [63] into four metabolites. On the basis of nuclear magnetic resonance and high-resolution mass spectral analyses, their structures (Figure 28.21) were identified as 2α,3β,7β-trihydroxy-olean-12-en-28-oic acid (**83**), 2α,3β,15α-trihydroxy-olean-12-en-28-oic acid (**84**), 2α,3β,7β,15α-tetrahydroxy-olean-12-en-28-oic acid (**85**), and 2α,3β,7β,13β-tetrahydroxy-olean-11-en-28-oic acid (**86**).

More recently, the incubation of maslinic acid (**82**) with *R. miehei* produced new derivatives [59]: an olean-11-en-28,13β-olide derivative (**87**), a metabolite hydroxylated at C-29 (**88**), an 11-oxo derivative (**89**), and two metabolites (**90–91**) with an 11α,12α-epoxy group, hydroxylated or not at C-30. Their structures (Figure 28.22) could be elucidated by extensive analyses of their spectroscopic data and by chemical correlations.

Glycyrrhetinic acid (**83**), also known as glycyrrhetic acid or 18β-glycyrrhetinic acid (3β-hydroxy-11-oxo-18β-olean-12-ene-29-oic acid), is the aglycone of saponosides isolated from roots of several *Glycyrrhiza* species (liquorice). This triterpenoid is one of the major constituents of a traditional Chinese medicine. The roots of *Glycyrrhiza uralensis* have several pharmacological activities, such as anti-inflammatory, anti-ulcer, and anticarcinogenic activities. Glycyrrhetinic acid and its derivatives have recently received some attention as potential therapeutic agents for several viral diseases including chronic hepatitis and AIDS.

The sweet saponoside of glycyrrhetinic acid, glycyrrhizin, has been transformed by several *Aspergillus* sp. first into its aglycone **83**, by hydrolysis of the diglucuronide oligosaccharide (see Section 28.5), then into the corresponding 3-oxo-glycyrrhetinic acid [64].

The major product of the microbial transformations of glycyrrhetinic acid (**83**) by *Curvularia lunata* ATCC 13432 [65], *Mucor spinosus* AS 3.3450 [16, 66], and *Mucor*

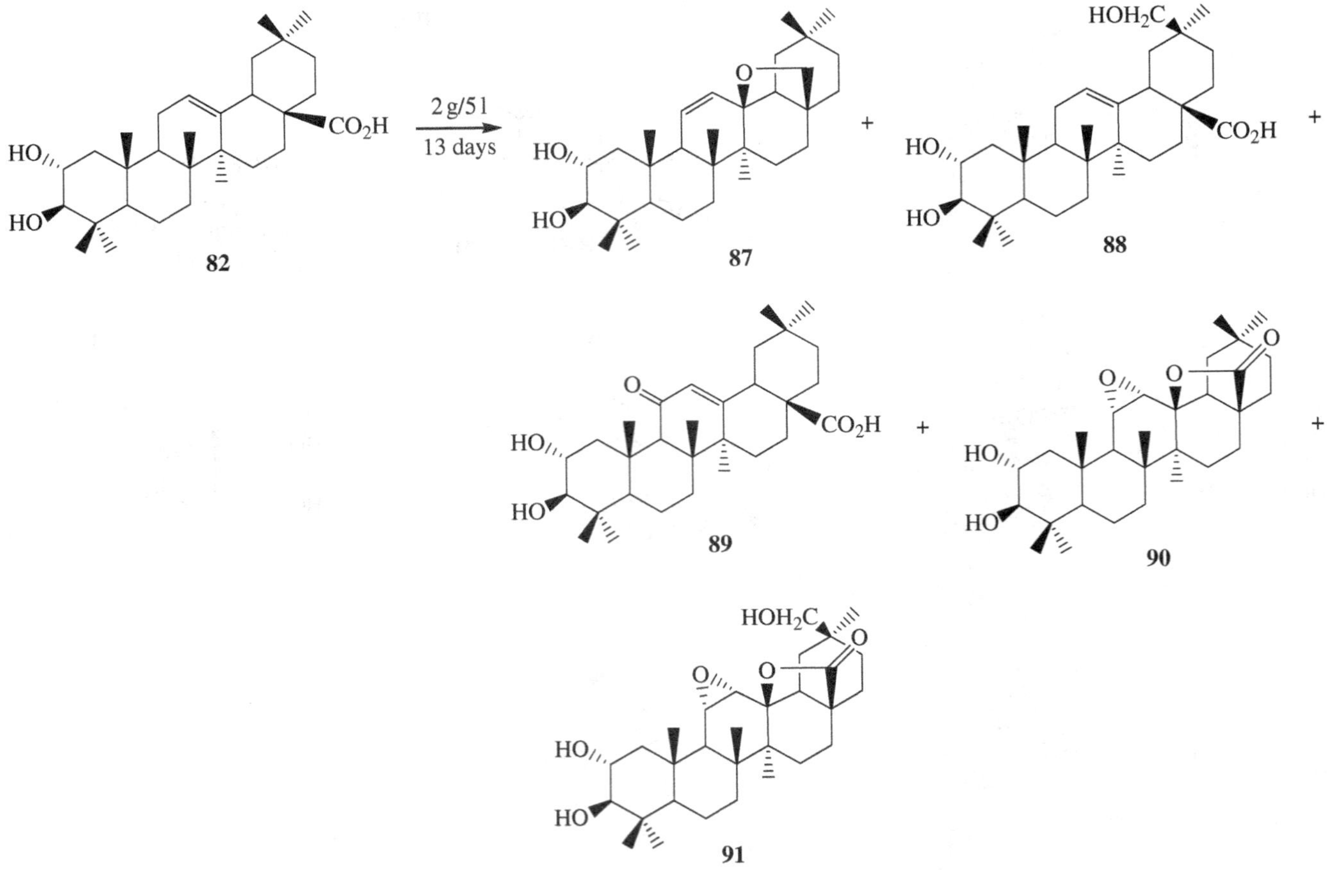

FIGURE 28.22
Microbial transformations of maslinic acid (**82**) by *Rhizomucor miehei* [60].

polymorphosporus [67] was identified as 7β-hydroxy-glycyrrhetinic acid (**84**) along with 15α-hydroxy-glycyrrhetinic acid (**85**) and other minor metabolites hydroxylated in positions 23 (**86**), 6β- (**87**), 7α- (**88**), 3-oxo-7α-hydroxy-glycyrrhetinic acid (**89**), and 3-oxo-15α-hydroxy-glycyrrhetinic acid (**90**) (Figure 28.23).

The main hydroxylating activity of *T. roseum* ATCC 8685 on glycyrrhetinic acid (**83**) was also directed to C-7 and C-15, yielding 7β- (**84**), 15α-hydroxy (**85**), and 7β,15α-dihydroxy (**91**) derivatives, which were also methylated at the carboxylic acid group by the microorganism [68]. The formation of a methyl glycyrrhetinate was also detected in the incubation of glycyrrhetinic acid (**83**) with *Nocardia* sp. NRRL 5646 [42].

When incubated in buffer with the mycelium of *Streptomyces* sp. (G-20), glycyrrhetinic acid (**83**) gave three conversion products (Figure 28.24). The major metabolite is a 22α-hydroxy derivative (**92**), and the two minors, a 22α,24- (**93**) and a 22α,23-dihydroxy (**94**) derivative [69, 70]. In addition, *Chainia antibiotica* was shown to cleave the ring A cycle of glycyrrhetinic acid (**71**) and 22α-hydroxy-glycyrrhetinic acid (**92**) to the corresponding 3,4-*seco*-compounds, probably via a Baeyer–Villiger-like reaction of the 3-oxo derivatives [69, 70].

It was also shown that the microbial transformation of glycyrrhetinic acid (**83**) by *Sphingomonas paucimobilis* (strain G5) [71, 72] produced a new metabolite identified as 3β-hydroxy-11-oxo-olean-12-ene-23,29-dioic acid (**95**) (Figure 28.25).

The microbial transformation of glycyrrhetinic acid (**83**) by *C. elegans* afforded one metabolite, 3β,7β-dihydroxy-11-oxo-olean-12-en-29-oic acid (**84**), while fermentation of **83** with *F. lini* afforded 3,11-dioxo-olean-12-en-29-oic acid (**96**) (Figure 28.26) which exhibited a potent lipoxygenase inhibitory activity [73].

An epimer at C-18 of glycyrrhetinic acid, 18α-glycyrrhetinic acid (3β-hydroxy-11-oxo-18α-olean-12-ene-29-oic acid, **97**), obtained by acidic treatment of glycyrrhetinic

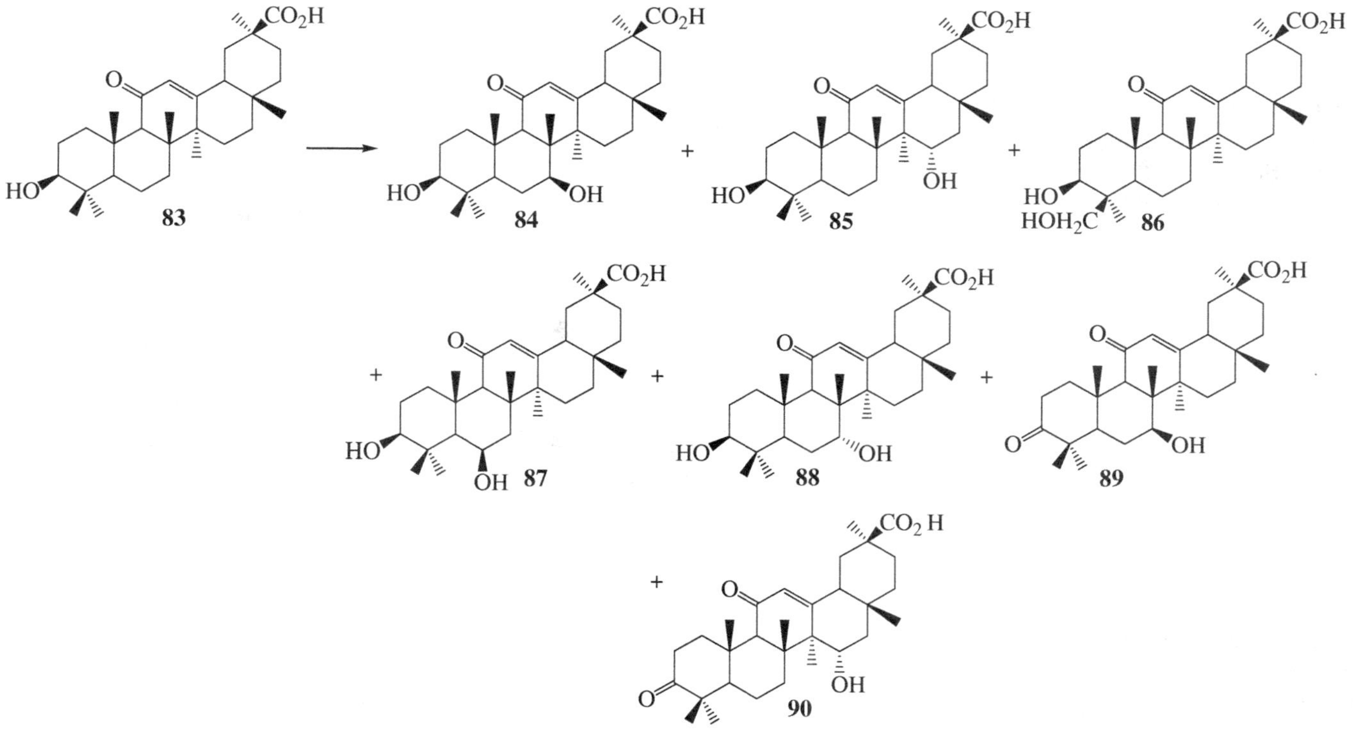

FIGURE 28.23

Microbial transformation of glycyrrhetinic acid (**83**) by *Mucor polymorphosporus* [67].

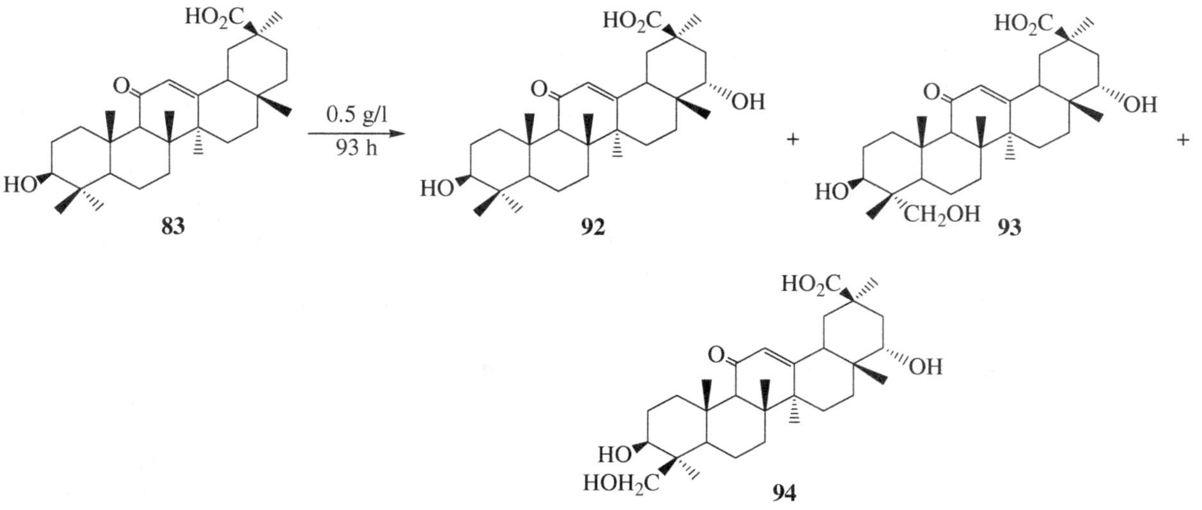

FIGURE 28.24

Microbial transformation of glycyrrhetinic acid (**83**) by *Streptomyces* sp. (G-20) [69, 70].

acid [65], was converted by *C. lunata* ATCC 13432, *T. roseum* ATCC 8685, *Cunninghamella* sp. ATCC 3229, *Mucor griseocyanus* ATCC 1207a, or *Helicostylum piriforme* ATCC 8992 into its 7β-hydroxy, 15α-hydroxy, or 7β,15α-dihydroxy derivatives (**98–100**, respectively) [74] (Figure 28.27). Similar results were obtained [74] when using as substrate liquiritic acid (3β-hydroxy-11-oxo-18β-olean-12-ene-30-oic acid, **101**), an epimer at C-20 isolated from the mother liquor of glycyrrhetinic acid purification [75]. Another epimer at C-18, obtained by acidic treatment of liquiritic acid [75], named 18α-liquiritic acid (3β-hydroxy-11-oxo-18β-olean-12-ene-30-oic acid, **105**), was transformed by *H. piriforme* and *Cunninghamella* sp. into its 7β-hydroxy derivative (**106**) and into its 7β-15α-dihydroxy derivative (**107**) by *T. roseum* [74] (Figure 28.27).

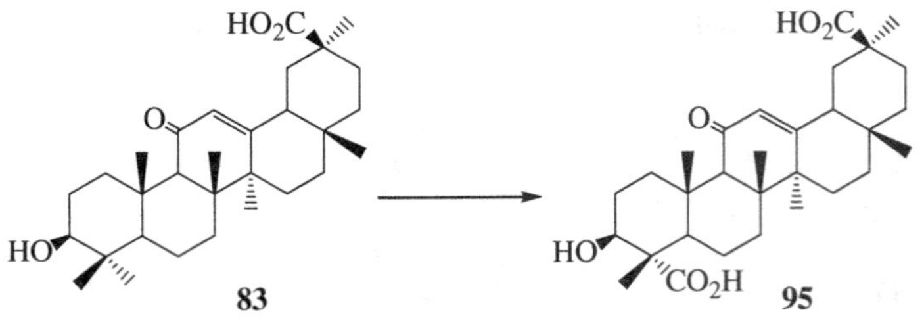

FIGURE 28.25

Microbial transformation of glycyrrhetinic acid (**83**) by *Sphingomonas paucimobilis* (strain G5) [71].

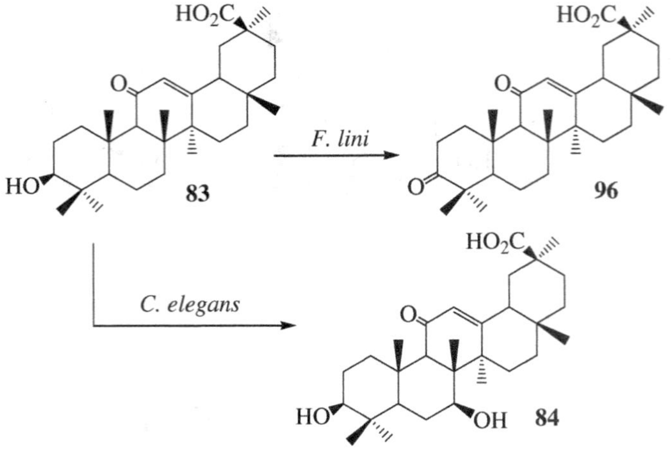

FIGURE 28.26

Microbial transformation of glycyrrhetinic acid (**83**) by *Fusarium lini* and *Cunninghamella elegans* [73].

In summary, microbial hydroxylation of β-amyrin and oleanolic acids may lead to new hydroxylations mainly in positions −7β, −15α but also frequently in other positions: −11β, −11α, −2α, −29, −30, −13β, −21β, −16α, −23, −24, −1β, −6β, and −7α (in order of priority, see Figure 28.28). With already hydroxylated olean derivatives such as echinocystic acid, phytolaccagenin, maslinic acid, or glycyrrhetinic acid as substrates, the corresponding complementary positions can be hydroxylated, obtaining a wide combinatorial variety of new multihydroxy derivatives, which should be difficult to achieve by classical chemical means and, potentially possess new or increased biological activities. Several of them have been previously isolated in plants as natural products. A summary of these results and yields is presented in Table 28.2.

There are only a few examples of the microbial hydroxylation of saponosides themselves because deglycosylation generally occurs before hydroxylation of the aglycone part. An example with quinovosides is described later (see Section 28.4). As another example, an endophytic fungal strain of the *Xylariaceae* family, isolated from the fruit pericarp of *Sapindus saponaria* and highly tolerant to the abundant endogenous saponins [76], is able to hydroxylate an exogenous saponoside (**108**) of hederagenin (23-hydroxy-oleanolic acid), without deglycosylation, to give a 22α-hydroxylated derivative (**109**) in 18% yield [77] (Figure 28.29). It should be interesting to test this activity on other saponosides of the olean or ursan families.

28.4 TYPICAL BIOTRANSFORMATIONS IN THE URSANE FAMILY

α-Amyrin has been isolated from various sources, most notably plant resins. Considerable amounts (up to grams per kilogram) of this triterpene are available in the resins of *Bursera* and *Protium* species of the *Burseraceae* family. Other known sources of α-amyrin include Mexican copal (up to 5 g/kg) [78], *Cassia obtusifolia* (140 mg/kg), and the resin of *Commiphora holtziana* (200 mg/kg) [79]

Ursolic acid (**87**), the simplest carboxy derivative of the ursane family, is widely distributed in various plants in both aglycone and glycosidic forms, and traditional uses of plants containing **87** in folk medicine are abundantly documented. Recent

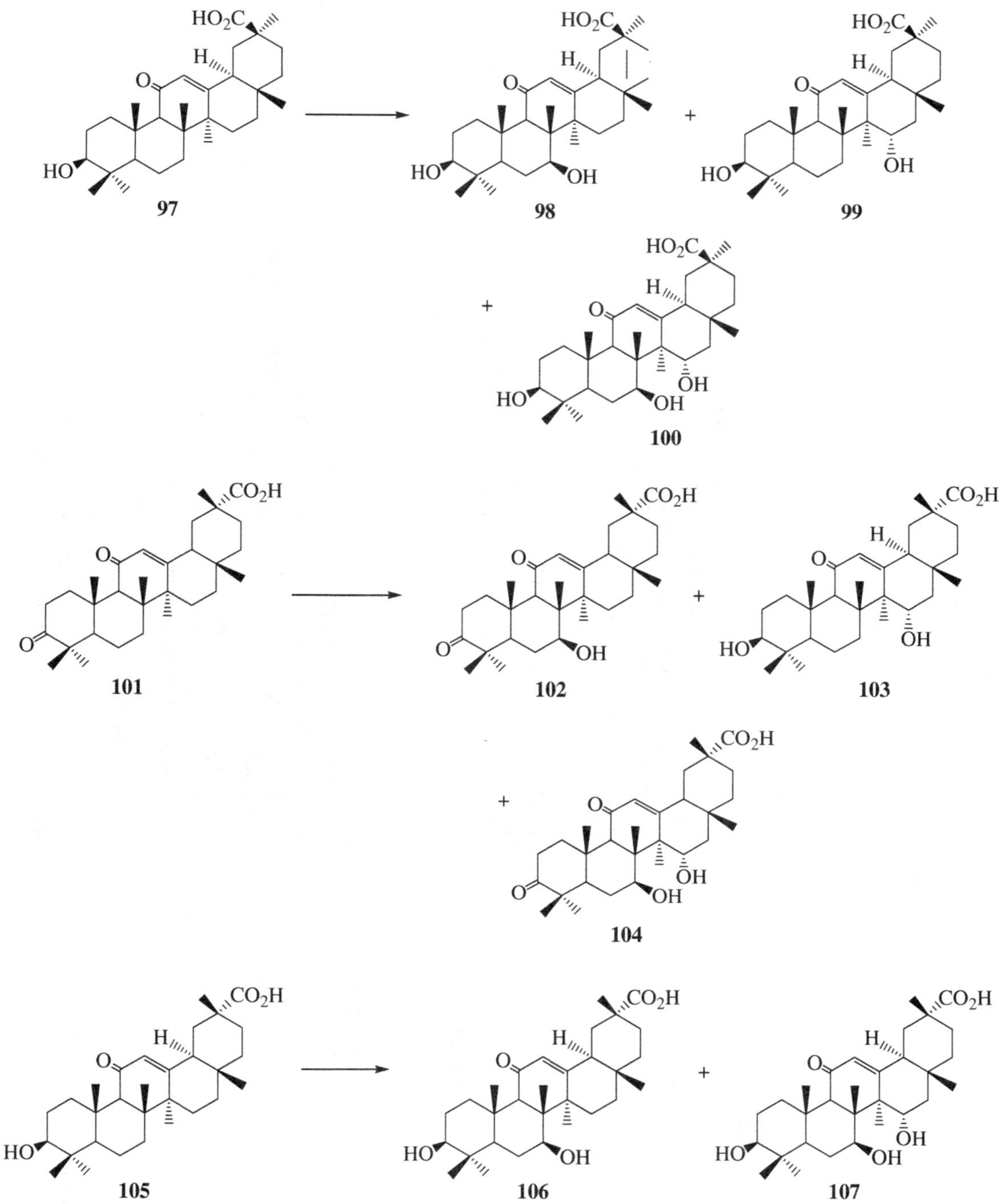

FIGURE 28.27

Microbial transformations of glycyrretinic epimers at C-18 (18α-glycyrretic acid, **97**) or at C-20 (liquiritic acid, **101**) or at C-18 and C-20 (18α-liquiritic acid, **105**) by various fungal strains [74].

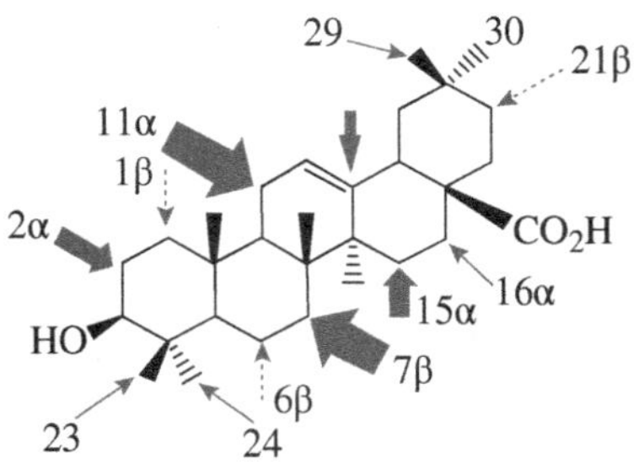

FIGURE 28.28

Main hydroxylation sites observed in the microbial transformations of olean derivatives.

TABLE 28.2 Microbial Transformation Products and Yields Obtained from Olean-Family PTs

Substrate PTs	Microorganisms	Products	Yields (%)
β-Amyrin (24)	*Lecanicillium muscarinium* [47]	3β-Hydroxy-11-oxo-olean-12-ene (**28**)	6.5
		3β-Hydroxy-11α,12α-epoxy-oleane-14-ene (**29**)	—
β-Amyrin acetate (25)	*Rhodobacter sphaeroides* [46]	β-Amyrin (**24**)	10
		3β,24-Dihydroxy-olean-12-ene (**26**)	9
		3β,28-Dihydroxy-olean-12-ene (**27**)	4
Oleanolic acid (30)	*Nocardia sp.* NRRL5646 [42]	3β-Hydroxy-olean-12-en-28-oic acid methyl ester	—
		3β,6β-Dihydroxy-olean-12-en-28-oic acid (**35**)	—
	Colletotrichum phomoides [48–51]	3β,6β,21β-Trihydroxy-olean-12-en-28-oic acid (**36**)	—
		3β-Hydroxy-21-oxo-olean-12-en-28-oic acid (**37**)	—
		3β-Hydroxy-11-oxo-olean-12-en-28-oic acid (**31**)	—
	Fusarium lini [48–51]	3β,15α-Dihydroxy-olean-12-en-28-oic acid (**32**)	—
		3β,12β-Dihydroxy-olean-11-en-28(13)-olide (**33**)	—
		3β-Hydroxy-olean-11,13(18)-dien-28-oic acid (**34**)	—
		3β,6β,21β-Trihydroxy-olean-12-en-28-oic acid (**44**)	—
	Penicillium chrysogenum [48–51]	3β-Hydroxy-21-oxo-olean-12-en-28-oic acid (**37**)	—
		3-Oxo-21β-hydroxy-olean-12-en-28-oic acid (**44**)	—
		3β-Hydroxy-olean-11,13(18)-dien-28-oic acid (**34**)	—
	Cunninghamella blakesleana [48–51]	3β,7β-Dihydroxy-olean-12-en-28-oic acid (**39**)	—
		3β-Hydroxy-11-oxo-olean-13(18)-en-28-oic acid (**40**)	—
		3β,7β-Dihydroxy-olean-11-en-28(13)-olide (**41**)	—
		3β,7β-Dihydroxy-olean-11,13(18)-dien-28-oic acid (**42**)	—
		1β,3β-Dihydroxy-olean-12-en-28-oic acid (**43**)	—
		2α,3β-Dihydroxy-olean-12-en-28-oic acid (**50**)	4
	F. lini [53]	2α,3β,11β-Trihydroxy-olean-12-en-28-oic acid (**51**)	3
		3β-Hydroxy-olean-12-en-28-oic acid-28-*O*-β-ᴅ-glucoside (**59b**)	4.5
	Alternaria longipes AS 3.2875 [55]	3β-Hydroxy-olean-12-en-28-oic acid-3-*O*-β-ᴅ-glucoside (**59a**)	0.3
		2α,3β,19α-Trihydroxy-olean-12-en-28-oic acid-28-*O*-β-ᴅ-glucoside (**60**)	0.1
		3β-Hydroxy-olean-12-en-28-oic acid-3,28-*O*-β-ᴅ-diglucoside (**59c**)	0.2
		2α,3β,19α-Trihydroxy-ursan-12-en-28-oic acid-28-*O*-β-ᴅ-glucoside (**151**)	0.2
	Penicillium adametzi AS 3.447 [55]	2α,3α,19α-Trihydroxy-ursan-12-en-28-oic acid-28-*O*-β-ᴅ-glucoside (**150**)	3.1
		3β,21β-Dihydroxy-olean-12-en-28-oic acid (**61**)	2.9
		3β,7β,21β-Trihydroxy-olean-12-en-28-oic acid (**62**)	0.2
		3β,7β,21β-Trihydroxy-olean-12-en-28-oic acid-28-*O*-β-ᴅ-glucoside	2.6
	Mucor rouxii [58]	3β,7β,21β-Trihydroxy-olean-12-en-28-oic acid (**68**)	1.2
		3-Oxo-7β-hydroxy-olean-12-en-28-oic acid (**67**)	3.2
		3-Oxo-21β-hydroxy-olean-12-en-28-oic acid (**70**)	6.3
		3-Oxo-7β,21β-dihydroxy-olean-12-en-28-oic acid (**69**)	5
	Rhizomucor miehei [60]	3β,29-Dihydroxy-olean-12-en-28-oic acid (**73**)	6
		3β,7β-29-Trihydroxy-olean-12-en-28-oic acid (**74**)	5
		1β,3β-29-Trihydroxy-olean-12-en-28-oic acid (**75**)	6
	Trichothecium roseum [59]	3-Oxo-15α-hydroxy-olean-12-en-28-oic acid (**71**)	7.5
		3-Oxo-7β,15α-dihydroxy-olean-12-en-28-oic acid (**72**)	40

	Streptomyces griseus ATCC 13273 [54]	3β-Hydroxy-olean-12-en-28,30-dioic acid (**54**)	90
		3β,23-Dihydroxy-olean-12-en-28,30-dioic acid (**55**)	0.75
		1β,11α-Dihydroxy-3-oxo-olean-11-en-28(13)-olide	
3-Oxo-oleanolic acid (52)	*Nocardia* sp. NRRL5646 [42]	3-Oxo-olean-12-en-28-oic acid methyl ester	—
		1β,11α-Dihydroxy-3-oxo-olean-12-en-28-oic acid (**64**)	2.3
	Absidia glauca [56]	1β-Hydroxy-3-oxo-olean-11-en-13,28-olide (**65**)	0.5
		1β,11α,21β-Trihydroxy-3-oxo-olean-12-en-28-oic acid (**66**)	0.25
		3-Oxo-olean-12-en-28-oic acid-28-*O*-β-ᴅ-glucoside (**57**)	50
	Aspergillus ochraceus CICC 40330 [54]	3β-Hydroxy-olean-12-en-28-oic acid-*O*-β-ᴅ-glucoside (**58**)	45
		3,4-*seco*-4β-Hydroxy-olean-12-en-28-oic acid	30
	Chaetomium longirostre [57]	3,4-*seco*-4,22β-Dihydroxy-olean-12-en-28-oic acid	28
		3-Oxo-olean-12-en-28,30-dioic acid (**56**)	30
	S. griseus ATCC 13273 [54]	3β,23-Dihydroxy-olean-12-en-28,30-dioic acid (**55**)	40
Esculentoside A (53)	*S. griseus* ATCC 13273 [54]	2β,3β-24-Trihydroxy-olean-12-en-28,29-dioic acid 29-methyl ester 3-*O*-β-xyloside	12
		2β,3β-24-Trihydroxy-olean-12-en-28,29-dioic acid 29-methyl ester	25
	A. ochraceus CICC 40330 [54]	2β,3β-24-Trihydroxy-olean-12-en-28,29-dioic acid 29-methyl ester 3-*O*-β-xyloside	50
Echinocystic acid (76)	*Nocardia corallina* CGMCC 4.1037 [61]	3-Oxo-16α-hydroxy-olean-12-en-28-oic acid (**77**)	23
		3β,16α-Dihydroxy-olean-12-en-28-oic acid-28-*O*-β-ᴅ-glucoside (**78**)	0.5
		3-Oxo-16α,21β-dihydroxy-olean-12-en-28-oic acid-28-*O*-β-ᴅ-glucoside (**79**)	0.4
Phytolaccagenin (80)	*S. griseus* ATCC 13273 [54, 62]	2β,3β-24,30-Tetrahydroxy-olean-12-en-28,29-dioic acid 29-methyl ester (**81**)	64
Maslinic acid (82)	*C. blakesleana* CGMCC 3.910 [64]	2α,3β,7β-Trihydroxy-olean-12-en-28-oic acid (**83**)	37.4
		2α,3β,15α-Trihydroxy-olean-12-en-28-oic acid (**85**)	1.7
		2α,3β,7β,15α-Tetrahydroxy-olean-12-en-28-oic acid (**84**)	3.2
		2α,3β,7β,13β-Tetrahydroxy-olean-11-en-28-oic acid (**86**)	3.4
	R. miehei [60]	2α,3β,29-Trihydroxy-olean-12-en-28-oic acid (**88**)	12.2
		2α,3β-Dihydroxy-11-oxo-olean-12-en-28-oic acid (**89**)	1
		2α,3β-Dihydroxy-olean-11-en-28(13)-olide (**87**)	0.8
		2α,3β-Dihydroxy-11α,12α-epoxy-olean-13,28-olide (**90**)	0.2
		2α,3β,29-Trihydroxy-11α,12α-epoxy-olean-13,28-olide (**91**)	0.9
Glycyrrhetinic acid (83)	*Cunninghamella elegans* [73]	3β,7β-Dihydroxy-11-oxo-olean-12-en-29-oic acid (**72**)	—
	Curvularia lunata ATCC 13432 [66]	3β,7β-Dihydroxy-11-oxo-olean-12-en-29-oic acid (**72**)	—
	Mucor spinosus AS 3.3450 [55]	3β,7β-Dihydroxy-11-oxo-olean-12-en-29-oic acid (**72**)	—
	F. lini [73]	3,11-Dioxo-olean-12-en-29-oic acid (**84**)	27
	Mucor polymorphosporus AS 3.3443 [67]	3β,7β-Dihydroxy-11-oxo-olean-12-en-29-oic acid (**72**)	18
		3β,15α-Dihydroxy-11-oxo-olean-12-en-29-oic acid (**73**)	2.7
		3β,24-Dihydroxy-11-oxo-olean-12-en-29-oic acid (**74**)	1.3
		3β,6β-Dihydroxy-11-oxo-olean-12-en-29-oic acid (**75**)	1.4
		3β,7α-Dihydroxy-11-oxo-olean-12-en-29-oic acid (**76**)	1.1
		7β-Hydroxy-3,11-dioxo-olean-12-en-29-oic acid (**77**)	0.9
		15α-Hydroxy-3,11-dioxo-olean-12-en-29-oic acid (**78**)	—
	Nocardia sp. NRRL 5646 [38]	3β-Hydroxy-11-oxo-olean-12-en-29-oic acid methyl ester	—
		3β-Hydroxy-11-oxo-olean-12-en-29-oic acid (**80**)	64

(Continued)

TABLE 28.2 (*Continued*)

Substrate PTs	Microorganisms	Products	Yields (%)
	Streptomyces sp. (G20) [69, 70]	3β,22α,24-Trihydroxy-11-oxo-olean-12-en-29-oic acid (**81**)	6.5
		3β,22α,23-Trihydroxy-11-oxo-olean-12-en-29-oic acid (**82**)	9.5
		3,4-*seco*-Glycyrrhetinic acid	60[a]
		3,4-*seco*-7β-Hydroxy-glycyrrhetinic acid	40[a]
	Chainia antibiotica I [69, 70]	3β-Hydroxy-11-oxo-olean-12-en-24,29-dioic acid (**83**)	2.4
	Sphingomonas paucimobilis (G5) [71, 72]	3,4-*seco*-4,23,24-Trinor-11-oxo-12-oleane-3,28,30-trioic acid	—
		3β,7β-Dihydroxy-11-oxo-olean-12-en-29-oic acid (**72**)	—
	T. roseum [68]	3β,15α-Dihydroxy-11-oxo-olean-12-en-29-oic acid (**73**)	—
		3β,7β,15α-Trihydroxy-11-oxo-olean-12-en-29-oic acid (**79**)	—
22α-Glycyrrhetinic acid (92)	*C. antibiotica* [70]	3,4-*seco*-22α-Hydroxy-glycyrrhetinic acid	40
18α-Glycyrrhetinic acid (97)	*C. lunata* ATCC 13432 [74]	3β,7β-Dihydroxy-11-oxo-18-*epi*-olean-12-en-29-oic acid (**98**)	—
		3β,15α-Dihydroxy-11-oxo-18-*epi*-olean-12-en-29-oic acid (**99**)	—
		3β,7β,15α-Trihydroxy-11-oxo-18-*epi*-olean-12-en-29-oic acid (**100**)	—
	T. roseum ATCC 8685 [74]	3β,7β-Dihydroxy-11-oxo-18-*epi*-olean-12-en-29-oic acid (**98**)	—
		3β,15α-Dihydroxy-11-oxo-18-*epi*-olean-12-en-29-oic acid (**99**)	—
		3β,7β,15α-Trihydroxy-11-oxo-18-*epi*-olean-12-en-29-oic acid (**100**)	—
	Cunninghamella ATCC 3229 [74]	3β,7β-Dihydroxy-11-oxo-18-*epi*-olean-12-en-29-oic acid (**98**)	—
		3β,15α-Dihydroxy-11-oxo-18-*epi*-olean-12-en-29-oic acid (**99**)	—
		3β,7β,15α-Trihydroxy-11-oxo-18-*epi*-olean-12-en-29-oic acid (**100**)	—
	Mucor griseocyanus ATCC 1207a [74]	3β,7β-Dihydroxy-11-oxo-18-*epi*-olean-12-en-29-oic acid (**98**)	—
		3β,15α-Dihydroxy-11-oxo-18-*epi*-olean-12-en-29-oic acid (**99**)	—
		3β,7β,15α-Trihydroxy-11-oxo-18-*epi*-olean-12-en-29-oic acid (**100**)	—
	Helicostylum piriforme ATCC 8992 [74]	3β,7β-Dihydroxy-11-oxo-18-*epi*-olean-12-en-29-oic acid (**98**)	—
		3β,15α-Dihydroxy-11-oxo-18-*epi*-olean-12-en-29-oic acid (**99**)	—
		3β,7β,15α-Trihydroxy-11-oxo-18-*epi*-olean-12-en-29-oic acid (**100**)	—
Liquiritic acid (20-*epi*-glycyrrhetinic acid) (101)	*C. lunata* ATCC 13432 [74]	3β,7β-Dihydroxy-11-oxo-20-*epi*-olean-12-en-30-oic acid (**102**)	—
		3β,15α-Dihydroxy-11-oxo-20-*epi*-olean-12-en-30-oic acid (**103**)	—
		3β,7β,15α-Trihydroxy-11-oxo-20-*epi*-olean-12-en-30-oic acid (**104**)	—
	T. roseum ATCC 8685 [74]	3β,7β-Dihydroxy-11-oxo-20-*epi*-olean-12-en-30-oic acid (**102**)	—
		3β,15α-Dihydroxy-11-oxo-20-*epi*-olean-12-en-30-oic acid (**103**)	—
		3β,7β,15α-Trihydroxy-11-oxo-20-*epi*-olean-12-en-30-oic acid (**104**)	—
	Cunninghamella ATCC 3229 [74]	3β,7β-Dihydroxy-11-oxo-20-*epi*-olean-12-en-30-oic acid (**102**)	—
		3β,15α-Dihydroxy-11-oxo-20-*epi*-olean-12-en-30-oic acid (**103**)	—
		3β,7β,15α-Trihydroxy-11-oxo-20-*epi*-olean-12-en-30-oic acid (**104**)	—
	M. griseocyanus ATCC 1207a [74]	3β,7β-Dihydroxy-11-oxo-20-*epi*-olean-12-en-30-oic acid (**102**)	—
		3β,15α-Dihydroxy-11-oxo-20-*epi*-olean-12-en-30-oic acid (**103**)	—
		3β,7β,15α-Trihydroxy-11-oxo-20-*epi*-olean-12-en-30-oic acid (**104**)	—
	H. piriforme ATCC 8992 [74]	3β,7β-Dihydroxy-11-oxo-20-*epi*-olean-12-en-30-oic acid (**102**)	—
		3β,15α-Dihydroxy-11-oxo-20-*epi*-olean-12-en-30-oic acid (**103**)	—
		3β,7β,15α-Trihydroxy-11-oxo-20-*epi*-olean-12-en-30-oic acid (**104**)	—
18α-Liquiritic acid (18-*epi*-liquiritic acid) (105)	*Cunninghamella* ATCC 3229 [74]	3β,7β-Dihydroxy-11-oxo-18-*epi*-olean-12-en-30-oic acid (**106**)	—
	T. roseum ATCC 8685 [74]	3β,7β,15α-Trihydroxy-11-oxo-18-*epi*-olean-12-en-30-oic acid (**107**)	—

[a] Yields obtained in different experimental conditions.

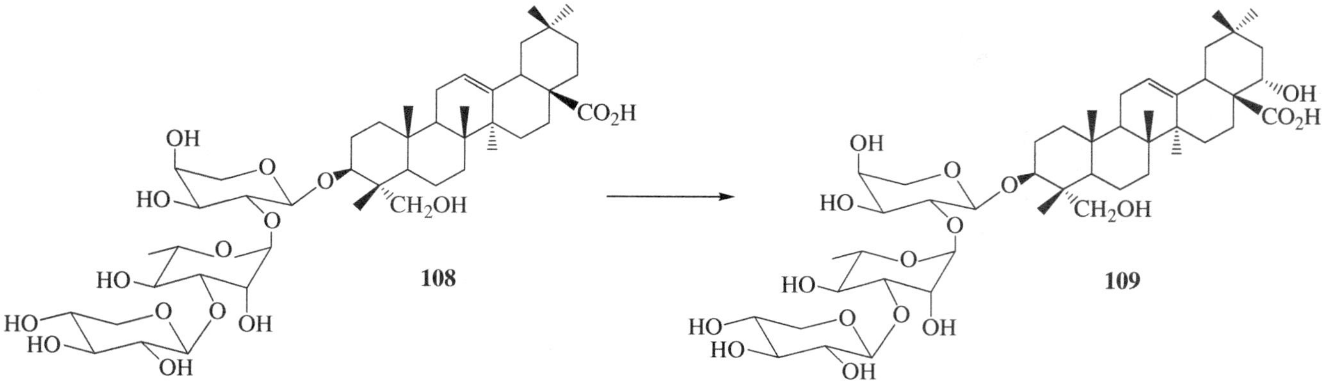

FIGURE 28.29
Microbial transformation of an hederagenin-derived saponoside (**108**) by a xylariaceous endophyte isolated from *Sapindus saponaria* [77].

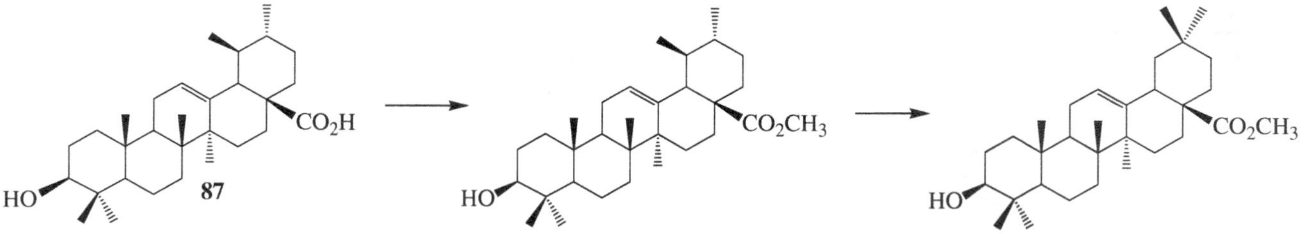

FIGURE 28.30
Biotransformation of ursolic acid to oleanolic acid methyl ester by *Nocardia* sp. NRRL 5646 [42].

studies have shown that ursolic acid possesses many biological effects, such as antioxidative, anti-inflammatory, antitumor, and hepatoprotective activity. The diverse effects of ursolic acid were reviewed by Ikeda *et al.* [7], and more recently by Salvador *et al.* [12]. Mazumdar [80] recently summarized the structure–activity relationships on cancer cells of ursolic acid and several ursane-type PTs obtained from natural sources and by chemical derivatization. Derivatization of ursolic acid by biomimetic oxidation with dioxoruthenium(VI) tetraphenylporphyrins was investigated. Although they have the same ursane-type triterpene core, the positions and numbers of hydroxyles on the terpene structures significantly affected the activity and the selectivity toward the tested cell lines.

Five PT acids, ursolic acid, oleanolic acid, betulinic acid, 23-hydroxybetulinic acid, and glycyrrhetinic acid were metabolized by *Nocardia* sp. NRRL 5646 to selectively furnish their corresponding 28-methyl esters. Notably, ursolic acid (**87**) was converted to oleanolic acid methyl ester via ursolic acid methyl ester as intermediate (Figure 28.30). The oleanolic ester is formed by participation of a "retrobiosynthetic" methyl migration from C-19 to C-20 [42].

From the screening of 41 microorganisms strains, only one (*Aspergillus flavus* ATCC 9170) was able to transform ursolic acid, resulting in the formation of ursonic acid (3-oxo-ursolic acid, **103**) in low yield [81].

When incubated with ursolic acid (**87**), an endophytic fungus *Pestalotiopsis microspora*, isolated from the medicinal plant *Huperzia serrata*, afforded four new metabolites [82]: 3-oxo-15α,30-dihydroxy-urs-12-en-28-oic acid (**88**), 3β,15α-dihydroxy-urs-12-en-28-oic acid (**89**), 3β,15α,30-trihydroxy-urs-12-en-28-oic acid (**90**), and 3,4-*seco*-ursan-4,30-dihydroxy-12-en-3,28-dioic acid (**91**) (Figure 28.31).

Another endophytic fungus isolated from *H. serrata* and identified as *Umbelopsis isabellina* transformed ursolic acid (**87**) into three products [83]: 3β-hydroxy-urs-11-en-28,13-olide (**92**), 3β,7β-dihydroxy-urs-11-en-28,13-olide (**93**), and 1β,3β-dihydroxy-urs-11-en-28,13-olide (**94**) (Figure 28.32). Although product **92** was a known

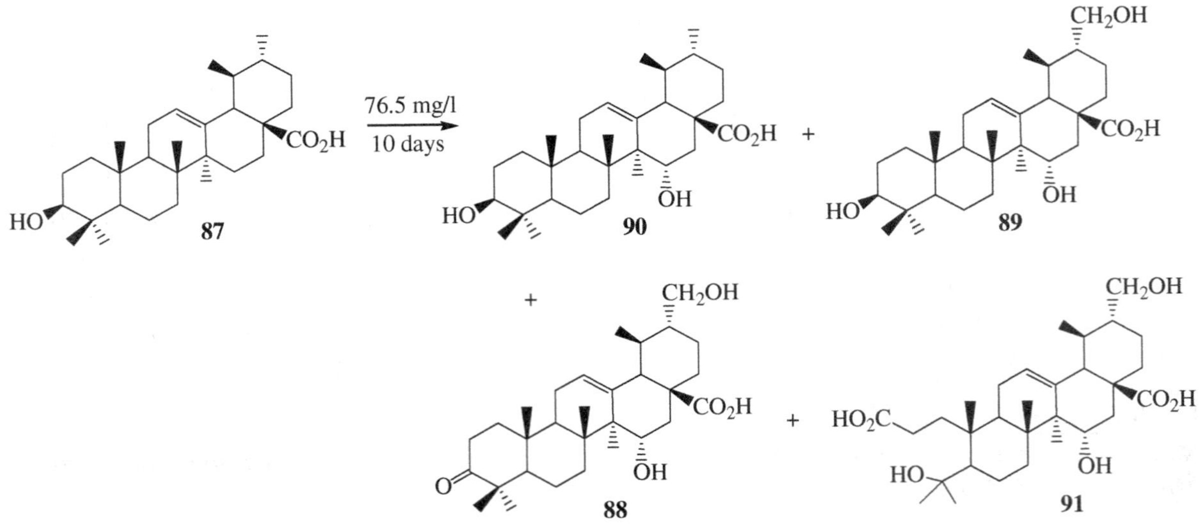

FIGURE 28.31

Microbial transformation of ursolic acid (**87**) by the endophyte *Pestalotiopsis microspora* [82].

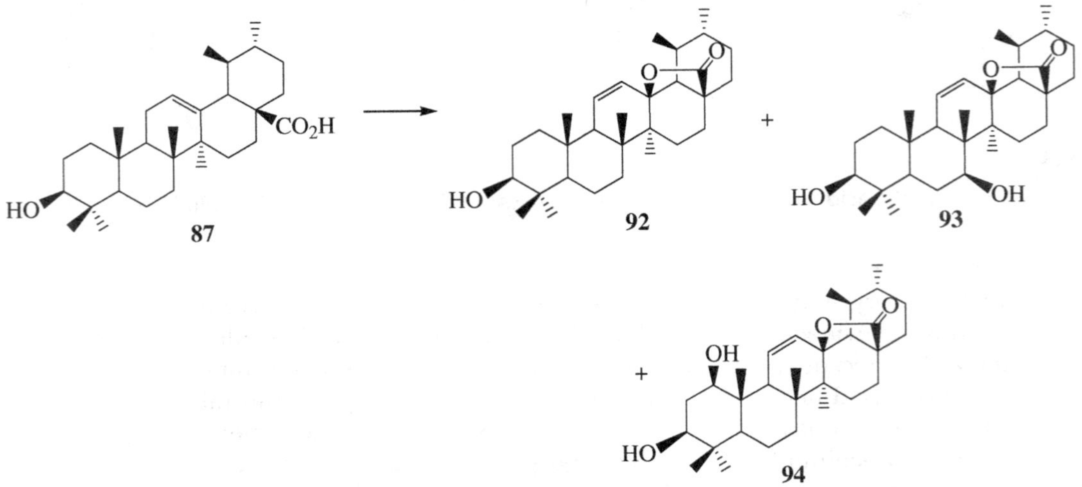

FIGURE 28.32

Microbial transformation of ursolic acid (**87**) by the endophyte *Umbelopsis isabellina* [83].

compound, it was obtained for the first time by microbial transformation. Products **93** and **94** were new compounds. The endophytic fungus *U. isabellina* can hydroxylate the C12–C13 double bond at position 13 of ursolic acid **87** and forms a five-member lactone. In the meantime, this fungus can also introduce the hydroxyl group at C-1 or C-7 of ursolic acid.

The bioconversion of ursolic acid (**87**) by *Alternaria alternata* was investigated, and eight metabolites were isolated and identified (Figure 28.33). The chemical structures of these metabolites were found to be 2α,3β-dihydroxy-urs-12-en-28-oic acid (corosolic acid, **95**), urs-12-en-2α,3β,28-triol (**96**), 3β,24-dihydroxy-urs-12-en-28-oic acid (**97**), 2α,3β,24-trihydroxy-urs-12-en-28-oic acid (**98**), 2α,3β,23,24-tetrahydroxyurs-12-en-28-oic acid (**99**), 3β,28-dihydroxy-12-ursene (uvaol, **100**), urs-12-en-3β-ol (**101**), and urs-12-en-2α,3β-diol (**102**), as determined by chemical and spectroscopic analyses [84]. Among these metabolites, **95** was found to be the main bioconversion product, with a yield of 77.6%. The proposed biosynthetic pathways of ursolic acids by *A. alternata* are presented in this chapter. Metabolites **95**, **98**, and **99** showed potent antiproliferative activities against human breast cancer, human colon cancer (Caco-2), and human liver cancer (HepG2) cell lines.

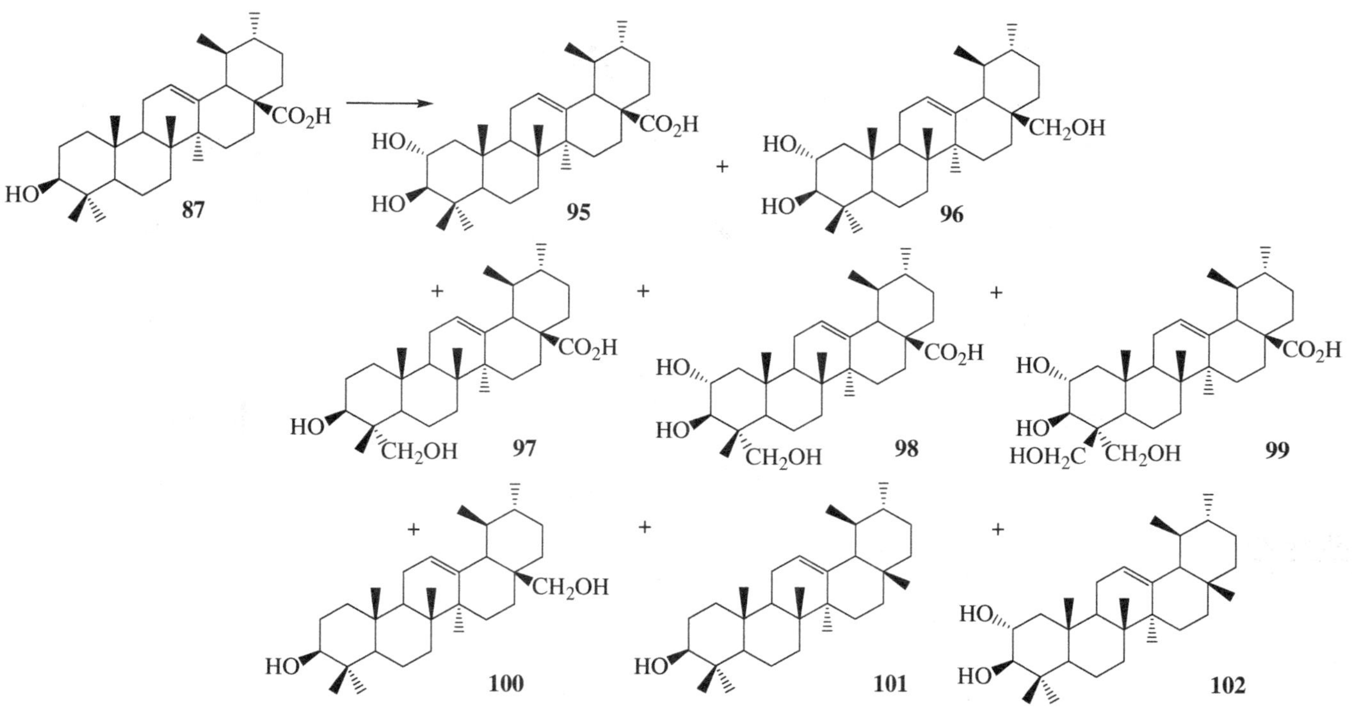

FIGURE 28.33
Microbial transformation of ursolic acid (**87**) by *Alternaria alternata* [84].

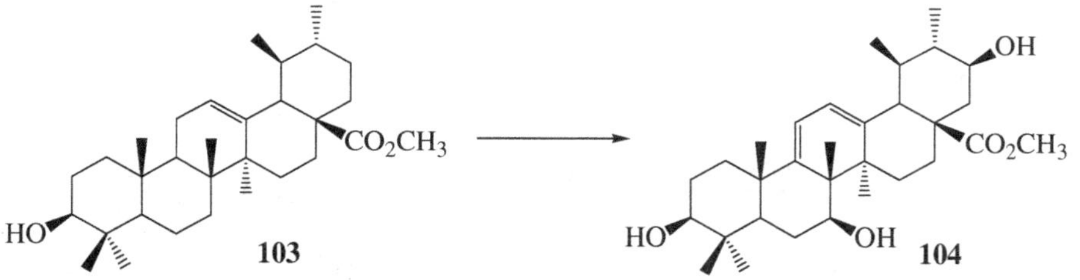

FIGURE 28.34
Microbial transformation of methyl ursolate by *Mucor plumbeus* ATCC 4740 [85].

Methyl ursolate, 3β,28-dihydroxy-urs-12-ene (**100**), and the 3β,28-bis (dimethylcarbamoxy) derivative of **100** were incubated with *Mucor plumbeus* ATCC 4740 by Collins *et al.* [85]. Only methyl ursolate underwent hydroxylation to give the novel compound 3β,7β,21β-trihydroxyursa-9(11),12-dien-28-oic acid methyl ester as the sole product (Figure 28.34).

Nocardia strains were used by Leipold *et al.* [86] to generate the following ursolic acid derivatives (Figure 28.35): *Nocardia* sp. 45077 synthesized ursonic acid (**103**) and 3-oxoursa-1,12-dien-28-oic acid (**104**). *Nocardia* sp. 46002 produced only ursonic acid and *Nocardia* sp. 43069 showed no metabolism at all. *Nocardia* sp. NRRL 5646 (previously used for the methylation of lupane and olean families PTs [42, 43]), *Nocardia* sp. 44822, and *Nocardia* sp. 44000 generated ursolic acid methyl ester (**105**), ursonic acid (**103**) and its methyl ester (**106**), and 3-oxoursa-1,12-dien-28-oic acid (**104**) and its methyl ester (**107**).

The microbial transformation of ursolic acid (**87**) by, filamentous fungus *Syncephalastrum racemosum* CGMCC 3.2500 produced, in low yields, five metabolites [87]: 3β,7β,21β-trihydroxy-urs-12-en-28-oic acid (**108**), 3β,21β-dihydroxy-urs-11-en-28-oic acid-13-lactone (**111**), 1β,3β,21β-trihydroxy-urs-12-en-28-oic acid (**109**), 3β,7β,21β-trihydroxy-urs-1-en-28-oic acid-13-lactone (**112**), and 21-oxo-1β,3β-dihydroxy-urs-12-en-28-oic acid (**110**) (Figure 28.36). In addition, the anti-HCV activity of compounds **108–112** was evaluated and shown to have no potential in the treatment of HCV.

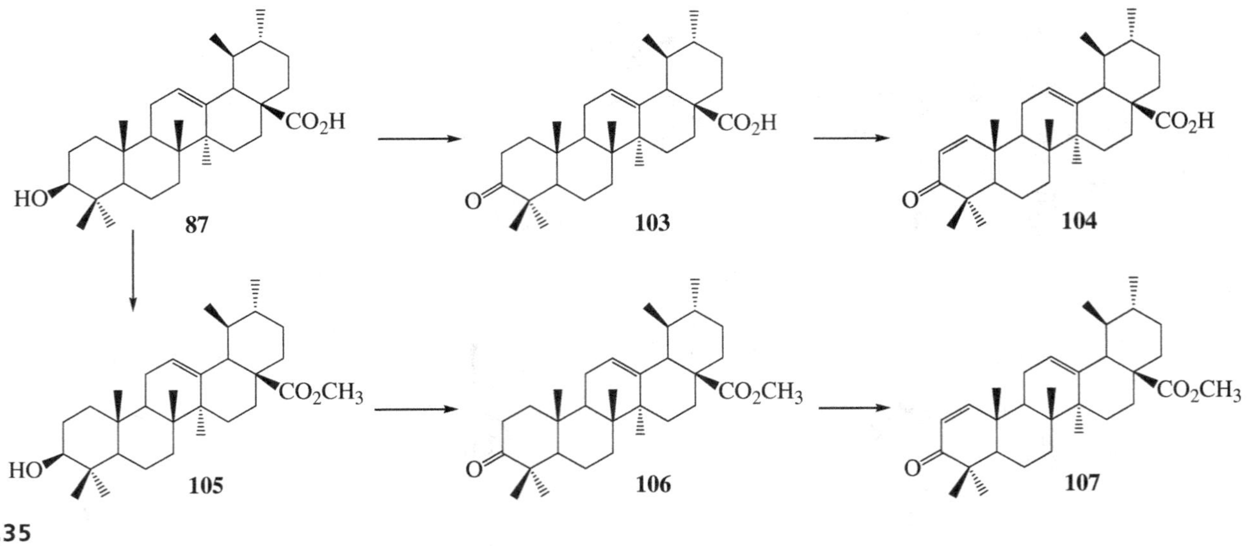

FIGURE 28.35
Microbial transformation of ursolic acid (**87**) by *Nocardia* sp. strains [86].

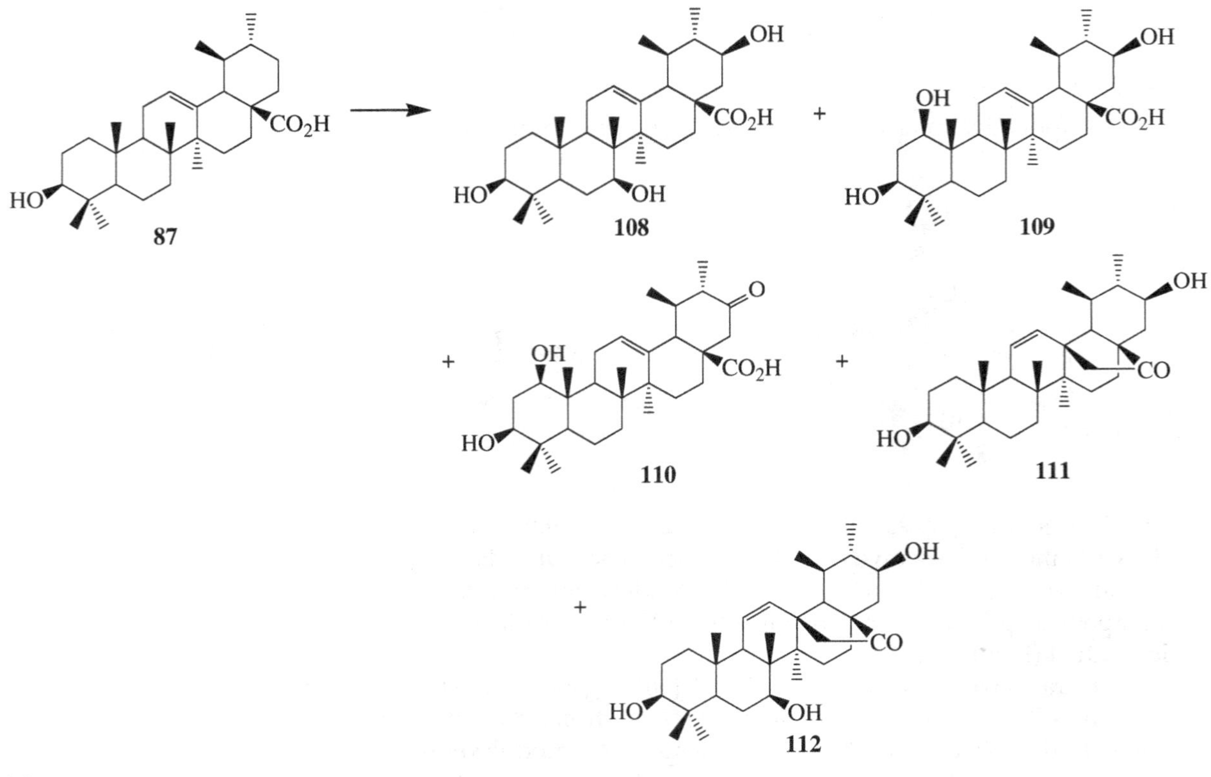

FIGURE 28.36
Microbial transformation of ursolic acid (**87**) by *Syncephalastrum racemosum* [87].

Asiatic acid (**113**) and its 6β-hydroxylated derivative madecassic acid (**114**) (Figure 28.37) are found in several plants but have been isolated essentially from *Centella asiatica* as free acids and glycosides (up to 2–3% of dry powdered leaves) [88, 89]. Preparations of *C. asiatica* ("gotu kola") are used in traditional and alternative medicine due to their wide spectrum of medicinal and cosmetic activities, such as collagen enhancement or burn wound healing, associated with these secondary metabolites [90, 91]. In addition, madecassoside, the natural glycosylated derivative of madecassic acid (**114**), has been recently investigated

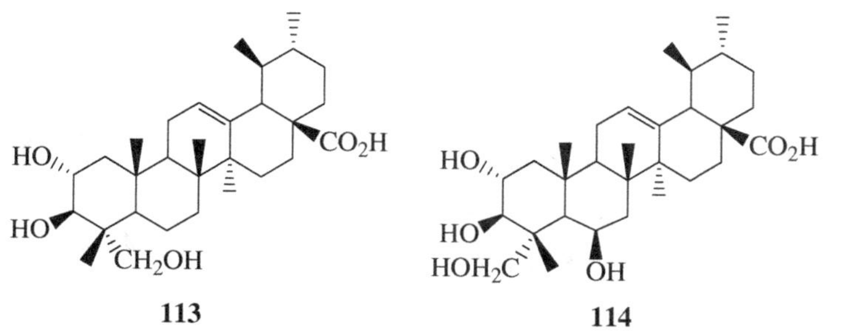

FIGURE 28.37

Structures of asiatic acid (**113**) and madecassic acid (**114**).

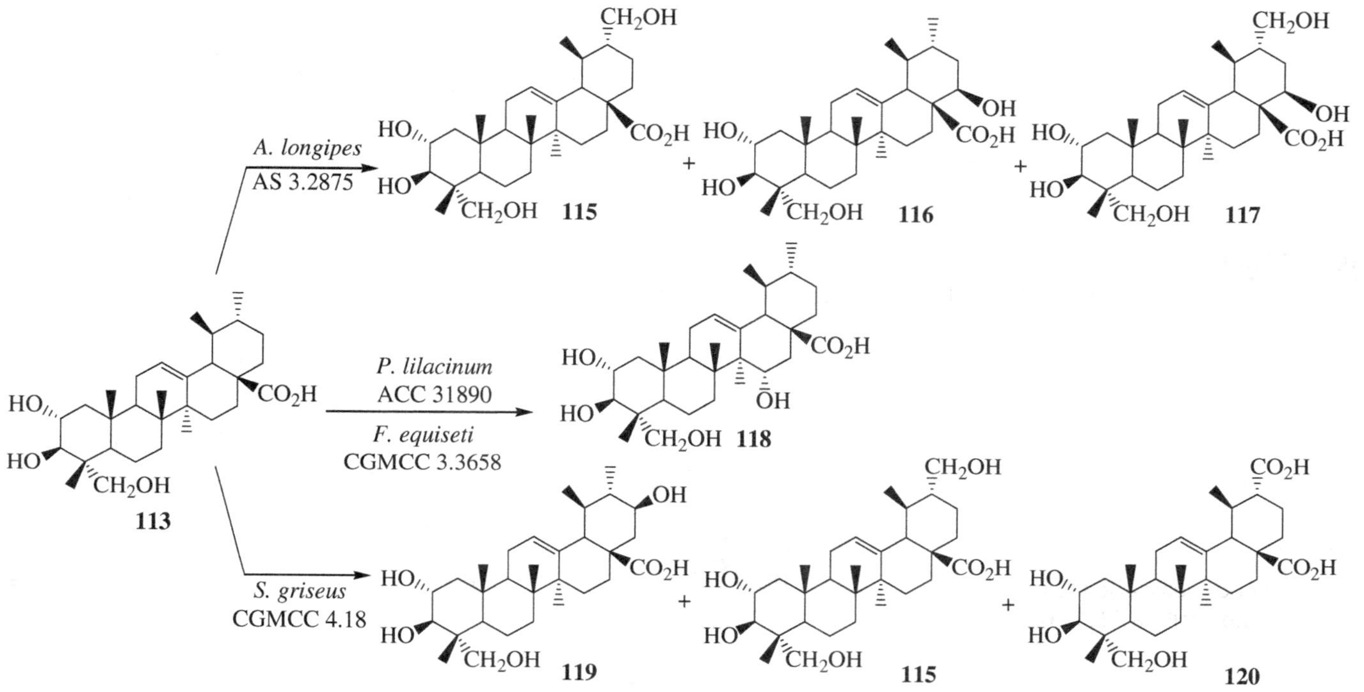

FIGURE 28.38

Microbial transformation of asiatic acid (**113**) by *Alternaria longipes* [93], *Penicillium lilacinum*, *Fusarium equiseti* or *Streptomyces griseus* [94].

for its neuroprotective effects in a rat model of the early phase of Parkinsonism and shown to significantly attenuate the induced reduction of dopamine in the striatum [92].

Asiatic acid (**113**) was converted by *A. longipes* AS 3.2875 to three compounds [93]: 2α,3β,24,30-tetrahydroxyurs-12-en-28-oic acid (**115**), 2α,3β,22β,24-tetrahydroxy-urs-12-ene-28-oic acid (**116**), and 2α,3β,22β,24,30-pentahydroxyurs-12-ene-28-oic acid (**117**) (Figure 28.38). Asiatic acid (**113**), incubated with *Penicillium lilacinum* ACCC31890 and *Fusarium equiseti* CGMCC 3.3658, gave an identical product, 2α,3β,15α,24-tetrahydroxyurs-12-en-28-oic acid (**118**). *Streptomyces griseus* CGMCC 4.18 afforded three derivatives [94]: 2α,3β,21β,24-tetrahydroxyurs-12-en-28-oic acid (**119**), 2α,3β,24,30-tetrahydroxyurs-12-en-28-oic acid (**115**), and 2α,3β,24-trihydroxy-urs-12-en-28,30-dioic acid (**120**). These modifications (Figure 28.38) at C-21, C-15, and C-30 dramatically decreased the cytotoxicity of asiatic acid on several human cancer cell lines.

New oxidation products, 2-keto,3β,24-dihydroxy-urs-12-en-28-oic acid (**121**), 2-keto,3β,15α-24-trihydroxy-urs-12-en-28-oic acid (**122**), and 2α,3β,15α,24-tetrahydr oxy-urs-12-en-28-oic acid (**123**), were obtained from the biotransformation of asiatic acid by the fungus *Fusarium avenaceum* AS 3.4594 (Figure 28.39) selected from among 25 fungal strains [95].

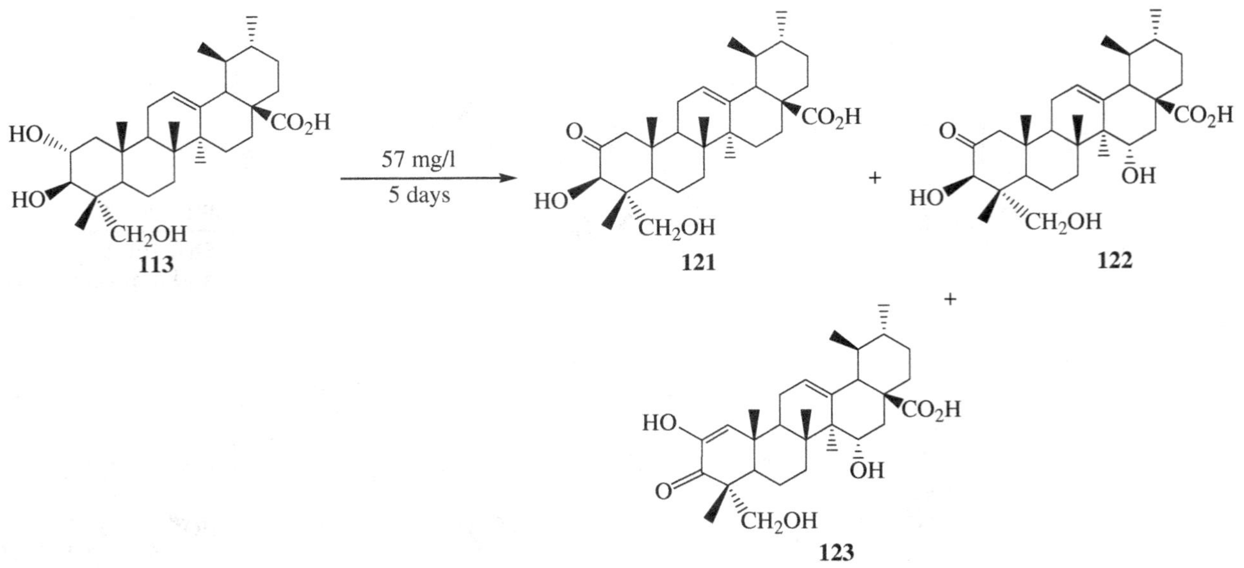

FIGURE 28.39
Microbial transformation of asiatic acid (**113**) by *Fusarium avenaceum* [95].

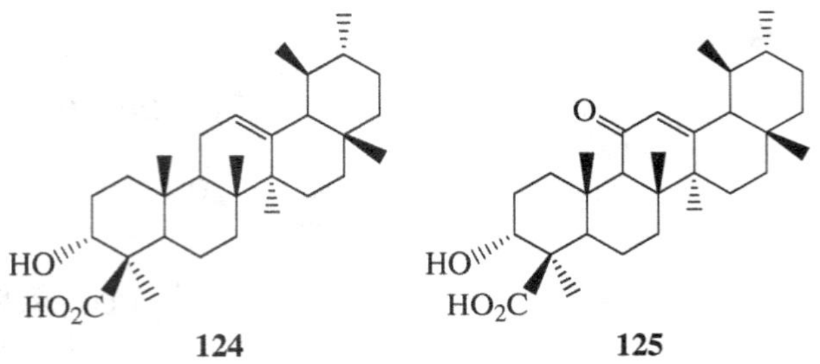

FIGURE 28.40
Structures of boswellic acid (**124**) and 11-keto-boswellic acid (**125**).

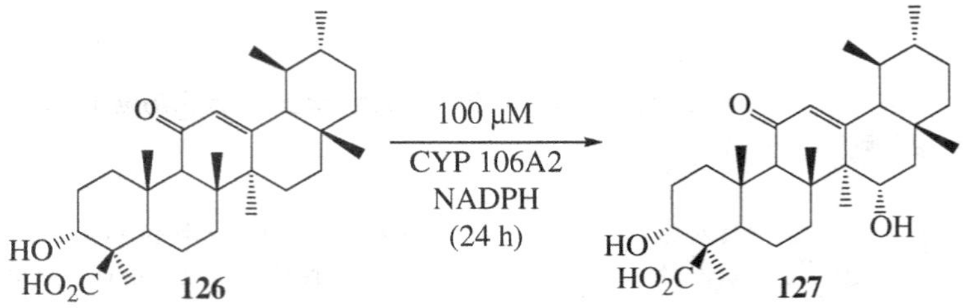

FIGURE 28.41
CYP106A2-catalyzed hydroxylation of 11-keto-β-boswellic acid [99].

Boswellic acids **124** and **125** (Figure 28.40), other ursane-type triterpenoids, are the major constituents of the gum derived from the plant *Boswellia serrata* (family *Burseraceae*, syn. *B. Glabra*), commonly known as "salai guggal," "white guggal," or "Indian olibanum" (incense, oliban). The gum exudate or the resin obtained from the bark of the tree has been widely used for several medical applications, such as arthritis, asthma, ulcers, and skin diseases [96]. Boswellic acids are putative 5-lipoxygenase inhibitors with a role in the management of various inflammatory conditions and control of cerebral edema in brain tumor patients [97, 98]. The antitumor activities of boswellic acids have also been reported [12].

Bleif *et al.* [99] developed a recombinant expression system for cytochrome P450 in *B. megaterium*, which was successfully applied for the whole-cell conversion of 11-keto-β-boswellic acid (**126**) to its 15α-hydroxy derivative **127** (Figure 28.41). The productivity was increased 15-fold in the heterologous expression system of *B. megaterium* MS941 (560.7 mg/l/day) compared with a relative productivity of 37.6 mg/l/day when using the natural *B. megaterium* strain ATCC 13368.

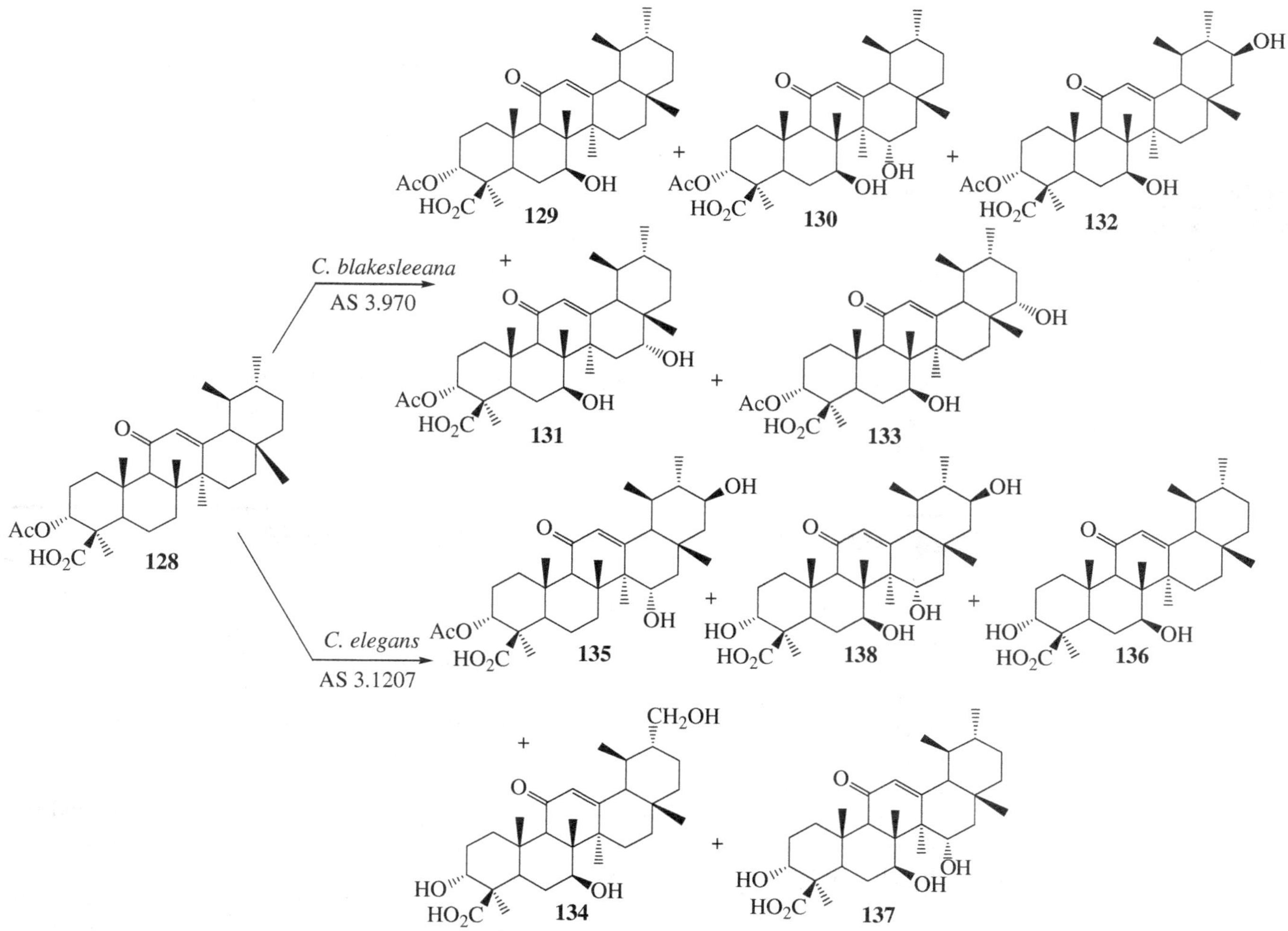

FIGURE 28.42

Microbial transformation of acetyl-11-keto-β-boswellic acid (**128**) by *Cunninghamella blakesleeana* [100] and *Cunninghamella elegans* [100, 101].

Twenty strains of fungi were screened for the transformation of 3-*O*-acetyl-11-keto-β-boswellic acid (**128**) [100, 101]. *Cunninghamella blakesleeana* AS 3.970 and *C. elegans* AS 3.1207 afforded several different metabolites acetylated or deacetylated in the −3 position (Figure 28.42): depending on the authors, *C. blakesleeana* gave acetylated or deacetylated 7β-hydroxy-11-keto-boswellic acid (**129, 136**), 7β,15α-dihydroxy-11-keto-boswellic acid (**130, 137**), 7β,16α-dihydroxy-11-keto-boswellic acid (**131**), 7β,21β-dihydroxy-11-keto-boswellic acid (**132**), 7β,22α-dihydroxy-11-keto-boswellic acid (**133**), and 7β,30-dihydroxy-11-keto-boswellic acid (**134**), whereas *C. elegans* gave 15α,21β-dihydroxy-11-keto-boswellic acid (**135**) and 7β,15α,21β-trihydroxy-11-keto-boswellic acid (**138**). Some of them exerted an inhibition of NO production in the 1–10 mM range, better or comparable with that of 3β-acetyl-11-keto-boswellic acid [100].

In summary, microbial transformations of α-amyrin and derived acids lead to new hydroxylations not only the major positions −15α and −7β but also frequently in other positions: −30, −2α, −13β, −21β, −24, −22β, −1β, −16α, −23, and −2β (in order of priority, see Figure 28.43). Already hydroxylated or keto-ursan(ursen)-derivatives such as ursolic acids, asiatic acid, and boswellic acids can be already hydroxylated at the corresponding complementary positions, affording a wide diversity of new multi-oxidized derivatives, difficult to achieve by classical chemical means and potentially

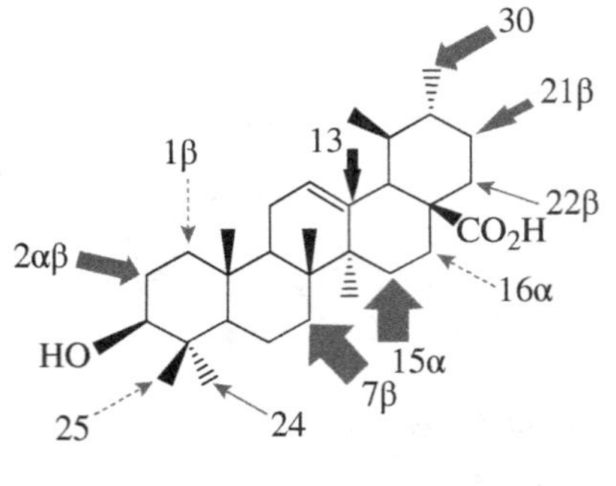

FIGURE 28.43
Main hydroxylation sites observed in the microbial transformations of ursan derivatives.

exhibiting new or increased biological activities. Several of them have been previously isolated in plants as natural products. A summary of these results and yields is presented in Table 28.3.

28.5 MICROBIAL TRANSFORMATIONS OF OTHER PTs

As an example of microbial transformations of less usual PTs, quinovic acid 3-*O*-β-D-6-deoxyglucoside (**138**), from the bark of *Mitragyna inermis* (Rubiaceae) and several glycosyl analogues (**139–141**), were deglycosylated by *Nocardia* sp. NRRL 5646 to their aglycone quinovic acid (**142**) and its rearranged 29,30-dimethyl counterpart, cincholic acid (**143**), via a carbon skeleton rearrangement involving a methyl group migration (Figure 28.44). The structures of the metabolites were established by ESI-LC/MS and 2D NMR techniques [102].

The microbial transformation of **138** by *S. griseus* ATCC13273 [103] allowed the hydroxylation of the 30-methyl of the aglycone in 57% yield, without hydrolysis of the glycosyl part (Figure 28.44).

Another atypical triterpene skeleton is the caenothane family, resulting from the ring contraction of ring A and isolated from the genera *Ziziphus* [104, 105]. Caenothic acid (**144**) is converted by *Mycobacterium* sp. (NRRL B-3805) to a 1,28-dimethyl ester, whereas transformation of the 3-oxo-1,28-dimethyl ester derivative resulted in the epimerization of the 1-carboxyester group [16, 104] (Figure 28.45).

28.6 GLYCOSYLATIONS AND DEGLYCOSYLATIONS

Soyasapogenols, oleane-derived aglycones of soyasaponins, can be produced from crude soybean saponin extract by acidic or enzymatic hydrolysis. Soyasapogenol B (**147**) is known to have hepatoprotective, antimutagenic, antivirus, and anti-inflammatory activities. Hydrolysis of soybean saponin with HCl in methanol gave three soyasapogenols, namely, soyasapogenol D (**150**), soyasapogenol B1 (**149**), and soyasapogenol A (**148**) (Figure 28.46).

In this context, several *Aspergillus* strains were screened for their saponin-hydrolyzing abilities to produce soyasapogenol B from soybean saponins added to the culture medium [106]. The microbial hydrolysis of soybean saponin extract by *Aspergillus terreus* led to the isolation of soyasapogenol B (**147**) as a major product (Figure 28.47). *Aspergillus terreus* II was shown to be the most potent microorganism in the hydrolysis of soybean saponins, leading to promising amounts of soyasapogenol B (about 20.5% of the added saponins, 24 mg/50 ml soyasapogenol B).

Maximum production of soyasapogenol B (about 152.3 mg/50 ml) was obtained using 1.5% (w/v) soybean saponin and 1.5% (w/v) glucose, at 32 °C after 72 h at pH 7 using phosphate buffer. Under these optimal conditions, the bioconversion efficiency increased from 20.5 to 85.3%. The isolation of soyasapogenols

TABLE 28.3 Microbial Transformation Products and Yields from Ursan-Family PTs

Substrate PTs	Microorganisms	Products	Yields (%)
Ursolic acid (87)	*Aspergillus flavus* ATCC 9170 [81]	3-Oxo-ursan-12-en-28-oic acid (**103**)	10.3
	Nocardia sp. NRRL 5646 [42, 86]	3β-Hydroxy-ursan-12-en-28-oic acid methyl ester (**105**)	35
		3β-Hydroxy-olean-12-en-28-oic acid methyl ester	
		3-Oxo-ursan-12-en-28-oic acid methyl ester (**106**)	5
	Nocardia sp. NRRL 45077 [86]	3-Oxo-ursan-12-en-28-oic acid (**103**)	50
		3-Oxo-ursan-1,12-dien-28-oic acid (**104**)	—
	Nocardia sp. NRRL 46002 [86]	3-Oxo-ursan-12-en-28-oic acid (**103**)	80
	Nocardia sp. NRRL 44822 [86]	3-Oxo-ursan-12-en-28-oic acid methyl ester (**106**)	15
		3β-Hydroxy-ursan-12-en-28-oic acid methyl ester (**105**)	15
		3-Oxo-ursan-1,12-dien-28-oic acid (**104**)	—
		3-Oxo-ursan-1,12-dien-28-oic acid methyl ester (**107**)	—
	Nocardia sp. NRRL 44000 [86]	3-Oxo-ursan-12-en-28-oic acid methyl ester (**106**)	30
		3β-Hydroxy-ursan-12-en-28-oic acid methyl ester (**105**)	30
		3-Oxo-ursan-1,12-dien-28-oic acid (**104**)	—
		3-Oxo-ursan-1,12-dien-28-oic acid methyl ester (**107**)	—
	Pestalotiopsis microspora [83]	3β,15α-Dihydroxy-ursan-12-en-28-oic acid (**90**)	1.2
		3β,15α,30-Trihydroxy-ursan-12-en-28-oic acid (**88**)	1.6
		3-Oxo-15α,30-dihydroxy-ursan-12-en-28-oic acid (**89**)	8.4
		3,4-*seco*-3-Oxo-15α,30-dihydroxy-ursan-12-en-28-oic acid (**91**)	1.2
	Umbelopsis isabellina [69]	3β-Hydroxy-ursan-11-en-13,28-olide (**92**)	0.3
		3β,7β-Dihydroxy-ursan-11-en-13,28-olide (**93**)	0.25
		1β,3β-Dihydroxy-ursan-11-en-13,28-olide (**94**)	0.45
	Alternaria alternata [84]	3β-Hydroxy-ursan-12-ene (**101**)	0.05
		2α,3β-Dihydroxy-ursan-12-ene (**102**)	0.02
		3β,28β-Dihydroxy-ursan-12-ene (**100**)	0.03
		2α,3β,28β-Trihydroxy-ursan-12-ene (**96**)	6.8
		3β,24-Dihydroxy-ursan-12-en-28-oic acid (**97**)	21.5
		2α,3β-Dihydroxy-ursan-12-en-28-oic acid (**95**)	77.6
		2α,3β,24-Trihydroxy-ursan-12-en-28-oic acid (**98**)	12.6
		2α,3β,23,24-Tetrahydroxy-ursan-12-ene (**99**)	6.1

(*Continued*)

TABLE 28.3 (*Continued*)

Substrate PTs	Microorganisms	Products	Yields (%)
	Syncephalastrum racemosum CGMCC 3.2500 [87]	1β,3β,21β-Trihydroxy-ursan-12-en-28-oic acid (**109**)	0.8
		1β,3β,21-Oxo-dihydroxy-ursan-12-en-28-oic acid (**110**)	1.5
		3β,7β,21β-Trihydroxy-ursan-12-en-28-oic acid (**108**)	1.5
		3β,21β-Dihydroxy-ursan-11-en-28,13-olide (**111**)	0.7
		3β,7β,21β-Trihydroxy-ursan-11-en-28,13-olide (**112**)	0.6
Ursolic acid methyl ester	*Mucor plumbeus* ATCC 4740 [85]	3β,7β,21β-Trihydroxy-ursan-9(11),12-dien-28-oic acid methyl ester	
Asiatic acid (113)	*Alternaria longipes* AS 3.2875 [93]	2α,3β,24,30-Tetrahydroxy-ursan-12-en-28-oic acid (**115**)	5
		2α,3β,22β,24-Tetrahydroxy-ursan-12-en-28-oic acid (**116**)	11
		2α,3β,22β,24,30-Pentahydroxy-ursan-12-en-28-oic acid (**117**)	2.5
	Fusarium avenaceum AS 3.4594 [95]	2-Oxo-3β,24-dihydroxy-ursan-12-en-28-oic acid (**121**)	3
		2-Oxo-3β,15α,24-trihydroxy-ursan-12-en-28-oic acid (**122**)	20
		2,3β,15α,24-Tetrahydroxy-ursan-1,12-dien-28-oic acid (**123**)	7.5
	Fusarium equiseti CGMCC 3.3658 [94]	2α,3β,15α,24-Tetrahydroxy-ursan-12-en-28-oic acid (**118**)	34
	Penicillium lilacinum ACCC 31890 [94]	2α,3β,15α,24-Tetrahydroxy-ursan-12-en-28-oic acid (**118**)	51
	Streptomyces griseus CGMCC 4.18 [94]	2α,3β,21β,24-Tetrahydroxy-ursan-12-en-28-oic acid (**119**)	7.6
		2α,3β,24,30-Tetrahydroxy-ursan-12-en-28-oic acid (**115**)	46
		2α,3β,24-Trihydroxy-ursan-12-en-28,30-dioic acid (**120**)	23
11-Oxo-boswellic acid (126)	*Bacillus megaterium* MS941 [99]	3α,15α-Dihydroxy-11-oxo-ursan-12-en-23-oic acid (**127**)	80
3-Acetoxy-11-oxo-boswellic acid (128)	*Cunninghamella blakesleana* AS 3.970 [100]	3α-Acetoxy,7β,15α-dihydroxy-11-oxo-ursan-12-en-23-oic acid (**130**)	0.4
		3α-Acetoxy,7β,21β-dihydroxy-11-oxo-ursan-12-en-23-oic acid (**132**)	3.3
		3α-Acetoxy,7β,22α-dihydroxy-11-oxo-ursan-12-en-23-oic acid (**133**)	0.9
		3α-Acetoxy,7β-hydroxy-11-oxo-ursan-12-en-23-oic acid (**129**)	1.6
		3α-Acetoxy,7β,16α-dihydroxy-11-oxo-ursan-12-en-23-oic acid (**131**)	0.8
	Cunninghamella elegans AS 3.1207 [100, 101]	3α-Acetoxy,15α,21β-dihydroxy-11-oxo-ursan-12-en-23-oic acid (**135**)	3.1
		3α-Acetoxy,7β,16α-dihydroxy-11-oxo-ursan-12-en-23-oic acid (**131**)	4.4
		3α,7β,30-Trihydroxy-11-oxo-ursan-12-en-23-oic acid (**134**)	—
		3α,7β-Dihydroxy-11-oxo-ursan-12-en-23-oic acid (**136**)	—
		3α,7β,15α-Trihydroxy-11-oxo-ursan-12-en-23-oic acid (**137**)	—
		3α,7β,15α,21β-Tetrahydroxy-11-oxo-ursan-12-en-23-oic acid (**138**)	—

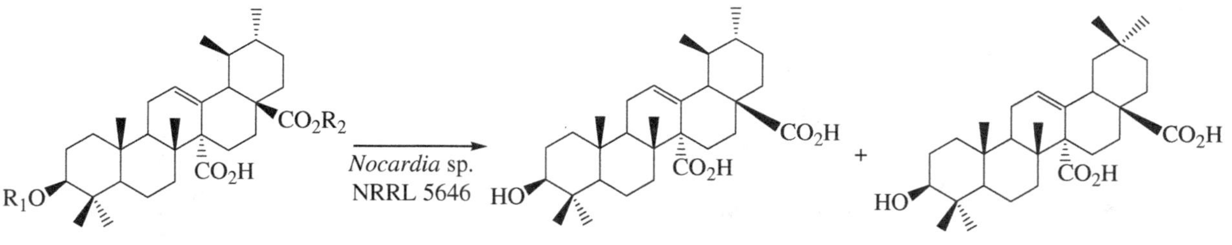

138: R_1 = β-D-6-deoxyglucose R_2 = H
139: R_1 = β-D-glucose R_2 = H
140: R_1 = β-D-glucose R_2 = β-D-glucose
141: R_1 = β-D-6-deoxyglucose R_2 = β-D-glucose

142 **143**

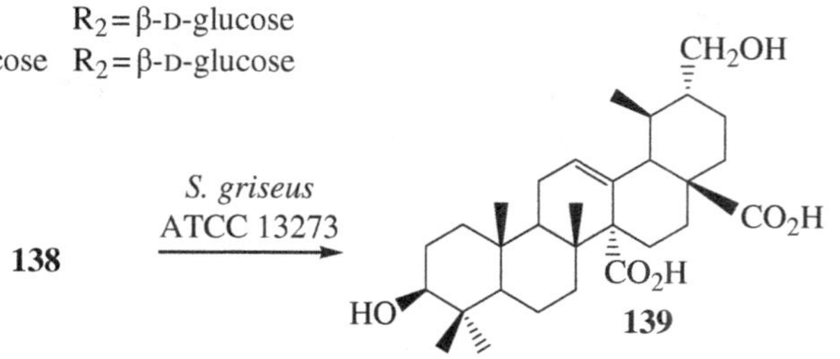

FIGURE 28.44

Microbial transformations of quinovic acid glycosides (**138–141**) by *Nocardia* sp. [102] or *Streptomyces griseus* [103].

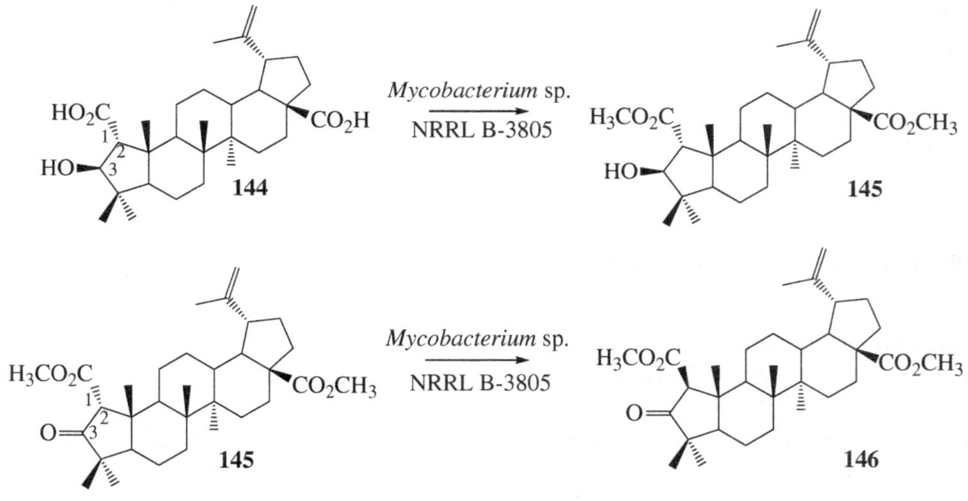

FIGURE 28.45

Microbial transformations of caenothic acid (**144**) by *Mycobacterium* sp. [104].

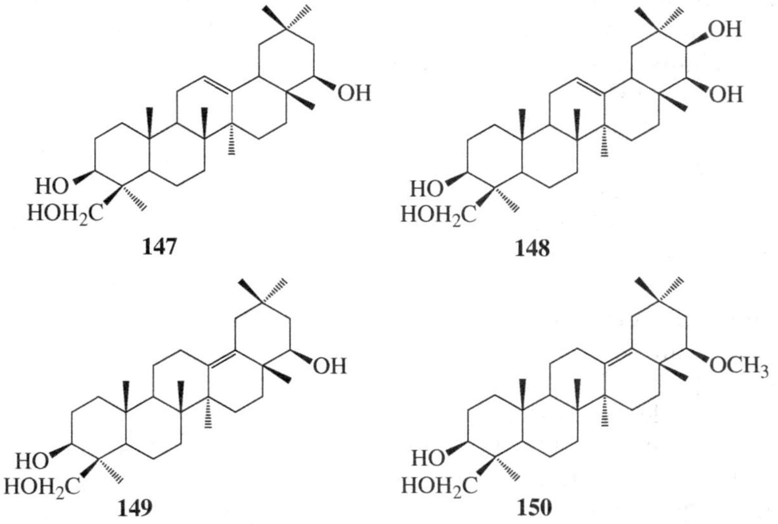

FIGURE 28.46

Main sapogenols obtained in the acidic hydrolysis of soybean saponins [106].

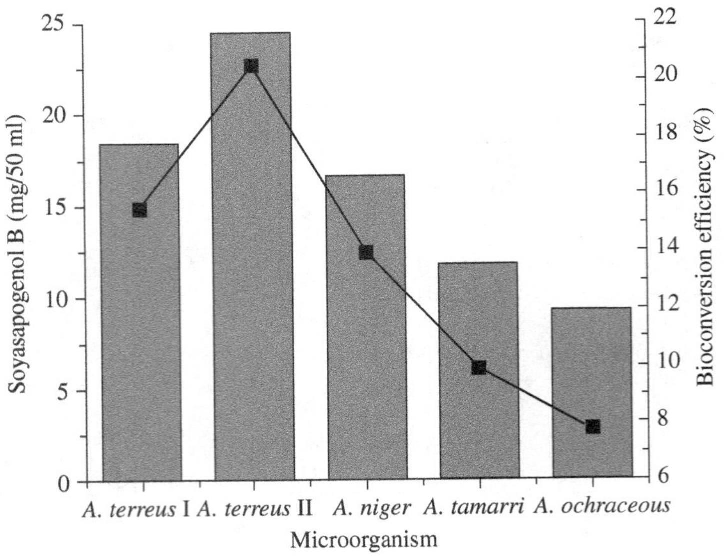

FIGURE 28.47
Bioconversion of soybean saponin by different *Aspergillus* strains. By courtesy of Amin *et al.* [107], © 2011 Informa Healthcare.

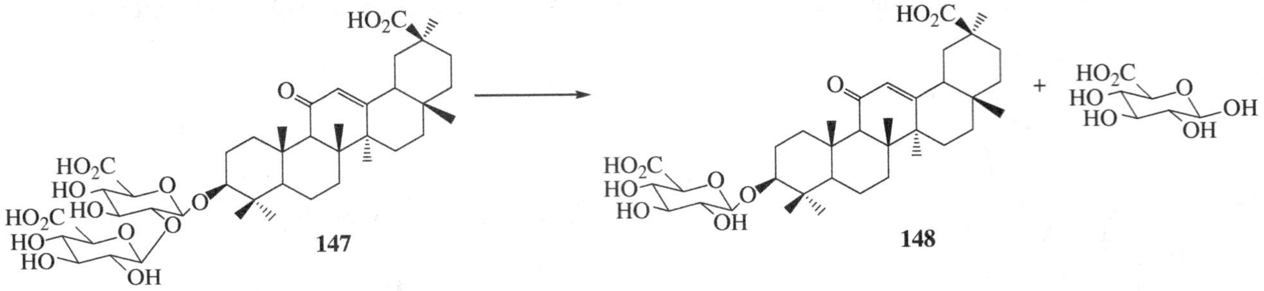

FIGURE 28.48
Selective hydrolysis of glycyrrhizin (**151**) to the monoglucuronide (**152**) by *Cryptococcus magnus* [108].

was performed using chromatographic methods and their structures identified on the basis of spectroscopic tools.

Among several other *Aspergillus* strains (see Section 28.2), *A. niger* GRM-3 selectively hydrolyses the 3-*O*-β-D-glucuronide linkage of the sugar chain of glycyrrhizin (**147**), liberating the aglycone glycyrrhetinic acid (**71**) and glucuronobiose. *Cryptococcus magnus*, a yeast isolated from soil, is able to selectively hydrolyze the terminal β-glucuronide linkage of glycyrrhizin to yield glycyrrhetinic acid 3-*O*-β-D-monoglucuronide (**148**), which is an excellent sweetener (nearly 1000-fold) comparatively to glycyrrhizin (Figure 28.48). Optimization of the culture and incubation conditions allowed a 95% yield to be obtained [106].

Similarly, the sequential hydrolysis of the oleanoside esculentoside (**53**) (see Figure 28.13) by *A. ochraceus* or *S. griseus* afforded successively the 3-*O*-β-xylosyl monosaccharide then the free aglycone [53] (see Section 28.2).

The weak hydrosolubility of PTs aglycones hampers the development of these natural anticancer agents. To circumvent this problem and to enhance their pharmacological properties, several glycosides have been synthesized by chemical or biological methods and tested for cytotoxicity against cancerous cell lines (for an example of chemical synthesis of glycosides of lupan-type triterpenes, see [108]).

Preparative scale biotransformation of betulinic acid (**1**) with resting-cell suspensions of *Cunninghamella* sp. NRRL 5695 resulted in the production in low yield of an acyl glucoside metabolite (Figure 28.49), which has been characterized as 28-*O*-β-D-glucopyranosyl-3β-hydroxy-lup-20(29)-en-28-oate (**149**) based on spectral and enzymatic

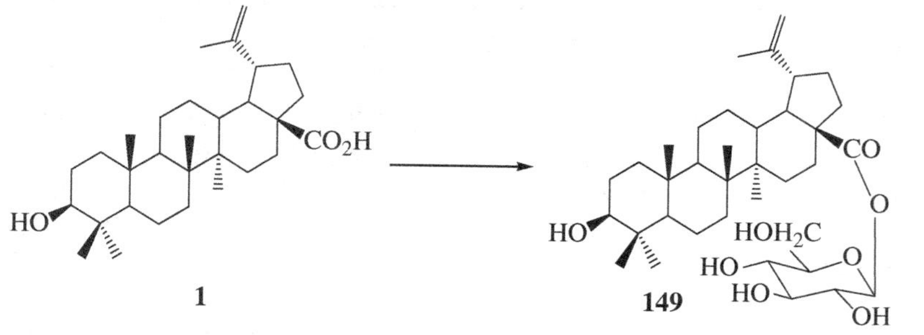

FIGURE 28.49
Microbial glycosylation of betulinic acid (**1**) by *Cunninghamella* sp. NRRL 5695 [109].

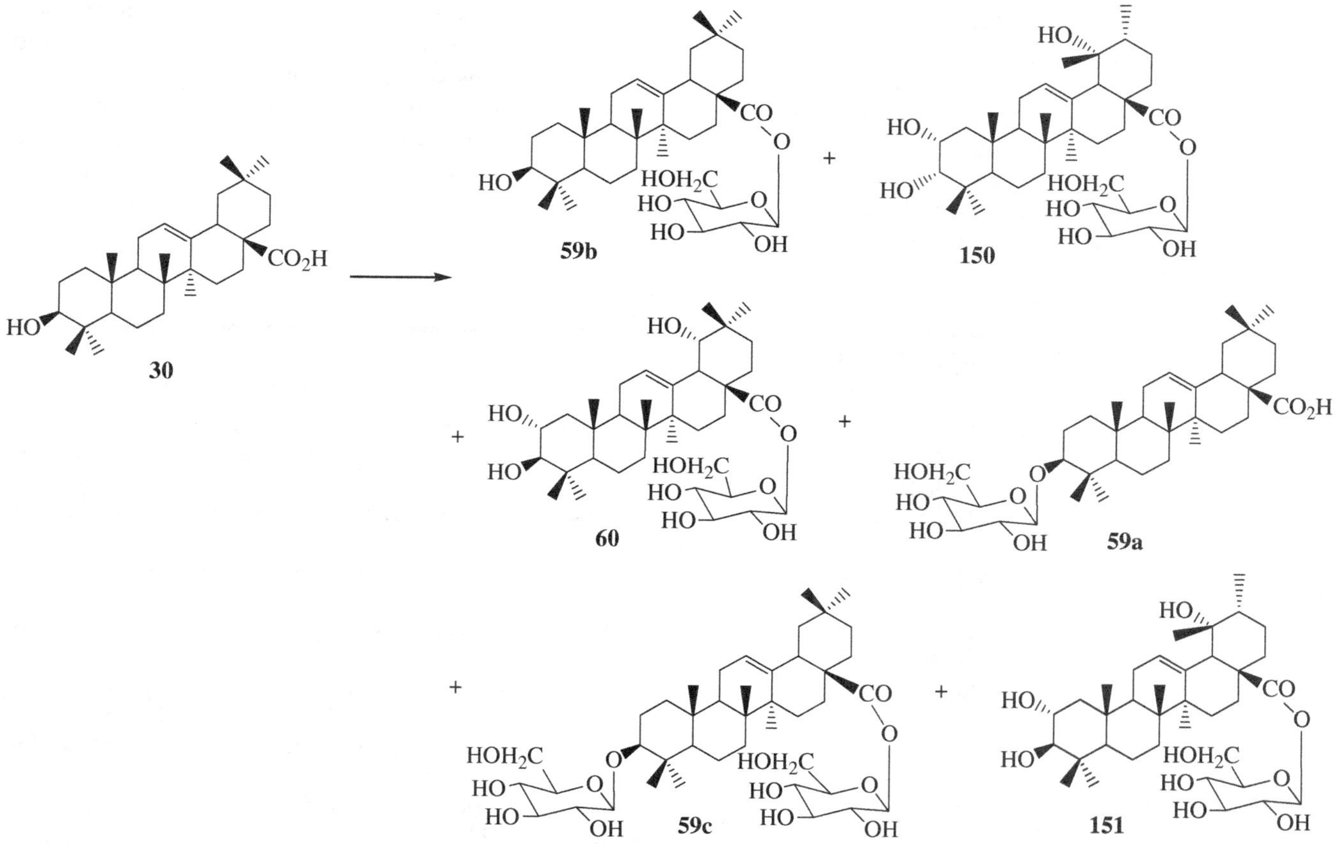

FIGURE 28.50
Microbial glycosylations and transformations of oleanolic acid (**30**) by *Alternaria longipes* [55].

hydrolysis data [109]. The *in vitro* cytotoxicity assay of metabolite **149** revealed no activity against several human melanoma cell lines.

In the microbial transformation of oleanolic acid (**30**) by *A. longipes*, several transformed products were isolated, most of them as acyl glycoconjugates [54]: oleanolic acid 28-*O*-β-D-glucopyranosyl ester (**59b**), 2α,3α,19α-trihydroxy-ursolic acid-28-*O*-β-D-glucopyranosyl ester (**150**), 2α,3β,19α-trihydroxy-oleanolic acid-28-*O*-β-D-glucopyranosyl ester (**60**), oleanolic acid-3-*O*-β-D-glucopyranoside (**59a**), 3-*O*-(β-D-glucopyranosyl)-oleanolic acid-28-*O*-β-D-glucopyranosyl ester (**59c**), and 2α,3β,19α-trihydroxy-ursolic acid-28-*O*-β-D-glucopyranosyl ester (**151**) (Figure 28.50). At the same time, the skeletons of compounds **60** and **150** were changed from olean to ursan (**150–151**).

Another similarly transformed product, 7β,21β-hydroxy-oleanolic acid-28-*O*-β-D-glucopyranosyl ester (**63**) was obtained [54] from **30** in an incubation with *P. adametzi*, together with 21β-hydroxy-oleanolic acid (**62**) and 7α,21β-dihydroxy-oleanolic acid (**61**) (see Figure 28.14). Compounds **63** and **151** exhibited stronger cytotoxic activities against HeLa cell lines than oleanolic acid.

28.7 CONCLUSION AND PERSPECTIVES

Whole-cell microbial transformations and particularly regio- and stereoselective oxygen introduction in several positions of the rings are frequently observed using sapogenins as substrates, resulting in multihydroxylated derivatives, sometimes already existing in plants, but difficult to obtain by chemical synthesis. These compounds are more hydrophilic than the original sapogenins and frequently exhibit similar or better pharmacological properties.

The direct transformation of saponosides (other than deglycosylation) is much less easy and has been rarely reported. That may be due to their decreased hydrophobicity and antibacterial effect on microorganisms, prohibiting any entry into the bacterial cell and thus any CYP450-dependent metabolism. The use of overexpressed more or less purified monooxygenase systems should probably overcome this problem, but with only weak possibilities for preparative-scale experiments, compared to microbial cultures. The search for endophytic strains housed in plants and able to realize those transformations, which have possibly gained resistance to saponosides and hydroxylation capabilities through horizontal gene transfer should also possibly solve this problem.

REFERENCES

[1] P. Burger, Caracterisation moléculaire de résines végétales archéologiques et actuelles: Etude de résines des Dipterocarpaceae, Ph.D. thesis, Université Louis Pasteur, Strasbourg, 2008.

[2] J. Patocka, Biologically active pentacyclic triterpenes and their current medicine signification, *J. Appl. Biomed.* 2003. **1**, 7–12.

[3] M. N. Laszczyk, Pentacyclic triterpenes of the lupane, oleanane and ursane group as tools in cancer therapy, *Planta Med.* 2009. **75**, 1549–1560.

[4] J. A. R. Salvador (Ed.) *Pentacyclic Triterpenes as Promising Agents in Cancer*. Nova Science Publishers, Inc., New York, 2010.

[5] L. H. Vazquez, J. Palazon, A. Navarro-Ocana. In *Phytochemicals: A Global Perspective of Their Role in Nutrition and Health* (Ed.: V. Rao). InTech Publisher, Rijeka, Croatia, 2012, pp. 487–502.

[6] H. Safayhi, E. R. Sailer, Anti-inflammatory actions of pentacyclic triterpenes, *Planta Med.* 1997. **63**, 487–493.

[7] Y. Ikeda, A. H. Murakami, H. Ohigashi, Ursolic acid: An anti- and pro-inflammatory triterpenoid, *Mol. Nutr. Food Res.* 2008. **52**, 26–42.

[8] S.-R. Wang, W. S. Fang, Pentacyclic triterpenoids and their saponins with apoptosis-inducing activity, *Curr. Top. Med. Chem.* 2009. **9**, 1581–1596.

[9] B. Siewert, E. Pianowski, R. Csuk, Esters and amides of maslinic acid trigger apoptosis in human tumor cells and alter their mode of action with respect to the substitution pattern at C-28, *Eur. J. Med. Chem.* 2013. **70**, 259–272.

[10] H. Sheng, H. Sun, Synthesis, biology and clinical significance of pentacyclic triterpenes: A multi-target approach to prevention and treatment of metabolic and vascular diseases, *Nat. Prod. Rep.* 2011. **28**, 543–593.

[11] S. Fulda, Betulinic acid: A natural product with anticancer activity, *Mol. Nutr. Food Res.* 2009. **53**, 140–146.

[12] J. A. R. Salvador, V. M. Moreira, B. M. F. Goncalves, A. S. Leal, Y. Jing, Ursane-type pentacyclic triterpenoids as useful platforms to discover anticancer drugs, *Nat. Prod. Rep.* 2012. **29**, 1463–1479.

[13] J. M. Castellano, A. Guinda, T. Delgado, M. Rada, J. A. Cayuela, Biochemical basis of the antidiabetic activity of oleanolic acid and related pentacyclic triterpenes, *Diabetes* 2013. **62**, 1791–1799.

[14] A. Alqahtani, K. Hamid, A. Kam, K. H. Wong, Z. Abdelhak, V. Razmovski-Naumovski, K. Chan, K. Li, P. W. Groundwater, G. Q. Li, The pentacyclic triterpenoids in herbal medicines and their pharmacological activities in diabetes and diabetic complications, *Curr. Med. Chem.* 2013. **20**, 908–931.

[15] S. Jäger, H. Trojan, T. Kopp, M. N. Laszczyk, A. Scheffler, Pentacyclic triterpene distribution in various plants—rich sources for a new group of multi-potent plant extracts, *Molecules* 2009. **14**, 2016–2031.

[16] A. Parra, F. Rivas, A. Garcia-Granados, A. Martinez, Microbial transformation of triterpenoids, *Mini-Rev. Org. Chem.* 2009. **6**, 307–320.

[17] K. Muffler, D. Leipold, M.-C. Scheller, C. Haas, J. Steingroewer, T. Bley, H. E. Neuhaus, M. A. Mirata, J. Schrader, R. Ulber, Biotransformation of triterpenes, *Process Biochem.* 2011. **46**, 1–15.

[18] M. Malinowska, E. Sikora, J. Ogonowski, Production of triterpenoids with cell and tissue cultures, *Acta Biochim. Pol.* 2013. **60**, 731–735.

[19] H. N. Bhatti, R. A. Khera, Biotransformations of diterpenoids and triterpenoids: A review, *J. Asian Nat. Prod. Res.* 2014. **16**, 70–104.

[20] A. Cano-Flores, Biotransformación de triterpenos con diferentes microorganismos, *Rev. Mex. Cienc. Farm.* 2013. **44**, 7–16.

[21] B. Dinda, S. Debnath, B. C. Mohanta, Y. Harigaya, Naturally occurring triterpenoid saponins, *Chem. Biodivers.* 2010. **7**, 2327–2580.

[22] M. Saleem, Lupeol, a novel anti-inflammatory and anti-cancer dietary triterpene, *Cancer Lett.* 2009. **285**, 109–115.

[23] T. C. de Carvalho, A. M. Polizeli, I. C. C. Turatti, M. E. Severiano, C. E. de Carvalho, S. R. Ambrósio, A. E. M. Crotti, U. S. de Figueiredo, P. C. Vieira, N. A. J. C. Furtado, Screening of filamentous fungi to identify biocatalysts for lupeol biotransformation, *Molecules* 2010. **15**, 6140–6151.

[24] P. Yogeeswari, D. Sriram, Betulinic acid and its derivatives: A review on their biological properties, *Curr. Med. Chem.* 2005. **12**, 657–666.

[25] G. A. Tolstikov, O. B. Flekhter, E. E. Shultz, L. A. Baltina, A. G. Tolstikov, Betulin and its derivatives. Chemistry and biological activity, *Chem. Sust. Dev.* 2005. **13**, 1–29.

[26] I.-C. Sun, H.-K. Wang, Y. Kashiwada, J. K. Shen, L. M. Cosentino, C. H. Chen, L. M. Yang, K. H. Lee, Anti-AIDS agents. 34. Synthesis and structure-activity relationships of betulin derivatives as anti-HIV agents, *J. Med. Chem.* 1998. **41**, 4648–4657.

[27] L. Baratto, M. Porsani, I. Pimentel, A. Pereira Netto, R. Paschke, B. Oliveira, Preparation of betulinic acid derivatives by chemical and biotransformation methods and determination of cytotoxicity against selected cancer cell lines, *Eur. J. Med. Chem.* 2013. **68**, 121–131.

[28] V. Zuco, R. Supino, S. C. Righetti, L. Cleris, E. Marchesi, C. Gambacorti-Passerini, F. Formelli, Selective cytotoxicity of betulinic acid on tumor cell lines, but not on normal cells, *Cancer Lett.* 2002. **175**, 17–25.

[29] Q.-H. Chen, J. Liu, H.-F. Zhang, G.-Q. He, M.-L. Fu, The betulinic acid production from betulin through biotransformation by fungi, *Enzyme Microb. Technol.* 2009. **45**, 175–180.

[30] Y. Feng, M. Li, J. Liu, T.-Y. Xu, R.-S. Fang, Q.-H. Chen, G.-Q. He, A novel one-step microbial transformation of betulin to betulinic acid catalysed by *Cunninghamella blakesleeana*, *Food Chem.* 2013. **136**, 73–79.

[31] M. S. Qazi, N. Sultana, M. Kamal, Biotransformation of pentacyclic terpene isolated from *Alstonia scholaris* (R.BR.), *Biocatal. Biotransform.* 2013. **31**, 148–152.

[32] D.-B. Mao, Y.-Q. Feng, Y.-H. Bai, C.-P. Xu, Novel biotransformation of betulin to produce betulone by *Rhodotorula mucilaginosa*, *J. Taiwan Inst. Chem. Eng.* 2012. **43**, 825–829.

[33] H. Liu, X.-L. Lei, N. Li, M.-H. Zong, Highly regioselective synthesis of betulone from betulin by growing cultures of marine fungus *Dothideomycete* sp. HQ 316564, *J. Mol. Catal. B: Enzym.* 2013. **88**, 32–35.

[34] J. Liu, M. L. Fu, Q. H. Chen, Biotransformation optimization of betulin into betulinic acid production catalysed by cultured *Armillaria luteo-virens* Sacc ZJUQH100-6 cells, *J. Appl. Microbiol.* 2011. **110**, 90–97.

[35] R. Azerad. In *Advances in Biochemical Engineering/Biotechnology (Biotransformations)*, Vol. **63** (Eds: K. Faber and T. Scheper). Springer-Verlag, Berlin-Heidelberg, 1999, pp. 169–218.

[36] E. A. Abourashed, A. M. Clark, C. Hufford, Microbial models of mammalian metabolism of xenobiotics: An updated review, *Curr. Med. Chem.* 1999. **6**, 359–374.

[37] R. Venisetty, V. Ciddi, Application of microbial biotransformation for the new drug discovery using natural drugs as substrates, *Curr. Pharm. Biotechnol.* 2003. **4**, 153–167.

[38] S. Kouzi, P. Chatterjee, J. Pezzuto, M. Hamann, Microbial transformations of the antimelanoma agent betulinic acid, *J. Nat. Prod.* 2000. **63**, 1653–1657.

[39] P. Chatterjee, S. A. Kouzi, J. M. Pezzuto, M. T. Hamann, Biotransformation of the antimelanoma agent betulinic acid by *Bacillus megaterium* ATCC 13368, *Appl. Environ. Microbiol.* 2000. **66**, 3850–3855.

[40] D. Z. L. Bastos, I. C. Pimentel, D. A. de Jesus, B. H. de Oliveira, Biotransformation of betulinic and betulonic acids by fungi, *Phytochemistry* 2007. **68**, 834–839.
[41] T. Akihisa, Y. Takamine, K. Yoshizumi, H. Tokuda, Y. Kimura, M. Ukiya, T. Nakahara, T. Yokochi, E. Ichiishi, H. Nishino, Microbial transformations of two lupane-type triterpenes and anti-tumor-promoting effects of the transformation products, *J. Nat. Prod.* 2002. **65**, 278–282.
[42] J. Zhang, Z.-H. Cheng, B.-Y. Yu, G. Cordell, S. Qiu, Novel biotransformation of pentacyclic triterpenoid acids by *Nocardia* sp. NRRL 5646, *Tetrahedron Lett.* 2005. **46**, 2337–2340.
[43] L.-W. Qian, J. Zhang, J.-H. Liu, B.-Y. Yu, Direct microbial-catalyzed asymmetric α-hydroxylation of betulonic acid by *Nocardia* sp. NRRL 5646, *Tetrahedron Lett.* 2009. **50**, 2193–2195.
[44] X. Xu, J. Dong, X. Mu, L. Sun, Supercritical CO2 extraction of oil, carotenoids, squalene and sterols from lotus (*Nelumbo nucifera* Gaertn) bee pollen, *Food Bioprod. Process.* 2011. **89**, 47–52.
[45] H. Rosas-Acevedo, T. Terrazas, M. E. González-Trujano, Y. Guzmán, M. Soto-Hernández, Anti-ulcer activity of *Cyrtocarpa procera* analogous to that of *Amphipterygium adstringens*, both assayed on the experimental gastric injury in rats, *J. Ethnopharmacol.* 2011. **134**, 67–73.
[46] G.-E. Yang, Z. Zhang, H. Bai, J. Gong, Y. Wang, B. Li, J. Li, Biotransformation of β-amyrin acetate by *Rhodobacter sphaeroides*, *J. Biosci. Bioeng.* 2008. **105**, 558–561.
[47] L. R. Martins, J. A. Takahashi, Rearrangement and oxidation of β-amyrin promoted by growing cells of *Lecanicillium muscarinium*, *Nat. Prod. Res.* 2010. **24**, 767–774.
[48] H. Hikino, S. Nabetani, T. Takemoto, Microbial transformation of oleanolic acid. 1, *Yakugaku Zasshi* 1969. **89**, 809–813.
[49] H. Hikino, S. Nabetani, T. Takemoto, Microbial transformation of oleanolic acid. 2, *Yakugaku Zasshi* 1971. **91**, 637–640.
[50] H. Hikino, S. Nabetani, T. Takemoto, Microbial transformation of oleanolic acid. 4, *Yakugaku Zasshi* 1972. **92**, 1528–1533.
[51] H. Hikino, S. Nabetani, T. Takemoto, Microbial transformation of oleanolic acid. 3, *Yakugaku Zasshi* 1972. **92**, 891–895.
[52] M. I. Choudhary, I. Batool, S. N. Khan, N. Sultana, S. A. A. Shah, A. Ur-Rahman, Microbial transformation of oleanolic acid by *Fusarium lini* and α-glucosidase inhibitory activity of its transformed products, *Nat. Prod. Res.* 2008. **22**, 489–494.
[53] Y.-Y. Zhu, L.-W. Qian, J. Zhang, J.-H. Liu, B.-Y. Yu, New approaches to the structural modification of olean-type pentacyclic triterpenes via microbial oxidation and glycosylation, *Tetrahedron* 2011. **67**, 4206–4211.
[54] D.-L. Liu, Y. Liu, F. Qiu, Y. Gao, J.-Z. Zhang, Biotransformation of oleanolic acid by *Alternaria longipes* and *Penicillium adametzi*, *J. Asian Nat. Prod. Res.* 2011. **13**, 160–167.
[55] N. Guo, Y. Zhao, W. S. Fang, Biotransformation of 3-oxo-oleanolic acid by *Absidia glauca*, *Planta Med.* 2010. **76**, 1904–1907.
[56] N. Shirane, Y. Hashimoto, K. Ueda, H. Takbnaka, K. Katoh, Ring-A cleavage of 3-oxo-olean-12-en-28-oic acid by the fungus *Chaetomium longirostre*, *Phytochemistry* 1996. **43**, 99–104.
[57] C. S. Capel, A. C. D. de Souza, T. C. de Carvalho, J. P. de Sousa, S. R. Ambrósio, C. H. Martins, W. R. Cunha, R. H. Galán, N. A. Furtado, Biotransformation using *Mucor rouxii* for the production of oleanolic acid derivatives and their antimicrobial activity against oral pathogens, *J. Ind. Microbiol.* 2011. **38**, 1493–1498.
[58] T. Gong, L. Zheng, X. Zhen, H.-X. He, H.-X. Zhu, P. Zhu, Microbial transformation of oleanolic acid by *Trichothecium roseum*, *J. Asian Nat. Prod. Res.* 2014. **16**, 383–386.
[59] A. Martinez, F. Rivas, A. Perojil, A. Parra, A. Garcia-Granados, A. Fernandez-Vivas, Biotransformation of oleanolic and maslinic acids by *Rhizomucor miehei*, *Phytochemistry* 2013. **94**, 229–237.
[60] X. Feng, Z. Zou, S. Fu, L. Sun, Z. Su, D.-A. Sun, Microbial oxidation and glucosidation of echinocystic acid by *Nocardia corallina*, *J. Mol. Catal. B: Enzym.* 2010. **66**, 219–223.
[61] L. Qian, J. Zhang, J. Liu, B. Yu, Regio-specific microbial hydroxylation of phytolaccagenin by *Streptomyces griseus* ATCC 13273, *Chin. J. Chem.* 2009. **27**, 1434–1436.
[62] A. Garcia-Granados, A. Martinez, J. N. Moliz, A. Parra, F. Rivas, 2-α,3-β-Dihydroxyolean-12-en-28-oic acid (maslinic acid), *Molecules* 1998. **3**(7), M88.
[63] X. Feng, J. Luan, F. F. Guo, D. P. Li, Z. Y. Chu, Microbial transformation of maslinic acid by *Cunninghamella blakesleana*, *J. Mol. Catal. B: Enzym.* 2012. **82**, 127–130.
[64] Y. Yamada, A. Nakamura, K. Yamamoto, H. Kikuzaki, Transformation of glycyrrhizic acid by *Aspergillus* sp., *Biosci. Biotechnol. Biochem.* 1994. **58**, 436–437.

[65] L. Canonica, G. Jommi, U. M. Pagnoni, F. Pelizzoni, B. M. Ranzi, C. Scolastico, Microbiological oxidation of triterpenoids. I. 7β—Hydroxyglycyrrhetic acid, *Gazz. Chim. Ital.* 1966. **96**, 820–831.

[66] J. Ma, X. L. Xin, W. Zhang, R. Lan, J. Y. Liu, Microbial transformation of glycyrrhetinic acid by *Mucor spinosus*, *Beijing Daxue Xuebao, Ziran Kexueban* 2008. **35**, 79–82.

[67] X. Xin, Y. Liu, M. Ye, H. Guo, D. Guo, Microbial transformation of glycyrrhetinic acid by *Mucor polymorphosporus*, *Planta Med.* 2006. **72**, 156–161.

[68] L. Canonica, M. Ferrari, G. Jommi, U. M. Pagnoni, F. Pelizzoni, B. M. Ranzi, S. Maroni, G. Nencini, T. Salvatori, Microbiological oxidation of triterpenoids. II. 15α-Hydroxyglycyrrhetic and 7β,15α-dihydroxyglycyrrhetic acids, *Gazz. Chim. Ital.* 1967. **97**, 1032–1051.

[69] K. Sakano, M. Ohshima, Structures of conversion products formed from 18β-glycyrrhetinic acid by *Streptomyces* sp. G-20, *Agric. Biol. Chem.* 1986. **50**, 763–766.

[70] K. Sakano, M. Ohshima, Microbial conversion of 18β-glycyrrhetinic acid and 22α-hydroxy-18β-glycyrrhetinic acid by *Chainia antibiotica*, *Agric. Biol. Chem.* 1986. **50**, 1239–1245.

[71] K. Yoshida, K. Furihata, H. Habe, H. Yamane, T. Omori, Microbial transformation of 18-glycyrrhetinic acid by *Sphingomonas paucimobilis* strain G5, *Biotechnol. Lett.* 2001. **23**, 1619–1624.

[72] K. Yoshida, K. Furihata, H. Yamane, T. Omori, Metabolism of 18β-glycyrrhetinic acid in *Sphingomonas paucimobilis* strain G5, *Biotechnol. Lett.* 2001. **23**, 253–258.

[73] M. I. Choudhary, Z. A. Siddiqui, S. Ahmed Nawaz, A. Ur-Rahman, Microbial transformation of 18β-glycyrrhetinic acid by *Cunninghamella elegans* and *Fusarium lini*, and lipoxygenase inhibitory activity of transformed products, *Nat. Prod. Res.* 2009. **23**, 507–513.

[74] M. Ferrari, U. M. Pagnoni, F. Pelizzoni, B. M. Ranzi, T. Salvatori, Microbiological oxidation of triterpenoids. III. Behavior of α-glycyrrhetic, liquiritic, and α-liquiritic acid, *Gazz. Chim. Ital.* 1969. **99**, 848–862.

[75] L. Canonica, G. Russo, E. Bombardelli, Triterpenes of *glycyrrhiza glabra*. II. Liquiritic acid, *Gazz. Chim. Ital.* 1966. **96**, 833–842.

[76] L. S. Amaral, M. Murgu, E. Rodriguez-Filho, A. Q. L. de Souza, M. I. de Moura Sarquis, A saponin tolerant and glycoside producer xylaraceous fungus isolated from fruits of *Sapindus saponaria*, *World J. Microbiol. Biotechnol.* 2008. **24**, 1341–1348.

[77] M. Murgu, L. F. Santos, G. D. de Souza, C. Daolio, B. Schneider, A. Ferreira, E. Rodriguez-Filho, Hydroxylation of a hederagenin derived saponin by a xylareaceous fungus found in fruits of *Sapindus saponaria*, *J. Braz. Chem. Soc.* 2008. **19**, 831–835.

[78] L. Hernández-Vázquez, S. Mangas, J. Palazón, A. Navarro-Ocana, Valuable medicinal plants and resins: Commercial phytochemicals with bioactive properties, *Ind. Crops Prod.* 2010. **31**, 476–480.

[79] L. O. A. Manguro, S. A. Opiyo, E. Herdtweck, P. Lemmen, Triterpenes of *Commiphora holtziana* oleo-gum resin, *Can. J. Chem.* 2009. **87**, 1173–1179.

[80] K. Mazumder, K. Tanaka, K. Fukase, Cytotoxic activity of ursolic acid derivatives obtained by isolation and oxidative derivatization by isolation and oxidative derivatization, *Molecules* 2013. **18**, 8929–8944.

[81] A. Ibrahim, S. Khalifa, I. Khafagi, Microbial metabolism of biologically active secondary metabolites from *Nerium oleander* L., *Chem. Pharm. Bull.* 2008. **56**, 1253–1258.

[82] S.-B. Fu, J.-S. Yang, J.-L. Cui, Q.-F. Meng, X. Feng, D.-A. Sun, Multihydroxylation of ursolic acid by *Pestalotiopsis microspora* isolated from the medicinal plant *Huperzia serrata*, *Fitoterapia* 2011. **82**, 1057–1061.

[83] S.-B. Fu, J.-S. Yang, J.-L. Cui, X. Feng, D.-A. Sun, Biotransformation of ursolic acid by an endophytic fungus from medicinal plant *Huperzia serrata*, *Chem. Pharm. Bull.* 2011. **59**, 1180–1182.

[84] Y. Wang, L. Xiang, M. Chen, Z.-X. Zhang, X. He, Substrate specificity for the 2α-hydroxylation of ursolic acid by *Alternaria alternata* and the antitumor activities of those metabolites, *J. Mol. Catal. B: Enzym.* 2012. **83**, 51–56.

[85] D. O. Collins, P. L. D. Ruddock, J. C. de Grasse, W. F. Reynolds, P. B. Reese, Microbial transformation of cadina-4,10(15)-dien-3-one, aromadendr-1(10)-en-9-one and methyl ursolate by *Mucor plumbeus* ATCC 4740, *Phytochemistry* 2002. **59**, 479–488.

[86] D. Leipold, G. Wünsch, M. Schmidt, H.-J. Bart, T. Bley, H. E. Neuhaus, H. Bergmann, E. Richling, K. Muffler, R. Ulber, Biosynthesis of ursolic acid derivatives by microbial metabolism of ursolic acid with *Nocardia* sp. strains—Proposal of new biosynthetic pathways, *Process Biochem.* 2010. **45**, 1043–1051.

[87] S. Fu, J. Yang, J. Cui, D.-A. Sun, Biotransformation of ursolic acid by *Syncephalastrum racemosum* CGMCC 3.2500 and anti-HCV activity, *Fitoterapia* 2013. **86**, 123–128.
[88] J. T. James, I. A. Dubery, Pentacyclic triterpenoids from the medicinal herb, *Centella asiatica* (L.) urban, *Molecules* 2009. **14**, 3922–3941.
[89] P. Puttarak, P. Panichayupakaranant, Factors affecting the content of pentacyclic triterpenes in *Centella asiatica* raw materials, *Pharm. Biol.* 2012. **50**, 1508–1512.
[90] M. Haftek, S. Mac-Mary, M.-A. Le Bitoux, P. Creidi, S. Seité, A. Rougier, P. Humbert, Clinical, biometric and structural evaluation of the long-term effects of a topical treatment with ascorbic acid and madecassoside in photoaged human skin, *Exp. Dermatol.* 2008. **17**, 946–952.
[91] F. Wu, D. Bian, Y. Xia, Z. Gong, Q. Tan, J. Chen, Y. Dai, Identification of major active ingredients responsible for burn wound healing of *Centella asiatica* herbs, *Evid. Based Complement. Alternat. Med.* 2012. **2012**, 848093.
[92] C.-L. Xu, R. Qu, J. Zhang, L.-F. Li, S.-P. Ma, Neuroprotective effects of madecassoside in early stage of Parkinson's disease induced by MPTP in rats, *Fitoterapia* 2013. **90**, 112–118.
[93] W.-N. He, J.-G. Dai, M. Ye, L.-J. Wu, D.-A. Guo, Microbial transformation of asiatic acid by *Alternaria longipes*, *J. Asian Nat. Prod. Res.* 2010. **12**, 760–764.
[94] F.-F. Guo, X. Feng, Z.-Y. Chu, D.-P. Li, L. Zhang, Z.-S. Zhang, Microbial transformation of asiatic acid, *J. Asian Nat. Prod. Res.* 2013. **15**, 15–21.
[95] F.-X. Huang, X.-H. Lin, W. Song, M. Ye, W.-Z. Yang, D.-A. Guo, Two new oxidation products obtained from the biotransformation of asiatic acid by the fungus *Fusarium avenaceum* AS 3.4594, *J. Asian Nat. Prod. Res.* 2012. **14**, 1039–1045.
[96] B. A. Shah, G. N. Qazi, S. C. Taneja, Boswellic acids: A group of medicinally important compounds, *Nat. Prod. Rep.* 2009. **26**, 72–89.
[97] H. Safayhi, E.-R. Sailer, H. P. T. Ammon, Mechanism of 5-lipoxygenase inhibition by acetyl-11-keto-β-boswellic acid, *Mol. Pharmacol.* 1995. **47**, 1212–1216.
[98] E.-R. Sailer, S. Schweizer, S. E. Boden, H. P. T. Ammon, H. Safayhi, Characterization of an acetyl-11-keto-â-boswellic acid and arachidonate-binding regulatory site of 5-lipoxygenase using photoaffinity labeling, *Eur. J. Biochem.* 1998. **256**, 364–368.
[99] S. Bleif, F. Hannemann, J. Zapp, D. Hartmann, J. Jauch, R. Bernhardt, A new *Bacillus megaterium* whole-cell catalyst for the hydroxylation of the pentacyclic triterpene 11-keto-β-boswellic acid (KBA) based on a recombinant cytochrome P450 system, *Appl. Microbiol. Biotechnol.* 2012. **93**, 1135–1146.
[100] Y. Sun, D. Liu, R.-G. Xi, X. Wang, Y. Wang, J. Hou, B. Zhang, C. Wang, K. Liu, X. Ma, Microbial transformation of acetyl-11-keto-boswellic acid and their inhibitory activity on LPS-induced NO production, *Bioorg. Med. Chem. Lett.* 2013. **23**, 1338–1342.
[101] X.-L. Xin, H. Huo, L. Chen, J. Li, J.-H. Sun, P.-W. Zheng, Y. Sun, Z.-M. Wu, Y.-H. Xiong, Microbial transformation of acetyl-11-keto-boswellic acid by *Cunninghamella elegans*, *J. Asian Nat. Prod. Res.* 2013. **15**, 1173–1178.
[102] Z.-H. Cheng, B.-Y. Yu, G. A. Cordell, S.-X. Qiu, Biotransformation of quinovic acid glycosides by microbes: Direct conversion of the ursane to the oleanane triterpene skeleton by *Nocardia* sp. NRRL 5646, *Org. Lett.* 2004. **6**, 3163–3165.
[103] Z.-H. Cheng, B.-Y. Yu, Y. Guo, S.-X. Qiu, Microbial hydroxylation of quinovic acid glycosides by *Streptomyces griseus* ATCC 13273, *Chin. J. Chem.* 2006. **24**, 95–98.
[104] S.-S. Lee, C.-J. Lin, K. C. Liu, Two triterpenes from *Paliurus ramosissimus*, *J. Nat. Prod.* 1992. **55**, 602–606.
[105] S. Guo, J.-A. Duan, Y.-P. Tang, N.-Y. Yang, D.-W. Qian, S.-L. Su, E.-X. Shang, Characterization of triterpenic acids in fruits of ziziphus species by HPLC-ELSD-MS, *J. Agric. Food Chem.* 2010. **58**, 6285–6289.
[106] T. Kuramoto, Y. Ito, M. Oda, Y. Tamura, S. Kitahata, Microbial production of glycyrrhetic acid 3-*O*-mono-β-D-glucuronide from glycyrrhizin by *Cryptococcus magnus* MG-27, *Biosci. Biotechnol. Biochem.* 1994. **58**, 455–458.
[107] H. A. S. Amin, A. G. Hanna, S. S. Mohamed, Comparative studies of acidic and enzymatic hydrolysis for production of soyasapogenols from soybean saponin, *Biocatal. Biotransform.* 2011. **29**, 311–319.
[108] C. Gauthier, J. Legault, M. Lebrun, P. Dufour, A. Pichette, Glycosidation of lupane-type triterpenoids as potent in vitro cytotoxic agents, *Bioorg. Med. Chem.* 2006. **14**, 6713–6725.
[109] P. Chatterjee, J. M. Pezzuto, S. A. Kouzi, Glucosidation of betulinic acid by *Cunninghamella* species, *J. Nat. Prod.* 1999. **62**, 761–763.

CHAPTER 29

Transaminases and their Applications

Sarah-Marie Dold, Christoph Syldatk, and Jens Rudat

Institute of Process Engineering in Life Sciences, Section II: Technical Biology, Karlsruhe Institute of Technology (KIT), Karlsruhe, Germany

29.1 INTRODUCTION

Transaminases are important enzymes in the synthesis of chiral amines, amino acids, and amino alcohols. In this chapter the properties of transaminases, the reaction mechanisms, and their selectivity and substrate specificity are presented. The synthesis of chiral building blocks for pharmaceutically relevant substances and fine chemicals with transaminases as biocatalysts is discussed. Enzymatic asymmetric synthesis and dynamic resolution are discussed using transaminases. Protein engineering by directed evolution as well as rational design of transaminases under process condition is presented to develop efficient bioprocesses.

29.2 GENERAL PROPERTIES OF TRANSAMINASES

Since their discovery, first published by Needham *et al.*, transaminases or aminotransferases (EC2.6.1.X) have received much attention as biocatalysts for the transformation of a keto acid to the corresponding amino acid/amine or vice versa [1]. Transaminases play an important role in amino acid metabolism and are ubiquitous in microbes and eukaryotic cells. They are pyridoxal-5′-phosphate (PLP)-dependent enzymes and are qualified as biocatalysts, due to their wide substrate scope, high enantio- and regioselectivity, high reaction rates, and stability [2]. As pictured in Scheme 29.1, the amino group of the amino donor is transferred to the carbonyl group of the amino acceptor.

Due to the aforementioned benefits of transaminases, they are utilized for the synthesis of optically pure amines, amino acids, and amino alcohols. These products are used as building blocks for fine chemicals and pharmaceutical agents or for food and agriculture. Furthermore, unusual amino acids, for example, β-amino acids, deployed in peptidomimetics, offer higher stability against proteases. For the synthesis of unnatural amino acids or bulky amines, novel ω-transaminases are used [3–7]. Several synthesis strategies were established and optimized for transaminases, facing the challenges of product and substrate inhibition and the need to shift the equilibrium toward the product. The increased importance of transaminases in the last decades has been reflected in ~107 transaminases listed in the BRENDA enzyme database (http://www.brenda-enzymes.org/) in 2014, compared to 77 in 2005.

Green Biocatalysis, First Edition. Edited by Ramesh N. Patel.

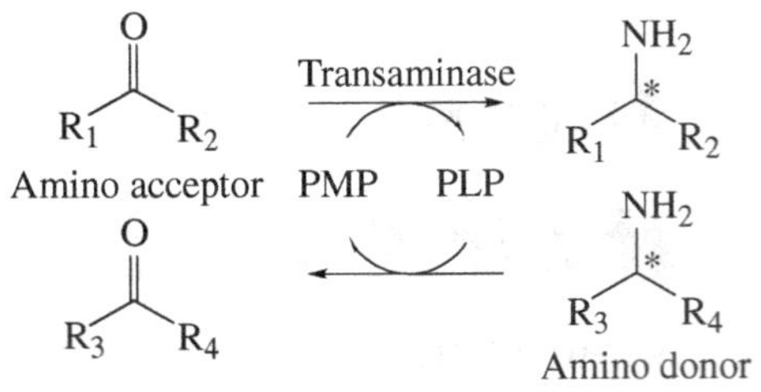

SCHEME 29.1
Transaminase-catalyzed reaction.

29.2.1 Classification as Pyridoxal-5′-Phosphate-Dependent Enzymes

Aminotransferases (AT) or transaminases (TA) comprise a major group of enzymes with different substrate spectra correlating with the sequence and the fold types of the different enzymes. Transaminases belong to the PLP-dependent enzymes, which consist of seven fold types of enzymes on the basis of amino acid sequence comparisons (latest update: http://bioinformatics.unipr.it/B6db) [8]. Transaminases can be found within the fold types I and IV (Table 29.1) [9–11]. According to the Pfam database (a database of protein families) where families correlate sharing high sequence similarity of defined protein regions, transaminases are divided into six subgroups [12]. The typical transaminases in fold type I are the well-known and described, aspartate aminotransferase and other (*S*)-selective transaminases. Recently a branched-chain aminotransferase from the thermophilic archaeon *Thermococcus* sp. CKU-1 has been discovered and classified as fold type I due to its high homology with the fold type I transaminases known so far [13]. Interestingly this enzyme reaches its maximum activity not until 95 °C. Furthermore it is noteworthy that this transaminase is active toward isoleucine and valine, which are not described for other transaminases in this fold type. In the fold type IV transaminases, lately an (*R*)-selective ω-transaminase from *Aspergillus terreus* has been described [14], as well as branched-chain aminotransferases.

Enzymes of fold type I/IV are active as homodimers with two active sites: a larger domain, with a seven-stranded β-sheet in the center, and a smaller domain, containing the C-terminus of the amino acid chain. PLP is covalently bound to the larger domain via the ε-amino group of a lysine residue. The difference between the two fold types is based on the mirror inverted binding of the phosphate in the active site. For fold type I classified enzymes the *re* face of the enzyme is exposed to the solvent rather than the *si* side of enzymes of fold type IV [11, 15].

29.2.2 Classification Based on Substrate Scope

Another possible classification of transaminases is based on their differing substrate scope [16, 17]. The amino donor for the transamination, catalyzed by transaminases, can be classified in three different chemical groups, differing in the presence and position of the functional groups, usually a negatively charged carboxylate group.

TABLE 29.1 Classification of Pyridoxal-5′-Phosphate-Dependent Transaminases Based on their Evolutionary Relationship

Pfam Subgroup	Fold Type	Enzyme
I/II	I	Aspartate AT Aromatic AT
III	I	ω-TA
IV	IV	Branched-chain AT (*R*)-Selective AT
V	I	Phosphoserine AT
VI (DegT_DnrJ_EryC1)	I	Sugar AT

Thus transaminases are divided into α-transaminases, ω-transaminases, and amine transaminases. Since α-transaminases need substrates with the α-carboxylic group adjacent to the amino, carbonyl, or keto-substituted α-carbon atom, these enzymes usually catalyze the transamination of proteinogenic amino acids in the metabolism and are industrially established for the production of canonic amino acids. The next group of transaminases—the ω-transaminases—accepts substrates with several carbon atoms between the carbonyl and the carboxylate group and substrates with the ketone or aldehyde function on the terminal carbon atom. Therefore the aminotransferases of this group are high in demand for the synthesis of various building blocks for pharmaceuticals, like chiral amines. Amine transaminases convert ketones into amines and do not require substrates with any carboxylic group. This is a huge benefit for the production of sterically challenging amines like rivastigmine or sitagliptin, among other amines. A very important difference between α-transaminases and amine transaminases is concerning the reaction equilibrium. The equilibrium of α-transaminase-catalyzed reactions is close to unity, whereas the production of alanine is strongly favored using amine transaminase as a biocatalyst, indicating the need to shift the equilibrium toward the product side (several solutions to these challenges are shown in Section 29.3.1).

29.2.3 Reaction Mechanism

Due to structure determination of the aspartate aminotransferase, the reaction mechanism of transaminases is well understood and examined in detail [10, 11, 18–20]. The transfer of the amino group is supported by the external cofactor PLP forming a Schiff base with the ε-amino group of the lysine on the active side of the enzyme (Lys-E), called internal aldimine. The internal aldimine keeps the PLP in a highly reactive condition, due to the positively charged nitrogen of the protonated imine, which is far more electrophilic than the aldehyde or ketone. Further conversion leads to pyridoxamine-5′-phosphate (PMP) reacting as an intermediate via the ping-pong bi–bi reaction mechanism [15]. The reversible reaction is accomplished in two reaction steps. In the first step PLP reacts with the L-aspartate to give PMP and oxaloacetate. In the second half-reaction, the amino acceptor, for example, α-ketoglutarate, regenerates the PLP, and the product L-glutamic acid is created (details are shown in Scheme 29.2). More precisely, the internal aldimine undergoes transamination with the amino group of the substrate to create the external aldimine. But it is important that either the internal aldimine is protonated and the amino group is not or the internal aldimine is deprotonated and the amino group is protonated for the extra proton that can be transferred between the amino group of the substrate and the imine nitrogen of the aldimine [19]. The Michaelis–Menten complex formed possesses the proton on the imine nitrogen and the free amino group of the substrate. This leads to a rapid attack on the free amino group and a geminal diamine mediated by 3′-oxygen of PLP. The diamine collapses into the external aldimine intermediate and the free amino residue of the lysine on the active side of the enzyme as the leaving group. The step after the external aldimine is the formation of the corresponding α-carbanionic intermediate (quinonoid) by loss of the substrate's α-hydrogen bond through deprotonation catalyzed by the free base of the lysine of the active side (Lys-E). This quinonoid possesses the electron pair from the α-hydrogen bond delocalized onto the pyridine nitrogen. Protonating the C4′ leads to the ketimine intermediate. The ketimine reacts to the carbinolamide intermediate by addition of water to the Cα atom catalyzed by Lys-E. Finally, deprotonation of the carbinolamide leads to the PMP/oxaloacetate Michaelis complex, which dissociates to the free enzyme and the product. Reversing the same steps with α-ketoglutarate as substrate leads to L-glutamic acid as amino acid product.

The influence of the PLP cofactor on the activity and stability of transaminases and their catalyzed reactions is shown for ω-transaminases as well [21–23].

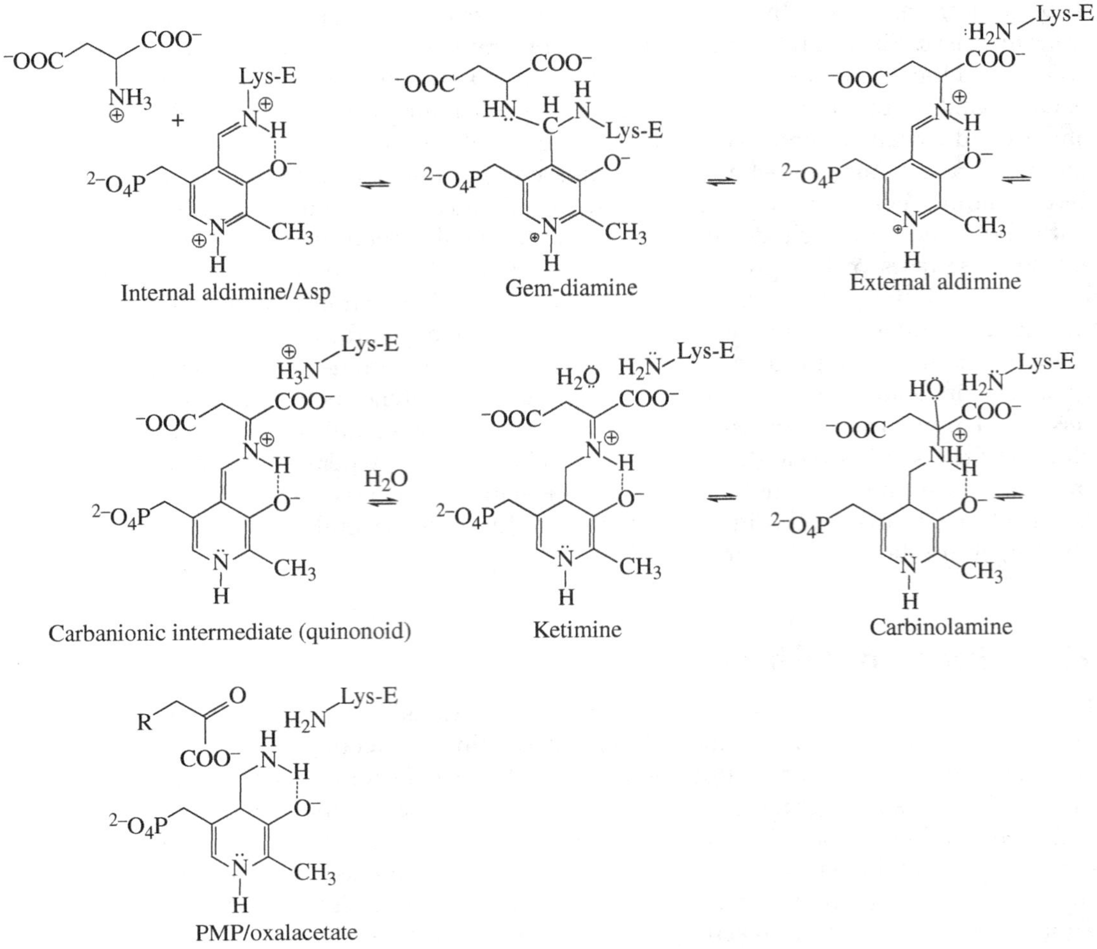

SCHEME 29.2
Reaction mechanism of transaminases with pyridoxal-5′-phosphate (PLP) as external cofactor via a ping-pong bi–bi mechanism. A two-step reaction, starting with an internal aldimine creating an external aldimine to pyridoxamine-5′-phosphate (PMP).

The main limitation in the described transamination reaction is inhibition by the substrates utilized and the formed products. It is noteworthy that substrate inhibition results from both amine enantiomers. For example, the (*R*)-enantiomer can form a Michaelis–Menten complex with PLP leading to dead-end complexes, whereas the (*S*)-enantiomer is able to convert the PLP to PMP [24]. This indicates that the substrate inhibition and therefore the wide substrate range are caused by the promiscuous binding pockets of the enzyme. Conversely, it is possible that the product forms a Michaelis–Menten complex with PLP of the free enzyme. Thus, the place for the substrate is taken and fast conversion rates are inhibited. Another limitation for amination with transaminases is the equilibrium of the reaction. Due to the reversibility of all the steps in the reaction mechanism, it is necessary to shift the equilibrium toward the product side to reach high conversion yields. Several possibilities to shift the equilibrium are described in Section 29.3.

29.2.4 Enantioselectivity of Transaminases

The most frequently described transaminases show high enantioselectivity for the chiral carbon atom with the carboxylic group at the α-position or the chiral center at the β-position. To determine the mechanism toward enantioselectivity of transaminases, a structural analysis is necessary. The crucial factor is the architecture of the active site of the enzyme, which gives the substrate the required orientation into the binding pockets [25]. Since the discovery of transaminases, numerous (*S*)-selective enzymes have been isolated. To expand the applications for biocatalysis, other (*R*)-selective transaminases are in demand. The most convenient tool to reveal (*R*)-selective transaminases is computational design or protein engineering (described in Sections 29.4.1 and 29.4.2) [26]. In the last years various transaminases with (*R*)-enantiopreference have been discovered. A study has established the structure of an (*R*)-selective amine–pyruvate aminotransferase from *Nectria haematococca*, which has the ability to convert (*R*)-methylbenzylamine and several (*R*)-amines and ketones [27]. Furthermore (*R*)-selective amine transaminases from *A. terreus* and *Aspergillus fumigatus* have been characterized [14, 28]. For the β-selectivity of aminotransferases, no binding mode has been reported. Wybenga *et al.* showed that the active site of the aminotransferase from *Mesorhizobium* sp. evolved specifically to receive both β- and α-amino acids [25]. This aminotransferase is enantioselective toward (*S*)-β-phenylalanine and toward (*R*)-3-amino-5-methylhexanoic acid and (*R*)-3-aminobutyric acid, due to their stereoconfiguration of the functional groups on the C-β-atom. The architecture of the active sides showed that the orientation of the substrates is forced to bind the carboxylic group at the arginine in the P-pocket and the side chain in the O-pocket and the addition of the amino group on the *si* face of the β-carbon of the keto acid. Moreover the O-pocket binding the aliphatic and hydrophilic side chains of β-amino acids is also capable of binding the α-carboxylic groups of α-amino acids.

Consequently, it is important to gain further knowledge about the structure and the mechanism of the relevant enzymes to facilitate in protein engineering in order to expand the substrate and thus the product range, leading to the desired optically pure enantiomers.

29.3 SYNTHESIS STRATEGIES WITH TRANSAMINASES

Transaminases are well established as biocatalysts in the production of a wide variety of chiral amines and amino acids. Their use as building blocks in pharmaceuticals, like in antiarrhythmic, cancer- or diabetes-treating drugs, and agrochemicals, makes it particularly important to find enzymatic approaches for their production and to overcome the chemical synthesis of chiral amines, which requires protective groups, toxic transition metal catalysts, harsh reaction conditions, and multireactions [29–32]. Reactions catalyzed by transaminases offer advantages like a promiscuous substrate spectrum, no need for external cofactors, high enantio- and stereoselectivity, mild reactions conditions, and good reaction yields. Only catalytic amounts of PLP, as described in Section 29.2.4, are needed as a cofactor to stabilize and to form the active dimer [4, 6, 29, 33].

Transaminases can either be utilized in kinetic resolution or asymmetric synthesis (Scheme 29.3). Asymmetric synthesis, starting with a prochiral ketone substrate, can theoretically lead to 100% conversion and is usually the preferred route to chiral products (Scheme 29.3a). Furthermore high enantiomeric purity is not dependent on conversion rates, whereas a kinetic resolution (Scheme 29.3b) needs 50% conversion for a high enantiomeric excess (ee). But kinetic resolution is thermodynamically favored, if pyruvate is the amino acceptor, compared to asymmetric synthesis where the equilibrium lies on the substrate side [5, 34]. To achieve 100% conversion, dynamic kinetic resolution serves as an alternative with spontaneous deracemization or the initiation of a suitable racemate for enantiomerically pure substrates (Scheme 29.3c). Deracemization in a one-pot two-step reaction with an (*S*)-and (*R*)-selective transaminase, respectively, is a method of choice, but unfortunately two enantiocomplementary enzymes are needed (Scheme 29.3d) [35]. Therefore deracemization with a dehydrogenase in the kinetic resolution step and a transaminase in the following step

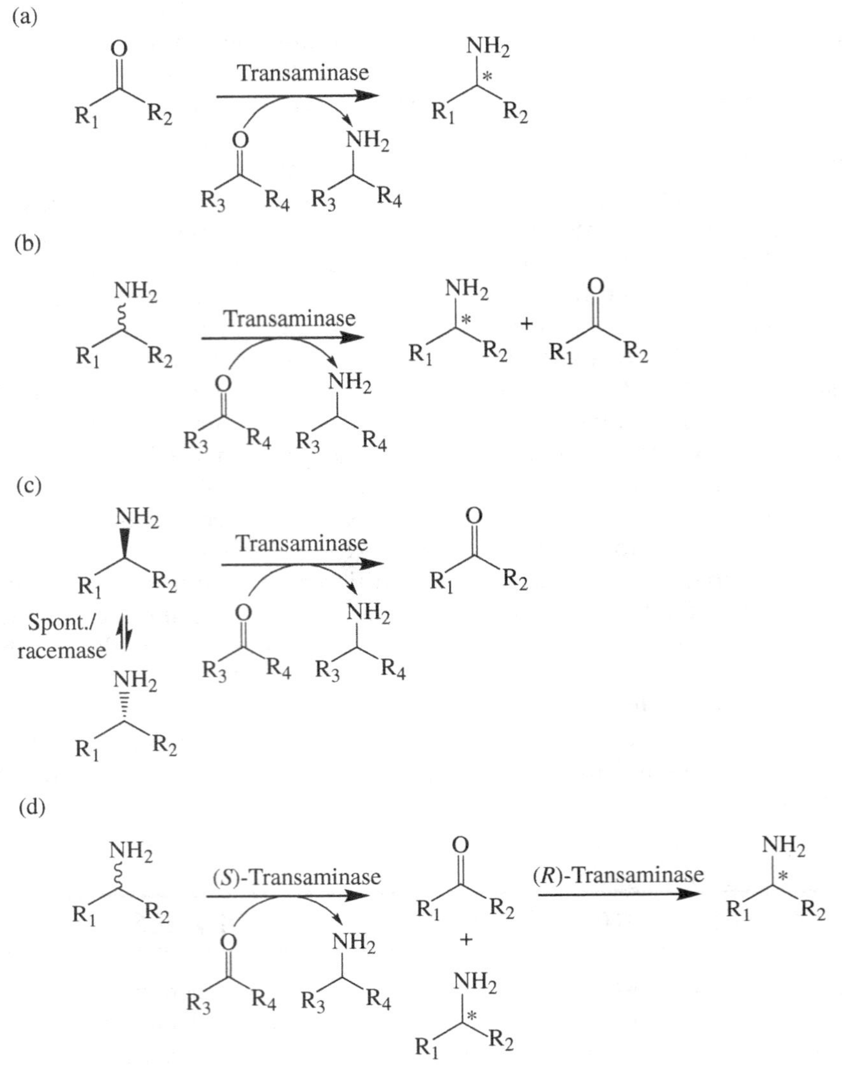

SCHEME 29.3

General synthesis strategies for transaminase-catalyzed reactions. (a) Asymmetric synthesis with transaminase. (b) Kinetic resolution with transaminase. (c) Dynamic kinetic resolution with transaminase. (d) One-pot two-step deracemization with transaminase.

is an option [36]. A variety of chiral amines and amino alcohols produced with asymmetric synthesis, dynamic kinetic resolution, and deracemization are given in Table 29.2. But overall the main problems of transaminase-catalyzed reactions are still product and substrate inhibition for each of the synthetic strategies described and the requirement to shift the reaction equilibrium for high reaction yields.

29.3.1 Synthesis of Chiral Amines

There is a high demand for the synthesis of chiral amines due to their application as chiral building blocks for more complex structures like pharmaceutical intermediates but also as fine chemicals or in agriculture. They are used in calcimimetic agents to treat hyperparathyroidism, in antiarrhythmic/antidiabetic drugs, as cholinesterase inhibitor, and as drugs for the treatment of Parkinson's and Alzheimer's disease [30, 39, 54]. Furthermore γ-aminobutyric acid (GABA), another chiral amine, is important in the nervous system and is used in its enantiopure form as a medication for several psychiatric disorders [37]. Ephedrine, a precursor of the illegal methamphetamine, is used as an anorectic agent, as a decongestant, or as hypertensive agent. It possesses two chiral centers, which affect the pharmacological application, due to its binding at different receptors [55]. Another chiral amine, active as an antidiabetic compound, is sitagliptin in its (*R*)-conformation [56, 57]. Consequently an enantiopure amine with

TABLE 29.2 A Variety of Chiral Amines and Amino Alcohols Produced with Asymmetric Synthesis (AS) and Dynamic Kinetic Resolution (DKR)/Deracemization (DE)

Product	Enzyme	Substrate	Synthesis Strategy	Reference
Pyrrolidine	(*R*)/(*S*)-ω-TA (*Vibrio fluvialis*, *Chromobacterium violaceum*, ATA-117)	(*rac*)-4-Oxo-3-phenylbutyric acid + L/D-alanine	DKR	[37]
Mexiletine	ω-TA (*C. violaceum*)/ATA-117	(*rac*)-Mexiletine + alanine	DE	[38]
Rivastigmine	ω-TA (*Paracoccus denitrificans*)	3-Acetylphenyl ethyl(methyl)carbamate + L-alanine	AS	[39]
α-Aminosteroids	Ω-TA (*Arthrobacter* sp.)	Steroid precursor + isopropylamine	AS	[40]
1-Phenoxypropane-2-amine	ω-TA (*Pseudomonas aeruginosa*, *Aspergillus terreus*)	1-Phenoxy-2-propane + D/L-alanine	AS	[41]
Nor(pseudo)ephedrine	Acetohydroxyacid synthase I/(*S*)/(*R*)-ω-TA (*C. violaceum*/*A. terreus*)	Pyruvate + benzaldehyde ⟶ (*R*)-phenylacetylcarbinol + D/L-alanine	DE	[42]
Methylbenzylamine	ω-TA (*Arthrobacter citreus*)	Acetophenone + isopropylamine	AS	[43]
Aminotetralin	ω-TA (*Arthrobacter* sp., *C. violaceum*)	1-/2-Tetralone + isopropylamine/alanine	AS	[44]
Amino alcohol				
2-Amino-1,3,4-butanetriol	Transketolase/β-alanine–pyruvate aminotransferase (*P. aeruginosa*)	Glycolaldehyde/β-hydroxypyruvate + methylbenzylamine		[45]
Amino acids				
L-2-Aminobutyric acid	ω-TA (*V. fluvialis*)	2-Oxobutyric acid + benzylamine	AS	[46]
(*R*)-Fluoroalanine	ω-TA (*V. fluvialis*)	3-Fluoropyruvate + methylbenzylamine		[47]
L-Homoalanine	(*S*)/(*R*)-ω-TA (*V. fluvialis*, *Arthrobacter* sp.)	α-Keto acid + isopropylamine	AS	[48]
L-Homoalanine	D-Amino acid oxidase + ω-TA (*V. fluvialis*)	(*rac*)-D-Homoalanine ⟶ 2-ketobutyric acid + benzylamine	DKR	[49]
2-Naphthylamine	D-Amino acid oxidase (*Rhodotorula gracilis*), L-aspartate aminotransferase (*Escherichia coli*)	(*rac*)-2-Naphtylamine + cysteine sulfinic acid	DE	[50]
β-Phenylalanine	Lipase (*Candida rugosa*) + ω-transaminase (*Mesorhizobium* sp.)	β-Keto acid ester ⟶ β-keto acid + 3-aminobutyric acid	AS	[51]
D-Amino-*N*-butyric acid	ω-Amino acid–pyruvate TA (*Alcaligenes denitrificans*)	Pyruvate + (*rac*)-β-amino-*N*-butyric acid	KR	[52]
L-Isoleucine	L-Alanine TA	2-Keto-3-methyl-valerate + L-alanine	AS	[53]

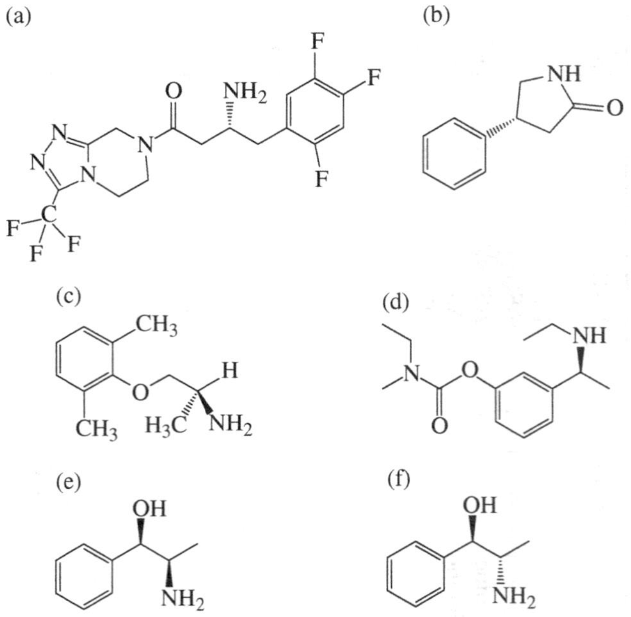

SCHEME 29.4

Industrial important halogenated chiral amines functioning as antiarrhythmic or antidiabetic drugs. (a) (*R*)-Sitagliptin. (b) (*R*)-4-Aminopyrrolidine-2-one. (c) (*S*)-Mexiletine. (d) (*S*)-Rivastigmine. (e) (1*R*,2*R*)-norpseudoephedrine (NPE). (f) (1*R*,2*S*)-Norephedrine (NE).

the correct conformation has to be produced. Recently, the asymmetric synthesis of prochiral diketones to 3-substituted cyclohexylamine derivatives serving as a building block for the multidrug-resistant protein MRP1 has been introduced [58]. As a consequence new synthesis strategies had to be developed, leading to transaminases with the benefits of a wide substrate range and a high enantioselectivity, giving the opportunity to produce chiral amines of the desired enantiomers. Some industrially important halogenated chiral amines are shown in Scheme 29.4.

Asymmetric Synthesis of Chiral Amines

As mentioned in Section 29.3 asymmetric synthesis has many advantages; however, the reversible reaction includes the conversion of a 2-oxo acid to an amino acid leading to equilibrium constants of ~1. To overcome this unfavorable equilibrium, large amounts of amino donors can be applied, but as a consequence this leads to substrate inhibition and may complicate the purification of the desired product. Various approaches to shift the equilibrium toward the product side were developed: the employment of an amino donor leading to a volatile by-product like acetone [59, 60] or precipitation due to the low solubility of the product compared to the keto acid [61, 62]. Another way to remove the product, leading to a shift of the equilibrium, is the use of an ion-exchange resin or spontaneous cyclization for product removal *in situ* [63] or via multienzyme networks briefly described in Section 29.3.1 and extensively reviewed by Simon *et al.* [64].

Equilibrium Shift by Transformation of the Keto By-Product

Höhne *et al.* discussed several solutions for the decomposition of a keto acid by product, efficiently for the efficient shifting the equilibrium to obtain higher product yields [17]. A novel single enzyme reaction with an ω-transaminase from *Chromobacterium violaceum* and 3-aminocyclohexa-1,5-dienecarboxylic acid as the amino donor was examined [65]. As shown in Scheme 29.5, a prochiral ketone (a) and the chosen amino donor (b) were leading to chiral amines. The ketone by-product (c) spontaneously tautomerizes, avoiding the reaction equilibrium and hence preventing product inhibition. Conversions up to 99% and a high ee were achieved with a ratio of prochiral ketone to amino donor of 1:1.05. The possible reverse reaction was also examined by applying the tautomerized ketone (c). But no significant amounts of

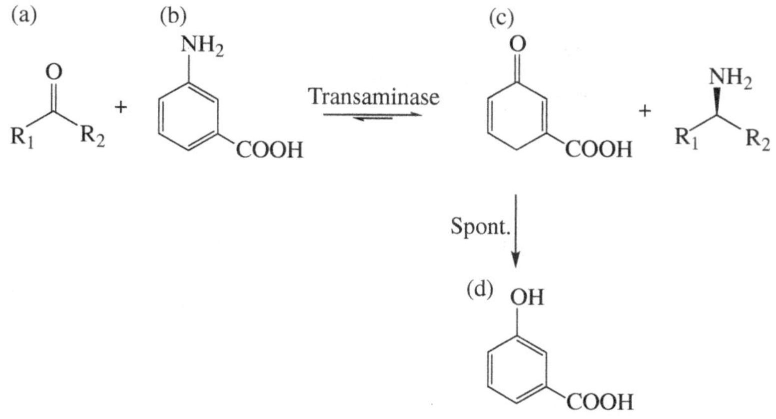

SCHEME 29.5
Equilibrium shift by spontaneously tautomerization of the ketone by-product. The prochiral ketone (a) and the chosen amino donor (b) are leading to the chiral amine by tautomerization of the ketone byproduct (c) to (d).

acetophenone were detected, which is apparently the reason for the high conversion rate in terms of minimal amino donor quantities.

Dealing with substrate inhibition caused by application of higher amounts of reaction educts needs further investigations in terms of ideal reaction conditions and amino donor/acceptor ratio. One crucial step for successful transamination is the choice of a suitable amino donor. Fesko *et al.* evaluated the influence of several amino donors and substrates to investigate a one-enzyme system [41]. The influence of the amino donor on the ee was shown by examination of reactions catalyzed by the ω-transaminase from *Paracoccus denitrificans*. Usually low ee values were obtained with alanine as the amino donor. The best results were acquired with 1-phenoxy-2-propanone and 1-phenylethylamine in sodium phosphate buffer pH 7.5 at 30 °C. To get higher substrate loading, it is necessary to add cosolvents due to the low solubility of the substrates. In this study, cyclohexane and sodium dodecyl sulfate (SDS) were best cosolvents. For the preparation of (*R*)- and (*S*)-phenoxypropane-2-amine in a similar yield, three enzymes were used for the synthesis [66]; *rac*-2-butylamine as amino donor usage was increased by tenfold (500 mM).

The evolved ω-transaminase from *Arthrobacter* sp., which is able to catalyze amination of sterically challenging keto amides, was utilized for the synthesis of the α-amino steroids 17-amino-1,3,5(10)-estratrien-3-ol and 17-amino-5α-androstan-3β-ol [40].

The transaminase was used in organic solvents (35% dimethylformamid) as a cosolvent to increase substrate concentration due to the low solubility in water. A 100-fold molar excess of 2-propylamine as the amino donor (1 M) was used, leading to conversion rates of 85–89% and ee of 90–96%.

Nowadays the focus on transaminase-catalyzed reactions includes overcoming traditional chemical synthetic methods, for example, halogenated or heteroatomic amines with the high enantiomeric purity and high yields needed for pharmacological approaches [67]. If the best reaction conditions for the contemplated reaction were established, the scale-up of the biocatalytic transamination and substrate concentration could be investigated. Several ortho-halogenated ketones were scaled up to 100 mM substrate concentrations with conversions of about 87–96%. Furthermore, brominated ketones were obtained with 250 mM concentration with an enzyme load of 20 mg. This effort leads to conversions of 85%.

Equilibrium Shift via By Product Removal
Lactate Dehydrogenase

In an enzymatic reaction catalyzed by ω-transaminases, ketones were converted into the desired chiral amine with the formation of the coproduct pyruvate from alanine. To remove pyruvate from the equilibrium, it was transformed into lactate by a lactate

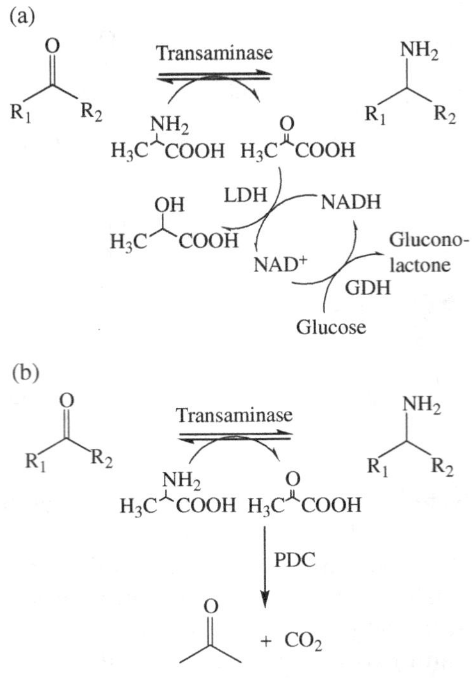

SCHEME 29.6 An overview of multienzyme networks established to shift the equilibrium of asymmetric synthesis toward the product side. (a) Lactate dehydrogenase (LDH)/glucose dehydrogenase (GDH). (b) Pyruvate decarboxylase (PDC).

dehydrogenase. The regeneration of the cofactor NADH needed by the dehydrogenase was achieved with glucose dehydrogenase shown in Scheme 29.6a [66]. Reaching higher conversion yields of 83% with this enzyme cascade combination compared to, for example, alanine dehydrogenase in cascade with formate dehydrogenase (FDH) (conversion rate of 29%) has lately been reported for the asymmetric synthesis of (*S*)-rivastigmine [39].

Pyruvate Decarboxylase

A drawback of using lactate dehydrogenase as a biocatalyst to remove pyruvate from the reaction equilibrium is the need for the NADH cofactor. Another possibility to eliminate the coproduct is the application of a pyruvate decarboxylase (Scheme 29.6b). A cofactor is not required, and the resulting products of pyruvate decarboxylation, acetaldehyde, and CO_2 are highly volatile, shifting the equilibrium toward the product [68]. Several pyruvate decarboxylases from yeast and bacteria are commercially available and are active at the same pH value as the transaminase required for the asymmetric synthesis of chiral amines.

Yeast Alcohol Dehydrogenase

A narrow substrate spectrum has been described for the yeast alcohol dehydrogenase (YADH) from *Saccharomyces cerevisiae*, making it a suitable biocatalyst only for molecules like methanol, ethanol, or in some cases acetone. Unfortunately the cofactor NADH is needed, which is regenerated by FDH (Scheme 29.6c). For the asymmetric synthesis of (*S*)-phenylethylamine with isopropylamine (IPA) as amino donor, acetone was converted to isopropyl alcohol catalyzed by YADH. The effectiveness of this method was compared to the reaction without YADH/FDH. A conversion yield of 99% was achieved with the YADH/FDH system while a conversion yield of 63–89% was obtained without YADH/FDH [69].

Alanine Dehydrogenase

An alternative for removing pyruvate from the reaction for higher conversion rates is alanine dehydrogenase (Scheme 29.6d). Here pyruvate is converted into D- or L-alanine with ammonia, and the cofactor NADH is regenerated by the FDH in the presence of NAD^+ [70]. This is described for the synthesis of aminotetralin based on the transamination of (*S*)-7-methoxy-2-tetralone with conversion rates and ee of 99% within 24h reaction time [44].

Equilibrium Shift via Supported Liquid Membrane

A new way of *in situ* product removal, shifting the equilibrium for the asymmetric synthesis of (*S*)-α-methylbenzylamine (MBA), was presented. A supported liquid membrane (SLM) system combined with a packed bed reactor enables continuous extraction of the amine product [43]. A conversion of 98% was reached with the extraction step using SLM, compared to 50% without the extraction step with SLM. The utilized biocatalyst was the ω-transaminase from *Arthrobacter citreus* in *Escherichia coli*, immobilized on chitosan beads imbedded in the reactor. An alkaline pH close to the pK_a value of MBA was selected to keep the reaction product deprotonated and therefore able to diffuse into the hydrophobic membrane. Reversely the stripping phase was kept at a basic pH; the product was protonated and cannot partition into the membrane as shown in Scheme 29.7. Thus, the pH difference is the driving force of the extraction, and back-extraction of the protonated MBA is prevented. The main limitation of the process was the minor selectivity between the amino donor and the product. Examination of the concentration of MBA and IPA in the reactor reservoir displayed a faster increase in MBA concentration. Two possible reasons were suggested. Firstly, MBA (logP value of 1.8) is more hydrophobic than IPA (logP value of 0.39) and was able to partition into the hydrophobic membrane. The second reason was based on the difference in the pK_a values between the two amines (9.54 MBA and 10.73 IPA). Hence, at pH 9 a higher amount of uncharged MBA was present. Another aspect should be considered; however, the concentration of IPA has to be maintained at a high level to assure the high conversion yields for MBA production and avoid excessive levels to limit the undesired extraction of IPA, which was constantly present in a ratio of 1:1 in the product reservoir.

To increase the product concentration to an economically viable level, usually of 50–100 g/l [71], a higher concentration of acetophenone (200 mM) was added. Here, a very important factor was the maintenance of the latter pH value throughout the entire reaction, which was achieved by higher buffer concentrations to control the pH in the substrate and product reservoir. It was also necessary to keep the membrane soaked with undecane and a specified amount of PLP in the substrate reservoir. Because of its hydrophilic nature, PLP was not able to partition into the membrane. High product levels (55 g/l) can be reached by fed-batch addition of

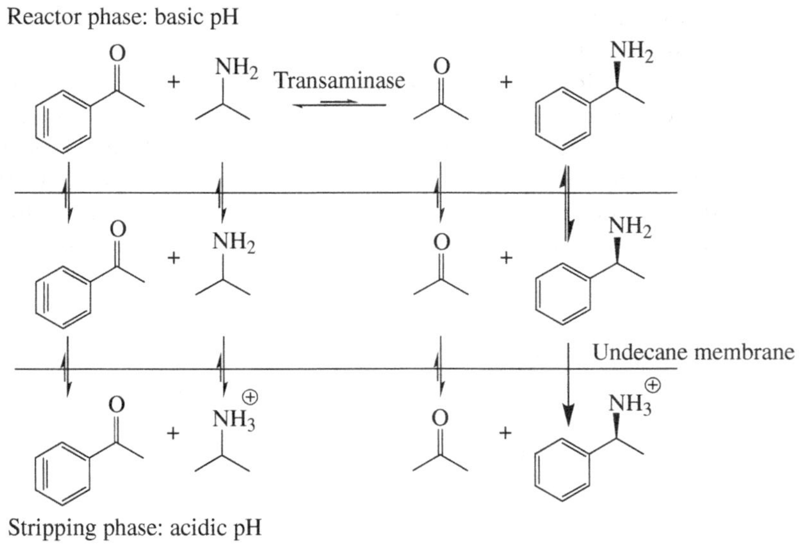

SCHEME 29.7

Supported liquid membrane system with a packed bed reactor for product extraction after asymmetric synthesis of chiral amines. No back-extraction is possible due to the different pH value separated in reservoirs by a hydrophobic undecane membrane.

acetophenone. After 80 h, the concentrations of IPA, MBA, acetone, and acetophenone were 460 mM, 480 mM, 78 mM, and 23 mM, respectively.

This extraction process for product removal and shifting of the equilibrium is the method of choice to obtain high yields of chiral amines, but is not suitable for the production of amino acids due to their zwitterionic character, only allowing poor partition into the membrane.

In summary, asymmetric synthesis is the method of choice for the production of enantiopure amines for pharmaceutically relevant molecules. Several possibilities to shift the equilibrium to the favored side of the reactions are described, leading to high conversion rates and high ee. Thus, compared to chemically synthesized amines, the enzymate route possesses considerable advantages for the economy and ecology.

Kinetic Resolution of Chiral Amines

Studies of transaminases in kinetic resolution elucidated several benefits. Compared to asymmetric synthesis the equilibrium favors product formation if pyruvate is used as an amino acceptor. To get enantiopure amines, kinetic resolution is an acceptable choice with yields of 50%. Various (*R*)- and (*S*)-selective transaminases are well established nowadays, leading to enantiopure (*S*)- and (*R*)-amines, with high ee. Unfortunately product inhibition is one major disadvantage of kinetic resolution. If a critical concentration of product is achieved, the maximum conversion is prevented. Based on kinetic modeling in a previous study from Shin and Kim, the inhibitory effects were based on the strong binding of the product to the PLP cofactor. Consequently the binding of the amino acceptor is hindered, and conversion of the substrate is not possible [72].

Several possibilities to manage this challenge have been explored recently. As mentioned before, the equilibrium toward the product side was favored if pyruvate was added. To elide the product inhibition, but to apply the necessary amount of amino acceptor, a continuous recycling of pyruvate was required. Truppo *et al.* developed an enzyme cascade with a transaminase combined with an amino acid oxidase (AAO) to shift the equilibrium constantly to the product side as shown in Scheme 29.8 [73].

Furthermore, an enzyme microreactor with a membrane contactor was determined to overcome the problem with a two-liquid phase reaction system *in situ* [74]. The organic phase served as both an extractor for acetophenone and a reservoir for the amino donor. Unfortunately solvent toxicity to the enzyme and limitations in the back-extraction of the water-soluble product with organic solvent were the main limitations. An elegant solution was found for water-soluble ketones generated by the kinetic resolution of *sec*-butylamine [34]. The reaction took place under reduced pressure (0.199 bar) harnessing the volatility of the ketone by-product 2-butanone. Further reactions were performed with whole cells due to the lower substrate and product inhibition effects. Thus, under reduced pressure (0.199 bar), 400 mM *sec*-butylamine converted into (*R*)-*sec*-butylamine with an ee of 98% and conversion rates up to 53%. A major drawback of this, however, was the evaporation of the reaction media. Since ω-transaminases are stable and active in organic solvents, ionic liquids (1-butyl-3-methylimidazolium hexafluorophosphate) were

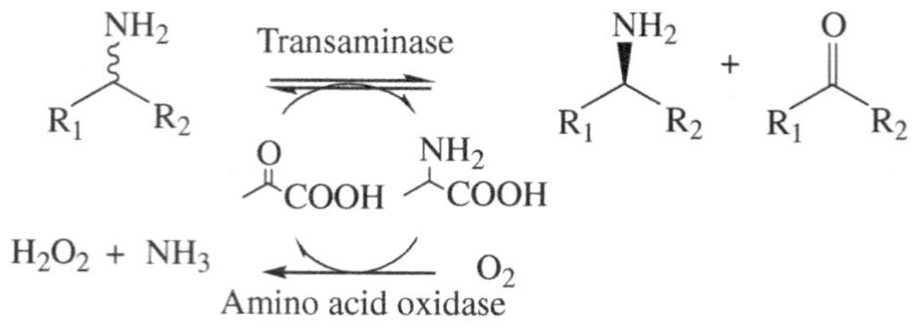

SCHEME 29.8

Overcoming product inhibition for an effective kinetic resolution using amino acid oxidase to recycle pyruvate.

investigated as reaction media, and the enzyme showed about 0.1–0.5% of its former activity. These results were leading to the conclusion that after further selection of ionic liquids, higher activities of the enzyme can be achieved.

Recently an ω-transaminase from *Ochrobactrum anthropi* has been isolated and surprisingly showed no product inhibition by acetophenone [75]. Various ketones shown in Table 29.3 were examined. No reaction limitation for the transaminase from *O. anthropi* and a residual enzyme activity up to 100% were detected, compared to the transaminase from *P. denitrificans* showing no residual activity at the same product concentrations.

In addition the transaminase from *O. anthropi* showed no substrate inhibition up to 500 mM of (*S*)-α-methylbenzylamine in any cases. For the kinetic resolution with 500 mM (*S*)-α-methylbenzylamine, 300 mM pyruvate, and 75 U/ml of *O. anthropi* transaminase, reaction yield of 95.3% and ee values of >99% were obtained after 3 h compared to the transaminase of *P. denitrificans* with no detectable conversion and the ee values of <22% obtained after 10 h.

This makes the ω-transaminase of *O. anthropi* an excellent biocatalyst for kinetic resolution of several chiral amines and a promising candidate for asymmetric synthesis.

Dynamic Kinetic Resolution for the Synthesis of Chiral Amines

Dynamic Kinetic Resolution

For the first time, dynamic kinetic resolution was mentioned for the production of 4-arylpyrrolidin-2-ones and their acyclic GABA analogs (Scheme 29.9) [37].

The transformation of racemic 4-oxo-3-phenylbutyric acid ethyl ester to the desired (*R*)-enantiomer was catalyzed by an (*R*)-ω-transaminase [76]. For a complete theoretical conversion, the coproduct pyruvate was converted to lactate by a lactate dehydrogenase. Under the tested reaction conditions (with 100 mg racemic substrate, 30 mg of the enzyme at 30 °C, and pH values between 6.5 and 9.0), spontaneous

TABLE 29.3 Ketone Inhibitor tested with the ω-Transaminase from *Ochrobactrum anthropi*

Ketone Inhibitor	Substrates	Residual Activity (%)
Acetophenone	(*S*)-α-Methylbenzylamine + 2-oxobutyrate	97 ± 6
Acetophenone	L-Alanine + 2-oxobutyrate	104 ± 7
Propiophenone	(*S*)-α-Ethylbenzylamine + pyruvate	97 ± 17
1-Indanone	(*S*)-1-Aminoindan + pyruvate	98 ± 23
2-Butanone	(*S*)-*sec*-Butylamine + pyruvate	93 ± 12
Cyclopropyl methyl ketone	(*S*)-Cyclopropylethylamine + pyruvate	105 ± 8

The concentration of the substrates was 20 mM and 0–10 mM of the ketone product, respectively.

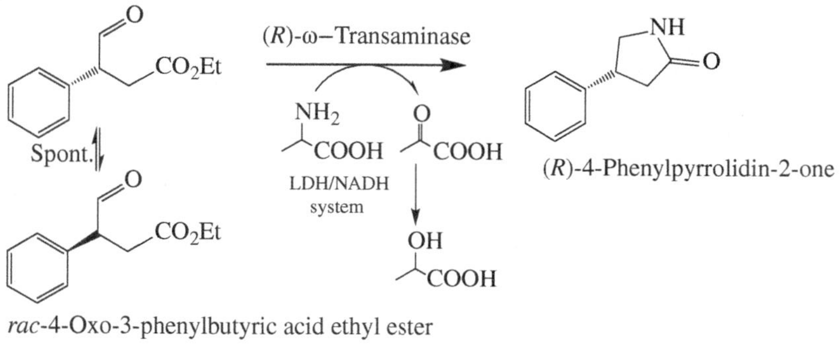

SCHEME 29.9 Dynamic kinetic resolution of 4-phenylpyrrolidin-2-one.

racemization of the substrate occurred, and the transamination yield was 91–93%, whereas the ee value remained between 42% (pH 9) and 69% (pH 6.5). To enhance the solubility of the substrate, 15% (v/v) DMSO was added to the reaction solution.

Another approach has recently been described by Cuetos *et al.* for the dynamic kinetic resolution of α-substituted β-amino esters [77]. Several ω-transaminases and acyclic alkyl-β-keto esters were tested, leading to high conversion rates and high ee and de values. With this result in mind it seems feasible that novel transaminases will be designed to create enantiomeric pure molecules under mild and economically feasible conditions.

Deracemization by a One-Pot Two-Step Procedures

For the deracemization of racemic amines by a one-pot, two-step procedures, two ω-transaminases with opposite enantiopreferences were needed. Koszelewski *et al.* examined the ω-transaminase-catalyzed transamination of several primary amines, obtaining enantiopure products with reaction yields of 100% [35]. For one, the (*R*)-enantiomer of mexiletine (shown in Scheme 29.4c in Section 29.3), which is pharmaceutically used as antiarrhythmic agent, is showing higher activity and better selective binding to a cardiac sodium channel compared to the (*S*)-enantiomer [38]. Therefore methods to prepare enantiopure mexiletine are desirable. A one-pot two-step process for deracemization requiring a kinetic resolution step followed by the reductive amination step was carried out. Complete removal of the undesired enantiomer was favored for the kinetic resolution step. Thus, the equilibrium had to be shifted to the product side by recycling the cosubstrate pyruvate via L-AAO at the expense of molecular oxygen to get complete conversion under the final reaction conditions. The peroxide produced was removed by a catalase [78]. It was beneficial that no interference from the emerging by-product occurred, due to the opposite configuration of alanine in the amination step (Scheme 29.10).

The best conversion yield in the kinetic resolution step, up to 55%, and ee values of >99% were achieved with 50 mM *rac*-mexiletine. After defining the best reaction conditions, the two reaction steps were combined by adding the ω-transaminase with opposite enantiopreference. Unfortunately it was shown that the ω-transaminase used in the first step also catalyzed the reductive amination reaction. To overcome this problem, prior to the addition of the transaminase with the opposite enantiopreference, the transaminase used in the first step was kept at 75 °C for 30 min. The combination of the ω-transaminase from *C. violaceum* and the transaminase ATA-117 achieved excellent conversion yield and ee values of >99% for the kinetic resolution and the reductive amination step, respectively.

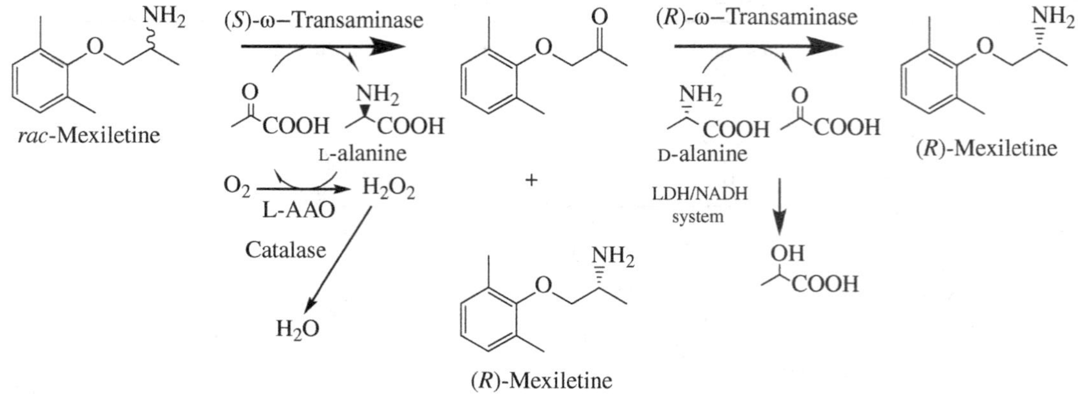

SCHEME 29.10
One-pot two-step deracemization of *rac*-mexiletine to get enantiopure (*R*)-mexiletine.

Another example of the one-pot, two-step enzyme cascade is the synthesis of enantiopure (1*R*,2*S*)-norephedrine (NE) or (1*R*,2*R*)-norpseudoephedrine (NPE) shown in Scheme 29.2e–f [42]. The first step of this multienzyme network was the carboligation reaction with intrinsic pyruvate decarboxylation. Pyruvate was decarboxylated and ligated to benzaldehyde, producing (*R*)-phenylacetylcarbinol (PAC) (ee > 98%) catalyzed by the thiamine diphosphate (ThDP)-dependent acetohydroxyacid synthase I (AHAS-I). The second step was the reductive amination of (*R*)-phenylacetylcarbinol by (*R*)-and (*S*)-selective transaminases, requiring an amino donor as cosubstrate. A smart cosubstrate recycling system was invented: with alanine as the cosubstrate for the second step resulting in pyruvate as coproduct serving as the substrate for the pyruvate decarboxylation catalyzed by AHAS-I in the first step. The bottleneck of this reaction was the reductive amination, necessitating further optimization of reaction parameters. Therefore, for the enzyme combination *Cv*-(*S*)-selective ω-transaminase/AHAS-I (leading to (1*R*,2*S*)-NE), the reaction conditions were 100 mM HEPES buffer (pH 7.5), 1 mg/ml ω-transaminase, 0.5 mg/ml AHAS-I, and a ratio 5 : 1 of alanine/PAC, at 25 °C. To improve the conversion rates the second enzyme was added consecutively after the complete conversion of benzaldehyde to PAC. Hence, the conversion of (1*R*,2*S*)-NE increased to 78%, whereas in the case of (1*R*,2*R*)-NPE catalyzed by the (*R*)-selective ω-transaminase from *A. terreus*, an even conversion yield of >96% was achieved.

In many cases the racemic amine is more easily available than the corresponding ketone. Thus, dynamic kinetic resolution or deracemization has gained increased attention over the last few years. Furthermore they are avoiding the limitation of kinetic resolution, leading to a high yield of optically pure amines, starting with racemic amines. The recycling mode of the coproducts is also a good way to reduce overall waste production.

29.3.2 Synthesis of Canonical and Noncanonical Amino Acids

The synthesis of all kinds of amino acids has been studied extensively in the last decades. Due to enhanced enzymatic and pharmacodynamic stability, as well as their diverse structures and biochemical properties, amino acids were maintained as chiral building blocks in numerous peptidomimetics, in single-enantiomer drugs, and also in various fields of agriculture [79, 80]. Due to the wide substrate specificity of transaminases, these enzymes are suitable as biocatalysts for the amination of keto acids or deracemization of racemic amines to produce enantiopure amino acids [2, 54, 81].

Some nonproteinogenic amino acids, produced enzymatically, are shown in Table 29.4; for example, L-homophenylalanine, a key intermediate of levetiracetam and brivaracetam applicable as antiepileptic drugs, or D-fluoroalanine, a key intermediate of antibiotics inactivating the bacterial D-alanine transaminase. In Table 29.4 a selection of proteinogenic and nonproteinogenic amino acids, the used biocatalysts, synthesis strategies, and the substrates used are listed.

Asymmetric Synthesis of Noncanonical Amino Acids

The asymmetric synthesis via α-transaminases was described for L-phenylalanine and L-homophenylalanine in several reports and reviews [17, 62, 86]. Herein the focus is on ω-transaminases that catalyzed the asymmetric synthesis of optically pure nonproteinogenic amino acids as building blocks for peptidomimetic and other pharmaceutical compounds. The overall advantage of ω-transaminase-catalyzed reactions is the ability to use achiral amino donors like benzylamine, which thermodynamically favors the equilibrium toward the product side.

For the synthesis of L-2-aminobutyric acid, the ω-transaminase from *Vibrio fluvialis* with 2-oxobutyric acid and benzylamine as substrates was established [46].

TABLE 29.4 An Overview of Some Industrial and Pharmacologically Relevant Amino Acids

Therapeutic Use/ Key Intermediate	Amino Acid	Structure	Reference
Antiepileptic drug	L-Homophenylalanine	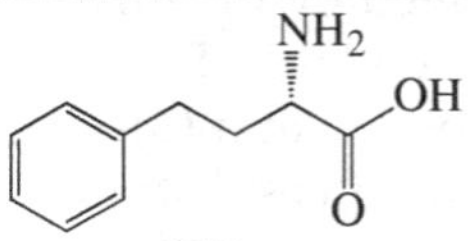	[82]
β-Lactam antibiotics	D-Phenylglycine	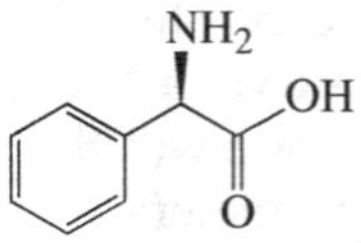	[83]
Antibiotic	D-Fluoroalanine	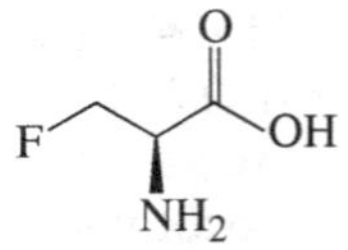	[84]
Pyrethroid insecticides	D-Valine	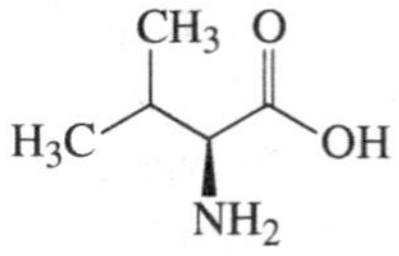	[85]
Receptor antagonist	L-2-Aminobutyric acid	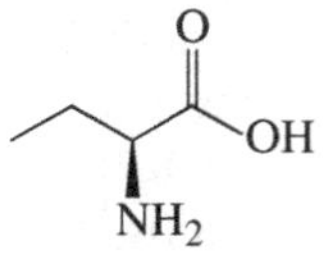	[46]
	L-2-Naphthyl alanine	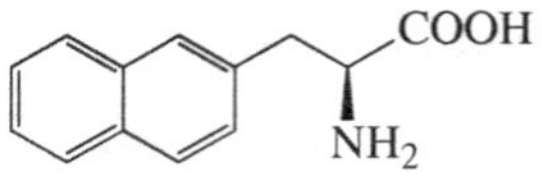	[50]
	β-Phenylalanine	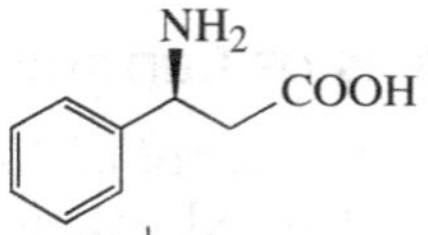	[51]
	L-β-Aminobutyric acid	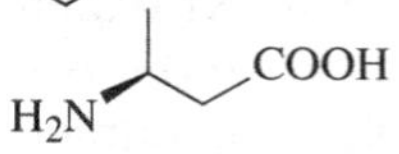	[52]

To overcome the product inhibition caused by benzaldehyde, an extractive biphasic reaction system was applied with hexane used as an extractor. Thus, considerable enhancements of reaction yield, up to 96%, were achieved compared to the aqueous system with conversion of about 39%.

An approach for the asymmetric synthesis of (*R*)-fluoroalanine using a biphasic reaction to overcome product inhibition caused by acetophenone was developed by Bea *et al.* [47]. The substrate fluoropyruvate and the product (*R*)-fluoroalanine remained in the aqueous phase, whereas acetophenone was extracted by isooctane, the selected organic solvent. The reaction yield for the conversion in the biphasic system was about 95–31% without acetophenone extraction.

A recently published study for the asymmetric synthesis of L-homoalanine also dealt with the problem of byproduct inhibition. Therefore the focus was on the choice of the most suitable amino donor. Starting from α-keto acids with IPA (giving acetone as coproduct) as amino donor, L-homoalanine was asymmetrically synthesized with ω-transaminase [48].

The asymmetric synthesis of D-amino acids is carried out by an (*R*)-selective ω-transaminase [87]. Several achiral ketones were examined, leading to D-configured amino acids like D-homoalanine, D-serine, D-fluoroalanine, D-alanine, and D-norvaline. A reaction yield of >99% and enantiopurity of >99.7% were achieved, except for D-norvaline. Remarkably, no product inhibition by acetophenone was observed at concentrations of 40 mM. The (*R*)-selective ω-transaminase displays a substrate spectrum very similar to the (*S*)-selective transaminase: ω-transaminase from *V. fluvialis*.

Deracemization and Dynamic Kinetic Resolution of Canonical and Noncanonical Amino Acids

For the preparation of 2-naphthyl alanine, a one-pot, two-step enzyme cascade was invented with D-AAO from *Rhodotorula gracilis* and an L-aminotransferase from *E. coli* [50]. Detail information about the deracemization steps was provided in Section 29.3.1 and shown in Scheme 29.3. Starting with the racemic 2-naphthyl alanine, the AAO generated the corresponding α-keto acid, which served as the substrate in the reductive amination step to enantiopure 2-naphthyl alanine. An irreversible amino donor, cysteine sulfinic acid, was used.

Recently Deracemization of racemic homoalanine to its optically pure enantiomer was recently examined [49]. The D-AAO used was catalyzing the oxidation of *rac*-homoalanine to 2-ketobutyrate followed by the conversion to L-homoalanine with ω-transaminase using benzylamine as an amino donor. The D-AAO was fused with *Vitreoscilla* hemoglobin, enhancing its activity and stability in the process. A catalase was used to catalyze the decomposition of hydrogen peroxide, the by-product of the AAO reaction. To deal with coproduct inhibition caused by benzaldehyde, a biphasic extraction system using isooctane as organic solvent was investigated. In the presence of the biphasic extraction system, a conversion of 98.8 and 99% ee was achieved after 24 h reaction with substrate concentrations of 100 and 200 mM of racemic homoalanine and benzylamine, respectively. As for the requirement of the cofactor FAD in the kinetic resolution step, the whole cells expressing both enzymes were used for the preparation of enantiopure homoalanine. After a 4 h reaction the complete conversion of the D-homoalanine was observed even without adding catalase and FAD. Using whole cells containing the transaminase and 500 mM *rac*-homoalanine and 500 mM benzylamine and 10 ml isooctane, the conversion yield of 97% and ee of >99% were obtained within 36 h.

29.3.3 Synthesis of β-Amino Acids

Considering the growing importance of β-amino acids as building blocks for peptidomimetics and bioactive compounds, there is the need for new synthesis strategies for their production in enantiopure β-amino acids. As shown in Scheme 29.11, the amino group of β-amino acids is attached to the β-carbon atom compared to α-amino acids. There are β^3-, β^2-, and $\beta^{2,3}$-substituted amino acids classified by the position and number of their residues [79, 80, 88].

Several β-amino acids occur naturally as free metabolites in metabolic pathways or as key intermediates in biosynthetic products. β-Alanine is the simplest β-amino acid that appears in pantothenic acid, a precursor of the coenzyme A. Further examples are (2*R*,3*S*)-*N*-benzoyl-3-phenylisoserine derived from (*R*)-β-phenylalanine, a compound in the antitumor agent paclitaxel from *Taxus brevifolia* [89], or as building blocks for β-lactam antibiotics [90] and in jasplakinolide, an antifungal compound [91] (Scheme 29.12).

Additionally, β-amino acids possess the ability to form β-peptide analogues to α-peptides. These secondary structures are highly stable against cleavage of proteolytic

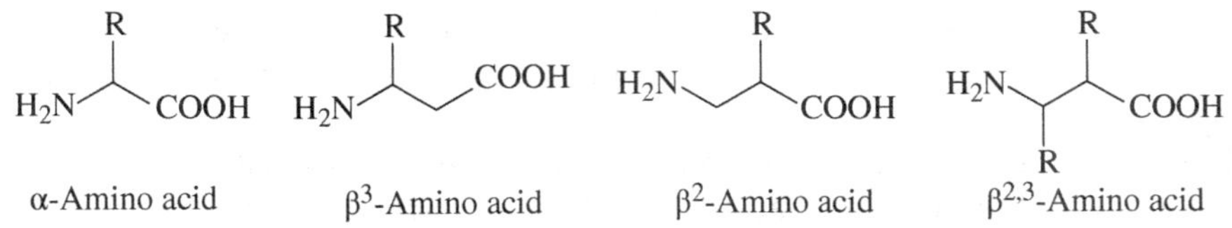

SCHEME 29.11 The structures of α- and β-amino acids.

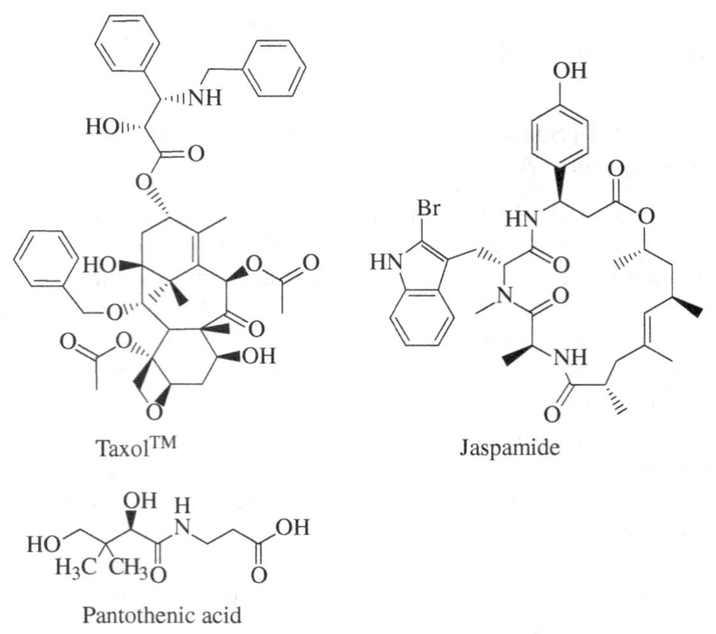

SCHEME 29.12

The structures of possible products containing β-amino acids as intermediates.

enzymes [92, 93]. Furthermore the mixed peptides containing α- and β-amino acids seemed to be more protected against the proteolytic digestion [94].

In the last few years several approaches to the enzymatic synthesis of enantiopure β-amino acids were pursued [6, 95]. Starting from *N*-acetylated ethyl esters, lipases [92], aminomutases [96, 97], acylases [98], monooxygenases [99], and amidases [100] were used for the synthesis of β-amino acids.

Due to benefits like the wide substrate scope and their role in the biodegradation process of β-amino acids by microbes, the focus for the synthesis of optically pure β-amino acids has been relied recently on transaminases.

Asymmetric Synthesis of β-Amino Acids

For the quite challenging asymmetric synthesis of β-amino acids, caused by the unstable prochiral β-ketone acid substrate, a coupled reaction combining the hydrolysis of the β-keto acid ester by a lipase from *Candida rugosa* and the transamination with a β-transaminase isolated from *Mesorhizobium* sp. was established [51]. With a conversion yield of 20% and >99% ee for the (*S*)-enantiomer of 3-amino-3-phenylpropionic acid, this coupled reaction was obtained. The equilibrium was favored due to 3-aminobutyric acid as an amino donor, converting into the volatile acetone with acetaldehyde as an intermediate. At first, the β-keto acid ester was considered to be an amino donor for the transamination reaction, but no formation of the corresponding β-amino acid ester could be observed leading to the assumption that transaminases possess no ability for the amination of β-keto acid ester.

Kinetic Resolution of β-Amino Acids

The advantage of the kinetic resolution of β-amino acids is the shift of the equilibrium toward the product side, due to the spontaneous decarboxylation of the β-keto acid by-product. For the synthesis of D-amino-*N*-butyric acid with an ω-transaminase from *Alcaligenes denitrificans* and pyruvate as amino acceptor, conversion yields of 53% and >99% ee were obtained [52]. The kinetic resolution of racemic β-phenylalanine catalyzed by the ω-transaminase from *Burkholderia phytofirmans* was examined recently [101].

Various ω-transaminases have been found to be suitable for asymmetric synthesis and kinetic resolution for enzymatic synthesis of amino acids with a high conversion yield and high ee, often in combination with other enzymes to shift the equilibrium. A huge benefit of the later enzymes is the wide substrate range; thus even amino acids with bulky side chains could be converted.

29.3.4 Synthesis of Amino Alcohols

Enantiopure amino alcohols are valuable building blocks for many pharmaceuticals [102]. One example is the asymmetric synthesis of 2-amino-1,3,4-butanetriol as a precursor for *syn*-amino alcohols and other bioactive compounds [45]. Two enzymes, a transketolase and a transaminase, were used in series. Hydroxypyruvate and glycolaldehyde were converted by the transketolase forming L-erythrulose; after adding the transaminase the conversion of 2-amino-1,3,4-butanetriol was observed with reaction yield of 21%. Unfortunately, the reaction rate for the transaminase-catalyzed step was 3000-fold below than that of the transketolase step, requiring the screening and/or mutagenesis for transaminases with increased activity toward L-erythrulose.

Recently an approach for the amination of *sec*-alcohols has developed [103]. Until now no microorganisms have been found capable of transforming *sec*-alcohols directly into amines, indicating the need of a two-enzyme cascade. As shown in Scheme 29.13, an alcohol dehydrogenase catalyzes the oxidation of the alcohol to the corresponding ketone, which was aminated by a transaminase. To shift the equilibrium, the pyruvate was removed via a lactate dehydrogenase, which was used for the recycling of NADPH as well.

29.3.5 Transaminase-Catalyzed Reactions with Whole Cells

In recent years, an upcoming approach for the synthesis of chiral amines and amino acids has been the use of whole cells in biocatalysis. Compared to reactions with isolated enzymes, no previous purification or immobilization steps plus cofactor regeneration *in vivo* are needed. But besides these benefits, whole-cell biotransformation shows some limitations. High enzyme activity may limit the cofactor regeneration, consequently leading to lower conversion yield [104]. High product levels can also have an effect on biotransformation activity [105], as well as the import of substrate and export of the product.

Synthesis of L-Isoleucine in Whole Cells

L-Isoleucine is a generally used amino acid and is an essential amino acid in feedstuffs and in infusion solutions, with annual production yielding amounts of 400 t [106]. A new way to use reductive amination in whole cells was recently invented [53]. A coupled enzyme system consisting of an L-alanine-dependent transaminase and an L-alanine dehydrogenase for the cofactor regeneration is needed to recycle the amino donor L-alanine as shown in Scheme 29.14. Both enzymes were expressed in *E. coli*.

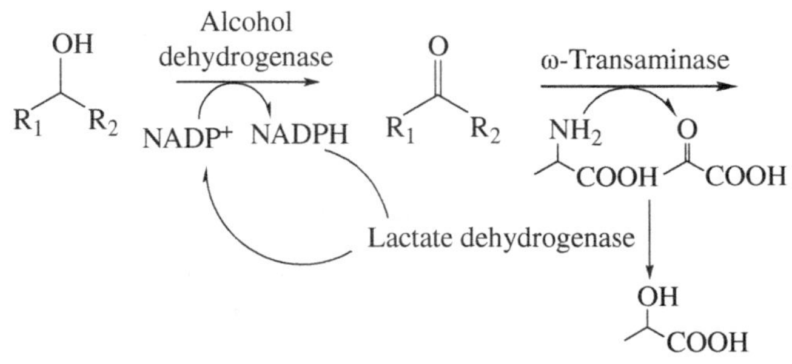

SCHEME 29.13
Basic reaction scheme for the amination of *sec*-alcohols.

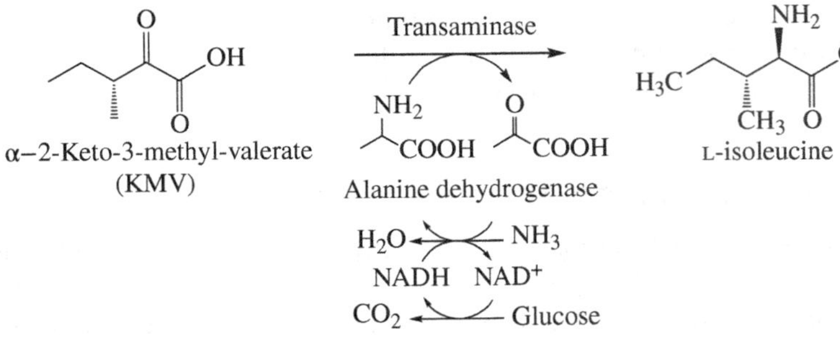

SCHEME 29.14

Reaction scheme for the synthesis of L-isoleucine with whole cells.

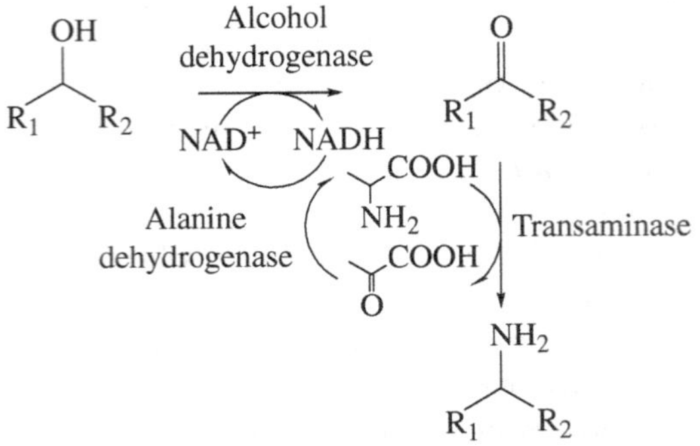

SCHEME 29.15

Reaction cascade for the synthesis of chiral amino alcohols in whole cells.

The reaction is driven by ammonium and glucose feed to the media. The NADH was regenerated by cellular metabolism from a carbon source like glucose.

2-Keto-3-methyl-valerate (KMV) was the least inhibitory substrate up to 40 mM and was therefore chosen for the synthesis of L-isoleucine. The best reaction yields and almost complete conversion of 100 mM KMV were obtained with 50 mM MOPS, 100 mM $(NH_4)_2SO_4$, and a glucose/KMV ratio of 1:2. In this reaction the overall yield of 0.6 g/g h was obtained.

Synthesis of Chiral Amino Alcohols and Amino Acids in Whole Cells

Given that transaminases are not able to aminate alcohols, one possibility for the synthesis of chiral amino alcohols is an enzyme cascade reaction carried out with whole cells [36]. Three enzymes cascaded in series were expressed in *E. coli*: first the alcohol dehydrogenase oxidized the alcohol to the corresponding aldehyde, which is converted into the amine by the transaminase as shown in Scheme 29.15. The recycling of pyruvate and cofactor regeneration were achieved by the alanine dehydrogenase.

Several cyclic aromatic alcohols and diols were tested. It was shown that conversion rates were dependent on chain length and primary and secondary hydroxyl groups. On the one hand, primary linear alcohols with short chains, such as 1-hexanol, achieved excellent product yields. However, the linear diols like 1,10-decanediol showed higher product yields with substrates with longer chain length. The aromatic benzyl alcohol showed excellent conversion rates. The best reaction conditions for 10 mM 1,10-decanediol was demonstrated with 50 mM HEPES pH 6.0, 250 mM alanine, and 275 mM $(NH_4)_2SO_4$. The best reaction yields of 91% were obtained with cyclohexanol; the highest ee value was reached with 2-octanol but with a lower conversion yield of 64%.

Transaminase Expression on the Surface of Escherichia coli *Cells*

The aforementioned disadvantage of whole-cell biotransformation with mass transfer limitation can be avoided by expression of the transaminase on the surface of the cells [107]. But this effort uncovered several drawbacks, like the need for the cofactor PLP and the required dimerization of protein on the surface to form the active enzyme. The surface expression of proteins has been reviewed for Gram-negative bacteria [108]. A frequently used autotransporter, in this case adhesin (AIDA-I), involved in diffuse adherence, formed the pore to translocate the protein to the outer membrane. The passenger enzyme was flanked by a His- (outside the enzyme) and a Myc- (inside the enzyme) tag. To gain active dimers, two correctly expressed transaminase–AIDA complexes must be close by and adjusted to the outside to be accessible for PLP. Transaminase activity was obtained in the outer membrane, but unfortunately no activity was detected in the whole-cell fraction. A possible explanation for that absence of activity could be the N-terminal proteolytic damage of the protein complex after translocation to the periplasm, due to the fact that an intact N-terminus was required for the secretion. Furthermore, it was noticed that larger recombinant protein complexes remaining longer in the unfolded conformation seemed to be more resistant to proteolytic degradation. However, the first step to surface-expressed transaminases has been made, with the future focus on transaminase activity not only on the outer membrane but within whole cells.

Saccharomyces cerevisiae *in Whole-Cell Transamination*

A promising new tool for the kinetic resolution of chiral amines has recently been demonstrated. Baker's yeast, *S. cerevisiae*, was applied in whole-cell biocatalysis for the kinetic resolution of (*R*)-1-phenylethylamine (PEA) by expressing the ω-transaminase from *Capsicum chinense* [109]. Racemic 1-phenylethylamine was converted to the enantiopure chiral amine with acetophenone being generated as coproduct. As glucose is the preferred carbon source for yeast cells, concentrations of 400 mM were used, internally generating the pyruvate that acts as the amino acceptor. Comparing to the performance of the yeast with externally added pyruvate as amino acceptor, glucose led to two times higher conversion without adding any external PLP cofactor. Nevertheless, adding PLP in amounts of 0.2 mM resulted in higher conversion yield (17–25% with an ee of 30%). Examinations of substrate and product inhibition toward the cell growth showed lower growth rates at 30 mM acetophenone and the highest reaction rates with 25 mM *rac*-1-phenylethylamine with resting cells. Unfortunately, in terms of industrial settings, further investigations have to be done, due to the still low conversion rates and further requirement of costly PLP for the reaction. Nevertheless, the yeast *S. cerevisiae* has the ability to express the ω-transaminase for the conversion of (*S*)-1-phenylethylamine with glucose as the sole carbon source, no need for external addition of the cofactor PLP, conversion yield up to 25% with resting cells with high stability/selectivity, and no by-product formation.

29.4 APPROACHES TO OPTIMIZE THE TRANSAMINASE-CATALYZED REACTIONS

An issue that has recently emerged was a microscale method to assay bioprocess options to improve bioconversion reactions yields by overcoming thermodynamic and kinetic limitations [110]. In this study they observed that the choice of the amino donor and the equilibrium shift via second enzyme reactions are the best options to increase the product yield.

Nevertheless, to customize industrial needs and to overcome the deficiencies of enzyme reactions, protein engineering makes it feasible to modify enzymes for certain processes [4, 68, 73, 111, 112].

A new approach to protein engineering is computational design. This method provides a wider range of gaining new active enzymes via the access to structures.

It is apparent that computational design improves the chances of finding active enzymes within a clone library yet the variety of the catalyzed reactions by the designed enzymes is still restricted [113]. The latest achievements in protein engineering have been reviewed extensively [114–116].

29.4.1 Protein Engineering by Rational Enzyme Design

Even though transaminases have broad substrate specificity, there are some limitations in using substrates with aliphatic or bulky side chains. Rational enzyme design, comprising homology modeling and site-directed mutagenesis (among other things), can to shift the substrate specificity and the enantiopreference of enzymes [117]. Therefore substantial knowledge of the enzyme structure is required. To elucidate the key determinants of the active site of a transaminase with unknown structure, homology modeling is the method of choice, demanding high sequence similarity with known protein structures.

An aromatic L-amino acid transaminase from *Enterobacter* sp. with unknown structure and no substrate specificity toward aromatic amino acid analogs was used to synthesize the L-phenylalanine analog L-diphenylalanine. Based on the interaction of the substrate with the active site of the enzyme, site-directed mutagenesis can alter the substrate specificity of the transaminase toward substrates with bulky side chains [118].

For the synthesis of chiral aliphatic amines with the ω-transaminase from *V. fluvialis*, homology modeling displayed a large substrate binding site and another smaller one. It has been shown that there is a negative correlation between the aromatic amines and the accessible surface. This might be caused by steric hindrances of two tryptophans in the larger binding site of the transaminase. Changing the amino acid from tryptophan to glycine reduces the hydrophobic interaction and allows binding of aromatic groups or aliphatic chains [119].

ω-Transaminases catalyze the synthesis of β-amino acids. To increase the relative activity to β-phenylalanine (3-amino-3-phenylpropionic acid), Hwang *et al.* redesigned the ω-transaminase from *Caulobacter crescentus* [120]. With site-directed mutagenesis, they mutated valine to glycine and asparagine into alanine to reduce steric hindrances and to enlarge the smaller pocket. The relative activity of the mutants increased 11-fold and 3-fold, respectively.

Rational enzyme design has the potential to shift the enantiopreference of a ω-transaminase by a single-point mutation. Svedendahl *et al.* described the mutation of the (*S*)-selective ω-transaminase from *A. citreus* to an (*R*)-selective transaminase converting 4-fluorophenylacetone [121]. Homology modeling displayed an active site loop close to the phenyl group of the substrate (4-fluorophenylacetone). This loop is involved in the binding of PLP. Mutation of one of the amino acids (valine) present in this loop created more space for the (pro-*R*) quinonoid (intermediate of the amination process; Section 29.2.3). This resulted in reversed enantiopreference toward this substrate.

To change the enantiopreference of transaminases, Höhne *et al.* searched *in silico* databases and took a closer look at D-selective and branched-chain transaminases [26]. After cloning the synthetic genes the new enzymes were investigated for selective activity and desired enantiopreference. The ee of the newly identified (*R*)-selective transaminase was >99.6%, and its activity was found to be in the same range as the known (*S*)-selective transaminase, which makes it suitable for the asymmetric synthesis of (*R*)-amines.

29.4.2 Protein Engineering by Directed Evolution

Another method used in protein engineering is directed evolution. Unlike rational enzyme design, this strategy requires neither structural information nor the knowledge of basic enzyme functions. But there is a need for screening or selection tests (like high-throughput screenings (HTS) and iterative cycle improvements) to detect the mutants possessing the desired properties [114]. Successful applications like gene shuffling and random mutagenesis create enzymes with higher substrate specificity, higher thermostability, and improved heterologous expression [122].

Directed Evolution for Higher Substrate Specificity

In 1998, Yano *et al.* modified an aspartate transaminase to higher substrate specificity toward β-branched substrates and 2-oxo acids by DNA shuffling, resulting in an ~10^5-fold higher catalytic efficiency (k_{cat}/K_M) toward 2-oxovaline, 2-oxoisoleucin, valine, and isoleucine [123]. The screening system for the mutants consisted of an auxotroph *E. coli* mutant, growing on a minimal plate without valine, isoleucine, and leucine, if carrying the plasmid with the gene *ilvE* for the aspartate transaminase capable of converting β-branched amino acids.

In an award-winning project (Green Chemistry Award 2010), an (*R*)-selective transaminase was engineered for the asymmetric synthesis of the antidiabetic compound sitagliptin [57]. Currently the synthesis of (*R*)-sitagliptin involves asymmetric hydrogenation at high pressure with a rhodium catalyst [56]. This process gave lower stereoselectivity and required further purification steps to get rid of the metal catalyst. Thus a transaminase had to be engineered with activity toward the prositagliptin ketone. Because of steric interference in the small binding pocket and potentially undesired interactions in the large binding pocket, transaminases are not able to bind prositagliptin. Thus, the large binding pocket of the enzyme was engineered after screening for an (*R*)-selective ω-transaminase with truncated substrates. After that the (*R*)-selective enzyme was mutated to be active with prositagliptin as substrate. Due to the low solubility of the prositagliptin ketone, not only the substrate specificity of the evolved transaminase but also the performance of the enzyme in organic solvents (here DMSO) had to be enhanced. To shift the equilibrium toward the product side, by using IPA as amino donor leading to acetone as the volatile keto by-product, the activity of the enzyme at higher reaction temperatures (<40 °C) had to be improved as well. The best results for the transformation of 200 g/l prositagliptin with an ee value of >99.95% under optimal reaction conditions were obtained with 6 g/l enzyme and 50% DMSO at pH 8 and 40 °C. Compared to the chemical synthesis with a rhodium catalyst, the overall yield increased from 10 to 13%, and the productivity (kilogram per liter a day) increased 53%. Furthermore, 19% of overall waste production was reduced, heavy metals were eliminated, and there was a reduction in manufacturing costs.

Recently, Midelfort *et al.* evaluated several engineering techniques, like homology modeling, bioinformatics, machine learning, crystal structure, and site-directed mutagenesis of specific sites, to evolve the transaminase from *V. fluvialis* in selectivity and activity for synthesis of (3*S*,5*R*)-ethyl-3-amino-5-methyloctanoate, an intermediate of the imagabalin synthesis [124]. Imagabalin is a candidate for the treatment of anxiety disorder. Up to now the synthesis of imagabalin has been based on chemical catalysis: however, the high costs make this process inadequate for industrial applications [125, 126]. With these combined engineering techniques, eight mutations were detected, of which five clearly provided more space and interaction with the substrate. The activity of the mutated enzyme was increased 60-fold toward (3*S*,5*R*)-ethyl-3-amino-5-methyloctanoate compared to the wild type and there was an increase in the de value of >95%.

Directed Evolution for Reduced Product Inhibition

To relieve product inhibition for higher product yields, Yun *et al.* created a mutant library of a transaminase from *V. fluvialis* by error-prone PCR [127]. The amino donor substrate was 2-aminoheptane and the inhibitory ketone was 2-butanone. In comparison with the wild-type ω-transaminase, the mutant had a twofold higher activity toward aliphatic ketones. Nevertheless the activity was similar toward short-chain aliphatic amines and aromatic ketones. For kinetic resolution of racemic 2-aminoheptane, the pH optimum was at 9.0. Crude cell extract of the mutant with 20 U/ml activity containing 100 mM *rac*-2-aminoheptane, 100 mM pyruvate, and 0.5 mM PLP at 37 °C gave an ee value of 99% for (*R*)-2-aminoheptane after 3 h reaction time. The greater conversion and the higher yield of the product compared to the wild-type enzyme decreased product inhibition.

Another approach in directed evolution was the evolution of an aspartate transaminase to an enzyme possessing the properties of the closely related tyrosine transaminase [128]. Eight rounds of DNA shuffling led to mutants with 100- to 270-fold increase of k_{cat}/K_M for phenylalanine and a 40- to 150-fold increase for tyrosine.

Directed Evolution for Improved Thermostability

The modification of (*S*)-aminotransferase from *A. citreus* for higher thermostability and increased tolerance for high amine concentration for the synthesis of substituted (*S*)-aminotetralin was achieved by error-prone PCR [59]. These properties are necessary to shift the equilibrium toward the product side by removing the volatile acetone at higher temperatures and to make the process economically competitive. Martin *et al.* demonstrated an increase in the specific activity of the aminotransferase for the mutant of the fifth generation at 55 °C compared to the wild type at 30 °C after five rounds of mutation. Therefore a 15-fold reduction of the enzyme load and a 5-fold reduction in reaction time compared to the wild type were achieved. The overall improvement of the specific activity is 280-fold higher, whereas K_M remained at 10 mM. The mutant screening system consisted of a colorimetric assay. The colonies grew overnight on an LB agar plate and were transferred on nitrocellulose paper which was placed on filter paper with the screening solution. The substituted (*S*)-aminotetralin and sodium pyruvate were used as substrates. The reaction with tetralone, a colored product, was formed upon exposure to air.

This colorimetric assay was used in a study by Matcham *et al.* for detection of mutants with higher tolerance toward (*S*)-methoxyisopropylamine ((*S*)-MOIPA) and higher thermal and chemical stability [60]. Due to a slightly increased product concentration after adding higher amounts of enzyme, it was necessary to improve the product tolerance of the enzyme. Error-prone PCR and several rounds of enzyme modification created a mutant with the required properties. At the end of the reaction the solution contained not only the desired product, (*S*)-MOIPA, but also acetone. To shift the equilibrium toward the product side by evaporating acetone by-product, a transaminase mutant thermally stable up to 50 °C was found after five rounds of mutation and used in this process. The final yield of (*S*)-MOIPA about 2 mM was obtained after 7 h reaction time with 5 g/l of the transaminase at 50 °C under vacuum.

Though naturally occurring enzymes scarcely show the desired manufacturing properties, these results demonstrated that it is possible to overcome certain limitations like substrate hindrances or deficient thermostability and supply transaminases for industrial processes.

29.4.3 Immobilization of Transaminases

In the previous sections, transaminases were referred as a suitable biocatalysts for the synthesis of chiral amines and amino acids. Their benefits are their wide substrate scope, high activity, and high enantio- and stereoselectivity. But some limitations in transaminase-catalyzed reactions occur as well, like product and substrate inhibition. To overcome these limitations and to improve or alter the enzyme's properties, enzyme immobilization can be the method of choice. Furthermore, to use enzymes in industrial processes, enzyme immobilization can help to solve certain challenges, such as enzyme recovery and reusability or enzyme stability. The characteristics of the carrier material are, besides the methods of immobilization, of particular importance to the performance of the immobilized enzymes. Specifically the physical properties of the matrices like particle diameter, swelling behavior, mechanical strength, and compression behavior are of major importance. The supporting matrices are classified as inorganic (e.g., natural minerals, bentonite) and organic (e.g., chitin, alginate, agarose). Most applications were performed with organic supporting material, due to their hydrophobic character and high porosity. The versatile methods of attaching the enzyme to the matrices are subdivided into reversible and irreversible. Irreversible methods are covalent coupling, entrapment, or cross-linking, whereas reversible immobilization is carried out via disulfide bonds or adsorption [129, 130]. Eventually immobilization of enzymes is an appropriate way to improve enzyme stability/activity and to reduce product inhibition [131].

Immobilization on Chitosan Beads

Chitosan is a main derivative of chitin gained by *N*-deacetylation by treating with 40–50% NaOH, followed by several purification steps. The reactive amino and hydroxyl groups of chitosan make it soluble in aqueous solution at $pH < 6.5$. The ability of chitosan as an immobilization substrate lies in its remarkable affinity to proteins, physiological inertness, nontoxicity, biocompatibility, and biodegradability [132].

The applicability of chitosan beads to immobilize ω-transaminases from *V. fluvialis* compared to immobilization on beads of different types was examined on the basis of the enzyme activity [133]. Higher residual activity on chitosan beads and a shift of the temperature optimum to 45 °C instead of 37 °C compared to free enzyme were detected. Furthermore, after five cycles of usage, 77% activity, retained compared to the activity on Eupergit™ beads at 19%. Eupergit consists of macroporous beads, made by copolymerization of *N,N'*-methylenebis(methacrylamide), glycidyl methacrylate, allyl glycidyl ether, and methacrylamide. The advantage of this type of beads is the outstanding chemical and mechanical stability, over pH ranges of 0–14 [134].

Recently several protocols have been reported for the immobilization on chitosan beads using glutaraldehyde as a linker: for the (*R*)-selective transaminase from *Gibberella zeae*, *Neosartorya fischeri* and *A. fumigatus* and for two (*S*)-selective transaminases from *Ruegeria pomeroyi* and *Rhodobacter sphaeroides* [135, 136]. Although the transaminases belong to different fold types, it was shown that they all could be immobilized with the same protocol, resulting in higher thermal stability, a slight shift of optimum pH, and optimal storage stabilities, and they could be used in asymmetric synthesis with IPA as the amino donor.

Another alternative for ω-transaminase immobilization on chitosan beads is the immobilization on catechol-chitosan–iron oxide nanoparticles. Ni *et al.* established a protocol for this composition carrying adhesive moieties for surface linkages [137]. After immobilization the enzyme showed higher thermal and pH stability compared to the free enzyme, and after 15 cycles of usage 50% of its activity remained.

Immobilization on a Sol–Gel Matrix

Entrapment of enzymes in a sol–gel matrix for immobilization is a useful method due to the fact that the gel can be tailored to a large range of network structures and processing conditions. Conditions like pH, gelation time, or hydrophobicity can be attuned to the special needs of a certain enzyme. The trapped enzymes showed higher activity and longer lifetimes compared to free enzymes. Within the silica cage the enzyme's mobility is restricted, and unfolding is prevented even under harsh conditions [138, 139]. The most studied material for gel preparation is SiO_2. Usually a silane precursor like sodium metasilicate (Na_2SiO_3) reacts under acidic or basic conditions to SiO_2 [140]. Furthermore every enzyme seems to prefer a different combination of sol–gel precursors [141]. Immobilizing ω-transaminases leads to higher activity over a broader range of pH and temperature. Improved results were obtained after the entrapment of an (*R*)-selective ω-transaminase from *Arthrobacter* sp. in a sol–gel matrix by Reetz *et al.* [142] and also with Celite™ 545 as an additive [143]. Celite, also called diatomaceous earth, consists of silica (SiO_2) and forms an inert porous structure. Additionally a one-pot two-step enzyme cascade was utilized whereby the immobilized enzyme of the first reaction step could be removed with filtration or centrifugation before initiation of the second step.

For the encapsulation of the ω-transaminase from *Arthrobacter* sp., a 1:5 ratio of silane tetramethoxysilane to methyltrialkoxysilane was found to be best with the washing step of the maturated sol–gel matrix also being crucial [144]. The highest conversion of the model compound was reached using IPA as the final washing agent. For sufficient substrate and product solubility, the addition of 10% v/v DMSO was necessary as well as for higher long-term stability of the enzyme.

Immobilization on a Nickel-Nitriloacetic Acid Agarose Beads

Another well established immobilization technique is via recombinant tags like polyhistidine tags. The idea came from immobilized metal affinity chromatography (IMAC) for the purification of proteins tagged with polyhistidine [145]. This method allows immobilization without prepurification steps and shows no interference with the enzyme structure or function, and does not affect the secretion or folding of fusion proteins within cells [146].

For the *in vitro* screening of enzyme pairs and enzymatic pathways, immobilized enzyme microreactor systems (IEMR) were established. This includes the reversible immobilization of His_6-tagged enzymes via nickel-nitriloacetic acid (Ni-NTA) linkage on the surface-derivatized silica. First of all the kinetic parameters for a new enzymatic pathway have to be defined via a transketolase and a ω-transaminase in a continuous flow system for a two-step asymmetric synthesis of chiral amino alcohols [147]. The apparent K_M value is found to be flow rate dependent with different flow rates ranging from 2 to 30 μl/min. The turnover rate k_{cat} was also found to be lower than that in solution phase possibly caused by steric hindrances, imperfect immobilization, or altered enzyme structure. The prototype for IEMR consists of two tubes cascaded in series with immobilized transketolase for the production of L-erythrulose and immobilized ω-transaminase for synthesis of 2-amino-1,3,4-butanetriol (Scheme 29.16).

For the ω-transaminase performing with a much lower reaction rate compared to the transketolase, a sevenfold higher amount of protein was loaded. About 83% conversion was obtained for the coupled reaction with initial concentrations of 60 mM of

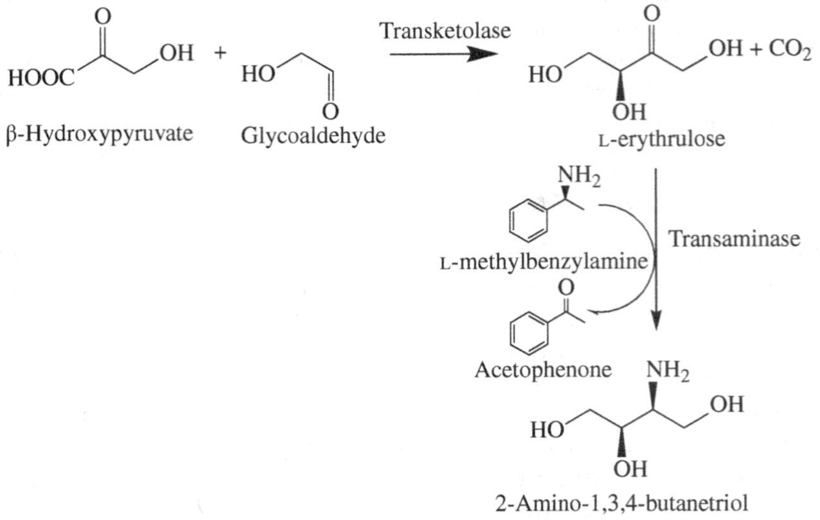

SCHEME 29.16

Synthesis of 2-amino-1,3,4-butanetriol by a transketolase/transaminase enzyme cascade via continuous flow dual reactor. The enzymes were immobilized inside the silica microreactors separately.

β-hydroxypyruvate and glycolaldehyde and 6 mM of L-methylbenzylamine within 20 min, at a lowest flow rate of 2 μl/min.

Immobilization of transaminases has been shown to be an excellent way to overcome some limitations occurring in the reactions, as well as to improve the reaction yield and the stability of the enzymes. Immobilization leads to a simple reaction process, due to the easy removal of the enzymes from the reaction solution. Furthermore the enzymes showed higher thermal and storage stability and a longer lifespan, and in many cases only a slightly decrease in activity occurred after several rounds of use. Concerning ecologically influences, the beads used for the immobilization are partly biodegradable and nontoxic and can be tailored in various network structures.

29.4.4 Process Development: A Fast Way to Identify Appropriate Transaminases

High-Throughput Assay to Determine Transaminase Activity

The development of rapid HTS assays is important to test the substrate scope, suitable amino donors/acceptors, and the stability under different reaction conditions, like temperature, pH, different solvents, and immobilization methods. Furthermore the rapid progress in protein engineering like directed evolution requires fast selection methods. This subject was extensively reviewed by Mathew *et al.* [148]. In the following a colorimetric, photometric, and kinetic assay for rapid transaminase activity screening is described and illustrated in Scheme 29.17.

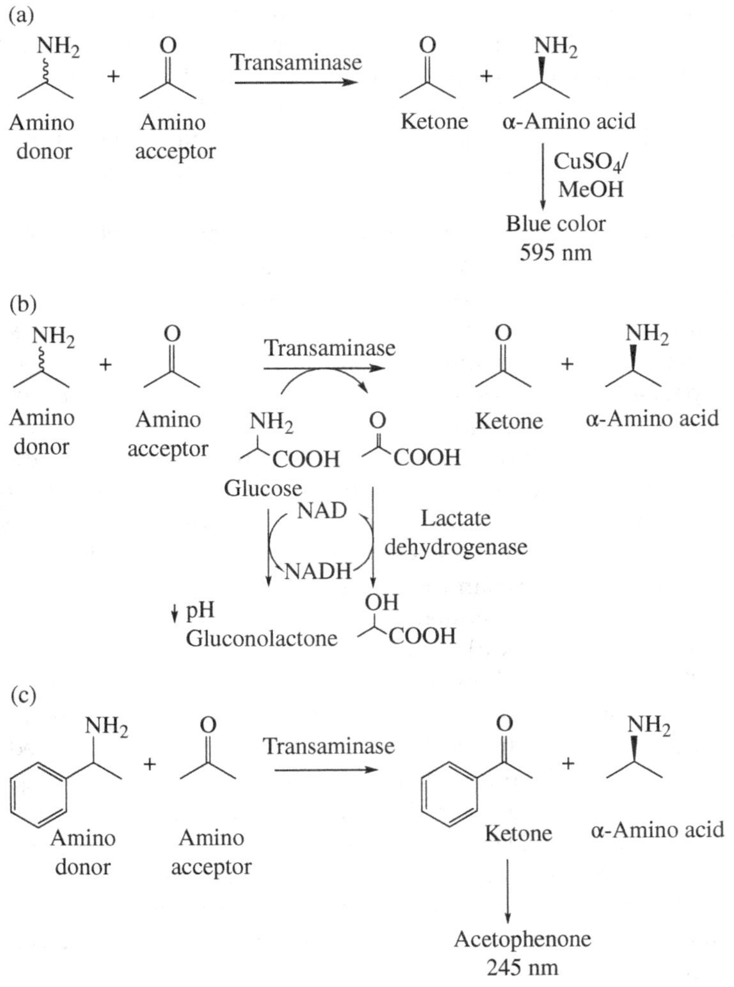

SCHEME 29.17
High-throughput screening assays to detect transaminase activity. (a) Colometric assay. (b) pH-based assay. (c) Photometric assay.

Kim *et al.* published an assay based on UV–Vis spectrometry [149]. As shown in Scheme 29.17a the α-amino acid produced during the reaction was stained with a solution containing $CuSO_4$/MeOH, forming a blue-colored copper ion complex, which can be quantified at 595 nm, whereas other substrates like amines, β-amino acids, α-keto acids, or ketones showed no color after adding the $CuSO_4$/MeOH solution. A major drawback of this method is the color complex, formed by α-amino acids in a phosphate buffer with cupric sulfate also absorbing at 595 nm. Hence, sufficient dialysis of the enzyme or the crude cell extracts has to be conducted.

Another approach is based on the reduction of the pH caused by gluconic acid, a coproduct of the transamination reaction, indicated with phenol red [111]. The coproduct pyruvate was removed by a lactate dehydrogenase to lactate to shift the equilibrium of the entire reaction toward the product side. The glucose dehydrogenase deployed to regenerate the needed cofactor NADH generates gluconic acid and leads to a decrease in the pH value (Scheme 29.17b).

Schätzle *et al.* published a photometric assay based on the high absorbance of acetophenone at 245 nm, a by-product formed in the transaminase reaction (Scheme 29.17c) [150]. Other compounds like pyruvate, alanine, and ketones absorb at the same wavelength but only with 5% of the absorbance compared to acetophenone. So real-time monitoring of the increasing absorbance of the by-product during the transaminase reaction is possible.

29.4.5 ω-Transaminases in Organic Solvents

Most enzymes require an aqueous environment for their best performances. This is in conflict with the current need for transaminases with a wide substrate specificity including poorly water-soluble molecules. To enhance the solubility of these substrates, cosolvents like DMSO have to be added. Besides this, organic solvents are deployed for product extraction to avoid product inhibition. Using transaminases in neat organic solvents does not require continuous pH control and additional product extraction.

Only a few studies have been published, examining the stability and activity of transaminases in organic solvents to overcome product inhibition and to expand the substrate spectrum of transaminases [44, 57, 151, 152].

Recently immobilized transaminases in neat organic solvents have been examined for the asymmetric synthesis of amines [153]. The immobilization of the ω-transaminase with Sepabeads was carried out by incubation of the transaminase with the beads in buffer for 48 h. By means of the enzyme activity, the screening of several organic solvents was carried out among which isopropyl acetate was the most suitable solvent. The immobilized enzyme showed high activity and stability under the chosen reaction conditions, as well as good yields of various tested amines.

But not only immobilized ω-transaminases showed activity in organic solvents. Several ω-transaminases, for example, from *A. citreus* or *C. violaceum*, were tested in methyl *tert*-butyl ether with a defined amount of water ($a_w = 0.6$) [154]. The optimum amount of water content was in a range from 0 to 0.7%, with significantly higher activity at 0.6%. All enzymes showed higher activity. In addition, 2-propylamine can be used in the reaction as an amino donor, where as could not be utilized in aqueous solutions. Furthermore no change in enantioselectivity was observed, remaining at >99% ee in all cases. In a comparison of the enzyme's stability in aqueous buffer and organic solvents, the catalytic residual activity was 4% after 5 min in buffer and 38% after 1 h and 22% after 24 h in water-free ethyl acetate, respectively. This clearly demonstrated that higher stability of ω-transaminases could be achieved in organic solvents than in aqueous buffer. But the most significant benefit for the reaction in organic solvents is that no substrate inhibition could be detected at higher substrate concentration.

29.5 CONCLUSION

Transaminases are most powerful tools for the synthesis of chiral amines, amino acids, and amino alcohols. In this chapter several approaches for the preparation of fine chemicals or building blocks for pharmaceuticals were discussed, like asymmetric synthesis or kinetic resolution. The main limitations of transaminase-catalyzed reactions are the need to shift the equilibrium to the product side and substrate and product inhibition. Some solutions to overcome such inhibition were presented here; for example, multienzyme cascades or biphasic extraction of the product. Protein engineering by directed evolution or rational enzyme design is a promising option to find transaminases with different substrate specificities and enantiopreferences. This is becoming more and more important for the pharmaceutical industry. Furthermore, it is a way to alter enzyme properties known so far, like thermostability and solvent and pH stability. Protein engineering has been assisted by the recently solved structures of certain transaminases.

REFERENCES

[1] Needham, D. M., *Biochemical Journal*, 1930, **24**(1), 208–227.

[2] Taylor, P. P., Pantaleone, D. P., *et al.*, *Trends in Biotechnology*, 1998, **16**(10), 412–418.

[3] Koszelewski, D., Tauber, K., *et al.*, *Trends in Biotechnology*, 2010, **28**(6), 324–332.

[4] Tufvesson, P., Lima-Ramos, J., *et al.*, *Biotechnology and Bioengineering*, 2011, **108**(7), 1479–1493.

[5] Hwang, B. Y., Cho, B. K., *et al.*, *Journal of Molecular Catalysis B: Enzymatic*, 2005, **37**(1–6), 47–55.

[6] Rudat, J., Brucher, B. R., and Syldatk, C., *AMB Express*, 2012, **2**(1), 11.

[7] Mathew, S. Y., *American Chemical Society*, 2012, **2**, 993–1001.

[8] Percudani, R., Peracchi, A., *et al.*, *BMC Bioinformatics*, 2009, **10**, 273–280.

[9] Schneider, G., Kack, H., and Lindqvist, Y., *Structure with Folding & Design*, 2000, **8**(1), R1–R6.

[10] Jansonius, J. N., *Current Opinion in Structural Biology*, 1998, **8**(6), 759–769.

[11] Eliot, A. C. and Kirsch, J. F., *Annual Review of Biochemistry*, 2004, **73**, 383–415.

[12] Finn, R. D. and Bateman, A., *Nucleic Acids Research*, 2014, **42**(Database issue), D222–D230.

[13] Uchida, Y., Hayashi, H., *et al.*, *Extremophiles*, 2014, **18**, 589–602.

[14] Lyskowski, A., Gruber, C., *et al.*, *PLoS One*, 2014, **9**(1), e87350.

[15] Kirby, A. J., Coenzymes I: Organic coenzymes. In: *Enzymatic Reaction Mechanisms*. P. A. Frey and A. D. Hegeman, Eds., Oxford University Press New York/Oxford, Vol. **118**, 2007, p. 148.

[16] Thomsen, M. and Skalden, L., *Acta Crystallographica Section F: Strucural Biology and Crystallization Communication*, 2013, **69**(12), 1415–1417.

[17] Höhne, M. and Bornscheuer, U., Application of transaminases in organic synthesis. In: *Enzyme Catalysis in Organic Synthesis*. K. Drauz, H. Gröger, and O. May, Eds., Wiley-VCH Verlag GmbH & Co. KGaA, Weinheim, Germany, 3rd Edition, 2012, pp. 779–820.

[18] Toney, M. D., *Archives of Biochemistry and Biophysics*, 2014, **544**, 119–127.

[19] Hayashi, H., Mizuguchi, H., and Kagamiyama, H., *Journal of Biological Chemistry*, 2003, **278**(11), 9481–9488.

[20] Goto, M., Miyahara, I., *et al.*, *Biochemistry*, 2003, **42**(13), 3725–3733.

[21] Humble, M. S., Cassimjee, K. E., *et al.*, *FEBS Journal*, 2012, **279**(5), 779–792.

[22] Cassimjee, K. E., Humble, M. S., *et al.*, *Organic & Biomolecular Chemistry*, 2012, **10**(28), 5466–5470.

[23] Shin, J. S., Yun, H., *et al.*, *Applied Microbiology and Biotechnology*, 2003, **61**(5–6), 463–471.

[24] Shin, J. S. and Kim, B. G., *Biotechnology and Bioengineering*, 2002, **77**(7), 832–837.

[25] Wybenga, G. G., Crismaru, C. G., *et al.*, *Journal of Biological Chemistry*, 2012, **287**(34), 28495–28502.

[26] Höhne, M., Schätzle, S., *et al.*, *Nature Chemical Biology*, 2010, **6**(11), 807–813.

[27] Sayer, C., Martinez-Torres, R. J., *et al.*, *FEBS Journal*, 2014, **281**(9), 2240–2253.

[28] Thomsen, M., Skalden, L., *et al.*, *Acta Crystallographica Section D: Biological Crystallography*, 2014, **70**(4), 1086–1093.
[29] Höhne, M. and Bornscheuer, U. T., *ChemCatChem*, 2009, **1**(1), 42–51.
[30] Nugent, T. C. and El-Shazly, M., *Advanced Synthesis & Catalysis*, 2010, **352**(5), 753–819.
[31] Uenishi, J., Hamada, M., *et al.*, *Journal of Organic Chemistry*, 2004, **69**(20), 6781–6789.
[32] Chelucci, G. and Baldino, S., *Tetrahedron: Asymmetry*, 2006, **17**(10), 1529–1536.
[33] Koszelewski, D., Goritzer, M., *et al.*, *ChemCatChem*, 2010, **2**(1), 73–77.
[34] Yun, H., Cho, B. K., and Kim, B. G., *Biotechnology and Bioengineering*, 2004, **87**(6), 772–778.
[35] Koszelewski, D., Clay, D., *et al.*, *European Journal of Organic Chemistry*, 2009, **2009**(14), 2289–2292.
[36] Klatte, S. and Wendisch, V. F., *Bioorganic and Medicinal Chemistry*, 2014, **22**(20), 5578–5585.
[37] Koszelewski, D., Clay, D., *et al.*, *Journal of Molecular Catalysis B: Enzymatic*, 2009, **60**(3–4), 191–194.
[38] Koszelewski, D., Pressnitz, D., *et al.*, *Organic Letters*, 2009, **11**(21), 4810–4812.
[39] Fuchs, M., Koszelewski, D., *et al.*, *Tetrahedron*, 2012, **68**(37), 7691–7694.
[40] Richter, N., Simon, R. C., *et al.*, *Chemical Communications*, 2014, **50**, 6098–6100.
[41] Fesko, K., Steiner, K., *et al.*, *Journal of Molecular Catalysis B: Enzymatic*, 2013, **96**(1), 103–110.
[42] Sehl, T., Hailes, H. C., *et al.*, *Angewandte Chemie International Edition*, 2013, **52**(26), 6772–6775.
[43] Rehn, G., Adlercreutz, P., and Grey, C., *Journal of Biotechnology*, 2014, **179**, 50–55.
[44] Pressnitz, D., Fuchs, C. S., *et al.*, *ACS Catalysis*, 2013, **3**(4), 555–559.
[45] Ingram, C. U., Bommer, M., *et al.*, *Biotechnology and Bioengineering*, 2007, **96**(3), 559–569.
[46] Shin, J. S., Kim, B. G., *et al.*, *Biotechnology Letters*, 2009, **31**(10), 1595–1599.
[47] Bea, H. S., Lee, S. H., and Yun, H., *Biotechnology and Bioprocess Engineering*, 2011, **16**(2), 291–296.
[48] Park, E. S., Dong, J. Y., and Shin, J. S., *Organic & Biomolecular Chemistry*, 2013, **11**(40), 6929–6933.
[49] Seo, Y. M., Mathew, S., *et al.*, *Organic & Biomolecular Chemistry*, 2012, **10**(12), 2482–2485.
[50] Caligiuri, A., D'Arrigo, P., *et al.*, *Biocatalysis and Biotransformation*, 2006, **24**(6), 409–413.
[51] Kim, J., Kyung, D., *et al.*, *Applied and Environmental Microbiology*, 2007, **73**(6), 1772–1782.
[52] Yun, H., Lim, S., *et al.*, *Applied and Environmental Microbiology*, 2004, **70**(4), 2529–2534.
[53] Lorenz, E., Klatte, S., and Wendisch, V. F. *Journal of Biotechnology*, 2013, **168**(3), 289–294.
[54] Breuer, M., Ditrich, K., *et al.*, *Angewandte Chemie International Edition*, 2004, **43**(7), 788–824.
[55] Rothman, R. B., Vu, N., *et al.*, *Journal of Pharmacology and Experimental Therapeutics*, 2003, **307**(1), 138–145.
[56] Hansen, K. B., Hsiao, Y., *et al.*, *Journal of the American Chemical Society*, 2009, **131**(25), 8798–8804.
[57] Savile, C. K., Janey, J. M., *et al.*, *Science*, 2010, **329**(5989), 305–309.
[58] Siirola, E., Mutti, F. G., *et al.*, *Advanced Synthesis & Catalysis*, 2013, **355**(9), 1703–1708.
[59] Martin, A. R., DiSanto, R., *et al.*, *Biochemical Engineering Journal*, 2007, **37**(3), 246–255.
[60] Matcham, G., Bhatia, M., *et al.*, *Chimia*, 1999, **53**(12), 584–589.
[61] Hanzawa, S., Oe, S., *et al.*, *Biotechnology Letters*, 2001, **23**(8), 589–591.
[62] Cho, B. K., Seo, J. H., *et al.*, *Biotechnology and Bioengineering*, 2003, **83**(2), 226–234.
[63] Truppo, M. D., Rozzell, J. D., and Turner, N. J., *Organic Process Research & Development*, 2010, **14**(1), 234–237.
[64] Simon, R. C., Richter, N., *et al.*, *ACS Catalysis*, 2014, **4**(1), 129–143.
[65] Wang, B., Land, H., and Berglund, P., *Chemical Communications*, 2013, **49**(2), 161–163.
[66] Koszelewski, D., Lavandera, I., *et al.*, *Advanced Synthesis & Catalysis*, 2008, **350**(17), 2761–2766.
[67] Paul, C. E. and Rodriguez-Mata, M., *Organic Process Research & Development*, 2014, **18**, 788–792.
[68] Höhne, M., Kuhl, S., *et al.*, *ChemBioChem*, 2008, **9**(3), 363–365.
[69] Cassimjee, K. E., Branneby, C., *et al.*, *Chemical Communications*, 2010, **46**(30), 5569–5571.
[70] Koszelewski, D., Lavandera, I., *et al.*, *Angewandte Chemie International Edition in English*, 2008, **47**(48), 9337–9340.
[71] Pollard, D. J. and Woodley, J. M., *Trends in Biotechnology*, 2007, **25**(2), 66–73.
[72] Shin, J. S. and Kim, B. G., *Biotechnology and Bioengineering*, 1998, **60**(5), 534–540.
[73] Truppo, M. D., Turner, N. J., and Rozzell, J. D., *Chemical Communications*, 2009, **2009**(16), 2127–2129.

[74] Shin, J. S., Kim, B. G., *et al.*, *Biotechnology and Bioengineering*, 2001, **73**(3), 179–187.
[75] Park, E. S. and Shin, J. S., *Applied and Environmental Microbiology*, 2013, **79**(13), 4141–4144.
[76] Allan, R. D., Bates, M. C., *et al.*, *Tetrahedron*, 1990, **46**(7), 2511–2524.
[77] Cuetos, A., Lavandera, I., and Gotor, V., *Chemical Communications*, 2013, **49**, 10688–10690.
[78] Nakajima, N., Esaki, N., and Soda, K., *Journal of the Chemical Society, Chemical Communications*, 1990, (13), 947–948.
[79] Hintermann, T. and Seebach, D., *Synlett*, 1997, 437–438.
[80] Hintermann, T. and Seebach, D., *Chimia*, 1997, **50**, 244–247.
[81] Bommarius, A. S., Schwarm, M., *et al.*, *Chimia*, 2001, **55**(1–2), 50–59.
[82] Sasa, M., *Journal of Pharmacological Sciences*, 2006, **100**(5), 487–494.
[83] Leuchtenberger, W., Huthmacher, K. and Drauz, K. *Applied Microbiology and Biotechnology*, 2005, **69**(1), 1–8.
[84] Goncalves, L. P. B., Antunes, O. A. C., *et al.*, *Organic Process Research & Development*, 2006, **10**(3), 673–677.
[85] Kamphuis, J., Boesten, W. H., *et al.*, *Advances in Biochemical Engineering/Biotechnology*, 1990, **42**, 133–186.
[86] Chao, Y. P., Lai, Z. J., *et al.*, *Biotechnology Progress*, 1999, **15**(3), 453–458.
[87] Park, E. S. Dong, J. Y., and Shin, J. S., *Applied Microbiology and Biotechnology*, 2014, **98**(2), 651–660.
[88] Seebach, D., Beck, A. K., and Bierbaum, D. J., *Chemistry and Biodiversity*, 2004, **1**(8), 1111–1239.
[89] Mekhail, T. M. and Markman, M., *Expert Opinion on Pharmacotherapy*, 2002, **3**(6), 755–766.
[90] Ojima, I., Lin, S., and Wang, T., *Current Medicinal Chemistry*, 1999, **6**, 927–954.
[91] Crews, P., Manes, L. V., and Boehler, M., *Tetrahedron Letters*, 1986, **27**(25), 2797–2800.
[92] Frackenpohl, J., Arvidsson, P. I., *et al.*, *ChemBioChem*, 2001, **2**(6), 445–455.
[93] Seebach, D., Gademann, K., *et al.*, *Helvetica Chimica Acta*, 1997, **80**, 2033–2038.
[94] Arvidsson, P. I., Frackenpohl, J., *et al.*, *ChemBioChem*, 2001, **2**(10), 771–773.
[95] Weiner, B., Szymanski, W., *et al.*, *Chemical Society Reviews*, 2010, **39**(5), 1656–1691.
[96] Verkuijl, B. J., Szymanski, W., *et al.*, *Chemical Communications*, 2010, **46**(6), 901–903.
[97] Wu, B., Szymanski, W., *et al.*, *Trends in Biotechnology*, 2011, **29**(7), 352–362.
[98] Groger, H., Trauthwein, H., *et al.*, *Organic & Biomolecular Chemistry*, 2004, **2**(14), 1977–1978.
[99] Rehdorf, J., Mihovilovic, M. D., *et al.*, *Chemistry*, 2010, **16**(31), 9525–9535.
[100] Engel, U., Syldatk, C., and Rudat, J., *Applied Microbiology and Biotechnology*, 2012, **94**(5), 1221–1231.
[101] Brucher, B., Syldatk, C., and Rudat, J., *Chemie Ingenieur Technik*, 2010, **82**(1–2), 155–160.
[102] Juaristi, E., *Anales De Quimica*, 1997, **93**(2), 135–142.
[103] Tauber, K., Fuchs, M., *et al.*, *Chemistry*, 2013, **19**(12), 4030–4035.
[104] Buhler, B., Park, J. B., *et al.*, *Applied and Environmental Microbiology*, 2008, **74**(5), 1436–1446.
[105] Kuhn, D., Fritzsch, F. S. O., *et al.*, *Journal of Biotechnology*, 2013, **163**(2), 194–203.
[106] Eggeling, L., Morbach, S., and Sahm, H., *Journal of Biotechnology*, 1997, **56**(3), 167–182.
[107] Gustavsson, M., Muraleedharan, M. N., and Larsson, G., *Applied and Environmental Microbiology*, 2014, **80**(7), 2293–2298.
[108] Leyton, D. L., Rossiter, A. E., *et al.*, *Nature Reviews Microbiology*, 2012, **10**(3), 213–225.
[109] Weber, N., Gorwa-Grauslund, M., *et al.*, *Applied Microbiology and Biotechnology*, 2014, **98**(10), 4615–4624.
[110] Halim, M., Rios-Solis, L., *et al.*, *Bioprocess and Biosystems Engineering*, 2014, **37**(5), 931–941.
[111] Truppo, M. D., Rozzell, J. D., *et al.*, *Organic & Biomolecular Chemistry*, 2009, **7**(2), 395–398.
[112] Woodley, J. M., *Current Opinion in Chemical Biology*, 2013, **17**(2), 310–316.
[113] Kries, H., Blomberg, R., *et al.*, *Current Opinion in Chemical Biology*, 2013, **17**(2), 221–228.
[114] Davids, T., Schmidt, M., *et al.*, *Current Opinion in Chemical Biology*, 2013, **17**(2), 215–220.
[115] Turanli-Yildiz, B., Protein engineering methods and applications. In: *Protein Engineering*. P. Kaumaya, Ed., InTechOpen, Rijeka, 2012, pp. 33–58.
[116] Nordwald, E. M., Garst, A., *et al.*, *Current Opinion in Biotechnology*, 2013, **24**(6), 1017–1022.
[117] Bommarius, A. S., Blum, J. K., and Abrahamson, M. J., *Current Opinion in Chemical Biology*, 2011, **15**(2), 194–200.
[118] Cho, B. K., Seo, J. H., *et al.*, *Biotechnology and Bioengineering*, 2006, **94**(5), 842–850.
[119] Cho, B. K., Park, H. Y., *et al.*, *Biotechnology and Bioengineering*, 2008, **99**(2), 275–284.

[120] Hwang, B. Y., Ko, S. H., *et al.*, *Journal of Microbiology and Biotechnology*, 2008, **18**(1), 48–54.
[121] Svedendahl, M., Branneby, C., *et al.*, *ChemCatChem*, 2010, **2**(8), 976–980.
[122] Petrounia, I. P., Arnold, F. H., *et al.*, *Current Opinion in Biotechnology*, 2000, **11**(4), 325–330.
[123] Yano, T., Oue, S., and Kagamiyama, H., *Proceedings of the National Academy of Sciences of the United States of America*, 1998, **95**(10), 5511–5515.
[124] Midelfort, K. S., Kumar, R., *et al.*, *Protein Engineering Design and Selection*, 2013, **26**(1), 25–33.
[125] Birch, M., Challenger, S., *et al.*, *Organic Process Research & Development*, 2011, **15**(6), 1358–1364.
[126] Birch, M., Challenger, S., *et al.*, *Organic Process Research & Development*, 2011, **15**(5), 1172–1177.
[127] Yun, H., Hwang, B. Y., *et al.*, *Applied and Environmental Microbiology*, 2005, **71**(8), 4220–4224.
[128] Rothman, S. C. and Kirsch, J. F., *Journal of Molecular Biology*, 2003, **327**(3), 593–608.
[129] Guisan, J. M., *Methods in Molecular Biology*, 2013, **1051**, 1–13.
[130] Brena, B., Gonzalez-Pombo, P., and Batista-Viera, F., *Methods in Molecular Biology*, 2013, **1051**, 15–31.
[131] Mateo, C., Palomo, J. M., *et al.*, *Enzyme and Microbial Technology*, 2007, **40**(6), 1451–1463.
[132] Krajewska, B., *Enzyme and Microbial Technology*, 2004, **35**(2–3), 126–139.
[133] Yi, S. S., Lee, C. W., *et al.*, *Process Biochemistry*, 2007, **42**(5), 895–898.
[134] Katchalski-Katzir, E. and Kraemer, D. M., *Journal of Molecular Catalysis B: Enzymatic*, 2000, **10**(1–3), 157–176.
[135] Mallin, H., Menyes, U., *et al.*, *ChemCatChem*, 2013, **5**(2), 588–593.
[136] Mallin, H., Höhne, M., and Bornscheuer, U. T., *Journal of Biotechnology*, 2014, **191**, 32–37.
[137] Ni, K., Zhou, X., *et al.*, *PLoS One*, 2012, **7**(7), e41101.
[138] Pierre, A. C., *Biocatalysis and Biotransformation*, 2004, **22**(3), 145–170.
[139] Avnir, D., Coradin, T., *et al.*, *Journal of Materials Chemistry*, 2006, **16**(11), 1013–1030.
[140] Brinker, C. J. and Scherer, G. W., *Sol-Gel Science: The Physics and Chemistry of Sol-Gel Processing*, Gulf Professional Publishing, Houston, TX, 1990.
[141] Lee, C. W., Yi, S. S., *et al.*, *Biotechnology and Bioprocess Engineering*, 2006, **11**(4), 277–281.
[142] Reetz, M. T., Tielmann, P., *et al.*, *Advanced Synthesis & Catalysis*, 2003, **345**(6–7), 717–728.
[143] Koszelewski, D., Müller, N., *et al.*, *Journal of Molecular Catalysis B: Enzymatic*, 2010, **63**(1–2), 39–44.
[144] Paivio, M. and Kanerva, L. T., *Process Biochemistry*, 2013, **48**(10), 1488–1494.
[145] Cheung, R. C., Wong, J. H., and Ng, T. B., *Applied Microbiology and Biotechnology*, 2012, **96**(6), 1411–1420.
[146] Cha, T., Guo, A., *et al.*, *Proteomics*, 2004, **4**(7), 1965–1976.
[147] Matosevic, S., Lye, G. J., and Baganz, F., *Journal of Biotechnology*, 2011, **155**(3), 320–329.
[148] Mathew, S., Shin, G., *et al.*, *Biotechnology and Bioprocess Engineering*, 2013, **18**(1), 1–7.
[149] Bum-Yeol Hwang, B.-G. K., *Enzyme and Microbial Technology*, 2004, **34**, 429–436.
[150] Schätzle, S., Höhne, M., *et al.*, *Analytical Chemistry*, 2009, **81**(19), 8244–8248.
[151] Shin, J. S. and Kim, B. G., *Biotechnology and Bioengineering*, 1997, **55**(2), 348–358.
[152] Yun, H., Kim, J., *et al.*, *Biotechnology and Bioengineering*, 2006, **93**(2), 391–395.
[153] Truppo, M. D., Strotman, H., and Hughes, G., *ChemCatChem*, 2012, **4**(8), 1071–1074.
[154] Mutti, F. G. and Kroutil, W., *Advanced Synthesis & Catalysis*, 2012, **354**(18), 3409–3413.

Index

Green Biocatalysis, First Edition. Edited by Ramesh N. Patel.